THE SELECTED WORKS OF
GEORGE E ANDREWS
(With Commentary)

ICP Selected Papers **ISSN: 2045-0990**

Published

Vol. 1 The Selected Papers of Sir Alan Fersht: Development of Protein Engineering
edited by Alan Fersht and Qinghua Wang

Vol. 2 The Selected Papers of Denis Noble CBE FRS: A Journey in Physiology Towards Enlightenment
edited by Denis Noble, Zhu Chen, Charles Auffray and Eric Werner

Vol. 3 The Selected Works of George E. Andrews (With Commentary)
by George E. Andrews and edited by Andrew V. Sills

Vol. 4 The Selected Papers of William N. Lipscomb, Jr.: A Legacy in Structure–Function Relationships
edited by Jianpeng Ma

ICP Selected Papers — Vol. 3

THE SELECTED WORKS OF GEORGE E ANDREWS

(With Commentary)

George E Andrews

Pennsylvania State University, USA

Edited by

Andrew V Sills

Georgia Southern University, USA

Imperial College Press

ICP

Published by

Imperial College Press
57 Shelton Street
Covent Garden
London WC2H 9HE

Distributed by

World Scientific Publishing Co. Pte. Ltd.
5 Toh Tuck Link, Singapore 596224
USA office: 27 Warren Street, Suite 401-402, Hackensack, NJ 07601
UK office: 57 Shelton Street, Covent Garden, London WC2H 9HE

British Library Cataloguing-in-Publication Data
A catalogue record for this book is available from the British Library.

Cover image reprinted with permission of the American Mathematical Society

ICP Selected Papers — Vol. 3
THE SELECTED WORKS OF GEORGE E ANDREWS
(With Commentary)

ISBN 978-1-84816-666-0

Typeset by Stallion Press
Email: enquiries@stallionpress.com

Printed in Singapore by B & Jo Enterprise Pte Ltd

To Joy and Leia

Author's Introduction

Since most of my thoughts and comments on the papers in this volume are interspersed throughout the chapters, I will make this introduction brief.

Mathematics has been a great joy to me. I owe much to countless people who have assisted me along the way. I have had the privilege of collaborating with more that 70 mathematicians, and I thank them for many exciting adventures. Pennsylvania State University has been my base since 1964; it has been a wonderful place to work, with friendly and helpful colleagues in both the faculty and administration. A more extensive account of my mathematical obligations was given in "Some Debts I Owe" [213] in the *Andrews Festschrift.*

Since 1995, Barbara Baum has been my staff assistant. Her professionalism, high standards, and keen insight have been of immeasurable value to me. This current volume would not have been possible without her thoughtful management of the many details involved in its preparation.

Serge Ballif typed my commentary for each chapter, and I am grateful both for his skill in the preparation, and his keen eye in spotting typos, infelicities, and egregious errors.

Ever since I met my student, Drew Sills, I have been vividly aware that he is as captivated by the theory of partitions as I am. Some time ago, he remarked that he would be pleased to assist in the preparation of my selected works, if I ever chose to undertake such a project. When the opportunity arose, he enthusiastically responded with tireless and painstaking efforts to edit this volume.

I would also like to thank Heng Huat Chan for suggesting to World Scientific that they publish my *Selected Works.* Siew Lian Tan, and her successor, Rok Ting Tan, at World Scientific have been most gracious and helpful in bringing this project to fruition.

Finally, I thank my wife, Joy, who has provided me with love, support and serenity for more than half a century. None of this would have been possible without her.

George E. Andrews
University Park, Pennsylvania
June 19, 2010

Editor's Preface

Personal Introduction

I first became aware of George E. Andrews as a result of seeing him interviewed on the television program *The Man Who Loved Numbers*, a biographical essay on Ramanujan, which aired in the United States as part of the *Nova* series on public television, in March 1988. Little did I suspect at the time that I was experiencing perhaps the most significant crossroads of my professional life. Just over twenty years later, George would ask me to edit his *Selected Works*. I can hardly express what an honor this has been. In the intervening two decades, I have had the pleasure of taking two year-long graduate number theory courses with George; completing an M.A. and a Ph.D. degree under his supervision; attending many of his talks at conferences, not to mention regular contact via visits, phone calls, and email. While the experience of interacting with George in person is unsurpassable, the wonderful papers contained in this volume nonetheless communicate the beauty of his mathematics with much the same clarity and enthusiasm as his lectures.

The Compilation of this Volume

Of course, when compiling a Selected Works, it is of prime importance to decide which papers are to be included, or, viewed from the opposite perspective, one must face the painful process of deciding what should be omitted. Back in the Fall of 2008, I began by consulting MathSciNet to see which of George's papers were the most cited, and by sending out an email to about one hundred mathematicians, including George's collaborators and former students, asking which papers they would like to see in such a collection. George and I also produced our own lists of favorites. We received a variety of responses to the email inquiry. Some respondents indicated one or two favorite papers; one indicated fifty "must haves"; several expressed a sentiment that I will paraphrase as follows: "only a *complete* works of George Andrews is acceptable; therefore as a matter of principle, I cannot participate in this survey." While I sympathize wholeheartedly with this latter sentiment,

I rationalize my participation in the project by hoping that the publication of this volume might help pave the way to the future publication of the *Complete Papers of George E. Andrews.*

Using the data I had collected, I compiled a spreadsheet and produced a ranked list of George's papers. George then made his (semi-)final selections using this list as a guide, whilst bearing in mind the page limitations prescribed by the publisher. I then pled the case for several papers that had not made the cut over some that had. In most cases, George agreed with my suggestions. One important, subjective guiding principle in our selection process was consideration of the current and the likely future availability of papers elsewhere. For instance, no paper published in the *Memoirs of the American Mathematical Society* was selected due to its length, and because the AMS has promised to keep such papers available. Papers in conference proceedings, particularly those that are out of print, were given special consideration. One such conference proceedings paper for which I lobbied, but ultimately was not included, was [174]: "Schur's theorem, Capparelli's conjecture and q-trinomial coefficients." I was hoping George would take the opportunity to include in his commentary the story surrounding the paper's main result. Given that this did not happen, I point the intrigued reader to §3 of my paper [Sil10], where I relate the story, based on my recollections of a conversation with George some years ago.

To the reader, you have in your hands a volume containing some of the most original and exciting mathematics in the areas of number theory, combinatorics, and q-series ever discovered. I have no doubt that readers, both contemporary and in the future, will be inspired by these pages to learn and discover new and beautiful mathematics, and that they will have a marvelously enjoyable time doing so.

Acknowledgments

I cannot even begin to express my debt of gratitude to George E. Andrews for being my teacher, mentor, role model, and inspiration.

Thanks are also due to the many people who shaped this volume by providing feedback on the papers they would like to see included. Special thanks are due to George's hardworking, creative, and kind staff assistant Barbara Baum. I cannot imagine a way that this project could have been completed without her tireless efforts.

I thank my parents, Ellen and Ray, for videotaping the *Nova* broadcast that introduced me to George E. Andrews. I thank my wife of twelve years, Leia, and my children, Eden, Hayden-Elijah and Fiona, for being my partners in this life. You have my unwavering love and gratitude.

Andrew V. Sills
Statesboro, Georgia
May 12, 2012

Contents

Curriculum Vitae

Born: December 4, 1938, Salem, Oregon, USA

Education

- B.S., M.A., Oregon State University, June, 1960
- Ph.D., University of Pennsylvania, May, 1964. Thesis advisor: Hans Rademacher

Employment Member of the Faculty, Pennsylvania State University

- 1964–1967 Assistant Professor
- 1967–1970 Associate Professor
- 1970–1981 Professor
- 1981– Evan Pugh Professor

Visiting Positions

- Visiting Professor, Massachusetts Institute of Technology, 1970–1971
- Visiting Lecturer, University of Erlangen, 1975
- Visiting Professor, University of Wisconsin, 1975–1976
- Visiting Lecturer, National University of Mexico, 1977
- Visiting Professor, University of New South Wales, 1978–1979
- Visiting Lecturer, New Zealand Mathematical Society, 1979
- Adjunct Professor, University of Waterloo, 1982–1992
- Visiting Scientist, Australian National University, 1983
- Visiting Professor, University of Strasbourg, 1983
- Visiting Ordway Professor, University of Minnesota, 1988
- Visiting Scientist, T. J. Watson Research Center, IBM, 1990–1991
- Visiting Fellow, University of Melbourne, 1997–1998
- Visiting Researcher, University of the Witwatersrand, Johannesburg, 1998
- Visiting Professor, Johannes Kepler University, Linz, Austria, 1998
- Visiting Research Fellow, School of Mathematical Sciences, University of Sussex, Brighton, UK, 1998
- Visiting Professor, University of Florida, 2005–present

Selected Honors

- Fulbright Scholar, University of Cambridge, 1960–1961
- Invited one-hour addresses at national American Mathematics Society-Mathematical Association of America meetings in 1972, 1977, 1993, and 1994
- Hedrick Lecturer, Mathematical Association of America, 1980
- Guggenheim Fellow, 1982–1983
- CBMS Regional Conference Principal Lecturer, Arizona State University, 1985
- J. S. Frame Lecturer, 1993
- Allegheny Region Distinguished Teaching Award, Mathematical Association of America
- Invited Lectures, American Association for the Advancement of Science in 1987 and 1996
- Elected Member, American Academy of Arts and Sciences, 1997
- Honorary Degree in Physics, University of Parma, May, 1998
- Invited participant, debate on teaching of first year university mathematics courses, International Conference of Mathematicians, Berlin, August, 1998
- Meeting of the 42nd Séminaire Lotharingien de Combinatoire, in honor of my sixtieth birthday, Maratea, Italy, September, 1998
- Two-day celebration of my sixtieth birthday at Pennsylvania State University (in conjunction with a regional AMS meeting), October, 1998
- Centennial Award from the Department of Mathematics, the University of Pennsylvania, "in recognition of contributions to pure mathematics and... mathematical education," October, 1999
- Honorary Degree, University of Florida, 2002
- Elected to National Academy of Sciences, 2003
- Sixty-fifth Birthday Conference at Pennsylvania State University, April, 2004
- Honorary Degree, University of Waterloo, 2004
- MAA Polya Lecturer, 2007–2008 and 2008–2009
- Elected President of AMS, 2007
- Honorary Professorship, Nankai University, 2008
- Seventieth Birthday Conference at Pennsylvania State University, December, 2008
- Selected Fellow of Society for Industrial and Applied Mathematics, 2009

Publications of George E. Andrews

Articles

[1] An asymptotic expression for the number of solutions of a general class of Diophantine equations. *Trans. Amer. Math. Soc.* 99:272–277 (1961)

[2] A lower bound for the volume of strictly convex bodies with many boundary lattice points. *Trans. Amer. Math. Soc.* 106:270–279 (1963)

[3] On estimates in number theory. *Amer. Math. Monthly* 70:1063–1065 (1963)

[4] A simple proof of Jacobi's triple product identity. *Proc. Amer. Math. Soc.* 16:333–334 (1965)

[5] On basic hypergeometric series, mock theta functions, and partitions, I. *Quart. J. Math.* 17:64–80 (1966)

[6] On basic hypergeometric series, mock theta functions, and partitions, II. *Quart. J. Math.* 17:132–143 (1966)

[7] q-identities of Auluck, Carlitz, and Rogers. *Duke Math. J.* 33:575–582 (1966)

[8] On the theorems of Watson and Dragonette for Ramanujan's mock theta functions. *Amer. J. Math.* 88:454–490 (1966)

[9] On generalizations of Euler's partition theorem. *Mich. Math. J.* 13:491–498 (1966)

[10] An analytical proof of the Rogers–Ramanujan-Gordon identities. *Amer. J. Math.* 88:844–846 (1966)

[11] Enumerative proofs of certain q-identities. *Glasgow Math. J.* 8:33–40 (1967)

[12] A generalization of a partition theorem of MacMahon. *J. Comb. Theory* 3:100–101 (1967)

[13] On Schur's second partition theorem. *Glasgow Math. J.* 9:127–132 (1967)

[14] Partition theorems related to the Rogers–Ramanujan identities. *J. Comb. Theory* 2:422–430 (1967)

[15] Some new partition theorems. *J. Comb. Theory* 2:431–436 (1967)

[16] A generalization of the Göllnitz–Gordon partition theorems. *Proc. Amer. Math. Soc.* 8:945–952 (1967)

[17] On the geometry of numbers in elementary number theory. *Amer. Math. Monthly* 74:1124–1125 (1967)

[18] On partition functions related to Schur's second partition theorem. *Proc. Amer. Math. Soc.* 18:441–444 (1968)

[19] Some formulae for the Fibonacci sequence with generalizations. *Fibonacci Quart.* 7:113–130 (1969)

[20] A general theorem on partitions with difference conditions. *Amer J. Math* 91:18–24 (1969)

[21] A new generalization of Schur's second partition theorem. *Acta Arith.* 14:429–434 (1968)

[22] A partition theorem of Göllnitz and related formulae. *Journal für die reine und angewandte Mathematik* 236:37–42 (1969)

[23] On q-difference equations for certain well–poised basic hypergeometric series. *Quart. J. Math.* 19:433–447 (1968)

[24] A polynomial identity which implies the Rogers–Ramanujan identities. *Scripta Math.* 28:297–305 (1970)

[25] On radix representation and the Euclidean algorithm. *Amer. Math. Monthly* 76:66–68 (1969)

[26] Two theorems of Euler and a general partition theorem. *Proc. Amer. Math. Soc.* 20:499–502 (1969)

[27] Some new partition theorems. II. *J. Comb. Theory* 7:262–263 (1969)

[28] A new property of partitions with applications to the Rogers–Ramanujan identities. *J. Comb. Theory* 10:266–270 (1971)

[29] Note on a partition theorem. *Glasgow Math. J.* 11:108–109 (1970)

[30] A note on the Bombieri–Selberg formula for algebraic curves. *Portugaliae Math.* 27:75–81 (1968)

[31] A generalization of the classical partition theorems. *Trans. Amer. Math. Soc.* 145:205–221 (1968)

[32] On Ramanujan's summation of ${}_1\Psi_1(a; b; z)$. *Proc. Amer. Math. Soc.* 22:552–553 (1969)

[33] On a calculus of partition functions. *Pacific J. Math.* 31:555–562 (1969)

[34] On a partition problem of J. J. Sylvester. *J. London Math. Soc.* 2(2):571–576 (1970)

[35] On a transformation of bilateral series with applications. *Proc. Amer. Math. Soc.* 25:554–558 (1970)

[36] On the Rogers–Ramanujan identities and partial q-difference equations. *Illinois J. Math.* 16:270–275 (1972)

[37] Generalizations of the Durfee square. *J. London. Math. Soc.* 3(2):563–570 (1971)

[38] The use of computers in search of identites of the Rogers–Ramanujan type. *Computers in Number Theory*, A. O. L. Atkin and B. J. Birch, eds., Academic Press, New York, pp. 377–387 (1971).

[39] A combinatorial proof of a partition function limit. *Amer. Math. Monthly* 78:276–278 (1971)

[40] On a theorem of Schur and Gleissberg. *Arch. Math.* 22:165–167 (1971)

[41] On a conjecture of Guinand for the plane partition function. *Proc. Edinburgh Math. Soc.*, Series II, 17:275–276 (1971)

[42] Congruence properties of the m-ary partition function. *J. Number Theory* 3:104–110 (1971)

[43] On a partition problem of H. L. Alder. *Pacific J. Math.* 36:279–284 (1971)

[44] Lambert series, false theta functions, and partitions. *Rocky Mt. J. Math.* 2:49–56 (1972)

[45] Summations and transformations for basic Appell series. *J. London Math. Soc.* 4(2):618–622 (1972)

[46] (with M. V. Subbarao and M. Vidyasagar). A family of combinatorial identities. *Canadian Math. Bull.* 15:11–18 (1972)

[47] On identities implying the Rogers–Ramanujan identities. *Houston Math. J.* 2(3):289–298 (1976)

[48] Two theorems of Gauss and allied identities proved arithmetically. *Pacific J. Math.* 41:563–578 (1972)

[49] Sieves for theorems of Euler, Rogers and Ramanujan. *Lecture Notes in Mathematics #251*, Springer-Verlag, New York, pp. 1–20 (1972)

[50] Sieves in the theory of partitions. *Amer. J. Math.* 94:1214–1230 (1972)

[51] Partition identities. *Advances in Math.* 9:10–51 (1972)

[52] On the foundations of combinatorial theory, V: Eulerian differential operators. *Studies in Applied Math.* 50:345–375 (1971)

[53] On the Alder polynomials and new generalization of the Rogers–Ramanujan identities. *Trans. Amer. Math. Soc.* 204:40–64 (1975)

[54] Combinatorial analysis and Fibonacci numbers. *Fibonacci Quart.* 12(2): 141–146 (1974)

[55] On the q-analog of Kummer's theorem and applications. *Duke Math. J.* 40(3):525–528 (1973)

[56] Partially ordered sets and the Rogers–Ramanujan identities. *Aequationes Math.* 12(1):94–107 (1975)

[57] A general theory of identities of the Rogers–Ramanujan type. *Bull. Amer. Math. Soc.* 80(6):1033–1052 (1974)

[58] On the general Rogers–Ramanujan theorem. *Memoir. Amer. Math. Soc.* 152, 86 pp. (1974)

[59] Applications of basic hypergeometric functions. *SIAM Rev.* 16(4):441–484 (1974)

[60] On Rogers–Ramanujan type identities related to the modulus 11. *Proc. London Math. Soc.* 30:330–346 (1975)

[61] An analytic generalization of the Rogers–Ramanujan identities for odd moduli. *Proc. Nat. Acad. Sci.* 71(10):4082–4085 (1974)

[62] A theorem on reciprocal polynomials with applications to permutations and compositions. *Amer. Math. Monthly* 82(8):830–833 (1975)

[63] Identities in combinatorics, I: On sorting two ordered sets. *Discrete Math.* 11(2):97–106 (1975)

[64] The theory of compositions, I: The ordered factorization of n and a conjecture of C. Long. *Canadian Math. Bull.* 18(4):479–484 (1975)

[65] The theory of compositions, II: Simon Newcomb's problem. *Utilitas Math.* 7:35–54 (1975)

[66] Identities in combinatorics, II: A q-analog of the Lagrange inversion theorem. *Proc. Amer. Math. Soc.* 53(1):240–245 (1975)

[67] q-analogs of the Watson and Whipple summations. *SIAM J. Math. Anal.* 7(3):332–336 (1976)

[68] MacMahon's prime numbers of measurement. *Amer. Math. Monthly* 82(9): 922–923 (1975)

[69] Partition ideals of order 1, the Rogers–Ramanujan identities and computers. *Proc. Séminaire Dubreil (algèbre) 19éme année* 20:1–16 (1975)

[70] Problems and prospects for basic hypergeometric functions. *Theory and Application of Special Functions*, R. Askey, ed., Academic Press, New York, pp. 191–214 (1975)

[71] The theory of compositions, III: MacMahon's formula and the Stanton–Cowan numbers. *Utilitas Math.* 9:283–290 (1976)

[72] MacMahon's conjecture on symmetric plane partitions. *Proc. Nat. Acad. Sci.* 74(2):426–429 (1977)

[73] (with R. Askey). Enumeration of partitions: The role of Eulerian series and q-orthogonal polynomials. *Higher Combinatorics*, M. Aigner, ed., Reidell Publ. Co., Dordrecht, pp. 3–26 (1977)

[74] An extension of Carlitz's bipartition identity. *Proc. Amer. Math. Soc.* 64(1):180–184 (1977)

[75] Plane partitions, II: The equivalence of the Bender-Knuth and MacMahon conjectures. *Pacific J. Math.* 72(2):283–291 (1977)

[76] Partitions, q-series and the Lusztig–Macdonald–Wall conjectures. *Invent. Math.* 41:91–102 (1977)

[77] (with I. Gessel). Divisibility properties of q-tangent numbers. *Proc. Amer. Math. Soc.* 68(1):380–384 (1978)

[78] An introduction to Ramanujan's "lost" notebook. *Amer. Math. Monthly* 86(2) 89–108 (1979)

[79] Plane partitions, I: The MacMahon conjecture. *Adv. in Math. Suppl. Studies* 1:131–150 (1979)

[80] Partitions and Durfee dissection. *Amer. J. Math.* 101:735–742 (1979)

[81] (with D. M. Bressoud). Vanishing coefficients in infinite product expansions. *J. Austral. Math. Soc.*, Series A, 27:199–202 (1979)

[82] Connection coefficient problems and partitions. *Proc. Symp. in Pure Math.* 34:1–24 (1979)

[83] Ramanujan and his "lost" notebook. *Vinculum* 16(3):91–94 (1979)

[84] Plane partitions, III: The weak Macdonald conjecture. *Invent. Math.* 53:193–225 (1979)

[85] An incredible formula of Ramanujan. *Austral. Math. Soc. Gaz.* 6:80–89 (1979)

[86] A note on partitions and triangles with integer sides. *Amer. Math. Monthly* 86(6):477–478 (1979)

[87] Notes on the Dyson conjecture. *SIAM J. Math. Anal.* 11(5):787–792 (1980)

[88] (with R. Askey). A simple proof of Ramanujan's ${}_1\Psi_1$ summation. *Aequa. Math.* 18:333–337 (1978)

[89] Implications of the MacMahon conjecture. *Lecture Notes in Mathematics #579*, Springer-Verlag, New York, pp. 287–296 (1977)

[90] Ramanujan's "lost" notebook, I: partial theta functions. *Adv. Math.* 41:137–172 (1981)

[91] Gap-frequency partitions and the Rogers–Selberg identities. *Ars Comb.* 9:201–210 (1980)

[92] Ramanujan's "lost" notebook, II: θ function expansions. *Adv. Math.* 41:173–185 (1981)

[93] Ramanujan's "lost" notebook, III: The Rogers–Ramanujan continued fraction. *Adv. Math.* 41:186–208 (1981)

[94] Multiple q-series identities. *Houston Math. J.* 7:11–22 (1981)

[95] Macdonald's conjecture and descending plane partitions. *Combinatorics, Representation Theory and Statistical Methods in Groups*, T. V. Narayana, ed., Marcel Dekker, New York, pp. 91–106 (1980)

[96] The Rogers–Ramanujan reciprocal and Minc's partition function. *Pacific J. Math.* 95:251–256 (1981)

[97] L. J. Rogers and the Rogers–Ramanujan identities. *Math. Chronicle* 11:1–15 (1982)

[98] (with R. Askey). Another q-extension of the β-function. *Proc. Amer. Math. Soc.* 81(1):97–100 (1981)

[99] (with D. Foata). Congruences for the q-secant numbers. *European J. Comb.* 1:283–297 (1980)

[100] The Mordell integrals and Ramanujan's "lost" notebook. *Lecture Notes in Mathematics #899*, Springer-Verlag, New York, pp. 10–48 (1980).

[101] The hard-hexagon model and the Rogers–Ramanujan type identities. *Proc. Natl. Acad. Sci.* 78:5290–5292 (1981)

[102] Ramanujan's "lost" notebook, IV: Stacks and alternating parity in partitions. *Adv. Math.* 53:55–74 (1984)

[103] On the Wall polynomials and the L-M-W conjectures. *Australian J. Math.* 37:17–26 (1984)

[104] Hecke modular forms and the Kac–Peterson identities. *Trans. Amer. Math. Soc.* 283(2):451–458 (1984)

[105] Euler's pentagonal number theorem. *Math. Mag.* 56(5):279–284 (1983) Reprinted in *The Genius of Euler: Reflections on his Life and Work*, W. Dunham, ed., MAA, Washington, pp. 225–232 (2007)

[106] Use and extension of Frobenius' representation of partitions. *Enumeration and Design*, Academic Press, New York, pp. 51–65 (1984)

[107] Multiple series Rogers–Ramanujan type identities. *Pac. J. Math.* 114(2): 267–283 (1984)

[108] (with E. Onofri). Lattice gauge theory, orthogonal polynomials and q-hypergeometric series. *Special Functions: Group Theoretical Aspects and Applications*, R. A. Askey *et al.*, eds., Reidel Pub. Co., Dordrecht pp. 163–188 (1984)

[109] Generalized Frobenius Partitions. *Memoir. Amer. Math. Soc. 49* (301) (1984)

[110] Ramanujan and SCRATCHPAD. *Proc. 1984 MACSYMA Users' Conf.*, General Electric, Schenectady, pp. 383–408 (1984)

[111] Physics, Ramanujan and computer algebra. *Computer Algebra*, D. V. Chudnovsky and R. Jenks, eds., M. Dekker, New York, pp. 97–109 (1989)

[112] (with R. J. Baxter and P. J. Forrester). Eight-vertex SOS model and generalized Rogers–Ramanujan-type identities. *J. of Stat. Phys.* 35:193–266 (1984)

[113] (with D. Bressoud). Identities in combinatorics, III: Further aspects of ordered set sorting. *Discrete Math.* 49:223–236 (1984)

[114] On a partition theorem of N. J. Fine. *J. Natl. Acad. Math.*, India, 1:105–107 (1983)

[115] Combinatorics and Ramanujan's "lost" notebook. *Surveys in Combinatorics 1985*, London Math. Soc. Lecture Note Series, I. Anderson, ed., Cambridge University Press, London, 103:1–23 (1985)

[116] The fifth and seventh order mock theta functions. *Trans. Amer. Math. Soc.* 293(1):113–134 (1986)

[117] Reciprocal polynomials and quadratic transformations. *Utilitas Math.* 28: 265–269 (1985)

[118] (with D. Bressoud). On the Burge correspondence between partitions and binary words. *Rocky Mt. J. Math.* 15:225–233 (1985)

[119] (with K. Uchimura). Identities in combinatorics, IV: Differentiation and harmonic numbers. *Utilitas Math.* 28:265–269 (1985)

[120] (with R. J. Baxter). Lattice gas generalization of the hard hexagon model, I: Star triangle relation and local densities. *J. Stat. Phys.* 44(1/2): 259–271 (1986)

[121] The Rogers–Ramanujan identities without Jacobi's triple product. *Rocky Mtn. J. Math.* 17:659–672 (1987)

[122] (with I. Goulden and D. M. Jackson). Shanks' convergence acceleration transform, Pade approximants and partitions. *J. Comb. Th., Series A.* 43(1):70–84 (1986)

[123] (with D. M. Bressoud, R. J. Baxter, W. Burge, P. J. Forrester and G. Viennot). Partitions with prescribed hook differences. *European J. Math.* 8:341–350 (1987)

[124] Ramanujan's "lost" notebook, V: Euler's partition identity. *Advances in Math.* 61(2):156–164 (1986)

[125] (with R. Askey). Classical orthogonal polynomials. *Proc. Conf. on Orthog. Polys. at Bar–le duc*, Lecture Notes in Math., Springer, Berlin, 1171: 36–62 (1985)

[126] (with A. K. Agarwal). Hook differences and lattice paths. *J. Statistical Planning and Inference* 14:5–14 (1986)

[127] ΕΥΡΗΚΑ! num $= \Delta + \Delta + \Delta$. *J. Num. Theory* 23(3):285–293 (1986)

[128] (with R. J. Baxter). Lattice gas generalization of the hard hexagon model. II. The local densities as elliptic functions. *J. Statisical Physics* 44:713–728 (1986)

[129] Questions and conjectures in partition theory. *Amer. Math. Monthly* 93:708–711 (1986)

[130] (with P. J. Forrester). Height probabilities in solid-on-solid models: I (letter to the editor) *J. Phys. A.: Math. Gen.* 19:L923–L926 (1986)

[131] (with A. K. Agarwal). Rogers–Ramanujan identities for partitions with "N copies of N." *J. Comb. Theory, Series A*, 45(1):40–49 (1987)

[132] (with R. J. Baxter). Lattice gas generalization of the hard hexagon model. III. q-trinomial coefficients. *J. of Stat. Physics* 47(3/4):297–330 (1987)

[133] Further problems on partitions. *Amer. Math. Monthly* 94(5):437–439 (1987)

[134] Catalan numbers, q-Catalan numbers and hypergeometric series. *J. Comb. Theory, Series A*, 44(2):267–273

[135] Plane partitions, IV: A conjecture of Mills–Robbins–Rumsey. *Aequationes Mathematicae* 33:230–250 (1987)

[136] (with A. K. Agarwal). On Asai's polynomials related to the twisting operators of finite classical groups. *J. Ramanujan Math. Soc.* 1:16–45 (1986)

[137] (with F. J. Dyson and D. R. Hickerson). Partitions and indefinite quadratic forms. *Invent. Math.* 91:391–407 (1988)

[138] (with F. G. Garvan). Ramanujan's "lost" notebook, VI: The mock theta conjectures. *Advances in Math.* 73:242–255 (1989)

[139] (with F. G. Garvan). Dyson's crank of a partition. *Bull. Amer. Math. Soc.* 18:167–171 (1988)

[140] (with R. J. Baxter). A motivated proof of the Rogers–Ramanujan identities. *Amer. Math. Monthly* 96:401–409 (1989)

[141] (with R. J. Baxter). SCRATCHPAD explorations for elliptic theta functions. *Computers in Mathematics, Lecture Notes in Pure and Appl. Math. 125*, D. V. Chudnovsky and R. D. Jenks, eds., M. Dekker, New York, 17–33 (1990)

[142] Ramanujan's fifth order mock theta functions as constant terms. *Ramanujan Revisited,* G. Andrews *et al.*, eds., Academic Press, New York, pp. 47–56 (1988)

[143] (with I. P. Goulden and D. M. Jackson). Generalizations of Cauchy's summation theorem for Schur functions. *Trans. Amer. Math. Soc.* 310:805–820 (1988)

[144] (with P. J. Forrester). Height probabilities in solid-on-solid models: II *J. Physics A: Math. Gen.* 20:4465–4481 (1987)

[145] (with A. K. Agarwal and D. M. Bressoud). The Bailey lattice. *J. Indian Math. Soc.* 51:57–73 (1987)

[146] Rogers–Ramanujan identities for two-color partitions. *Indian J. of Math.* 29:117–125 (1987)

[147] Introduction to *Srinivasa Ramanujan: the Lost Notebook and Other Unpublished Papers*, Narosa, New Delhi (1987)

[148] J. J. Sylvester, Johns Hopkins and partitions. *A Century of Mathematics in America, Part I*, P. Duren, ed., Amer. Math. Soc., Providence (1988)

[149] An identity of Sylvester and the Rogers–Ramanujan identities. *Number Theory, Madras 1987, Lecture Notes in Math. 1395*, Springer, Berlin, pp. 64–72 (1989)

[150] Mock theta functions. *Proc. Symposia in Pure Math.* 49:283–298 (1989)

[151] A formula of Ramanujan from the "Lost" Notebook. *Ramanujan International Symposium on Analysis*, N. K. Thakare, ed., Macmillan India, Delhi (1989)

[152] On the proofs of the Rogers–Ramanujan identities. From *q-Series and Partitions, IMA Volumes in Math. and its Application*, D. Stanton, ed., Springer, New York, 18:1–14 (1989)

[153] (with Y. Kitaoka). A result on q-series and its application to quadratic forms. *J. Number Th.* 34:54–62 (1990)

[154] Euler's "Exemplum Memorabile Inductionis Fallacis" and q-trinomial coefficients. *J. Amer. Math. Soc.* 3:653–669 (1990)

[155] Partitions and the Gaussian sum. *The Mathematical Heritage of C. F. Gauss*, G. Rassias, ed., World Sc. Pub., Singapore (1990)

[156] q-trinomial coefficients and Rogers–Ramanujan type identities. *Analytic Number Theory*, B. Berndt *et al.*, eds., Birkhäuser, Boston, pp. 1–11 (1990)

[157] (with B. Berndt, L. Jacobsen, and R. L. Lamphere). Variations on the Rogers–Ramanujan continued fraction. *Number Theory, Madras (1987) Lecture Notes in Math., No. 1395*, Springer, Berlin, pp. 73–83 (1989)

[158] (with J. B. Olsson). Partition identities with an application to group representation theory. *Journal für die reine und angewandte Mathematik* 413:198–212 (1991)

[159] (with D. R. Hickerson). Ramanujan's "lost" notebook, VII: The sixth order mock theta functions. *Advances in Math.* 89:60–105 (1991)

[160] (with D. M. Jackson). An algebraically derived q-analog of a character sum associated with a class of semiregular permutations. *Pac. J. Math.* 144(2):207–218 (1990)

[161] A page from Ramanujan's lost notebook. *Indian J. of Math.* 32:207–216 (1990).

[162] (with G. Almkvist). A Hardy–Ramanujan formula for restricted partitions. *J. Number Th.* 38:135–144 (1991)

[163] On the difference of successive Gaussian polynomials. *J. Planning and Stat. Inf.* 34:19–22 (1993)

[164] On Ramanujan's empirical calculation for the Rogers–Ramanujan identities. *Contemporary Math.* 143:27–35 (1993).

[165] Bailey chains and generalized Lambert series, I: Four identities of Ramanujan. *Illinois J. Math.* 36:251–274 (1992)

[166] Plane partitions, V: The T.S.S.C.P.P. conjecture. *J. Combinatorial Th. (A)* 66(1):28–39 (1994)

[167] (with B. C. Berndt, L. Jacobsen, and R. L. Lamphere). The continued fractions found in the unorganized portions of Ramanujan's notebooks. *Memoir. Amer. Math. Soc.*, No. 477, 99:vi+71 pp. (1992)

[168] Application of SCRATCHPAD to problems in special functions and combinatorics. In *Trends in Computer Algebra*, R. Janssen, ed., Springer Lecture Notes in Comp. Sci., No. 296, 159–166 (1988)

[169] Three aspects of partitions. *Proc. of 25th Lotharingian Combinatorics Conf.*, Sept 1990

[170] The reasonable and unreasonable effectiveness of number theory in statistical mechanics. *Proc. Symp. in Appl. Math.* 46:21–34 (1992)

[171] On the Greene–Krammer theorem and related identities. *Ganita* 43:69–73 (1992)

[172] (with S. B. Ekhad and D. Zeilberger). A short proof of Jacobi's formula for the number of representations of an integer as a sum of four squares. *Amer. Math. Monthly*, 100:274–276 (1993)

[173] (with W. H. Burge). Determinant identities. *Pac. J. Math.*, 158:1–14 (1993)

[174] Schur's theorem, Capparelli's conjecture and q-trinomial coefficients. *Contemporary Math.* 166:141–151 (1994)

[175] (with D. M. Jackson and T. Visentin). A hypergeometric analysis of the genus series for a class of 2-cell embeddings in orientable surfaces. *S.I.A.M. J. Math. Analysis*, 25:243–255 (1994)

[176] (with C. Bessenrodt and J. B. Olsson). Partition identities and labels for some modular characters. *Trans. Amer. Math. Soc.* 344:597–615 (1994)

[177] (with K. Alladi and B. Gordon). Generalizations and refinements of a partition theorem of Gollnitz. *Journal für die reine und angewandte Mathematik* 460:165–188 (1995)

[178] (with K. Alladi and B. Gordon). Refinements and generalizations of Capparelli's conjecture on partitions. *J. of Algebra* 174:636–658 (1995)

[179] Modular equations and Ramanujan's chapter 16, entry 29. *Proc. Indian Acad. Sci. (Math.)*, 104:225–235 (1994)

[180] (with P. Paule). Some questions concerning computer-generated proofs of a binomial double-sum identity. *J. Symbolic Comp.* 16:147–153 (1993)

[181] (with D. Crippa and K. Simon). q-series arising from the study of random graphs. *S.I.A.M. J. Discrete Math.* 10(1):41–56 (1997)

[182] (with D. Bowman). A full extension of the Rogers–Ramanujan continued fraction. *Proc. Amer. Math. Soc.* 123(11):3343–3350 (1995)

[183] On a conjecture of Peter Borwein. *J. Symbolic Computation* 20:487–501 (1995)

[184] Pfaff's Method (I): The Mills–Robbins–Rumsey determinant. *Discrete Math.* 193:43–60 (1998)

[185] Pfaff's Method (II): Diverse applications. *J. Computational & Appl. Math.* 68: 15–23 (1996)

[186] Pfaff's Method (III): Comparison with the WZ method. *Electronic J. Combin.*, (Foata Festschrift) 3(2):R21, 18 pp., (1996). Also in *Proceedings Organic Math Workshop* at `http://www.cecm.sfu.ca/projects/OMP/`; *Organic Mathematics Canadian and Mathematical Society Proceedings*, J. Borwein, P. Borwein, L. Jorgenson and R. Corless, eds., Vol. 20, pp. 19–33,

AMS, Providence (1997); and *The Foata Festschrift* a reprint of the *Electronic J. of Combinatorics*, J. Désarménien, A. Kerber and V. Strehl (eds.), Vol 3 (no. 2), pp. 517–534. Imprimerie Louis-Jean, Gap (1996)

[187] (with D. Stanton). Determinants in Plane Partitions Enumeration. *Europ J. Combinatorics*, 19:273–282 (1998)

[188] (with J. P. O. Santos). Rogers–Ramanujan type identities for partitions with attached odd parts. *Ramanujan Journal* 1(1):91–99 (1997)

[189] (with C. Bessenrodt and J. B. Olsson). A refinement of a partition identity and blocks of some modular characters. *Arch. Math.* 66:101–113 (1996)

[190] Rogers–Ramanujan polynomials for modulus 6. In *Analytic Number Theory: Proceedings of a Conference in Honor of Heini Halberstam*, B. C. Berndt, H. G. Diamond and A. J. Hildebrand, eds., vol I, Birkhäuser, Boston, pp. 17–30 (1996)

[191] The well-poised thread. *Ramanujan Journal* 1(1):7–23 (1997)

[192] (with K. Alladi). A new key identity for Göllnitz's (Big) partition theorem. In *Contemporary Mathematics 210 Number Theory*, Proceedings of a Conference of The Ramanujan Mathematical Society, January 3–6, 1996, V. K. Murty and M. Waldschmidt, eds., AMS, Providence, pp. 229–241 (1998)

[193] MacMahon's partition analysis: I. the lecture hall partition theorem. In *Mathematical Essays in Honor of Gian-Carlo Rota*, B. E. Sagan and R. P. Stanley, eds, Birkhäuser, Boston, pp. 1–22 (1998)

[194] (with R. Roy). Ramanujan's method in q-series congruences. *Electronic J. Combin.*, (Wilf Festschrift) 4(2):R2, 7 pp. (1997)

[195] (with K. Alladi). A quartic key identity for a partition theorem of Göllnitz. *J. Number Theory* 75:220–236, (1999)

[196] q-analogs of the binomial coefficient congruences of Babbage, Wolstenholme and Glaisher. *Discrete Mathematics* 204:15–25, (1999)

[197] (with A. Berkovich) A trinomial analogue of Bailey's lemma and N = 2 superconformal invariance. *Comm. in Math. Physics* 192:245–260 (1998)

[198] Simplicity and surprise in Ramanujan's "Lost Notebook." *Amer. Math. Monthly* 104(10):918–925 (1997)

[199] (with D. Bowman) The Bailey transform and D. B. Sears. *Quaestiones Mathematicae*, 22(1):19–26 (1999)

[200] Schur's theorem, partitions with odd parts and the Al-Salam–Carlitz polynomials. *AMS Contemporary Mathematics*, 254:45–53 (2000)

[201] (with A. and J. Knopfmacher) Engel expansions and the Rogers–Ramanujan identities. *J. Number Theory* 80(2):273–290 (2000). Also available online at http://www.idealibrary.com/

[202] (with R. Lewis) The ranks and cranks of partitions moduli 2, 3 and 4. *J. Number Theory*, 85:74–84 (2000)

[203] (with O. Warnaar and A. Schilling) An A_2 Bailey lemma and Rogers–Ramanujan-type identities. *J. Amer. Math. Soc.* 12:677–702 (1999)

[204] (with R. Lewis) An algebraic identity of F. H. Jackson and its implications for partitions. *Discrete Math.* 232:77–83 (2001)

[205] (with R. Lewis) Restricted bipartitions *Discrete Math.* 232(1–3):85–89 (2001)

[206] MacMahon's partition analysis: II. Fundamental theorems. *Annals of Combinatorics* 4:327–338 (2000)

[207] (with P. Paule and A. Riese) MacMahon's partition analysis: The omega package. *Europ J. Combinatorics* 22(7):887–904 (2001)

[208] (with P. Paule) MacMahon's partition analysis IV: Hypergeometric multisums. Paper B42i, Issue 42, *Séminaire Lotharingien de Combinatoire* (electronic journal) (Issue 42 consists of papers presented at meeting held in my honor at Maratea, Italy, September, 1998). Also in *The Andrews Festschrift: Seventeen Papers on Classical Number Theory and Combinatorics*, eds. D. Foata and G-N. Han, Springer, Berlin pp. 189–208 (2001)

[209] Stacked lattice boxes. *Annals of Combinatorics* 3:115–130 (1999)

[210] (with A. Knopfmacher) An algorithmic approach to discovering and proving q-series identity. *Algorithmica* 29:34–43 (2001)

[211] Three-quadrant Ferrers graphs. *Indian J. Math.* 42(1):1–7 (2000)

[212] (with P. Paule, A. Riese, and V. Strehl) MacMahon's partition analysis V: Bijections, recursions and magic squares. *Algebraic Combinatorics and Applications*, proceedings of Euroconference Alcoma 1999, September 12–19, 1999, Goessweistein, Germany, A. Betten, A. Kohnert, R. Laue and A. Wassermann, eds., Springer, Berlin, pp. 1–39 (2001)

[213] Some debts I owe. Paper B42a, Issue 42, Séminaire Lotharingien de Combinatoire (electronic journal) (Issue 42 consists of papers presented at meeting held in my honor at Maratea, Italy, September, 1998) Also in *The Andrews Festschrift: Seventeen Papers on Classical Number Theory and Combinatorics*, eds. D. Foata and G-N. Han, Springer, Berlin, pp. 1–16 (2001)

[214] (with A. Knopfmacher and P. Paule) An infinite family of Engel expansions of Rogers–Ramanujan type. *Advances in Applied Math* 25:2–11 (2000)

[215] (with C. Krattenthaler, L. Orsina, and P. Papi) *ad*–nilpotent b–ideals in sl(n) having a fixed class of nilpotence: Combinatorics and enumeration. *Trans. Amer. Math. Soc.*, 354(10):3835–3853 (2002)

[216] (with R. Lewis and Z-G. Liu) An identity relating a theta function to a sum of Lambert series. *Bull. London Math. Soc.*, 33:25–31 (2001)

[217] (with L. Guo, W. Keigher and K. Ono) Baxter algebras and Hopf algebras. *Trans. Amer. Math. Soc.* 355(11):4639–4656 (2003)

[218] (with A. Knopfmacher and B. Zimmermann) On the number of distinct values of multinomial coefficients. *Journal of Number Theory* 118:15–30 (2006)

[219] (with R. Lewis and J. Lovejoy) Partitions with designated summands. *Acta Arithmetica* 105(1):51–66 (2002)

[220] Umbral calculus, Bailey chains and pentagonal number theorems. *JCT(A)* 91:464–475 (2000)

[221] The Friedman–Joichi–Stanton monotonicity conjecture at primes. *DIMACS Series in Discrete Mathematics and Theoretical Computer Science*, 64, M. B. Nathanson, ed., AMS, Providence, pp. 9–15 (2004)

[222] (with K. Ono and J. Jimenez Urroz) q-series identities and values of certain L-functions. *Duke Math. Journal* 108(3):395–419 (2001)

[223] (with A. Knopfmacher, P. Paule and B. Zimmermann) Engel expansions of q-series by computer algebra. *Symbolic Computation, Number Theory, Special Functions, Physics and Combinatorics*, proceedings of Conference on Symbolic Computation, Number Theory, Special Functions, Physics and Combinatorics, University of Florida, November, 1999, Frank Garvan and Mourad Ismail, eds., Kluwer, Dordrecht, pp. 33–57 (2001)

[224] (with K. Alladi and A. Berkovich) A four parameter generalization of Göllnitz's (big) partition theorem. *DIMACS Series in Discrete Mathematics*

and Theoretical Computer Science, 64, M. B. Nathanson, ed., AMS, Providence, pp. 1–7 (2004)

[225] Bailey's transform, lemma, chain and tree. *Special Functions 2000*, eds. J. Bustoz *et al.*, Kluwer, Dordrecht, pp. 1–22 (2001)

[226] (with J. Wimp) Some q-orthogonal polynomials and related Hankel determinants. *Rocky Mt. Journal of Math.* 32(2):429–442 (2002)

[227] Partitions: At the interface of q-series and modular forms. *Ramanujan Journal* 7(1,2,3):385–400 (2003)

[228] (with P. Paule and A. Riese) MacMahon's partition analysis VI: A new reduction algorithm. *Annals Combin.* 5:251–270 (2002)

[229] (with P. Paule and A. Riese) MacMahon's partition analysis VII: Constrained compositions. *AMS Contemporary Mathematics* 291:11–27 (2001)

[230] (with P. Paule and A. Riese) MacMahon's partition analysis VIII: Plane partition diamonds. *Advances in Appl. Math.* 27:231–242 (2001)

[231] A Bailey lemma from the quintuple product. In *Number Theory & Discrete Mathematics*, A. Agarwal *et al.* eds., Hindustan Book Agency, New Delhi, pp. 37–46 (2002)

[232] (with P. Paule and A. Riese) MacMahon's partition analysis IX: k-gon partitions. *Bull. Austral. Math. Soc.* 64:321–329 (2001)

[233] (with A. Knopfmacher, P. Paule & H. Prodinger) q-Engel series expansions and Slater's identities. *Quaestiones Mathematicae* 24(30):403–416 (2001)

[234] (with A. Berkovich) The WP-Bailey tree and its implications. *J. London Math. Soc.* (2) 66:529–549 (2002)

[235] (with B. Berndt, J. Sohn, A.J. Yee and A. Zaharescu) Continued fractions with three limit points. *Advances in Math* 192:231–258 (2005)

[236] (with B. Berndt, J. Sohn, A. J. Yee and A. Zaharescu) On Ramanujan's continued fraction for $(q^2;q^3)_\infty/(q;q^3)_\infty$. *Trans. Amer. Math. Soc.* 355(6):2397–2411 (2003)

[237] (with R. A. Pèrez) A new partition identity arising from complex dynamics. *Annals of Combinatorics* 9:245–257 (2005)

[238] Fibonacci numbers and the Rogers–Ramanujan identities. *Fibonacci Quarterly* 42(1):3–19 (2004)

[239] (with K. Alladi and A. Berkovich) A new four parameter q-series identity and its partition implications. *Invention. Math.* 153(2):231–260 (2003)

[240] (with I. P. Goulden and D. M. Jackson) Determinants of random matrices and Jack polynomials of rectangular shape. *Studies Applied Math* 110:377–390 (2003)

[241] a-Gaussian polynomials and finite Rogers–Ramanujan identities. In *Theory and Applications of Special Functions: A Volume Dedicated to Mizan Rahman*, M. Ismail and E. Koelink, eds., Springer, New York, pp. 39–60 (2005).

[242] Ramanujan's "Lost Notebook" VIII: the entire Rogers–Ramanujan function. *Advances in Math* 191:393–407 (2005)

[243] On a partition function of Richard Stanley. Stanley Festschrift, *Electronic Journal of Combinatorics*, Vol. 11 (2), `http://www.combinatorics.org/`

[244] Ramanujan's "Lost Notebook." IX the partial theta function as an entire function. *Advances in Math* 191:408–422 (2005)

[245] (with P. Freitas) Extension of Abel's lemma with q-series implications. *Ramanujan Journal* 10:137–152 (2005)

[246] Carlitz and the general ${}_3\phi_2$. Askey Special Issues, *Ramanujan Journal* 13, #1/2/3:311–318 (2007)

[247] The meaning of Ramanujan. *Proceedings of International Conference Number Theory for Secure Communications*, SASTRA University, December 2003, 1–14

[248] (with P. Paule and C. Schneider) Plane partitions VI: Stembridge's TSPP theorem. *Advances in Appl. Math.* 34:709–739 (2005)

[249] (with P. Paule and A. Riese) MacMahon's partition analysis X: Plane partitions with diagonals. *Southeast Asian Journal Math. and Sci.* 3(1) (in honor of R. P. Agarwal) 3–14 (2004)

[250] (with P. Paule and A. Riese) MacMahon's partition analysis XI: Broken diamonds and modular forms. *Acta Arith.* 126.3:281–294 (2007)

[251] Ramanujan and partial fractions. In *Contributions to the History of Indian Mathematics*, vol. 3 in series Culture and History of Mathematics, G. Emch, R. Sridharan, and M. Srinivas, eds., Hindustan Book Agency, New Delhi, pp. 251–260 (2005)

[252] Partitions with short sequences and mock theta functions. *Proc. Nat. Acad. Sci.* 102(13):4666–4671 (2005)

[253] (with K. Alladi, K. Ono, and R. McIntosh) On the work of Basil Gordon. *JCT(A)* 113:21–38 (2006)

[254] (with P. Paule) Partition analysis XII: Plane partitions (accepted, *J. London Math Soc.*)

[255] (with O. Warnaar) The product of partial theta functions. *Adv. Appl. Math.* 39:116–120 (2007)

[256] A Fine dream. *International Journal of Number Theory* 3(3):325–334 (2007)

[257] (with P. Paule) MacMahon's Dream. In *Developments in Math: Partitions, q-series, and Modular Forms, Proc. 2008 Gainesville Conference*, K. Alladi and F. Garvan, eds., 23, Springer, New York, pp. 1–12 (2012).

[258] (with J. Sellers) On Sloane's generalization of non-squashing stacks of boxes. *Discrete Mathematics* 307:1185–1190 (2007)

[259] (with H. Eriksson, F. Petrov, and D. Romik) Integrals, partitions and MacMahon's theorem. *JCT(A)* 114:545–554 (2007)

[260] (with B. Berndt) Your hit parade: The top ten most fascinating formulas from Ramanujan's Lost Notebook. *AMS Notices* 55(1):18–30 (2008)

[261] Partitions with initial repetitions. (accepted, *Acta Math. Sinica*)

[262] (with O. Warnaar) The Bailey transform and false theta functions. *Ramanujan J.* 14:173–188 (2007)

[263] Partitions, Durfee symbols, and the Atkin–Garvan moments of ranks. *Inventiones Math.* 169:37–73 (2007)

[264] A survey of multipartitions: Congruences and identities. In *Surveys in Number Theory*, K. Alladi, ed., Developments in Mathematics, vol. 17, Springer, New York, pp. 1–19 (2008)

[265] Euler's "De Partitio Numerorum." *Bulletin AMS* 44(4):561–573 (2007)

[266] The number of smallest parts in the partitions of n. *Journal für die reine und angewandte Mathamatik* 624:133–142 (2008)

[267] The theory of compositions, IV: Multicompositions. *The Mathematics Student* (Special Centenary Volume):25–31 (2007)

[268] (with S. Corteel and C. D. Savage) On q-series identities arising from lecture hall partitions. *International J. Number Theory* 5(2):327–337 (2009)

[269] The finite Heine transformation. *Combinatorial Number Theory: Proceedings of Integers Conference 2007*, De Gruyter Proceedings in Mathematics, B. Landman *et al.*, eds., pp. 1–6. De Gruyter, (2009) Carrouton, Georgia. (Will also appear in online volume of *Integers*)

[270] Parity in partition identities. *Ramanujan Journal* 23:45–90 (2010)

[271] Partitions with distinct evens. *Advances in Combinatorial Mathematics. Proceedings of the Waterloo Workshop in Computer Algebra 2008.* I. Kotsireas and E. zima, eds., Springer, New York, pp. 31–37 (2009)

[272] (with L. Littlejohn) A combinatorial interpretation of the Legendre-Stirling numbers, *Proc. Amer. Math. Soc.* 137(8):2581–2590 (2009)

[273] (with M.D. Hirschhorn and J. A. Sellers) Arithmetic properties of partitions with even parts distinct. *Ramanujan Journal* 23:169–181 (2010)

[274] The meaning of Ramnujam now and for the future. *Ramanujan Journal* 20: 257–273 (2009)

[275] q-Catalan identities. In *The Legacy of Alladi Ramakrishnan in the Mathematical Sciences*, K. Alladi, J. R. Klauder, and C. R. Rao, eds., Springer, New York, pp. 183–190 (2010)

[276] On Shapiro's Catalan convolution. *Advances in Appl. Math.* 46:15–24 (2011)

[277] Concave compositions. *Electronic J. of Combinatorics* 18(2): P6 (2011) Also available online at http://www.combinatorics.org/Volume_18/Abstracts/v18i2p6.html

[278] q-Orthogonal polynomials, Rogers-Ramnujan identities, and mock theta functions. *Proceedings of the Steklov Institute Dedicated to the 75th birthday of A. A. Karatsuba* 276, 21–32 (2012)

[279] (with J. Bell) Euler's pentagonal number theorem and the Rogers-Fine identity. (accepted, *Annals of Combin.*)

[280] Partitions with early conditions. (submitted)

[281] (with S. Rose) MacMahon's sum-of-divisors functions, Chebyshev polynomials, and quasi-modular forms. (accepted, *Journal für die reine und angewandte Mathematik*)

[282] Differences of partitiion functions: The anti-telescoping method. (accepted, *Ramanujan J.*)

[283] (with W. Gawronski and L. Littlejohn) The Legendre-Stirling numbers. *Discrete Math.* 311:1255–1272 (2011)

[284] Concave and convex compositions. (accepted, *Ramanujan J.*)

[285] Basis partitions and the Rogers-Ramanujan identities. (accepted, *Ramanujan J.*)

[286] (with C. Savage and H. Wilf) Hypergeometric identities associated with statistics on words. (in preparation)

[287] (with F. Garvan and J. Liang) Combinatorial interpretations of congruences for the spt function. (accepted, *Ramanujan J.*)

[288] (with M. Merca) The truncated pentagonal number theorem. (submitted)

[289] (with F. Garvan and J. Liang) Self-conjugate vector partitions and the parity of the spt function. (submitted)

[290] (with R. Rhoades and S. Zwegers) Modularity of the concave composition generating function. (in preparation)

[291] (with E. Egge, W. Gawronski and L. Littlejohn) The Jacobi-Stirling numbers. (in preparation)

[292] (with S. H. Chan and B. Kim) The odd moments of ranks and cranks. (in preparation)

Books (Author)

[Book1] *Number Theory*, W. B. Saunders Co., Philadelphia, A, 259 pp. (1971), Reissued, Dover, New York, (1995).

[Book2] *The Theory of Partitions*, The Encyclopedia of Mathematics and its Applications Series, Addison-Wesley Pub. Co., New York, 300 pp. (1976). Reissued, Cambridge University Press, New York, (1998)

[Book3] *Partitions: Yesterday and Today*, New Zealand Mathematical Society, Wellington, New Zealand, 56 pp. (1979)

[Book4] *q-Series: Their Development and Application in Analysis, Number Theory, Combinatorics, Physics and Computer Algebra*, C.B.M.S. Regional Conference Series in Math, No. 66, American Math. Soc., Providence (1986)

[Book5] *Special Functions*, Research Monograph (with R. Askey, Univ. of Wisconsin and R. Roy, Beloit College), Cambridge University Press, Cambridge, UK, 300 pp. (1998)

[Book6] *Integer Partitions* (with Kimmo Eriksson), Cambridge University Press, Cambridge, UK, 152 pp. (2004)

[Book7] *Ramanujan's Lost Notebook, Part I* (with Bruce Berndt, University of Illinois), Springer, New York, 437 pp. (2005)

[Book8] *Ramanujan's Lost Notebook, Part II* (with Bruce Berndt), Springer, New York, 420 pp. (2009)

Books (Editor)

[Book9] *The Collected Papers of P. A. MacMahon, Vol. I*, MIT Press, Cambridge, MA, 1464 pp. (1979)

[Book10] *The Collected Papers of P. A. MacMahon, Vol. II*, MIT Press, Cambridge, MA, 952 pp. (1986)

[Book11] *Chapter Notes for N. J. Fine's Basic Hypergeometric Series*, Surveys in Math., No. 27, Amer. Math. Soc., Providence, pp. 561–573 (1988)

[Book12] *Ramanujan Revisited*, Academic Press, New York, (1988)

[Book13] *The Rademacher Legacy*, Contemp. Math., Vol. 166, (1994)

[Book14] *Askey-Olver Volume of S.I.A.M. J. Math.*, Vol. 25, (1994)

[Book15] *Topics in Number Theory in Honor of B. Gordon and S. Chowla, Mathematics and its Applications*, Vol. 467 (with S. Ahlgren and K. Ono). Kluwer, Boston, 258 pp. (1999)

Mathematical Reviews, Prefaces, and Obituaries

[Rev1] (book review) *The Collected Papers of Alfred Young, 1873–1940*, edited by G. Robinson, Univ. of Toronto Press, Toronto, 684 pp. (1977). Bull. Amer. Math. Soc. 1(16):980–997 (1979)

[Rev2] (book review) *q-Hypergeometric Functions and Applications*, H. Exton, E. Horwood. *Series Mathematics and its Applications*, Halsted Press, Chichester, 347 pp. (1983). Bull. London Math. Soc. 16:332–335 (1984)

[Rev3] (book review) *Enumerative Combinatorics, Vol. I*, Richard P. Stanley, Wadsworth and Brooks/Cole, Monterey, xi + 306 pp. (1986). Bull. Amer. Math. Soc. 17:360–365 (1987)

[Rev4] (book review) *Pi and the A.G.M.*, J. M. and P. B. Borwein, Wiley, New York, xi + 414 pp. (1987). Bull. Amer. Math. Soc. 22:198–201 (1990)

[Rev5] (book review) *Basic Hypergeometric Series*, G. Gasper and M. Rahman, Cambridge University Press, Cambridge, xx + 287 pp. (1990). Amer. Math. Monthly, 98:282–285 (1991)

[Rev6] (book review) *Ramanujan's Notebooks, Part III*, Bruce C. Berndt, Springer-Verlag, New York, 510 pp. (1991). Math. Intelligencer 16(3):69–71 (1994)

[Rev7] (book review) *generatingfunctionology*, Herbert S. Wilf, Academic Press, Boston, 228 pp., (1994). S.I.A.M. Discrete Math Newsletter 5(1): 1–2 (1994)

[Rev8] (book review) *Mathematical Thinking and Problem Solving*, Alan H. Schoenfeld, Erlbaum Associates, Hillsdale, New Jersey, 339 pp. (1994). Amer. Scientist 84:302–303 (1996)

[Rev9] (book review) *Ramanujan: Letters and Commentary*, Bruce C. Berndt and Robert A. Rankin, American Math. Society, London Math. Society, Providence, xiv + 347 pp. (1995). Current Science 75(10):1079–1080 (1998)

[Rev10] (book review) *Elements of the Theory of Numbers*, J. B. Dence and T. P. Dence, Academic Press, San Diego, (1999), 517 pp. Missouri Journal of Mathematical Sciences 12(1):48–49 (2000)

[Rev11] (book review) *Enumerative Combinatorics, Vol. 2*, Richard P. Stanley, Cambridge University Press, Cambridge, 581 pp. (1999). Bull. London Math. Soc. 32:622–624 (2000)

[Rev12] (book review) *Mathematics Education Research: A Guide for the Research Mathematician*, C. McKnight, A. Magid, T. J. Murphy, and M. McKnight, AMS, 106 pp. Amer. Math. Monthly, 108:281–285 (2001)

[Rev13] (preface) *Infinite families of exact sums of squares formulas*, Jacobi elliptic functions, continued fractions, and Schur functions by S. Milne. Ramanujan Journal 6(1):5–6 (2002)

[Rev14] (preface) Teaching in the MASS program. *MASS Selecta Teaching and Learning Advanced Undergraduate Mathematics*, eds., S. Katok, A. Sossinsky and S. Tabachnikov, 7–8. AMS, Providence (2003)

[Rev15] (commentary) (with K. Ono) *Ramanujan's congruences and Dyson's crank.* Proc. Nat, Acad. Sci. 102(43), 15277 (2005)

[Rev16] (foreword) (in Hindi) *Great Mathematician: Srinivasa Ramanujan* by Narenda Kumar Govil and Bhudev Sharma. Prabhat Prakashan, New Delhi, pp. 9–10 (2005)

[Rev17] (preface) (with G. Gasper and S. Suslov) *Special issue of the Ramanujan Journal in honor of R. Askey's 70th birthday.* Ramanujan Journal 13:5–6 (2007)

[Rev18] (obituary) (with R. C. Corless, D. J. Jeffrey, A. R. Pears, and J. Sheehan) EDWARD MAITLAND WRIGHT 1906–2005. Bull. London Math. Soc. 39:857–865 (2007)

Mathematical Essays

[Essay1] (essay) The Death of Proof? Semi-Rigorous Mathematics? You've Got to Be Kidding! *Math. Intelligencer* **16** (4):16–18 (1994)

[Essay2] (essay) The Irrelevance of Calculus Reform: Ruminations of a Sage-on-the-Stage. *UME Trends* **6** (6):17 and 23 (1995). Reprinted (with permission) *Virginia Mathematics Teacher* **22** (3):5–6 (1996)

[Essay3] (essay) Commentary on Assessment Standards for School Mathematics. *Notices of AMS* **44** (4):458–462 (1997)

[Essay4] (essay) Mathematical Education: A Case for Balance. *The College Mathematics Journal* **27** (5):341–355 (1996)

[Essay5] (note) The Geometric Series in Calculus. *Amer. Math. Monthly* **105** (1): 36–40 (1998)

[Essay6] (essay) Remarks on Calculus Reform and Calculators. *Mathematics & Technology in the Classroom* **4** (1) (1997)

[Essay7] (essay) Calculators and Reform: Up Close and Personal. *Rendezvous Idaho State University Journal of Arts and Letters* **38** (1) (2003)

[Essay8] (essay) Drowning in the data deluge, *AMS Notices* **59** (7): 933–941 (2012)

Biographical Notes

George W. Eyre Andrews was born on December 4, 1938 and grew up on a farm ten miles outside Salem, Oregon. For a time, he had contemplated a career in law or engineering, but graduated from Oregon State College in June, 1960 with simultaneous B.S. and M.A. degrees in mathematics. He spent the 1960–1961 academic year at the University of Cambridge in England, supported by a Fulbright Fellowship.

Andrews returned to the U.S. in 1961, and enrolled in the graduate program in mathematics at the University of Pennsylvania where 1961–1962 was a special year in number theory. Since he was planning on studying prime numbers, Andrews decided Penn would be his best choice for graduate school. However, after learning about the enigmatic, self-taught Indian mathematical genius Srinivasa Ramanujan, Andrews' primary interest shifted to another branch of analytic number theory. He wrote his doctoral dissertation on the subject of Ramanujan's third-order mock theta functions under the direction of the great Hans Rademacher. It was also Rademacher who introduced Andrews to the theory of partitions. Andrews was Rademacher's last Ph.D. student as Rademacher retired from the Faculty of Pennsylvania in 1962. Andrews completed his Ph.D. in 1964.

Andrews was then offered faculty positions at a couple of universities; he decided to accept the offer from Pennsylvania State, in part due to his desire to work with Pennsylvania State number theorists Sarvadaman Chowla and Nathan Fine. He was appointed as an Assistant Professor of Mathematics in Fall 1964; received tenure and promotion to the rank of Associate Professor in 1967, and became Full Professor in 1970. In 1981, he was appointed Evan Pugh Professor of Mathematics. Other than periodic sabbaticals and leave (1971–1972 at MIT, 1975–1976 at Wisconsin, 1978–1979 at the University of New South Wales, 1990–1991 at IBM, and 1996–1997 at the University of Melbourne), Andrews has spent his entire career in the faculty of Pennsylvania State. Since 2006, he has also spent part of each spring semester as a Distinguished Visiting Professor of Mathematics at the University of Florida.

Andrews' mathematical research over the past four decades has been very diverse. Although primarily considered a number theorist, his work has overlapped with other areas, including mathematical analysis, computer science, and theoretical physics. In fact, the honorary doctorate he received in 1998 from the University of Parma was in Physics, not Mathematics, and conferred in recognition of his work on statistical mechanics.

Despite this diversity, one thread is common in much of Andrews' work: the influence of Srinivasa Ramanujan, the great mathematician who was born in India in 1887. From an early age, Ramanujan showed exceptional talent in mathematics. With no appropriate teacher available for so brilliant a student, it fell to Ramanujan to teach himself from whatever books he could manage to borrow. He devoted virtually all of his time to mathematics and neglected his other subjects. Locked within the confines of a rigid system, Ramanujan failed to qualify for higher education. When he was 26, Ramanujan wrote to the eminent English mathematician G. H. Hardy of Trinity College at the University of Cambridge, including some of his results (but no proofs or indications of how he derived them). Hardy was extremely impressed, and a correspondence between the two men followed. Eventually, Ramanujan was able to accept Hardy's invitation to Cambridge, to study and collaborate. Ramanujan became ill during his stay in England, and returned to his native India in 1919. Tragically, he died at the age of 32 in 1920. Despite his physical deterioration during 1919–1920, Ramanujan remained at the height of his mathematical powers. In his last letter to Hardy, Ramanujan described a new collection of functions that he had discovered and named "mock theta functions." These were the same functions that Andrews would begin studying in his Ph.D. thesis years later.

In 1976, while visiting England, Andrews arranged to see papers from the estate of the mathematician G. N. Watson which were deposited in the Trinity College Library. To his surprise and amazement, he found in a box, a collection of about 100 sheets of paper, on which he recognized Ramanujan's handwriting. Soon he realized that in his hands was Ramanujan's work on the mock theta functions, which had been unknown to the mathematical community, aside from the one letter Hardy had received in 1919. In this way, Andrews discovered what is usually referred to as Ramanujan's "Lost" Notebook.

Later, Andrews assessed that, whilst about 20 percent of the results contained in the "Lost" Notebook had been discovered by other mathematicians in the ensuing five decades, the vast majority of results were new to the mathematical community. Ramanujan had only presented results, and as with his first letter to Hardy, neither gave an as to how he indication discovered them, nor provided any proofs. Since unearthing had recognized the "Lost" Notebook, Andrews, and his colleague Bruce Berndt of the University of Illinois, have managed to provide proofs for many of Ramanujan's formulas, as have a number of Berndt's Ph.D. students. Berndt and Andrews are currently preparing an edited version of the "Lost" Notebook. The first two of an estimated four volumes have appeared at the time of writing.

The preceding tale of Andrews' discovery of the "Lost" Notebook, while perhaps one of the more colorful episodes in his mathematical life, is only one of his myriad mathematical accomplishments. To date, Andrews has published over 275 papers and written eight books. He has edited eight others, including the weighty two-volume collection of the works of P. A. MacMahon. Andrews is one of the world's leading authorities on partitions, q-series, and hypergeometric functions, and has remained creative and active into his seventies. He has served as thesis advisor for 24 Ph.D. students and 18 master's degree students.

In addition, Andrews has served on journal editorial boards; numerous departmental, college, university, and AMS committees, advisory panels for the National Science Foundation and for analogous granting agencies abroad, and has also been chairman of the Department of Mathematics twice, despite a professed lack of affinity for administration. (During his second "sentence" to department chairmanship, he was instrumental in bringing into being Penn State's unique and very successful MASS program.) From 2009–2010 he served as the sixtieth president of the American Mathematical Society.

He has also been an active critic of some recent trends in mathematics education, particularly certain aspects of calculus reform.

However, there are many facets to George E. Andrews to which a formal curriculum vitae cannot do justice. His students report that as a classroom lecturer, Andrews' clarity is unsurpassed. He always welcomes them, whether during office hours, or during an unscheduled "drop by", with a friendly, "Come in!", "Have a chair!", and an inviting, cheerful grin.

Ph.D. Students of George E. Andrews Through 2011

(1) David J. DeVries, Ph.D., Pennsylvania State University, August 1969. *Lattices of Equivalence Relations and Generalized Primes.*
(2) Edward Parberry, Ph.D., Pennsylvania State University, November 1969. *On a Partition Function Related to the Theorem of Schur.*
(3) Ming-Po Chen, Ph.D., Pennsylvania State University, May 1972. *Asymptotic Theorems in Partition Theory.*
(4) Doris Jean Mitchell, Ph.D., Pennsylvania State University, May 1972. *Generating Functions for Various Sets of Solid Partitions.*
(5) Willard Connor, Ph.D., Pennsylvania State University, August 1972. *Partition Theorems Related to Some Identities of Rogers and Watson.*
(6) Margret Kothmann, Ph.D., Pennsylvania State University, May 1977. *On the Ranks of Partitions.*
(7) Kevin W. J. Kadell, Ph.D., Pennsylvania State University, November 1979. *Generalizations of Basic Hypergeometric Series.*
(8) Michael D. Hirschhorn, Ph.D., University of New South Wales, April 1980. *Developments in the Theory of Partitions.* Co-advised by George E. Andrews and John Loxton.
(9) Louis Worthy Kolitsch, Ph.D., Pennsylvania State University, May 1985. *Some Analytic and Arithmetic Properties of Generalized Frobenius Partitions.*
(10) Francis G. Garvan, Ph.D., Pennsylvania State University, January 1986. *Generalizations of Dyson's Rank.*
(11) Scott Stevens, Ph.D., Pennsylvania State University, December 1987. *Group-Action Graphs and Ramsey Graph Theory: Investigating the Ramsey Numbers in $R(K_{1,n}, K_{k,m})$ and $R(K_{1,n}, B_{k,m})$.* Co-advised by Torrence D. Parsons and George E. Andrews.
(12) Nicolas Santa-Gadea, Ph.D., Pennsylvania State University, December 1990 *On the Rank and the Crank Moduli 8, 9, and 12.*
(13) José Plínio de Oliveira Santos, Ph.D., Pennsylvania State University, May 1991. *Computer Algebra and Identities of the Rogers–Ramanujan Type.*

(14) Kuo-Jye Chen, Ph.D., Pennsylvania State University, August 1992. *An Asymptotic Formula in Partition Theory.*

(15) Garth Payne, Ph.D., Pennsylvania State University, December 1997. *Multivariate Hypergeometric Terms.*

(16) Jeremy Lovejoy, Ph.D., Pennsylvania State University, August 2000. *Arithmetic of Partition Functions.* Co-advised by George E. Andrews and Ken Ono.

(17) Andrew Victor Sills, Ph.D., University of Kentucky, May 2002. *Computer Assisted Explorations of Rogers–Ramanujan Type Identities.* Co-advised by George E. Andrews and Avinash Sathaye.

(18) Augustine O. Munagi, Ph.D., University of Lagos (Nigeria), December 2005. *Restricted Partition Formulas, q-Partial Fractions and the Rademacher Conjecture.* Co-advised by George E. Andrews and Monsur A. Kenu.

(19) William J. Keith, Ph.D., Pennsylvania State University, August 2007. *Rank of partitions and degree symbols.*

(20) Michael J. Rowell, Ph.D., Pennsylvania State University, August 2007. *The Bailey transform and conjugate Bailey pairs.* Co-advised by Ae Ja Yee and George E. Andrews.

(21) Kagan Kursungoz, Ph.D., Pennsylvania State University, May 2009. *Parity considerations in Andrews-Gordon identities and the K-marked Durfee symbols.*

(22) J. Brandt Kronholm, Ph.D., SUNY at Albany, May 2010. *On Congruence Properties of* $p(n, m)$. Co-advised by George E. Andrews and Antun Milas.

(23) Shishuo Fu, Ph.D., Pennsylvania State University, August 2011. *Bijective Methods and combinatorial studies of problems in partition theory and related areas.*

(24) Heiko Toth, Ph.D., Pennsylvania State University, December 2011. *Asymptotics in Partition Functions.*

Papers Reproduced

1. Andrews, G. E., A lower bound volume of strictly convex bodies with many boundary lattice points. Trans. Amer. Math. Soc. (2) 106, 270–279 (1963)
2. Andrews, G. E., On the theorems of Watson and Dragonette for Ramanujan's mock theta functions. Amer. J. Math. (2) 88, 454–490 (1966)
3. Andrews, G. E., A simple proof of Jacobi's triple product identity. Proc. Amer. Math. Soc. (2) 16, 333–334. (1965)
4. Andrews, G. E. & Askey, R., Classical orthogonal polynomials. Orthogonal Polynomials and Applications, Bar-le-Duc (1984). Repr. in Lecture Notes in Maths 1171, Springer, 36–62 (1985)
5. Andrews, G. E., On q-difference equations for certain well-poised basic hypergeometric series. Quart. J. Math. (2) 19, 433–447 (1968)
6. Andrews, G. E., An analytic generalization of the Rogers–Ramanujan identities for odd moduli. Proc. Nat. Acad. Sci. (10) 71, 4082–4085 (1974)
7. Andrews, G. E., Multiple series Rogers–Ramanujan type identities. Pac. J. Math. (2) 14, 267–283 (1984)
8. Andrews, G. E., The fifth and seventh order mock theta functions. Trans. Amer. Math. Soc. (1) 293, 113–134 (1986)
9. Andrews, G. E., Dyson, F. J. & Hickerson, D. R., Partitions and indefinite quadratic forms. Invent. Math. 91, 391–407 (1988)
10. Andrews, G. E., Umbral calculus. Bailey chains and pentagonal number theorems. JCT (A) 91, 464–475 (2000)
11. Andrews, G. E., Baileys transform, lemma, chain and tree. Special functions 2000, Kluwer Academic, 1–22 (2001)
12. Andrews, G. E., Euler's "Exemplum Memorabile Inductionis Fallacis" and q-trinomial coefficients. J. Amer. Math. Soc. (3) 3, 65–669 (1990)
13. Andrews, G. E., q-Trinomial coefficients and Rogers–Ramanujan type identities. Analytic Number Theory, Springer, 1–11 (1990)
14. Andrews, G. E., On a conjecture of Peter Borwein. J. Symbolic Computation 20, 487–501 (1995)
15. Andrews, G. E., An introduction to Ramanujan's "Lost" Notebook. Amer, Math. Monthly (2) 86, 89–108 (1979)

16. Andrews, G. E., Baxter, R. J. & Forrester, P. J., Eight-vertex SOS model and generalized Rogers–Ramanujan-type identities. J. of Stat. Phys. (3/4) 35, 193–266 (1984)
17. Andrews, G. E. & Baxter, R. J., Lattice gas generalization of the hard hexagon model, III. q-Trinomial Coefficients. J. of Stat. Phys. (3/4) 47, 297–330 (1987)
18. Andrews, G. E., Pfaff's method (II): diverse applications. J. Computational Appl. Math. 68, 15–23 (1996)
19. Andrews, G. E., MacMahon's partition analysis: II fundamental theorems. Annals of Combinatorics 4, 327–338 (2000)
20. Andrews, G. E., An analytic proof of the Rogers–Ramanujan–Gordon identities. Amer. J. Math. (4) 88, 844–846 (1966)
21. Andrews, G. E., Some new partition theorems. J. Comb. Theory (4) 2, 431–436 (1967)
22. Andrews, G. E., A generalization of the Göllnitz-Gordon partition theorems. Proc. Amer. Math. Soc. (5) 18, 945–952 (1967)
23. Andrews, G. E., Sieves for theorems of Euler, Rogers, and Ramanujan. Lecture Notes in Maths 1171, Springer, 1–20 (1982)
24. Andrews, G. E., Combinatorics and Ramanujan's "lost" notebook. Surveys in Combinatorics 1985, Invited Papers for the Tenth British Combinatorial Conference, Cambridge University Press, 1–23 (1985)
25. Andrews, G. E., On a partition problem of H. L. Alder. Pacific J. Math. (2) 36, 279–284 (1971)
26. Andrews, G. E., Partitions and Durfee dissection. Amer. J. Math. (3) 101, 735–742 (1979)
27. Andrews, G. E., Use and extension of Frobenius' representation of partitions. Enumeration and Design, Academic Press, 51–65 (1984)
28. Andrews, G. E. & Garvan, F. G., Dyson's crank of a partition. Bull. Amer. Math. Soc. (2) 18, 167–171 (1988)
29. Andrews, G. E. & Olsson, J. B., Partition identities with an application to group representation theory. Journal für die reine und angewandte Mathematik 413, 198–212 (1991)
30. Alladi, K., Andrews, G. E. & Gordon, B., Generalizations and refinements of a partition theorem of Göllnitz. Journal für die reine und angewandte Mathematik 460, 165–188 (1995)
31. Andrews, G. E. & Lewis, R., An algebraic identity of F. H. Jackson and its implications for partitions. Discrete Math. 232, 77–83 (2001)
32. Andrews, G. E., Partitions with short sequences and mock theta functions. Proc. Nat. Acad. Sci. USA 102, 4666–4671 (2005)
33. Andrews, G. E., Partitions, Durfee symbols, and the Atkin–Garvan moments of ranks. Invent. Math. 169, 37–73 (2007)
34. Andrews, G. E., The number of smallest parts in the partitions of n. Journal für die reine und angewandte Mathematik 624, 133–142 (2008)

35. Andrews, G. E., On a partition theorem of N. J. Fine. J. Nat. Acad. Math. 1, India, 105–107 (1983)
36. Andrews, G. E., The Friedman–Joichi–Stanton monotonicity conjecture at primes. DIMACS Series in Discrete Mathematics and Theoretical Computer Science 64, AMS, 1–23 (2004)
37. Andrews, G. E., The theory of compositions IV: multicompositions. The Mathematics Student, 25–31 (2007)
38. Andrews, G. E,, MacMahon's conjecture on symmetric plane partitions. Proc. Nat. Acad. Sci. (2) 74, 426–429 (1977)
39. Andrews, G. E., Plane partitions (III): the weak Macdonald conjecture, Invent. Math. 53,193–225 (1979)
40. Andrews, G. E., Plane partitions IV: a conjecture of Mills–Robbins–Rumsey. Aequationes Mathematicae 33, 230–250 (1987)
41. Andrews, G. E. & Burge, W. H., Determinant identities. Pac. J. Math. (1) 158, 1–14 (1993)
42. Andrews, G. E., Catalan numbers, q-Catatan numbers and hypergeometric series. J. Comb. Theory, Series A (2) 44, 267–273 (1987)
43. Andrews, G. E., Fibonacci numbers and the Rogers–Ramanujan identities. Fibonacci Quarterly (1) 42, 3–19 (2004)
44. Andrews, G. E. & Baxter, R. J., SCRATCHPAD explorations for elliptic theta functions. Computers in Mathematics, Lecture Notes in Pure and Appl. Math. 125, 17–33 (1990)
45. Andrews, G. E., A note on the Bombieri–Selberg formula for algebraic curves. Portugaliae Mathematica (2) 27, 75–81 (1968)
46. Andrews, G. E., MacMahon's prime numbers of measurement. Amer. Math. Monthly (9) 82, 922–923 (1975)
47. Andrews, G. E., EYPHKA! num $= \Delta + \Delta + \Delta$. J. Num. Theory (3) 23, 285–293 (1986)
48. Andrews, G. E. & Roy, R., Ramanujan's method in q-series congruences. Electronic J. Combin. (2) 4 (1997)
49. Andrews, G. E., q-Analogs of the binomial coefficient congruences of Babbage, Wolstenholme and Glaisher. Discrete Math 204, 15–25 (1999)
50. Andrews, G. E., Knopfmacher, A. and Zimmermann, One the number of distinct values of multinomial coefficients. Journal of Number Theory 118, 15–30 (2006)
51. Andrews, G. E., Partition identities. Advances in Math (1) 9, 10–51 (1972)
52. Andrews, G. E., A general theory of identities of the Rogers–Ramanujan type. Bull. Amer. Math. Soc. (6) 80, 1033–1052 (1974)
53. Andrews, G. E., Partitions: at the interface of q-series and modular forms. Ramanujan Journal 7, 385–400 (2003)
54. Andrews, G. E., Applications of basic hypergeometric functions. SIAM Rev (4) 16, 441–484 (1974)
55. Andrews, G. E., Problems and prospects for basic hypergeometric functions. Theory and Application of Special Functions, Academic Press, 191–224 (1975)

56. Andrews, G. E., The well-poised thread: an organized chronicle of some amazing summations and their implications. Ramanujan Journal 1, 7–23 (1997)
57. Andrews, G. E., L. J. Rogers and the Rogers–Ramanujan identities. Math. Chronicle (Part 2) 11, 1–15 (1982)
58. Andrews, G. E., Mathematical education: a case for balance. The College Mathematics Journal (5) 27, 341–355 (1996)

Photo 1 Dominique Foata and George Andrews, February 18, 2008

Photo 2 Krishna Alladi, John Thompson, George Andrews, Richard Askey, and Alladi Ramakrishna

Photo 3 George and Joy Andrews

Photo 4 George Andrews and Bruce Berndt

Photo 5 Some of George Andrews' students in front of the Octacube. From left to right: Shishuo Fu, Heiko Todt, William Keith, Kagan Kursungoz, Louis Kolitsch, Jeremy Lovejoy (behind George Andrews), George Andrews, Mike Hirschhorn, Brandt Kronholm, Mike Rowell, Frank Garvan, Drew Sills, and Plinio Santos

Photo 6 George Andrews teaching a class of future teachers

Photo 7 James Sellers, George Andrews, Ae Ja Yee, and Krishna Alladi

Photo 8 Barbara Baum, Paul Baum, and Ole Warnaar at George Andrews' 70th Birthday Dinner, December 6, 2008

Photo 9 George Andrews and Rajiv Gandhi, December 22, 1987

Photo 10 Claudio Procesi, Ian Macdonald, Xavier Viennot, Richard Stanley, George Andrews, Gian-Carlo Rota, Curtis Greene, Anders Björner, and Adriano Garsia, at conference in honor of Adriano Garsia in Taormina, Sicily, July 1994

Photo 11 George Andrews signing the National Academy of Sciences membership book, Ron Graham in attendance, April, 2004

Photo 12 George Andrews, Peter Paule, and William Chen

Chapter 1

The Geometry of Numbers

When I was an undergraduate student at Oregon State University, I took a number of advanced courses from Helmut Groemer. This amazing young professor could, it seemed, teach anything: mathematical logic, statistics, probability, and differential geometry for example. During my junior year, my family had some financial difficulties, and I was told that I should find new ways to finance my last year in college. My solution was to complete my B.S. the following summer and do an M.A. in what would have been my senior year. Oregon State University required a research thesis for the M.A. So, thinking I might become a logician, I approached Groemer for a logic thesis problem. He replied that he had no logic problems, but perhaps I might consider a problem in the geometry of numbers. This turn of events sent me down the number theory road, something for which I am eternally grateful.

The result of my M.A. thesis was to prove the main theorem in [1]:

Theorem 1.1. *We are given a closed, strictly convex body C with N lattice points (i.e. points with integer coordinates) on its surface. If $S(C)$ denotes the surface content of the boundary of C, then there exists a constant $k(n)$ depending only on n such that*

$$S(C) > k(n)N^{(n+1)/n},$$

where n denotes the dimensionality of space.

While I was carrying out this research, it became clear to me that there was a corresponding inequality related to volume, but I was unable to find a proof.

During my M.A. year, I received a Fulbright Grant to study at the University of Cambridge. Winning this grant was a complete obsession of mine because the lady who has been my wife for the last 50 years had told me that we should probably delay our marriage to a time when it would be more convenient for each of us, except, of course, if I did win the fellowship to Cambridge, as this would be too wonderful an opportunity to pass up.

My advisor at Cambridge was J. W. S. Cassels, and I spent this glorious first year of marriage taking courses from all the Cambridge greats: Cassels, Davenport, Ingham, Birch and Swinnerton-Dyer. Also, during that year, I got the idea for how to prove the main result in [2], the paper that is included in this volume.

Theorem 1.2. *If C is an n-dimensional strictly convex body with N noncoplanar lattice points on its boundary, then*

$$V(C) > \kappa(n)N^{(n+1)/(n-1)}$$

where $\kappa(n) > 0$ depends only on n.

I finished the proof of this theorem in the Fall of 1961. I had chosen to attend the University of Pennsylvania after my Fulbright Grant year because the University made 1961–1962 the year of number theory to honor Hans Rademacher's seventieth birthday. I knew Paul Bateman would be there, and I thought I might ask to be his student. As it turned out, Rademacher taught a year long course (his last at Penn) on additive number theory. I became his last student and was hooked.

I did subsequently illustrate some elementary applications of [2] in [3]. The conclusion of [3] suggests that the result in [2] might be improved if further constraints are applied to the convex body. I was never able to go on with this, but Swinnerton-Dyer responded to my challenge in [Swi74]. He dealt with counting lattice points on convex curves undergoing magnification by a factor of N. I had hoped one could prove that there were at most $O(N^{\epsilon})$ lattice points on the curve. Swinnerton-Dyer constructed lovely counterexamples to show that this was impossible if the curve was not infinitely differentiable.

My final contribution to this area was to use the geometry of numbers to prove the theorem determining the solutions of the linear diophantine equation [17]. This paper was reprinted by the MAA in the volume Selected Papers in Algebra [Mon77].

Although I have not worked in this area since the 1960s, I have been most pleased to see that the main theorem in [2] has been greatly extended by others. The two-dimensional result in [2] was improved by Diviš [Div76], where he improved the constant in the inequality. In 2008, there was a Bulletin of the AMS survey of this general area by Bárány [Bár08], where he begins section 13 "... by quoting a beautiful theorem of G. E. Andrews [2] from 1963." Bárány cites five subsequent proofs of this result.

1.1 A Lower Bound for the Volume of Strict Convex Bodies with Many Boundary Lattice Points

A LOWER BOUND FOR THE VOLUME OF STRICTLY CONVEX BODIES WITH MANY BOUNDARY LATTICE POINTS

BY

GEORGE E. ANDREWS(1)

In a recent paper it was shown that if a strictly convex body C in n-dimensional space contains N noncoplanar lattice points (i.e., points with integer coordinates) on its boundary, then

$$S(C) > k(n)N^{(n+1)/n}$$

where $S(C)$ denotes the surface area of C and $k(n) > 0$ is a constant depending only on n [1]. If $V(C)$ denotes the volume of C and

$$V(C) \geqq c'(n)[S(C)]^{n/(n-1)}$$

where $c'(n) > 0$ is a constant depending only on n (which is true if C is an n-dimensional sphere for example), then the above theorem implies that

$$V(C) > k'(n)N^{(n+1)/(n-1)}$$

where $k'(n) > 0$ is a constant depending only on n. The object of this paper is to show that the restriction to

$$V(C) \geqq c'(n)[S(C)]^{n/(n-1)}$$

is unnecessary. The main result will be the following theorem.

THEOREM. *If C is an n-dimensional strictly convex body with N noncoplanar lattice points on its boundary, then*

$$V(C) > \kappa(n)N^{(n+1)/(n-1)}$$

where $\kappa(n) > 0$ depends only on n.

This paper will be divided into two sections. In the first section we shall obtain certain elementary but necessary results. In the second section we shall prove the above theorem.

Received by the editors January 25, 1962.

(1) The research on this paper was begun at the University of Cambridge under a Fulbright grant and completed at the University of Pennsylvania under a National Science Foundation Graduate Fellowship.

I wish to thank Dr. H. Groemer for suggesting this problem to me when I was a student at Oregon State University.

I. In this section we shall be primarily concerned with certain elementary results in n-dimensional Euclidean geometry. The following notations and conventions will be used throughout this paper.

At all times the bodies we shall deal with will be considered closed and bounded; however, the cones we shall deal with will of course be unbounded.

We shall call a linear $(n-1)$-dimensional boundary element of an n-dimensional polytope or cone a "face."

Given the point $(a_1, \cdots, a_n)$ and the direction $(A_1 : \cdots : A_n)$, the "ray" emanating from $(a_1, \cdots, a_n)$ in the direction $(A_1 : \cdots : A_n)$ is given by the set of points $(a_1 + A_1 t, \cdots, a_n + A_n t)$ and $t \geqq 0$.

If H is a supporting hyperplane of C, a convex body, then the "exterior normal vector" of H is the one on the opposite side of H from C. We define an "exterior normal ray" as a ray emanating from some given point of H in the direction of the exterior normal vector.

The symbol O will always denote the origin.

At all times, $\sigma(n) = \pi^{n/2}/\Gamma(n/2+1)$; $\pi^{n/2}/\Gamma(n/2+1)$ is the volume of the n-dimensional unit sphere [7, p. 136].

By $(Po)_n$, $(Po)_n^a$, $(Po)_n^b, \cdots$, we shall denote n-dimensional, convex polytopes [7, p. 96].

By Θ_i we shall denote an n-dimensional solid angle(2). This symbol will be used to denote both the angle and the measure of the angle, but this should cause no confusion.

We let $\Delta_i(\Theta_i)$ be an n-dimensional pyramid [7, p. 111] with vertex $\boldsymbol{\alpha}_i$, vertex angle Θ_i, and an $(n-1)$-dimensional convex polytope $(Po)_{n-1}^{\Delta^i}$ as base. We shall call the collection of all faces of $\Delta_i(\Theta_i)$ other than $(Po)_{n-1}^{\Delta^i}$ the "sides" of $\Delta_i(\Theta_i)$. To $\Delta_i(\Theta_i)$ we associate a similar pyramid $\delta_i(\Theta_i)$ which has vertex $\boldsymbol{\alpha}_i$, vertex angle Θ_i, and has its base parallel to $(Po)_{n-1}^{\Delta^i}$ but is such that the total area of its sides is equal to 1. Also associated with $\Delta_i(\Theta_i)$ is a cone $\mathscr{C}_i$ which is the collection of all rays emanating from $\boldsymbol{\alpha}_i$ and intersecting $(Po)_{n-1}^{\Delta^i}$. We designate by $\tilde{\mathscr{C}}_i$ the set of all points $\mathbf{y}$ for which the angle $\angle\, \mathbf{y}\boldsymbol{\alpha}_i\mathbf{x}$ is at least $\pi/2$ if $x \in \mathscr{C}_i$, and $\tilde{\Theta}_i$ will denote the n-dimensional angle defined by $\tilde{\mathscr{C}}_i$.

If Θ is an n-dimensional solid angle defined by n rays (i.e., these rays make up the edges of Θ), then

$$\sin \Theta =_{\text{df.}} \left| \det (c_j^i) \right|$$

where c_j^i is the jth direction cosine of the ith ray [7, p. 122].

LEMMA 1. $\sin \Theta \leqq (n-1)!\,\Theta$.

Proof. If the point $\boldsymbol{\alpha}$ is the vertex of Θ, then $(1/n!) \sin \Theta$ is the volume of a simplex with edges of unit length emanating from $\boldsymbol{\alpha}$. Hence the whole of this simplex is contained in the body that the cone defined by Θ cuts out of the unit

(2) The interiors of the solid angles we shall consider will always be contained in half-spaces.

sphere with center at $\boldsymbol{\alpha}$. The measure of the volume common to this cone and the unit sphere is Θ/n. Therefore

$$\frac{1}{n!}\sin\Theta \leqq \frac{\Theta}{n},$$

or

$$\sin\Theta \leqq (n-1)!\,\Theta. \qquad \text{q.e.d.}$$

The following lemma is most likely due to Mahler [6] and is very similar to Theorem V of Dr. Cassels' book [3, p. 117]. Unfortunately the author was not able to find this result in readily available literature, and it is thus included for completeness.

LEMMA 2. *Let C be a closed convex body with volume $V(C) > 0$. Then there exist points $\mathbf{x}_0, \cdots, \mathbf{x}_n \in C$ such that C contains the simplex $\mathscr{D}_n$ with $\mathbf{x}_0, \cdots, \mathbf{x}_n$ as vertices and C is contained in the simplex $\mathscr{B}_n$ with vertices*

$$\mathbf{x}_j + \sum_{i=0;\, i\neq j}^{n} (\mathbf{x}_i - \mathbf{x}_j) \qquad j = 0, \cdots, n.$$

Proof. Pick the points $\mathbf{x}_0, \cdots, \mathbf{x}_n$ so that $V(\mathscr{D}_n)$ is a maximum. Such a choice is possible since C is closed and bounded. By convexity $\mathscr{D}_n$ is contained in C. We note that $\mathscr{B}_n$ is a simplex bounded by n hyperplanes each passing through a vertex of $\mathscr{D}_n$ parallel to the opposite face of $\mathscr{D}_n$. If C were not contained in $\mathscr{B}_n$, then for some face $S_i(n-1)$ (containing x_i) of $\mathscr{B}_n$ there would be a point $\mathbf{y}$ of C which was on the opposite side of $S_i(n-1)$ from $\mathscr{D}_n$. Then $\mathbf{x}_0, \cdots, \mathbf{x}_{i-1}, \mathbf{y}, \mathbf{x}_{i+1}, \cdots, \mathbf{x}_n$ would define a simplex contained in C with a larger volume than $V(\mathscr{D}_n)$, a contradiction. Therefore $C \subset \mathscr{B}_n$. q.e.d.

It is easily verified that $\mathscr{B}_n$ and $\mathscr{D}_n$ are similar simplexes the ratio of similitude being $(n:1)$; both $\mathscr{B}$ and $\mathscr{D}_n$ have the same centroid, and $V(\mathscr{D}_n) = n^{-n}V(\mathscr{B}_n)$.

COROLLARY 1. $V(\mathscr{D}_n) \leqq V(C) \leqq n^n V(\mathscr{D}_n)$; $V(\mathscr{B}_n) \geqq V(C) \geqq n^{-n}V(\mathscr{B}_n)$.

Proof. Clear from Lemma 2 and the fact that $V(\mathscr{D}_n) = n^{-n}V(\mathscr{B}_n)$. q.e.d.

LEMMA 3. *If $\mathscr{C}_1 \subset \mathscr{C}_2$ and both have $\boldsymbol{\alpha}$ as vertex, then $\tilde{\mathscr{C}}_2 \subset \tilde{\mathscr{C}}_1$.*

Proof. If $\mathbf{x} \in \tilde{\mathscr{C}}_2$ and $\mathbf{y} \in \mathscr{C}_1$, then $\mathbf{y} \in \mathscr{C}_2$, and hence $\angle \mathbf{x}\boldsymbol{\alpha}\mathbf{y} \geqq \pi/2$. Hence $\mathbf{x} \in \tilde{\mathscr{C}}_1$. q.e.d.

LEMMA 4. *If an n-dimensional cone $\mathscr{C}$ is defined by n rays (i.e., $\mathscr{C}$ is the convex cover of n linearly independent rays emanating from a point $\boldsymbol{\alpha}$), then $\tilde{\mathscr{C}}$ is the convex cover of the n rays emanating from $\boldsymbol{\alpha}$ each ray being an exterior normal ray to a face of $\mathscr{C}$.*

Proof. Let $\mathscr{C}^*$ denote the convex cover of the above mentioned n exterior normal rays of $\mathscr{C}$. Without loss of generality we let $\boldsymbol{\alpha} = \mathbf{O}$. Let the defining rays of $\mathscr{C}$ have direction numbers $(a_1^i : \cdots : a_n^i)$ $i = 1, \cdots, n$. Then the defining rays of $\mathscr{C}^*$ have direction numbers $\varepsilon_i(A_1^i : \cdots : A_n^i)$ $i = 1, \cdots, n$ where A_j^i is the cofactor of a^i in $\det(a_j^i)$ and $\varepsilon_i = \pm 1$ is chosen so that $\varepsilon_i(A_1^i, \cdots, A_n^i)$ is an exterior normal vector to $\mathscr{C}$. We let $\mathbf{A}^i = \varepsilon_i(A_1^i, \cdots, A_n^i)$ and $\mathbf{a}^i = (a_1^i, \cdots, a_n^i)$.

if $\mathbf{x} \in \mathscr{C}^*$, then $\mathbf{x} = \lambda_1 \mathbf{A}^1 + \cdots + \lambda_n \mathbf{A}^n$ where all $\lambda_i \geqq 0$. If $\mathbf{y} \in \mathscr{C}$, then $\mathbf{y} \cdot \mathbf{A}^i \leqq 0$; hence $\mathbf{y} \cdot \mathbf{x} \leqq 0$. Thus $\mathbf{x} \in \tilde{\mathscr{C}}$.

If $\mathbf{x} \notin \mathscr{C}^*$ then $\mathbf{x} = \lambda_1' \mathbf{A}^1 + \cdots + \lambda_n' \mathbf{A}^n$ where at least one of the λ_i' is negative, say λ_k'. Then $\mathbf{a}^k \cdot \mathbf{A}^i = 0$ if $i \neq k$ and $\mathbf{a}^k \cdot \mathbf{A}^k = -|\det(a_j^i)| < 0$ since $\mathbf{a}^k$ and $\mathbf{A}^k$ must lie on opposite sides of the face defining $\mathbf{A}^k$. Hence $\mathbf{x} \cdot \mathbf{a}^k = \lambda_k' \mathbf{a}^k \cdot \mathbf{A}^k > 0$, and $\mathbf{a}^k$ is a point of $\mathscr{C}$ for which $\angle \mathbf{x} \mathrm{O} \mathbf{a}^k < \pi/2$. Thus $\mathbf{x} \notin \tilde{\mathscr{C}}$.

Therefore $\mathscr{C}^* = \tilde{\mathscr{C}}$. q.e.d.

LEMMA 5. *If $\mathscr{C}$ is defined by n rays as in Lemma 4 and Θ is the vertex angle of $\mathscr{C}$, then*

$$V[\delta(\Theta)] \leqq \left(\frac{n!}{n^{2n}}\right)^{1/(n-1)} \sin^{1/(n-1)} \tilde{\Theta}.$$

Proof. In this case we see that $\delta(\Theta)$ is an n-dimensional simplex. For convenience we shall consider the origin O as the vertex of $\mathscr{C}$. The edges of $\delta(\Theta)$ emanating from O are given by the vectors $(a_1^i, \cdots, a_n^i)$, $i = 1, \cdots n$. Now

$$V[\delta(\Theta)] = \frac{1}{n!} |\det(a_j^i)|,$$

and by Lemma 4 we know that $\tilde{\Theta}$ is defined by the convex cover of the rays emanating from O normal to the faces of $\mathscr{C}$; thus

$$\sin \tilde{\Theta} = \left| \det \left(\frac{A_j^i}{(A_1^{i2} + \cdots + A_n^{i2})^{1/2}} \right) \right| = \left\{ \prod_{i=1}^{n} \left(A_1^{i^2} + \cdots + A_n^{i^2} \right)^{-1/2} \right\} \cdot \left| \det(A_j^i) \right|$$

where A_j^i is the cofactor of a_j^i in $\det(a_j^i)$. By the definition of $\delta(\Theta)$, the area of all of its sides (i.e., excluding its base) equals 1. Now $[1/(n-1)!](A_1^{i^2} + \cdots + A_n^{i^2})^{1/2}$ gives the area of the ith side of $\delta(\Theta)$. Hence

$$\left[\frac{1}{(n-1)!}\right]^n \prod_{i=1}^{n} \left(A_1^{i^2} + \cdots + A_n^{i^2} \right)^{1/2} \leqq n^{-n}$$

since $\sum_{i=1}^n |\xi_i| = 1$ implies that $\max \prod_{i=1}^n |\xi_i| = n^{-n}$. Thus

$$\prod_{i=1}^{n} \left(A_1^{i^2} + \cdots + A_n^{i^2} \right)^{-1/2} \geqq \left[\frac{n}{(n-1)!}\right]^n.$$

Now

$$|\det(a_j^i)| \cdot |\det(A_j^i)| = |\det(a_j^i)|^n.$$

Hence

$$\sin \tilde{\Theta} \geqq \left(\frac{n}{(n-1)!}\right)^n |\det(A_j^i)| = \left(\frac{n}{(n-1)!}\right)^n |\det(a_j^i)|^{n-1}$$

$$= \left(\frac{n}{(n-1)!}\right)^n (n!)^{n-1} (V[\delta(\Theta)])^{n-1}$$

$$= \frac{n^{2n}}{n!} \left(V[\delta(\Theta)] \right)^{n-1}.$$

Therefore

$$V[\delta(\Theta)] \leqq \left(\frac{n!}{n^{2n}}\right)^{1/(n-1)} \sin^{1/(n-1)} \tilde{\Theta}. \qquad \text{q.e.d.}$$

LEMMA 6. *$V[\delta(\Theta)] \leqq \xi(n)\cdot\tilde{\Theta}^{1/(n-1)}$ where $\xi(n)$ depends only on n.*

Proof. In the manner of Lemma 2 we circumscribe the base polytope of $\delta(\Theta)$ with an $(n-1)$-dimensional simplex $\mathscr{B}_{n-1}$, and we inscribe an $(n-1)$-dimensional simplex $\mathscr{D}_{n-1}$. Now $\mathscr{B}_{n-1}$ forms an n-dimensional pyramid $\Delta(\Theta')$ with $\boldsymbol{\alpha}$ the vertex of Θ, and $\Theta' \supset \Theta$.

Let us for a moment restrict our attention to H, the hyperplane containing $\mathscr{D}_{n-1}$, $\mathscr{B}_{n-1}$, and the base of $\delta(\Theta)$. As we noted after Lemma 2, $\mathscr{B}_{n-1}$ and $\mathscr{D}_{n-1}$ have a common centroid. We now inscribe a $\mathscr{B}'_{n-1}$ inside $\mathscr{D}_{n-1}$ in exactly the same manner that $\mathscr{D}_{n-1}$ is inscribed in $\mathscr{B}_{n-1}$. Thus in this case $\mathscr{B}_{n-1}$, $\mathscr{D}_{n-1}$, and $\mathscr{B}'_{n-1}$ all have a common centroid say $\mathbf{c}$. If we shrink each linear dimension of H by a factor of $(n-1)^{-2}$ about the point $\mathbf{c}$, then $\mathscr{B}_{n-1}$ will be transformed into $\mathscr{B}'_{n-1}$.

If we now shrink each linear dimension of n-dimensional space by a factor of $(n-1)^{-2}$ about the point $\mathbf{c}$, then as above $\mathscr{B}_{n-1}$ will be transformed into $\mathscr{B}'_{n-1}$, and $\boldsymbol{\alpha}$ becomes $\mathbf{c}+(n-1)^{-2}(\boldsymbol{\alpha}-\mathbf{c})$ which was originally an interior point of $\delta(\Theta)$. Thus the n-dimensional simplex $S'(n)$ defined by $\mathscr{B}'_{n-1}$ and $\mathbf{c} + (n-1)^{-2}(\boldsymbol{\alpha}-\mathbf{c})$ is contained in $\delta(\Theta)$, and $\Delta(\Theta')$ is similar to $S'(n)$ the ratio of similitude being $((n-1)^{+2}:1)$. Since $S'(n) \subset \delta(\Theta)$, the area of the sides of $S'(n)$ must be $\leqq 1$. Thus

$$V[\Delta(\Theta')] = (n-1)^{2n}V[S'(n)] \leqq (n-1)^{2n}V[\delta(\Theta')].$$

By Lemma 5,

$$V[\delta(\Theta')] \leqq \left(\frac{n!}{n^{2n}}\right)^{1/(n-1)} \sin^{1/(n-1)}\tilde{\Theta}'.$$

By Lemmas 1 and 3,

$$\sin\tilde{\Theta}' \leqq (n-1)!\,\tilde{\Theta}' \leqq (n-1)!\,\tilde{\Theta}.$$

Hence

$$\begin{aligned} V[\delta(\Theta)] &\leqq V[\Delta(\Theta')] \\ &\leqq (n-1)^{2n}\left(\frac{n!}{n^{2n}}\right)^{1/(n-1)} ((n-1)!)^{1/(n-1)}\tilde{\Theta}^{1/(n-1)} \\ &= \xi(n)\,\tilde{\Theta}^{1/(n-1)}. \qquad \text{q.e.d.} \end{aligned}$$

LEMMA 7. *Given any bounded convex body C in n-dimensional space with volume $V(C) > 0$, there exists an affine transformation T of determinant ± 1 such that $T: C \to C'$ and*

$$V(C) = V(C') \geqq \gamma(n)\,[S(C')]^{n/(n-1)}$$

where $\gamma(n)$ depends only on n.

Proof. As in Lemma 2, we inscribe a simplex $\mathcal{D}_n$ in C and circumscribe a simplex $\mathcal{B}_n$ with the properties listed in that lemma. We choose our transformation T so that it sends $\mathcal{D}_n$ into a simplex $\mathcal{D}'_n$ of the same volume but with equal edges. Thus since

$$V(\mathcal{D}_n) = V(\mathcal{D}'_n) = |\det T|\, V(\mathcal{D}_n),$$

we have $|\det T| = 1$ [4, p. 6]. Since parallelism and volume are preserved under this transformation, we see that $\mathcal{B}_n$ is also transformed into a simplex with equal edges say $\mathcal{B}'_n$. If a is the length of an edge of $\mathcal{D}'_n$, then

$$V(\mathcal{D}'_n) = \frac{a^n}{n!}\left(\frac{n+1}{2^n}\right)^{1/2}$$

[7, p. 126], and thus

$$S(\mathcal{D}'_n) = (n+1)\left(\frac{a^{n-1}}{(n-1)!}\left(\frac{n}{2^{n-1}}\right)^{1/2}\right).$$

Since $\mathcal{B}'$ is similar to $\mathcal{D}'_n$,

$$V(\mathcal{B}'_n) = \frac{(na)^n}{n!}\left(\frac{n+1}{2^n}\right)^{1/2},$$

and

$$S(\mathcal{B}'_n) = (n+1)\left(\frac{(na)^{n-1}}{(n-1)!}\left(\frac{n}{2^{n-1}}\right)^{1/2}\right).$$

Since $S(\mathcal{B}'_n) \geqq S(C') \geqq S(\mathcal{D}'_n)$ by [2, p. 47], and since

$$V(\mathcal{B}') = \frac{1}{n!}(n+1)^{1/2}\left(\frac{(n-1)!}{(n-1)\sqrt{n}}\right)^{n/(n-1)}(S(\mathcal{B}'_n))^{n/(n-1)} = c_1(n)\,(S(\mathcal{B}'_n))^{n/(n-1)},$$

we obtain

$$\begin{aligned} V(C) = V(C') \geqq V(\mathcal{D}'_n) &= n^{-n}V(\mathcal{B}'_n) = n^{-n}c_1(n)(S(\mathcal{B}'_n))^{n/(n-1)} \\ &\geqq n^{-n}c_1(n)(S(C'))^{n/(n-1)} \\ &= \gamma(n)(S(C'))^{n/(n-1)} \end{aligned}$$

q.e.d.

One might be tempted to believe that Lemma 7 and the result contained in [1] together imply the theorem to be proved in § II; however, the transformation T given above does not necessarily have integral coefficients which it would have to have if the desired implication were to hold.

The following result is undoubtedly well known, but the author was unable to find this result completely proved elsewhere.

LEMMA 8(3). *Given a polytope $(Po)_n$ with vertices $\alpha_1, \cdots, \alpha_M$, then*

$$\sum_{i=1}^{M} \tilde{\Theta}_i = n\sigma(n) = \frac{2\pi^{n/2}}{\Gamma(n/2)}$$

where $\tilde{\Theta}_i$ is the n-dimensional exterior angle at α_i.

Proof. Our problem is to show that the exterior angles (i.e. $\tilde{\Theta}_1, \cdots, \tilde{\Theta}_M$) of $(Po)_n$ completely fill the unit sphere. Let us think of the unit sphere as the totality of line segments of length 2 with midpoint in the origin. First, we show that by translating the exterior angles of $(Po)_n$ to the origin we obtain all such segments. This follows almost directly from Theorem IV of [3, p. 115] which states (among other things) that if $\mathscr{K}$ is a bounded convex body then there are precisely two supporting hyperplanes to $\mathscr{K}$ parallel to any given hyperplane H.

We now pick a direction $(A_1 : \cdots : A_n)$. There are two supporting hyperplanes H_1 and H_2 to $(Po)_n$ with exterior unit normals

$$\vec{n}_1 = (A_1^2 + \cdots + A_n^2)^{-1/2}(A_1, \cdots, A_n) \text{ and } \vec{n}_2 = (A_1^2 + \cdots + A_n)^{-1/2}(-A_1, \cdots, -A_n)$$

respectively. Now each of these supporting hyperplanes contains at least one vertex of $(Po)_n$ say $\alpha_i \subset H_1$ and $\alpha_j \subset H_2$ (since a supporting hyperplane to a closed bounded convex set contains at least one extreme point of the set [4, p. 24]). If we consider $\vec{n}_1$ to be emanating from α_i, we have $\vec{n}_1 \subset \tilde{\Theta}_i$ and similarly $\vec{n}_2 \subset \tilde{\Theta}_j$. Thus after translating the exterior angles of $(Po)_n$ so that their vertices coincide with O, we obtain at least one pair of vectors $(A_1^2 + \cdots + A_n^2)^{-1/2} (A_1, \cdots, A_n)$ and $(A_1^2 + \cdots + A_n^2)^{-1/2} (-A_1, \cdots, -A_n)$ which gives the required segment for the direction $(A_1 : \cdots : A_n)$. Hence

$$\sum_{i=1}^{M} \tilde{\Theta}_i \geqq \frac{2\pi^{n/2}}{\Gamma(n/2)} = n\sigma(n).$$

On the other hand, if a certain ray ρ emanating from α_i is completely in the interior of $\tilde{\Theta}_i$, then $\angle \mathbf{x}\alpha_i\mathbf{y} > \pi/2$ if $\mathbf{y} \in \Theta_i$ and $x \in \rho$. Hence the hyperplane H through α_i which has this ray as an exterior normal ray is such that $H \cap (Po)_n = \alpha_i$. By [3, p. 115], no other supporting hyperplane to $(Po)_n$ has an exterior normal ray in this direction. Thus after translating the exterior angles of $(Po)_n$ so that their vertices coincide with O, none of the interiors of the various angles intersect. Hence

$$\sum_{i=1}^{M} \tilde{\Theta}_i \leqq \frac{2\pi^{n/2}}{\Gamma(n/2)} = n\sigma(n).$$

Therefore

$$\sum_{i=1}^{M} \tilde{\Theta}_i = \frac{2\pi^{n/2}}{\Gamma(n/2)} = n\sigma(n). \qquad \text{q.e.d.}$$

(3) For $n=2$, this is simply Gauss's theorem on the curvature integral for polygons.

II. Having established the necessary properties of polytopes in § I, we now proceed to the general theorem. We shall restate it here for convenience.

THEOREM. *If C is an n-dimensional strictly convex body with N noncoplanar lattice points on its boundary, then*

$$V(C) > \kappa(n)N^{(n+1)/(n-1)}$$

where $\kappa(n) > 0$ *depends only on n.*

Proof. Call the set of N boundary lattice points $B_a(N)$. The members of $B_a(N)$ are the vertices of a convex polytope entirely in the interior of C [1, Lemma 1, p. 273]. Call this polytope $(Po)_n^a$. Let us now multiply each dimension of space by 3. This transforms $(Po)_n^a$ into $(Po)_n^b$ with $V[(Po)_n^a] = 3^{-n}V[(Po)_n^b]$. Call the set of vertices of $(Po)_n^b$, $B_b(N)$. Clearly if $\mathbf{p} = (p_1, \cdots, p_n)$ is in $B_b(N)$, then $p_i \equiv 0 \pmod 3$ for $i = 1, \cdots, n$. Suppose $\mathbf{p}_i$ is a specific vertex of $(Po)_n^b$. Note that the convex cover of the set of lattice points $\{\mathbf{p}_i + (\mathbf{p}_j - \mathbf{p}_i)/3, i \neq j, i = 1, \cdots, N, j = 1, \cdots, N\}$ is a convex polytope $(Po)_n^c$ contained in $(Po)_n^b$. To each vertex $\mathbf{p}_i$ of $(Po)_n^b$ there is a corresponding face $(Po)_{n-1}^i$ of $(Po)_n^c$ lying on a hyperplane H_i such that $\mathbf{p}_i$ lies on the opposite side of H_i from all other members of $B_b(N)$, and if $i \neq j$, then $(Po)_{n-1}^i \neq (Po)_{n-1}^j$ [1, Proof of Lemma 3, p. 274]. Thus H_i truncates $(Po)_n^b$ at the vertex $\mathbf{p}_i$. A pyramid $\Delta_i(\Theta_i)$ is thus formed at each vertex $\mathbf{p}_i$ of $(Po)_n^b$ where $\mathbf{p}_i$ is the vertex and $H \cap (Po)_n^b$ is the base. These pyramids are all disjoint by construction, and the sides of each pyramid are part of the surface (or boundary) of $(Po)_n^b$. By construction, $\Delta_i(\Theta_i)$ contains in its base an $(n-1)$-dimensional convex polytope with lattice point vertices $(q_{s1}, \cdots, q_{sn})$ $s = 1, \cdots m \geqq n$. Thus $\Delta_i(\Theta_i)$ must at least contain an n-dimensional simplex $S(n)$ with lattice point vertices since $\mathbf{p}_i = (p_{i1}, \cdots, p_{in})$ is a lattice point, and thus

$$V[\Delta_i(\Theta_i)] \geqq V(S(n)) = \frac{1}{n!}\left|\det(q_{st} - p_{it})\right| \geqq \frac{1}{n!}.$$

We now apply Lemma 7 to $(Po)_n^b$ and transform $(Po)_n^b$ into $(Po)_n^d$ with

$$V[(Po)_n^b] = V[(Po)_n^d] = \gamma(n)[S[(Po)_n^d]]^{n/(n-1)}.$$

The pyramids dealt with above are also transformed into new pyramids $\Delta_i'(\Theta_i')$, and still $V[\Delta_i'(\Theta_i')] = V[\Delta_i(\Theta_i)] \geqq 1/n!$. Let $\mathscr{S}_i$ denote the surface area of the sides of $\Delta_i'(\Theta_i')$. Then clearly

$$\mathscr{S}_i^{n/(n-1)} \cdot V[\delta_i'(\Theta_i')] = V[\Delta_i'(\Theta_i')]$$

since $\delta_i'(\Theta_i')$ is the pyramid similar to $\Delta_i'(\Theta_i')$ but such that the total area of its sides is equal to 1.

Hence

$$\left(\frac{V[(Po)_n^d]}{\gamma(n)}\right)^{(n-1)/n} \geqq S[(Po)_n^d]$$

$$\geqq \sum_{i=1}^{N} \mathscr{S}_i$$

$$= \sum_{i=1}^{N} \left[\{\mathscr{S}_i \cdot (V(\delta_i'(\Theta_i')))^{(n-1)/n}\} \frac{1}{(V(\delta_i'(\Theta_i')))^{(n-1)/n}}\right]$$

$$\geqq \left[\sum_{i=1}^{N} \{\mathscr{S}_i \cdot (V(\delta_i'(\Theta_i')))^{(n-1)/n}\}^{n/(n+1)}\right]^{(n+1)/n}$$

$$\cdot \left[\sum_{i=1}^{N} \left\{\frac{1}{(V(\delta_i'(\Theta_i')))^{(n-1)/n}}\right\}^{-n}\right]^{-1/n}$$

(by Hölder's inequality [5, p. 24])

$$= \left[\sum_{i=1}^{N} \{V(\Delta_i'(\Theta_i'))\}^{(n-1)/(n+1)}\right]^{(n+1)/n} \left[\sum_{i=1}^{N} (V(\delta_i'(\Theta_i')))^{n-1}\right]^{-1/n}$$

$$\geqq \left[\sum_{i=1}^{N} (n!)^{-(n-1)/(n+1)}\right]^{(n+1)/n} \left[\sum_{i=1}^{N} (\xi(n))^{n-1}\tilde{\Theta}_i'\right]^{-1/n} \quad \text{(by Lemma 6)}$$

$$= n!)^{-(n-1)/n} N^{(n+1)/n} (\xi(n))^{-(n-1)/n} \left(\sum_{i=1}^{N} \tilde{\Theta}_i'\right)^{-1/n}$$

$$= (n!)^{-(n-1)/n} N^{(n+1)/n} (\xi(n))^{-(n-1)/n} \left(\frac{2\pi^{n/2}}{\Gamma(n/2)}\right)^{-1/n} \quad \text{(by Lemma 8)}$$

$$= \kappa'(n) N^{(n+1)/n}.$$

We now combine these results.

$$V(C) \geqq V[(Po)_n^a]$$

$$= 3^{-n} V[(Po)_n^b]$$

$$= 3^{-n} V[(Po)_n^d]$$

$$\geqq 3^{-n}\gamma(n)(\kappa'(n))^{n/(n-1)} N^{(n+1)/(n-1)},$$

and letting $\kappa(n) = (\frac{1}{2})\, 3^{-n}\, \gamma(n)(\kappa'(n))^{n/n-1}$, we obtain

$$V(C) > \kappa(n) N^{(n+1)/(n-1)}. \quad \text{q.e.d.}$$

We now have a new proof for the main result in [1, p. 272].

COROLLARY. *If C is an n-dimensional strictly convex body with N noncoplaner lattice points on its boundary, then*

$$S(C) > c(n)N^{(n+1)/n}.$$

Proof. By the isoperimetric inequality [4, p. 104],

$$\begin{aligned} S(C) &\geqq n(\sigma(n))^{1/n}[V(C)]^{(n-1)/n} \\ &> n(\sigma(n))^{1/n}(\kappa(n))^{(n-1)/n}N^{(n+1)/n} \\ &= c(n)N^{(n+1)/n}. \end{aligned}$$

q.e.d.

BIBLIOGRAPHY

1. G. E. Andrews, *An asymptotic expression for the number of solutions of a general class of Diophantine equations*, Trans. Amer. Math. Soc. **99** (1961), 272–277.

2. T. Bonnesen and W. Fenchel, *Theorie der konvexen Körper*, Chelsea, New York, 1948.

3. J. W. S. Cassels, *An introduction to the geometry of numbers*, Springer-Verlag, Berlin, 1959.

4. H. G. Eggleston, *Convexity*, Cambridge Tracts in Mathematics and Mathematical Physics, No. 47, Cambridge Univ. Press, Cambridge, 1958.

5. G. H. Hardy, J. E. Littlewood and G. Pólya, *Inequalities*, Cambridge Univ. Press, Cambridge, 1952.

6. K. Mahler, *The geometry of numbers*, Duplicated lectures, Boulder, Colorado, 1950.

7. D. M. Y. Sommerville, *An introduction to the geometry of n dimensions*, Dover, New York, 1958.

UNIVERSITY OF PENNSYLVANIA,
PHILADELPHIA, PENNSYLVANIA

1.1.1 Errata

p. 274, top equation: $1/(-1) \to 1/(n-1)$.
p. 277, penultimate displayed equation: $/(n-) \to n/(n-1)$.
p. 278, lines 10 and 11: $-1/ \to -1/n$.

Chapter 2

q-Series

Instead of writing ten chapters on different aspects of q-series, I have chosen to divide one chapter into ten subsections.

2.1 Ph.D. thesis
2.2 Heine's Method
2.3 The Askey Influence
2.4 Well-poised series
2.5 Bailey Chains
2.6 q-Trinomials and the Borwein Conjecture
2.7 Ramanujan's "Lost" Notebook
2.8 Statistical Mechanics
2.9 The Pfaff Trilogy
2.10 MacMahon's "Partition Analysis"

At the beginning of each subsection, I shall provide a commentary.

2.1 Ph.D. Thesis

As described in Chapter 1, my early work through my M.A. (at Oregon State University), my Fulbright grant year at Cambridge and into the Fall of 1961, was all on the geometry of numbers. However, during 1962–1964, the magic of partitions and q-series performed by Fine and Rademacher engrossed me.

I asked Rademacher if I could write my thesis under his direction. He agreed and almost immediately suggested that I improve the work in Leila Dragonette's thesis on mock theta functions [Dra52]. Rademacher suggested that I utilize the Poisson summation formula to deal most effectively with the transformations of the third-order mock theta functions. I stubbornly resisted his recommendation for more than a year. Eventually, I came to realize that Watson had been using a disguised version of the Poisson summation in his famous "Final Problem" paper [Wat36], and I can well remember my delight on seeing the transformations of $f(q)$ under the modular group fall naturally onto the page. I think that Rademacher was most pleased by the fact that I had identified the sums of roots of unity in the asymptotic series with those in the Hardy–Ramanujan formula for $p(n)$.

The natural question left open in my thesis was: Is the formula for the coefficients, in fact, exact?

This exceedingly difficult question (posed again in [227]) was answered positively in a stunningly brilliant work by Ono and Bringmann [Bri-O08]. Garthwaite provided a comparable treatment of $\omega(q)$ [Gar08].

2.1.1 *On the theorems of Watson and Dragonette for Ramanujan's mock theta functions*

ON THE THEOREMS OF WATSON AND DRAGONETTE FOR RAMANUJAN'S MOCK THETA FUNCTIONS.

By GEORGE E. ANDREWS.[1]

I. Introduction. The mock theta functions were first studied by Ramanujan; G. H. Hardy in [9; p. 354] defines such functions as follows

> . . . a "mock θ-function" is a function defined by a q-series convergent for $|q| < 1$, for which we can calculate asymptotic formulae, when q tends to a "rational point" $e^{2r\pi i/s}$, of the same degree of precision as those furnished, for the ordinary θ-functions, by the theory of linear transformation.

The mock theta functions we shall study are the third order functions given below:

$$f(q) = \sum_{n=0}^{\infty} \frac{q^{n^2}}{(1+q)^2(1+q^2)^2 \cdots (1+q^n)^2}$$

$$\phi(q) = \sum_{n=0}^{\infty} \frac{q^{n^2}}{(1+q^2)(1+q^4) \cdots (1+q^{2n})}$$

$$\psi(q) = \sum_{n=0}^{\infty} \frac{q^{n^2}}{(1-q)(1-q^3) \cdots (1-q^{2n-1})}$$

$$\omega(q) = \sum_{n=0}^{\infty} \frac{q^{2n(n+1)}}{(1-q)^2(1-q^3)^2 \cdots (1-q^{2n+1})^2}$$

$$\upsilon(q) = \sum_{n=0}^{\infty} \frac{q^{n(n+1)}}{(1+q)(1+q^3) \cdots (1+q^{2n+1})}$$

There are two other third order functions $\chi(q)$ and $\rho(q)$ [12; p. 62]. They satisfy the following equations [12; p. 63]

$$\chi(q) = \tfrac{1}{4}f(q) + \tfrac{3}{4}\theta_4{}^2(0, q^3) \prod_{r=1}^{\infty} (1-q^r)^{-1}$$

$$\rho(q) = -\tfrac{1}{2}\omega(q) + \tfrac{3}{2}[\tfrac{1}{2}q^{-3/8}\theta_2(0, q^{3/2})]^2 \prod_{r=1}^{\infty} (1-q^{2r})^{-1}.$$

Received November 16, 1964.

[1] Presented to the faculty of the Graduate School of the University of Pennsylvania in partial fulfillment of the requirements for the degree of Doctor of Philosophy, 1964.

This research was partially completed under a National Science Foundation Graduate Fellowship and partially under NSF grant GP1640.

We shall not study them; however, their behavior can be derived from these equations once the behavior of f and ω is known.

In the paper [2], Dragonette derived general linear transformation formulae for the third order mock theta functions and used her results to obtain very precise asymptotic formulae for the power series coefficients of these functions. G. N. Watson in [12] had already studied the third order mock theta functions under the specific transformations; $S: \tau \to \tau + 1$ and $T: \tau \to 1/\tau$. Since these transformations generate the modular group, Dragonette was able to derive the general modular transformation formulae by the process of iteration. Her method, however, left certain roots of unity unevaluated and also gave only estimates for the "error" terms arising. In this paper, we shall generalize Watson's method [12; pp. 73-76] and shall thus obtain precise linear transformation formulae with all roots of unity and "error" terms evaluated. This will carry us through Sections II and III. In Section IV, we shall show that all sums of roots of unity that are to be considered are actually generalized Kloosterman sums and therefore subject to the results of Salié [10] and Lehmer [6]. We shall show, in fact, that the exponential sums appearing in the asymptotic expansions of Dragonette [2; Theorems II and IV] are actually special cases of the exponential sum $A_k(n)$ associated with the partition function (cf. [6]). In Section V, we shall apply our results to the problem of improving the error terms in the asymptotic expansion theorems of Dragonette. For example, Dragonette has proved the following [2; p. 478]

THEOREM. *The coefficient of* q^n *in the series* $f(q) = \sum_{n=0}^{\infty} A(n) q^n$ *is*

$$A(n) = \sum_{0<k\leq n^{\frac{1}{2}}} \frac{\lambda(k)\exp\{\pi(n-1/24)^{\frac{1}{2}}/k\sqrt{6}\}}{k^{\frac{1}{2}}(n-1/24)^{\frac{1}{4}}} + O(n^{\frac{1}{2}}\,\mathrm{og}\, n)$$

where

$$\lambda(1) = \tfrac{1}{2}(-1)^{n-1}$$

and

$$\lambda(k) = \tfrac{1}{2} \sum_{\substack{h \bmod 2k \\ \text{g.c.d.}(h,2k)=1}} \exp(-\pi i h n/k)\,\epsilon_{h,k}{}^{(f)} \qquad (k = 2, 3, \cdots, [n^{\frac{1}{2}}]).$$

We shall improve Dragonette's result to the following

THEOREM 5.1. *The coefficient of* q^n *in the series* $f(q) = \sum_{n=0}^{\infty} A(n)$ *is*

$$A(n) = \sum_{k=1}^{[n^{\frac{1}{2}}]} \frac{\lambda(k)\exp\{\pi(n-1/24)^{\frac{1}{2}}/k\sqrt{6}\}}{k^{\frac{1}{2}}(n-1/24)^{\frac{1}{4}}} + O(n^{\epsilon})$$

where

$$\lambda(k) = \begin{cases} \frac{1}{2}(-1)^{\frac{1}{2}(k+1)} A_{2k}(n) & k \text{ odd} \\ \frac{1}{2}(-1)^{\frac{1}{2}k} A_{2k}(n-\frac{1}{2}k) & k \text{ even,} \end{cases}$$

and ϵ *is positive and arbitrarily small.* $A_k(n)$ *is the exponential sum appearing in the Hardy-Ramanujan formula for* $p(n)$ [9; p. 284].

It seems likely that a closer study of the asymptotic expansion for $A(n)$ would show that the above series would diverge if extended to (since $\lambda(k) = O(k^{\frac{1}{2}+\epsilon})$ and probably $|\lambda(k)| > c \cdot k^{\frac{1}{2}}$ for some $c > 0$ and an infinite number of k). If, however, e^x is replaced by $2\sinh x$ in the above expansion, the problem becomes deeper. I would conjecture that the new series is not absolutely convergent (for probably $|\lambda(p)| > c \cdot p^{\frac{1}{2}}$ for some $c > 0$ and a positive proportion of all primes p), but the extreme accuracy of Dragonette's numerical results [2; p. 494] makes plausible the conjecture that the new series is conditionally convergent and that it converges to $A(n)$.

The author wishes to express his deep indebtedness to Professor Hans Rademacher who suggested this study and gave much advice and encouragement throughout the course of the research.

II. The modular transformation formulae of f and ω. Our first two theorems of this section improve the results in Dragonette's Theorem I [2; p. 477]. The next two theorems improve Dragonette's results for ω in Theorem III [2; p. 499, line (7.7)]. We complete this section by simplifying the expressions for the roots of unity we have obtained.

Throughout the course of this paper we shall endeavor to keep our notation consistent with that of Dragonette. The following well-known theorem together with its notation and conventions will be used over and over.

If $P(q) = \prod_{r=1}^{\infty} (1-q^r)^{-1}$, then

$$P(\exp\{2\pi i(h+iz)/k\}) = \omega_{h,k} z^{\frac{1}{2}} \exp\{\pi(z^{-1}-z)/12k\} \cdot P(\exp\{2\pi i(h'+iz^{-1})/k\}) \tag{2.0.1}$$

where $\operatorname{Re} z > 0$ with the principal branch of $z^{\frac{1}{2}}$ taken, $hh' \equiv -1 \pmod{k}$, and $\omega_{h,k}$ a $24k$-th root of unity [9; pp. 290-291]. We remark that $P(q)$ is essentially the modular form $\eta(\tau)$, and (2.0.1) is the transformation formula for $\eta(\tau)$ in thin disguise.

THEOREM 2.1. *Let* g.c.d.$(h,k) = 1$, $hh' + kk' = -1$, h *be even, and* $\operatorname{Re} z > 0$ *with the principal branch of* $z^{\frac{1}{2}}$ *taken. Then*

$$f(\exp\{\pi i(h+iz)/k\}) = \eta_{h,k}^{(f)} 2^{\frac{3}{2}} z^{-\frac{1}{2}} \exp(-\pi(4z^{-1}+z/8)/3k) \cdot \omega(\exp\{2\pi i(h'+iz^{-1})/k\}) + E_1^{(f)}(h,k;z)$$

where

$$\eta_{h,k}^{(f)} = \exp\{\tfrac{1}{2}\pi i(2h'+k'-3-3h'k'+3h'/k)\}\omega_{\frac{1}{2}h,k},$$

and

$$E_1^{(f)}(h,k;z) = 2^{\frac{1}{2}} k^{-1} \exp(-\pi z/24k) \sum_{\nu \bmod k} \delta_1^{(f)}(\nu;\tfrac{1}{2}h;2h';k) z^{\frac{1}{2}} J_1^{(f)}(\nu;k;z)$$

with

$$\delta_1^{(f)}(\nu;h;h';k) = (-1)^{\nu}\omega_{h,k}\exp\{\pi i h'(-3\nu^2+\nu)/k\}$$

and

$$J_1^{(f)}(\nu;k;z) = \int_{-\infty}^{\infty} \frac{\exp(-3\pi z x^2/2k)}{\cosh(\pi i(\nu-\frac{1}{6})/k - \pi z x/2k)}\, dx.$$

Finally if $1 \leqq \nu \leqq k$, *we have*

$$|z^{\frac{1}{2}} J_1^{(f)}(\nu;k;z)| \leqq 3^{-\frac{1}{2}} 2^{\frac{3}{2}} N^{\frac{1}{2}} k\, |k-2(\nu-\tfrac{1}{6})|^{-1}$$

provided $z = kN^{-2} - ik\phi$ *with* $|\phi| \leqq k^{-1}N^{-1}$ *and* $1 \leqq k \leqq N$.

THEOREM 2.2. *Let* g.c.d.$(h,k)=1$, $hh''+2kk''=-1$, h *be odd, and* $\operatorname{Re} z > 0$ *with the principal branch of* $z^{\frac{1}{2}}$ *taken. Then*

$$f(\exp\{\pi i(h+iz)/k\}) = \epsilon_{h,k}^{(f)} z^{-\frac{1}{2}} \exp(-\pi(z+z^{-1})/24k) f(\exp\{\pi i(h''+iz^{-1})/k\}) + E_2^{(f)}(h,k;z)$$

where

$$\epsilon_{h,k}^{(f)} = (-1)^{k+\frac{1}{2}(hh''-1)} \exp(\pi i h''/2 + 3\pi i h h''^2 k/2)\omega_{h,2k},$$

and

$$E_2^{(f)}(h,k;z) = E_1^{(f)}(2h,2k;2z) = \exp(-\pi z/24k) k^{-1} \sum_{\nu \bmod k} \delta_1^{(f)}(\nu;h;h'';2k) z^{\frac{1}{2}} J_1^{(f)}(\nu;2k;2z)$$

with $\delta_1^{(f)}(\nu;h;h'';2k)$ *and* $J_1^{(f)}(\nu;2k;2z)$ *as in Theorem* 2.1.

THEOREM 2.3. *Let* g.c.d.$(h,k)=1$, $hh'+kk'=-1$, h *be even, and* $\operatorname{Re} z > 0$ *with the principal branch of* $z^{\frac{1}{2}}$ *taken. Then*

$$\omega(\exp\{\pi i(h+iz)/k\}) = \tfrac{1}{2} z^{-\frac{1}{2}} \epsilon_{h,k}^{(\omega)} \exp(2\pi z/3k + \pi z^{-1}/12k) \cdot f(\exp\{2\pi i(h'+iz^{-1})/k\}) + E_1^{(\omega)}(h,k;z)$$

where

$$\epsilon_{h,k}^{(\omega)} = (-1)^{\frac{1}{2}(k-1)} \exp(3\pi ihk/4 - 3\pi ih/4k)\,\omega_{h,k}$$

and

$$E_1^{(\omega)}(h,k;z) = k^{-1}\exp(2\pi z/3k) \sum_{\nu \bmod k} \delta_1^{(\omega)}(\epsilon;\nu;h;h';k)\, z^{\frac{1}{2}} J_1^{(\omega)}(\nu;k;z)$$

with

$$\delta_1^{(\omega)}(\epsilon;\nu;h;h';k) = i\epsilon_{h,k}^{(\omega)} \exp(-3\pi ih'\nu^2/k + \pi ih'/k)$$

and

$$J_1^{(\omega)}(\nu;k;z) = \int_{-\infty}^{\infty} \frac{\exp(-3\pi zx^2/k)}{\tanh\{\pi i(\nu - \frac{1}{6})/k - \pi zx/k\}}\,dx.$$

Finally if $-\frac{1}{2}(k+1) \leqq \nu \leqq \frac{1}{2}(k-1)$, *we have*

$$|z^{\frac{1}{2}} J_1^{(\omega)}(\nu;k;z)| \leqq 2^{-\frac{1}{4}} 3^{-\frac{1}{2}} \exp(\pi/6) N^{\frac{1}{2}} k\, |\nu - \tfrac{1}{6}|^{-1}$$

provided $z = kN^{-2} - ik\phi$ *with* $|\phi| \leqq (kN)^{-1}$ *and* $1 \leqq k \leqq N$.

THEOREM 2.4. *Let* g.c.d.$(h,k) = 1$, $hh'' + 2kk'' = -1$, *h be odd, and* $\operatorname{Re} z > 0$ *with the principal branch of* $z^{\frac{1}{2}}$ *taken. Then*

$$\omega(\exp\{\pi i(h+iz)/k\}) = \eta_{h,k}^{(\omega)} z^{-\frac{1}{2}} \exp(2\pi z/3k - 2\pi z^{-1}/3k) \cdot \omega(\exp\{\pi i(h''+iz^{-1})/k\}) + E_2^{(\omega)}(h,k;z)$$

where

$$\eta_{h,k}^{(\omega)} = i(-1)^{\frac{1}{2}(h''+1)} \exp\{-3\pi ih''k''/2 - 3\pi i(h''-h)/4k\}\,\omega_{h,k}$$

and

$$E_2^{(\omega)}(h,k;z) = k^{-1}\exp(2\pi z/3k) \sum_{\nu \bmod k} (-1)^{\nu} \exp(-3\pi ih''/4k) \cdot \delta_1^{(\omega)}(\eta;\mu;h;h'';k)\, z^{\frac{1}{2}} J_1^{(\omega)}(\mu;k;z)$$

with $\delta_1^{(\omega)}(\eta;\mu;h;h'';k)$ and $J_1^{(\omega)}(\mu;k;z)$ *as in Theorem* 2.3 *and* $\mu = \nu + \frac{1}{2}$ *written for simplicity of notation.*

We now prove these four theorems. The proof of Theorem 2.1 is given in detail; since the proofs of all four theorems are similar, we omit many of the details in the remaining theorems. Also, it should be noted that Theorem 2.3 may be derived directly from Theorem 2.1; however, such a derivation would be complicated and would not likely shorten our exposition.

Proof of Theorem 2.1. We shall freely utilize the results in [12]. We let

$$F(q) = f(q)/P(q)$$

(2.1.1)
$$= f(q) \prod_{r=1}^{\infty} (1 - q^r)$$

$$= 1 + 4 \sum_{n=1}^{\infty} (-1)^n q^{\frac{1}{2}n(3n+1)} (1 + q^n)^{-1} \quad [12;\ \text{p. } 64]$$

$$= 1 + 4 \sum_{n=1}^{\infty} (-1)^n q^{3n^2/2} (q^{-\frac{1}{2}n} + q^{\frac{1}{2}n})^{-1}.$$

If $q = \exp\{\pi i(h + iz)/k\}$, then

$$F(q) = 1 + 2 \sum_{n=1}^{\infty} \frac{(-1)^n \exp\{3n^2\pi i(h + iz)/2k\}}{\cosh(n\pi i(h + iz)/2k)}$$

$$= \sum_{n=-\infty}^{\infty} \frac{(-1)^n \exp(3\pi i h n^2/2k - 3\pi z n^2/2k)}{\cosh(\pi i h n/2k - \pi z n/2k)}, \text{ letting } n = mk + \nu$$

(2.1.3)
$$= \sum_{\nu=0}^{k-1} \sum_{m=-\infty}^{\infty} \frac{(-1)^{mk+\nu} \exp\{3\pi i h(m^2k^2 + 2mk\nu + \nu^2)/2k - 3\pi z(mk + \nu)^2/2k\}}{\cosh\{\pi i h m/2 + \pi i h \nu/2k - \pi z(mk + \nu)/2k\}}$$

$$= \sum_{\nu=0}^{k-1} (-1)^\nu \exp(3\pi i h \nu^2/2k) \sum_{m=-\infty}^{\infty} \frac{(-1)^m \exp\{-3\pi z(mk + \nu)^2/2k\}}{\cosh\{\pi i h \nu/2k - \pi z(mk + \nu)/2k\}};$$

by the Poisson summation formula [1; pp. 7-9] we obtain

$$F(q) = \sum_{\nu=0}^{k-1} (-1)^\nu \exp(3\pi i h \nu^2/2k) \sum_{n=-\infty}^{\infty} \int_{-\infty}^{\infty} \frac{\exp\{-3\pi z(kx + \nu)^2/2k + \pi i x + 2\pi i n x\}}{\cosh\{\pi i h \nu/2k - \pi z(kx + \nu)/2k\}}\, dx;$$

making the change of variable $kx + \nu \to x$, we obtain

$$F(q) = k^{-1} \sum_{\nu=0}^{k-1} (-1)^\nu \exp(3\pi i h \nu^2/2k) \sum_{n=-\infty}^{\infty} \int_{-\infty}^{\infty} \frac{\exp\{(2n + 1)\pi i(x - \nu)/k - 3\pi z x^2/2k\}}{\cosh(\pi i h \nu/2k - \pi z x/2k)}\, dx.$$

Notice now that if ν is replaced by $\nu \pm k$ the change introduced is

$$(-1)^{\pm k} \exp\{3\pi i h(\pm 2k + k^2)/2k\} (-1)^{(2n+1) - \frac{1}{2}h} = 1.$$

Hence we have

$$F(q) = k^{-1} \sum_{\nu \bmod k} (-1)^\nu \exp(3\pi i h \nu^2/2k) \sum_{n=-\infty}^{\infty} \int_{-\infty}^{\infty} \frac{\exp\{(2n + 1)\pi i(x - \nu)/k - 3\pi z x^2/2k\}}{\cosh(\pi i h \nu/2k -)zx/2k)}\, dx.$$

Now

$$\sum_{n=-\infty}^{\infty} = \sum_{n=0}^{\infty} + \sum_{n=-1}^{-\infty} = \sum_{n=0}^{\infty} + \sum_{n=1}^{\infty} \int_{-\infty}^{\infty} \frac{\exp\{-(2n - 1)\pi i(x - \nu)/k - 3\pi z x^2/2k\}}{\cosh(\pi i h \nu/2k -)zx/2k)}\, dx;$$

sending $x \to -x$ and $n \to n + 1$ we derive

$$\sum_{n=-\infty}^{\infty} = \sum_{n=0}^{\infty} + \sum_{n=0}^{\infty} \int_{-\infty}^{\infty} \frac{\exp\{(2n + 1)\pi i(x + \nu)/k - 3\pi z x^2/2k\}}{\cosh(-\pi i h \nu/2k - \pi z x/2k)}\, dx.$$

Replacing ν by $-\nu$ in the second sum we see that

$$(2.1.4)\quad F(q)=2k^{-1}\sum_{\nu \bmod k}(-1)^{\nu}\exp(3\pi ih\nu^2/2k)\cdot\sum_{n=0}^{\infty}\int_{-\infty}^{\infty}\frac{\exp\{(2n+1)\pi i(x-\nu)/k-3\pi zx^2/2k\}}{\cosh(\pi ih\nu/2k-\pi zx/2k)}\,dx.$$

We now define $C_k(x)=\dfrac{\cosh 3x}{\cosh\,(x/k)}$.

Clearly $C_k(x)$ is an entire function of x if k is an odd integer. Hence if

$$\begin{aligned}\mathfrak{F}_n(x)&=\frac{\exp\{(2n+1)\pi i(x-\nu)/k-3\pi zx^2/2k\}}{\cosh(\pi ih\nu/2k-\pi zx/2k)}\\&=\frac{\exp\{(2n+1)\pi i(x-\nu)/k-3\pi zx^2/2k\}}{\cosh\{3(\pi ih\nu-\pi zx)/2\}}C_k(\{\pi ih\nu-\pi zx\}/2)\\&=\frac{(-1)^{\frac{1}{2}h\nu}\exp\{(2n+1)\pi i(x-\nu)/k-3\pi zx^2/2k\}}{\cosh(3\pi zx/2)}C_k(\{\pi ih\nu-\pi zx\}/2)\end{aligned}$$

then $\mathfrak{F}_n(x)$ has (at most) simple poles at

$$x_m=(2m+1)i/3z,\ m\ \text{any integer}.$$

The residue at x_m is

$$\lambda_{n,m}=2(3\pi z)^{-1}(-1)^{\frac{1}{2}h\nu-m}\exp\{(2n+1)\pi i(x_m-\nu)/k-3\pi zx_m{}^2/2k\}C_k(\{\pi ih\nu-\pi zx_m\}/2).$$

Now by Cauchy's theorem,

$$\left\{\int_{-\infty}^{\infty}-P\int_{-\infty+(2n+1)i/3z}^{\infty+(2n+1)i/3z}\right\}\mathfrak{F}_n(x)\,dx=2\pi i(\lambda_{n,0}+\lambda_{n,1}+\cdots+\tfrac{1}{2}\lambda_{n,n}).$$

Hence

$$\begin{aligned}F(q)&=2k^{-1}\sum_{\nu \bmod k}(-1)^{\nu}\exp(3\pi ih\nu^2/2k)2\pi i\sum_{n=0}^{\infty}(\lambda_{n,0}+\lambda_{n,1}+\cdots+\tfrac{1}{2}\lambda_{n,n})\\&\quad+2k^{-1}\sum_{\nu \bmod k}(-1)^{\nu}\exp(3\pi ih\nu^2/2k)\sum_{n=0}^{\infty}P\int_{-\infty+x_n}^{\infty+x_n}\mathfrak{F}_n(x)\,dx\\&=\Sigma_1+\Sigma_2.\end{aligned}$$

It will turn out that Σ_1 will yield the main term in the transformation formula involving $\omega(q)$, and Σ_2 will yield the "error" term.

In Σ_1, we have

$$\sum_{n=0}^{\infty}\{\lambda_{n,0}+\lambda_{n,1}+\cdots+\tfrac{1}{2}\lambda_{n,n}\}=\sum_{n=0}^{\infty}\{\tfrac{1}{2}\lambda_{m,m}+\lambda_{m+1,m}+\lambda_{m+2,m}+\cdots\},$$

by rearrangement of the first series,

$$=\sum_{n=0}^{\infty}\lambda_{m,m}\{\tfrac{1}{2}+\sum_{l=1}^{\infty}\exp\{2\pi i(x_m-\nu)l/k\}$$

$$\text{since } \lambda_{m+l,m}=\exp\{2\pi i(x_m-\nu)l/k\}\cdot\lambda_{m,m}$$

(2.1.6)

$$=\tfrac{1}{2}\sum_{n=0}^{\infty}\lambda_{m,m}\frac{1+\exp\{2\pi i(x_m-\nu)/k\}}{1-\exp\{2\pi i(x_m-\nu)/k\}}.$$

We now take a closer look at $\lambda_{m,m}$. Since $\lambda_{m,m}$ is a residue of $\mathfrak{F}_m(x)$, we note that $\lambda_{m,m}\neq 0$ if and only if $\cosh(\pi ih\nu/2k-\pi zx_m/2k)=0$, i.e. $\cos\{\tfrac{1}{2}\pi(h\nu/k-(2m+1)/3k\}=0$. Thus we must have $3h\nu-(2m+1)\equiv 0 \pmod{3k}$.

If $2m+1\not\equiv 0 \pmod 3$, then the above congruence is impossible. Hence $2m+1=6j+3$. Therefore,

$$h\nu-(2j+1)\equiv 0 \pmod k, \text{ or}$$

$$\nu\equiv -h'(2j+1) \pmod k$$

$$\text{provided } hh'\equiv -1 \pmod k.$$

Thus when considering $\sum_{\nu \bmod k}$ we see that the only non-zero term is the term $\nu\equiv -h'(2j+1) \pmod k$.

We now evaluate $\lambda_{m,m}$ when $m=3j+1$ and $\nu\equiv -h'(2j+1) \pmod k$.

$$\lambda_{m,m}=\lim_{x\to x_m}(x-x_m)\mathfrak{F}_m(x)$$

$$=\frac{\exp\{(2m+1)\pi i(x_m-\nu)/k-3\pi zx_m{}^2/2k\}}{-\tfrac{1}{2}\pi zk^{-1}\sinh(\pi ih\nu/2k-\pi zx_m/2k)}$$

$$=\frac{\exp\{3(2j+1)\pi i[(2j+1)iz^{-1}+h'(2j+1)]/k+3\pi(2j+1)^2/2kz\}}{-\tfrac{1}{2}\pi zk^{-1}\sinh\{-\pi ihh'(2j+1)/2k-\pi i(2j+1)/2k\}}$$

(2.1.7)

$$=\frac{-2k\cdot\exp\{3(2j+1)^2\pi i(h'+iz^{-1})/k+3\pi(2j+1)^2/2kz\}}{i\pi z\sin\{\pi(2j+1)kk'/2k\}}$$

$$=-2k(\pi iz)^{-1}(-1)^{k'j+\frac{1}{2}(k'-1)}\exp\{3(2j+1)^2\pi i(h'+iz^{-1})/k$$

$$+3\pi(2j+1)^2/2kz\}.$$

Hence

$$\Sigma_1=2k^{-1}2\pi i\tfrac{1}{2}\sum_{j=0}^{\infty}(-1)^{-h'(2j+1)}\exp\{3\pi ih(h'(2j+1))^2\}$$

$$\cdot(-2k)(\pi iz)^{-1}(-1)^{k'j+\frac{1}{2}(k'-1)}$$

$$\cdot \exp\{3(2j+1)^2\pi i(h'+iz^{-1})/k+3\pi(2j+1)^2/2kz\}$$

$$\frac{1+\exp\{2\pi i[(2j+1)i/z+h'(2j+1)]/k\}}{1-\exp\{2\pi i[(2j+1)i/z+h'(2j+1)]/k\}}.$$

$$(2.1.8)\quad = 4z^{-1}(-1)^{h'-1+\frac{1}{2}(k'-1)}\sum_{j=0}^{\infty}(-1)^j\exp\{3\pi ih'(2j+1)^2(-1-kk')/2k\}$$

$$\cdot \exp\{3(2j+1)^2\pi i(h'+iz^{-1})/k+3\pi(2j+1)^2/2kz\}$$

$$\frac{1+\exp\{2\pi i[(2j+1)i/z+h'(2j+1)]/k\}}{1-\exp\{2\pi i[(2j+1)i/z+h'(2j+1)]/k\}}.$$

$$= 4z^{-1}(-1)^{h'-1-\frac{1}{2}(k'-1)}\exp\{-3\pi ih'k'/2+3\pi i(h'+iz^{-1})/2k\}$$

$$\cdot \sum_{j=0}^{\infty}(-1)^j\exp\{3j(j+1)2\pi i(h'+iz^{-1})/k\}$$

$$\frac{1+\exp\{2\pi i(2j+1)(h'+iz^{-1})/k\}}{1-\exp\{2\pi i(2j+1)(h'+iz^{-1})/k\}}.$$

$$= 4z^{-1}(-1)^{h'-1-\frac{1}{2}(k'-1)}\exp\{-3\pi ih'k'/2+3\pi ih'/2k-3\pi/2kz\}$$

$$\frac{\omega(\exp[2\pi i(h'+iz^{-1})/k])}{P(\exp[4\pi i(h'+iz^{-1})/k])}\quad [12;\ \text{p. } 66].$$

We now evaluate Σ_2. We recall that

$$\Sigma_2 = 2k^{-1}\sum_{\nu \bmod k}(-1)^\nu \exp(3\pi ih\nu/2k)\sum_{n=0}^{\infty}P\int_{-\infty+x_n}^{\infty+x_n}\mathfrak{F}_n(x)\,dx,$$

where $x=(2n+1)i/3z$.

Now

$$2\sum_{n=0}^{\infty} = \sum_{n=0}^{\infty}+\sum_{n=0}^{-\infty}P\int_{-\infty-(2n+1)i/3z}^{\infty+(2n+1)i/3z}\frac{\exp\{-(2n-1)\pi i(x-\nu)/k-3\pi zx^2/2k\}}{\cosh(\pi ih\nu/2k-\pi zx/2k)}dx$$

$$= \sum_{n=0}^{\infty}+\sum_{n=1}^{\infty-}P\int_{-\infty+(-2n+1)i/3z}^{\infty+(-2n+1)i/3z}\frac{\exp\{(2n+1)\pi i(x-\nu)/k-3\pi zx^2/2k\}}{\cosh(\pi ih\nu/2k-\pi zx/2k)}\,dx$$

upon replacing ν by $-\nu$ and sending $x\to -x$; thus

$$\Sigma_2 = k^{-1}\sum_{\nu \bmod k}(-1)^\nu\exp(3\pi ih\nu^2/2k)\sum_{n=-\infty}^{\infty}P\int_{-\infty+x_n}^{\infty+x_n}\mathfrak{F}_n(x)\,dx.$$

Now

$$P\int_{-\infty+x_n}^{\infty+x_n}\mathfrak{F}_n(x)\,dx = P\int_{-\infty}^{\infty}\mathfrak{F}_n(x+x_n)\,dx,$$

and

$$\exp\{(2n+1)\pi i(x+x_n-\nu)/k-3\pi z(x+x_n)^2/2k\}$$

$$= \exp\{(2n+1)\pi i x/\mathrm{k} - (2n+1)^2\pi/3kz - (2n+1)\pi i\nu/k$$
$$- 3\pi z x^2/2k - \pi x(2n+1)i/k\}\cdot\exp\{\pi(2n+1)^2/6kz\}$$
$$= \exp\{-(2n+1)\pi i\nu/k - (2n+1)^2\pi/6kz - 3\pi z x^2/2k\}$$

We return to Σ_2 letting $n = 3p + \delta$,

$$\Sigma_2 = k^{-1}\sum_{\nu \bmod k}(-1)^\nu \exp(3\pi i h\nu^2/2k)\sum_{\delta=-1}^{1}\sum_{p=-\infty}^{\infty}\exp\{-(2n+1)\pi i\nu/\mathrm{k} - (2n+1)^2\pi/6kz\}$$

(2.1.9)

$$\cdot P\int_{-\infty}^{\infty}\frac{\exp(-3\pi z x^2/2k)}{\cosh\{\pi i h\nu/2k - \pi z x/2k - \pi i p/k - \pi i(2\delta+1)/6k\}}dx$$

If we replace $\frac{1}{2}h - p$ by ν, i. e. $\nu \to -2h'\nu - 2ph'$, then

$$\cosh\{\pi i(\tfrac{1}{2}h\nu - p)/k - \pi z x/2k - \pi i(2\delta+1)/6k\}$$
$$\to \cosh\{-\pi i p/k - \pi i(1+kk')(\nu+p)/k - \pi z x/2k - \pi i(2\delta+1)/6k\}$$
$$= (-1)^{\nu+p}\cosh\{\pi i\nu/k - \pi z x/2k - \pi i(2\delta+1)/6k\},$$

and

$$(-1)^\nu\exp\{3\pi i h\nu^2/2k - (2n+1)\pi i\nu/k\} \to \exp\{3\pi i h 4h'^2(\nu+p)^2$$
$$+ 2(6p+2\delta+1)\pi i h'(\nu+p)/k\}$$
$$= \exp\{-6\pi i h'(\nu^2+2\nu p+p^2)/k + 12p\pi i h'(\nu+p)/k$$
$$+ 2(2\delta+1)\pi i h' p/k + 2(2\delta+1)\pi i h'\nu/k\}$$
$$= \exp\{6\pi i h' p^2/k + (2\delta+1)\pi i 2h' p/k$$
$$- 6\pi i h'\nu^2/k + (2\delta+1)\pi i 2h'\nu/k\}.$$

Thus in Σ_2 we now have $\sum_{p=-\infty}^{\infty}$ free of x and ν. We define

$$U_\delta(z) = \sum_{p=-\infty}^{\infty}(-1)^p\exp\{6\pi i h' p^2/k + (2\delta+1)\pi i h' p/k - (6p+2\delta+1)^2\pi/6kz\}$$

(2.1.10)

$$= \exp\{-(2\delta+1)^2\pi/6kz\}\sum_{p=-\infty}^{\infty}(-1)^p\exp\{2\pi i(h'+iz^{-1})\cdot(3p^2+(2\delta+1)p)/k\}.$$

Letting $\tau' = (h'+iz^{-1})/k$, we have for $\delta = 1$

$$U_1(z) = \exp(-3\pi/2kz)\sum_{p=-\infty}^{\infty}(-1)^p\exp\{2\pi i\tau' 3p(p+1)\}.$$

In this series the p-th term and the $(-p-1)$-st term cancel each other, hence $U_1(z)=0$.

On the other hand,

$$
\begin{aligned}
U_0(z) &= \exp(-\pi/6kz)\sum_{p=-\infty}^{\infty}(-1)^p\exp\{2\pi i\tau' p(3p+1)\} \\
&= \exp(-\pi/6kz)\sum_{p=-\infty}^{\infty}(-1)^p\exp\{2\pi i\tau' p(3p-1)\} \\
&= U_{-1}(z) \qquad (2.1.11) \\
&= \frac{\exp(-\pi/6kz)}{P(\exp[4\pi i(h'+iz^{-1})/k])} \qquad [4;\ \text{p. } 284].
\end{aligned}
$$

Hence

$$
\begin{aligned}
\Sigma_2 = \frac{k^{-1}\exp(-\pi/6kz)}{P[4\pi i(h'+iz^{-1})/k]}\sum_{\delta=-1}^{0}\sum_{\nu \bmod k}(-1)^\nu\exp\{&-6\pi i h'\nu^2/k \\
&+(2\delta+1)\pi i 2h'\nu/k\} \\
\cdot\int_{-\infty}^{\infty}\frac{\exp(-3\pi z x^2/2k)}{\cosh\{\pi i\nu/k-\pi z x/2k-\pi i(2\delta+1)/6k\}}\,dx.
\end{aligned}
$$

In the case $\delta=-1$, if we replace ν by $-\nu$ and send $x\to -x$ we get the case $\delta=0$, hence

$$
\begin{aligned}
\Sigma_2 &= \frac{k^{-1}2\cdot\exp(-\pi/6kz)}{P[4\pi i(h'+iz^{-1})/k]}\sum_{\nu \bmod k}(-1)^\nu\exp\{-6\pi i h'\nu^2/k+2\pi i h'\nu/k\} \qquad (2.1.12)\\
&\quad\cdot\int_{-\infty}^{\infty}\frac{\exp(-3\pi z x^2/2k)}{\cosh\{\pi i(\nu-\frac{1}{6})/k-\pi z x/2k\}}\,dx.
\end{aligned}
$$

Since h is even, we may write (2.0.1) as

$$
\begin{aligned}
P(\exp\{2\pi i(\tfrac{1}{2}h+i\tfrac{1}{2}z)/k\}) = \omega_{\frac{1}{2}h,k}2^{-\frac{1}{2}}z^{\frac{1}{2}}\exp(\pi/6kz \\
-\pi z/24k)P(\exp\{4\pi i(h'+iz^{-1})/k\});
\end{aligned}
$$

thus since we have established

$$
\frac{f(\exp\{\pi i(h+iz)/k\})}{P(\exp\{\pi i(h+iz)/k\})}=\Sigma_1+\Sigma_2,
$$

we have

$$
\begin{aligned}
&f(\exp[\pi i(h+iz)/k]) \\
&\quad=(-1)^{h'-1+\frac{1}{2}(k'-1)}\exp\{-3\pi i h'k'/2+3\pi i h'/2k\}\omega_{\frac{1}{2}h,k}2^{\frac{3}{2}}z^{-\frac{1}{2}} \\
&\qquad\cdot\exp(-4\pi/3kz-\pi z/24k)\omega(\exp[2\pi i(h'+iz^{-1})/k])
\end{aligned}
$$

$$(2.1.13) \qquad + 2^{\frac{1}{2}} z^{\frac{1}{2}} k^{-1} \exp(-\pi z/24k) \omega_{\frac{1}{2}h,k} \sum_{\nu \bmod k} (-1)^{\nu} \exp(-6\pi i h'\nu^2/k - 2\pi i h'\nu/k)$$
$$\cdot \int_{-\infty}^{\infty} \frac{\exp(-3\pi z x^2/2k)}{\cosh\{\pi i(\nu - \frac{1}{6})/k - \pi z x/2k\}} dx.$$

This is the required transformation formula; thus when h is even, the general linear transformation sends f into ω plus a correction $E_1^{(f)}(h, k; z)$.

We proceed to the final part of the theorem in which we estimate the integral involved in $E_1^{(f)}(h, k; z)$.

$$\begin{aligned} z^{\frac{1}{2}} J_1^{(f)}(\nu; k; z) &= z^{\frac{1}{2}} \int_{-\infty}^{\infty} \frac{\exp(-3\pi z x^2/2k)}{\cosh\{\pi i(\nu - \frac{1}{6})/k - \pi z k/2k\}} dx \\ &= 2k\pi^{-1} z^{-\frac{1}{2}} \int_{-\infty}^{\infty} \frac{\exp(-6kt^2/\pi z)}{\cosh\{\pi i(\nu - \frac{1}{6})/k - t\}} dt \end{aligned}$$

by Cauchy's theorem. Now with $z = kN^{-2} - ik\phi$ and $1 \leqq \nu \leqq k$, we have

$$| z^{\frac{1}{2}} J_1^{(f)}(\nu; k; z) | \leqq 2k\pi^{-1} | z |^{-\frac{1}{2}} \int_{-\infty}^{\infty} \frac{\exp\{-6k(\operatorname{Re} z^{-1})t^2/\pi\}}{| \cosh\{\pi i(\nu - \frac{1}{6})/k - t\} |} dt$$

$$\leqq 2k\pi^{-1} | z |^{-\frac{1}{2}} | \cos\{\pi(\nu - \tfrac{1}{6})/k\} |^{-1} \int_{-\infty}^{\infty} \exp\{-6k(\operatorname{Re} z^{-1})t^2/\pi\} dt$$

$$\leqq 2k\pi^{-1} | z |^{-\frac{1}{2}} | 1 - 2k^{-1}(\nu - \tfrac{1}{6}) |^{-1} \pi^{\frac{1}{2}} \{12k(\operatorname{Re} z^{-1})\}^{-\frac{1}{2}} \int_{-\infty}^{\infty} \exp(-u^2/2) du$$

$$\leqq k^{\frac{3}{2}} (3\pi)^{-\frac{1}{2}} | k - 2(\nu - \tfrac{1}{6}) |^{-1} | z |^{-\frac{1}{2}} (\operatorname{Re} z^{-1})^{-\frac{1}{2}} (2\pi)^{\frac{1}{2}}$$
$$= k^{\frac{3}{2}} (\tfrac{2}{3})^{\frac{1}{2}} | k - 2(\nu - \tfrac{1}{6}) |^{-1} | z |^{-\frac{1}{2}} \{(\operatorname{Re} z)/| z |^2\}^{-\frac{1}{2}}.$$

Now $\operatorname{Re} z = kN^{-2}$, and

$$\begin{aligned} | z | &= (k^2 N^{-4} + k^2 \phi^2)^{\frac{1}{2}} \\ &\leqq (k^2 N^{-4} + k^2 k^{-2} N^{-2})^{\frac{1}{2}} \\ &= \{(k^2 + N^2) N^{-4}\}^{\frac{1}{2}} \\ &\leqq 2^{\frac{1}{2}} N^{-1}. \end{aligned}$$

Hence

$$| z |^{\frac{1}{2}} (\operatorname{Re} z)^{-\frac{1}{2}} \leqq 2^{\frac{1}{4}} N^{-\frac{1}{2}} N k^{-\frac{1}{2}} = 2^{\frac{1}{4}} N^{\frac{1}{2}} k^{-\frac{1}{2}}.$$

Thus

$$| z^{\frac{1}{2}} J_1^{(f)}(\nu; k; z) | \leqq 2^{\frac{3}{4}} N^{\frac{1}{2}} k 3^{-\frac{1}{2}} | k - 2(\nu - \tfrac{1}{6}) |^{-1}$$

This concludes the proof of Theorem 2.1.

Proof of Theorem 2.2. As before

(2.2.1) $$F(q) = f(q)/P(q)$$

(2.2.2) $$= 1 + 4\sum_{n=1}^{\infty} \frac{(-1)^n q^{\frac{1}{2}n(3n+1)}}{1+q^n}$$

(2.2.3) $$= \Sigma(-1)^\nu \exp(3\pi i h\nu^2/2) \sum_{\nu=0}^{2k-1} \frac{(-1)^m \exp\{-3\pi z(2mk+\nu)^2/2k\}}{\cosh\{\pi i h\nu/2k - \pi z(2mk+\nu)/2k\}}$$

by the Poisson summation formula [1; pp. 7-9] we obtain

(2.2.4) $$F(q) = k^{-1} \sum_{\nu \bmod 2k} (-1)^\nu \cdot \exp(3\pi i h\nu^2/2k) \sum_{n=0}^{\infty} \int_{-\infty}^{\infty} \frac{\exp\{(2n+1)\pi i(x-\nu)/2k - 3\pi z x^2/2k\}}{\cosh(\pi i h\nu/2k - \pi z x/2k)} dx.$$

We now define

$$S_{2k}(x) = \frac{\sinh 3x}{\cosh(x/2k)}.$$

Clearly $S_{2k}(x)$ is an entire function of x since k is an integer. Hence if

$$\mathfrak{F}_n(x) = \frac{\exp\{(2n+1)\pi i(x-\nu)/2k - 3\pi z x^2/2k\}}{\cosh(\pi i h\nu/2k - \pi z x/2k)}$$
$$= \frac{-(-1)^\nu \exp\{(2n+1)\pi i(x-\nu)/2k - 3\pi z x^2/2k\} S_{2k}(\pi i h\nu - \pi z x)}{\sinh 3\pi z x},$$

then $\mathfrak{F}_n(x)$ has (at most) simple poles at

$$x_m = 2mi/6z = mi/3z.$$

The residue at x_m is

$$\lambda_{n,m} = -(-1)^{\nu+m}(3\pi z)^{-1} \exp\{(2n+1)\pi i(x_m - \nu)/2k - 3\pi z x_m^2/2k\} S_{2k}(\pi i h\nu - \pi z x_m).$$

Now by Cauchy's theorem,

$$\left\{\int_{-\infty}^{\infty} - \int_{-\infty+(2n+1)i/6z}^{\infty+(2n+1)i/6z}\right\} \mathfrak{F}_n(x)dx = 2\pi i\{\tfrac{1}{2}\lambda_{n,0} + \lambda_{n,1} + \cdots + \lambda_{n,n}\}.$$

Hence

(2.2.5) $$F(q) = k^{-1} \sum_{\nu \bmod 2k} (-1)^\nu \exp(3\pi i h\nu^2/2k) \sum_{n=0}^{\infty} 2\pi i\{\tfrac{1}{2}\lambda_{n,0} + \lambda_{n,1} + \cdots + \lambda_{n,n}\}$$
$$+ k^{-1} \sum_{\nu \bmod 2k} (-1)^\nu \exp(3\pi i h\nu^2/2k) \sum_{n=0}^{\infty} \int_{-\infty+(2n+1)i/6z}^{\infty+(2n+1)i/6z} \mathfrak{F}_n(x)\,dx$$
$$= \Sigma_1 + \Sigma_2.$$

In this instance, Σ_1 will yield the main term in the transformation formula involving $f(q)$, and Σ_2 will again yield the "error" term.

In Σ_1, we have

$$\sum_{n=0}^{\infty}\{\tfrac{1}{2}\lambda_{n,0}+\lambda_{n,1}+\cdots+\lambda_{n,n}\}$$

(2.2.6)

$$=\tfrac{1}{2}\frac{\lambda_{0,0}}{1-\exp\{\pi i(x_m-\nu)/k\}}+\sum_{m=1}^{\infty}\frac{\lambda_{m,m}}{1-\exp\{\pi i(x_m-\nu)/k\}}.$$

We thus obtain

$$\text{(2.2.6a)}\quad \Sigma_1=\pi i k^{-1}\sum_{\nu \bmod 2k}(-1)^{\nu}\exp(3\pi i h\nu^2/2k)\sum_{m=-\infty}^{\infty}\frac{\lambda_{m,m}}{1-\exp\{\pi i(x_m-\nu)/k\}}.$$

We now take a closer look at $\lambda_{m,m}$. Since $\lambda_{m,m}$ was a residue of $\mathfrak{F}_m(x)$, we note that $\lambda_{m,m}\neq 0$ if and only if $\cosh(\pi i h\nu/2k-\pi z x_m/2k)=0$, i.e. $\cos\{\pi(h\nu/k-m/3k)/2\}=0$. Thus we must have $3h\nu-m\equiv 3k \pmod{6k}$. If $m\not\equiv 0 \pmod 3$, then this last congruence is impossible. Hence $m=3j$.

$$h\nu-j\equiv k \pmod{2k}$$

or

$$\nu\equiv -h''(k+j) \pmod{2k}$$

provided $hh''\equiv -1 \pmod{2k}$.

Thus when considering $\sum_{\nu \bmod 2k}$, we see that the only non-zero term is the term $\nu\equiv -h''(k+j) \pmod{2k}$.

We now evaluate $\lambda_{m,m}$ with $m=3j$ and $\nu\equiv -h''(k+j) \pmod{2k}$.

$$\lambda_{m,m}=\lim_{x\to x_m}(x-x_m)\mathfrak{F}_m(x)$$

(2.2.7)

$$=2k(\pi i z)^{-1}(-1)^{jk''+\frac{1}{2}(hh''-1)+j}\exp\{\pi i h''/2 + (6j+1)\pi i j(h''+iz^{-1})/2k+3\pi j^2/2kz\}$$

where $hh''+2kk''=-1$. Hence

(2.2.8)

$$\Sigma_1=z^{-1}(-1)^{k+\frac{1}{2}(hh''-1)}\exp(\pi i h''/2 + 3\pi i h h''^2 k/2)\frac{f(\exp\{\pi i(h''+iz^{-1})/k\})}{P(\exp\{\pi i(h''+iz^{-1})/k\})}$$ by [12; p. 64].

We now evaluate Σ_2.

$$\Sigma_2=k^{-1}\sum_{\nu \bmod 2k}(-1)^{\nu}\exp(3\pi i h\nu^2/2k)\sum_{n=0}^{\infty}\int_{-\infty+(2n+1)i/6z}^{\infty+(2n+1)i/6z}\mathfrak{F}_n(x)\,dx$$

$$=\tfrac{1}{2}k^{-1}\sum_{\nu \bmod 2k}(-1)^{\nu}\exp(3\pi i h\nu^2/2k)\sum_{n=-\infty}^{\infty}\int_{-\infty+(2n+1)i/6z}^{\infty+(2n+1)i/6z}\mathfrak{F}_n(x)\,dx,$$

as in Theorem 2.1.

Now

$$\int_{-\infty+(2n+1)i/6z}^{\infty+(2n+1)i/6z} \mathcal{F}_n(x)\,dx = \int_{\infty}^{-\infty} \mathcal{F}_n(x(2n+1)i/6z)\,dx,$$

and in Σ_2 letting $n = 3p + \delta$, we have

$$\begin{aligned}\Sigma_2 = \tfrac{1}{2}k^{-1} &\sum_{h \bmod 2k} (-1)^{\nu} \\ &\cdot \exp(3\pi i h\nu^2/2k) \sum_{\delta=-1}^{1} \sum_{p=-\infty}^{\infty} \exp\{-(2n+1)\pi i\nu/2k\} \\ &\cdot \exp\{-(2n+1)^2\pi/24kz\} \\ &\int_{-\infty}^{\infty} \frac{\exp(-3\pi z x^2/2k)}{\cosh\{\pi i h\nu/2k - \pi z x/2k - \pi i(6p+2\delta+1)/12k\}}dx \end{aligned} \tag{2.2.9}$$

If we replace $h\nu - p$ by ν i.e. $\nu \to -h''\nu - h''p$, then in Σ_2 we now have $\sum_{p=-\infty}^{\infty}$ free of x and ν. We define

$$\begin{aligned} U_\delta(z) - \sum_{p=-\infty}^{\infty} (-1)^p \exp\{&3\pi i h'' p^2/2k \\ &+ (2\delta+1)\pi i h'' p/2k - (6p+2\delta+1)^2\pi/24kz\} \end{aligned} \tag{2.2.10}$$

$$= \begin{cases} 0 \text{ if } \delta = 1 \\ \exp(-\pi/24kz) \dfrac{1}{P(\exp\{\pi i(h''+iz^{-1})/k\})} \end{cases} \tag{2.2.11}$$

as in Theorem 2.1. Hence

$$\begin{aligned}\Sigma_2 = \frac{k^{-1}\exp(-\pi/24kz)}{P(\exp\{\pi i(h''+iz^{-1})/k\})} &\sum_{\nu \bmod 2k} (-1)^{\nu} \\ &\cdot \exp(-3\pi i h''\nu^2/2k + \pi i h''\nu/2k) \\ &\cdot \int_{-\infty}^{\infty} \frac{\exp(-3\pi z x/2k)}{\cosh(\pi i h\nu/2k - \pi z x/2k - \pi i/12k)}\,dx. \end{aligned} \tag{2.2.12}$$

Since h is odd, we may write (2.0.1) as

$$\begin{aligned} &P(\exp\{2\pi i(h+iz)/2k\}) \\ &\quad = \omega_{h,2k} z^{\frac{1}{2}} \exp(\pi/24kz - \pi z/24k) P(\exp\{2\pi i(h''+iz^{-1})/2k\}); \end{aligned}$$

thus since we have established

$$\frac{f(\exp\{\pi i(h+iz)/k\})}{P(\exp\{\pi i(h+iz)/k\})} = \Sigma_1 + \Sigma_2,$$

we have

$$
\begin{aligned}
&f(\exp\{\pi i(h+iz)/k\}) \\
&\quad=(-1)^{k+\frac{1}{2}(hh''-1)}\exp(\pi ih''/2+3\pi ihh''^2k/2)\omega_{h,2k}z^{-\frac{1}{2}} \\
&\qquad\cdot\exp(\pi/24kz-\pi z/24k)f(\exp\{\pi i(h''+iz^{-1})/k\}) \\
&\qquad+k^{-1}\omega_{h,2k}z^{\frac{1}{2}}\exp(-\pi z/24k)\sum_{\nu \bmod 2k}(-1)^{\nu} \\
&\qquad\cdot\exp(-3\pi ih''\nu^2/2k+\pi ih''\nu/2k) \\
&\qquad\cdot\int_{-\infty}^{\infty}\frac{\exp(-3\pi zx^2/2k)}{\cosh(\pi i\nu/2k-\pi zx/2k-\pi i/12k)}dx.
\end{aligned}
\tag{2.2.13}
$$

Thus when h is odd, the general linear transformation sends f into f plus a correction $E_2^{(f)}(h,k;z)$; this completes Theorem 2.2.

Proof of Theorem 2.3. We let

$$\Omega(q)=\omega(q)/P(q^2) \tag{2.3.1}$$

$$=q^{-\frac{3}{4}}\sum_{n=0}^{\infty}(-1)^n q^{3(n+\frac{1}{2})^2}\frac{1+q^{2n+1}}{1-q^{2n+1}},\quad [12;\ \text{p. } 66], \tag{2.3.2}$$

and if $q=\exp\{\pi i(h+iz)/k\}$

$$
\begin{aligned}
&q^{\frac{3}{4}}\Omega(q) \\
&=-\tfrac{1}{2}\sum_{\nu=0}^{k=1}\sum_{m=-\infty}^{\infty}\frac{(-1)^{\nu+m}\exp\{3(\nu+\frac{1}{2})^2\pi ih/k-3\pi z(nk+\nu+\frac{1}{2})^2/k\}}{\tanh\{(\nu+\frac{1}{2})\pi ih/k-(mk+\nu+\frac{1}{2})\pi z/k\}};
\end{aligned}
\tag{2.3.3}
$$

by the Poisson summation formula [1; pp. 7-9] we obtain

$$
\begin{aligned}
q^{\frac{3}{4}}\Omega(q)&=-k^{-1}\sum_{\nu \bmod k}(-1)^{\nu}\exp(3\pi ih\mu^2/k)\sum_{n=0}^{\infty} \\
&\qquad\cdot\int_{-\infty}^{\infty}\frac{\exp\{(2n+1)\pi i(x-\mu)/k-3\pi zx^2/k\}}{\tanh(\pi ih\mu/k-\pi zx/k)}dx
\end{aligned}
\tag{2.3.4}
$$

as in the previous theorems with $\mu=\nu+\frac{1}{2}$. Hence if $\mathcal{S}_k(x)=\coth(x/k)\cdot\sinh 3x$, then we have $\mathcal{S}_k(x)$ is an entire function of x since k is an integer, and

$$
\begin{aligned}
\mathcal{F}_n(x)&=\frac{\exp\{(2n+1)\pi i(x-\mu)/k-3\pi zx^2/k\}}{\tanh(\pi ih\mu/k-\pi zx/k)} \\
&=\frac{-\exp\{(2n+1)\pi i(x-\mu)/k-3\pi zx^2/k\}\mathcal{S}_k(\pi ih\mu-\pi zx)}{(-1)^{\frac{3}{2}h}\sinh 3\pi zx}.
\end{aligned}
$$

Thus $\mathcal{F}_n(x)$ has (at most) simple poles at

$$x_m=mi/3z=2mi/6z.$$

The residue at x_m is

$$\lambda_{n,m} = -(-1)^{m+\frac{1}{2}h}(3\pi z)^{-1}\exp\{(2n+1)\pi i(x_m - \mu)/k - 3\pi z x_m{}^2/k\mathcal{S}_k(\pi i h\mu - \pi z x_m),$$

and

$$\lambda_{m+l,m} = \exp\{2\pi i l(x_m - \mu)/k\}\lambda_{m,m}.$$

Now by Cauchy's theorem,

$$\left\{\int_{-\infty}^{\infty} - \int_{-\infty+(2n+1)i/6z}^{\infty+(2n+1)i/6z}\right\}\mathfrak{F}_n(x)dx = 2\pi i\{\tfrac{1}{2}\lambda_{n,0} + \lambda_{n,1} + \cdots + \lambda_{n,n}\}.$$

Hence

$$\begin{aligned} q^{\frac{3}{8}}\Omega(q) &= -2\pi i k^{-1}\sum_{\nu \bmod k}(-1)^\nu \\ &\quad\cdot\exp(3\pi i h\mu^2/k)\sum_{n=0}^{\infty}\{\tfrac{1}{2}\lambda_{n,0} + \lambda_{n,1} + \cdots + \lambda_{n,n}\} \\ &\quad - k^{-1}\sum_{\nu \bmod k}(-1)^\nu\exp(3\pi i h\mu^2/k)\sum_{n=0}^{\infty}\int_{-\infty+(2n+1)i/6z}^{\infty+(2n+1)i/6z}\mathfrak{F}_n(x)\,dx. \\ &= \Sigma_1 + \Sigma_2. \end{aligned} \tag{2.3.5}$$

Again Σ_1 yields the main term of the transformation this time involving $f(q)$, and Σ_2 as always yields the "error" term.

In Σ_1 we have

$$\begin{aligned} &\sum_{n=0}^{\infty}\{\lambda_{n,0} + \cdots + \lambda_{n,n}\} \\ &= \tfrac{1}{2}\frac{\lambda_{0,0}}{1 - \exp\{2\pi i(x_m - \mu)/k\}} + \sum_{m=1}^{\infty}\frac{\lambda_{m,m}}{1 - \exp\{2\pi i(x_m - \mu)/k\}}; \end{aligned} \tag{2.3.6}$$

so

$$\begin{aligned} \Sigma_1 &= -\pi i k^{-1}\sum_{\nu \bmod k}(-1)^\nu \\ &\quad\cdot\exp(3\pi i h\mu^2/k)\sum_{m=-\infty}^{\infty}\frac{\lambda_{m,m}}{1 - \exp\{2\pi i(x_m - \mu)/k\}}. \end{aligned} \tag{2.3.6a}$$

We now take a closer look at $\lambda_{m,m}$. Since $\lambda_{m,m}$ was a residue of $\mathfrak{F}_m(x)$ we note that $\lambda_{m,m} \neq 0$ if and only if $\tanh(\pi i h\mu/k - \pi z x/k) = 0$, i.e. $\sin\{h\mu/k - m/3k\} = 0$. Thus we must have

$$3h(2\nu + 1) - 2m \equiv 0 \pmod{6k}$$

since μ is the abbreviation for $\nu + \frac{1}{2}$. If $m \not\equiv 0 \pmod 3$, then the above congruence is impossible. Hence $m = 3j$.

$$\tfrac{1}{2}h(2\nu+1)-j\equiv 0 \pmod{k}$$

or

$$2\nu+1\equiv -2h'j \pmod{k}$$

or

$$\nu\equiv \tfrac{1}{2}(k+1)(-2h'j-1) \pmod{k}$$

$$\equiv -h'j-\tfrac{1}{2}(k+1) \pmod{k};$$

provided $hh'\equiv -1 \pmod{k}$.

Thus when considering $\sum_{\nu \bmod k}$ we see that the only non-zero term is the term $\nu\equiv -h'j-\frac{1}{2}(k+1) \pmod{k}$.

We now evaluate $\lambda_{m,m}$ when $m=3j$ and $\nu\equiv -h'j-\frac{1}{2}(k+1) \pmod{k}$.

(2.3.7)
$$\lambda_{m,m}=\lim_{x\to x_m}(x-x_m)\mathcal{F}_m(x)$$
$$=-ik(\pi z)^{-1}\exp\{\pi ij(h'+iz^{-1})/\mathrm{k}-3\pi j^2/kz+6\pi ih'j^2/k\}.$$

Hence

(2.3.8) $$\Sigma_1=(2z)^{-1}(-1)^{\frac{1}{2}(k-1)}\exp(3\pi ihk/4)F(\exp\{2\pi i(h'+iz^{-1})/k\}),$$

where $F(q)$ is defined in the first line of the proof of Theorem 2.1.

We now evaluate Σ_2. With $n=3p+\delta$ and $\mu=\nu+\frac{1}{2}$, we have

(2.3.9)
$$\Sigma_2=-(2k)^{-1}\sum(-1)^{\nu}\exp(3\pi ih\mu^2/k)\sum_{\delta=-1}^{1}\sum_{p=-\infty}^{\infty}$$
$$\cdot\int_{-\infty}^{\infty}\frac{\exp\{-(2n+1)\pi i\mu/k-(2n+1)^2\pi/12kz-3\pi zx^2/k\}}{\tanh\{\pi ih\mu/k-\pi zx/k-\pi i(6p+2\delta+1)/6k\}}dx.$$

If we replace $h\mu-p$ by ν, i.e. $\nu\to -h'(\nu+p)-\frac{1}{2}(k+1)$, then we have $\sum_{p=-\infty}^{\infty}$ free of x and ν. We define

(2.3.10)
$$U_\delta(z)=\sum_{p=-\infty}^{\infty}(-1)^p\exp\{\pi ih'(3p^2+(2\delta+1)p)/k$$
$$-(6p+2\delta+1)^2\pi/12kz\}.$$

As before

$$U_1(z)=0$$

(2.3.11) $$U_{-1}(z)=U_0(z)=\frac{\exp(-\pi/12kz)}{P(\exp\{2\pi i(h'+iz^{-1})/k\})}.$$

Thus

$$\Sigma_2=\frac{ik^{-1}(-1)^{\frac{1}{2}(k-1)}\exp(3\pi ihk/4-\pi/12kz)}{P(\exp\{2\pi i(h'+iz^{-1})/k\})}\sum_{\nu \bmod k}exp(-3\pi ih'\nu^2/k)$$

(2.3.12)

$$\cdot \exp(\pi i h'\nu/k) \int_{-\infty}^{\infty} \frac{\exp(-3\pi z x^2/k)}{\tanh(\pi i\nu/k - \pi z x/k - \pi i/6k)} dx.$$

We have established

$$\exp(3\pi i(h+iz)/4k) \frac{\omega(\exp\{\pi i(b+iz)/k\})}{P(\exp\{2\pi i(h+iz)/k\})} = \Sigma_1 + \Sigma_2,$$

hence by (2.0.1)

$$\begin{aligned} &\omega(\exp\{\pi i(h+iz)/k\}) \\ (2.3.13) \quad &= \tfrac{1}{2} z^{-\frac{1}{2}}(-1)^{\frac{1}{2}(k-1)} \exp(3\pi i h k/4 - 3\pi i h/4k)\omega_{h,k} \exp(\pi/12kz + 2\pi z/3k) \\ &\qquad \cdot f(\exp\{2\pi i(h' + iz^{-1})/k\}) \\ &\quad + i(-1)^{\frac{1}{2}(k-1)} \exp(3\pi i h k/4 - 3\pi i h/4k + 2\pi z/3k) k^{-1} z^{\frac{1}{2}} \omega_{h,k} \\ &\quad \cdot \sum_{\nu \bmod k} \exp(-3\pi i h'\nu^2/k + \pi i h'\nu/k) \int_{-\infty}^{\infty} \frac{\exp(-3\pi z x^2/k)}{\tanh(\pi i(\nu - \frac{1}{6})/k - \pi z x/k)} dx. \end{aligned}$$

This is the required transformation formula.

We proced to the final part of the theorem in which we estimate the integral appearing in $E_1^{(\omega)}(h, k; z)$.

$$\begin{aligned} z^{\frac{1}{2}} J_1^{(\omega)}(\nu; k; z) &= z^{\frac{1}{2}} \int_{-\infty}^{\infty} \frac{\exp(-3\pi z x^2/k)}{\tanh(\pi i(\nu - \frac{1}{6})/k - \pi z x/k)} dx \\ &= k(\pi z^{\frac{1}{2}})^{-1} \int_{-\infty}^{\infty} \frac{\exp(-3kt^2/\pi z)}{\tanh(\pi i(\nu - \frac{1}{6})/k - t)} dt \end{aligned}$$

by Cauchy's theorem. Now with $z = kN^{-2} - ik\phi$ and $-\frac{1}{2}(k+1) \leqq \nu \leqq \frac{1}{2}(k-1)$, we have

$$| z^{\frac{1}{2}} J_1^{(\omega)}(\nu; k; z) | \leqq k\pi^{-1} | z |^{-\frac{1}{2}}$$

$$\int_{-\infty}^{\infty} \frac{\exp(-3k(\operatorname{Re} z^{-1})t^2/\pi) | \cosh\{\pi i(\nu - \frac{1}{6})/k - t\} |}{| \sinh\{\pi i(\nu - \frac{1}{6})/k - t\} |} dt.$$

Now

$$| \cosh(x + iy) | = (\cos^2 y \cosh^2 x + \sin^2 y \sinh^2 x)^{\frac{1}{2}} \leqq \sqrt{2} \cosh x,$$

and

$$| \sinh(x + iy) | = (\sinh^2 x \cos^2 y + \cosh^2 x \sin^2 y)^{\frac{1}{2}} \geqq | \sin y |.$$

Thus

$$| z^{\frac{1}{2}} J_1^{(\omega)}(\nu; k; z) | \leqq k\sqrt{2}\, \pi^{-1} | z |^{-\frac{1}{2}} | \sin\{\pi(\nu - \tfrac{1}{6})/k\} |^{-1} \int_{-\infty}^{\infty}$$

$$\cdot \exp(-3k(\operatorname{Re} z^{-1})t^2/\pi) \cosh t \, dt$$

$$\leqq k^2\sqrt{2}\,\pi^{-1}\,|z|^{-\frac{1}{2}}\,|2(\nu-\tfrac{1}{6})|^{-1}\int_{-\infty}^{\infty}\exp(-3k(\operatorname{Re} z^{-1})t^2/\pi+t)\,dt$$

since $-\frac{1}{2}(k+1)\leqq\nu\leqq\frac{1}{2}(k-1)$;

$$=k^2\pi^{-1}(2z)^{-\frac{1}{2}}\,|\nu-\tfrac{1}{6}|^{-1}\exp(\pi/12k(\operatorname{Re} z^{-1}))\int_{-\infty}^{\infty}\exp(-3k(\operatorname{Re} z^{-1})t^2/\pi)\,dt.$$

Now

$$\begin{aligned}k(\operatorname{Re} z^{-1})&=k^2N^{-2}\,|z|^{-2}\\&\geqq k^2N^{-2}(k^2N^{-4}+k^2/k^2N^2)^{-1}\\&\geqq\tfrac{1}{2}k^2\geqq\tfrac{1}{2}.\end{aligned}$$

Hence

$$\begin{aligned}|z^{\frac{1}{2}}J_1^{(\omega)}(\nu;k;z)|&\leqq k^2(2\,|z|)^{-\frac{1}{2}}\pi^{-1}\,|\nu-\tfrac{1}{6}|^{-1}\exp(\pi/6)\\&\qquad\cdot\int_{-\infty}^{\infty}\exp(-6k(\operatorname{Re} z^{-1})t^2/\pi)\sqrt{2}\,dt\\&\leqq k^2\,|z|^{-\frac{1}{2}}\sqrt{2}\,|\nu-\tfrac{1}{6}|^{-1}\exp(\pi/6)\,(12k(\operatorname{Re} z^{-1}))^{-\frac{1}{2}}\end{aligned}$$

as in Theorem 2.1,

$$\leqq 2^{-\frac{1}{2}}3^{-\frac{1}{2}}\,|\nu-\tfrac{1}{6}|^{-1}\exp(\pi/6)N^{\frac{1}{2}}k.$$

This concludes the proof of Theorem 2.3.

Proof of Theorem 2.4. As before

$$\Omega(q)=\frac{\omega(q)}{P(q)^2} \tag{2.4.1}$$

$$=q^{-\frac{2}{3}}\sum_{n=0}^{\infty}(-1)^nq^{3(n+\frac{1}{2})^2}\frac{1+q^{2n+1}}{1-q^{2n+1}},\ [12;\ \text{p. } 66], \tag{2.4.2}$$

and if $q=\exp\{\pi i(h+iz)/k\}$,

(2.4.3)
$$q^{\frac{2}{3}}\Omega(q)=-\tfrac{1}{2}\sum_{\nu=0}^{2k-1}\sum_{m=-\infty}^{\infty}\frac{(-1)^\nu\exp\{3(\nu+\frac{1}{2})^2\pi ih/k-3(2mk+\nu+\frac{1}{2})^2\pi z/k\}}{\tanh\{(\nu+\frac{1}{2})\pi ih/k-(2mk+\nu+\frac{1}{2})\pi z/k\}};$$

by the Poisson summation formula [1; pp. 7-9] we obtain

(2.4.4)
$$\begin{aligned}q^{\frac{2}{3}}\Omega(q)=-(2k)^{-1}&\sum_{\nu \bmod 2k}(-1)^\nu\exp(3\pi ih\mu^2/k)\\&\cdot\sum_{n=0}^{\infty}{}^{*}\int_{-\infty}^{\infty}\frac{\exp\{2n\pi i(x-\mu)/2k-3\pi zx^2/k\}}{\tanh(\pi ih\mu/k-\pi zx/k)}\,dx\end{aligned}$$

where $\sum_{n=0}^{\infty}{}^{*}a_n=\frac{1}{2}a_0+\sum_{n=0}^{\infty}a_n$ and $\mu=\nu+\frac{1}{2}$.

Note that if ν is replaced by $\nu + k$, the change introduced is

$$(-1)^k \exp\{3\pi ih(k(\nu+k+1)+k\nu)/k - \pi in\} = (-1)^{2\nu+2k+1+n} = (-1)^{n+1}.$$

Hence if n is even the terms ν and $\nu + k$ cancel each other, and if n is odd then these terms are identical. Thus

$$q^{\frac{3}{8}}\Omega(q) = -k^{-1} \sum_{\nu \bmod k} (-1)^\nu \cdot \exp(3\pi ih\mu^2/k) \sum_{n=0}^{\infty} \int_{-\infty}^{\infty} \frac{\exp\{(2n+1)\pi i(x-\mu)/k - 3\pi zx^2/k\}}{\tanh(\pi ih\mu/k - \pi zx/k)} dx. \tag{2.4.4a}$$

As in Theorem 2.3, we define $\mathcal{S}_k(x) = \coth(x/k)\sinh(3x)$, then if

$$\mathcal{F}_n(x) = \frac{\exp\{(2n+1)\pi i(x-\mu)/k - 3\pi zx^2/k\}}{\tanh(\pi ih\mu/k - \pi zx/k)} = \frac{(-1)^{\nu+\frac{1}{2}(3h-1)}\exp\{(2n+1)\pi i(x-\mu)/k - 3\pi zx^2/k\}\mathcal{S}_k(\pi ih\mu - \pi zx)}{i\cosh(3\pi zx)}$$

$\mathcal{F}_n(x)$ *has* (at most) simple poles at $x_m = (2m+1)i/6z$. The residue at x_m is

$$\lambda_{n,m} = \frac{(-1)^{m+\nu+\frac{1}{2}(3h-1)}\exp\{(2n+1)\pi i(x-\mu)/k - 3\pi zx^2/k\}\mathcal{S}_k(\pi ih\mu - \pi zx_m)}{3\pi zi},$$

and

$$\lambda_{m+l,m} = \exp\{2\pi il(x_m - \mu)/k\}\lambda_{m,m}.$$

Now by Cauchy's theorem,

$$\left\{\int_{-\infty}^{\infty} - P\int_{-\infty+x_n}^{\infty+x_n}\right\}\mathcal{F}_n(x)\,dx = 2\pi i\{\lambda_{n,0} + \lambda_{n,1} + \cdots + \tfrac{1}{2}\lambda_{n,n}\}.$$

Hence

$$\begin{aligned} q^{\frac{3}{8}}\Omega(q) &= -2\pi ik^{-1}\sum_{\nu \bmod k}(-1)^\nu \cdot \exp(3\pi ih\mu^2/k)\sum_{n=0}^{\infty}\{\lambda_{n,0} + \lambda_{n,1} + \cdots + \tfrac{1}{2}\lambda_{n,n}\} \\ &\quad - k^{-1}\sum_{\nu \bmod k}(-1)^\nu \exp(3\pi ih\mu^2/k)\sum_{n=0}^{\infty} P\int_{-\infty+x_n}^{\infty+x_n}\mathcal{F}_n(x)\,dx \\ &= \Sigma_1 + \Sigma_2. \end{aligned} \tag{2.4.5}$$

Σ_1 yields the main term of the transformation which involves $\omega(q)$ this time, and Σ_2 yields the "error" term.

In Σ_1 we have

$$(2.4.6)\quad \sum_{m=0}^{\infty}\{\lambda_{n,0}+\lambda_{n,1}+\cdots+\tfrac{1}{2}\lambda_{n,n}\}=\tfrac{1}{2}\sum_{m=0}^{\infty}\lambda_{m,m}\frac{1+\exp\{2\pi i(x_m-\mu)/k\}}{1-\exp\{2\pi i(x_m-\mu)/k\}}$$

We now take a closer look at $\lambda_{m,m}$. Since $\lambda_{m,m}$ was a residue of $\mathfrak{F}_m(x)$, we note that $\lambda_{m,m}\neq 0$ if and only if

$$\tanh(\pi ih\mu/\mathrm{k}-\pi zx_m/k)=0,\ \text{i.e.}\ \sin\{\pi(h\mu/k-(2m+1)/6k)=0.$$

Thus we must have

$$3h(2\nu+1)-(2m+1)\equiv 0\ (\mathrm{mod}\ 6k)$$

If $2m+1\not\equiv 0\ (\mathrm{mod}\ 3)$, then the above congruence is impossible. Thus $2m+1=6j+3$, and

$$h(2\nu+1)-(2j+1)\equiv 0\ (\mathrm{mod}\ 2k),$$

or

$$2\nu+1\equiv -h''(2j+1)\ (\mathrm{mod}\ 2k),$$

or

$$\nu\equiv-\tfrac{1}{2}\{h''(2j+1)+1\}\ (\mathrm{mod}\ k).$$

Thus when considering $\sum_{\nu \bmod k}$ we see that the only non-zero term is the term $\nu\equiv-\frac{1}{2}\{h''(2j-1)-1\}\ (\mathrm{mod}\ k)$.

We now evaluate $\lambda_{m,m}$ when $m=3j+1$ and $\nu=-\frac{1}{2}\{h''(2j+1)+1\}$.

$$\lambda_{m,m}=\lim_{x\to x_m}\mathfrak{F}_m(x)$$

$$(2.4.7)\qquad =-k(\pi z)^{-1}\exp\{-3\pi(2j+1)^2/4kz+\tfrac{1}{2}3(2j+1)^2\pi ih''/k\}.$$

Hence

$$(2.4.8)\quad \Sigma_1=iz^{-1}(-1)^{\frac{1}{2}(h''+1)}\exp\{-3\pi ih''k''/2+3\pi i(h''+iz^{-1}(/4k\}\cdot\Omega(\exp\{\pi i(h''+iz^{-1})/k\}).$$

We now evaluate Σ_2. In Σ_2 if $n=3p+\delta$, we have

$$(2.4.9)\quad \begin{aligned}\Sigma_2=-(2k)^{-1}\sum_{\delta=-1}^{1}\sum_{p=-\infty}^{\infty}(-1)^{\nu}\exp\{3\pi ih\mu^2/k\\ -\pi(2n+1)^2/12kz-(2n+1)\pi i\mu/k\}\\ \cdot\int_{-\infty}^{\infty}\frac{\exp(-3\pi zx^2/k)}{\tanh\{\pi ih\mu/k-\pi zx/k-\pi i(6p+2\delta+1)/6k\}}dx.\end{aligned}$$

If we replace $(2\nu+1)h-2p$ by $(2\nu+1) \pmod{2k}$, i.e.

$$\nu \to -\tfrac{1}{2}\{h''(2\nu+1+2p)+1\} \pmod{k},$$

then we have $\sum_{p=-\infty}^{\infty}$ free of x and ν. We define

$$U_\delta(z) = \sum_{p=-\infty}^{\infty} (-1)^p \exp\{\pi i h''(3p^2(2\delta+1)p)/k - \pi(6p+2\delta+1)^2/12kz\} \tag{2.4.10}$$

$$= \begin{cases} 0 \text{ if } \delta = 1 \\ \dfrac{\exp(-\pi/12kz)}{P(\exp\{2\pi i(h''+iz^{-1})/k\})} \end{cases} \quad \text{if } \delta = -1 \text{ or } 0, \text{ as before.} \tag{2.4.11}$$

Thus

(2.4.12)

$$\Sigma_2 = \frac{-k^{-1}(-1)^{\frac{1}{2}(h''+1)}\exp(-\pi/12kz - 3\pi i h''k''/2)}{P(\exp\{2\pi i(h''+iz^{-1})/k\})} \sum_{\nu \bmod k} (-1)^\nu$$

$$\cdot \exp(-3\pi i h''\mu^2/k + \pi i h''\mu/k)$$

$$\int_{-\infty}^{\infty} \frac{\exp(-3\pi z x^2/k)}{\tanh(\pi i h\mu/k - \pi z x/k - \pi i/6k)}\, dx.$$

We have established

$$\exp\{3\pi i(h+iz)/4k\} \frac{\omega(\exp\{\pi i(h+iz)/k\})}{P(\exp\{2\pi i(h+iz)/k\})} = \Sigma_1 + \Sigma_2;$$

hence by (2.0.1)

$$\begin{aligned} &\omega(\exp\{\pi i(h+iz)/k\}) \\ &= i(-1)^{\frac{1}{2}(h''+1)}\exp(-3\pi i h''k''/2 + 3\pi i h''/4k - 3\pi i h/4k)\omega_{h,k} \\ &\quad \cdot z^{-\frac{1}{2}}\exp(-2\pi z^{-1}/3k + 2\pi z/3k)\omega(\exp\{\pi i(h''+iz^{-1})/k\}) \\ &\quad - k^{-1}(-1)^{\frac{1}{2}(h''+1)}\exp(-3\pi i h''k''/2 - 3\pi i h/4k)\omega_{h,k} z^{\frac{1}{2}}\exp(2\pi z/3k) \end{aligned} \tag{2.4.13}$$

$$\cdot \sum_{\nu \bmod k} (-1)^\nu \exp(-3\pi i h''\mu^2/k + \pi i h''\mu/k)$$

$$\cdot \int_{-\infty}^{\infty} \frac{\exp(-3\pi z x^2/k)}{\tanh(\pi i h\mu/k - \pi z x/k - \pi i/6k)}\, dx.$$

Thus when h is odd, the general linear transformation sends ω into ω plus a correction $E_2^{(\omega)}(h,k;z)$; this concludes the proof of Theorem 2.4.

We now proceed to the task of simplifying the expressions for $\epsilon_{h,k}^{(f)}$, $\eta_{h,k}^{(f)}$, $\epsilon_{h,k}^{(\omega)}$, $\eta_{h,k}^{(\omega)}$.

COROLLARY 2.5. *We have*

$$\epsilon_{h,k}^{(f)} = \begin{cases} (-1)^{\frac{1}{2}(k+1)}\omega_{h,2k} & k \text{ odd} \qquad (2.5.1a) \\ i(-1)^{\frac{1}{2}(h-1+k)}\omega_{h,2k} & k \text{ even} \qquad (2.5.1b) \end{cases}$$

$$\eta_{h,k}^{(f)} = (-1)^{\frac{1}{2}(k-1)}\exp\{3(k+1)^2\pi ih'/2k\}\omega_{\frac{1}{2}h,k} \tag{2.5.2}$$

$$\epsilon_{h,k}^{(\omega)} = (-1)^{\frac{1}{2}(k-1)}\exp\{2\pi ih3(k^2-1)/8k\}\omega_{h,k} \tag{2.5.3}$$

$$\eta_{h,k}^{(\omega)} = \begin{cases} i(-1)^{\frac{1}{2}(h'''+1)}\exp\{3\pi i(h'''-h)/4k\}\omega_{h,k}, \quad k \text{ even}, \; hh''' \equiv -1 \pmod{8k}; & (2.5.4a) \\ (-1)^{\frac{1}{2}(k+1)}\exp\{2\pi ih'(5k^2+3)/8k + 2\pi ih3(k^2-1)/8k\}\omega_{h,k} \quad k \text{ odd}. & (2.5.4b) \end{cases}$$

Remark. Our results agree with those of Dragonette insofar as is possible to check; however, it appears she has made a slight error in line (7.8) [2; p. 499]. She gives the incorrect evaluation of $\epsilon_{h,k}^{(\omega)}$; if one uses line (7.2) of her paper, one easily derives a result on agreement with ours.

Proof. The proofs of these results involve routine simplifications of the expressions already obtained. For (2.5.1), we use the fact that h is odd, $hh'' = -1 - 2kk''$, and $\exp(\pi ih''/2) = -\exp(\pi ih/2)$. For (2.5.2), we use the fact that h is even, k is odd, and $k' \equiv -k - khh' \pmod 8$. (2.5.3) is deduced directly. For (2.5.4a), we use the fact that h is odd, k is even, and $h'' \equiv h''' \pmod 4$. For (2.5.4b), we use the fact that h is odd, k is odd, and $k'' \equiv -\frac{1}{2}(-k - khh'') \pmod 8$.

III. The modular transformation formulae of ϕ, ψ, and ν. Having obtained the transformation formulae for f and ω, we obtain the corresponding formulae for ϕ, ψ, and ν by means of certain identities involving the mock theta functions [2; p. 475, (1.2(a)), (1.2(d))].

LEMMA 3.1. *If* $\theta_2(0,q) = \sum_{n=-\infty}^{\infty} q^{(n+\frac{1}{2})^2}$, *and* $\theta_4(0,q) = \sum_{n=-\infty}^{\infty}(-1)^n q^{n^2}$ *with* $hh' + kk' = -1$ *and* g.c.d.$(h,k) = 1$, *then we have for* $(\operatorname{Re} z) > 0$

$$\theta_2^2(0, \exp\{\pi i(h+iz)/k\} = iz^{-1}(-1)^{\frac{1}{2}(h-1)}\exp(\pi ihk'/2)$$

$$\cdot\, \theta_2^2(0, \exp\{\pi i(h'+iz^{-1})/k\}) \tag{3.1.1}$$

h odd, k even, h' odd;

$$\theta_4^2(0, \exp\{\pi i(h+iz)/k\} = z^{-1}(-1)^{\frac{1}{2}(k-1)}\exp(-\pi ih'k/2)$$

$$\cdot\, \theta_2^2(0, \exp\{\pi i(h'+iz^{-1})/k\}) \tag{3.1.2}$$

h even, k odd, k' odd;

$$\theta_4^2(0, \exp\{\pi i(h+iz)/k\}) = iz^{-1}(-1)^{\frac{1}{2}(h-1)}\exp(-\pi ihk/2)$$

$$\cdot\, \theta_4^2(0, \exp\{\pi i(h'+iz^{-1})/k\}) \tag{3.1.3}$$

h odd, k' even, h' odd;

$$\theta_2^2(0, \exp\{\pi i(h+iz)/k\}) = z^{-1}(-1)^{\frac{1}{2}(k-1)}\exp(\pi ihk/2)$$

$$\cdot\, \theta_4^2(0, \exp\{\pi i(h'+iz^{-1})/k\}) \tag{3.1.4}$$

k odd, k' odd, h' even;

Proof. These formulae are obtained directly from [11] using Table XX on page 241 and Table XLII on page 262. In Table XX, we let $-h=a$, $k=b$, $k'=c$, $h'=d$, $\tau=(h+iz)/k$, $T=(h'+iz^{-1})/k$. We obtain

(3.1.1) with k' even from line 1 of Table XX, and lines (2), (6), (7), (8) of XLII;

(3.1.1) with k' odd from line 2 of Table XX, and lines (2), (6), (7), (8) of XLII;

(3.1.2) with h' odd from line 6 of Table XX, and lines (2), (6), (7), (8) of XLII;

(3.1.2) with h' even from line 5 of Table XX, and lines (2), (6), (7), (8) of XLII;

(3.1.3) with k even from line 1 of Table XX, and lines (4), (6), (7), (8) of XLII;

(3.1.3) with k odd from line 3 of Table XX, and lines (4), (6), (7), (8) of XLII;

(3.1.4) with h odd from line 4 of Table XX, and lines (4), (6), (7), (8) of XLII;

(3.1.4) with h even from line 5 of Table XX, and lines (4), (6), (7), (8) of XLII.

In the next two lemmas, we study two theta functions closely related to the mock theta functions.

LEMMA 3.2. *If* $G_1(q) = \theta_4(0,q)\prod_{r=1}^{\infty}(1+q^r)^{-1}$ *and*

$$G_2(q) = \tfrac{1}{2}q^{-\frac{1}{4}}\theta_2(0,q)\prod_{r=1}^{\infty}(1+q^{2r}),$$

then

$$G_1(\exp\{\pi i(h+iz)/k\}) = \begin{cases} 2^{\frac{3}{2}}z^{-\frac{1}{2}}(-1)^{h'}\exp(-\pi ih'/k)\eta_{h,k}{}^{(f)}\exp(-\pi/3kz \\ \qquad -\pi z/24k)G_2(\exp\{\pi i(h'+iz^{-1})/k\}) & h \text{ even} \\ z^{-\frac{1}{2}}(-1)^k\epsilon_{h,k}{}^{(f)}\exp(\pi/24kz \\ \qquad -\pi z/24k)G_1(\exp\{\pi i(h''+iz^{-1})/k\}) & h \text{ odd} \end{cases}$$

where in the top line $hh' \equiv -1 \pmod{k}$ and in the bottom line $hh'' \equiv -1 \pmod{2k}$. $\epsilon_{h,k}{}^{(f)}$ appears in Theorem 2.2, and $\eta_{h,k}{}^{(f)}$ appears in Theorem 2.1.

Proof. We note that

$$\begin{aligned} G_1(q) &= \theta_4(0,q)\prod_{r=1}^{\infty}(1+q^r)^{-1} \\ &= \theta_4(0,q)\prod_{r=1}^{\infty}(1+q^r)^{-1}\prod_{r=1}^{\infty}(1-q^r)\prod_{r=1}^{\infty}(1-q^r)^{-1} \\ &= \theta_4{}^2(0,q)P(q) \qquad [5;\ \text{p. } 185] \end{aligned}$$

$$\begin{aligned} G_2(q) &= \tfrac{1}{2}q^{-\frac{1}{4}}\theta_2(0,q)\prod_{r=1}^{\infty}(1+q^{2r}) \\ &= \tfrac{1}{2}q^{-\frac{1}{4}}\theta_2(0,q)\prod_{r=1}^{\infty}(1+q^{2r})\prod_{r=1}^{\infty}(1-q^{4r})\prod_{r=1}^{\infty}(1-q^{4r})^{-1} \\ &= \tfrac{1}{4}q^{-\frac{1}{2}}\theta_2{}^2(0,q)P(q^4) \qquad [5;\ \text{p. } 185]. \end{aligned}$$

The remainder of the proof is routine. Since $G_1(q) = \theta_4{}^2(0,q)P(q)$, we apply Lemma 3.1 to $\theta_4{}^2(0,q)$, and (2.0.1) to $P(q)$; the desired result follows directly using the expressions for $\epsilon_{h,k}{}^{(f)}$ and $\eta_{h,k}{}^{(f)}$ found in Corollary 2.5.

LEMMA 3.3. *If* G_1 *and* G_2 *are as in the previous lemma, then*

$$G_2(\exp\{\pi i(h+iz)/k\})=\begin{cases}(2z)^{-\frac{1}{2}}(-1)^h\exp(\pi ih/k)\epsilon_{2h,k}{}^{(\omega)}\exp(\pi z/3k \\ \quad +\pi z^{-1}/24k)G_1[(-1)^{h'}\exp\{\pi i(h'+iz^{-1})/k\}] & k \text{ odd}\\ z^{-\frac{1}{2}}(-1)^{k'}\exp\{\pi i(h-h')/k\}\eta_{h,\frac{1}{2}k}{}^{(\omega)}\exp(\pi z/3k \\ \quad -\pi z^{-1}/3k)G_2(\exp\{\pi i(h'+iz^{-1})/k\}) & k \text{ even}\end{cases}$$

where $hh'+kk'=-1$.

Proof. The proof of this lemma is exactly like that of Lemma 3.2. Since $G_2(q)=\frac{1}{4}q^{-\frac{1}{2}}\theta_2{}^2(0,q)P(q^4)$, we apply Lemma 3.1 to $\theta_2{}^2(0,q)$ and (2.0.1) to $P(q^4)$; the desired result follows directly using the expressions $\epsilon_{h,k}{}^{(\omega)}$ and $\eta_{h,k}{}^{(\omega)}$ found in Corollary 2.5.

To obtain the linear transformation formulae for ϕ, ψ, and v we utilize the following identities partially stated by Ramanujan [9; p. 354] and proved by Watson [12; p. 63].

$$2\phi(q)=f(-q)+G_1(-q) \tag{A}$$

$$4\psi(q)=-f(-q)+G_1(-q) \tag{B}$$

$$\mp q^{-1}v(\pm q)=\omega(q^2)\mp q^{-1}G_2(q) \tag{C}$$

where G_1 and G_2 are given in Lemma 3.2,

THEOREM 3.4. *Let* g.c.d.$(h,k)=1$, $hh'+kk'=-1$, *also if h is odd* $hh''+2kk''=-1$, *and* $\operatorname{Re} z>0$ *with the principal branch of $z^{\frac{1}{2}}$ taken. Then*

$$\phi[\exp\{\pi i(h+iz)/k\}]=\begin{cases} z^{-\frac{1}{2}}\epsilon_{h,k}{}^{(\phi)}\exp(-\pi z/24k+\pi z^{-1}/24k)\,\cdot \\ \quad \cdot\,\phi(\exp\{\pi i(h''+iz^{-1})/k\}) \\ \quad +\frac{1}{2}E_2{}^{(f)}(h+k,k;z) \quad (k \text{ even}) & (3.4.1)\\ 2^{\frac{1}{2}}z^{-\frac{1}{2}}\eta_{h,k}{}^{(\phi)}\exp(-\pi z/24k-\pi z^{-1}/3k) \\ \quad \cdot\, v[(-1)^{h'-1}\exp\{\pi i(h'+iz^{-1})/k\}] \\ \quad +\frac{1}{2}E_1{}^{(f)}(h+k,k;z) \quad (k \text{ odd},\ h \text{ odd}) & (3.4.2)\\ 2z^{-\frac{1}{2}}\sigma_{h,k}{}^{(\phi)}\exp(-\pi z/24k+\pi z^{-1}/24k) \\ \quad \cdot\,\psi[(-1)^{h'-1}\exp\{\pi i(h'+iz^{-1})/k\}] \\ \quad +\frac{1}{2}E_2{}^{(f)}(h+k,k;z) \quad (h \text{ even}) & (3.4.3)\end{cases}$$

$$\psi[\exp\{\pi i(h+iz)/k\}]=\begin{cases} \frac{1}{2}z^{-\frac{1}{2}}\epsilon_{h,k}{}^{(\psi)}\exp(-\pi z/24k+\pi z^{-1}/24k) \\ \quad \cdot\,\phi[(-1)^{h'}\exp\{\pi i(h'+iz^{-1})/k\}] \\ \quad -\frac{1}{4}E_2{}^{(f)}(h+k,k;z) \quad (h \text{ even}) & (3.4.4)\\ (2z)^{-\frac{1}{2}}\eta_{h,k}{}^{(\psi)}\exp(-\pi z/24k-\pi z^{-1}/3k) \\ \quad \cdot\, v[(-1)^{h'}\exp\{\pi i(h'+iz^{-1})/k\}] \\ \quad -\frac{1}{4}E_1{}^{(f)}(h+k,k;z) \quad (h \text{ odd},\ k \text{ odd}) & (3.4.5)\\ z^{-\frac{1}{2}}\sigma_{h,k}{}^{(\psi)}\exp(-\pi z/24k+\pi z^{-1}/24k) \\ \quad \cdot\,\psi(\exp\{\pi i(h''+iz^{-1})/k\}) \\ \quad -\frac{1}{4}E_2{}^{(f)}(h+k,k;z) \quad (k \text{ even}) & (3.4.6)\end{cases}$$

$$(3.4.7)\qquad (2z)^{-\frac{1}{2}}\epsilon_{h,k}{}^{(v)}\exp(\pi z/3k+\pi z^{-1}/3k)\cdot\phi[(-1)^{h'-1}\exp\{\pi i(h'+iz^{-1})/k\}]-\exp\{\pi i(h+iz)/k\}E_1{}^{(\omega)}(2h,k;2z)\quad (h\text{ odd, k odd})$$

$$(3.4.8)\quad v(\exp\{\pi i(h+iz)/k\})=\begin{cases} z^{-\frac{1}{2}}\eta_{h,k}{}^{(v)}\exp(\pi z/3k-\pi z^{-1}/3k)\cdot v[(-1)^{k'}\exp\{\pi i(h'+iz^{-1})/k\}]-\exp\{\pi i(h+iz)/k\}E_2{}^{(\omega)}(h,\tfrac{1}{2}k;z) & (k\text{ even})\end{cases}$$

$$(3.4.9)\qquad 2^{\frac{1}{2}}z^{-\frac{1}{2}}\sigma_{h,k}{}^{(v)}\exp(\pi z/3k+\pi z^{-1}/24k)\cdot\psi[(-1)^{h'-1}\exp\{\pi i(h'+iz^{-1})k]-\exp\{\pi i(h+iz)/k\}E_1{}^{(\omega)}(2h+2k,k;2z)\quad (h\text{ even})$$

with

$$\epsilon_{h,k}{}^{(\phi)}=\epsilon_{h+k,k}{}^{(f)}\quad (k\text{ even})$$
$$\eta_{h,k}{}^{(\phi)}=(-1)^{h'}\exp(-\pi ih'/k)\eta_{h+k,k}{}^{(f)}\quad (h+k\text{ even})$$
$$\sigma_{h,k}{}^{(\phi)}=-\epsilon_{h+k,k}{}^{(f)}\quad (h\text{ even})$$
$$\epsilon_{h,k}{}^{(\psi)}=-\epsilon_{h+k,k}{}^{(f)}\quad (h\text{ even})$$
$$\eta_{h,k}{}^{(\psi)}=(-1)^{h'}\exp(-\pi ih'/k)\eta_{h+k,k}{}^{(f)}\quad (h+k\text{ even})$$
$$\sigma_{h,k}{}^{(\psi)}=\epsilon_{h+k,k}{}^{(f)}\quad (k\text{ even})$$
$$\epsilon_{h,k}{}^{(v)}=-\exp(\pi ih/k)\epsilon_{2h,k}{}^{(\omega)}\quad (h+k\text{ even})$$
$$\eta_{h,k}{}^{(v)}=(-1)^{k'}\exp\{\pi i(h-h')/k\}\eta_{h,\frac{1}{2}k}{}^{(\omega)}\quad (k\text{ even})$$
$$\sigma_{h,k}{}^{(v)}=\exp(\pi ih/k)\epsilon_{2h,k}{}^{(\omega)}\quad (h\text{ even})$$

where $\epsilon_{h,k}{}^{(f)}$, $\eta_{h,k}{}^{(f)}$, $\epsilon_{h,k}{}^{(\omega)}$, $\eta_{h,k}{}^{(\omega)}$ *and* $E_1{}^{(f)}(h,k;z)$, $E_2{}^{(f)}(h,k;z)$, $E_1{}^{(\omega)}(h,k;z)$, $E_2{}^{(\omega)}(h,k;z)$ *are given in Theorems* 2.1, 2.2, 2.3 *and* 2.4 *respectively.*

Proof. We shall derive (3.4.1) only. The remaining formulae are proved in a similar manner.

$$(3.4.10)\qquad 2\phi(\exp\{\pi i(h+iz)/k\}) =f(\exp\{\pi i[(h+k)+iz]/k\}+G_1(\exp\{\pi i[(h+k)+iz]/k\}).$$

Now for k even, we use the fact that $(h+k)''\equiv h''+k \pmod{2k}$ since $(h+k)(h''+k)=hh''+k(h+h''+k)\equiv -1 \pmod{2k}$. Hence by (A)

$$(3.4.11)\qquad 2\phi(-\exp\{\pi i[(h+k)''+iz^{-1}]/k\}) =f(\exp\{\pi i[(h+k)''+iz^{-1}]/k\})+G_1(\exp\{\pi i[(h+k)''+iz^{-1}]/k\}).$$

Multiplying (3.4.11) by $z^{-\frac{1}{2}}\epsilon_{h+k,k}^{(f)}\exp(-\pi z/24k+\pi z^{-1}/24k)$, then subtracting from (3.4.10) and using Theorem 2.2 and Lemma 3.2, we get

$$\begin{aligned}&\phi(\exp\{\pi i(h+iz)/k\})\\ (3.4.12)\qquad &=z^{-\frac{1}{2}}\epsilon_{h+k,k}^{(f)}\exp(-\pi z/24k+\pi z^{-1}/24k)\\ &\quad\cdot\phi(\exp\{\pi i(h''+iz^{-1})/k\})+\tfrac{1}{2}E_2^{(f)}(h+k,k;z)\end{aligned}$$

which is (3.4.1).

IV. The exponential sums. Before evaluating the exponential sums that arise for the mock theta functions, we give the following lemma which will be useful in the next section.

LEMMA 4.1. *If* g.c.d.$(D,k)=1$, $d\mid 24$, *and* $0\leqq\sigma_1<\sigma_2\leqq k$,

$$A_k(D;d;F,G;\sigma_1;\sigma_2)=\sum_{\substack{h \bmod dk\\ \sigma_1\leqq Dh^*\leqq\sigma_2\\ \text{g.c.d.}(h,dk)=1}}\omega_{h,k}\exp(-2\pi iFh/dk\ 2\pi iGh^*/dk),$$

where $\omega_{h,k}$ *appears in* (2.0.1), *and* $hh^*\equiv-1 \pmod{dk}$, *then*

$A_k(D;d;F,G;\sigma_1;\sigma_2)=O(((24F+d)/d,k)^{\frac{1}{2}}k^{\frac{1}{2}+\epsilon})$, *as* $k\to+\infty$, *where the constant implied by the O-symbol is absolute.*

Proof. Many papers contain theorems like this for sums related to the standard Kloosterman sum (e.g. [3], [7]). However, by [9; p. 284] we have

$$\omega_{h,k}=-i\left(\frac{-k}{h}\right)\exp\{2\pi i[(4k^2+3k-1)h-(k^2-1)\bar{h}]/24k\}$$

if k is even, g.c.d.$(h,24)=\Theta$ and $h\bar{h}\equiv-1\ (\bmod 24k/\Theta)$;

$$\omega_{h,k}=\exp\{-\pi i(k-1)/4\}\left(\frac{-h}{k}\right)\exp\{2\pi i[(k^2-1)h-(k^2-1)\bar{h}]/24k\}$$

if k is odd, g.c.d.$(h,24)=\Theta$ and $h\bar{h}\equiv-1\ (\bmod 24k/\Theta)$. Thus we see that the sum we are considering is in fact related to a generalized Kloosterman sum; such sums have recently been studied in detail in [13]. A generalized Klosterman sum is of the form

$$\sum_{\substack{h \bmod k \\ \text{g.c.d.}(h,k)=1}} \chi(h)\exp\{2\pi i(ah - h')/k\}$$

where $\chi(h)$ is a quadratic character. This sum satisfies the same estimates as the original Kloosterman sum; one may easily prove that such sums are multiplicative, and one has

$$\Big| \sum_{\substack{h \bmod p^\alpha \\ \text{g.c.d.}(h,p)=1}} \chi(h)\exp\{2\pi i(ah - h')/p^\alpha\} \Big| \leqq 2p^{\frac{1}{2}\alpha}, \qquad \alpha > 1$$

by [6; p. 283] and [10; p. 102, line (54)].

This information gives by the method of Lehner [7; § 10]

$$\begin{aligned} | A_k(D; d; F, G; \sigma_1, \sigma_2) &= \begin{cases} O\{(-24F/d + 4k^2 + 3k - 1, 24k)^{\frac{1}{2}}k^{\frac{1}{2}+\epsilon}\} & (k \text{ even}) \\ O\{(-24F/d + k^2 - 1, 24k)^{\frac{1}{2}}k^{\frac{1}{2}+\epsilon}\} & (k \text{ odd}) \end{cases} \\ &= O\{(24F/d + 1, k)^{\frac{1}{2}}k^{\frac{1}{2}+\epsilon}\}. \end{aligned}$$

We now go to the exponential sums in question. The following equations are direct consequences of Corolary 2. 5. We shall only exhibit the exponential sums necessary to deal with Theorem 5. 1. Examination of Theorem 3. 4 shows that analogous results hold for the exponential sums relevant in the proof of Theorem 5. 2. If we let

$$\lambda_k^{(f)}(\epsilon; \sigma_1; \sigma_2; n) = \sum_{\substack{h \bmod 2k \\ \text{g.c.d.}(h,2k)=1 \\ \sigma_1 \leqq h' \leqq \sigma_2}} \epsilon_{h,k}^{(f)}\exp(-\pi i h n/k), \tag{4. 2}$$

then

$$\lambda_k^{(f)}(\epsilon; \sigma_1; \sigma_2; n) = \begin{cases} (-1)^{\frac{1}{2}(k+1)} \sum_{j=0}^{1} A_{2k}(1; 1; n, 0; \sigma_1 - jk, \sigma_2 - jk) & (k \text{ odd}) \\ (-1)^{\frac{1}{2}k} \sum_{j=0}^{1} \mathrm{A}_{2k}(1, 1; n - \frac{1}{2}k, 0; \sigma_1 - jk, \sigma_2 - jk) & (k \text{ even}) \end{cases}$$

If we let

$$\lambda_k^{(f)}(\eta; \sigma_1; \sigma_2; n) = \sum_{\substack{h \bmod 2k \\ \text{g.c.d.}(h,k)=1 \\ 2 \mid h \\ \sigma_1 \leqq h' \leqq \sigma_2}} \eta_{h,k}^{(f)}\exp(-\pi i h n/k), \tag{4. 3}$$

then

$$\lambda_k^{(f)}(\eta; \sigma_1; \sigma_2; n) = (-1)^{\frac{1}{2}(k-1)} A_k(\tfrac{1}{2}(k+1); 1; n, 3(k+1)^3/8; \sigma_1; \sigma_2).$$

The following two exponential sums will also arise when we estimate the error terms in the asymptotic expansions. If we let

$$(4.4)\quad \Lambda_{k,1}^{(f)}(\delta;\sigma_1,\sigma_2;n)=\sum_{\substack{h \bmod 2k\\ \text{g.c.d.}(h,2k)=1\\ 2\mid h\\ \sigma_1\leqq h'\leqq\sigma_2\\ \sigma_1\leqq Dh^*\leqq\sigma_2}} \delta_1^{(f)}(\nu;\tfrac{1}{2}h;2h';k)\exp(-\quad ihn/k),$$

then

$$\Lambda_{k,1}^{(f)}(\delta;\sigma_1,\sigma_2;n)=(-1)^{\nu}\sum_{\substack{h \bmod k\\ \text{g.c.d.}(2h,k)=1\\ \sigma_1\leqq\frac{1}{2}(k+1)h'\leqq\sigma_2}} \delta_1^{(f)}(\nu;h;h';k)\exp(-2\pi ihn/k)$$

$$=(-1)^{\nu}A_k(\tfrac{1}{2}(k+1);1;n,\tfrac{1}{2}(\nu-3\nu^2);\sigma_1,\sigma_2).$$

If we let

$$(4.5)\quad \Lambda_{k,2}^{(f)}(\delta;\sigma_1,\sigma_2;n)=\sum_{\substack{h \bmod 2k\\ \text{g.c.d.}(h,2k)=1\\ \sigma_1\leqq h'\leqq\sigma_2}} \delta_1^{(f)}(\nu;h;h'';2k)\exp(-\pi ihn/k),$$

then

$$\Lambda_{k,2}^{(f)}(\delta;\sigma_1,\sigma_2;n)$$

$$=(-1)^{\nu}\sum_{j=0}^{1}A_{2k}(1;1;n,\tfrac{1}{2}(\nu-3\nu^2);\sigma_1-jk;\sigma_2-jk).$$

We now treat the exponential sums which actually appear in the asymptotic expansions.

THEOREM 4.6. *If*

$$(4.6.1)\qquad \lambda(k)=\tfrac{1}{2}\sum_{\substack{h \bmod 2k\\ \text{g.c.d.}(h,2k)=1}} \epsilon_{h,k}^{(f)}\exp(-\pi ihn/k);$$

$$(4.6.2)\qquad \lambda^{(\phi)}(k)=\tfrac{1}{2}\sum_{\substack{h \bmod 2k\\ \text{g.c.d.}(h,2k)=1\\ 2\mid k}} \epsilon_{h,k}^{(\phi)}\exp(-\pi ihn/k);$$

$$(4.6.3)\qquad \lambda^{(\psi)}(k)=\tfrac{1}{2}\sum_{\substack{h \bmod 2k\\ \text{g.c.d.}(h,k)=1\\ 2\mid h}} \epsilon_{h,k}^{(\psi)}\exp(-\pi ihn/k);$$

$$(4.6.4)\qquad \lambda^{(\upsilon)}(k)=\tfrac{1}{2}\sum_{\substack{h \bmod 2k\\ \text{g.c.d.}(h,k)=1\\ 2\nmid h,2\nmid k}} \epsilon_{h,k}^{(\upsilon)}\exp(-\pi ihn/k);$$

(4.6.5) $\lambda^{(\omega)}(k)=\frac{1}{2}\sum_{\substack{h \bmod 2k \\ \text{g.c.d.}(h,k)=1 \\ 2\mid h}} \epsilon_{h,k}{}^{(\omega)}\exp(-\pi ihn/k)$; *then*

$$\lambda(k)=\begin{cases}\frac{1}{2}(-1)^{\frac{1}{2}(k+1)}A_{2k}(n) & (k \text{ odd}) \\ \frac{1}{2}(-1)^{\frac{1}{2}k}A_{2k}(n-\frac{1}{2}k) & (k \text{ even}),\end{cases} \tag{4.6.6}$$

$$\lambda^{(\phi)}(k)=\tfrac{1}{2}(-1)^{n+\frac{1}{2}k}A_{2k}(n-\tfrac{1}{2}k) \tag{4.6.7}$$

$$\lambda^{(\psi)}(k)=\tfrac{1}{2}(-1)^{n+\frac{1}{2}(k-1)}A_{2k}(n) \tag{4.6.8}$$

$$\lambda^{(\nu)}(k)=\tfrac{1}{2}(-1)^{n+\frac{1}{2}(k-1)}A_{k}\{(n-1)[\tfrac{1}{2}(k+1)]^2-3(k^2-1)/8\} \tag{4.6.9}$$

$$\lambda^{(\omega)}(k)=\tfrac{1}{2}(-1)^{\frac{1}{2}(k-1)}A_{k}\{\tfrac{1}{2}n(k+1)-3(k^2-1)/8\} \tag{4.6.10}$$

where $A_k(n)=\sum_{\substack{h \bmod k \\ \text{g.c.d.}(h,k)=1}} \omega_{h,k}\exp(-2\pi ihn/k)$ *and* $\omega_{h,k}$ *appears in* (2.0.1).

Proof. (4.6.6) is a special case of (4.2). (4.6.7) follows immediately from the definition of $\epsilon_{h,k}{}^{(\phi)}$ in Theorem 3.4. (4.6.8) also follows immediately from the definition of $\epsilon_{h,k}{}^{(\psi)}$ in Theorem 3.4. We deal with (4.6.10) next. We have

$$\lambda^{(\omega)}(k)=\frac{1}{2}\sum_{\substack{h \bmod 2k \\ \text{g.c.d.}(h,k)=1 \\ 2\mid h}} \epsilon_{h,k}{}^{(\omega)}\exp(-\pi ihn/k)$$

$$=\frac{1}{4}(-1)^{\frac{1}{2}(k-1)}\sum_{j=0}^{1}\sum_{\substack{h \bmod 2k \\ \text{g.c.d.}(h,k)=1}} \omega_{h,k}\exp\{\pi ihj+2\pi ih3(k^2-1)/8k-\pi ihn/k\}$$

[by Corollary 2.5]

$$=\frac{1}{4}(-1)^{\frac{1}{2}(k-1)}\sum_{j=0}^{1}\sum_{\substack{h \bmod k \\ \text{g.c.d.}(h,k)=1}} \omega_{h,k}\exp\{\pi ihj+2\pi ih3(k^2-1)/8k-\pi ihn/k\}\cdot(1+(-1)^{jk+n})$$

$$=\frac{1}{2}(-1)^{\frac{1}{2}(k-1)}\sum_{\substack{h \bmod k \\ \text{g.c.d.}(h,k)=1}} \omega_{h,k}\exp(-2\pi ih\{\tfrac{1}{2}n(k+1)-3(k^2-1/8\}/k)$$

$$=\tfrac{1}{2}(-1)^{\frac{1}{2}(k-1)}A_k\{\tfrac{1}{2}n(k+1)-3(k^2-1)/8\}.$$

$$\lambda^{(\nu)}(k)=-\frac{1}{2}\sum_{\substack{h \bmod 2k \\ \text{g.c.d.}(h,k)=1 \\ 2\nmid h,\, 2\nmid k}} \epsilon_{2h,k}{}^{(\omega)}\exp(\pi ih/k-\pi ihn/k) \quad \text{[by Theorem 3.4]}$$

$$=\frac{1}{2}(-1)^{n}\sum_{\substack{h \bmod 2k \\ \text{g.c.d.}(2h,k)=1}} \epsilon_{2h,k}{}^{(\omega)}\exp\{-\pi ih(n-1)/k\}$$

$$= \tfrac{1}{4}(-1)^n \sum_{j=0}^{1} \sum_{\substack{h \bmod 2k \\ \text{g.c.d.}(h,k)=1}} \epsilon_{2h,k}^{(\omega)} \exp(-\pi i h(n-1)/k + \pi i h j)$$

$$= \tfrac{1}{2}(-1)^n \sum_{\substack{h \bmod k \\ \text{g.c.d.}(h,k)=1}} \epsilon_{2h,k}^{(\omega)} \exp\{-\pi i h(n-1)(k+1)/k\}$$

$$= \tfrac{1}{2}(-1)^n \sum_{\substack{h \bmod 2k \\ \text{g.c.d.}(h,k)=1 \\ 2|h}} \epsilon_{h,k}^{(\omega)} \exp\{-\pi i h(n-1)\tfrac{1}{2}(k+1)\}$$

$$= \tfrac{1}{2}(-1)^{n+\frac{1}{2}(k-1)} A_k\{(n-1)[\tfrac{1}{2}(k+1)]^2 - 3(k^2-1)/8\} \text{ by (4.6.10).}$$

This concludes Theorem 4.6.

V. The asymptotic formulae. In [2], Dragonette has given asymptotic expansions for the power series coefficients of f, ω, ϕ, ψ, and v (as discussed in § I). In this section we shall improve her results.

Since Dragoneete obtained the main terms for the expansion of $A(n)$ using the Hardy-Ramanujan-Rademacher method, we shall refer the reader to her paper for many of the details. Our method centers upon the use of estimates of generalized Kloosterman sums together with our precise knowledge of the error terms involved in the modular transformation formulae. We now prove the theorem stated in the introduction.

Proof of Theorem 5.1. For the initial Farey dissection we refer the reader to [2; pp. 489-90]. Dragonette obtains there

$$(5.1.1)\quad A(n) = \tfrac{1}{2}\sum_{k=1}^{N} \sum_{\substack{h \bmod 2k \\ \text{g.c.d.}(h,2k)=1}} \exp(-\pi i h n/k) \int_{-\theta'_{h,k}}^{\theta''_{h,k}} \{z^{-\frac{1}{2}} \epsilon_{h,k}^{(f)}$$

$$\cdot \exp(\pi z n/k - \pi z/24k + \pi z^{-1}/24k) f(\exp\{\pi i(h'' + iz^{-1})/k\}) + E_2^{(f)}(h,k;z)\} d\phi$$

$$+ \tfrac{1}{2}\sum_{k=1}^{N} \sum_{\substack{h \bmod 2k \\ \text{g.c.d.}(h,k)=1 \\ 2|h}} \exp(-\pi i h n/k) \int_{-\theta'_{h,k}}^{\theta''_{h,k}} \{2^{\frac{3}{2}} z^{-\frac{1}{2}} \eta_{h,k}^{(f)}$$

$$\cdot \exp(\pi z n/k - \pi z/24k - 4\pi z^{-1}/3k)\, \omega(\exp\{2\pi i(h' + iz^{-1})/k\}) + E_1^{(f)}(h,k;z)\} d\phi$$

$$(5.1.2)\qquad = \sum_{k=1}^{N} \frac{\lambda(k)\exp\{\pi(n-1/24)^{\frac{1}{2}}/k6^{\frac{1}{2}}\}}{k^{\frac{1}{2}}(n-1/24)^{\frac{1}{2}}} + O(\sum_{k=1}^{N} (n,k)^{\frac{1}{2}} k^{\epsilon-1}) + O(n^{\epsilon})$$

$$+\tfrac{1}{2}\sum_{k=1}^{N}\sum_{\substack{h \bmod 2k\\ \text{g.c.d.}(h,2k)=1}} \exp(-\pi i h n/k)\int_{-\theta'_{h,k}}^{\theta''_{h,k}} E_2^{(f)}(h,k;z)\,d\phi$$

$$+\tfrac{1}{2}\sum_{k=1}^{N}\sum_{\substack{h \bmod 2k\\ \text{g.c.d.}(h,2k)=1}} \exp(-\pi i h n/k)\int_{-\theta'_{h,k}}^{\theta''_{h,k}} E_1^{(f)}(h,k;z)\,d\phi$$

$$(5.1.3)\qquad = \Sigma_1 + O(n^{\epsilon}) + \Sigma_2 + \Sigma_3,$$

We remark that (5.1.3) is obtained from (5.1.1) by almost the same method that Dragonette uses in obtaining [2; p. 493, line (5.5)] from [2; p. 490, line (4.5)]. The main alteration in the proof is our use of estimates for truncated generalized Kloosterman sums given by Lemma 4.1 and equations (4.2) and (4.3); this is done in exactly the way that Rademacher modifies the Hardy-Ramanujan method to obtain Fourier coefficients for the modular invariant, $J(\tau)$, [8]. Also we exhibit the contributions of $E_1^{(f)}(h,k;z)$ and $E_2^{(f)}(h,k;z)$ whereas Dragonette uses the estimate

$$|E_j^{(f)}(h,k;z)| = O(k\log k) \qquad j=1,2.$$

We shall show in fact that

$$|\Sigma_j| = O(n^{\epsilon}) \qquad j=2,3.$$

For reference we give the following resumè of the notation used above.

$$\theta'_{h,k} = k^{-1}(k_1+k)^{-1}, \qquad \theta''_{h,k} = k^{-1}(k_2+k)^{-1}$$

where $h_1/k_1 < h/k < h_2/k_2$ are adjacent Farey fractions in the Farey series of order N. As a simple consequence of the theory of Farey series we have

$$k_1 \equiv -h' \pmod{k}, \qquad k_2 \equiv h' \pmod{k}, \text{ and}$$
$$N-k < k_1 \leqq N, \qquad N-k < k_2 \leqq N.$$
$$z = kN^{-2} - ik\phi.$$

Also throughout our work here we take $N=[n^{\frac{1}{2}}]$. Using the notation of Theorems 2.1 and 2.2, we have

$$\Sigma_2 = \tfrac{1}{2}\sum_{k=1}^{N} k^{-1}\sum_{\substack{h \bmod 2k\\ \text{g.c.d.}(h,2k)=1}} \exp(-\pi i h n/k)\sum_{\nu \bmod k}\delta_1^{(f)}(\nu;h;h'';2k)$$
$$\cdot\int_{-\theta'_{h,k}}^{\theta''_{h,k}} z^{\frac{1}{2}} J_1^{(f)}(\nu;2k;4z)\,d\phi$$

$$= \tfrac{1}{2}\sum_{k=1}^{N} k^{-1}\sum_{\nu=1}^{k}\sum_{\substack{h \bmod 2k \\ \text{g.c.d.}(h,2k)=1}} \delta_1^{(f)}(\nu; h; h''; 2k)\exp(-\pi i h n/k)$$

$$\cdot\left\{\int_{-k^{-1}(N+k)^{-1}}^{k^{-1}(N+k)^{-1}} + \int_{-k^{-1}(k_1+k)^{-1}}^{-k^{-1}(N+k)^{-1}} + \int_{k^{-1}(N+k)^{-1}}^{k^{-1}(k_2+k)^{-1}}\right\}$$

$$= \Sigma_{21} + \Sigma_{22} + \Sigma_{23}.$$

$$|\Sigma_{21}| = O\Big(\sum_{k=1}^{N} k^{-1}(n,k)^{\frac{1}{2}}k^{\frac{1}{2}+\epsilon'}(kN)^{-1}\sum_{\nu=1}^{k} N^{\frac{1}{2}}k\,|k-2(\nu-\tfrac{1}{6})|^{-1}\Big)$$

(by Theorem 2.1 and (4.5))

$$= O\Big(N^{-\frac{1}{2}}\sum_{k=1}^{N}(n,k)^{\frac{1}{2}}k^{-\frac{1}{2}-2\epsilon''}\Big)$$

$$= O\Big(N^{-\frac{1}{2}}\sum_{\substack{d\mid n \\ d\leq n^{\frac{1}{2}}}} d^{\frac{1}{2}}\sum_{\substack{k=1 \\ \text{g.c.d.}(n,k)=1}}^{n^{\frac{1}{2}}} k^{-\frac{1}{2}+2\epsilon''}\Big)$$

$$= O\Big(N^{-\frac{1}{2}}\sum_{d\mid n} d^{\frac{1}{2}}\sum_{\substack{k'=1 \\ d\leq n^{\frac{1}{2}}}}^{n^{\frac{1}{2}}/d} d^{-\frac{1}{2}+2\epsilon''}k'^{-\frac{1}{2}+2\epsilon''}\Big)$$

$$= O\Big(N^{-\frac{1}{2}}\sum_{\substack{d\mid n \\ d\leq n^{\frac{1}{2}}}} d^{2\epsilon''}n^{\frac{1}{4}+\epsilon''}d^{-\frac{1}{2}-2\epsilon''}\Big)$$

$$= O\Big(n^{\epsilon''}\sum_{\substack{d\mid n \\ d\leq n^{\frac{1}{2}}}} d^{-\frac{1}{2}}\Big)$$

$$= O(n^{\epsilon}).$$

Σ_{22} and Σ_{23} are of the same structure so we treat only Σ_{23}.

$$\Sigma_{23} = \tfrac{1}{2}\sum_{k=1}^{N} k^{-1}\sum_{\nu=1}^{k}\sum_{\substack{h \bmod 2k \\ \text{g.c.d.}(h,2k)=1}} \delta_1^{(f)}(\nu; h; h''; 2k)\exp(-\pi i h n/k)$$

$$\cdot\sum_{l=k_2+k}^{N+k-1}\int_{k^{-1}(l+1)^{-1}}^{k^{-1}l^{-1}} z^{\frac{1}{2}}J_1^{(f)}(\nu; 2k; 4z)\,d\phi$$

$$= \tfrac{1}{2}\sum_{k=1}^{N} k^{-1}\sum_{\nu=1}^{k}\sum_{l=N+1}^{N+k-1}\int_{k^{-1}(l+1)^{-1}}^{k^{-1}l^{-1}} z^{\frac{1}{2}}J_1^{(f)}(\nu; 2k; 4z)\,d\phi$$

$$\cdot\sum_{\substack{h \bmod 2k \\ \text{g.c.d.}(h,2k)=1 \\ N<k_2+k\leq l}} \delta_1^{(f)}(\nu; h; h''; 2k)\exp(-\pi i h n/k).$$

The inner sum is now a truncated generalized Kloosterman sum by (4.5) and subject to the estimate of Lemma 4.1. Hence

$$|\Sigma_{23}| = O\Big(\sum_{k=1}^{N} k^{-1} \sum_{\nu=1}^{k} N^{\frac{1}{2}} k \,|\, k - 2(\nu - \tfrac{1}{6})|^{-1} (kN)^{-1} (n,k)^{\frac{1}{2}} k^{\frac{1}{2}+\epsilon'}\Big)$$

$$= O\Big(N^{-\frac{1}{2}} \sum_{k=1}^{N} (n,k)^{\frac{1}{2}} k^{-\frac{1}{2}+\epsilon''} \log k\Big)$$

$$= O(N^{\epsilon'''}) \qquad (\text{just as for } \Sigma_{21})$$

$$= O(n^{\epsilon}).$$

Thus we have $|\Sigma_2| = O(n^{\epsilon})$ for any $\epsilon > 0$. Σ_3 is handled in exactly the same manner. Hence

$$A(n) = \sum_{k=1}^{[n^{\frac{1}{2}}]} \frac{\lambda(k)\exp\{\pi(n - 1/24)^{\frac{1}{2}}/k6^{\frac{1}{2}}\}}{k^{\frac{1}{2}}(n - 1/24)^{\frac{1}{4}}} + O(n^{\epsilon}).$$

Coupling this with (4.6.6), we obtain the required result.

In an exactly analogous manner one may obtain the following

THEOREM 5.1. *Let $M(q)$ be one of the functions ϕ, ψ, ν, ω. Then if*

$$M(q) = \sum_{n=0}^{\infty} A^{(M)}(n) q^n,$$

$$A^{(\phi)}(n) = \sum_{\substack{k=1 \\ 2\,|\,k}}^{[n^{\frac{1}{2}}]} \frac{\lambda^{(\phi)}(k)\exp\{\pi(n - 1/24)^{\frac{1}{2}}/k6^{\frac{1}{2}}\}}{k^{\frac{1}{2}}(n - 1/24)^{\frac{1}{4}}} + O(n^{\epsilon}),$$

$$\lambda^{(\phi)}(k) = \tfrac{1}{2}(-1)^{n+\frac{1}{2}k} A_{2k}(n - \tfrac{1}{2}k) \qquad (\text{by } (4.6.7))$$

$$A^{(\psi)}(n) = \sum_{\substack{k=1 \\ (k,2)=1}}^{[n^{\frac{1}{2}}]} \frac{\lambda^{(\psi)}(k)\exp\{\pi(n - 1/24)^{\frac{1}{2}}/k6^{\frac{1}{2}}\}}{2k^{\frac{1}{2}}(n - 1/24)^{\frac{1}{4}}} + O(n^{\epsilon}),$$

$$\lambda^{(\psi)}(k) = \tfrac{1}{2}(-1)^{n+\frac{1}{2}(k-1)} A_{2k}(n) \qquad (\text{by } (4.6.8))$$

$$A^{(\nu)}(n) = \sum_{\substack{k=1 \\ (k,2)=1}}^{[n^{\frac{1}{2}}]} \frac{\lambda^{(\nu)}(k)\exp\{\pi(n - 1/3)^{\frac{1}{2}}/k6^{\frac{1}{2}}\}}{(2k)^{\frac{1}{2}}(n + 1/3)^{\frac{1}{4}}} + O(n^{\epsilon}),$$

$$\lambda^{(\nu)}(k) = \tfrac{1}{2}(-1)^{n+\frac{1}{2}(k-1)} A_k\{(n-1)[\tfrac{1}{2}(k+1)]^2 - 3(k^2-1)/8\} \qquad (\text{by } 4.6.9))$$

$$A^{(\omega)}(n) = \sum_{\substack{k=1 \\ (k,2)=1}}^{[n^{\frac{1}{2}}]} \frac{\lambda^{(\omega)}(k)\exp\{\pi(n+2/3)^{\frac{1}{2}}/k3^{\frac{1}{2}}\}}{k^{\frac{1}{2}}(n+2/3)^{\frac{1}{4}}} + O(n^{\epsilon}),$$

$$\lambda^{(\omega)}(k) = \tfrac{1}{2}(-1)^{\frac{1}{2}(k-1)} A_k\{n\tfrac{1}{2}(k+1) - 3(k^2-1)/8\} \quad \text{(by 4.6.10)).}$$

THE PENNSYLVANIA STATE UNIVERSITY,
UNIVERSITY PARK, PENNSYLVANIA.

REFERENCES.

[1] R. Bellman, *A brief introduction to theta functions*, Holt, Rinehart and Winston, New York, 1961.

[2] L. A. Dragonette, "Some asymptotic formulae for the mock theta series of Ramanujan," *Transactions of the American Mathematical Society*, vol. 72 (1952), pp. 474-500.

[3] T. Estermann, "Vereinfachter Beweis eines Satzes von Kloosterman," *Abhandlungen Hamburg Mathematisches Seminar*, vol. 7 (1929), pp. 82-98.

[4] G. H. Hardy and E. M. Wright, *An introduction to the theory of numbers*, 4th ed., Oxford, Oxford University Press, 1960, 421 pp.

[5] K. G. J. Jacobi, *Fundamenta nova theoriae functionum ellipticarum*, Köningsberg, 1829.

[6] D. H. Lehmer, "The series for the partition function," *Transactions of the American Mathematical Society*, vol. 43 (1938), pp. 271-295.

[7] J. Lehner, "A partition function connected with the modulus five," *Duke Journal of Mathematics*, vol. 8 (1941), pp. 631-655.

[8] H. Rademacher, "The Fourier coefficients of the modular invariant $J(\tau)$," *American Journal of Mathematics*, vol. 60 (1938), 501-512.

[9] S. Ramanujan, *Collected papers of Srinivasa Ramanujan*, Cambridge, Cambridge University Press, 1927, 355 pp.

[10] H. Salié, "Uber die Kloostermanschen Summen $S(u,v;q)$ (" *Mathematik Zeitschrift*, vol. 34 (1931), pp. 91-109.

[11] J. Tannery and J. Molk, *Elements de la Theorie des Fonctions Elliptiques*, Paris, 1896, vol. 2, 299 pp.

[12] G. N. Watson, "The final problem: an account of the mock theta functions," *Journal of the London Mathematical Society*, vol. 11 (1936), pp. 55-80.

[13] M. I. Knopp and J. R. Smart, "On Kloosterman sums connected with modular forms of half-integral dimension," *Illinois Journal of Mathematics*, vol. 8 (1964), pp. 480-487.

2.1.2 Erratum

p. 490, formula for $\lambda^{(\omega)}(n)$: As pointed out by Garthwaite [Gar08], the initial factor on the right-hand side should be $\frac{1}{4}$ rather than $\frac{1}{2}$.

2.2 Heine's Method

I have restricted this section to two short papers [4] and [55]. Paper [4] is the first one I published using Heine's method. This short proof of the Jacobi Triple product identity was independently discovered at the same time by Menon [Men65]. I wrote this paper in response to a challenge posed by Richard Bellman in his charming book: [Bel61] *Introduction to Theta Functions*, and he wrote me a gracious congratulatory note after [4] appeared.

Put succinctly, Heine's method consists of multiple well-chosen applications of the q-binomial series. I first encountered it in [Wat37] when Fine asked me to report on it in his 1963 class on basic hypergeometric functions. Papers [5], [6], [7], [32], [35], [45], [46], [81], and [98] fit more or less directly in this category. It should be noted that while I am grateful to Fine for making me report on [Wat37], this was, nonetheless, responsible for me writing [5], [6], [7] (and most unfortunately [23]) in Watson's idiosyncratic notation (see §2.4.1.1).

Paper [45] is, on the surface, a surprise in that it reduces the first q-Appell function to a ${}_3\phi_2$. This was probably the paper that engendered a correspondence between Richard Askey and me.

Paper [52] was written during my first sabbatical at MIT, and at Gian-Carlo Rota's invitation, included in his celebrated "On the Foundations of Combinatorial Theory" series. The operator methods used are at least related to Heine through the observations of Rogers [Rog1893]. Paper [55] (also reprinted in this section) has turned out to be useful in other contexts: [268], [Cor-L09], [McL-S10], [Sav-S10].

Papers [77], [87], [88], [171], and [216], are, at least, tangentially related to the theme of this section. Paper [87] inspired Richard Askey to write [Ask80], which set off a variety of further research by Habsieger [Hab88], Kadell [Kad88], Macdonald [Macd82], and others.

Paper [76] (and the related [103] and [136]) dealt with problems arising in group theory starting with the Lusztig–Macdonald–Wall conjecture (*cf.* [Lus77]). While [76] fits well in this section, I include [103] and [136] because of the related topic, even though the methods are much more in line with those in [50] and [123].

2.2.1 *A simple proof of Jacobi's triple product identity*

A SIMPLE PROOF OF JACOBI'S TRIPLE PRODUCT IDENTITY

GEORGE E. ANDREWS

Bellman remarks in [1, p. 42] that there are no simple proofs known of the complete triple product identity

$$\prod_{n=0}^{\infty}\{(1-x^{2n+2})(1+x^{2n+1}z)(1+x^{2n+1}z^{-1})\} = \sum_{n=-\infty}^{\infty} x^{n} z^{n}$$

with $z\neq 0$ and $|x|<1$.

However the two identities of Euler,

$$\text{(E1)}\qquad \prod_{n=0}^{\infty}(1+x^n z) = \sum_{n=0}^{\infty}\frac{x^{n(n-1)/2}z^n}{(1-x)\cdots(1-x^n)}, \qquad |x|<1,$$

and

$$\text{(E2)}\qquad \prod_{n=0}^{\infty}(1+x^n z)^{-1} = \sum_{n=0}^{\infty}\frac{(-1)^n z^n}{(1-x)\cdots(1-x^n)}, \quad |x|<1, |z|<1,$$

are rather easily established [1, p. 49]. It does not seem to have been noticed that Jacobi's triple product identity follows simply from Euler's identities.

$$\prod_{n=0}^{\infty}(1+x^{2n+1}z) = \sum_{n=0}^{\infty}\frac{x^{n}z^{n}}{(1-x^2)\cdots(1-x^{2n})} \qquad \text{(by (E1))}$$

$$= \sum_{n=0}^{\infty}\frac{x^{n^2}z^n \prod_{j=0}^{\infty}(1-x^{2n+2+2j})}{\prod_{j=0}^{\infty}(1-x^{2j+2})}$$

$$= \frac{1}{\prod_{j=0}^{\infty}(1-x^{2j+2})}\sum_{n=-\infty}^{\infty} x^{n^2}z^n \prod_{j=0}^{\infty}(1-x^{2n+2+2j})$$

(all terms of the sum with negative n are zero)

Received by the editors September 2, 1964.

$$= \frac{1}{\prod_{j=0}^{\infty}(1-x^{2j+2})} \sum_{n=-\infty}^{\infty} x^{n^2} z^n \sum_{m=0}^{\infty} \frac{(-1)^m x^{m^2+m+2nm}}{(1-x^2)\cdots(1-x^{2m})} \quad \text{(by (E1))}$$

$$= \frac{1}{\prod_{j=0}^{\infty}(1-x^{2j+2})} \sum_{m=0}^{\infty} \frac{(-1)^m (xz^{-1})^m}{(1-x^2)\cdots(1-x^{2m})} \sum_{n=-\infty}^{\infty} x^{(n+m)^2} z^{n+m}$$

$$= \frac{1}{\prod_{j=0}^{\infty}(1-x^{2j+2})} \prod_{j=0}^{\infty}(1+x^{2j+1}z^{-1})^{-1} \sum_{n=-\infty}^{\infty} x^{n^2} z^n$$

(using (E2) and replacing $n+m$ by n in the inner sum).

The above argument is valid provided $|x| < |z|$. The complete result for all nonzero z follows by analytic continuation.

Bibliography

1. R. Bellman, *A brief introduction to theta functions*, Holt, Rinehart and Winston, New York, 1961.

The Pennsylvania State University

2.2.2 On the q-analog of Kummer's theorem and applications

ON THE q-ANALOG OF KUMMER'S THEOREM AND APPLICATIONS

GEORGE E. ANDREWS

1. Introduction. The q-analogs for Gauss's summation of ${}_2F_1[a, b; c; 1]$ and Saalschutz's summation of ${}_3F_2[a, b, -n; c, a+b-c-n+1; 1]$ are well known, namely, E. Heine [8; p. 107, Equation (6)] showed that

$$ {}_2\phi_1\begin{bmatrix} a,\ b;\ q,\ c/ab \\ c \end{bmatrix} = \frac{(c/a)_\infty (c/b)_\infty}{(c)_\infty (c/ab)_\infty} \tag{1.1} $$

where

$$ {}_m\phi_n\begin{bmatrix} a_1, \cdots, a_m;\ q, z \\ b_1, \cdots, b_n \end{bmatrix} = \sum_{j=0}^{\infty} \frac{(a_1)_j \cdots (a_m)_j z^j}{(q)_j (b_1)_j \cdots (b_n)_j}, $$

and $(a)_n = (a;q)_n = (1-a)(1-aq)\cdots(1-aq^{n-1})$, $(a)_\infty = (a;q)_\infty = \lim_{n\to\infty}(a)_n$. (See also [12; p. 97, Equation (3.3.2.2)].) F. H. Jackson [9; p. 145] showed that

$$ {}_3\phi_2\begin{bmatrix} a,\ b,\ q^{-n};\ q,\ q \\ c,\ abq/cq^n \end{bmatrix} = \frac{(c/a)_n (c/b)_n}{(c)_n (c/ab)_n}. \tag{1.2} $$

The q-analog of Dixon's summation of ${}_3F_2[a, b, c; 1+a-b, 1+a-c; 1]$ was more difficult to find, and indeed only a partial analog is true; namely, W. N. Bailey [5] and F. H. Jackson [10; p. 167, Equation (2)] proved that if $a = q^{-2n}$ where n is a positive integer, then

$$ {}_3\phi_2\begin{bmatrix} a,\ b,\ c;\ q,\ \frac{q^2 a^{\frac12}}{bc} \\ \frac{aq}{b},\ \frac{aq}{c} \end{bmatrix} = \frac{(b/a)_\infty (c/a)_\infty (qa^{\frac12})_\infty (bca^{-\frac12})_\infty}{(ba^{-\frac12})_\infty (ca^{-\frac12})_\infty (a^{-1}qa)_\infty (bca^{-1})_\infty}. \tag{1.3} $$

There are three other well-known summations for the ${}_2F_1$ series, namely, Kummer's theorem [12; p. 243, Equation (III. 5)]

$$ {}_2F_1[a, b; 1+a-b; -1] = \frac{\Gamma(1+a-b)\Gamma\left(1+\frac{a}{2}\right)}{\Gamma\left(\frac12+\frac{a}{2}\right)\Gamma\left(1+\frac{a}{2}-b\right)}, \tag{1.4} $$

Gauss's second theorem [12; p. 243, Equation III. 6)]

Received December 23, 1972. The author was partially supported by National Science Foundation Grant GP-23774.

$$ {}_2F_1\left[a, b; \frac{1}{2}+\frac{a}{2}+\frac{b}{2}; \frac{1}{2}\right] = \frac{\Gamma\left(\frac{1}{2}\right)\Gamma\left(\frac{1}{2}+\frac{a}{2}+\frac{b}{2}\right)}{\Gamma\left(\frac{1}{2}+\frac{a}{2}\right)\Gamma\left(\frac{1}{2}+\frac{b}{2}\right)}, \tag{1.5} $$

and Bailey's theorem [12; p. 243, Equation (III. 7)]

$$ {}_2F_1\left[a, 1-a; c; \frac{1}{2}\right] = \frac{\Gamma\left(\frac{c}{2}\right)\Gamma\left(\frac{1}{2}+\frac{c}{2}\right)}{\Gamma\left(\frac{c}{2}+\frac{a}{2}\right)\Gamma\left(\frac{1}{2}+\frac{c}{2}-\frac{a}{2}\right)}. \tag{1.6} $$

Of these three, only a q-analog of Kummer's theorem is known, namely, [6; p. 711] (see also [5; p. 173])

$$ {}_2\phi_1\left[\begin{matrix} a, b; q, -\frac{q}{b} \\ \\ \frac{qa}{b} \end{matrix}\right] = \frac{(aq;q^2)_\infty(-q)_\infty(q^2a/b^2;q^2)_\infty}{(qa/b)_\infty(-q/b)_\infty}. \tag{1.7} $$

The only known proof of (1.7) consists of a specialization of parameters in Jackson's summation of the well-poised ${}_6\phi_5$.

Our object here is to provide a very simple proof of (1.7) and to show that the following q-analogs of Gauss's second theorem and Bailey's theorem hold:

$$ \sum_{n=0}^{\infty} \frac{(a)_n(b)_n q^{\frac{1}{2}n(n+1)}}{(q)_n(qab;q^2)_n} = \frac{(-q)_\infty(aq;q^2)_\infty(bq;q^2)_\infty}{(qab;q^2)_\infty}, \tag{1.8} $$

$$ \sum_{n=0}^{\infty} \frac{(b)_n(q/b)_n c^n q^{\frac{1}{2}n(n-1)}}{(q^2;q^2)_n(c)_n} = \frac{(qc/b;q^2)_\infty(bc;q^2)_\infty}{(c)_\infty}. \tag{1.9} $$

2. The q-analog of Kummer's theorem. Here we utilize the summation technique that was successfully employed in [1], [2], [3] and [4]. We shall need the elementary summation [12; p. 92, Equation (3.2.2.11)]

$$ \sum_{n=0}^{\infty} \frac{(A)_n T^n}{(q)_n} \equiv {}_1\phi_0[A; q, T] = (AT)_\infty/(T)_\infty. \tag{2.1} $$

Therefore

$$ \begin{aligned} {}_2\phi_1\left[\begin{matrix} a, b; q, -q/b \\ qa/b \end{matrix}\right] &= \sum_{n=0}^{\infty} \frac{(a)_n(b)_n(-q/b)^n}{(q)_n(aq/b)_n} \\ &= \frac{(a)_\infty}{(aq/b)_\infty} \sum_{n=0}^{\infty} \frac{(b)_n(-q/b)^n(aq^{n+1}/b)_\infty}{(q)_n(aq^n)_\infty} \\ &= \frac{(a)_\infty}{(aq/b)_\infty} \sum_{n=0}^{\infty} \frac{(b)_n(-q/b)^n}{(q)_n} \sum_{m=0}^{\infty} \frac{(q/b)_m a^m q^{nm}}{(q)_m} \end{aligned} $$

$$= \frac{(a)_\infty}{(aq/b)_\infty} \sum_{m=0}^{\infty} \frac{(q/b)_m a^m (-q^{m+1})_\infty}{(q)_m (-q^{m+1}/b)_\infty}$$

$$= \frac{(a)_\infty (-q)_\infty}{(aq/b)_\infty (-q/b)_\infty} \sum_{m=0}^{\infty} \frac{(q^2/b^2;\, q^2)_m a^m}{(q^2;\, q^2)_m}$$

$$= \frac{(a)_\infty (-q)_\infty (aq^2/b^2;\, q^2)_\infty}{(aq/b)_\infty (-q/b)(a;\, q^2)_\infty}$$

$$= \frac{(aq;\, q^2)_\infty (-q)_\infty (aq^2/b^2;\, q^2)_\infty}{(aq/b)_\infty (-q/b)_\infty},$$

and so (1.7) is established. Technically what we have done is to set $\alpha = b$, $\beta = a$, $\gamma = qa/b$, and $\tau = -q/b$ in [3; Equation (I 1)]; we then have observed that the resulting ${}_2\phi_1$ with base q reduces to a ${}_1\phi_0$ with base q^2.

3. The q-analogs of Gauss's second theorem and Bailey's theorem. As is well-known [12; p. 32], both (1.5) and (1.6) may be deduced by the application of Kummer's theorem to the following identity:

$$(3.1) \qquad (1 - z)^{-a} {}_2F_1\left[\begin{matrix} a,\, b;\, -z/(1-z) \\ c \end{matrix}\right] = {}_2F_1\left[\begin{matrix} a,\, c-b;\, z \\ c \end{matrix}\right].$$

The following lemma is the q-analog of (3.1), and from it we may deduce (1.8) and (1.9) utilizing the q-analog of Kummer's theorem. Actually this lemma was given by F. H. Jackson [9; p. 145, Equation (4)]; his proof entails the development of a q-analog of Euler's transformation of power series. We include a short proof that shows this result to be a limiting case of an identity of N. Hall [7].

LEMMA. $$\sum_{n=0}^{\infty} \frac{(\beta)_n (\alpha)_n (-1)^n q^{\binom{n}{2}} (x\gamma/\beta)^n}{(q)_n (\gamma)_n (x\alpha)_n} = \frac{(x)_\infty}{(x\alpha)_\infty} \sum_{n=0}^{\infty} \frac{(\gamma/\beta)_n (\alpha)_n x^n}{(q)_n (\gamma)_n}.$$

Proof. N. Hall [7] (see also [11; p. 174, Equation (10.1)]) has proved the result

$${}_3\phi_2\left[\begin{matrix} a_1\,,\, a_2\,,\, a_3\,;\, q,\, b_1 b_2/a_1 a_2 a_3 \\ b_1\,,\, b_2 \end{matrix}\right]$$

$$= \frac{(b_2/a_3)_\infty (b_1 b_2/a_1 a_2)_\infty}{(b_2)_\infty (b_1 b_2/a_1 a_2 a_3)_\infty} {}_3\phi_2\left[\begin{matrix} b_1/a_1\,,\, b_1/a_2\,,\, a_3\,;\, q,\, b_2/a_3 \\ b_1\,,\, b_1 b_2/a_1 a_2 \end{matrix}\right].$$

Our lemma follows directly by the substitutions $a_1 = \beta$, $a_3 = \alpha$, $b_1 = \gamma$, $b_2 = x\alpha$ if we then let $a_2 \to \infty$.

To obtain (1.8), we set $\alpha = a$, $\beta = b$, $\gamma = q^{\frac12} a^{\frac12} b^{\frac12}$ and $x = -q^{\frac12} b^{\frac12} a^{-\frac12}$ in the lemma. Hence

$$\sum_{n=0}^{\infty} \frac{(a)_n(b)_n q^{\frac{1}{2}n(n+1)}}{(q)_n(qab;\, q^2)_n} = \frac{(-q^{\frac{1}{2}}b^{\frac{1}{2}}a^{-\frac{1}{2}})_\infty}{(-q^{\frac{1}{2}}b^{\frac{1}{2}}a^{\frac{1}{2}})_\infty}\, {}_2\phi_1\left[\begin{matrix} a,\ q^{\frac{1}{2}}a^{\frac{1}{2}}b^{-\frac{1}{2}};\ q,\ -q^{\frac{1}{2}}b^{\frac{1}{2}}a^{-\frac{1}{2}} \\ q^{\frac{1}{2}}a^{\frac{1}{2}}b^{\frac{1}{2}} \end{matrix}\right]$$

$$= \frac{(-q^{\frac{1}{2}}b^{\frac{1}{2}}a^{-\frac{1}{2}})_\infty}{(-q^{\frac{1}{2}}b^{\frac{1}{2}}a^{\frac{1}{2}})_\infty} \frac{(aq;\, q^2)_\infty(-q)_\infty(bq;\, q^2)_\infty}{(q^{\frac{1}{2}}b^{\frac{1}{2}}a^{\frac{1}{2}})_\infty(-q^{\frac{1}{2}}b^{\frac{1}{2}}a^{-\frac{1}{2}})_\infty} \qquad \text{(by (1.7))}$$

$$= \frac{(-q)_\infty(aq;\, q^2)_\infty(bq;\, q^2)_\infty}{(abq;\, q^2)_\infty}$$

which is (1.8).

To obtain (1.9), we set $\alpha = q/b$, $\beta = b$, $\gamma = c$ and $x = -b$ in the lemma. Hence

$$\sum_{n=0}^{\infty} \frac{(b)_n(q/b)_n c^n q^{\frac{1}{2}n(n-1)}}{(q)_n(c)_n(-q)_n} = \sum_{n=0}^{\infty} \frac{(b)_n(q/b)_n c^n q^{\frac{1}{2}n(n-1)}}{(q^2;\, q^2)_n(c)_n}$$

$$= \frac{(-b)_\infty}{(-q)_\infty}\, {}_2\phi_1\left[\begin{matrix} c/b,\ q/b;\ q,\ -b \\ c \end{matrix}\right]$$

$$= \frac{(-b)_\infty(qc/b;\, q^2)_\infty(-q)_\infty(bc;\, q^2)_\infty}{(-q)_\infty(c)_\infty(-b)_\infty}$$

$$= \frac{(qc/b;\, q^2)_\infty(bc;\, q^2)_\infty}{(c)_\infty}.$$

References

1. George E. Andrews, *On basic hypergeometric series, mock theta functions, and partitions, I*, Quart. J. Math., vol. 17(1966), pp. 64–80.
2. ———, *On basic hypergeometric series, mock theta functions, and partitions, II*, Quart. J. Math., vol. 17(1966), pp. 132–143.
3. ———, *q-identities of Auluck, Carlitz, and Rogers*, Duke Math. J., vol. 33(1966), pp. 575–581.
4. ———, *Summations and transformations for basic Appell series*, J. London Math. Soc., vol. 4(1972), pp. 618–622.
5. W. N. Bailey, *A note on certain q-identitities*, Quart. J. Math., vol. 12(1941), pp. 173–175.
6. J. A. Daum, *The basic analogue of Kummer's theorem*, Bull. Amer. Math. Soc., vol. 48(1942), pp. 711–713.
7. Newman A. Hall, *An algebraic identity*, J. London Math. Soc., vol. 11(1936), p. 276.
8. Eduard Heine, *Theorie der Kugelfunctionen*, Berlin, Reimer, 1878.
9. F. H. Jackson, *Transformations of q-series*, Mess. Math., vol. 39(1910), pp. 145–153.
10. ———, *Certain q-identities*, Quart. J. Math., vol. 12(1941), pp. 167–172.
11. D. B. Sears, *On the transformation theory of basic hypergeometric functions*, Proc. London Math. Soc., vol. 53(1951), pp. 158–180,
12. Lucy J. Slater, *Generalized Hypergeometric Functions*, Cambridge, Cambridge Univ. Press, 1966.

Department of Mathematics, Pennsylvania State University, University Park, Pennsylvania 16802

2.3 The Askey Influence

I spent the 1975–1976 academic year at the University of Wisconsin at the invitation of Richard Askey. We devoted much of a seminar that year to the work of Hahn [Hah49] on q-analogs of orthogonal polynomials. Askey [Ask01] published a paper on my year in Madison and our collaboration in "The Work of George Andrews: A Madison Perspective" in the *Andrews Festschrift*. Both that year, and the interactions leading up to it, greatly broadened the methods I developed in studying q-hypergeometric series.

Here, I have chosen to include just one paper [125] to provide a sample of this variety. There were a number of papers with Askey [73], [88], [98], [125]; each in some way revealing the interaction of orthogonal polynomials (Askey) and q-series methods (Andrews). Other papers have a direct connection to Askey, for example, [87], [94], [117], [226]. Paper [140] grew from my desire to highlight Rodney Baxter's proof of the Rogers–Ramanujan identities. It should be noted that only three people independently discovered proofs of the Rogers–Ramanujan identities: Rogers [Rog1893], Schur [Schu17], and Baxter.

2.3.1 Classical orthogonal polynomials (with R. Askey)

CLASSICAL ORTHOGONAL POLYNOMIALS

George E. Andrews[(1)] and Richard Askey[(2)]

Abstract.

There have been a number of definitions of the classical orthogonal polynomials, but each definition has left out some important orthogonal polynomials which have enough nice properties to justify including them in the category of classical orthogonal polynomials. We summarize some of the previous work on classical orthogonal polynomials, state our definition, and give a few new orthogonality relations for some of the classical orthogonal polynomials.

1. Introduction.

The classical orthogonal polynomials are sometimes described as those of Hermite, Laguerre and Jacobi, and including as special cases of the latter the polynomials of Legendre and Tchebychef. This is too narrow a definition, and when widened there are many other interesting polynomials contained in this class, but not so many that we cannot find all of them. Actually there are a number of different places to put the boundary of the classical polynomials. We think we have the right class, but others in the past have also felt that and their definitions were narrower than the one we will give here.

Jacobi polynomials, $P_n^{(\alpha,\beta)}(x)$, the most general of the three sets of polynomials mentioned above, can be defined by

$$(1-x)^{\alpha}(1+x)^{\beta}P_n^{(\alpha,\beta)}(x) = \frac{(-1)^n}{2^n n!}\frac{d^n}{dx^n}[(1-x)^{n+\alpha}(1+x)^{n+\beta}] \tag{1.1}$$

A formula of this type is called a Rodrigues formula. When $\alpha,\beta > -1$ they are orthogonal with respect to the beta distribution on $[-1,1]$.

$$\int_{-1}^{1} P_n^{(\alpha,\beta)}(x)P_m^{(\alpha,\beta)}(x)(1-x)^{\alpha}(1+x)^{\beta}\,dx = 0, \qquad m \neq n$$

$$= \frac{2^{\alpha+\beta+1}\Gamma(n+\alpha+1)\Gamma(n+\beta+1)}{(2n+\alpha+\beta+1)\Gamma(n+1)\Gamma(n+\alpha+\beta+1)} = h_n^{\alpha,\beta} \qquad m = n \tag{1.2}$$

(1) Supported in part by NSF grant MCS-8201733
(2) Supported in part by NSF grant DMS-840071.

All formulas for orthogonal polynomials which are given without a reference can be found in [26] or [52].

A set of polynomials $\{P_n(x)\}_{n=0}^{N}$, $N \le \infty$, is orthogonal if there is a positive measure $d\alpha(x)$ so that

$$\begin{aligned}\int_{-\infty}^{\infty} P_n(x)P_m(x)\,d\alpha(x) &= 0, \qquad m \ne n \le N\\ &= h_n > 0, \qquad m = n \le N.\end{aligned}$$

From [1.1] it is possible to obtain a number of explicit series representations for Jacobi polynomials. The standard one which will be extended below is

$$P_n^{(\alpha,\beta)}(x) = \frac{(\alpha+1)_n}{n!}\, {}_2F_1\left(\begin{matrix}-n, n+\alpha+\beta+1\\ \alpha+1\end{matrix}; \frac{1-x}{2}\right) \qquad [1.3]$$

The series on the right hand side of [1.3] is a hypergeometric series. Hypergeometric series are series $\Sigma\, c_n$ with c_{n+1}/c_n a rational function of n. They are usually written as

$${}_pF_q\left(\begin{matrix}a_1,\dots,a_p\\ b_1,\dots,b_q\end{matrix}; t\right) = \sum_{k=0}^{\infty} \frac{(a_1)_k \cdots (a_p)_k t^k}{(b_1)_k \cdots (b_q)_k k!} \qquad [1.4]$$

where the shifted factorial $(a)_k$ is defined by

$$(a)_k = \Gamma(k+a)/\Gamma(a) = a(a+1)\cdots(a+k-1). \qquad [1.5]$$

Since

$$\frac{d}{dt}\, {}_pF_q\left(\begin{matrix}a_1,\dots,a_p\\ b_1,\dots,b_q\end{matrix}; t\right) = \frac{a_1\cdots a_p}{b_1\cdots b_q}\, {}_pF_q\left(\begin{matrix}a_1+1,\dots,a_p+1\\ b_1+1,\dots,b_q+1\end{matrix}; t\right)$$

it is easy to see that

$$\frac{d}{dx} P_n^{(\alpha,\beta)}(x) = \frac{(n+\alpha+\beta+1)}{2} P_{n-1}^{(\alpha+1,\beta+1)}(x). \qquad [1.6]$$

When [1.1] is differentiated the result can be written as

$$\frac{d}{dx}(1-x)^{\alpha}(1+x)^{\beta}P_n^{(\alpha,\beta)}(x) = -2(n+1)(1-x)^{\alpha-1}(1+x)^{\beta-1}P_{n+1}^{(\alpha-1,\beta-1)}(x). \qquad [1.7]$$

Formulas [1.6] and [1.7] can be combined to give

$$(1-x^2)y'' + [\beta-\alpha-(\alpha+\beta+2)x]y' + n(n+\alpha+\beta+1)y = 0 \qquad [1.8]$$

when $y = P_n^{(\alpha,\beta)}(x)$.

There are formulas like [1.1], [1.6] and [1.8] for Laguerre and Hermite polynomials. Laguerre polynomials, $L_n^{\alpha}(x)$, are defined by

$$L_n^\alpha(x) = \frac{(\alpha+1)_n}{n!}\ {}_1F_1\left(\begin{matrix}-n\\ \alpha+1\end{matrix};\ x\right) \qquad [1.9]$$

or

$$x^\alpha e^{-x} L_n^\alpha(x) = \frac{1}{n!}\frac{d^n}{dx^n}[x^{n+\alpha}e^{-x}]. \qquad [1.10]$$

When $\alpha > -1$ they are orthogonal. This orthogonality is

$$\begin{aligned}\int_0^\infty L_n^\alpha(x)L_m^\alpha(x)x^\alpha e^{-x}\,dx &= 0, && m \neq n,\\ &= \frac{\Gamma(n+\alpha+1)}{n!} && m = n.\end{aligned} \qquad [1.11]$$

Hermite polynomials, $H_n(x)$, can be defined by

$$e^{-x^2}H_n(x) = (-1)^n \frac{d^n}{dx^n} e^{-x^2}. \qquad [1.12]$$

Their hypergeometric representation is a bit more complicated.

$$H_n(x) = (2x)^n\ {}_2F_0\left(\begin{matrix}-n/2, (1-n)/2\\ -\end{matrix};\ \frac{-1}{x^2}\right) \qquad [1.13]$$

and the orthogonality is

$$\begin{aligned}\int_{-\infty}^\infty H_n(x)H_m(x)e^{-x^2}\,dx &= 0, && m \neq n\\ &= 2^n n!\sqrt{\pi}, && m = n.\end{aligned} \qquad [1.14]$$

Sonine [47] considered the problem of finding all sets of orthogonal polynomials $\{P_n(x)\}_0^\infty$ whose derivatives $\{P'_{n+1}(x)\}_{n=0}^\infty$ are also orthogonal. From [1.6], Jacobi polynomials have this property. Laguerre and Hermite polynomials also do, since

$$\frac{d}{dx} L_n^\alpha(x) = -L_{n-1}^{\alpha+1}(x) \qquad [1.15]$$

and

$$\frac{d}{dx} H_n(x) = 2nH_{n-1}(x) \qquad [1.16]$$

Sonine [47] showed that up to a linear change of variables these are the only polynomials that have this property. Hahn [29] rediscovered this theorem. This was the first of a number of theorems that said that a set of orthogonal polynomials that satisfies a certain property is either the Jacobi, Laguerre or Hermite polynomials. The next theorem was found by Bochner [21]. He showed that if

$$A(x)y'' + B(x)y' + \lambda_n y = 0, \qquad n = 0,1,\ldots, \qquad [1.17]$$

where $A(x)$ and $B(x)$ are independent of n, λ_n is independent of x, and $y = p_n(x)$ is a polynomial of degree n which form a set of orthogonal polynomials, then $\{p_n(x)\}$ are either Jacobi polynomials, Laguerre polynomials or Hermite polynomials. By this we mean as before that there is a linear change of variables that gives one of these polynomial sets as defined above.

A third theorem was found by Tricomi [58]. Let a set of orthogonal polynomials $\{p_n(x)\}_0^\infty$ be given by

$$p_n(x) = \frac{K_n}{w(x)} \frac{d^n}{dx^n} [w(x) [T(x)]^n] \tag{1.18}$$

where K_n is a constant, $T(x)$ is a polynomial in x whose coefficients are independent of n, and $w(x)$ is independent of n. Then $\{p_n(x)\}$ are either Jacobi, Laguerre or Hermite polynomials.

All of these theorems have often been interpreted to say that the only orthogonal polynomials that have many nice properties are Jacobi, Laguerre and Hermite polynomials. Fortunately before any of the above theorems were found, Tchebychef discovered another very important set of orthogonal polynomials that has nice extensions of [1.1], [1.6] and [1.8].

2. Further classical hypergeometric orthogonal polynomials.

Tchebychef [53] introduced polynomials orthogonal with respect to the uniform distribution on an equally spaced set of points. Rather than use his notation we will use the notation which is now standard. In a later paper [54], he gave a Rodrigues type formula, using a finite difference operator instead of a derivative. From this it is easy to find an explicit representation as a hypergeometric polynomial. Still later [55] he found a more general set of orthogonal polynomials. These polynomials are called Hahn polynomials and are now defined by

$$Q_n(x;\ \alpha,\beta,N) = {}_3F_2\left(\begin{matrix}-n,n+\alpha+\beta+1,-x\\ \alpha+1,-N\end{matrix};\ 1\right) \tag{2.1}$$

Their orthogonality relation is

$$\sum_{x=0}^{N} Q_n(x)Q_m(x)\frac{(\alpha+1)_x(\beta+1)_{N-x}}{x!(N-x)!} = 0, \qquad m \neq n \leq N \tag{2.2}$$

If

$$\Delta f(x) := f(x+1) - f(x) \tag{2.3}$$

then

40

$$\Delta Q_n(x;\alpha,\beta,N) = -\frac{n(n+\alpha+\beta+1)}{(\alpha+1)N} Q_{n-1}(x;\ \alpha+1,\beta+1,N-1). \quad [2.4]$$

The Rodrigues type formula is

$$\binom{x+\alpha}{x}\binom{N-x+\beta}{N-x}\binom{N}{n} Q_n(x;\ \alpha,\beta,N) = \binom{n+\beta}{n}\Delta^n\left[\binom{x+\alpha}{\alpha+n}\binom{N-x+\beta+n}{\beta+n}\right] \quad [2.5]$$

The Hahn polynomials contain Jacobi polynomials as limits.

$$\lim_{N\to\infty} Q_n(xN;\alpha,\beta,N) = \frac{P_n^{(\alpha,\beta)}(1-2x)}{P_n^{(\alpha,\beta)}(1)} \quad [2.6]$$

They also contain three other important sets of orthogonal polynomials as limits

$$\lim_{\alpha\to\infty} Q_n(x;\ \alpha,\alpha(1-p)/p,N) = {}_2F_1(-n,-x;\ -N;\ p^{-1}) =: K_n(x;\ p,N) \quad [2.7]$$

$$\lim_{N\to\infty} Q_n(x;\ \beta-1,(1-c)N/c,N) = {}_2F_1(-n,-x;\ \beta;\ 1-c^{-1}) =: M_n(x;\ \beta,c) \quad [2.8]$$

$$\lim_{N\to\infty} Q_n(x;\ N-1,\ N^2a^{-1},N) = {}_2F_0(-n,-x;\ -\,;\ -a^{-1}) =: C_n(x;\ a) \quad [2.9]$$

These polynomials are called Krawtchouk, Meixner and Charlier polynomials respectively and their orthogonality relations are

$$\sum_{x=0}^{N} K_n(x;\ p,N)K_m(x;\ p,N)\binom{N}{x}p^x(1-p)^{N-x} = 0, \quad m\neq n\leq N,\ 0<p<1, \quad [2.10]$$

$$\sum_{x=0}^{\infty} M_n(x;\ \beta,c)M_m(x;\ \beta,c)\frac{(\beta)_x}{x!}c^x = 0, \quad m\neq n,\ 0<c<1,\ \beta>0, \quad [2.11]$$

$$\sum_{x=0}^{\infty} C_n(x;\ a)C_m(x;\ a)a^x/x! = 0, \quad m\neq n,\ a>0. \quad [2.12]$$

There is one further set of orthogonal polynomials of this type. These polynomials were found by Meixner [38] when he solved the following problem.

Given two power series

$$A(z) = \sum_{n=0}^{\infty} a_n z^n, \qquad a_0 \neq 0,$$

$$B(z) = \sum_{n=1}^{\infty} b_n z^n, \qquad b_1 \neq 0,$$

find all sets of orthogonal polynomials $\{P_n(x)\}_{n=0}^{N}$ that come from the generating function

$$A(z)e^{xB(z)} = \sum_{n=0}^{N} P_n(x)z^n, \qquad N \le \infty. \qquad [2.13]$$

Laguerre polynomials come from

$$(1-z)^{-\alpha-1}\exp(-xz/(1-z)) = \sum_{n=0}^{\infty} L_n^{\alpha}(x)z^n$$

Similarly Meixner, Krawtchouk, Charlier and Hermite polynomials have generating functions of the form [2.13]. Meixner found one further set of orthogonal polynomials, and stated the orthogonality relation. This set of polynomials was rediscovered and generalized by Pollaczek [40]. We call these Meixner-Pollaczek polynomials. They are given as

$$P_n^{(a)}(x;\varphi) = e^{in\varphi}\,{}_2F_1\left(\begin{matrix}-n, a+ix\\ 2a\end{matrix}; 1-e^{-2i\varphi}\right), \quad a>0, \quad 0<\varphi<\pi. \qquad [2.14]$$

Their orthogonality is

$$\int_{-\infty}^{\infty} P_n^{(a)}(x;\varphi)P_m^{(a)}(x;\varphi)e^{(2\varphi-\pi)x}|\Gamma(a+ix)|^2\,dx = 0, \quad m\neq n, \quad 0<\varphi<\pi, \qquad [2.15]$$

Some statisticians have called the weight function

$$w(x) = e^{(2\varphi-\pi)x}|\Gamma(a+ix)|^2$$

the Meixner hypergeometric distribution [35]. This is an inappropriate name. A better name is the continuous binomial distribution. Krawtchouk polynomials, Meixner polynomials and Meixner-Pollaczek polynomials are really the same polynomials, with different parameters, and a linear change of variables for the Meixner-Pollaczek polynomials. Krawtchouk polynomials are orthogonal with respect to the binomial distribution, Meixner polynomials are orthogonal with respect to the negative binomial distribution, so Meixner-Pollaczek polynomials should be orthogonal with respect to some type of binomial distribution. Since the distribution is absolutely continuous, the name "the continuous binomial distribution" is natural.

Toscano [57] found a Rodrigues type formula for the Meixner-Pollaczek polynomials using the operator

$$\delta f(x) = f(x+i/2) - f(x-i/2). \qquad [2.16]$$

Stieltjes [48] found another set of orthogonal polynomials from a continued fraction, and Carlitz extended them and gave an explicit orthogonality relation [22]. The polynomials are

$$p_n(x) = i^n \, {}_3F_2\left(\begin{matrix}-n, n+1, \gamma-ix \\ 1, 2\gamma\end{matrix}; 1\right), \quad 0 < \gamma < 1, \tag{2.17}$$

and the orthogonality is

$$\int_{-\infty}^{\infty} \frac{p_n(x)\,p_m(x)}{[\sin^2\pi\gamma \ \cosh^2\pi x + \cos^2\pi\gamma \ \sinh^2\pi x]}\,dx = 0, \quad m\neq n. \tag{2.18}$$

These polynomials were extended by Askey and Wilson [17] to

$$p_n(x) = i^n \, {}_3F_2\left(\begin{matrix}-n, n+2\alpha+2\gamma-1, \gamma-ix \\ \alpha+\gamma, 2\gamma\end{matrix}; 1\right) \tag{2.19}$$

and their orthogonality relation is

$$\int_{-\infty}^{\infty} p_n(x)\,p_m(x)\,|\Gamma(\alpha+ix)\Gamma(\gamma+ix)|^2\,dx = 0, \quad m \neq n \tag{2.20}$$

$\alpha, \gamma > 0$. These polynomials have a Rodrigues formula, and they are the end of the polynomials that have a Rodrigues formula using a difference operator. However this is not the end of the classical polynomials. Two different extensions will be given in the next two sections.

3. Basic hypergeometric classical orthogonal polynomials.

A basic hypergeometric series is a series Σc_n with c_{n+1}/c_n a rational function of q^n with a fixed q. Define

$$(a;q)_n = (1-a)(1-aq)\cdots(1-aq^{n-1}), \quad n = 1,2,\ldots,$$
$$= 1, \quad n = 0. \tag{3.1}$$

If $|q| < 1$ then

$$(a;q)_\infty = \prod_{k=0}^{\infty}(1-aq^k) \tag{3.2}$$

The basic hypergeometric series is best written as

$${}_{p+1}\varphi_{p+r}\left(\begin{matrix}a_0, \ldots a_p \\ b_1, \ldots, b_p\end{matrix}; q, t\right)$$
$$= \sum_{k=0}^{\infty} \frac{(a_1;q)_k \cdots (a_p;q)_k}{(b_1;q)_k \cdots (b_p;q)_k} \frac{(-1)^{kr} q^{rk(k-1)/2}\, t^k}{(q;q)_k} \tag{3.3}$$

Basic hypergeometric extensions of all of the polynomials in sections 1 and 2 exist. Sometimes there is more than one extension.

The first extension was found by Markov [37], and is an extension of the case $\alpha = \beta = 0$ of [2.1] that Tchebychef found in [53] and [54]. Rather than use Markov's notation we will use the one which is now standard.

$$Q_n(t;\ 1,1,N;\ q) = {}_3\phi_2\left(\begin{matrix} q^{-n},\ q^{n+1},\ t;\ q,q \\ q;\ q^{-N} \end{matrix}\right) \quad [3.4]$$

The orthogonality is

$$\sum_{x=0}^{N} Q_n(q^{-x})Q_m(q^{-x})q^{-x} = 0 \qquad m \neq n \leq N. \quad [3.5]$$

The recurrence relation is a special case of a more general one which will be given in the next section.

The next two examples were published in the same year. Rogers [42] considered the polynomials whose recurrence relation is

$$2xH_n(x|q) = H_{n+1}(x|q) + (1-q^n)H_{n-1}(x|q) \quad [3.6]$$

He used these polynomials to find the Rogers-Ramanujan identities

$$\frac{1}{(q;q^5)_\infty(q^4;q^5)_\infty} = \sum_{n=0}^{\infty}\frac{q^{n^2}}{(q;q)_n} \quad [3.7]$$

$$\frac{1}{(q^2;q^5)_\infty(q^3;q^5)_\infty} = \sum_{n=0}^{\infty}\frac{q^{n^2+n}}{(q;q)_n} \quad [3.8]$$

and a number of other interesting results, but he was not aware that these polynomials are orthogonal. The orthogonality when $-1 < q < 1$ was stated by Szegö [51]. It is

$$\int_{-1}^{1} H_n(x|q)H_m(x|q)\prod_{k=0}^{\infty}[1 - 2(2x^2-1)q^k + q^{2k}](1-x^2)^{-\frac{1}{2}}dx = 0,\ m\neq n \quad [3.9]$$

In the next year Rogers [43] considered an extension. These polynomials satisfy

$$\begin{aligned} 2x(1-\beta q^n)C_n(x;\ \beta|q) &= (1-q^{n+1})C_{n+1}(x;\ \beta|q) \\ &\quad + (1-\beta^2q^{n-1})C_{n-1}(x;\ \beta|q). \end{aligned} \quad [3.10]$$

Rogers knew these polynomials extend the symmetric Jacobi polynomials $P_n^{(\alpha,\alpha)}(x)$ and found a number of very important results for them. The orthogonality was found in the late 1970's by Askey and Wilson [18], and other proofs have been given by Askey and Ismail [11], [12]. When $-1 < q, \beta < 1$ the orthogonality is

$$\int_{-1}^{1} C_n(x;\ \beta|q)C_m(x;\ \beta|q) \prod_{k=0}^{\infty} \left[\frac{1-2(2x^2-1)q^k+q^{2k}}{1-2(2x^2-1)\beta q^k+\beta^2 q^{2k}}\right] \frac{dx}{\sqrt{1-x^2}} = 0$$

$$m \neq n. \qquad [3.11]$$

Also in 1894 Stieltjes [49] found a continued fraction that gives an interesting set of orthogonal polynomials. Wigert [59] generalized these polynomials (Stieltjes had them for one value of q, Wigert for general q), found an explicit representation as a series, and gave one orthogonality relation. These polynomials can be defined by

$$\begin{aligned} p_n(x) &= \sum_{k=0}^{n} \frac{(q^{-n};q)_k}{(q;q)_k} q^{k^2/2}(q^{n+1}x)^k \\ &= {}_1\varphi_1\left(\begin{matrix} q^{-n} \\ 0 \end{matrix};\ q, -q^{n+\frac{3}{2}}x\right) \end{aligned} \qquad [3.12]$$

and Wigert's orthogonality is

$$\int_0^{\infty} p_n(x)p_m(x)w(x)\,dx = 0, \qquad m \neq n,$$

$$w(x) = \exp(-k^2 \log^2 x), \qquad q = \exp(-(2k^2)^{-1}) \qquad [3.13]$$

The moment problem is indeterminate, so there are many different measures which give the orthogonality relation [3.13]. Two more are

$$w(x) = [(-q^{\frac{1}{2}}x;\ q)_{\infty}(-q^{\frac{1}{2}}x^{-1};\ q)_{\infty}]^{-1}$$

with respect to Lebesgue measure, and the same function with respect to the measure $d_q x$ defined by

$$\int_0^{\infty} f(x)\,d_q x = (1-q) \sum_{n=-\infty}^{\infty} f(q^n)q^n$$

See [8].

Geronimus [28] considered the following problem. Given real points $x_1, x_2, \ldots,$ form

$$w_k(x) = \prod_{j=1}^{k} (x-x_j)$$

Define a set of polynomials $\{p_n(x)\}$ by

$$p_n(x) = \sum_{k=0}^{n} a_{n-k}b_k w_k(x) \qquad [3.14]$$

with $b_n \neq 0$, $a_0 \neq 0$. He found necessary and sufficient conditions on the coefficients a_k and b_k for $\{p_n(x)\}$ to be a set of orthogonal polynomials. These conditions were not simple enough for him to find explicitly all the polynomials, but he found many examples. These include the Stieltjes-Wigert polynomials, a different q-extension of Hermite polynomials than that of Rogers, and a set of polynomials which could be called q-Charlier polynomials.

Probably the most important characterization theorem was found by Hahn [30]. It was so important because he found a number of new polynomials and facts about them. He considered a q-extension of the derivative as

$$D_q f(x) = \frac{f(qx) - f(x)}{(q-1)x} \qquad [3.15]$$

and looked for all sets of orthogonal polynomials that satisfy one of five conditions. The first is that a set of orthogonal polynomials $\{p_n(x)\}$ has associated to it another set of orthogonal polynomials $\{q_n(x)\}$ that satisfies

$$D_q p_n(x) = c_n q_{n-1}(x) \qquad [3.16]$$

for a sequence of coefficients c_n.

The second condition is that a set of orthogonal polynomials satisfies a second order q-difference equation:

$$(a_{11}x^2+a_{12}x+a_{13})D_q^2 y + (a_{21}x+a_{22})D_q y + a_3 y = 0 \qquad [3.17]$$

The third condition is that there is a Rodrigues type formula. This time it takes the form

$$\pi(x)p_n(x) = D_q^n[f_1(x)f_2(x)\dots f_n(x)\pi(x)], \qquad f_i(x) = f_{i+1}(qx) \qquad [3.18]$$

where $\pi(x)$ and $f_1(x)$ are independent of n.

Hahn introduced two other conditions which were stated correctly in the body of the paper but in too abbreviated a form in the introduction. The first is that the orthogonal polynomials have a series representation of the form

$$p_n(x) = \sum_{k=0}^{n} a_{n,k}\varphi_k(x) \qquad [3.19]$$

where $\varphi_k(x) = x^k$ or $\varphi_k(x) = (x;q)_k$ and the coefficients $a_{n,k}$ satisfy

$$h_1(n,k)a_{n,k} = h_2(n,k)a_{n,k-1}$$

with h_1 and h_2 polynomials in q^n and q^k.

The last condition Hahn introduced dealt with moments. If the orthogonality relation is

$$\int_{-\infty}^{\infty} p_n(x)\,p_m(x)\,d\alpha(x) = 0, \qquad m \neq n,$$

and the moments M_n are defined by

$$\int_{-\infty}^{\infty} \varphi_n(x)\,d\alpha(x) = M_n, \qquad \varphi_n(x) = x^n \quad \text{or} \quad \varphi_n(x) = (x;q)_n \qquad [3.20]$$

then

$$\frac{M_n}{M_{n-1}} = \frac{a+bq^n}{c+dq^n}, \qquad ad - bc \neq 0, \qquad n = 1,2,\ldots,N, \qquad N \leq \infty. \qquad [3.21]$$

All of these conditions lead to the same set of orthogonal polynomials. The most general one is

$$Q_n(x;a,b,N:q) = {}_3\varphi_2\left(\begin{matrix} q^{-n},\ q^{n+1}ab,x \\ aq,q^{-N} \end{matrix};\ q,q\right), \qquad n = 0,1,\ldots,N, \ [3.22]$$

or

$$p_n(x;a,b,c:q) = {}_3\varphi_2\left(\begin{matrix} q^{-n},\ q^{n+1}ab,x \\ aq,cq \end{matrix};\ q,q\right), \qquad n = 0,1,\ldots. \quad [3.23]$$

The orthogonality for [3.22] is

$$\sum_{x=0}^{N} Q_n(q^{-x})\,Q_m(q^{-x})\frac{(aq;q)_x(bq;q)_{N-x}}{(q;q)_x(q;q)_{N-x}}\,\frac{1}{(aq)^x} = 0, \qquad m \neq n \leq N, \qquad [3.24]$$

as can be seen easily by setting $d = 0$ in the more general orthogonality relation [4.14], which was proved in [16] and will be commented on in the next section. When $a = b = 1$ this reduces to the orthogonality [3.5] Markov found [37].

We worked out the orthogonality relation for [3.23] in 1976 but have not published it yet. This seems to be an appropriate place to give it. To state it in a way that is relatively easy to remember recall the q-integral which was introduced independently by Thomae [56] and Jackson [33].

$$\int_0^a f(x)\,d_qx = a(1-q)\sum_{n=0}^{\infty} f(aq^n)\,q^n \qquad [3.25]$$

and

$$\int_{-d}^{c} f(x)\,d_qx = \int_{0}^{c} f(x)\,d_qx - \int_{0}^{-d} f(x)\,d_qx \qquad [3.26]$$

A q-beta integral put on the interval $[-d,c]$ is

$$\int_{-d}^{c} \frac{(-qx/d;q)_\infty (qx/c;q)_\infty}{(-q^{\beta+1}x/d;q)_\infty (q^{\alpha+1}x/c;q)_\infty} d_qx$$

$$= \frac{cd(1-q)(q;q)_\infty (q^{\alpha+\beta+2};q)_\infty (-d/c;q)_\infty (-c/d;q)_\infty}{(c+d)(q^{\alpha+1};q)_\infty (q^{\beta+1};q)_\infty (-q^{\alpha+1}d/c;q)_\infty (-q^{\beta+1}c/d;q)_\infty} = M \qquad [3.27]$$

See [5]. This is also a special case of an identity of Sears [44], as pointed out by Al-Salam and Verma [4].

The orthogonality of the polynomials $p_n(x)$ defined by [3.23] can be obtained as follows. One wants to have the variable in [3.27] normalized so it can be absorbed in part of the weight function. We will take it as $xq^{\alpha+1}/c$. It is not clear what the best normalization for the polynomials is, so we will take one that is similar to the standard one for Jacobi polynomials.

$$P_n^{(\alpha,\beta)}(x;c,d:q) = c^n q^{-(\alpha+1)n} \frac{(q^{\alpha+1};q)_n (-q^{\alpha+1}d/c;q)_n}{(q;q)_n (-q;q)_n}$$

$$\qquad {}_3\varphi_2\left(\begin{matrix} q^{-n}, q^{n+\alpha+\beta+1}, xq^{\alpha+1}/c \\ q^{\alpha+1}, -q^{\alpha+1}d/c \end{matrix}; q,q\right) \qquad [3.28]$$

Then

$$\int_{-d}^{c} P_n^{(\alpha,\beta)}(x;c,d:q) \left(\frac{-qx^{\beta+1}}{d};q\right)_k \frac{(qx/c;q)_\infty (-qx/d;q)_\infty}{(q^{\alpha+1}x/c;q)_\infty (-q^{\beta+1}x/d;q)_\infty} d_qx$$

$$= \frac{c^n q^{-(\alpha+1)n}(q^{\alpha+1};q)_n (-q^{\alpha+1}d/c;q)_n}{(q;q)_n (-q;q)_n} \sum_{j=0}^{n} \frac{(q^{-n};q)_j (q^{n+\alpha+\beta+1};q)_j q^j}{(q^{\alpha+1};q)_j (-q^{\alpha+1}d/c;q)_j (q;q)_j}$$

$$\int_{-d}^{c} \frac{(qx/c;q)_\infty (-qx/d;q)_\infty}{(q^{\alpha+j+1}x/c;q)_\infty (-q^{\beta+k+1}x/d;q)_\infty} d_qx$$

$$= \frac{c^n q^{-(\alpha+1)n} cd(-d/c;q)_\infty (-c/d;q)_\infty (1-q)(q;q)_\infty (q^{\alpha+\beta+k+2};q)_\infty}{(q;q)_\infty (-q;q)_\infty (c+d)(q^{\alpha+n+1};q)_\infty (q^{\beta+k+1};q)_\infty (-q^{\alpha+n+1}d/c;q)_\infty (-q^{\beta+k+1}c/d;q)_\infty}$$

$$\cdot \sum_{j=0}^{n} \frac{(q^{-n};q)_j (q^{n+\alpha+\beta+1};q)_j}{(q^{k+\alpha+\beta+2};q)_j (q;q)_j} q^j .$$

This series can be summed, since

$$ {}_2\varphi_1\left(\begin{matrix}q^{-n},a\\c\end{matrix};q,q\right) = \frac{(\frac{c}{a};q)_n}{(c;q)_n}a^n, \quad [3.29] $$

and the resulting value for the integral is a multiple of $(q^{1+k-n};q)_n$ which vanishes when $k = 0,1,\ldots,n-1$. When all the constants are carried through we see that

$$ \int_{-d}^{c} P_n^{(\alpha,\beta)}(x;c,d:q)P_m^{(\alpha,\beta)}(x;c,d:q)\frac{(qx/c;q)_\infty(-qx/d;q)_\infty}{(q^{\alpha+1}x/c;q)_\infty(-q^{\beta+1}x/d;q)_\infty}d_qx = 0 \quad m \neq n $$

$$ = \frac{(cd)^n q^{\binom{n}{2}}(q^{\alpha+1};q)_n(q^{\beta+1};q)_n(1-q^{\alpha+\beta+1})(-q^{\beta+1}c/d;q)_n(-q^{\alpha+1}d/c;q)_n}{(q^{\alpha+\beta+1};q)_n(q;q)_n(1-q^{2n+\alpha+\beta+1})(-q;q)_n(-q;q)_n} M, \quad m=n \quad [3.30] $$

where M is defined in [3.27].

If $w(x;\alpha,\beta;c,d)$ denotes the weight function,

$$ w(-x;\alpha;\beta;c,d) = w(x;\beta,\alpha;d,c) \quad [3.31] $$

so there is a symmetry relation forced on $P_n^{(\alpha,\beta)}(x;c,d:q)$. It is

$$ P_n^{(\alpha,\beta)}(-x;c,d:q) = (-1)^n P_n^{(\beta,\alpha)}(x;d,c:q) \quad [3.32] $$

When $c = d = 1$ we have

$$ \lim_{q\to 1} P_n^{(\alpha,\beta)}(x;c,d:q) = P_n^{(\alpha,\beta)}(x). \quad [3.33] $$

The one set of polynomials Hahn treated in some detail were a different set of q-Jacobi polynomials. Define

$$ p_n^{(\alpha,\beta)}(x:q) = \frac{(q^{\alpha+1};q)_n}{(q;q)_n}\,{}_2\varphi_1\left(\begin{matrix}q^{-n},\ q^{n+\alpha+\beta+1}\\q^{\alpha+1}\end{matrix};q,qx\right). \quad [3.34] $$

Then the same type of argument shows that

$$ \int_0^1 p_n^{(\alpha,\beta)}(x:q)\ p_m^{(\alpha,\beta)}(x:q)x^\alpha\frac{(xq;q)_\infty}{(xq^{\beta+1};q)_\infty}d_qx = 0, \quad m \neq n,\ \alpha > -1. \quad [3.35] $$

From the explicit formulas [3.28] and [3.34] and the symmetry relation [3.32] we have

$$ P_n^{(\alpha,\beta)}(x;1,0:q) = $$

$$= (-1)^n \frac{q^{n(n-1)/2}(q^{\beta+1};q)_n}{(q;q)_n(-q;q)_n} \; {}_2\varphi_1\left(\begin{matrix} q^{-n}, q^{n+\alpha+\beta+1} \\ q^{\beta+1} \end{matrix}; q, qx\right)$$

$$= \frac{(-1)^n q^{n(n-1)/2}}{(-q;q)_n} p_n^{(\beta,\alpha)}(x:q) \qquad [3.36]$$

When we worked out the orthogonality relation [3.30] we did not know about a set of q-Jacobi polynomials whose weight function is absolutely continuous, and so named these two classes of Jacobi polynomials the big q-Jacobi polynomials (for $P_n^{(\alpha,\beta)}(x:q)$) and the little q-Jacobi polynomials (for $p_n^{(\alpha,\beta)}(x:q)$). We do not have a better name, but are open to suggestions.

There is another connection between some of these two classes of polynomials. The ultraspherical or symmetric Jacobi polynomials are an important subclass of Jacobi polynomials. These are $P_n^{(\alpha,\alpha)}(x)$, or under a different normalization

$$C_n^\lambda(x) = \frac{(2\lambda)_n}{(\lambda+\frac{1}{2})_n} P_n^{(\lambda-\frac{1}{2}\lambda-\frac{1}{2})}(x) \qquad [3.27]$$

Probably the most important q extension of $C_n^\lambda(x)$ is the one found by Rogers [43] (see [3.10]), so we will not renormalize the symmetric big q-Jacobi polynomials. There are a pair of quadratic transformations connecting some Jacobi polynomials.

$$P_{2n}^{(\alpha,\alpha)}(x) = \frac{(\alpha+1)_{2n}n!(-1)^n}{(2n)!(\alpha+1)_n} P_n^{(-\frac{1}{2},\alpha)}(1-2x^2) \qquad [3.38]$$

$$P_{2n+1}^{(\alpha,\alpha)}(x) = \frac{(\alpha+1)_{2n+1}n!(-1)^n}{(2n+1)!(\alpha+1)_n} xP_n^{(\frac{1}{2},\alpha)}(1-2x^2) \qquad [3.39]$$

A q-extension is

$$P_{2n}^{(\alpha,\alpha)}(x;1,1:q) = P_{2n}^{(\alpha,\alpha)}(x;q) = c_n p_n^{(-\frac{1}{2},\alpha)}(x^2:q^2) \qquad [3.40]$$

and

$$P_{2n+1}^{(\alpha,\alpha)}(x;1,1:q) = P_{2n+1}^{(\alpha,\alpha)}(x;q) = d_n x p_n^{(\frac{1}{2},\alpha)}(x^2:q^2) \qquad [3.41]$$

where the coefficients c_n and d_n can be found by setting $x = 1$.

A q-extension of Meixner's class of polynomials was found by Al-Salam and Chihara [3]. Recall that Meixner found all sets of orthogonal polynomials generated by

$$f(r)e^{xg(r)} = \sum_{n=0}^{\infty} p_n(x)r^n \qquad [3.42]$$

50

with $f(r)$ and $g(r)$ functions analytic in a neighborhood of $r = 0$, $f(0) \neq 0$, $g(0) = 0$, $g'(0) \neq 0$. It is not clear how to find a q-extension of the function on the left hand side, but after all such orthogonal polynomials have been found one can look for other properties they have in common. One is the following. These polynomials come in five different classes. One is Hermite polynomials, and the generating function for them is

$$\sum_{n=0}^{\infty} \frac{H_n(x)}{n!} r^n = e^{2xr-r^2} \qquad [3.43]$$

From this it is easy to see that

$$\sum_{k=0}^{n} \binom{n}{k} H_k(x) H_{n-k}(y) = 2^{-\frac{n}{2}} H_n((x+y)2^{-\frac{1}{2}})$$

The other classes have a generating function of the form

$$\sum_{n=0}^{\infty} p_n^{\lambda}(x) r^n = [h(r)]^{\lambda} e^{xg(r)} \qquad [3.44]$$

and so satisfy

$$\sum_{k=0}^{n} p_k^{\lambda}(x) p_{n-k}^{\mu}(y) = p_n^{\lambda+\mu}(x+y) \qquad [3.45]$$

Al-Salam and Chihara solved the following problem. Given two sets of orthogonal polynomials $\{p_n(x)\}$ and $\{q_n(x)\}$ form

$$\sum_{k=0}^{n} p_k(x) q_{n-k}(y) = Q_n(x,y). \qquad [3.46]$$

They found all such sets with $\{Q_n(x,y)\}$ a set of orthogonal polynomials in x for an infinite number of y. The polynomials can be normalized so their generating function is

$$\sum_{n=0}^{\infty} p_n(x) r^n = \prod_{n=0}^{\infty} \frac{(1-arq^n+br^2q^{2n})}{(1-xrq^n+cr^2q^{2n})} \qquad [3.47]$$

when $|q| < 1$, and a similar infinite product with q replaced by q^{-1} when $|q| > 1$. Descriptions of these polynomials will be given in section 4.

After the generating function [3.47] has been found it is clear what generating function to consider as an extension of Meixner's series [3.44]. We state the following conjecture.

Conjecture. If $f(r)$ and $h(r)$ are functions analytic in a circle about zero, $f(0) \neq 0$, $h(0) = 0$, $h'(0) \neq 0$, and if polynomials $\{p_n(x)\}$ are defined by

$$f(r) \prod_{n=0}^{\infty} (1-xh(q^n r))^{-1} = \sum_{n=0}^{\infty} p_n(x) r^n \qquad [3.48]$$

for a q, $-1 < q < 1$, they are orthogonal if and only if $f(r)$ and $h(r)$ can be renormalized to be

$$f(r) = \prod_{n=0}^{\infty} (1 - arq^n + br^2q^{2n})/(1 + cr^2q^{2n})$$

and

$$h(r) = r(1 + cr^2)^{-1}$$

for appropriate real parameters a, b and c.

Some other examples of classical orthogonal polynomials which are basic hypergeometric functions have been found.

Hahn [31] found a set dual to [3.22] and [3.23], but did not give the orthogonality. Al-Salam and Carlitz [2] found two sets of orthogonal polynomials. These arose again in the following problem solved by Chihara [23]. He looked at the problem of Geronimus when the x_i are all equal. Thus he wanted to find all sets of orthogonal polynomials of the form

$$p_n(x) = \sum_{k=0}^{n} a_{n-k} b_k x^k , \qquad [3.49]$$

or equivalently all polynomials which have a generating function of the form

$$A(r)B(xr) = \sum_{n=0}^{\infty} p_n(x) r^n \qquad [3.50]$$

with $A(r)$ and $B(r)$ analytic in a neighborhood of $r = 0$ and $A(0) \neq 0$, $B(0) \neq 0$. See [24, Chapter 5.5] for a description of these polynomials. Many of these polynomials are "classical", and the rest are closely related to classical polynomials.

4. The final sets of classical orthogonal polynomials.

This section is titled in a very strong way, and we hope that someone will come along and prove we are wrong, just as has happened to everyone else who has tried to characterize the classical polynomials.

52

They have always had too small a class. We could pull the most general sets of classical polynomials out of the air, as they seemed to arise in [16] and [18]. Another way would be to start with Leonard's characterization theorem, which is one good reason for the title of this section. However this seems to be a reasonable place to record how these polynomials were discovered. We have been writing a book on this material since 1976, and in an early draft of one chapter one of us was looking for a representation of the polynomials $Q_n(x;\alpha,\alpha,N)$ which would be obviously symmetric, since their weight function $\binom{x+\alpha}{x}\binom{N-x+\alpha}{N-x}$ is symmetric about $x = N/2$. He integrated a ${}_2F_1$ quadratic transformation to obtain such a formula, and the resulting series was a ${}_4F_3$ rather than a ${}_3F_2$. He gave this to a student, J. Wilson, with the suggestion that something more could be found at the ${}_4F_3$ level. Wilson [60] then found the following two sets of orthogonal polynomials.

$$R_n(\lambda(x);\alpha,\beta,\gamma,\delta) = {}_4F_3\left(\begin{matrix}-n,\ n+\alpha+\beta+1,\ -x,\ x+\gamma+\delta+1\\ \alpha+1,\ \beta+\delta+1,\ \gamma+1\end{matrix};1\right) \qquad [4.1]$$

When $\beta+\delta+1 = -N$, N a positive integer, then

$$\sum_{x=0}^{N} R_n(\lambda(x))R_m(\lambda(x))\frac{(\gamma+\delta+1)_x\left(\frac{\gamma+\delta+3}{2}\right)_x(\alpha+1)_x(-N)_x(\gamma+1)_x}{x!\left(\frac{\gamma+\delta+1}{2}\right)_x(\gamma+\delta-\alpha+1)_x(N+\gamma+\delta+2)_x(\delta+1)_x} = 0, \qquad [4.2]$$

$m \neq n \leq N$. Here $\lambda(x) = x(x+\gamma+\delta+1)$. $R_n(\lambda(x))$ is a polynomial of degree $2n$ in x, but of degree n in $\lambda(x)$.

$$\frac{W_n(x^2;a,b,c,d)}{(a+b)_n(a+c)_n(a+d)_n} = {}_4F_3\left(\begin{matrix}-n,\ n+a+b+c+d-1,\ a+ix,\ a-ix\\ a+b,\ a+c,\ a+d\end{matrix};1\right) \qquad [4.3]$$

When $a,b,c,d > 0$ or if any of the parameters are complex they occur in conjugate pairs and have positive real parts, then

$$\int_0^\infty W_n(x^2)W_m(x^2)\left|\frac{\Gamma(a+ix)\Gamma(b+ix)\Gamma(c+ix)\Gamma(d+ix)}{\Gamma(2ix)}\right|^2 dx = 0, \quad m\neq n. \qquad [4.4]$$

See Wilson [61]. After these relations were found, and a more general orthogonality which contains both but loses the positivity of the weight function, L. Durand asked what was the connection between [4.2] and Racah's orthogonality of 6-j symbols of angular momentum theory. It turns out they are the same, except Racah's representation of the 6-j symbols has to be transformed using the following transformation formula of Whipple to obtain a polynomial times part of the weight function. Whipple's symmetry for balanced ${}_4F_3$ can be obtained from [4.3] and [4.4]. The polynomials defined in [4.3] are obviously symmetric in b, c and d, while the weight function in [4.4] is symmetric in a, b, c and d. Thus the polynomials in [4.3] are also symmetric in a and b, at least up to a factor independent of x. The extra factors $(a+b)_n(a+c)_n(a+d)_n$ were chosen to make this constant one.

After these polynomials were found it was a simple task to find a basic hypergeometric extension of [4.1], [4.2] and [4.3], and the only problem in extending [4.4] was to show that

$$\frac{1}{2\pi}\int_0^{\pi}\left|\frac{(e^{2i\theta};q)_\infty}{(ae^{i\theta};q)_\infty(be^{i\theta};q)_\infty(ce^{i\theta};q)(de^{i\theta};q)_\infty}\right|^2 d\theta$$

$$= \frac{(abcd;q)_\infty}{(q;q)_\infty(ab;q)_\infty(ac;q)_\infty(ad;q)_\infty(bc;q)_\infty(bd;q)_\infty(cd;q)} \qquad [4.5]$$

when a, b, c and d are real and in absolute value less than one, and a more general contour integral in the general case. This was surprisingly hard, and it has taken over five years before relatively simple ways of evaluating this integral were found. Wilson's very ingenious elliptic function argument, which was the only hard part of the first derivation, is in [18]; and later proofs are given in [7] and [41]. The easiest evaluation was just found by Ismail and Stanton [32].

Before stating Leonard's theorem, we mention a classical theorem that was used to obtain many of the characterization theorems. Every set of orthogonal polynomials satisfies a three term recurrence relation of the form

$$xp_n(x) = A_n p_{n+1}(x) + B_n p_n(x) + C_n p_{n-1}(x), \qquad [4.6]$$

with A_n, B_n, C_n real and $A_{n-1}C_n > 0$. Conversely if [4.6] holds, $p_{-1}(x) = 0$, $p_0(x) = 1$, A_n, B_n and C_n are real and $A_{n-1}C_n > 0$ then $\{p_n(x)\}$ is a set of orthogonal polynomials. There is at least one positive measure $d\alpha(x)$ so that

$$\int_{-\infty}^{\infty} p_n(x)p_m(x)\,d\alpha(x) = 0, \quad m \neq n. \qquad [4.7]$$

If $A_{n-1}C_n > 0$ for $n = 1,2,\ldots,$ then [4.7] holds for $m,n = 0,1,\ldots$. If $A_{n-1}C_n > 0$ for $n = 1,2,\ldots,N,$ then [4.7] holds for $m,n = 0,1,\ldots,N$. There are some very important examples where only the first $N+1$ polynomials are orthogonal with respect to a positive measure.

The above result is often attributed to Favard, but it was published earlier. See references in [6]. This is an easy theorem, and many mathematicians know everything that is needed to prove it. The three term recurrence relation is a discrete analogue of a Sturm-Liouville two point boundary value problem, with the boundary values $p_{-1}(x) = 0$, $p_{N+1}(x) = 0$. This could be extended to a radiation condition, but it is the appropriate condition for discrete classical orthogonal polynomials. Just as in the classical Sturm-Liouville theory, the eigenfunctions are orthogonal. In this case the orthogonality is

$$\sum_{n=0}^{N} p_n(x_{k,n})p_n(x_{j,w})k_n = \delta_{j,k}/w_k \qquad [4.8]$$

where $x_{j,N}$ are the zeros of $p_{N+1}(x_{j,N})$ and k_n and w_n can be found, k_n easily from the recurrence relation, w_n with a bit more difficulty. Then

$$\{p_n(x_{j,N})k_j^{\frac{1}{2}} w_n^{\frac{1}{2}}\}_{j,n=0}^{N}$$

is an orthogonal matrix, so

$$\sum_{j=0}^{N} p_n(x_{j,N})p_m(x_{j,n})w_j = \delta_{m,n}/k_n \qquad [4.9]$$

Then $w_j > 0$, and $\sum w_j = k_0^{-1}$, so there is a subsequence N_r and a positive measure $d\alpha(x)$ with

$$\int_{-\infty}^{\infty} p_n(x)p_m(x)d\alpha(x) = \lim_{N_r\to\infty} \sum_{k=0}^{N_r} p_n(x_{k,N_r})p_m(x_{k,N_r})w_k = 0, \ m\neq n. \qquad [4.10]$$

Observe that the first orthogonality relation obtained is usually not for a set of polynomials, for there is no reason for $p_n(x_{k,n})$ to be a polynomial of degree k. However there are a number of important examples where it is. One example is the Krawtchouk polynomial [2.7]

$$K_n(x;p,N) = {}_2F_1(-n, -x; -N; p^{-1}).$$

This is a polynomial of degree n in x, and of degree x in n. In fact these polynomials are symmetric:

$$K_n(x;p,N) = K_x(n;p,N), \quad x,n = 0,1,\ldots,N.$$

As we remarked, Hahn [30] found the polynomials $Q_n(x;\alpha,\beta;N:q)$ [3.22], but did not find their orthogonality. If he had worked out the orthogonality he could have found the orthogonality for

$$R_n(\lambda(x);\gamma,\delta,N:q) = {}_3\varphi_2\left(\begin{matrix} q^{-n}, q^{-x}, q^{x+\gamma+\delta+1} \\ q^{\gamma+1}, q^{-N} \end{matrix} ; q,q\right) \qquad [4.11]$$

$\lambda(x) = q^{-x} + q^{x+\gamma+\delta+1}$, since

$$R_n(\lambda(x);\gamma,\delta,N:q) = Q_x(q^{-n};\gamma,\delta,N:q) \qquad [4.12]$$

and the orthogonality for $R_n(\lambda(x))$ follows from that of $Q_n(q^{-x})$ via the argument above. Hahn discovered the polynomials $R_n(\lambda(x))$ [31], but did not find their orthogonality relation.

One can ask what orthogonal polynomials $p_n(\lambda(x))$ have the property that they are polynomials of degree x in a variable $\mu(n)$. It is necessary to introduce $\lambda(x)$ and $\mu(n)$ to include [3.22] and [4.11]. Leonard [36] proved that if the orthogonality is on at least nine points then these polynomials are the q-Racah polynomials (the q-extensions of [4.1]) or one of their limiting or special cases. Another proof is in [20]. These polynomials are

$$R_n(\lambda(x);a,b,c,d;q) = {}_4\varphi_3\left(\begin{matrix} q^{-n}, q^{n+1}ab, q^{-x}, q^{x+1}cd \\ aq, bdq, cq \end{matrix} ; q,q\right). \qquad [4.13]$$

$\lambda(x) = q^{-x} + q^{x+1}cd$ and $bdq = q^{-N}$. The orthogonality is

$$\sum_{x=0}^{N} R_n(\lambda(x))R_m(\lambda(x))w(x) = \delta_{m,n}/k_n, \quad n,m = 0,1,\ldots,N, \qquad [4.14]$$

$$w(x) = \frac{(cdq;q)_x(1-cdq^{2x+1})(aq;q)_x(bdq;q)_x(cq;q)_x}{(q;q)_x(1-cdq)(cdqa^{-1};q)_x(cqb^{-1};q)_x(dq;q)_x}(abq)^{-x} \qquad [4.15]$$

$$k_n = \frac{(abq;q)_n(1-abq^{2n+1})(cq;q)_n(bdq;q)_n(aq;q)_n(cdq)^{-n}}{(q;q)_n(1-abq)(abqc^{-1};q)_n(aqd^{-1};q)_n(bq;q)_n} \qquad [4.16]$$

$$\frac{(a^{-1}cdq;q)_\infty(b^{-1}cq;q)_\infty(dq;q)_\infty(a^{-1}b^{-1}q^{-1};q)_\infty}{(cdq^2;q)_\infty(a^{-1}b^{-1}c;q)_\infty(a^{-1}d;q)_\infty(b^{-1};q)_\infty}$$

The infinite products in k_n can also be written as

$$\frac{(bq;q)_N(abq/c;q)_N}{(b/c;q)_N(abq^2;q)_N} \quad \text{when} \quad d = q^{-N-1}b^{-1}.$$

This orthogonality relation is in [16]. The recurrence relation given there is

$$-(1-q^{-x})(1-q^{x+1}cd)R_n(\lambda(x)) = A_nR_{n+1}(\lambda(x)) - (A_n + C_n)R_n(\lambda(x)) + C_nR_{n-1}(\lambda(x)) \quad [4.17]$$

$$A_n = \frac{(1-abq^{n+1})(1-aq^{n+1})(1-bdq^{n+1})(1-cq^{n+1})}{(1-abq^{2n+1})(1-abq^{2n+2})}$$

$$C_n = \frac{q(1-q^n)(1-bq^n)(c-abq^n)(d-aq^n)}{(1-abq^{2n})(1-abq^{2n+1})}$$

A nice use of these polynomials is given by Perlstadt [39]. The absolutely continuous version of these polynomials is

$$\frac{a^nW_n(x;a,b,c,d|q)}{(ab;q)_n(ac;q)_n(ad;q)_n} = {}_4\varphi_3\left(\begin{matrix} q^{-n},\ q^{n-1}abcd;\ ae^{i\theta},\ ae^{-i\theta} \\ ab,\ ac,\ ad \end{matrix};q,q\right) \quad [4.18]$$

$x = \cos\theta$, and when $-1 < a,b,c,d,q < 1$ the orthogonality is

$$\int_{-1}^{1} W_n(x)W_m(x)\,\frac{h(x,1)h(x,q^{\frac{1}{2}})h(x,-1)h(x,-q^{\frac{1}{2}})\,dx}{h(x,a)h(x,b)h(x,c)h(x,d)(1-x^2)^{\frac{1}{2}}} = \delta_{m,n}/k_n \quad [4.19]$$

and $$h(x,a) = (ae^{i\theta};q)_\infty(ae^{-i\theta};q)_\infty$$

These polynomials are classical polynomials since they have a Rodrigues type formula, an appropriate divided difference operator acting on them gives a set of orthogonal polynomials, and they satisfy a second order difference equation in x which is of Sturm-Liouville type. These are all given for the polynomials $W_n(x)$ in [18]. Wilson [60] found these results for the ${}_4F_3$ orthogonal polynomials in his thesis but has not published them yet. The connections between 6-j symbols and orthogonal polynomials, including a Rodrigues formula, are starting to be rediscovered by physicists. See [46] and [50].

Askey and Wilson closed [18] with a chart of the classical orthogonal polynomials which can be given as hypergeometric series. It is too early to give the corresponding chart or charts for the basic hypergeometric classical orthogonal polynomials, but there are some gaps in these still undrawn charts that can be filled. One concerns the dual q-Hahn polynomials when $q > 1$. These polynomials can be given as

$$Q_n(\mu(x)) = Q_n(\mu(x);\ a,c,d;q) = {}_3\varphi_2\left(\begin{matrix} q^{-n},\ q^{-x},\ q^{x+1}d \\ aq,\ cq \end{matrix};\ q,q\right). \quad [4.20]$$

$\mu(x) = q^{-x} + q^{x+1}d$ and their recurrence relation is

$$-(1-q^{-x})(1-q^{x+1}d)Q_n(\mu(x)) = A_nQ_{n+1}(\mu(x)) - (A_n+C_n)Q_n(\mu(x)) + C_nQ_{n-1}(\mu(x)) \quad [4.21]$$

$$A_n = (1-cq^{n+1})(1-aq^{n+1})$$

$$C_n = q(1-q^n)(d-acq^n).$$

One orthogonality relation is

$$\sum_{x=0}^{\infty} Q_n(\mu(x))Q_m(\mu(x))w(x) = 0, \qquad m \neq n$$

$$w(x) = \frac{(dq;q)_x(1-dq^{2x+1})(aq;q)_x(cq;q)_x(-1)^x q^{-\binom{x}{2}}(acq^2)^{-x}}{(q;q)_x(1-dq)(\frac{dq}{a};q)_x(\frac{dq}{c};q)_x}$$

Both of these follow from the above results for the q-Wilson polynomials when $b \to 0$ and d is replaced by d/c to give symmetry in a and c. Some of the special cases of this when $c = 0$ were treated in [13, sections 3.12 and 3.13]. The case when the moment problem is uniquely determined was treated there. In the general case when $a \neq 0$ and $c \neq 0$ Chihara has shown that the moment problem is always indeterminate. This along with other details will appear separately.

We close this section with our definition of the classical orthogonal polynomials.

Definition. _A set of orthogonal polynomials is classical if it is a special case or a limiting case of the_ ${}_4\varphi_3$ _orthogonal polynomials_ given by [4.13] or [4.18].

5. A few limiting cases.

There are quite a few sets of orthogonal polynomials that arise from the ${}_4\varphi_3$ polynomials by appropriate specializations. One surprising type occurs when q is a root of unity. In [12] it was pointed out that the case $q \to -1$ leads to polynomials orthogonal with respect to $|x|^{\alpha-1}(1-x^2)^{\alpha/2}$ on $[-1,1]$. From the ${}_4\varphi_3$ polynomials it is easy to obtain the polynomials orthogonal with respect to $|x|^{\alpha}(1-x^2)^{\beta}$, again by letting $q \to -1$. When $q \to e^{2\pi i/k}$ there are eight classes of polynomials that arise, depending on whether k is even or odd and whether the factors $(1-x^2)^{-\frac{1}{2}}$, $(1-x^2)^{\frac{1}{2}}$, $(1-x)^{\frac{1}{2}}(1-x)^{-\frac{1}{2}}$ or $(1-x)^{-\frac{1}{2}}(1+x)^{\frac{1}{2}}$ occur. These cases include the discrete orthogonality of $\cos n\theta$, $\sin(n+1)\theta$, $\sin(n+\frac{1}{2})\theta$ and $\cos(n+\frac{1}{2})\theta$ $0 \le n \le N$ on roots of unity as limiting cases. The present paper is already too long, so we refer the reader to [1] and [15] for these polynomials. Much more needs to be done with them. One fascinating problem is to try to find the second order differential equations they satisfy. The existence of these equations was proved by Atkinson and Everitt [19], and earlier by Shohat [45]. As Shohat wrote, the existence of these differential equations was really discovered by Laguerre [34].

6. Applications.

The real reason we care about the classical orthogonal polynomials is their usefulness. Books could be written on some of the applications, so the best we can do here is list a few references. Quite a few applications are given in papers in [14] and others are mentioned in the preface to this book. A recent application of Jacobi polynomials was a surprise to everyone. L. deBranges [25] reduced the Bieberbach-Milin conjecture to proving

$$ {}_3F_2\left(\begin{matrix}-n,\ n+\alpha+2,\ (\alpha+1)/2\\ \alpha+1,\ (\alpha+3/2)\end{matrix};t\right) > 0, \quad 0 \le t < 1, \quad \alpha = 2,4,\ldots, \qquad [6.1] $$

Earlier Askey and Gasper [10] had shown that

$$ \sum_{k=0}^{n} P_k^{(\alpha,0)}(x) = \frac{(\alpha+2)_n}{n!}\ {}_3F_2\left(\begin{matrix}-n,\ n+\alpha+2,\ (\alpha+1)/2\\ \alpha+1,\ (\alpha+3)/2\end{matrix};\frac{1-x}{2}\right) \qquad [6.2] $$

and that this ${}_3F_2$ is positive when $\alpha > -2$, $-1 < x \le 1$. Gasper [27] has proven the much deeper result that

$$ \sum_{k=0}^{n} \frac{P_k^{(\alpha,-\frac12)}(x)}{P_k^{(-\frac12,\alpha)}(1)} > 0, \quad -1 < x \le 1, \quad \alpha > \tfrac12. $$

When these inequalitites were proved Askey and Gasper had no idea they would complete a proof of the Bieberbach conjecture, but they knew these were deep results which could be used to prove some interesting general facts.

Finally, a number of problems about the classical orthogonal polynomials are given in [9].

References

[1] W. Al-Salam, W. Allaway and R. Askey, Sieved ultraspherical polynomials, Trans. Amer. Math. Soc. 284 (1984), 39-55.

[2] W. Al-Salam and L. Carlitz, Some orthogonal q-polynomials, Math. Nach. 30 (1965), 47-61.

[3] W. Al-Salam and T. S. Chihara, Convolutions of orthogonal polynomials, SIAM J. Math. Anal. 7 (1976), 16-28.

[4] W. Al-Salam and A. Verma, Some remarks on q-beta integral, Proc. Amer. Math. Soc. 85 (1982), 360-362.

[5] G. E. Andrews and R. Askey, Another q-extension of the beta function, Proc. Amer. Math. Soc. 81 (1981), 97-100.

[6] R. Askey, Comment to [68-1], G. Szegö, Collected Papers, vol. 3, Birkhäuser Boston, 1982, 866-869.

[7] R. Askey, An elementary evaluation of a beta type integral, Indian J. Pure Appl. Math. 14 (1983), 892-895.

[8] R. Askey, Limits of some q-Laguerre polynomials, J. Approx. Th., to appear.

[9] R. Askey, Some problems about special functions and computations, Rendiconti Semin. Mate. Univ. e Polit. di Torino, to appear.

[10] R. Askey and G. Gasper, Positive Jacobi polynomial sums. II, Amer. J. Math. 98 (1976), 709-737.

[11] R. Askey and M. Ismail, The Rogers q-ultraspherical polynomials, Approximation Theory III, ed. E. W. Cheney, Academic Press, New York, 1980, 175-182.

[12] R. Askey and M. Ismail, A generalization of ultraspherical polynomials, in Studies in Pure Mathematics, ed. P. Erdös, Birkhäuser, Basel, 1983, 55-78.

[13] R. Askey and M. Ismail, Recurrence relations, continued fractions and orthogonal polynomials, Memoirs Amer. Math. Soc., 300, 1984.

[14] R. Askey, T. Koornwinder and W. Schempp (editors), Special Functions: Group Theoretical Aspects and Applications, Reidel, Dordrecht, Boston, Lancaster, 1984.

[15] R. Askey and D. P. Shukla, Sieved Jacobi polynomials, to appear.

[16] R. Askey and J. Wilson, A set of orthogonal polynomials that generalize the Racah coefficients or 6-j symbols, SIAM J. Math. Anal. 10 (1979), 1008-1016.

[17] R. Askey and J. Wilson, A set of hypergeometric orthogonal polynomials, SIAM J. Math. Anal. 13 (1982), 651-655.

[18] R. Askey and J. Wilson, Some basic hypergeometric orthogonal polynomials that generalize Jacobi polynomials, Memoirs Amer. Math. Soc. 1985.

[19] F. V. Atkinson and W. N. Everitt, Orthogonal polynomials which satisfy second order differential equations, in E. B. Christoffel, ed. P. L. Butzer and F. Fehér, Birkhäuser, Basel, 1981, 173-181.

[20] E. Bannai and T. Ito, Algebraic Combinatorics I: Association Schemes, Benjamin/Cummins, Menlo Park, CA, 1984.

[21] S. Bochner, Über Sturm-Liouvillesche Polynomsysteme, Math. Zeit., 29 (1929), 730-736.

[22] L. Carlitz, Bernoulli and Euler numbers and orthogonal polynomials, Duke Math. J., 26 (1959), 1-15.

[23] T. S. Chihara, Orthogonal polynomials with Brenke type generating functions, Duke Math. J. 35 (1968), 505-518.

[24] T. S. Chihara, An Introduction to Orthogonal Polynomials, Gordon and Breach, New York, London, Paris, 1978.

[25] L. de Branges, A proof of the Bieberbach conjecture, Acta Math.

[26] A. Erdélyi et. al., Higher Transcendental Functions, vol. 2, McGraw Hill, New York, 1952, reprinted Krieger, Malabar, Florida, 1981.

[27] G. Gasper, Positive sums of the classical orthogonal polynomials, SIAM J. Math. Anal. 8 (1977), 423-447.

[28] J. Geronimus, The orthogonality of some systems of polynomials, Duke Math. J., 14 (1947), 503-510.

[29] W. Hahn, Über die Jacobischen Polynome und zwei verwandte Polynomklassen, Math. Zeit. 39 (1935), 634-638.

[30] W. Hahn, Über Orthogonalpolynome die q-Differenzengleichungen genügen, Math. Nath. 2 (1949), 4-34.

[31] W. Hahn, Über Polynome, die gleichzeitig zwei verschiedenen Orthogonalsystemen angehören, Math. Nach. 2 (1949), 263-278,

[32] M. Ismail and D. Stanton, paper on q-Hermite polynomials, to appear.

[33] F. H. Jackson, On q-definite integrals, Quart. J. Pure Appl. Math., 41 (1910), 193-203.

[34] E. Laguerre, Sur la réduction en fractions continues d'une fraction qui satisfait a une équation différentielle linéaire du premier ordre dont les coefficients sont rationnels, J. math. pure appl. (4)1, 1885, 135-165, Oeuvres de Laguerre, second edition, Tome II, Chelsea, New York, 1972, 685-711.

[35] C. D. Lai. A survey of Meixner's hypergeometric distribution, Mathematical Chronicle, 6 (1977), 6-20.

[36] D. Leonard, Orthogonal polynomials, duality and association schemes, SIAM J. Math. Anal., 13 (1982), 656-663.

[37] A. Markoff, On some applications of algebraic continued fractions (in Russian), Thesis, St. Petersburg, 1884, 131 pp.

[38] J. Meixner, Orthogonale Polynomsysteme mit einer besonderen Gestalt der erzeugenden Funktion, J. London Math. Soc. 9 (1934), 6-13.

[39] M. Perlstadt, A property of orthogonal polynomial families with polynomial duals, SIAM J. Math. Anal. 15 (1984), 1043-1054.

[40] F. Pollaczek, Sur une famille de polynomes orthogonaux qui contient les polynomes d'Hermite et de Laguerre comme cas limites, C. R. Acad. Sci., Paris 230 (1950), 1563-1565.

[41] M. Rahman, A simple evaluation of Askey and Wilson's q-beta integral, Proc. Amer. Math. Soc., 92 (1984), 413-417.

61

[42] L. J. Rogers, Second memoir on the expansion of certain infinite products, Proc. London Math. Soc., 25 (1894), 318-343.

[43] L. J. Rogers, Third memoir on the expansion of certain infinite products, Proc. London Math. Soc., 26 (1895), 15-32.

[44] D. B. Sears, Transformation of basic hypergeometric functions of special type, Proc. London Math. Soc. 52 (1951), 467-483.

[45] J. J. Chokhate (J. Shohat), Sur une classe étendue de fractions continues algébriques et sur les polynomes de Tchebycheff correspondants, C. R. Acad. Sci., Paris, 191 (1930), 989-990.

[46] Ya. A. Smorodinskiǐ and S. K. Suslov, 6-j symbols and orthogonal polynomials, Yad. Fiz. 36 (1982), 1066-1071, translation, Sov. J. Nucl. Phys. 36 (1982), 623-625.

[47] N. Ja. Sonine, Über die angenäherte Berechnung der bestimmten Integrale und über die dabei vorkommenden ganzen Functionen, Warsaw Univ. Izv. 18 (1887), 1-76 (Russian). Summary in Jbuch. Fortschritte Math. 19, 282.

[48] T. J. Stieltjes, Sur quelques intégrales definies et leur développement en fractions continues, Quart. J. Math. 24 (1890), 370-382; Oeuvres, T. 2, Noordhoff, Groningen, 1918, 378-394.

[49] T. J. Stieltjes, Recherches sur les fractions continues, Annales de la Faculté des Sciences de Toulouse, 8 (1894), J 1-122; 9 (1895), A1-47; Oeuvres, T. 2, 398-566.

[50] S. K. Suslow, Rodrigues formula for the Racah coefficients, Yad. Fiz. 37 (1983), 795-796, translation, Sov. J. Nucl. Phys. 37 (1983), 472-473.

[51] G. Szegö, Ein Beitrag zur Theorie der Thetafunktionen, Sitz. Preuss. Akad. Wiss. Phys. Math. Kl., XIX (1926), 242-252, Collected Papers, Vol. I, Birkhäuser Boston, 1982, 795-805.

[52] G. Szegö, Orthogonal Polynomials, Amer. Math. Soc. Colloq. Publ. 23, Amer. Math. Soc. Providence, RI, 1975.

[53] P. L. Tchebychef, Sur les fractions continues, Oeuvres, T. I., Chelsea, New York, 203-230.

[54] P. L. Tchebychef, Sur une nouvelle série, Oeuvres, T. I., Chelsea, New York, 381-384.

[55] P. L. Tchebychef, Sur l'interpolation des valeurs équidistantes, Oeuvres, II, Chelsea, New York, 1961, 219-242.

[56] J. Thomae, Beiträge zur Theorie der durch die Heinesche Reihe; $1 + ((1 - q^{\alpha})(1-q^{\beta})/(1-q)(1-q^{\gamma}))x + \cdots$ darstellbaren Functionen, J. reine und angew. Math. 70 (1869), 258-281.

[57] L. Toscono, I polinomi ipergeometrici nel calcolo delle differenze finite, Boll. Un. Mat. Ital. (3) 4 (1949), 398-409.

[58] F. Tricomi, Serie Ortogonali di Funzioni, Torino, 1948.

[59] S. Wigert, Sur les polynomes orthogonaux et l'approximation des fonctions continues, Arkiv för Matem., Astron. och Fysik. 17 (1923), no. 18, 15 pp.

[60] J. Wilson, Hypergeometric series, recurrence relations and some new orthogonal functions, Ph.D. thesis, Univ. Wisconsin, Madison, 1978.

[61] J. Wilson, Some hypergeometric orthogonal polynomials, SIAM J. Math. Anal. 11 (1980), 690-701.

Pennsylvania State University
University of Wisconsin-Madison

2.4 Well-poised series

The work in this section really started with my efforts to understand Göllnitz's proof of what are now called the Göllnitz–Gordon identities [Göl60]. This led inexorably to [16], and subsequently to the paper included in this section [23] (which suffers from Watson's non-standard notation, as well as too much condensed notation). Indeed, a significant side value of [60] is that it restates many of the results of [23], and [23] was subsequently important in the theory of partitions and q-series (for example, [31], [67], [58], [102], [149], [176], [188]).

Especially noteworthy is [73], which is a collaboration with Askey, based on Askey's previously successful effort to link orthogonal polynomials with well-poised series.

Paper [121] uses well-poised series to provide a proof of the Rogers–Ramanujan identities without invoking Jacobi's triple product. This is a paper arising purely from a challenge set by Doron Zeilberger at Oberwolfach in 1984.

2.4.1 On q-difference equations for certain well-poised basic hypergeometric series

ON q-DIFFERENCE EQUATIONS FOR CERTAIN WELL-POISED BASIC HYPERGEOMETRIC SERIES

By GEORGE E. ANDREWS†

[Received 11 October 1967]

1. Introduction

IN this paper we shall study the following q-series.

$$C_{k,i}(a_1,\ldots,a_\lambda;x;q) = C_{k,i}((a);x;q)_\lambda = \sum_{n=0}^{\infty} \frac{M_n(x)q^{-in}(1-x^iq^{2ni})\Pi_n(-x,q)P_n(1;a_1,\ldots,a_\lambda)}{(1-x)\Pi_n(-q,q)P_n(xq;a_1^{-1},\ldots,a_\lambda^{-1})}; \tag{1.1}$$

$$D_{k,i}(a_1,\ldots,a_\lambda;x;q) = D_{k,i}((a);x;q)_\lambda = \sum_{n=0}^{\infty} \frac{M_n(xq)q^{-in}\Pi_n(-xq,q)P_n(1;a_1,\ldots,a_\lambda)}{\Pi_n(-q,q)P_n(xq;a_1^{-1},\ldots,a_\lambda^{-1})}\times$$
$$\times\left\{1+\frac{(-1)^{\lambda+1}x^iq^{(2n+1)i-\lambda n}(a_1\ldots a_\lambda)^{-1}p_{n+1}(1;a_1,\ldots,a_\lambda)}{p_{n+1}(xq;a_1^{-1},\ldots,a_\lambda^{-1})}\right\}; \tag{1.2}$$

$$H_{k,i}(a_1,\ldots,a_\lambda;x;q) = H_{k,i}((a);x;q)_\lambda = P_\infty(xq;a_1^{-1},\ldots,a_\lambda^{-1})C_{k,i}((a);x;q)_\lambda/\Pi_\infty(-xq,q); \tag{1.3}$$

$$J_{k,i}(a_1,\ldots,a_\lambda;x;q) = J_{k,i}((a);x;q)_\lambda = P_\infty(xq;a_1^{-1},\ldots,a_\lambda^{-1})D_{k,i}((a);x;q)_\lambda/\Pi_\infty(-xq,q), \tag{1.4}$$

where

$$\Pi_n(b,q) = \prod_{j=0}^{n-1}(1+bq^j), \qquad \Pi_\infty(b,q) = \prod_{j=0}^{\infty}(1+bq^j),$$

$$p_n(b;y_1,\ldots,y_\lambda) = \prod_{j=1}^{\lambda}(1-by_jq^{n-1}),$$

$$P_n(b;y_1,\ldots,y_\lambda) = \prod_{j=1}^{n}p_j(b;y_1,\ldots,y_\lambda),$$

$$P_\infty(b;y_1,\ldots,y_\lambda) = \prod_{j=1}^{\infty}p_j(b;y_1,\ldots,y_\lambda),$$

$$M_n(x) = M_n((a);\lambda;k;x;q) = (-1)^{n(\lambda+1)}x^{kn}q^{\frac{1}{2}(2k-\lambda+1)n^2+\frac{1}{2}(\lambda+1)n}(a_1\ldots a_\lambda)^{-n}.$$

$C_{k,i}((a);x;q)_\lambda$ generalizes a certain well-poised basic hypergeometric

† Partially supported by National Science Foundation grant GP 6663.

series that appears in several famous theorems: namely,

$$C_{k,1}((a);x;q)_\lambda$$

$$= {}_{\lambda+3}\Phi_{\lambda+2}\left[\begin{matrix} x, qx^{\frac{1}{2}}, -qx^{\frac{1}{2}}, a_1,\ldots, a_\lambda \\ x^{\frac{1}{2}}, -x^{\frac{1}{2}}, xq/a_1,\ldots, xq/a_\lambda \end{matrix}; q, \frac{q^{\frac{1}{2}(2k-\lambda+1)n+\frac{1}{2}(\lambda-1)}x^k(-1)^{\lambda+1}}{a_1 a_2 \ldots a_\lambda}\right]. \quad (1.5)$$

The notation is that of Slater [(**10**) 91]. In particular, when $\lambda = 3, k = 1$, this function appears in the limiting form of Jackson's theorem [(**10**) 96, equation (3.3.1.3)]; when $\lambda = 5$, $k = 2$, the resulting series appears in Watson's q-analogue of Whipple's theorem [(**10**) 100, equation (3.4.1.5)]. Selberg (**9**) has studied these series in the case $\lambda = 0$; the case $\lambda = 1$ was treated in (**2**).

Our main object is to obtain q-difference equations for these series. Originally these results were to appear in a future work on partitions (**1**); however, once the results were derived, it was found that, apart from their intrinsic interest, they were quite useful in proving theorems on infinite products, continued fractions, and q-series. Hence it was felt that a separate exhibition of these theorems would be worth while.

In § 3, we shall derive some theorems on infinite products. For example,

$$_4\Phi_3\left[\begin{matrix} x, qx^{\frac{1}{2}}, -qx^{\frac{1}{2}}, a; q, x^{\frac{1}{2}}q^{\frac{1}{2}n}a^{-1} \\ x^{\frac{1}{2}}, -x^{\frac{1}{2}}, xqa^{-1} \end{matrix}\right] = \frac{\Pi_\infty(-xq,q)\Pi_\infty((xq)^{\frac{1}{2}}a^{-1},q)}{\Pi_\infty(-xqa^{-1},q)\Pi_\infty((xq)^{\frac{1}{2}},q)}. \quad (1.6)$$

We shall also obtain known results like the limiting form of Jackson's theorem [(**10**) 96, equation (3.3.1.3)].

In § 4, we obtain new theorems on continued fractions. The most interesting result here seems to be Theorem 6, which may well be the 'general theorem' on continued fractions referred to by Ramanujan in one of his letters to Hardy [(**7**) xxviii].

THEOREM 6. *If* $a_1^{-1}+a_2^{-1} = -b$, *and* $a_1^{-1}a_2^{-1} = -a$, *then*

$$\frac{H_{2,1}(a_1,a_2;x;q)}{H_{2,1}(a_1,a_2;xq;q)} = 1+bxq+\frac{xq(1+axq^2)}{1+bxq^2+}\,\frac{xq^2(1+axq^3)}{1+bxq^3+}\cdots.$$

As we shall see, this theorem contains many of the most elegant known results on continued fractions of the Rogers–Ramanujan type.

In § 5, we shall apply our results to the study of q-series. Much work has been done on problems of this type by Starcher (**12**) and Fine (**5**) for q-difference equations of first and second order. We shall therefore restrict ourselves to equations of orders 3 and 4.

2. Main theorems

Theorem 1 will contain the principal results; in Theorem 2, we shall collect some useful information for later use.

Theorem 1. *Let k and i be real with $2k+1 \geqslant \lambda \geqslant 0$ where λ is an integer; $|q| < 1$. Furthermore, if $2k+1 = \lambda$, then*

$$|x^k(a_1 \ldots a_\lambda)^{-1} q^{\frac{1}{2}(\lambda+1)-i}| < 1.$$

Under these conditions,

$$H_{k,i}((a);x;q)_\lambda - H_{k,i-1}((a);x;q)_\lambda = x^{i-1} J_{k,k-i+1}((a);x;q)_\lambda; \tag{2.1}$$

$$J_{k,i}((a);x;q)_\lambda = \sum_{j=0}^{\lambda} (-1)^j \sigma_j(a_1^{-1},\ldots,a_\lambda^{-1}) x^j q^j H_{k,i-j}((a);xq;q)_\lambda; \tag{2.2}$$

$$H_{k,-i}((a);x;q)_\lambda = -x^{-i} H_{k,i}((a);x;q)_\lambda; \tag{2.3}$$

$$H_{k,0}((a);x;q)_\lambda = 0. \tag{2.4}$$

$\sigma_j(y_1,\ldots,y_\lambda)$ is the j-th elementary symmetric function of $y_1,\ldots,y_\lambda$; $\sigma_0(y_1,\ldots,y_\lambda) = 1$.

Proof. We start with (2.1).

$$\begin{aligned} &C_{k,i}((a);x;q)_\lambda - C_{k,i-1}((a);x;q)_\lambda \\ &\quad = \sum_{n=0}^{\infty} M_n(x)(1-x)^{-1}\{q^{-in}(1-q^n) + x^{i-1}q^{n(i-1)}(1-xq^n)\} \times \\ &\qquad\qquad \times \frac{\Pi_n(-x,q)\,P_n(1;a_1,\ldots,a_\lambda)}{\Pi_n(-q,q)\,P_n(xq;a_1^{-1},\ldots,a_\lambda^{-1})}. \end{aligned}$$

Consequently splitting the series into two parts by means of the terms enclosed in $\{\}$, we have

$$\begin{aligned} &C_{k,i}((a);x;q)_\lambda - C_{k,i-1}((a);x;q)_\lambda \\ &= \sum_{n=1}^{\infty} M_n(x) q^{-in} \frac{\Pi_{n-1}(-xq,q)\,P_n(1;a_1,\ldots,a_\lambda)}{\Pi_{n-1}(-q,q)\,P_n(xq;a_1^{-1},\ldots,a_\lambda^{-1})} + \\ &\qquad + x^{i-1} \sum_{n=0}^{\infty} M_n(x) q^{n(i-1)} \frac{\Pi_n(-xq,q)\,P_n(1;a_1,\ldots,a_\lambda)}{\Pi_n(-q,q)\,P_n(xq;a_1^{-1},\ldots,a_\lambda^{-1})} \\ &= (-1)^{\lambda+1} x^k q^{k+1-i} (a_1 \ldots a_\lambda)^{-1} \sum_{n=0}^{\infty} M_n(x) q^{(2k-\lambda-i+1)n} \times \\ &\qquad \times \frac{\Pi_n(-xq,q)\,P_{n+1}(1;a_1,\ldots,a_\lambda)}{\Pi_n(-q,q)\,P_{n+1}(xq;a_1^{-1},\ldots,a_\lambda^{-1})} + \\ &\qquad + x^{i-1} \sum_{n=0}^{\infty} M_n(x) q^{n(i-1)} \frac{\Pi_n(-xq,q)\,P_n(1;a_1,\ldots,a_\lambda)}{\Pi_n(-q,q)\,P_n(xq;a_1^{-1},\ldots,a_\lambda^{-1})} \\ &= x^{i-1} \sum_{n=0}^{\infty} M_n(xq) q^{-(k-i+1)n} \frac{\Pi_n(-xq,q)\,P_n(1;a_1,\ldots,a_\lambda)}{\Pi_n(-q,q)\,P_n(xq;a_1^{-1},\ldots,a_\lambda^{-1})} \times \\ &\qquad \times \left\{1 + \frac{(-1)^{\lambda+1}(xq)^{k-i+1} q^{2n(k-i+1)-\lambda n}(a_1 \ldots a_\lambda)^{-1} p_{n+1}(1;a_1,\ldots,a_\lambda)}{p_{n+1}(xq;a_1^{-1},\ldots,a_\lambda^{-1})}\right\} \\ &= x^{i-1} D_{k,k-i+1}((a);x;q)_\lambda. \end{aligned}$$

We now deduce (2.1) utilizing (1.3) and (1.4).

We prove (2.2) as follows. In (1.2) we combine fractions in the term enclosed in $\{\}$. This yields

$$D_{k,i}((a);x;q)_\lambda = \sum_{n=0}^{\infty} M_n(xq)q^{-in}\frac{\Pi_n(-xq,q)P_n(1;a_1,\dots,a_\lambda)}{\Pi_n(-q,q)P_{n+1}(xq;a_1^{-1},\dots,a_\lambda^{-1})}\times$$

$$\times\Big\{\sum_{j=0}^{\lambda}(-1)^j x^j\sigma_j(a_1^{-1},\dots,a_\lambda^{-1})q^{j(n+1)}+$$

$$+(-1)^{\lambda+1}x^i q^{(2n+1)i-\lambda n}\sum_{j=0}^{\lambda}(-1)^{-j}\sigma_j(a_1^{-1},\dots,a^{-1})q^{n(\lambda-j)}\Big\}$$

$$= \sum_{n=0}^{\infty} M_n(xq)q^{-in}\frac{\Pi_n(-xq,q)P_n(1;a_1,\dots,a_\lambda)}{\Pi_n(-q,q)P_{n+1}(xq;a_1^{-1},\dots,a_\lambda^{-1})}\times$$

$$\times\sum_{j=0}^{\lambda}(-1)^j\sigma_j(a_1^{-1},\dots,a_\lambda^{-1})x^jq^{j(n+1)}(1-x^{i-j}q^{(2n+1)(i-j)})$$

$$= (1-xq)(p_1(xq;a_1^{-1},\dots,a_\lambda^{-1}))^{-1}\sum_{j=0}^{\lambda}(-1)^j\sigma_j(a_1^{-1},\dots,a_\lambda^{-1})x^jq^j\times$$

$$\times\sum_{n=0}^{\infty}M_n(xq)q^{-(i-j)n}(1-xq)^{-1}(1-(xq)^{i-j}q^{2n(i-j)})\times$$

$$\times\frac{\Pi_n(-(xq),q)P_n(1;a_1,\dots,a_\lambda)}{\Pi_n(-q,q)P_n((xq)q;a_1^{-1},\dots,a_\lambda^{-1})}$$

$$= (1-xq)(p_1(xq;a_1^{-1},\dots,a_\lambda^{-1}))^{-1}\sum_{j=0}^{\lambda}(-1)^j\sigma_j(a_1^{-1},\dots,a_\lambda^{-1})x^jq^j\times$$

$$\times C_{k,i-j}((a);xq;q)_\lambda.$$

We now deduce (2.2), utilizing (1.3) and (1.4).

(2.3) follows directly from the identity

$$-x^{-i}q^{-in}(1-x^iq^{2ni}) \equiv q^{-n(-i)}(1-x^{-i}q^{2n(-i)}).$$

(2.4) follows directly from the fact that $1-x^iq^{2ni}=0$ for all n when $i=0$.

We now proceed to Theorem 2.

THEOREM 2.

$$C_{k,1}(a_1,\dots,a_\lambda,-x;x^2;q) = (1+x)C_{k-\frac{1}{2},\frac{1}{2}}(a_1,\dots,a_\lambda;x^2;q); \tag{2.5}$$

$$\lim_{a_\lambda\to\infty} H_{k,i}(a_1,\dots,a_\lambda;x;q) = H_{k,i}(a_1,\dots,a_{\lambda-1};x;q); \tag{2.6}$$

$$\lim_{a_\lambda\to\infty} J_{k,i}(a_1,\dots,a_\lambda;x;q) = J_{k,i}(a_1,\dots,a_{\lambda-1};x;q); \tag{2.7}$$

$$H_{k,1}((a);x;q)_\lambda = J_{k,k}((a);x;q)_\lambda = J_{k,k+1}((a);x;q)_\lambda; \tag{2.8}$$

if the two sets $\{a_1,\dots,a_\lambda\}$ and $\{qa_1^{-1},\dots,qa_\lambda^{-1}\}$ are equal, then

$$D_{k,i}((a);1;q)_\lambda = \Pi_\infty(-(-1)^{c-\lambda}q^{2k+1-i-\frac{1}{2}\lambda},q^{2k-\lambda+1})\times$$

$$\times\Pi_\infty(-(-1)^{c-\lambda}q^{i-\frac{1}{2}\lambda},q^{2k-\lambda+1})\Pi_\infty(-q^{2k-\lambda+1},q^{2k-\lambda+1}), \tag{2.9}$$

where $a_1\dots a_\lambda = (-1)^c q^{\frac{1}{2}\lambda}$.

Proof. We start with (2.5).

$$C_{k,1}(a_1,\ldots,a_\lambda,-x;x^2;q)=\sum_{n=0}^{\infty}(-1)^{n(\lambda+1)}x^{2(k-\frac{1}{2})n}(a_1\ldots a_\lambda)^{-n}q^{\frac{1}{2}(2(k-\frac{1}{2})-\lambda+1)n^2}\times$$

$$\times q^{\frac{1}{2}\lambda n}\frac{(1-xq^n)(1+xq^n)}{(1-x^2)}\frac{\Pi_n(-x^2,q)P_n(1;a_1,\ldots,a_\lambda)}{\Pi_n(-q,q)P_n(x^2q;a_1^{-1},\ldots,a_\lambda^{-1})}\frac{(1+x)}{(1+xq^n)}$$

$$=(1+x)C_{k-\frac{1}{2},\frac{1}{2}}(a_1,\ldots,a_\lambda;x^2;q).$$

(2.6) and (2.7) are deduced from an appeal to Tannery's theorem [(**14**) 371] and the fact that

$$\lim_{a_\lambda\to\infty}a_\lambda^{-n}\Pi_n(-a_\lambda,q)=\lim_{a_\lambda\to\infty}(a_\lambda^{-1}-1)(a_\lambda^{-1}-q)\ldots(a_\lambda^{-1}-q^{n-1})$$

$$=(-1)^nq^{\frac{1}{2}n(n-1)}.$$

We obtain the first equation in (2.8) from (2.1) in the case $i=1$ utilizing (2.4). To obtain the remainder of (2.8), we set $i=0$ in (2.1) and then apply (2.3) and (2.4).

Finally we treat (2.9). Under the conditions stated there, we have immediately that $a_1a_2\ldots a_\lambda=(qa_1^{-1})(qa_2^{-1})\ldots(qa)_\lambda^{-1}$ and therefore

$$(a_1a_2\ldots a_\lambda)^2=q^\lambda.$$

Consequently $a_1a_2\ldots a_\lambda=(-1)^cq^{\frac{1}{2}\lambda}$. Thus

$$D_{k,i}((a);1;q)_\lambda$$

$$=\sum_{n=0}^{\infty}(-1)^{n(\lambda-c+1)}q^{\frac{1}{2}(2k-\lambda+1)n^2+\frac{1}{2}n+(k-i)n}\{1+(-1)^{\lambda-c+1}q^{(2n+1)i-\lambda n-\frac{1}{2}\lambda}\}$$

$$=\sum_{n=-\infty}^{\infty}(-1)^{n(\lambda-c+1)}q^{\frac{1}{2}(2k-\lambda+1)n^2+\frac{1}{2}(2k-2i+1)n}.$$

(2.9) now follows from Jacobi's identity [(**10**) 105, equation (3.5.8)].

3. Infinite products

By means of equations (2.1)–(2.4) and (2.8), we may derive first-order difference equations for $H_{k,1}((a);x;q)_\lambda$ in the cases $k=0$, $\lambda=1$; $k=\frac{1}{2}$, $\lambda=1$; $k=1$, $\lambda=3$. Formulae for infinite products are then easily deduced.

THEOREM 3.

$$_4\Phi_3\begin{bmatrix}x,qx^{\frac{1}{2}},-qx^{\frac{1}{2}},a;q,a^{-1}\\ x^{\frac{1}{2}},-x^{\frac{1}{2}},xq/a\end{bmatrix}=0.$$

Proof.

$$H_{0,1}(a;x;q)=J_{0,0}(a;x;q)=J_{0,1}(a;x;q)\quad(\text{by (2.8) } k=0).$$

$$J_{0,1}(a;x;q)=H_{0,1}(a;xq;q)\quad(\text{by (2.2) } k=0,\ i=1,\text{ and (2.4)}).$$

Therefore $$H_{0,1}(a;x;q)=H_{0,1}(a;xq;q).$$

This implies $$H_{0,1}(a;x;q) = H_{0,1}(a;xq^N;q)$$

for any integer $N \geqslant 0$. Consequently since $|q| < 1$ and $H_{0,1}(a;x;q)$ is continuous in x at $x = 0$, we have, by (**10**) 92, equation (3.2.2.11),

$$H_{0,1}(a;x;q) = \sum_{n=0}^{\infty} \frac{\Pi_n(-a,q)a^{-n}}{\Pi_n(-q,q)} = 0.$$

Therefore, by (1.3), $$C_{0,1}(a;x;q) = 0.$$

This reduces to Theorem 3 by (1.5) with $k = 0$, $\lambda = 1$.

THEOREM 4. *Equation* (1.6) *is valid.*

Proof.

$$H_{\frac{1}{2},1}(a;x;q) = J_{\frac{1}{2},\frac{1}{2}}(a;x;q) = J_{\frac{1}{2},\frac{3}{2}}(a;x;q) \quad (\text{by } (2.8)\ k = \tfrac{1}{2}).$$

$$(1+x^{\frac{1}{2}})H_{\frac{1}{2},\frac{1}{2}}(a;x;q) = J_{\frac{1}{2},1}(a;x;q) \quad (\text{by } (2.1)\ k = \tfrac{1}{2}, \text{ and } (2.3)).$$

$$J_{\frac{1}{2},1}(a;x;q) = H_{\frac{1}{2},1}(a;xq;q) \quad (\text{by } (2.2)\ k = \tfrac{1}{2},\ i = 1,\ \lambda = 1, \text{ and } (2.4)).$$

$$J_{\frac{1}{2},\frac{1}{2}}(a;x;q) = (1+(xq)^{\frac{1}{2}}a^{-1})H_{\frac{1}{2},\frac{1}{2}}(a;xq;q)$$
$$(\text{by } (2.2)\ k = \tfrac{1}{2},\ i = \tfrac{1}{2},\ \lambda = 1,\ (2.3), \text{ and } (2.4)).$$

Hence

$$\begin{aligned} H_{\frac{1}{2},1}(a;x;q) &= (1+(xq)^{\frac{1}{2}}a^{-1})H_{\frac{1}{2},\frac{1}{2}}(a;xq;q) \\ &= (1+(xq)^{\frac{1}{2}}a^{-1})(1+(xq)^{\frac{1}{2}})^{-1}J_{\frac{1}{2},1}(a;xq;q) \\ &= (1+(xq)^{\frac{1}{2}}a^{-1})(1+(xq)^{\frac{1}{2}})^{-1}H_{\frac{1}{2},1}(a;xq^2;q). \end{aligned}$$

We may now deduce

$$H_{\frac{1}{2},1}(a;x;q) = \frac{\Pi_\infty((xq)^{\frac{1}{2}}/a,q)}{\Pi_\infty((xq)^{\frac{1}{2}},q)}$$

by the technique employed in Theorem 3.

Consequently, by (1.3),

$$C_{\frac{1}{2},1}(a;x;q) = \frac{\Pi_\infty(-xq,q)\Pi_\infty((xq)^{\frac{1}{2}}/a,q)}{\Pi_\infty(-xq/a,q)\Pi_\infty((xq)^{\frac{1}{2}},q)}.$$

This reduces to Theorem 4 by (1.5) with $k = \frac{1}{2}$, $\lambda = 1$.

THEOREM 5.

$${}_6\Phi_5\begin{bmatrix} x, qx^{\frac{1}{2}}, -qx^{\frac{1}{2}}, a_1, a_2, a_3;\ q, xq/a_1a_2a_3 \\ x^{\frac{1}{2}}, -x^{\frac{1}{2}}, xq/a_1, xq/a_2, xq/a_3 \end{bmatrix}$$
$$= \frac{\Pi_\infty(-xq,q)\Pi_\infty(-xq/a_1a_2,q)\Pi_\infty(-xq/a_1a_3,q)\Pi_\infty(-xq/a_2a_3,q)}{\Pi_\infty(-xq/a_1,q)\Pi_\infty(-xq/a_2,q)\Pi_\infty(-xq/a_3,q)\Pi_\infty(-xq/a_1a_2a_3,q)}.$$

Remark. This theorem is known as the limiting form of Jackson's theorem [(**10**) 96, equation (3.3.1.3)].

Proof. We let $\sigma_j = \sigma_j(a_1^{-1}, a_2^{-1}, a_3^{-1})$.

$$H_{1,1}((a);x;q)_3 = J_{1,1}((a);x;q)_3 = J_{1,2}((a);x;q)_3 \quad \text{(by (2.8) } k = 1\text{)}. \tag{3.1}$$

$$J_{1,1}((a);x;q)_3 = (1-\sigma_2 xq)H_{1,1}((a);xq;q)_3 + \sigma_3 xqH_{1,2}((a);xq;q)_3$$
$$\text{(by (2.2) } k = 1,\ i = 1,\ \lambda = 3, \text{(2.3), and (2.4))}. \tag{3.2}$$

$$J_{1,2}((a);x;q)_3 = H_{1,2}((a);xq;q)_3 - xq(\sigma_1 - xq\sigma_3)H_{1,1}((a);xq;q)_3$$
$$\text{(by (2.2) } k = 1,\ i = 2,\ \lambda = 3, \text{(2.3), and (2.4))}. \tag{3.3}$$

Hence, by (3.3) and (3.1),

$$H_{1,2}((a);xq;q)_3 = H_{1,1}((a);x;q)_3 + xq(\sigma_1 - xq\sigma_3)H_{1,1}((a);xq;q)_3. \tag{3.4}$$

Substituting (3.4) into (3.2) and utilizing (3.1), we obtain

$$(1-xq\sigma_3)H_{1,1}((a);x;q)_3 = (1-\sigma_2 xq + x^2q^2\sigma_1\sigma_3 - x^3q^3\sigma_3^2)H_{1,1}((a);xq;q)_3. \tag{3.5}$$

We may rewrite this last formula as

$$H_{1,1}((a);x;q)_3 = \frac{(1-xq/a_1a_2)(1-xq/a_1a_3)(1-xq/a_2a_3)H_{1,1}((a);xq;q)_3}{(1-xq/a_1a_2a_3)}. \tag{3.6}$$

We may now deduce

$$H_{1,1}((a);x;q)_3 = \frac{\Pi_\infty(-xq/a_1a_2, q)\Pi_\infty(-xq/a_1a_3, q)\Pi_\infty(-xq/a_2a_3, q)}{\Pi_\infty(-xq/a_1a_2a_3, q)} \tag{3.7}$$

by the technique employed in Theorem 3.

Consequently, by (1.3),

$$C_{1,1}((a);x;q)_3$$
$$= \frac{\Pi_\infty(-xq, q)\Pi_\infty(-xq/a_1a_2, q)\Pi_\infty(-xq/a_1a_3, q)\Pi_\infty(-xq/a_2a_3, q)}{\Pi_\infty(-xq/a_1, q)\Pi_\infty(-xq/a_2, q)\Pi_\infty(-xq/a_3, q)\Pi_\infty(-xq/a_1a_2a_3, q)}. \tag{3.8}$$

This reduces to Theorem 5 by (1.5) with $k = 1$, $\lambda = 3$.

4. Continued fractions of the Rogers–Ramanujan type

Our results here depend for the most part on the derivation of second-order q-difference equations. We shall study the cases $k = 2$, $i = 1$, $\lambda = 2$; $k = 1$, $i = \frac{1}{2}$, $\lambda = 1$.

We start with the result stated in § 1, as Theorem 6.

Proof of Theorem 6. Here we let $\sigma_1(a_1^{-1}, a_2^{-1}) = -b$, $\sigma_2(a_1^{-1}, a_2^{-1}) = -a$.

$$H_{2,2}((a);x;q)_2 - H_{2,1}((a);x;q)_2 = xJ_{2,1}((a);x;q)_2 \quad \text{(by (2.1) } k = i = 2\text{)}. \tag{4.1}$$

$$H_{2,1}((a);x;q)_2 = J_{2,2}((a);x;q)_2 = J_{2,3}((a);x;q)_2 \quad \text{(by (2.8) } k = 2\text{)}. \tag{4.2}$$

$$J_{2,2}((a);x;q)_2 = H_{2,2}((a);xq;q)_2 + bxqH_{2,1}((a);xq;q)_2 \quad \text{(by (2.2) } k=i=\lambda=2\text{)}. \tag{4.3}$$

$$J_{2,1}((a);x;q)_2 = (1+axq)H_{2,1}((a);xq;q)_2 \quad \text{(by (2.2) } k=2,\ i=1,\ \lambda=2\text{, (2.3), and (2.4))}. \tag{4.4}$$

Now (4.2) and (4.3) imply that

$$H_{2,1}((a);x;q)_2 - bxqH_{2,1}((a);xq;q)_2 = H_{2,2}((a);xq;q)_2. \tag{4.5}$$

(4.1), (4.4), and (4.5) imply that

$$H_{2,1}((a);x;q)_2 - (bxq+1)H_{2,1}((a);xq;q)_2 = xq(1+axq^2)H_{2,1}((a);xq^2;q)_2. \tag{4.6}$$

Consequently, rewriting (4.6), we obtain

$$H_{2,1}((a);x;q)_2 = (1+bxq)H_{2,1}((a);xq;q)_2 + xq(1+axq^2)H_{2,1}((a);xq^2;q)_2. \tag{4.7}$$

Thus

$$\frac{H_{2,1}((a);x;q)_2}{H_{2,1}((a);xq;q)_2} = 1+bxq+\frac{xq(1+axq^2)}{H_{2,1}((a);xq;q)_2/H_{2,1}((a);xq^2;q)_2}. \tag{4.8}$$

By iterating (4.8) we obtain Theorem 6.

COROLLARY 1. (Ramanujan [(**7**) 215])

$$\frac{\Pi_\infty(-q^2,q^5)\Pi_\infty(-q^3,q^5)}{\Pi_\infty(-q,q^5)\Pi_\infty(-q^4,q^5)} = 1+\frac{q}{1+}\frac{q^2}{1+}\frac{q^3}{1+}\cdots.$$

Proof. In Theorem 6, let $a_1 \to \infty$, $a_2 \to \infty$. This yields $b = a = 0$. Utilizing (4.2), (2.6), (2.9), and (2.2) for $\lambda = 0$, we find

$$H_{2,1}(\ ;1;q) = J_{2,2}(\ ;1;q) = \frac{\Pi_\infty(-q^2,q^5)\Pi_\infty(-q^3,q^5)\Pi_\infty(-q^5,q^5)}{\Pi_\infty(-q,q)};$$

$$H_{2,1}(\ ;q;q) = J_{2,1}(\ ;1;q) = \frac{\Pi_\infty(-q,q^5)\Pi_\infty(-q^4,q^5)\Pi_\infty(-q^5,q^5)}{\Pi_\infty(-q,q)}.$$

The desired result now follows directly from Theorem 6.

The next result is equivalent to one stated by Ramanujan [(**7**) xxviii] and was proved by G. N. Watson [(**13**) 236].

COROLLARY 2.

$$\frac{\Pi_\infty(-q^3,q^6)\Pi_\infty(-q^3,q^6)}{\Pi_\infty(-q,q^6)\Pi_\infty(-q^5,q^6)} = 1+\frac{q(1+q)}{1+}\frac{q^2(1+q^2)}{1+}\frac{q^3(1+q^3)}{1+}\cdots.$$

Proof. In Theorem 6, let $a_1 = -a_2 = q^{\frac{1}{2}}$. Hence $b = 0$, $a = q^{-1}$. The desired result now follows as in Corollary 1.

COROLLARY 3. (Gordon [(6) 741])

$$\frac{\Pi_\infty(-q^3,q^8)\Pi_\infty(-q^5,q^8)}{\Pi_\infty(-q,q^8)\Pi_\infty(-q^7,q^8)} = 1+q+\frac{q^2}{1+q^3+}\,\frac{q^4}{1+q^5+}\,\frac{q^6}{1+q^7+}\cdots.$$

Proof. In Theorem 6, first replace q by q^2, then set $a_1 = -q$, and let $a_2 \to \infty$. Hence $a = 0$, $b = q^{-1}$. The desired result now follows as in Corollary 1.

Our next theorem generalizes a continued fraction expansion of A. Selberg [(9) 19].

THEOREM 7. *Let* $a_1^{-1} = -a$. *Then*

$$\frac{(1+x)H_{1,\frac12}(a_1;x^2;q^2)}{H_{1,\frac12}(a_1;x^2q^2;q^2)} = 1+ax^2q^2+\frac{xq(1-axq^2)(1+xq)}{1+ax^2q^4+}\,\frac{xq^2(1-axq^3)(1+xq^2)}{1+ax^2q^6+}\cdots.$$

Proof. Here $\sigma_1(a_1^{-1}) = -a$.

$$H_{1,\frac12}(a_1;x;q)(1+x^{\frac12}) = J_{1,\frac32}(a_1;x;q) \quad \text{(by (2.1) with } k=1,\ i=\tfrac12). \tag{4.9}$$

$$H_{1,\frac32}(a_1;x;q)-H_{1,\frac12}(a_1;x;q) = x^{\frac12}J_{1,\frac12}(a_1;x;q) \quad \text{(by (2.1) with } k=1,\ i=\tfrac32). \tag{4.10}$$

$$J_{1,\frac12}(a_1;x;q) = H_{1,\frac12}(a_1;xq;q)-ax^{\frac12}q^{\frac12}H_{1,\frac12}(a_1;xq;q) \quad \text{(by (2.2) with } k=1,\ i=\tfrac12\text{, and (2.3))}. \tag{4.11}$$

$$J_{1,\frac32}(a_1;x;q) = H_{1,\frac32}(a_1;xq;q)+axqH_{1,\frac12}(a_1;xq;q) \quad \text{(by (2.2) with } k=1,\ i=\tfrac32). \tag{4.12}$$

Hence, by (4.9), (4.10), and (4.12),

$$(1+x^{\frac12})H_{1,\frac12}(a_1;x;q) = (1+axq)H_{1,\frac12}(a_1;xq;q)+x^{\frac12}q^{\frac12}J_{1,\frac12}(a_1;xq;q). \tag{4.13}$$

Thus, by (4.11) and (4.13),

$$(1+x^{\frac12})H_{1,\frac12}(a_1;x;q) = (1+axq)H_{1,\frac12}(a_1;xq;q)+x^{\frac12}q^{\frac12}(1-ax^{\frac12}q)H_{1,\frac12}(a_1;xq^2;q). \tag{4.14}$$

If we let $$f(x) = \Pi_\infty(x,q)H_{1,\frac12}(a_1;x^2;q^2),$$

then $$f(x) = (1+ax^2q^2)f(xq)+xq(1-axq^2)(1+xq)f(xq^2).$$

Consequently

$$\frac{f(x)}{f(xq)} = 1+ax^2q^2+\frac{xq(1-axq^2)(1+xq)}{f(xq)/f(xq^2)}. \tag{4.15}$$

By iterating (4.15), we obtain Theorem 7.

5. q-series identities

Intricate q-series identities seem to be somewhat difficult to obtain by the method of q-difference equations. Although the series form of the Rogers–Ramanujan identities is relatively easily derived by the method of q-difference equations [(9) 11], Selberg deduced the corresponding identities of modulus 7 from the related q-difference equation by involved and ingenious use of certain recurrence formulae. We shall show in Theorem 8 that a much simpler, though possibly less revealing, derivation of the identities of modulus 7 may be given. In Theorem 9, we shall give a similar proof of Bailey's identities related to the modulus 9.

LEMMA 1. *Suppose, for* $|q| < 1$, *there exists a function* $f(x)$ *which is analytic at* $x = 0$, *and satisfies*

$$\sum_{i=0}^{s} p_i(x,q) f(xq^i) = 0, \qquad f(0) = 1 \tag{5.1}$$

$\Big($*where* $p_i(x,q)$ *is a polynomial in* x *and* q, $p_i(x,q) = \sum_{j=0}^{d_i} \xi_j^{(i)}(q) x^j\Big)$, *and that* $\sum_{i=0}^{s} \xi_0^{(i)}(q) y^i \neq 0$ *for* $|y| < 1$. *Then* $f(x)$ *is the only function analytic at* 0 *for which* (5.1) *is fulfilled.*

Proof. Suppose $g(x)$ is analytic at 0 and fulfils (5.1). Let

$$g(x) = \sum_{n=0}^{\infty} B_n(q) x^n; \qquad f(x) = \sum_{n=0}^{\infty} A_n(q) x^n.$$

Now $B_0(q) = A_0(q) = 1$, since $g(0) = f(0) = 1$. Assume $B_n(q) = A_n(q)$ for $n < N$. Then, by (5.1),

$$0 = \sum_{i=0}^{s} \sum_{j=0}^{d_i} \xi_j^{(i)}(q) x^j \sum_{n=0}^{\infty} A_n(q) x^n q^{in} = \sum_{m=0}^{\infty} x^m \textstyle\sum_1 \xi_j^{(i)}(q) A_n(q) q^{in},$$

where $\sum_1$ is over all j and n such that $j+n = m$, $0 \leqslant n$, $0 \leqslant j \leqslant d_i$, $0 \leqslant i \leqslant s$. Hence the inner sum in the last series is zero for all m, and the same is true if $A_n(q)$ is replaced by $B_n(q)$.

Thus

$$A_N(q) \sum_{i=0}^{s} \xi_0^{(i)}(q) q^{in} + \sum_{\substack{j+n=N \\ 0 \leqslant n < N,\, 0 < j \leqslant d_i \\ 0 \leqslant i \leqslant s}} \xi_j^{(i)}(q) A_n(q) q^{in} = 0. \tag{5.2}$$

But, by hypothesis, $\sum_{i=0}^{s} \xi_0^{(i)}(q) q^{in} \neq 0$. Hence

$$\begin{aligned} A_N(q) &= \Big(\sum_{i=0}^{s} \xi_0^{(i)}(q) q^{in}\Big)^{-1} \big(-\textstyle\sum_2 \xi_j^{(i)}(q) A_n(q) q^{in}\big) \\ &= \Big(\sum_{i=0}^{s} \xi_0^{(i)}(q) q^{in}\Big)^{-1} \big(-\textstyle\sum_2 \xi_j^{(i)}(q) B_n(q) q^{in}\big) \\ &= B_N(q), \end{aligned}$$

where $\sum_2$ is over all j, n, and i such that

$$j+n = N,\ 0 \leqslant n < N,\ 0 < j \leqslant d_i,\ 0 \leqslant i \leqslant s.$$

Hence by induction, $A_n(q) = B_n(q)$ for all n. Therefore $f(x) = g(x)$.

The next theorem is originally due to Rogers [(**8**) 331]; however, it was rediscovered by Selberg [(**9**) 15], and was subsequently proved by Dyson [(**4**) 35], Bailey [(**3**) 421], and Slater [(**10**) 155].

THEOREM 8.

$$\prod_{\substack{n=1\\ n\not\equiv 0,\ \pm 3 \pmod 7}}^{\infty} (1-q^n)^{-1} = \Pi_\infty(q,q) \sum_{n=0}^{\infty} \frac{q^{2n^2}}{\Pi_n(-q^2,q^2)\Pi_{2n}(q,q)}; \quad (5.3)$$

$$\prod_{\substack{n=1\\ n\not\equiv 0,\ \pm 2 \pmod 7}}^{\infty} (1-q^n)^{-1} = \Pi_\infty(q,q) \sum_{n=0}^{\infty} \frac{q^{2n^2+2n}}{\Pi_n(-q^2,q^2)\Pi_{2n}(q,q)}; \quad (5.4)$$

$$\prod_{\substack{n=1\\ n\not\equiv 0,\ \pm 1 \pmod 7}}^{\infty} (1-q^n)^{-1} = \Pi_\infty(q,q) \sum_{n=0}^{\infty} \frac{q^{2n^2+2n}}{\Pi_n(-q^2,q^2)\Pi_{2n+1}(q,q)}. \quad (5.5)$$

Proof. From (2.1) and (2.2), we see that if

$$Q_i(x) = J_{3,i}(\ ;x;q),$$

then

$$Q_3(x)-Q_2(x) = x^2q^2Q_1(xq); \quad (5.6)$$

$$Q_2(x)-Q_1(x) = xqQ_2(xq); \quad (5.7)$$

$$Q_1(x) = Q_3(xq). \quad (5.8)$$

These equations may be solved for $Q_3(x)$ to yield

$$Q_3(x)-(1+xq)Q_3(xq)-x^2q^2Q_3(xq^2)+x^3q^5Q_3(xq^3) = 0. \quad (5.9)$$

Hence by Lemma 1, (5.9) has a unique solution (since it has one solution) with $Q_3(0) = 1$. Therefore the system (5.6)–(5.8) has unique solutions with $Q_1(0) = Q_2(0) = Q_3(0) = 1$.

Consider now

$$R_3(x) = \sum_{m=0}^{\infty} \frac{x^{2m}q^{2m^2}\Pi_\infty(xq^{2m+1},q)}{\Pi_m(-q^2,q^2)}; \quad (5.10)$$

$$R_2(x) = \sum_{m=0}^{\infty} \frac{x^{2m}q^{2m^2+2m}\Pi_\infty(xq^{2m+1},q)}{\Pi_m(-q^2,q^2)}; \quad (5.11)$$

$$R_1(x) = \sum_{m=0}^{\infty} \frac{x^{2m}q^{2m^2+2m}\Pi_\infty(xq^{2m+2},q)}{\Pi_m(-q^2,q^2)}. \quad (5.12)$$

Clearly

$$R_1(x) = R_3(xq).$$

Also

$$\begin{aligned} R_3(x)-R_2(x) &= \sum_{m=0}^{\infty} \frac{x^{2m}q^{2m^2}\Pi_\infty(xq^{2m+1},q)(1-q^{2m})}{\Pi_m(-q^2,q^2)} \\ &= \sum_{m=0}^{\infty} \frac{x^{2(m+1)}q^{2(m+1)^2}\Pi_\infty(xq^{2m+3},q)}{\Pi_m(-q^2,q^2)} \\ &= x^2q^2 \sum_{m=0}^{\infty} \frac{(xq)^{2m}q^{2m^2+2m}\Pi_\infty((xq)q^{2m+2},q)}{\Pi_m(-q^2,q^2)} \\ &= x^2q^2R_1(xq). \end{aligned}$$

Finally

$$\begin{aligned} R_2(x)-R_1(x) &= \sum_{m=0}^{\infty} \frac{x^{2m}q^{2m^2+2m}\Pi_\infty(xq^{2m+2},q)(1+xq^{2m+1}-1)}{\Pi_m(-q^2,q^2)} \\ &= xq \sum_{m=0}^{\infty} \frac{(xq)^{2m}q^{2m^2+2m}\Pi_\infty((xq)q^{2m+1},q)}{\Pi_m(-q^2,q^2)} \\ &= xqR_2(xq). \end{aligned}$$

Thus we see that the functions $R_a(x)$ $(a = 1, 2, 3)$ fulfil (5.6), (5.7), and (5.8); therefore by the remark following (5.9), $R_a(x) = Q_a(x)$ $(a = 1, 2, 3)$. Taking $x = 1$, $a = 1, 2, 3$, and utilizing (2.9), we obtain (5.3), (5.4), and (5.5).

Our last theorem is originally due to Bailey [(**3**) 422]; it has subsequently been proved by L. J. Slater [(**11**) 156].

THEOREM 9.

$$\prod_{\substack{n=1 \\ n\not\equiv 0,\ \pm 4 \pmod 9}}^{\infty} (1-q^n)^{-1} = \frac{\Pi_\infty(-q^3,q^3)}{\Pi_\infty(-q,q)} \sum_{m=0}^{\infty} \frac{q^{3m^2}\Pi_{3m}(-q,q)}{\Pi_m(-q^3,q^3)\Pi_{2m}(-q^3,q^3)}; \tag{5.13}$$

$$\prod_{\substack{n=1 \\ n\not\equiv 0,\ \pm 2 \pmod 9}}^{\infty} (1-q^n)^{-1} = \frac{\Pi_\infty(-q^3,q^3)}{\Pi_\infty(-q,q)} \sum_{m=0}^{\infty} \frac{q^{3m^2+3m}\Pi_{3m}(-q,q)(1-q^{3m+2})}{\Pi_m(-q^3,q^3)\Pi_{2m+1}(-q^3,q^3)}; \tag{5.14}$$

$$\prod_{\substack{n=1 \\ n\not\equiv 0,\ \pm 1 \pmod 9}}^{\infty} (1-q^n)^{-1} = \frac{\Pi_\infty(-q^3,q^3)}{\Pi_\infty(-q,q)} \sum_{m=0}^{\infty} \frac{q^{3m^2+3m}\Pi_{3m+1}(-q,q)}{\Pi_m(-q^3,q^3)\Pi_{2m+1}(-q^3,q^3)}. \tag{5.15}$$

Proof. From (2.1) and (2.2), we see that if

$$S_i(x) = J_{4,i}(\ ;x;q),$$

then

$$S_4(x)-S_3(x) = x^3q^3S_1(xq); \tag{5.16}$$

$$S_3(x)-S_2(x) = x^2q^2S_2(xq); \tag{5.17}$$

$$S_2(x)-S_1(x) = xqS_3(xq); \tag{5.18}$$

$$S_1(x) = S_4(xq). \tag{5.19}$$

These equations may be solved for $S_4(x)$ to yield

$$S_4(x)-(1+xq)S_4(xq)-x^2q^2(1+xq+xq^2)S_4(xq^2)+ \\ +x^4q^7S_4(xq^3)+x^6q^{13}S_4(xq^4) = 0. \tag{5.20}$$

Hence by Lemma 1, (5.20) has a unique solution (since it has one) with $S_4(0) = 1$. Therefore the system (5.16)–(5.19) has unique solutions with $S_1(0) = S_2(0) = S_3(0) = S_4(0) = 1$.

Consider now

$$T_4(x) = \sum_{m=0}^{\infty} \frac{x^{3m}q^{3m^2}\Pi_\infty(-x^3q^{6m+3},q^3)}{\Pi_m(-q^3,q^3)\Pi_\infty(-xq^{3m+1},q)}; \tag{5.21}$$

$$T_3(x) = \sum_{m=0}^{\infty} \frac{x^{3m}q^{3m^2+3m}(1-x)\Pi_\infty(-x^3q^{6m+3},q^3)}{\Pi_m(-q^3,q^3)\Pi_\infty(-xq^{3m},q)}; \tag{5.22}$$

$$T_2(x) = \sum_{m=0}^{\infty} \frac{x^{3m}q^{3m^2+3m}(1-x^2q^{3m+2})\Pi_\infty(-x^3q^{6m+6},q^3)}{\Pi_m(-q^3,q^3)\Pi_\infty(-xq^{3m+1},q)}; \tag{5.23}$$

$$T_1(x) = \sum_{m=0}^{\infty} \frac{x^{3m}q^{3m^2+3m}\Pi_\infty(-x^3q^{6m+6},q^3)}{\Pi_m(-q^3,q^3)\Pi_\infty(-xq^{3m+2},q)}. \tag{5.24}$$

Clearly

$$T_1(x) = T_4(xq).$$

Furthermore

$$\begin{aligned} T_2(x)-T_1(x) &= \sum_{m=0}^{\infty} \frac{x^{3m}q^{3m^2+3m}\Pi_\infty(-x^3q^{6m+6},q^3)(1-x^2q^{3m+2}-1+xq^{3m+1})}{\Pi_m(-q^3,q^3)\Pi_\infty(-xq^{3m+1},q)} \\ &= \sum_{m=0}^{\infty} \frac{x^{3m}q^{3m^2+3m}\Pi_\infty(-x^3q^{6m+6},q^3)xq^{3m+1}(1-xq)}{\Pi_m(-q^3,q^3)\Pi_\infty(-xq^{3m+1},q)} \\ &= xq\sum_{m=0}^{\infty} \frac{(xq)^{3m}q^{3m^2+3m}\Pi_\infty(-(xq)^3q^{6m+3},q^3)(1-xq)}{\Pi_m(-q^3,q^3)\Pi_\infty(-(xq)q^{3m},q)} \\ &= xqT_3(xq). \end{aligned}$$

Also

$$\begin{aligned} T_4(x)-T_3(x) &= \sum_{m=0}^{\infty} \frac{x^{3m}q^{3m^2}\Pi_\infty(-x^3q^{6m+3},q^3)(1-xq^{3m}-q^{3m}+xq^{3m})}{\Pi_m(-q^3,q^3)\Pi_\infty(-xq^{3m},q)} \\ &= x^3q^3\sum_{m=0}^{\infty} \frac{(xq)^{3m}q^{3m^2+3m}\Pi_\infty(-(xq)^3q^{6m+6},q^3)}{\Pi_m(-q^3,q^3)\Pi_\infty(-(xq)q^{3m+2},q)} \\ &= x^3q^3T_1(xq). \end{aligned}$$

In order to verify (5.17) for the $T_a(x)$, we must make use of the following auxiliary function, $h(x)$.

$$h(x) = \sum_{m=0}^{\infty} \frac{x^{3m+4}q^{3(m+1)(m+2)}\prod_\infty(-x^3q^{6m+9}, q^3)}{\prod_m(-q^3, q^3)\prod_\infty(-xq^{3m+3}, q)}$$
$$= \sum_{m=0}^{\infty} \frac{x^{3m+1}q^{3m(m+1)}\prod_\infty(-x^3q^{6m+3}, q^3)(1-q^{3m})}{\prod_m(-q^3, q^3)\prod_\infty(-xq^{3m}, q)}.$$

Thus

$$T_3(x)-T_2(x)+h(x)$$
$$= \sum_{m=0}^{\infty} \left\{\frac{x^{3m}q^{3m^2+3m}\prod_\infty(-x^3q^{6m+3}, q^3)(1-x)}{\prod_m(-q^3, q^3)\prod_\infty(-xq^{3m}, q)} - \right.$$
$$-\frac{x^{3m}q^{3m^2+3m}(1-x^2q^{3m+2})\prod_\infty(-x^3q^{6m+6}, q^3)}{\prod_m(-q^3, q^3)\prod_\infty(-xq^{3m+1}, q)} +$$
$$\left. + \frac{x^{3m+1}q^{3m^2+3m}\prod_\infty(-x^3q^{6m+3}, q^3)(1-q^{3m})}{\prod_m(-q^3, q^3)\prod_\infty(-xq^{3m}, q)}\right\}$$
$$= \sum_{m=0}^{\infty} \frac{x^{3m}q^{3m^2+3m}\prod_\infty(-x^3q^{6m+6}, q^3)}{\prod_m(-q^3, q^3)\prod_\infty(-xq^{3m}, q)} \times$$
$$\times\{(1-x)(1-x^3q^{6m+3})-(1-x^2q^{3m+2})(1-xq^{3m})+x(1-q^{3m})(1-x^3q^{6m+3})\}$$
$$= \sum_{m=0}^{\infty} \frac{x^{3m}q^{3m^2+3m}\prod_\infty(-x^3q^{6m+6}, q^3)}{\prod_m(-q^3, q^3)\prod_\infty(-xq^{3m}, q)} x^2q^{3m+2}(1-xq^{3m})(1-xq^{3m+1})$$
$$= \sum_{m=0}^{\infty} \frac{x^2q^2x^{3m}q^{3m^2+6m}\prod_\infty(-x^3q^{6m+9}, q^3)(1-x^3q^{6m+6})}{\prod_m(-q^3, q^3)\prod_\infty(-xq^{3m+2}, q)}$$
$$= \sum_{m=0}^{\infty} \frac{x^2q^2x^{3m}q^{3m^2+6m}\prod_\infty(-x^3q^{6m+9}, q^3)(1-x^2q^{3m+4}+x^2q^{3m+4}(1-xq^{3m+2}))}{\prod_m(-q^3, q^3)\prod_\infty(-xq^{3m+2}, q)}$$
$$= x^2q^2 \sum_{m=0}^{\infty} \frac{(xq)^{3m}q^{3m^2+3m}\prod_\infty(-(xq)^3q^{6m+6}, q^3)(1-(xq)^2q^{3m+2})}{\prod_m(-q^3, q^3)\prod_\infty(-xq^{3m+2}, q)} +$$
$$+ \sum_{m=0}^{\infty} \frac{x^{3m+4}q^{3(m+1)(m+2)}\prod_\infty(-x^3q^{6m+9}, q^3)}{\prod_m(-q^3, q^3)\prod_\infty(-xq^{3m-3}, q^3)}$$
$$= x^2q^2T_2(xq)+h(x).$$

Therefore $$T_3(x)-T_2(x) = x^2q^2T_2(xq).$$

Thus we see that the functions $T_a(x)$ ($a = 1, 2, 3, 4$) fulfil (5.16)–(5.19); therefore by the remark following (5.20), $T_a(x) = S_a(x)$ ($a = 1, 2, 3, 4$). Taking $x = 1$, $a = 1, 2, 4$ and utilizing (2.9), we obtain (5.13), (5.14), and (5.15).

REFERENCES

1. G. E. Andrews, 'A generalization of the classical partition theorems' (to appear).

2. —— 'A generalization of the Göllnitz–Gordon partition theorems', *Proc. Amer. Math. Soc.* 18 (1967) 945–52.

3. W. N. Bailey, 'Some identities in combinatory analysis', *Proc. London Math. Soc.* (2) 49 (1947) 421–35.

4. F. J. Dyson, 'Three identities in combinatory analysis', *J. London Math. Soc.* 18 (1943) 35–39.

5. N. J. Fine, 'Some basic hypergeometric series and applications' (unpublished monograph).

6. B. Gordon, 'Some continued fractions of the Rogers–Ramanujan type', *Duke Math. J.* 32 (1965) 741–8.

7. S. Ramanujan, *Collected Papers* (Cambridge, 1927).

8. L. J. Rogers, 'On two theorems of combinatory analysis and some allied identities', *Proc. London Math. Soc.* (2) 16 (1917) 315–36.

9. A. Selberg, 'Über einige arithmetische Identitäten', *Avh. norske Vidensk-Akad.* (1936) 1–23 (No. 8).

10. L. J. Slater, *Generalized hypergeometric functions* (Cambridge, 1966).

11. —— 'Further identities of the Rogers–Ramanujan type', *Proc. London Math. Soc.* (2) 54 (1952) 147–67.

12. G. W. Starcher, 'On identities arising from solutions of q-difference equations and some interpretations in number theory', *Amer. J. of Math.* 53 (1930) 801–16.

13. G. N. Watson, 'Theorems stated by Ramanujan IX: Two continued fractions', *J. London Math. Soc.* 4 (1929) 231–7.

14. T. M. MacRobert, *Functions of a complex variable* (MacMillan, New York, 1958).

The Pennsylvania State University
University Park, Pennsylvania 16801

Remark (added in proof). Theorem 3 is a limiting case of an identity due to R. P. Agarwal, *Proc. Camb. Phil. Soc.* 49 (1953) 441–5, last equation.

2.4.1.1 Notes on notation used in "On q-difference equations for certain well-poised basic hypergeometric series"

The papers [5], [6], [7], and [23] make use of the notation employed by Watson [Wat37]:

$$\prod_n (a,q) = \prod_{j=0}^{n-1} (1+aq^j).$$

The standard notation is:

$$(a;q)_n = \prod_{j=0}^{n-1} (1-aq^j);$$

so that

$$\prod_n (a,q) = (-a;q)_n.$$

Of course,

$$\prod_\infty (a,q) = \lim_{n\to\infty} \prod_n (a,q) = (-a;q)_\infty.$$

Along with the Watsonian notation, another nonstandard notation employed in [23] appears as Eq. (1.6) on page 434:

$$ {}_4\Phi_3 \left[\begin{matrix} x, qx^{\frac{1}{2}}, -qx^{\frac{1}{2}}, a \\ x^{\frac{1}{2}}, -x^{\frac{1}{2}}, xqa^{-1} \end{matrix} ; q, x^{\frac{1}{2}} q^{\frac{1}{2}n} a^{-1} \right] = \frac{\prod_\infty(-xq,q)\prod_\infty((xq)^{\frac{1}{2}}a^{-1},q)}{\prod_\infty(-xqa^{-1},q)\prod_\infty((xq)^{\frac{1}{2}},q)}. \quad (2.1)$$

The precedent for the confusing use of n in the left-hand side of (2.1) occurs at the top of page 91 in [Sla66]. What is intended is:

$$\sum_{n=0}^{\infty} \frac{(x;q)_n(1-xq^{2n})(a;q)_n}{(q;q)_n(1-x)(xq/a;q)_n} a^{-n}x^{n/2}q^{n^2/2} = \frac{(xq;q)_\infty(-\sqrt{xq}/a;q)_\infty}{(xq/a;q)_\infty(-\sqrt{xq};q)_\infty}.$$

2.4.1.2 Erratum

A letter sent by Ken Stolarsky, dated February 4, 1968, contained an enthusiastic response to a preprint version of [23]. In this letter, Dr. Stolarsky kindly pointed out a number of misprints in the preprint. All but one of these misprints were corrected in the published version. Many thanks to Barbara Baum for finding Dr. Stolarsky's letter in the Andrews files.

p. 446, line -7: Denominator factor $\Pi_\infty(-xq^{3m-3},q^3)$ should be $\Pi_\infty(-xq^{3m-3},q)$.

2.5 Bailey Chains

Rogers' second proof of the Rogers–Ramanujan identities [Rog17] was perfected by Bailey in two papers [Bai47], and [Bai49]. The Bailey transform (as named by Lucy Slater [Sla66]) has its first statement in [Bai49].

In the following paragraph from [Bai49], Bailey describes how he could produce finite versions of his theorems:

"4. If, instead of using the analogue of Gauss's theorem, we use the analogue of Saalschütz's theorem, we can take

$$\delta_r = \frac{(\rho_1)_r(\rho_2)_r(x^{-N})_r}{(\rho_1\rho_2 x^{-N}/a)_r} x^r,$$

where N is a positive integer, and we find that

$$\gamma_n = \frac{(ax/\rho_1)_N(ax/\rho_2)_N}{(ax)_N(ax/\rho_1\rho_2)_N}\frac{(-1)^n(\rho_1)_n(\rho_2)_n(x^{-N})_n}{(ax/\rho_1)_n(ax/\rho_2)_n(ax^{N+1})_n}\left(\frac{ax}{\rho_1\rho_2}\right)^n x^{\frac{1}{2}n(2N-n+1)}.$$

These values of δ_r, γ_n reduce to the previous values when $N \to \infty$, and lead to more general results involving only terminating basic series. We are, however, more concerned with identities of the Rogers–Ramanujan type in this paper, as the most general formulae for basic series (apart from those already given) are too involved to be any great interest."

What Bailey missed by his dismissal of finite versions is a powerful iterative method: the "Bailey chain." The discovery of this fact arose when I was trying to find another way of integrating orthogonal polynomials into a proof of the Rogers–Ramanujan identities (*cf.* Problem 74–12, "The Rogers–Ramanujan identities," S.I.A.M. Rev., 16 (1974), 390).

The first paper I have included in this section is [61], which reveals for the first time that there are elegant analytic generalizations of the Rogers–Ramanujan identities. Often this result:

$$\sum_{n_1,n_2,\ldots,n_{k-1}\geqq 0} \frac{q^{N_1^2+N_2^2+\cdots+N_{k-1}^2+N_i+N_{i+1}+\cdots+N_{k-1}}}{(q)_{n_1}(q)_{n_2}\cdots(q)_{n_{k-1}}} = \prod_{\substack{n=1 \\ n\not\equiv 0,\pm i \pmod{2k+1}}}^{\infty} (1-q^n)^{-1}, \tag{2.2}$$

where $N_j = n_j + n_{j+1} + \cdots + n_{k-1}$ is referred to as the Andrews–Gordon identity because of Gordon's parallel combinatorial theorem in [Gor61]. However, there is no hint in [Gor61] that (2.2) might exist. Actually, I had found the $k = 3$ case of (2.2) several years earlier. I had cast it aside because I was looking for a new proof of the Rogers–Selberg theorems [Sel36], and the double series I found was not what I was looking for. Thus [61] had to wait until I was open to finding this theorem. Here is an instance of an adage that has guided me: "Try to hear what the mathematics is trying to tell you!"

Paper [107] is where the terms "Bailey pair," "Bailey chain" and "Bailey's lemma" originate, and so it is the central paper included in this section. It should be noted that the iterative nature of multiple series identities first arose in [70] (included in Chapter 6) and independently by Peter Paule [Paule87].

Paper [116] (also included) uses the Bailey chain to find new representations of the fifth- and seventh-order mock theta functions. These representations were essential in the subsequent, explosive study of mock theta functions. Paper [142] uses [116] to view the fifth-order mock theta functions as "constant terms" in certain infinite products. Dean Hickerson [Hic88a], [Hic88b], building on and greatly extending these ideas, was able to prove the Mock Theta Conjectures from [138]. Strange formulas from Ramanujan's "Lost" Notebook, proved in [124], led to conjectures about some of the related q-series [129]. Paper [137] answers some of the questions posed in [129], and again utilizes the power of the Bailey chain. Hence it is included in this section. Five different mathematicians had discovered the remarkable multiplicative properties of the coefficients of:

$$\sum_{n=0}^{\infty} \frac{q^{n(n+1)/2}}{(-q;q)_n};$$

however, none had determined the use of the Bailey chain in establishing this multiplicativity.

It is also the case that Bailey pairs can be extended to multiple series. Thus I have included [220], which builds on the work of Berkovich, McCoy, and Schilling [Berk-MS96], and also Paule [Paule87]. Perhaps my favorite identity from this paper is the tri-pentagonal number theorem:

$$\sum_{i,j,k \geqq 0} \frac{q^{i^2+j^2+k^2}}{(q;q)_{i+j-k}(q;q)_{i+k-j}(q;q)_{j+k-i}}$$
$$= \frac{1}{(q;q)_\infty^3} \sum_{i,j,k=-\infty}^{\infty} (-1)^{i+j+k} q^{i(3i-1)/2+j(3j-1)+k(3k-1)+ij+ik+jk}.$$

Berkovich [Berk10] has found the most elegant proof of this result. A combinatorial interpretation is given in [264].

The final paper included in this subsection is [225]. Again, buried in Bailey's paper [Bai49], was the basic idea for extending Bailey chains in a way that would account for the q-analog of Dougall's theorem. Berkovich and I followed up on this in [234], and Spiridonov [Spi02] saw how to extend these more general Bailey pairs to the newly developing field of elliptic hypergeometric functions.

Paper [152] contrasts the various proofs of the Rogers–Ramanujan identities, especially as related to the Bailey chain. Bowman and I looked at what are now called "conjugate Bailey pairs" in [199].

Paper [203], which is mostly due to the insights of Schilling and Warnaar, extends Bailey chains very effectively, where the underlying classical group is A_2. Paper [231] looks at the Bailey lemma arising from the quintuple product identity.

2.5.1 An analytic generalization of the Rogers–Ramanujan identities for odd moduli

Reprinted from
Proc. Nat. Acad. Sci. USA
Vol. 71, No. 10, pp. 4082–4085, October 1974

An Analytic Generalization of the Rogers-Ramanujan Identities for Odd Moduli

(basic hypergeometric series/partitions/combinational identities/partition identities)

GEORGE E. ANDREWS

410 McAllister Building, Pennsylvania State University, University Park, Pa. 16802

Communicated by Mark Kac, July 12, 1974

ABSTRACT A $(k-1)$-fold Eulerian series expansion is given for $\Pi(1-q^n)^{-1}$, where the product runs over all positive integers n that are not congruent to $0, i$ or $-i$ modulo $2k+1$. The Rogers-Ramanujan identities are the cases $k = i = 2$ and $k = i + 1 = 2$.

1. Introduction

The Rogers-Ramanujan identities have a long and interesting history. In 1894, L. J. Rogers proved the following two identities which have subsequently become known as the Rogers-Ramanujan or Rogers-Ramanujan-Schur identities (see ref. 1; pp. 90–99, for the early history of these identities):

$$\sum_{n\geqq 0}\frac{q^{n^2}}{(q)_n} = \prod_{\substack{n=1\\ n\not\equiv 0,\pm 2 (\mathrm{mod}\ 5)}}^{\infty} (1-q^n)^{-1}, \qquad [1.1]$$

and

$$\sum_{n\geqq 0}\frac{q^{n^2+n}}{(q)_n} = \prod_{\substack{n=1\\ n\not\equiv 0,\pm 1 (\mathrm{mod}\ 5)}}^{\infty} (1-q^n)^{-1}, \qquad [1.2]$$

where $(a)_0 = 1$, $(a)_n = (a;q)_n = (1-a)(1-aq)\cdots(1-aq^{n-1})$, and $|q| < 1$.

The search for identities of a similar character has gone on intermittently ever since; however, no simple general identities of this nature have ever been discovered.

V. N. Singh's formula (ref. 2, Eq. 3.6) comes the closest to this objective; however it contains a rather complicated representation of the Alder polynomial.

L. J. Rogers (ref. 3, p. 331, Eq. 6) also discovered further identities related to the modulus 7; for example, he proved that

$$\prod_{m=1}^{\infty}(1+q^m)\sum_{n\geqq 0}\frac{q^{2n^2}}{(q^2;q^2)_n(-q)_{2n}} = \prod_{\substack{n=1\\ n\not\equiv 0,\pm 3 (\mathrm{mod}\ 7)}}^{\infty} (1-q^n)^{-1}. \qquad [1.3]$$

In this same vein, W. N. Bailey (ref. 4, p. 421) proved three identities related to the modulus 9; for example

$$\prod_{m=1}^{\infty}(1+q^m+q^{2m})\sum_{m\geqq 0}\frac{q^{3m^2}(q)_{3m}}{(q^3;q^3)_m(q^3;q^3)_{2m}} = \prod_{\substack{n=1\\ n\not\equiv 0,\pm 4 (\mathrm{mod}\ 9)}}^{\infty} (1-q^n)^{-1}. \qquad [1.4]$$

A large amount of work was done on identities of this type, and a collection of some 130 such identities was published by L. J. Slater (5). In each of the identities given by Slater, the modulus arising in the infinite product is always $\leqq 64$ and always has its prime factor among $\{2,3,5,7\}$.

Recently (6) five double series identities were obtained related to the modulus 11; thus

$$\prod_{m=1}^{\infty}(1+q^m+q^{2m}+q^{3m})\sum_{n\geqq 0}\sum_{j\geqq 0} \times \frac{q^{4n^2+12nj+8j^2+j}(-1)^j(q)_{4n+2j}}{(q^4;q^4)_n(q^2;q^2)_j(q^4;q^4)_{2n+2j}} = \prod_{\substack{n=1\\ n\not\equiv 0,\pm 5 (\mathrm{mod}\ 11)}}^{\infty} \times (1-q^n)^{-1}. \qquad [1.5]$$

Unfortunately the methods used to prove [1.5] (as well as those used by Rogers, Bailey, and Slater) seem limited to special cases and do not provide general theorems for an infinite family of moduli.

In attempting to generalize fully [1.1] and [1.2], H. L. Alder (7) was able to show that there exist polynomials $G_{k,i}(n;q)$ such that

$$\sum_{n\geqq 0}\frac{G_{k,i}(n;q)}{(q)_n} = \prod_{\substack{n=1\\ n\not\equiv 0,\pm i (\mathrm{mod}\ 2k+1)}}^{\infty} (1-q^n)^{-1}; \qquad [1.6]$$

however neither his work nor subsequent papers (refs. 7; 8, Sec. 6; 9; 10; 2; 11; 12) on the Alder polynomials yielded analytic identities of the elegance and simplicity of [1.1] and [1.2].

The object of this paper is to prove the following generalization of the Rogers-Ramanujan identities:

THEOREM 1. Let $1 \leqq i \leqq k$ be integers; then

$$\sum_{n_1,n_2,\cdots,n_{k-1}\geqq 0}\frac{q^{N_1^2+N_2^2+\cdots+N_{k-1}^2+N_i+N_{i+1}+\cdots+N_{k-1}}}{(q)_{n_1}(q)_{n_2}\cdots(q)_{n_{k-1}}} = \prod_{\substack{n=1\\ n\not\equiv 0,\pm i (\mathrm{mod}\ 2k+1)}}^{\infty} (1-q^n)^{-1}, \qquad [1.7]$$

where $N_j = n_j + n_{j+1} + \cdots + n_{k-1}$.

We remark that this result reduces to [1.1] for $k = i = 2$, and to [1.2] for $k = 2$, $i = 1$; otherwise, to our knowledge, no special case of this result has appeared in the literature. We should point out that the work in Section 3 here shows that V. N. Singh's formula (ref. 2,

Eq. 3.6) can be transformed into *Theorem 1* in the case $k = i$ by suitable rearrangements of series. For example, with $k = i = 3$, we obtain

$$\sum_{n,m\geq 0}^{\infty} \frac{q^{2n^2+2nm+m^2}}{(q)_n(q)_m} = \prod_{\substack{n=1\\ n\not\equiv 0,\pm 3 \pmod 7}}^{\infty} (1-q^n)^{-1} \quad [\mathbf{1.8}]$$

instead of [**1.3**].

It is important to note that the technique we shall use to prove *Theorem 1* is applicable to a number of problems in the theory of partitions, and we hope in time to apply it to some of the questions raised in (ref. 13; Sec. 3) concerning partition identities.

After proving *Theorem 1* in Section 2, we shall utilize our results to obtain a new formula for the Alder polynomials.

2. Proof of Theorem 1

We begin with a function that was originally introduced by L. J. Rogers (14) and was considered independently by A. Selberg (15).

$$Q_{k,i}(x) = \sum_{n\geq 0} \times \frac{(-1)^n x^{kn} q^{1/2(2k+1)n(n+1)-in}(1 - x^i q^{(2n+1)i})}{(q)_n (xq^{n+1})_\infty} \quad [\mathbf{2.1}]$$

where $(a)_\infty = (a;q)_\infty = \lim_{n\to\infty}(a;q)_n$.

For completeness we include a proof of the following functional equation that was independently discovered by Rogers (14) and Selberg (15).

$$Q_{k,i}(x) - Q_{k,i-1}(x) = (xq)^{i-1} Q_{k,k-i+1}(xq). \quad [\mathbf{2.2}]$$

This follows from the fact that

$$Q_{k,i}(x) - Q_{k,i-1}(x) = \sum_{n\geq 0} \frac{(-1)^n x^{kn} q^{1/2(2k+1)n(n+1)}}{(q)_n (xq^{n+1})_\infty} (q^{-in} - x^i q^{(n+1)i} - q^{(-i+1)n} + x^{i-1} q^{(n+1)(i-1)})$$
$$= \sum_{n\geq 0} \frac{(-1)^n x^{kn} q^{1/2(2k+1)n(n+1)}}{(q)_n (xq^{n+1})_\infty} (q^{-in}(1-q^n) + x^{i-1}q^{(n+1)(i-1)}(1 - xq^{n+1})) = \sum_{n\geq 1} \frac{(-1)^n x^{kn} q^{1/2(2k+1)n(n+1)-in}}{(q)_{n-1}(xq^{n+1})_\infty}$$
$$+ (xq)^{i-1} \sum_{n\geq 0} \frac{(-1)^n x^{kn} q^{1/2(2k+1)n(n+1)+n(i-1)}}{(q)_n (xq^{n+2})_\infty} = -x^k q^{2k+1-i} \sum_{n\geq 0} \frac{(-1)^n x^{kn} q^{1/2(2k+1)n(n+1)+(2k+1-i)n}}{(q)_n (xq^{n+2})_\infty} + (xq)^{i-1}$$
$$\times \sum_{n\geq 0} \frac{(-1)^n x^{kn} q^{1/2(2k+1)n(n+1)+n(i-1)}}{(q)_n (xq^{n+2})_\infty} = (xq)^{i-1} \sum_{n\geq 0} \frac{(-1)^n (xq)^{kn} q^{1/2(2k+1)n(n+1)-(k-i+1)n}(1 - (xq)^{k-i+1} q^{(2n+1)(k-i+1)})}{(q)_n (xq^{n+2})_\infty}$$
$$= (xq)^{i-1} Q_{k,k-i+1}(xq).$$

Since $Q_{k,0}(x) = 0$ by [**2.1**], we see that by [**2.2**]

$$Q_{1,1}(x) = Q_{1,1}(xq) = Q_{1,1}(xq^2) = \cdots$$
$$= \lim_{n\to\infty} Q_{1,1}(xq^n) = Q_{1,1}(0) = 1. \quad [\mathbf{2.3}]$$

The next formula is new and is the key to the proof of *Theorem 1*.

$$Q_{k,i}(x) = \sum_{n\geq 0} \frac{x^{(k-1)n} q^{(k-1)n^2+(k-i)n}}{(q)_n} Q_{k-1,i}(xq^{2n}), \quad [\mathbf{2.4}]$$

for $1 \leq i \leq k$, $k \geq 2$.

To establish [**2.4**] we begin by observing that if we define $Q_{k,i}(x) = \sum_{M,N\geq 0} C_{k,i}(M,N) x^M q^N$, then by [**2.1**] and [2.2]: (*i*) $C_{k,0}(M,N) = 0$ for all M and N, (*ii*) $C_{k,i}(M,N) = 0$ if either M or N is nonpositive and $M^2 + N^2 \neq 0$, (*iii*) $C_{k,i}(0,0) = 1$ for $1 \leq i \leq k$, (*iv*) $C_{k,i}(M,N) - C_{k,i-1}(M,N) = C_{k,k-i+1}(M - i + 1, N - M)$ for $1 \leq i \leq k$. Now these four conditions uniquely determine the $C_{k,i}(M,N)$; this is easily established by a double induction first on N and then on i. Hence we see that the $Q_{k,i}(x)$ $(0 \leq i \leq k)$ are the only functions analytic in x and q around (0,0) that satisfy the functional equation [2.2] for $1 \leq i \leq k$ and the boundary conditions $Q_{k,0}(x) \equiv 0$, $Q_{k,i}(0) = 1$, $1 \leq i \leq k$. To prove [**2.4**] then, we need only show that the right hand side of this identity also fulfills these defining conditions. We define for $k \geq 2$, $0 \leq i \leq k$,

$$R_{k,i}(x) = \sum_{n\geq 0} \frac{x^{(k-1)n} q^{(k-1)n^2+(k-i)n}}{(q)_n} Q_{k-1,i}(xq^{2n}).$$

Immediately we see that $R_{k,0}(x) = 0$ and $R_{k,i}(0) = 1$ for $1 \leq i \leq k$. Furthermore for $1 \leq i \leq k$

$$R_{k,i}(x) - R_{k,i-1}(x) = \sum_{n\geq 0} \frac{x^{(k-1)n} q^{(k-1)n^2+(k-i)n}}{(q)_n}$$
$$\times (Q_{k-1,i}(xq^{2n}) - q^n Q_{k-1,i-1}(xq^{2n}))$$
$$= \sum_{n\geq 0} \frac{x^{(k-1)n} q^{(k-1)n^2+(k-i)n}}{(q)_n} (Q_{k-1,i-1}(xq^{2n})$$
$$+ (xq^{2n+1})^{i-1} Q_{k-1,k-i}(xq^{2n+1}) - q^n Q_{k-1,i-1}(xq^{2n}))$$
$$= \sum_{n\geq 0} \frac{x^{(k-1)n} q^{(k-1)n^2+(k-i)n}}{(q)_n} (1 - q^n) Q_{k-1,i-1}(xq^{2n})$$
$$+ (xq)^{i-1} \sum_{n\geq 0} \frac{x^{(k-1)n} q^{(k-1)n^2+(k+i-2)n}}{(q)_n}$$
$$\times (Q_{k-1,k-i+1}(xq^{2n+1}) - (xq^{2n+2})^{k-i} Q_{k-1,i-1}(xq^{2n+2}))$$
$$= \sum_{n\geq 0} \frac{x^{(k-1)n+(k-1)} q^{(k-1)n^2+(3k-i-2)n+(2k-i-1)}}{(q)_n}$$
$$\times Q_{k-1,i-1}(xq^{2n+2}) + (xq)^{i-1} R_{k,k-i+1}(xq)$$
$$- \sum_{n\geq 0} \frac{x^{(k-1)n+k-1} q^{(k-1)n^2+(3k-i-2)n+(2k-i-1)}}{(q)_n}$$
$$\times Q_{k-1,i-1}(xq^{2n+2}) = (xq)^{i-1} R_{k,k-i+1}(xq).$$

Thus, since the $R_{k,i}(x)$ fulfill all the defining conditions, we see that $R_{k,i}(x) = Q_{k,i}(x)$ for $0 \leqq i \leqq k$, $k \geqq 2$. Hence [**2.4**] is established

Next we wish to show that for $1 \leqq i \leqq k$, $k \geqq 2$,

$$Q_{k,i}(x) = \sum_{n_1,n_2,\cdots,n_{k-1}\geqq 0} \frac{x^{N_1+N_2+\cdots+N_{k-1}}q^{N_1^2+N_2^2+\cdots+N^2_{k-1}+N_i+\cdots+N_{k-1}}}{(q)_{n_1}(q)_{n_2}\cdots(q)_{n_{k-1}}}, \qquad [2.5]$$

where $N_j = n_j + n_{j+1} + \cdots + n_{k-1}$.

When $k = 2$, we see by [**2.4**] and [**2.3**] that

$$Q_{2,1}(x) = \sum_{n\geqq 0} \frac{x^n q^{n^2+n}}{(q)_n},$$

which is just [**2.5**] for $k = 2$, $i = 1$. Furthermore by [**2.2**]

$$Q_{2,2}(x) = Q_{2,1}(xq^{-1}) = \sum_{n\geqq 0} \frac{x^n q^{n^2}}{(q)_n},$$

which is [**2.5**] for $k = i = 2$. Hence [**2.5**] is valid when $k = 2$.

We now assume [**2.5**] is valid for $k - 1$. Then for $1 \leqq i \leqq k - 1$, we see by [**2.4**] that

$$Q_{k,i}(x) = \sum_{n_{k-1}\geqq 0} \frac{x^{(k-1)n_{k-1}}q^{(k-1)n^2_{k-1}+(k-i)n_{k-1}}}{(q)_{n_{k-1}}} \cdot \sum_{n_1,\cdots,n_{k-2}\geqq 0} \frac{(xq^{2n_{k-1}})^{\nu_1+\nu_2+\cdots+\nu_{k-2}}q^{\nu_1^2+\nu_2^2+\cdots+\nu^2_{k-2}+\nu_i+\cdots+\nu_{k-2}}}{(q)_{n_1}(q)_{n_2}\cdots(q)_{n_{k-1}}},$$

where $\nu_j = n_j + n_{j+1} + \cdots + n_{k-2}$. If we write $N_j = \nu_j + n_{k-1}$, $N_{k-1} = n_{k-1}$, then

$$Q_{k,i}(x) = \sum_{n_1,\cdots,n_{k-1}\geqq 0} \frac{x^{N_1+\cdots+N_{k-1}}q^{N_1^2+N_2^2+\cdots+N^2_{k-1}+N_i+N_{i+1}+\cdots+N_{k-1}}}{(q)_{n_1}(q)_{n_2}\cdots(q)_{n_{k-1}}},$$

which is [**2.5**] for $1 \leqq i < k$. When $i = k$, by [**2.2**]

$$\begin{aligned} Q_{k,k}(x) &= Q_{k,1}(xq^{-1}) \\ &= \sum_{n_1,\cdots,n_{k-1}\geqq 0} \frac{x^{N_1+\cdots+N_{k-1}}q^{N_1^2+N_2^2+\cdots+N^2_{k-1}}}{(q)_{n_1}(q)_{n_2}\cdots(q)_{n_{k-1}}}, \end{aligned}$$

which is [**2.5**] for $i = k$.

We now easily deduce *Theorem 1* from [**2.1**] and [**2.5**].

$$\begin{aligned} &\prod_{\substack{n=1 \\ n\not\equiv 0,\pm i (\mathrm{mod}\ 2k+1)}}^{\infty} (1-q^n)^{-1} \\ &= \frac{\sum_{n=0}^{\infty} (-1)^n q^{\frac{1}{2}(2k+1)n(n+1)-in}(1-q^{(2n+1)i})}{(q)_\infty} \\ &\qquad \text{(by Jacobi's identity (ref. 16, pp. 169–170)} \\ &= Q_{k,i}(1) \qquad \text{(by [2.1])} \\ &= \sum_{n_1,\cdots,n_{k-1}\geqq 0} \frac{q^{N_1^2+N_2^2+\cdots+N^2_{k-1}+N_i+N_{i+1}+\cdots+N_{k-1}}}{(q)_{n_1}(q)_{n_2}\cdots(q)_{n_{k-1}}} \\ &\qquad \text{(by [2.5]).} \end{aligned}$$

3. The Alder polynomials

Equation [2.5] allows us to deduce easily the following representation for the Alder polynomials, $G_{k,i}(n;q)$, defined by

$$Q_{k,i}(x) = \sum_{n\geqq 0} \frac{G_{k,i}(n;q)x^n}{(q)_n}. \qquad [3.1]$$

THEOREM 2. For $1 \leqq i \leqq k$,

$$\begin{aligned} G_{k,i}(n;q) = &\sum_{\substack{n_1+2n_2+\cdots+(k-1)n_{k-1}=n \\ n_1,\cdots,n_{k-2}\geqq 0}} \begin{bmatrix} N_{k-2} \\ n_{k-2} \end{bmatrix} \\ &\cdot \begin{bmatrix} N_{k-3} \\ n_{k-3} \end{bmatrix} \cdots \begin{bmatrix} N_1 \\ n_1 \end{bmatrix} (q^{N_1+1})_{N_2+\cdots+N_{k-1}} \\ &\cdot q^{N_1^2+\cdots+N^2_{k-1}+N_i+\cdots+N_{k-1}}, \end{aligned} \qquad [3.2]$$

where $N_j - n_j + n_{j+1} + \cdots + n_{k-1}$, and $\begin{bmatrix} m \\ n \end{bmatrix} = (q)_m(q)_n^{-1}(q)_{n-m}^{-1}$, the Gaussian polynomial.

Proof. Comparing [**3.1**] with [**2.5**], we see that

$$\begin{aligned} G_{k,i}(n;q) &= (q)_n \sum_{\substack{N_1+\cdots+N_{k-1}=n \\ n_1,\cdots,n_{k-1}\geqq 0}} \frac{q^{N_1^2+\cdots+N^2_{k-1}+N_i+\cdots+N_{k-1}}}{(q)_{n_1}(q)_{n_2}\cdots(q)_{n_{k-1}}} \\ &= \sum_{\substack{n_1+2n_2+\cdots+(k-1)n_{k-1}=n \\ n_1,\cdots,n_{k-1}\geqq 0}} \frac{(q)_{n_{k-2}+n_{k-1}}}{(q)_{n_{k-2}}(q)_{n_{k-1}}} \\ &\quad \cdot \frac{(q)_{n_{k-3}+n_{k-2}+n_{k-1}}}{(q)_{n_{k-2}+n_{k-1}}(q)_{n_{k-3}}} \cdot \cdots \cdot \frac{(q)_{n_1+\cdots+n_{k-1}}}{(q)_{n_2+\cdots+n_{k-1}}(q)_{n_1}} \\ &\quad \cdot (q^{n_1+\cdots+n_{k-1}+1})_{n_2+2n_3+\cdots+(k-2)n_{k-1}} \\ &\quad \cdot q^{N_1^2+\cdots+N^2_{k-1}+N_i+\cdots+N_{k-1}} \\ &= \sum_{\substack{n_1+2n_2+\cdots+(k-1)n_{k-1}=n \\ n_1,\cdots,n_{k-1}\geqq 0}} \begin{bmatrix} N_{k-2} \\ n_{k-2} \end{bmatrix}\begin{bmatrix} N_{k-3} \\ n_{k-3} \end{bmatrix} \cdots \begin{bmatrix} N_1 \\ n_1 \end{bmatrix} \\ &\quad \cdot (q^{N_1+1})_{N_2+\cdots+N_{k-1}} \cdot q^{N_1^2+\cdots+N^2_{k-1}+N_i+\cdots+N_{k-1}}, \end{aligned} \qquad [3.3]$$

as desired.

Of all the previous representations of the Alder polynomials, [**3.2**] most closely resembles that of V. N. Singh (ref. 12, Eq. 3.7). This is probably due to the fact that both approaches use an iterative process. In fact, we may, by the following series of algebraic manipulations, transform Singh's representation into ours:

$$G_{k,k}(n;q) = q^{n^2} \sum_{t_1,\cdots,t_{k-2}\geqq 0} \frac{(q^{n-2t_1+1})_{2t_1}(q^{t_1-2t_2+1})_{2t_2}\cdots(q^{t_{k-3}-2t_{k-2}+1})_{2t_{k-2}}}{(q)_{t_1}(q)_{t_2}\cdots(q)_{t_{k-2}}} \cdot \frac{q^{-2t_1(n-t_1)-2t_2(t_1-t_2)-\cdots-2t_{k-2}(t_{k-3}-t_{k-2})}}{(q^{n-2t_1+1})_{t_2}(q^{t_1-2t_2+1})_{t_3}\cdots(q^{t_{k-4}-2t_{k-3}+1})_{t_{k-2}}}$$

$$= q^{n^2} \sum_{t_1,\cdots,t_{k-2}\geqq 0} \frac{(q)_n}{(q)_{n-2t_1}} \cdot \frac{(q)_{t_1}}{(q)_{t_1-2t_2}} \cdot \cdots \cdot \frac{(q)_{t_{k-3}}}{(q)_{t_{k-3}-2t_{k-2}}(q)_{t_1}(q)_{t_2}\cdots(q)_{t_{k-2}}} \cdot \frac{q^{-2t_1(n-t_1)-2t_2(t_1-t_2)-2t_3(t_2-t_3)-\cdots-2t_{k-2}(t_{k-3}-t_{k-2})}}{(q^{n-2t_1+1})_{t_2}(q^{t_1-2t_2+1})_{t_3}\cdots(q^{t_{k-4}-2t_{k-3}+1})_{t_{k-2}}}$$

$$= q^{n^2} \sum_{t_1,\cdots,t_{k-2}\geqq 0} \frac{(q)_n q^{-2t_1(n-t_1)-2t_2(t_1-t_2)-\cdots-2t_{k-2}(t_{k-3}-t_{k-2})}}{(q)_{n-2t_1+t_2}(q)_{t_1-2t_2+t_3}\cdots(q)_{t_{k-4}-2t_{k-3}+t_{k-2}}(q)_{t_{k-3}-2t_{k-2}}(q)_{t_{k-2}}}. \quad [3.4]$$

Now we let $t_{k-2} = n_{k-1}$, $t_{k-3} = n_{k-2} + 2n_{k-1}, \cdots$, $t_{k-j-1} = n_{k-j} + 2n_{k-j+1} + \cdots + jn_{k-1}$ for $1 \leqq j \leqq k-2$. Thus $t_{k-2} = n_{k-1}$, $t_{k-3} - 2t_{k-2} = n_{k-2}$, $t_{k-j-1} - 2t_{k-j} + t_{k-j+1} = n_{k-j}$, and $t_{k-j-1} - t_{k-j} = n_{k+j} + \cdots + n_{k-j} \equiv N_{k-j}$. Therefore,

$$\begin{aligned} &n^2 - 2t_1(n-t_1) - 2t_2(t_1-t_2) - \cdots \\ &\quad - 2t_{k-2}(t_{k-3}-t_{k-2}) \\ &= n^2 - 2(N_2 + \cdots + N_{k-1})(n - N_2 - \cdots - N_{k-1}) \\ &\quad - 2(N_3 + \cdots + N_{k-1})N_2 \\ &\quad - 2(N_4 + \cdots + N_{k-1})N_3 \\ &\quad \vdots \\ &\quad - 2N_{k-1}\cdot N_{k-2} \\ &= n^2 - 2(N_2 + \cdots + N_{k-1})n + 2N_2^2 + \cdots \\ &\quad + 2N^2_{k-1} + 2\sum_{2\leqq i<j\leqq k-1} N_iN_j \\ &= (n - N_2 - \cdots - N_{k-1})^2 + N_2^2 + \cdots + N^2_{k-1}. \end{aligned}$$

Hence the above shifts in summation (i.e., $t_{k-j-1} = \sum_{l=1}^{j} ln_{k-j-1+l}$) transform [3.4] into [3.3].

4. Conclusion

Apart from the fact that we now have a reasonable analytic generalization of the Rogers-Ramanujan identities, there is at least one other significant feature related to the present work. Namely, we now have good reason to hope that the iterative use of q-difference equations (such as was used to prove [2.4]) will be instrumental in providing at least a partial answer to Question 2 of (ref. 13, Sec. 3):

"What finite linear q-difference equations with polynomial coefficients···have solutions that can be represented by 'higher dimentional' q-series?"

This work was partially supported by National Science Foundation Grant GP-23774.

1. Hardy, G. H. (1940) *Ramanujan* (Cambridge University Press, London) (Reprinted: Chelsea, New York).
2. Singh, V. N. (1957) "Certain generalized hypergeometric identities of the Rogers-Ramanujan type," *Pac. J. Math.* **7**, 1011–1014.
3. Rogers, L. J. (1917) "On two theorems of combinatory analysis and some allied identities," *Proc. London Math. Soc.* **16**, 315–336.
4. Bailey, W. N. (1947) "Some identities in combinatory analysis," *Proc. London Math. Soc.* **49**, 421–435.
5. Slater, L. J. (1952) "Further identities of the Rogers-Ramanujan type," *Proc. London Math. Soc.* **54**, 147–167.
6. Andrews, G. E. (1974) "On Rogers-Ramanujan type identities related to the modulus 11," *Proc. London Math. Soc.*, in press.
7. Alder, H. L. (1954) "Generalizations of the Rogers-Ramanujan identities," *Pac. J. Math.* **4**, 161–168.
8. Alder, H. L. (1969) "Partition identities—from Euler to the present," *Amer. Math. Monthly* **76**, 733–746.
9. Andrews, G. E. (1974) "On the Alder polynomials and a new generalization of the Rogers-Ramanujan identities," *Trans. Amer. Math. Soc.*, in press.
10. Carlitz, L. (1960) "Note on Alder's polynomials," *Pac. J. Math.* **10**, 517–519.
11. Singh, V. N. (1957) "Certain generalized hypergeometric identities of the Rogers-Ramanujan type (II)," *Pac. J. Math.* **7**, 1691–1699.
12. Singh, V. N. (1959) "A note on the computation of Alder's polynomials," *Pac. J. Math.* **9**, 271–275.
13. Andrews, G. E. (1974) "A general theory of identities of the Rogers-Ramanujan type," *Bull. Amer. Math. Soc.*, in press.
14. Ramanujan, S. & Rogers, L. J. (1919) "Proof of certain identities in combinatory analysis," *Proc. Cambridge Phil. Soc.* **19**, 211–216.
15. Selberg, A. (1936) "Über einige arithmetische Identitäten," *Avh. N. Vidensk. Akad. Oslo Mat. Naturvidensk. Kl.* **8**.
16. Andrews, G. E. (1971) *Number Theory* (W. B. Saunders, Philadelphia, Pa.).

2.5.2 Multiple series Rogers–Ramanujan type identities

PACIFIC JOURNAL OF MATHEMATICS
Vol. 114, No. 2, 1984

MULTIPLE SERIES ROGERS-RAMANUJAN TYPE IDENTITIES

GEORGE E. ANDREWS

It is shown how each of the classical identities of Rogers-Ramanujan type can be embedded in an infinite family of multiple series identities. The method of construction is applied to four of L. J. Rogers' elegant series related to the quintuple product identity. Other applications are also presented.

1. Introduction. The Rogers-Ramanujan identities [**6**; Ch. 7] are given analytically as follows:

$$1+\sum_{n=1}^{\infty}\frac{q^{n^2}}{(1-q)(1-q^2)\cdots(1-q^n)} = \prod_{n=0}^{\infty}\frac{1}{(1-q^{5n+1})(1-q^{5n+4})}, \tag{1.1}$$

$$1+\sum_{n=1}^{\infty}\frac{q^{n^2+n}}{(1-q)(1-q^2)\cdots(1-q^n)} = \prod_{n=0}^{\infty}\frac{1}{(1-q^{5n+2})(1-q^{5n+3})}. \tag{1.2}$$

Numerous authors [**18**], [**19**], [**16**], [**12**], [**13**], [**22**], [**23**] in the first half of this century found related results connecting q-series resembling those in (1.1) and (1.2) with various modular forms and functions. The culmination of their efforts may be found in the two papers of L. J. Slater [**22**], [**23**] wherein over 130 such identities are cataloged.

Within the last decade it has been observed that if one extends the q-series allowed to multiple series then infinite families of Rogers-Ramanujan type identities can be found [**4**], [**5**], [**14**], [**17**], [**25**], [**26**]. For example [**4**],

$$\sum_{n_{k-1}\geq\cdots\geq n_1\geq 0}\frac{q^{n_1^2+n_2^2+\cdots+n_{k-1}^2}}{(q)_{n_{k-1}-n_{k-2}}(q)_{n_{k-2}-n_{k-3}}\cdots(q)_{n_1}} = \prod_{\substack{n=1\\ n\not\equiv 0,\pm k \pmod{2k+1}}}^{\infty}(1-q^n)^{-1}, \tag{1.3}$$

where $(q)_n = (1-q)(1-q^2)\cdots(1-q^n)$. Such series have opened several new combinatorial vistas related to such identities [**8**], [**14**].

However other identities originally found by Rogers have not been fully generalized to infinite families (work in this direction is given in [**17**]). The following four results of L. J. Rogers (for a history and combinatorial applications see [**11**]) are perhaps the best examples of "isolated" Rogers-Ramanujan type identities:

$$(1.4)\quad 1+\sum_{n=1}^{\infty}\frac{q^{n^2}}{(1-q)(1-q^2)\cdots(1-q^{2n})}$$
$$=\prod_{n=0}^{\infty}\frac{1}{(1-q^{2n+1})(1-q^{20n+4})(1-q^{20n+8})}$$
$$=\prod_{n=0}^{\infty}\frac{(1-q^{10n+10})(1-q^{10n+2})(1-q^{10n+8})(1-q^{20n+6})(1-q^{20n+14})}{(1-q^{n+1})}$$

$$(1.5)\quad \sum_{n=0}^{\infty}\frac{q^{n^2+2n}}{(1-q)(1-q^2)\cdots(1-q^{2n+1})}$$
$$=\prod_{n=0}^{\infty}\frac{1}{(1-q^{2n+1})(1-q^{20n+8})(1-q^{20n+12})}$$
$$=\prod_{n=0}^{\infty}\frac{(1-q^{10n+10})(1-q^{10n+4})(1-q^{10n+6})(1-q^{20n+2})(1-q^{20n+18})}{(1-q^{n+1})}$$

$$(1.6)\quad \sum_{n=0}^{\infty}\frac{q^{n^2+n}}{(1-q)(1-q^2)\cdots(1-q^{2n})}=\prod_{\substack{n=1\\ n\not\equiv 0,\pm1,\pm8,\pm9,10\ (\mathrm{mod}\ 20)}}^{\infty}(1-q^n)^{-1}$$
$$=\prod_{n=0}^{\infty}\frac{(1-q^{20n+10})(1-q^{10n+1})(1-q^{10n+9})(1-q^{20n+8})(1-q^{20n+12})}{(1-q^{n+1})}$$

$$(1.7)\quad \sum_{n=0}^{\infty}\frac{q^{n^2+n}}{(1-q)(1-q^2)\cdots(1-q^{2n+1})}=\prod_{\substack{n=1\\ n\not\equiv 0,\pm3,\pm4,\pm7,10\ (\mathrm{mod}\ 20)}}^{\infty}(1-q^n)^{-1}$$
$$=\prod_{n=0}^{\infty}\frac{(1-q^{10n+10})(1-q^{10n+3})(1-q^{10n+7})(1-q^{20n+16})(1-q^{20n+4})}{(1-q^{n+1})}.$$

One reason these four are considered noteworthy is because of the appearance of the Quintuple Product Identity [**3**; p. 466] in each second line. This is especially intriguing in light of the fact that all previous infinite family extensions [**4**], [**5**], [**14**] have involved only the Triple Product of Jacobi [**6**; p. 21].

The object of this paper is to show that *all* of the 130 identities given in Slater's compendium [**21**] can be embedded in infinite families of results. In particular, identities (1.4)–(1.7) can be generalized to infinite families. For example, the generalization of (1.4) is

$$\sum_{n_k \geq n_{k-1} \geq \cdots \geq n_1 \geq 0} \frac{q^{n_1^2+n_2^2+\cdots+n_k^2}}{(q)_{n_k-n_{k-1}}(q)_{n_{k-1}-n_{k-2}}\cdots(q)_{n_2-n_1}(q)_{2n_1}} = \prod_{\substack{n=1 \\ n \not\equiv 0,\, \pm(k+1) \pmod{6k+4} \\ n \not\equiv \pm(4k+2) \pmod{12k+8}}}^{\infty} (1-q^n)^{-1}. \tag{1.8}$$

In fact the extensions of (1.4)–(1.7) presented in this paper all rely on the Quintuple Product Identity (see (3.5), (3.9), (3.13)).

In §2, we shall prove Theorem 1 which provides the means for these extensions. In §3 we prove the generalizations of (1.4)–(1.7). In a very natural way certain questions concerning the "reducibility" of Rogers-Ramanujan type identities arise; some initial observations on this topic are provided in §4. In §5 we explore how Theorem 1 can be applied to several identities that seem to fall naturally outside of the domain effectively outlined by L. J. Slater in [**22**] and [**23**].

The results we prove are most closely related to S. Milne's striking and elegant multiple series expansion for the general Rogers-Selberg function [**17**]. Indeed Milne sketches applications using the quintuple product identity [**17**; pp. 641–642], and presumably the specializations he gives in [**17**; eq. (3.12) abcd] make his function ${}_{(a)}\phi_{1,1}(x;\, q)$ summable to yield our (1.3), (3.5), (3.9) and (3.13). The primary advantage of our approach is that it allows us to avoid having to sum special cases of ${}_{(a)}\phi_{1,1}(x;\, q)$ and instead lets us quote known summations from Slater's extended tables [**22**]. Indeed one might say that Milne's work represents the full generalization of Rogers' second proof of (1.1) and (1.2) [**20**] while the results in this paper fully generalize Rogers' first proof [**18**].

2. The main theorem. The germ of our work lies in W. N. Bailey's fundamental papers [**12**], [**13**] on Rogers-Ramanujan type identities. In particular, the very brief §4 of Bailey's paper [**13**] provides the mechanism for producing an infinite family of identities out of one identity. To give Bailey's results we require the following standard notation [**24**; p. 89]:

$$(a;\, q)_\infty = (a)_\infty = \prod_{n=0}^{\infty}(1-aq^n), \tag{2.1}$$

$$(a;\, q)_n = (a)_n = (a;\, q)_\infty/(aq^n;\, q)_\infty \tag{2.2}$$

$$\left(= (1-a)(1-aq)\cdots(1-aq^{n-1}) \text{ for } n \text{ a nonnegative integer}\right).$$

The assertion by Bailey in [**13**; §4] is:

BAILEY'S LEMMA. *If for* $n \geq 0$

$$\beta_n = \sum_{r=0}^{n} \frac{\alpha_r}{(q)_{n-r}(aq)_{n+r}}, \tag{2.3}$$

then

$$\sum_{n\geq 0} \left(\frac{(\rho_1)_n(\rho_2)_n(aq/\rho_1\rho_2)^n \alpha_n}{(aq/\rho_1)_n(aq/\rho_2)_n} \right) \frac{1}{(q)_{N-n}(aq)_{N+n}} \tag{2.4}$$
$$= \sum_{n\geq 0} \frac{(\rho_1)_n(\rho_2)_n(aq/\rho_1\rho_2)^n \beta_n (aq/\rho_1\rho_2)_{N-n}}{(q)_{N-n}(aq/\rho_1)_N(aq/\rho_2)_N}.$$

To obtain the above result one merely takes the results from [**13**; §4] and fills them into the identity in [**13**; §2]. Finally algebraic simplification provides the above form of the result.

The important observation is that (2.4) is again an instance of (2.3) with now

$$\alpha_r' = \frac{(\rho_1)_r(\rho_2)_r(aq/\rho_1\rho_2)^r \alpha_r}{(aq/\rho_1)_r(aq/\rho_2)_r}, \tag{2.5}$$

and

$$\beta_N' = \sum_{n\geq 0} \frac{(aq/\rho_1\rho_2)_{N-n}(\rho_1)_n(\rho_2)_n(aq/\rho_1\rho_2)^n \beta_n}{(q)_{N-n}(aq/\rho_1)_N(aq/\rho_2)_N}. \tag{2.6}$$

Thus Bailey's Lemma can be iterated ad infinitum, and this iteration is what is made explicit in the following result.

THEOREM 1. *Let the sequences* α_n *and* β_n *be related by* (2.3), *then*

$$\sum_{n\geq 0} \frac{(b_1)_n(c_1)_n(b_2)_n(c_2)_n \cdots (b_k)_n(c_k)_n(q^{-N})_n}{(aq/b_1)_n(aq/c_1)_n(aq/b_2)_n(aq/c_2)_n \cdots (aq/b_k)_n(aq/c_k)_n(aq^{N+1})_n} \tag{2.7}$$
$$\times \left(\frac{a^k q^{k+N}}{b_1c_1 \cdots b_kc_k} \right)^n q^{-\binom{n}{2}}(-1)^n \alpha_n$$
$$= \frac{(aq)_N(aq/b_kc_k)_N}{(aq/b_k)_N(aq/c_k)_N} \sum_{n_k \geq n_{k-1} \geq \cdots \geq n_1 \geq 0} \frac{(b_k)_{n_k}(c_k)_{n_k} \cdots (b_1)_{n_1}(c_1)_{n_1}}{(q)_{n_k-n_{k-1}}(q)_{n_{k-1}-n_{k-2}} \cdots (q)_{n_2-n_1}}$$
$$\times \frac{(q^{-N})_{n_k}(aq/b_{k-1}c_{k-1})_{n_k-n_{k-1}} \cdots (aq/b_1c_1)_{n_2-n_1}}{(b_kc_kq^{-N}/a)_{n_k}(aq/b_{k-1})_{n_k}(aq/c_{k-1})_{n_k} \cdots (aq/b_1)_{n_2}(aq/c_1)_{n_2}}$$
$$\times q^{n_1+n_2+\cdots+n_k} a^{n_1+\cdots+n_{k-1}}(b_{k-1}c_{k-1})^{-n_{k-1}} \cdots (b_1c_1)^{-n_1}\beta_{n_1}.$$

Proof. We proceed by mathematical induction. When $k = 1$, (2.7) reduces to

$$(2.8)\quad \sum_{n\geq 0} \frac{(b_1)_n(c_1)_n(q^{-N})_n}{(aq/b_1)_n(aq/c_1)_n(aq^{N+1})_n}\left(\frac{aq^{1+N}}{b_1c_1}\right)^n(-1)^n q^{-\binom{n}{2}}\alpha_n$$

$$= \frac{(aq)_N(aq/b_1c_1)_N}{(aq/b_1)_N(aq/c_1)_N}\sum_{n_1\geq 0}\frac{(b_1)_{n_1}(c_1)_{n_1}(q^{-N})_{n_1}q^{n_1}\beta_{n_1}}{(b_1c_1q^{-N}/a)_{n_1}},$$

which is (2.4) (with $\rho_1 = b_1$, $\rho_2 = c_1$) multiplied by $(q)_N(aq)_N$ and simplified. Hence the case $k = 1$ is merely Bailey's Lemma.

Now assume the result true up to but not including a particular k. Then the case $k - 1$ is merely a new instance of (2.3) with

$$(2.9)\quad \alpha_n^* = \frac{(b_1)_n(c_1)_n\cdots(b_{k-1})_n(c_{k-1})_n}{(aq/b_1)_n(aq/c_1)_n\cdots(aq/b_{k-1})_n(aq/c_{k-1})_n}$$

$$\times\left(\frac{a^{k-1}q^{k-1}}{b_1c_1\cdots b_{k-1}c_{k-1}}\right)^n\alpha_n,$$

$$(2.10)\quad \beta_n^* = \frac{(aq/b_{k-1}c_{k-1})_n}{(q)_n(aq/b_{k-1})_n(aq/c_{k-1})_n}$$

$$\times\sum_{n_{k-1}\geq\cdots\geq n_1\geq 0}\frac{(b_{k-1})_{n_{k-1}}(c_{k-1})_{n_{k-1}}\cdots(b_1)_{n_1}(c_1)_{n_1}}{(q)_{n_{k-1}-n_{k-2}}\cdots(q)_{n_2-n_1}}$$

$$\times\frac{(q^{-n})_{n_{k-1}}(ab/b_{k-2}c_{k-2})_{n_{k-1}-n_{k-2}}\cdots(aq/b_1c_1)_{n_2-n_1}}{(b_{k-1}c_{k-1}q^{-n}/a)_{n_{k-1}}(aq/b_{k-2})_{n_{k-1}}(aq/c_{k-2})_{n_{k-1}}\cdots(aq/b_1)_{n_2}(aq/c_1)_{n_2}}$$

$$\times q^{n_1+n_2+\cdots+n_{k-1}}a^{n_1+\cdots+n_{k-2}}(b_{k-2}c_{k-2})^{-n_{k-2}}\cdots(b_1c_1)^{-n_1}\beta_{n_1}.$$

Hence we may substitute α_n^* and β_n^* into (2.4) with $\rho_1 = b_k$, $\rho_2 = c_k$ and then multiply through by $(q)_N(aq)_N$. The resulting lefthand side of (2.4) is obviously the lefthand side of (2.7). The resulting righthand side is

$$(2.11)\quad (q)_N(aq)_N\sum_{n_k\geq 0}\frac{(b_k)_{n_k}(c_k)_{n_k}(aq/b_kc_k)^{n_k}\beta_{n_k}^*(aq/b_kc_k)_{N-n_k}}{(q)_{N-n_k}(aq/b_k)_N(aq/c_k)_N}$$

$$= \frac{(aq)_N(aq/b_kc_k)_N}{(aq/b_k)_N(aq/c_k)_N}\sum_{n_k\geq 0}\frac{(b_k)_{n_k}(c_k)_{n_k}(q^{-N})_{n_k}q^{n_k}\beta_{n_k}^*}{(b_kc_kq^{-N}/a)_{n_k}}$$

$$= \frac{(aq)_N (aq/b_k c_k)_N}{(aq/b_k)_N (aq/c_k)_N} \sum_{n_k \geq 0} \frac{(aq/b_{k-1}c_{k-1})_{n_k} q^{n_k}}{(q)_{n_k} (aq/b_{k-1})_{n_k} (aq/c_{k-1})_{n_k}}$$

$$\times \sum_{n_{k-1} \geq \cdots \geq n_1 \geq 0} \frac{(b_k)_{n_k} (c_k)_{n_k} \cdots (b_1)_{n_1} (c_1)_{n_1}}{(q)_{n_{k-1}-n_{k-2}} \cdots (q)_{n_2-n_1}}$$

$$\times \frac{(q^{-N})_{n_k} (q)_{n_k} (aq/b_{k-1}c_{k-1})_{n_k - n_{k-1}}}{(b_k c_k q^{-N}/a)_{n_k} (q)_{n_k - n_{k-1}} (aq/b_{k-1}c_{k-1})_{n_k}} \left(\frac{a}{b_{k-1}c_{k-1}} \right)^{n_{k-1}}$$

$$\times \frac{(aq/b_{k-2}c_{k-2})_{n_{k-1}-n_{k-2}} \cdots (aq/b_1 c_1)_{n_2-n_1}}{(aq/b_{k-2})_{n_{k-1}} (aq/c_{k-2})_{n_{k-1}} \cdots (aq/b_1)_{n_2} (aq/c_1)_{n_2}}$$

$$\times q^{n_1 + \cdots + n_{k-1}} a^{n_1 + \cdots + n_{k-2}} (b_{k-2}c_{k-2})^{-n_{k-2}} \cdots (b_1 c_1)^{-n_1} \beta_{n_1}$$

$$= \frac{(aq)_N (aq/b_k c_k)_N}{(aq/b_k)_N (aq/c_k)_N} \sum_{n_k \geq n_{k-1} \geq \cdots \geq n_1 \geq 0} \frac{(b_k)_{n_k} (c_k)_{n_k} \cdots (b_1)_{n_1} (c_1)_{n_1}}{(q)_{n_k - n_{k-1}} \cdots (q)_{n_2 - n_1}}$$

$$\times \frac{(q^{-N})_{n_k} (aq/b_{k-1}c_{k-1})_{n_k - n_{k-1}} \cdots (aq/b_1 c_1)_{n_2 - n_1}}{(b_k c_k q^{-N}/a)_{n_k} (aq/b_{k-1})_{n_k} (aq/c_{k-1})_{n_k} \cdots (aq/b_1)_{n_2} (aq/c_1)_{n_2}}$$

$$\times q^{n_1 + \cdots + n_k} a^{n_1 + \cdots + n_{k-1}} (b_{k-1}c_{k-1})^{-n_{k-1}} \cdots (b_1 c_1)^{-n_1} \beta_{n_1},$$

which is the desired form of the right hand side of (2.7). Thus Theorem 1 is proved. □

While Theorem 1 looks cumbersome it has in fact a number of famous and important specializations. The most important occurs for

$$\alpha_n = \frac{(1 - aq^{2n})(a)_n (-1)^n q^{\binom{n}{2}}}{(1-a)(q)_n}, \tag{2.12}$$

$$\beta_n = \delta_{n,0} = \begin{cases} 1 & \text{if} \quad n = 0, \\ 0 & \quad\quad n > 0. \end{cases} \tag{2.13}$$

The fact that α_n and β_n satisfy (2.3) is immediate from the following result (with $M = n$) which is easily proved by mathematical induction on M.

$$\sum_{j=0}^{M} \frac{(1 - aq^{2j})(q^{-n})_j (a)_j q^{nj}}{(1-a)(aq^{n+1})_j (q)_j} = \frac{(aq)_M q^{nM} (q^{1-n})_M}{(q)_M (aq^{n+1})_M}, \tag{2.14}$$

a result derived originally by R. P. Agarwal [**1**; p. 444, final equation].

The case $k = 1$ of Theorem 1 with α_n given by (2.12) and β_n given by (2.13) is the terminating case of the limiting form of Jackson's theorem [**24**; p. 96]. The case $k = 2$ is Watson's celebrated q-analog of Whipple's Theorem [**28**] which implies (among many other results) the Rogers-Ramanujan identities (equations (1.1) and (1.2)).

3. q-Series extensions of (1.3)–(1.7). We have chosen four identities from among Slater's list of 130. However it should be emphasized that *Slater derives all* 130 *identities from* (α_n, β_n) *pairs* applied to (2.4) with $N \to \infty$. We may therefore derive for each (α_n, β_n) pair given by Slater in **[22]** an infinite family of identities corresponding to $k = 1, 2, 3, \ldots$ in Theorem 1. We have chosen to generalize (1.4)–(1.7) as striking examples of how this works. To simplify our proceedings we consider a special limiting case of Theorem 1.

THEOREM 2. *Let α_n and β_n be related by* (2.3). *Then*

$$\frac{1}{(aq)_\infty} \sum_{n \geq 0} q^{kn^2} a^{kn} \alpha_n \tag{3.1}$$

$$= \sum_{n_k \geq n_{k-1} \geq \cdots \geq n_1 \geq 0} \frac{a^{n_1 + \cdots + n_k} q^{n_1^2 + n_2^2 + \cdots + n_k^2} \beta_{n_1}}{(q)_{n_k - n_{k-1}} (q)_{n_{k-1} - n_{k-2}} \cdots (q)_{n_2 - n_1}}.$$

Proof. Let $N, b_1, b_2, \ldots, b_k, c_1, c_2, \ldots, c_k$ all tend to infinity in Theorem 1. □

THEOREM 3. *Identity* (1.8) *is valid.*

REMARK. All we need do for this result and the remaining results in this section is to choose the appropriate (α_n, β_n) pair given by Slater **[22]** and fill it in to (3.1).

Proof. Since (1.4) appears as identity (98) in Slater **[23]** we see that the requisite (α_n, β_n) pair in **[22]** is given by her $A(1)$, i.e.

$$\alpha_m = \begin{cases} -q^{6n^2 - 5n + 1}, & m = 3n - 1 > 0, \\ q^{6n^2 - n} + q^{6n^2 + n}, & m = 3n > 0, \\ -q^{6n^2 + 5n + 1}, & m = 3n + 1 > 0, \\ 1, & m = 0, \end{cases} \tag{3.2}$$

$$\beta_n = 1/(q)_{2n}, \tag{3.3}$$

and furthermore $a = 1$ (Slater's x is our aq^{-1}). Hence by Theorem 2 (with $a = 1$),

$$\sum_{n_k \geq \cdots \geq n_1 \geq 0} \frac{q^{n_1^2+n_2^2+\cdots+n_k^2}}{(q)_{n_k-n_{k-1}}(q)_{n_{k-1}-n_{k-2}}\cdots(q)_{n_2-n_1}(q)_{2n_1}} \tag{3.4}$$

$$= \frac{1}{(q)_\infty}\left(\sum_{n=1}^{\infty} q^{k(3n-1)^2}\left(-q^{6n^2-5n+1}\right) + 1 + \sum_{n=1}^{\infty} q^{k(3n)^2}\left(q^{6n^2-n}+q^{6n^2+n}\right)\right.$$
$$\left. + \sum_{n=0}^{\infty} q^{k(3n+1)^2}\left(-q^{6n^2+5n+1}\right)\right)$$
$$= \frac{1}{(q)_\infty} \sum_{n=-\infty}^{\infty} q^{(9k+6)n^2+n} - q^{k+1}\sum_{n=-\infty}^{\infty} q^{(9k+6)n^2+(6k+5)n}$$
$$= \frac{1}{(q)_\infty} \prod_{n=0}^{\infty} (1-q^{(6k+4)(n+1)})(1-q^{k+1+(6k+4)n})(1-q^{5k+3+(6k+4)n})$$
$$\times(1-q^{4k+2+(12k+8)n})(1-q^{8k+6+(12k+8)n})$$

(by [**6**; p. 21, Th. 2.8])

$$= \prod_{\substack{n=1 \\ n\not\equiv 0,\, \pm(k+1)\ (\mathrm{mod}\ 6k+4) \\ n\not\equiv \pm(4k+2)\ (\mathrm{mod}\ 12k+k)}}^{\infty} (1-q^n)^{-1},$$

as desired. □

THEOREM 4.

$$\sum_{n_k \geq n_{k-1} \geq \cdots \geq n_1 \geq 0} \frac{q^{n_1^2+n_2^2+\cdots+n_k^2+2n_1+n_2+\cdots+n_k}}{(q)_{n_k-n_{k-1}}(q)_{n_{k-1}-n_{k-2}}\cdots(q)_{n_2-n_1}(q)_{2n_1+1}} \tag{3.5}$$
$$= \prod_{\substack{n=1 \\ n\not\equiv 0,\, \pm(2k+2)\ (\mathrm{mod}\ 6k+4) \\ n\not\equiv \pm 2k\ (\mathrm{mod}\ 12k+8)}}^{\infty} (1-q^n)^{-1}.$$

Proof. This result reduces to (1.5) when $k = 1$, and since (1.5) appears as identity (96) in Slater [**23**] we see that the requisite (α_n, β_n) pair in [**22**] is given by her $A(4)$, i.e.

$$\alpha_m = \begin{cases} q^{6n^2-4n}, & m = 3n-1 > 0, \\ q^{6n^2+4n}, & m = 3n > 0, \\ -q^{6n^2+8n+2} - q^{6n^2+4n}, & m = 3n+1 > 0, \\ 1, & m = 0, \end{cases} \tag{3.6}$$

$$\beta_n = q^n/(q^2;q)_{2n}, \tag{3.7}$$

and furthermore $a = q$.

Hence by Theorem 2, with $a = q$ and both sides multiplied by $(1-q)^{-1}$,

$$\sum_{n_k \geq n_{k-1} \geq \cdots \geq n_1 \geq 0} \frac{q^{n_1^2+n_2^2+\cdots+n_k^2+2n_1+n_2+\cdots+n_k}}{(q)_{n_k-n_{k-1}}(q)_{n_{k-1}-n_{k-2}}\cdots(q)_{n_2-n_1}(q)_{2n_1+1}} \tag{3.8}$$

$$= \frac{1}{(q)_\infty}\left(\sum_{n=1}^{\infty} q^{k(3n-1)^2+k(3n-1)}\left(q^{6n^2-4n}\right) + 1 + \sum_{n=1}^{\infty} q^{k(3n)^2+3kn}\left(q^{6n^2+4n}\right)\right.$$

$$\left. + \sum_{n=0}^{\infty} q^{k(3n+1)^2+k(3n+1)}\left(-q^{6n^2+8n+2} - q^{6n^2+4n}\right)\right)$$

$$= \frac{1}{(q)_\infty}\left(\sum_{n=-\infty}^{\infty} q^{(9k+6)n^2+(3k+4)n} - q^{2k}\sum_{n=-\infty}^{\infty} q^{(9k+6)n^2+(9k+4)n}\right)$$

$$= \frac{1}{(q)_\infty}\left(\sum_{n=-\infty}^{\infty} q^{(9k+6)n^2+(3k+4)n} - q^{2k+2}\sum_{n=-\infty}^{\infty} q^{(9k+6)n^2+(9k+8)n}\right)$$

(replacing n by $-n-1$ in the second sum)

$$= \frac{1}{(q)_\infty}\prod_{n=0}^{\infty}\left(1-q^{(6k+4)(n+1)}\right)\left(1-q^{2k+2+(6k+4)n}\right)\left(1-q^{4k+2+(6k+4)n}\right)$$

$$\times\left(1-q^{10k+8+(12k+8)n}\right)\left(1-q^{2k+(12k+8)n}\right)$$

(by [**6**; p. 21, Th. 2.8])

$$= \prod_{\substack{n=1 \\ n\not\equiv 0,\, \pm(2k+2) \pmod{6k+4} \\ n\not\equiv \pm 2k \pmod{12k+8}}}^{\infty} (1-q^n)^{-1},$$

as desired. □

THEOREM 5.

$$\sum_{n_k \geq n_{k-1} \geq \cdots \geq n_1 \geq 0} \frac{q^{n_1^2+n_2^2+\cdots+n_k^2+n_1}}{(q)_{n_k-n_{k-1}}(q)_{n_{k-1}-n_{k-2}}\cdots(q)_{n_2-n_1}(q)_{2n_1}} \tag{3.9}$$

$$= \prod_{\substack{n=1 \\ n\not\equiv 0,\, \pm k \pmod{6k+4} \\ n\not\equiv \pm(4k+4) \pmod{12k+8}}}^{\infty} (1-q^n)^{-1}$$

Proof. This result reduces to (1.6) when $k = 1$, and since (1.6) appears as identity (99) in Slater [**23**] we see that the requisite (α_n, β_n) pair in [**22**] is given by her $A(3)$, i.e.

$$\alpha_m = \begin{cases} -q^{6n^2-2n}, & m = 3n-1>0, \\ q^{6n^2-2n} + q^{6n^2+2n}, & m = 3n > 0, \\ -q^{6n^2+2n}, & m = 3n+1>0, \\ 1, & m = 0, \end{cases} \tag{3.10}$$

$$\beta_n = q^n/(q)_{2n}, \tag{3.11}$$

and furthermore $a = 1$.

Hence by Theorem 2, with $a = 1$

$$\sum_{n_k \geq n_{k-1} \geq \cdots \geq n_1 \geq 0} \frac{q^{n_1^2+n_2^2+\cdots+n_k^2+n_1}}{(q)_{n_k-n_{k-1}}(q)_{n_{k-1}-n_{k-2}}\cdots(q)_{n_2-n_1}(q)_{2n_1}} \tag{3.12}$$

$$= \frac{1}{(q)_\infty}\left(\sum_{n=1}^{\infty} q^{k(3n-1)^2}\left(-q^{6n^2-2n}\right) + 1 \right.$$

$$\left. + \sum_{n=1}^{\infty} q^{k(3n)^2}\left(q^{6n^2-2n} + q^{6n^2+2n}\right) + \sum_{n=0}^{\infty} q^{k(3n+1)^2}\left(-q^{6n^2+2n}\right)\right)$$

$$= \frac{1}{(q)_\infty}\left(\sum_{n=-\infty}^{\infty} q^{(9k+6)n^2-2n} - q^k \sum_{n=-\infty}^{\infty} q^{(ik+6)n^2+(6k+2)n}\right)$$

$$= \frac{1}{(q)_\infty} \prod_{n=0}^{\infty} \left(1-q^{(6k+4)(n+1)}\right)\left(1-q^{k+(6k+4)n}\right)\left(1-q^{(5k+4)+(6k+4)n}\right)$$

$$\times\left(1-q^{4k+4+(12k+8)n}\right)\left(1-q^{8k+4+(12k+8)n}\right)$$

(by [**6**; p. 21, Th. 2.8])

$$= \prod_{\substack{n=1 \\ n \not\equiv 0, \pm k \ (\mathrm{mod}\ 6k+4) \\ n \not\equiv \pm(4k+4)\ (\mathrm{mod}\ 12k+8)}}^{\infty} (1-q^n)^{-1},$$

as desired. □

Finally we have

THEOREM 5.

$$\sum_{n_k \geq n_{k-1} \geq \cdots \geq n_1 \geq 0} \frac{q^{n_1^2+n_2^2+\cdots+n_k^2+n_1+n_2+\cdots+n_k}}{(q)_{n_k-n_{k-1}}(q)_{n_{k-1}-n_{k-2}}\cdots(q)_{n_2-n_1}(q)_{2n_1+1}} = \prod_{\substack{n=1 \\ n \not\equiv 0, \pm(2k+1)\ (\mathrm{mod}\ 6k+4) \\ n \not\equiv \pm(2k+2)\ (\mathrm{mod}\ 12k+8)}}^{\infty} (1-q^n)^{-1}. \tag{3.13}$$

Proof. This result reduces to (1.7) when $k = 1$, and since (1.7) appears as identity (94) in Slater [**23**] we see that the requisite (α_n, β_n) pair in [**22**] is given by her $A(2)$, i.e.

$$\alpha_m = \begin{cases} q^{6n^2-n}, & m = 3n-1 > 0, \\ q^{6n^2+n}, & m = 3n \geq 0, \\ -q^{6n^2+5n+1} - q^{6n^2+7n+2}, & m = 3n+1 > 0, \end{cases} \tag{3.13}$$

$$\beta_n = 1/(q^2; q)_{2n}, \tag{3.14}$$

and here $a = q$.

Hence by Theorem 2 with $a = q$ and both sides multiplied by $(1-q)^{-1}$

$$\sum_{n_k \geq n_{k-1} \geq \cdots \geq n_1 \geq 0} \frac{q^{n_1^2+n_2^2+\cdots+n_k^2+n_1+n_2+\cdots+n_k}}{(q)_{n_k-n_{k-1}}(q)_{n_{k-1}-n_{k-2}}\cdots(q)_{n_2-n_1}(q)_{2n_1+1}} \tag{3.15}$$

$$\begin{aligned}
&= \frac{1}{(q)_\infty}\left(\sum_{n=1}^{\infty} q^{k(3n-1)^2+k(3n-1)}\left(q^{6n^2-n}\right) + \sum_{n=0}^{\infty} q^{k(3n)^2+k3n}\left(q^{6n^2+n}\right)\right. \\
&\qquad \left. + \sum_{n=0}^{\infty} q^{k(3n+1)^2+k(3n+1)}\left(-q^{6n^2+5n+1} - q^{6n^2+7n+2}\right)\right) \\
&= \frac{1}{(q)_\infty}\left(\sum_{n=-\infty}^{\infty} q^{(9k+6)n^2+3kn+n} - q^{2k+1}\sum_{n=-\infty}^{\infty} q^{(9k+6)n^2+(9k+5)n}\right) \\
&= \frac{1}{(q)_\infty}\prod_{n=0}^{\infty}(1-q^{(6k+4)(n+1)})(1-q^{2k+1+(6k+4)n})(1-q^{4k+3+(6k+4)n}) \\
&\qquad \times(1-q^{2k+2+(12k+8)n})(1-q^{10k+6+(12k+8)n}) \\
&= \prod_{\substack{n=1 \\ n\not\equiv 0,\, \pm(2k+1)\ (\mathrm{mod}\ 6k+4) \\ n\not\equiv \pm(2k+2)\ (\mathrm{mod}\ 12k+8)}}^{\infty} (1-q^n)^{-1},
\end{aligned}$$

as desired. □

As is obvious from the above, each pair (α_n, β_n) given by Slater in [**22**] can be inserted in Theorem 2 or Theorem 1. The four examples we have chosen constitute applications to probably the most elegant results considered by Rogers et al. beyond the original Rogers-Ramanujan identities.

4. The pairs of sequences (α_n, β_n). Often in this subject one may be given only one of the sequences α_n or β_n. If the α_n are given then the β_n are determined by (2.3). Less well known but still of importance is the "inversion" of (2.3). Namely [7; p. 8, Lemma 3]

$$\alpha_n = (1 - aq^{2n}) \sum_{j=0}^{n} \frac{(aq)_{n+j-1}(-1)^{n-j} q^{\binom{n-j}{2}}}{(q)_{n-j}} \beta_j. \tag{4.1}$$

Also of interest is the creation of (α'_n, β'_n) given by Bailey's Lemma. Furthermore if (α'_n, β'_n) are given then one can immediately determine α_n from (2.5), and one can solve for $\beta_0, \beta_1, \ldots, \beta_N$ from the diagonal system of equations (2.6). Thus the operation of forming the sequence of pairs:

$$(\alpha_n, \beta_n) \to (\alpha'_n, \beta'_n) \to (\alpha''_n, \beta''_n) \to \tag{4.2}$$

which was codified explicitly in Theorem 1 can, in fact, be extended to the left as well. In actual fact, if we consider $\alpha_n = \alpha_n^0$, $\beta_n = \beta_n^0$, then we can form

$$\begin{aligned} \cdots \to (\alpha_n^{(-2)}, \beta_n^{(-2)}) \to (\alpha_n^{(-1)}, \beta_n^{(-1)}) \to (\alpha_n^0, \beta_n^0) \\ \to (\alpha_n^1, \beta_n^1) \to \cdots \end{aligned} \tag{4.3}$$

where each pair is related to the next through instances of (2.5) and (2.6). In the following, we shall refer to a pair of sequences (α_n, β_n) related by (2.3) to be a *Bailey pair*. We shall call (4.2) an *ordinary Bailey chain*, and (4.3) a *bilateral Bailey chain*. For any given Bailey chain one would like to designate a "simplest" Bailey pair as the "reduced" element of the chain. At the current state of knowledge it does not appear to be fruitful (or perhaps possible) to provide a rigorous definition of "reduced". Suffice it to say that the Bailey pair given by (2.12) and (2.13) are the obvious "reduced" pair in their family.

Another important way to produce new Bailey pairs (α_n, β_n) from previous ones is through the replacements $a \to a^{-1}$, $q \to q^{-1}$. Thus if $\alpha_n = \alpha_n(a, q)$, $\beta_n = \beta_n(a, q)$,

$$A_k(a, q) = A_k = a^k q^{k^2} \alpha_k(a^{-1}, q^{-1}),$$

$$B_k(a, q) = B_k = a^{-k} q^{-k^2-k} \beta_k(a^{-1}, q^{-1}),$$

then the replacement of a by a^{-1} and q by q^{-1} in (2.3) yields

$$B_n = \sum_{r=0}^{n} \frac{A_r}{(q)_{n-r}(aq)_{n+r}}. \tag{4.4}$$

Thus the substitution $a \to a^{-1}$, $q \to q^{-1}$ yields a new pair satisfying (2.3). We call (A_n, B_n) the *dual Bailey pair* of (α_n, β_n).

For example, the Bailey pair (α_n, β_n) given by (2.12) and (2.13) is self-dual. On the other hand Slater's A-Table **[22]** of eight pairs (α_n, β_n) consists of the four pairs we utilized in §3 and the 4 duals of these pairs.

5. New applications. The q-series identities that have previously appeared primarily have their genesis in the work of L. J. Rogers **[18]**, **[19]**. Indeed the majority of Bailey pairs given by Slater in **[22]** arise, as she notes, from Rogers' pathbreaking papers. In fact, the Bailey pairs used in the classical development of this subject have always taken α_n as zero or plus or minus a power of q. The β_n have always been simple finite q-products.

Let us start by delving into the work of I. Schur on the Rogers-Ramanujan type identities. Schur in his famous initial paper **[21]** on this topic provided three polynomial identities that fit the Bailey pair criterion. Namely he proved (where $[{A \atop B}] = (q)_A/((q)_B(q)_{A-B})$ and $[x]$ is the greatest integer in x)

$$\sum_{\lambda=-\infty}^{\infty} (-1)^\lambda q^{\lambda(3\lambda+1)/2} \begin{bmatrix} n \\ \left[\frac{n-3\lambda}{2}\right] \end{bmatrix} = 1, \tag{5.1}$$

$$\sum_{\lambda=-\infty}^{\infty} (-1)^\lambda q^{\lambda(5\lambda+1)/2-2\alpha\lambda} \begin{bmatrix} n \\ \left[\frac{n-5\lambda}{2}\right]+\alpha \end{bmatrix} = \sum_{j\geq 0} q^{j^2+\alpha j} \begin{bmatrix} n-\alpha-j \\ j \end{bmatrix}, \tag{5.2}$$

where in (5.2) $\alpha = 0$ or 1. Actually (5.2) first appears in **[2]**; however, as noted there, the essential observations for this identity go back to Schur. From (5.1) and (5.2) one can obtain 12 Bailey pairs: the first 6 come from n even or odd, $\alpha = 0$ or 1. The last 6 are the dual Bailey pairs of the first 6. In fact, Rogers had proved (5.1) in **[18]** since when n is even (5.1) is Bailey pair $A(1)$ of **[22]** and when n is odd (5.1) is Bailey pair $A(2)$ of **[22]**. As mentioned previously, Rogers also had the dual Bailey pairs of $A(1)$ and $A(2)$. Of the remaining eight possible Bailey pairs implied by (5.2), we choose the case n even, $\alpha = 0$. In this instance the Bailey pair is

$$\alpha_n = \begin{cases} q^{m(10m-1)} + q^{m(10m+1)}, & n = 5m > 0, \\ 1, & n = 0, \\ -q^{10m^2+11m+3}, & n = 5m+3 > 0, \\ -q^{10m^2-11m+3}, & n = 5m-3 > 0. \end{cases} \tag{5.3}$$

$$\beta_n = \frac{1}{(q)_{2n}} \sum_{j=0}^{n} q^{j^2} \begin{bmatrix} 2n-j \\ j \end{bmatrix}, \tag{5.4}$$

where $a = 1$.

When this pair is used in conjunction with Theorem 2, the resulting identity is

$$\begin{aligned} \frac{1}{(q)_\infty} \Bigg(& \prod_{m=0}^{\infty} (1 - q^{(m+1)(50k+20)})(1 + q^{(25k+9)+(50k+20)m})(1 + q^{(25k+11)+(50k+20)m}) \\ & - q^{4k+2} \prod_{m=0}^{\infty} (1 - q^{(m+1)(50k+20)})(1 + q^{5k+1+(50k+20)m})(1 + q^{45k+19+(50k+20)m}) \Bigg) \\ = & \sum_{n_k \geq \cdots \geq n_1 \geq n_0 \geq 0} \frac{q^{n_k^2 + \cdots + n_1^2 + n_0^2} (q)_{2n_1 - n_0}}{(q)_{n_k - n_{k-1}} \cdots (q)_{n_2 - n_1} (q)_{2n_1 - 2n_0} (q)_{2n_1} (q)_{n_0}}. \end{aligned} \tag{5.5}$$

D. Bressoud [**15**], in a study building on Rogers' work, has found several other new Bailey pairs. As an example, we list

$$\alpha_n = \begin{cases} (-1)^m (q^{m(5m+1)/2} + q^{m(5m-1)/2}), & n = 2m, \\ 1, & n = 0, \\ 0, & n \text{ odd}, \end{cases} \tag{5.6}$$

$$\beta_n = \frac{1}{(q)_{2n}} \sum_{j=0}^{n} \begin{bmatrix} n \\ j \end{bmatrix} q^{j^2}, \tag{5.7}$$

where $a = 1$.

Hence by Theorem 2,

$$\begin{aligned} & \sum_{n_k \geq \cdots \geq n_1 \geq n_0 \geq 0} \frac{q^{n_k^2 + \cdots + n_1^2 + n_0^2} (q)_{n_1}}{(q)_{n_k - n_{k-1}} \cdots (q)_{n_2 - n_1} (q)_{n_1 - n_0} (q)_{n_0} (q)_{2n_1}} \\ & = \prod_{\substack{n=1 \\ n \not\equiv 0, \pm(4k+2) \pmod{8k+5}}}^{\infty} (1 - q^n)^{-1}, \end{aligned} \tag{5.8}$$

which, of course, reduces to (1.1) for $k = 0$.

Finally, and most off the beaten track, we shall see that the Wall polynomials [**27**] from group theory fit into the Bailey pair scheme.

The Wall polynomials were first defined in [**27**] in order to determine the generating functions for the numbers of conjugacy classes in the symplectic and orthogonal groups over finite fields of characteristic 2. They are defined by the following recurrences:

$$X_{2n+1} = X_{2n} + q^{2n+1} X_{2n-1}, \tag{5.9}$$

$$X_{2n+2} = X_{2n+1} + q^{n+1}(1 + q^{n+1})\left(X_{2n+1} + (1 - q^{2n+1})X_{2n-1}\right), \tag{5.10}$$

where $X_{-1} = X_{-1}(a, b, q) = a$ and $X_0 = X_0(a, b, q) = b$.

Recently it has been shown that **[10]** among other results

$$X_{2n-1}(1,1,q) = \sum_{j=-\infty}^{\infty} q^{j^2} \begin{bmatrix} 2n \\ n+2j \end{bmatrix}, \tag{5.11}$$

$$X_{2n-1}(0,1,q) = \sum_{j\geq 0} q^{j^2+j} \begin{bmatrix} 2n \\ n-2j-1 \end{bmatrix}. \tag{5.12}$$

Hence $\beta_n = X_{2n-1}(1,1,q)/(q)_{2n}$ forms a Bailey pair with

$$\alpha_n = \begin{cases} 2q^{j^2}, & n = 2j, \\ 1, & n = 0, \\ 0, & n \text{ odd}, \end{cases} \tag{5.13}$$

and $a = 1$.

Thus by Theorem 2

$$\sum_{n_k \geq \cdots \geq n_1 \geq 0} \frac{q^{n_k^2 + \cdots + n_1^2} X_{2n_1-1}(1,1,q)}{(q)_{n_k - n_{k-1}} \cdots (q)_{n_2-n_1}(q)_{2n_1}} = \frac{1}{(q)_\infty} \sum_{j=-\infty}^{\infty} q^{(4k+1)j^2}. \tag{5.14}$$

Finally $\beta_n = X_{2n-1}(0,1,q)/(q)_{2n}$ forms a Bailey pair with

$$\alpha_n = \begin{cases} q^{j^2+j}, & n = 2j+1, \\ 0, & n \text{ even}. \end{cases} \tag{5.15}$$

with $a = 1$.

Hence by Theorem 2

$$\sum_{n_k \geq \cdots \geq n_1 \geq 0} \frac{q^{n_k^2 + \cdots + n_1^2} X_{2n_1-1}(0,1,q)}{(q)_{n_k - n_{k-1}} \cdots (q)_{n_2-n_1}(q)_{2n_1}} = \frac{\sum_{j=0}^{\infty} q^{k(2j+1)^2+j^2+j}}{(q)_\infty}. \tag{5.16}$$

6. Conclusion. Our object has not been to add heaps of further Rogers-Ramanujan identities to the already extended lists. Rather it has been to exhibit clearly mechanisms for their construction. The problems of additive number theory **[8]**, combinatorics and special function **[5]**, and physics **[9]** will presumably dictate which Bailey pairs and Bailey chains will be of most interest. Presumably we have here provided the framework for their construction.

References

[1] R. P. Agarwal, *On the partial sums of series of hypergeometric type*, Proc. Cambridge Phil. Soc., **49** (1953), 441–445.

[2] G. E. Andrews, *A polynomial identity which implies the Rogers-Ramanujan identities*, Scripta Math., **28** (1970), 297–305.

[3] ______, *Applications of basic hypergeometric functions*, S.I.A.M. Review, **16** (1974), 441–484.

[4] ______, *An analytic generalization of the Rogers-Ramanujan identities for odd moduli*, Proc. Nat. Acad. Sci. U.S.A., **71** (1974), 4082–4085.

[5] ______, *Problems and Prospects for Basic Hypergeometric Functions*, *The Theory and Applications of Special Functions*, (R. Askey, ed.), Academic Press, New York, 1975, pp. 191–224.

[6] ______, *The Theory of Partitions*, Encyclopedia of Mathematics and Its Applications, Vol. 2, Addison-Wesley, 1976.

[7] ______, *Connection coefficient problems and partitions*, *Relations Between Combinatorics and Other Parts of Mathematics*, D. K. Ray-Chaudhuri ed., Proc. Symp. Pure Math., **34** (1979), 1–24.

[8] ______, *Partitions and Durfee dissection*, Amer. J. Math., **101** (1979), 735–742.

[9] ______, *The hard-hexagon model and Rogers-Ramanujan type identities*, Proc. Nat. Acad. Sci. U.S.A., **78** (1981), 5290–5292.

[10] ______, *On the Wall polynomials and the L-M-W conjectures*, J. Austral. Math. Soc. (to appear).

[11] G. E. Andrews and R. Askey, *Enumeration of partitions: the role of Eulerian series and q-orthogonal polynomials*, *from Higher Combinatorics*, M. Aigner ed., Reidel, Dordrecht, 1977, pp. 3–26.

[12] W. N. Bailey, *Some identities in combinatory analysis*, Proc. London Math. Soc., (2), **49** (1947), 421–435.

[13] ______, *Identities of the Rogers-Ramanujan type*, Proc. London Math. Soc., (2), **50** (1949), 1–10.

[14] D. M. Bressoud, *Analytic and combinatorial generalizations of the Rogers-Ramanujan identities*, Memoirs Amer. Math. Soc., **24** (1980), No. 227, 54 pp.

[15] ______, *Some identities for terminating q-series*, Math. Proc. Camb. Phil. Soc., **89** (1981), 211–223.

[16] F. H. Jackson, *Examples of a generalization of Euler's transformation for power series*, Messenger of Math., **57** (1928), 169–187.

[17] S. Milne, *A generalization of Andrews' reduction formula for the Rogers-Selberg functions*, Amer. J. Math., **104** (1982), 635–643.

[18] L. J. Rogers, *Second memoir on the expansion of certain infinite products*, Proc. London Math. Soc., **25** (1894), 318–343.

[19] ______, *On two theorems of combinatory analysis and some allied identities*, Proc. London Math. Soc., (2), **16** (1917), 315–336.

[20] L. J. Rogers and S. Ramanujan, *Proof of certain identities in combinatory analysis*, Proc. Camb. Phil. Soc., **19** (1919), 211–214.

[21] I. Schur, *Ein Beitrag zur additiven Zahlentheorie und zur Theorie der Kettenbrüche*, Sitz.-Ber. Preuss. Akad. Wiss. Phys.-Math. Kl., 1917, pp. 302–321. (Reprinted in I. Schur, Gesammelte Abhandlungen, Vol. 2, pp. 117–136, Springer, Berlin, 1973).

[22] L. J. Slater, *A new proof of Rogers's transformations of infinite series*, Proc. London Math. Soc. (2), **53** (1951), 460–475.

[23] _____, *Further identities of the Rogers-Ramanujan type*, Proc. London Math. Soc. (2), **54** (1952), 147–167.

[24] _____, *Generalized Hypergeometric Functions*, Cambridge University Press, London and New York, 1966.

[25] A. Verma and V. K. Jain, *Transformations between basic hypergeometric series on different bases and identities of Rogers-Ramanujan type*, J. Math. Analysis and Appl., **76** (1980), 230–269.

[26] _____, *Transformations of non-terminating basic hypergeometric series, their contour integrals and applications to Rogers-Ramanujan identities*, J. Math. Analysis and Appl., **87** (1982), 9–44.

[27] G. E. Wall, *On the conjugacy classes in the unitary, symplectic and orthogonal groups*, J. Australian Math. Soc., **3** (1963), 1–62.

[28] G. N. Watson, *A new proof of the Rogers-Ramanujan identities*, J. London Math. Soc., **4** (1929), 4–9.

Received October 27, 1982 and in revised form January 4, 1983. Partially supported by National Science Foundation Grant MCS 8201733, and by a Guggenheim Fellowship.

The Pennsylvania State University
University Park, PA 16802

2.5.3 The fifth- and seventh-order mock theta functions

TRANSACTIONS OF THE
AMERICAN MATHEMATICAL SOCIETY
Volume 293, Number 1, January 1986

THE FIFTH AND SEVENTH ORDER MOCK THETA FUNCTIONS

BY
GEORGE E. ANDREWS[1]

ABSTRACT. The theory of Bailey chains is extended to yield identities for Hecke type modular forms and related generalizations. The extended results allow us to produce Hecke type series for the fifth and seventh order mock theta functions. New results on the generating function for sums of three squares also follow, and a new proof that every integer is the sum of three triangular numbers is given.

1. Introduction. The mock theta functions were named and first studied by S. Ramanujan. Four months before he died, he sketched his results in a letter to G. H. Hardy [**14**, pp. 354–355]. The results in this letter formed the basis on which the intermittent study of these functions has proceeded [**1–3**, **11**, **19**, **20**]. Ramanujan included in his letter [**14**, pp. 354–355] four separate classes of mock theta functions: one class of third order, two of fifth order, and one of seventh order.

There have been a number of deep results obtained about the third order functions [**3**, **11**, **19**]. This is because Watson was able to find representations of them which allowed him to study their behaviour under the fundamental transformations of the modular group [**19**]. For example,

$$f(q) = 1 + \sum_{n=1}^{\infty} \frac{q^{n^2}}{(1+q)^2(1+q^2)^2\cdots(1+q^n)^2} \tag{1.1}$$

is one of the third order functions. Watson showed that

$$f(q)\prod_{n=1}^{\infty}(1-q^n) = 1 + 4\sum_{n=1}^{\infty}\frac{(-1)^n q^{n(3n+1)/2}}{1+q^n}, \tag{1.2}$$

and he then adroitly used the Poisson summation formula on the right-hand side of (1.2) to obtain the modular transformations of $f(q)$.

In 1976, Ramanujan's "Lost" Notebook came to light [**5**]. In it we find many results for the mock theta functions beyond those contained in Ramanujan's last letter. In particular, while (1.2) was not included in the last letter, it and generalizations of it are found in the "Lost" Notebook [**5**, p. 97, (3.3)].

However the fifth and seventh order functions have been more of a problem. No formula like (1.2) has ever been found for any of them, and none exists in Ramanujan's "Lost" Notebook. Indeed in his paper on the fifth order functions, Watson states [**20**, p. 274]: "I have failed to construct a complete and exact

Received by the editors August 6, 1984 and, in revised form, December 10, 1984.
1980 *Mathematics Subject Classification*. Primary 33A30, 10J05, 33A25, 05A19.
[1]Partially supported by National Science Foundation Grant MCS 8201733.

transformation theory of the functions, on the lines of the transformation theory of the functions of the third order, and, in view of the complexity of all the series which are involved, I am becoming somewhat skeptical concerning the existence of an exact transformation theory for functions of the fifth order."

The object of this paper is to provide the counterparts of (1.2) for the fifth and seventh order mock theta functions. This is, I believe, the necessary first step in finding the transformation theory whose existence is doubted by Watson. As an example, let us consider

$$f_0(q) = 1 + \sum_{n=1}^{\infty} \frac{q^{n^2}}{(1+q)(1+q^2)\cdots(1+q^n)}, \tag{1.3}$$

one of the fifth order mock theta functions [20, p. 277]. We shall show in §5 that

$$f_0(q)\prod_{n=1}^{\infty}(1-q^n) = \sum_{j=-\infty}^{\infty}\sum_{n\geq|j|}(-1)^j q^{n(5n+1)/2-j^2}(1-q^{4n+2}). \tag{1.4}$$

Note the resemblance of the expression on the right-hand side of (1.4) to certain identities for modular forms due to Hecke [12] and Rogers [15] (see §4). Presumably this resemblance can be exploited to obtain the transformation theory alluded to by Watson.

The next three sections describe the necessary background for our work. §5 is a slight digression since we are able to prove certain Hecke type identities directly from our work as well as a formula related to sums of three squares. Also we obtain the new identity

$$\begin{aligned}\left(\sum_{n=0}^{\infty} q^{\binom{n+1}{2}}\right)^3 &= \sum_{n=0}^{\infty}\sum_{j=0}^{2n} \frac{q^{2n^2+2n-\binom{j+1}{2}}\left(1+q^{2n+1}\right)}{\left(1-q^{2n+1}\right)} \\ &= \sum_{n=0}^{\infty}\sum_{j=0}^{2n} \frac{q^{n+j(4n+1-j)/2}\left(1+q^{2n+1}\right)}{\left(1-q^{2n+1}\right)},\end{aligned} \tag{1.5}$$

from which follows immediately Gauss's classic result that every natural number is the sum of three triangular numbers. §§6 and 7 contain our main results on the mock theta functions.

2. Bailey chains. In [8], we presented a comprehensive treatment of Rogers-Ramanujan type identities based on a little known result of W. N. Bailey [9, §4]. For the statement of Bailey's Lemma we need the following standard notation:

$$(a;q)_\infty = (a)_\infty = \prod_{n=0}^{\infty}(1-aq^n), \tag{2.1}$$

$$(a;q)_n = (a)_n = (a;q)_\infty/(aq^n;q)_\infty \tag{2.2}$$

$$\left(= (1-a)(1-aq)\cdots(1-aq^{n-1}) \text{ for } n \text{ a nonnegative integer}\right).$$

BAILEY'S LEMMA. *If for $n \geq 0$ the sequences $\{\alpha_n\}$ and $\{\beta_n\}$ are related by*

$$\beta_n = \sum_{r=0}^{n} \frac{\alpha_r}{(q)_{n-r}(aq)_{n+r}}, \tag{2.3}$$

then for $n \geq 0$

$$\beta'_n = \sum_{r=0}^{n} \frac{\alpha'_r}{(q)_{n-r}(aq)_{n+r}}, \tag{2.4}$$

where

$$\beta'_n = \frac{1}{(aq/\rho_1)_n(aq/\rho_2)_n} \sum_{j=0}^{n} \frac{(\rho_1)_j(\rho_2)_j(aq/\rho_1\rho_2)_{n-j}(aq/\rho_1\rho_2)^j\beta_j}{(q)_{n-j}} \tag{2.5}$$

and

$$\alpha'_r = \frac{(\rho_1)_r(\rho_2)_r(aq/\rho_1\rho_2)^r\alpha_r}{(aq/\rho_1)_r(aq/\rho_2)_r}. \tag{2.6}$$

The above formulation is not at all the way Bailey stated this result [**9**, §4]. However, formulated as above it turns out to be incredibly powerful in obtaining and understanding Rogers-Ramanujan type identities. Pairs α_n, β_n can be substituted into identities like (3.1) to yield directly Rogers-Ramanujan type identities. The point is that once you find a pair of sequences α_n, β_n that satisfies (2.3) you can produce a new pair α'_n, β'_n that satisfies the same identity. Thus an infinite family

$$(\alpha_n, \beta_n) \to (\alpha'_n, \beta'_n) \to (\alpha''_n, \beta''_n) \to \cdots$$

of such "Bailey pairs" can be obtained merely by iterating Bailey's Lemma. Furthermore, if only the α_n sequence is given, then the β_n sequence is completely determined by (2.3). If only the β_n sequence is given, then (2.3) may be inverted to yield [**5**, Lemma 3]

$$\alpha_n = (1 - aq^{2n}) \sum_{j=0}^{n} \frac{(aq)_{n+j-1}(-1)^{n-j}q^{\binom{n-j}{2}}}{(q)_{n-j}}\beta_j. \tag{2.7}$$

Thus if only the β_n sequence is given, then the α_n sequence is completely determined by (2.7).

Furthermore, the sequence may be extended to the left as well:

$$\cdots \to (\alpha_n^{(-2)}, \beta_n^{(-2)}) \to (\alpha_n^{(-1)}, \beta_n^{(-1)}) \to (\alpha_n, \beta_n) \to (\alpha_n, \beta'_n) \to \cdots.$$

Obviously from (2.6)

$$\alpha_r^{(-1)} = \frac{(aq/\rho_1)_r(aq/\rho_2)_r(\rho_1\rho_2/aq)^r\alpha_r}{(\rho_1)_r(\rho_2)_r}. \tag{2.8}$$

To back up in the chain of β's is a little trickier. The relation is

$$\beta_n^{(-1)} = \frac{1}{(\rho_1)_n(\rho_2)_n} \sum_{j=0}^{n} \frac{(aq/\rho_1)_j(aq/\rho_2)_j(\rho_1\rho_2/aq)_{n-j}(\rho_1\rho_2/aj)^{2n-j}\beta_j}{(q)_{n-j}}. \tag{2.9}$$

To see this, let us define

$$B_n = \left(\frac{aq}{\rho_1}\right)_n\left(\frac{aq}{\rho_2}\right)_n\beta_n, \tag{2.10}$$

(2.11) $$B_n^{(-1)} = (\rho_1)_n(\rho_2)_n\left(\frac{aq}{\rho_1\rho_2}\right)^n \beta_n^{(-1)},$$

(2.12) $$b(t) = \sum_{n=0}^{\infty} B_n t^n$$

and

(2.13) $$b_{-1}(t) = \sum_{n=0}^{\infty} B_n^{(-1)} t^n.$$

Then by the q-binomial series [**4**, p. 17, (2.2.1)], we see that (2.5) is equivalent to

(2.14) $$b(t) = \frac{(aqt/\rho_1\rho_2)_\infty b_{(-1)}(t)}{(t)_\infty}.$$

Obviously (2.14) is equivalent to

(2.15) $$b_{-1}(t) = \frac{(t)_\infty}{(aqt/\rho_1\rho_2)_\infty} b(t),$$

and (2.15) yields (2.8) by invocation again of the q-binomial series.

We now have the four identities ((2.3), (2.7)–(2.9)) necessary so that we may start with either sequence of a Bailey pair, obtain the other sequence and then move either direction in the Bailey chain.

3. The role of SCRATCHPAD. How can one gain a foothold in studying the mock theta functions? Our approach was to implement the study of Bailey chains on SCRATCHPAD, IBM's symbolic algebra package. To keep things as simple as possible we considered the case where ρ_1 and $\rho_2 \to \infty$ and $a = 1$. Thus from Bailey's Lemma, we see that

(3.1) $$\sum_{n=0}^{\infty} q^{n^2}\beta_n = (q)_\infty \sum_{n=0}^{\infty} q^{n^2}\alpha_n,$$

where by (2.7)

(3.2) $$\alpha_n = (1 - q^{2n}) \sum_{j=0}^{n} \frac{(q)_{n+j-1}(-1)^{n-j} q^{\binom{n-j}{2}}\beta_j}{(q)_{n-j}}.$$

Suppose we want the left side of (3.1) to coincide with (1.1). Then we must take

(3.3) $$\beta_n = \frac{1}{(1+q)^2(1+q^2)^2\cdots(1+q^n)^2}.$$

We now define α_n and β_n explicitly in SCRATCHPAD and ask for the first four values of α_n. SCRATCHPAD responds with

$$1$$
$$\frac{-4q}{q+1}$$
$$\frac{4q^3}{q^2+1}$$
$$\frac{-4q^6}{q^3+1}.$$

The pattern suggested is quite clear; namely it is reasonable to conjecture that α_n is $(-1)^n 4q^{n(n+1)/2}/(1+q^n)$ for $n \geq 1$, $\alpha_0 = 1$. The insertion of this conjecture in (3.1) coincides, not surprisingly, with the known fact (1.2).

Suppose now we redefine β_n by

$$\beta_n = \frac{1}{(1+q)(1+q^2)\cdots(1+q^n)}. \tag{3.4}$$

This then makes the left-side of (3.1) identical with the series in (1.3) for the fifth order mock theta function $f_0(q)$. The pattern for the related α_n arising from (3.2) is now somewhat complicated so we ask SCRATCHPAD for the first nine values:

$$\begin{aligned}
&1\\
&q^2 - 3q\\
&q^7 - 2q^6 - q^5 + 2q^4 + 2q^3\\
&q^{15} - 2q^{14} - q^{12} + 4q^{11} - 2q^8 - 2q^6\\
&q^{26} - 2q^{25} + q^{22} + 2q^{21} - 2q^{18} - 2q^{17} + 2q^{13} + 2q^{10}\\
&\vdots\\
&q^{100} - 2q^{99} + 2q^{96} - q^{92} - 2q^{88} + 2q^{84} + 2q^{83}\\
&\quad -2q^{76} - 2q^{75} + 2q^{67} + 2q^{64} - 2q^{56} - 2q^{51} + 2q^{43} + 2q^{36}.
\end{aligned}$$

A careful look at α_8 (and, in the actual discovery, α_7 and α_9) suggests that a reasonable conjecture for α_n is

$$q^{n(3n+1)/2} \sum_{j=-n}^{n} (-1)^j q^{-j^2} - q^{n(3n-1)/2} \sum_{j=-n+1}^{n-1} (-1)^j q^{-j^2}. \tag{3.5}$$

If we insert this conjectured formula for α_n into (3.1) we obtain (1.4) almost directly. Thus SCRATCHPAD provides a powerful tool for the empirical analysis of Bailey chains. It should be emphasized that identity (1.4) appears nowhere in any of Ramanujan's known writings.

4. Extensions of Shanks' formulas. Once the truncated theta series are observed arising in (3.5), one immediately recalls the two elegant papers by D. Shanks on truncated theta series [**16, 17**]. Presumably, if Shanks' results can be embedded in the hierarchy of q-hypergeometric function identities, then adequate generalizations of his results should allow us to derive (3.5) and related formulae for the other fifth order mock theta functions. It turns out that if we move to the left one place in the Bailey chain for the fifth order mock theta functions, our work is greatly simplified. The following two lemmas provide us adequate q-hypergeometric series machinery for the appropriate new Bailey pairs given in Theorems 3 and 4.

LEMMA 1. *The sequences A_n, B_n form a Bailey pair where*

$$B_n = \frac{(-1)^n (b)_n q^{-n(n-1)/2} a^{-n}}{(q)_n (bq)_n}, \tag{4.1}$$

$$A_n = \frac{(-1)^n (aq)_{n-1}(1-aq^{2n})q^{n(n-1)/2}}{(bq)_n} \sum_{j=0}^{n} \frac{(b)_j q^{j(1-n)} a^{-j}}{(q)_j}. \tag{4.2}$$

PROOF. All that is necessary is to evaluate (2.7) with β_n replaced by B_n:

$$\begin{aligned}
A_n &= (1-aq^{2n}) \sum_{j=0}^{n} \frac{(aq)_{n+j-1}(-1)^{n-j} q^{\binom{n-j}{2}} B_j}{(q)_{n-j}} \\
&= \frac{(-1)^n (aq)_{n-1}(1-aq^{2n}) q^{\binom{n}{2}}}{(q)_n} \sum_{j=0}^{n} \frac{(aq^n)_j (q)_n (b)_j a^{-j} q^{j(1-n)}}{(q)_j (q)_{n-j} (bq)_j} \\
&= \frac{(-1)^n (aq)_{n-1}(1-aq^{2n}) q^{\binom{n}{2}}}{(q)_n} \sum_{j=0}^{n} \frac{(q^{-n})_j (b)_j (aq^n)_j q^{j-\binom{j}{2}} a^{-j}}{(q)_j (bq)_j} \\
&= \frac{(-1)^n (aq)_{n-1}(1-aq^{2n}) q^{\binom{n}{2}}}{(q)_n} \lim_{t\to 0} \sum_{j=0}^{n} \frac{(q^{-n})_j (b)_j (aq^n)_j (q/at)^j}{(q)_j (bq)_j (t^{-1})_j} \\
&= \frac{(-1)^n (aq)_{n-1}(1-aq^{2n}) q^{\binom{n}{2}}}{(q)_n} \cdot \frac{(q)_n}{(bq)_n} \sum_{j=0}^{n} \frac{(q^{-n})_j (b)_j a^{-j} q^{j(1-n)}}{(q)_j (q^{-n})_j}
\end{aligned}$$

(by [22, p. 175, (10.2)] with $a = q^{-n}, c = aq^n, e = bq, f = t^{-1}, p = q$)

$$= \frac{(-1)^n (aq)_{n-1}(1-aq^{2n}) q^{\binom{n}{2}}}{(bq)_n} \sum_{j=0}^{n} \frac{(b)_j a^{-j} q^{j(1-n)}}{(q)_j},$$

as desired. □

LEMMA 2.

$$1 + \sum_{j=1}^{n} \frac{(a)_j (1-aq^{2j})(b)_j a^j q^{j^2} b^{-j}}{(a)_j (1-a)(aq/b)_j} = \frac{(aq)_n}{(aq/b)_n} \sum_{j=0}^{n} \frac{(b)_j (aq^{n+1}/b)^j}{(q)_j}. \tag{4.3}$$

PROOF. This result is merely a limiting case of Watson's q-analog of Whipple's theorem [**18**, p. 100, (3.4.1.5)]. In fact, (4.3) follows immediately from the substitutions $g = q^{-n}, e = b, c = aq^{n+1}, f \to \infty, d \to \infty$. □

We remark that if in (4.3) we take $a = 1$ and let $b \to \infty$, we obtain Shank's finite version of Euler's pentagonal number theorem [**16**, p. 747, (2)]. Shank's finite version of Gauss's theorem [**17**, p. 609, (3′)] follows by replacing q by q^2 in (4.3) and then setting $a = 1, b = q$.

THEOREM 3. *The sequences A_n, B_n form a Bailey pair where*

$$B_n = \frac{(-1)^n (b)_n q^{-n(n-1)/2} a^{-n}}{(q)_n (bq)_n}, \tag{4.4}$$

$$(4.5)\qquad A_n = \frac{(-1)^n (aq)_{n-1}(b)_n a^{-n} q^{-\binom{n}{2}}(1-aq^{2n})}{(bq)_n (q)_n} + \frac{(-1)^n q^{\binom{n}{2}} b^{n-1}(aq/b)_{n-1}(1-aq^{2n})}{(bq)_n} \cdot \left(1 + \sum_{j=1}^{n-1} \frac{(aq)_{j-1}(1-aq^{2j})(b)_j a^{-j} q^{-j^2} b^{-j}}{(q)_j (aq/b)_j}\right).$$

PROOF. Since the B_n sequence given by (4.4) is the same as the one in Lemma 1, equation (4.1), we see that we only need to identify the expression in (4.2) with the right-hand side of (4.5). In Lemma 2 replace a by a^{-1}, b by b^{-1}, q by q^{-1} and n by $n-1$; hence

$$(4.6)\qquad 1 + \sum_{j=1}^{n-1} \frac{(aq)_{j-1}(1-aq^{2j})(b)_j a^{-j} b^{-j} q^{-j^2}}{(q)_j (aq/b)_j} = \frac{b^{1-n}(aq)_{n-1}}{(aq/b)_{n-1}} \sum_{j=0}^{n-1} \frac{(b)_j a^{-j} q^{j(1-n)}}{(q)_j}.$$

Now let us examine the sum in (4.2). We split off the term $j = n$ which yields the first summand on the right-hand side of (4.5); the remainder of the sum in (4.2) is identical with the sum of the right-hand side of (4.6). Hence multiplying both sides of (4.6) by $b^{n-1}(aq/b)_{n-1}/(aq)_{n-1}$ and substituting the resulting left-hand side into (4.2), we obtain the second term in (4.5). □

THEOREM 4. *The sequences A'_n, B'_n form a Bailey pair where*

$$(4.7)\qquad B'_n = \frac{1}{(bq)_n},$$

$$(4.8)\qquad A'_n = \frac{(-1)^n (aq)_{n-1}(b)_n q^{\binom{n+1}{2}}(1-aq^{2n})}{(bq)_n (q)_n} + \frac{(-1)^n a^n q^{n(3n-1)/2} b^{n-1}(aq/b)_{n-1}(1-aq^{2n})}{(bq)_n} \cdot \left(1 + \sum_{j=1}^{n-1} \frac{(aq)_{j-1}(1-aq^{2j})(b)_j a^{-j} q^{-j^2} b^{-j}}{(q)_j (aq/b)_j}\right).$$

PROOF. We apply Bailey's Lemma (with $\rho_1, \rho_2 \to \infty$) to the Bailey pair in Theorem 3 and the resulting $A'_n = a^n q^{n^2} A_n$ which is precisely (4.8). Now by (2.5)

$$(4.9)\qquad B'_n = \sum_{j=0}^{n} \frac{a^j q^{j^2}}{(q)_{n-j}} \frac{(-1)^j (b)_j q^{-j(j-1)/2} a^{-j}}{(q)_j (bq)_j} = \frac{1}{(q)_n} \sum_{j=0}^{n} \frac{(q^{-n})_j (b)_j}{(q)_j (bq)_j} q^{(n+1)j} = \frac{1}{(bq)_n} \quad \text{by } [\mathbf{18}, \text{p. } 97, (3.3.2.6)],$$

as desired. □

5. Hecke modular form identities. E. Hecke [**12**] made an extensive study of double theta type series involving an indefinite quadratic form. For example, he showed that

$$(5.1)\qquad (q)_\infty^2 = \sum_{m=-\infty}^{\infty} \sum_{n \geq 2|m|} (-1)^{n+m} q^{\binom{n+1}{2} - m(3m-1)/2}.$$

Actually as D. Bressoud points out [**10**], identity (5.1) was discovered originally by L. J. Rogers [**15**]. Subsequent studies of these types of identities have been made by Kac and Peterson [**13**], Bressoud [**10**], and the author [**7**].

The Bailey pair in Theorem 3 leads naturally to Theorem 5, an infinite product expansion which implies several Hecke modular form type results.

THEOREM 5. *Let A_n be defined by* (4.5); *then*

$$(5.2)\qquad \frac{(q)_\infty (bq/y)_\infty (aq)_\infty}{(bq)_\infty (q/y)_\infty (aq/y)_\infty} = \sum_{n=0}^{\infty} \frac{(y)_n (-a/y)^n q^{\binom{n+1}{2}} A_n}{(aq/y)_n}.$$

PROOF. We take the Bailey pair from Theorem 3 and substitute into (2.4) with $\rho_1 = y$, $\rho_2 \to \infty$, $n \to \infty$. This yields

$$(5.3)\qquad \frac{1}{(aq/y)_\infty} \sum_{n=0}^{\infty} \frac{(y)_n (b)_n}{(q)_n (bq)_n} \left(\frac{q}{y}\right)^n = \frac{1}{(aq)_\infty} \sum_{r=0}^{\infty} \frac{(y)_r}{(aq/y)_r} \left(-\frac{a}{y}\right)^r q^{\binom{r+1}{2}} A_r.$$

Now

$$(5.4)\qquad \sum_{n=0}^{\infty} \frac{(y)_n (b)_n (q/y)^n}{(q)_n (bq)_n} = \frac{(q)_\infty (bq/y)_\infty}{(bq)_\infty (q/y)_\infty}$$

by [**18**, p. 97, (3.3.2.5)]. Substituting the right-hand side of (5.4) into (5.3) we obtain (5.2). □

To apply (5.2) with ease to Hecke modular form identities, we prove three lemmas.

LEMMA 6. *Let $A_n(a, b, q)$ denote the A_n in* (4.5). *Then*

$$(5.5)\qquad A_n(aq, b, q) = \frac{(-1)^n (1 - aq^{2n+1}) q^{\binom{n}{2}} b^n (aq/b)_n}{(1 - aq)(bq)_n} \cdot \left(1 + \sum_{j=1}^{n} \frac{(aq)_{j-1} (1 - aq^{2j}) (b)_j a^{-j} q^{-j^2} b^{-j}}{(q)_j (aq/b)_j}\right).$$

PROOF. By (4.2)

$$A_n(aq, b, q) = \frac{(-1)^n (aq^2)_{n-1} (1 - aq^{2n+1}) q^{\binom{n}{2}}}{(bq)_n} \cdot \sum_{j=0}^{n} \frac{(b)_j q^{-jn} a^{-j}}{(q)_j}$$

$$= \frac{(-1)^n (aq)_n (1 - aq^{2n+1}) q^{\binom{n}{2}}}{(1 - aq)(bq)_n} \frac{b^n (aq/b)_n}{(aq)_n}$$

(equation continues)

$$\cdot\left(1+\sum_{j=1}^{n}\frac{(aq)_{j-1}(1-aq^{2j})(b)_j a^{-j}q^{-j^2}b^{-j}}{(q)_j(aq/b)_j}\right)$$

(by (4.6) with n replaced by $n+1$)

$$=\frac{(-1)^n(1-aq^{2n+1})q^{\binom{n}{2}}b^n(aq/b)_n}{(1-aq)(bq)_n}$$

$$\cdot\left(1+\sum_{j=1}^{n}\frac{(aq)_{j-1}(1-aq^{2j})(b)_j a^{-j}q^{-j^2}b^{-j}}{(q)_j(aq/b)_j}\right),$$

as desired. □

Lemma 7. *Let $A_n(a,b,q)$ denote the A_n in (4.5). Then*

$$A_n(1,-1,q)=q^{\binom{n+1}{2}}\sum_{j=-n}^{n}(-1)^j q^{-j^2}-q^{\binom{n}{2}}\sum_{j=-n+1}^{n-1}(-1)^j q^{-j^2}, \tag{5.6}$$

$$A_n(q,-1,q)=\frac{(1-q^{2n+1})q^{\binom{n}{2}}}{(1-q)}\left(1+2\sum_{j=1}^{n}(-1)^j q^{-j^2}\right), \tag{5.7}$$

$$A_n(1,q^{-1},q^2)=(-1)^n\left(q^{n^2+2n}\sum_{j=0}^{2n}q^{-\binom{j+1}{2}}-q^{n^2-2n}\sum_{j=0}^{2n-2}q^{-\binom{j+1}{2}}\right), \tag{5.8}$$

$$A_n(q^2,q,q^2)=\frac{(-1)^n(1+q^{2n+1})q^{n^2}}{1+q}\sum_{j=0}^{2n}q^{-\binom{j+1}{2}}, \tag{5.9}$$

$$A_n(1,0,q)=q^{n^2+n}\sum_{j=-n}^{n}(-1)^j q^{-j(3j+1)/2}-q^{n^2-n}\sum_{j=-n+1}^{n-1}(-1)^j q^{-j(3j+1)/2}, \tag{5.10}$$

$$A_n(q,0,q)=\frac{q^{n^2}(1-q^{2n+1})}{(1-q)}\sum_{j=-n}^{n}(-1)^j q^{-j(3j+1)/2}. \tag{5.11}$$

Proof. By (4.5),

$$A_n(1,-1,q)=(-1)^n 2q^{-\binom{n}{2}}-q^{\binom{n}{2}}(1-q^n)\left(1+2\sum_{j=1}^{n-1}(-1)^j q^{-j^2}\right)$$

$$=q^{\binom{n+1}{2}}\sum_{j=-n}^{n}(-1)^j q^{-j^2}-q^{\binom{n}{2}}\sum_{j=-n+1}^{n-1}(-1)^j q^{-j^2},$$

as asserted in (5.6).

By (5.5),

$$A_n(q,-1,q)=\frac{(1-q^{2n+1})q^{\binom{n}{2}}}{(1-q)}\left(1+2\sum_{j=1}^{n}(-1)^j q^{-j^2}\right),$$

as asserted in (5.7).

By (4.5),

$$A_n(1, q^{-1}, q^2) = \frac{(-1)^n(1+q^{2n})q^{n-n^2}(1-q^{-1})}{(1-q^{2n-1})} + \frac{(-1)^n q^{(n-1)^2}(1-q^{4n})}{(1-q)}$$

$$\cdot\left(1 + \sum_{j=1}^{n-1} \frac{(1+q^{2j})(1-q^{-1})(1-q)q^{j-2j^2}}{(1-q^{2j-1})(1-q^{2j+1})}\right)$$

$$= \frac{(-1)^n(1+q^{2n})q^{n-n^2}(1-q^{-1})}{(1-q^{2n-1})}$$

$$+ \frac{(-1)^n q^{(n-1)^2}(1-q^{4n})}{(1-q)} \sum_{j=-n+1}^{n-1} \frac{(1-q^{-1})(1-q)q^{j-2j^2}}{(1-q^{2j-1})(1-q^{2j+1})}$$

$$= \frac{(-1)^n(1+q^{2n})q^{n-n^2}(1-q^{-1})}{(1-q^{2n-1})} + \frac{(-1)^n q^{(n-1)^2}(1-q^{4n})(1-q^{-1})}{q^{-1}(1-q^2)}$$

$$\cdot \sum_{j=-n+1}^{n-1} q^{j-2j^2}\left(\frac{q^{-1}}{1-q^{2j-1}} - \frac{q}{1-q^{2j+1}}\right)$$

$$= \frac{(-1)^n(1+q^{2n})q^{n-n^2}(1-q^{-1})}{(1-q^{2n-1})}$$

$$+ \frac{(-1)^n q^{(n-1)^2}(1-q^{4n})}{(1+q)} \sum_{j=-n+1}^{n-1} \frac{q^{j-2j^2}(1+q)}{(1-q^{2j+1})}$$

(where we have replaced j by $-j$ in the first sum)

$$= \frac{(-1)^n(1+q^{2n})q^{n-n^2}(1-q^{-1})}{(1-q^{2n-1})} - \frac{(-1)^n q^{(n-1)^2-n-2n^2}(1-q^{4n})}{(1-q^{-2n+1})}$$

$$+(-1)^n q^{(n-1)^2}(1-q^{4n}) \sum_{j=0}^{n-1}\left(\frac{q^{j-2j^2}}{1-q^{2j+1}} + \frac{q^{-j-1-2(j+1)^2}}{1-q^{-2j-1}}\right)$$

$$= (-1)^n q^{-n^2-n}(1+q^{2n})$$

$$-(-1)^n q^{(n-1)^2}(1-q^{4n}) \sum_{j=0}^{n-1} \frac{q^{-2j^2-3j-2}(1-q^{4j+2})}{(1-q^{2j+1})}$$

$$= (-1)^n q^{-n^2-n}(1+q^{2n})$$

$$-(-1)^n q^{n^2-2n}(1-q^{4n}) \sum_{j=0}^{n-1}\left(q^{-\binom{2j+2}{2}} + q^{-\binom{2j+1}{2}}\right)$$

$$= (-1)^n q^{-n^2-n}(1+q^{2n}) - (-1)^n q^{n^2-2n}(1-q^{4n}) \sum_{j=0}^{2n-1} q^{-\binom{j+1}{2}}$$

$$= (-1)^n q^{n^2+2n} \sum_{j=0}^{2n} q^{-\binom{j+1}{2}} - (-1)^n q^{n^2-2n} \sum_{j=0}^{2n-2} q^{-\binom{j+1}{2}},$$

as asserted in (5.8).

Next by (5.5)

$$A_n(q^2,q,q^2) = \frac{(-1)^n(1+q^{2n+1})q^{n^2}}{1+q}\left(1+\sum_{j=1}^{n}(1+q^{2j})q^{-2j^2-j}\right)$$
$$= \frac{(-1)^n(1+q^{2n+1})q^{n^2}}{1+q}\sum_{j=0}^{2n}q^{-\binom{j+1}{2}},$$

as asserted in (5.9).

By (4.5),

$$A_n(1,0,q) = (-1)^n q^{-\binom{n}{2}}(1+q^n)$$
$$-q^{n^2-n}(1-q^{2n})\left(1+\sum_{j=1}^{n-1}(1+q^j)(-1)^j q^{-j(3j+1)/2}\right)$$
$$= (-1)^n q^{-\binom{n}{2}}(1+q^n) - q^{n^2-n}(1-q^{2n})\sum_{j=-n+1}^{n-1}(-1)^j q^{-j(3j+1)/2}$$
$$= q^{n^2+n}\sum_{j=-n}^{n}(-1)^j q^{-j(3j+1)/2} - q^{n^2-n}\sum_{j=-n+1}^{n-1}(-1)^j q^{-j(3j+1)/2},$$

as asserted in (5.10).

Finally by (5.5)

$$A_n(q,0,q) = \frac{q^{n^2}(1-q^{2n+1})}{(1-q)}\left(1+\sum_{j=1}^{n}(1+q^j)q^{-j^2-\binom{j+1}{2}}(-1)^j\right)$$
$$= \frac{q^{n^2}(1-q^{2n+1})}{(1-q)}\sum_{j=-n}^{n}(-1)^j q^{-j(3j+1)/2},$$

which is (5.11). □

LEMMA 8. *Let $A_n = A_n(a, b, q)$ be defined by* (4.5). *Then*

$$\frac{(q)_\infty(aq)_\infty}{(bq)_\infty} = \sum_{n=0}^{\infty} a^n q^{n^2} A_n, \tag{5.12}$$

$$\frac{(q)_\infty^2(-bq)_\infty}{(-q)_\infty^2(bq)_\infty} = 1 + 2\sum_{n=1}^{\infty}\frac{q^{n(n+1)/2}}{1+q^n}A_n(1,b,q), \tag{5.13}$$

$$\frac{(q^2;q^2)_\infty^2(bq;q^2)_\infty}{(q;q^2)_\infty^2(bq^2;q^2)_\infty} = \sum_{n=0}^{\infty}(-1)^n q^{n^2}A_n(1,b,q^2). \tag{5.14}$$

PROOF. These identities are immediate from Theorem 5 under the substitutions $y \to \infty$; $a = 1, y = -1$; finally replace q by q^2, then set $a = 1, y = q$. □

It is now possible to combine Lemmas 7 and 8 to obtain 18 different identities of the Hecke modular form type. We note only:

$$(5.15)\quad (q)_\infty^2(q;q^2)_\infty = 1 + \sum_{\substack{n=1\\|j|\le n}}^{\infty} (-1)^j q^{n(3n+1)/2-j^2} - \sum_{\substack{n=1\\|j|<n}}^{\infty} (-1)^j q^{n(3n-1)/2-j^2}$$

$$= \sum_{\substack{n=0\\|j|\le n}}^{\infty} (-1)^j q^{n(3n+1)/2-j^2}(1-q^{2n+1}).$$

This result follows from (5.12) with $a = 1$, $b = -1$ and (5.6). We can also find a new representation for the generating function of $r_3(n)$, the number of representations of n as a sum of three squares:

$$(5.16)\quad \sum_{n=0}^{\infty} (-1)^n r_3(n) q^n = \left(\sum_{n=-\infty}^{\infty} (-1)^n q^{n^2}\right)^3 = \left(\frac{(q)_\infty}{(-q)_\infty}\right)^3$$

$$= 1 + 4\sum_{n=1}^{\infty} \frac{(-1)^n q^n}{1+q^n} - 2\sum_{\substack{n=1\\|j|<n}}^{\infty} \frac{q^{n^2-j^2}(1-q^n)(-1)^j}{1+q^n}.$$

The first two equalities are classical [**4**, p. 23, (2.2.12)]; the last follows from (5.13) with $b = -1$ and (5.6).

Finally we consider (5.2) in the case $q \to q^2$. Then $a = q^2$, $b = y = q$. This yields

$$(5.17)\quad \left(\sum_{n=0}^{\infty} q^{\binom{n+1}{2}}\right)^3 = \left(\frac{(q^2;q^2)_\infty}{(q;q^2)_\infty}\right)^3$$

$$= \sum_{n=0}^{\infty}\sum_{j=0}^{2n} \frac{q^{2n^2+2n-\binom{j+1}{2}}(1+q^{2n+1})}{(1-q^{2n+1})}$$

$$= \sum_{n=0}^{\infty}\sum_{j=0}^{2n} \frac{q^{n+j(4n+1-j)/2}(1+q^{2n+1})}{(1-q^{2n+1})}.$$

Clearly the left-hand side of (5.17) is the generating function for the number of representations of an integer as a sum of three triangular numbers. Inspecting the last entry in (5.7) we see that in this series the nth term contributes at least 1 to the coefficient of q^n; hence all coefficients are positive. Thus we have just proved Gauss's famous result that every integer is the sum of three triangular numbers.

6. The fifth order mock theta functions. In the following theorem, eight of the ten fifth order mock theta functions of Ramanujan [**14**, pp. 354–355] will be related to double sums of the Hecke modular form type. We are not able to give such formulas for either $\chi_0(q)$ or $\chi_1(q)$ [**20**, pp. 277–278]. However, since $\chi_0(q) = 2F_0(q) - \phi_0(-q)$ and $\chi_1(q) = 2F_1(q) + q^{-1}\phi_1(-q)$ [**20**, pp. 277, 279], we see that each of $\chi_0(q)$ and $\chi_1(q)$ is a linear combination of functions that do appear in our main result. Each equation in Theorem 9 will provide both the definition and the appropriate Hecke representation of the fifth order mock theta function under consideration.

THEOREM 9.

$$f_0(q) \equiv \sum_{n=0}^{\infty} \frac{q^{n^2}}{(-q)_n} = \frac{1}{(q)_\infty} \sum_{\substack{n=0 \\ |j| \le n}}^{\infty} (-1)^j q^{n(5n+1)/2 - j^2} (1 - q^{4n+2}), \tag{6.1}$$

$$F_0(q) \equiv \sum_{n=0}^{\infty} \frac{q^{2n^2}}{(q; q^2)_n} \tag{6.2}$$

$$= \frac{1}{(q^2; q^2)_\infty} \sum_{n=0}^{\infty} \sum_{j=0}^{2n} (-1)^n q^{5n^2+2n-\binom{j+1}{2}} (1 + q^{6n+3}),$$

$$1 + 2\psi_0(q) \equiv \sum_{n=0}^{\infty} (-1; q)_n q^{\binom{n+1}{2}} \tag{6.3}$$

$$= \frac{(-q)_\infty}{(q)_\infty} \left(1 + 2 \sum_{n=1}^{\infty} (-1)^n q^{2n^2+n} - 2 \sum_{\substack{n=1 \\ |j|<n}}^{\infty} (-1)^j q^{n(5n-1)/2 - j(3j+1)/2} (1 - q^n) \right),$$

$$\phi_0(q) \equiv \sum_{n=0}^{\infty} (-q; q^2)_n q^{n^2} \tag{6.4}$$

$$= \frac{(-q; q^2)}{(q^2; q^2)} \sum_{\substack{n=0 \\ |j| \le n}}^{\infty} (-1)^j q^{5n^2+2n-3j^2-j} (1 - q^{6n+3}),$$

$$f_1(q) \equiv \sum_{n=0}^{\infty} \frac{q^{n^2+n}}{(-q)_n} = \frac{1}{(q)_\infty} \sum_{\substack{n=0 \\ |j| \le n}}^{\infty} (-1)^j q^{n(5n+3)/2 - j^2} (1 - q^{2n+1}), \tag{6.5}$$

$$F_1(q) = \sum_{n=0}^{\infty} \frac{q^{2n^2+2n}}{(q; q^2)_{n+1}} \tag{6.6}$$

$$= \frac{1}{(q^2; q^2)_\infty} \sum_{n=0}^{\infty} \sum_{j=0}^{2n} (-1)^n q^{5n^2+4n-\binom{j+1}{2}} (1 + q^{2n+1}),$$

$$\psi_1(q) = \sum_{n=0}^{\infty} (-q)_n q^{\binom{n+1}{2}} \tag{6.7}$$

$$= \frac{(-q)_\infty}{(q)_\infty} \sum_{\substack{n=0 \\ |j| \le n}}^{\infty} (-1)^j q^{n(5n+3)/2 - j(3j+1)/2} (1 - q^{2n+1}),$$

$$\phi_1(q) = \sum_{n=0}^{\infty} (-q; q^2)_n q^{(n+1)^2} \tag{6.8}$$

$$= \frac{(-q; q^2)_\infty}{(q^2; q^2)_\infty} \sum_{\substack{n=0 \\ |j| \le n}}^{\infty} (-1)^j q^{5n^2+4n-3j^2-j} (1 + q^{2n+1}).$$

PROOF. We shall rely on many of the results in §§4 and 5. Our jumping off point is Theorem 4 and the observation that $A_n'(a,b,q) = a^n q^{n^2} A_n(a,b,q)$. Thus all our formulae for the $A_n(a,b,q)$ can be applied in evaluating $A_n'(a,b,q)$.

We begin with (6.1), and let $B_n' = B_n'(b,q)$.

$$\begin{aligned}
f_0(q) &= \sum_{n=0}^{\infty} q^{n^2} B_n'(-1,q) \quad (\text{by (4.7)}) \\
&= \frac{1}{(q)_\infty} \sum_{n=0}^{\infty} q^{n^2} A_n'(1,-1,q) \quad (\text{by (3.1)}) \\
&= \frac{1}{(q)_\infty} \sum_{n=0}^{\infty} q^{2n^2} A_n(1,-1,q) \quad (\text{by (2.6)}) \\
&= \frac{1}{(q)_\infty} \left(\sum_{\substack{n=0 \\ |j|\le n}}^{\infty} q^{n(5n+1)/2-j^2}(-1)^j - \sum_{\substack{n=0 \\ |j|<n}}^{\infty} q^{n(5n-1)/2-j^2}(-1)^j \right) \quad (\text{by (5.6)}) \\
&= \frac{1}{(q)_\infty} \sum_{\substack{n=0 \\ |j|\le n}}^{\infty} q^{n(5n+1)/2-j^2}(-1)^j (1-q^{4n+2}),
\end{aligned}$$

which is (6.1).

Next

$$\begin{aligned}
F_0(q) &= \sum_{n=0}^{\infty} q^{2n^2} B_n'(q^{-1},q^2) \quad (\text{by (4.7)}) \\
&= \frac{1}{(q^2;q^2)_\infty} \sum_{n=0}^{\infty} q^{2n^2} A_n'(1,q^{-1},q^2) \quad (\text{by (3.1)}) \\
&= \frac{1}{(q^2;q^2)} \sum_{n=0}^{\infty} q^{4n^2} A_n(1,q^{-1},q^2) \quad (\text{by (2.6)}) \\
&= \frac{1}{(q^2;q^2)_\infty} \left(\sum_{n=0}^{\infty} \sum_{j=0}^{2n} (-1)^n q^{5n^2+2n-\binom{j+1}{2}} \right. \\
&\qquad \left. - \sum_{n=0}^{\infty} \sum_{j=0}^{2n-2} (-1)^n q^{5n^2-2n-\binom{j+1}{2}} \right) \quad (\text{by (5.8)}) \\
&= \frac{1}{(q^2;q^2)} \sum_{n=0}^{\infty} \sum_{j=0}^{2n} (-1)^n q^{5n^2+2n-\binom{j+1}{2}} (1+q^{6n+3}),
\end{aligned}$$

as asserted in (6.2).

$$\begin{aligned}
1 + 2\psi_0(q) &= \sum_{n=0}^{\infty} (-1;q)_n q^{\binom{n+1}{2}} B_n'(0,q) \\
&= \frac{(-q)_\infty}{(q)_\infty} \sum_{n=0}^{\infty} \frac{(-1;q)_n q^{\binom{n+1}{2}} A_n'(1,0,q)}{(-q;q)_n} \\
&\qquad (\text{by (2.4)}, n\to\infty, \rho_1\to\infty, \rho_2=-1, a=1)
\end{aligned}$$

(equation continues)

$$= \frac{(-q)_\infty}{(q)_\infty}\left(1 + 2\sum_{n=1}^{\infty} \frac{q^{\binom{n+1}{2}+n^2}}{1+q^n} A_n(1,0,q)\right) \quad \text{(by (2.6))}$$

$$= \frac{(-q)_\infty}{(q)_\infty}\left(1 + 2\sum_{n=1}^{\infty} \frac{q^{n(3n+1)/2}}{1+q^n}\left((-1)^n q^{\binom{n+1}{2}}(1+q^n) - q^{n^2-n}(1-q^{2n}) \sum_{|j|<n} (-1)^j q^{-j(3j+1)/2}\right)\right)$$

$$\text{(by (5.10))}$$

$$= \frac{(-q)_\infty}{(q)_\infty}\left(1 + 2\sum_{n=1}^{\infty} (-1)^n q^{2n^2+n} - 2\sum_{\substack{n=1 \\ |j|<n}}^{\infty} (-1)^j q^{n(5n-1)/2 - j(3j+1)/2}(1-q^n)\right),$$

as desired.

Next

$$\phi_0(q) = \sum_{n=0}^{\infty} q^{n^2}(-q;q^2)_n B_n'(0,q^2)$$

$$= \frac{(-q;q^2)_\infty}{(q^2;q^2)_\infty} \sum_{n=0}^{\infty} q^{n^2} A_n'(1,0,q^2)$$

$$\left(\text{by (2.4)}, q \to q^2, n \to \infty, \rho_1 \to \infty, \rho_2 = -q, a = 1\right)$$

$$= \frac{(-q;q^2)_\infty}{(q^2;q^2)_\infty} \sum_{n=0}^{\infty} q^{3n^2} A_n(1,0,q^2) \quad \text{(by (2.6))}$$

$$= \frac{(-q;q^2)_\infty}{(q^2;q^2)_\infty} \sum_{n=0}^{\infty} q^{3n^2}\left(q^{2n^2+2n} \sum_{|j|\le n} (-1)^j q^{-3j^2-j} - q^{2n^2-2n} \sum_{|j|<n} (-1)^j q^{-3j^2-j}\right)$$

$$= \frac{(-q;q^2)_\infty}{(q^2;q^2)_\infty} \sum_{\substack{n=0 \\ |j|\le n}}^{\infty} (-1)^j q^{5n^2+2n-3j^2-j}(1-q^{6n+3}),$$

as asserted in (6.4).

The proofs of the remaining four identities in Theorem 9 are exactly parallel to the four just given. The fundamental change concerns the replacement of a by q (or by q^2 when q has been replaced by q^2). We therefore omit the details. □

7. The seventh order mock theta functions. These three functions are perhaps the most mysterious in all of Ramanujan's work. They are defined in the last three lines of Ramanujan's *Collected Papers* [**14**, p. 355]. His only assertion about them is that they are not related to each other. G. N. Watson makes cryptic reference to them in [**19**, p. 80]; his short paper on the dilogarithm [**21**] was clearly motivated by a less than successful study of the asymptotics of the seventh order functions as $q \to 1$. To my knowledge nothing else about them appears in the literature.

We are able to produce a counterpart of Theorem 9 for them; however the scope of our results is much more limited than the generality of results related to the fifth order functions. Clearly much remains unknown about the seventh order functions; I hope that these results will facilitate further study.

Our work relies on the Bailey chain theory of §2. We begin with the lemmas necessary to establish the appropriate Bailey pairs. The polynomials to be considered are best expressed in terms of the Gaussian polynomials [**4**, p. 35]:

$$\begin{bmatrix} n \\ m \end{bmatrix} = \begin{bmatrix} n \\ m \end{bmatrix}_q = \frac{(q)_n}{(q)_m (q)_{n-m}}. \tag{7.1}$$

These polynomials satisfy the recurrences [**4**, p. 35]:

$$\begin{bmatrix} n \\ m \end{bmatrix} = \begin{bmatrix} n-1 \\ m-1 \end{bmatrix} + q^m \begin{bmatrix} n-1 \\ m \end{bmatrix} \tag{7.2}$$

$$= \begin{bmatrix} n-1 \\ m \end{bmatrix} + q^{n-m} \begin{bmatrix} n-1 \\ m-1 \end{bmatrix}. \tag{7.3}$$

LEMMA 10. *Let sequences of polynomials (in q) U_n, V_n, W_n be defined by*

$$U_n = \sum_{j=1}^{n} \begin{bmatrix} n+j-1 \\ 2j-1 \end{bmatrix} (q)_{j-1} (-1)^{n-j} q^{\binom{n-j}{2}}, \tag{7.4}$$

$$V_n = (1 - q^{2n}) \sum_{j=0}^{n} \begin{bmatrix} n+j \\ n-j \end{bmatrix} \frac{(q)_j}{1 - q^{n+j}} (-1)^{n-j} q^{\binom{n-j}{2}}, \tag{7.5}$$

$$W_n = \sum_{j=0}^{n} \begin{bmatrix} n+j \\ n-j \end{bmatrix} (q)_j (-1)^{n-j} q^{\binom{n-j}{2}}. \tag{7.6}$$

Then

$$V_n = W_n - q^{2n-1} W_{n-1}, \tag{7.7}$$

$$V_n = -(1 - q^n) q^{2n-2} U_{n-1} + (-1)^n 2 q^{\binom{n+1}{2}}, \tag{7.8}$$

$$W_n = -q^n U_n - q^{2n-2} U_{n-1}. \tag{7.9}$$

PROOF. Since $1 - q^{2n} = 1 - q^{n+j} + q^{n+j}(1 - q^{n-j})$, we may rewrite V_n as

$$\begin{aligned} V_n &= \sum_{j=0}^{n} \left(\begin{bmatrix} n+j \\ n-j \end{bmatrix} + q^{n+j} \begin{bmatrix} n+j-1 \\ n-j-1 \end{bmatrix} \right) (q)_j (-1)^{n-j} q^{\binom{n-j}{2}} \\ &= W_n - q^{2n-1} \sum_{j=0}^{n-1} \begin{bmatrix} n-1+j \\ n-1-j \end{bmatrix} (q)_j (-1)^{n-1-j} q^{\binom{n-1-j}{2}} \\ &= W_n - q^{2n-1} W_{n-1}, \end{aligned}$$

which is (7.7).

$$
\begin{aligned}
\frac{V_n-(-1)^n 2q^{\binom{n+1}{2}}}{1-q^n} &= (-1)^n q^{\binom{n}{2}} + \sum_{j=1}^{n} \begin{bmatrix} n+j \\ n-j \end{bmatrix} \frac{(1+q^n)}{(1-q^{n-j})} (q)_j (-1)^{n-j} q^{\binom{n-j}{2}} \\
&\qquad \text{(by (7.5))} \\
&= (-1)^n q^{\binom{n}{2}} + \sum_{j=1}^{n} \begin{bmatrix} n+j-1 \\ n-j-1 \end{bmatrix} (q)_j (-1)^{n-j} q^{\binom{n-j}{2}} \\
&\quad + \sum_{j=1}^{n} \begin{bmatrix} n+j-1 \\ n-1 \end{bmatrix} (q)_{j-1} (-1)^{n-j} q^{\binom{n-j+1}{2}} \\
&\quad \left(\text{where we have used } 1+q^n = (1-q^{n-j}) + q^{n-j}(1+q^j)\right) \\
&= \sum_{j=0}^{n} \begin{bmatrix} n+j-1 \\ n-j-1 \end{bmatrix} (q)_j (-1)^{n-j} q^{\binom{n-j}{2}} \\
&\quad - \sum_{j=0}^{n-1} \begin{bmatrix} n+j \\ n-j-1 \end{bmatrix} (q)_j (-1)^{n-j} q^{\binom{n-j}{2}} \\
&= -\sum_{j=0}^{n-2} q^{2j+1} \begin{bmatrix} n+j-1 \\ n-j-2 \end{bmatrix} (q)_j (-1)^{n-j} q^{\binom{n-j}{2}} \quad \text{(by (7.3))} \\
&= -q^{2n-2} \sum_{j=1}^{n-1} \begin{bmatrix} n-1+j-1 \\ n-1-j \end{bmatrix} (q)_{j-1} (-1)^{n-j-1} q^{\binom{n-1-j}{2}} \\
&= -q^{2n-2} U_{n-1},
\end{aligned}
$$

which is equivalent to (7.8).

Finally,

$$
\begin{aligned}
W_n + q^n U_n &= \sum_{j\ge 0} \begin{bmatrix} n+j \\ n-j \end{bmatrix} (q)_j (-1)^{n-j} q^{\binom{n-j}{2}} \\
&\quad + q^n \sum_{j=1}^{n} \begin{bmatrix} n+j-1 \\ n-j \end{bmatrix} (q)_{j-1} (-1)^{n-j} q^{\binom{n-j}{2}} \\
&= (-1)^n q^{\binom{n}{2}} + \sum_{j=1}^{n} (q)_{j-1} (-1)^{n-j} q^{\binom{n-j}{2}} \left(\begin{bmatrix} n+j \\ n-j \end{bmatrix} (1-q^j) + q^n \begin{bmatrix} n+j-1 \\ n-j \end{bmatrix} \right) \\
&= (-1)^n q^{\binom{n}{2}} + (1+q^n) \sum_{j=1}^{n} (q)_j (-1)^{n-j} q^{\binom{n-j}{2}} \frac{(q)_{n+j-1}}{(q)_{2j}(q)_{n-j}} \\
&= (-1)^n q^{\binom{n}{2}} + \sum_{j\ge 1} (q)_j (-1)^{n-j} q^{\binom{n-j}{2}} \begin{bmatrix} n+j-1 \\ n-j-1 \end{bmatrix} \\
&\quad + \sum_{j\ge 1} (q)_{j-1} (-1)^{n-j} q^{\binom{n-j+1}{2}} \begin{bmatrix} n+j-1 \\ n-j \end{bmatrix} \\
&\qquad \left(\text{where we have used } 1+q^n = (1-q^{n-j}) + q^{n-j}(1+q^j)\right)
\end{aligned}
$$

(equation continues)

$$= \sum_{j\ge 0} (q)_j(-1)^{n-j}q^{\binom{n-j}{2}}\begin{bmatrix} n+j-1 \\ n-j-1 \end{bmatrix}$$

$$+ \sum_{j\ge 1} (q)_{j-1}(-1)^{n-j}q^{\binom{n-j+1}{2}}\left(q^{2j-1}\begin{bmatrix} n+j-2 \\ n-j-1 \end{bmatrix} + \begin{bmatrix} n+j-2 \\ n-j \end{bmatrix}\right)$$

(by (7.3))

$$= \sum_{j=0}^{n-1} (q)_j(-1)^{n-j}q^{\binom{n-j}{2}}\begin{bmatrix} n+j-1 \\ n-j-1 \end{bmatrix} - q^{2n-2}U_{n-1}$$

$$+ \sum_{j=1}^{n} (q)_{j-1}(-1)^{n-j}q^{\binom{n-j+1}{2}}\begin{bmatrix} n+j-2 \\ n-j \end{bmatrix}$$

$$= -q^{2n-2}U_{n-1},$$

since the third term is the same as the first once j is replaced by $j+1$. Thus (7.9) is estabished. □

LEMMA 11. *For the U_n defined by* (7.4) *we have*

$$U_{2n} = -2q^{3n^2-2n}\sum_{j=0}^{n-1} q^{-j^2-j}, \tag{7.10}$$

$$U_{2n+1} = q^{3n^2+n}\sum_{|j|\le n} q^{-j^2}. \tag{7.11}$$

PROOF. Equations (7.10) and (7.11) follow directly by induction from the initial values $U_0 = 0$, $U_1 = 1$, $U_2 = -2q$, and for $n \ge 3$

$$U_n - q^{3n-5}U_{n-2} = -(-1)^n 2q^{\binom{n}{2}}. \tag{7.12}$$

The initial values follow immediately from (7.4). Now by (7.7)–(7.9)

$$\left(-q^nU_n - q^{2n-2}U_{n-1}\right) - q^{2n-1}\left(-q^{n-1}U_{n-1} - q^{2n-4}U_{n-2}\right)$$

$$= W_n - q^{2n-1}W_{n-1} = V_n = -(1-q^n)q^{2n-2}U_{n-1} + (-1)^n 2q^{\binom{n+1}{2}};$$

simplifying the extremes of this equation we find

$$-q^nU_n + q^{4n-5}U_{n-2} = (-1)^n 2q^{\binom{n+1}{2}},$$

and multiplying through by $-q^{-n}$ we obtain (7.12). □

We are now prepared to derive the Bailey pairs for the seventh order mock theta functions.

Define

$$\mathscr{B}_n(0) = \frac{1}{(q^{n+1})_n}, \tag{7.13}$$

$$\mathscr{B}_n(1) = \frac{1}{(q^n)_n},\quad n > 0, \qquad \mathscr{B}_0(1) = 0, \tag{7.14}$$

$$\mathscr{B}_n(2) = \frac{1}{(q^{n+1})_{n+1}}, \tag{7.15}$$

and

$$\mathcal{A}_{2n}(0) = q^{3n^2+n} \sum_{|j| \le n} q^{-j^2} - q^{3n^2-n} \sum_{|j|<n} q^{-j^2}, \tag{7.16}$$

$$\mathcal{A}_{2n+1}(0) = -2q^{3n^2+4n+1} \sum_{j=0}^{n} q^{-j^2-j} + 2q^{3n^2+2n} \sum_{j=0}^{n-1} q^{-j^2-j}, \tag{7.17}$$

$$\mathcal{A}_{2n}(1) = -2q^{3n^2-2n}(1-q^{4n}) \sum_{j=0}^{n-1} q^{-j^2-j}, \tag{7.18}$$

$$\mathcal{A}_{2n+1}(1) = q^{3n^2+n}(1-q^{4n+2}) \sum_{|j|\le n} q^{-j^2}, \tag{7.19}$$

$$\mathcal{A}_{2n}(2) = \frac{1}{1-q}\left(q^{3n^2+n} \sum_{|j|\le n} q^{-j^2} + 2q^{3n^2+2n} \sum_{j=0}^{n-1} q^{-j^2-j}\right), \tag{7.20}$$

$$\mathcal{A}_{2n+1}(2) = \frac{-1}{1-q}\left(2q^{3n^2+4n+1} \sum_{j=0}^{n} q^{-j^2-j} + q^{3n^2+5n+2} \sum_{|j|\le n} q^{-j^2}\right). \tag{7.21}$$

Lemma 12. *$(\mathcal{A}_n(i), \mathcal{B}_n(i))$ form a Bailey pair for $i = 0, 1$, $a = 1$ and for $i = 2$, $a = q$.*

Proof. By (2.7) with $a = 1$, the other half of the Bailey pair for $\mathcal{B}_n(0)$ (with $a = 1$) is

$$\begin{aligned}
&(1-q^{2n}) \sum_{j=0}^{n} \frac{(q)_{n+j-1}}{(q)_{n-j}} (-1)^{n-j} q^{\binom{n-j}{2}} \mathcal{B}_j(0) \\
&= (1-q^{2n}) \sum_{j=0}^{n} \begin{bmatrix} n+j \\ n-j \end{bmatrix} \frac{(q)_j}{(1-q^{n+j})} (-1)^{n-j} q^{\binom{n-j}{2}} \\
&= V_n \quad (\text{by } (7.5)) \\
&= (-1)^n 2q^{\binom{n+1}{2}} - (1-q^n) q^{2n-2} U_{n-1} \quad (\text{by } (7.8)) \\
&= \mathcal{A}_n(0).
\end{aligned}$$

The formulas for $\mathcal{A}_n(0)$ given by (7.16) and (7.17) follow immediately from Lemma 11. Hence the $i = 0$ case is established.

By (2.7) with $a = 1$, the other half of the Bailey pair for $\mathcal{B}_n(1)$ (with $a = 1$) is

$$\begin{aligned}
&(1-q^{2n}) \sum_{j=0}^{n} \frac{(q)_{n+j-1}}{(q)_{n-j}} (-1)^{n-j} q^{\binom{n-j}{2}} \mathcal{B}_j(1) \\
&= (1-q^{2n}) \sum_{j=1}^{n} \begin{bmatrix} n+j-1 \\ n-j \end{bmatrix} (q)_{j-1} (-1)^{n-j} q^{\binom{n-j}{2}} \\
&= (1-q^{2n}) U_n \quad (\text{by } (7.4)) \\
&= \mathcal{A}_n(1),
\end{aligned}$$

where the formulas for $\mathcal{A}_n(1)$ given by (7.18) and (7.19) follow directly from Lemma 11. Hence the case $i = 1$ is established.

Finally by (2.7) with $a = q$, the other half of the Bailey pair for $\mathscr{B}_n(2)$ (with $a = q$) is

$$\frac{(1-q^{2n+1})}{(1-q)}\sum_{j=0}^{n}\frac{(q)_{n+j}}{(q)_{n-j}}(-1)^{n-j}q^{\binom{n-j}{2}}\mathscr{B}_j(2)$$

$$=\frac{(1-q^{2n+1})}{(1-q)}\sum_{j=0}^{n}\begin{bmatrix}n+j+1\\ n-j\end{bmatrix}\frac{(q)_j}{1-q^{n+j+1}}(-1)^{n-j}q^{\binom{n-j}{2}}$$

$$=\frac{1}{1-q}\sum_{j=0}^{n}\begin{bmatrix}n+j+1\\ n-j\end{bmatrix}(q)_j\frac{(1-q^{n+j+1}+q^{n+j+1}(1-q^{n-j}))}{(1-q^{n+j+1})}(-1)^{n-j}q^{\binom{n-j}{2}}$$

$$=\frac{1}{1-q}\left(\sum_{j=0}^{n}\begin{bmatrix}n+j+1\\ n-j\end{bmatrix}(q)_j(-1)^{n-j}q^{\binom{n-j}{2}}\right.$$

$$\left.+\sum_{j=0}^{n}\begin{bmatrix}n+j\\ n-j-1\end{bmatrix}(q)_j(-1)^{n-j}q^{\binom{n-j}{2}+n+j+1}\right)$$

$$=\frac{1}{1-q}\left(U_{n+1}-q^{2n}U_n\right)=\mathscr{A}_n(2),$$

where the formulas for $\mathscr{A}_n(2)$ given by (7.20) and (7.21) follow directly from Lemma 11. Hence the case $i = 2$ is established. □

We are now set for our main result on the seventh order mock theta functions, which does indeed link these functions with Hecke type series involving the number 7.

THEOREM 13.

$$\mathscr{F}_0(q)\equiv\sum_{n=0}^{\infty}\frac{q^{n^2}}{(q^{n+1})_n} \tag{7.22}$$

$$=\frac{1}{(q)_\infty}\left(\sum_{\substack{n=0\\ |j|\le n}}q^{7n^2+n-j^2}(1-q^{12n+6})\right.$$

$$\left.-2q\sum_{n=0}^{\infty}\sum_{j=0}^{n}q^{7n^2+8n-j^2-j}(1-q^{12n+13})\right),$$

$$\mathscr{F}_1(q)\equiv\sum_{n=1}^{\infty}\frac{q^{n^2}}{(q^n)_n} \tag{7.23}$$

$$=\frac{1}{(q)_\infty}\left(-2\sum_{n=0}^{\infty}\sum_{j=0}^{n-1}q^{7n^2-2n-j^2-j}(1-q^{4n})\right.$$

$$\left.+\sum_{\substack{n=0\\ |j|\le n}}q^{7n^2+5n+1-j^2}(1-q^{4n+2})\right),$$

$$(7.24)\quad \mathscr{F}_2(q) \equiv \sum_{n=0}^{\infty} \frac{q^{n^2+n}}{(q^{n+1})_{n+1}}$$

$$= \frac{1}{(q)_\infty}\left(\sum_{\substack{n=0\\|j|\le n}}^{\infty} q^{7n^2+n-j^2}(1-q^{8n+3}) \right.$$

$$\left. -2q^2 \sum_{n=0}^{\infty}\sum_{j=0}^{n} q^{7n^2+8n-j^2-j}(1-q^{8n+7}) \right).$$

PROOF. For $i = 0, 1$, we have

$$\mathscr{F}_i(q) = \sum_{n=0}^{\infty} q^{n^2}\mathscr{B}_n(i) = \frac{1}{(q)} \sum_{n=0}^{\infty} q^{n^2}\mathscr{A}_n(i),$$

and equations (7.22) and (7.23) now follow by substituting the formulae for $\mathscr{A}_n(i)$ from (7.16)–(7.19) into this identity.

For $i = 2$, $a = q$, we have

$$\mathscr{F}_2(q) = \sum_{n=0}^{\infty} q^{n^2+n}\mathscr{B}_n(2) = \frac{1}{(q^2)_\infty} \sum_{n=0}^{\infty} q^{n^2+n}\mathscr{A}_n(2),$$

and (7.24) follows from (7.21) and (7.22). □

8. Conclusion. Clearly we are just beginning to understand Ramanujan's last brilliant creation: the Mock Theta Functions. It is my hope that these discoveries will assist in proving many of the results on fifth order functions that are given in his "Lost" Notebook **[5]**. Also is there significant arithmetic information about sums of three squares in (5.16)?

Finally we suggest a further study of the Bailey pair $(\alpha_n(b), \beta_n(b))$, where $\beta_n(b) = (bq)_n/(q)_{2n}$. The first few $\alpha_n(b)$ are

$$\alpha_0(b) = 1,$$
$$\alpha_1(b) = -bq - b,$$
$$\alpha_2(b) = b^2q^3 + bq^4 + bq^3 - q^2,$$
$$\alpha_3(b) = -b^3q^6 - b^2q^8 - b^2q^7 - b^2q^6 - bq^8 + bq^5 + q^7 + q^5.$$

Furthermore $\sum_{n\ge 0} q^{n^2}\beta_n(b)$ is an important theta series for $b = 0, q^{-1/2}, -1$, and is the first seventh order mock series for $b = 1$. Can a representation of $\alpha_n(b)$ be found that directly yields these facts as special cases?

REFERENCES

1. G. E. Andrews, *On basic hypergeometric series, mock theta functions and partitions*. I, Quart. J. Math. Oxford Ser. **17** (1966), 64–80.

2. ______, *On basic hypergeometric series, mock theta functions and partitions*. II, Quart. J. Math. Oxford Ser. **17** (1966), 132–143.

3. ______, *On the theorems of Watson and Dragonette for Ramanujan's mock-theta functions*, Amer. J. Math. **88** (1966), 454–490.

4. ______, *The theory of partitions*, Encyclopedia of Math. and Its Applications, Vol. 2 (G.-C. Rota, ed.), Addison-Wesley, Reading, Mass., 1976.

5. ______, *An introduction to Ramanujan's "lost" notebook*, Amer. Math. Monthly **86** (1979), 89–108.

6. ______, *Connection coefficient problems and partitions*, Proc. Sympos. Pure Math., vol. 34, Amer. Math. Soc., Providence, R. I., 1979, pp. 1–24.

7. ______, *Hecke modular forms and the Kac-Petersen identities*, Trans. Amer. Math. Soc. **283** (1984), 451–458.

8. ______, *Multiple series Rogers-Ramanujan type identities*, Pacific J. Math. **114** (1984), 267–283.

9. W. N. Bailey, *Identities of the Rogers-Ramanujan type*, Proc. London Math. Soc. (2) **50**(1949), 1–10.

10. D. M. Bressoud, *Hecke modular forms and q-Hermite polynomials* (to appear).

11. L. A. Dragonette, *Some asymptotic formulae for the mock theta series of Ramanujan*, Trans. Amer. Math. Soc. **72** (1952), 474–500.

12. E. Hecke, *Über einen Zusammenhang zwischen elliptischen Modulfunktionen und indefiniten quadratischen Formen*, Mathematische Werke, Vandenhoeck und Ruprecht, Göttingen, 1959, pp. 418–427.

13. V. G. Kac and D. H. Peterson, *Affine Lie algebras and Hecke modular forms*, Bull. Amer. Math. Soc. (N.S.) **3** (1980), 1057–1061.

14. S. Ramanujan, *Collected papers*, Cambridge Univ. Press, London, 1927; Reprinted, Chelsea, New York, 1962.

15. L. J. Rogers, *Second memoir on the expansion of certain infinite products*, Proc. London Math. Soc. **25** (1894), 318–343.

16. D. Shanks, *A short proof of an identity of Euler*, Proc. Amer. Math. Soc. **2** (1951), 747–749.

17. ______, *Two theorems of Gauss*, Pacific J. Math. **8** (1958), 609–612.

18. L. J. Slater, *Generalized hypergeometric functions*, Cambridge Univ. Press, London, 1966.

19. G. N. Watson, *The final problem: an account of the mock theta functions*, J. London Math. Soc. **11** (1936), 55–80.

20. ______, *The mock theta functions*. II, Proc. London Math. Soc. (2) **42** (1937), 272–304.

21. ______, *A note on Spence's logarithm transcendant*, Quart. J. Math. Oxford Ser. **8** (1937), 39–42.

22. D. B. Sears, *On the transformation theory of basic hypergeometric functions*, Proc. London Math. Soc. (2) **53** (1951), 158–180.

Department of Mathematics, The Pennsylvania State University, University Park, Pennsylvania 16802

2.5.4 Partitions and indefinite quadratic forms (with F.J. Dyson and D.R. Hickerson)

Invent. math. 91, 391–407 (1988)

Inventiones mathematicae

Partitions and indefinite quadratic forms

George E. Andrews [1,*], Freeman J. Dyson [2], and Dean Hickerson [1,**]

[1] Department of Mathematics, Pennsylvania State University, University Park, PA 16802, USA

[2] Institute for Advanced Study, Princeton, New Jersey 08540, USA

1. Introduction

In [3], the function

$$\sigma(q)=1+\sum_{n=1}^{\infty}\frac{q^{n(n+1)/2}}{(1+q)(1+q^2)\dots(1+q^n)} \tag{1.1}$$

appeared in connection with identities from Ramanujan's "Lost" Notebook. In [5] it was noted that if

$$\begin{aligned}\sigma(q)&=\sum_{n=0}^{\infty}S(n)\,q^n\\&=1+q-q^2+2q^3+\dots+4q^{45}+\dots+6q^{1609}+\dots\end{aligned} \tag{1.2}$$

then the following two conjectures are quite plausible.

Conjecture 1. $\limsup|S(n)|=+\infty$.

Conjecture 2. $S(n)=0$ for infinitely many n.

The function $\sigma(q)$ first arose to our knowledge in Ramanujan's "Lost" Notebook, and it is discussed at length in [3]. In [5] the uniqueness of $\sigma(q)$ as a power series arising from a simple q-series is highlighted.

In [5] it was noted that $S(n)$ is equal to the number of partitions of n into distinct parts with even rank minus the number with odd rank. The rank of a partition (see [1, p. 142], [7]) is the largest part minus the number of parts. Thus $S(8)$ is -2 because $6+2$ and $5+2+1$ have even rank while 8, $7+1$, $5+3$ and $4+3+1$ have odd rank.

In this paper we shall prove a number of results relating $S(n)$ to the arithmetic of $\mathbf{Q}(\sqrt{6})$. Our main result requires an arithmetic function $T(m)$ defined for

* Partially supported by National Science Foundation Grant DMS-8503324

** Partially supported by a Postdoctoral Research Grant from the University of California at Davis

all integers $\equiv 1 \pmod{24}$ as follows: For (positive or negative) integers $m \equiv 1 \pmod{24}$, consider Pell's equation

$$u^2 - 6v^2 = m. \tag{1.3}$$

Note that if (u, v) is a solution of this equation then $u \equiv \pm 1 \pmod 6$ and v is even. We call two solutions (u, v) and (u', v') equivalent if

$$u' + v'\sqrt{6} = \pm(5 + 2\sqrt{6})^r (u + v\sqrt{6}) \tag{1.4}$$

for some integer r. By induction on $|r|$, it is easy to show that if (u, v) and (u', v') are equivalent, then $u + 3v \equiv \pm(u' + 3v') \pmod{12}$. Let $T(m)$ be the excess of the number of inequivalent solutions of (1.3) with $u + 3v \equiv \pm 1 \pmod{12}$ over the number of them with $u + 3v \equiv \pm 5 \pmod{12}$.

Theorem 2. *For* $n \geqq 0$, $S(n) = T(24n+1)$.

By studying the arithmetic of $\mathbf{Q}(\sqrt{6})$ it is then possible to prove that $S(n)$ assumes all integer values and does so infinitely often. This result is much stronger than the above two conjectures; indeed it trivially implies them both.

In Section 2 we prove the generating function counterpart of Theorem 2:

Theorem 1.

$$\sigma(q) = \sum_{\substack{n \geqq 0 \\ |j| \leqq n}} (-1)^{n+j} q^{n(3n+1)/2 - j^2}(1 - q^{2n+1}). \tag{1.5}$$

We note in passing that (1.5) closely resembles [4; p. 124, eq. (5.15)]:

$$\prod_{n=1}^{\infty} (1-q^n)^2(1-q^{2n-1}) = \sum_{\substack{n \geqq 0 \\ |j| \leqq n}} (-1)^j q^{n(3n+1)/2 - j^2}(1 - q^{2n+1}). \tag{1.6}$$

Since (1.6) is so similar to (1.5), we found a proof for (1.5) by adapting the methods used in [4] to prove (1.6).

We devote Section 3 to enough arithmetic in $\mathbf{Q}(\sqrt{6})$ to obtain an explicit determination of $T(m)$ and thus prove Conjectures 1 and 2. In Section 4 we provide an alternative derivation of Theorem 1 which is short and reasonably self-contained. In Section 5 we give a partition-theoretic interpretation of $T(m)$ for $m < 0$. We conclude with a look at future projects.

We wish to emphasize both the power and the surprise of this interaction between the theory of partitions and algebraic number theory. We know of no other partition function in the literature which assumes all integral values as $S(n)$ does. This result would have been impossible without the analysis of $T(m)$ provided by the arithmetic in $\mathbf{Q}(\sqrt{6})$.

2. Proof of Theorems 1 and 2

We shall follow the Bailey chain technique developed in [2] and [4]. This approach requires a study of two sequences of rational functions of q:

$$b_n = \frac{(-q)^n}{(q^2;q^2)_n}, \tag{2.1}$$

and

$$c_n = q^{n^2+n} \sum_{j=-n}^{n} (-1)^j q^{-j^2} = q^{n^2+n}\left(1+2\sum_{j=1}^{n}(-1)^j q^{-j^2}\right), \tag{2.2}$$

where

$$(A)_n = (A;q)_n = (1-A)(1-Aq)\dots(1-Aq^{n-1}) = \prod_{j=0}^{\infty}\frac{(1-Aq^j)}{(1-Aq^{j+n})}. \tag{2.3}$$

We will derive, in succession, four linear equations relating the b_n and c_n. The last of these will easily imply Theorem 1.

Lemma 1. *For* $n \geqq 0$

$$c_n = \sum_{j=0}^{n} \frac{(q)_{n+j}(-1)^{n-j} q^{\binom{n-j+1}{2}} b_j}{(q)_{n-j}}. \tag{2.4}$$

Proof. We let $g_n(q)$ denote the right-hand side of (2.4). Then by (2.1)

$$\begin{aligned} g_n(q) &= (-1)^n \sum_{j=0}^{n} \frac{(q)_{n+j}\, q^{\binom{n-j+1}{2}+j}}{(q)_{n-j}(q^2;q^2)_j} \\ &= (-1)^n q^{\binom{n+1}{2}} \sum_{j=0}^{n} \frac{(q^{-n})_j (q^{n+1})_j (-q)^j}{(q)_j(-q)_j} \\ &= \frac{2(-1)^n q^n}{1+q^n} \sum_{j=0}^{n} \frac{(q^{-n})_j(-q^{1+n})^j}{(-q^{1-n})_j} \end{aligned} \tag{2.5}$$

(by [1; p. 38, last line] with $b=q^{-n}$, $a=q^{n+1}$, $t=c=-q$). Therefore

$$\begin{aligned} q^{n^2+n} g_n(q^{-1}) &= \frac{(q)_n}{(-q)_n} \sum_{j=0}^{n} \frac{(-1)_{n-j}\, q^{(n+1)(n-j)}(-1)^{n-j}}{(q)_{n-j}} \\ &= \frac{(q)_n}{(-q)_n} \sum_{j=0}^{n} \frac{(-1)_j\, q^{j(n+1)}(-1)^j}{(q)_j}. \end{aligned} \tag{2.6}$$

We now apply Lemma 2 of [4; p. 118] with $a=1$ and $b=-1$. (Note that [4] contains a misprint: In the first denominator, "$(a)_j$" should be "$(q)_j$".) We obtain

$$1+2\sum_{j=1}^{n}(-1)^j q^{j^2}=\frac{(q)_n}{(-q)_n}\sum_{j=0}^{n}\frac{(-1)_j(-1)^j q^{(n+1)j}}{(q)_j}. \tag{2.7}$$

Comparing (2.6) and (2.7), we find

$$q^{n^2+n}g_n(q^{-1})=1+2\sum_{j=1}^{n}(-1)^j q^{j^2}, \tag{2.8}$$

or equivalently

$$g_n(q)=q^{n^2+n}\left(1+2\sum_{j=1}^{n}(-1)^j q^{-j^2}\right), \tag{2.9}$$

as desired. □

Lemma 2. *Suppose α_n, β_n, γ_n are sequences of functions of a and q related as follows:*

$$\alpha_n=\gamma_n-\gamma_{n-1}, \qquad n\geqq 0, \qquad (\gamma_{-1}=0); \tag{2.10}$$

$$\gamma_n=\sum_{j=0}^{n}\frac{(aq)_{n+j}(-1)^{n-j}q^{\binom{n-j+1}{2}}\beta_j}{(q)_{n-j}}. \tag{2.11}$$

Then

$$\alpha_n=(1-aq^{2n})\sum_{j=0}^{n}\frac{(aq)_{n+j-1}(-1)^{n-j}q^{\binom{n-j}{2}}\beta_j}{(q)_{n-j}}. \tag{2.12}$$

Proof. By (2.10) and (2.11),

$$\begin{aligned}\alpha_n&=\gamma_n-\gamma_{n-1}\\&=\sum_{j=0}^{n}\beta_j\left(\frac{(aq)_{n+j}(-1)^{n-j}q^{\binom{n-j+1}{2}}}{(q)_{n-j}}-\frac{(aq)_{n+j-1}(-1)^{n-j-1}q^{\binom{n-j}{2}}}{(q)_{n-j-1}}\right)\\&=\sum_{j=0}^{n}\beta_j\frac{(aq)_{n+j-1}(-1)^{n-j}q^{\binom{n-j}{2}}}{(q)_{n-j}}((1-aq^{n+j})q^{n-j}+(1-q^{n-j}))\\&=(1-aq^{2n})\sum_{j=0}^{n}\frac{(aq)_{n+j-1}(-1)^{n-j}q^{\binom{n-j}{2}}\beta_j}{(q)_{n-j}},\end{aligned} \tag{2.13}$$

as desired. □

Proof of Theorem 1. We recall Bailey's Lemma from [2; p. 270]:
If for $n \geqq 0$

$$\beta_n = \sum_{r=0}^{n} \frac{\alpha_r}{(q)_{n-r}(aq)_{n+r}}, \tag{2.14}$$

then

$$\sum_{n\geqq 0} \left(\frac{(\rho_1)_n (\rho_2)_n (aq/\rho_1 \rho_2)^n \alpha_n}{(aq/\rho_1)_n (aq/\rho_2)_n} \right) \frac{1}{(q)_{N-n}(aq)_{N+n}}$$
$$= \sum_{n\geqq 0} \frac{(\rho_1)_n (\rho_2)_n (aq/\rho_1 \rho_2)^n \beta_n (aq/\rho_1 \rho_2)_{N-n}}{(q)_{N-n}(aq/\rho_1)_N (aq/\rho_2)_N}. \tag{2.15}$$

We multiply both sides of (2.15) by $\left(1 - \frac{aq}{\rho_1}\right)$ and then specialize the resulting equation (as well as (2.14)) to $a=1$, $\rho_1 = q$, $\rho_2 \to \infty$, $N \to \infty$. Observing that

$$(\rho_2)_n (1/\rho_2)^n = (1/\rho_2 - 1)(1/\rho_2 - q)\ldots(1/\rho_2 - q^{n-1}) \to (-1)^n q^{\binom{n}{2}}$$

as $\rho_2 \to \infty$ we obtain the following:
If for $n \geqq 0$

$$\beta_n = \sum_{r=0}^{n} \frac{\alpha_r}{(q)_{n-r}(q)_{n+r}}, \tag{2.16}$$

then

$$\sum_{n\geqq 0} (-1)^n q^{\binom{n}{2}} (1-q^n)\, \alpha_n = \sum_{n\geqq 0} (q)_n (-1)^n q^{\binom{n}{2}} \beta_n, \tag{2.17}$$

assuming convergence (which will be trivial in our application).
By Lemmas 1 and 2 taken together, we have for $n \geqq 1$

$$c_n - c_{n-1} = (1-q^{2n}) \sum_{j=0}^{n} \frac{(q)_{n+j}(-1)^{n-j} q^{\binom{n-j}{2}} b_j}{(1-q^{n+j})(q)_{n-j}}. \tag{2.18}$$

By the inversion formula for Bailey pairs [2; p. 278, Eq. (4.1)] we know

$$b_n = \sum_{r=0}^{n} \frac{(c_r - c_{r-1})}{(q)_{n-r}(q)_{n+r}}, \tag{2.19}$$

where $c_0 = 1$, $c_{-1} = 0$. Thus (2.19) is an instance of (2.16), and so by (2.17)

$$\sum_{n\geqq 0} ((-1)^n q^{\binom{n}{2}} + (-1)^{n+1} q^{\binom{n+1}{2}})(c_n - c_{n-1}) = \sum_{n\geqq 0} \frac{q^{\binom{n+1}{2}}}{(-q)_n} = \sigma(q). \tag{2.20}$$

Now assuming absolute convergence of all series involved, we see that

$$\begin{aligned}\sum_{n\geqq 0} (t_n+t_{n+1})(c_n-c_{n-1}) \\ = \sum_{n\geqq 0} t_n c_n + \sum_{n\geqq 0} t_{n+1} c_n - \sum_{n\geqq 0} t_n c_{n-1} - \sum_{n\geqq 0} t_{n+1} c_{n-1} \\ = \sum_{n\geqq 0} t_n c_n - \sum_{n\geqq 0} t_{n+2} c_n = \sum_{n\geqq 0} (t_n - t_{n+2}) c_n. \end{aligned} \tag{2.21}$$

Applying (2.21) to the left-hand side of (2.20), we find that

$$\sigma(q) = \sum_{\substack{n\geqq 0 \\ |j|\leqq n}} (-1)^{n+j} q^{n(3n+1)/2-j^2}(1-q^{2n+1}), \tag{2.22}$$

as desired. □

In preparation for Theorem 2 we must present some background on Pell's equation.

We remark that, in general, two solutions (u_1, v_2) and (u_2, v_2) of Pell's equation

$$u^2 - Dv^2 = m \tag{2.23}$$

are called equivalent if there exist (x, y) such that

$$x^2 - Dy^2 = 1 \tag{2.24}$$

and

$$u_1 + v_1\sqrt{D} = (x+y\sqrt{D})(u_2+v_2\sqrt{D}). \tag{2.25}$$

The following lemma enables us to select one solution from each equivalence class of solutions of (2.23).

Lemma 3. *Let (x_1, y_1) be the fundamental solution of $x^2 - Dy^2 = 1$; i.e. the solution in which x_1 and y_1 are minimal positive. If $m>0$, then each equivalence class of solutions of $u^2 - Dv^2 = m$ contains a unique (u, v) with $u>0$ and*

$$-\frac{y_1}{x_1+1} u < v \leqq \frac{y_1}{x_1+1} u.$$

If $m<0$, the corresponding conditions are $v>0$ and

$$-\frac{Dy_1}{x_1+1} v < u \leqq \frac{Dy_1}{x_1+1} v.$$

Proof. Since two solutions (u, v) and (u', v') of $u^2 - Dv^2 = m$ are equivalent if and only if $u'+v'\sqrt{D} = \pm(x_1+y_1\sqrt{D})^r(u+v\sqrt{D})$ for some integer r, it is clear that, for any positive number α, each equivalence class contains a unique (u, v) with $u+v\sqrt{D}$ in the half-open interval $(\alpha, \alpha(x_1+y_1\sqrt{D})]$. In particular, each class contains a unique (u, v) with

$$\sqrt{\frac{|m|}{x_1+y_1\sqrt{D}}}<u+v\sqrt{D}\leqq\sqrt{|m|(x_1+y_1\sqrt{D})}. \tag{2.26}$$

For $m>0$, we have $u\neq 0$, so

$$\frac{v}{u}=\frac{1}{\sqrt{D}}\frac{(u+v\sqrt{D})-(u-v\sqrt{D})}{(u+v\sqrt{D})+(u-v\sqrt{D})}=\frac{1}{\sqrt{D}}\frac{(u+v\sqrt{D})^2-m}{(u+v\sqrt{D})^2+m}$$
$$=\frac{1}{\sqrt{D}}\left[1-\frac{2m}{(u+v\sqrt{D})^2+m}\right]. \tag{2.27}$$

Since the right-hand side of (2.27) is an increasing function of $u+v\sqrt{D}$ for $u+v\sqrt{D}\geqq 0$, a straightforward calculation shows that (2.26) is equivalent to the condition stated in the theorem.

A similar proof works for $m<0$, using

$$\frac{u}{v}=\sqrt{D}\left[1+\frac{2m}{(u+v\sqrt{D})^2-m}\right]. \quad \square$$

We are now ready to treat:

Proof of Theorem 2

For the equation $u^2-6v^2=24n+1$ with $n\geqq 0$, we have $D=6$, $x_1=5$, and $y_1=2$, so each equivalence class of solutions contains a unique (u, v) with $u>0$ and $-u/3<v\leqq u/3$. Since $u\equiv\pm 1 \pmod 6$ and v is even, we can write $u=6i+\delta$ and $v=2j$, where δ is either 1 or 5. The inequalities then become $i\geqq 0$ and $|j|\leqq i$. Furthermore, $u+3v=6(i+j)+\delta$, so (u, v) is counted positively by $T(24n+1)$ if either $\delta=1$ and $i+j$ is even or $\delta=5$ and $i+j$ is odd, and negatively otherwise. Hence

$$T(24n+1)=\sum_{\substack{i\geqq 0\\ |j|\leqq i\\ (6i+1)^2-24j^2=24n+1}}(-1)^{i+j}-\sum_{\substack{i\geqq 0\\ |j|\leqq i\\ (6i+5)^2-24j^2=24n+1}}(-1)^{i+j}$$
$$=\sum_{\substack{i\geqq 0\\ |j|\leqq i\\ i(3i+1)/2-j^2=n}}(-1)^{i+j}-\sum_{\substack{i\geqq 0\\ |j|\leqq i\\ i(3i+1)/2+2i+1-j^2=n}}(-1)^{i+j},$$

so

$$\sum_{n\geqq 0}T(24n+1)\,q^n=\sum_{\substack{i\geqq 0\\ |j|\leqq i}}(-1)^{i+j}q^{i(3i+1)/2-j^2}-\sum_{\substack{i\geqq 0\\ |j|\leqq i}}(-1)^{i+j}q^{i(3i+1)/2+2i+1-j^2}$$
$$=\sum_{\substack{i\geqq 0\\ |j|\leqq i}}(-1)^{i+j}q^{i(3i+1)/2-j^2}(1-q^{2i+1}).$$

Since this equals the right-hand side of (1.5), Theorem 2 follows immediately. $\square$

3. A formula for $T(m)$

In this section we obtain a formula for $T(m)$ in terms of the prime factorization of m, and use it to prove Conjectures 1 and 2.

To evaluate $T(m)$, we will work in the ring of integers in $\mathbf{Q}(\sqrt{6})$. We need the following facts and definitions:

The set of algebraic integers in $\mathbf{Q}(\sqrt{6})$ is $\{u+v\sqrt{6}: u, v \in \mathbf{Z}\}$, which we denote $\mathbf{Z}[\sqrt{6}]$. This set is a unique factorization domain [10; Theorems 245 and 248]. If $\alpha = u+v\sqrt{6}$, then the conjugate of α, denoted $\bar{\alpha}$, is defined to be $u-v\sqrt{6}$. The norm of α is $N(\alpha)=\alpha\bar{\alpha}=u^2-6v^2$. A number ε in $\mathbf{Z}[\sqrt{6}]$ is called a unit if $1/\varepsilon$ is also in $\mathbf{Z}[\sqrt{6}]$. The group of units consists of all numbers of the form $\pm(5+2\sqrt{6})^r$ for integers r. Two numbers, α and α', are called associates if $\alpha'=\alpha\varepsilon$ for some unit ε; associate elements have the same norm. For each rational integer m, there are only finitely many nonassociate elements with norm m [8, Theorem 2.2.5]. If p is a rational prime $\equiv \pm 7$ or $\pm 11 \pmod{24}$, then p remains prime in $\mathbf{Z}[\sqrt{6}]$. If, instead, $p \equiv \pm 1$ or $\pm 5 \pmod{24}$, then p factors as the product of two conjugate but nonassociate primes, $p=\pi\bar{\pi}$, where $N(\pi)=N(\bar{\pi})=p$. (See [8, Theorem 3.8.1].)

We let R be the set of α in $\mathbf{Z}[\sqrt{6}]$ such that $N(\alpha)$ is relatively prime to 6. Note that R is closed under multiplication. If $\alpha=u+v\sqrt{6}$, then $N(\alpha) \equiv u^2 \pmod 6$, so $\alpha \in R$ if and only if u is relatively prime to 6. Hence, if $\alpha \in R$ then $u^2 \equiv 1 \pmod{24}$ so $N(\alpha) \equiv 1-6v^2 \equiv 1$ or $19 \pmod{24}$.

Throughout the remainder of this section, Greek letters will denote elements of R; this assumption will not always be stated.

Definition: For $\alpha = u+v\sqrt{6}$ in R,

$$f(\alpha)=\begin{cases} 1 \text{ if either } v \text{ is even and } u+3v \equiv \pm 1 \pmod{12} \\ \qquad \text{or } v \text{ is odd and } 2u+3v \equiv \pm 1 \pmod{12}, \\ -1 \text{ if either } v \text{ is even and } u+3v \equiv \pm 5 \pmod{12} \\ \qquad \text{or } v \text{ is odd and } 2u+3v \equiv \pm 5 \pmod{12}. \end{cases} \tag{3.1}$$

Note that $f(\alpha)$ is defined for each α in R. For, if v is even then

$$u+3v \equiv \pm 1 + 0 = \pm 1 \pmod 6;$$

if v is odd then

$$2u+3v \equiv \pm 2 + 3 \equiv \pm 1 \pmod 6.$$

The next lemma summarizes the facts we will use about f.

Lemma 4. (a) *The function f is completely multiplicative; that is, for any α and β in R, $f(\alpha\beta)=f(\alpha)f(\beta)$.*

(b) *If* α *and* α' *are associates then* $f(\alpha)=f(\alpha')$.

(c) $f(\bar{\alpha})=\begin{cases} f(\alpha) \text{ if } N(\alpha)\equiv 1 \pmod{24}, \\ -f(\alpha) \text{ if } N(\alpha)\equiv 19 \pmod{24}. \end{cases}$

Proof. (a) Let $\alpha=u+v\sqrt{6}$, $\beta=r+s\sqrt{6}$, and $\alpha\beta=x+y\sqrt{6}$. Note that $x=ur+6vs$ and $y=us+vr$. There are three cases, depending on the parities of v and s. (Recall that u and r are odd.)

Case 1: If v and s are even, then y is even and

$$\begin{aligned} x+3y&=(u+3v)(r+3s)-3vs \\ &\equiv(u+3v)(r+3s) \pmod{12}, \end{aligned}$$

since v and s are even. By considering the possibilities for $u+3v$ and $r+3s$ mod 12, it then follows that $f(\alpha\beta)=f(\alpha)f(\beta)$.

Case 2: If v and s are odd, then y is even. We proceed as in Case 1, using

$$\begin{aligned} x+3y&=(2u+3v)(2r+3s)-3(u+v)(r+s) \\ &\equiv(2u+3v)(2r+3s) \pmod{12}. \end{aligned}$$

Case 3: Suppose v and s are of opposite parity, say v is even and s is odd. We proceed as before, using the fact that y is odd and

$$\begin{aligned} 2x+3y&=(u+3v)(2r+3s)+3v(s-r) \\ &\equiv(u+3v)(2r+3s) \pmod{12}. \end{aligned}$$

(b) This follows from part (*a*), since $\alpha'=\pm(5+2\sqrt{6})^r\alpha$ for some r, and since $f(-1)=f(5+2\sqrt{6})=1$.

(c) By part (a), $f(\alpha)f(\bar{\alpha})=f(\alpha\bar{\alpha})=f(N(\alpha))$, so $f(\bar{\alpha})=f(N(\alpha))f(\alpha)^{-1}$ $=f(N(\alpha))f(\alpha)$, since $f(\alpha)=\pm 1$. But $f(N(\alpha))$ is 1 if $N(\alpha)\equiv 1$ (mod (24), and -1 if $N(\alpha)\equiv 19 \pmod{24}$. □

The relation between f and T is given by:

Lemma 5. *For* $m\equiv 1 \pmod{24}$,

$$T(m)=\sum_{N(\alpha)=m} f(\alpha), \tag{3.2}$$

where the summation is over a complete set of nonassociate elements α *with norm* m. (*By Lemma* 4 (b), *it does not matter which set we use.*)

Proof. Writing $\alpha=u+v\sqrt{6}$, we have $N(\alpha)=m$ if and only if

$$u^2-6v^2=m. \tag{3.3}$$

Further, α and $\alpha'=u'+v'\sqrt{6}$ are associates if and only if (u, v) and (u', v') are equivalent solutions of (3.3). Since $m\equiv 1 \pmod{24}$, (3.3) implies that v is even, so $f(\alpha)$ is 1 if $u+3v\equiv\pm 1 \pmod{12}$ and -1 if $u+3v\equiv\pm 5 \pmod{12}$. Hence the sum in (3.2) counts solutions of (3.3) with the same 'weights' as $T(m)$. □

We now take (3.2) as the definition of $T(m)$ for any $m \equiv 1 \pmod 6$. This extension of the definition turns out to be trivial, as we now show.

Lemma 6. *If* $m \equiv 1 \pmod 6$ *but* $m \not\equiv 1 \pmod{24}$, *then* $T(m)=0$.

Proof. For $m \equiv 7$ or $13 \pmod{24}$, the sum in (3.2) is empty, so suppose $m \equiv 19 \pmod{24}$. As α runs through a complete set of nonassociate elements of norm m, so does $\bar{\alpha}$, so

$$2T(m) = \sum_{N(\alpha)=m} f(\alpha) + \sum_{N(\alpha)=m} f(\bar{\alpha}) = 0,$$

by Lemma 4(c). □

Lemma 7. *T is multiplicative; that is, if m and n are relatively prime and* $m \equiv n \equiv 1 \pmod 6$, *then* $T(mn) = T(m)\,T(n)$.

Proof. Let $\{\alpha_1, \ldots, \alpha_r\}$ be a complete set of nonassociate elements of norm m, and let $\{\beta_1, \ldots, \beta_s\}$ be such a set for norm n. By unique factorization, it follows that $\{\alpha_i \beta_j : 1 \leqq i \leqq r, 1 \leqq j \leqq s\}$ is such a set for norm mn. Hence

$$T(mn) = \sum_{i,j} f(\alpha_i \beta_j) = \sum_i f(\alpha_i) \sum_j f(\beta_j) = T(m)\,T(n). \quad \square$$

We next evaluate $T(m)$ for prime powers. Since $T(m)$ is defined only for $m \equiv 1 \pmod 6$, it is more convenient to replace primes $\equiv 5 \pmod 6$ by their negatives.

Lemma 8. *Let p be a prime* $\equiv 1 \pmod 6$ *or the negative of a prime* $\equiv 5 \pmod 6$, *and let* $e \geqq 1$. *Then*:

(a) *If* $p \not\equiv 1 \pmod{24}$ *and e is odd, then* $T(p^e)=0$.

(b) *If* $p \not\equiv 1 \pmod{24}$ *and e is even, then*

$$T(p^e) = \begin{cases} 1 & \text{if } p \equiv 13 \text{ or } 19 \pmod{24}, \\ (-1)^{e/2} & \text{if } p \equiv 7 \pmod{24}. \end{cases}$$

(c) *If* $p \equiv 1 \pmod{24}$ *then either* $T(p)=2$ *or* $T(p)=-2$. *In the first case,* $T(p^e)=e+1$. *In the second case,* $T(p^e)=(-1)^e(e+1)$.

Proof. In part (a), $p^e \not\equiv 1 \pmod{24}$, so $T(p^e)=0$ by Lemma 6.

For parts (b) and (c), suppose that $N(\alpha)=p^e$. Write $\alpha = \pi_1 \pi_2, \ldots, \pi_r$, where the π_i's are primes in $\mathbf{Z}[\sqrt{6}]$. Then each π_i is a divisor of $N(\alpha)$, and is therefore a divisor of p.

For $p \equiv 7$ or $13 \pmod{24}$, this implies that each π_i is an associate of p, so $r=e/2$ and α is an associate of $p^{e/2}$. Thus the sum in (3.2) has only one term:

$$T(p^e) = f(p^{e/2}) = f(p)^{e/2} = \begin{cases} 1 & \text{if } p \equiv 13 \pmod{24}, \\ (-1)^{e/2} & \text{if } p \equiv 7 \pmod{24}. \end{cases}$$

For $p \equiv 1$ or $19 \pmod{24}$, each π_i must be an associate of either π or $\bar{\pi}$, where π is a fixed prime of norm p. Thus α is an associate of $\pi^{e-i}\bar{\pi}^i$ for some

$0 \leqq i \leqq e$. The sum in (3.2) has $e+1$ terms:

$$T(p^e) = \sum_{i=0}^{e} f(\pi^{e-i}\bar{\pi}^i) = \sum_{i=0}^{e} f(\pi)^{e-i} f(\bar{\pi})^i = \sum_{i=0}^{e} f(\pi)^{e-i} \varepsilon^i f(\pi)^i$$

$$= f(\pi)^e \sum_{i=0}^{e} \varepsilon^i,$$

where ε is 1 if $p \equiv 1 \pmod{24}$ and -1 if $p \equiv 19 \pmod{24}$.

For $p \equiv 19 \pmod{24}$ and e even, we have

$$T(p^e) = (\pm 1)^e \sum_{i=0}^{e} (-1)^i = 1.$$

For $p \equiv 1 \pmod{24}$, we have

$$T(p^e) = f(\pi)^e \sum_{i=0}^{e} 1^i = f(\pi)^e (e+1).$$

If $f(\pi) = 1$ then $T(p) = 2$ and $T(p^e) = e+1$. If $f(\pi) = -1$ then $T(p) = -2$ and $T(p^e) = (-1)^e(e+1)$. □

Note that every $m \equiv 1 \pmod 6$, $m \neq 1$, has a factorization of the form

$$m = p_1^{e_1} p_2^{e_2} \dots p_r^{e_r}, \tag{3.4}$$

where $r \geqq 1$, each p_i is either a prime $\equiv 1 \pmod 6$ or the negative of a prime $\equiv 5 \pmod 6$, the p_i's are distinct, and the e_i's are positvie integers. (Just change each prime $\equiv 5 \pmod 6$ to its negative in the prime factorization of $|m|$.) Putting this together with Lemmas 7 and 8, we have the following result, which gives almost a closed form for $T(m)$. We say "almost" because we know of no way to determine the sign of $T(p)$ for $p \equiv 1 \pmod{24}$ which is easier than solving the equation $u^2 - 6v^2 = p$.

Theorem 3. *Let $m \neq 1$ be an integer $\equiv 1 \pmod 6$. Suppose m has the factorization (3.4). Then $T(m) = T(p_1^{e_1}) \dots T(p_r^{e_r})$, where*

$$T(p^e) = \begin{cases} 0 & \text{if } p \not\equiv 1 \pmod{24} \text{ and } e \text{ is odd.} \\ 1 & \text{if } p \equiv 13 \text{ or } 19 \pmod{24} \text{ and } e \text{ is even,} \\ (-1)^{e/2} & \text{if } p \equiv 7 \pmod{24} \text{ and } e \text{ is even,} \\ e+1 & \text{if } p \equiv 1 \pmod{24} \text{ and } T(p) = 2. \\ (-1)^e(e+1) & \text{if } p \equiv 1 \pmod{24} \text{ and } T(p) = -2. \end{cases}$$

In particular, $T(m) = 0$ if and only if there is some i for which $p_i \not\equiv 1 \pmod{24}$ and e_i is odd. □

We now apply this result to the distribution of values of $S(n)$.

Theorem 4. *$S(n)$ is almost always 0; that is, the set of n for which $S(n) \neq 0$ has density 0. On the other hand, $S(n)$ takes on every integer value infinitely often.*

Proof. For given $x>0$, let P_x be the set of primes $\leq x$ and $\equiv 7 \pmod{24}$. By Theorem 3, if $S(n) \neq 0$ and $p \in P_x$, then the exponent of p in $24n+1$ is even. In particular, for each $p \in P_x$, either $p \nmid 24n+1$ or $p^2 | 24n+1$. The density of the set of such n is clearly

$$\prod_{p \in P_x}\left(1-\frac{1}{p}+\frac{1}{p^2}\right)<\prod_{p \in P_x}\left(1-\frac{1}{2p}\right)<\prod_{p \in P_x} e^{-1/2p}=\exp\left(-\frac{1}{2}\sum_{p \in P_x}\frac{1}{p}\right).$$

Letting $x \to \infty$ and using the fact that

$$\sum_{p \equiv 7 \pmod{24}} \frac{1}{p}$$

diverges [9; p. 121], we conclude that the set of n for which $S(n) \neq 0$ has density 0.

To prove the second claim, let k be a given integer. If $k=0$, then k occurs infinitely often, by the result just proved. So suppose $k \neq 0$. Note that $T(73)=2$, since every solution (u, v) of $u^2-6v^2=73$ is equivalent to either $(13, 4)$ or $(13, -4)$. Hence if k is positive, then

$$S\left(\frac{73^{k-1}p^2-1}{24}\right)=T(73^{k-1}p^2)=[(k-1)+1]\cdot 1=k,$$

for any prime $p \equiv 13$ or $19 \pmod{24}$. If k is negative, then

$$S\left(\frac{73^{-k-1}p^2-1}{24}\right)=T(73^{-k-1}p^2)=[(-k-1)+1]\cdot(-1)^{2/2}=k,$$

for any prime $p \equiv 7 \pmod{24}$. □

As noted earlier, Conjectures 1 and 2 follow immediately.

4. A short proof of Theorem 1

The proof of Theorem 1 given in Sect. 2 places that result clearly in the context of Bailey chains as developed in [2] and [4]; however it seems appropriate to include a short and nearly self-contained proof of Theorem 1. We shall only need as background

$$\sum_{n=0}^{\infty}\frac{(a)_n(1-aq^{2n})(y)_n q^{n^2}a^n y^{-n}}{(q)_n(1-a)(aq/y)_n}=\frac{(aq)_\infty}{(aq/y)_\infty}, \tag{4.1}$$

and

$$\sum_{n=0}^{\infty}\frac{(a)_n(b)_n(c/ab)^n}{(q)_n(c)_n}=\frac{(c/a)_\infty(c/b)_\infty}{(c)_\infty(c/ab)_\infty}. \tag{4.2}$$

These equations are Lemmas 1 and 3 of [6] and each is proved there is less than half a page.

We remark that the c_n given by (2.2) are uniquely defined by

$$c_0=1, \qquad c_n=q^{2n}c_{n-1}+2(-1)^n q^n, \qquad n\geqq 1. \tag{4.3}$$

We can therefore replace Lemma 1 by the following observation: If

$$\gamma_n=(-1)^n \sum_{j=0}^{n} \frac{(q)_{n+j}\, q^{\binom{n-j+1}{2}+j}}{(q)_{n-j}(q^2;q^2)_j}, \tag{4.4}$$

then clearly $\gamma_0=1$, and for $n\geqq 1$

$$\begin{aligned}
&\gamma_n-q^{2n}\gamma_{n-1}\\
&=(-1)^n \sum_{j=0}^{n} \frac{(q)_{n+j-1}\, q^{\binom{n-j}{2}+j}}{(q)_{n-j}(q^2;q^2)_j}(q^{n-j}(1-q^{n+j})+q^{2n}(1-q^{n-j}))\\
&=(-1)^n \sum_{j=0}^{n} \frac{(q)_{n+j-1}\, q^{\binom{n-j}{2}+j}}{(q)_{n-j}(q^2;q^2)_j}\, q^{n-j}(1-q^{2n})\\
&=\frac{(-1)^n q^{\binom{n+1}{2}}(1-q^{2n})(q)_{n-1}}{(q)_n} \sum_{j=0}^{n} \frac{(q^{-n})_j (q^n)_j (-q)^j}{(q)_j(-q)_j}\\
&=(-1)^n q^{\binom{n+1}{2}}(1+q^n)\frac{(-q^{1-n})_n}{(-q)_n}\\
&\qquad\qquad \text{(by (4.2), } a=q^n, b=q^{-n}, c=-q)\\
&=2(-1)^n q^n.
\end{aligned} \tag{4.5}$$

Hence $c_n=\gamma_n$ for $n\geqq 0$.

Therefore

$$\begin{aligned}
&\sum_{\substack{n\geqq 0\\ |j|\leqq n}} (-1)^{n+j} q^{n(3n+1)/2-j^2}(1-q^{2n+1})\\
&\quad=\sum_{n\geqq 0} (-1)^n q^{\binom{n}{2}}(1-q^{2n+1})c_n\\
&\quad=\sum_{n\geqq 0} q^{\binom{n}{2}}(1-q^{2n+1}) \sum_{j=0}^{n} \frac{(q)_{n+j}\, q^{\binom{n-j+1}{2}+j}}{(q)_{n-j}(q^2;q^2)_j}\\
&\qquad\qquad \text{(by (4.4) and (4.6))}\\
&\quad=\sum_{j\geqq 0} \frac{q^{\binom{j+1}{2}}}{(q^2;q^2)_j} \sum_{n\geqq j} \frac{q^{n^2-jn}(1-q^{2n+1})(q)_{n+j}}{(q)_{n-j}}\\
&\quad=\sum_{j\geqq 0} \frac{q^{\binom{j+1}{2}}}{(q^2;q^2)_j} \sum_{n\geqq 0} \frac{q^{n^2+jn}(1-q^{2n+2j+1})(q)_{n+2j}}{(q)_n}
\end{aligned}$$

$$= \sum_{j \geq 0} \frac{q^{\binom{j+1}{2}}}{(q^2;q^2)_j} (q)_{2j} \sum_{n \geq 0} \frac{q^{n^2+jn}(1-q^{2n+2j+1})(q^{2j+1})_n}{(q)_n}$$

$$= \sum_{j \geq 0} \frac{q^{\binom{j+1}{2}}(q)_{2j}}{(q^2;q^2)_j} \frac{(q^{2j+1})_\infty}{(q^{j+1})_\infty}$$

(by (4.1) with $a=q^{2j+1}, y=q^{j+1}$)

$$= \sum_{j \geq 0} \frac{q^{\binom{j+1}{2}}}{(-q)_j},$$

and thus Theorem 1 is again established.

5. The case $m<0$

Theorem 2 gave a partition-theoretic interpretation of $T(m)$ for $m>0$. It turns out that there is a similar result for $m<0$. Specifically, for $n \geq 1$, consider partitions of n into odd parts, with the property that if k occurs as a part, then all positive odd numbers less than k also occur. Let $S^*(n)$ be the excess of the number of such partitions with largest part $\equiv 3 \pmod 4$ over the number with largest part $\equiv 1 \pmod 4$.

Theorem 5. *For* $n \geq 1$, $2S^*(n) = T(1-24n)$.

It is easily seen that

$$\sum_{n \geq 1} S^*(n) q^n = \sum_{n \geq 1} \frac{(-1)^n q^{n^2}}{(q;q^2)_n}.$$

Also, an application of Lemma 3 shows that

$$\sum_{n \geq 1} T(1-24n) q^n = 2 \sum_{m \geq 0} \sum_{2k \geq 3m+1} (-1)^{m+k} q^{k^2-m(3m+1)/2}(1+q^{2(k-m)}). \quad (5.1)$$

Letting $m=2n-j-1$ and $k=3n-j-1$ the right-hand side of (5.1) becomes

$$2 \sum_{n \geq 1} (-1)^n q^{n(3n-1)}(1+q^{2n}) \sum_{j=0}^{2n-1} q^{-j(j+1)/2}.$$

Hence Theorem 5 is equivalent to

Theorem 6.

$$\sum_{n \geq 1} \frac{(-1)^n q^{n^2}}{(q;q^2)_n} = \sum_{n \geq 1} (-1)^n q^{n(3n-1)}(1+q^{2n}) \sum_{j=0}^{2n-1} q^{-j(j+1)/2}. \quad (5.2)$$

We will actually prove this with q replaced by $\sqrt{q}$, since it simplifies the argument. Our proof is very similar to the second proof of Theorem 1 so, for brevity, we will leave out many of the algebraic details.

We begin with an analogue of Lemma 1.

Lemma 9. *For* $n \geqq 0$,

$$\sum_{j=0}^{2n-1} q^{-j(j+1)/4} = (-1)^n q^{-\binom{n}{2}} \sum_{j=1}^{n} \frac{(-1)^j q^{j^2/2-nj}(q)_{n+j-1}}{(q)_{n-j}(q)_{j-1}(\sqrt{q})_j}. \tag{5.3}$$

Proof. Let d_n denote the right-hand side of (5.3). Then

$$\begin{aligned} d_n - d_{n-1} &= (-1)^n q^{-\binom{n}{2}} \sum_{j=1}^{n} \frac{(-1)^j q^{j^2/2-nj}(q)_{n+j-2}}{(q)_{n-j}(q)_{j-1}(\sqrt{q})_j} \\ &\quad \cdot((1-q^{n+j-1})+q^{n+j-1}(1-q^{n-j})) \\ &= (-1)^{n-1} q^{(1-n-n^2)/2} \frac{1-q^{2n-1}}{1-\sqrt{q}} \sum_{j=0}^{n-1} \frac{(q^{1-n})_j (q^n)_j (\sqrt{q})^j}{(q)_j (q^{3/2})_j} \\ &= q^{n/2-n^2}(1+q^{n-1/2}) = q^{-(2n-2)(2n-1)/4} + q^{-(2n-1)(2n)/4}, \end{aligned}$$

by (4.2) with $a=q^n$, $b=q^{1-n}$, $c=q^{3/2}$. The result now follows by induction on n. □

We therefore have

$$\begin{aligned} &\sum_{n \geqq 1} (-1)^n q^{n(3n-1)/2}(1+q^n) \sum_{j=0}^{2n-1} q^{-j(j+1)/4} \\ &= \sum_{n \geqq 1} q^{n^2}(1+q^n) \sum_{j=1}^{n} \frac{(-1)^j q^{j^2/2-nj}(q)_{n+j-1}}{(q)_{n-j}(q)_{j-1}(\sqrt{q})_j} \\ &= \sum_{j \geqq 1} \frac{(-1)^j q^{j^2/2}}{(q)_{j-1}(\sqrt{q})_j} \sum_{n \geqq j} \frac{q^{n(n-j)}(1+q^n)(q)_{n+j-1}}{(q)_{n-j}} \\ &= \sum_{j \geqq 1} \frac{(-1)^j q^{j^2/2}(q)_{2j-1}}{(q)_{j-1}(\sqrt{q})_j} \sum_{n \geqq 0} \frac{q^{n(n+j)}(1+q^{n+j})(q^{2j})_n}{(q)_n} \\ &= \sum_{j \geqq 1} \frac{(-1)^j q^{j^2/2}}{(\sqrt{q})_j}, \end{aligned}$$

by (4.1) with $a=q^{2j}$ and $y=q^j$. Theorem 6 now follows by replacing q by q^2. □

6. Conclusion

There appears to be a rich theory of identities of the sort considered here. Consider for example identity (1.6). The right hand side can be written as

$$\sum_{n \geqq 0} f(24n+1)\, q^n,$$

where $f(m)$ is the excess of the number of inequivalent solutions of $u^2-6v^2=m$, in which either $u>0$ and $2u+3v\equiv 2$ or $4 \pmod{12}$ or $u<0$ and $2u+3v\equiv 8$ or $10 \pmod{12}$ over the number in which either $u>0$ and $2u+3v\equiv 8$ or $10 \pmod{12}$ or $u<0$ and $2u+3v\equiv 2$ or $4 \pmod{12}$. But the left hand side equals

$$(q)_\infty (q;q^2)_\infty (q)_\infty = \sum_n (-1)^n q^{n^2} \sum_n (-1)^n q^{n(3n+1)/2} = \sum_{n\geqq 0} g(24n+1)\, q^n, \tag{6.1}$$

where $g(m)$ is the excess of the number of inequivalent solutions of $x^2+6y^2=m$ in which $x+3y\equiv \pm 1 \pmod{12}$ over the number in which $x+3y\equiv \pm 5 \pmod{12}$. Thus (1.6) implies that $f(m)=g(m)$ for $m\equiv 1 \pmod{24}$. This seems quite surprising, since $f(m)$ is related to arithmetic in $\mathbf{Q}(\sqrt{6})$, while $g(m)$ is related to $\mathbf{Q}(\sqrt{-6})$.

We have done some computing to see if other such relations exist, and it appears that they are quite common. In fact, it appears that for any two quadratic fields, there is some such equation. For example, for $\mathbf{Q}(\sqrt{-2})$ and $\mathbf{Q}(\sqrt{-3})$ it turns out that, for $m\equiv 1 \pmod{24}$, the excess of the number of solutions of $u^2+2v^2=m$ in which $u\equiv \pm 1 \pmod{12}$ over those in which $u\equiv \pm 5 \pmod{12}$ equals the excess of the number of solutions of $x^2+3y^2=m$ in which $x+3y\equiv \pm 1 \pmod 8$ over those in which $x+3y\equiv \pm 3 \pmod 8$. In terms of generating functions, this says that

$$\sum_n (-1)^n q^{n(3n+1)/2} \sum_n q^{3n^2} = \sum_n (-q)^{n(3n+1)/2} \sum_n (-1)^n q^{2n^2}, \tag{6.2}$$

which follows easily from the triple product identity. We have checked about 15 pairs of quadratic fields, and found such an identity in each case.

For our purposes, the most interesting identities are those relating $\mathbf{Q}(\sqrt{2})$, $\mathbf{Q}(\sqrt{3})$, and $\mathbf{Q}(\sqrt{6})$, since the conditions on solutions of $x^2-6y^2=m$ turn out to be those used to define $T(m)$. Specifically, for (positive or negative) $m\equiv 1 \pmod{24}$, $T(m)$ equals that excess of the number of solutions of $x^2-2y^2=m$ in which $y\equiv 0$ or $\pm 2 \pmod{12}$ over those in which $y\equiv 6$ or $\pm 4 \pmod{12}$, and $T(m)$ equals the excess of the number of solutions of $x^2-3y^2=m$ in which $x+3y\equiv \pm 1$ or $\pm 5 \pmod{24}$ over those in which $x+3y\equiv \pm 7$ or $\pm 11 \pmod{24}$. In terms of generating functions, these say (for $m\geqq 0$) that

$$\sigma(q) = \sum_{9|i|<|6n+1|} (-1)^i q^{n(3n+1)/2-3i^2} \tag{6.3}$$

and

$$\sigma(q) = \sum_{8|i|<|6n+1|} (-1)^{n+i} (-q)^{n(3n+1)/2-2i^2}, \tag{6.4}$$

respectively.

In a subsequent paper, H. Cohen considers a number of these questions and natural extensions of this work.

References

1. Andrews, G.E.: The theory of partitions. In: Rota, G.-C. (ed.) Encyclopedia of math. and its applications, vol. 2. Reading, Mass.: Addison-Wesley 1976
2. Andrews, G.E.: Multiple series Rogers-Ramanujan type identities. Pac. J. Math. **114**, 267–283 (1984)
3. Andrews, G.E.: Ramanujan's "lost" notebook *V*, Euler's partition identity. Adv. Math. **61**, 156–164 (1986)
4. Andrews, G.E.: The fifth and seventh order mock theta functions. Trans. Am. Math. Soc. **293**, 113–134 (1986)
5. Andrews, G.E.: Questions and conjectures in partition theory. Am. Math. Monthly **93**, 708–711 (1986)
6. Andrews, G.E.: ΕΥΡΗΚΑ! num $=\Delta+\Delta+\Delta$. J. Number Theory **23**, 285–293 (1986)
7. Atkin, A.O.L., Swinnerton-Dyer, P.: Some properties of partitions. Proc. London Math. Soc. (3) **4**, 84–106 (1954)
8. Borevich, Z.I., Shafarevich, I.R.: Number theory. New York London: Academic Press 1966
9. Chandrasekharan, K.: Introduction to analytic number theory. Die Grundlehren der mathematischen Wissenschaften, Band 148. Berlin Heidelberg New York: Springer 1968
10. Hardy, G.H., Wright, E.M.: An introduction to the theory of numbers, 4th edition. London: Oxford University Press 1968

Oblatum 12-XI-1987

2.5.5 Umbral calculus, Bailey chains, and pentagonal number theorems

Journal of Combinatorial Theory, Series A 91, 464–475 (2000)
Doi:10.1006/jcta.2000.3111, available online at http://idealibeary.com

Umbral Calculus, Bailey Chains, and Pentagonal Number Theorems

George E. Andrews[1]

Department of Mathematics, The Pennsylvania State University, University Park, Pennsylvania 16802
E-mail: andrews@math.psu.edu

Communicated by the Managing Editors

Received April 4, 2000

DEDICATED TO THE MEMORY OF GIAN-CARLO ROTA

We revisit the umbral methods used by L. J. Rogers in his second proof of the Rogers–Ramanujan identities. We shall study how subsequent methods such as the Bailey chains and their variants arise naturally from Rogers' insights. We conclude with the introduction of multi-dimensional Bailey chains and apply them to prove some new Pentagonal Number Theorems. © 2000 Academic Press

AMS subject classifications: 11P81, 33D15, 33D80.

1. INTRODUCTION

In the summer of 1998, I attended the conference, Combinatorics and Physics '98, held at Los Alamos and organized by Bill Chen and Jim Louck. One day, Gian-Carlo invited me and several others to lunch at a Mexican restaurant in a small town many miles from Los Alamos. During the drive there, he and I discussed at length Problem 4, A Unified Theory of Special Functions, in his paper Ten Mathematics Problems I Will Never Solve [11]. As we discussed this, I mentioned my belief that L. J. Rogers' second proof of the Rogers–Ramanujan identities [10; Sect. 1] can be viewed as an umbral calculus method that has its origins in the notorious "eighteen papers of Liouville" [5, Chap. XI] which were explained umbrally by Humbert [7]. Furthermore, I speculated, an analysis of subsequent work viewed through the umbral calculus lens should provide an overarching account of Bailey Chains [2] and their extensions (e.g., [4, 9]). He seemed intrigued and told me to be sure to write it up and send

[1] Partially supported by the National Science Foundation under Grant DMS-9206993.

it to him. I said that, of course I would. I put the project on a back burner, and it remained there until the spring of 1999 when the earthshaking news arrived: Rota had died. Tragic and unbelievable news!

The invitation to prepare a paper for this volume brought our Los Alamos conversation back to mind vividly. In Sections 2 and 3, I shall provide the details of the project I outlined in my conversation with Gian-Carlo.

The genesis of all this, of course, is (as I said above) Rogers' second proof of the Rogers–Ramanujan identities. Consequently, the work in Sections 2 and 3 will be justified only if we are led to something really new. Section 4 will describe some new results. To maintain reasonable brevity we shall restrict the discussion to some new Pentagonal Number Theorems.

Recall the famous Pentagonal Number Theorem of Euler [1; p. 11]:

$$1=\frac{\sum_{n=-\infty}^{\infty}(-1)^n q^{n(3n-1)/2}}{\prod_{n=1}^{\infty}(1-q^n)}. \tag{1.1}$$

In Section 4 we shall prove

$$\sum_{n=1}^{\infty}\frac{q^{2n^2}}{(q;q)_{2n}}=\frac{\sum_{n,m=-\infty}^{\infty}(-1)^{n+m}q^{n(3n-1)/2+m(3m-1)/2+nm}}{\prod_{n=1}^{\infty}(1-q^n)^2}, \tag{1.2}$$

and

$$\sum_{i,j,k\geq 0}\frac{q^{i^2+j^2+k^2}}{(q;q)_{i+j-k}(q;q)_{i+k-j}(q;q)_{j+k-i}}$$
$$=\frac{\sum_{n,m,p=-\infty}^{\infty}(-1)^{n+m+p}q^{n(3n-1)/2+m(3m-1)/2+p(3p-1)/2+nm+np+mp}}{\prod_{n=1}^{\infty}(1-q^n)^3}. \tag{1.3}$$

We conclude with a discussion of possible extensions of the ideas introduced here.

2. ROGERS' SECOND PROOF

We now provide a variation of L. J. Rogers' second proof of the Rogers–Ramanujan identities [10; Sec. 1]. The starting point is Jacobi's triple product identity [6; p. 239, Eq. (II.28)]

$$\sum_{n=-\infty}^{\infty}q^{n^2}z^n=(q^2;q^2)_\infty(-zq;q^2)_\infty(-z^{-1}q;q^2)_\infty, \tag{2.1}$$

where

$$(A; q)_N = \prod_{n=0}^{\infty} \frac{(1 - Aq^n)}{(1 - Aq^{n+N})}. \tag{2.2}$$

To (2.1) we shall apply Euler's identity [6; p. 236, Eq. (II.1)]

$$\sum_{n=0}^{\infty} \frac{q^{n^2} z^n}{(q^2; q^2)_n} = (-zq; q^2)_\infty. \tag{2.3}$$

Hence

$$\frac{\sum_{n=-\infty}^{\infty} q^{n^2} z^n}{(q^2; q^2)_\infty} = \sum_{r, s \geqq 0} \frac{z^{r-s} q^{r^2+s^2}}{(q^2; q^2)_r (q^2; q^2)_s}. \tag{2.4}$$

We now invoke an umbral mapping:

$$z^n \to \begin{cases} 0 & \text{if } n \text{ odd} \\ \alpha_{\frac{n}{2}} & \text{if } n \text{ even.} \end{cases} \tag{2.5}$$

Consequently

$$\frac{\sum_{n=-\infty}^{\infty} q^{4n^2} \alpha_n}{(q^2; q^2)_\infty} = \sum_{\substack{r, s \geqq 0 \\ r \equiv s \pmod 2}} \frac{\alpha_{\frac{r-s}{2}} q^{r^2+s^2}}{(q^2; q^2)_r (q^2; q^2)_s}. \tag{2.6}$$

The double sum on the right may be simplified if we rewrite the indices using $r = m + j$ and $s = m - j$ (which is legitimate because r and s have the same parity). If in addition we replace α_s by α_s / q^{2s^2} and then replace q by $q^{1/2}$, we find that

$$\frac{1}{(q; q)_\infty} \sum_{n=-\infty}^{\infty} q^{n^2} \alpha_n = \sum_{m \geqq 0} q^{m^2} \beta_m, \tag{2.7}$$

where

$$\beta_m = \sum_{j=-m}^{m} \frac{\alpha_j}{(q; q)_{m-j} (q; q)_{m+j}}. \tag{2.8}$$

The first Rogers–Ramanujan identity follows once one proves that if

$$\alpha_j = (-1)^j q^{j(3j-1)/2}, \tag{2.9}$$

then

$$\beta_m = \frac{1}{(q;q)_m}. \tag{2.10}$$

The fact that (2.8) is fulfilled for these particular α_j and β_m is proved by mathematical induction (cf. [3; pp. 22–23]), and we shall not repeat the proof here. Once this has been established one is allowed to conclude that (2.7) also holds; therefore

$$\frac{1}{(q;q)_\infty} \sum_{n=-\infty}^{\infty} (-1)^n q^{n(5n-1)/2} = \sum_{m \geqq 0} \frac{q^{m^2}}{(q;q)_m}. \tag{2.11}$$

By Jacobi's triple product (2.1), we directly deduce from (2.11) that

$$\frac{1}{(q;q^5)_\infty (q^4;q^5)_\infty} = \sum_{m \geqq 0} \frac{q^{m^2}}{(q;q)_m}, \tag{2.12}$$

which is the well-known first Rogers–Ramanujan identity [6; p. 36].

The above proof differs slightly from Rogers' version. He replaced z by $e^{i\theta}$ and wrote everything in terms of cosines. In this way the sum in (2.8) would have had $0 \leqq j \leqq m$ instead of $-m \leqq j \leqq m$.

We shall exploit the bilateral nature of (2.8) in our developments in Section 3. However, it is important to observe that several authors, notably Paule [9] followed by Berkovich, McCoy, and Schilling [4], have recognized that bilateral sums are quite valuable in extensions of the work of Rogers.

The most important idea to be grasped here is simply this. Rogers' second proof of the Rogers–Ramanujan identities consists purely of an umbral map applied to a classical identity from elliptic theta functions, in this case Jacobi's triple product identity.

There are plenty of other identities of this nature, e.g., the quintuple product [6; p. 134], Winquist's identity [13], the Macdonald identities [8], [12], etc. In the interests of brevity we follow one obvious path in the next section, namely the product of independent instances of Jacobi's triple product. This will set us up for the Pentagonal Number Theorems proved in Section 5.

3. MULTI-DIMENSIONAL BAILEY CHAINS

In this section we shall follow up on just one possibility suggested by the discussion at the end of Section 2. Namely, let us look at the same idea applied to the product of s independent copies of Jacobi's triple product:

$$\prod_{j=1}^{s} \sum_{n_j=-\infty}^{\infty} z_j^{n_j} q^{n_j^2} = \prod_{j=1}^{s} (q^2; q^2)_\infty (-z_j q; q^2)_\infty (-z_j^{-1} q; q^2)_\infty]. \tag{3.1}$$

Precisely s copies of the umbral treatment of the case $s=1$ can be carried out, and one will obtain an s-fold version of (2.7) and (2.8). Having made this initial observation, we can jump directly to a full s-fold extension of Bailey's lemma:

THEOREM 1. *If for* $n_1, n_2, ..., n_s \geqq 0$,

$$\beta_{n_1, n_2, ..., n_s} = \sum_{r_1=-\infty}^{n_1} \sum_{r_2=-\infty}^{n_2} \cdots \sum_{r_s=-\infty}^{n_s} \frac{\alpha_{r_1, r_2, ..., r_s}}{\prod_{j=1}^{s} (q;q)_{n_j - r_j} (a_j q; q)_{n_j + r_j}} \tag{3.2}$$

then

$$\beta'_{n_1, n_2, ..., n_s} = \sum_{r_1=-\infty}^{n_1} \sum_{r_2=-\infty}^{n_2} \cdots \sum_{r_s=-\infty}^{n_s} \frac{\alpha'_{r_1, r_2, ..., r_s}}{\prod_{j=1}^{s} (q;q)_{n_j - r_j} (a_j q; q)_{n_j + r_j}}, \tag{3.3}$$

where

$$\alpha'_{r_1, r_2, ..., r_s} = \left(\prod_{j=1}^{s} \frac{(\rho_j)_{r_j} (\sigma_j)_{r_j} (q_j q/(\rho_j \sigma_j))^{r_j}}{(a_j q/\rho_j)_{r_j} (a_j q/\sigma_j)_{r_j}} \right) \alpha_{r_1, r_2, ..., r_s} \tag{3.4}$$

and

$$\beta'_{n_1, n_2, ..., n_s} = \prod_{j=1}^{s} \sum_{m_j=-\infty}^{n_j} \frac{(\rho_j)_{m_j} (\sigma_j)_{m_j} (a_j; q/(\rho_j \sigma_j))_{n_j - m_j} (\frac{a_j; q}{\rho_j \sigma_j})^{m_j} \beta_{m_1, ..., m_s}}{(q)_{n_j - m_j} (a_j q/\rho_j)_{n_j} (a_j q/\sigma_j)_{n_j}}. \tag{3.5}$$

When $s=1$, this is the bilateral Bailey pair formula given in [4; p. 48]. The proof of this formula is precisely the proof of the $s=1$ case now repeated independently s times. Given this fact, we omit the details.

In the next section, we shall only be interested in the limiting case of Theorem 1, in which $a_1 = a_2 = \cdots = a_s = 1$, and each of $n_1, n_2, ..., n_s$, $\rho_1, \rho_2, ..., \rho_s$, $\sigma_1, \sigma_2, ..., \sigma_s \to +\infty$. This provides what might be termed the Weak Multi-dimensional Bailey Lemma. It is the natural multi-dimensional analog of (2.7) and (2.8).

COROLLARY 1. *If for* $n_1, n_2, ..., n_s \geqq 0$

$$\beta_{n_1, n_2, ..., n_s} = \sum_{r_1=-n_1}^{n_1} \sum_{r_2=-n_2}^{n_2} \cdots \sum_{r_s=-n_s}^{n_s} \frac{\alpha_{r_1, r_2, ..., r_s}}{\prod_{j=1}^{s} (q;q)_{n_j - r_j} (q;q)_{n_j + r_j}}, \tag{3.6}$$

then

$$\sum_{n_1, \ldots, n_s \geq 0} q^{n^2_1 + n^2_2 + \cdots + n^2_s} \beta_{n_1, n_2, \ldots, n_s}$$

$$= \frac{1}{(q; q)^s_\infty} \sum_{m_1, m_2, \ldots, m_s = -\infty}^{\infty} q^{m^2_1 + m^2_2 + \cdots + m^2_s} \alpha_{m_1, m_2, \ldots, m_s}. \quad (3.7)$$

4. PENTAGONAL NUMBER THEOREMS

Our object here is to provide some striking applications of our work in Section 2. We shall prove three results of increasing difficulty.

THEOREM 2. *Identity* (1.2) *is valid.*

Proof. We apply Corollary 1 with $s = 2$ and

$$\alpha_{m_1, m_2} = (-1)^{m_1 + m_2} q^{\binom{m_1 + m_2}{2}}. \quad (4.1)$$

Consequently

$$(q; q)_{2n_1} (q; q)_{2n_2} \beta_{n_1, n_2}$$

$$= \sum_{r_1 = -n_1}^{n_1} \sum_{r_2 = -n_2}^{n_2} \begin{bmatrix} 2n_1 \\ n_1 - r_1 \end{bmatrix} \begin{bmatrix} 2n_2 \\ n_2 - r_2 \end{bmatrix} (-1)^{r_1 + r_2} q^{\binom{r_1 + r_2}{2}}, \quad (4.2)$$

where

$$\begin{bmatrix} A \\ B \end{bmatrix} = \begin{cases} 0, & \text{if } B < 0 \text{ or } B > A \\ \frac{(q; q)_A}{(q; q)_B (q; q)_{A-B}} & \text{otherwise.} \end{cases} \quad (4.3)$$

By symmetry we may assume $n_2 \geq n_1$, and by the q-binomial theorem [1; p. 36, Eq. (3.3.6)], we deduce that

$$(q; q)_{2n_1} (q; q)_{2n_2} \beta_{n_1 n_2}$$

$$= (-1)^{n_2} \sum_{r_1 = -n_1}^{n_1} \begin{bmatrix} 2n_1 \\ n_1 - r_1 \end{bmatrix} (-1)^{r_1} q^{\binom{r_1}{2} - r_1 n_2 + \binom{n_2}{2}} (q^{r_1 - n_2}; q)_{2n_2}$$

$$= \begin{cases} 0 & \text{if } n_2 \neq n_1 \\ (q; q)_{2n_1} & \text{if } n_1 = n_2. \end{cases} \quad (4.4)$$

Consequently

$$\beta_{n_1, n_2} = \begin{cases} 0 & \text{if} \quad n_1 \neq n_2 \\ \frac{1}{(q;q)_{2n_1}} & \text{if} \quad n_1 = n_2. \end{cases} \tag{4.5}$$

Theorem 2 now follows immediately from (4.1), (4.5) and Corollary 1. ∎

Our next result is harder to prove than Theorem 2 primarily because we introduce a small asymmetry in the definition of α_{m_1, m_2}.

THEOREM 3.

$$\sum_{m, n \geq 0} \frac{(-1)^n q^{m^2+2mn+2n^2}(-1;q)_m}{(q;q^2)_{m+n}(q^2;q^2)_n(q;q)_m} = \frac{\sum_{i,j=-\infty}^{\infty} (-1)^j q^{i(3i-1)/2+j(3j-1)/2+ij}}{(q;q)_\infty^2}. \tag{4.6}$$

Proof. We apply Corollary 1 with $s = 2$ and

$$\alpha_{m_1, m_2} = (-1)^{m_2} q^{\binom{m_1+m_2}{2}}. \tag{4.7}$$

Consequently

$$(q;q)_{2n_1}(q;q)_{2n_2}\beta_{n_1,n_2} = \sum_{r_1=-n_1}^{n_1} \sum_{r_2=-n_2}^{n_2} \begin{bmatrix} 2n_1 \\ n_1+r_1 \end{bmatrix} \begin{bmatrix} 2n_2 \\ n_2+r_2 \end{bmatrix} (-1)^{r_2} q^{\binom{r_1+r_2}{2}}. \tag{4.8}$$

By the q-binomial theorem we find

$$(q;q)_{2n_1}(q;q)_{2n_2}\beta_{n_1,n_2} = \sum_{r_1=-n_1}^{n_1} \begin{bmatrix} 2n_1 \\ n_1+r_1 \end{bmatrix} q^{\binom{r_1}{2}}(q^{1-r_1};q)_{n_2}(q^{r_1};q)_{n_2}. \tag{4.9}$$

Now the only non-vanishing terms in (4.9) occur for $n_2 < r_1 \leq n_1$ and $-n_1 \leq r_1 \leq -n_2$. Therefore

$$\begin{aligned}(q;q)_{2n_1}&(q;q)_{2n_2}\beta_{n_1,n_2} \\ &= \sum_{r_1 > n_2} \begin{bmatrix} 2n_1 \\ n_1+r_1 \end{bmatrix} q^{\binom{r_1}{2}}(q^{1-r_1};q)_{n_2}(q^{r_1};q)_{n_2} \\ &\quad + \sum_{r_1 \geq n_2} \begin{bmatrix} 2n_1 \\ n_1-r_1 \end{bmatrix} q^{\binom{-r_1}{2}}(q^{r_1+1};q)_{n_2}(q^{-r_1};q)_{n_2}\end{aligned}$$

$$= -q^{n_2} \sum_{r_1 \geqq 0} \begin{bmatrix} 2n_1 \\ n_1 + r_1 \end{bmatrix} q^{\binom{r_1}{2} - r_1} (q^{1-r_1}; q)_{n_2 - 1} (q^{r_1}; q)_{n_2} (1 - q^{2r_1})$$

(by combining the two sums)

$$= -q^{n_2 - 1} \sum_{r_1 \geqq 0} \begin{bmatrix} 2n_1 \\ n_1 + r_1 + 1 \end{bmatrix} q^{\binom{r_1}{2}} (q^{-r_1}; q)_{n_2 - 1} (q^{r_1 + 1}; q)_{n_2} (1 - q^{2r_1 + 2})$$

(shifting r_1 to $r_1 + 1$)

$$= (-1)^{n_2} (q; q)_{2n_2} \begin{bmatrix} 2n_1 \\ n_1 + n_2 \end{bmatrix}$$

$$\times {}_6\phi_5 \left(\begin{matrix} q^{2n_2}, q^{n_2+1}, -q^{n_2+1}, q^{n_2-n_1}, q^{n_2+1/2}, -q^{n_2+1/2}; q, -q^{n_1-n_2} \\ q^{n_2}, -q^{n_2}, q^{n_2+n_1+1}, q^{n_2+1/2}, -q^{n_2+1/2} \end{matrix} \right)$$

(in the notation of [6; p. 4])

$$= (-1)^{n_2} \begin{bmatrix} 2n_1 \\ n_1 + n_2 \end{bmatrix} \frac{(q)_{n_1+n_2} (-1; q)_{n_1 - n_2}}{(q^{2n_2+1}; q^2)_{n_1 - n_2}}$$

(by [6; p. 238, Eq. (II.21)])

$$= (-1)^{n_2} (-q; q)_{n_1} (-1; q)_{n_1 - n_2} (q; q^2)_{n_2} (q^{n_1 - n_2 + 1}; q)_{n_2}. \tag{4.10}$$

Note that if $n_2 > n_1$, then β_{n_1, n_2} must be 0. Consequently by Corollary 1,

$$\sum_{n_1 \geqq n_2} \beta_{n_1, n_2} q^{n_1^2 + n_2^2} = \frac{\sum_{i,j=-\infty}^{\infty} (-1)^j q^{i^2 + j^2 + \binom{i+j}{2}}}{(q; q)_\infty^2}. \tag{4.11}$$

Shifting the index n_1 to $n_1 + n_2$ and replacing β_{n_1, n_2} by the expression found for it in (4.10) we obtain (4.6), thus proving Theorem 3. ∎

Our final theorem, the proof of (1.3), requires an initial inversion lemma.

LEMMA 1. *If two sequences* $\{x_n\}_{n=0}^{\infty}$ *and* $\{\rho_n\}_{n=0}^{\infty}$ *satisfy either of the following two relations for all n, then they satisfy both.*

$$x_n = \sum_{j=0}^{n} (-1)^j \begin{bmatrix} n+j \\ n-j \end{bmatrix} q^{\binom{j+1}{2} - nj} \rho_j, \tag{4.12}$$

and

$$\rho_n = \sum_{k=0}^{n} (-1)^k q^{\binom{k}{2}} \begin{bmatrix} 2n+1 \\ n+k+1 \end{bmatrix} \frac{(1 - q^{2k+1})}{(1 - q^{2n+1})} x_k. \tag{4.13}$$

Proof. This is merely a special case of the inversion theorem for Bailey pairs [2; p. 278] with $a=q$, $\beta_n=\rho_n/(q^2;q)_{2n-1}$, and $\alpha_n=(-1)^n q^{\binom{n}{2}}(1-q^{2n+1})\,x_n$. ▮

THEOREM 4. *Identity* (1.3) *is valid.*

Proof. We apply Corollary 1 with $s=3$ and

$$\alpha_{m_1,m_2,m_3}=(-1)^{m_1+m_2+m_3}\,q^{\binom{m_1+m_2+m_3}{2}}. \tag{4.14}$$

If we can prove that

$$\sum_{i=-n_1}^{n_1}\sum_{j=-n_2}^{n_2}\sum_{k=-n_3}^{n_3}(-1)^{i+j+k}\,q^{\binom{i+j+k}{2}}\begin{bmatrix}2n_1\\ n_1+i\end{bmatrix}\begin{bmatrix}2n_2\\ n_2+j\end{bmatrix}\begin{bmatrix}2n_3\\ n_3+k\end{bmatrix}$$
$$=\frac{(q;q)_{2n_1}(q;q)_{2n_2}(q;q)_{2n_3}}{(q)_{n_1+n_2-n_3}(q;q)_{n_1+n_3-n_2}(q;q)_{n_2+n_3-n_1}}, \tag{4.15}$$

then Corollary 1 implies directly that identity (1.3) holds. So to conclude our proof we must establish (4.15).

To this end, we define

$$S(n_1,n_2,k)$$
$$=\frac{1}{(q;q)_{2n_1}}\sum_{i=-n_1}^{n_1}\sum_{j=-n_2}^{n_2}(-1)^{i+j}\,q^{\binom{i+j}{2}+k(i+j)}\begin{bmatrix}2n_1\\ n_1+i\end{bmatrix}\begin{bmatrix}2n_2\\ n_2+j\end{bmatrix}, \tag{4.16}$$

and

$$T(n_1,n_2,n_3)=\frac{(q;q)_{2n_2}}{(q;q)_{n_1+n_2-n_3}}\begin{bmatrix}2n_3\\ n_1+n_3-n_2\end{bmatrix}. \tag{4.17}$$

Then we may rewrite (4.15) as

$$\sum_{k=-n_3}^{n_3}(-1)^k q^{\binom{k}{2}}\,S(n_1,n_2,k)\begin{bmatrix}2n_3\\ n_3+k\end{bmatrix}=T(n_1,n_2,n_3). \tag{4.18}$$

Next we note that by replacing i by $-i$ and j by $-j$ in (4.16), we have

$$S(n_1,n_2,k)=S(n_1,n_2,1-k). \tag{4.19}$$

Using (4.19) to combine positive and negative terms in (4.18), we may rewrite (4.18) as

$$q^{n_3}\sum_{k=0}^{n_3}(-1)^k q^{\binom{k}{2}} S(n_1,n_2,k+1)\begin{bmatrix}2n_3+1\\ n_3+k+1\end{bmatrix}\frac{(1-q^{2k+1})}{(1-q^{2n_3+1})}$$
$$=T(n_1,n_2,n_3). \tag{4.20}$$

We now apply our lemma with $n=n_3$, $\rho_n=q^{-n}T(n_1,n_2,n)$ and $x_k=S(n_1,n_2,k+1)$. So we see that (4.15) is actually equivalent to

$$S(n_1,n_2,n+1)=\sum_{j=0}^{n}(-1)^j\begin{bmatrix}n+j\\ n-j\end{bmatrix}q^{\binom{j+1}{2}-nj}q^{-j}\,T(n_1,n_2,j). \tag{4.21}$$

We have now reduced the proof of Theorem 4 to proving that (4.21) is true. Furthermore, owing to the symmetry in (1.3), we may without loss of generality assume that $n_1\geqq n_2\geqq n_3$. The right-hand side of (4.21) when reduced to q-hypergeometric notation [6; p. 4] is, in fact, equal to

$$\frac{(-1)^n(q;q)_{2n}(q;q)_{2n_2}q^{-\binom{n+1}{2}}}{(q;q)_{n_1+n_2-n}(q;q)_{n_1+n-n_2}(q;q)_{n_2+n-n_1}}\,{}_3\phi_2\left(\begin{matrix}q^{-n_1+n_2-n},q^{-n_2+n_1-n},0;q;q\\ q^{n_1+n_2-n+1},q^{-2n}\end{matrix}\right). \tag{4.22}$$

Now we can simplify the expression for $S(n_1,n_2,k)$ given in (4.16) by applying the q-binomial theorem [6; p. 236, Eq. (II.4)]

$$S(n_1,n_2,k)$$
$$=\frac{1}{(q;q)_{2n_1}}\sum_{j=-n_1-n_2}^{n_2-n_1}(-1)^j q^{\binom{j}{2}+kj}\begin{bmatrix}2n_2\\ n_2+n_1+j\end{bmatrix}(q^{k+j};q)_{2n_1}$$
$$=(-1)^{n_1+n_2}q^{\binom{n_1+n_2+1}{2}-k(n_1+n_2)}$$
$$\times\frac{1}{(q;q)_{2n_1}}\sum_{j\geqq 0}\frac{(q^{-2n_2},q)_j}{(q;q)_j}\frac{(q^{k-n_2-n_1};q)_{2n_1+j}}{(q^{k-n_2-n_1};q)_j}q^{j(k+n_2-n_1)}$$
$$=\frac{(-1)^{n_1+n_2}q^{\binom{k-n_2-n_1}{2}-\binom{k}{2}}(q^{k-n_2-n_1};q)_{2n_1}}{(q;q)_{2n_1}}$$
$$\times{}_2\phi_1\left(\begin{matrix}q^{-2n_2},q^{k-n_2+n_1};q,q^{k+n_2-n_1}\\ q^{k-n_2-n_1}\end{matrix}\right)\quad\text{(in the notation of [6; p. 4]).} \tag{4.23}$$

Therefore

$$S(n_1, n_2, n+1) = S(n_1, n_2, -n) \qquad \text{(by (4.19))}$$
$$= \frac{(-1)^{n_1+n_2} q^{\binom{n_1+n_2+n+1}{2}-\binom{n+1}{2}} (q^{-n_1-n_2-n}; q)_{2n_1}}{(q;q)_{2n_1}}$$
$${}_2\phi_1\left(\begin{matrix} q^{-2n_2}, q^{-n-n_2+n_1}; q, q^{n_2-n-n_1} \\ q^{-n_1-n_2-n} \end{matrix}\right)$$
$$= (-1)^{n_1+n_2} q^{\binom{n_1-n_2}{2}+n^2_2+nn_2-n^2_1+nn_1-n^2} \begin{bmatrix} 2n \\ n_2+n-n_1 \end{bmatrix}$$
$${}_2\phi_1\left(\begin{matrix} q^{-n-n_2+n_1}, q^{-n-n_2+n_1}; q; q^{-2n_1} \\ q^{-2n} \end{matrix}\right) \quad \text{(by [6; p. 240, Eq. (III.2)])} \tag{4.24}$$

Finally then, the proof of Theorem 4 is reduced to proving that the expression in (4.22) is equal to the right-hand side of (4.24). So what we must prove is

$${}_3\phi_2\left(\begin{matrix} q^{-n_1+n_2-n}, q^{-n_2+n_1-n}, 0; q, q \\ q^{n_1+n_2-n+1}, q^{-2n} \end{matrix}\right)$$
$$= \frac{(-1)^{n+n_1+n_2} q^{\binom{n_1-n_2}{2}+n_2^2+nn_2-n_1^2+nn_1-n^2+\binom{n+1}{2}} (q;q)_{n_1+n_2-n}}{(q;q)_{2n_2}}$$
$$\times {}_2\phi_1\left(\begin{matrix} q^{-n-n_2+n_1}, q^{-n-n_2+n_1}; q, q^{-2n_1} \\ q^{-2n} \end{matrix}\right). \tag{4.25}$$

But this is just [6; Eq. (III.11), p. 241] with $c=0$, $d=q^{-2n}$, $e=q^{n_1+n_2-n+1}$, $b=q^{-n_1+n_2-n}$, and the n of (III.11) replaced by $n+n_2-n_1$. Therefore since (4.25) is true we see that in fact Theorem 4 is proved. ▮

5. CONCLUSION

There are several important summary observations. First of all, it would be very nice to have a simpler proof of the polynomial identity (4.15). It is rather surprising that (4.15) is so difficult in light of the fact that the just slightly simpler (4.4) is very easy. In fact, using a q-version of K. Wegschaider's "MultiSum" package, P. Paula and A. Riese derived a recurrence proof of (4.15); see also http://www.risc.uni-linz.ac.at/research/combinat/risc/.

It is natural to ask for a nice q-series expansion for an s-dimensional Pentagonal Number series, namely

$$\frac{\sum_{n_1, n_2, \ldots, n_s} (-1)^{n_1+n_2+\cdots+n_2} q^{n_1^2+n_2^2+\cdots+n_s^2+\binom{n_1+n_2+\cdots+n_s}{2}}}{(q;q)_\infty^s}. \tag{5.1}$$

Finally and most important, the object here was to observe that the Bailey chain could be viewed as arising from the application of an umbral operator to a classical elliptic function identity, namely Jacobi's Triple Product Identity. There are many such identities, such as the Quintuple Product Identity [6; p. 134], Winquist's Identity [13], or the Macdonald identities [8, 12] in general. Whether comparable theorem applications exist related to these identities remains to be seen.

REFERENCES

1. G. E. Andrews, "The Theory of Partitions," Encyclopedia of Mathematics and Its Applications, Vol. 2 (G.-C. Rota, Ed.), Addison–Wesley, Reading, MA, 1976. Reprinted: Cambridge Univ. Press, Cambridge, UK, 1998.
2. G. E. Andrews, Multiple series Rogers–Ramanujan type identities, *Pac. J. Math.* **141** (1984), 267–283.
3. G. E. Andrews, "q-Series, Their Development and Application in Analysis, Number Theory, Combinatorics, Physics and Computer Algebra," CBMS Regional Conference Lecture Series, Vol. 66, Amer. Math. Soc., Providence, RI, 1986.
4. A. Berkovich, B. M. McCoy, and A. Schilling, $N=2$ supersymmetry and Bailey pairs, *Phys. A* **228** (1996), 33–62.
5. L. E. Dickson, "History of the Theory of Numbers," Vol. 2, Carnegie Inst. of Washington, Publ. 256, 1920. Reissued: Chelsea, New York, 1952.
6. G. Gasper and M. Rahman, "Basic Hypergeometric Series," Encyclopedia of Mathematics and Its Applications, Vol. 35 (G.-C. Rota, Ed.), Cambridge Univ. Press, Cambridge, UK, 1990.
7. G. Humbert, Démonstration analytique d'une formule de Liouville, *Bull. Sci. math. (2)* **34** (1910), 29–31.
8. I. G. Macdonald, Affine root systems and Dedekind's η-function, *Invent. Math.* **15** (1972), 91–143.
9. P. Paule, The concept of Bailey chains, *Sém. Lothar. Combin. B* **18f** (1987), 24.
10. L. J. Rogers, On two theorems of combinatory analysis and some allied identities, *Proc. London Math. Soc. (2)* **16** (1917), 315–336.
11. G. C. Rota, Ten mathematics problems I will never solve, *DMV Mitteilungen* **2** (1998), 45–52.
12. D. Stanton, An elementary approach to the Macdonald identities, *in* "q-Series and Partitions" (D. Stanton, Ed.), IMA Volumes in Mathematics and Its Applications, Vol. 18, Springer-Verlag, Berlin, 1989.
13. L. Winquist, An elementary proof of $p(11m+6) \equiv 0 \pmod{11}$, *J. Combin. Theory* **6** (1969), 56–59.

2.5.6 Bailey's transform, lemma, chains and tree

BAILEY'S TRANSFORM, LEMMA, CHAINS AND TREE

GEORGE E. ANDREWS
Department of Mathematics
The Pennsylvania State University
University Park, PA 16802 USA

Abstract. In this paper we shall provide a brief survey of the work begun by L. J. Rogers and W. N. Bailey which has led to an iterative method for producing infinite chains of q-series identities. Apart from providing the reader with leads to the study of previous accomplishments, we shall emphasize the importance of examination of the seminal works in order to discern topics open to further development. This will lead us directly to a new construct: the Bailey tree.

1. Introduction

Bailey chains have their genesis in the work of L. J. Rogers. He first proved the following identities in 1894 [30], and they have become known as the Rogers-Ramanujan identities

$$\sum_{n=0}^{\infty} \frac{q^{n^2}}{(q;q)_n} = \frac{1}{(q, q^4; q^5)_\infty} \tag{1.1}$$

and

$$\sum_{n=0}^{\infty} \frac{q^{n^2+n}}{(q;q)_n} = \frac{1}{(q^2, q^3; q^5)_\infty}, \tag{1.2}$$

where

$$(A)_n = (A;q)_n = \prod_{j=0}^{n-1} (1 - Aq^j), \tag{1.3}$$

J. Bustoz et al. (eds.), Special Functions 2000, 1–22.

and

$$(A_1, A_2, \ldots, A_r; q)_n = \prod_{j=1}^{r} (A_j; q)_n . \tag{1.4}$$

We shall provide a proof of (1.1) and (1.2) in Section 2. It is essentially Bressoud's simplification of Rogers' second proof [19]. This will provide us with the simplest possible example of a Bailey chain (cf. [5], [7; Ch.3]).

1.1. DEFINITION

A General Bailey Chain is an infinite sequence ($i \geqq 0$) of pairs of sequences of rational functions $\{\alpha_n^{(i)}\}_{n\geqq 0}$ and $\{\beta_n^{(i)}\}_{n\geqq 0}$ of two variables a and q and possibly others

$$(\alpha_n, \beta_n) \to (\alpha_n', \beta_n') \to (\alpha_n'', \beta_n'') \to \cdots . \tag{1.5}$$

There is an identity independent of i connecting $\alpha_n^{(i)}$ and $\beta_n^{(i)}$, say

$$\beta_n^{(i)} = F_n(\alpha_0^{(i)}, \alpha_1^{(i)}, \alpha_2^{(i)}, \ldots, \alpha_n^{(i)}) ; \qquad (i \geqq 0) \tag{1.6}$$

furthermore there are given rules of construction

$$\beta_n^{(i)} = G_n(\beta_0^{(i-1)}, \ldots, \beta_n^{(i-1)}) \qquad (i \geqq 1) \tag{1.7}$$

and

$$\alpha_n^{(i)} = H_n(\alpha_0^{(i-1)}, \ldots, \alpha_r^{(i-1)}) \qquad (i \geqq 1) , \tag{1.8}$$

again G_n and H_n are independent of i.

Put in this abstract setting, it is hard to discern the relationship of this iterative process with the Rogers-Ramanujan identities (1.1) and (1.2). This will become clear in Section 2. Section 3 will look at the overarching method dubbed by L. J. Slater, the Bailey transform [38; p. 58]. Section 4 will provide a summary of how this idea has been broadly applied. In Section 5, we consider the Milne-Lilly extensions [25]–[27], and other extensions [12]. We also return to a look at Rogers' second paper [31] with an eye to extensions as described in [8]. In Section 6, we reconsider Bailey's first papers on this topic; this leads us to new Bailey chains in Section 7 and eventually to the Bailey tree in Section 8.

2. The Bressoud-Rogers Proof

In this section we will provide the simplest known example of a non-trivial Bailey chain.

BAILEY'S TRANSFORM, LEMMA, CHAINS AND TREE 3

Theorem 1 *Let*

$$\alpha_n^{(0)} = \frac{(1-aq^{2n})(a;q)_n(-1)^n q^{\binom{n}{2}}}{(1-a)(q;q)_n} \tag{2.1}$$

and

$$\beta_n^{(0)} = \delta_{n,0}\,. \tag{2.2}$$

Furthermore for $i > 0$

$$\alpha_n^{(i)} = a^n q^{n^2} \alpha_n^{(i-1)} \tag{2.3}$$

and

$$\beta_n^{(i)} = \sum_{j=0}^{n} \frac{q^{j^2} a^j \beta_j^{(i-1)}}{(q;q)_{n-i}}\,. \tag{2.4}$$

Then $\{\alpha_n^{(i)}\}_{n\geqq 0}$, $\{\beta_n^{(i)}\}_{n\geqq 0}$ $(i \geqq 0)$ form a Bailey Chain.

2.1. PROOF

The G_n and H_n of (1.7) and (1.8) are clearly defined by (2.3) and (2.4). We shall show that

$$F_n(x_0, x_1, \ldots, x_n) = \sum_{j=0}^{n} \frac{x_j}{(q;q)_{n-j}(aq;q)_{n+j}}\,. \tag{2.5}$$

The treatment of the assertion that

$$\beta_n^{(0)} = \sum_{j=0}^{n} \frac{\alpha_j^{(0)}}{(q;q)_{n_j}(aq;q)_{n+j}} \tag{2.6}$$

follows immediately from the fact (proved easily by induction on M) that

$$\sum_{j=0}^{M} \frac{(a;q)_j(1-aq^{2j})(q^{-n};q);q^{nj}}{(q;q)_j(aq^{n+1};q)_j(1-a)} = \frac{(aq;q)_M q^{nM}(q^{1-n};q)_M}{(q;q)_M(aq^{n+1};q)_M}\,. \tag{2.7}$$

Now we assume that (1.6) holds for $i-1$ with F_n given by (2.5). Hence

$$\beta_n^{(i)} \;=\; \sum_{j=0}^{n} \frac{q^{j^2} a^j \beta_j^{(i-1)}}{(q;q)_{n-j}}$$

$$= \sum_{j=0}^{n} \frac{q^{j^2} a^j}{(q;q)_{n-j}} \sum_{r=0}^{j} \frac{\alpha_r^{(i-1)}}{(q;q)_{j-r}(aq;q)_{j+r}}$$

$$= \sum_{r=0}^{n} \sum_{j=r}^{n} \frac{q^{j^2} a^j \alpha_r^{(i-1)}}{(q;q)_{n-j}(q;q)_{j-r}(aq;q)_{j+r}}$$

$$= \sum_{r=0}^{n} \sum_{j=0}^{n-r} \frac{q^{j^2+2jr} a^j \alpha_r^{(i)}}{(q;q)_{n-j-r}(q;q)_j(aq;q)_{j+2r}}$$

$$= \sum_{r=0}^{n} \frac{\alpha_r^{(i)}}{(q;q)_{n-r}(aq;q)_{2r}}$$

$$\times \sum_{j\geqq 0} \frac{(q^{-n+r};q)_j(-1)^j q^{(n-r)j+\binom{j+1}{2}+2jr}}{(q;q)_j(aq^{2r+1};q)_j}$$

$$= \sum_{r=0}^{r} \frac{\alpha_r^{(i)}}{(q;q)_{n-r}(aq;q)_{2r}} \frac{1}{(aq^{2r+1};q)_{n-r}}$$

(by [23]; p. 11, eq (1.5.2) with $b \to \infty$)

$$= \sum_{r=0}^{n} \frac{\alpha_r^{(i)}}{(q;q)_{n-r}(aq;q)_{n+r}}, \tag{2.8}$$

which proves that (1.6) holds for all i. □

The first couple of instances of (1.6) reveal the power of Bailey Chains. When $i = 1$,

$$\beta_n^{(1)} = \frac{1}{(q;q)_n} \tag{2.9}$$

by (2.4), and

$$\alpha_n^{(1)} = \frac{(1-aq^{2n})(a;q)_n(-a)^n q^{n(3n-1)/2}}{(1-a)(q;q)_n}. \tag{2.10}$$

Next,

$$\beta_n^{(2)} = \sum_{j=0}^{n} \frac{q^{j^2} q^j}{(q)_{n-j}(q)_j}, \tag{2.11}$$

and

$$\alpha_n^{(2)} = \frac{(1-aq^{2n})(a;q)_n(-1)^n a^{2n} q^{n(5n-1)/2}}{(1-a)(q;q)_n}. \tag{2.12}$$

Now by Theorem 1, we know that

$$\beta_n^{(2)} = \sum_{r=0}^{n} \frac{\alpha_r^{(2)}}{(q;q)_{n-r}(aq;q)_{n+r}}, \tag{2.13}$$

and therefore

$$\beta_\infty^{(2)} = \frac{1}{(q;q)_\infty (aq;q)_\infty} \sum_{r=0}^{\infty} \alpha_r^{(2)}, \tag{2.14}$$

i.e. after multiplication by $(q;q)_\infty$, we obtain

$$\sum_{n=0}^{\infty} \frac{a^n q^{n^2}}{(q;q)_n} = \frac{1}{(aq;q)_\infty} \sum_{n=0}^{\infty} \frac{(a;q)_n (1-aq^{2n})(-1)^n a^{2n} q^{n(5n-1)/2}}{(q;q)_n (1-a)}. \tag{2.15}$$

From this familiar formula and Jacobi's Triple Product Indentity [23; p. 12, eq. (1.6.1)] we obtain (1.1) by setting $a = 1$, and (1.2) by setting $a - q$.

This process was essentially carried out by D. Bressoud in [19] (cf. [17; Sec. 3.4]). It is a limiting case of the proof given in Chapter 3 of [7]. We go through it here to illustrate as cleanly as possible precisely how a Bailey Chain works.

3. The Bailey Transform

The actual transformation that we performed in (2.8) was treated in a fully abstract manner by Bailey in [15]. L. J. Slater [38; p. 58] has christened this the Bailey Transform:

If

$$\beta_n = \sum_{r=0}^{n} \alpha_r u_{n-r} v_{n+r}, \tag{3.1}$$

and

$$\gamma_n = \sum_{r=n}^{\infty} \delta_r u_{r-n} v_{r+n}, \tag{3.2}$$

then

$$\sum_{n=0}^{\infty} \alpha_n \gamma_n = \sum_{n=0}^{\infty} \beta_n \delta_n. \tag{3.3}$$

As Bailey notes [15; p. 1] the proof is a mere series rearrangement.

$$\sum_{n=0}^{\infty}\alpha_n\gamma_n = \sum_{n=0}^{\infty}\sum_{r=n}^{\infty}\alpha_n\delta_r u_{r-n}v_{r+n} \tag{3.4}$$

$$= \sum_{r=0}^{\infty}\sum_{n=0}^{r}\delta_r\alpha_n u_{r-n}v_{r+n} \tag{3.5}$$

$$= \sum_{r=0}^{\infty}\delta_r\beta_r\,. \tag{3.6}$$

In (2.8), the Bailey Transform underlies the process in the instance $u_n = 1/(q;q)_n$, $v_n = 1/(aq;q)_n$, $\delta_r = a^r q^{r^2}$, $\gamma_n = a^n q^{n^2}/(aq;q)_\infty$.

In Section 4, we shall examine other instances of the Bailey Transform that lead to other Bailey Chains. Indeed, one of the reasons for a separate section on the Bailey Transform is to distinguish it from Bailey's Lemma (an instance of the Bailey Transform [cf. (4.1)-(4.4) in Section 4]) and Bailey Chains (as defined in Section 1).

Section 4 will only consider the case $u_n = 1/(q;q)_n$, $v_n = 1/(aq;q)_n$, and in Section 5 we shall restrict ourselves to natural multidimensional extensions of this instance. However, there are several quite distinct instances of Bailey's Transform in the literature, and we shall touch briefly on a few.

D. Bressoud [18] was perhaps the first to obtain really striking variations. He chose

$$u_n = \frac{(\beta;q)_n}{(q;q)_n},\quad v_n = \frac{(\alpha\beta;q)_n}{(q;q)_n},\quad \text{and}\quad \delta_n = r^n\,. \tag{3.7}$$

Perhaps his most appealing application is the discovery of two new polynomial identities which imply the Rogers-Ramanujan identities:

$$\sum_{m=0}^{N} q^{m^2}\begin{bmatrix}N\\m\end{bmatrix} = \sum_{m=-\infty}^{\infty}(-1)^m q^{m(5m+1)/2}\begin{bmatrix}2N\\N+2m\end{bmatrix} \tag{3.8}$$

and

$$(1-q^N)\sum_{n=0}^{\infty} q^{m^2+m}\begin{bmatrix}N-1\\m\end{bmatrix} = \sum_{m=-\infty}^{\infty}(-1)^m q^{m(5m+3)/2}\begin{bmatrix}2N\\N+2m+1\end{bmatrix}, \tag{3.9}$$

where

$$\begin{bmatrix}A\\B\end{bmatrix} = \begin{bmatrix}A\\B\end{bmatrix}_q = \begin{cases}\frac{(q;q)_A}{(q;q)_B(q;q)_{A-B}} & \text{if } 0 \leqq A \leqq B\\ 0 & \text{otherwise.}\end{cases} \tag{3.10}$$

In [10], it was pointed out that one of D. B. Sears' most useful transformations [35; p. 167, Th. 3] was the following instance of the Bailey Transform:

$$\begin{aligned} u_n &= \frac{(ab/e;q)_n}{(q;q)_n}, \quad v_n = 1, \quad \gamma_s = \sum_{t=0}^{\infty} \frac{(ab/e;q)_t \delta_{s+t}}{(q;q)_t} \\ \alpha_n &= \frac{(e/a;q)_n (e/b;q)_n \left(\frac{ab}{e}\right)^n}{(q;q)_n (e;q)_n}, \quad \beta_n = \frac{(a;q)_n (b;q)_n}{(q;q)_n (e;q)_n}. \end{aligned} \tag{3.11}$$

Following up on this [10], one can obtain an $(r-1)$-fold expansion of

$$\begin{aligned} &{}_{r+1}\phi_r \left(\begin{matrix} c_0, c_1, \ldots, c_r \\ f_1, \ldots, f_r \end{matrix} ; q, \frac{f_1 f_2 \cdots f_r}{c_0 c_1 \cdots c_r} \right) \\ &:= \sum_{n \geqq 0} \frac{(c_0, c_1, \ldots, c_r; q)_n}{(q, f_1, \ldots, f_r; q)_n} \left(\frac{f_1 f_2 \ldots f_r}{c_0 c_1 \ldots c_r} \right)^n . \end{aligned} \tag{3.12}$$

When $r = 2$, the result reduces to

$$\begin{aligned} &{}_3\phi_2 \left(\begin{matrix} c_0, c_1, c_2 \\ f_1, f_2 \end{matrix} ; q, \frac{f_1 f_2}{c_0 c_1 c_2} \right) \\ &= \frac{\left(\frac{f_2}{c_2}, \frac{f_1 f_2}{c_0 c_2}; q \right)_\infty}{\left(f_2, \frac{f_1 f_2}{c_0 c_1 c_2}; q \right)_\infty} {}_3\phi_2 \left(\begin{matrix} f_1/c_0, f_1/c_1, c_2 \\ f_1, \frac{f_1 f_2}{c_0 c_1} \end{matrix} ; q, f_2/c_2 \right) \end{aligned} \tag{3.13}$$

a familiar and useful formula [23, p. 62, eq. (3.2.7)].

Very much related to this is a series of papers by Schilling and Warnaar [32], [33], [34] in which they also concentrate on the γ_n and δ_n and consider multi-dimensional extensions thereof. Also a variation on γ_n and δ_n appears in Bailey's paper [15, Sec. 5].

In this same general area, but not explicitly following from the Bailey Transform is the work in [9] on a q-trinomial coefficient analog of Bailey's Lemma. In [39], O. Warnaar shows how this work is closely related to the classical Bailey Lemma.

4. One-Dimensional Bailey Chains

The proof of the Rogers-Ramanujan identities given in Section 2 is in fact a very special case of the following result known as Bailey's Lemma [5], [7; Ch. 3].

If for $n \geq 0$

$$\beta_n = \sum_{r=0}^{n} \frac{\alpha_r}{(q;q)_{n-r} (aq;q)_{n+r}} , \tag{4.1}$$

then

$$\beta'_n = \sum_{r=0}^{n} \frac{\alpha'_r}{(q;q)_{n-r}(aq;q)_{n+r}}, \tag{4.2}$$

where

$$\alpha'_r = \frac{(\rho_1;q)_r(\rho_2;q)_r(aq/\rho_1\rho_2)^r\alpha_r}{(aq/\rho_1;q)_r(aq/\rho_2;q)_r} \tag{4.3}$$

and

$$\beta'_n = \sum_{j\geq 0} \frac{(\rho_1;q)_j(\rho_2;q)_j(aq/\rho_1\rho_2;q)_{n-j}(aq/\rho_1\rho_2)^j\beta_j}{(q;q)_{n-j}(aq/\rho_1;q)_n(aq/\rho_2;q)_n}. \tag{4.4}$$

We shall suppress the proof of this; it is very similar to the proof of (2.8) (cf. [7]; Ch. 3). This, of course, can be iterated immediately to produce a Bailey Chain as described in Section 1.

This result as outlined (but not stated) by Bailey himself ([15]; Sec. 4) uses the Bailey Transform in the case where

$$\begin{aligned} u_n &= \frac{1}{(q;q)_n}, \qquad v_n = \frac{1}{(aq;q)_n}, \\ \delta_n &= \frac{(\rho_1,\rho_2,q^{-N};q)_n q^n}{(\rho_1\rho_2 q^{-N}/a;q)_n}, \\ \gamma_n &= \frac{(aq/\rho_1,aq/\rho_2;q)_N}{(aq,aq/(\rho_1\rho_2);q)_N}\ \frac{(\rho_1,\rho_2,q^{-N};q)_n\left(\frac{-aq}{\rho_1\rho_2}\right)^n q^{nN-\binom{n}{2}}}{(aq/\rho_1,aq/\rho_2,aq^{N+1};q)_n}. \end{aligned} \tag{4.5}$$

Subsequently this Bailey Chain has been applied to prove many disparate results. Its original application [3] (cf. [7]; Ch. 3) was to provide an analog of Watson's q-analog of Whipple's Theorem [41] for

$$ {}_{2k+r}\phi_{2k+3}\left(\begin{matrix} a, q\sqrt{a}, -q\sqrt{a}, b_1, c_1, b_2, c_2, \ldots, b_k, c_k, q^{-N} \\ \sqrt{a}, -\sqrt{a}, \frac{aq}{b_1}, \frac{aq}{c_1}, \frac{aq}{b_2}, \frac{aq}{c_2}, \ldots, \frac{aq}{b_k}, \frac{aq}{c_k}, aq^{N+1} \end{matrix}; q, \frac{a^k q^{k+N}}{b_1c_1\cdots b_kc_k}\right). \tag{4.6}$$

This in turn can be specialized to the natural extension of the first Rogers-Ramanujan identity [2], [4; Ch. 7]:

$$\sum_{n_1\geq n_2\geq\cdots n_{k-1}\geq 0} \frac{q^{n_1^2+n_2^2+\cdots+n_{k-1}^2}}{(q;q)_{n_1-n_2}(q;q)_{n_2-n_3}\cdots(q;q)_{n_{k-2}-n_{k-1}}(q;q)_{n_{k-1}}}$$

$$= \prod_{\substack{n=1 \\ n \not\equiv 0, \pm k \pmod{2k+1}}}^{\infty} \frac{1}{1-q^n}. \tag{4.7}$$

But other results also follow such as [11; p. 392]

$$\sum_{n=0}^{\infty} \frac{q^{\binom{n+1}{2}}}{(-q;q)_n} = \sum_{\substack{n \geqq 0 \\ |j| \leqq n}} (-1)^{n+j} q^{n(3n+1)/2-j^2}(1-q^{2n+1}), \tag{4.8}$$

or [11; p. 404]

$$\sum_{n \geqq 1} \frac{(-1)^n q^{n^2}}{(q;q^2)_n} = \sum_{n \geqq 1} \sum_{j=0}^{2n-1} (-1)^n q^{n(3n-1)-j(j+1)/2}(1+q^{2n}). \tag{4.9}$$

Further applications are given in [5], [6], [36] and [37].

5. Multidimensional Bailey Chains

Lilly and Milne were the first researchers [25], [26] and [27] to recognize that the concepts described in Sections 2–5 could be extended to higher dimensional series. In a substantial series of papers extending classical ordinary and q-hypergeometric functions to multiple series based on the unitary or symplectic groups, S. Milne and his collaborators laid the ground work for an extension of Bailey's Lemma and Bailey chains to these groups.

As an example of their achievement we state their generalization for the symplectic groups C_ℓ [26; pp. 494–495]:

Theorem 2 *Let $A = \{A_y\}$ and $B = \{B_y\}$ be sequences that satisfy*

$$B_N = \sum_{\substack{0 \le y_k \le N_k \\ k=1,2,\ldots,\ell}} \left\{ \prod_{r,s=1}^{\ell} \left[\left(q\, \frac{x_r}{x_s}\, q^{y_r-y_s} \right)_{N_r-y_r}^{-1} (q x_r x_s q^{y_r+y_s})_{N_r-y_r}^{-1} \right] A_y \right\} \tag{5.1}$$

for every $N_i \ge 0$, $i = 1, 2, \ldots, \ell$. (A and B form a C_ℓ Bailey pair). If we define

$$A'_N := \left\{ \prod_{k=1}^{\ell} \left[\frac{(\alpha x_k)_{N_k} (q x_k \beta^{-1})_{N_k}}{(\beta x_k)_{N_k} (q x_k \alpha^{-1})_{N_k}} \right] \left(\frac{\beta}{\alpha} \right)^{(N_1+\cdots+N_\ell)} A_N \right\} \tag{5.2}$$

and if we also define

$$
\begin{aligned}
B'_N \quad := \quad & \sum_{\substack{0 \le m_k \le N_k \\ k=1,2,\dots,\ell}} \left\{ \left(\frac{\beta}{\alpha}\right)_{(N_1+\cdots+N_\ell)-(m_1+\cdots+m_\ell)} \right. \\
\times \quad & \prod_{k=1}^{\ell} \left[\frac{(\alpha x_k)_{m_k} (q x_k \beta^{-1})_{m_k}}{(\beta x_k)_{N_k} (q x_k \alpha^{-1})_{N_k}} \right] \left(\frac{\beta}{\alpha}\right)^{(m_1+\cdots+m_\ell)} \\
\times \quad & \prod_{1 \le r < s \le \ell} [(q x_r x_s q^{m_r+m_s})^{-1} (q x_r x_3 q^{N_s - m_s})^{-1}_{N_r - m_r}] \\
\times \quad & \left. \prod_{r,s=1}^{\ell} \left(q \frac{x_r}{x_s} q^{m_r - m_s} \right)^{-1}_{N_r - m_r} B_m \right\}, \qquad (5.3)
\end{aligned}
$$

then $A' = \{A'_y\}$ *and* $B' = \{B'_y\}$ *also satisfy (5.1); i.e. they form a* C_ℓ *Bailey pair.*

In [27], they consider the applications of their extensions and provide relevant generalizations of the q-analog of Whipple's transformation, the q-Dougall summation and other results.

In [12], Schilling, Warnaar and I provide an alternative approach to multiple series. It is not nearly as broad in scope as the Lilly-Milne work, but it does yield a variety of new series-product identities. The principal ideas of [12] culminate in the following formulation.

5.1. DEFINITION

(A_2 Bailey pair of type I). Denote $\alpha_k = \alpha_{k_1,k_2,k_3}$ and $\beta_L = \beta_{L_1,L_2}$, and let $\alpha = \{\alpha_k\}_{\substack{k_1 \ge k_2 \ge k_3 \\ k_1+k_2+k_3=0}}$ and $\beta = \{\beta_l\}_{L_1,L_2 \ge 0}$ be a pair of sequences that satisfies

$$
\beta_L = \sum_{\substack{k_1 \ge k_2 \ge k_3 \\ k_1+k_2+k_3=0}} \alpha_k \sum_r \frac{q^{r_1 r_{23}}}{(q)_{r_1} (q)_{r_2} (aq)_{r_3} (q)_{r_{12}} (q)_{r_{13}} (q)_{r_{23}}}, \qquad (5.4)
$$

where $\sum_r$ denotes a sum over $r_1, \dots, r_{23}$ such that

$$
r_1 + r_{12} + r_{13} = L_2 - k_1, \quad r_2 + r_{12} + r_{23} = L_2 - k_2, \quad r_3 + r_{13} + r_{23} = L_2 - k_3
$$

and

$$
r_1 + r_2 + r_3 = 2L_1 - L_2, \quad r_{12} + r_{13} + r_{23} = 2L_2 - L_1.
$$

Then (α, β) forms an A_2 Bailey pair of type I relative to a.

5.2. DEFINITION

(A_2 Bailey pair of type II). Denote $\alpha_k = \alpha_{k_1,k_2,k_3}$ and $\beta_L = \beta_{L_1,L_2}$, and let $\alpha = \{\alpha_k\}_{\substack{k_1 \geq k_2 \geq k_3 \\ k_1+k_2+k_3=0}}$ and $\beta = \{\beta_L\}_{L_1,L_2 \geq 0}$ be a pair of sequences that satisfies

$$\beta_L = \sum_{\substack{k_1 \geq k_2 \geq k_3 \\ k_1+k_2+k_3=0}} \alpha_k \times \frac{(aq)_{L_1+L_2}}{(aq)_{L_1+k_1}(aq)_{L_1+k_2}(q)_{L_1+k_3}(q)_{L_2-k_1}(q)_{L_2-k_2}(aq)_{L_2-k_3}}. \quad (5.5)$$

Then (α, β) forms an A_2 Bailey pair of type II relative to a.

With these definitions, the A_2 version of Bailey's lemma reads

Theorem 3 [12] *Let(α, β) form an A_2 bailey pair of type $T = I, II$ relative to a. Then the pair (α', β') given by*

$$\begin{aligned} \alpha'_k &= a^{k_1+k_2} q^{\frac{1}{2}(k_1^2+k_2^2+k_3^2)} \alpha_k, \\ \beta'_L &= f_L^{(T)} \sum_{r_1=0}^{L_1} \sum_{r_2=0}^{L_2} \frac{a^{r_1} q^{r_1^2-r_1r_2+r_2^2}}{(q)_{L_1-r_1}(q)_{L_2-r_2}} \beta_r \end{aligned} \quad (5.6)$$

forms an A_2 Bailey pair of type II relative to a. Here $f_L^{(I)} = (aq)_{L_1+L_2}^{-1}$ and $f_L^{(II)} = 1$.

As examples of the series-product results that follow from this theorem, we include [12; Th. 5.2]

Theorem 4 *For $|q| < 1$,*

$$\sum_{r_1 r_2 \geq 0} \frac{q^{r_1^2-r_1r_2+r_2^2}}{(q)_{r_1}} \begin{bmatrix} 2r_1 \\ r_2 \end{bmatrix} = \prod_{n=1}^{\infty} \frac{1}{(1-q^{7n-1})^2(1-q^{7n-3})(1-q^{7n-4})(1-q^{7n-6})^2}, \quad (5.7)$$

$$\sum_{r_1 r_2 \geq 0} \frac{q^{r_1^2-r_1r_2+r_2^2+r_1+r_2}}{(q)_{r_1}} \begin{bmatrix} 2r_1 \\ r_2 \end{bmatrix}$$

$$= \prod_{n=1}^{\infty} \frac{1}{(1-q^{7n-2})(1-q^{7n-3})^2(1-q^{7n-4})^2(1-q^{7n-5})}, \quad (5.8)$$

and

$$\sum_{r_1 r_2 \geq 0} \frac{q^{r_1^2 - r_1 r_2 + r_2^2 + r_1}}{(q)_{r_1}} \begin{bmatrix} 2r_1 + 1 \\ r_2 \end{bmatrix}$$
$$= \sum_{r_1, r_2 \geq 0} \frac{q^{r_1^2 - r_1 r_2 + r_2^2 + r_2}}{(q)_{r_1}} \begin{bmatrix} 2r_1 \\ r_2 \end{bmatrix} \quad (5.9)$$
$$= \prod_{n=1}^{\infty} \frac{1}{(1-q^{7n-1})(1-q^{7n-2})(1-q^{7n-3})}$$
$$\times \frac{1}{(1-q^{7n-4})(1-q^{7n-5})(1-q^{7n-6})}.$$

(the symbol $\begin{bmatrix} A \\ B \end{bmatrix}$ *is defined in* (3.10)).

In [8], I undertook an examination of the work of L. J. Rogers which inspired all of Bailey's insights. The main result there might be described as a "direct product" generalization of Bailey's Lemma and Bailey Chains [7; Ch. 3]. Its bilateral nature is dealt with in the $s = 1$ case in [16] and [29].

Theorem 5 *If for* $n_1, n_2, \ldots, n_s \geqq 0$,

$$\beta_{n_1,n_2,\ldots,n_s} = \sum_{r_1=-\infty}^{n_1} \sum_{r_2=-\infty}^{n_2} \cdots \sum_{r_s=-\infty}^{n_s} \frac{\alpha_{r_1,r_2,\ldots,r_s}}{\prod_{j=1}^{s} (q;q)_{n_j - r_j} (a_j q; q)_{n_j + r_j}} \quad (5.10)$$

then

$$\beta'_{n_1,n_2,\ldots,n_s} = \sum_{r_1=-\infty}^{n_1} \sum_{r_2=-\infty}^{n_2} \cdots \sum_{r_s=-\infty}^{n_s} \frac{\alpha'_{r_1,r_2,\ldots,r_s}}{\prod_{j=1}^{s} (q;q)_{n_j - r_j} (a_j q; q)_{n_j + r_j}}, \quad (5.11)$$

where

$$\alpha'_{r_1,r_2,\ldots,r_s} = \left(\prod_{j=1}^{s} \frac{(\rho_j)_{r_j} (\sigma_j)_{r_j} (a_j q / (\rho_j \sigma_j))^{r_j}}{(a_j q/\rho_j)_{r_j} (a_j q/\sigma_j)_{r_j}} \right) \alpha_{r_1,r_2,\ldots,r_s} \quad (5.12)$$

and

$$\beta'_{n_1,n_2,\ldots,n_s} \tag{5.13}$$
$$= \prod_{j=1}^{s} \sum_{m_j=-\infty}^{n_j} \frac{(\rho_j)_{m_j}(\sigma_j)_{m_j}(a_jq/(\rho_j\sigma_j))_{n_j-m_j}\left(\frac{a_jq}{\rho_j\sigma_j}\right)^{m_j}\beta_{m_j,\ldots,m_s}}{(q)_{n_j-m_j}(a_jq/\rho_j)_{n_j}(a_jq/\sigma_j)_{n_j}}.$$

The instance of this theorem that is utilized in [8] is the

First Corollary *If for* $n_1, n_2, \ldots, n_s \geqq 0$

$$\beta_{n_1,n_2,\ldots,n_s} = \sum_{r_1=-n_1}^{n_1} \sum_{r_2=-n_2}^{n_2} \cdots \sum_{r_s=-n_s}^{n_s} \frac{\alpha_{r_1,r_2,\ldots,r_s}}{\prod_{j=1}^{s}(q;q)_{n_j-r_j}(q;q)_{n_j+r_j}}, \tag{5.14}$$

then

$$\sum_{n_1,\ldots,n_s \geqq 0} q^{n_1^2+n_2^2+\cdots+n_s^2}\beta_{n_1,n_2,\ldots,n_s} \tag{5.15}$$
$$= \frac{1}{(q;q)_\infty^s} \sum_{m_1,m_2,\ldots,m_s=-\infty} q^{m_1^2+m_2^2+\cdots+m_s^2}\alpha_{m_1,m_2,\ldots,m_s}.$$

To illustrate the implications of this result for q-series identities, new pentagonal number theorems are proved:

$$\sum_{n=1}^{\infty} \frac{q^{2n^2}}{(q;q)_{2n}} = \frac{\sum_{n,m=-\infty}^{\infty} (-1)^{n+m} q^{n(3n-1)/2+m(3m-1)/2+nm}}{\prod_{n-1}^{\infty}(1-q^n)^2}. \tag{5.16}$$

and

$$\sum_{i,j,k\geqq 0} \frac{q^{i^2+j^2+k^2}}{(q;q)_{i+j-k}(q;q)_{i+k-j}(q;q)_{j+k-i}} \tag{5.17}$$
$$= \frac{\sum_{n,m,p=-\infty}^{\infty} (-1)^{n+m+p} q^{n(3n-1)/2+m(3m-1)/2+p(3p-1)/2+nm+np+mp}}{\prod_{n=1}^{\infty}(1-q^n)^3}.$$

To further illustrate the application of these results we shall prove:

Theorem 6

$$\sum_{m,n\geqq 0}\frac{(-1)^m q^{2m^2+2n^2}(-q;q^2)_{n-m}}{(q;q^2)_{n-m}(q^4;q^4)_m(q^4;q^4)_m}=\frac{\sum\limits_{i,j=-\infty}^{\infty}(-1)^i q^{3i^2+3j^2+2ij}}{(q^2;q^2)_\infty^2}. \tag{5.18}$$

Proof. This result is a direct application of the First Corollary wherein we replace q by q^2, set $s=2$, and

$$\alpha_{i,j}=(-1)^i q^{(i+j)^2}, \tag{5.19}$$

$$\beta_{m,n}=\frac{(-1)^m(-q;q^2)_{n-m}}{(q;q^2)_{n-m}(q^4;q^4)_m(q^4;q^4)_n}. \tag{5.20}$$

This means that we must prove that

$$\sum_{i,j=-\infty}^{\infty}(-1)^i q^{(i+j)^2}\begin{bmatrix}2m\\m+i\end{bmatrix}_{q^2}\begin{bmatrix}2n\\n+j\end{bmatrix}_{q^2} = \frac{(-1)^m(-q;q^2)_{n-m}(q^2;q^4)_m(q^2;q^4)_n}{(q;q^2)_{n-m}}. \tag{5.21}$$

To see this, we note

$$\begin{aligned}
&\sum_{i,j}(-1)^i q^{(i+j)^2}\begin{bmatrix}2m\\m+i\end{bmatrix}_{q^2}\begin{bmatrix}2n\\n+j\end{bmatrix}_{q^2} \qquad (5.22)\\
&=\sum_j\begin{bmatrix}2n\\n+j\end{bmatrix}_{q^2} q^{j^2}\sum_i(-1)^k q^{i^2+2ij}\begin{bmatrix}2m\\m+i\end{bmatrix}_{q^2}\\
&=\sum_j\begin{bmatrix}2n\\n+j\end{bmatrix}_{q^2} q^{j^2}(q^{2j+1},q^2)_m(q^{1-2j};q^2)_m\\
&\qquad\qquad \text{(by [4; p. 49, Ex. 1])}\\
&=\frac{(-1)^m(q;q^2)_{m-n}q^{(n+m)^2}}{(q;q^2)_{-m-n}}\,{}_2\phi_1\left(\begin{matrix}q^{-4n},q^{2m-2n+1}\\q^{-2m-2n+1}\end{matrix};q^2,-q^{2n-2m+1}\right)\\
&=\frac{(-1)^m(-q;q^2)_{n-m}(q^2;q^4)_m(q^2;q^4)_r}{(q;q^2)_{n-m}}\\
&\qquad\qquad \text{(by [1; eq. (1.7)], [23; Sec. 1.8])},
\end{aligned}$$

which is the desired result. □

6. A New Bailey Chain, the WP-Bailey Chain

The work described in Section 4 yields (among many other results) transformations for the series given in (4.6). The most famous of these are the case $k = 1$, a summation due to F. H. Jackson [24], and $k = 2$, Watson's q-analog of Whipple's theorem.

However, there is a classical transformation due to Bailey [23; p. 39, eq. (2.9.1)]) which is not a direct corollary of the Bailey Chains considered in Section 4.

$$
{}_{10}\phi_9\left(\begin{matrix} a, qa^{\frac{1}{2}}, -qa^{\frac{1}{2}}, b, c, d, e, f, \lambda aq^{n+1}/ef, q^{-n} \\ a^{\frac{1}{2}}, -a^{\frac{1}{2}}, aq/b, aq/c, aq/d, aq/e, aq/f, efq^{-n}/\lambda, aq^{n+1} \end{matrix}; q, q\right)
$$
$$
= \frac{(aq, aq/ef, \lambda q/e, \lambda q/f; q)_n}{(aq/e, aq/f, \lambda q/ef, \lambda q; q)_n}\, {}_{10}\phi_9\left(\begin{matrix} \lambda, q\lambda^{\frac{1}{2}}, q\lambda^{\frac{1}{2}}, \lambda b/a, \lambda c/a, \lambda d/a, \\ \lambda^{\frac{1}{2}} - \lambda^{\frac{1}{2}}, aq/b, aq/c, aq/d, \end{matrix}\right.
$$
$$
\left.\begin{matrix} e, f, \lambda aq^{n+1}/ef, q^{-n}, \\ \lambda q/e, \lambda q/f, efq^{-n}/a, \lambda q^{n+1} \end{matrix}; q, q\right), \tag{6.1}
$$

where $\lambda = qa^2/bcd$.

If we examine the Bailey method for proving (6.1) (beautifully, presented in [23; Sec. 2.2 and Sec. 2.9]), we find that we can easily produce a new type of Bailey Chain. We shall call it a WP-Bailey Chain because both the series in (6.1) are examples of very well-poised basic hypergeometric series [23; p. 32].

Definition 6.1 *We say that two sequences $\alpha_n(a,k)$ and $\beta_n(a,k)$ form a WP-Bailey pair provided*

$$
\beta_n(a,k) = \sum_{i=0}^{n} \frac{(k/a;q)_{n-i}(k;q)_{n+i}}{(q;q)_{n-i}(aq;q)_{n+i}}\, \alpha_i(a,k)\,. \tag{6.2}
$$

Note that if $k = 0$ then we have a Bailey pair as considered in Section 4.

Theorem 7 *If $\alpha_n(a,k)$ and $\beta_n(a,k)$ form a WP-Bailey pair, then so do $\alpha'_n(a,k)$ and $\beta'_n(a,k)$, where*

$$
\alpha'_n(a,k) = \frac{(\rho_1;q)_n(\rho_2;q)_n(aq/\rho_1\rho_2)^n}{(aq/\rho_1;q)_n(aq/\rho_2;q)_n}\, \alpha_n\left(a, \frac{k\rho_1\rho_2}{aq}\right) \tag{6.3}
$$

and

$$
\beta'_n(a,k) = \frac{\left(\frac{k\rho_1}{a};q\right)_n \left(\frac{k\rho_2}{a}q\right)_n}{\left(\frac{aq}{\rho_1},q\right)_n \left(\frac{aq}{\rho_2},q\right)_n} \sum_{j=0}^{n} \frac{(\rho_1;q)_j(\rho_2;q)_j}{\left(\frac{k\rho_2}{a};q\right)_j \left(\frac{k\rho_1}{a};q\right)_j} \tag{6.4}
$$

$$\times \frac{\left(1-\frac{k\rho_1\rho_2 q^{2j-1}}{a}\right)\left(\frac{aq}{\rho_1\rho_2};q\right)_{n-j}(k;q)_{n+j}}{\left(1-\frac{k\rho_1\rho_2}{aq}\right)(q;q)_{n-j}\left(\frac{k\rho_1\rho_2}{a};q\right)_{n+j}}\left(\frac{aq}{\rho_1\rho_2}\right)^j$$

$$\times \quad \beta_j\left(a, \frac{k\rho_1\rho_2}{aq}\right).$$

Remark Note that if we at $k = 0$, this result reduces to the assertions in (4.1)–(4.4).

Proof. We must show that $\alpha'_n(a,k)$ and $\beta'_n(a,k)$ as given in (6.3) and (6.4) actually satisfy (6.2). By (6.4) and (6.2)

$$\begin{aligned}
\beta'_n(a,k) &= \frac{\left(\frac{k\rho_1}{a}, \frac{k\rho_2}{a};q\right)_n}{\left(\frac{aq}{\rho_1}, \frac{aq}{\rho_2};q\right)_n}\sum_{j=0}^{n}\frac{(\rho_1,\rho_2;q)_j}{\left(\frac{k\rho_2}{a}, \frac{k\rho_1}{a};q\right)_j} \\
&\times \frac{\left(1-\frac{k\rho_1\rho_2 q^{2j-1}}{a}\right)\left(\frac{aq}{\rho_1\rho_2};q\right)_{n-j}(k;q)_{n+j}}{\left(1-\frac{k\rho_1\rho_2}{aq}\right)(q;q)_{n-j}\left(\frac{k\rho_1\rho_2}{a};q\right)_{n+j}}\left(\frac{aq}{\rho_1\rho_2}\right)^j \\
&\times \sum_{i=0}^{j}\frac{\left(\frac{k\rho_1\rho_2}{a^2q};q\right)_{j-i}\left(\frac{k\rho_1\rho_2}{aq};q\right)_{j+1}\alpha_i\left(a,\frac{k\rho_1\rho_2}{aq}\right)}{(q;q)_{j-i}(aq;q)_{j+i}} \\
&= \frac{\left(\frac{k\rho_1}{a}, \frac{k\rho_2}{a};q\right)_n}{\left(\frac{aq}{\rho_1}, \frac{aq}{\rho_2};q\right)_n}\sum_{i=0}^{n}\sum_{j=0}^{n-i}\frac{(\rho_1,\rho_2;q)_{j+i}}{\left(\frac{k\rho_2}{a}, \frac{k\rho_1}{a};q\right)_{j+i}} \\
&\times \frac{\left(1-\frac{k\rho_1\rho_2 q^{2i+2j-1}}{a}\right)\left(\frac{aq}{\rho_1\rho_2};q\right)_{n-i-j}(k;q)_{n+i+j}\left(\frac{aq}{\rho_1\rho_2}\right)^{i+j}}{\left(1-\frac{k\rho_1\rho_2}{aq}\right)(q;q)_{n-i-j}\left(\frac{k\rho_1\rho_2}{a};q\right)_{n+i+j}} \\
&\times \frac{\left(\frac{k\rho_1\rho_2}{a^2q};q\right)_j\left(\frac{k\rho_1\rho_2}{aq};q\right)_{j+2i}\alpha_i\left(a,\frac{k\rho_1\rho_2}{aq}\right)}{(q;q)_j(aq;q)_{j+2i}} \\
&= \frac{\left(\frac{k\rho_1}{a}, \frac{k\rho_2}{a};q\right)_n}{\left(\frac{aq}{\rho_1}, \frac{aq}{\rho_2};q\right)_n}\sum_{i=0}^{n}\frac{(\rho_1,\rho_2;q)_i}{\left(\frac{k\rho_2}{a}, \frac{k\rho_1}{a};q\right)_i}\alpha_i\left(a,\frac{k\rho_1\rho_2}{aq}\right) \\
&\times \frac{\left(\frac{aq}{\rho_1\rho_2};q\right)_{n-i}(k;q)_{n+i}\left(\frac{aq}{\rho_1\rho_2}\right)^i\left(1-\frac{k\rho_1\rho_2 q^{2i-1}}{a}\right)\left(\frac{k\rho_1\rho_2}{aq};q\right)_{2i}}{(q;q)_{n-i}\left(\frac{k\rho_1\rho_2}{a};q\right)_{n+i}\left(1-\frac{k\rho_1\rho_2}{aq}\right)(aq;q)_{2i}}
\end{aligned}$$

$$\times \; {}_8\phi_7\left(\begin{matrix}\frac{k\rho_1\rho_2q^{2i-1}}{a}, q^{i+1}\sqrt{\frac{k\rho_1\rho_2}{aq}}, -q^{i+1}\sqrt{\frac{k\rho_1\rho_2}{aq}}, \rho_1q^i, \rho_2q^i, \\ q^i\sqrt{\frac{k\rho_1\rho_2}{aq}}, -q^i\sqrt{\frac{k\rho_1\rho_2}{aq}}, \frac{k\rho_2q^i}{a}, \frac{k\rho_1q^i}{a}, \end{matrix}\right.$$
$$\left.\begin{matrix}\frac{k\rho_1\rho_2}{a^2q}, kq^{n+i}, q^{-n+i} \\ aq^{2i+1}, \frac{\rho_1\rho_2q^{-n+i}}{a}, \frac{k\rho_1\rho_2q^{n+i}}{a}\end{matrix}; q, q\right)$$

$$= \frac{\left(\frac{k\rho_1}{a}, \frac{k\rho_2}{a}; q\right)_n}{\left(\frac{aq}{\rho_1}, \frac{aq}{\rho_2}; q\right)_n} \sum_{i=0}^{n} \frac{(\rho_1, \rho_2; q)_i}{\left(\frac{k\rho_2}{a}, \frac{k\rho_1}{a}; q\right)_i} \alpha_i\left(a, \frac{k\rho_1\rho_2}{aq}\right)$$
$$\times \frac{\left(\frac{aq}{\rho_1\rho_2}; q\right)_{n-i} (k;q)_{n+i} \left(\frac{aq}{\rho_1\rho_2}\right)^i \left(\frac{k\rho_1\rho_2}{a}; q\right)_{2i}}{(q;q)_{n-i}\left(\frac{k\rho_1\rho_2}{a}; q\right)_{n+i} (aq;q)_{2i}}$$
$$\times \frac{\left(\frac{k\rho_1\rho_2q^{2i}}{a}, \frac{k}{a}, \frac{\rho_1q^{-n}}{a}, \frac{\rho_2q^{-n}}{a}; q\right)_{n-i}}{\left(\frac{k\rho_2q^i}{a}, \frac{k\rho_1q^i}{a}, \frac{\rho_1\rho_2q^{-n+i}}{a}, \frac{q^{-n-i}}{a}; q\right)_{n-i}}$$

(by [23; p. 35, eq. (2.6.2)])

$$= \sum_{i=0}^{n} \frac{\left(\frac{k}{a}; q\right)_{n-i} (k;q)_{n+i}}{(q;q)_{n-i}(aq;q)_{n+i}} \frac{(\rho_1, \rho_2; q)_i}{\left(\frac{aq}{\rho_1}, \frac{aq}{\rho_2}; q\right)_i} \left(\frac{aq}{\rho_1\rho_2}\right)^i \alpha_i\left(a, \frac{k\rho_1\rho_2}{aq}\right)$$
$$= \sum_{i=0}^{n} \frac{\left(\frac{k}{a}; q\right)_{n-i} (k;q)_{n+i}}{(q;q)_{n-i}(aq;q)_{n+i}} \alpha_i'(a,k), \tag{6.5}$$

as desired. □

It is now a simple matter to begin a WP-Bailey chain by noting that

$$\beta_n(a,k) = \delta_{n,0}, \tag{6.6}$$

$$\alpha_n(a,k) = \frac{(a;q)_n(1-aq^{2n})\left(\frac{a}{k};q\right)_n}{(q;q)_n\,(1-a)\,(kq;q)_n}\left(\frac{k}{a}\right)^n \tag{6.7}$$

forms a WP-Bailey pair. This follows immediately by setting $m = n$ in the following identity

$$\sum_{n=0}^{m} \frac{\left(\frac{k}{a}; q\right)_{n-i} (k;q)_{n+i}\alpha_i(a,k)}{(q;q)_{n-i}(aq;q)_{n+i}.}$$

$$= \frac{\left(\frac{kq^{-m}}{a};q\right)_n (1-k)(kq^{m+1};q)_n(q^{1-n};q)_m q^{nm}}{(1-kq^n)(aq^{m+1};q)_n(q;q)_n(q;q)_m}. \tag{6.8}$$

Identity (6.8) follows immediately by mathematical induction on m.

Corollary (F. H. Jackson's q-analog of Dougall's theorem [24], [23; p. 35, eq. (2.6.2)])

$$ {}_8\phi_7\left(\begin{matrix} a, q\sqrt{a}, -q\sqrt{a}, kq^n, \rho_1, \rho_2, \dfrac{a^2q}{k\rho_1\rho_2}, q^{-n} \\ \sqrt{a}, -\sqrt{a}, \dfrac{aq^{1-n}}{k}, \dfrac{aq}{\rho_1}, \dfrac{aq}{\rho_2}, \dfrac{k\rho_1\rho_2}{a}, aq^{n+1} \end{matrix}; q, q\right) = \frac{\left(aq, \frac{aq^{1-n}}{\rho_1 k}, \frac{aq^{1-n}}{\rho_2 k}, \frac{aq}{\rho_1\rho_2}; q\right)_n}{\left(\frac{aq^{1-n}}{k}, \frac{aq}{\rho_1}, \frac{aq}{\rho_2}, \frac{aq^{1-n}}{k\rho_1\rho_2}; q\right)_n}. \tag{6.9}$$

Proof. This is just the assertion that the WP-Bailey pair $\alpha_n'(a,k)$ and $\beta_n'(a,k)$ defined from Theorem 7 satisfy (6.2) where $\alpha_n(a,k)$ and $\beta_n(a,k)$ are given by (6.6) and (6.7). □

Corollary (W. N. Bailey's very well-poised ${}_{10}\phi_9$ identity [23; p. 38, eq. (2.9.1)]) *Equation (6.1) is valid.*

Proof. This is (after changes of variable), the assertion that the WP-Bailey pair $\alpha_n''(a,k)$, $\beta_n''(a,k)$ (with new parameters ρ_3 and ρ_4) defined from (6.3) and (6.4) satisfy (6.2) where $\alpha_n'(a,k)$ and $\beta_n'(a,k)$ are given in the previous corollary. □

We close this section by remarking that the N-th entry in this WP-Bailey chain produces a representation of

$$ {}_{2N+4}\phi_{2N+3}\left(\begin{matrix} a, q\sqrt{a}, -a\sqrt{a}, \rho_1, \rho_2, \rho_3\rho_4, \dots, \rho_{2N-3}, \rho_{2N-2}, \lambda q^n, q^{-n} \\ \sqrt{a}, -\sqrt{a}, \frac{aq}{\rho_1}, \frac{aq}{\rho_2}\dots, \frac{aq}{\rho_{2N-3}}, \frac{aq}{\rho_{2N-2}}, \frac{aq^{1-n}}{\lambda}, aq^{n+1} \end{matrix}; q, q\right) \tag{6.10}$$

(where $a^Nq^{N-1} = \rho_1\rho_2\cdots\rho_{2N-2}\lambda$) as an $(N-2)$ fold summation. Thus when $N=4$ (the first new case) we find the very-well poised ${}_{12}\phi_{11}$ of the above type represented by a double sum.

7. An Alternative WP-Bailey Pair

In [13; Sec. 9] and [14], W. N. Bailey used ideas similar to those recounted up to this point in order to develop identities for nearly poised series. These are series in which one column of parameters does not satisfy the well-poised requirement.

When we translate Bailey's ideas into the world of Bailey chains, a most extraordinary thing happens. Namely, in contrast with Theorem 3, we find that from any given WP-Bailey pair we can construct a new WP-Bailey pair $\widetilde{\alpha}_n(a,k)$, $\widetilde{\beta}_n(a,k)$ which is not at all the pair given by (6.3) and (6.4).

Theorem 8 *If $\alpha_n(a,k)$ and $\beta_n(a,k)$ form a WP-Bailey pair, then so do $\widetilde{\alpha}_n(a,k)$ and $\widetilde{\beta}_n(a,k)$, where*

$$\widetilde{\alpha}_n(a,k) = \frac{\left(\frac{qa^2}{k};q\right)_{2n}\left(\frac{k^2}{qa^2}\right)^n}{(k;q)_{2n}}\alpha_n\left(a,\frac{qa^2}{k}\right), \tag{7.1}$$

$$\widetilde{\beta}_n(a,k) = \sum_{j=0}^{n}\frac{\left(\frac{k^2}{a^2q};q\right)_{n-j}}{(q;q)_{n-j}}\left(\frac{k^2}{a^2q}\right)^j\beta_j\left(a,\frac{aq^2}{k}\right). \tag{7.2}$$

Remark There is absolutely no intersection of the WP-Bailey pairs produced by Theorem 3 and those produced by Theorem 4.

Proof. We are given that the sequences $\alpha_n(a,k)$ and $\beta_n(a,k)$ satisfy (6.2). Hence

$$\begin{aligned}
\widetilde{\beta}_n(a,k) &= \sum_{j=0}^{n}\frac{\left(\frac{k^2}{a^2q};q\right)_{n-j}\left(\frac{k^2}{a^2q}\right)^j}{(q;q)_{n-j}}\sum_{i=0}^{j}\frac{\left(\frac{qa}{k};q\right)_{j-1}\left(\frac{qa^2}{k};q\right)_{j+i}}{(q;q)_{j-i}(aq;q)_{j+i}}\alpha_i\left(a,\frac{qa^2}{k}\right)\\
&= \sum_{i=0}^{n}\sum_{j=0}^{n-i}\frac{\left(\frac{k^2}{a^2q};q\right)_{n-i-j}\left(\frac{k^2}{a^2q}\right)^{j+i}\left(\frac{qa}{k};q\right)_j\left(\frac{aq^2}{k};q\right)_{j+2i}\alpha_i\left(a,\frac{qa^2}{k}\right)}{(q;q)_{n-i-j}(q;q)_j(aq;q)_{j+2i}}\\
&= \sum_{i=0}^{n}\frac{\left(\frac{k^2}{a^2q};q\right)_{n-i}\left(\frac{k^2}{a^2q}\right)^i\left(\frac{aq^2}{k};q\right)_{2i}\alpha_i\left(a,\frac{aq^2}{k}\right)}{(q;q)_{n-i}(aq;q)_{2i}}\\
&\quad\times\sum_{j=0}^{n-i}\frac{(q^{-n+i};q)_j}{\left(\frac{a^2q^{2-n+i}}{k^2};q\right)_j}q^j\frac{\left(\frac{aq}{k};q\right)_j\left(\frac{a^2q^{2i+1}}{k};q\right)_j}{(q;q)_j(aq^{2i+1};q)_j}\\
&= \sum_{i=0}^{n}\frac{\left(\frac{k^2}{a^2q};q\right)_{n-i}\left(\frac{k^2}{a^2q}\right)^i\left(\frac{qa^2}{k};q\right)_{2i}\alpha_i\left(a,\frac{qa^2}{k}\right)}{(q;q)_{n-i}(aq;q)_{2i}}\\
&\quad\times\frac{\left(\frac{k}{a};q\right)_{n-i}(kq^{2i};q)_{n-i}}{(aq^{2i+1};q)_{n-i}\left(\frac{k^2}{a^2q};q\right)_{n-i}}
\end{aligned}$$

(by [23; p. 13, eq. (1.7.2)])

$$= \sum_{i=0}^{n} \frac{\left(\frac{k}{a};q\right)_{n-i}(k;q)_{n+i}}{(q;q)_{n-i}(aq;q)_{n+i}};\frac{\left(\frac{qa^2}{k};q\right)_{2i}\left(\frac{k^2}{a^2q}\right)^i \alpha_i\left(a;\frac{qa^2}{k}\right)}{(k;q)_{2i}}$$

$$= \sum_{i=0}^{n} \frac{\left(\frac{k}{a};q\right)_{n-i}(k;q)_{n+i}}{(q;q)_{n-i}(aq:q)_{n+i}}\widetilde{\alpha}_i(a,k), \tag{7.3}$$

as desired. □

The most immediate application of Theorem 4 is the application in which the WP-Bailey pair $\alpha_n'(a,k)$ and $\beta_n'(a,k)$ of the first Corollary are taken as input. The resulting assertion that $\widetilde{\alpha}_n(a,k)$ and $\widetilde{\beta}_n(a,k)$ form a WP-Bailey pair is equivalent to the result of Bailey [13; p. 431]:

$$\begin{aligned} &{}_5\phi_4\left(\begin{matrix} a,\ b,\ c,\ d,\ q^{-N} \\ aq/b,\ aq/c,\ aq/d,\ a^2q^{-N}/k^2 \end{matrix};q,q\right) \\ &= \frac{(kq/a)_N(k^2q/a)_N}{(kq)_N(k^2q/a^2)_N} \\ &\times {}_{12}\phi_{11}\left(\begin{matrix} k,\ q\sqrt{k}, -q\sqrt{k}, kb/a,\ kc/a,\ kd/a, & \sqrt{a}, -\sqrt{a}, \\ \sqrt{k},\ -\sqrt{k},\ aq/b,\ aq/c,\ aq/d,\ kq/\sqrt{a}, & -kq/\sqrt{a}, \end{matrix}\right. \\ &\qquad \left.\begin{matrix} \sqrt{(aq)}, & -\sqrt{(aq)}, & k^2q^{N+1}/a, & q^{-N} \\ k\sqrt{(q/a)}, & -k\sqrt{(q/a)}, & aq^{-N}/k, & kq^{N+1} \end{matrix};q,q\right). \end{aligned} \tag{7.4}$$

8. Summary and the Bailey Tree

First it should be noted that this survey paper has shortchanged a number of papers that have made substantial contributions to Bailey chains. Among these are [16], [20], [21], [22], [29]. I would draw special attention to [40] wherein Warnaar gives an extended account of Bailey chains especially emphasizing their role in statistical mechanics.

Finally, I note again the surprise contained in Sections 7 and 8. The new, two parameter WP-Bailey pairs (which reduce to the original Bailey pairs when $k = 0$) generate new WP-Bailey pairs in two entirely different ways (Theorems 3 and 4). Thus at each step in the generation of a chain there are two offspring pairs. So instead of generating a chain, we are generating a binary tree. We shall wait to explore Bailey Trees subsequently.

The author was supported in part by NSF Grant DMS-9206993.

References

1. G. E.Andrews, *On the q-analog of Kummer's theorem and applications*, Duke Math. J. **40** (1973), 525–528.

2. G. E. Andrews, *An analytic generalization of the Rogers-Ramanujan identities for odd moduli*, Proc. Nat. Acad. Sci. USA **71** (1974), 4082–4085.
3. G. E. Andrews, *Problems and prospects for basic hypergeometric functions* in: "Theory and Applications of Special Functions", R. Askey, ed., Academic Press, New York, 1975, pp. 191–224.
4. G. E. Andrews, *The Theory of Partitions*, Encyl. of Math. and Its Appl., Vol. 2, Addison-Wesley, Reading, 1976; Reissued: Cambridge University Press, Cambridge, 1985 and 1998.
5. G. E. Andrews, *Multiple series Rogers-Ramanujan type identities*, Pacific J. Math. **114** (1984), 267–283.
6. G. E. Andrews, *The fifth and seventh order mock theta functions*, Trans. Amer. Math. Soc. **293** (1986), 113–134.
7. G. E. Andrews, *q-Series: Their Development*, CBMS Regional Conf. Lecture Series **66**, Amer. Math. Soc., Providence, 1986.
8. G. E. Andrews, *Umbral calculus, Bailey chains and pentagonal number theorems*, J. Comb. Theory, Ser. A, **91** (2000), 464–475.
9. G. E. Andrews and A. Berkovich, *A trinomial analogue of Bailey's lemma and $N = 2$ superconformal invariance*, Comm. in Math. Phys. **192** (1998), 245–260.
10. G. E. Andrews and D. Bowman, *The Bailey transform and D. B. Sears*, Quaest. Math. **22** (1999), 19–26.
11. G. E. Andrews, F. J. Dyson and D. Hickerson, *Partitions and indefinite quadratic forms*, Invent. Math. **91** (1988), 391–407.
12. G. E. Andrews, A. Schilling and S. O. Warnaar, *An A_2 Bailey Lemma and Rogers-Ramanujan-type identities*, J. Amer. Math. Soc. **12** (1999), 677–702.
13. W. N. Bailey, *Some identities in combinatory analysis*, Proc. London Math. Soc. (2) **49** (1947), 421–435.
14. W. N. Bailey, *A transformation of nearly-poised basic hypergeometric series*, J. London Math. Soc. **22** (1947), 237–240.
15. W. N. Bailey, *Identities of the Rogers-Ramanujan type*, Proc. London Math. Soc. (2) **50** (1949), 1–10.
16. A. Berkovich, B. M. McCoy and A. Schilling, *$N = 2$ supersymmetry and Bailey pairs*, Phys. A **228** (1996), 33–62.
17. J. M. and P. B. Borwein, *Pi and the AGM*, Wiley, New York, 1987.
18. D. M. Bressoud, *Some identities for terminating q-series*, Math. Proc. Camb. Phil. Soc. **81** (1981), 211–223.
19. D. M. Bressoud, *An easy proof of the Rogers-Ramanujan identities*, J. Number Theory **16** (1983), 235–241.
20. D. M. Bressoud, *A matrix inverse*, Proc. Amer. Math. Soc. **88** (1983), 446–448.
21. D. M. Bressoud, *The Bailey lattice: an introduction*, in: "Ramanujan Revisited", G. E. Andrews et al, eds., Academic Press, New York 1988, pp. 57–67.
22. O. Foda and V.-H. Iuano, *Virasoro character identities from the Andrews-Bailey construction*, Int. J. Mod. Phys. A **12** (1997), 1651–1676.
23. G. Gasper and M. Rahman, *Basic Hypergeometric Series*, Encyl. of Math and Its Appl., Vol. 35, Cambridge University Press, Cambridge, 1990.
24. F. H. Jackson, *Summation of q-hypergeometric series*, Messenger of Math. **50** (1921), 101–112.
25. G. M. Lilly and S. C. Milne, *The A_ℓ and C_ℓ Bailey transform and lemma*, Bull. Amer. Math. Soc. (NS) **26** (1992), 258–263.
26. G. M. Lilly and S. C. Milne, *The C_ℓ Bailey transform and Bailey lemma*, Constr. Approx. **9** (1993), 473–500.
27. G. M. Lilly and S. C. Milne, *Consequences of the A_ℓ and C_ℓ Bailey transform and Bailey lemma*, Discr. Math. **139** (1995), 319–346.
28. P. Paule, *On identities of the Rogers-Ramanujan type*, J. Math. Anal. and Appl. **107** (1985), 225–284.
29. P. Paule, *The concept of Bailey chains*, Sem. Lothar. Combin. B, 18f (1987), 24.

30. L. J. Rogers, *Second memoir on the expansion of certain infinite products*, Proc. London Math. Soc. **25** (1894), 318–343.
31. L. J. Rogers, *On two theorems of combinatory analysis and allied identities*, Proc. London Math. Soc. (2) **16** (1917), 315–336.
32. A. Schilling and S. O. Warnaar, *A higher-level Bailey lemma*, Int. J. Mod. Phys., **B11** (1997), 189–195.
33. A. Schilling and S. O. Warnaar, *A higher level Bailey lemma, proof and application*, Ramanujan journal **2** (1998), 327–349.
34. A. Schilling and S. O. Warnaar, *Conjugate Bailey pairs*, to appear.
35. D. B. Sears, *On the transformation theory of basic hypergeometric functions*, Proc. London Math. Soc. (2) **53** (1951), 158–180.
36. L. J. Slater, *A new proof of Rogers' transformations of infinite series*, Proc. London Math. Soc. (2) **53** (1951), 460–475.
37. L. J. Slater, *Further identities of the Rogers-Ramanujan type*, Proc. London Math. Soc. (2) **54** (1952), 147–167.
38. L. J. Slater, *Generalized Hypergeometric Functions*, Cambridge University Press, Cambridge, 1966.
39. O. Warnaar, *A note on the trinomial analogue of Bailey's lemma*, J. Combin. Theory A **81** (1998), 114–118.
40. O. Warnaar, *50 years of Bailey's lemma*, Kerber Festschrift, to appear.
41. G. N. Watson, *A new proof of the Rogers-Ramanujan identities*, J. London Math. Soc., **4** (1929) pp. 4–9.

2.6 *q*-Trinomials and the Borwein Conjecture

In [132], Baxter and I use a q-analog of the coefficients in:

$$\prod_{j=1}^{\infty}\left(1+x^{i}+x^{2j}\right)$$

to solve a model in statistical mechanics. This paper is included in Section 2.8. These q-trinomial coefficients turned out to be delightfully useful and their discovery is recounted in [141]: a paper included in Chapter 5.

The first use outside physics was to solve a mystery posed by Euler in 1765. This paper [154] is also included here. The q-trinomial coefficients can also be used to give new finite versions of the Rogers–Ramanujan identities, as well as the Göllnitz–Gordon identities, and this paper [156] is included. Paper [144] follows up on [156] in proving the Capparelli conjectures.

Finally, the Borwein conjecture was a problem I studied for five years before publishing my attempts (none of which succeeded) [183]. In light of the continuing interest in this problem, this paper has been included as well.

In [132], Baxter and I conclude by suggesting that valuable and interesting analogs of:

$$\prod_{j=1}^{n}\left(1+x^{j}+x^{2j}+\cdots+x^{(k-1)j}\right)$$

ought to exist. Schilling and Warnaar [Schi-W98] have indeed found such generalizations with appealing applications.

2.6.1 Euler's "Exemplum Memorabile Inductionis Fallacis" and q-trinomial coefficients

JOURNAL OF THE
AMERICAN MATHEMATICAL SOCIETY
Volume 3, Number 3, July 1990

EULER'S "EXEMPLUM MEMORABILE INDUCTIONIS FALLACIS" AND q-TRINOMIAL COEFFICIENTS

GEORGE E. ANDREWS

1. Introduction

In [5], R. J. Baxter and the author introduced q-analogs of the coefficients in the Laurent polynomial

$$(1 + x + x^{-1})^n = \sum_{j=-n}^{n} \binom{n}{j}_2 x^j . \tag{1.1}$$

The coefficients in (1.1) are called trinomial coefficients (although the same name is used for the coefficients arising in $(x + y + z)^n$). Now trinomial coefficients have a rather sparse literature. There are occasional references to them in combinatorics books (e.g., [11]; indeed Euler found them worthy of a 20 page account [12]). However they have seen little development perhaps for two reasons: (1) their elementary properties mimic closely those of binomial coefficients, and (2) there is no nice simple formula for them as there is for binomial coefficients: $\binom{n}{m} = n!/(m!(n-m)!)$. Indeed the simplest formulas for $\binom{n}{m}_2$ (easily derived from (1.1)) are

$$\binom{n}{m}_2 = \sum_{j \geq 0} \frac{n!}{j!(j+m)!(n-2j-m)!}, \tag{1.2}$$

and

$$\binom{n}{m}_2 = \sum_{j \geq 0} (-1)^j \binom{n}{j} \binom{2n-2j}{n-m-j}, \tag{1.3}$$

neither of which is especially attractive.

It came as a real surprise that q-analogs of these numbers were to play a crucial role in the solution of a model in statistical mechanics [5]. For example

Received by the editors October 26, 1989.

1980 *Mathematics Subject Classification* (1985 *Revision*). Primary 11P68; Secondary 05A10, 05A15, 05A30.

The author was partially supported by National Science Foundation Grant DMS-820269.

[5, §4], we required a representation of the polynomials defined by

$$Y_m(a, b; q) = q^{m(3-b)} \sum_{j=4-b}^{3} Y_{m-1}(a, j; q), \tag{1.4}$$

$$Y_0(a, b; q) = \begin{cases} 1 & \text{if } a = b, \\ 0 & \text{otherwise} \end{cases} \tag{1.5}$$

(where a and b are among $\{1, 2, 3\}$), so that, for example,

$$Y_\infty(3, 3; q) = \prod_{\substack{n=1 \\ n \not\equiv 0, \pm 3 \pmod 7}}^{\infty} (1 - q^n)^{-1} \tag{1.6}$$

would be an immediate corollary. The representation we found was [5, p. 319, equation (4.7), $j = k = 1$]

$$\begin{aligned} Y_m(3; q) = &\sum_{\mu=-\infty}^{\infty} q^{14\mu^2+\mu} \binom{m; 7\mu; q}{7\mu}_2 \\ &- \sum_{\mu=-\infty}^{\infty} q^{14\mu^2+13\mu+3} \binom{m; 7\mu+3; q}{7\mu+3}_2, \end{aligned} \tag{1.7}$$

where

$$\binom{m; B; q}{A}_2 = \sum_{j \geq 0} \frac{q^{j(j+B)}(q)_n}{(q)_j (q)_{j+A} (q)_{n-2j-A}}, \tag{1.8}$$

and

$$(A)_j = (A; q)_j = (1 - A)(1 - Aq) \cdots (1 - Aq^{j-1}). \tag{1.9}$$

As we have described in [6], consideration of these polynomials was literally thrust upon us. We should emphasize the similarity between the above and Schur's polynomial proof of the Rogers-Ramanujan identities [13, §4]. The polynomials Schur examined were given by $G_1(q) = G_2(q) = 1$ and $G_n(q) = G_{n-1}(q) + q^{n-2} G_{n-2}(q)$ for $n > 2$. Schur showed that

$$G_{n+1}(q) = \sum_{\lambda=-\infty}^{\infty} (-1)^\lambda q^{\lambda(5\lambda+1)/2} \begin{bmatrix} n \\ \left\lfloor \frac{n-5\lambda}{2} \right\rfloor \end{bmatrix}_q, \tag{1.10}$$

where $\lfloor x \rfloor$ is the greatest integer $\leq x$ and

$$\begin{bmatrix} A \\ B \end{bmatrix} = \begin{bmatrix} A \\ B \end{bmatrix}_q = \begin{cases} \dfrac{(1 - q^A)(1 - q^{A-1}) \cdots (1 - q^{A-B+1})}{(1 - q^B)(1 - q^{B-1}) \cdots (1 - q)}, & 0 \leq B \leq A, \\ 0 & \text{otherwise.} \end{cases} \tag{1.11}$$

Since generalizations of Schur's identity (1.10) have been utilized extensively in physics [3, 8] and additive number theory [1, 4, 7, 9, 10], one may naturally ask whether the q-trinomial coefficients, appearing in a similar context such as (1.7), may have diverse applications rivalling those of the q-binomial coefficients.

It would seem that to answer this question one would need to identify the q-trinomial coefficients as the generating functions of certain partition-like objects. While there are certain rather artificial interpretations (see §7), none has yielded any combinatorial explanation of (1.7) as the combinatorics of the q-binomial coefficients explains (1.10) [1]. This is surely the key aspect of these polynomials, and unfortunately it will not be resolved in this paper. However we shall consider related questions which we hope will provide further evidence for the power of q-trinomial coefficients and consequently for the importance of understanding them fully.

Obviously one way to proceed is to attempt to investigate any unresolved mysteries related to the trinomial coefficients themselves. As Richard Guy and Donald Knuth pointed out, Euler [12, pp. 54–55] presented such a mystery in two truly surprising pages devoted to Exemplum Memorabile Inductionis Fallacis. Euler first computed $\binom{m}{0}_2$ for $0 \le m \le 9$:

$$1, 1, 3, 7, 19, 51, 141, 393, 1107, 3139, \ldots .$$

He then triples each entry in a row shifted one to the right:

$$\begin{array}{l} 1, 1, 3, 7, 19, 51, 141, 393, 1107, 3139, \ldots \\ \quad 3, 3, 9, 21, 57, 153, 423, 1179, 3321, \ldots , \end{array}$$

and starting with the first two-entry column, he subtracted the first row from the second:

$$2, 0, 2, 2, 6, 12, 30, 72, 182, \ldots ,$$

each of which may be factored into two consecutive integers:

$$1\cdot 2, 0\cdot 1, 1\cdot 2, 1\cdot 2, 2\cdot 3, 3\cdot 4, 5\cdot 6, 8\cdot 9, 13\cdot 14, \ldots .$$

The first factors make up the Fibonacci sequence F_n defined by $F_{-1} = 1$, $F_0 = 0$, $F_n = F_{n-1} + F_{n-2}$ for $n > 0$.

Surprisingly, however, this marvelous rule

$$3\binom{m+1}{0}_2 - \binom{m+2}{0}_2 = F_m(F_m + 1), \qquad -1 \le m \le 7, \tag{1.12}$$

is *false* for $m > 7$. In order to understand (1.12) we define

$$E_m(a, b) = \sum_{\lambda=-\infty}^{\infty} \left(\binom{m}{10\lambda + a}_2 - \binom{m}{10\lambda + b}_2 \right). \tag{1.13}$$

As part of Theorem 2.1, we show that

$$2E_{m+1}(0, 1) = F_m(F_m + 1), \tag{1.14}$$

from which (1.12) follows by inspection (see Corollary 2.2). However our real concern lies with the q-analogs of the various formulae appearing in Theorem 2.1. What develops in this instance is not another proof of the Rogers-Ramanujan identities, but rather new identities for the even and odd parts of the Rogers-Ramanujan functions. To be more explicit, we recall the first Rogers-Ramanujan function:

$$G(q) = \sum_{n=0}^{\infty} \frac{q^{n^2}}{(1-q)(1-q^2)\cdots(1-q^n)} = \prod_{n=0}^{\infty} \frac{1}{(1-q^{5n+1})(1-q^{5n+4})}. \tag{1.15}$$

Then for the even part of $G(q)$ we have (see (6.22))

$$\frac{1}{2}(G(q)+G(-q)) = \frac{\sum_{\lambda=-\infty}^{\infty}(q^{60\lambda^2-4\lambda} - q^{60\lambda^2+44\lambda+8})}{\prod_{n=1}^{\infty}(1-q^{2n})}, \tag{1.16}$$

and for the odd part we have (see (6.22))

$$\frac{1}{2}(G(q)-G(-q)) = \frac{\sum_{\lambda=-\infty}^{\infty}(q^{60\lambda^2+16\lambda+1} - q^{60\lambda^2+64\lambda+17})}{\prod_{n=1}^{\infty}(1-q^{2n})}. \tag{1.17}$$

These identities, which can be proved by other means, are limiting cases of the following Schur polynomial identities (see (6.16) and (6.17)):

$$\begin{aligned} &\frac{1}{2}(G_{2m+1}(q^{1/2}) + G_{2m+1}(-q^{1/2})) \\ &\quad = \sum_{\lambda=-\infty}^{\infty} q^{30\lambda^2-2\lambda} \binom{m;\ 10\lambda;\ q}{10\lambda}_2 - \sum_{\lambda=-\infty}^{\infty} q^{30\lambda^2+22\lambda+4} \binom{m;\ 10\lambda+4;\ q}{10\lambda+4}_2, \end{aligned} \tag{1.18}$$

$$\begin{aligned} &\frac{q^{-1/2}}{2}(G_{2m+1}(q^{1/2}) - G_{2m+1}(-q^{1/2})) \\ &\quad = \sum_{\lambda=-\infty}^{\infty} q^{30\lambda^2+8\lambda} \binom{m;\ 10\lambda+1;\ q}{10\lambda+1}_2 - \sum_{\lambda=-\infty}^{\infty} q^{30\lambda^2+32\lambda+8} \binom{m;\ 10\lambda+5;\ q}{10\lambda+5}_2. \end{aligned} \tag{1.19}$$

By investigating Euler's Exemplum Memorabile Inductionis Fallacis we are led to results which raise more questions than they answer. In particular, the legacy of Schur's polynomials [1, 4, 8, 9, 10] suggests that q-trinomial coefficients can provide combinatorial explanations of (1.7), (1.18), and (1.19).

Just as Schur [13, equation (30)] proved the simpler

$$\sum_{\mu=-\infty}^{\infty} (-1)^{\mu} q^{\mu(3\mu+1)/2} \begin{bmatrix} n \\ \lfloor \frac{n-3\mu}{2} \rfloor \end{bmatrix} = 1 \tag{1.20}$$

(a finite version of Euler's Pentagonal Number Theorem), we shall also (in §§3 and 5) look at the much simpler case of (1.13) in which the 10 is replaced by 6.

In §2 we shall examine $E_m(a, b)$ and Euler's remarkable example of a misleading induction. Section 3 briefly treats the sequences of (1.13) when 10 is replaced by 6. In §4 we provide the necessary background and new lemmas for the q-trinomial coefficients. In §5 we provide q-analogs of the results in §3. In §6 we treat q-analogs of the $E_m(a, b)$; this yields, among other things, (1.18) and (1.19). Our work raises new questions and these are briefly described in the conclusion.

I wish to thank Ray and Christine Ayoub for translating Euler's Exemplum Memorabile Inductionis Fallacis for me.

2. The $E_m(a, b)$ and Euler's misleading induction

Actually the only hard part of understanding (1.12) is writing down (1.13). Once that has been done the following theorem is quite easy.

Theorem 2.1.

$$2E_m(1, 2) = 2E_{m-1}(0, 3) = 2E_{m+1}(0, 1) = F_m(F_m + 1), \tag{2.1}$$

$$E_m(1, 4) = E_{m+1}(2, 3) = F_{m+1}F_m, \tag{2.2}$$

$$2E_m(3, 4) = 2E_{m-1}(2, 5) = 2E_{m+1}(4, 5) = F_m(F_m - 1), \tag{2.3}$$

$$2E_m(1, 3) = F_{2m} + F_m, \tag{2.4}$$

$$2E_m(2, 4) = F_{2m} - F_m, \tag{2.5}$$

$$2E_m(1, 5) = F_{2m+1} - F_{m-1}, \tag{2.6}$$

$$2E_m(0, 4) = F_{2m+1} + F_{m-1}, \tag{2.7}$$

$$2E_m(0, 2) = F_{2m-1} + F_{m+1}, \tag{2.8}$$

$$2E_m(3, 5) = F_{2m-1} - F_{m+1}, \tag{2.9}$$

$$E_m(0, 5) = F_{2m-1} + F_mF_{m-1}. \tag{2.10}$$

Remark. This result gives us $E_m(a, b)$ for every pair of integers a, b because clearly

$$E_m(a, b) = -E_m(b, a), \qquad E_m(10r + a, 10s + b) = E_m(a, b),$$

and

$$E_m(10 - a, b) = E_m(a, b) = E_m(a, 10 - b).$$

Proof. We recall the Euler-Binet formula for the Fibonacci numbers [12, p. 54]

$$F_n = \frac{\phi^n - \bar{\phi}^n}{\sqrt{5}}, \tag{2.11}$$

where $\phi = (1 + \sqrt{5})/2$ and $\bar{\phi} = (1 - \sqrt{5})/2$ are roots of the equation

$$x^2 - x - 1 = 0. \tag{2.12}$$

Thus each extreme right-hand entry of (2.1)–(2.10) is of the form

$$A(\phi^2)^m + B(\bar{\phi}^2)^m + C\phi^m + D\bar{\phi}^m + E(-1)^m. \tag{2.13}$$

Therefore each of these sequences satisfies a fifth-order recurrence whose auxiliary equation [15, p. 153] is

$$(x^2 - x - 1)(x^2 - 3x + 1)(x + 1) = x^5 - 3x^4 - x^3 + 5x^2 + x - 1, \tag{2.14}$$

since ϕ^2 and $\bar{\phi}^2$ are the roots of $x^2 - 3x + 1 = 0$.

On the other hand, if $\xi = e^{2\pi i/10}$, then

(2.15)

$$\begin{aligned}
E_m(a, b) &= \frac{1}{10}\sum_{\mu=-\infty}^{\infty}\sum_{j=0}^{9}\binom{m}{\mu}_2(\xi^{j(\mu-a)} - \xi^{j(\mu-b)}) \\
&= \frac{1}{10}\sum_{j=0}^{9}(1+\xi^j+\xi^{-j})^m(\xi^{-aj}-\xi^{-bj}) \\
&= \frac{1}{5}\sum_{j=1}^{4}(1+2\cos(\pi j/5))^m(\cos(\pi aj/5)-\cos(\pi bj/5)) \\
&\qquad + \frac{1}{10}(-1)^m((-1)^a-(-1)^b) \\
&= \frac{1}{5}\sum_{j=1}^{2}\{(1+2\cos(\pi j/5))^m - (1-2\cos(\pi j/5))^m\} \\
&\qquad \times (\cos(\pi aj/5)-\cos(\pi bj/5)) + \frac{1}{10}(-1)^m((-1)^a-(-1)^b) \\
&= \alpha(\phi^2)^m + \beta(\overline{\phi}^2)^m + \gamma\phi^m + \delta\overline{\phi}^m + \varepsilon(-1)^m,
\end{aligned}$$

since $\phi^2 = 1 + 2\cos(\pi/5)$, $\overline{\phi}^2 = 1 - 2\cos(2\pi/5)$, $\phi = 1 + 2\cos(2\pi/5)$, and $\overline{\phi} = 1 - 2\cos(\pi/5)$.

Hence by (2.13) and (2.15) we see that every entry in (2.1)–(2.10) satisfies the fifth-order recurrence whose auxiliary polynomial is (2.14). Therefore our theorem follows by mathematical induction provided all the assertions are true for $m \le 6$, and this is easily verified by inspection. □

Corollary 2.2. *For* $m \le 7$, *Equation* (1.14) *holds.*

Proof. From (1.1) we see that

$$\binom{m}{j}_2 = \binom{m-1}{j-1}_2 + \binom{m-1}{j}_2 + \binom{m-1}{j+1}_2. \tag{2.16}$$

Hence

(2.17)

$$\begin{aligned}
3\binom{m+1}{0}_2 - \binom{m+2}{0}_2 &= 3\binom{m+1}{0}_2 - \binom{m+1}{-1}_2 - \binom{m+1}{0}_2 - \binom{m+1}{1}_2 \\
&= 2\binom{m+1}{0}_2 - 2\binom{m+1}{1}_2 \\
&= 2E_{m+1}(0, 1) \quad (\text{for } m \le 7) \\
&= F_m(F_m + 1) \quad (\text{by } (2.1)). \quad \square
\end{aligned}$$

Obviously similar corollaries can be produced from all the other assertions in Theorem 2.1; however none seems quite this dramatic.

Also, the reader may wonder why we have treated differences of two trinomial coefficients in our sums. The following formula (proved exactly like Theorem 2.1) makes clear how much simpler the differencing results turn out:

$$\text{(2.18)}\quad \sum_{\lambda=-\infty}^{\infty} \binom{m}{10\lambda+a}_2 = \begin{cases} \frac{1}{2}F_m^2 + \frac{1}{2}F_{m-1}^2 + \frac{1}{2}F_{m-1}F_m + \frac{2}{5}F_{m-1} + \frac{1}{5}F_m + \frac{3^m}{10} & \text{if } a = 0, \\ \frac{1}{2}F_{m-1}F_m + \frac{1}{2}F_m^2 - \frac{1}{10}F_{m-1} + \frac{1}{5}F_m + \frac{3^m}{10} & \text{if } a = 1, \\ \frac{1}{2}F_{m-1}F_m - \frac{1}{10}F_{m-1} - \frac{3}{10}F_m + \frac{3^m}{10} & \text{if } a = 2, \\ -\frac{1}{2}F_{m-1}F_m - \frac{1}{10}F_{m-1} - \frac{3}{10}F_m + \frac{3^m}{10} & \text{if } a = 3, \\ -\frac{1}{2}F_{m-1}F_m - \frac{1}{2}F_m^2 - \frac{1}{10}F_{m-1} + \frac{1}{5}F_m + \frac{3^m}{10} & \text{if } a = 4, \\ -\frac{1}{2}F_{m-1}^2 - \frac{1}{2}F_{m-1}F_m - \frac{1}{2}F_m^2 + \frac{2}{5}F_{m-1} + \frac{1}{5}F_m + \frac{3^m}{10} & \text{if } a = 5. \end{cases}$$

Additionally, we have been unable to find any reasonable q-analogs of any part of (2.18).

3. Recurrences related to 6

Obviously the sequences defined by (1.13) are special cases of

$$\text{(3.1)}\quad \mathscr{E}_m(k\,;a\,,b) = \sum_{\lambda=-\infty}^{\infty} \left(\binom{m}{k\lambda+a}_2 - \binom{m}{k\lambda+b}_2\right).$$

The case $k = 7$ is considered extensively in [5]. Elsewhere we shall treat $k = 4$ and 5, which have somewhat different q-analogs. For later work in this paper, we shall find it useful to treat the case $k = 6$.

Theorem 3.1.

$$\text{(3.2)}\quad \begin{aligned} \mathscr{E}_{m+1}(6\,;0\,,1) = \mathscr{E}_{m+1}(6\,;0\,,1) = \mathscr{E}_{m+1}(6\,;2\,,3) = \mathscr{E}_m(6\,;1\,,2) \\ = \mathscr{E}_{m-1}(6\,;0\,,3) = (2^m-(-1)^m)/3, \end{aligned}$$

$$\text{(3.3)}\quad \mathscr{E}_m(6\,;0\,,2) = \mathscr{E}_m(6\,;\,,1\,,3) = 2^{m-1}.$$

Proof. Now each entry satisfies a recurrence whose auxiliary polynomial is $x^2 - x - 2 = 0$. The rest is as before, only easier. □

4. q-trinomial coefficients

In [5], R. J. Baxter and the author studied six q-analogs of $\binom{m}{j}_2$. To our surprise, there are intricate facts and relationships among the q-analogs which we found quite refractory. We shall require a selection of these results including three q-analogs. First, the definitions [5, Equations (2.7), (2.8), (2.9), (2.13), and (2.14)]:

$$\text{(4.1)}\quad \binom{m\,;B\,;q}{A}_2 = \sum_{j\geq 0} \frac{q^{j(j+B)}(q)_m}{(q)_j(q)_{j+A}(q)_{m-2j-A}},$$

$$\text{(4.2)}\quad T_0(m\,,A\,,q) = \sum_{j=0}^{m}(-1)^j \begin{bmatrix} m \\ j \end{bmatrix}_{q^2} \begin{bmatrix} 2m-2j \\ m-A-j \end{bmatrix},$$

and

$$T_1(m, A, q) = \sum_{j=0}^{m}(-q)^j \begin{bmatrix} m \\ j \end{bmatrix}_{q^2} \begin{bmatrix} 2m-2j \\ m-A-j \end{bmatrix}, \tag{4.3}$$

where

$$\begin{bmatrix} A \\ B \end{bmatrix}_q = \begin{bmatrix} A \\ B \end{bmatrix} = \frac{(q)_A}{(q)_B(q)_{A-B}} \tag{4.4}$$

and

$$(A; q)_n = (A)_n = (1-A)(1-Aq)\cdots(1-Aq^{n-1}). \tag{4.5}$$

These are indeed q-analogs of the trinomial coefficients since [5, p. 299]

$$\binom{m; B; 1}{A}_2 = T_0(m, A, 1) = T_1(m, A, 1) = \binom{m}{A}_2. \tag{4.6}$$

There are easily established symmetry relations [5, p. 299, Equation (2.15)]

$$\binom{m; B; q}{-A}_2 = q^{A(A+B)}\binom{m; B+2A; q}{A}_2, \tag{4.7}$$

while, immediately from (4.4), we see that

$$T_i(m, A, q) = T_i(m, -A, q) \qquad (i = 0 \text{ or } 1). \tag{4.8}$$

We shall require only two of the several Pascal triangle recurrences [5, Equations (2.17) and (2.19)]:

$$\begin{aligned} T_1(m, A, q) = {}& T_1(m-1, A, q) + q^{m+A}T_0(m-1, A+1, q) \\ & + q^{m-A}T_0(m-1, A-1, q), \end{aligned} \tag{4.9}$$

$$\begin{aligned} T_0(m, A, q) = {}& T_0(m-1, A-1, q) + q^{m+A}T_1(m-1, A, q) \\ & + q^{2m+2A}T_0(m-1, A+1, q). \end{aligned} \tag{4.10}$$

Next we have an identity which reduces to a tautology when $q = 1$ [5, p. 301, Equation (2.20)]:

$$\begin{aligned} & T_1(m, A, q) - q^{m-A}T_0(m, A, q) - T_1(m, A+1, q) \\ & \quad + q^{m+A+1}T_0(m, A+1, q) = 0, \end{aligned} \tag{4.11}$$

and two important identities [5, p. 305, Equations (2.33_i) and (2.34_i)]:

$$\binom{m; A; q^2}{A}_2 = q^{m^2-A^2}T_0(m, A, q^{-1}) \tag{4.12}$$

and

$$\binom{m; A-1; q^2}{A}_2 = q^{m(m-1)-A(A-1)}T_1(m, A, q^{-1}). \tag{4.13}$$

Also, we need to consider a sum of the $T_0(m, A, q)$, so we define

$$U(m, A, q) = T_0(m, A, q) + T_0(m, A+1, q). \tag{4.14}$$

Limiting values of our q-analogs are also needed [5, pp. 309–310, Equations (2.48), (2.53), (2.54)]:

$$\lim_{m\to\infty}\binom{m;A;q}{A}_2=\frac{1}{(q)_\infty},\tag{4.15}$$

$$\lim_{m\to\infty}U(m,A,q)=\frac{(-q;q^2)_\infty}{(q^2;q^2)_\infty}.\tag{4.16}$$

In addition, two new lemmas are needed.

Lemma 4.1. *For* $m\geq 1$,

(4.17)

$$U(m,A,q)=(1+q^{2m-1})U(m-1,A,q)$$
$$+q^{m-A}T_1(m-1,A-1,q)+q^{m+A+1}T_1(m-1,A+2,q).$$

Proof.

$$\begin{aligned}
U(m,A,q)&-(1+q^{2m-1})U(m-1,A,q)\\
&=T_0(m,A,q)+T_0(m,A+1,q)\\
&\quad-(1+q^{2m-1})(T_0(m-1,A,q)+T_0(m-1,A+1,q))\\
&=T_0(m,A,q)-T_0(m-1,A+1,q)\\
&\quad+T_0(m,A+1,q)-T_0(m-1,A,q)\\
&\quad-q^{2m-1}T_0(m-1,A+1,q)-q^{2m-1}T_0(m-1,A,q)\\
&=(q^{m-A}T_1(m-1,A,q)+q^{2m-2A}T_0(m-1,A-1,q))\\
&\quad+(q^{m+A+1}T_1(m-1,A+1,q)+q^{2m+2A+2}T_0(m-1,A+2,q))\\
&\quad-q^{2m-1}T_0(m-1,A+1,q)-q^{2m-1}T_0(m-1,A,q)
\end{aligned}$$

(by two applications of equation (4.10)—with A replaced by $-A$ in the first and by $A+1$ in the second)

$$\begin{aligned}
&=q^{m-A}(T_1(m-1,A,q)+q^{m-A}T_0(m-1,A-1,q)\\
&\qquad\qquad-q^{m+A-1}T_0(m-1,A,q))\\
&\quad+q^{m+A+1}(T_1(m-1,A+1,q)+q^{m+A+1}T_0(m-1A+2,q)\\
&\qquad\qquad-q^{m-A-2}T_0(m-1,A+1,q))\\
&=q^{m-A}T_1(m-1,A-1,q)+q^{m+A+1}T_1(m-1,A+2,q)
\end{aligned}$$

(by two applications of (4.11)—with A replaced by $A-1$ in the first and by $A+1$ in the second, and both with m replaced by $m-1$). □

Lemma 4.2. *For* $m\geq 2$,

(4.18)

$$U(m,A,q)=(1+q+q^{2m-1})U(m-1,A,q)-qU(m-2,A,q)$$
$$+q^{2m-2A}T_0(m-1,A-2,q)+q^{2m+2A+2}T_0(m-2,A+3,q).$$

Proof. By Lemma 4.1,

$$\begin{aligned}
&U(m, A, q) - (1+q+q^{2m-1})U(m-1, A, q) + qU(m-2, A, q) \\
&= q^{m-A}T_1(m-1, A-1, q) + q^{m+A+1}T_1(m-1, A+2, q) \\
&\quad - qU(m-1, A, q) + qU(m-2, A, q) \\
&= q^{m-A}(T_1(m-2, A-1, q) + q^{m+A+2}T_0(m-2, A, q) \\
&\qquad + q^{m-A}T_0(m-2, A-2, q)) \\
&\quad + q^{m+A+1}(T_1(m-2, A+2, q) + q^{m+A+1}T_0(m-2, A+3, q) \\
&\qquad + q^{m-A-3}T_0(m-2, A+1, q)) \\
&\quad - q(T_0(m-2, A+1, q) + q^{m-1-A}T_1(m-2, A, q) \\
&\qquad + q^{2m-2A-2}T_0(m-2, A-1, q)) \\
&\quad - q(T_0(m-2, A, q) + q^{m+A}T_1(m-2, A+1, q) \\
&\qquad + q^{2m+2A}T_0(m-2, A+2, q)) \\
&\quad + qT_0(m-2, A, q) + qT_0(m-2, A+1, q)
\end{aligned}$$

(by two applications of (4.10) and two of (4.9))

$$\begin{aligned}
&= q^{2m-2A}T_0(m-2, A-2, q) + q^{2m+2A+2}T_0(m-2, A+3, q) \\
&\quad + q^{m-A}(T_1(m-2, A-1, q) - T_1(m-2, A, q) \\
&\qquad - q^{m-A-1}T_0(m-2, A-1, q) + q^{m+A-2}T_0(m-2, A, q)) \\
&\quad + q^{m+A+1}(T_1(m-2, A+2, q) - T_1(m-2, A+1, q) \\
&\qquad + q^{m-A-3}T_0(m-2, A+1, q) - q^{m+A}T_0(m-2, A+2, q)) \\
&\hspace{6cm}\text{(by cancelling and regrouping)} \\
&= q^{2m-2A}T_0(m-2, A-2, q) + q^{2m+2A+2}T_0(m-2, A+3, q)
\end{aligned}$$

(by two applications of (4.11)—with A replaced by $A-1$ in the first and by $-A-2$ in the second, and both with m replaced by $m-2$). □

5. The q-analog of §3

In this section we shall prove an identity, Theorem 5.1, which is, in fact, the q-analog of (3.4). From this we shall also deduce identities (83) and (86) in L. J. Slater's compendium [14] of Rogers-Ramanujan type identities.

Theorem 5.1.

$$\begin{aligned}
\prod_{j=1}^{m}(1+q^{2j-1}) &= \sum_{\lambda=-\infty}^{\infty} q^{12\lambda^2+2\lambda}(T_0(m, 6\lambda, q) + T_0(m, 6\lambda+1, q)) \\
&\quad - \sum_{\lambda=-\infty}^{\infty} q^{12\lambda^2+10\lambda+2}(T_0(m, 6\lambda+2, q) + T_0(m, 6\lambda+3, q)).
\end{aligned}$$

Proof. Let us denote the left-hand side of our theorem by L_m and the right-hand side of R_m. Clearly L_m is uniquely defined by the conditions

$$L_m = \begin{cases} 1 & \text{if } m = 0, \\ (1+q^{2m-1})L_{m-1} & \text{if } m > 0. \end{cases} \tag{5.1}$$

On the other hand, $R_0 = 1$ and

$$\begin{aligned} R_m &= \sum_{\lambda=-\infty}^{\infty} q^{12\lambda^2+2\lambda} U(m, 6\lambda, q) - \sum_{\lambda=-\infty}^{\infty} q^{12\lambda^2+10\lambda+2} U(m, 6\lambda+2, q) \\ &= (1+q^{2m-1})R_{m-1} + \sum_{\lambda=-\infty}^{\infty} q^{12\lambda^2+2\lambda+m-6\lambda} T_1(m-1, 6\lambda-1, q) \\ &\quad + \sum_{\lambda=-\infty}^{\infty} q^{12\lambda^2+2\lambda+m+6\lambda+1} T_1(m-1, 6\lambda+2, q) \\ &\quad - \sum_{\lambda=-\infty}^{\infty} q^{12\lambda^2+10\lambda+2+m-6\lambda-2} T_1(m-1, 6\lambda+1, q) \\ &\quad - \sum_{\lambda=-\infty}^{\infty} q^{12\lambda^2+10\lambda+2+m+6\lambda+3} T_1(m-1, 6\lambda+4, q) \\ &= (1+q^{2m-1})R_{m-1} \end{aligned}$$

since the first sum (with λ replaced by $-\lambda$) cancels the third, and the fourth sum (with λ replaced by $-\lambda-1$) cancels the second. □

Corollary 5.2.

$$\sum_{j\ge 0} q^{2j^2} \begin{bmatrix} m \\ 2j \end{bmatrix} = \sum_{\lambda=-\infty}^{\infty} q^{12\lambda^2-\lambda} \binom{m;\, 6\lambda;\, q}{6\lambda}_2 - \sum_{\lambda=-\infty}^{\infty} q^{12\lambda^2+7\lambda+1} \binom{m;\, 6\lambda+2;\, q}{6\lambda+2}_2, \tag{5.2}$$

$$\sum_{j\ge 0} q^{2j^2+2j} \begin{bmatrix} m \\ 2j+1 \end{bmatrix} = \sum_{\lambda=-\infty}^{\infty} q^{12\lambda^2+5\lambda} \binom{m;\, 6\lambda+1;\, q}{6\lambda+1}_2 - \sum_{\lambda=-\infty}^{\infty} q^{12\lambda^2+13\lambda+3} \binom{m;\, 6\lambda+3;\, q}{6\lambda+3}_2. \tag{5.3}$$

Proof. Equation (5.2) is the even portion of Theorem 5.1 once it has had q replaced by q^{-1} and the result multiplied by q^{m^2}; equation (5.3) is the odd portion. □

Finally we note that if $m \to \infty$ in (5.2), then we obtain (using (4.15) and the Jacobi triple product [2, p. 21])

$$\sum_{j=0}^{\infty} \frac{q^{2j^2}}{(q)_{2j}} = \prod_{n=1}^{\infty}(1+q^{24n-11})(1+q^{24n-13})(1-q^{24n}) - q\sum_{n=1}^{\infty}(1+q^{24n-5})(1+q^{24n-19})(1-q^{24n}),$$

which is Slater's equation (83) [14, p. 160], and from (5.3) in the same way we obtain

$$\sum_{j=0}^{\infty} \frac{q^{2j^2+2j}}{(q)_{2j+1}} = \prod_{n=1}^{\infty}(1+q^{24n-7})(1+q^{24n-17})(1-q^{24n}) - q^2\prod_{n=1}^{\infty}(1+q^{24n-1})(1+q^{24n-23})(1-q^{24n}),$$

which is Slater's equation (86) [14, p. 161].

6. Results related to the Rogers-Ramanujan identities

In order to provide q-analogs for some of the results in §2, we require the following polynomials:

$$G_n(q) = \begin{cases} 0 & \text{if } n=0, \\ 1 & \text{if } n=1, \\ G_{n-1}(q)+q^{n-2}G_{n-2} & \text{if } n \geq 2, \end{cases} \tag{6.1}$$

and

$$H_n(q) = \begin{cases} 0 & \text{if } n=0, \\ 1 & \text{if } n=1, \\ H_{n-1}(q)+q^{n-1}H_{n-2}(q) & \text{if } n \geq 2. \end{cases} \tag{6.2}$$

Obviously

$$G_n(1) = H_n(1) = F_n. \tag{6.3}$$

Furthermore, it is a simple exercise in mathematical induction to prove that

$$G_{2n}(-1) = H_{2n}(-1) = F_n \tag{6.4}$$

and

$$G_{2n-3}(-1) = H_{2n+3}(-1) = F_n. \tag{6.5}$$

In [2, p. 50, Example 10], it is shown that

$$G_n(q) = \sum_{j\geq 0} q^{j^2} \begin{bmatrix} n-1-j \\ j \end{bmatrix} \tag{6.6}$$

and

$$H_n(q) = \sum_{j\geq 0} q^{j^2+j} \begin{bmatrix} n-1-j \\ j \end{bmatrix}. \tag{6.7}$$

Reversing the order of summation in (6.6) and (6.7) and replacing q by q^{-1} we obtain related reciprocal polynomials:

$$R_m(q) \equiv q^{m^2} G_{2m+1}(q^{-1}) = \sum_{j=0}^{m} q^{j^2} \begin{bmatrix} m+j \\ 2j \end{bmatrix} \tag{6.8}$$

and

$$S_m(q) = q^{m^2-1} H_{2m}(q^{-1}) = \sum_{j=0}^{m} q^{j^2+2j} \begin{bmatrix} m+j \\ 2j+1 \end{bmatrix}. \tag{6.9}$$

Theorem 6.1.

$$\begin{aligned} R_m(q) = &\sum_{\lambda=-\infty}^{\infty} q^{40\lambda^2+4\lambda} U(m,\, 10\lambda,\, q) \\ &- \sum_{\lambda=-\infty}^{\infty} q^{40\lambda^2+36\lambda+8} U(m,\, 10\lambda+4,\, q), \end{aligned} \tag{6.10}$$

$$\begin{aligned} S_m(q) = &\sum_{\lambda=-\infty}^{\infty} q^{40\lambda^2+12\lambda} U(m,\, 10\lambda+1,\, q) \\ &- \sum_{\lambda=-\infty} q^{40\lambda^2+28\lambda+4} U(m,\, 10\lambda+3,\, q). \end{aligned} \tag{6.11}$$

Proof. From (6.8) and (6.1) we can derive a defining second-order recurrence for $R_m(q)$:

$$R_m(q) = \begin{cases} 1 & \text{if } m = 0, \\ 1+q & \text{if } m = 1, \\ (1+q+q^{2m-1})R_{m-1}(q) - qR_{m-2}(q) & \text{if } m \geq 2. \end{cases} \tag{6.12}$$

Hence to prove (6.10) we only need show that the right-hand side satisfies the same defining recurrence. Inspection shows the validity of (6.10) for $m = 0$ and 1. So if $\rho_m(q)$ denotes the right-hand side of (6.10), then by Lemma 4.2 we obtain

$$\begin{aligned} &\rho_m(q) - (1+q+q^{2m-1})\rho_{m-1}(q) + q\rho_{m-2}(q) \\ &= \sum_{\lambda=-\infty}^{\infty} a^{40\lambda^2+4\lambda}(q^{2m-20\lambda} T_0(m-2,\, 10\lambda-2,\, q) \\ &\qquad + q^{2m+20\lambda+2} T_0(m-2,\, 10\lambda+3,\, q)) \\ &\quad - \sum_{\lambda=-\infty}^{\infty} q^{40\lambda^2+36\lambda+8}(q^{2m-20\lambda-8} T_0(m-2,\, 10\lambda+2,\, q) \\ &\qquad + q^{2m+20\lambda+10} T_0(m-2,\, 10\lambda+7,\, q)) \\ &= 0, \end{aligned} \tag{6.13}$$

since the first portion of the first sum (with λ replaced by $-\lambda$) cancels the first portion of the second sum, and the second portion of the second sum (with λ replaced by $-\lambda-1$) cancels the second portion of the first sum.

Thus $\rho_m(q)$ satisfies (6.12) and consequently (6.10) is valid.

In the same way, it is easily verified that $S_m(q)$ satisfies

$$S_m(q)=\begin{cases}0 & \text{if } m=0\,,\\ 1 & \text{if } m=1\,,\\ (1+q+q^{22m-1})S_{m-1}(q)-qS_{m-2}(q) & \text{if } m\geq 2\,.\end{cases} \tag{6.14}$$

As before, we let $\sigma_m(q)$ denote the right-hand side of (6.11), and we need only show that $\sigma_m(q)$ satisfies (6.14). Now by Lemma 4.2 we obtain

$$\begin{aligned}\sigma_m(q)&-(1+q+q^{2m-1})\sigma_{m-1}(q)+q\sigma_{m-2}(q)\\ &=\sum_{\lambda=-\infty}^{\infty}q^{40\lambda^2+12\lambda}(q^{2m-20\lambda-2}T_0(m-2\,,\,10\lambda-1\,,\,q)\\ &\qquad\qquad+q^{2m+20\lambda+4}T_0(m-2\,,\,10\lambda+4\,,\,q))\\ &\quad-\sum_{\lambda=-\infty}^{\infty}q^{40\lambda^2+28\lambda+4}(q^{2m-20\lambda-6}T_0(m-2\,,\,10\lambda+1\,,\,q)\\ &\qquad\qquad+q^{2m+20\lambda+8}T_0(m-2\,,\,10\lambda+6\,,\,q))\\ &=0\,,\end{aligned} \tag{6.15}$$

since the first portion of the first sum (with λ replaced by $-\lambda$) cancels the first portion of the second sum, and the second sum (with λ replaced by $-\lambda-1$) cancels the second portion of the first sum. □

Corollary 6.2.

(6.16)

$$\begin{aligned}&\frac{1}{2}(G_{2m+1}(q^{1/2}+G_{2m+1}(-q^{1/2}))\\ &\quad=\sum_{\lambda=-\infty}^{\infty}q^{30\lambda^2-2\lambda}\begin{pmatrix}m\,;\,10\lambda\,;\,q\\ 10\lambda\end{pmatrix}_2-\sum_{\lambda=-\infty}^{\infty}q^{30\lambda^2+22\lambda+4}\begin{pmatrix}m\,;\,10\lambda+4\,;\,q\\ 10\lambda+4\end{pmatrix}_2,\end{aligned}$$

(6.17)

$$\begin{aligned}&\frac{q^{-1/2}}{2}(G_{2m+1}(q^{1/2})-G_{2m+1}(-q^{1/2}))\\ &\quad=\sum_{\lambda=-\infty}^{\infty}q^{30\lambda^2+8\lambda}\begin{pmatrix}m\,;\,10\lambda+1\,;\,q\\ 10\lambda+1\end{pmatrix}_2-\sum_{\lambda=-\infty}^{\infty}q^{30\lambda^2+32\lambda+8}\begin{pmatrix}m\,;\,10\lambda+5\,;\,q\\ 10\lambda+5\end{pmatrix}_2,\end{aligned}$$

(6.18)

$$\begin{aligned}&\frac{1}{2}(H_{2m}(q^{1/2})+H_{2m}(-q^{1/2})\\ &\quad=\sum_{\lambda=-\infty}^{\infty}q^{30\lambda^2+4\lambda}\begin{pmatrix}m\,;\,10\lambda+1\,;\,q\\ 10\lambda+1\end{pmatrix}_2-\sum_{\lambda=-\infty}^{\infty}q^{30\lambda^2+16\lambda+2}\begin{pmatrix}m\,;\,10\lambda+3\,;\,q\\ 10\lambda+3\end{pmatrix}_2,\end{aligned}$$

$$\begin{aligned}\frac{q^{-1/2}}{2}(H_{2m}(q^{1/2}) - H_{2m}(-q^{1/2})) &= \sum_{\lambda=-\infty}^{\infty} q^{30\lambda^2+14\lambda+1}\binom{m;\ 10\lambda+2;\ q}{10\lambda+2}_2 \\ &- \sum_{\lambda=-\infty}^{\infty} q^{30\lambda^2+26\lambda+5}\binom{m;\ 10\lambda+4;\ q}{10\lambda+4}_2 .\end{aligned} \tag{6.19}$$

Proof. These four identities arise by replacing q by q^{-1} in (6.10) and (6.11) and then extracting even functions and odd functions. □

Note that by (6.4) and (6.5) we see that Corollary 6.2 provides q-analogs of (2.7), (2.6), (2.4), and (2.5) respectively.

Finally we note that if $m \to \infty$ in (6.10), then we obtain (using (4.16) and Jacobi's triple product [2, p. 21])

$$\sum_{j=0}^{\infty} \frac{q^{j^2}}{(q)_{2j}} = \prod_{n=1}^{\infty} \frac{(1+q^{2n-1})(1-q^{20n-12})(1-q^{20n-8})(1-q^{20n})}{(1-q^{2n})}, \tag{6.20}$$

which is Slater's equation (79) [14, p. 160]. If $m \to \infty$ in (6.11) we obtain (using (4.16) and Jacobi's triple product [2, p. 21])

$$\sum_{j=0}^{\infty} \frac{q^{j^2+2j}}{(q)_{2j+1}} = \prod_{n=1}^{\infty} \frac{(1+q^{2n-1})(1-q^{20n-4})(1-q^{20n-16})(1-q^{20n})}{(1-q^{2n})}, \tag{6.21}$$

which is equivalent to Slater's equation (96) [14, p. 162].

To conclude this section we replace q by q^2 in both (6.16) and (6.17), and then multiply the second resulting equation by q and add it to the first; taking the limit as $m \to \infty$ we obtain

$$\begin{aligned}\sum_{j=0}^{\infty} \frac{q^{j^2}}{(q)_j} = \frac{1}{(q^2;q^2)_\infty}\Bigg\{ &\sum_{\lambda=-\infty}^{\infty} (q^{60\lambda^2-4\lambda} - q^{60\lambda^2+44\lambda+8}) \\ &+ q\sum_{\lambda=-\infty}^{\infty} (q^{60\lambda^2+16\lambda} - q^{60\lambda^2+64\lambda+16})\Bigg\}.\end{aligned} \tag{6.22}$$

This is the Rogers-Ramanujan series [2, p. 104]; however the right-hand side is split into even odd parts with each part a simple theta function quotient. This dissection is related clearly to Watson's representation of this series wherein even and odd parts can be readily discerned [16, p. 64, line 4].

Exactly the same treatment of (6.18) and (6.19) yields for the second Roger-Ramanujan series:

$$\sum_{j=0}^{\infty}\frac{q^{j^2+j}}{(q)_j}=\frac{1}{(q^2;q^2)_\infty}\left\{\sum_{\lambda=-\infty}^{\infty}(q^{60\lambda^2+8\lambda}-q^{60\lambda^2+32\lambda+4})+q\sum_{\lambda=-\infty}^{\infty}(q^{60\lambda^2+28\lambda+2}-q^{60\lambda^2+52\lambda+10})\right\}. \tag{6.23}$$

7. Conclusion

Having unravelled Euler's "remarkable example of misleading induction", we are left with deeper mysteries. For example, what is a combinatorial explanation of Theorems 5.1 or 6.1, or Corollary 6.2? Surely before we can answer such a question we must find a full combinatorial explanation of the q-trinomial coefficients themselves. Standard partition-theoretic arguments [2, Chapter 2] show that

$$\binom{m;0;q}{0}=\sum_{j\geq 0}q^{j^2}\begin{bmatrix}n\\j\end{bmatrix}\begin{bmatrix}n-j\\j\end{bmatrix} \tag{7.1}$$

is the generating function for all partitions in which the largest part is $\leq n$ and the number of parts plus the edge of the Durfee square is also $\leq n$. This interpretation can easily be generalized to $\binom{m;A;q}{A}_2$; however, it has yet to reveal anything significant about the various surprising identities involving q-trinomial coefficients. Indeed the combinatorics of the identities in §4 is still unknown.

References

1. G. E. Andrews, *Sieves in the theory of partitions*, Amer. J. Math. **94** (1972), 1214–1230.
2. ——, *The theory of partitions*, Encyclopedia of Mathematics and Its Applications (G.-C. Rota, ed.), vol. 2, Addison-Wesley, Reading, MA, 1976; reissued by Cambridge Univ. Press, London and New York, 1985.
3. ——, *The hard-hexagon model and Rogers-Ramanujan type identities*, Proc. Nat. Acad. Sci. U.S.A. **78** (1981), 5290–5292.
4. ——, *Uses and extensions of Frobenius' representation of partitions*, Enumeration and Design (D. M. Jackson and S. A. Vanstone, eds.), Academic Press, 1984, pp. 51–65.
5. G. E. Andrews and R. J. Baxter, *Lattice gas generalization of the hard hexagon model*. III. *q-trinomial coefficients*, J. Statist. Phys. **47** (1987), 297–330.
6. ——, *SCRATCHPAD explorations for elliptic theta functions* (Proc. 1986 Stanford Computers and Mathematics Conf.), Computers in Mathematics (D. Chudnovsky and R. Jenks, eds.), Marcel Dekker, New York (to appear).
7. G. E. Andrews, R. J. Baxter, D. M. Bressoud, W. H. Burge, P. J. Forrester, and G. Viennot, *Partitions with prescribed hook differences*, European J. Combin. **8** (1987), 341–350.
8. G. E. Andrews, R. J. Baxter, and P. J. Forrester, *Eight-vertex SOS model and generalized Rogers-Ramanujan-type identities*, J. Statist. Phys. **35** (1984), 193–266.
9. D. M. Bressoud, *Extension of the partition sieve*, J. Number Theory **12** (1980), 87–100.

10. W. H. Burge, *A correspondence between partitions related to generalizations of the Rogers-Ramanujan identities*, Discrete Math. **34** (1981), 9–15.

11. L. Comtet, *Advanced combinatorics*, Reidel, Dordrecht and Boston, 1974.

12. L. Euler, *Observations analyticae*, Novi Commentarii Academiae Scientarum Petropolitanae **11** (1765), 124–143; also in *Opera Omnia*, Series 1, vol. 15, Teubner, pp. 50–69.

13. I. Schur, *Ein Beitrag zur additiven Zahlentheorie und zur theorie der Kettenbrüche*, S.-B. Preuss. Akad. Wiss. Phys.-Math. Kl., 1917, pp. 302–321; reprinted in *Gesammelte Abhandlungen*, vol. 2, Springer, Berlin, 1973, pp. 117–136.

14. L. J. Slater, *Further identities of the Rogers-Ramanujan type*, Proc. London Math. Soc. (2) **54** (1952), 147–167.

15. M. Spiegel, *Finite differences and difference equations*, McGraw-Hill, New York, 1971.

16. G. N. Watson, *Proof of certain identities in combinatory analysis*, J. Indian Math. Soc. **20** (1934), 57–69.

ABSTRACT. The trinomial coefficients are defined centrally by $\sum_{j=-m}^{\infty} \binom{m}{j}_2 x^j = (1 + x + x^{-1})^m$. Euler observed that for $-1 \leq m \leq 7$, $3\binom{m+1}{0}_2 - \binom{m+2}{0}_2 = F_m(F_m + 1)$, where F_m is the mth Fibonacci number. The assertion is false for $m > 7$. We prove general identities—one of which reduces to Euler's assertion for $m \leq 7$. Our main object is to analyze q-analogs extending Euler's observation. Among other things we are led to finite versions of dissections of the Rogers-Ramanujan identities into even and odd parts.

DEPARTMENT OF MATHEMATICS, THE PENNSYLVANIA STATE UNIVERSITY, UNIVERSITY PARK, PENNSYLVANIA 16802

2.6.2 q-trinomial coefficients and Rogers–Ramanujan type identities

q-Trinomial Coefficients and Rogers-Ramanujan Type Identities

GEORGE E. ANDREWS

To my friend, Paul Bateman, on his seventieth birthday

1. Introduction

There are many proofs [4], [2, Ch. 7] of the celebrated Rogers-Ramanujan identities:

$$1+\sum_{n=1}^{\infty}\frac{q^{n^2}}{(1-q)(1-q^2)\cdots(1-q^n)}=\prod_{n=0}^{\infty}\frac{1}{(1-q^{5n+1})(1-q^{5n+4})}, \quad (1.1)$$

$$1+\sum_{n=1}^{\infty}\frac{q^{n^2+n}}{(1-q)(1-q^2)\cdots(1-q^n)}=\prod_{n=0}^{\infty}\frac{1}{(1-q^{5n+2})(1-q^{5n+3})}. \quad (1.2)$$

However one of the most useful proofs is due to Schur [16, §4]:

If $D_0(q) = D_1(q) = 1$ and $D_n(q) = D_{n-1}(q) + q^{n-1}D_{n-2}(q)$ for $n > 1$, then

$$D_n(q)=\sum_{\lambda=-\infty}^{\infty}(-1)^{\lambda}q^{\lambda(5\lambda+1)/2}\begin{bmatrix} n \\ \lfloor\frac{n-5\lambda}{2}\rfloor\end{bmatrix}_q, \quad (1.3)$$

where $\lfloor x\rfloor$ is the greatest integer $\le x$ and

$$\begin{bmatrix} A \\ B\end{bmatrix}_q=\begin{cases}\dfrac{(1-q^A)(1-q^{A-1})\cdots(1-q^{A-B+1})}{(1-q^B)(1-q^{B-1})\cdots(1-q)}, & 0\le B\le A,\\ 0, & \text{otherwise.}\end{cases} \quad (1.4)$$

Partially supported by National Science Foundation Grant DMS-870269

Similarly if $D_1^*(q) = D_2^*(q) = 1$ and $D_n^*(q) = D_{n-1}^*(q) + q^{n-1}D_{n-2}^*(q)$ for $n > 2$, then

$$D_n^*(q) = \sum_{\lambda=-\infty}^{\infty} (-1)^\lambda q^{\lambda(5\lambda-3)/2} \begin{bmatrix} n \\ \lfloor \frac{n-5\lambda}{2} \rfloor + 1 \end{bmatrix}_q . \tag{1.5}$$

A brief combinatorial argument reveals that (1.1) follows from (1.3) and (1.2) from (1.5) by letting $n \to \infty$.

There are several points of importance about Schur's proof. They have led to completely new generalizations of the Rogers-Ramanujan identities [**1**], [**3**], [**8**], [**10**], [**11**]. Also the fact that (1.3) and (1.5) are polynomial identities allows for treatment of duality in the Hard Hexagon model [**3**]; such polynomial representations are, at times, essential in these statistical mechanics models [**9**, §2.6].

In [**7**], Baxter and I required comparable polynomial representations related to the modulus 7 instance of B. Gordon's [**14**] generalization of the Rogers-Ramanujan identities. However instead of the Gaussian polynomials playing an essential role, now q-analogs of trinomial coefficients took their place. For example, suppose $Y_m(b;q)$ is the (polynomial) generating function for partitions $a_1 + a_2 + \cdots + a_j$ with $a_i \geq a_{i+1}$, $a_i - a_{i+2} \geq 2$, $a_1 \leq m$, and m appears as a part exactly $3 - b$ times ($b = 1, 2,$ or 3).

Then B. Gordon [**33**, Thm. 1, d=t=3] has proved that

$$Y_\infty(3;q) = \prod_{\substack{n=1 \\ n \not\equiv 0, \pm 3 \bmod 7}}^{\infty} (1-q^n)^{-1}. \tag{1.6}$$

Baxter and I found that [**7**, p.319, eq. (4.7), j=k=1]

$$\begin{aligned} Y_m(3;q) = &\sum_{\mu=-\infty}^{\infty} q^{14\mu^2+\mu} \begin{pmatrix} m; 7\mu; q \\ 7\mu \end{pmatrix}_2 \\ &- \sum_{\mu=-\infty}^{\infty} q^{14\mu^2+13\mu+3} \begin{pmatrix} m; 7\mu+3; q \\ 7\mu+3 \end{pmatrix}_2, \end{aligned} \tag{1.7}$$

where

$$\begin{pmatrix} n; B; q \\ A \end{pmatrix}_2 = \sum_{j \geq 0} \frac{q^{j(j+B)}(q)_n}{(q)_j (q)_{j+A} (q)_{n-2j-A}}, \tag{1.8}$$

and

$$(A)_j = (A;q)_j = (1-A)(1-Aq)\cdots(1-Aq^{j-1}). \tag{1.9}$$

q-TRINOMIAL COEFFICIENTS 3

Now a double application of the binomial theorem shows that

$$(1+x+x^{-1})^n = \sum_{k=-n}^{n} \binom{n}{k}_2 x^k$$
$$= \sum_{k=-n}^{n} \left(\sum_{j\geq 0} \frac{n!}{j!(j+k)!(n-2j-k)!} \right) x^k. \quad (1.10)$$

Thus the polynomials in (1.8) are q-analogs of trinomial coefficients. (Note: these are *not* to be confused with the coefficients in $(x+y+z)^n$ which are also often called trinomial coefficients.)

Our object here is to explore other applications of q-trinomial coefficients to polynomials familiar in additive number theory. Perhaps most surprising and intriguing is the fact that the Schur polynomials $D_n(q)$ andd $D_n^*(q)$ can be related to the q-trinomial coefficients also. Namely

$$D_n(q) = \sum_{\mu=-\infty}^{\infty} q^{10\mu^2+\mu} \binom{n; 5\mu; q}{5\mu}_2 - \sum_{\mu=-\infty}^{\infty} q^{10\mu^2-9\mu+2} \binom{n; 5\mu-2; q}{5\mu-2}_2, \quad (1.11)$$

and

$$D_n^*(q) = \sum_{\mu=-\infty}^{\infty} q^{10\mu^2+3\mu} \binom{n; 5\mu+1; q}{5\mu+1}_2 - \sum_{\mu=-\infty}^{\infty} q^{10\mu^2-7\mu+1} \binom{n; 5\mu-2; q}{5\mu-2}_2. \quad (1.12)$$

In Section 2 we shall collect the necessary background on q-trinomial coefficients. In Section 3 we shall prove the above formulas. In Section 4 we consider related partition polynomials due to B. Gordon [**15**] arising from the Göllnitz–Gordon identities [**2**, §7.4]. We conclude with some discussion of the possible application of such identities.

2. Background

In [**7**], q-trinomial coefficients are extensively developed. In this section we shall present without proof those formulae necessary for our subsequent work. All these results are proved fully in [**7**].

We shall need three q-analogs of the trinomial coefficients. Recalling (1.8):

$$\binom{m; B; q}{A}_2 = \sum_{j\geq 0} \frac{q^{j(j+B)}(q)_m}{(q)_j(q)_{j+A}(q)_{m-2j-A}}; \quad (2.1)$$

in addition [**7**, eqs (2.8) and (2.9)]

$$T_0(m,A,q)=\sum_{j=0}^{m}(-1)^j\begin{bmatrix} m\\ j\end{bmatrix}_{q^2}\begin{bmatrix} 2m-2j\\ m-A-j\end{bmatrix}_q, \tag{2.2}$$

and

$$T_1(m,A,q)=\sum_{j=0}^{m}(-q)^j\begin{bmatrix} m\\ j\end{bmatrix}_{q^2}\begin{bmatrix} 2m-2j\\ m-A-j\end{bmatrix}_q. \tag{2.3}$$

These polynomials possess the following symmetry relations [**7**, eq. (2.15)]:

$$\binom{m;B;q}{-A}_2=q^{A(A+B)}\binom{m;B+2A;q}{A}_2, \tag{2.4}$$

$$T_i(m,-A,q)=T_i(m,A,q)\qquad (i=0,1). \tag{2.5}$$

Next there are four Pascal triangle type formulas that are necessary [**7**, eqs. (2.16), (2.25), (2.28), (2.29)]:

$$\begin{aligned}T_1(m,A,q)&=T_1(m-1,A,q)\\&\quad+q^{m+A}T_0(m-1,A+1,q)+q^{m-A}T_0(m-1,A-1,q),\end{aligned} \tag{2.6}$$

$$\begin{aligned}\binom{m;A-1;q}{A}_2&=q^{m-1}\binom{m-1;A-1;q}{A}_2\\&\quad+q^A\binom{m-1;A+1;q}{A+1}_2+\binom{m-1;A-1;q}{A-1}_2,\end{aligned} \tag{2.7}$$

$$\begin{aligned}\binom{m;B;q}{A}_2&=\binom{m-1;B;q}{A}_2\\&\quad+q^{m-A-1+B}\binom{m-1;B;q}{A+1}_2+q^{m-A}\binom{m-1;B-1;q}{A-1}_2,\end{aligned} \tag{2.8}$$

$$\begin{aligned}\binom{m;B;q}{A}_2&=\binom{m-1;B;q}{A}_2\\&\quad+q^{m-A}\binom{m-1;B-2;q}{A-1}_2+q^{m+B}\binom{m-1;B+1;q}{A+1}_2.\end{aligned} \tag{2.9}$$

Additionally there are two further identities needed [**7**, eq. (2.20]:

$$\begin{aligned}&T_1(m,A,q)-q^{m-A}T_0(m,A,q)\\&\qquad-T_1(m,A+1,q)+q^{m+A+1}T_0(m,A+1,q)=0\end{aligned} \tag{2.10}$$

and [**7**, eq. (2.27) corrected]

$$\begin{aligned}&\binom{m;A;q}{A}_2+q^m\binom{m;A;q}{A+1}_2\\&\qquad-\binom{m;A+1;q}{A+1}_2-q^{m-A}\binom{m;A-1;q}{A}_2=0.\end{aligned} \tag{2.11}$$

In closing this section, we note that [**7**, paragraph following (2.14)]

$$\binom{m; B; 1}{A}_2 = T_0(m, A, 1) = T_1(m, A, 1) = \binom{m}{A}_2. \tag{2.12}$$

Thus these polynomials are indeed q-analogs of the trinomial coefficients $\binom{m}{A}_2$.

3. Schur's polynomials

Our object here is the proof of (1.11) and (1.12). In the remarks just prior to (1.3) we find the defining recurrence for $D_n(q)$. Hence we only need to show that the right-hand side of (1.11), which we shall denote $d_n(q)$, satisfies the same conditions.

Clearly

$$d_0(q) = d_1(q) = 1. \tag{3.1}$$

Now

$$\begin{aligned} d_n(q) - d_{n-1}(q) & \qquad\qquad (3.2) \\ = \sum_{\mu=-\infty}^{\infty} q^{10\mu^2+\mu} & \left(\binom{n; 5\mu; q}{5\mu}_2 - \binom{n-1; 5\mu; q}{5\mu}_2 \right) \\ - \sum_{\mu=-\infty}^{\infty} q^{10\mu^2-9\mu+2} & \left(\binom{n; 5\mu-2; q}{5\mu-2}_2 - \binom{n-1; 5\mu-2; q}{5\mu-2}_2 \right) \\ = \sum_{\mu=-\infty}^{\infty} q^{10\mu^2+\mu} & \left(q^{n-5\mu} \binom{n-1; 5\mu-1; q}{5\mu-1}_2 + q^{n-1} \binom{n-1; 5\mu; q}{5\mu+1}_2 \right) \\ - \sum_{\mu=-\infty}^{\infty} q^{10\mu^2-9\mu+2} & \left(q^{n+5\mu-2} \binom{n-1; 5\mu-1; q}{5\mu-1}_2 \right. \\ & \left. + q^{n-5\mu+2} \binom{n-1; 5\mu-4; q}{5\mu-3}_2 \right) \end{aligned}$$

(by (2.8) applied to the first sum and (2.9) applied to the second)

$$\begin{aligned} = q^{n-1} \Bigg(& \sum_{\mu=-\infty}^{\infty} q^{10\mu^2+\mu} \binom{n-1; 5\mu; q}{5\mu+1}_2 \\ & - \sum_{\mu=-\infty}^{\infty} q^{10\mu^2-14\mu+5} \binom{n-1; 5\mu-4; q}{5\mu-3}_2 \Bigg) \end{aligned}$$

$$
\begin{aligned}
&= q^{n-1}\Bigg(\sum_{\mu=-\infty}^{\infty} q^{10\mu^2+\mu}\left(\binom{n-2;5\mu;q}{5\mu}_2\right.\\
&\qquad \left.+q^{n-2}\binom{n-2;5\mu;q}{5\mu+1}_2+q^{5\mu+1}\binom{n-2;5\mu+2;q}{5\mu+2}_2\right)\\
&\quad -\sum_{\mu=-\infty}^{\infty} q^{10\mu^2-14\mu+5}\left(q^{5\mu-3}\binom{n-2;5\mu-2;q}{5\mu-2}_2\right.\\
&\qquad \left.+q^{n-2}\binom{n-2;5\mu-4;q}{5\mu-3}_2+\binom{n-2;5\mu-4;q}{5\mu-4}_2\right)\Bigg)
\end{aligned}
$$

(by (2.7) applied to both sums)

$$
\begin{aligned}
&= q^{n-1}d_{n-1}(q)\\
&\quad+\sum_{\mu=-\infty}^{\infty} q^{10\mu^2+6\mu+1}\left(\binom{n-2;5\mu+2;q}{5\mu+2}_2-\binom{n-2;5\mu+1;q}{5\mu+1}_2\right.\\
&\qquad \left.+q^{n-5\mu-3}\binom{n-2;5\mu;q}{5\mu+1}_2-q^{n-2}\binom{n-2;5\mu+1;q}{5\mu+2}_2\right)\\
&= q^{n-1}d_{n-2}(q) \qquad \text{(by (2.11))}.
\end{aligned}
$$

Hence $d_n(q)$ fulfills the defining conditions for $D_n(q)$; consequently $d_n(q) = D_n(q)$ and equation (1.11) is established.

Equation (1.12) is proved in almost exactly the same way. The only change is in the first step where (2.9) is applied to the first sum and (2.8) to the second. We therefore omit the details.

4. Göllnitz–Gordon polynomials

In [**15**, p. 741], B. Gordon proves the following two partition theorems.

First Göllnitz–Gordon Identity. *The number of partitions of any positive integer n into parts $\equiv 1$, 4, or 7, (mod 8) is equal to the number of partitions of the form $n = n_1 + n_2 + \cdots + n_k$, where $n_i \geq n_{i+1} + 2$ and $n_i \geq n_{i+1} + 3$ if n_i is even $(1 \leq i \leq k-1)$.*

Second Göllnitz–Gordon Identity. *The number of partitions of n into parts $\equiv 3$, 4, or 5 (mod 8) is equal to the number of partitions $n = n_1 + n_2 + \cdots + n_k$ satisfying $n_k \geq 3$ in addition to the inequalities listed in the first Göllnitz–Gordon identity.*

H. Göllnitz [**12**] had found these theorems previously; however he first published them in [**13**].

Both Göllnitz and Gordon base their proof on the following two identities due to Lucy Slater [**17**, eqs. (36) and (34)]:

$$\sum_{n=0}^{\infty} \frac{q^{n^2}(-q;q^2)_n}{(q^2;q^2)_n} = \prod_{n=0}^{\infty} \frac{1}{(1-q^{8n+1})(1-q^{8n+4})(1-q^{8n+7})}, \tag{4.1}$$

and

$$\sum_{n=0}^{\infty} \frac{q^{n^2+2n}(-q;q^2)_n}{(q^2;q^2)_n} = \prod_{n=0}^{\infty} \frac{1}{(1-q^{8n+3})(1-q^{8n+4})(1-q^{8n+5})}. \tag{4.2}$$

In his proof, Gordon [**15**, p. 744] examines the polynomials $S_0(q) = 1+q$, $S_1(q) = 1+q+q^2+q^3+q^4$, and for $n > 1$,

$$S_n(q) = (1+q^{2n+1})S_{n-1}(q) + q^{2n}S_{n-2}(q). \tag{4.3}$$

Gordon [**15**, pp. 744–745] shows that $S_n(q)$ is the generating function for the second type of partitions considered in the first Göllnitz–Gordon identity with the added condition that all parts are $\leq 2n+1$.

To prove the first Göllnitz–Gordon identity one need only show that $S_\infty(q)$ is equal to the right-hand side of (4.1). However no representations of $S_n(q)$ are known which yield the infinite product on the right-hand side of (4.1) directly in the limit. It turns out, however, that

$$S_{n-1}(q) = \sum_{\mu=-\infty}^{\infty} (-1)^\mu q^{4\mu^2+\mu}(T_0(n,4\mu,q) + T_0(n,4\mu+1,q)). \tag{4.4}$$

To prove (4.4), it is convenient to define

$$U(m,A,q) = T_0(m,A,q) + T_0(m,A+1,q). \tag{4.5}$$

Next we note that Gordon's polynomials are equally well-defined by taking $S_{-1}(q) = 1$, $S_0(q) = 1+q$, and then requiring (4.3) to hold for all $n \geq 1$. Let us denote the right-hand side of (4.4) by $s_{n-1}(q)$. Then clearly $s_{-1}(q) = 1$ and $s_0(q) = T_0(1,0,q) + T_0(1,1,q) = 1+q$.

In order to prove the recurrence (4.3), we must use

$$\begin{aligned} &U(m,A,q) - (1+q^{2m-1})U(m-1,A,q) \\ &\quad = q^{m-A}T_1(m-1,A-1,q) + q^{m+A+1}T_1(m-1,A+2,q), \end{aligned} \tag{4.6}$$

a result easily derived from the formulae given in Section 2 and proved in full in [**6**, Lemma 1].

Hence by (4.6) with $n \geq 2$:

$$
\begin{aligned}
&s_{n-1}(q) - (1+q^{2n-1})s_{n-2}(q) - q^{2n-2}s_{n-3}(q) \qquad (4.7)\\
&= \sum_{\mu=-\infty}^{\infty} (-1)^\mu q^{4\mu^2+\mu}\Big(q^{n-4\mu}T_1(n-1,4\mu-1,q)\\
&\qquad + q^{n+4\mu+1}T_1(n-1,4\mu+2,q)\Big)\\
&\quad - \sum_{\mu=-\infty}^{\infty} (-1)^\mu q^{4\mu^2+\mu+2n-2}\Big(T_0(n-2,4\mu,q) + T_0(n-2,4\mu+1,q)\Big)\\
&= \sum_{\mu=-\infty}^{\infty} (-1)^\mu q^{4\mu^2-3\mu+n}\Big(T_1(n-2,4\mu-1,q)\\
&\qquad + q^{n+4\mu-2}T_0(n-2,4\mu,q) + q^{n-4\mu}T_0(n-2,4\mu-2,q)\Big)\\
&\quad + \sum_{\mu=-\infty}^{\infty} (-1)^\mu q^{4\mu^2+5\mu+n+1}\Big(T_1(n-2,4\mu+2,q)\\
&\qquad + q^{n+4\mu+1}T_0(n-2,4\mu+3,q) + q^{n-4\mu-3}T_0(n-2,4\mu+1,q)\Big)\\
&\quad - \sum_{\mu=-\infty}^{\infty} (-1)^\mu q^{4\mu^2+\mu+2n-2}\Big(T_0(n-2,4\mu,q) + T_0(n-2,4\mu+1,q)\Big)\\
&= \sum_{\mu=-\infty}^{\infty} (-1)^\mu q^{4\mu^2+\mu}\Big(q^{n-4\mu}T_1(n-2,4\mu-1,q)\\
&\qquad + q^{2n-8\mu}T_0(n-2,4\mu-2,q) + q^{n+4\mu+1}T_1(n-2,4\mu+2,q)\\
&\qquad + q^{2n+8\mu+2}T_0(n-2,4\mu+3,q)\Big).
\end{aligned}
$$

Treat this last expression as four separate sums and shift μ to $\mu-1$ in the third and fourth sums. Hence

$$
\begin{aligned}
&s_{n-1}(q) - (1+q^{2n-1})s_{n-2}(q) - q^{2n-2}s_{n-3}(q)\\
&= q^n \sum_{\mu=-\infty}^{\infty} (-1)^\mu q^{4\mu^2-3\mu}\Big(T_1(n-2,4\mu-1,q) + q^{n-4\mu}T_0(n-2,4\mu-2,q)\\
&\qquad - T_1(n-2,4\mu-2,q) - q^{n+4\mu-3}T_0(n-2,4\mu-1,q)\Big)\\
&= 0 \qquad \text{(by (2.10))}. \qquad (4.8)
\end{aligned}
$$

Thus (4.4) is established.

An exactly similar result holds for the second Göllnitz–Gordon identity. In this case [**15**, p. 745], Gordon defines a sequence of polynomials $T_n(q)$, given by $T_{-1}(q) = 0$, $T_0(q) = 1$, and for $n > 0$

$$T_n(q) = (1 + q^{2n+1})T_{n-1}(q) + q^{2n}T_{n-2}(q). \tag{4.9}$$

In the same manner as above, it is possible to show that

$$T_{n-1}(q) = \sum_{\mu=-\infty}^{\infty} (-1)^{\mu} q^{4\mu^2+3\mu} \Big(T_0(n, 4\mu+1, q) + T_0(n, 4\mu+2, 1) \Big). \tag{4.10}$$

To close this section we note that the limiting cases of these polynomials are easily seen to be the respective infinite products in the identities (4.1) and (4.2).

In [**6**, eq. (4.16)], we note the limit

$$\lim_{m\to\infty} U(m, A, q) = \prod_{n=1}^{\infty} \frac{(1+q^{2n-1})}{(1-q^{2n})}. \tag{4.11}$$

Hence by (4.4),

$$\begin{aligned}
\lim_{n\to\infty} S_{n-1}(q) &= \prod_{n=1}^{\infty} \frac{(1+q^{2n-1})}{(1-q^{2n})} \sum_{\mu=-\infty}^{\infty} (-1)^{\mu} q^{4\mu^2+\mu} \\
&= \prod_{n=1}^{\infty} \frac{(1-q^{8n-2})(1-q^{8n-6})(1-q^{8n})(1-q^{8n-5})(1-q^{8n-3})}{(1-q^n)} \\
&\qquad \text{(by Jacobi's triple product [2, p. 22, Cor. 2.9])} \\
&= \prod_{n=1}^{\infty} \frac{1}{(1-q^{8n-1})(1-q^{8n-4})(1-q^{8n-7})},
\end{aligned} \tag{4.12}$$

and by (4.10),

$$\begin{aligned}
\lim_{n\to\infty} T_{n-1}(q) &= \prod_{n=1}^{\infty} \frac{(1+q^{2n-1})}{(1-q^{2n})} \sum_{\mu=-\infty}^{\infty} (-1)^{\mu} q^{4\mu^2+3\mu} \\
&= \prod_{n=1}^{\infty} \frac{(1-q^{8n-2})(1-q^{8n-6})(1-q^{8n})(1-q^{8n-7})(1-q^{8n-1})}{(1-q^n)} \\
&\qquad \text{(by Jacobi's triple product [2, p. 22, Cor. 2.9])} \\
&= \prod_{n=1}^{\infty} \frac{1}{(1-q^{8n-3})(1-q^{8n-4})(1-q^{8n-5})}.
\end{aligned}$$

5. Conclusion

In a way, this paper is, I hope, the mere beginning of further study of q-trinomial coefficients. The most intriguing question to me is this:

Is there a nice combinatorial explanation of (1.11), (1.12), (4.4) and (4.10)? As was noted in the introduction, the combinatorial extensions of (1.3) have been quite fruitful. The beginning of the explanation of (1.3) requires that we know that $\begin{bmatrix} A \\ B \end{bmatrix}_q$ is the generating function for partitions into at most B parts, each $\leq A - B$. Then a sieve is introduced concerning successive ranks **[1]**. Thus the simplest question is: what are natural partition-theoretic interpretations of any and all the q-trinomial coefficients that would support a sieve-theoretic interpretation of (1.11), (1.12), (4.4) and (4.10)?

REFERENCES

[1] G. E. Andrews, Sieves in the theory of partitions, Amer. J. Math. **94** (1972), 1214–1230.

[2] G. E. Andrews, *The Theory of Partitions*, Vol. 2, Encyclopedia of Mathematics and Its Applications (G.-C. Rota, ed.), Addison–Wesley, Reading, 1976; reissued: Cambridge University Press, London and New York, 1984.

[3] G. E. Andrews, The hard-hexagon model and the Rogers-Ramanujan type identities, Proc. Nat. Acad. Sci. U. S. A. **78** (1981), 5290–5292.

[4] G. E. Andrews, Uses and extensions of Frobenius' representations of partitions, in *Enumeration and Design* (D. M. Jackson and S. A. Vanstone, eds.), Academic Press 1984, pp. 51–65.

[5] G. E. Andrews, On the proofs of the Rogers-Ramanujan identities, in *q-Series and Partitions* (Dennis Stanton, ed.), IMA Volumes in Mathematics and its Applications, Springer–Verlag, New York, 1989, pp. 1–14.

[6] G. E. Andrews, Euler's "Exemplum Memorabile Inductionis Fallacis" and q-trinomial coefficients (to appear).

[7] G. E. Andrews and R. J. Baxter, Lattice gas generalization of the hard hexagon model. III. q-trinomial coefficients, J. Stat. Phys. **47** (1987), 297–330.

[8] G. E. Andrews, R. J. Baxter, D. M. Bressoud, W. H. Burge, P. J. Forrester, and G. Viennot, Partitions with prescribed hook differences, Europ. J. Combinatorics **8** (1987), 341–350.

[9] G. E. Andrews, R. J. Baxter, and P. J. Forrester, Eight-vertex SOS model and generalized Rogers-Ramanujan-type identities, J. Stat. Phys. **35** (1984), 193–266.

[10] D. M. Bresssoud, Extension of the partition sieve, J. Number Th. **12** (1980), 87–100.

[11] W. H. Burge, A correspondence between partitions related to generalizations of the Rogers-Ramanujan identities, Discrete Math. **34** (1981), 9–15.

[12] H. Göllnitz, *Einfache Partitionen* (unpublished), Diplomarbeit W. S. 1960, Göttingen, 65 pp..

[13] H. Göllnitz, Partitionen mit Differenzenbedingungen, J. Reine Angew. Math. **225** (1967), 154–190.

[14] B. Gordon, A combinatorial generalization of the Rogers-Ramanujan identities, Amer. J. Math. **83** (1961), 393–399.

[15] B. Gordon, Some continued fractions of the Rogers-Ramanujan type, Duke Math. J. **31** (1965), 741–748.

[16] I. Schur, Ein Beitrag zur additiven Zahlentheorie und zur Theorie der Kettenbrüche, S.-B. Preuss. Akad. Wiss. Phys.-Math. Kl., 302–321; reprinted in: *I. Schur, Gesammelte Abhandlungen, Vol. 2,* Springer–Verlag, Berlin 1973.

[17] L. J. Slater, Further identities of the Rogers–Ramanujan type, Proc. London Math. Soc. (2) **54** (1952), 147–167.

George E. Andrews
Dept. of Mathematics
The Pennsylvania State University
University Park, PA 16802

2.6.3 On a conjecture of Peter Borwein

J. Symbolic Computation (1995) **20**, 487–501

On a Conjecture of Peter Borwein

GEORGE E. ANDREWS[†]

The Pennsylvania State University, University Park, Pennsylvania 16802

(Received 3 July 1994)

We consider several conjectures by Peter Borwein concerning products of the form $\prod_{n=1}^{pN}(1-q^n)/\prod_{n=1}^{N}(1-q^{pn})$, $p=3,5$.

1. Introduction

Peter Borwein (1990) has recently uncovered several charming and intriguing mysteries of great simplicity. The most easily stated is the

FIRST BORWEIN CONJECTURE (Polynomial). For the polynomials $A_n(q)$, $B_n(q)$ and $C_n(q)$ defined by

$$\prod_{j=1}^{n}(1-q^{3j-2})(1-q^{3j-1}) = A_n(q^3) - qB_n(q^3) - q^2C_n(q^3), \tag{1.1}$$

each has non-negative coefficients.

The natural setting for this problem is the theory of partitions.

FIRST BORWEIN CONJECTURE (Partitions). Let $B_e(n,N)$ (respectively $B_0(n,N)$) denote the number of partitions of N into an even (respectively odd) number of distinct non-multiples of 3 each $< 3n$. Then

$$B_e(n,N) - B_0(n,N) \begin{cases} \geq 0 & \text{if } 3|N \\ \leqq 0 & \text{otherwise.} \end{cases} \tag{1.2}$$

This is the sort of intriguing simply stated problem that devotees of the theory of partitions love. It should be stated at the beginning that we are unable to prove these conjectures. We proceed by listing the remaining ones.

SECOND BORWEIN CONJECTURE. For the polynomials $\alpha_n(q)$, $\beta_n(q)$ and $\gamma_n(q)$ defined by

$$\prod_{j=1}^{n}(1-q^{3j-2})^2(1-q^{3j-1})^2 = \alpha_n(q^3) - q\beta_n(q^3) - q^2\gamma_n(q^3), \tag{1.3}$$

each has non-negative coefficients.

[†] Partially supported by National Science Foundation Grant DMS 8702695-04.

THIRD BORWEIN CONJECTURE. For the polynomials $\nu_n(q)$, $\phi_n(q)$, $\chi_n(q)$, $\psi_n(q)$ and $\omega_n(q)$ defined by

$$\prod_{j=1}^{n}(1-q^{5j-4})(1-q^{5j-3})(1-q^{5j-2})(1-q^{5j-1})$$
$$=\nu_n(q^5)-q\phi_n(q^5)-q^2\chi_n(q^5)-q^3\psi_n(q^5)-q^4\omega_n(q^5), \tag{1.4}$$

each has non-negative coefficients.

Our focus in this paper will be the First and Second Conjectures. Indeed we consider the following:

REFINED CONJECTURE. For the polynomials $a(m,n,t,q)$, $b(m,n,t,q)$ and $c(m,n,t,q)$ defined by

$$\prod_{j=1}^{m}(1-q^{3j-2})(1-q^{3j-1})\prod_{k=1}^{n}(1-zq^{3j-2})(1-zq^{3j-1})$$
$$=\sum_{t=0}^{2n} z^t(a(m,n,t,q^3)-q\,b(m,n,t,q^3)-q^2c(m,n,t,q^3)), \tag{1.5}$$

each has non-negative coefficients provided $m \geq 1$.

We see immediately that the Refined Conjecture implies the First and Second Borwein Conjectures. This is because $A_n(q) = a(n,0,0,q)$, $B_n(q) = b(n,0,0,q)$, $C_n(q) = c(n,0,0,q)$ while $\alpha_n(q) = \sum_{t=0}^{2n} a(n,n,t,q)$, $\beta_n(q) = \sum_{t=0}^{2n} b(n,n,t,q)$ and $\gamma_n(q) = \sum_{t=0}^{2n} c(n,n,t,q)$.

Our study will proceed as follows. In Section 2, we shall examine the original question considered by P. Borwein which led to the conjectures described above. In Sections 3 and 4, we consider the First Borwein Conjecture and link it to the mod 9 Rogers–Ramanujan identities of Bailey (1947). In Section 5 we examine the Refined Conjecture and provide a few elementary representations of the a, b, and c polynomials, and we prove limiting cases of the Refined Conjecture. We conclude with a discussion of relevant computer algebra questions.

2. The Original Theorem

In a study of modular forms, P. Borwein (1993) considered

$$\prod_{n=1}^{\infty}\frac{(1-q^n)}{(1-q^{pn})}=\sum_{n=0}^{\infty}c_p(n)q^n, \tag{2.1}$$

where p is a prime.

THEOREM 2.1. *For all primes p, $c_p(n)$ and $c_p(n+p)$ have the same sign for each $n \geqq 0$ (i.e., $c_p(n)c_p(n+p) \geqq 0$ for each $n \geqq 0$).*

REMARK. F. Garvan and P. Borwein (1990) have also proved this result although I understand by somewhat different means relying on the work of Rankin.

PROOF. If $p = 2$ then we see immediately that $(-1)^n c_2(n)$ is the number of partitions of n into distinct odd parts. Consequently, $c_2(n) \geqq 0$ if n is even and $c_2(n) \leqq 0$ if n is odd thus confirming the theorem for $p = 2$.

To make succinct the remainder of our argument we introduce the standard notation:

$$(A;q)_n = (A)_n = \prod_{j=0}^{n-1}(1-Aq^j), \tag{2.2}$$

$$(A;q)_\infty = (A)_\infty = \prod_{j=0}^{\infty}(1-Aq^j), \tag{2.3}$$

$$(A_1,A_2,\ldots,A_r;q)_\infty = (A_1,A_2,\ldots,A_r)_\infty = (A_1)_\infty(A_2)_\infty\cdots(A_r)_\infty. \tag{2.4}$$

For p odd, we note (see Andrews, 1976; p. 11, equation (1.3.1))

$$\begin{aligned}\frac{(q)_\infty}{(q^p;q^p)_\infty} &= \frac{\sum_{j=-\infty}^{\infty}(-1)^j q^{j(3j+1)/2}}{(q^p;q^p)_\infty} \qquad (2.5)\\
&= \sum_{r=-\frac{p-1}{2}}^{\frac{p-1}{2}}\ \sum_{k=-\infty}^{\infty}(-1)^{kp+r}q^{(kp+r)(3kp+3r+1)/2}\Big/(q^p;q^p)_\infty\\
&= \sum_{r=\frac{1-p}{2}}^{\frac{p-1}{2}}(-1)^r q^{r(3r+1)/2}\sum_{k=-\infty}^{\infty}(-1)^k q^{(3p^2k^2+kp(6r+1))/2}\Big/(q^p;q^p)_\infty\\
&= \sum_{r=\frac{1-p}{2}}^{\frac{p-1}{2}}(-1)^r q^{r(3r+1)/2}\frac{(q^{3p^2},q^{p(3p+1)/2+3pr},q^{p(3p-1)/2-3pr};q^{3p^2})_\infty}{(q^p;q^p)_\infty}\end{aligned}$$

(by Andrews, 1976; p. 21, equation (2.2.10))

$$\begin{aligned}&= Q_{\frac{3p-1}{2},\frac{3p-1}{2}}(1;q^p) + \sum_{r=1}^{\frac{p-1}{2}}(-1)^r q^{r(3r-1)/2}\big(Q_{\frac{3p-1}{2},\frac{3p+1}{2}-3r}(1;q^p)\\
&+ q^r Q_{\frac{3p-1}{2},\frac{3p-1}{2}-3r}(1;q^p)\big),\end{aligned}$$

where we are using the functions (see Andrews, 1976; Section 7.3) from B. Gordon's Theorem (see Gordon, 1961)

$$Q_{k,i}(1;q) = \sum_{n\geqq 0}A_{k,i}(n)q^n = \sum_{n\geqq 0}B_{k,i}(n)q^n, \tag{2.6}$$

where

$$A_{k,i}(n) = \text{the number of partitions of } n \text{ into parts} \not\equiv 0, \pm i \pmod{2k+1}, \tag{2.7}$$

and

$$\begin{aligned}B_{k,i}(n) = {}& \text{the number of partitions of } n \text{ of the form } n = b_1+b_2+\cdots+b_j\\
& \text{wherein } b_h \geqq b_{h+1},\ b_h - b_{h+k-1} \geqq 2, \text{ and at most } i-1 \text{ of the } b_h \text{ equal } 1.\end{aligned} \tag{2.8}$$

Thus we have broken our series up into terms of the form

$$(-1)^r q^{r(3r\pm 1)/2}Q_{\frac{3p-1}{2},j}(1;q^p). \tag{2.9}$$

Clearly this expression has $(-1)^r$ as the sign of all of its coefficients. Hence the only way to get variations of sign between $c_p(n)$ and $c_p(n+p)$ is if two or more such terms combine together to produce this effect. So under what circumstances with $-\infty < r_0, r_1 < \infty$ can we have

$$r_0(3r_0+1)/2 \equiv r_1(3r_1+1)/2 \pmod{p}? \tag{2.10}$$

Or equivalently since p is odd

$$(r_0 - r_1)(3r_0 + 3r_1 + 1) \equiv 0 \pmod{p}. \tag{2.11}$$

Since p is prime we have only two solutions namely

$$r_0 \equiv r_1 \text{ or } r_0 \equiv -r_1 - 1/3 \pmod{p}. \tag{2.12}$$

This means that for any given residue class mod p, at most two terms in (2.5) (of the form (2.9)) contribute to the same residue class. Note that if there are less than two terms we are done by the comment preceding (2.10). Furthermore, because we are considering only one residue system mod p we see that $r_0 \not\equiv r_1$ mod p. Thus the only way we obtain coefficients from a particular residue class for the subscripts is with, say r, and $s \equiv -r - 1/3 \pmod{p}$. In other words, for a given residue class we have

$$\begin{aligned}&(-1)^r q^{r(3r\pm1)/2} Q_{\frac{3p-1}{2}, \frac{3p\pm1}{2}-3r}(1; q^p),\\ &+(-1)^s q^{s(3s\pm1)/2} Q_{\frac{3p-1}{2}, \frac{3p\pm1}{2}-3s}(1; q^p).\end{aligned} \tag{2.13}$$

Where by the use of $\pm$ we may assume $0 \leqq r \leqq (p-1)/2$, $0 \leqq s \leqq (p-1)/2$ and not both r and s zero. By symmetry we may also assume $r \leqq s$, and indeed $r < s$ because the pentagonal numbers $r(3r+1)/2$ $(-\infty < r < \infty)$ are all distinct and $r(3r-1)/2 \equiv r(3r+1)/2 \pmod{p}$ implies $r \equiv 0 \pmod{p}$. Therefore we may write our two terms in (2.13) as

$$(-1)^r q^{r(3r\pm1)/2} \left(Q_{\frac{3p-1}{2}, \frac{3p\pm1}{2}-3r}(1; q^p) \pm q^{\lambda p} Q_{\frac{3p-1}{2}, \frac{3p\pm1}{2}-3s}(1; q^p)\right) \tag{2.14}$$

where $\lambda \geqq 1$. Now if the plus sign holds before the second Q in (2.14) then the constancy of sign for this residue class is established. Furthermore, since $s > r$

$$\begin{aligned}\frac{3p\pm1}{2} - 3r &\geqq \frac{3p-1}{2} - 3r > \frac{3p+1}{2} - 3s\\ &\geqq \frac{3p\pm1}{2} - 3s.\end{aligned} \tag{2.15}$$

Hence in the only case that could contribute variations in sign we have an expression of the form

$$(-1)^r q^{r(3r\pm1)/2} \left(Q_{\frac{3p-1}{2}, i}(1; q^p) - q^{\lambda p} Q_{\frac{3p-1}{2}, j}(1; q^p)\right) \tag{2.16}$$

where $\frac{3p-1}{2} \geqq i > j > 0$. But

$$Q_{\frac{3p-1}{2}, i}(1; q) - q^{\lambda} Q_{\frac{3p-1}{2}, j}(1; q) = \sum_{n=0}^{\infty} \left(B_{\frac{3p-1}{2}, i}(n) - B_{\frac{3p-1}{2}, j}(n-\lambda)\right) q^n, \tag{2.17}$$

and with $R = (3p-1)/2$,

$$\begin{aligned}B_{R,i}(n) - B_{R,j}(n-\lambda) &\geqq B_{R,i}(n) - B_{R,i}(n-\lambda) \qquad \text{(by equation (2.8))}\\ &= A_{R,i}(n) - A_{R,i}(n-\lambda) \qquad \text{(by equation (2.6))}\\ &\geqq 0 \quad \text{(by equation (2.7), see also Andrews, 1981; pp. 191–192).}\end{aligned}$$

Hence $Q_{R,i}(1, q) - q^{\lambda} Q_{R,j}(1, q)$ for $R \geqq i > j > 0$ has non-negative coefficients.

Therefore for any $n \geqq 0$,

$$c_p(n) \quad \text{and} \quad c_p(n+p)$$

have the same sign (i.e., $c_p(n)c_p(n+p) \geqq 0$). □

3. The First Conjecture

In this section we shall present the easily derived representations and recurrences for $A_n(q)$, $B_n(q)$ and $C_n(q)$.

THEOREM 3.1. *$A_0(q) = 1$, $B_0(q) = C_0(q) = 0$. For $n \geqq 1$*

$$A_n(q) = (1+q^{2n-1})A_{n-1}(q) + q^n B_{n-1}(q) + q^n C_{n-1}(q) \tag{3.1}$$

$$B_n(q) = (1+q^{2n-1})B_{n-1}(q) + q^{n-1} A_{n-1}(q) - q^n C_{n-1}(q) \tag{3.2}$$

$$C_n(q) = (1+q^{2n-1})C_{n-1}(q) + q^{n-1} A_{n-1}(q) - q^{n-1} B_{n-1}(q) \tag{3.3}$$

$$A_n(q) = \sum_{\lambda=-\infty}^{\infty} (-1)^\lambda q^{\lambda(9\lambda+1)/2} \begin{bmatrix} 2n \\ n+3\lambda \end{bmatrix} \tag{3.4}$$

$$B_n(q) = \sum_{\lambda=-\infty}^{\infty} (-1)^\lambda q^{\lambda(9\lambda-5)/2} \begin{bmatrix} 2n \\ n+3\lambda-1 \end{bmatrix} \tag{3.5}$$

$$C_n(q) = \sum_{\lambda=-\infty}^{\infty} (-1)^\lambda q^{\lambda(9\lambda+7)/2} \begin{bmatrix} 2n \\ n+3\lambda+1 \end{bmatrix} \tag{3.6}$$

where

$$\begin{bmatrix} A \\ B \end{bmatrix} = \begin{cases} 0 & \text{if } B < 0 \text{ or } B > A \\ \frac{(q)_A}{(q)_B (q)_{A-B}} & \text{otherwise.} \end{cases} \tag{3.7}$$

PROOF. Recurrences (3.1)–(3.3) are immediately deduced from (1.1), namely

$$\begin{aligned} &A_n(q^3) - qB_n(q^3) - q^2 C_n(q^3) \\ &= (1-q^{3n-2})(1-q^{3n-1})(A_{n-1}(q^3) - qB_{n-1}(q^3) - q^2 C_{n-1}(q^3)). \end{aligned} \tag{3.8}$$

As for (3.4)–(3.6) we can deduce these from the q-binomial theorem (Andrews, 1976; p. 36, equation (3.3.6)):

$$\begin{aligned} \prod_{j=1}^{n}(1-q^{3j-2})(1-q^{3j-1}) &= \sum_{j=-n}^{n} (-1)^j q^{j(3j+1)/2} \begin{bmatrix} 2n \\ n+j \end{bmatrix}_{q^3} \\ &= \sum_{\mu=-1}^{1} \sum_{\nu=-\infty}^{\infty} (-1)^{3\nu+\mu} q^{(3\nu+\mu)(9\nu+3\mu+1)/2} \begin{bmatrix} 2n \\ n+3\nu+\mu \end{bmatrix}_{q^3} \\ &= -q \sum_{\nu=-\infty}^{\infty} (-1)^\nu q^{(27\nu^2-15\nu)/2} \begin{bmatrix} 2n \\ n+3\nu-1 \end{bmatrix}_{q^3} \\ &\quad + \sum_{\nu=-\infty}^{\infty} (-1)^\nu q^{(27\nu^2+3\nu)/2} \begin{bmatrix} 2n \\ n+3\nu \end{bmatrix}_{q^3} \\ &\quad - q^2 \sum_{\nu=-\infty}^{\infty} (-1)^\nu q^{(27\nu^2+21\nu)/2} \begin{bmatrix} 2n \\ n+3\nu+1 \end{bmatrix}_{q^3} \end{aligned}$$

$$= -qB_n(q^3) + A_n(q^3) - q^2C_n(q^3). \tag{3.9}$$

□

4. Bailey's Identities and q-Quadratic Transformations

Sears (1951) and Carlitz (1969) (see also Gasper and Rahman, 1990; Ex. 2.26, p. 55) have independently discovered the following identity which is a terminating q-analog of a quadratic transformation due to Whipple: if $a = q^{-n}$ where n is a non-negative integer,

$$\sum_{j=0}^{n} \frac{(a)_j(b)_j(c)_j\left(\frac{aqx}{bc}\right)^j}{(q)_j\left(\frac{aq}{b}\right)_j\left(\frac{aq}{c}\right)_j} = (xq^{-n})_n \sum_{0 \leqq 2j \leqq n} \frac{(a)_{2j}\left(\frac{aq}{bc}\right)_j q^j}{(q)_j\left(\frac{aq}{b}\right)_j\left(\frac{aq}{c}\right)_j(ax)_j(q/x)_j}. \tag{4.1}$$

Seemingly unrelated to (4.1) are three identities discovered by Bailey (1947, p. 422) which are effectively Rogers–Ramanujan type identities for the modulus 9:

$$\prod_{\substack{n=1 \\ n \not\equiv 0, \pm 4 \pmod 9}}^{\infty} (1-q^n)^{-1} = \frac{(q^3;q^3)_\infty}{(q)_\infty} \sum_{n=0}^{\infty} \frac{q^{3n^2}(q)_{3n}}{(q^3;q^3)_n(q^3;q^3)_{2n}} \tag{4.2}$$

$$\prod_{\substack{n=1 \\ n \not\equiv 0, \pm 2 \pmod 9}}^{\infty} (1-q^n)^{-1} = \frac{(q^3;q^3)_\infty}{(q)_\infty} \sum_{n=0}^{\infty} \frac{q^{3n^2+3n}(q)_{3n}(1-q^{3n+2})}{(q^3;q^3)_n(q^3;q^3)_{2n+1}} \tag{4.3}$$

$$\prod_{n \not\equiv , \pm 1 \pmod 9} (1-q^n)^{-1} = \frac{(q^3;q^3)_\infty}{(q)_\infty} \sum_{n=0}^{\infty} \frac{q^{3n^2+3n}(q)_{3n+1}}{(q^3;q^3)_n(q^3;q^3)_{2n+1}}. \tag{4.4}$$

In the remainder of this section we shall show that identity (4.1) may be specialized to yield identities for $A_n(q)$, $B_n(q)$ and $C_n(q)$ and that these identities are just finite versions of (4.2)–(4.4) which yield these latter three in the limit.

THEOREM 4.1. *For* $n > 0$,

$$A_n(q) = \sum_{0 \leqq 3j \leqq n} \frac{(q^3;q^3)_{n-j-1}(1-q^{2n})(q)_{3j}q^{3j^2}}{(q)_{n-3j}(q^3;q^3)_{2j}(q^3;q^3)_j}, \tag{4.5}$$

$$B_n(q) = \sum_{0 \leqq 3j \leqq n-1} \frac{(q^3;q^3)_{n-j-1}(1-q^{3j+2}-q^{n+3j+2}+q^{n+1})(q)_{3j}q^{3j^2+3j}}{(q)_{n-3j-1}(q^3;q^3)_{2j+1}(q^3;q^3)_j}, \tag{4.6}$$

$$C_n(q) = \sum_{0 \leqq 3j \leqq n-1} \frac{(q^3;q^3)_{n-j-1}(1-q^{3j+1}-q^{n+3j+2}+q^{n})(q)_{3j}q^{3j^2+3j}}{(q)_{n-3j-1}(q^3;q^3)_{2j+1}(q^3;q^3)_j}. \tag{4.7}$$

PROOF. In order to prove these assertions, we must not only use (4.1) as it stands, but we must also use the following double application of it:

$$\sum_{j=0}^{n} \frac{(a)_j(b)_j(c)_j\left(\frac{aqx}{bc}\right)^j}{(q)_j\left(\frac{aq}{b}\right)_j\left(\frac{a}{c}\right)_j} = \frac{1}{1-\frac{a}{c}} \sum_{j=0}^{n} \frac{(a)_j(b)_j(c)_j}{(q)_j\left(\frac{aq}{b}\right)_j\left(\frac{aq}{c}\right)_j}\left(\frac{aqx}{bc}\right)^j$$

$$+ \frac{1}{1-\frac{c}{a}} \sum_{j=0}^{n} \frac{(a)_j(b)_j(c)_j\left(\frac{aq^2x}{bc}\right)^j}{(q)_j\left(\frac{aq}{b}\right)_j\left(\frac{aq}{c}\right)_j} \qquad \text{(where } a = q^{-n}\text{)}$$

$$= \sum_{j \geqq 0} \frac{(a)_{2j} \left(\frac{aq}{bc}\right)_j q^j (axq^{j+1})_{n-j-1}(1 - xaq^j + \frac{xa}{c} - \frac{aq^j}{c})}{(q)_j \left(\frac{aq}{b}\right)_j \left(\frac{a}{c}\right)_{j+1} \left(\frac{q}{x}\right)_j}. \tag{4.8}$$

First we note that (4.7) is immediate from (4.6) because by (3.5) and (3.6)

$$q^{n^2-1} B_n\left(\frac{1}{q}\right) = C_n(q). \tag{4.9}$$

This leaves us three cases each of (4.5) and (4.6) depending on the residue class of $n \bmod 3$.

For example,

$$\begin{aligned}
A_{3n}(q) &= \sum_{\lambda=-\infty}^{\infty} (-1)^\lambda q^{\lambda(9\lambda+1)/2} \begin{bmatrix} 6n \\ 3n+3\lambda \end{bmatrix} \\
&= \sum_{\lambda=0}^{2n} (-1)^{\lambda-n} q^{(\lambda-n)(9\lambda-9n+1)/2} \begin{bmatrix} 6n \\ 3\lambda \end{bmatrix} \\
&= (-1)^n q^{n(9n-1)/2} \sum_{\lambda=0}^{2n} \frac{(q^{-6n};q^3)_\lambda (q^{-6n+1};q^3)_\lambda (q^{-6n+2};q^3)_\lambda q^{(9n+2)\lambda}}{(q^3;q^3)_\lambda (q^2;q^3)_\lambda (q;q^3)_\lambda} \\
&= \sum_{0 \leqq j \leqq n} \frac{(q^3;q^3)_{3n-j-1}(1-q^{6n})(q)_{3j} q^{3j^2}}{(q)_{3n-3j}(q^3;q^3)_{2j}(q^3;q^3)_j},
\end{aligned} \tag{4.10}$$

where to obtain the last line we have replaced q by q^3 in (4.1), then replaced n by $2n$, b by q^{-6n+1}, c by q^{-6n+2}, x by q^{3n+2}, and finally we have reversed the order of summation.

The remaining cases of (4.5) are done in exactly the same way. Also, (4.6) is done in this manner except that for the final step one invokes (4.8) instead of (4.1). □

Apart from yielding (4.2)–(4.4) as limiting results, Theorem 4.1 can be viewed as representations of the polynomials $A_n(q)$, $B_n(q)$ and $C_n(q)$ which are "closer" to showing that these polynomials have non-negative coefficients. To make this comment clear let us examine $A_6(q)$ term-by-term in (3.4) and then again in (4.5). By (3.4),

$$\begin{aligned}
A_6(q) \; = \; & \\
&(1 + q + 2q^2 + 3q^3 + 5q^4 + 7q^5 + 11q^6 + 13q^7 + 18q^8 + 22q^9 + 28q^{10} \\
&\quad +32q^{11} + 39q^{12} + 42q^{13} + 48q^{14} + 51q^{15} + 55q^{16} + 55q^{17} \\
&\quad +58q^{18} + 55q^{19} + 55q^{20} + 51q^{21} + 48q^{22} + 42q^{23} + 39q^{24} \\
&\quad +32q^{25} + 28q^{26} + 22q^{27} + 18q^{28} + 13q^{29} + 11q^{30} + 7q^{31} \\
&\quad +5q^{32} + 3q^{33} + 2q^{34} + q^{35} + q^{36}) - (q^4 + q^5) \\
&(1 + q + 2q^2 + 3q^3 + 4q^4 + 5q^5 + 7q^6 + 8q^7 + 10q^8 + 12q^9 + 13q^{10} + 14q^{11} \\
&\quad +15q^{12} + 15q^{13} + 15q^{14} + 15q^{15} + 14q^{16} + 13q^{17} + 12q^{18} + 10q^{19} + 8q^{20} \\
&\quad +7q^{21} + 5q^{22} + 4q^{23} + 3q^{24} + 2q^{25} + q^{26} + q^{27}) + q^{17} + q^{19}.
\end{aligned} \tag{4.11}$$

While by (4.5),

$$\begin{aligned}A_6(q) &= \\ &(1+q+2q^2+2q^3+4q^4+5q^5+7q^6+8q^7+11q^8+12q^9+16q^{10}\\ &+17q^{11}+20q^{12}+21q^{13}+24q^{14}+25q^{15}+27q^{16}+26q^{17}\\ &+28q^{18}+26q^{19}+27q^{20}+25q^{21}+24q^{22}+21q^{23}+20q^{24}\\ &+17q^{25}+16q^{26}+12q^{27}+11q^{28}+8q^{29}+7q^{30}+5q^{31}\\ &+4q^{32}+2q^{33}+2q^{34}+q^{35}+q^{36})+(q^3+q^6+q^9-2q^{15}\\ &-2q^{18}-2q^{21}+q^{27}+q^{30}+q^{33})+(q^{12}-q^{13}-q^{14}\\ &+q^{15}-q^{16}+2q^{18}-q^{20}+q^{21}-q^{22}-q^{23}+q^{24}).\end{aligned} \tag{4.12}$$

Thus in (4.11) we find the coefficient of q^{18} is $(58-15-15)$ while in (4.12) it is $(28-2+2)$. Inspection makes it clear that the "inclusion–exclusion" effect in (3.4) is much heavier than in (4.5). Hence it might be possible to prove that $A_n(q)$ has positive coefficients by establishing sufficiently tight bounds on the coefficients that arise term-by-term in (4.5).

5. The Refined Conjecture

In this section we present the counterpart of Theorem 3.1 for the polynomials given in (1.5). Unfortunately the complexity of our results has so far prevented us from finding an analog of Theorem 4.1.

THEOREM 5.1.

$$a(m,0,t,q)=\begin{cases}A_m(q) & \text{if } t=0\\ 0 & \text{if } t>0,\end{cases} \tag{5.1}$$

$$b(m,0,t,q)=\begin{cases}B_m(q) & \text{if } t=0\\ 0 & \text{if } t>0,\end{cases} \tag{5.2}$$

$$c(m,0,t,q)=\begin{cases}C_m(q) & \text{if } t=0\\ 0 & \text{if } t>0,\end{cases} \tag{5.3}$$

$$\begin{aligned}a(m,n,t,q) &= (1+q^{2m-1})a(m-1,n,t,q)\\ &\quad + q^m b(m-1,n,t,q)+q^m c(m-1,n,t,q),\end{aligned} \tag{5.4}$$

$$\begin{aligned}b(m,n,t,q) &= (1+q^{2m-1})b(m-1,n,t,q)\\ &\quad + q^{m-1}a(m-1,n,t,q)-q^m c(m-1,n,t,q),\end{aligned} \tag{5.5}$$

$$\begin{aligned}c(m,n,t,q) &= (1+q^{2m-1})c(m-1,n,t,q)\\ &\quad + q^{m-1}a(m-1,n,t,q)+q^{m-1}b(m-1,n,t,q),\end{aligned} \tag{5.6}$$

$$\begin{aligned}a(m,n,t,q) &= a(m,n-1,t,q)+q^{2n-1}a(m,n-1,t-2,q)\\ &\quad + q^n b(m,n-1,t-1,q)+q^n c(m,n-1,t-1,q),\end{aligned} \tag{5.7}$$

$$\begin{aligned}b(m,n,t,q) &= b(m,n-1,t,q)+q^{2n-1}b(m,n-1,t-2,q)\\ &\quad + q^{n-1}a(m,n-1,t-1,q)-q^n c(m,n-1,t-1,q),\end{aligned} \tag{5.8}$$

$$\begin{aligned} c(m,n,t,q) &= c(m,n-1,t,q) + q^{2n-1}c(m,n-1,t-2,q) \\ &\quad + q^{n-1}a(m,n-1,t-1,q) - q^{n-1}b(m,n-1,t-1,q), \end{aligned} \tag{5.9}$$

$$\begin{aligned} a(m,n,t,q) &= q^t a(m,n-1,t,q) + q^{t-1}a(m,n-1,t-2,q) \\ &\quad + q^t b(m,n-1,t-1,q) + q^t c(m,n-1,t-1,q), \end{aligned} \tag{5.10}$$

$$\begin{aligned} b(m,n,t,q) &= q^t b(m,n-1,t,q) + q^{t-1}b(m,n-1,t-2,q) \\ &\quad + q^{t-1}a(m,n-1,t-1,q) - q^t c(m,n-1,t-1,q), \end{aligned} \tag{5.11}$$

$$\begin{aligned} c(m,n,t,q) &= q^t c(m,n-1,t,q) + q^{t-1}c(m,n-1,t-2,q) \\ &\quad + q^{t-1}a(m,n-1,t-1,q) - q^{t-1}b(m,n-1,t-1,q), \end{aligned} \tag{5.12}$$

where $m \geqq 1$, $n \geqq 0$ *in (5.4)–(5.6);* $m \geqq 1$, $n \geqq 1$ *in (5.7)–(5.9) and* $m \geqq 1$, $n \geqq 1$ *in (5.10)–(5.12).*

PROOF. Identities (5.1)–(5.3) follow from

$$\prod_{j=1}^{m}(1-q^{3j-1})(1-q^{3j-2}) = a(m,0,0,q^3) - q\,b(m,0,0,q^3) - q^2c(m,0,0,q^3); \tag{5.13}$$

identities (5.4)–(5.6) follow from

$$\begin{aligned} &(1-q^{3m-2}-q^{3m-1}+q^{6m-3}) \\ &\times \sum_{t=0}^{2n} z^t(a(m-1,n,t,q^3) - q\,b(m-1,n,t,q^3) - q^2c(m-1,n,t,q^3)) \\ &= \sum_{t=0}^{2n} z^t(a(m,n,t,q^3) - q\,b(m,n,t,q^3) - q^2c(m,n,t,q^3)); \end{aligned} \tag{5.14}$$

identities (5.7)–(5.9) follow from

$$\begin{aligned} &(1-zq^{3n-2}-zq^{3n-1}+z^2q^{6n-3}) \\ &\times \sum_{t=0}^{2n} z^t(a(m,n-1,t,q^3) - q\,b(m,n-1,t,q^3) - q^2c(m,n-1,t,q^3)) \\ &= \sum_{t=0}^{2n} z^t(a(m,n,t,q^3) - q\,b(m,n,t,q^3) - q^2c(m,n,t,q^3)); \end{aligned} \tag{5.15}$$

and finally identities (5.10)–(5.12) follow from

$$\begin{aligned} &(1-zq-zq^2+z^2q^3) \\ &\times \sum_{t=0}^{2n-2} z^tq^{3t}(a(m,n-1,t,q^3) - q\,b(m,n-1,t,q^3) - q^2c(m,n-1,t,q^3)) \\ &= \sum_{t=0}^{2n} z^t(a(m,n,t,q^3) - q\,b(m,n,t,q^3) - q^2c(m,n,t,q^3)). \end{aligned} \tag{5.16}$$

□

We now note that these polynomials are new q-analogs of the trinomial coefficients in Andrews and Baxter (1987, 1990).

The trinomial coefficients are centrally defined by

$$(1+t+t^{-1})^n = \sum_{j=-n}^{n} \binom{n}{j}_2 t^j. \tag{5.17}$$

COROLLARY 5.1. *For* $m \geqq 1$, $n \geqq 0$

$$a(m,n,t,1) = 2\cdot 3^{m-1}\binom{n}{t-n}_2, \tag{5.18}$$

$$b(m,n,t,1) = 3^{m-1}\binom{n}{t-n}_2, \tag{5.19}$$

$$c(m,n,t,1) = 3^{m-1}\binom{n}{t-n}_2. \tag{5.20}$$

PROOF. We proceed by induction. First we easily deduce that $A_m(1) = 2\cdot 3^{m-1}$, $B_m(1) = 3^{m-1}$ and $C_m(1) = 3^{m-1}$ from (3.1)–(3.3). Hence by (5.1)–(5.3) we see that Corollary 5.1 is true for $n = 0$. The induction step is secured by setting $q = 1$ in the recurrences (5.7)–(5.9) and invoking the recurrence (directly derived from (5.17))

$$\binom{n}{j}_2 = \binom{n-1}{j-1}_2 + \binom{n-1}{j}_2 + \binom{n-1}{j+1}_2. \tag{5.21}$$

□

COROLLARY 5.2.

$$\lim_{t\to\infty} q^{-3t^2}(a(\infty,\infty,2t,q^3) - q\,b(\infty,\infty,2t;q^3) - q^2c(\infty,\infty,2t,q^3)) = \frac{(q)_\infty(-q^2;q^2)_\infty(q^6;q^6)_\infty}{(q^3;q^3)^3_\infty(-q^6;q^6)_\infty}, \tag{5.22}$$

$$\lim_{t\to\infty} q^{-3t^2-3t-1}(a(\infty,\infty,2t+1,q^3) - q\,b(\infty,\infty,2t+1;q^3) - q^2c(\infty,\infty,2t+1,q^3)) = \frac{-(q)_\infty(-q;q^2)_\infty(q^6;q^6)_\infty}{(q^3;q^3)^3_\infty(-q^3;q^6)_\infty}. \tag{5.23}$$

PROOF. By letting m and $n\to\infty$ in (1.5), we see that

$$\sum_{t=0}^{\infty} z^t(a(\infty,\infty,t,q^3) - q\,b(\infty,\infty,t,q^3) - q^2\,c(\infty,\infty,t,q^3)) \tag{5.24}$$

$$= \frac{(q)_\infty}{(q^3;q^3)_\infty}(zq;q^3)_\infty(zq^2;q^3)_\infty$$

$$= \frac{(q)_\infty}{(q^3;q^3)_\infty}\sum_{m,n\geqq 0}\frac{(-1)^{m+n}z^{m+n}q^{3\binom{m}{2}+3\binom{n}{2}+m+2n}}{(q^3;q^3)_n(q^3;q^3)_m}$$

(by Andrews, 1967; p. 19, equation (2.2.6))

$$= \frac{(q)_\infty}{(q^3;q^3)_\infty}\left\{\sum_{t\geqq 0} q^{2t}z^{2t}\sum_{n=-\infty}^{\infty}\frac{q^{3\binom{t-n}{2}+3\binom{n+t}{2}+n+t}}{(q^3;q^3)_{n+t}(q^3;q^3)_{t-n}}\right.$$

$$\left.-\sum_{t\geqq 0} q^{2t+1}z^{2t+1}\sum_{n=-\infty}^{\infty}\frac{q^{3\binom{t+1-n}{2}+3\binom{n+t}{2}+n+t}}{(q^3;q^3)_{n+t}(q^3;q^3)_{t+1-n}}\right\}$$

$$= \frac{(q)_\infty}{(q^3;q^3)_\infty}\left\{\sum_{t\geqq 0} q^{3t^2}z^{2t}\sum_{n=-\infty}^{\infty}\frac{q^{3n^2+n}}{(q^3;q^3)_{n+t}(q^3;q^3)_{t-n}}\right.$$

$$\left.-\sum_{t\geqq 0} q^{3t^2+3t+1}z^{2t+1}\sum_{n=-\infty}^{\infty}\frac{q^{3n^2-2n}}{(q^3;q^3)_{n+t}(q^3;q^3)_{t+1-n}}\right\}. \tag{5.25}$$

Hence

$$\lim_{t\to\infty} q^{-3t^2}(a(\infty,\infty,2t,q^3) - q\,b(\infty,\infty,2t,q^3) - q^2c(\infty,\infty,2t,q^3))$$

$$= \frac{(q)_\infty}{(q^3;q^3)^3_\infty}\sum_{n=-\infty}^{\infty} q^{3n^2+n}$$

$$= \frac{(q)_\infty}{(q^3;q^3)^3_\infty}(q^6;q^6)_\infty(-q^2;q^6)_\infty(-q^4;q^6)_\infty$$

(by Andrews, 1976; p. 21, Theorem 2.8)

$$= \frac{(q)_\infty(-q^2;q^2)_\infty(q^6;q^6)_\infty}{(q^3;q^3)^3_\infty(-q^6;q^6)_\infty},$$

and

$$\lim_{t\to\infty} q^{-3t^2-3t-1}(a(\infty,\infty,2t+1,q^3) - q\,b(\infty,\infty,2t+1,q^3) - q^2c(\infty,\infty,2t+1,q^3))$$

$$= -\frac{(q)_\infty}{(q^3;q^3)^3_\infty}\sum_{n=-\infty}^{\infty} q^{3n^2-2n}$$

$$= \frac{-(q)_\infty(q^6;q^6)_\infty(-q;q^6)_\infty(-q^5;q^6)_\infty}{(q^3;q^3)^3_\infty}$$

$$= \frac{-(q)_\infty(-q;q^2)_\infty(q^6;q^6)_\infty}{(q^3;q^3)^3_\infty(-q^3;q^6)_\infty}.$$

□

Before we look further at the limiting cases of these polynomials, we note elementary reciprocal relations.

Corollary 5.3.

$$q^{m^2+n^2}a(m,n,2n-t,q^{-1}) = a(m,n,t,q) \tag{5.26}$$

$$q^{m^2+n^2-1}b(m,n,2n-t,q^{-1}) = c(m,n,t,q,) \tag{5.27}$$

$$q^{m^2+n^2-1}c(m,n,2n-t,q^{-1}) = b(m,n,t,q). \tag{5.28}$$

PROOF. Equation (5.28) is just (5.27) rewritten with t replaced by $2n-t$ and q replaced by q^{-1} and is included for completeness. In fact, all the assertions follow from the immediate identity

$$q^{3m^2+3n^2}z^{2n}\prod_{j=1}^{m}(1-q^{-3j+1})(1-q^{-3j+2})\prod_{h=1}^{n}(1-z^{-1}q^{-3h+1})(1-z^{-1}q^{-3h+2})$$
$$=\prod_{j=1}^{m}(1-q^{3j-1})(1-q^{3j-2})\prod_{h=1}^{n}(1-zq^{3h-1})(1-zq^{3h-2}).$$

□

In light of Corollary 5.2, we now define the obvious limiting functions

$$\mathcal{A}_e(q)=\lim_{t\to\infty}q^{-t^2}a(\infty,\infty,2t,q),\tag{5.29}$$

$$\mathcal{A}_0(q)=\lim_{t\to\infty}q^{-t^2-t-1}a(\infty,\infty,2t+1,q),\tag{5.30}$$

$$\mathcal{B}_e(q)=\lim_{t\to\infty}q^{-t^2}b(\infty,\infty,2t,q),\tag{5.31}$$

$$\mathcal{B}_0(q)=\lim_{t\to\infty}q^{-t^2-t}b(\infty,\infty,2t+1,q).\tag{5.32}$$

COROLLARY 5.4.

$$\mathcal{A}_e(q)=\frac{(q^{12};q^{12})_\infty(q^2;q^2)_\infty(q^2;q^4)_\infty}{(q)_\infty^3}\times\left((q^5;q^{12})_\infty(q^7;q^{12})_\infty-q(q;q^{12})_\infty(q^{11};q^{12})_\infty\right),\tag{5.33}$$

$$\mathcal{B}_e(q)=\frac{(q^2;q^2)_\infty(q^2;q^4)_\infty(q^{12};q^{12})_\infty(q^3;q^{12})_\infty(q^9;q^{12})_\infty}{(q)_\infty^3},\tag{5.34}$$

$$\mathcal{A}_0(q)=\frac{2(q^{12};q^{12})_\infty(q^2;q^{12})_\infty(q^{10};q^{12})_\infty}{(q)_\infty^2(q^2;q^4)_\infty},\tag{5.35}$$

$$\mathcal{B}_0(q)=\frac{(-q^2;q^2)_\infty(q^6;q^{12})_\infty(q^6;q^6)_\infty}{(q)_\infty^2}.\tag{5.36}$$

PROOF. These identities are derivable from Corollary 5.2 as follows: by (5.22)

$$\lim_{t\to\infty}q^{-3t^2}(a(\infty,\infty,2t;q^3)-q\,b(\infty,\infty,2t,q^3)-q^2c(\infty,\infty,2t,q^3))$$
$$=\frac{(q^6;q^6)_\infty}{(q^3;q^3)_\infty^3(-q^6;q^6)_\infty}\sum_{n=-\infty}^{\infty}(-1)^nq^{2n^2-n}$$

(by Andrews, 1976; p. 21, equation (2.2.10))

$$=\frac{(q^6;q^6)_\infty(q^6;q^{12})_\infty}{(q^3;q^3)_\infty^3}\left(\sum_{n=-\infty}^{\infty}(-1)^nq^{18n^2-3n}\right.$$
$$\left.-q\sum_{n=-\infty}^{\infty}(-1)^nq^{18n^2+9n}-q^3\sum_{n=-\infty}^{\infty}(-1)^nq^{18n^2-15n}\right).\tag{5.37}$$

Identities (5.33) and (5.34) are now obtained from (5.37) by isolating the appropriate powers of q classified modulo 3 and simplifying the result into infinite products using Jacobi's triple product (see Andrews, 1976; p. 21, Theorem 2.8).

Finally, by (5.23)

$$\lim_{t\to\infty} q^{-3t^2-3t}(a(\infty,\infty,2t+1,q^3) - q\,b(\infty,\infty,2t+1,q^3) - q^2c(\infty,\infty,2t+1,q^3))$$

$$= \frac{-q}{(q^3;q^3)^2_\infty(q^6;q^{12})_\infty} \sum_{n=-\infty}^{\infty} (-1)^n q^{2n^2}$$

$$= \frac{(-q^6;q^6)_\infty}{(q^3;q^3)^2_\infty}\left(-q\sum_{n=-\infty}^{\infty}(-1)^n q^{18n^2} + 2q^3\sum_{n=-\infty}^{\infty}(-1)^n q^{18n^2+12n}\right) \tag{5.38}$$

and just as (5.37) implies (5.33) and (5.34), so too (5.38) implies (5.35) and (5.36). □

COROLLARY 5.5. *Each of $\mathcal{A}_e(q)$, $\mathcal{A}_0(q)$, $\mathcal{B}_e(q)$, $\mathcal{B}_0(q)$ has non-negative power series coefficients.*

PROOF. First we note that any function of the form

$$\prod_{n=1}^{\infty}(1-q^n)^{-\lambda_n} \tag{5.39}$$

(where each λ_n is a non-negative integer) has non-negative power series coefficients. For each of $\mathcal{B}_e(q)$, $\mathcal{B}_0(q)$ and $\frac{1}{2}\mathcal{A}_0(q)$ it is easily checked that Corollary 5.4 provides precisely such infinite products. Thus we need only seriously examine (5.33).

To handle (5.33) we must recall the Generalized Göllnitz–Gordon partition theorem (see Andrews, 1967; p. 945, Theorem 1):

Let a and k be integers with $0 < a \leqq k$. Let $C_{k,a}(n)$ denote the number of partitions of n into parts which are neither $\equiv 2 \pmod 4$ nor $\equiv 0, \pm(2a-1) \pmod{4k}$. Let $D_{k,a}(n)$ denote the number of partitions of n of the form $n = \sum_{i=1}^{\infty} f_i \cdot i$ (f_i is the number of times the part i appears in the partition) with $f_1 + f_2 \leqq a-1$ and for each $i \geqq 1$, $f_{2i-1} \leqq 1$ and $f_{2i} + f_{2i+1} + f_{2i+2} \leqq k-1$. Then

$$C_{k,a}(n) = D_{k,a}(n). \tag{5.40}$$

From this theorem and the definitions of $C_{k,a}(n)$ and $D_{k,a}(n)$ contained therein, we see immediately that

$$C_{k,a}(n) \geqq C_{k,a}(n-1) \qquad \text{if } a > 1 \tag{5.41}$$

(because $C_{k,a}(n) - C_{k,a}(n-1)$ is the number of partitions among those enumerated by $C_{k,a}(n)$ that do not have 1 as a part), and

$$D_{k,a}(n) \geqq D_{k,a'}(n) \qquad \text{if } a \geqq a' \tag{5.42}$$

(because the only difference in the sets partitions enumerated by the two partition functions is that for the first one $f_1 + f_2 \leqq a-1$, while for the second a more stringent condition holds, $f_1 + f_2 \leqq a' - 1$).

Hence by (5.33)

$$\begin{aligned}\mathcal{A}_e(q) &= \frac{(q^2;q^2)_\infty}{(q)_\infty^2}\Big(\sum_{n\geqq 0} C_{3,3}(n)q^n - q\sum_{n\geqq 0} C_{3,1}(n)q^n\Big)\\ &= \frac{(q^2;q^2)_\infty}{(q)_\infty^2}\sum_{n\geqq 0}(C_{3,3}(n) - C_{3,1}(n-1))q^n. \qquad (5.43)\end{aligned}$$

By the comments following (5.39), it is clear that $(q^2;q^2)_\infty/(q)_\infty^2$ has non-negative coefficients. Also,

$$\begin{aligned}C_{3,3}(n) - C_{3,1}(n-1) &\geqq C_{3,3}(n-1) - C_{3,1}(n-1) && \text{(by equation (5.41))}\\ &= D_{3,3}(n-1) - D_{3,1}(n-1) && \text{(by equation (5.40))}\\ &\geqq 0 && \text{(by equation (5.42)).}\end{aligned}$$

Thus in (5.43) we see that $\mathcal{A}_e(q)$ is the product of two expressions each with non-negative power series coefficients. Hence $\mathcal{A}_e(q)$ has non-negative power series coefficients. □

6. The Conjectures and Computer Algebra

The first point to make is that the Refined Conjecture is certainly something I would have been much less likely to discover without AXIOM. It was my hope to find a generalization of the original conjectures that would be easier to prove than the originals.

I half suspect that the Refined Conjecture is harder to prove than the original; however the connection between (5.33) (and, in fact, (5.34) also) and the Generalized Göllnitz–Gordon Theorem suggests the possibility of a significant underlying combinatorial structure relative to $a(m,n,t,q)$ and $b(m,n,t,q)$.

I have spent a large amount of time using AXIOM trying to obtain representations of any (or all) of the polynomials described here that would yield the conjectures directly from the representations. The best any of these searches did was Theorem 4.1 which I erroneously initially thought would yield the First Conjecture.

However, I do believe that such a search still has great merit. It is difficult to shake the belief that there must be sets of combinatorial objects enumerated by each of the polynomials considered here. If such objects exist they are surely elusive (at least to me). This leads one to speculate that they are not close relatives of the more familiar combinatorial objects historically connected with q-series (e.g., partitions with difference conditions or finite vector spaces).

In any event, the search itself may add to the various algorithms and strategies that have evolved from such studies in the past.

References

Andrews, G.E. (1967). A generalization of the Göllnitz–Gordon partition theorems, *Proc. Amer. Math. Soc.* **18**, 945–952.

Andrews, G.E. (1976). *The Theory of Partitions, Encyclopedia of Mathematics and Its Appl.*, Vol. 2, Addison-Wesley, Reading, (Cambridge: Cambridge University Press, Reissued 1985).

Andrews, G.E. (1981). Ramanujan's "lost" notebook III. The Rogers–Ramanujan continued fraction, *Adv. in Math.* **41**, 186–208.

Andrews, G.E. (1986). q-Series. Their Development and Application in Analysis, Number Theory, Combinatorics, Physics, and Computer Algebra, CBMS Regional Conference Lecture Series, No. 66, *Amer. Math. Soc.*, Providence.

Andrews, G.E., Baxter, R.J. (1987). Lattice gas generalization of the hard hexagon model. III q-trinomial coefficients. *J. Stat. Phys.* **47**, 297–330.

Andrews, G.E., Baxter, R.J. (1990). SCRATCHPAD explorations for elliptic theta functions, from Computers in Mathematics. In (Chudnovsky, D.V. and Jenks, R.D. eds) *Lecture Notes in Pure and Appl. Math.*, No. 125, pp. 17–33. New York: Dekker.

Bailey, W.N. (1947). Some identities in combinatory analysis, *Proc. London Math. Soc.* **49**(2):421–435.

Borwein, P. (1990). (Personal communication.)

Borwein, P. (1993). Some restricted partition functions, *J. Number Th.* **45**, 228–240.

Carlitz, L. (1969). Some formulas of F. H. Jackson, *Monatsh. für Math.* **73**, 193–198.

Gasper, G., Rahman, M. (1990). Basic Hypergeometric series, *Encyclopedia of Math. and Its Appl.*, Vol. 35. Cambridge: Cambridge University Press.

Gordon, B. (1961). A combinatorial generalization of the Rogers–Ramanujan identities, *Amer. J. Math.* **83**, 393–399.

Sears, D.B. (1951). Transformations of basic hypergeometric functions of special type, *Proc. London Math. Soc.* **52**(2):467–483.

2.7 Ramanujan's "Lost" Notebook

Bruce Berndt and I are providing an edited version of the "Lost" Notebook in five volumes (*cf.* [Book7], and [Book8]). I have published more than twenty papers on the "Lost" Notebook: [78], [83], [90], [92], [93], [100], [102], [110], [115], [116], [124], [137], [138], [142], [147], [151], [159], [161], [198], [235], [236], [242], [244], [260].

In addition, papers [85], [164], [165], [167], [179], [247], and [251] are related to problems and identities considered in Ramanujan's famous notebooks which have been beautifully elucidated in five volumes by Bruce Berndt [Bern85-98].

Since almost everything in the above papers will appear in the five-volume work Berndt and I are editing, I have chosen to include here only [78] which introduced the "Lost" Notebook.

2.7.1 An introduction to Ramanujan's "Lost" Notebook

AN INTRODUCTION TO RAMANUJAN'S "LOST" NOTEBOOK

GEORGE E. ANDREWS

1. Introduction. In the spring of 1976, I visited the Trinity College Library at Cambridge University. Dr. Lucy Slater had suggested to me that there were materials deposited there from the estate of the late G. N. Watson which might contain some work on q-series. In one box of materials from Watson's estate I found a number of items written by the famous Indian mathematician S. Ramanujan (1887–1920). The most interesting item in this box was a manuscript of more than one hundred pages in Ramanujan's distinctive handwriting which contains over six hundred mathematical formulae listed one after the other without proof. *It is my contention that this manuscript, or notebook, was written during the last year of Ramanujan's life, after his return to India from England.* My evidence (given in Section 3) for this assertion is all indirect; in the words of Stephen Leacock, "It is what we call circumstantial evidence—the same thing that people are hanged for."

The fascinating story of Ramanujan's short life and brilliant career has been told several times, and the interested reader is referred to the accounts in Ramanujan's *Collected Papers* **[18]**, in *Ramanujan* by G. H. Hardy **[14]**, and in *Ramanujan the Man and the Mathematician* by S. R. Ranganathan **[20]**. There are three famous notebooks written by Ramanujan **[19]**. During the past 60 years, these have formed the basis for numerous papers by many mathematicians, including G. H. Hardy, G. N. Watson, L. J. Rogers, and many others. G. N. Watson **[26]** presented a nice survey of the notebooks in 1931, and B. Berndt **[10]** has written a new survey giving an up-to-date account of recent work related to the notebooks. Watson **[26**, p. 138] suggests that Ramanujan's work on the three famous notebooks concluded around 1913. A fully edited version of them was never completed, although G. N. Watson and B. M. Wilson initiated such a project. In 1957, the Tata Trust brought out a photostat edition of the three notebooks **[19]**.

We shall consider the origin of this "lost" notebook later. Before going further, let us state a few of the marvelous identities that appear in this work:

$$1+\sum_{n=1}^{\infty}\frac{q^j}{\prod_{j=1}^{n}(1+aq^j)\left(1+\frac{q^j}{a}\right)} = (1+a)\sum_{n=0}^{\infty}a^{3n}q^{\frac{1}{2}n(3n+1)}(1-a^2q^{2n+1}) - a\frac{\sum_{n=0}^{\infty}(-1)^n a^{2n}q^{n(n+1)/2}}{\prod_{j=1}^{\infty}(1+aq^j)\left(1+\frac{q^j}{a}\right)}; \tag{1.1}$$

$$\sum_{n=0}^{\infty}\frac{q^n}{(1+q)(1+q^3)\cdots(1+q^{2n+1})} = \sum_{n=0}^{\infty}(-1)^n q^{6n^2+4n}(1+q^{4n+2}); \tag{1.2}$$

Professor Andrews received his Ph.D. at the University of Pennsylvania in 1964. He was the late Professor Hans Rademacher's last student. He was a Fulbright scholar at Cambridge University in 1960–61, and has spent leaves of absence from Pennsylvania State University at M.I.T. and Wisconsin. He has published extensively on the theory of partitions and basic hypergeometric functions. His books are *Number Theory* (1971) and *The Theory of Partitions* (1976). He has recently edited *The Collected Papers of P. A. MacMahon.—Editors*

$$\left[\cfrac{1}{1+\cfrac{q}{1+\cfrac{q^2}{1+\cfrac{q^3}{1+\cfrac{q^4}{\vdots}}}}}\right]^3 = \frac{\sum_{n=0}^{\infty} q^{5n^2+4n}\frac{1+q^{5n+2}}{1-q^{5n+2}} - \sum_{n=0}^{\infty} q^{5n^2+6n+1}\frac{1+q^{5n+3}}{1-q^{5n+3}}}{\sum_{n=0}^{\infty} q^{5n^2+2n}\frac{1+q^{5n+1}}{1-q^{5n+1}} - \sum_{n=0}^{\infty} q^{5n^2+8n+3}\frac{1+q^{5n+4}}{1-q^{5n+4}}}; \tag{1.3}$$

$$\frac{G(aq,\lambda q)}{G(a,\lambda)} = \cfrac{1}{1+\cfrac{aq+\lambda q}{1+\cfrac{bq+\lambda q^2}{1+\cfrac{aq^2+\lambda q^3}{1+\cfrac{bq^2+\lambda q^4}{\vdots}}}}}, \tag{1.4}$$

where

$$G(a,\lambda)=G(a,\lambda;b;q)=1+\sum_{n=1}^{\infty}\frac{q^{n(n+1)/2}(a+\lambda)(a+\lambda q)\cdots(a+\lambda q^{n-1})}{(1-q)(1-q^2)\cdots(1-q^n)(1+bq)(1+bq^2)\cdots(1+bq^n)};$$

$$\frac{1}{1-\alpha}+\sum_{n=1}^{\infty}\frac{\beta^n}{(1-\alpha x^n)(1-\alpha x^{n-1}y)(1-\alpha x^{n-2}y^2)\cdots(1-\alpha y^n)}$$

$$=\frac{1}{1-\beta}+\sum_{n=1}^{\infty}\frac{\alpha^n}{(1-\beta x^n)(1-\beta x^{n-1}y)(1-\beta x^{n-2}y^2)\cdots(1-\beta y^n)}. \tag{1.5}$$

To my knowledge none of these five identities appears in the literature. Identity (1.1) was rediscovered by N. J. Fine in the early 1950's; however, he never published his proof.

Identity (1.2) is a "false" theta series identity. Results like (1.2) were studied by L. J. Rogers [**22**]; however, this elegant result appears to have escaped him. As we shall see, identity (1.2) implies a partition identity like that deduced from Euler's Pentagonal Number Theorem [**5**, p. 10].

Identity (1.3) is related to the famous Rogers–Ramanujan continued fraction [**5**, p. 104]:

$$\cfrac{1}{1+\cfrac{q}{1+\cfrac{q^2}{1+\cfrac{q^3}{\vdots}}}} = \prod_{n=0}^{\infty}\frac{(1-q^{5n+1})(1-q^{5n+4})}{(1-q^{5n+2})(1-q^{5n+3})}$$

$$=\frac{\sum_{n=0}^{\infty}(-1)^n q^{(5n^2+3n)/2}-\sum_{n=0}^{\infty}(-1)^n q^{(5n^2+7n+2)/2}}{\sum_{n=0}^{\infty}(-1)^n q^{(5n^2+n)/2}-\sum_{n=0}^{\infty}(-1)^n q^{(5n^2+9n+4)/2}}; \tag{1.6}$$

however, it lies somewhat deeper.

Identity (1.4) is a new Rogers–Ramanujan type continued fraction from which Ramanujan deduces five corollaries (our equations (7.10)–(7.14)) in which continued fractions are represented by infinite products. The special case $a=0$ was given in Chapter 16 of Ramanujan's older notebooks [**19**, Vol. 2, Eq. (13), p. 195]. In the older notebooks Ramanujan gives a hint for proving the continued fraction identity by giving explicit formulae for the convergents. The proof we shall give relies on some q-series transformations due to Rogers.

Identity (1.5) looks disarmingly simple (and is). Its analytic proof is quite mundane; however, there is an amusing combinatorial proof.

I have chosen these five identities to give some flavor of Ramanujan's achievements in this "lost" notebook. Since there were over 600 to choose from, these results cannot really be called an accurate sample, but merely a tantalizing introduction. I plan to write a series of papers in which I shall organize these formulae into somewhat sensible groupings. I shall prove as many of Ramanujan's formulae as I can, and the rest I shall present for the consideration of others.

2. The mathematical setting for the "lost" notebook. The vast majority of the formulae in the lost notebook (including the results we have chosen) concern q-series and related theta functions. For example, we meet again and again the series

$$\sum_{n=0}^{\infty}\frac{(1-a)(1-aq)\cdots(1-aq^{n-1})(1-b)(1-bq)\cdots(1-bq^{n-1})t^n}{(1-q)(1-q^2)\cdots(1-q^n)(1-c)(1-cq)\cdots(1-cq^{n-1})}. \tag{2.1}$$

This is called the q-analog of the famous hypergeometric series studied by Euler, Gauss, and others:

$$\sum_{n=0}^{\infty}\frac{a(a+1)\cdots(a+n-1)b(b+1)\cdots(b+n-1)t^n}{n!\;\;c(c+1)\cdots(c+n-1)}. \tag{2.2}$$

It is called the q-analog because the rising factorial $a(a+1)\cdots(a+n-1)$ of the hypergeometric series has been uniformly replaced by the rising q-factorial $(1-a)(1-aq)\cdots(1-aq^{n-1})$. There are many, many results known about the hypergeometric series (2.2) (see [**25**, Chap. 1] for a nice introduction). For example, when $b=c$ in (2.2) we have the famous binomial series theorem:

$$\sum_{n=0}^{\infty}\frac{a(a+1)\cdots(a+n-1)t^n}{n!}=(1-t)^{-a},\qquad |t|<1. \tag{2.3}$$

In parallel we have the q-binomial series theorem if $b=c$ in (2.1):

$$\sum_{n=0}^{\infty}\frac{(1-a)(1-aq)\cdots(1-aq^{n-1})t^n}{(1-q)(1-q^2)\cdots(1-q^n)}=\prod_{n=0}^{\infty}\frac{(1-atq^n)}{(1-tq^n)}. \tag{2.4}$$

There is also an integral representation of the hypergeometric function (2.2) [**25**, p. 20]:

$$\sum_{n=0}^{\infty}\frac{a(a+1)\cdots(a+n-1)b(b+1)\cdots(b+n-1)t^n}{n!\;\;c(c+1)\cdots(c+n-1)}$$

$$=\frac{\Gamma(c)}{\Gamma(b)\Gamma(c-b)}\int_0^1 x^{b-1}(1-x)^{c-b-1}(1-tx)^{-a}\,dx. \tag{2.5}$$

The investigations of q-series have shown that the natural q-analog of (2.5) is not an integral representation for the series in (2.1), but rather an identity (due to Heine) between two such series given in this paper as identity (4.1). L. J. Rogers [**21**] noted that (4.1) could be applied to itself, and as a result he easily deduced the new and significant identities (4.6) and (7.1). Heine [**15**, p. 325] originally discovered (7.1), but he proved it in a more complicated way. Actually, (6.1) is a q-analog of Euler's famous identity for the hypergeometric function [**24**, p. 10].

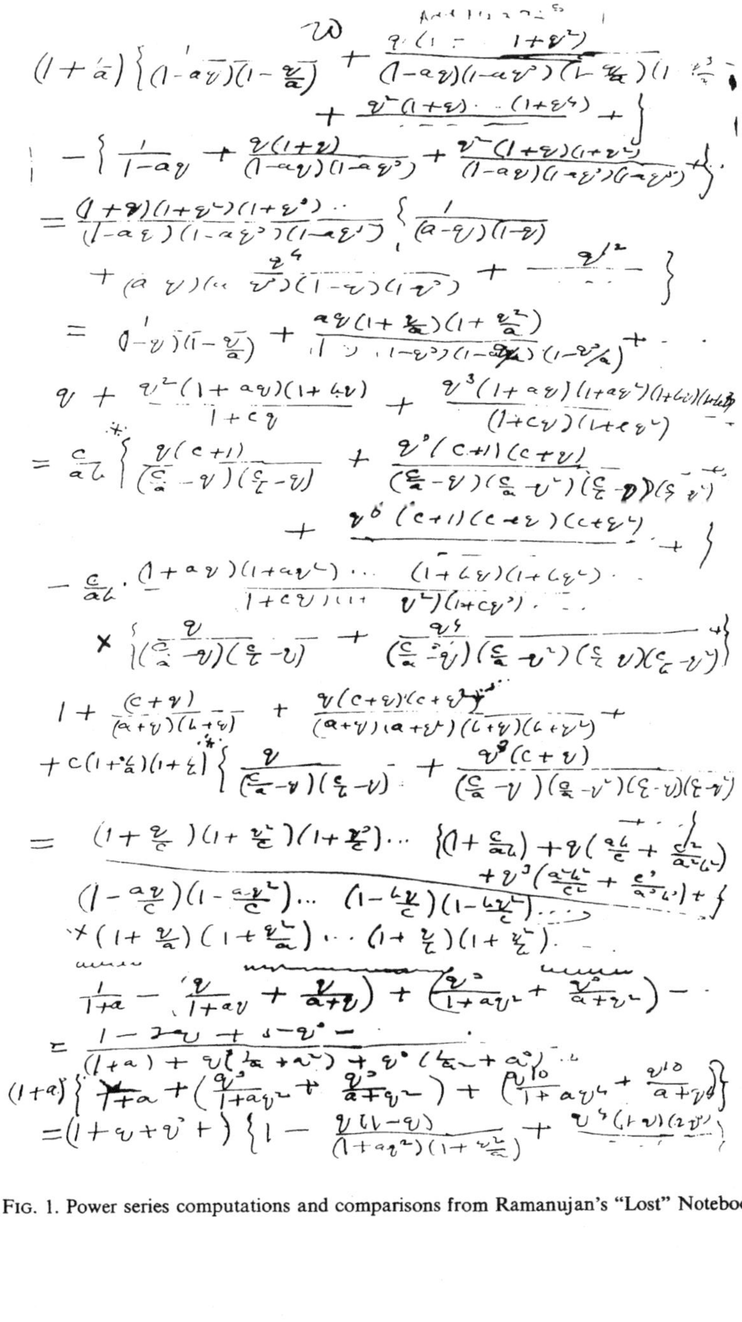

FIG. 1. Power series computations and comparisons from Ramanujan's "Lost" Notebook.

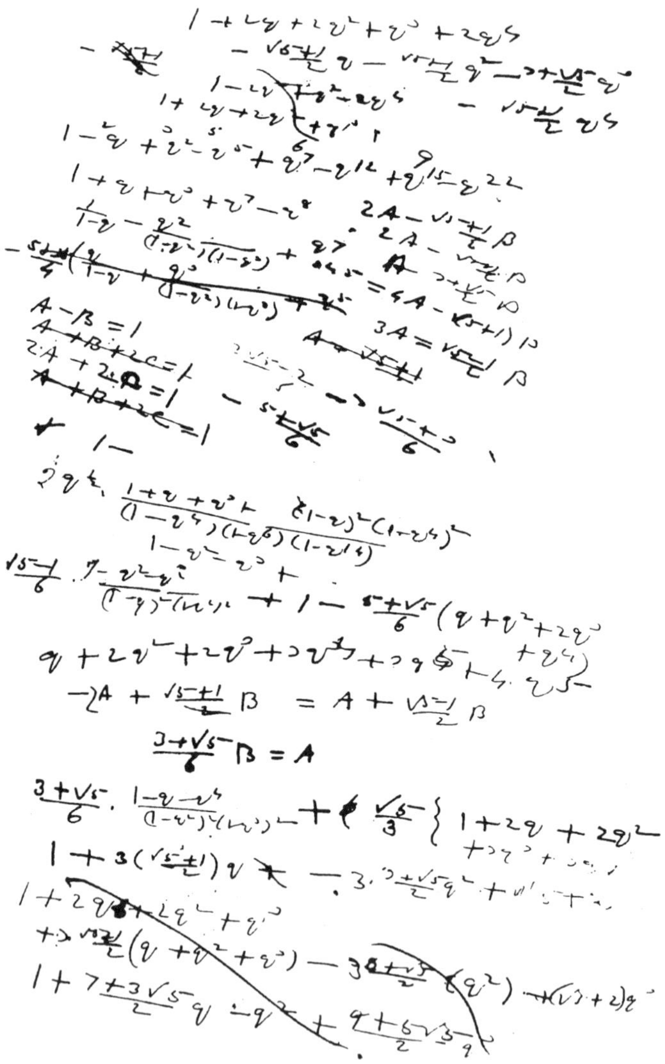

FIG. 2. A page of formulas from Ramanujan's "Lost" Notebook.

$$\sum_{n=0}^{\infty} \frac{a(a+1)\cdots(a+n-1)b(b+1)\cdots(b+n-1)t^n}{n!\ \ c(c+1)\cdots(c+n-1)}$$

$$=(1-t)^{c-a-b}\sum_{n=0}^{\infty} \frac{(c-a)(c-a+1)\cdots(c-a+n-1)(c-b)(c-b+1)\cdots(c-b+n-1)t^n}{n!\quad c(c+1)\cdots(c+n-1)}. \tag{2.6}$$

To understand fully Ramanujan's use of the terms "theta series," "false theta series," and "mock theta series," we quote from Ramanujan's last letter to Hardy [**27**, pp. 57–61] (also mentioned in Section 3 below). After the quotation we shall make a few comments to clarify further some of Ramanujan's ideas.

If we consider a ϑ-function in the transformed Eulerian form, e.g.,

$$1+\frac{q}{(1-q)^2}+\frac{q^4}{(1-q)^2(1-q^2)^2}+\frac{q^9}{(1-q)^2(1-q^2)^2(1-q^3)^2}+\cdots, \tag{A}$$

$$1+\frac{q}{1-q}+\frac{q^4}{(1-q)(1-q^2)}+\frac{q^9}{(1-q)(1-q^2)(1-q^3)}+\cdots, \tag{B}$$

and determine the nature of the singularities at the points

$$q=1, q^2=1, q^3=1, q^4=1, q^5=1, \ldots,$$

we know how beautifully the asymptotic form of the function can be expressed in a very neat and closed exponential form. For instance, when $q=e^{-t}$ and $t \to 0$,

$$(\mathrm{A})=\sqrt{\left(\frac{t}{2\pi}\right)}\exp\left(\frac{\pi^2}{6t}-\frac{t}{24}\right)+O(1),$$

$$(\mathrm{B})=\sqrt{\left(\frac{2}{5-\sqrt{5}}\right)}\exp\left(\frac{\pi^2}{15t}-\frac{t}{60}\right)+O(1),$$

and similar results at other singularities.

If we take a number of functions like (A) and (B), it is only in a limited number of cases the terms close as above; but in the majority of cases they never close as above. For instance, when $q=e^{-t}$ and $t \to 0$,

$$1+\frac{q}{(1-q)^2}+\frac{q^3}{(1-q)^2(1-q^2)^2}+\frac{q^6}{(1-q)^2(1-q^2)^2(1-q^3)^2}+\cdots$$

$$=\sqrt{\left(\frac{t}{2\pi\sqrt{5}}\right)}\exp\left[\frac{\pi^2}{5t}+a_1t+a_2t^2+\cdots+O(a_kt^k)\right], \tag{C}$$

where $a_1=1/8\sqrt{5}$, and so on. The function (C) is a simple example of a function behaving in an unclosed form at the singularities.

Now a very interesting question arises. Is the converse of the statements concerning the forms (A) and (B) true? That is to say: Suppose there is a function in the Eulerian form and suppose that all or an infinity of points are exponential singularities, and also suppose that at these points the asymptotic form of the function closes as neatly as in the cases of (A) and (B). The question is: Is the function taken the sum of two functions one of which is an ordinary ϑ-function and the other a (trivial) function which is $O(1)$ at all the points $e^{2m\pi i/n}$? The answer is it is not necessarily so. When it is not so, I call the function a Mock ϑ-function. I have not proved rigorously that it is not necessarily so. But I have constructed a number of examples in which it is inconceivable to construct a ϑ-function to cut out the singularities of the original function. Also I have shown that if it is necessarily so then it leads to the following assertion: viz., it is possible to construct two power series in x, namely, $\Sigma a_n x^n$ and $\Sigma b_n x^n$, both of which have essential singularities on the unit circle, are convergent when $|x|<1$, and tend to finite limits at every point $x=e^{2r\pi i/s}$, and at the same time the limit of $\Sigma a_n x^n$ at the point $x=e^{2r\pi i/s}$ is equal to the limit of $\Sigma b_n x^n$ at the point $x=e^{-2r\pi i/s}$.

This assertion seems to be untrue. Anyhow, we shall go to the examples and see how far our assertions are true. I have proved that, if

$$f(q)=1+\frac{q}{(1+q)^2}+\frac{q^4}{(1+q)^2(1+q^2)^2}+\cdots,$$

then

$$f(q)+(1-q)(1-q^3)(1-q^5)\cdots(1-2q+2q^4-2q^9+\cdots)=O(1)$$

at all the points $q=-1, q^3=-1, q^5=-1, q^7=-1, \ldots$; and at the same time

$$f(q)-(1-q)(1-q^3)(1-q^5)\cdots(1-2q+2q^4-2q^9+\cdots)=O(1)$$

at all the points $q^2=-1, q^4=-1, q^6=-1, \ldots$. Also, obviously, $f(q)=O(1)$ at all the points $q=1, q^3=1, q^5=1, \ldots$. And so $f(q)$ is a Mock ϑ-function.

When $q=-e^{-t}$ and $t\to 0$,

$$f(q)+\sqrt{\left(\frac{\pi}{t}\right)}\exp\left(\frac{\pi^2}{24t}-\frac{t}{24}\right)\to 4.$$

The coefficient of q^n in $f(q)$ is

$$(-1)^{n-1}\frac{\exp\left\{\pi\sqrt{\left(\frac{1}{6}n-\frac{1}{144}\right)}\right\}}{2\sqrt{\left(n-\frac{1}{24}\right)}}+O\left[\frac{\exp\left\{\frac{1}{2}\pi\sqrt{\left(\frac{1}{6}n-\frac{1}{144}\right)}\right\}}{\sqrt{\left(n-\frac{1}{24}\right)}}\right].$$

It is inconceivable that a single ϑ-function could be found to cut out the singularities of $f(q)$.

When Ramanujan refers to a ϑ-function, he apparently means sums, products, and quotients of series of the form

$$\sum_{n=-\infty}^{\infty}(-1)^{\epsilon n}q^{an^2+bn},$$

where $\epsilon=0$ or 1. Note that the full assertions connected with his series (A) and (B) are:

$$1+\frac{q}{(1-q)^2}+\frac{q^4}{(1-q)^2(1-q^2)^2}+\cdots=\frac{1}{\sum_{n=-\infty}^{\infty}(-1)^n q^{n(3n-1)/2}} \qquad \text{(A)}'$$

(an identity of Euler) [5, Chaps. 1 and 2], and

$$1+\frac{q}{1-q}+\frac{q^4}{(1-q)(1-q^2)}+\cdots=\frac{\sum_{n=-\infty}^{\infty}(-1)^n q^{n(5n-1)/2}}{\sum_{n=-\infty}^{\infty}(-1)^n q^{n(3n-1)/2}} \qquad \text{(B)}'$$

(the first Rogers–Ramanujan identity [5, Chap. 7]). The mock theta function $f(q)$ described by Ramanujan is also discussed in Section 3.

Theta functions have had considerable impact on various branches of mathematics, extending from mathematical physics to number theory. R. Bellman [9] has given a charming introduction to the many facets of theta functions. The mock theta functions are related to problems in additive number theory (see [8] and [7]).

Finally there are the "false theta functions." These are simply theta series with the "wrong" signs. For example, both

$$\sum_{n=-\infty}^{\infty}(-1)^n q^{6n^2+4n}=1-q^2-q^{10}+q^{16}+q^{32}-\cdots$$

and

$$\sum_{n=-\infty}^{\infty} q^{6n^2+4n}=1+q^2+q^{10}+q^{16}+q^{32}+\cdots$$

are theta series; however, the series on the right-hand side of (1.2) is not, in that

$$\sum_{n=0}^{\infty}(-1)^n q^{6n^2+4n}(1+q^{4n+2})=1+q^2-q^{10}-q^{16}+q^{32}+\cdots.$$

These "false" theta series do not seem to possess the analytic interest that Ramanujan describes for the ϑ-series and mock ϑ-series; however, they do crop up in some elegant identities such as (1.2).

There are other results from the general theory of q-series that we shall require in our treatment of the "lost" notebook. However, the ideas covered in this section constitute the fundamental tools in our treatment.

3. The origin of the "lost" notebook. There is no introduction or covering letter with this manuscript. Indeed there are only a few words scattered here and there throughout the manuscript. Concerning this notebook, Miss Rosemary Graham, Manuscript Cataloguer of the Trinity College Library, says: "... the notebook and other material was discovered among Watson's papers by Dr. J. M. Whittaker, who wrote the obituary of Professor Watson for the Royal Society. He passed the papers to Professor R. A. Rankin of Glasgow University, who, in December 1968, offered them to Trinity College so that they might join the other Ramanujan manuscripts already given to us by Professor Rankin on behalf of Professor Watson's widow."

Ramanujan's wife gives the following description of Ramanujan's last year (April 1919–April 1920) before his death: "He returned from England only to die, as the saying goes. He lived for less than a year. Throughout this period I lived with him without break. He was only skin and bones. He often complained of severe pain. In spite of it he was always busy doing his Mathematics. That evidently helped him to forget the pain. I used to gather the sheets of paper which he filled up. I would also give the slate whenever he asked for it." (Quoted from [**20**, p. 91].)

Of this intense mathematical activity, we have up to now only known of the mock theta functions. These functions were described in Ramanujan's last letter to Hardy, written from the University of Madras and dated January 12, 1920 [**18**, p. xxxi]: "I am extremely sorry for not writing you a single letter up to now...I discovered very interesting functions recently which I call 'Mock' ϑ-functions. Unlike the 'False' ϑ-functions (studied partially by Prof. Rogers in his interesting paper) they enter into mathematics as beautifully as the ordinary theta functions. I am sending you with this letter some examples." Besides the material quoted in Section 2, Ramanujan also defines four third-order mock theta functions, ten fifth-order functions and three seventh-order functions. He also includes three identities satisfied by the third-order functions and five identities satisfied by his first five fifth-order functions. He states that the other five fifth-order functions also satisfy similar identities.

Subsequent authors ([**27**], [**28**], [**12**], [**1**], [**2**], [**3**], [**7**]) have studied the mock theta functions extensively. All of these papers study the work that Ramanujan described in his last letter. Now the "lost" notebook contains all of the formulae for the third- and fifth-order mock theta functions given in Ramanujan's last letter. Furthermore, it contains the five identities for the second family of fifth-order functions that were only mentioned but not stated in the letter. The "lost" notebook also contains one-parameter generalizations of the third-order identities that were rediscovered independently in 1965 [**2**].

It appears that either Watson did not possess the "lost" notebook in the late 1930's when he worked on mock theta functions or he had filed it away and forgotten it. In any event, Watson [**27**, p. 61] says that he believes Ramanujan did not possess transformation formulae for the third-order mock theta functions such as

$$f(q)\prod_{n=1}^{\infty}(1-q^n)=1+4\sum_{n=1}^{\infty}\frac{(-1)^n q^{n(3n+1)/2}}{1+q^n}, \tag{3.1}$$

where $f(q)$ is one of Ramanujan's third-order mock theta functions given by

$$f(q)=1+\sum_{n=1}^{\infty}\frac{q^{n^2}}{(1+q)^2(1+q^2)^2\cdots(1+q^n)^2}. \tag{3.2}$$

However in the "lost" notebook we find

$$\prod_{n=1}^{\infty}(1-abq^n)\left(1+\sum_{n=1}^{\infty}\frac{q^{n^2}}{(1+aq)(1+aq^2)\cdots(1+aq^n)(1+bq)(1+bq^2)\cdots(1+bq^n)}\right)$$

$$=(1+a)(1+b)\sum_{n=0}^{\infty}\frac{(-1)^n a^n b^n q^{n(3n+1)/2}(1-abq^{2n})}{(1+aq^n)(1+bq^n)(1-ab)}\prod_{j=1}^{n}\frac{(1-abq^{j-1})}{(1-q^j)}, \tag{3.3}$$

which reduces to (3.1) for $a\to 1$, $b\to 1$. Indeed the master identity used by Watson to construct all of his mock theta function transformations is obtained directly from (3.3) by the substitutions $a\to e^{i\theta}$, $b\to e^{-i\theta}$.

Watson rightly suggests the importance of (3.1) by pointing out that he was only able to prove that $f(q)$ really was a mock theta function by utilizing (3.1). In a second paper [**28**], Watson studied the fifth-order mock theta functions and proved the five identities associated with each family. While Watson did succeed in proving that the fifth-order mock theta functions do indeed behave like theta functions [**28**, §6], he was unable to prove that they are not themselves theta functions. This was because he did not know a result comparable to (3.1) for the fifth-order functions. However, the "lost" notebook contains several such results. For example, $\phi_0(q)$ defined by

$$\phi_0(q)=1+\sum_{n=1}^{\infty}q^{n^2}(1+q)(1+q^3)\cdots(1+q^{2n-1}) \tag{3.4}$$

is one of the fifth-order functions, and in the "lost" notebook we find a result equivalent to

$$\phi_0(-q)=\prod_{n=0}^{\infty}\frac{(1-q^{5n+5})(1+q^{5n+2})(1+q^{5n+3})}{(1-q^{10n+2})(1-q^{10n+8})}$$

$$+1-\sum_{n=0}^{\infty}\frac{q^{5n^2}}{(1-q)(1-q^6)\cdots(1-q^{5n+1})(1-q^4)(1-q^9)\cdots(1-q^{5n-1})}. \tag{3.5}$$

Hence by (3.3) with $q\to q^5$ and then $a\to -q$, $b\to -q^{-1}$, we see that

$$\phi_0(-q)=\prod_{n=0}^{\infty}\frac{(1-q^{5n+5})(1+q^{5n+2})(1+q^{5n+3})}{(1-q^{10n+2})(1-q^{10n+8})}$$

$$+1-\prod_{n=0}^{\infty}(1-q^{5n+5})^{-1}\left\{\frac{1}{1-q}+(1-q^{-1})\sum_{n=1}^{\infty}\frac{(-1)^n q^{n(15n+5)/2}(1+q^{5n})}{(1-q^{5n+1})(1-q^{5n-1})}\right\}. \tag{3.6}$$

Now Watson was obviously totally unaware of (3.5) or he surely would have found (3.6). If he had found (3.6), the methods he used to prove that $f(q)$ is a mock theta function should have led him to a proof that the fifth-order functions are indeed true mock theta functions.

In a later paper I hope to examine all of the formulae in the "lost" notebook for the mock theta-functions. For now, I hope I have made the case for my assertion that this notebook was composed during the last year of Ramanujan's life, when, by his own words, he discovered the mock theta functions. I should add that while only a fraction (perhaps 5%) of the notebook is on the mock theta functions themselves, much of the rest of it involves related q-series expansions of theta functions and false theta functions. Thus it is not unreasonable to assume that the entire notebook was composed during Ramanujan's last year, especially since the results on the mock theta functions are scattered through it. Finally, the fact that its existence was never mentioned by anyone for over 55 years leads me to call it "lost." B. Birch [11] has found some other notes of Ramanujan's in the library of the Oxford Mathematical Institute; however, these notes comprise only 33 pages and, for the most part, apparently treat different formulae from those found in the "lost" notebook.

4. Proof of identity (1.1). In order to prove (1.1), we shall prove the following stronger result, which is also found in the "lost" notebook:

$$1+\sum_{n=1}^{\infty}\frac{q^n}{(1+aq)(1+aq^2)\cdots(1+aq^n)(1+bq)(1+bq^2)\cdots(1+bq^n)}$$

$$=(1+a^{-1})\left(1+\sum_{n=1}^{\infty}\frac{(-1)^n q^{n(n+1)/2}b^n a^{-n}}{(1+bq)(1+bq^2)\cdots(1+bq^n)}\right)-\frac{a^{-1}\sum_{n=0}^{\infty}(-1)^n q^{n(n+1)/2}b^n a^{-n}}{\prod_{j=1}^{\infty}(1+aq^j)(1+bq^j)}. \tag{4.1}$$

To obtain (1.1) from (4.1), first interchange a and b in (4.1) and then set $b=a^{-1}$; identity (1.1) follows once we recall the identity of L. J. Rogers [22, p. 335, Eq. (3)]:

$$1+\sum_{n=1}^{\infty}\frac{(-1)^n y^{2n} q^{n(n+1)/2}}{(1-yq)(1-yq^2)\cdots(1-yq^n)}=\sum_{n=0}^{\infty}(-1)^n y^{3n} q^{n(3n+1)/2}(1-y^2q^{2n+1}). \tag{4.2}$$

To simplify working with these series, we introduce the following standard notation:

$$(a)_n=(a;q)_n=(1-a)(1-aq)\cdots(1-aq^{n-1}). \tag{4.3}$$

The symbol $(a)_n$ is defined for all real n by the relation

$$(a)_n=(a;q)_n=\prod_{m=0}^{\infty}\frac{(1-aq^m)}{(1-aq^{m+n})},\qquad |q|<1.$$

Finally, we define

$$(a)_\infty=(a;q)_\infty=\prod_{m=0}^{\infty}(1-aq^m),\qquad |q|<1. \tag{4.4}$$

To prove (4.1) let us recall an identity of Euler [5, p. 19]:

$$\sum_{m=0}^{\infty}\frac{A^m q^{m(m+1)/2}}{(q)_m}=(-Aq)_\infty, \tag{4.5}$$

and an identity of L. J. Rogers [21, p. 171]:

$$\sum_{n=0}^{\infty}\frac{(\alpha)_n(\beta)_n\tau^n}{(q)_n(\gamma)_n}=\frac{(\gamma/\beta)_\infty(\beta\tau)_\infty}{(\gamma)_\infty(\tau)_\infty}\sum_{n=0}^{\infty}\frac{(\beta)_m(\alpha\beta\tau/\gamma)_m\gamma^m\beta^{-m}}{(q)_m(\beta\tau)_m}. \tag{4.6}$$

Now multiply both sides of (3.1) by $(-bq)_\infty$ and apply (3.5) with $A=bq^n$:

$$\sum_{n=0}^{\infty}\frac{q^n}{(-aq)_n}\sum_{m=0}^{\infty}\frac{q^{mn+m(m+1)/2}b^m}{(q)_m}$$

$$=(1+a^{-1})\sum_{n=0}^{\infty}(-1)^n q^{n(n+1)/2}b^n a^{-n}\sum_{n=0}^{\infty}\frac{q^{mn+m(m+1)/2}b^m}{(q)_m}-\frac{a^{-1}\sum_{n=0}^{\infty}(-1)^n q^{n(n+1)/2}b^n a^{-n}}{(-aq)_\infty}. \tag{4.7}$$

If we compare coefficients of b^N on both sides of (4.7) and multiply the resulting identity by $q^{-N(N+1)/2}$, we find that (4.1) is equivalent to

$$\frac{1}{(q)_N}\sum_{n=0}^{\infty}\frac{q^{(N+1)n}}{(-aq)_n}=a^{-N}(1+a^{-1})\sum_{m=0}^{N}\frac{(-1)^{N+m}a^m}{(q)_m}-\frac{(-1)^N a^{-N-1}}{(-aq)_\infty}. \tag{4.8}$$

To prove (4.8) we set $\beta=q,\alpha=0,\gamma=-aq$, and $\tau=q^{N+1}$ in (4.6); hence after dividing by $(q)_N$ we find

$$\begin{aligned}\frac{1}{(q)_N}\sum_{n=0}^{\infty}\frac{q^{(N+1)n}}{(-aq)_n}&=\frac{1}{(q)_N}\frac{(q^{N+2})_\infty(1+a)}{(q^{N+1})_\infty}\sum_{n=0}^{\infty}\frac{(-a)^n}{(q^{N+2})_n}\\&=(1+a)(-a)^{-N-1}\sum_{n=0}^{\infty}\frac{(-a)^{n+N+1}}{(q)_{n+N+1}}\\&=(1+a)(-a)^{-N-1}\left\{\sum_{n=0}^{\infty}\frac{(-a)^n}{(q)_n}-\sum_{n=0}^{N}\frac{(-a)^n}{(q)_n}\right\}\\&=(1+a)(-a)^{-N-1}\left\{\frac{1}{(-a)_\infty}-\sum_{n=0}^{N}\frac{(-a)^n}{(q)_n}\right\}\\&=\frac{-(-1)^N a^{-N-1}}{(-aq)_\infty}+a^{-N}(1+a^{-1})\sum_{n=0}^{N}\frac{(-1)^{n+N}a^n}{(q)_n},\end{aligned} \tag{4.9}$$

which is (4.8). Thus (4.1) is established. Note that in the penultimate step in (4.9) we used a second identity of Euler [**5**, p. 19]:

$$\sum_{n=0}^{\infty}\frac{A^n}{(q)_n}=\frac{1}{(A)_\infty}. \tag{4.10}$$

There exist generalizations of (4.1), and we shall discuss these later in this series of papers.

5. Proof of identity (1.2). To prove this result we shall require (4.2) and (4.6) as well as the fundamental transformation of E. Heine [**15**, p. 306]:

$$\sum_{n=0}^{\infty}\frac{(\alpha)_n(\beta)_n\tau^n}{(q)_n(\gamma)_n}=\frac{(\beta)_\infty(\alpha\tau)_\infty}{(\gamma)_\infty(\tau)_\infty}\sum_{m=0}^{\infty}\frac{(\gamma/\beta)_m(\tau)_m\beta^m}{(q)_m(\alpha\tau)_m}. \tag{5.1}$$

To begin we replace q by q^2 in (5.1) and then set $\alpha=0,\beta=q^2,\gamma=q^3$ and $\tau=-q$; after dividing both sides of the resulting identity by $(1-q)$ we find that

$$\sum_{n=0}^{\infty}\frac{(-1)^n q^n}{(q;q^2)_{n+1}}=\frac{(q^2;q^2)_\infty}{(q;q^2)_\infty(-q;q^2)_\infty}\sum_{m=0}^{\infty}\frac{(q;q^2)_m(-q;q^2)_m q^{2m}}{(q^2;q^2)_m}$$

$$
\begin{aligned}
&= (q^2;q^2)_\infty \sum_{m=0}^{\infty} \frac{q^{2m}}{(q^2;q^2)_m} \frac{1}{(q^{4m+2};q^4)_\infty} \\
&= (q^2;q^2)_\infty \sum_{m=0}^{\infty} \frac{q^{2m}}{(q^2;q^2)_m} \sum_{n=0}^{\infty} \frac{q^{n(4m+2)}}{(q^4;q^4)_n} \qquad \text{(by (4.10))} \\
&= (q^2;q^2)_\infty \sum_{n=0}^{\infty} \frac{q^{2n}}{(q^4;q^4)_n} \frac{1}{(q^{4n+2};q^2)_\infty} \qquad \text{(by (4.10))} \\
&= \sum_{n=0}^{\infty} \frac{(q^2;q^2)_{2n} q^{2n}}{(q^4;q^4)_n} \\
&= \sum_{n=0}^{\infty} (q^2;q^4)_n q^{2n}.
\end{aligned} \tag{5.2}
$$

Now in (4.6) replace q by q^4, then set $\beta = q^4, \alpha = q^2, \tau = q^2$, and let $c \to 0$; this yields

$$
\begin{aligned}
\sum_{n=0}^{\infty} (q^2;q^4)_n q^{2n} &= \sum_{m=0}^{\infty} \frac{(-1)^m q^{2m^2+2m}}{(q^2;q^4)_{m+1}} \\
&= -\sum_{m=1}^{\infty} \frac{(-1)^m q^{2m^2-2m}}{(q^2;q^4)_m} \\
&= 1 - \sum_{n=0}^{\infty} (-1)^n q^{6n^2-4n}(1-q^{8n}) \quad \text{by (4.2)} \\
&= \sum_{n=0}^{\infty} (-1)^n q^{6n+4n}(1+q^{4n+2}).
\end{aligned} \tag{5.3}
$$

Now if we equate the first member of (5.2) with the final member of (5.3), we deduce (1.2) upon replacing q by $-q$.

The following partition theorem is easily deduced from (1.2); it would be nice to have a combinatorial proof of this result:

THEOREM. *Let $R_i(n)$ denote the number of partitions of n into odd parts wherein the largest part is congruent to i* (mod 4) *and appears an odd number of times while all other parts appear an even number of times. Then*

$$
R_1(n) - R_3(n) = \begin{cases} (-1)^j & \text{if } n = 12j^2+8j+1 \quad \text{or} \quad 12j^2+16j+5 \\ 0 & \text{otherwise.} \end{cases} \tag{5.4}
$$

Proof. We note that

$$
1 + \sum_{n=1}^{\infty} (R_1(n) - R_3(n)) q^n = \sum_{n=0}^{\infty} \frac{(-1)^n q^{2n+1}}{(1-q^2)(1-q^6)\cdots(1-q^{4n+2})}. \tag{5.5}
$$

since

$$
\frac{1}{(1-q^{1+1})(1-q^{3+3})\cdots(1-q^{(2n+1)+(2n+1)})} \tag{5.6}
$$

is the generating function for partitions with odd parts each $\leqslant 2n+1$ and each appearing an even number of times. Our theorem now follows if we replace q by $-q^2$ in (1.2) and multiply the resulting identity by q.

For example, when $n=5$ the partitions enumerated by $R_1(5)$ are 5 and $1+1+1+1+1$ while that enumerated by $R_3(5)$ is $3+1+1$; hence $R_1(5)-R_3(5)=2-1=1$ as predicted. When $n=11$, $R_1(11)$ enumerates $9+1+1, 5+3+3, 5+1+1+1+1+1+1$ and $1+1+1+1+1+1+1+1+1+1+1$ while $R_3(11)$ enumerates $11, 7+1+1+1+1, 3+3+3+1+1, 3+1+1+1+1+1+1+1+1$; hence $R_1(11)-R_3(11)=4-4=0$ as predicted.

Later in this series we shall present further partition theorems implied by Ramanujan's work in this "lost" notebook.

6. Proof of identity (1.3). This continued fraction identity depends both on the well-known Rogers–Ramanujan continued fraction (1.6) and on what has become known as Ramanujan's ${}_1\psi_1$ summation:

$$
\begin{aligned}
{}_1\psi_1\left[\begin{matrix} a; q, t \\ b \end{matrix}\right] &= \sum_{n=-\infty}^{\infty} \frac{(a)_n}{(b)_n} t^n \\
&= \frac{(b/a)_\infty (az)_\infty (q/az)_\infty (q)_\infty}{(q/a)_\infty (b/az)_\infty (b)_\infty (z)_\infty}. \qquad (6.1)
\end{aligned}
$$

Proofs of (6.1) can be found in **[4]**, **[6]**, **[7]**.

We start on (1.3) by proving an elementary series identity: for $i=1, 2, 3$, or 4

$$
\sum_{n=0}^{\infty} q^{5n^2+2in} \frac{1+q^{5n+i}}{1-q^{5n+i}} = \sum_{n=0}^{\infty} \frac{q^{in}}{1-q^{5n+i}}. \qquad (6.2)
$$

This identity holds because

$$
\begin{aligned}
\sum_{n=0}^{\infty} \frac{q^{in}}{1-q^{5n+i}} &= \sum_{n=0}^{\infty} \sum_{n=0}^{\infty} q^{in+5nm+mi} \\
&= \sum_{n=0}^{\infty} \left(\sum_{m=0}^{n} + \sum_{m=n+1}^{\infty} \right) q^{in+5nm+im} \\
&= \sum_{m=0}^{\infty} \sum_{n=0}^{\infty} q^{i(n+m)+5(n+m)n+im} \\
&\quad + \sum_{n=0}^{\infty} \sum_{m=0}^{\infty} q^{in+5n(m+n+1)+i(m+n+1)} \\
&= \sum_{m=0}^{\infty} \frac{q^{5m^2+2im}}{1-q^{5m+i}} + \sum_{n=0}^{\infty} \frac{q^{5n^2+5n+2in+i}}{1-q^{5n+i}} \\
&= \sum_{n=0}^{\infty} q^{5n^2+2in} \frac{1+q^{5n+i}}{1-q^{5n+i}}. \qquad (6.3)
\end{aligned}
$$

Next we see that by (6.1) for $i=1, 2, 3$, or 4

$$
\begin{aligned}
\sum_{n=-\infty}^{\infty} \frac{q^{in}}{1-q^{5n+i}} &= \frac{1}{1-q^i} {}_1\psi_1\left[\begin{matrix} q^i; q^5, q^i \\ q^{5+i} \end{matrix}\right] \\
&= \frac{(q^5;q^5)_\infty^2 (q^{2i};q^5)_\infty (q^{5-2i};q^5)_\infty}{(a^{5-i};q^5)_\infty^2 (q^i;q^5)_\infty^2}, \qquad (6.4)
\end{aligned}
$$

Hence

$$\frac{\sum_{n=0}^{\infty} q^{5n^2+4n}\frac{1+q^{5n+2}}{1-q^{5n+2}}-\sum_{n=0}^{\infty} q^{5n^2+6n+1}\frac{1+q^{5n+3}}{1-q^{5n+3}}}{\sum_{n=0}^{\infty} q^{5n^2+2n}\frac{1+q^{5n+1}}{1-q^{5n+1}}-\sum_{n=0}^{\infty} q^{5n^2+8n+3}\frac{1+q^{5n+4}}{1-q^{5n+4}}}$$

$$=\frac{\sum_{n=0}^{\infty}\frac{q^{2n}}{1-q^{5n+2}}-\sum_{n=0}^{\infty}\frac{q^{3n+1}}{1-q^{5n+3}}}{\sum_{n=0}^{\infty}\frac{q^{n}}{1-q^{5n+1}}-\sum_{n=0}^{\infty}\frac{q^{4n+3}}{1-q^{5n+4}}}\qquad\text{(by (6.3))}$$

$$=\frac{\sum_{n=-\infty}^{\infty}\frac{q^{2n}}{1-q^{5n+2}}}{\sum_{n=-\infty}^{\infty}\frac{q^{n}}{1-q^{5n+1}}}$$

$$=\frac{(q^5;q^5)_\infty^2(q^4;q^5)_\infty(q;q^5)_\infty}{(q^3;q^5)_\infty^2(q^2;q^5)_\infty^2}\cdot\frac{(q^4;q^5)_\infty^2(q;q^5)_\infty^2}{(q^5;q^5)_\infty^2(q^2;q^5)_\infty(q^3;q^5)_\infty}$$

$$=\left(\frac{(q^4;q^5)_\infty(q;q^5)_\infty}{(q^2;q^5)_\infty(q^3;q^5)_\infty}\right)^3$$

$$=\left[\cfrac{1}{1+\cfrac{q}{1+\cfrac{q^2}{1+\cfrac{q^3}{\ddots}}}}\right]^3\qquad\text{(by (1.6)).}$$

Also in the "lost" notebook are more than a dozen identities deduced from special cases of (6.4) combined with (1.6).

7. Proof of identity (1.4). Instead of determining the convergents of this continued fraction, we deduce this identity from a family of relations. To obtain this family, we require a further identity of E. Heine [**15**, p. 325] (see also L. J. Rogers [**21**, p. 171]):

$$\sum_{n=0}^{\infty}\frac{(\alpha)_n(\beta)_n\tau^n}{(q)_n(\gamma)_n}=\frac{(\alpha\beta\tau/\gamma)_\infty}{(\tau)_\infty}\sum_{n=0}^{\infty}\frac{(\gamma/\alpha)_n(\gamma/\beta)_n}{(q)_n(\gamma)_n}\left(\frac{\alpha\beta\tau}{\gamma}\right)^n. \tag{7.1}$$

Next we define for nonnegative integral i:

$$F_i(a,\lambda;b;q)=\sum_{n=0}^{\infty}\frac{q^{(n+2i+1)n/2}(-\lambda a^{-1}q^i)_n a^n}{(q)_n(-bq)_{n+i}}, \tag{7.2}$$

$$H_i(a,\lambda;b;q)=\sum_{n=0}^{\infty}\frac{q^{(n+2i+3)n/2}(-\lambda a^{-1}q^i)_n a^n}{(q)_n(-bq)_{n+i}}. \tag{7.3}$$

Immediately from (7.2) and (7.3) we obtain our first relation:

$$F_i(a,\lambda;b;q)-H_i(a,\lambda;b;q) = \sum_{n=0}^{\infty} \frac{q^{(n+2i+1)n/2}(-\lambda a^{-1}q^i)_n a^n}{(q)_n(-bq)_{n+i}}(1-q^n)$$

$$= \sum_{n=0}^{\infty} \frac{q^{(n+2i+2)(n+1)/2}(-\lambda a^{-1}q^i)_{n+1}a^{n+1}}{(q)_n(-bq)_{n+i+1}}$$

$$=(aq^{i+1}+\lambda q^{2i+1})\sum_{n=0}^{\infty} \frac{q^{(n+2(i+1)+1)n/2}(-\lambda a^{-1}q^{i+1})_n a^n}{(q)_n(-bq)_{n+i+1}}$$

$$=(aq^{i+1}+\lambda q^{2i+1})F_{i+1}(a,\lambda;b;q). \tag{7.4}$$

Next we transform $F_i(a,\lambda;b;q)$ and $H_i(a,\lambda;b;q)$ utilizing (6.1). First in (7.1) we set

$$\gamma=-bq^{i+1}, \quad \tau=-\frac{q^{i+2}a}{\alpha}, \quad \beta=-\gamma a^{-1}q^i$$

and let $\alpha\to\infty$:

$$H_i(a,\lambda;b;q)=(-\lambda b^{-1}q^{i+1})_\infty \sum_{n=0}^{\infty} \frac{(ba\lambda^{-1})_n(-\lambda b^{-1}q^{i+1})^n}{(q)_n(-bq)_{n+i}}; \tag{7.5}$$

next we set

$$\gamma=-bq^{i+1}, \tau=-\frac{q^{i+1}a}{\alpha}, \beta=-\lambda a^{-1}q^i$$

and let $\alpha\to\infty$:

$$F_i(a,\lambda;b;q)=(-\lambda b^{-1}q^i)_\infty \sum_{n=0}^{\infty} \frac{(ba\lambda^{-1})_n(-\lambda b^{-1}q^i)^n}{(q)_n(-bq)_{n+i}}. \tag{7.6}$$

From (7.5) and (7.6) we obtain our second relation:
$H_i(a,\lambda;b;q)-F_{i+1}(a,\lambda;b;q)$

$$=(-\lambda b^{-1}q^{i+1})_\infty \sum_{n=0}^{\infty} \frac{(ba\lambda^{-1})_n(-\lambda b^{-1}q^{i+1})^n}{(q)_n(-bq)_{n+i+1}}(1+bq^{n+i+1}-1)$$

$$=(bq^{i+1}+\lambda q^{2i+2})(-\lambda b^{-1}q^{i+2})_\infty \sum_{n=0}^{\infty} \frac{(ba\lambda^{-1})_n(-\lambda b^{-1}q^{i+2})^n}{(q)_n(-bq)_{n+i+1}}$$

$$=(bq^{i+1}+\lambda q^{2i+2})H_{i+1}(a,\lambda;b;q). \tag{7.7}$$

Finally we note that from the definition of $G(a,\lambda)$ given after (1.4), we have

$$G(a,\lambda)=F_0(a,\lambda;b;q), \tag{7.8}$$

and

$$G(aq,\lambda q)=H_0(a,\lambda;b;q). \tag{7.9}$$

We now deduce (1.4) from iterated application of (7.7) and (7.4):

$$\frac{G(aq,\lambda q)}{G(a,\lambda)}=\frac{1}{\dfrac{F_0(a,\lambda;b;q)}{H_0(a,\lambda;b;q)}} \quad \text{(by (7.8) and (7.9))}$$

$$= \cfrac{1}{1+\cfrac{aq+\lambda q}{\cfrac{H_0(a,\lambda;b;q)}{F_1(a,\lambda;b;q)}}} \quad \text{(by (7.4))}$$

$$= \cfrac{1}{1+\cfrac{aq+\lambda q}{1+\cfrac{bq+\lambda q^2}{\cfrac{F_1(a,\lambda;b;q)}{H_1(a,\lambda;b;q)}}}} \quad \text{(by (7.7))}$$

$$= \cfrac{1}{1+\cfrac{aq+\lambda q}{1+\cfrac{bq+\lambda q^2}{1+\cfrac{aq^2+\lambda q^3}{\cfrac{H_1(a,\lambda;b;q)}{F_2(a,\lambda;b;q)}}}}} \quad \text{(by (7.4))}$$

$$\vdots \quad .$$

Convergence of the continued fraction is assured by the fact that both H_i/F_{i+1} and F_i/H_i are analytic in $|q|<1$ with each of the form $1+O(q^{i+1})$.

Having proved (1.4), let us now examine the corollaries discovered by Ramanujan:

$$\prod_{n=0}^{\infty} \frac{(1-q^{2n+1})}{(1-q^{4n+2})^2} = \cfrac{1}{1+\cfrac{q}{1+\cfrac{q^2+q}{1+\cfrac{q^3}{1+\cfrac{q^4+q^2}{1+\vdots}}}}} \quad . \tag{7.10}$$

Identity (7.10) follows if we set $a=0$, $b=1$, $\lambda=1$ in (1.4) and recall (4.5), which implies

$$G(0,\lambda) = \sum_{n=0}^{\infty} \frac{q^{n^2}\lambda^n}{(q^2;q^2)_n} = (-\lambda q;q^2)_\infty,$$

so

$$\frac{G(0,q)}{G(0,1)} = \frac{(-q^2;q^2)_\infty}{(-q;q^2)_\infty} = \frac{1}{(q^2;q^4)_\infty(-q;q^2)_\infty} \quad \text{(by [5, p. 5, Eq. (1.2.5)])} \quad = \frac{(q;q^2)_\infty}{(q^2;q^4)_\infty^2}.$$

Next we have an identity originally discovered by Eisenstein [13, p. 36]:

$$\sum_{n=0}^{\infty} (-1)^n q^{n(n+1)/2} = \cfrac{1}{1+\cfrac{q}{1+\cfrac{q^2-q}{1+\cfrac{q^3}{1+\cfrac{q^4-q^2}{\vdots}}}}} \quad . \tag{7.11}$$

This result follows by our setting $a=0$, $\lambda=1$, $b=-1$ in (1.4) and noting

$$\sum_{n=0}^{\infty} \frac{q^{n^2-n}\tau^n}{(q)_n(\gamma)_n} = \frac{1}{(\gamma)_\infty} \sum_{m=0}^{\infty} \frac{(\tau/\gamma)_m}{(q)_m}(-1)^m \gamma^m q^{m(m-1)/2},$$

which follows from (3.6) if we replace τ by $\tau\alpha^{-1}\beta^{-1}$ and let α and β tend to infinity. Hence

$$\frac{G(0,q;-1;q)}{G(0,1;-1;q)} = \frac{\sum_{m=0}^{\infty} \frac{q^{m^2+m}}{(q)_m^2}}{\sum_{m=0}^{\infty} \frac{q^{m^2}}{(q)_m^2}} = \frac{\frac{1}{(q)_\infty} \sum_{m=0}^{\infty} (-1)^m q^{m(m+1)/2}}{1/(q)_\infty} = \sum_{m=0}^{\infty} (-1)^m q^{m(m+1)/2}.$$

$$\prod_{n=0}^{\infty} \frac{(1-q^{6n+1})(1-q^{6n+5})}{(1-q^{6n+3})^2} = \cfrac{1}{1+\cfrac{q+q^2}{1+\cfrac{q^2+q^4}{1+\cfrac{q^3+q^6}{1+\vdots}}}}\quad . \tag{7.12}$$

To obtain this result we replace q by q^2 in (1.4), then we set $a=q^{-1}$, $b=1$, $\lambda=1$. Then we observe

$$G(q^{-1},1;1;q^2) = \sum_{n=0}^{\infty} \frac{q^{n^2}(-q;q^2)_n}{(q^4;q^4)_n} = \frac{(-q;q^2)_\infty(q^3;q^3)_\infty(q^3;q^6)_\infty}{(q^2;q^2)_\infty},$$

a result given by Slater [24, p. 154, Eq. (25)], and finally

$$G(q,q^2;1;q^2) = \sum_{n=0}^{\infty} \frac{q^{n^2+2n}(-q;q^2)_n}{(q^4;q^4)_n}$$

$$= \frac{(-q;q^2)_\infty(q^6;q^6)_\infty(q;q^6)_\infty(q^5;q^6)_\infty}{(q^2;q^2)_\infty}.$$

This last formula has apparently not been stated in the literature before; however, it is easily deduced if we set $y=-q^{1/2}$ and let $z\to\infty$ in Slater's identity E(4) [23, p. 469].

$$\prod_{n=0}^{\infty} \frac{(1-q^{8n+1})(1-q^{8n+7})}{(1-q^{8n+3})(1-q^{8n+5})} = \cfrac{1}{1+\cfrac{q^2+q}{1+\cfrac{q^4}{1+\cfrac{q^6+q^3}{\vdots}}}}\quad . \tag{7.13}$$

Here we replace q by q^2 in (1.4); then we set $a=q^{-1}$, $b=0$, $\lambda=1$. Under these substitutions

$$\frac{G(q,q^2;0;q^2)}{G(q^{-1},1;0;q^2)}=\frac{\sum_{n=0}^{\infty}\frac{q^{n^2+2n}(-q;q^2)_n}{(q^2;q^2)_n}}{\sum_{n=0}^{\infty}\frac{q^{n^2}(-q;q^2)_n}{(q^2;q^2)_n}}$$

$$=\frac{(q;q^8)_\infty(q^7;q^8)_\infty}{(q^3;q^8)_\infty(q^5;q^8)_\infty},$$

where the last equation follows from two identities of Slater [**24**, p. 155, Eqs. (34) and (36)].

Besides the case $a=b=0,\lambda=1$ which yields (1.6), Ramanujan presented one other corollary which involves a false theta function.

$$\sum_{n=0}^{\infty}(-1)^n q^{3n^2+2n}(1+q^{2n+1})=\cfrac{1}{1+\cfrac{q^2-q}{1+\cfrac{q^4-q^2}{1+\cfrac{q^6-q^3}{\vdots}}}}\quad.\tag{7.14}$$

To obtain this result we replace q by q^2 in (1.4); then we set $a=-q^{-1}$, $b=-1$, $\lambda=1$. Thus

$$\frac{G(-q,q^2;-1;q^2)}{G(-q^{-1},1;-1;q^2)}=\lim_{\tau\to\infty}\frac{\sum_{n=0}^{\infty}\frac{(q/\tau;q^2)_n(q;q^2)_n\tau^n q^{2n}}{(q^2;q^2)_n(q^2;q^2)_n}}{\sum_{n=0}^{\infty}\frac{(q/\tau;q^2)_n(q;q^2)_n\tau^n}{(q^2;q^2)_n(q^2;q^2)_n}}$$

$$=\frac{\frac{(q^3;q^2)}{(q^2;q^2)}\sum_{n=0}^{\infty}\frac{(-1)^n q^{n^2+n}}{(q^3;q^2)_n}}{\frac{(q;q^2)_\infty}{(q^2;q^2)_\infty}}\qquad\text{(by (4.6))}$$

$$=\sum_{n=0}^{\infty}(-1)^n\frac{q^{n^2+n}}{(q;q^2)_{n+1}}$$

$$=\sum_{n=0}^{\infty}(-1)^n q^{3n^2+2n}(1+q^{2n+1}),$$

where the last equation is due to L. J. Rogers [**22**, p. 333, Eq. (4)].

8. Proof of identity (1.5). This result is a good deal easier than the others we have considered. First we give an analytic proof. For this proof we require a special case of the q-analog of the binomial series (2.4) [**5**, p. 36, Eq. (3.37)]:

$$\frac{1}{(z)_{n+1}}=\sum_{m=0}^{\infty}\frac{(q)_{n+m}}{(q)_n(q)_m}z^m.\tag{8.1}$$

Thus the left-hand side of (1.5) is

$$\sum_{n=0}^{\infty} \frac{\beta^n}{(\alpha x^n)_{n+1}}$$

where $q = y/x$. Hence

$$\sum_{n=0}^{\infty} \frac{\beta^n}{(\alpha x^n)_{n+1}} = \sum_{n=0}^{\infty} \sum_{m=0}^{\infty} \beta^n \alpha^m \frac{(q)_{n+m}}{(q)_n (q)_m} x^{nm},$$

and the symmetry in α and β is now obvious. Hence (1.5) is established.

Combinatorially we see that the coefficient of $\alpha^m \beta^n$ in

$$\sum_{n=0}^{\infty} \frac{\beta^n}{(1-\alpha x^n)(1-\alpha x^{n-1}y)\cdots(1-\alpha y^n)} \tag{8.2}$$

is a homogeneous polynomial in x and y of degree nm. In this polynomial the coefficient of $y^j x^{nm-j}$ is the number of partitions of j into at most m parts each at most n, say $j = \alpha_1 + \cdots + \alpha_r$. Then the exponent on x is

$$nm - j = (n-\alpha_1) + (n-\alpha_2) + \cdots + (n-\alpha_r) + \underbrace{n + \cdots + n}_{m-r \text{ terms}}$$

which is another partition of at most m parts each $\leqslant n$ and it is uniquely determined by $\alpha_1 + \cdots + \alpha_r$, n and m. Now if we conjugate all the partitions under consideration we merely interchange the roles of n and m. Hence (8.2) is symmetric in α and β.

9. Conclusion. This is the first of several papers that will treat Ramanujan's "lost" notebook. I hope that the results discussed here indicate that Ramanujan discovered many results of interest in the last year of his life. I wish to close by quoting the last two paragraphs of G. N. Watson's presidential address to the London Mathematical Society in 1935 [**27**, p. 80]. In this address Watson has just finished an extensive discussion of the mock theta functions based on the last letter of Ramanujan, which we quoted in Sections 2 and 3. Watson concludes as follows:

> The study of Ramanujan's work and of the problems to which it gives rise inevitably recalls to mind Lamé's remark that, when reading Hermite's papers on modular functions, "on a la chair de poule." I would express my own attitude with more prolixity by saying that such a formula as
>
> $$\int_0^{\infty} e^{-3\pi x^2} \frac{\sinh \pi x}{\sinh 3\pi x}\, dx = \frac{1}{e^{2\pi/3}\sqrt{3}} \sum_{n=0}^{\infty} \frac{e^{-2n(n+1)\pi}}{(1+e^{-\pi})^2(1+e^{-3\pi})^2 \cdots (1+e^{-(2n+1)\pi})^2}$$
>
> gives me a thrill which is indistinguishable from the thrill which I feel when I enter the Sagrestia Nuova of the Capelle Medicee and see before me the austere beauty of the four statues representing Day, Night, Evening, and Dawn which Michelangelo has set over the tombs of Guiliano de'Medici and Lorenzo de'Medici.
>
> Ramanujan's discovery of the mock theta functions makes it obvious that his skill and ingenuity did not desert him at the oncoming of his untimely end. As much as any of his earlier work, the mock theta functions are an achievement sufficient to cause his name to be held in lasting remembrance. To his students such discoveries will be a source of delight and wonder until the time shall come when we too shall make our journey to that Garden of Proserpine where
>
> "Pale, beyond porch and portal,
> Crowned with calm leaves, she stands
> Who gathers all things mortal
> With cold immortal hands."

Acknowledgment. The author wishes to thank the Master and Fellows of Trinity College, Cambridge, for supplying photographs of the two pages from Ramanujan's "Lost" Notebook herein reproduced. (See Figures 1 and 2.)

Partially supported by National Science Foundation Grant MPS 74-07282.

References

1. G. E. Andrews, On the theorems of Watson and Dragonette for Ramanujan's mock theta functions, Amer. J. Math., 88 (1966) 454–490.

2.———, On basic hypergeometric series, mock theta functions and partitions I, Quart. J. Math. Oxford Ser., 17 (1966) 64–80.

3.———, On basic hypergeometric series, mock theta functions and partitions II, Quart. J. Math. Oxford Ser., 17 (1966) 132–143.

4.———, On Ramanujan's summation of ${}_1\psi_1(a;b;z)$, Proc. Amer. Math. Soc., 22 (1969) 552–553.

5.———, The Theory of Partitions, Encyclopedia of Math. and Its Applications, Vol. 2, Addison-Wesley, Reading, 1976.

6. G. E. Andrews and R. Askey, A simple proof of Ramanujan's summation of the ${}_1\psi_1$, Aequationes Math. (to appear).

7.———, The classical and discrete orthogonal polynomials and their q-analogs (to appear).

8. A. O. L. Atkin and H. P. F. Swinnerton-Dyer, Some properties of partitions, Proc. London Math. Soc. (3), 4 (1953) 84–106.

9. R. Bellman, A Brief Introduction to Theta Functions, Holt, Rinehart and Winston, New York, 1961.

10. B. C. Berndt, Ramanujan's notebooks (to appear).

11. B. J. Birch, A look back at Ramanujan's notebooks, Proc. Cambridge Phil. Soc., 78 (1975) 73–79.

12. L. A. Dragonette, Some asymptotic formulae for the mock theta series of Ramanujan, Trans. Amer. Math. Soc., 72 (1952) 474–500.

13. G. Eisenstein, Mathematische Werke, Vol. 1, Chelsea, New York, 1975.

14. G. H. Hardy, Ramanujan, Cambridge University Press, London and New York, 1940; reprinted by Chelsea, New York.

15. E. Heine, Untersuchungen über die Reihe..., J. Reine Angew. Math., 34 (1847) 285–328.

16. M. Ismail, A simple proof of Ramanujan's ${}_1\psi_1$ sum, Proc. Amer. Math. Soc., 63 (1977) 185–186.

17. M. Kothmann, On the ranks of partitions, Ph.D. Thesis, The Pennsylvania State University, 1977.

18. S. Ramanujan, Collected Papers, Cambridge University Press, London and New York, 1927; reprinted Chelsea, New York.

19.———, Notebooks of Srinivasa Ramanujan (2 volumes), Tata Institute of Fundamental Research, Bombay, 1957.

20. S. R. Ranganathan, Ramanujan the Man and the Mathematician, Asia Publishing House, Bombay, 1967.

21. L. J. Rogers, On a threefold symmetry in the elements of Heine's series, Proc. London Math. Soc., 24 (1893) 171–179.

22.———, On two theorems of combinatory analysis and some allied identities, Proc. London Math. Soc. (2), 16 (1916) 315–336.

23. L. J. Slater, A new proof of Roger's transformations of infinite series, Proc. London Math. Soc. (2), 53 (1951) 460–475.

24.———, Further identities of the Rogers–Ramanujan type, Proc. London Math. Soc. (2), 54 (1952) 147–167.

25.———, Generalized Hypergeometric Functions, Cambridge University Press, Cambridge, 1966.

26. G. N. Watson, Ramanujan's Note Books, J. London Math. Soc., 6 (1931) 137–153.

27.———, The final problem: An account of the mock theta functions, J. London Math. Soc., 11 (1936) 55–80.

28.———, The mock theta functions II, Proc. London Math. Soc. (2) 42 (1937) 272–304.

Department of Mathematics, The Pennsylvania State University, University Park, PA 16802.

2.8 Statistical Mechanics

I have recounted in detail in [Book3] my initial contact with Baxter concerning the hard hexagon model. Thus followed seven papers joint with Baxter and others: [112], [120], [123], [128], [132], [140], [141]. I have included [112] and [132] as the most significant representatives of this collaboration. I was deeply honored in 1978 to recieve an honorary dectorate in physics from the University of Parma, one of the world's oldest universities.

Apart from the work with Baxter and Forrester, I also have joint physics papers with Onofri [108], with Forrester alone [130], [144] and with Berkovich [197].

Paper [101] is my only solo-authored paper related to the hard hexagon model. A further account of my physics work, including a leisurely account of [101], occurs in Chapter 8 of [Book3].

2.8.1 *Eight-vertex SOS model and generalized Rogers–Ramanujan-type identities (with R.J. Baxter and P.J. Forrester)*

Journal of Statistical Physics, Vol. 35, Nos. 3/4, 1984

Eight-Vertex SOS Model and Generalized Rogers–Ramanujan-Type Identities

George E. Andrews,[1,2] R. J. Baxter,[3] and P. J. Forrester[3]

Received December 21, 1983

The eight-vertex model is equivalent to a "solid-on-solid" (SOS) model, in which an integer height l_i is associated with each site i of the square lattice. The Boltzmann weights of the model are expressed in terms of elliptic functions of period $2K$, and involve a variable parameter η. Here we begin by showing that the hard hexagon model is a special case of this eight-vertex SOS model, in which $\eta = K/5$ and the heights are restricted to the range $1 \leqslant l_i \leqslant 4$. We remark that the calculation of the sublattice densities of the hard hexagon model involves the Rogers–Ramanujan and related identities. We then go on to consider a more general eight-vertex SOS model, with $\eta = K/r$ (r an integer) and $1 \leqslant l_i \leqslant r-1$. We evaluate the local height probabilities (which are the analogs of the sublattice densities) of this model, and are automatically led to generalizations of the Rogers–Ramanujan and similar identities. The results are put into a form suitable for examining critical behavior, and exponents $\beta, \alpha, \bar{\alpha}$ are obtained.

KEY WORDS: Statistical mechanics; lattice statistics; number theory; eight-vertex model; solid-on-solid model; hard hexagon model; Rogers–Ramanujan identities.

1. THE LOCAL HEIGHT PROBABILITIES P_a

1.1. Introduction

Many of the exactly solved two-dimensional models in statistical mechanics are equivalent to special cases of the eight-vertex model.[1–3] For instance,

[1] Department of Mathematics, McAllister Building, Pennsylvania State University, University Park, Pennsylvania 16802.

[2] Supported by the Guggenheim Foundation and in part by the National Science Foundation, grant MCS 8201733.

[3] Research School of Physical Sciences, The Australian National University, G.P.O. Box 4, Canberra 2601, Australia.

the critical Potts model is equivalent to a six-vertex model,[4] which is the eight-vertex model with two-vertex configurations given zero weight. (One point that should be made is that the six-vertex model can be solved in an electric field,[5] which is not true of the eight-vertex model. Here we shall consider only the "zero-field" eight-vertex model.)

The triangular three-spin model was originally proposed and solved[6] quite independently of the eight-vertex model: three years went by before it was realized[7] that it can be transformed into a special eight-vertex model.

It seems that history must repeat itself: it is now a little over three years since the hard hexagon model was solved,[8,9] and until now it was believed to be distinct from the eight-vertex model (p. 453 of Ref. 10). However, one way of solving the eight-vertex model[11–14] is to convert it to a "solid-on-solid" (SOS) model (Section 3 of Ref. 12, and Section 9.1.2 of Ref. 14). We begin this paper by showing that the hard hexagon model is in fact a special case of this SOS model. For this case, the eight-vertex parameter η has the value

$$\eta = K/5 \tag{1.1.1}$$

where $2K$ and $2iK'$ are the periods of the elliptic functions that naturally occur.

One intriguing feature of the hard hexagon model is that the Rogers–Ramanujan and related identities occur very naturally in the calculation of the local densities.[9] These involve functions such as

$$\begin{aligned} G(x) &= \prod_{m=1}^{\infty} \left[(1 - x^{5m-4})(1 - x^{5m-1})\right]^{-1} \\ H(x) &= \prod_{m=1}^{\infty} \left[(1 - x^{5m-3})(1 - x^{5m-2})\right]^{-1} \end{aligned} \tag{1.1.2}$$

where the powers of x in the infinite products increase in intervals of 5.

In this paper we generalize the hard hexagon model by considering the corresponding eight-vertex SOS model with

$$\eta = K/r \tag{1.1.3}$$

r being a positive integer which is sometimes (but not always) restricted to odd values. We then use the corner-transfer-matrix technique to calculate the local height probabilities of this model (these are the analogs of the hard hexagon densities) and are led (as we hoped) to various generalizations of the Rogers–Ramanujan identities. These are almost certainly closely related to Gordon's generalization,[15,16] and involve the functions

$$G_{ri}(x) = \prod_{\substack{m=1 \\ m \not\equiv 0, \pm i \pmod r}}^{\infty} (1 - x^m)^{-1} \tag{1.1.4}$$

which are natural generalizations of $H(x), G(x)$, reducing to them when $r = 5$ and $i = 1, 2$.

1.2. Eight-Vertex SOS Model

Consider a square lattice $\mathscr{L}$, wound on a cylinder so that the last column is followed by the first. With each site i associate an integer "height" l_i. With each face associate a weight $W(l_i, l_j \mid l_m, l_n)$, where i, j, n, m are the four surrounding sites, ordered anticlockwise from the lower left, as in Fig. 1. Then the partition function is

$$Z = \sum \prod_{(i,j,n,m)} W(l_i, l_j \mid l_m, l_n) \qquad (1.2.1)$$

where the sum is over all the allowed arrangements of heights on the lattice, and the product is over all faces (i, j, n, m) of the lattice.

Require that the heights of adjacent sites must differ by 1: this is equivalent to taking $W(l, m' \mid l', m)$ to be zero unless

$$|l - m'| = |m' - m| = |m - l'| = |l' - l| = 1 \qquad (1.2.2)$$

To within a uniform additive shift of l, m', l', m, there are six possible ways of satisfying (1.2.2), as shown in Fig. 2.

Now consider the usual zero-field eight-vertex model, with weights a, b, c, d defined as in Eqs. (1)–(4) of Ref. 11. Define k, η, v, ρ by Eq. (8) of

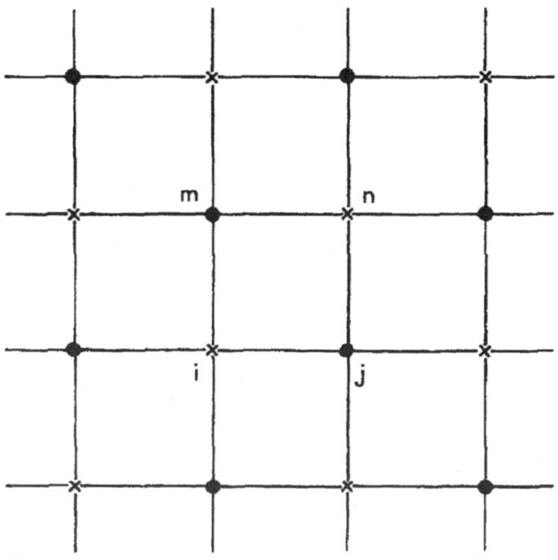

Fig. 1. The square lattice $\mathscr{L}$, showing a typical face (i, j, n, m) and the two sublattices X and Y (denoted, respectively, by crosses and circles).

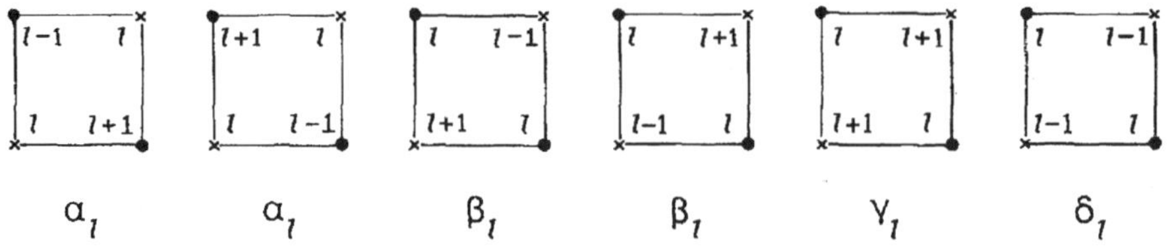

Fig. 2. The six possible arrangements of heights round a face of the lattice. The location of the sublattices is irrelevant here: crosses and circles may be interchanged.

Ref. 11, i.e., by

$$\begin{aligned} a &= \rho\Theta(2\eta)\Theta(v-\eta)H(v+\eta) \\ b &= \rho\Theta(2\eta)H(v-\eta)\Theta(v+\eta) \\ c &= \rho H(2\eta)\Theta(v-\eta)\Theta(v+\eta) \\ d &= \rho H(2\eta)H(v-\eta)H(v+\eta) \end{aligned} \tag{1.2.3}$$

where $H(u), \Theta(u)$ are the elliptic theta functions of argument u and modulus k. Let $h(u)$ be the function

$$h(u) = H(u)\Theta(u) \tag{1.2.4}$$

and let

$$\rho' = \rho\Theta(0), \qquad w_l = w_0 + 2l\eta \tag{1.2.5}$$

where w_0 is a constant, as yet arbitrary. Then it is shown in Refs. 11 and 12 that this eight-vertex model on $\mathscr{L}$ is equivalent to an SOS model on $\mathscr{L}$ with weights

$$\begin{aligned} W(l,l+1\,|\,l-1,l) &= \rho' h(v+\eta) \\ W(l,l-1\,|\,l+1,l) &= \rho' h(v+\eta)h(w_{l+1})/h(w_l) \\ W(l+l,l\,|\,l,l-1) &= \rho' h(v-\eta)h(w_{l-1})/h(w_l) \\ W(l-1,l\,|\,l,l+1) &= \rho' h(v-\eta) \\ W(l+1,l\,|\,l,l+1) &= \rho' h(2\eta)h(w_l+\eta-v)/\left[h(w_l)h(w_{l+1})\right] \\ W(l-1,l\,|\,l,l-1) &= \rho' h(2\eta)h(w_l-\eta+v) \end{aligned} \tag{1.2.6}$$

(taking $\mathscr{L}$ to be the dual of the lattice used in Ref. 12, where the heights are associated with faces rather than sites).

From the usual definitions[17,18] of the elliptic theta functions,

$$h(u) = 2p^{1/4}\sin\frac{\pi u}{2K}\prod_{n=1}^{\infty}\left(1-2p^n\cos\frac{\pi u}{K}+p^{2n}\right)(1-p^{2n})^2 \tag{1.2.7}$$

where K and K' are the complete elliptic integrals of the first kind, and p is the "nome"

$$p = e^{-\pi K'/K} \tag{1.2.8}$$

Noting that K is a function only of p, from (1.2.5)–(1.2.7) we see that we have five parameters at our disposal, which we can take to be p, $p^{1/4}\rho'$, η, w_0, and v. Here we shall regard the first four as real constants, and v as a variable (possibly complex). In particular, p must lie in the interval

$$-1 < p < 1 \tag{1.2.9}$$

which implies that K is real and positive.

In Refs. 11 and 12, particular attention is paid to the case when there exists an integer L such that $2L\eta$ is a period of the elliptic functions. This means that we can regard two heights l_i in the SOS model as being equal if they differ only by an integer multiple of L: it is then natural to interpret the adjacency restrictions (1.2.2) to modulo L, and this can be convenient in setting up a Bethe ansatz with which to solve the SOS model. [Incidentally, Pegg(19) has obtained the general conditions under which an SOS model satisfying (1.2.2) can be solved by a Bethe ansatz.]

Here we shall not adopt this "modulo L" viewpoint, but will interpret (1.2.2) strictly as written. This means that if we move from left to right along a row of $\mathscr{L}$, returning ultimately (because $\mathscr{L}$ is wound on a cylinder) to our starting point, then we must see as many increases in height (between adjacent sites) as decreases. (In Bethe ansatz terms, there are as many "up arrows" as "down arrows" per row.) It follows that we can in fact ignore the restriction (9) of Ref. 11, as well as the renormalization (10), and take $H(u)$, $\Theta(u)$ to be the usual elliptic theta functions.(17,18) In (1.2.3) we have used the fact that $\Theta(u)$ is an even function, while $H(u)$ is odd.

When we say, before (1.2.6), that the eight-vertex and SOS models are "equivalent," we mean that the eigenvalues of the transfer matrix of the former are also eigenvalues of the transfer matrix of the latter. If the SOS model has no eigenvalues larger than the maximum eight-vertex eigenvalues, then it follows that the two models have the same partition function per site in the limit when the number of rows of $\mathscr{L}$ becomes infinite.

The eigenvectors and eigenvalues of the SOS model transfer matrix are given by Eqs. (1.16)–(1.23) or Ref. 13, with $n = N/2$. [Actually these equations are over-restrictive: all the eight-vertex eigenvalues are certainly given by them(2) but the SOS model has other eigenvectors and eigenvalues obtained by multiplying the RH sides of (1.16) and (1.23) by z^l and z^2, respectively, while also multiplying the two additive terms on the right-hand side of (1.21) by z and z^{-1}. This z can be any complex number, and this extra degree of freedom reflects the fact that we are at present allowing the heights to take all integer values from $-\infty$ to ∞, with no "height boundary condition." When they are restricted to a finite set of values, boundary constraints are imposed on the eigenvectors which fix z. For the hard hexagon case discussed below, it appears that $z^{10}=1$. The eigenvectors are in general linear combinations of those given in Ref. 13 that correspond to degenerate eigenvalues.]

The eigenvectors of the SOS model are independent of v (reflecting the fact that eight-vertex transfer matrices with the same values of k and η, but different values of v, commute[(2)]). The eigenvalues are independent of the parameter w_0.

A trivial variation of the model is to associate extra weights $F(l_i,l_j)$, $F^{-1}(l_i,l_j)$ with the faces above and below each horizontal edge (i,j). Similarly, weights $G(l_i,l_m), G^{-1}(l_i,l_m)$ can be associated with the faces to the right and left of each vertical edge (i,m). Clearly such weights cancel out of the partition function, but they multiply the weight function $W(l,m'\mid l',m)$ by

$$F(l,m')G(l,l')/\left[F(l',m)G(m',m)\right] \tag{1.2.10}$$

One particular such transformation is to multiply W by $x(l)/x(m)$, for any function $x(l)$.

Performing both, taking

$$\begin{aligned} F(l,l\pm 1) = G(l+1,l) &= \left[G(l-1,l)\right]^{-1} \\ &= \left[h(w_l)\right]^{1/2} \end{aligned} \tag{1.2.11}$$

and $x(l)=i^l$, the weights (1.10) become

$$\begin{aligned} &W(l,l+1\mid l-1,l) = W(l,l-1\mid l+1,l) = \alpha_l \\ &W(l+1,l\mid l,l-1) = W(l-1,l\mid l,l+1) = \beta_l \\ &W(l+1,l\mid l,l+1) = \gamma_l, \qquad W(l-1,l\mid l,l-1) = \delta_l \end{aligned} \tag{1.2.12a}$$

where $\alpha_l,\alpha_l,\beta_l,\beta_l,\gamma_l,\delta_l$ are the weights of the six possible configurations, as shown in Fig. 2, and are given by

$$\begin{aligned} \alpha_l &= \rho' h(v+\eta) \\ \beta_l &= \rho' h(\eta - v)\left[h(w_{l-1})h(w_{l+1})\right]^{1/2}/h(w_l) \\ \gamma_l &= \rho' h(2\eta)h(w_l+\eta-v)/h(w_l) \\ \delta_l &= \rho' h(2\eta)h(w_l-\eta+v)/h(w_l) \end{aligned} \tag{1.2.12b}$$

From now on we shall take W to be given by (1.2.12) rather than (1.2.6). This is a more symmetric form, as it has the property that

$$W(l,m'\mid l',m) = W(l,l'\mid m',m) = W(m,m'\mid l',l) \tag{1.2.13}$$

The only further transformation of the type (1.2.10) that can be applied without violating these symmetries is to multiply $W(l, m' \mid l', m)$ by

$$x^{l-m} g_l g_m / (g_{l'} g_{m'}) \tag{1.2.14}$$

where $x^4 = 1$ and g_l is arbitrary. When $x = i$ and $g_l = 1$, this transformation merely negates β_l.

Note that if the height of one site is even, then the heights of all its neighbors must be odd. If we divide the square lattice into two sublattices X and Y (the crosses and circles of Fig. 1), such that any site of X has neighbors only in Y, and conversely, then it follows that Z is the sum of two individual partition functions Z_1 and Z_2. In Z_1, the only allowed configurations are those with odd heights on X sites, and even heights on Y sites. Conversely for Z_2. Thus in this sense the SOS model splits into two separate, but equivalent, models. We shall refer to them as "sub-SOS" models.

1.3. Hard Hexagon Model

The hard hexagon model[(8)] is a special case of the hard square model with diagonal interactions. With each site i of the square lattice $\mathscr{L}$ associate an occupation number σ_i, such that $\sigma_i = 0$ or 1 according to whether the site is empty or contains a particle. As with the SOS model, associate a weight $W_H(\sigma_i, \sigma_j \mid \sigma_m, \sigma_n)$ with each face (i, j, n, m). No two particles can be adjacent, which is equivalent to saying that $\sigma_i \sigma_j = 0$ for all edges (i, j) of $\mathscr{L}$.

This model can be solved provided a certain restriction (Eq. 24 of Ref. 8) is satisfied (the hard hexagon model is included in this solvable case). The weights W_H can then naturally be parametrized in terms of elliptic functions. Using the parametrization of Eq. (2.12) of Ref. 20, or equivalently Eq. (14.2.39) or Ref. 10, replacing u therein by $\pi(\eta - v)/(2K)$ and noting that $\theta(u)$ and $\theta_1(u)$ therein are proportional to $h(2Ku/\pi)$ (with appropriate elliptic moduli), we find that the nonzero values of W_H are

$$\begin{aligned}
&W_H(00 \mid 00) = h(5\eta - v)/h(4\eta) \\
&W_H(10 \mid 00) = W_H(00 \mid 01) = h(\eta - v)/\left[h(2\eta)h(4\eta)\right]^{1/2} \\
&W_H(01 \mid 00) = W_H(00 \mid 10) = h(\eta + v)/h(2\eta) \\
&W_H(10 \mid 01) = h(3\eta + v)/h(4\eta) \\
&W_H(01 \mid 10) = h(3\eta - v)/h(2\eta)
\end{aligned} \tag{1.3.1}$$

where

$$\eta = K/5 \tag{1.3.2}$$

We can convert this to an SOS model satisfying (1.2.2). For each occupation number σ_i define a height l_i by

$$l_i = 3 - 2\sigma_i, \qquad \text{if } i \text{ is on sublattice } X$$
$$= 2\sigma_i + 2, \qquad \text{if } i \text{ is on sublattice } Y \tag{1.3.3}$$

It then follows that the restrictions $\sigma_i, \sigma_j = 0, 1$ and $\sigma_i\sigma_j = 0$ are equivalent to

$$1 \leqslant l_i, l_j \leqslant 4, \qquad |l_i - l_j| = 1 \tag{1.3.4}$$

for all pairs (i, j) of adjacent sites. Further, defining

$$W(l_i, l_j \mid l_m, l_n) = W_H(\sigma_i, \sigma_j \mid \sigma_m, \sigma_n) \tag{1.3.5}$$

and using the relations

$$h(-u) = -h(u) = h(2K + u) \tag{1.3.6}$$

we find that this function W is precisely the same as that given by (1.2.12), with

$$\rho' = \left[h(2\eta)\right]^{-1}, \qquad w_0 = 0 \tag{1.3.7}$$

The definitions (1.3.3) ensure that l_i is odd on an X site, even on a Y site, so we have a sub-SOS model. However, we could just as easily have interchanged the X and Y sublattices, and obtained the other SOS model [with the same weights, and still satisfying (1.3.4)]. Altogether, it follows that

$$Z_{\text{SOS}} = 2Z_{HH} \tag{1.3.8}$$

where Z_{HH} is the hard-hexagon partition function and Z_{SOS} is the eight-vertex SOS partition function, given by (1.2.1), (1.2.12) and (1.3.7), with each height l_i restricted to the values $1, 2, 3, 4$.

Restricted and Unrestricted SOS Models. Let us call this SOS model, where $1 \leqslant l_i \leqslant 4$, the "restricted" SOS model. It is not the same as the unrestricted model discussed in Section 1.2, where $-\infty < l_i < \infty$, but it is closely related.

Consider the unrestricted model. In particular, consider two successive rows R and R' of $\mathscr{L}$, and the product P of the weights of the intervening faces. Suppose that all the heights on R lie between 1 and 4. Then $P = 0$ unless the heights on R' also all lie between 1 and 4. To see this, note that if R' is the upper row, then it can only contain a height 0 or 5 via the configurations shown in Fig. 3. These contain weights $W(21 \mid 10)$, $W(34 \mid 45)$, respectively, and from (1.2.12) these vanish because

$$h(w_0) = h(0) = 0, \qquad h(w_5) = h(10\eta) = h(2K) = 0 \tag{1.3.9}$$

Similar considerations apply if R' is the lower row.

If all the weights were finite, this would mean that the transfer matrix of the unrestricted SOS model broke up into diagonal blocks, one of which

Fig. 3. Configurations containing heights 0 and 5 in the upper row, and only heights 1, 2, 3, 4 in the lower row.

would be the restricted SOS model transfer matrix. More precisely, this would happen if $\eta = K/5$ and w_0 is allowed to tend to zero.

In fact the situation is more complicated than this, since some of the other transfer matrix elements [e.g., those containing a factor $W(4,5\,|\,5,4)$] then become infinite. Even so, we have still been able to verify for a two-column lattice that the relevant eigenvalues and eigenvector elements of the unrestricted model become those of the restricted one. This suggests that the restricted and unrestricted SOS models may have at least some transfer matrix eigenvalues in common. This would be consistent with the fact that the hard hexagon model (in its regime I, III, and IV) has the same free energy per site as the corresponding eight-vertex model.[21]

Another argument in favor of this assertion is to compare the equations that determine the transfer matrix eigenvalues of the unrestricted SOS model and the hard hexagon model. Each eigenvalue Λ is an entire function of the variable v. For the SOS model, it follows from Eq. (1.23) of Ref. 13 that each $\Lambda(v)$ satisfies the functional equation

$$\Lambda(v)\,Q(v) = z\phi(v-\eta)\,Q(v+2\eta) + z^{-1}\phi(v+\eta)\,Q(v-2\eta) \quad (1.3.10)$$

where z is the parameter discussed in Section 1.2,

$$\phi(v) = \left[\rho\Theta(0)h(v)\right]^N \quad (1.3.11)$$

N is the (even) number of columns of the lattice $\mathscr{L}$, and

$$Q(v) = \prod_{j=1}^{N/2} h(v-u_j) \quad (1.3.12)$$

$u_1, \ldots, u_{N/2}$ being constants.

Writing (1.3.10) with v replaced by $v+2k\eta$, where $k=0,1,\ldots,4$, noting that

$$Q(v+10\eta) = (-1)^{N/2}\,Q(v) \quad (1.3.13)$$

we obtain five homogeneous linear equations for $Q(v+2k\eta)$, $k=0, 1, \ldots, 4$. Eliminating these gives a 5th-degree equation for the function

$\Lambda(v)$, and it can be verified that this equation is automatically satisfied if

$$\Lambda(v)\Lambda(v-2\eta)=\phi(v+\eta)\phi(v-3\eta)+\phi(v-\eta)\Lambda(v+4\eta) \quad (1.3.14)$$

provided

$$z^5=-(-1)^{N/2} \quad (1.3.15)$$

(In fact the 5×5 matrix of coefficients then has rank of only 3.)

However, (1.3.14) is precisely the equation satisfied by the eigenvalues of the transfer matrix of the hard hexagon model [Eq. (3.3) of Ref. 20, using (1.3.7) and the transformations mentioned prior to Eq. (1.3.1) herein]. Thus there may be (and almost certainly are) eigenvalues satisfying both (1.3.10) and (1.3.14), provided z in the unrestricted SOS model is given by (1.3.15).

1.4. Restricted SOS Model with $\eta = K/r$

From now on we consider in this paper an SOS model satisfying (1.2.2), (1.2.5), and (1.2.12), with

$$\eta = K/r \quad (1.4.1)$$

$$w_0 = 0 \quad (1.4.2)$$

and each height l_i restricted to the integer values

$$l_i = 1,2,\ldots,r-1 \quad (1.4.3)$$

Here r is an arbitrary integer not less than 3.

For $r=5$ we regain the restricted SOS model which we have shown to be equivalent to the hard hexagon model [more precisely, to the hard square model with diagonal interactions that has weights given by (1.3.1)]. Thus the present model is a natural generalization of the hard hexagon model. It has the property that $h(w_l)$ is nonzero for $l=1,\ldots,r-1$, while

$$h(w_0)=h(w_r)=0 \quad (1.4.4)$$

This means that the weights (1.2.12) are finite (and in general nonzero) provided the restriction (1.4.3) is satisfied.

As with the $r=5$ case, it is likely that this model is in some sense equivalent to the unrestricted model with $-\infty < l_i < \infty$, and hence to the eight-vertex model. However, we shall not need this equivalence. All we need is the fact that the restricted model, like the unrestricted one, satisfies the star–triangle relation [Eq. (13.3.6) of Ref. 10, (2.49) of Ref. 22]:

$$\sum_g W(b,c\,|\,a,g)W'(a,g\,|\,f,e)W''(g,c\,|\,e,d)$$

$$=\sum_g W''(a,b\,|\,f,g)W'(b,c\,|\,g,d)W(g,d\,|\,f,e) \quad (1.4.5)$$

Here W is the weight function defined by (1.2.2), (1.2.5), (1.2.12); W' and

W'' are defined similarly, but with v replaced by v' and v'', respectively, where

$$v' = v + v'' - \eta \tag{1.4.6}$$

[If we set $u = \eta - v$, then we regain the usual relation $u' = u + u''$: Eq. (13.3.10) of Ref. 10.]

The equation (1.4.5) is to be true for all values of the six heights a, b, c, d, e, f. Since W, W', W'' each vanish unless their arguments satisfy (1.6), we must have $|a-b| = |b-c| = |c-d| = |d-e| = |e-f| = |f-a| = 1$. Apart from a uniform additive shift of all the heights, this means that there are 20 cases to consider. These occur in pairs, one being obtained from the other by interchanging a with d, b with e, and c with f, which merely interchanges the two sides of the equation (1.4.5).

It follows that (1.4.5) consists of ten distinct equations. Taking $\alpha_l', \ldots, \delta_l'$ to be the values of $\alpha_l, \ldots, \delta_l$ in (1.2.12) when W, v are replaced by W', v'; and similarly for $\alpha_l'', \ldots, \delta_l''$; seven of the equations are

$$\begin{aligned}
\beta_l\alpha_l'\beta_l'' + \gamma_l\delta_{l+1}'\gamma_l'' &= \delta_{l+1}\gamma_l'\delta_{l+1}'' + \beta_{l+1}\alpha_{l+1}'\beta_{l+1}'' \\
\beta_l\gamma_{l-1}'\beta_l'' + \gamma_l\alpha_l'\gamma_l'' &= \alpha_l\gamma_l'\alpha_l'' \\
\beta_l\alpha_l'\delta_l'' + \gamma_l\delta_{l+1}'\beta_l'' &= \alpha_l\beta_l'\delta_{l+1}'' \\
\delta_l\alpha_l'\delta_l'' + \beta_l\delta_{l+1}'\beta_l'' &= \alpha_l\delta_l'\alpha_l'' \\
\delta_l\gamma_{l-1}'\beta_l'' + \beta_l\alpha_l'\gamma_l'' &= \alpha_l\beta_l'\gamma_{l-1}'' \\
\alpha_{l+1}\alpha_l'\alpha_{l+1}'' &= \alpha_l\alpha_{l+1}'\alpha_l'' \\
\alpha_{l+1}\beta_{l+1}'\beta_l'' &= \alpha_l\beta_l'\beta_{l+1}''
\end{aligned} \tag{1.4.7}$$

The other three equations can be obtained by interchanging the unprimed and double-primed symbols.

For the unrestricted model, l in (1.4.7) takes all integer values and we can verify directly, using (1.2.12b), that the equations are satisfied.

For the restricted model, each of $a, b, \ldots, f$ must lie in the interval $1, \ldots, r-1$. This means that l takes the values $1, \ldots, r-2$ in the first equation in (1.4.7), the values of $2, \ldots, r-2$ in the next four equations, and the values $2, \ldots, r-3$ in the last two. Further, the height g in (1.4.5) must also lie in the interval $1, \ldots, r-1$: this means that in the first equation the terms

$$\beta_1\alpha_1'\beta_1'', \qquad \beta_{r-1}\alpha_{r-1}'\beta_{r-1}'' \tag{1.4.8}$$

should be deleted. (They occur for $l = 1$ and $l = r-2$, and correspond to $g = 0, r$, respectively.)

However, from (1.2.12b) and (1.4.4) it is clear that the terms (1.4.8) vanish for the unrestricted model. It follows that (1.4.7), and hence the star–triangle relation (1.4.5), is indeed satisfied by the restricted SOS model.

The Case When r Is Odd. The model has some special features when r is odd, i.e., when there exists an integer n such that

$$r = 2n + 1 \tag{1.4.9}$$

(Note that the hard hexagon model is such a case, with $n = 2$.) In this case, there are as many even heights in the allowed range (1.4.3) as there are odd heights. Each of the two sub-SOS models can then naturally be expressed as a "lattice gas" generalization of the hard hexagon model. For the one with odd heights on X sites and even heights on Y sites, define σ_i by

$$\begin{aligned} \sigma_i &= \tfrac{1}{2}(2n - 1 - l_i), && \text{if} \quad i \in X \\ &= \tfrac{1}{2}(l_i - 2), && \text{if} \quad i \in Y \end{aligned} \tag{1.4.10}$$

Then the σ_i are integers with permitted values $0, 1, \ldots, n-1$, satisfying

$$\sigma_i + \sigma_j = n - 2 \quad \text{or} \quad n - 1 \tag{1.4.11}$$

for all adjacent pairs (i, j) of sites. We can regard these σ_i as "occupation numbers" and use these to define the states of the system.

The SOS model weight function W is the same for all faces of the lattice $\mathscr{L}$, but the mapping (1.4.10) is not: it depends on whether the face has an X or a Y site at its lower-left corner (if $\mathscr{L}$ were a checkerboard it would depend on whether the face were black or white). It follows that W, when regarded as a function of the occupation numbers σ_i, may similarly depend on the face under consideration. In fact this does not occur, because the function W defined by (1.2.12), (1.4.1), and (1.4.2) has the symmetry property

$$W(r - l, r - m' \mid r - l', r - m) = W(l, m' \mid l', m) \tag{1.4.12}$$

Thus when (1.4.9) is satisfied, each sub-SOS model is equivalent to a uniform lattice gas in which there must be either $n - 2$ or $n - 1$ particles on every pair of adjacent sites.

1.5. Expressions for the Local Height Probabilities

The Rogers–Ramanujan identities naturally enter the calculation of the sublattice densities of the hard hexagon model.[9] For the more general case of our restricted SOS model, the analogs of these densities are the local height probabilities:

$$P_a = Z^{-1} \sum \delta(l_1, a) \prod_{(i,j,n,m)} W(l_i, l_j \mid l_m, l_n) \tag{1.5.1}$$

where Z is the partition function defined by (1.2.1) and the sum and product have the same meanings as therein, l_1 is the height of the center site

1 of the lattice, and a is an integer between 1 and $r-1$. Thus P_a is the probability that site 1 has height a, and it follows from (1.2.1) and (1.5.1) that

$$\sum_{a=1}^{r-1} P_a = 1 \tag{1.5.2}$$

In evaluating the probability P_a, we suppose initially that the lattice $\mathscr{L}$ is planar (i.e., *not* wound on a cylinder or torus) and finite, and fix the boundary heights to have the values they would assume in a particular ground state configuration [i.e., a set of values of the heights that maximizes the summand in (1.2.1)]. Finally, we shall take the limit when the lattice becomes infinitely large, all boundary sites being infinitely far from the center site 1.

For definiteness, let us suppose that site 1 lies on the X sublattice, and refer to sub-SOS models, states, and boundary conditions as "even" if they have even heights on X sites, odd heights on Y sites. If the converse, then "odd." For every even ground state, an odd one can be formed by interchanging the X and Y sublattices. Thus the ground states of the total SOS model can be divided into equal numbers of even and odd ground states.

P_a will depend on the particular ground state that is chosen for the boundary conditions. One significant but trivial dependence is that if the ground state is even, clearly

$$P_a = 0, \qquad \text{unless } a \text{ is even} \tag{1.5.3}$$

Thus in this case the summation in (1.5.2) can be restricted to even values of a. (In the other case it can be restricted to odd values.)

If there is only one even (and one odd) ground state, then we expect the system to be "disordered." This means that in the limit of a large lattice we expect P_a to be independent of the boundary conditions, so long as they are even. If there is more than one even ground state, then the system is "ordered" and P_a depends on the boundary conditions.

As with the hard hexagon model, we can calculate P_a by using corner transfer matrices.[10,23] This is done in Appendix A, attention being focused on the cases when v is real and $-\eta < v < 3\eta$. These are the cases when the weight function W is positive, or can be made so by an appropriate choice of ρ' and by applying a transformation of the type (1.2.14) with $x = i$, i.e., by negating β_l.

We have to distinguish the cases when v is greater or less than η, and when p in (1.2.7) is positive or negative. Altogether this gives us four "regimes" to consider. Labeling them analogously to those of the hard hexagon model, they are

$$\begin{array}{lll} \text{regime I:} & -1<p<0, & \eta<v<3\eta \\ \text{regime II:} & 0<p<1, & \eta<v<3\eta \\ \text{regime III:} & 0<p<1, & -\eta<v<\eta \\ \text{regime IV:} & -1<p<0, & -\eta<v<\eta \end{array} \tag{1.5.4}$$

It is convenient to define an integer t such that

$$\begin{aligned} t &= 2-r, && \text{in regimes I and II} \\ &= 2, && \text{in regimes III and IV} \end{aligned} \tag{1.5.5}$$

and to define a function $E(z,x)$ by

$$E(z,x)=\prod_{n=1}^{\infty}(1-x^{n-1}z)(1-x^{n}z^{-1})(1-x^{n}) \tag{1.5.6}$$

for all complex numbers z, x such that $|x|<1$. This is basically an elliptic theta function, and by Jacobi's triple product identity (Theorem 2.8 of Ref. 16), it has the series expansion:

$$E(z,x)=\sum_{n=-\infty}^{\infty}(-1)^{n}x^{n(n-1)/2}z^{n} \tag{1.5.7}$$

Regimes II and III. The parameter p is positive in regimes II and III, so we can define ϵ and x (both real and positive) so that

$$p=e^{-\epsilon}, \qquad x=\exp\left[-4\pi^{2}/(r\epsilon)\right] \tag{1.5.8}$$

We find in Appendix A that

$$P_a=S^{-1}u_aX_m(a,b,c;x^{t}) \tag{1.5.9}$$

for $a=1,\ldots,r-1$, where

$$u_a=x^{(2-t)(2a-r)^2/(16r)}E(x^{a},x^{r}) \tag{1.5.10}$$

$$X_m(a,b,c;q)=\sum_{l_2,\ldots,l_m}q^{\phi(\mathbf{l})} \tag{1.5.11}$$

$$\phi(\mathbf{l})=\sum_{j=1}^{m}j\,|l_{j+2}-l_j|/4 \tag{1.5.12}$$

$$S=\sum_{1\leqslant a<r}u_aX_m(a,b,c;x^{t}) \tag{1.5.13}$$

Here $\mathbf{l}=\{l_1,l_2,\ldots,l_{m+2}\}$ is a set of integer heights satisfying the restrictions

$$1\leqslant l_j\leqslant r-1, \qquad |l_{j+1}-l_j|=1 \tag{1.5.14a}$$

for $j\geqslant 1$. The summation in (1.5.11) is over all allowed values of

$l_2, \ldots, l_m$, the end heights l_1, l_{m+1}, l_{m+2} being fixed at the values

$$l_1 = a, \qquad l_{m+1} = b, \qquad l_{m+2} = c \tag{1.5.14b}$$

Thus a, b, c must all lie in the range $1, 2, \ldots, r-1$, and must satisfy

$$|b - c| = 1, \qquad m + a - b = \text{even} \tag{1.5.15}$$

We can verify the symmetry properties

$$u_{r-a} = u_a, \qquad X_m(r-a, r-b, r-c; q) = X_m(a, b, c; q) \tag{1.5.16}$$

which are consequences of the symmetry (1.4.12) of W.

The heights $l_1, \ldots, l_{m+2}$ correspond to the heights of sites on the center row of $\mathcal{L}$, starting at the center site 1 and moving rightwards to the boundary. Thus l_{m+1} and l_{m+2}, i.e., b and c, are boundary heights, and are to be fixed according to the ground state chosen for the boundary conditions.

In a ground state, the values of $l_1, \ldots, l_{m+2}$ are such as to maximize the summand in (1.5.11). In regime III, t is positive, so $|q| < 1$ and the ground states are obtained by minimizing $\phi(\mathbf{l})$. Plainly this is achieved by taking

$$\begin{aligned} l_j &= l_X, \qquad \text{if } j \text{ is odd} \\ &= l_Y, \qquad \text{if } j \text{ is even} \end{aligned} \tag{1.5.17}$$

where l_X, l_Y are any two integers satisfying $|l_X - l_Y| = i$, $1 \leqslant l_X, l_Y \leqslant r-1$. There are $2r-4$ such values of l_X and l_Y, so there are $2r-4$ ground states, of which $r-2$ are even and $r-2$ are odd. (For the hard hexagon model, with $r = 5$, this gives $5 - 2 = 3$ ground states, in agreement with the observation of Huse[(24)].)

We remark in Appendix A that all ground states (for all regimes) are invariant under uniform shifts in the southwest to northeast diagonal direction. We can use this to obtain the ground state heights on all rows of $\mathcal{L}$, given the heights $l_1, \ldots, l_{m+2}$ on the center row. From (1.5.17) it follows that in regime III a ground state has

$$\begin{aligned} l_j &= L_X, \qquad \text{if } j \in X \\ &= l_Y, \qquad \text{if } j \in Y \end{aligned} \tag{1.5.18}$$

for all sites j of the lattice.

In regime II, t is negative and the ground states maximize $\phi(\mathbf{l})$. This is achieved by taking the l_j (for the linear sequence $l_1, \ldots, l_{m+2}$) to increase in unit steps to $r-1$, then decrease to 1, then increase again to $r-1$, and so on. More precisely, let j_0 be any integer from 1 to $2r-4$ and, for given j, let

$$k = j - j_0, \qquad \operatorname{mod}(2r-4) \tag{1.5.19a}$$

(thus $0 \leqslant k < 2r - 4$). Then

$$l_j = 1 + |k - r + 2| \tag{1.5.19b}$$

There are $2r - 4$ such sequences, depending on the value of j_0, so there are $2r - 4$ ground states, as in regime III.

In these regimes II and III there are as many ground states as there are permitted values of b and c, with $|b - c| = 1$. Thus we can either select a ground state and then evaluate b and c from (1.5.14b), or we can choose b and c and then determine the corresponding ground state.

As remarked above, we ultimately want to calculate P_a in the limit of an infinitely large lattice. This corresponds to letting $m \to \infty$ in (1.5.9)–(1.5.15), while keeping the ground state [i.e., l_X and l_Y in (1.5.17), or j_0 in (1.5.19)] fixed.

Regimes I and IV. In these regimes p is negative. We now define ϵ and x (both real and positive) by

$$p = -e^{-\epsilon}, \qquad x = \exp(-2\pi^2/r\epsilon) \tag{1.5.20}$$

(For the hard hexagon case, this x is the square of that used in Refs. 8, 9, and 20.) We find that

$$P_a = T^{-1} v_a Y_m(a,b,c;x^{-t}) \tag{1.5.21}$$

for $a = 1, \ldots, r-1$, where

$$v_a = x^{[(t-1)r-(t-2)a]a/2r} E(x^a, -x^{r/2}) \tag{1.5.22}$$

$$Y_m(a,b,c;q) = \sum_{l_2,\ldots,l_m} q^{\psi(\mathbf{l})} \tag{1.5.23}$$

$$\psi(\mathbf{l}) = \sum_{j=1}^{m} jH(l_j, l_{j+1}, l_{j+2}) \tag{1.5.24}$$

$$T = \sum_{1 \leqslant a < r} v_a Y_m(a,b,c;x^{-t}) \tag{1.5.25}$$

Here $\mathbf{l}$ and the $l_2, \ldots, l_m$ summation have the same meaning as in (1.5.14). The function $H(l,l',l'')$ is defined by

$$H(l,l',l'') = 0, \quad \text{if} \quad l \neq l'' \tag{1.5.26a}$$

$$\begin{aligned} H(l,l+1,l) &= 0, \quad \text{if} \quad l \leqslant r/2 \\ &= 1, \quad \text{if} \quad l > r/2 \end{aligned} \tag{1.5.26b}$$

$$\begin{aligned} H(l,l-1,l) &= 1, \quad \text{if} \quad l \leqslant r/2 \\ &= 0, \quad \text{if} \quad l > r/2 \end{aligned} \tag{1.5.26c}$$

These formulas (1.5.20)–(1.5.26) apply regardless of whether r is even or odd, but only when r is odd do we have both the symmetries

$$v_{r-a} = v_a, \qquad Y_m(r-a, r-b, r-c; q) = Y_m(a, b, c; q) \quad (1.5.27)$$

[which are analogous to (1.5.16) and consequences of (1.4.12)]. Further, there are complications in discussing the ground states when r is even. (Both these difficulties arise because the corner transfer matrices are not diagonal in the limit $x \to 0$ when r is even: the necessary diagonalization means that if $l_{j-1} = l_{j+1} = r/2$, then l_j is merely an eigenvalue label and does not correspond to a height on the lattice.)

For r odd, the ground states are then of the form given by (1.5.17) and (1.5.18). For regime I we want to minimize $\psi(\mathbf{l})$, which corresponds to taking

$$(l_X, l_Y) = (n, n+1) \quad \text{or} \quad (n+1, n), \qquad \text{where} \quad n = (r-1)/2 \quad (1.5.28)$$

Thus there are just two ground states, one for each sub-SOS model, and we expect the system to be disordered.

For regime IV we want to maximize $\psi(\mathbf{l})$, which corresponds to taking l_X, l_Y to be any pair of integers in the range $1, \ldots, r-1$ satisfying $|l_X - l_Y| = 1$, except for the values (1.5.29). Thus there are $2r-6$ ground states in regime IV.

2. EVALUATION OF THE PROBABILITIES P_a

2.1. General Comments

We have obtained the expressions (1.5.9), (1.5.21) for the local height probabilities P_a, involving the m-fold sums X_m and Y_m defined by (1.5.11) and (1.5.23). Our object in this part is to obtain more tractable expressions for X_m and Y_m, considered as functions of their argument q; and to study the limit $m \to \infty$. In this limit we find that X_m and Y_m are actually modular forms (p. 114 of Ref. 25). Indeed the results in regimes I, III, and IV are sums of at most two simple quotients of elliptic theta functions (Ref. 18, Chap. 21). Regime II yields multidimensional theta series.

Our approach is to first keep m finite, and show that X_m and Y_m can be written as sums of Gaussian polynomials. This treatment has its genesis in Ref. 26, which in turn was suggested by I. Schur's original treatment of the Rogers–Ramanujan–Schur identities.[(27)] While this approach may be more cumbersome than the one used for regimes I, III, and IV of the original hard hexagon model (Ref. 10, pp. 432–443), it does seem essential in this more general setting. In particular the classical theory of q-difference equations and q series[(28,29)] appears barely adequate (see Appendix B) to handle regime II via polynomials. Unfortunately we do not know what the

Rogers–Ramanujan-type "series product" identities are for either regime II or IV except when $r = 5$ [Eqs. (14.5.22) and (14.5.50) of Ref. 10]. We hope to discuss the Rogers–Ramanujan aspects of regimes I and III in a subsequent paper; indeed the appropriate identities for regime I are given in Ref. 30 and in Eq. (7.3.7) of Ref. 16.

2.2. Gaussian Polynomials

The Gaussian polynomials (which we shall use for calculating X_m, Y_m) are defined as

$$\begin{aligned}\begin{bmatrix} N \\ M \end{bmatrix} \equiv \begin{bmatrix} N \\ M \end{bmatrix}_q &= \prod_{j=1}^{M} \frac{1-q^{N-M+j}}{1-q^j}, \qquad 0 \leqslant M \leqslant N \\ &= 0 \qquad\qquad \text{otherwise} \end{aligned} \tag{2.2.1}$$

We shall use the subscript q only when confusion might otherwise arise.

It is easy to prove (by mathematical induction on N) that these are indeed polynomials in q. One merely utilizes either of the fundamental recurrences (p. 35 of Ref. 16)

$$\begin{bmatrix} N \\ M \end{bmatrix} - \begin{bmatrix} N-1 \\ M \end{bmatrix} = q^{N-M} \begin{bmatrix} N-1 \\ M-1 \end{bmatrix} \tag{2.2.2}$$

$$\begin{bmatrix} N \\ M \end{bmatrix} - \begin{bmatrix} N-1 \\ M-1 \end{bmatrix} = q^{M} \begin{bmatrix} N-1 \\ M \end{bmatrix} \tag{2.2.3}$$

In addition to the recurrences, we need that

$$\begin{bmatrix} N \\ M \end{bmatrix} = \begin{bmatrix} N \\ N-M \end{bmatrix} \tag{2.2.4}$$

$$\begin{bmatrix} N \\ M \end{bmatrix}_{q^{-1}} = q^{-M(N-M)} \begin{bmatrix} N \\ M \end{bmatrix}_q \tag{2.2.5}$$

and for $|q| < 1$,

$$\lim_{M,N\to\infty} \begin{bmatrix} N+M \\ M \end{bmatrix} = Q(q)^{-1} \tag{2.2.6}$$

where

$$Q(q) = \prod_{n=1}^{\infty} (1-q^n) \tag{2.2.7}$$

We shall also use the notations

$$(A)_n = (A;q)_n = \prod_{j=0}^{n-1} (1-q^j A). \tag{2.2.8}$$

Thus in particular

$$(q)_\infty = \prod_{j=0}^{\infty}(1-q^{j+1}) = Q(q) \tag{2.2.9}$$

Two simple identities that we shall use are

$$(A)_{m+n} = (A)_m(q^mA)_n \tag{2.2.10}$$

$$(A)_n = (-A)^n q^{(1/2)n(n-1)}(q^{1-n}A^{-1})_n \tag{2.2.11}$$

2.3. Regime III

We shall consider the regimes in order of difficulty: regime III is the easiest to handle and provides a good prototype for subsequent work.

The function $X_m(a,b,c;q)$, or simply $X_m(a,b,c)$, is defined by (1.5.11), (1.5.12), and (1.5.14). By examining admissible sequences of $\{l_2,\ldots,l_m\}$ and splitting them into two classes according to whether $l_m = b-1$ or $l_m = b+1$, we see that the $X_m(a,b,c)$ are totally defined by the following recurrences and initial values ($m > 0$, $1 \leqslant a \leqslant r-1$):

$$X_m(a,b,b+1) = X_{m-1}(a,b+1,b) + q^{m/2}X_{m-1}(a,b-1,b),$$
$$1 \leqslant b \leqslant r-2 \quad (2.3.1)$$

$$X_m(a,b,b-1) = X_{m-1}(a,b-1,b) + q^{m/2}X_{m-1}(a,b+1,b),$$
$$2 \leqslant b \leqslant r-1 \quad (2.3.2)$$

$$X_m(a,0,1) = X_m(a,r,r-1) = 0 \tag{2.3.3}$$

$$\begin{aligned} X_0(a,b,c) &= 1, \qquad \text{if} \quad a=b \quad \text{and} \quad c=b\pm 1 \\ &= 0, \qquad \text{otherwise} \end{aligned} \tag{2.3.4}$$

Our main object is to show that as $m\to\infty$, X_m converges to a difference of two theta series divided by $Q(q)$.

Theorem 2.3.1. For $m \geqslant 0$, $1 \leqslant a,b,c < r$, $c = b\pm 1$, $m+a-b$ an even integer,

$$X_m(a,b,c) = q^{(a^2-a)/4}\{F_m(a,b,c) - F_m(-a,b,c)\} \tag{2.3.5}$$

where

$$F_m(a,b,c) = \sum_{\lambda=-\infty}^{\infty} q^{(r-1)\lambda(r\lambda-a)+[bc+(2r\lambda-a)(b+c-1)]/4}\begin{bmatrix} m \\ \frac{1}{2}(m+a-b)-r\lambda \end{bmatrix} \tag{2.3.6}$$

[Because of the definition (2.2.1), for given m, a, b there are only a finite number of nonzero terms in the λ summation.]

Proof. Since (2.3.1)–(2.3.4) uniquely define the $X_m(a,b,c)$, we need only show that the right-hand side of (2.3.5) satisfies the same conditions. Now

$$F_m(a,b,b+1) - F_{m-1}(a,b+1,b)$$
$$= \sum_{\lambda=-\infty}^{\infty} q^{(r-1)\lambda(r\lambda-a)+b(b+1+4r\lambda-2a)/4}$$
$$\times \left\{ \begin{bmatrix} m \\ \frac{1}{2}(m+a-b)-r\lambda \end{bmatrix} - \begin{bmatrix} m-1 \\ \frac{1}{2}(m+a-b)-1-r\lambda \end{bmatrix} \right\} \quad (2.3.7)$$

Using (2.2.3) to simplify the bracketted difference of Gaussian polynomials, it follows at once that

$$F_m(a,b,b+1) - F_{m-1}(a,b+1,b) = q^{m/2}F_{m-1}(a,b-1,b) \quad (2.3.8)$$

Similarly,

$$F_m(a,b,b-1) - F_{m-1}(a,b-1,b)$$
$$= \sum_{\lambda=-\infty}^{\infty} q^{(r-1)\lambda(r\lambda-a)+(b-1)(b+4r\lambda-2a)/4}$$
$$\times \left\{ \begin{bmatrix} m \\ \frac{1}{2}(m+a-b)-r\lambda \end{bmatrix} - \begin{bmatrix} m-1 \\ \frac{1}{2}(m+a-b)-r\lambda \end{bmatrix} \right\} \quad (2.3.9)$$

and on using (2.2.2) we obtain

$$F_m(a,b,b-1) - F_{m-1}(a,b-1,b) = q^{m/2}F_{m-1}(a,b+1,b) \quad (2.3.10)$$

Thus $F_m(a,b,c)$ satisfies the linear recurrence relations (2.3.1) and (2.3.2). Since these are independent of a, so does $F_m(-a,b,c)$, and hence the expression (2.3.5) for $X_m(a,b,c)$.

It remains to verify that (2.3.3) and (2.3.4) are satisfied. From (2.3.6)

$$F_m(a,0,1) = \sum_{\lambda=-\infty}^{\infty} q^{(r-1)\lambda(r\lambda-a)} \begin{bmatrix} m \\ \frac{1}{2}(m+a)-r\lambda \end{bmatrix} \quad (2.3.11)$$

$$F_m(a,r,r-1) = \sum_{\lambda=-\infty}^{\infty} q^{(r-1)(\lambda+1/2)[r(\lambda+1/2)-a]} \begin{bmatrix} m \\ \frac{1}{2}(m+a)-r(\lambda+1/2) \end{bmatrix} \quad (2.3.12)$$

Using (2.2.4) and replacing λ by $-\lambda, -1-\lambda$, respectively, in these two equations, we find that

$$F_m(a,0,1) = F_m(-a,0,1), \qquad F_m(a,r,r-1) = F_m(-a,r,r-1) \quad (2.3.13)$$

From (2.3.5) it follows immediately that (2.3.3) is satisfied.

Finally, from (2.2.1) and (2.3.6), $F_0(a,b,c)$ will be zero unless $a \equiv b \pmod{2r}$. Remembering that $1 \leqslant a, b < r$, from (2.3.5) we obtain

$$X_0(a,b,c) = q^{(a^2-a)/4} F_0(a,a,c)\delta_{a,b} \tag{2.3.14}$$

where, as in part I,

$$\begin{aligned} \delta_{a,b} \equiv \delta(a,b) &= 1, \qquad \text{if} \quad a = b \\ &= 0, \qquad \text{otherwise} \end{aligned} \tag{2.3.15}$$

For $F_0(a,b,c)$, the only nonzero term in the summation in (2.3.6) is that with $\lambda = 0$; hence

$$F_0(a,a,c) = q^{(a-a^2)/4} \tag{2.3.16}$$

It follows that (2.3.4) is also satisfied, which completes the proof of Theorem 2.3.1. ■

We now go on to consider the limit $m \to \infty$. From (1.5.9), the argument q of the function $X_m(a,b,c;q)$ is $q = x^t$. From (1.5.5), $t = 2$ in regime III, so here we have

$$|q| < 1 \tag{2.3.17}$$

We shall need the function $\Delta(a,d;q)$, defined for $1 \leqslant a \leqslant r-1$, $1 \leqslant d \leqslant r-2$ by

$$\begin{aligned} \Delta(a,d;q) = \sum_{\lambda=-\infty}^{\infty} & q^{r(r-1)\lambda^2 + rd\lambda + a(a-1)/4} \\ & \times \left\{ q^{-(r-1)a\lambda - (1/2)ad} - q^{(r-1)a\lambda + (1/2)ad} \right\} \end{aligned} \tag{2.3.18}$$

i.e., using (1.5.7),

$$\begin{aligned} \Delta(a,d;q) = q^{a(a-1)/4} \{ & q^{-(1/2)ad} E\left[-q^{(r-a)(r-1)+rd}, q^{2r(r-1)} \right] \\ & - q^{(1/2)ad} E\left[-q^{(r+a)(r-1)+rd}, q^{2r(r-1)} \right] \} \end{aligned} \tag{2.3.19}$$

Theorem 2.3.2. For $1 \leqslant a,b,c < r$, $c = b \pm 1$,

$$\lim_{m\to\infty} X_m(a,b,c) = Q(q)^{-1} q^{bc/4} \Delta(a, \tfrac{1}{2}(b+c-1); q) \tag{2.3.20}$$

where the limit is taken through values of m with same parity as $a - b$.

Proof. This is a direct consequence of Theorem 2.3.1. One merely applies (2.2.6) to the Gaussian polynomial in (2.3.6), then uses (2.3.5) and (1.5.7). The passage to a limit inside the summation in (2.3.6) is easily justified due to the uniform boundedness of the Gaussian polynomial inside $|q| \leqslant 1 - \epsilon$ and the rapid absolute convergence of the λ series caused by the exponent on q being quadratic in λ.

We must distinguish the case in which the large-m limit is taken through even values of $m - b$ from that in which it is taken through odd values. In the former case a must be an even integer (we have the "even sub-SOS model" discussed in Section 1); in the latter case a is an odd integer (the "odd sub-SOS model").

Substituting (2.3.18) into (1.5.9) and (1.5.13), using (1.5.5), we find that the local height probability P_a tends to a limit as $m \to \infty$, being given by

$$P_a = u_a \Delta(a, d; x^2) \Big/ \sum_{1 \leqslant a' \leqslant r-1} u_{a'} \Delta(a', d; x^2) \qquad (2.3.21)$$

where

$$d = \tfrac{1}{2}(b + c - 1), \qquad 1 \leqslant d \leqslant r - 2 \qquad (2.3.22)$$

and a, a' are either both even integers satisfying $1 \leqslant a, a' \leqslant r - 1$ [this is the "even" case (1.5.3)], or are both odd integers satisfying $1 \leqslant a, a' \leqslant r - 1$.

Allowing for these even and odd cases, and noting that the integer d takes $r - 2$ values, we see that there are $2r - 4$ different functions P_a given by (2.3.21). These correspond to the $2r - 4$ ground states (1.5.17). ■

2.4. Regime I

The results for this regime closely parallel those of regime III. Our work is slightly more complicated owing to the intricacy of the summands in the definition (1.5.23)–(1.5.26) of the relevant polynomials $Y_m(a, b, c; q)$.

As before, we examine admissible sequences of $l_2, \ldots, l_m$ satisfying (1.5.14), and split them into two classes according to whether $l_m = b - 1$ or $l_m = b + 1$. Writing the polynomials $Y_m(a, b, c; q)$ simply as $Y_m(a, b, c)$ we find that they are completely defined by the following recurrences and initial values ($m > 0$, $1 \leqslant a < r$):

$$Y_m(a, b, b+1) = Y_{m-1}(a, b-1, b) + q^{m\chi(b+1)} Y_{m-1}(a, b+1, b), \qquad 1 \leqslant b \leqslant r-2 \qquad (2.4.1)$$

$$Y_m(a, b, b-1) = Y_{m-1}(a, b+1, b) + q^{m[1-\chi(b-1)]} Y_{m-1}(a, b-1, b), \qquad 2 \leqslant b \leqslant r-1 \qquad (2.4.2)$$

$$Y_m(a, 0, 1) = Y_m(a, r, r-1) = 0 \qquad (2.4.3)$$

$$Y_0(a, b, b \pm 1) = \delta_{a,b} \qquad (2.4.4)$$

where $\chi(i)$ is the characteristic function of all numbers $\leqslant n$:

$$\chi(i) = 1, \quad \text{if } i \leqslant n$$
$$= 0, \quad \text{if } i > n \qquad (2.4.5)$$

and n is the integer part of $r/2$, i.e.,

$$\begin{aligned} n &= (r-1)/2, \qquad \text{if } r \text{ odd} \\ &= r/2, \qquad \text{if } r \text{ even} \end{aligned} \tag{2.4.6}$$

We shall in fact find that we can solve (2.4.1)–(2.4.5) without using (2.4.6); so in this section and the next we shall ignore (2.4.6) and allow n to be any integer in the interval $1 \leqslant n \leqslant r-2$.

Note that $\chi(b+1) = 1 - \chi(b-1) = 0$ for $b = n$ and for $b = n+1$. For these values of b it follows that the right-hand sides of (2.4.1) and (2.4.2) are the same and hence

$$Y_m(a,b,b-1) = Y_m(a,b,b+1), \qquad b = n, n+1 \tag{2.4.7}$$

We shall need the following two functions:

$$\alpha(a,b,c) = -2a + \tfrac{1}{2}r(b-c+3) - r\chi(c) \tag{2.4.8}$$

$$\beta(m;a,b,c) = \tfrac{1}{4}(b-c+1)(m+b-a) - \tfrac{1}{2}a + \tfrac{1}{2}\chi(c)\left[m(c-b)+a-b\right] \tag{2.4.9}$$

Lemma 2.4.1. For a, b integers,

$$\alpha(a,b,b+1) = \alpha(a,b-1,b), \qquad b \neq n \tag{2.4.10}$$

$$\beta(m;a,b,b+1) = \beta(m-1;a,b-1,b), \qquad b \neq n \tag{2.4.11}$$

$$\alpha(a,b,b-1) = \alpha(a,b+1,b), \qquad b \neq n+1 \tag{2.4.12}$$

$$\beta(m;a,b,b-1) = \beta(m-1;a,b+1,b), \qquad b \neq n+1 \tag{2.4.13}$$

$$\alpha(a,b-1,b) = \alpha(a,b+1,b), -r \tag{2.4.14}$$

$$\beta(m-1;a,b-1,b) = \beta(m-1;a,b+1,b), + \tfrac{1}{2}(a-b-m) + m\chi(b) \tag{2.4.15}$$

Also, for a an integer and $b = n$ or $n+1$,

$$\alpha(a,b,b-1) = \alpha(a,b,b+1) \tag{2.4.16}$$

$$\beta(m;a,b,b-1) = \beta(m;a,b,b+1) \tag{2.4.17}$$

These lemmas follow directly from the definitions (2.4.8) and (2.4.9).

The following five lemmas contain the results necessary to provide expressions for the $Y_m(a,b,c)$; they each concern the polynomial

$G_m(a,b,c)$, defined by ($m - a - b$ an even integer)

$$G_m(a,b,c) = \sum_{\lambda=-\infty}^{\infty} q^{2r\lambda^2+\alpha(a,b,c)\lambda+\beta(m;a,b,c)} \begin{bmatrix} m \\ \frac{1}{2}(m+a-b)-r\lambda \end{bmatrix} \tag{2.4.18}$$

Lemma 2.4.2. For a and b integers, with $b \neq n$,

$$G_m(a,b,b+1) = G_{m-1}(a,b-1,b) + q^{m\chi(b+1)} G_{m-1}(a,b+1,b) \tag{2.4.19}$$

Proof. Using (2.4.10) and (2.4.11),

$$G_m(a,b,b+1) - G_{m-1}(a,b-1,b)$$

$$= \sum_{\lambda=-\infty}^{\infty} q^{2r\lambda^2+\alpha(a,b-1,b)\lambda+\beta(m-1;a,b-1,b)}$$

$$\times \left\{ \begin{bmatrix} m \\ \frac{1}{2}(m+a-b)-r\lambda \end{bmatrix} - \begin{bmatrix} m-1 \\ \frac{1}{2}(m+a-b)-r\lambda \end{bmatrix} \right\} \tag{2.4.20}$$

Using (2.2.2) to simplify the bracketted difference of Gaussian polynomials, then using (2.4.14) and (2.4.15) and comparing the result with (2.4.18), we find that

$$G_m(a,b,b+1) - G_{m-1}(a,b-1,b) = q^{m\chi(b)} G_{m-1}(a,b+1,b) \tag{2.4.21}$$

Noting that $\chi(b) = \chi(b+1)$ for $b \neq n$, we obtain (2.4.19). ■

Lemma 2.4.3. For a and b integers, with $b \neq n+1$,

$$G_m(a,b,b-1) - G_{m-1}(a,b+1,b) = q^{m[1-\chi(b-1)]} G_{m-1}(a,b-1,b) \tag{2.4.22}$$

Proof. The proof parallels that of the preceding lemma. We use (2.4.12) and (2.4.13) to write the left-hand side of (2.4.22) as a sum involving the difference of two Gaussian polynomials then use (2.2.3) to express this difference as a single polynomial, and finally use (2.4.14), (2.4.15) and the fact that $\chi(b) = \chi(b-1)$ for $b \neq n+1$. ■

Lemma 2.4.4. For $b = n$ and $b = n+1$,

$$G_m(a,b,b-1) = G_m(a,b,b+1) \tag{2.4.23}$$

This follows at once from (2.4.16) and (2.4.17).

Lemma 2.4.5. For any integer a,

$$G_m(a,0,1) = G_m(-a,0,1) \tag{2.4.24}$$

$$G_m(a,r,r-1) = G_m(-a,r,r-1) \tag{2.4.25}$$

Proof. From (2.4.8), (2.4.9), and (2.4.18),

$$G_m(a,0,1) = \sum_{\lambda=-\infty}^{\infty} q^{2r\lambda^2 - 2a\lambda + \frac{1}{2}m} \begin{bmatrix} m \\ \frac{1}{2}(m+a) - r\lambda \end{bmatrix} \tag{2.4.26}$$

$$G_m(a,r,r-1) = \sum_{\lambda=-\infty}^{\infty} q^{2r(\lambda+1/2)^2 - 2a(\lambda+1/2) + (1/2)m} \begin{bmatrix} m \\ \frac{1}{2}(m+a) - r(\lambda+\frac{1}{2}) \end{bmatrix} \tag{2.4.27}$$

Using (2.2.4), we observe that the right-hand side of (2.4.26) is unchanged by negating a and λ. Similarly, the right-hand side of (2.4.27) is unchanged by negating a and replacing λ by $-\lambda - 1$. The symmetries (2.4.24) and (2.4.25) immediately follow. ■

Lemma 2.4.6. For $1 \leqslant |a|, b < r$, and $c = b \pm 1$,

$$G_0(a,b,c) = q^{-a/2}\delta_{a,b} \tag{2.4.28}$$

Proof. From (2.2.1), when $m = 0$ the Gaussian polynomial in (2.4.18) vanishes unless $a - b = 2r\lambda$, which can happen only when $a = b$ and $\lambda = 0$. Using (2.4.8) and (2.4.9), the summand in (2.4.18) then has value $q^{-a/2}$. The result (2.4.28) follows. ■

We are now in a position to state and establish our result for the function $Y_m(a,b,c)$.

Theorem 2.4.7. For $m \geqslant 0$, $1 \leqslant a,b,c < r$, $c = b \pm 1$, and $b - a - m$ an even integer,

$$Y_m(a,b,c) = q^{a/2}[G_m(a,b,c) - G_m(-a,b,c)] \tag{2.4.29}$$

Proof. Since $Y_m(a,b,c)$ is completely defined by the recurrences and boundary conditions (2.4.1)–(2.4.4), we need only show that they are satisfied by (2.4.29). From Lemma 2.4.2, the recurrence relation (2.4.1) is satisfied (for $b \neq n$) by $Y_m(a,b,c) = G_m(a,b,c)$. Since the relation is unchanged by negating a, it is also satisfied by $Y_m(a,b,c) = G_m(-a,b,c)$. Since it is linear and homogeneous, it is also satisfied by (2.4.29).

Similarly, Lemma 2.4.3 implies that Eq. (2.4.2) is satisfied by (2.4.29) for $b \neq n+1$, while Lemma 2.4.4 implies that Eq. (2.4.7) is satisfied for $b = n, n+1$.

Thus when $b = n$ we have proved that (2.4.2) and (2.4.7) are satisfied. Since (2.4.7) was obtained by eliminating the right-hand sides of (2.4.1) and (2.4.2), it follows that (2.4.1) is also satisfied for $b = n$. Similarly, (2.4.2) is satisfied for $b = n+1$. (In fact, Lemmas 2.4.2 and 2.4.3 are valid for $b = n$ and $b = n+1$, but their proofs are then different from those given above. We have introduced (2.4.7) to avoid the need to give the proofs for these special cases.)

Thus the expression (2.4.29) for $Y_m(a,b,c)$ satisfies the recurrences (2.4.1) and (2.4.2) for all allowed values of b. From Lemmas 2.4.5 and 2.4.6 it is easily seen that the boundary conditions (2.4.3) and (2.4.4) are also satisfied. This completes the proof of Theorem 2.4.7. ■

The Limit $m \to \infty$. We now want to take the limit $m \to \infty$. From (1.5.5) we see that $t < 0$ in regime I, so from (1.5.21) the argument q of our function $Y_m(a,b,c;q)$ is numerically less than 1:

$$|q| < 1 \tag{2.4.30}$$

We have essentially four cases to consider: $c = b \pm 1$, $\chi(c) = 0$, or 1. In order to avoid digressions during the proofs of the main results, we first prove the following lemma, using the definitions (2.2.1), (2.2.8), and (2.2.9).

Lemma 2.4.8.

$$\lim_{m\to\infty} q^{-m/2}\left\{\begin{bmatrix} m \\ \frac{1}{2}m+B \end{bmatrix} - \begin{bmatrix} m \\ \frac{1}{2}m+B+b \end{bmatrix}\right\}$$
$$= (1-q^b)(q^{-B-b+1}-q^{B+1})/[(1-q)(q)_\infty] \tag{2.4.31}$$

Proof. Set $m = 2k$ and note that

$$q^{-k}\left\{\begin{bmatrix} 2k \\ k+B \end{bmatrix} - \begin{bmatrix} 2k \\ k+B-b \end{bmatrix}\right\}$$
$$= q^{-k}\sum_{j=0}^{b-1}\left\{\begin{bmatrix} 2k \\ k+B+j \end{bmatrix} - \begin{bmatrix} 2k \\ k+B+j+1 \end{bmatrix}\right\}$$
$$= \sum_{j=0}^{b-1} \frac{(q)_{2k}(q^{-B-j}-q^{B+j+1})}{(q)_{k+B+j+1}(q)_{k-B-j}} \tag{2.4.32}$$

using the definitions (2.2.1) and (2.2.8). Taking the limit $k \to \infty$, the right-hand side becomes

$$\sum_{j=0}^{b-1} (q^{-B-j} - q^{B+j+1})/(q)_\infty \tag{2.4.33}$$

Performing the j summation, we obtain (2.4.31), as desired.

We shall save space in the subsequent developments if we define "$\lim_{m\to\infty}$" to mean the limit restricted to those values of m of the same parity as $a - b$.

Theorem 2.4.9. Let a and b be integers, $1 \leqslant a < r$, $1 \leqslant b < n$. Then

$$\begin{aligned}\lim_{m\to\infty} q^{-m} Y_m(a,b,b+1) &= \frac{q^{1-b}(1-q^b)}{(1-q)(q)_\infty} \sum_{\mu=-\infty}^{\infty} (-1)^\mu q^{(1/2)r\mu(\mu+1)-a\mu} \\ &= q^{1-b}(1-q^b)E(q^a,q^r)/[(1-q)(q)_\infty]\end{aligned} \tag{2.4.34}$$

where the sum is over all integers μ, and the function $E(x,z)$ is given by (1.5.7).

Proof. We begin by noting that for $1 \leqslant b < n$,

$$\alpha(a,b,b+1) = -2a, \qquad \beta(m;a,b,b+1) = \tfrac{1}{2}(m-b) \tag{2.4.35}$$

Using these expressions in (2.4.18) and (2.4.29), negating λ in $G_m(-a,b,b+1)$ and using (2.2.4), we obtain

$$\begin{aligned}Y_m(a,b,b+1) = \sum_{\lambda=-\infty}^{\infty} & q^{2r\lambda^2 - 2a\lambda + \frac{1}{2}(m-b+a)} \\ & \times \left\{ \begin{bmatrix} m \\ \frac{1}{2}(m+a-b) - r\lambda \end{bmatrix} - \begin{bmatrix} m \\ \frac{1}{2}(m+a+b) - r\lambda \end{bmatrix} \right\}\end{aligned} \tag{2.4.36}$$

Taking the limit $m\to\infty$, using (2.4.31), we obtain

$$\begin{aligned}&\lim_{m\to\infty} q^{-m} Y_m(a,b,b+1) \\ &\quad = \frac{q^{1-b}(1-q^b)}{(1-q)(q)_\infty} \sum_{\lambda=-\infty}^{\infty} q^{2r\lambda^2}\left[q^{(r-2a)\lambda} - q^{a-(r+2a)\lambda}\right]\end{aligned} \tag{2.4.37}$$

Splitting the summation into two series, taking $\lambda = \mu/2$, (μ even) in the first, and taking $\lambda = (\mu+1)/2$ (μ odd) in the second, we obtain the desired result (2.4.34). ■

Theorem 2.4.10. Let a and b be integers, $1 \leqslant a < r$, $n \leqslant b \leqslant r-2$. Then

$$\lim_{m\to\infty} Y_m(a,b,b+1) = [(q)_\infty]^{-1} \sum_{\mu=-\infty}^{\infty} (-1)^\mu q^{(1/2)r\mu(\mu+1)-a\mu}$$
$$= E(q^a, a^r)/(q)_\infty \qquad (2.4.38)$$

Proof. This case is simpler than the previous, in that we can use the elementary formula (2.2.6), rather than the more sophisticated Lemma 2.4.8.

Since $n \leqslant b \leqslant r-2$, we have

$$\alpha(a,b,b+1) = r-2a, \qquad \beta(m;a,b,b+1) = -\tfrac{1}{2}a \qquad (2.4.39)$$

From (2.4.18) and (2.4.29), letting $m \to \infty$ and using (2.2.6), we obtain

$$\lim_{m\to\infty} Y_m(a,b,b+1) = [(q)_\infty]^{-1} \sum_{\lambda=-\infty}^{\infty} q^{2r\lambda^2+r\lambda}(q^{-2a\lambda} - q^{2a\lambda+a}) \qquad (2.4.40)$$

Splitting this series into two, setting $\lambda = \mu/2$ in the first, and $(-\mu-1)/2$ in the second, we obtain the desired result (2.4.38).

Theorem 2.4.11. Let a and b be integers, $1 \leqslant a < r$, $2 \leqslant b \leqslant n+1$. Then

$$\lim_{m\to\infty} Y_m(a,b,b-1) = E(q^a,q^r)/(q)_\infty \qquad (2.4.41)$$

Proof. We merely note from (2.4.8) and (2.4.9) that

$$\alpha(a,b,b-1) = r-2a, \qquad \beta(m;a,b,b-1) = -\tfrac{1}{2}a \qquad (2.4.42)$$

These are the same expressions as those in (2.4.39), so the proof proceeds exactly as for Theorem 2.4.10. ■

Theorem 2.4.12. Let a and b be integers, $1 \leqslant a < r$, $n+2 \leqslant b < r$. Then

$$\lim_{m\to\infty} q^{-m} Y_m(a,b,b-1) = q^{1-r+b}(1-q^{r-b})E(q^a,q^r)/[(1-q)(q)_\infty] \qquad (2.4.43)$$

Proof. In this case

$$\alpha(a,b,b-1) = 2r-2a, \qquad \beta(m;a,b,b-1) = \tfrac{1}{2}(m+b-2a) \qquad (2.4.44)$$

Using (2.4.18) and (2.4.29); negating λ in $G_m(a,b,b-1)$ and using (2.2.4), while replacing λ in $G_m(-a,b,b-1)$ by $\lambda - 1$, we find that

$$Y_m(a,b,b-1) = \sum_{\lambda=-\infty}^{\infty} q^{2r\lambda^2 - 2(r-a)\lambda + \frac{1}{2}(m+b-a)} \times \left\{ \begin{bmatrix} m \\ \frac{1}{2}(m-a+b) - r\lambda \end{bmatrix} - \begin{bmatrix} m \\ \frac{1}{2}(m+2r-a-b) - r\lambda \end{bmatrix} \right\} \qquad (2.4.45)$$

The right-hand side of this equation is the same as that of (2.4.36), but with a, b replaced by $r-a, r-b$. Making these replacements throughout Theorem 2.4.9, and using the identity

$$E(z,x) = E(x/z,x) \qquad (2.4.46)$$

we at once obtain our desired result (2.4.43). ■

The last four theorems give the limiting behavior of $Y_m(a,b,c)$ as m becomes large, for all integers a,b,c such that $1 \leqslant a,b,c \leqslant r-1$, $c = b \pm 1$. Note that in every case the appropriate limit factors into the form

$$E(q^a, q^r) \times [\text{function of } b,c,q] \qquad (2.4.47)$$

Substituting the appropriate limiting forms of $Y_m(a,b,c)$ into (1.5.21) and (1.5.25), it follows that P_a tends to a limit as $m \to \infty$, and that the function of b,c,q in (2.4.47) cancels out of the final expression for P_a, leaving

$$P_a = v_a E(q^a, q^r) \Big/ \sum_{1 \leqslant a' \leqslant r-1} v_{a'} E(q^{a'}, q^r), \qquad (2.4.48)$$

where, as in regime III (and indeed in all regimes), a and a' are either both even integers, or are both odd. From (1.5.21) and (1.5.5), the argument q of $Y_m(a,b,c;q)$ is here related to the parameter x by

$$q = x^{r-2} \qquad (2.4.49)$$

We see that the limit (2.4.48) is indepedent of the values b,c of the boundary heights l_{m+1}, l_{m+2}. [It is even independent of the integer n, which as we mentioned after (2.4.6) can be regarded in this section as an independent integer.] This indicates that the model is disordered in regime I, and is consistent with the fact that there are then just the two ground states (1.5.29).

2.5. Regime IV

We now come to our first exploitation of the duality principle that was utilized for treating the Rogers–Ramanujan-type identities occurring in the original hard hexagon model.[26] In this regime we see from (1.5.5) that $t > 0$, so from (1.5.21) the argument q of the polynomial $Y_m(a,b,c;q)$ is now greater than 1. Our interest therefore now centers on the polynomials reciprocal to those of regime I.

Throughout this section we shall write $y_m(a,b,c)$ for $Y_m(a,b,c)$ with q replaced by q^{-1}, and $g_m(a,b,c)$ for $G_m(a,b,c)$ with q replaced by q^{-1}. As before $b-a-m$ is even and $c = b \pm 1$.

It is convenient to define

$$\gamma(m;a,b,c) = \tfrac{1}{4}(a-b)^2 - \beta(m;a,b,c) \tag{2.5.1}$$

Lemma 2.5.1.

$$q^{m^2/4} g_m(a,b,c) = q^{\gamma(m;a,b,c)} \sum_{\lambda=-\infty}^{\infty} q^{r(r-2)\lambda^2 - [\alpha(a,b,c)+ra-rb]\lambda} \times \begin{bmatrix} m \\ \frac{1}{2}(m+a-b) - r\lambda \end{bmatrix} \tag{2.5.2}$$

Proof. This follows immediately by applying (2.2.5) to (2.4.18). ■

As in section 2.4, we shall denote by $\lim_{m\to\infty}$ the limit taken through those integers m of the same parity as $a-b$.

Lemma 2.5.2.

$$\lim_{m\to\infty} q^{m^2/4} y_m(a,b,c) = Q(q)^{-1} \sum_{\lambda=-\infty}^{\infty} q^{r(r-2)\lambda^2 + rb\lambda - \frac{1}{2}a} \times \left\{ q^{\gamma(m;a,b,c) - [\alpha(a,b,c)+ra]\lambda} - q^{\gamma(m;-a,b,c) - [\alpha(-a,b,c)-ra]\lambda} \right\} \tag{2.5.3}$$

Proof. This follows from applying (2.2.6) to (2.5.2), and using (2.4.29) with q replaced by q^{-1}. ■

The result (2.5.3) can be simplified by defining two further functions:

$$\tau(b,c) = \tfrac{1}{2}(c-b-1) + \chi(c) \tag{2.5.4}$$

for $1 \leqslant b, c \leqslant r-1$, $c = b \pm 1$; then

$$\Gamma(a,d;q) = \sum_{\lambda=-\infty}^{\infty} q^{r(r-2)\lambda^2 + rd\lambda + (a-d-1)^2/4} \{ q^{-(r-2)a\lambda} - q^{(r-2)a\lambda + ad} \} \tag{2.5.5}$$

for $1 \leqslant a \leqslant r-1$ and $1 \leqslant d \leqslant r-3$. Using (1.5.7), we can write (2.5.5) as

$$\Gamma(a,d;q) = q^{(a-d-1)^2/4}\{E[-q^{r(r-2+d)-(r-2)a}, q^{2r(r-2)}] - q^{ad}E[-q^{r(r-2+d)+(r-2)a}, q^{2r(r-2)}]\} \quad (2.5.6)$$

The function $\tau(b,c)$ is an integer, with values $-1, 0, 1$.

Lemma 2.5.3.

$$\lim_{m\to\infty} q^{[m+|\tau(b,c)|]^2/4} y_m(a,b,c) = \Gamma[a, b+\tau(b,c)-1; q]/Q(q) \quad (2.5.7)$$

Proof. This is just a restatement of Lemma 2.5.2, as is most easily verified by explicitly considering the four cases $c = b \pm 1$, $c \leqslant n$ or $c > n$. ■

From (1.5.21), (1.5.25), (1.5.5), it follows that P_a tends to a limit as $m \to \infty$, given by

$$P_a = v_a\Gamma(a,d;x^2) \Big/ \sum_{1 \leqslant a' \leqslant r-1} v_{a'}\Gamma(a',d;x^2) \quad (2.5.8)$$

where

$$d = b + \tau(b,c) - 1 \quad (2.5.9)$$

and a, a' are either both even integers (the "even" sub-SOS model), or are both odd integers (the "odd" sub-SOS model).

We note that P_a depends on d. From (2.5.5) and (2.4.5),

$$\begin{aligned} d &= b, && \text{if} \quad b+1 = c \leqslant n \\ &= b-1, && \text{if} \quad b+1 = c > n \\ &= b-1, && \text{if} \quad b-1 = c \leqslant n \\ &= b-2, && \text{if} \quad b-1 = c > n \end{aligned} \quad (2.5.10)$$

Since $1 \leqslant b, c \leqslant r-1$, we see that d is an integer satisfying

$$1 \leqslant d \leqslant r-3 \quad (2.5.11)$$

Thus d takes only $r-3$ values. Allowing for the two possible sets of values of a (even or odd), it follows that there are $2(r-3)$ functions P_a, corresponding to the $2r-6$ ground states in regime IV.

2.6. Regime II

This is by far the most complicated of the regimes. From (1.5.5) we see that $t < 0$, so from (1.5.9) the argument q of the function $X_m(a,b,c;q)$ is greater than 1. We are therefore interested in the polynomials (in $q^{1/2}$) that are reciprocal to those of regime III, which we can define as

$$x_m(a,b,c) = x_m(a,b,c;q) = q^{m(m+1)/4}X_m(a,b,c;q^{-1}) \quad (2.6.1)$$

From (2.3.1) and (2.3.2) we see that the $x_m(a,b,c)$ satisfy the recurrences ($m > 0$, $1 \leqslant a < r$):

$$x_m(a,b,b+1) = q^{m/2}x_{m-1}(a,b+1,b) + x_{m-1}(a,b-1,b), \qquad 1 \leqslant b \leqslant r-2 \quad (2.6.2)$$

$$x_m(a,b,b-1) = q^{m/2}x_{m-1}(a,b-1,b) + x_{m-1}(a,b+1,b), \qquad 2 \leqslant b \leqslant r-1 \quad (2.6.3)$$

Lemma 2.6.1. For $m \geqslant 0$, $1 \leqslant a,b,c < r$, $c = b \pm 1$, $m + a - b$ an even integer

$$x_m(a,b,c) = q^{(m+a-b)/4}\{f_m(a,b,c) - f_m(-a,b,c)\} \quad (2.6.4)$$

where

$$f_m(a,b,c) = \sum_{\lambda=-\infty}^{\infty} q^{r\lambda^2 - a\lambda + (b+1-c)(2r\lambda+b-a)/4}\begin{bmatrix} m \\ \frac{1}{2}(m+a-b) - r\lambda \end{bmatrix} \quad (2.6.5)$$

Proof. This follows immediately on replacing q by q^{-1} in Theorem 2.3.1 and using (2.6.1). ■

Remarks on Procedure. Using (2.2.6), we can easily verify that $f_m(a,b,c)$ tends to a limit as $m \to \infty$. Unfortunately this limit is an even function of a (this follows by replacing λ by $-\lambda$, or by $-\lambda - 1$), so the expression $\{f_m(a,b,c) - f_m(-a,b,c)\}$ in (2.6.4) vanishes.

This is still true even if we expand the Gaussian polynomial in (2.6.5) in increasing powers of $q^{m/2}$ and keep a finite number of terms in the expansion, which is basically the method we employed in regime I. (We shall in fact find for large m that $\{f_m(a,b,c) - f_m(-a,b,c)\}$ is q raised to the power of a *quadratic* form in m, which accounts for the failure of this method.)

We therefore need a more sophisticated procedure to determine the large-m behavior of $x_m(a,b,c)$, and it is this that makes regime II so difficult. We shall begin by showing that $x_m(a,2,1)$ can be expressed in terms of the function

$$\rho_N(q,w,s;d_1,\ldots,d_2) = \sum_{j=0}^{N} q^{j^2+j-Nj}w^{-j}$$

$$\times(1-q^{N-2j}w)\prod_{i=1}^{s}\left[(d_i)_j(wd_i)_{N-j}\right]^{-1} \quad (2.6.6)$$

where N, s are nonnegative integers and we are using the notation (2.2.8).

We shall further show that one of $d_1, \ldots, d_s$ is equal to the argument q, while another is equal to q/w. As is shown in Appendix B, the function ρ_N is then proportional to a special case of a "well-poised q-hypergeometric series".[28,29,32] Its behavior in the limit $N \to \infty$ is determined in Appendix B, using a standard theorem for such series (Theorem 4 of Ref. 29).

Using this result (Theorem B5), we can evaluate $x_m(a,2,1)$ in the limit of m large. We then obtain the limits of all the $x_m(a,b,c)$ by sequentially solving the recurrence relations (2.6.2), (2.6.3) for $x_m(a,1,2)$, $x_m(a,3,2)$, $x_m(a,2,3), \ldots, x_m(a,r-2,r-1)$. Finally we identify our results when $r=5$ with those of the original hard hexagon model.[9,10]

Evaluation of $x_m(a,2,1)$ in the Limit of m Large. We begin by expressing $x_m(a,2,1)$, for $m+a$ an even integer, in terms of the function ρ_N. We shall need the integers $u, v, \lambda_0, \lambda_1, N$, defined by

$$\tfrac{1}{2}(m-a)+1 = r\lambda_0 + u, \qquad 0 \leqslant u < r \quad (2.6.7)$$

$$\tfrac{1}{2}(m+a)+1 = r\lambda_1 + v, \qquad 0 \leqslant v < r \quad (2.6.8)$$

$$N = \lambda_0 + \lambda_1 \quad (2.6.9)$$

Adding (2.6.7) and (2.6.8), we see that $m, N, u+v$ are related by

$$m+2 = rN + u + v \quad (2.6.10)$$

Theorem 2.6.2. For $1 \leqslant a < r$, $m \geqslant 0$, $m+a$ even,

$$x_m(a,2,1) = L_m(a)\rho_N(q^r, q^{v-u}, r; q^{u+1}, q^{u+2}, \ldots, q^{u+r}) \quad (2.6.11)$$

where

$$L_m(a) = \frac{q^{r\lambda_0^2+(a-r)\lambda_0+(m+2-a)/4}(q)_{m+1}}{(q)_u(q)_v} \quad (2.6.12)$$

Proof. Using Lemma 2.6.1, negating λ in $f_m(-a,2,1,)$,

$$x_m(a,2,1) = q^{(m+2-a)/4} \sum_{\lambda=-\infty}^{\infty} q^{r\lambda^2+(r-a)\lambda} \times \left\{ \begin{bmatrix} m \\ \frac{1}{2}(m+a)-1-r\lambda \end{bmatrix} - q^{a-2\lambda} \begin{bmatrix} m \\ \frac{1}{2}(m+a)+1-r\lambda \end{bmatrix} \right\} \tag{2.6.13}$$

The summand in (2.6.13) is zero unless

$$-\tfrac{1}{2}(m-a)-1 \leqslant r\lambda \leqslant \tfrac{1}{2}(m+a)+1 \tag{2.6.14}$$

The definitions (2.6.7) and (2.6.8) are arranged so that $-\lambda_0$ is the smallest value of λ in the range (2.6.14), while λ_1 is the largest. Thus we can restrict the summation in (2.6.13) to

$$-\lambda_0 \leqslant \lambda \leqslant \lambda_1 \tag{2.6.15}$$

From (2.2.1) it is readily verified, for $-1 \leqslant k \leqslant m+1$, that

$$\begin{bmatrix} m \\ k-1 \end{bmatrix} - q^{2k-m} \begin{bmatrix} m \\ k+1 \end{bmatrix} = \frac{(1-q^{2k-m})(q)_{m+1}}{(q)_{k+1}(q)_{m-k+l}} \tag{2.6.16}$$

Using this identity, (2.6.13) can be written as

$$x_m(a,2,1) = q^{(m+2-a)/4} \sum_{\lambda=-\lambda_0}^{\lambda_1} q^{r\lambda^2+(r-a)\lambda} \times \frac{(1-q^{a-2r\lambda})(q)_{m+1}}{(q)_{(1/2)(m+a)+1-r\lambda}(q)_{(1/2)(m-a)+1+r\lambda}} \tag{2.6.17}$$

Now we replace λ by $j-\lambda_0$, and use (2.6.7)–(2.6.9), (2.6.12), and (2.2.10) to obtain

$$x_m(a,2,1) = L_m(a) \sum_{j=0}^{N} q^{rj^2+(r-rN+u-v)j} \frac{1-q^{rN-2rj+v-u}}{(q^{u+1})_{rj}(q^{v+1})_{rN-rj}} \tag{2.6.18}$$

Now we note that

$$(q^{u+1})_{rj} = (q^{u+1};q^r)_j(q^{u+2};q^r)_j \cdots (q^{u+r};q^r)_j \tag{2.6.19}$$

and similarly for $(q^{v+1})_{r(N-j)}$. Comparing the summation in (2.6.18) with that in (2.6.6), we obtain the desired result (2.6.11). ■

Now we want to let $m \to \infty$. We shall find it convenient, both here and in Appendix B, to adopt the convention that by

$$\lim_{n\to\infty} g_n = f_n \tag{2.6.20a}$$

we mean

$$\lim_{n\to\infty} (g_n/f_n) = 1 \tag{2.6.20b}$$

We shall need the function

$$\Phi_a(z) = \frac{E(q^a,q^r)E(-z,q)[Q(q^r)]^3}{E(-z,q^r)E(-q^a z,q^r)[Q(q)]^2} \tag{2.6.21}$$

This is an analytic function of z in the domain $0 < |z| < \infty$, so has a Laurent expansion:

$$\Phi_a(z) = \sum_{j=-\infty}^{\infty} \eta_{a,j} z^j \tag{2.6.22}$$

We shall find that we can express the $x_m(a,b,c)$ in terms of these coefficients $\eta_{a,j}$.

Theorem 2.6.3. For $1 \leqslant a < r$, $m + a$ even,

$$\lim_{m\to\infty} x_m(a,2,1) = (-1)^N q^{r\lambda_0^2+(a-r)\lambda_0+(m+2-a)/4-(1/2)u(u+1)} \eta_{v-u,-N-u} \tag{2.6.23}$$

Proof. When $m\to\infty$ we see from (2.6.7)–(2.6.10) that u and v remain between 0 and $r-1$, while $N\to\infty$. Also, one of the arguments $q^{u+1},\ldots,q^{u+r}$ of ρ_N in Theorem 2.6.2 must be equal to q^r, while another must be equal to q^{r+u-v}. These two special arguments must be distinct, since if $u = v$ then (2.6.7) and (2.6.8) give $a = r(\lambda_1 - \lambda_0)$, i.e., $a \equiv 0 \pmod r$, which are not allowed values of a.

We can therefore use Theorem B5 to obtain the limiting behavior of ρ_N in Theorem 2.6.2. This gives that $x_m(a,2,1)$ is the coefficient of z^{-N} in the Laurent expansion of

$$\frac{L_m(a)(q^r;q^r)_\infty^3 E(q^{v-u},q^r)}{E(z,q^r)E(q^{v-u}z,q^r)} \prod_{i=1}^{r} \frac{E(q^{u+i}z^{-1},q^r)}{(q^r;q^r)_\infty (q^{u+i};q^r)_\infty (q^{v+i};q^r)_\infty} \tag{2.6.24}$$

Now we use the triple product form (1.5.6) of $E(z,x)$ to note that

$$E(q^{u+i}z^{-1},q^r)/(q^r;q^r)_\infty = (q^{u+i}z^{-1};q^r)_\infty (q^{r-u-i}z;q^r)_\infty \tag{2.6.25}$$

Using the identity (2.6.19), it follows that the product over i in (2.6.24) is simply

$$(q^{u+1}z^{-1})_\infty (q^{-u}z)_\infty / [(q^{u+1})_\infty (q^{v+1})_\infty] \tag{2.6.26}$$

Substituting this, and using the definition (2.6.12) of $L_m(a)$, with $(q)_{m+1}$ replaced by $(q)_\infty$, the function (2.6.24) becomes

$$\frac{q^{r\lambda_0^2+(a-r)\lambda_0+(m+2-a)/4}(q^r;q^r)_\infty^3 E(q^{v-u},q^r)E(q^{-u}z,q)}{E(z,q^r)E(q^{v-u}z,q^r)(q)_\infty^2} \tag{2.6.27}$$

From (1.5.6) we can readily verify that for all integers u,

$$E(q^{-u}z,q) = (-z)^u q^{-(1/2)u(u+1)}E(z,q) \tag{2.6.28}$$

so from (2.6.21) and (2.2.9) we see that (2.6.27) is equal to

$$q^{r\lambda_0^2+(a-r)\lambda_0+(m+2-a)/4-(1/2)u(u+1)}(-z)^u\Phi_{v-u}(-z) \tag{2.6.29}$$

Recalling that $x_m(a,2,1)$ is the coefficient of z^{-N} in the Laurent expansion of this function (2.6.29), we obtain the desired result (2.6.23). ■

The form for $x_m(a,2,1)$ given by the preceding theorem involves the integers λ_0, v, u, N defined by (2.6.7)–(2.6.9). It turns out that we can eliminate them in favor of our original variables a and m. To do this we need certain recurrence and symmetry properties of $\Phi_a(z)$ and $\eta_{a,j}$.

Lemma 2.6.4. For all integers a, and nonzero complex numbers

$$\Phi_{a+r}(z) = -z\Phi_a(z) \tag{2.6.30}$$

$$\Phi_a(q^r z) = z^{2-r}q^{a-(1/2)r(r-1)}\Phi_a(z) \tag{2.6.31}$$

$$\Phi_a(q^{-a}z^{-1}) = z^{1-a}q^{(1/2)a(1-a)}\Phi_a(z) \tag{2.6.32}$$

Proof. These three properties all follow directly from the definition (2.6.21), the recurrence formula (2.6.28), and the identity [readily verifiable from (1.5.6)]:

$$E(z^{-1},q) = -z^{-1}E(z,q) \quad ■ \tag{2.6.33}$$

Lemma 2.6.5. For all integers a and j,

$$\eta_{a+r,j} = -\eta_{a,j-1} \tag{2.6.34}$$

$$\eta_{a,j+r-2} = q^{rj-a+(1/2)r(r-1)}\eta_{a,j} \tag{2.6.35}$$

$$\eta_{a,a-1-j} = q^{(1/2)a(a-2j-1)}\eta_{a,j} \tag{2.6.36}$$

Proof. Substituting the series form (2.6.22) of $\Phi_a(z)$ into Lemma 2.6.4 and equating coefficients in the resulting expansions, we obtain (2.6.34)–(2.6.36). Repeated application of (2.6.34) and (2.6.35) yields the corollaries:

$$\eta_{a+rk,j} = (-1)^k \eta_{a,j-k} \tag{2.6.37}$$

$$\eta_{a,j+(r-2)k} = q^{k[rj-a+(1/2)r+(1/2)r(r-2)k]}\eta_{a,j} \tag{2.6.38}$$

for all integers a, j, k. ■

Theorem 2.6.6. For $1 \leqslant a < r$, $m + a$ even,

$$\lim_{m\to\infty} x_m(a,2,1) = q^{-(m+2-a)^2/8}\eta_{a,(1/2)(a-m-2)} \tag{2.6.39}$$

Proof. From (2.6.7) and (2.6.8),

$$v - u = a + r(\lambda_0 - \lambda_1) \tag{2.6.40}$$

From (2.6.37), with $k = \lambda_0 - \lambda_1$, it follows that

$$\begin{aligned}(-1)^N \eta_{v-u,-N-u} &= (-1)^{N+\lambda_0-\lambda_1}\eta_{a,-N-u-\lambda_0+\lambda_1} \\ &= \eta_{a,-u-2\lambda_0}\end{aligned} \tag{2.6.41}$$

after using (2.6.9).

Now we use (2.6.38), with $j = -u - r\lambda_0$ and $k = \lambda_0$, to obtain

$$\eta_{a,-u-2\lambda_0} = q^{\lambda_0[(1/2)r-a-ru-(1/2)r(r+2)\lambda_0]}\eta_{a,-u-r\lambda_0} \tag{2.6.42}$$

so from this and (2.6.41), Theorem 2.6.3 becomes

$$\lim_{m\to\infty} x_m(a,2,1) = q^{-(1/2)(u+r\lambda_0)(u+r\lambda_0+1)+(m+2-a)/4}\eta_{a,-u-r\lambda_0} \tag{2.6.43}$$

Note that u, λ_0 enter this expression only in the combination $u + r\lambda_0$, which from (2.6.7) is $\frac{1}{2}(m - a + 2)$. The result (2.6.39) follows. ■

Evaluation of $x_m(a,b,c)$ in the Limit of m Large. Now that we have $x_m(a,2,1)$, we can use the recurrence relations (2.6.2) and (2.6.3) to obtain $x_m(a,b,c)$ in general. We first need a lemma concerning the large-j behavior of $\eta_{a,j}$.

Lemma 2.6.7. For all integers a, j, define $\hat{\eta}_{a,j}$ by

$$\eta_{a,j} = q^{[(1/2)rj(j+1)-aj]/(r-2)}\hat{\eta}_{a,j} \tag{2.6.44}$$

Then

$$\hat{\eta}_{a,j+r-2} = \hat{\eta}_{a,j} \tag{2.6.45}$$

Proof. This follows immediately from (2.6.35). Since $\hat{\eta}_{a,j}$ is periodic in j, it is bounded, so this lemma gives the large-j behavior of $\eta_{a,j}$.

Theorem 2.6.8. For $1 \leqslant a < r$, $m+a-b$ even,

$$\lim_{m\to\infty} x_m(a,b,b+1) = q^{-(m+a-b)^2/8}\eta_{a,(1/2)(m+a-b)}, \qquad 1 \leqslant b \leqslant r-2 \tag{2.6.46a}$$

$$\lim_{m\to\infty} x_m(a,b,b-1) = q^{-(m-a+b)^2/8}\eta_{a,(1/2)(a-m-b)}, \qquad 2 \leqslant b \leqslant r-1 \tag{2.6.46b}$$

Proof. From (2.3.3) and (2.6.1), $x_m(a,0,1)$ in the recurrence relation (2.6.2) is to be interpreted as zero. Setting $b=1$ therein, we therefore obtain

$$x_m(a,1,2) = x_{m-1}(a,2,1) \tag{2.6.47}$$

Using Theorem 2.6.6 and Eq. (2.6.36), we can at once verify that (2.6.46a) is satisfied for $b=1$, and (2.6.46b) is satisfied for $b=2$.

We now proceed by induction. Suppose that (2.6.46a) is satisfied for $1 \leqslant b \leqslant b_0 - 1$, and (2.6.46b) is satisfied for $2 \leqslant b \leqslant b_0$, where $2 \leqslant b_0 \leqslant r-2$. (We have just shown this is so for $b_0 = 2$.) Setting $b = b_0$ in the recurrence relation (2.6.3) and rearranging, we obtain

$$x_{m-1}(a,b_0+1,b_0) = x_m(a,b_0,b_0-1) - q^{m/2}x_{m-1}(a,b_0-1,b_0) \tag{2.6.48}$$

In the limit $m\to\infty$, both terms on the right-hand side are known to be given by (2.6.46). Thus their ratio is, using Lemma 2.6.7 and temporarily dropping the suffix on b_0,

$$\frac{q^{m/2}x_{m-1}(a,b-1,b)}{x_m(a,b,b-1)} = \frac{q^{(1/2)m(b-a+1)}\eta_{a,(1/2)(m+a-b)}}{\eta_{a,(1/2)(a-b-m)}} = \frac{q^{m(r-1-b)/(r-2)}\hat{\eta}_{a,(1/2)(m+a-b)}}{\hat{\eta}_{a,(1/2)(a-b-m)}} \tag{2.6.49}$$

Since $\hat{\eta}_{a,j}$ is bounded and $b < r-1$, this ratio tends to zero as $m\to\infty$. Thus we can ignore the last term in (2.6.48). Using (2.6.46b) to evaluate $x_m(a,b_0,b_0-1)$, it follows at once that $x_{m-1}(a,b_0+1,b_0)$ is also given, for m large, by (2.6.46b). Thus (2.6.46b) is true for $b=b_0+1$.

The second step of the inductive proof is to set $b=b_0$ in (2.6.2). Both terms on the right-hand side are now known to be given by (2.6.46), so (temporarily dropping the suffix on b_0)

$$\frac{q^{m/2}x_{m-1}(a,b+1,b)}{x_{m-1}(a,b-1,b)} = \frac{q^{(1/2)m(a-b+1)}\eta_{a,(1/2)(a-m-b)}}{\eta_{a,(1/2)(m+a-b)}} = \frac{q^{m(b-1)/(r-2)}\hat{\eta}_{a,(1/2)(a-m-b)}}{\hat{\eta}_{a,(1/2)(m+a-b)}} \tag{2.6.50}$$

Since $b > 1$, this ratio tends to zero as $m \to \infty$, so we can ignore the first term on the right-hand side of (2.6.2). Using (2.6.46a) to evaluate the second term, we at once find that $x_m(a, b_0, b_0 + 1)$ on the left-hand side is also given by (2.6.46a). Thus (2.6.46a) is true for $b = b_0$.

This completes the inductive loop. Taking $b_0 = 2, 3, \ldots, r - 2$, we establish the results (2.6.46). ■

Throughout the rest of this section, let

$$j = -b + (c - b)m \tag{2.6.51}$$

Since $c = b \pm 1$, j is either $m - b$ or $-m - b$. It is an integer, and has the same parity as a.

We shall use the function $\hat{x}_m(a, b, c)$, defined in terms of $x_m(a, b, c)$ by

$$x_m(a, b, c) = q^{j(j+r)/(4r-8)} \hat{x}_m(a, b, c) \tag{2.6.52}$$

This function $\hat{x}_m$ contains fractional powers of q, but its large-m behavior is simpler to discuss than that of x_m itself, as is evident from the following theorem.

Theorem 2.6.9. For $1 \leqslant a < b,\ c < r,\ c = b \pm 1,\ m + a - b$ even,

$$\lim_{m \to \infty} \hat{x}_m(a, b, c) = q^{a(r-a)/(4r-8)} \hat{\eta}_{a,(1/2)(a+j)} \tag{2.6.53}$$

Proof. This is just a restatement of Theorem 2.6.8, using the definitions (2.6.44), (2.6.51), (2.6.52). ■

Note that m, b, c enter the right-hand side of (2.6.53) only via the integer j, and that $a + j$ must be even. Further, from (2.6.45), $\hat{\eta}_{a,i}$ is a periodic function of i, of period $r - 2$. Thus as m is increased (in increments of two), $\hat{x}_m(a, b, c)$ sequentially takes $r - 2$ different values, returning after $r - 2$ increments to its original value. Varying b and c merely changes the starting point of this cycle.

From (1.5.5.), (1.5.9), (1.5.13), (2.6.1), and (2.6.52), the local height probabilities are given by

$$P_a = u_a \hat{x}_m(a, b, c; q) \Big/ \sum_{1 \leqslant a' \leqslant r-1} u_{a'} \hat{x}_m(a', b, c; q) \tag{2.6.54}$$

where a, a' are either confined to the set of even integers (the "even" sub-SOS model), or to the set of odd integers (the "odd" model). The argument q is related to the parameter x of (1.5.8) (which is not to be confused with the functions $x_m(a, b, c; q)$ and $\hat{x}_m(a, b, c; q)$ by

$$q = x^{r-2} \tag{2.6.55}$$

For m large and a fixed, there are just $r - 2$ different values of $\hat{x}_m(a, b, c; q)$ that can be obtained by varying m, b, and c, i.e., by varying the boundary conditions on the lattice. From (2.6.54) it follows that for a given parity of a, there are just $r-2$ different functions P_a that can be formed by varying the boundary conditions. Altogether this gives us $2r-4$ functions, corresponding to the $2r-4$ ground states (1.5.19).

The Case $r=5$. As was shown in Sections 1.3 and 1.4, when $r=5$ we regain the origin hard hexagon model. Thus the results (2.6.46)–(2.6.55) of this section should then reduce to those for regime II of the hard hexagon model.[9,10,26] From their definitions, the quantities $F_k(0), F_k(1)$ in Refs. 9, 10, and 26 are related to our functions $\hat{x}_m(a,b,c)$ by

$$\begin{aligned} F_k(\sigma) &= q^{(1/2)(\sigma-1)-(2-k)(3-k)/6}\hat{x}_{6i-2k+b-2}(2+2\sigma,b,b+1), \qquad 1\leqslant b\leqslant 3 \\ &= q^{(1/2)(\sigma-1)-(2-k)(3-k)/6}\hat{x}_{6i-2k-b}(2+2\sigma,b,b-1), \qquad 2\leqslant b\leqslant 4 \end{aligned} \tag{2.6.56}$$

where $k=1,2,3$, $\sigma=0,1$, and the right-hand side is to be evaluated in the limit $i\to\infty$.

We can take the limit $i\to\infty$ by using Theorem 2.6.9, which gives

$$\begin{aligned} F_k(\sigma) &= q^{[2\sigma-(2-k)(3-k)]/6}\hat{\eta}_{2+2\sigma,\sigma-k+3i} \\ &= q^{[2\sigma-(2-k)(3-k)]/6}\hat{\eta}_{2+2\sigma,\sigma+k-3i+1} \end{aligned} \tag{2.6.57}$$

We see that both expressions in (2.6.57) are independent of b, so it is irrelevant what value we took in (2.6.56). Also, from (2.6.36) and (2.6.44), $\hat{\eta}_{a,a-1-j}=\hat{\eta}_{a,j}$ for all integers a,j. It follows that the two expressions in (2.6.57), and hence in (2.6.56), are equivalent.

The integer i still appears in (2.6.57), but from the periodicity relation (2.6.45) it plays no role, the expressions in (2.6.57) being independent of i. We can therefore now set $i=0$ and use the first expression. Then from (2.6.44) we obtain

$$F_k(\sigma) = q^{\sigma k-1+k-k^2}\eta_{2+2\sigma,\sigma-k} \tag{2.6.58}$$

for $k=1,2,3$ and $\sigma=0,1$.

If we now use the definitions (2.6.21), (2.6.22) of $\eta_{a,j}$, the resulting expressions for the $F_k(\sigma)$ are not of the same form as those obtained originally.[9,10] To establish the equivalence we need the following theorem. As throughout this paper, $E(z,x)$ is the elliptic theta function defined by (1.5.6), or equivalently (1.5.7).

Theorem 2.6.10. For $|q|<1$, and all nonzero complex numbers z,w,

$$\begin{aligned} &E(z,q)E(w,q)E(zw,q)E(w/z,q) \\ &= [Q(q)]^2\{E(w^3,q^3)[E(qz^3,q^3)-zE(qz^{-3},q^3)] \\ &\qquad -(w/z)E(z^3,q^3)[E(qw^3,q^3)-wE(qw^{-3},q^3)]\} \end{aligned} \tag{2.6.59}$$

Proof. Like all elementary theta function identities this can be proved using Liouville's theorem (Section 15.3 of Ref. 10). Regard q, w as constants and z as a variable. Let

$$f(z) = \frac{\text{right-hand side of (2.6.59)}}{\text{left-hand side of (2.6.59)}} \tag{2.6.60}$$

Then by using (2.6.28) it is readily verified that $f(z)$ has the periodicity property $f(qz) = f(z)$. Since $E(z,q)$ is analytic for $0 < |z| < \infty$, the only possible singularities of $f(z)$ in this domain are simple poles, occurring when the left-hand side of (2.6.59) vanishes, in particular when $z = 1$, w, or w^{-1}. However, using the obvious identities $E(1,q) = 0$, $E(w^{-3},q) = -w^{-3}E(w^3,q)$, we can immediately verify that the right-hand side of (2.6.59) then also vanishes, so the poles have zero residue and $f(z)$ is in fact analytic at $z = 1, w, w^{-1}$. Since $f(qz) = f(z)$, it is therefore analytic at $z = q^n, q^n w, q^n w^{-1}$ for all integers n, which exhausts all the zeros of the left-hand side of (2.6.59).

It follows that $f(z)$ is analytic in $0 < |z| < \infty$, so is bounded in $q \leqslant |z| \leqslant 1$. Again using the periodicity property $f(qz) = f(z)$, the function $f(z)$ must be bounded everywhere. This means that $f(z)$ must be analytic at $z = 0$ and ∞, so by Liouville's theorem it must be a constant.

From the infinite product expansions (1.5.6) and (2.2.7), it is readily established that if $\omega = e^{2\pi i/3}$, then (for all complex numbers z)

$$E(z,q)E(\omega z,q)E(\omega^{-1}z,q) = \left[\,Q(q)\right]^3 E(z^3,q^3)/Q(q^3) \tag{2.6.61}$$

$$E(q,q^3) = Q(q) \tag{2.6.62}$$

$$E(\omega,q^3) = (1-\omega)Q(q^3) \tag{2.6.63}$$

Setting $z = \omega$ in (2.6.59) and (2.6.60), and using these identities, it follows that

$$f(\omega) = 1 \tag{2.6.64}$$

Since $f(z)$ is a constant, this implies that $f(z) \equiv 1$, which establishes the theorem. ■

Lemma 2.6.11. For $|q| < 1$ and all complex numbers z,

$$E(z,q) = Q(q)\prod_{i=1}^{r}\left[E(q^{i-1}z,q^r)/Q(q^r)\right] \tag{2.6.65}$$

Proof. Like (2.6.19), this follows directly from the infinite product expansions (1.5.6) and (2.2.7). ■

We can use the preceding lemma and theorem to write $\Phi_a(z)$ in a form from which the original hard hexagon results can be regained. First we define

$$\zeta_a = q^{(1/2)(a-1)(3a-4)}\left[E(q^{3a+5},q^{15}) - q^a E(q^{5-3a},q^{15})\right]/Q(q) \quad (2.6.66)$$

$$\zeta'_a = q^{(1/2)(a-1)(3a-4)}E(q^{3a},q^{15})/Q(q) \quad (2.6.67)$$

Theorem 2.6.12. For $r=5$, $1 \leqslant a \leqslant 4$ and all complex numbers z,

$$\begin{aligned}\Phi_a(z) &= \zeta_a z^{1-a}E(-q^{6a}z^{-3},q^{15}) \\ &\quad + \zeta'_a z^{-a}\left[q^a E(-q^{6a+5}z^{-3},q^{15}) + q^{3a}z^{-1}E(-q^{5-6a}z^3,q^{15})\right]\end{aligned} \quad (2.6.68)$$

Proof. Negate z in Lemma 2.6.11 and substitute the resulting expression for $E(-z,q)$ into (2.6.21). This gives

$$\Phi_a(z) = \frac{E(q^a,q^5)E(-q^iz,q^5)E(-q^jz,q^5)E(-q^kz,q^5)}{Q(q)Q(q^5)^2} \quad (2.6.69)$$

where i,j,k are the integers $1,2,3,4$ excluding a. Using the quasi-periodic property (2.6.68) of the E function, together with the simple identity $E(z,x)=E(x/z,x)$, the result (2.6.69) can be written as

$$\begin{aligned}\Phi_a(z) &= q^{(1/2)(a-1)(3a-4)}z^{1-a}E(q^a,q^5)E(-q^{3a}z^{-1},q^5)E(-q^{2a}z^{-1},q^5) \\ &\quad \times E(-q^a z^{-1},q^5)/\left[Q(q)Q(q^5)^2\right]\end{aligned} \quad (2.6.70)$$

This expression contains a product of four E functions, which is the same as that on the left-hand side of Theorem 2.6.10, provided that z,w,q in that theorem are replaced by $q^a, -q^{2a}z^{-1}, q^5$. We can therefore apply Theorem 2.6.10 to (2.6.70). Doing so, we obtain the desired result (2.6.68). ■

From (2.6.22) and (1.5.7), an immediate *corollary* of this theorem is that, for $1 \leqslant a \leqslant 4$ and all integers i,

$$\begin{aligned}\eta_{a,3i+1-a} &= \zeta_a q^{15i(i+1)/2-6ai} \\ \eta_{a,3i-a} &= \zeta'_a q^{15i(i+1)/2-(6a+5)i+a} \\ \eta_{a,3i-a-1} &= \zeta'_a q^{15i(i-1)/2-(6a-5)i+3a}\end{aligned} \quad (2.6.71)$$

[These expressions satisfy the periodicity and symmetry relations (2.6.34)–(2.6.36).]

Using these expressions (2.6.71) for $\eta_{a,j}$, from (2.6.58) it follows that

$$\begin{aligned} F_1(0) &= q^{-1}\zeta_2 = \left[E(q^4,q^{15}) + qE(q,q^{15})\right]/Q(q) \\ F_1(1) &= q^{-9}\zeta_4 = \left[-qE(q^2,q^{15}) + E(q^7,q^{15})\right]/Q(q) \\ F_2(0) &= F_3(0) = q^{-1}\zeta_2' = E(q^6,q^{15})/Q(q) \\ F_2(1) &= F_3(1) = q^{-11}\zeta_4' = qE(q^{12},q^{15})/Q(q) \end{aligned} \tag{2.6.72}$$

which are the results originally obtained for the hard hexagon model [Eq. 64 of Ref. 9, (14.5.22) of Ref. 10, Theorems 1–6 of Ref. 26].

3. NORMALIZATION OF THE PROBABILITIES P_a

3.1. The Normalization Factor M

The probabilities P_a are given, for the four regimes, by (2.3.21), (2.4.48), (2.5.8), and (2.6.54). Each of these equations is of the form

$$P_a = \mu_a/M \tag{3.1.1}$$

where

$$M = \sum_{1 \leqslant a \leqslant r}^{*} \mu_a \tag{3.1.2}$$

the * indicating that the sum is either confined to the set of even integers or to the set of odd integers. It is sometimes convenient to write μ_a in the form

$$\mu_a = \mu_a' + \mu_{-a}' \tag{3.1.3}$$

Using (2.3.19), (2.5.6), (2.6.53), and the definitions (1.5.5), (1.5.10), (1.5.22) of u_a, v_a, we find that for the four regimes, μ_a or μ_a' is given by

$$\text{I} \qquad \mu_a = x^{(1/2)a(a+1-r)}E(x^a, -x^{r/2})E\left[x^{(r-2)a}, x^{r(r-2)}\right] \tag{3.1.4a}$$

$$\text{II} \qquad \mu_a = E(x^a, x^r)\hat{\eta}_{a,(1/2)(a+j)} \tag{3.1.4b}$$

$$\text{III} \qquad \mu_a' = x^{(1/2)a(a-1-2d)}E(x^a, x^r)E\left[-x^{2(r-a)(r-1)+2rd}, x^{4r(r-1)}\right] \tag{3.1.4c}$$

$$\text{IV} \qquad \mu_a' = x^{(1/2)a(a-1-2d)}E(x^a, x^{r/2})E\left[-x^{2(r-a)(r-2)+2rd}, x^{4r(r-2)}\right] \tag{3.1.4d}$$

We have omitted factors that are independent of a, and hence cancel out of (3.1.1). The function $\hat{\eta}_{a,i}$ is defined by (2.6.21), (2.6.22), and (2.6.44) with $q = x^{r-2}$. It satisfies the relations

$$\hat{\eta}_{a,i} = \hat{\eta}_{a,i+r-2} = x^{-2a-r}\hat{\eta}_{a+2r,i+r} = -x^{-a}\hat{\eta}_{-a,i-a} \tag{3.1.5}$$

for all integers a, i.

In (3.1.4), a, j, d are integers satisfying the restrictions

$$a + j = \text{even}, \qquad 1 \leqslant d \leqslant d_{\max} \tag{3.1.6}$$

where $d_{\max} = r - 2, r - 3$ for regimes III, IV, respectively. The integers j and d are to be regarded as fixed (i.e., independent of a): they determine the phase of the system in the ordered regimes.

The function μ_a is needed only for $1 \leqslant a < r$, and μ'_a for $1 \leqslant |a| < r$, but it can be convenient to extend the definitions (3.1.3) and (3.1.4) to all integers a. Using (2.6.28), (2.6.33), and (3.1.5), we can then verify that

$$\mu_{-a} = \mu_a = \mu_{a+2r}, \qquad \mu_0 = \mu_{\pm r} = 0 \tag{3.1.7}$$

Thus (3.1.2) can be put into the form

$$M = \frac{1}{2} \sum_{-r \leqslant a < r}^{*} \mu_a \tag{3.1.8}$$

It is clear from (3.1.1) and (3.1.2) that M is a normalization factor whose presence in (3.1.1) ensures that the local height probabilities satisfy the condition

$$\sum_{1 \leqslant a < r}^{*} P_a = 1 \tag{3.1.9}$$

This means that M plays a similar role to the partition function: in fact if we trace back its origin we see that M is proportional to S and T in (1.5.13) and (1.5.25), to the denominators in (A26) and (A1), and hence to the partition function (1.2.1) of the eight-vertex SOS model.

We also see from (3.1.7) and (3.1.8) that M has a simple mathematical significance: it is the sum of $\frac{1}{2}\mu_a$ over all equal-parity values of a within a complete period $2r$. For those cases (at present regimes III and IV) where μ'_a is defined, it is also true that

$$\mu'_a = \mu'_{a+2r}, \qquad \mu'_0 = \mu'_{\pm r} = 0 \tag{3.1.10}$$

so we can, using (3.1.3), write (3.1.8) as

$$M = \sum_{-r \leqslant a < r}^{*} \mu'_a \tag{3.1.11}$$

3.2. Theta-Function "Sums-of-Products" Identities

The Case $r = 5$. For the original hard hexagon model, which is when $r = 5$, it has been noted that the normalization factor M always turns out to be a simple product of theta functions. (There are nine different possibilities, depending on the regime and the phase.) Further, this simplification is always a corollary of one or more of the 40 "sums-of-products" identities[34] listed by Ramanujan and subsequently proved by Rogers,[35] Darling,[36] Mordell,[37] and Watson.[38] These identities involve the function $Q(x)$ defined by (2.2.7), as well as the function $P(x), G(x), H(x)$ defined by

$$P(x) = \prod_{n=1}^{\infty} (1 - x^{2n-1}) \tag{3.2.1}$$

$$G(x) = \prod_{n=1}^{\infty} \left[(1 - x^{5n-4})(1 - x^{5n-1})\right]^{-1} \tag{3.2.2}$$

$$H(x) = \prod_{n=1}^{\infty} \left[(1 - x^{5n-3})(1 - x^{5n-2})\right]^{-1} \tag{3.2.3}$$

For instance, consider regime I with $r = 5$ and a even. Then from (3.1.2) and (3.1.4a), setting $x_1 = -x^{1/2}$.

$$\begin{aligned} M &= \mu_2 + \mu_4 \\ &= x^{-2}E(x^2, -x^{2.5})E(x^6, x^{15}) + E(x^4, -x^{2.5})E(x^{12}, x^{15}) \\ &= x_1^{-4}Q(x_1)\,Q(x_1^6)\left\{H(x_1)G(x_1^6) - x_1G(x_1)H(x_1^6)\right\} \end{aligned} \tag{3.2.4}$$

However, Eq. 8 in Birch's list[34] of Ramanujan's identities is

$$G(x^6)H(x) - xG(x)H(x^6) = P(x)/P(x^3) \tag{3.2.5}$$

so we deduce at once that

$$\begin{aligned} M &= x_1^{-4}Q(x_1)\,Q(x_1^6)P(x_1)/P(x_1^3) \\ &= x_1^{-4}E(x_1, x_1^2)E(-x_1^3, x_1^{12}) \end{aligned} \tag{3.2.6}$$

The other $r = 5$ cases involve Eqs. 2, 5, 6, and 23 of Birch's[34] list, and are given in Refs. 9, 10, and 20.

Do these identities generalize to arbitrary values of r, so that M is always a simple product of theta functions? It turns out that the answer is yes, as is shown in the remainder of this section. As usual, regime II has to be treated separately; for regimes I, III, and IV there is one general identity (Theorem 3.2.1) that covers all cases.

Regimes I, III, and IV. The general identity that we shall need is the following.

Theorem 3.2.1. Let x, y be real numbers such that $|x| < 1$ and

$$y^{m/2} = -\epsilon x^r, \qquad \epsilon = \pm 1 \tag{3.2.7}$$

where m, r are positive integers and $1 \leqslant m < 2r$; then for all complex numbers z

$$\sum_{-r<a<r}^{*} x^{(1/2)a(a-1)} z^a E(x^a, y) E[\epsilon^{m-1} x^{(2r-m)(r+a)} z^{2r}, x^{2r(2r-m)}]$$
$$= \tfrac{1}{2}\{E(-z,x)E(z^{-1}, y/x) \pm E(z,x)E(-z^{-1}, y/x)\} \quad (3.2.8)$$

where the upper (plus) choice of the $\pm$ sign in (3.2.8) is to be made if the summation is restricted to even values of a; the lower (minus) sign applies if the sum is restricted to odd values.

Proof. First consider the quantity

$$J = \sum_{-\infty<a<\infty}^{*} \sigma(a) \quad (3.2.9)$$

where

$$\sigma(a) = x^{(1/2)a(a-1)} z^a E(x^a, y) \quad (3.2.10)$$

From (3.2.7) and the quasiperiodicity property (2.6.28) of $E(z,q)$, we can verify that (for all integers k)

$$\sigma(a+2kr) = x^{2kra+2k^2r^2-kr} z^{2kr} (-x^a)^{-km} y^{-km(km-1)/2} \sigma(a)$$
$$= (-\epsilon^{m-1})^k x^{(2r-m)k(a+kr)} z^{2kr} \sigma(a) \quad (3.2.11)$$

For any integer a, there are unique integers a_0, k such that

$$a = a_0 + 2kr, \qquad -r \leqslant a_0 < r \quad (3.2.12)$$

and a_0 has the same parity as a. Thus (3.2.9) can be written

$$J = \sum_{-r\leqslant a_0<r}^{*} \sum_{k=-\infty}^{\infty} \sigma(a_0 + 2kr) \quad (3.2.13)$$

Dropping the suffix on a_0 and substituting the expression (3.2.11) for $\sigma(a+2kr)$ into (3.2.13), the k summation can be performed by using (1.5.7). This gives

$$J = \sum_{-r\leqslant a<r}^{*} \sigma(a) E[\epsilon^{m-1} x^{(2r-m)(r+a)} z^{2r}, x^{2r(2r-m)}] \quad (3.2.14)$$

Using (3.2.10) and noting that $\sigma(-r) = 0$, we see that J is equal to the left-hand side of (3.2.8).

Now we go back to the definition (3.2.9)–(3.2.10) of J. Let J_e and J_0 be the values of J when the summation in (3.2.9) is restricted, respectively, to even and odd values of a. Then

$$J_e + \tau J_0 = \sum_{a=-\infty}^{\infty} \tau^a \sigma(a) \quad (3.2.15)$$

where $\tau = \pm 1$ and the summation is now over *all* integers a. Using (3.2.10) and (1.5.7), it follows that

$$J_e + \tau J_0 = \sum_{a=-\infty}^{\infty} \tau^a x^{(1/2)a(a-1)} z^a \sum_{k=-\infty}^{\infty} (-1)^k x^{ak} y^{(1/2)k(k-1)}$$

$$= \sum_{k=-\infty}^{\infty} (-\tau)^k z^{-k} (y/x)^{(1/2)k(k-1)} \sum_{j=-\infty}^{\infty} \tau^j z^j x^{(1/2)j(j-1)} \quad (3.2.16)$$

where we have interchanged the a, k summations and set $a = j - k$. Since the summand in (3.2.16) factors into a function of k times a function of j, the sum is a product of two E functions. Using (1.5.7), we obtain

$$J_e + \tau J_0 = E(-\tau z, x) E(\tau z^{-1}, y/x) \quad (3.2.17)$$

Taking sums and differences of this equation for $\tau = +1$ and $\tau = -1$, we can evaluate J_e and J_0. Remembering that the left-hand side of (3.2.8) is either J_e or J_0, we obtain (3.2.8), as desired. ■

Simple special cases of Theorem 3.2.1 occur when

$$z = x^l \quad \text{or} \quad z = -(y/x)^l \quad (3.2.18)$$

where l is an integer. The last term in (3.2.8) then vanishes, so the right-hand side reduces to the single product

$$\tfrac{1}{2} E(-z, x) E(z^{-1}, y/x) \quad (3.2.19)$$

and is independent of whether the summation is restricted to even or to odd values of a.

The particular identities that we shall need are all such simple special cases of Theorem 3.2.1. We shall also need the symmetry and periodicity properties (2.4.46) and (2.6.28) of $E(z,q)$, as well as the elementary identities (true for all integers d and complex numbers z, q with $|q| < 1$)

$$E(z,q) = E(-qz^2, q^4) - zE(-qz^{-2}, q^4) \quad (3.2.20)$$

$$E(-1,q) = 2E(-q, q^4) \quad (3.2.21)$$

$$E(-q^{-d}, q) = 2q^{-(1/2)d(d+1)} E(-q, q^4) \quad (3.2.22)$$

[The first follows from (1.5.7) by splitting the sum therein into two parts, one with n even and the other with n odd; the second is obtained from the first by taking $z = -1$; the third follows from the second and from (2.6.28).]

Taking $m = 2$, $\epsilon = -1$, $y = x^r$, and $z = x^{-d}$ (d an integer) in Theorem 3.2.1, using (2.4.46) and (3.2.22), we obtain

$$\sum_{-r<a<r}^{*} x^{(1/2)a(a-1-2d)} E(x^a, x^r) E\left[-x^{2(r-1)(r-a)+2rd}, x^{4r(r-1)}\right]$$

$$= x^{-(1/2)d(d+1)} E(-x, x^4) E(x^d, x^{r-1}) \quad (3.2.23)$$

Similarly, taking $m = 4$, $\epsilon = -1$, $y = -x^{r/2}$, and $z = x^{-d}$ (d an integer), we obtain

$$\sum_{-r<a<r}^{*} x^{(1/2)a(a-1-2d)}E(x^{a}, -x^{r/2})E[-x^{2(r-2)(r-a)+2rd}, x^{4r(r-2)}]$$
$$= x^{-(1/2)d(d+1)}E(-x, x^{4})E[x^{d}, -x^{(r-2)/2}] \tag{3.2.24}$$

Identities (3.2.32) and (3.2.24) are relevant to regimes III and IV, respectively. For regime I we take $m = 4$, $\epsilon = -1$, $y = -x^{r/2}$, and $z = w^{-(r-2)/2}$: this ensures that $z = -x/y$, so the second possibility in (3.2.18) is satisfied (for r even, so is the first). We then group the $(a, -a)$ terms together in (3.2.8) and use (2.6.33), (3.2.20), (2.6.28), and (3.2.21) to obtain

$$\sum_{1\leqslant a<r}^{*} x^{(1/2)a(a+1-r)}E(x^{a}, -x^{r/2})E[x^{(r-2)a}, x^{r(r-2)}]$$
$$= x^{(2r-r^{2}-\nu)/8}E(-x^{\nu/2}, x)E(x^{(r-2)/2}, x^{2r-4}) \tag{3.2.25}$$

where

$$\begin{aligned} \nu &= 0, \quad \text{if } r \text{ is even} \\ &= 1, \quad \text{if } r \text{ is odd} \end{aligned} \tag{3.2.26}$$

[For r even, (3.2.25) is a special case of (3.2.24), with $d = (r-2)/2$.]

Regime II. For regime II, the identity we want to establish is

$$\sum_{1\leqslant a<r}^{*} E(x^{a}, x^{r})\hat{\eta}_{a,(1/2)(a+j)} = Q(q) \tag{3.2.27}$$

for all integers j, the sum being over integers a with the same parity as j. The function $\hat{\eta}_{a,i}$ is defined by (2.6.21), (2.6.22), and (2.6.44) with $q = x^{r-2}$. From (3.1.2) and (3.1.4b), M is equal to the left-hand side of (3.2.27), so it follows that

$$M = Q(q) \tag{3.2.28}$$

One way to prove (3.2.27) is to first establish the following theorem.

Theorem 3.2.2. Let $\Phi_a(z)$ be defined by (2.6.21), with $q = x^{r-2}$. Then for all complex numbers z,

$$\sum_{1\leqslant a<r}^{*} x^{ra(a-2)/8}z^{-a/2}E(x^{a}, x^{r})\Phi_{a}(q^{-a/2}z)$$
$$= x^{-r\alpha/8}z^{(1/2)\alpha-1}Q(q)E(-x^{(1/2)r\alpha}z, x^{r}) \tag{3.2.29}$$

where $\alpha = 0$ if the sum is restricted to even values of a, $\alpha = 1$ if the sum is restricted to odd values.

Proof. We regard z as a complex variable and use Liouville's theorem, rather as we did in Theorem 2.6.10. First we verify from (2.6.21) and

(1.5.6) that

$$\lim_{z\to -1} \Phi_a(z) = Q(q)$$
$$\lim_{z\to -1} \Phi_a(q^{-a}z) = (-1)^{a-1} q^{-(1/2)a(a-1)} Q(q) \tag{3.2.30}$$

Then we consider (3.2.29) when $z = -q^{b/2}$, where b is an integer of the same parity as a and α. If $1 \leqslant |b| < r$, then all terms in the sum on the left-hand side vanish except the one with $a = |b|$. Using (3.2.30) and (2.6.28), we can verify that (3.2.29) is satisfied. If $b = 0$ or r (and has the same parity as a and α), then all terms on the left-hand side and right-hand side vanish, so (3.2.29) is trivially satisfied.

Now let

$$f(z) = (\text{left-hand side} - \text{right-hand side}) \text{ of } (3.2.29) \tag{3.2.31}$$

This function is analytic in the domain $0 < |z| < \infty$, and by using (2.6.28) we can verify that

$$f(q^r z) = z^{2-r} q^{-(1/2)r(r-1)} f(z) \tag{3.2.32}$$

(This quasiperiodicity relation is satisfied by each term in the a summation.) From the above remarks it follows that

$$f(-q^{b/2}) = 0 \tag{3.2.33}$$

for all integers b of the same parity as a and α. This means that the function

$$g(z) = f(z)/E(-q^{-\alpha/2}z, q) \tag{3.2.34}$$

is also analytic in $0 < |z| < \infty$ [the denominator has only simple zeros, at which $f(z)$ vanishes].

Finally, consider the function

$$G(z) = z^{\alpha/2} g(z) E(-q^{\alpha/2}z, q^r) E(-q^{-\alpha/2}z, q^r) \tag{3.2.35}$$

[which is proportional to $f(z)$ divided by the $a = \alpha$ term of the sum in (3.2.29)]. This is analytic in $0 < |z| < \infty$ and is periodic:

$$G(q^r z) = G(z) \tag{3.2.36}$$

It is therefore bounded in $q^r \leqslant |z| \leqslant 1$, and hence in $0 < |z| < \infty$. This means that it must be analytic at $z = 0$ and ∞, so by Liouville's theorem it must be a constant. It vanishes when $z = -q^{\alpha/2}$, so the constant must be zero. Hence $G(z)$, $g(z)$, and $f(z)$ vanish identically, which proves the theorem. ■

The required identity (3.2.27) can now be established by Laurent expanding both sides of (3.2.29) in powers of z and equating coefficients, using (2.6.22), (2.6.44), and (1.5.7).

There is an alternative way to establish (3.2.28). Note that $\hat{\eta}$ enters

(3.1.4b) only because of the result (2.6.53) of Theorem 2.6.9. If we go back to the original functions $\hat{x}_m$ and, using (2.6.52) to x_m, we see that (3.1.2) is equivalent to

$$M = \lim_{m\to\infty} M_m(b,c) \tag{3.2.37}$$

where, for $1 \leqslant b < c < r$ and $c = b \pm 1$,

$$M_m(b,c) = \sum_{1\leqslant a<r}{}^{*}\, x^{(a+j)(a-j-r)/4} E(x^a, x^r) x_m(a,b,c) \tag{3.2.38}$$

the integer j being defined by (2.6.51); the summation is over integers a with the same parity as j.

In Section 2.6 we evaluated the large-m limit of $x_m(a,b,c)$. The derivation was very complicated, involving (in Appendix B) the theory of very well-poised q-hypergeometric series. By contrast, the large-m limit of $M_m(b,c)$ can be obtained quite simply and directly: the key is the following theorem.

Theorem 3.2.3. Let $M_n(b,c)$ be defined by (3.2.38), (2.6.51) and (2.6.4)–(2.6.5), with $q = x^{r-2}$. Define ξ (an integer or half an integer) by

$$\xi = (c-b)(\tfrac{1}{2}r - b) \tag{3.2.39}$$

Then for $|x| < 1$, $1 \leqslant b$, $c < r$ and $c = b \pm 1$,

$$M_m(b,c) = \sum_{t=-\infty}^{\infty} (-1)^t x^{(1/2)rt^2-(m+\xi)t} (q^{1-t};q)_m \tag{3.2.40}$$

where we are using the notation (2.2.8).

Proof. From (2.6.4) and (2.6.33) we can verify that the summand in (3.2.38) is an even function of a, and that

$$M_m(b,c) = \sum_{-r\leqslant a<r}{}^{*}\, E(x^a, x^r)\sigma(a) \tag{3.2.41}$$

where

$$\sigma(a) = x^{(a+j)(a-j-r)/4} q^{(m+a-b)/4} f_m(a,b,c) \tag{3.2.42}$$

and we have used the fact that $\sigma(0) = \sigma(-r) = 0$.

We now proceed similarly to the first part of the proof of Theorem 3.2.1. Incrementing a by $2r$, and replacing λ in (2.6.5) by $\lambda + 1$, we find that

$$\sigma(a+2r) = x^{2a+r}\sigma(a) \tag{3.2.43}$$

and hence, by recurrence,

$$\sigma(a+2kr) = x^{2ka+rk(2k-1)}\sigma(a) \tag{3.2.44}$$

for all integers k.

If we define

$$J = \sum_{-\infty<a<\infty}^{*} \sigma(a)(1-x^a) \tag{3.2.45}$$

then by proceeding as in (3.2.12) and (3.2.13) we obtain, using (1.5.7) and (3.2.41),

$$\begin{aligned} J &= \sum_{-r\leqslant a_0<r}^{*} \sigma(a_0) \sum_{k=-\infty}^{\infty} \left[x^{2ka_0+rk(2k-1)} - x^{(2k+1)a_0+rk(2k+1)} \right] \\ &= \sum_{-r\leqslant a<r}^{*} \sigma(a) \sum_{l=-\infty}^{\infty} (-1)^l x^{la+(1/2)rl(l-1)} \\ &= \sum_{-r\leqslant a<r}^{*} \sigma(a) E(x^a, x^r) \\ &= M_m(b,c) \end{aligned} \tag{3.2.46}$$

Thus J is equal to $M_m(b,c)$. On the other hand, substituting (2.6.5) into (3.2.42) and (3.2.45), interchanging the a, λ summations and setting

$$a = -j + 2(r-2)\lambda + 2t \tag{3.2.47}$$

(t must be an integer), we obtain

$$\begin{aligned} J = \sum_{\lambda=-\infty}^{\infty} \sum_{t=-\infty}^{\infty} (1 - q^{2\lambda}x^{2t-j}) q^{\lambda(2\lambda+b-c)-(1/2)(b+1-c)t} \\ \times x^{t(t-j-1)} \begin{bmatrix} m \\ \frac{1}{2}(1+b-c)m + t - 2\lambda \end{bmatrix} \end{aligned} \tag{3.2.48}$$

Let $F(\lambda, t)$ be the summand in (3.2.48) with the factor $(1 - q^{2\lambda}x^{2t-j})$ omitted. Then one can verify that

$$F(\lambda + \tfrac{1}{2}, t+1) = q^{2\lambda}x^{2t-j} F(\lambda, t) \tag{3.2.49}$$

so (3.2.48) can be written as

$$J = \sum_{\lambda=-\infty}^{\infty} \sum_{t=-\infty}^{\infty} \left[F(\lambda, t) - F(\lambda + \tfrac{1}{2}, t+1) \right] \tag{3.2.50}$$

The two terms in (3.2.50) can be grouped together by replacing (λ, t) by $(\frac{1}{2}\mu, t)$ in the first, and by $(\frac{1}{2}\mu - \frac{1}{2}, t-1)$ in the second. Then μ is even for the first term, odd for the second: combining them we get

$$J = \sum_{\mu=-\infty}^{\infty} \sum_{t=-\infty}^{\infty} (-1)^{\mu} F(\tfrac{1}{2}\mu, t) \tag{3.2.51}$$

the sums being over all integers μ, t.

Next we interchange the μ, t summations (such interchanges are justified by the rapid convergence of the series due to the quadratic exponents in the summand), and set

$$\mu = t + (b - c)k \tag{3.2.52}$$

(Since $|b - c| = 1$, k is an integer.) Then, after substituting the explicit form of the summand, (3.2.51) becomes

$$J = \sum_{t=-\infty}^{\infty} (-1)^t x^{(1/2)rt(t-1)-jt} \sum_{k=0}^{m} (-1)^k q^{(1/2)k(k+1)+(b-c)kt} \begin{bmatrix} m \\ k \end{bmatrix} \tag{3.2.53}$$

where we have used (2.2.4), and the fact that $|b - c| = 1$. The k summation can be restricted to $0 \leqslant k \leqslant m$ because of (2.2.1).

The k summation can now be performed by using the identity

$$\sum_{k=0}^{m} (-1)^k q^{(1/2)k(k-1)} z^k \begin{bmatrix} m \\ k \end{bmatrix} = (z)_m \tag{3.2.54}$$

(Theorem 3.3, p. 36 of Ref. 16). This gives

$$j = \sum_{t=-\infty}^{\infty} (-1)^t x^{(1/2)rt(t-1)-jt} (q^{1+(b-c)t}; q)_m \tag{3.2.55}$$

Replacing t by $(c - b)t$, using (2.6.51) and remembering that $J = M_m(b, c)$, we obtain the desired result (3.2.40). This completes the proof of Theorem 3.2.3. ■

It is instructive to explicitly expand the product in the summand of (3.2.40):

$$(q^{1-t}; q)_m = (1 - q^{1-t})(1 - q^{2-t}) \cdots (1 - q^{m-t}) \tag{3.2.56}$$

We see that this vanishes if $t = 1, 2, \ldots, m$, so we can restrict the t summation to the two regions $t \leqslant 0$ and $t > m$. Setting $t = -s$ in the first, $t = s + m + 1$ in the second, and using (2.2.11), we can write (3.2.40) as

$$M_m(b, c) = \sum_{s=0}^{\infty} (-1)^s x^{(1/2)rs^2 + ms} \left[x^{\xi s} - x^{-\xi s + [(1/2)r - \xi](m+1)} \right] (q^{s+1}; q)_m \tag{3.2.57}$$

It is now easy to take the limit $m \to \infty$. From (3.2.39), $|\xi| < \frac{1}{2}r$, so as $m \to \infty$ all terms in the summation in (3.2.57) tend uniformly to zero, except for the first part of the $s = 0$ term. Thus, using (3.2.37),

$$M = \lim_{m \to \infty} M_m(b, c) = (q; q)_\infty = Q(q) \tag{3.2.58}$$

We have thus rederived (3.2.28), without using the working of Appendix B and Section 2.6, and without introducing the functions η and $\hat{\eta}$ (which are multidimensional theta series).

3.3. Final Results

We can now write down explicit expressions for the local height probabilities. They are given by (3.1.1)–(3.1.4), and for regimes I, II, III, IV, we can write M as a simple product of theta functions by using (3.2.25), (3.2.28), (3.2.23), (3.2.24), respectively. We obtain

$$\text{I:}\quad P_a = x^{(1/2)a(a+1-r)+(r^2-2r+\nu)/8}E(x^a, -x^{r/2}) \times \frac{E[x^{(r-2)a}, x^{r(r-2)}]}{\{E(-x^{\nu/2}, x)E(x^{(r-2)/2}, x^{2r-4})\}} \tag{3.3.1a}$$

$$\text{II:}\quad P_a = E(x^a, x^r)\hat{\eta}_{a,(1/2)(a+j)}/Q(q) \tag{3.3.1b}$$

III and IV:

$$P_a = x^{(1/2)a(a-1)+(1/2)d(d+1)}E(x^a, y) \times \frac{\{x^{-ad}E[-x^{2l(r-a)+2rd}, x^{4rl}] - x^{ad}E[-x^{2l(r+a)+2rd}, x^{4rl}]\}}{[E(-x, x^4)E(x^d, y/x)]} \tag{3.3.1c}$$

where ν is defined by (3.2.26); $y = x^r$, $l = r - 1$ in regime III; while in regime IV $y = -x^{r/2}$, $l = r - 2$. In all cases x lies in the interval (0, 1).

The function $\hat{\eta}$ is defined by (2.6.21), (2.6.22), and (2.6.44), with $q = x^{r-2}$. Setting $j = j_0 + (r-2)k$ in (2.6.22), where j_0 and k are integers such that $0 \leqslant j_0 \leqslant r-3$, we can perform the k summation by using the periodicity property (2.6.45). [This procedure parallels that in (3.2.9)–(3.2.14).] We obtain the identity

$$\Phi_a(z) = \sum_{j=0}^{r-3} x^{(1/2)rj(j+1)-aj}\hat{\eta}_{a,j}z^jE[-q^{(1/2)r(r-1)+rj-a}z^{r-2}, q^{r(r-2)}] \tag{3.3.2}$$

where $q = x^{r-2}$. We can regard the $\hat{\eta}_{a,j}$ as being defined by this identity and the periodicity relation (2.6.45).

We can readily use (1.5.6) or (1.5.7) to expand all these results in increasing powers of x, so these forms are convenient for examining the behavior when x is small, which is when the eight-vertex SOS model is completely disordered (regime I) or completely ordered (regime II, III, and IV): the local height probabilities P_a take their ground state values.

Critical Behavior. The other extreme is when $x \to 1$. From (1.5.8) and (1.5.20), ϵ is then large and p in (1.2.3)–(1.2.9) is numerically small. In the limit when $x = 1$ and $p = 0$ the eight-vertex weights a, b, c, d in (1.2.3) satisfy

$$d = 0, \qquad |(a^2 + b^2 - c^2)/(2ab)| < 1 \tag{3.3.3}$$

The model is then critical,[10] so it is of particular interest to consider this case and to be able to develop perturbation expansions in powers of p.[24] This can be done by transforming to elliptic functions of conjugate modulus. [Examples of such transformations have already occurred in (A30) and (A51).] The results are conventionally expressed in terms of the standard theta functions,[17,18] defined for all complex numbers u, q, with $|q| < 1$, by

$$\theta_1(u,q^2) = 2|q|^{1/4}\sin u \prod_{n=1}^{\infty}(1 - 2q^{2n}\cos 2u + q^{4n})(1 - q^{2n}) \quad (3.3.4a)$$

$$\theta_4(u,q^2) = \prod_{n=1}^{\infty}(1 - 2q^{2n-1}\cos 2u + q^{4n-2})(1 - q^{2n}) \quad (3.3.4b)$$

$$\theta_2(u,q^2) = \theta_1\left(\frac{\pi}{2} + u, q^2\right), \qquad \theta_3(u,q^2) = \theta_4\left(\frac{\pi}{2} + u, q^2\right) \quad (3.3.4c)$$

In regimes I and IV we encounter θ_1 functions with q^2 negative, which is why we take q^2 rather than q to be the second argument. Note that θ_1 remains real even when q^2 is negative. Sometimes we shall find it convenient to remove the $2|q|^{1/4}$ factor and to work with the functions

$$\tilde{\theta}_i(u,q^2) = \tfrac{1}{2}|q|^{-1/4}\theta_i(u,q^2), \qquad i = 1,2 \quad (3.3.5)$$

When $q = 0$, these functions $\tilde{\theta}_1$ and $\tilde{\theta}_2$ reduce to $\sin u$ and $\cos u$.

The "conjugate modulus" identities that we shall need are [for all complex numbers u, ϵ such that $\mathrm{Re}(\epsilon) > 0$]

$$\theta_1(u,e^{-\epsilon}) = \rho(u,\epsilon)E(e^{-4\pi u/\epsilon}, e^{-4\pi^2/\epsilon}) \quad (3.3.6a)$$

$$\theta_4(u,e^{-\epsilon}) = \rho(u,\epsilon)E(-e^{-4\pi u/\epsilon}, e^{-4\pi^2/\epsilon}) \quad (3.3.6b)$$

$$\theta_1(\tfrac{1}{2}u, -e^{-\epsilon/4}) = 2^{1/2}\rho(u,\epsilon)E(e^{-4\pi u/\epsilon}, -e^{-4\pi^2/\epsilon}) \quad (3.3.6c)$$

where

$$\rho(u,\epsilon) = (2\pi/\epsilon)^{1/2}\exp\left[(2\pi u - 2u^2 - \tfrac{1}{2}\pi^2)/\epsilon\right] \quad (3.3.6d)$$

These are Eqs. (14.2.42) and (14.6.1) of Ref. 10, after allowing for differences of notation. Note that if ϵ is large then the second argument of the function E is numerically close to 1, which is what happens near criticality in our results (3.3.1). The product or series expansions (1.5.6) and (1.5.7) then converge very slowly. On the other hand, the second argument of the functions θ_1 and θ_4 is then small, so the product expansions (3.3.4) converge rapidly. This is why it is desirable near criticality to use the transformation (3.3.6).

First apply the transformation to the definition (3.3.2) of the function $\hat{\eta}_{a,j}$ that occurs in regime II. The parameters x, p, and ϵ are then related by

(1.5.8). Using (2.6.21) and setting

$$\lambda_{a,j} = [(r-2)/r]^{1/2} \exp\left[-\frac{\epsilon}{8} + \frac{r\epsilon}{24(r-2)} - \frac{(r-2)\pi^2}{6r\epsilon} + \frac{(r-a)^2\pi^2}{2r^2\epsilon} \right] \hat{\eta}_{a,j} \tag{3.3.7}$$

$$z = \exp[-4\pi(r-2)u/\epsilon] \tag{3.3.8}$$

$$t = p^{1/(r-2)} = e^{-\epsilon/(r-2)} \tag{3.3.9}$$

$$F_a(u) = \frac{Q(t)^3 \tilde{\theta}_1(\pi a/r, t)\theta_4(ru, t^r)}{Q(t^r)^2 \theta_4(u,t)\theta_4(u+\pi a/r, t)} \tag{3.3.10}$$

the identity (3.3.2) transforms to

$$F_a(u) = \frac{r}{4(r-2)} \sum_{j=0}^{r-3} \lambda_{a,j} \theta_4 \left[u + \frac{\pi(r-1)}{2(r-2)} + \frac{\pi j}{r-2} - \frac{\pi a}{r(r-2)}, t^{1/(r-2)} \right] \tag{3.3.11}$$

true for all complex numbers u. This $F_a(u)$ is an entire function of u, periodic of period π.

From (2.6.45) and (3.3.7)

$$\lambda_{a,j+r-2} = \lambda_{a,j} \tag{3.3.12}$$

We can regard the $\lambda_{a,j}$ as defined by (3.3.8)–(3.3.12). We can make this more explicit by introducing the Fourier coefficients $f_{a,n}$ of $F_a(u)$:

$$F_a(u) = \sum_{n=-\infty}^{\infty} f_{a,n} e^{2inu} \tag{3.3.13}$$

which satisfy the quasiperiodicity relation

$$f_{a,n+r-2} = -t^{n+(1/2)(r-2)} e^{-2\pi i a/r} f_{a,n} \tag{3.3.14}$$

Using the series expansion of $\theta_4(u)$ (Eq. 8.192.1 of Ref. 17), we find that the $\lambda_{a,j}$ are given by

$$\lambda_{a,j} = \frac{4}{r} \sum_{n=0}^{r-3} \exp\left[\frac{-2\pi i n(j+\frac{1}{2}-a/r)}{r-2} \right] t^{-n^2/(2r-4)} f_{a,n} \tag{3.3.15}$$

for all integers a, j.

Now we apply the transformations (3.3.6) to our results (3.3.1), using the relations (1.5.8) (between x, ϵ, and p) in regimes II and III, and the relations (1.5.20) in regimes I and IV. It is convenient to define

$$s = p^{1/(4r-r)}, \qquad t = |p|^{1/(r-2)} \tag{3.3.16}$$

$$R_a = r/\theta_1(\pi a/r, p) \tag{3.3.17}$$

[This is consistent with (3.3.9).] After a fair bit of cumbersome manipulation we obtain, for the four regimes I–IV,

$$\text{I:}\quad P_a = \frac{2\theta_1(\pi a/r, t^2)}{R_a\theta_4(\pi v/2, p^{2r})\theta_2(0, -t^r)} \tag{3.3.18a}$$

$$\text{II:}\quad P_a = \lambda_{a,(1/2)(a+j)}\tilde{\theta}_1(\pi a/r, p)/Q(t^r) \tag{3.3.18b}$$

$$\text{III:}\quad P_a = \frac{\theta_3(\pi a/2r - \pi d/(2r-2), s) - \theta_3(\pi a/2r + \pi d/(2r-2), s)}{R_a\theta_4(0, p^r)\theta_1(\pi d/(r-1), s^{4r})} \tag{3.3.18c}$$

$$\text{IV:}\quad P_a = \frac{\theta_3(\pi a/2r - \pi d/(2r-4), t^{1/2}) - \theta_3(\pi a/2r + \pi d/(2r-4), t^{1/2})}{R_a\theta_4(0, p^{2r})\theta_1(\pi d/(r-2), -t^r)} \tag{3.3.18d}$$

As before, v is defined by (3.2.26), $Q(q)$ by (2.2.7). The integers j and d determine the phase of the system in the ordered regimes. They must satisfy (3.1.6) and are defined in terms of the boundary conditions by (2.6.51), (2.3.18)–(2.3.20), and (2.5.10), for regimes II, III, IV, respectively. [Thus $d = \frac{1}{2}(b + c - 1)$ in regime III.]

These results are expressed in terms of the original parameter p that entered (1.2.3)–(1.2.9), which is negative for regimes I and IV, positive for II and III. For fixed values of w_0/K, v/K, and η/K [from (1.4.1), $\eta/K = r^{-1}$], the SOS model weights W are real functions of p, analytic for $|p| < 1$. At $p = 0$ we are on the interface between regimes I and II (for $\eta < v < 3\eta$), or between III and IV (for $-\eta < v < \eta$). The model is then critical, so we can naturally regard p as a "deviation from criticality" variable.

At $p = 0$ we find from (3.3.18) that for all regimes

$$P_a = P_a^{(c)} = 4r^{-1}\sin^2(\pi a/r) \tag{3.3.19}$$

The (c) denoting the critical value of P_a. Note that this result is independent of j and d, i.e., is independent of boundary conditions and is the same for all phases. This is what we expect to happen at criticality: long-range order has disappeared.

We can expand the P_a in increasing powers of p (or s, or t). For regime II we first need to note from (3.3.10) and (3.3.13) that

$$f_{a,n} = t^{n/2}e^{in\pi a/r}\sin\frac{(n+1)\pi a}{r}\{1 + 0(t^2)\}, \qquad 0 \leqslant n \leqslant r-2 \tag{3.3.20}$$

$$f_{a,n} = \sin(\pi a/r) + t^2\sin(3\pi a/r) + \cdots \tag{3.3.21}$$

We then find that

$$\text{I:}\quad P_a = P_a^{(c)}\{1-(-p)^{2/(r-2)}[1+2\cos(2\pi a/r)] + O[p, p^{6/(r-2)}]\} \tag{3.3.22a}$$

$$\text{II:}\quad P_a = P_a^{(c)}\left\{1+4p^{(1/2)(r-3)/(r-2)^2}\cos\frac{\pi a}{r}\cos\frac{\pi(j+1)}{r-2} + O[p, p^{(r-4)/(r-2)^2}]\right\} \tag{3.3.22b}$$

$$\text{III:}\quad P_a = P_a^{(c)}\left\{1+4p^{3/(8r-8)}\cos\frac{\pi a}{r}\cos\frac{\pi d}{r-1} + O[p, p^{1/(r-1)}]\right\} \tag{3.3.22c}$$

$$\text{IV:}\quad P_a = P_a^{(c)}\left\{1+4(-p)^{3/(4r-8)}\cos\frac{\pi a}{r}\cos\frac{\pi d}{r-2} + O[p, p^{2/(r-2)}]\right\} \tag{3.3.22d}$$

[The terms of order $p^{(r-4)/(r-2)^2}$ in (3.3.22b) occur only for $r \geqslant 6$.]

For the ordered regimes II, III, IV, the leading deviations from criticality are proportional to the differences between the P_a for different phases, and hence to the long-range-order parameter. Their exponents are therefore the critical exponent β,[10] so

$$\beta = \frac{r-3}{2(r-2)^2}, \qquad \frac{3}{8(r-1)}, \qquad \frac{3}{4(r-2)} \tag{3.3.23}$$

for regimes II, III, IV, respectively.

The probabilities P_a are in general local properties of the lattice: they depend on the site to which the height l_1 in (1.5.1) refers. For definiteness we have up to now taken this to be the center site, but it could be any site deep within the lattice.

In regime I the P_a depend only on whether the site lies on the X or the Y sublattice. (We can think of these sublattices as the centers of the black and white faces of a checkerboard.) For a given phase, the same is true in regimes III and IV. For regime II, there are $r-2$ different functions P_a, depending on the position of the site, corresponding to the $r-2$ distinct (modulo $2r-4$) equal-parity values of j that are allowed in (3.3.1b) and (3.3.18b). Thus in regime II we can define an average function $\tilde{P}_a$ obtained by averaging over these values of j. From (3.3.18b), (3.3.15), and (3.3.21), this is given by

$$\begin{aligned}\tilde{P}_a &= 4r^{-1}f_{a,0}\tilde{\theta}_1(\pi a/r, p)/Q(t^r)\\ &= P_a^{(c)}\{1+p^{2/(r-2)}[1+2\cos(2\pi a/r)]+\cdots\}\end{aligned} \tag{3.3.24}$$

In regime I, the corresponding average $\bar{P}_a$ is equal to P_a. For both regimes, $\bar{P}_a$ is a nonlocal property of the model, independent of boundary conditions and phase. (For the hard hexagon model $\bar{P}_a$ corresponds to the mean density.) From (3.3.22a) and (3.3.24) we see that for regimes I and II,

$$\bar{P}_a = P_a^{(c)}\{1 + \text{sgn}(p)|p|^{1-\bar{\alpha}}[1 + 2\cos(2\pi a/r)] + \cdots\} \qquad (3.3.25a)$$

where the critical exponent $\bar{\alpha}$ is given by

$$1 - \bar{\alpha} = 2/(r-2) \qquad (3.3.25b)$$

Free Energy. To complete the discussion of the critical behavior, we should consider the free energy of the SOS model, with heights restricted by (1.4.3). This can be obtained by the "inversion relation" method,[21] using (A11), (A12), and (A13). It is the same as the free energy of the unrestricted model, and of the original eight-vertex model with weights given by (1.2.3), except that in regime II the free energy is that of the "SDP-like" eight-vertex model.[21]

Let ρ', η, v, K, p be the parameters entering the definitions (1.2.3)–(1.2.9) of the Boltzmann weights of the model, and $h(u)$ the function defined by (1.2.4) and (1.2.7). In regimes II and III define

$$\begin{aligned} \rho_0 &= \rho' h(2\eta)\exp[-\pi(v^2 - \eta^2)/(2KK')] \\ \tau &= 2\pi K/K', \qquad \lambda = 2\pi\eta/K' \\ u &= \pi(\eta + v)/K', \qquad q = e^{-\tau} \end{aligned} \qquad (3.3.26)$$

Then we can verify that the definition (1.2.3) of the eight-vertex model weights a, b, c, d is equivalent to that of Eqs. (6.1)–(6.5) and (6.11) of Ref. 21 (apart from negating and permuting some of $w_1, \ldots, w_4$ therein: such transformations leave the partition function unchanged[2]). It follows that κ, the partition function per site, is given by Eqs. (6.30) and (6.32) of Ref. 21. Using (1.5.4), we see that $-\eta < v < \eta$, $0 < u < \lambda$ in regime III, so Eq. (6.32a) is applicable; in regime II, $\eta < v < \kappa - \eta$, $\lambda < u < \frac{1}{2}\tau$, so (6.32b) applies.

For regime III we can use the original eight-vertex model results.[2] The $q, \pi/\mu$ of Eq. (E10) of Ref. 2 correspond to $p^{1/2}, K/\eta$ herein, so we find that the dominant critical singularity in κ is

$$\text{III:} \quad \kappa_{\text{sing}} = p^{K/(2\eta)} \qquad (3.3.27)$$

This is multiplied by a factor that in general is a nonzero analytic function of p at $p = 0$. Exceptions occur when $r = K/\eta$ is an integer, which from (1.4.1) is precisely the case we are considering. If r is even (3.3.27) should be multiplied by a factor $\ln p$; if r is odd the singular part of κ disappears altogether, as has been observed for the hard hexagon model.[8,10]

For regime II, a similar calculation to that of Appendix E of Ref. 2 yields

$$\text{II:} \quad \kappa_{\text{sing}} = p^{K/(K-2\eta)} \tag{3.3.28}$$

In regimes I and IV the parameter p is negative. Define L'/K by (A50), i.e.,

$$p = -e^{-\pi L'/K} \tag{3.3.29}$$

and define

$$\begin{gathered} \rho_0 = \rho' h(2\eta)\exp\left[-\pi(v^2-\eta^2)/2KL'\right] \\ q^2 = -e^{-\pi K/L'}, \qquad \lambda = \pi\eta/L' \\ u = \pi(\eta + v)/2L' \end{gathered} \tag{3.3.30}$$

These definitions ensure that (1.2.3) is equivalent to Eqs. (6.1)–(6.5) of Ref. 21. It is now Eqs. (6.9a) and (6.9b) that are appropriate, the former applying in regime IV ($-\eta < v < \eta$), the latter in regime I, ($\eta < v < K - \eta$). From them we find that the singular part of κ is

$$\text{IV:} \quad \kappa_{\text{sing}} = (-p)^{K/2\eta} \tag{3.3.31}$$

$$\text{I:} \quad \kappa_{\text{sing}} = (-p)^{K/(K-2\eta)} \tag{3.3.32}$$

These free energy results (3.3.26)–(3.3.32) apply for all values of η in the interval $(0, K/2)$. Setting $r = K/\eta$ as in (1.4.1), we see that

$$\kappa_{\text{sing}} = |p|^{2-\alpha} \tag{3.3.33}$$

where

$$\begin{aligned} 2 - \alpha &= r/(r-2), && \text{in regimes I and II} \\ &= r/2, && \text{in regimes III and IV} \end{aligned} \tag{3.3.34}$$

For the hard hexagon case, when $r = 5$, these values of α agree with those previously obtained (Eq. 14.6.10 of Ref. 10), with the proviso, noted after (3.3.27), that κ then has no singular part in regime III.

We note also that in regimes I and II the free energy exponent α is the same as the average height probability exponent $\bar{\alpha}$ given by (3.3.25). In general it is not obvious why this should be so, since differentiating the logarithm of the partition function with respect to p gives not only terms proportional to $\bar{P}_a$, but also terms proportional to the correlation of four heights round a face of the lattice. (For the hard hexagon case, with $r = 5$

and $v = 3\eta$, such multiple-height correlations do not occur, so then we must have $\bar{\alpha} = \alpha$.)

4. SUMMARY

We have considered the SOS model with weights given by (1.2.6), or equivalently (1.2.12b), which is known[11,12] to have the same partition function as the eight-vertex model. We have shown that the hard hexagon model is equivalent to a special case of this SOS model, in which $\eta = K/5$ and each height l_i is restricted to the range $1 \leqslant l_i \leqslant 4$.

One intriguing feature of the hard hexagon model is that the Rogers–Ramanujan[39,40] identities, and many related identities, occur naturally in the calculation of the sublattice densities.[9,20,26] In the SOS model these densities correspond to the probability that the height at a particular site has a given value. We have therefore considered a more general SOS model, in which $\eta = K/r$ (r an integer), each height being restricted to the range $1 \leqslant l_i \leqslant r-1$. Sure enough, when we calculate the local height probabilities we are led to generalizations of the Rogers–Ramanujan identities.

There are two main sorts of such identities. In Section 2 we evaluate the $(m-1)$-fold sums (1.5.11) and (1.5.23), and show that in the limit $m \to \infty$ they are modular forms. (In regimes I, III, and IV they are sums of at most two simple quotients of elliptic theta functions.) These identities are generalizations of the Rogers–Ramanujan identities, as well as of some similar identities listed by Slater.[31] They are presumably related to Gordon's generalization.[15,16]

The other sort of identity occurs in Section 3, where we calculate the normalization factors M occurring in the definition of the local height probabilities P_a. [These factors are proportional to (1.5.13) or (1.5.25), and ensure that the necessary condition (1.5.2) is satisfied.] For regimes I, III, and IV these identities involve sums of products of theta functions of different nomes. They are generalizations of some of the 40 such identities listed by Ramanujan,[34] and are derivable from Theorem 3.2.1.

Regime II is much more difficult than the other regimes. Its solution (in the limit $m \to \infty$) involves the theory of very well-poised q – hypergeometric series (see Appendix B), and introduces the multi-dimensional theta series $\eta_{a,j}$ and $\hat{\eta}_{a,j}$. (Curiously, the normalization factor M can be evaluated without these complications, as is shown in Theorem 3.2.3.)

In Section 3.3. we present our final results for the P_a, put them into a form suitable for examining critical behavior, and obtain some of the critical exponents $\beta, \alpha, \bar{\alpha}$ of the restricted SOS model.

For definiteness, we have focused our attention on the case $\eta = K/r$ (r an integer), which is the obvious immediate generalization of the hard hexagon case $\eta = K/5$. However, the working can be further generalized to the case $\eta = sK/r$ (r and s integers, $1 \leqslant s < r$), as is being done by one of us (PJF).

APPENDIX A: CORNER TRANSFER MATRICES

Let A, B, C, D be the corner transfer matrices (CTMs) corresponding to the lower-right, upper-right, upper-left, and lower-left quadrants of the lattice, as on p. 366 of Ref. 10. Each will break up into $r-1$ diagonal blocks, one for each value of the center spin l_1. Let S_a be the corresponding diagonal matrix whose diagonal entries are unity for the block with $l_1 = a$, all other elements being zero. Then (1.5.1) is equivalent to

$$P_a = \text{Trace}\, S_a ABCD / \text{Trace}\, ABCD \tag{A1}$$

The matrices A, B, C, D are products of local "face transfer matrices" U_j and V_j that merely add one face at a time to the lattice [Eq. (13.2.4) of Ref. 10, Eq. (9) of Ref. 23]. These matrices have elements

$$(U_j)_{\mathbf{l}\mathbf{l}'} = W(l_j, l_{j+1} \mid l_{j-1}, l_j') \prod_{\substack{k=1 \\ \neq j}}^{m} \delta(l_k, l_k') \tag{A2}$$

$$(V_j)_{\mathbf{l}\mathbf{l}'} = W(l_{j-1}, l_j \mid l_j', l_{j+1}) \prod_{\substack{k=1 \\ \neq j}}^{m} \delta(l_k, l_k') \tag{A3}$$

Here $\mathbf{l}$ denotes the set of heights $\{l_1, l_2, \ldots, l_m\}$, $\mathbf{l}'$ denotes $\{l_1', l_2', \ldots, l_m'\}$, and $j = 2, 3, \ldots, m+1$. The heights $l_1, l_2, \ldots$ correspond to those an observer would see if he started at the center of the lattice and walked outwards along the edges. Thus they (and $l_1', l_2', \ldots$) must satisfy the adjacency condition (1.2.2), i.e.,

$$|l_{j+1} - l_j| = |l_{j+1}' - l_j'| = 1, \qquad j \geqslant 1 \tag{A4}$$

The boundary heights of the lattice are to be fixed at their ground state values (there may be more than one ground state, in which case we have to select a particular one). The heights $l_{m+1}, l_{m+1}', l_{m+2}$ that occur in the definitions of $U_m, V_m, U_{m+1}, V_{m+1}$ are to be given these boundary values, which will in general mean that these boundary face transfer matrices depend on the face to which they refer.

Special Properties of the U_j, V_j. From (1.2.12), all the matrices U_j, V_j, A, B, C, D are functions of v, so we can for instance write U_j as

$U_j(v)$. When $v = \eta$, we see from (1.2.12) that, for l, m, n, p satisfying (1.2.2),

$$W(l,m \mid n,p) = \rho' h(2\eta)\delta(l,p) \tag{A5}$$

By substituting this result into (A2), it follows that

$$U_j(\eta) = \rho' h(2\eta) I \tag{A6}$$

where I is the identity matrix.

Similarly, when $v = -\eta$ we find from (1.2.12) that

$$W(l,m \mid n,p) = \rho' h(2\eta)\left[\frac{h(w_l)h(w_p)}{h(w_m)h(w_n)}\right]^{1/2} \delta(m,n) \tag{A7}$$

Using (A3), it follows that

$$V_j(-\eta) = \rho' h(2\eta) R_{j-1} R_{j+1} R_j^{-2} \tag{A8}$$

where R_j is the diagonal matrix with entries

$$(R_j)_{ll'} = \left[h(w_{l_j})\right]^{1/2} \prod_{k=1}^{m} \delta(l_k, l_k') \tag{A9}$$

For general values of v, the matrices U_j and V_j are very sparse, breaking up into one-by-one and two-by-two blocks. They can therefore easily be inverted. Defining

$$\xi(v) = \left[\rho'^2 h(2\eta + v) h(2\eta - v)\right]^{-1} \tag{A10}$$

we find that the inverse of U_j is also given by (A2), but with W multiplied by $\xi(v-\eta)$ and v in (1.2.12) replaced by $2\eta - v$. Thus

$$U_j^{-1}(v) = \xi(v-\eta) U_j(2\eta - v) \tag{A11}$$

Similarly, the inverse of V_j is given by (A3), but with $W(l,m \mid n,p)$ multiplied by $\xi(v+\eta)h(w_m)h(w_n)/[h(w_l)h(w_p)]$, and v replaced by $-2\eta - v$. It follows that

$$V_j^{-1}(v) = \xi(v+\eta) R_{j-1}^{-2} R_j^2 V_j(-2\eta - v) R_j^2 R_{j+1}^{-2} \tag{A12}$$

These inversion relations (A11), (A12) are consistent with (A6) and (A8), being satisfied by them when $v = \eta, -\eta$, respectively.

Actually, we shall use neither (A11) nor (A12), but a simple corollary of (A12) that follows from (1.4.1) and the fact that W is negated if v is increased by $2K$:

$$V_j^{-1}(v) = -\xi(v+\eta) R_{j-1}^{-2} R_j^2 V_j\left[(2r-2)\eta - v\right] R_j^2 R_{j+1}^{-2} \tag{A13}$$

Properties of A, B, C, D. The matrix A, i.e., $A(v)$, is defined as

$$A(v) = F_2 F_3 \ldots F_{m+1} \tag{A14a}$$

where

$$F_j = U_{m+1}^{(j)} U_m^{(j)} U_{m-1} \cdots U_j \tag{A14b}$$

the superscripts (j) on U_{m+1} and U_m denoting the fact that these matrices (but not $U_2, \ldots, U_{m-1}$) depend on the boundary heights and hence on the value of j in (A14b).

The matrices B, C, D are defined similarly, but with each U_j replaced by V_j, U_j^T, V_j^T, respectively. In fact U_j and V_j are symmetric matrices, but it does not necessarily follow that $C = A$ or $D = B$, since the boundary heights for C (D) may differ from those of A (B).

The ground states [i.e., the values of the heights of all lattice sites that maximize the summand in (1.2.1)], do themselves depend upon v, but only in the sense that they have fixed values in one domain of the complex v plane, then change discontinuously to other fixed values as v crosses from this domain to another. We find that we have two domains to consider:

$$\begin{aligned} &\mathscr{D}_1: \quad -1 < \mathrm{Re}(v/\eta) < 1 \\ &\mathscr{D}_2: \quad 1 < \mathrm{Re}(v/\eta) < r-1 \end{aligned} \tag{A15}$$

We can allow v to reach, or even briefly cross, the domain boundaries, so long as we interpret our results as the appropriate analytic continuations of the intradomain values.

In both cases we can therefore let $v = \eta$. From (A6) and (1.14) it follows that (to within irrelevant scalar factors)

$$A(\eta) = C(\eta) = I \tag{A16}$$

This result is consistent with the fact that the ground states are unchanged by uniform shifts in the SW–NE direction [i.e., the height on site (x, y) is the same as that on site $(x+1, y+1)$]. Since U_j, V_j are symmetric, it follows that

$$\begin{gathered} A^T(v) = A(v), \qquad C^T(v) = C(v) \\ D^T(v) = B(v) \end{gathered} \tag{A17}$$

In the domain $\mathscr{D}_1$, we can also let $v = -\eta$. Substituting the form (A8) of $V_j(-\eta)$ into the appropriate analogs of (A14), we find that $R_2, R_3 \ldots$ all cancel out, leaving (to within irrelevant scalar factors)

$$B(-\eta) = D(-\eta) = R_1 \tag{A18}$$

Thus in domain $\mathscr{D}_1$

$$A(\eta)B(-\eta)C(\eta)D(-\eta) = R_1^2 \tag{A19}$$

In the domain $\mathscr{D}_2$ we can allow v to be close to the value $(r-1)\eta$ (this is a "virtual inversion point"[(21)]), and use (A13). Together with (A14) and the symmetry property of the ground states, this implies that

$$[B(v)]^{-1} = R_1^{-2} D[(2r-2)\eta - v] \tag{A20}$$

[There is a problem in deriving this, in that (A13) is not necessarily satisfied for $j = m+1$. However, it seems that we can ignore this boundary difficulty in the limit of m large.]

The matrix R_1, like $S_1, \ldots, S_{r-1}$, commutes with all the corner transfer matrices, so in domain $\mathscr{D}_2$ we have the relation

$$A(\eta)B(v)C(\eta)D[(2r-2)\eta - v] = R_1^2 \tag{A21}$$

provided v is sufficiently close to $(r-1)\eta$.

A, B, C, D **as Exponentials in** v. The star–triangle relation (1.33), (1.34) implies[10,22] that in the limit of m large the corner transfer matrices have the form (to within irrelevant scalar factors)

$$\begin{aligned} A(v) &= Q_1 M_1 e^{v\mathscr{H}} Q_2^{-1} \\ B(v) &= Q_2 M_2 e^{-v\mathscr{H}} Q_3^{-1} \\ C(v) &= Q_3 M_3 e^{v\mathscr{H}} Q_4^{-1} \\ D(v) &= Q_4 M_4 e^{-v\mathscr{H}} Q_1^{-1} \end{aligned} \tag{A22}$$

where $\mathscr{H}, Q_1, \ldots, Q_4, M_1, \ldots, M_4$ are matrices that are independent of v and commute with $R_1, S_1, \ldots, S_{r-1}$, and $\mathscr{H}, M_1, \ldots, M_4$ are diagonal. Substituting these forms into (A1), we obtain

$$P_a = \operatorname{Trace} S_a M_1 M_2 M_3 M_4 / \operatorname{Trace} M_1 M_2 M_3 M_4 \tag{A23}$$

Thus we need to calculate the product $M_1 M_2 M_3 M_4$. First substitute the forms (A22) into (A19) or (A21) depending on whether v is in domain $\mathscr{D}_1$ or $\mathscr{D}_2$. We obtain

$$M_1 M_2 M_3 M_4 e^{2t\eta\mathscr{H}} = R_1^2 \tag{A24}$$

where

$$t = 2, 2-r, \qquad \text{in domains } \mathscr{D}_1, \mathscr{D}_2 \tag{A25}$$

respectively. Thus (A23) becomes

$$P_a = \operatorname{Trace} S_a R_1^2 e^{-2t\eta\mathscr{H}} / \operatorname{Trace} R_1^2 e^{-2t\eta\mathscr{H}} \tag{A26}$$

This simplifies the problem to the calculation of $\mathscr{H}$. However, setting $v = \eta$ in the first of the equations (A22) and using (A16), we find that

$$Q_1 M_1 = Q_2 e^{-\eta\mathscr{H}} \tag{A27}$$

and hence

$$A(v) = Q_2 e^{(v-\eta)\mathscr{H}} Q_2^{-1} \tag{A28}$$

Thus $\exp[(v-\eta)\mathscr{H}]$ is simply the diagonal form of the single corner transfer matrix $A(v)$. We can calculate this by using certain quasiperiodic properties of $A(v)$ (corresponding to incrementing v by iK', or $2iK'$): these imply that the elements of $\mathscr{H}$ are integer multiples of π/K', and we can calculate these integers from special limiting cases.

To do this, we have to distinguish whether the nome p in the definition (1.2.7) of the elliptic function $h(u)$ is positive or negative. Together with the division (A15) of the complex v plane, this gives us four cases to consider. If we restrict v to be real and less than 3η (which ensures that the Boltzmann weights can be chosen to be real and nonnegative), then these four cases are the four "regimes" in (1.5.4). However, here we shall still find it convenient to regard v as a complex variable, and extend the allowed values of v to

$$\begin{aligned} v &\in \mathscr{D}_2, && \text{in regimes I and II} \\ &\in \mathscr{D}_1, && \text{in regimes III and IV} \end{aligned} \tag{A29}$$

From (A25), we see that the values of t in the four regimes are given by (1.5.5).

Regimes II and III. When $0 < p < 1$ (regimes II and III), we can make a conjugate modulus transformation[10,17] so as to write the definition (1.2.7) of the function $h(u)$ as

$$h(u) = \tau \exp\left[-\pi(u-K)^2/(2KK')\right] E(e^{-2\pi u/K'}, y) \tag{A30}$$

where

$$y = \exp(-4\pi K/K') \tag{A31}$$

$$\tau = \frac{K}{K'} \prod_{n=1}^{\infty} \frac{1 - y^{n/2}}{1 + y^{n/2}} \tag{A32}$$

and the function $E(z, y)$ is defined by (1.5.6). Note that y, like p, lies in the interval (0, 1), but is close to one when p is small, and vice-versa.

Using the form (A30) for $h(u)$, and defining

$$x = e^{-4\pi\eta/K'}, \qquad w = e^{-2\pi(\eta - v)/K'} \tag{A33}$$

$$g_l = \exp\left[\pi(v-\eta)(w_l - K)^2/8\eta KK'\right] \tag{A34}$$

$$\nu = \rho' \tau E(x, y) \exp\left[\pi(4K\eta - K^2 - v^2 - 3\eta^2)/2KK'\right] \tag{A35}$$

$$\mu_l = \exp(-2\pi w_l/K'), \qquad E_l = E(\mu_l, y) \tag{A36}$$

then, using (1.2.5) but not (1.4.1) nor (1.4.2), we can write (1.2.12b) as

$$\begin{aligned}
\alpha_l &= \nu\left[g_l^2/(g_{l-1}g_{l+1})\right]w^{1/2}E(xw^{-1})/E(x) \\
\beta_l &= \nu(g_{l-1}g_{l+1}/g_l^2)\left[xE_{l-1}E_{l+1}/(wE_l^2)\right]^{1/2}E(w)/E(x) \\
\gamma_l &= \nu(g_{l+1}/g_l)^2E(\mu_l w)/E(\mu_l) \\
\delta_l &= \nu(g_{l-1}/g_l)^2E(\mu_l w^{-1})/E(\mu_l)
\end{aligned} \tag{A37}$$

where here we have written $E(z, y)$ simply as $E(z)$.

[These reduce to the hard hexagon weights in the case $\eta = K/5$, provided that here we replace ν, g_l by unity, and take r in Eq. (28b) of Ref. 8 to be $w^{1/2}$.] Now using (1.4.1) and (1.4.2), we can verify that the x defined by (A33) is the same as that defined by (1.5.8), and that

$$y = x^r, \qquad \mu_l = x^l \tag{A38}$$

When calculating A, we can ignore the parameter ν in (A37), since it merely contributes an irrelevant scalar factor. The terms involving g_l have the effect of multiplying $W(l,m'\mid l',m)$ by $g_l g_m/g_{l'}g_{m'}$ and this in turn simply divides $A_{\mathbf{l}\mathbf{l}'}$ by g_{l_1}. The remaining terms in (A37) involve v only via integer or half-integer powers of w, so are periodic functions of v, of period $2iK'$. Within the appropriate domain $\mathscr{D}_1$ or $\mathscr{D}_2$, defined in (A15), it appears that both A and its diagonal form are analytic functions of v, so from (A28) it follows that the elements of $\exp[(v-\eta)\mathscr{H}]$ are

$$\left[e^{(v-\eta)\mathscr{H}}\right]_{\mathbf{l}\mathbf{l}'} = g_{l_1}^{-1}w^{N(\mathbf{l})/2}\delta(\mathbf{l},\mathbf{l}') \tag{A39}$$

where $N(\mathbf{l})$ is an integer function of $\mathbf{l}$, and we use the notation

$$\delta(\mathbf{l},\mathbf{l}') \equiv \prod_{k=1}^{m}\delta(l_k, l_k') \tag{A40}$$

Assuming (as seems perfectly reasonable) that $\mathscr{H}$ does not change discontinuously with p, the integers $N(\mathbf{l})$ must be independent of p. We can therefore obtain them by studying the simple case when p and w are close to one, i.e., when η/K' and K/K' are large, $-1 < \mathrm{Re}(v/\eta) < 3$ and

$$x \ll |x| \ll x^{-1} \tag{A41}$$

It then follows from (A37) that

$$|\beta_l^2/(\gamma_l\delta_l)| \ll 1 \tag{A42}$$

for $l = 2, \ldots, r-2$, which implies that the matrices U_j, and hence A, are near diagonal (in the sense that their off-diagonal elements give a negligible contribution to their eigenvalues). In the limit when x is small, we can take

β_l to be zero: from (1.2.12a) and (A37) we then obtain

$$W(l,m' \mid l',m) = \nu[\, g_l g_m/(g_{l'} g_{m'})\,] w^{|l'-m'|/4} \delta_{lm} \tag{A43}$$

The corner transfer matrix A is then diagonal, with entries

$$\begin{aligned} A_{\mathbf{l}\mathbf{l}'} &= \prod_{j=1}^{m} [\, W(l_{j+1}, l_{j+2} \mid l_j, l_{j+1})\,]^j \delta(\mathbf{l}, \mathbf{l}') \\ &= g_{l_1}^{-1} w^{\phi(\mathbf{l})} \delta(\mathbf{l}, \mathbf{l}') \end{aligned} \tag{A44}$$

where $\phi(\mathbf{l})$ is defined by (1.5.12) and we have ignored some $\mathbf{l}$-independent (i.e., scalar) factors in A. The boundary heights l_{m+1}, l_{m+2} that occur here must be fixed at their values for the particular ground state under consideration.

From (A28), $\exp[(v-\eta)\mathscr{H}]$ is the diagonal form of A, so in this small-x limit it is given by (A44). From (A39) it follows that

$$N(\mathbf{l}) = 2\phi(\mathbf{l}) \tag{A45}$$

This is indeed an integer function; using the continuity argument mentioned above, $N(\mathbf{l})$ must be given by (A45) for all p in the interval $(0,1)$. From (A39), (A33), and (A34) (with $\eta - v$ replaced by $2t\eta$), it follows that

$$[\, e^{-2t\eta\mathscr{H}}\,]_{\mathbf{l}\mathbf{l}'} = x^{[-t(2l_1-r)^2/16r + t\phi(\mathbf{l})]} \delta(\mathbf{l}, \mathbf{l}') \tag{A46}$$

Also, from (A9) and (A30),

$$(R_1^2)_{\mathbf{l}\mathbf{l}'} = \tau x^{(2l_1-r)^2/8r} E(x^{l_1}, y) \delta(\mathbf{l}, \mathbf{l}') \tag{A47}$$

Substituting these results into (A26), cancelling the scalar factor τ, and remembering that

$$(S_a)_{\mathbf{l}\mathbf{l}'} = \delta(l_1, a) \delta(\mathbf{l}, \mathbf{l}') \tag{A48}$$

we obtain the result (1.5.9)–(1.5.14).

Regimes I and IV. When $-1 < p < 0$ (regimes I and IV), the elliptic integral K is real, but K' is complex, being of the form

$$K' = L' + iK \tag{A49}$$

where L' is real and positive, and

$$p = -e^{-\pi L'/K} \tag{A50}$$

We can write (1.2.7) in the "conjugate modulus" form[10]

$$h(u) = \tau \exp[\, \pi u(K-u)/2KL'\,] E(e^{-\pi u/L'}, y) \tag{A51}$$

where the function $E(z, y)$ is again defined by (1.5.6), and in these regimes

I and IV we define y and τ by

$$y = -\exp(-\pi K/L') \tag{A52}$$

$$\tau = y^{1/4}\frac{K}{L'}\prod_{n=1}^{\infty}\frac{1-y^{2n}}{1-y^{2n-1}} \tag{A53}$$

Since y is negative, the constant τ is complex. This is just a trivial complication due to the standard definitions of the elliptic theta functions. It is readily compensated by an appropriate definition of ρ' in (1.2.12b). In any case, τ will cancel out of our final result for P_a.

Substituting this form for $h(u)$ into (1.2.12b), defining

$$x = e^{-2\pi\eta/L'}, \qquad w = e^{-\pi(\eta-v)/L'} \tag{A54}$$

$$g_l = \exp\left[\pi(v-\eta)(w_l - K)^2/8\eta KL'\right] \tag{A55}$$

$$\nu = \rho'\tau E(x, y)\exp\left[\pi(2K\eta - 3\eta^2 - v^2)/2KL'\right] \tag{A56}$$

$$\mu_l = \exp(-\pi w_l/L'), \qquad E_l = E(\mu_l, y) \tag{A57}$$

we find that α_l, β_l are again given by (A37), while the expressions for γ_l, δ_l now contain extra factors $w^{1/2}, w^{-1/2}$, respectively. Thus

$$\begin{aligned}\gamma_l &= \nu(g_{l+1}/g_l)^2 w^{1/2} E(\mu_l w)/E(\mu_l)\\ \delta_l &= \nu(g_{l-1}/g_l)^2 w^{-1/2} E(\mu_l w^{-1})/E(\mu_l)\end{aligned} \tag{A58}$$

[For the case $\eta = K/5$, the hard hexagon weights, Eq. (28a) of Ref. 8, can be obtained by replacing x, w, g_l, ν herein by $x^2, w^{-1}, w^{-l/2}, w^{1/2}$.]

From (1.4.1) and (1.4.2),

$$y = -x^{r/2}, \qquad \mu_l = x^l \tag{A59}$$

As in regimes II and III, we expect the elements of the diagonal form of A to be integer or half-integer powers of w, divided by g_{l_1}, i.e., we expect (A39) to be valid. To obtain these powers, we focus attention on the case when x is small and

$$x^{1/2} \ll |w| \ll x^{-1/2} \tag{A60}$$

Provided $r \neq 2l$, we find that (A42) is again satisfied, so in the limit of x small we can take $\beta_l = 0$. Provided l' and m' are not both equal to $r/2$, it follows that

$$W(l, m' \mid l', m) = \nu w^{1/2}(g_l g_m/g_{l'}g_{m'})w^{-H(l', l, m')}\delta_{lm} \tag{A61}$$

where the function $H(l, l', l'')$ is defined by (1.5.26).

For odd values of r, it follows immediately that A is diagonal in the limit $x \to 0$, with elements

$$A_{\mathbf{ll}'} = g_{l_1}^{-1} w^{-\psi(\mathbf{l})} \delta(\mathbf{l}, \mathbf{l}') \tag{A62}$$

where $\psi(\mathbf{l})$ is defined by (1.5.24) and we have ignored l-independent factors. As in regimes II and III, the boundary heights l_{m+1}, l_{m+2} are to be fixed at their ground state values.

The argument is a little more complicated if r is even, since (A42) then fails for $l = r/2$. This means that in the small-x limit most of the off-diagonal elements $A_{\mathbf{ll}'}$ of A are zero, but nonzero elements occur when

$$l_j \neq l_j' \quad \text{and} \quad l_{j-1} = l_{j+1} = l_{j-1}' = l_{j+1}' = r/2 \tag{A63}$$

It follows that A is block diagonal, with blocks that are direct products of one-by-one or two-by-two matrices. It can therefore readily be diagonalized: the eigenvectors are found to be independent of w [as (A28) implies], while the diagonal form is again given by (A62), (1.5.24), and (1.5.26).

From (A28), (A39), and (A62), we see that in the small-x limit

$$N(\mathbf{l}) = -2\psi(\mathbf{l}) \tag{A64}$$

This is indeed an integer function, so should be valid for all x in the interval $0 < x < 1$, i.e., for all p in $(-1, 0)$. Replacing $\eta - v$ by $2t\eta$ in (A39), (A54), and (A55), it follows that

$$\left[e^{-2t\eta\mathscr{H}^{\mathscr{P}}} \right]_{\mathbf{ll}'} = x^{[-t(2l_1 - r)^2/8r - t\psi(\mathbf{l})]} \delta(\mathbf{l}, \mathbf{l}') \tag{A65}$$

while from (A9) and (A51),

$$\left(R_1^2\right)_{\mathbf{ll}'} = \tau x^{l_1(2l_1 - r)/2r} E(x^{l_1}, y) \delta(\mathbf{l}, \mathbf{l}') \tag{A66}$$

Substituting these expressions into (A26), using (A48) and (A59), and canceling a factor $\tau x^{-tr/8}$, we obtain the result (1.5.21)–(1.5.26).

APPENDIX B: q-HYPERGEOMETRIC SERIES

The study of q-hypergeometric series has an extensive literature.[28,29,32] We shall consider just those developments necessary for the treatment of regime II. We shall use the notation (2.2.8) and its generalization

$$(A_1, A_2, \ldots, A_r; q)_n = (A_1)_n (A_2)_n \cdots (A_r)_n \tag{B1}$$

The *ordinary q-hypergeometric series* is defined by

$${}_n\phi_s\begin{pmatrix} \alpha_1, \ldots, \alpha_n; q, t \\ \beta_1, \ldots, \beta_s \end{pmatrix} = \sum_{j=0}^{\infty} \frac{(\alpha_1, \ldots, \alpha_n; q)_j t^j}{(q, \beta_1, \ldots, \beta_s; q)_j} \tag{B2}$$

This series terminates after finitely many terms if one of the α_i is a nonpositive integer power of q: this is the case we shall be considering.

The series is said to be "well poised" if $s = n - 1$ and

$$\beta_{j-1}\alpha_j = qa, \qquad j = 1, \ldots, n \tag{B3}$$

where $a = \alpha_1$ and we take β_0 to be the fixed entry q in (B2). It is "very well poised" if it is also true that

$$\alpha_2, \alpha_3, \beta_1, \beta_2 = qa^{1/2}, -qa^{1/2}, a^{1/2}, -a^{1/2} \tag{B4}$$

One can readily verify that in this case the entries $\alpha_2, \alpha_3, \beta_1, \beta_2$ in (B2) merely contribute a combined factor

$$\frac{(qa^{1/2})_j(-qa^{1/2})}{(a^{1/2})_j(-a^{1/2})_j} = \frac{1 - aq^{2j}}{1 - a} \tag{B5}$$

to the summand.

We shall restrict our attention to such very well poised q-hypergeometric series, and shall take n to be even, $n \geqslant 6$, and α_n to be the nonpositive integer power of q. Then we can write $\alpha_1, \ldots, \alpha_n$ as

$$\alpha_1, \ldots, \alpha_n = a, qa^{1/2}, -qa^{1/2}, b_1, c_1, b_2, c_2, \ldots, b_k, c_k, q^{-N} \tag{B6}$$

where

$$n = 2k + 4, \qquad s = 2k + 3 \tag{B7}$$

and N is a nonnegative integer. We take t to be

$$t = a^k q^{k+N}/(b_1 \ldots b_k c_1 \ldots c_k) \tag{B8}$$

The series (B2) can then be written as a $(k - 1)$-dimensional summation (Theorem 4 of Ref. 29). At first sight this may not appear to be progress, but for $k = 2$ the identity yields Watson's[(33)] proof of the Rogers–Ramanujan identities. We shall in fact find that it enables us to handle the tricky $m \to \infty$ limit for regime II.

We do not need the full theorem, but can restrict our attention to the limit when

$$c_1, \ldots, c_k \to 0 \tag{B9}$$

(Since the series is a rational function of each c_i, there is no difficulty in letting them become zero.) Then Theorem 4 of Ref. 29 becomes the following.

Theorem B1. For N, k integers, $N \geqslant 0$, $k \geqslant 1$,

$$\sum_{j=0}^{N} q^{(1/2)kj(1-j)} a^j \frac{1 - q^{2j}a}{1 - a} \prod_{i=1}^{k+2} \frac{(-b_i)^{-j}(b_i)_j}{(aq/b_i)_j}$$

$$= q^{(1/2)N(N-1)}(aq)_N \sum_{\{m\}} \prod_{i=1}^{k} \frac{(b_{i+1})_{M_i} q^{(1/2)(m_i + M_i)(m_i - M_i + 1)}}{(-b_i)^{M_i}(q)_{m_i}(aq/b_i)_{M_i}} \tag{B10}$$

where

$$b_{k+1} = q^{-N}, \qquad b_{k+2} = a \tag{B11}$$

$$M_i = m_1 + m_2 + \cdots + m_i \tag{B12}$$

and the $\{m\}$ summation is over all integer values of $m_1, \ldots, m_k$ such that

$$m_i \geqslant 0, \quad i = 1, \ldots, k; \qquad m_1 + m_2 + \cdots + m_k = N \tag{B13}$$

Remarks. The left-hand side of (B10) is the q-hypergeometric series (B2) with the substitutions (B3)–(B9). This theorem is precisely the same as Theorem 4 of Ref. 29 with $c_1, \ldots, c_k$ set equal to zero: we have merely condensed the notation somewhat by using the definitions (B11)–(B13) and the properties (2.2.10), (2.2.11).

Theorem B2. Let the function ρ_N be defined by (2.6.6), and set

$$d_{k+1} = q/w, \qquad d_{k+2} = q \tag{B14}$$

Then for N, k integers, $N \geqslant 0$, $k \geqslant 1$:

$$\rho_N(q, w, k+2; d_1, d_2, \ldots, d_{k+2})$$
$$= (1-w) \sum_{\{m\}} \prod_{i=1}^{k} \frac{q^{(1/2)m_i(m_i-1)}(-d_i)^{m_i}}{(q)_{m_i}(d_i)_{M_i}(wd_i)_{N-M_{i-1}}} \tag{B15}$$

where the M_i are defined by (B12) (wht $M_0 = 0$) and the summation is over all integers $m_1, \ldots, m_k$ satisfying (B13).

Proof. This is merely a restatement of Theorem B1, in which we have set

$$a = q^{-N}w^{-1}, \qquad b_i = aq/d_i, \qquad i = 1, \ldots, k+2 \tag{B16}$$

Doing this, using (2.2.10) and (2.2.11), and multiplying both sides of (B10) by

$$(1 - q^N w)/\left[(wd_1)_N (wd_2)_N \cdots (wd_{k+2})_N\right] \tag{B17}$$

we obtain the desired result (B15). It also follows that the function ρ_N in (B15) is proportional to the very well-poised series in (B10).

We shall be interested in the limit when $N \to \infty$, while $q, w, k, d_1, \ldots, d_k$ remain fixed. We must exclude the cases when the function ρ_N is infinite, so for the rest of this Appendix it is to be understood that, for all integers i and all integers n such that $n \geqslant 0$,

$$d_i \neq q^{-n}, \qquad wd_i \neq q^{-n} \tag{B18}$$

We also restrict our attention to the case when

$$|q| < 1 \tag{B19}$$

Theorem B3. For $k \geqslant 1$,

$$\lim_{N\to\infty} \rho_N(q, w, k+2; d_1, \ldots, d_k, qw^{-1}, q)$$

$$= (1-w) \sum_{\{m\}} \prod_{i=1}^{k} \frac{q^{(1/2)m_i(m_i-1)}(-d_i)^{m_i}}{(q)_\infty (d_i)_\infty (wd_i)_\infty} \tag{B20}$$

where now the $\{m\}$ summation is over all integers m_i such that

$$-\infty < m_i < \infty, \quad i = 1, \ldots, k; \qquad m_1 + m_2 + \cdots + m_k = N \tag{B21}$$

we are using the convention (2.6.20).

Proof. The denominator in (B15) is uniformly bounded. The numerator is maximized when $m_i = k^{-1}N + O(1)$, $i = 1, \ldots, k$. Let f be some number which is large compared with unity but much less than N/k, e.g., $N/2k$. Split the $\{m\}$ summation in Theorem B2 into two parts A and B, A having all m_i greater than f, B containing all other values of $m_1, \ldots, m_k$. Then the power $\frac{1}{2}\sum m_i(m_i - 1)$ of q ensures that for N large B is negligible compared with A. In part A the suffixes $m_i, M_i, N - M_{i-1}$ (for $i = 1, \ldots, k$) all tend to infinity with N, so the terms in the denominator of (B15) can all be replaced by their respective limits. The $\{m\}$ summation can then be extended to include negative values of the m_i, since the extra terms thus introduced are also negligible compared with A.

Theorem B4. For $k \geqslant 1$,

$$\lim_{N\to\infty} \rho_N(q, w, k+2; d_1, \ldots, d_k, qw^{-1}, q)$$

is equal to the coefficient of z^{-N} in the Laurent expansion in powers of z of

$$(1-w) \prod_{i=1}^{k} \frac{E(d_i/z, q)}{(q)_\infty (d_i)_\infty (wd_i)_\infty} \tag{B22}$$

where $E(z, x)$ is the elliptic function given by (1.5.7).

Proof. Substituting the series form (1.5.7) of the function $E(z, x)$ into (B22) and collecting terms in the k-fold product that are proportional to z^{-N}, we obtain the right-hand side of (B20).

Theorem B5. For $s \geqslant 3$, one of $d_1, \ldots, d_s$ equal to q/w and another equal to q,

$$\lim_{N\to\infty} \rho_N(q, w, s; d_1, \ldots, d_s)$$

is equal to the coefficient of z^{-N} in the Laurent expansion in powers of z of

$$\frac{(q)_\infty^3 E(w, q)}{E(z, q) E(zw, q)} \prod_{i=1}^{s} \frac{E(d_i/z, q)}{(q)_\infty (d_i)_\infty (wd_i)_\infty} \tag{B23}$$

Proof. Using the definition (2.6.6), both sides of (B23) are symmetric functions of $d_1, \ldots, d_s$, so it is sufficient to prove the relation for some particular ordering of $d_1, \ldots, d_s$. Choose

$$d_{s-1} = qw^{-1}, \qquad d_s = q \tag{B24}$$

Substituting these values into (B23), setting $s = k + 2$ and noting from (1.5.6) that $E(q/z, q) = E(z, q)$ and

$$E(w,q) = (1 - w)(q)_\infty(qw)_\infty(qw^{-1})_\infty \tag{B25}$$

NOTE ADDED IN PROOF

Many of the identities of Secs. 2 and 3 were suggested and/or checked by partially expanding them in powers of the variable q or x, using a computer and (in Sec. 2) IBM's symbolic manipulation language "Scratchpad." For r odd, the identity (3.2.25) has been obtained by Bressoud (Eq. 5.3 of Ref. 41, with $k = (r - 1)/2$, and $a - x^n$ therein corrected to $1 + x^n$).

REFERENCES

1. C. Fan and F. Y. Wu, *Phys. Rev. B* **2**:723 (1970).
2. R. J. Baxter, *Ann Phys. (N.Y.)* **70**:193 (1972).
3. F. Y. Wu, *Phys. Rev. B* **4**:2312 (1971); L. P. Kadanoff and F. J. Wegner, *Phys. Rev. B* **4**:3989 (1971).
4. R. J. Baxter, *J. Phys. C* **6**:L445 (1973).
5. E. H. Lieb and F. Y. Wu, in *Phase Transitions and Critical Phenomena*, Vol. 1, C. Domb and M. S. Green, eds. (Academic, London, 1972), pp. 321–490.
6. R. J. Baxter and F. Y. Wu, *Austral. J. Phys.* **27**:357 and 369 (1974).
7. R. J. Baxter and I. G. Enting, *J. Phys. A* **9**:L149 (1976).
8. R. J. Baxter, *J. Phys. A*, **13**:L61 (1980).
9. R. J. Baxter, *J. Stat. Phys.* **26**:427 (1981).
10. R. J. Baxter, *Exactly Solved Models in Statistical Mechanics* (Academic, London, 1982).
11. R. J. Baxter, *Ann. Phys. (N.Y.)*, **76**:1 (1973).
12. R. J. Baxter, *Ann. Phys. (N.Y.)*, **76**:25 (1973).
13. R. J. Baxter, *Ann. Phys. (N.Y.)*, **76**:48 (1973).
14. M. Gaudin, *La fonction d'onde de Bethe* (Masson, Paris, 1983).
15. B. Gordon, *Am. J. Math.* **83**:393 (1961).
16. G. E. Andrews, *The Theory of Partitions* (Addison-Wesley, Reading, Massachusetts, 1976).
17. I. S. Gradshteyn and I. M. Ryzhik, *Table of Integrals, Series and Products* (Academic, New York, 1965).
18. E. T. Whittaker and G. N. Watson, *A Course of Modern Analysis* (Cambridge University Press, Cambridge, 1950).
19. N. E. Pegg, *J. Phys. A* **15**:L549 (1982).
20. R. J. Baxter and P. A. Pearce, *J. Phys. A* **15**:897 (1982); **16**:2239 (1983).
21. R. J. Baxter, *J. Stat. Phys.* **28**:1 (1982).

22. R. J. Baxter, in *Fundamental Problems in Statistical Mechanics V*, E. G. D. Cohen, ed. (North-Holland, Amsterdam, 1980).
23. R. J. Baxter, *Physica* **106A**:18 (1981).
24. D. A. Huse, *Phys. Rev. Lett.* **49**:1121 (1982).
25. T. M. Apostol, *Modular Functions and Dirichlet Series in Number Theory* (Springer, New York, 1976).
26. G. E. Andrews, *Proc. Natl. Acad. Sci. (U.S.A.)* **78**:5290 (1981).
27. I. Schur, *Sitzungsber. Akad. Wissensch. Berlin Phys.-Math. Kl.* p. 488 (1926).
28. C. R. Adams, *Bull. Am. Math. Soc.* **37**:361 (1931).
29. G. E. Andrews, in *Theory and Application of Special Functions*, R. Askey, ed. (Academic, New York, 1975).
30. G. E. Andrews, *Proc. Natl. Acad. Sci. (U.S.A.)* **71**:4082 (1974).
31. L. J. Slater, *Proc. London Math. Soc. (2)* **54**:147 (1952).
32. L. J. Slater, *Generalized Hypergeometric Functions* (Cambridge University Press, Cambridge, 1966).
33. G. N. Watson, *J. London Math. Soc.* **4**:4 (1929).
34. B. J. Birch, *Math. Proc. Cambridge Philos. Soc.* **78**:73 (1975).
35. L. J. Rogers, *Proc. London Math. Soc. (2)*, **19**:387 (1921).
36. H. B. C. Darling, *Proc. London Math. Soc. (2)* **19**:350 (1921).
37. L. J. Mordell, *Proc. London Math. Soc. (2)* **20**:408 (1922).
38. G. N. Watson, *J. Indian Math. Soc.* **20**:57 (1933).
39. L. J. Rogers, *Proc. London Math. Soc.* **25**:318 (1894).
40. S. Ramanujan, *Proc. Cambridge Philos. Soc.* **19**:214 (1919).
41. D. M. Bressoud, *J. London Math. Soc. (2)* **16**:9 (1977).

2.8.2 Lattice gas generalization of the hard hexagon model. III. q-trinomial coefficients (with R.J. Baxter)

Reprinted from JOURNAL OF STATISTICAL PHYSICS
Vol. 47, Nos. 3/4, May 1987

Lattice Gas Generalization of the Hard Hexagon Model. III. q-Trinomial Coefficients

George E. Andrews[1] **and R. J. Baxter**[2]

Received December 12, 1986

In the first two papers in this series we considered an extension of the hard hexagon model to a solvable two-dimensional lattice gas with at most two particles per pair of adjacent sites, and we described the local densities in terms of elliptic theta functions. Here we present the mathematical theory behind our derivation of the local densities. Our work centers on q-analogs of trinomial coefficients.

KEY WORDS: Statistical mechanics; lattice statistics; number theory; hard hexagon model; Rogers–Ramanujan identities; trinomial coefficients; q-series.

1. INTRODUCTION

This is the third paper in our series[(1,2)] on lattice gas generalizations of the hard hexagon model. In the first paper we noted that it might be possible to find solvable square-lattice statistical mechanics models corresponding to Gordon's[(3,4)] generalization of the Rogers–Ramanujan identities. This idea was strongly suggested by the intimate connection between the Rogers–Ramanujan identities[(5,6)] and the original solution of the hard hexagon model.[(7)]

In the first paper[(1)] (subsequently referred to as I) we obtained the solution of the star–triangle relation in the case of two particles per site and we found the local densities for each of the four regimes as multiple sums. We note that Kuniba *et al.* independently obtained this solution of the star–triangle relation, and that it has now been further generalized,[(8–12)] in particular, to an arbitrary number of particles per site. In the second

[1] Department of Mathematics, Pennsylvania State University, University Park, Pennsylvania 16802.
[2] Research School of Physical Science, Australian National University, Canberra 2601, Australia.

paper[(2)] (subsequently called II), we presented the local densities for the four regimes in terms of elliptic theta functions and deduced the critical behavior. Here we give the mathematical derivation of the results in II.

Surprisingly, the extensive literature on q-series and elliptic theta functions failed to provide the mathematical tools necessary for our derivations. Briefly, what is needed is a q-analog of trinomial coefficients, i.e., the coefficients of x^j in $(1+x+x^2)^n$. The literature is sparse on trinomial coefficients perhaps because they lack both depth and elegance. It was consequently a great surprise to us that q-analogs of trinomial coefficients were both the key to the mathematics of our model and fairly complicated mathematical objects.

Very recently in a brilliant tour de force, Date *et al.*[(21)] have generalized the problem yet further, to a double hierarchy of models specified by two-integer L and N, where $L-3 \geqslant N \geqslant 0$. They have shown that the star-triangle relation is satisfied, and derived (by a method different from ours) the local densities.

2. q-TRINOMIAL COEFFICIENTS

2.1. Fundamentals

We begin by considering the polynomial $(1+x+x^2)^n$. The coefficients of the expanded form of this expression are called trinomial coefficients.[(13)] We shall not follow Comtet's notation exactly, since a variation seems more natural for our purposes:

$$(1+x+x^2)^n = \sum_{j=-n}^{n} \binom{n}{j}_2 x^{j+n} \tag{2.1}$$

By double applications of the binomial theorem to the second and third expressions in

$$(1+x+x^2)^n = [1+x(1+x)]^n = [(1+x)^2 - x]^n \tag{2.2}$$

we find by coefficient comparison that

$$\binom{n}{j}_2 = \sum_{h \geqslant 0} \frac{n!}{h!\,(h+j)!\,(n-j-2h)!} \tag{2.3}$$

$$= \sum_{h \geqslant 0} (-1)^h \binom{n}{h} \binom{2n-2h}{n-j-h} \tag{2.4}$$

Furthermore, we easily deduce from (2.1) that

$$\binom{n}{j}_2 = \binom{n}{-j}_2 \tag{2.5}$$

and

$$\binom{n}{j}_2 = \binom{n-1}{j-1}_2 + \binom{n-1}{j}_2 + \binom{n-1}{j+1}_2 \tag{2.6}$$

Equations (2.3) and (2.4) provide two representations of $\binom{n}{j}_2$, and these lead to six apparently distinct q-analogs:

$$\binom{m;\,B;\,q}{A}_2 = \sum_{j\geqslant 0} \frac{q^{j(j+B)}(q)_m}{(q)_j(q)_{j+A}(q)_{m-2j-A}} \tag{2.7}$$

$$T_0(m, A, q) = \sum_{j=0}^{m} (-1)^j \begin{bmatrix} m \\ j \end{bmatrix}_{q^2} \begin{bmatrix} 2m-2j \\ m-A-j \end{bmatrix} \tag{2.8}$$

$$T_1(m, A, q) = \sum_{j=0} (-q)^j \begin{bmatrix} m \\ j \end{bmatrix}_{q^2} \begin{bmatrix} 2m-2j \\ m-A-j \end{bmatrix} \tag{2.9}$$

$$t_0(m, A, q) = \sum_{j=0}^{m} (-1)^j q^{j^2} \begin{bmatrix} m \\ j \end{bmatrix}_{q^2} \begin{bmatrix} 2m-2j \\ m-A-j \end{bmatrix} \tag{2.10}$$

$$t_1(m, A, q) = \sum_{j=0}^{m} (-1)^j q^{j^2-j} \begin{bmatrix} m \\ j \end{bmatrix}_{q^2} \begin{bmatrix} 2m-2j \\ m-A-j \end{bmatrix} \tag{2.11}$$

$$\tau_0(m, A, q) = \sum_{j=0}^{m} (-1)^j q^{mj-\binom{j}{2}} \begin{bmatrix} m \\ j \end{bmatrix} \begin{bmatrix} 2m-2j \\ m-A-j \end{bmatrix} \tag{2.12}$$

where

$$\begin{bmatrix} A \\ B \end{bmatrix}_q = \begin{bmatrix} A \\ B \end{bmatrix} = \prod_{j=1}^{B} \frac{(1-q^{A+1-j})}{(1-q^j)} \tag{2.13}$$

$$(A;q)_n = (A)_n = (1-A)(1-Aq)\cdots(1-Aq^{n-1}) \tag{2.14}$$

If $q=1$, then each of the above polynomials is equal to $\binom{n}{j}_2$ by (2.3) or (2.4). We next establish the analogs of (2.5) and (2.6) that will be important to us.

First we note that each of (2.8)–(2.12) is symmetric in A and $-A$. As for (2.7), we note

$$\begin{aligned} \binom{m;\,B;\,q}{-A}_2 &= \sum_{j\geqslant 0} \frac{q^{j(j+B)}(q)_m}{(q)_j(q)_{j-A}(q)_{m-2j+A}} \\ &= \sum_{j\geqslant 0} \frac{q^{(j+A)(j+A+B)}(q)_m}{(q)_{j+A}(q)_j(q)_{m-2j-A}} \\ &= q^{A(A+B)} \binom{m;\,B+2A;\,q}{A}_2 \end{aligned} \tag{2.15}$$

2.2. Recurrences

There are numerous analogs of (2.6). We present several. First,

$$T_1(m, A, q) = T_1(m-1, A, q) + q^{m+A}T_0(m-1, A+1, q) + q^{m-A}T_0(m-1, A-1, q) \tag{2.16}$$

The proof of (2.1) is somewhat intricate:

$$\begin{aligned} &T_1(m-1, A, q) + q^{m+A}T_0(m-1, A+1, q) \\ &= \sum_{j\geq 0} (-q)^j \begin{bmatrix} m-1 \\ j \end{bmatrix}_{q^2} \left(\begin{bmatrix} 2m-2j-2 \\ m+A-j-1 \end{bmatrix} + q^{m+A-j} \begin{bmatrix} 2m-2j-2 \\ m+A-j \end{bmatrix} \right) \\ &= \sum_{j\geq 0} (-q)^j \begin{bmatrix} m-1 \\ j \end{bmatrix}_{q^2} \begin{bmatrix} 2m-2j-1 \\ m+A-j \end{bmatrix} \end{aligned} \tag{2.17}$$

[by Ref. 4, p. 35, Eq. (3.3.4)]

Hence

$$\begin{aligned} &T_1(m, A, q) - T_1(m-1, A, q) - q^{m+A}T_0(m-1, A+1, q) \\ &= \sum_{j\geq 0} (-q)^j \begin{bmatrix} m \\ j \end{bmatrix}_{q^2} \begin{bmatrix} 2m-2j \\ m+A-j \end{bmatrix} \\ &\quad - \sum_{j\geq 0} (-q)^j \begin{bmatrix} m-1 \\ j \end{bmatrix}_{q^2} \begin{bmatrix} 2m-2j-1 \\ m+A-j \end{bmatrix} \\ &= \sum_{j\geq 0} (-1)^j \begin{bmatrix} m-1 \\ j \end{bmatrix}_{q^2} \frac{(q)_{2m-2j-1}[(1-q^{2m}) - (1-q^{m-A-j})]}{(q)_{m+A-j}(q)_{m-A-j}} \\ &= \sum_{j\geq 0} (-q)^j \begin{bmatrix} m-1 \\ j \end{bmatrix}_{q^2} \frac{(q)_{2m-2j-1} q^{m-A-j}[1-q^{m+A+j}]}{(q)_{m+A-j}(q)_{m-A-j}} \\ &= q^{m-A} \sum_{j\geq 0} (-1)^j \begin{bmatrix} m-1 \\ j \end{bmatrix}_{q^2} \\ &\quad \times \frac{(q)_{2m-2j-1}[(1-q^{m+A-j}) + q^{m+A-j}(1-q^{2j})]}{(q)_{m+A-j}(q)_{m-A-j}} \\ &= q^{m-A} \left\{ \sum_{j\geq 0} (-1)^j \begin{bmatrix} m-1 \\ j \end{bmatrix}_{q^2} \begin{bmatrix} 2m-2j-1 \\ m+A-j-1 \end{bmatrix} \right. \\ &\quad \left. + \sum_{j\geq 0} (-1)^j \begin{bmatrix} m-1 \\ j-1 \end{bmatrix}_{q^2} \begin{bmatrix} 2m-2j \\ m+A-j \end{bmatrix} \right\} \end{aligned}$$

$$= q^{m-A}\left\{\sum_{j\geq 0}(-1)^j\begin{bmatrix} m-1 \\ j \end{bmatrix}_{q^2}\begin{bmatrix} 2m-2j-1 \\ m+A-j-1 \end{bmatrix}\right.$$
$$\left. - \sum_{j\geq 0}(-1)^j q^{m+A-j-1}\begin{bmatrix} m-1 \\ j \end{bmatrix}_{q^2}\begin{bmatrix} 2m-2j-2 \\ m+A-j-1 \end{bmatrix}\right\}$$
$$= q^{m-A}\sum_{j\geq 0}(-1)^j\begin{bmatrix} m-1 \\ j \end{bmatrix}_{q^2}\begin{bmatrix} 2m-2j-2 \\ m+A-j-2 \end{bmatrix}$$
$$= q^{m-A}T_0(m-1, A-1, q) \qquad \text{[by Ref. 4, p. 35, Eq. (3.3.4)]} \qquad (2.18)$$

as desired.

A second q-analog of (2.6) is

$$T_0(m, A, q) = T_0(m-1, A-1, q) + q^{m+A}T_1(m-1, A, q)$$
$$+ q^{2m+2A}T_0(m-1, A+1, q) \qquad (2.19)$$

The proof of (2.19) relies on the proof of (2.16); namely, by (2.17)

$$T_0(m, A, q) - q^{m+A}[T_1(m-1, A, q) + q^{m+A}T(m-1, A+1, q)]$$
$$= \sum_{j\geq 0}(-1)^j\begin{bmatrix} m \\ j \end{bmatrix}_{q^2}\begin{bmatrix} 2m-2j \\ m+A-j \end{bmatrix}$$
$$- \sum_{j\geq 0}(-1)^j q^{m+A+j}\begin{bmatrix} m-1 \\ j \end{bmatrix}_{q^2}\begin{bmatrix} 2m-2j-1 \\ m+A-j \end{bmatrix}$$
$$= \sum_{j\geq 0}(-1)^j\begin{bmatrix} m-1 \\ j \end{bmatrix}_{q^2}\frac{(q)_{2m-2j-1}[(1-q^{2m}) - q^{m+A+j}(1-q^{m-A-j})]}{(q)_{m+A-j}(q)_{m-A-j}}$$
$$= \sum_{j\geq 0}(-1)^j\begin{bmatrix} m-1 \\ j \end{bmatrix}_{q^2}\frac{(q)_{2m-2j-1}(1-q^{m+A+j})}{(q)_{m+A-j}(q)_{m-A-j}}$$
$$= q^{-m+A}[q^{m-A}T_0(m-1, A-1, q)]$$

[by the last six equations in (2.18)]

Next we consider some recurrences that reduce to tautologies when $q = 1$. First,

$$T_1(m, A, q) - q^{m-A}T_0(m, A, q) - T_1(m, A+1, q)$$
$$+ q^{m+A+1}T_0(m, A+1, q) = 0 \qquad (2.20)$$

We prove this in two steps:

$$\begin{aligned}
&T_1(m, A, q) - q^{m-A}T_0(m, A, q)\\
&\quad = \sum_{j\geq 0} (-q)^j \begin{bmatrix} m \\ j \end{bmatrix}_{q^2} \begin{bmatrix} 2m-2j \\ m-A-j \end{bmatrix} (1-q^{m-A-j})\\
&\quad = (1-q^{2m}) \sum_{j\geq 0} (-q)^j \begin{bmatrix} m-1 \\ j \end{bmatrix}_{q^2} \begin{bmatrix} 2m-2j-1 \\ m-A-j-1 \end{bmatrix}\\
&\quad = (1-q^{2m}) \sum_{j\geq 0} (-q)^j \begin{bmatrix} m-1 \\ j \end{bmatrix}_{q^2} \begin{bmatrix} 2m-2j-1 \\ m+A-j \end{bmatrix}
\end{aligned} \tag{2.21}$$

Also,

$$\begin{aligned}
&T_1(m, A+1, q) - q^{m+A+1}T_0(m, A+1, q)\\
&\quad = \sum_{j\geq 0} (-q)^j \begin{bmatrix} m \\ j \end{bmatrix}_{q^2} \begin{bmatrix} 2m-2j \\ m+A+1-j \end{bmatrix} (1-q^{m+A+1-j})\\
&\quad = (1-q^{2m}) \sum_{j\geq 0} (-q)^j \begin{bmatrix} m-1 \\ j \end{bmatrix}_{q^2} \begin{bmatrix} 2m-2j-1 \\ m+A-j \end{bmatrix}
\end{aligned} \tag{2.22}$$

Equation (2.20) now follows by substracting (2.22) from (2.21).

Next we have two identities for $\binom{n;B;q}{A}_2$:

$$\binom{m;\, B-1;\, q}{A}_2 = \binom{m;\, B;\, q}{A}_2 + q^B(1-q^m)\binom{m-1;\, B+1;\, q}{A+1}_2 \tag{2.23}$$

$$\binom{m;\, B;\, q}{A}_2 = q^{m-A}\binom{m;\, B-2;\, q}{A}_2 + (1-q^m)\binom{m-1;\, B;\, q}{A}_2 \tag{2.24}$$

Each is easily proved. First (2.23):

$$\begin{aligned}
&\binom{m;\, B-1;\, q}{A}_2 - \binom{m;\, B;\, q}{A}_2\\
&\quad = \sum_{j\geq 0} \frac{q^{j(j+B-1)}(q)_m(1-q^j)}{(q)_j\,(q)_{j+A}(q)_{m-2j-A}}\\
&\quad = \sum_{j\geq 0} \frac{q^{(j+1)(j+B)}(q)_m}{(q)_j\,(q)_{j+A+1}(q)_{m-2j-A-2}}\\
&\quad = q^B(1-q^m)\binom{m-1;\, B+1;\, q}{A+1}_2
\end{aligned}$$

Now, by (2.24)

$$\begin{aligned}
&\binom{m;\,B;\,q}{A}_2 - q^{m-A}\binom{m;\,B-2;\,q}{A}_2 \\
&= \sum_{j\geq 0} \frac{q^{j(j+B)}(q)_m(1-q^{m-2j-A})}{(q)_j\,(q)_{j+A}\,(q)_{m-2j-A}} \\
&= (1-q^m)\sum_{j\geq 0} \frac{q^{j(j+B)}\,(q)_{m-1}}{(q)_j\,(q)_{j+A}\,(q)_{m-2j-A-1}} \\
&= (1-q^m)\binom{m-1;\,B;\,q}{A}_2
\end{aligned}$$

We close this subsection with restatements of (2.16), (2.19), and (2.20) for the $\binom{m;B;q}{A}_2$ and we add two further recurrences:

$$\begin{aligned}
\binom{m;\,A-1;\,q}{A}_2 &= q^{m-1}\binom{m-1;\,A-1;\,q}{A}_2 \\
&\quad + q^A\binom{m-1;\,A+1;\,q}{A+1}_2 + \binom{m-1;\,A-1;\,q}{A-1}_2
\end{aligned} \tag{2.25}$$

$$\begin{aligned}
\binom{m;\,A;\,q}{A}_2 &= q^{m-A}\binom{m-1;\,A-1;\,q}{A-1}_2 \\
&\quad + q^{m-A-1}\binom{m-1;\,A-1;\,q}{A}_2 + \binom{m-1;\,A+1;\,q}{A+1}_2
\end{aligned} \tag{2.26}$$

$$\binom{m;\,A;\,q}{A}_2 + q^m\binom{m;\,A;\,q}{A+1}_2 - \binom{m;\,A+1;\,q}{A+1}_2 - q^{m-A}\binom{m-1;\,A-1;\,q}{A}_2 = 0 \tag{2.27}$$

$$\begin{aligned}
\binom{m;\,B;\,q}{A}_2 &= \binom{m-1;\,B;\,q}{A}_2 + q^{m-A-1+B}\binom{m-1;\,B;\,q}{A+1}_2 \\
&\quad + q^{m-A}\binom{m-1;\,B-1;\,q}{A-1}_2
\end{aligned} \tag{2.28}$$

$$\begin{aligned}
\binom{m;\,B;\,q}{A}_2 &= \binom{m-1;\,B;\,q}{A}_2 + q^{m-A}\binom{m-1;\,B-2;\,q}{A-1}_2 \\
&\quad + q^{m+B}\binom{m-1;\,B+1;\,q}{A+1}_2
\end{aligned} \tag{2.29}$$

Now (2.29) is merely (2.28) with A replaced by $-A$, B by $B-2A$, and (2.15) applied. As for (2.28), we see that

$$\begin{aligned}
&\binom{m;\,B;\,q}{A}_2-\binom{m-1;\,B;\,q}{A}_2\\
&=\sum_{j\geqslant 0}\frac{q^{j(j+B)}(q)_{m-1}\,[(1-q^m)-(1-q^{m-2j-A})]}{(q)_j\,(q)_{j+A}\,(q)_{m-2j-A}}\\
&=\sum_{j\geqslant 0}\frac{q^{j(j+B)}(q)_{m-1}\,q^{m-2j-A}[(1-q^j)+q^j(1-q^{j+A})]}{(q)_j\,(q)_{j+A}\,(q)_{m-2j-A}}\\
&=\sum_{j\geqslant 0}\frac{q^{(j+1)(j+1+B)+m-2(j+1)-A}(q)_{m-1}}{(q)_j\,(q)_{j+A+1}\,(q)_{m-1-2j-(A+1)}}\\
&\quad+\sum_{j\geqslant 0}\frac{q^{j(j+B-1)+m-A}(q)_{m-1}}{(q)_j\,(q)_{j+A-1}\,(q)_{m-1-2j-(A-1)}}\\
&=q^{m-A-1+B}\binom{m-1;\,B;\,q}{A+1}_2+q^{m-A}\binom{m-1;\,B-1;\,q}{A-1}_2
\end{aligned}$$

which proves (2.28).

2.3. Generating Functions

The generating function for $\binom{n}{j}_2$ is quite simple and is given by (2.1). The related generating functions for the q-analogs are not so simple; however, they prove useful in applications.

Recall the q-hypergeometric function (Ref. 14, p. 65) given by

$$_r\phi_s\binom{a_1,\,a_2,\ldots,\,a_r;\,q,\,t}{b_1,\ldots,\,b_s}=\sum_{n\geqslant 0}\frac{(a_1)_n\,(a_2)_n\cdots(a_r)_n\,t^n}{(q)_n\,(b_1)_n\,(b_2)_n\cdots(b_s)_n}\tag{2.30}$$

Then

$$\sum_{A=-n}^{n}x^{n+A}q^{\binom{A}{2}}\binom{n;\,A-r;\,q}{A}_2=q^{\binom{n+1}{2}-nr}\,{}_2\phi_1\binom{q^{-n},\,-x;\,q,\,-xq^r}{0}\tag{2.31}$$

To see this, we observe that

$$\begin{aligned}
&\sum_{A=-n}^{n}x^{n+A}q^{\binom{A}{2}}\binom{n;\,A-r;\,q}{A}_2\\
&=\sum_{j\geqslant 0}q^{j(j-r)}\begin{bmatrix}n\\ j\end{bmatrix}x^n\sum_{A=-n}^{n}x^Aq^{\binom{A}{2}+Aj}\begin{bmatrix}n-j\\ j+A\end{bmatrix}
\end{aligned}$$

[by (2.7)]

$$= \sum_{j \geqslant 0} q^{j(j-r)} \begin{bmatrix} n \\ j \end{bmatrix} x^n \sum_{A=-\infty}^{\infty} x^{A-j} q^{\binom{A-j}{2}+(A-j)j} \begin{bmatrix} n-j \\ A \end{bmatrix}$$

$$= \sum_{j \geqslant 0} q^{\binom{j+1}{2}-jr} \begin{bmatrix} n \\ j \end{bmatrix} x^{n-j} (-x)_{n-j}$$

[by Ref. 4, Theorem 3.3, p. 36]

$$= \sum_{j \geqslant 0} q^{\binom{n-j+1}{2}-(n-j)r} \begin{bmatrix} n \\ j \end{bmatrix} x^{j} (-x)_{j}$$

$$= q^{\binom{n+1}{2}-nr} {}_2\phi_1 \begin{pmatrix} q^{-n}, -x; q, -xq^r \\ 0 \end{pmatrix}$$

We also have

$$\sum_{A=-n}^{n} x^{n+A} q^{\binom{A}{2}} \tau_0(n, A, q)$$

$$= q^{\binom{n+1}{2}} \sum_{A=-n}^{n} x^{n+A} q^{\binom{A}{2}} \sum_{j \geqslant 0} (-1)^{n-j} q^{-\binom{-j}{2}} \begin{bmatrix} n \\ j \end{bmatrix} \begin{bmatrix} 2j \\ A+j \end{bmatrix}$$

$$= q^{\binom{n+1}{2}} \sum_{j \geqslant 0} (-1)^{n-j} q^{-\binom{j+1}{2}} \begin{bmatrix} n \\ j \end{bmatrix} x^n \sum_{A=-\infty}^{\infty} x^{A-j} q^{\binom{A-j}{2}} \begin{bmatrix} 2j \\ A \end{bmatrix}$$

$$= q^{\binom{n+1}{2}} \sum_{j \geqslant 0} (-1)^{n-j} x^{n-j} \begin{bmatrix} n \\ j \end{bmatrix} (-xq^{-j})_{2j}$$

$$= (-x)^n q^{\binom{n+1}{2}} \sum_{j \geqslant 0} \frac{(q^{-n})_j (-x)_j (-x^{-1}q)_j q^{j(n-j)}}{(q)_j}$$

$$= (-x)^n q^{\binom{n+1}{2}} \lim_{\tau \to 0} {}_3\phi_2 \begin{pmatrix} q^{-n}, -x, -x^{-1}q; q, \tau^{-2}q^n \\ \tau^{-1}, \tau^{-1}q \end{pmatrix} \qquad (2.32)$$

2.4. Identities

There are really only two families of polynomials of interest in Eqs. (2.7)–(2.12). The apparently different polynomials listed there are related by the following identities:

$$T_i(m, A, q^{-1}) = q^{A^2-m^2} t_i(m, A, q), \qquad i = 0, 1 \qquad (2.33_i)$$

$$\begin{pmatrix} m; A-i; q^2 \\ A \end{pmatrix}_2 = q^{i(A-m)} t_i(m, A, q), \qquad i = 0, 1 \qquad (2.34_i)$$

$$\tau_0(m, A, q) = \begin{pmatrix} m; A; q \\ A \end{pmatrix}_2 \qquad (2.35)$$

Identity (2.33) is quite easy. One merely replaces q by q^{-1} in (2.8) and (2.9) and then simplifies by using the trivial identity

$$\begin{bmatrix} A \\ B \end{bmatrix}_{q^{-1}} = q^{-B(A-B)} \begin{bmatrix} A \\ B \end{bmatrix} \tag{2.36}$$

To prove (2.34) it is necessary to prove a third identity also. Define

$$t_2(m, A, q) = \sum_{j \geqslant 0} (-1)^j q^{j^2 - j} \begin{bmatrix} m \\ j \end{bmatrix}_{q^2} \begin{bmatrix} 2m-2j \\ m-j-A \end{bmatrix} \times \frac{(1 + q^{2A} - q^{m+A+j-1} - q^{m+A-j})}{(1 - q^{2m-2j-1})} \tag{2.37}$$

We shall prove (2.34) simultaneously with

$$\binom{m;\, A-2;\, q^2}{A}_2 = q^{A-m} t_2(m, A, q) \tag{2.38}$$

We note directly from the relevant definition that

$$\binom{m;\, B;\, q^2}{A}_2 = t_0(m, m, q) = t_1(m, m, q) = t_2(m, m, q) = 1$$

and

$$\binom{m;\, -m;\, q^2}{-m}_2 = t_0(m, -m, q) = 1$$

$$\binom{m;\, -m-1;\, q^2}{-m} = q^{-2m} = q^{(-m)-m} t_1(m, -m, q)$$

$$\binom{m;\, -m-2;\, q^2}{-m}_2 = q^{-4m} = q^{-2m} \frac{(1 + q^{-2m} - q^{-1} - 1)}{(1 - q^{2m-1})}$$

$$= q^{(-m)-m} t_2(m, -m, q)$$

Hence (2.34_0), (2.34_1), and (2.38) are valid for $A = \pm m$, and for $|A| > m$ they are the tautology "$0 = 0$."

We proceed by induction on m. If $m = 0$, then either $A = 0$ or $|A| > 0$ and both these cases have been established above. By (2.23) with $B = A$,

$$\binom{m;\, A-1;\, q^2}{A}_2 = \binom{m;\, A;\, q^2}{A}_2 + q^{2A}(1 - q^{2m}) \binom{m-1;\, A+1;\, q^2}{A+1}_2 \tag{2.39}$$

By (2.23) with $B = A - 1$

$$\binom{m;\, A-2;\, q^2}{A}_2 = \binom{m;\, A-1;\, q^2}{A}_2 + q^{2(A-1)}(1-q^{2m})\binom{m-1;\, A;\, q^2}{A+1}_2 \tag{2.40}$$

And by (2.24) with $B = A$

$$\binom{m;\, A;\, q^2}{A}_2 = q^{2(m-A)}\binom{m;\, A-2;\, q^2}{A}_2 + (1-q^{2m})\binom{m-1;\, A;\, q^2}{A}_2 \tag{2.41}$$

Now Eqs. (2.39)–(2.41) can be viewed as recurrences that relate

$$\binom{m;\, A-i;\, q^2}{A}_2 \qquad (i=0, 1, 2)$$

to the same q-trinomial coefficients with m replaced by $m-1$. Furthermore, if we regard (2.39)–(2.41) as a linear system in

$$\binom{m;\, A-i;\, q^2}{A}_2 \qquad (i=0, 1, 2)$$

then the determinant of the system is

$$\begin{vmatrix} -1 & 1 & 0 \\ 0 & -1 & 1 \\ 1 & 0 & -q^{2m-2A} \end{vmatrix} = 1 - q^{2m-2A} \tag{2.42}$$

and this expression is nonzero unless $A = m$; however, we have already disposed of the case $A = m$. Therefore, if we can establish that the right-hand sides of (2.34_0), (2.34_1), and (2.38) satisfy (2.39)–(2.41), then we will have proved the necessary and sufficient recurrences for the induction step passing from $m-1$ to m. We have

$$\begin{aligned} &q^{A-m}t_1(m, A, q) - t_0(m, A, q) \\ &\quad = q^{A-m} \sum_{j\geq 0} (-1)^j q^{j^2-j} \begin{bmatrix} m \\ j \end{bmatrix}_{q^2} \\ &\qquad \times \begin{bmatrix} 2m-2j \\ m-A-j \end{bmatrix} [(1-q^{2j}) + q^{2j}(1-q^{m-A-j})] \\ &\quad = q^{A-m}(1-q^{2m}) \sum_{j\geq 0} (-1)^j q^{j^2-j} \begin{bmatrix} m-1 \\ j-1 \end{bmatrix}_{q^2} \begin{bmatrix} 2m-2j \\ m-j-A \end{bmatrix} \\ &\qquad + q^{A-m}(1-q^{2m}) \sum_{j\geq 0} (-1)^j q^{j^2+j} \begin{bmatrix} m-1 \\ j \end{bmatrix}_{q^2} \begin{bmatrix} 2m-2j-1 \\ m-j-A-1 \end{bmatrix} \end{aligned}$$

$$= q^{A-m}(1-q^{2m})\left(-\sum_{j\geqslant 0}(-1)^j q^{j^2+j}\begin{bmatrix} m-1 \\ j \end{bmatrix}_{q^2}\begin{bmatrix} 2m-2j-1 \\ m-j-A-1 \end{bmatrix}\right.$$

$$\left.+\sum_{j\geqslant 0}(-1)^j q^{j^2+j}\begin{bmatrix} m-1 \\ j \end{bmatrix}_{q^2}\begin{bmatrix} 2m-2j-1 \\ m-j-A-1 \end{bmatrix}\right)$$

$$= q^{A-m}(1-q^{2m})\sum_{j\geqslant 0}(-1)^j q^{j^2+j}\begin{bmatrix} m-1 \\ j \end{bmatrix}_{q^2}\begin{bmatrix} 2m-2j-2 \\ m-j-A-2 \end{bmatrix} q^{m-j+A}$$

[by Ref. 4, p. 35, Eq. (3.3.3)]

$$= q^{2A}(1-q^{2m})\, t_0(m-1, A+1, q) \tag{2.43}$$

which is (2.39)

Next we treat (2.41):

$$t_0(m, A, q) - (1-q^{2m})\, t_0(m-1, A, q)$$

$$= \sum_{j\geqslant 0}(-1)^j q^{j^2}\begin{bmatrix} m \\ j \end{bmatrix}_{q^2}\begin{bmatrix} 2m-2j \\ m-j-A \end{bmatrix}$$

$$-\sum_{j\geqslant 0}(-1)^j q^{j^2}\begin{bmatrix} m \\ j \end{bmatrix}_{q^2}(1-q^{2m-2j})\begin{bmatrix} 2m-2j-2 \\ m-j-A-1 \end{bmatrix}$$

$$= \sum_{j\geqslant 0}(-1)^j q^{j^2}\begin{bmatrix} m \\ j \end{bmatrix}_{q^2}\frac{(q)_{2m-2j-2}(1-q^{2m-2j})}{(q)_{m-j-A}\,(q)_{m-j+A}}$$

$$\times[(1-q^{2m-2j-1})-(1-q^{m-j-A})(1-q^{m-j+A})]$$

$$= q^{m-A}\sum_{j\geqslant 0}(-1)^j q^{j^2-j}\begin{bmatrix} m \\ j \end{bmatrix}_{q^2}$$

$$\times\begin{bmatrix} 2m-2j \\ m-j-A \end{bmatrix}\frac{(1+q^{2A}-q^{m+A-j-1}-q^{m+A-j})}{(1-q^{2m-2j-1})}$$

$$= q^{2m-2A}[q^{A-m}t_2(N, A, q)] \tag{2.44}$$

which is (2.41).

Finally, we consider (2.40):

$$q^{A-m}t_1(m, A, q) - q^{A-m}t_2(m, A, q)$$

$$= q^{A-m}\left\{\sum_{j\geqslant 0}(-1)^j q^{j^2-j}\begin{bmatrix} m \\ j \end{bmatrix}_{q^2}\begin{bmatrix} 2m-2j \\ m-j-A \end{bmatrix}\right.$$

$$-\sum_{j\geqslant 0}(-1)^j q^{j^2-j}\begin{bmatrix} m \\ j \end{bmatrix}_{q^2}$$

$$\left.\times\begin{bmatrix} 2m-2j \\ m-j-A \end{bmatrix}\frac{(1+q^{2A}-q^{m+A-j-1}-q^{m+A-j})}{(1-q^{2m-2j-1})}\right\}$$

$$= q^{A-m} \sum_{j\geqslant 0} (-1)^j q^{j^2-j} \begin{bmatrix} m \\ j \end{bmatrix}_{q^2} \begin{bmatrix} 2m-2j \\ m-j-A \end{bmatrix} \frac{1}{(1-q^{2m-2j-1})}$$
$$\times (1-q^{2m-2j-1}-1-q^{2A}+q^{m+A-j-1}+q^{m+A-j})$$
$$= q^{A-m} \sum_{j\geqslant 0} (-1)^j q^{j^2-j} \begin{bmatrix} m \\ j \end{bmatrix}_{q^2}$$
$$\times \begin{bmatrix} 2m-2j \\ m-j-A \end{bmatrix} \frac{(1-q^{m-A-j})(-q^{2A}+q^{m+A-j-1})}{(1-q^{2m-2j-1})}$$
$$= -q^{3A-m}(1-q^{2m}) \sum_{j\geqslant 0} (-)^j q^{j^2-j} \begin{bmatrix} m-1 \\ j \end{bmatrix}_{q^2} \begin{bmatrix} 2m-2j-2 \\ m-j-A-2 \end{bmatrix}$$
$$= -q^{2(A-1)}(1-q^{2m})\, q^{(A+1)-(m-1)} t_1(m-1, A+1, q) \tag{2.45}$$

which is (2.40).

The necessary recurrences have been established to complete our proof of (2.34_0), (2.34_1), and (2.38) by mathematical induction.

We now turn to (2.35). By (2.31) and (2.32), we see that to establish (2.35) we need only prove

$$q^{-nr}\, {}_2\phi_1 \begin{pmatrix} q^{-n}, -x; q, -xq^r \\ 0 \end{pmatrix}$$
$$= (-x)^n \lim_{\tau\to 0} {}_3\phi_2 \begin{pmatrix} q^{-n}, -x, -x^{-1}q; q, \tau^{-2}q^n \\ \tau^{-1}, \tau^{-1}q \end{pmatrix} \tag{2.46}$$

This is merely a limiting case of [Ref. 15; Eq. (10.2)]

$${}_3\phi_2 \begin{pmatrix} q^{-n}, b, c; q, efq^n/bc \\ e, f \end{pmatrix} = \frac{(e/b)_n}{(e)_n}\, {}_3\phi_2 \begin{pmatrix} q^{-n}, b, f/c; q, q \\ q^{1-n}b/e, f \end{pmatrix} \tag{2.47}$$

by taking $b=-x$, $c=-x^{-1}q$, $e=\tau^{-1}$, $f=\tau^{-1}q$.

2.5. Asymptotics

In subsequent sections it will be very important to know what happens to the q-trinomial coefficients as $m\to\infty$:

$$\lim_{m\to\infty} \begin{pmatrix} m; A; q \\ A \end{pmatrix}_2 = \lim_{m\to\infty} \tau_0(m, A, q) = \frac{1}{(q)_\infty} \tag{2.48}$$

$$\lim_{m\to\infty} \begin{pmatrix} m; A-1; q \\ A \end{pmatrix}_2 = \frac{1+q^A}{(q)_\infty} \tag{2.49}$$

$$\lim_{m\to\infty} t_0(m, A, q) = \frac{1}{(q^2; q^2)_\infty} \tag{2.50}$$

$$\lim_{m\to\infty} T_1(m, A, q) = \frac{(-q^2; q^2)_\infty}{(q^2; q^2)_\infty} \tag{2.51}$$

$$\lim_{m\to\infty} q^{-m} t_1(m, A, q) = \frac{q^{-A} + q^A}{(q^2; q^2)_\infty} \tag{2.52}$$

$$\lim_{\substack{m\to\infty \\ m-A \text{ even}}} T_0(m, A, q) = \frac{\alpha(q^2)}{(q^2; q^2)_\infty} \tag{2.53}$$

$$\lim_{\substack{m\to\infty \\ m-A \text{ odd}}} T_0(m, A, q) = \frac{q\beta(q^2)}{(q^2; q^2)_\infty} \tag{2.54}$$

where

$$\begin{aligned}\alpha(q) &= (q; q^8)_\infty (q^7; q^8)_\infty (q^8; q^8)_\infty (q^6; q^{16})_\infty (q^{10}; q^{16})_\infty (q)_\infty^{-1} \\ &= [(-q^{11}: q^{24})_\infty (-q^{13}; q^{24})_\infty (q^{24}; q^{24})_\infty \\ &\quad - q(-q^5; q^{24})_\infty (-q^{19}; q^{24})_\infty (q^{24}; q^{24})_\infty](q)_\infty^{-1} \\ &= \tfrac{1}{2}[(-q^{1/2}; q)_\infty + (q^{1/2}; q)_\infty]\end{aligned} \tag{2.55}$$

and

$$\begin{aligned}\beta(q) &= (q^3; q^8)_\infty (q^5; q^8)_\infty (q^8; q^8)_\infty (q^2; q^{16})_\infty (q^{14}; q^{16})_\infty (q)_\infty^{-1} \\ &= [(-q^7; q^{24})_\infty (-q^{17}; q^{24})_\infty (q^{24}; q^{24})_\infty \\ &\quad - q^2(-q; q^{24})_\infty (-q^{23}; q^{24})_\infty (q^{24}; q^{24})_\infty](q)_\infty^{-1} \\ &= q^{-1/2}[(-q^{1/2}; q)_\infty - (q^{1/2}; q)_\infty]\end{aligned} \tag{2.56}$$

Note that the limits in (2.53) and (2.54) are not taken as m tends to infinity through all integral values, but only those values of m of the indicated parity are allowed.

To obtain (2.48) and (2.49), we note

$$\begin{aligned}\lim_{m\to\infty} \begin{pmatrix} m; A; q \\ A \end{pmatrix}_2 &= \sum_{j\geqslant 0} \frac{q^{j(j+B)}}{(q)_j (q)_{j+A}} \\ &= \frac{1}{(q)_\infty} \sum_{j=0}^{A-B} \begin{bmatrix} A-B \\ j \end{bmatrix} q^{j^2+jB}\end{aligned} \tag{2.57}$$

[by Ref. 14, p. 576, Eq. (12), $\alpha = \beta = 1/t$,

$\tau = t^2 q^{1+B}$, $c = q^{A+1}$, $t \to 0$]

Taking (2.35) into account, we see that (2.48) is the case $A = B$ of (2.57) and (2.49) is the case $A - 1 = B$ of (2.57).

Next we treat (2.50):

$$\lim_{m\to\infty} t_0(m, A, q) = \lim_{m\to\infty} \sum_{j\geqslant 0} (-1)^j q^{j^2} \begin{bmatrix} m \\ j \end{bmatrix}_{q^2} \begin{bmatrix} 2m-2j \\ m-A-j \end{bmatrix}$$

$$= \frac{1}{(q)_\infty} \sum_{j=0}^{\infty} \frac{(-1)^j q^{j^2}}{(q^2; q^2)_j} = \frac{(q; q^2)_\infty}{(q)_\infty} = \frac{1}{(q^2; q^2)_\infty}$$

[by Ref. 4, p. 19, Eq. (2.2.6)]

For (2.51) we see that

$$\lim_{m\to\infty} T_1(m, A, q) = \lim_{m\to\infty} \sum_{j=0}^{m} (-q)^j \begin{bmatrix} m \\ j \end{bmatrix}_{q^2} \begin{bmatrix} 2m-2j \\ m-A-j \end{bmatrix}$$

$$= \frac{1}{(q)_\infty} \sum_{j=0}^{\infty} \frac{(-q)^j}{(q^2; q^2)_j} = \frac{1}{(-q; q^2)_\infty (q)_\infty} = \frac{(-q^2; q^2)_\infty}{(q^2; q^2)_\infty}$$

[by Ref. 4, p. 19, Eq. (2.2.5) and p. 5, Eq. (1.2.5)]

Now, for (2.52)

$$\lim_{m\to\infty} q^{-m} t_1(m, A, q) = q^{-A} \lim_{m\to\infty} \binom{m; A-1; q}{A}_2$$

$$= \frac{q^{-A}(1 + q^{2A})}{(q^2; q^2)_\infty}$$

by (2.49).

The limits in (2.53) and (2.54) are the toughest. To obtain them, we need the following identities, which are easily obtained from (2.7) by reversing the index of summation:

If $m - A$ is even,

$$q^{(m-A)(m+B)/2} \binom{m; B; q^{-1}}{A}_2 = \sum_{j\geqslant 0} \frac{q^{2j^2 + (B-A)j} (q)_m}{(q)_{2j} (q)_{(m-A-2j)/2} (q)_{(m+A-2j)/2}} \tag{2.58}$$

If $m - A$ is odd,

$$q^{(m-A)(m+B)/2 + (A-B-1)/2} \binom{m; B; q^{-1}}{A}_2$$

$$= \sum_{j\geqslant 0} \frac{q^{2j^2 + (B-A+2)j} (q)_m}{(q)_{2j+1} (q)_{(m-A-1-2j)/2} (q)_{(m+A-1-2j)/2}} \tag{2.59}$$

We now prove (2.53):

$$\lim_{\substack{m\to\infty\\ m-A\text{ even}}} T_0(m, A, q)$$

$$= \lim_{\substack{m\to\infty\\ m-A\text{ even}}} q^{m^2-A^2} t_0(m, A, q^{-1}) \qquad [\text{by } (2.33_0)]$$

$$= \lim_{\substack{m\to\infty\\ m-A\text{ even}}} q^{m^2-A^2} \begin{pmatrix} m; A; q^{-2} \\ A \end{pmatrix}_2 \qquad [\text{by } (2.34_0)]$$

$$= \lim_{m\to\infty} \sum_{j\geqslant 0} \frac{q^{4j^2}(q^2; q^2)_m}{(q^2; q^2)_{2j}\,(q^2; q^2)_{(m-A-2j)/2}\,(q^2; q^2)_{(m+A-2j)/2}}$$

$[\text{by } (2.58)]$

$$= \frac{1}{(q^2; q^2)_\infty} \sum_{j=0}^{\infty} \frac{q^{4j^2}}{(q^2; q^2)_{2j}}$$

$$= \frac{\alpha(q^2)}{(q^2; q^2)_\infty} \qquad [\text{by Ref. 17, p. 160, Eq. (83)}] \tag{2.60}$$

We conclude this section with (2.54):

$$\lim_{\substack{m\to\infty\\ m-A\text{ odd}}} T_0(m, A, q)$$

$$= \lim_{\substack{m\to\infty\\ m-A\text{ odd}}} q^{m^2-A^2} t_0(m, A, q^{-1}) \qquad [\text{by } (2.33_0)]$$

$$= \lim_{\substack{m\to\infty\\ m-A\text{ odd}}} q^{m^2-A^2} \begin{pmatrix} m; A; q^{-2} \\ A \end{pmatrix}_2 \qquad [\text{by } (2.34_0)]$$

$$= \lim_{m\to\infty} q \sum_{j\geqslant 0} \frac{q^{4j^2+4j}\,(q^2; q^2)_m}{(q^2; q^2)_{2j+1}\,(q^2; q^2)_{(m-A-1-2j)/2}\,(q^2; q^2)_{(m+A-1-2j)/2}}$$

$[\text{by } (2.59)]$

$$= \frac{q}{(q^2; q^2)_\infty} \sum_{j=0}^{\infty} \frac{q^{4j^2+4j}}{(q^2; q^2)_{2j+1}}$$

$$= \frac{q\beta(q^2)}{(q^2; q^2)_\infty} \qquad [\text{by Ref. 17, p. 161, Eq. (86)}] \tag{2.61}$$

3. REGIME III

In this section we establish the formulas (3.16a)–(3.16d) of II. Our approach is to represent the polynomials $X_m(a, b, c, q^{-1})$ given by (2.6) in II in terms of the q-trinomial coefficients given in Section 2. We begin by defining

$$\begin{aligned} F_m(j, k, l; q)_\varepsilon &= F_m(j, k, l)_\varepsilon \\ &= \sum_{\mu=-\infty}^{\infty} q^{35\mu^2 + (7k+7l-10j-5)\mu + (j-k)(j-l)} T_\varepsilon(m, 7\mu + k - j, q) \end{aligned} \tag{3.1}$$

where $\varepsilon = 0$ or 1. We then set

$$F_m(j, k, l) = F_m(j, k, l)_{\delta(k,l)} \tag{3.2}$$

where $\delta(k, k) = 1$, $\delta(k, l) = 0$ if $k \neq l$, and we define

$$\rho_0 = 1, \qquad \rho_1 = 3, \qquad \rho_2 = 2 \tag{3.3}$$

Then under conditions (2.5a) and (2.6) of II with $0 \leqslant j, k, l \leqslant 2$:

$$X_m(\rho_j, \rho_k, \rho_l; q^{-1}) = F_m(j, k, l) - q^{-2j-1} F_m(-j-1, k, l) \tag{3.4}$$

To prove (3.4), we need only establish that the right-hand side satisfies the following defining recurrences and initial conditions:

$$X_0(a, b, c; q^{-1}) = \begin{cases} 1 & \text{for } a = b \\ 0 & \text{otherwise} \end{cases} \tag{3.5}$$

$$X_m(a, b, c; q^{-1}) = \sum_{h=4-b}^{3} q^{|\rho^{-1}(h) - \rho^{-1}(c)|m} X_{m-1}(a, h, b; q^{-1}) \tag{3.6}$$

where, by (3.3), $\rho^{-1}(1) = 0$, $\rho^{-1}(3) = 1$, $\rho^{-1}(2) = 2$, and throughout $b + c \geqslant 4$, $1 \leqslant a, b, c \leqslant 3$.

We remark that (3.6) has been apparently altered from the natural recurrences arising from (2.6) of II; however, inspection shows that it yields exactly the same recurrences that the function $H(a, b, c)$ yields. Furthermore, the ρ-function seems to be the natural one to use in expressing our regime III polynomials succinctly.

To establish (3.5), we note that by (2.8) and (2.9),

$$F_0(j, k, l) = \begin{cases} 1 & \text{if } j = k \\ 0 & j \neq k \end{cases} \tag{3.7}$$

and this shows that the right-hand side of (3.4) satisfies (3.5).

We now translate term by term the identities (2.16), (2.19), and (2.20) into recurrences for the $F_m(j, k, l)_\varepsilon$:

$$\begin{aligned} F_m(j, k, l)_1 &\\ &= F_{m-1}(j, k, l)_1 \\ &\quad + q^{m+k-l} F_{m-1}(j, k+1, l)_0 + q^{m+l-k} F_{m-1}(j, k-1, l)_0 \end{aligned} \tag{3.8}$$

$$\begin{aligned} F_m(j, k, l)_0 &\\ &= q^m F_{m-1}(j, k, l+1) \\ &\quad + q^{2m+k-l-1} F_{m-1}(j, k+1, l+1)_0 \\ &\quad + q^{l-k+1} F_{m-1}(j, k-1, l+1)_0 \end{aligned} \tag{3.9a}$$

$$\begin{aligned} F_m(j, k, l)_0 &\\ &= q^{k-l+1} F_{m-1}(j, k+1, l-1)_0 + q^m F_{m-1}(j, k, l-1)_1 \\ &\quad + q^{2m+l-k-1} F_{m-1}(j, k-1, l-1)_0 \\ &\qquad [\text{by (2.19) with } -A \text{ replacing } A] \end{aligned} \tag{3.9b}$$

$$\begin{aligned} &F_m(j, k, l)_1 - q^m F_m(j, k, l-1)_0 \\ &\quad q^{k-l+1} F_m(j, k+1, l-1)_1 + q^{m+k-l+1} F_m(j, k+1, l)_0 = 0 \end{aligned} \tag{3.10a}$$

$$\begin{aligned} &F_m(j, k, l)_1 - q^m F_m(j, k, l+1)_0 - q^{l+1-k} F_m(j, k-1, l+1)_1 \\ &\quad q^{m+l+1-k} F_m(j, k-1, l)_0 = 0 \\ &\quad [\text{by (2.20) with } -A \text{ replacing } A] \end{aligned} \tag{3.10b}$$

Next we consider (3.1) with μ replaced by $-\mu$:

$$F_m(j, k, l)_\varepsilon = q^{2k-2j} F_m(-j-1, -k-1, -l+1)_\varepsilon \tag{3.11}$$

Finally we take μ into $\mu + 1$ in (3.1):

$$F_m(j, k, l)_\varepsilon = q^{4k+9} F_m(j, k+7, l+3)_\varepsilon \tag{3.12}$$

There are six pairs b, c that are admissible in (3.6). We shall use (3.8)–(3.12) to show that the right-hand side of (3.4) also satisfies (3.6) in these instances. Let $X_m^*(j, k, l; q^{-1})$ denote the right-hand side of (3.4). For $b = c = 3$, we have

$$
\begin{aligned}
&X_m^*(\rho_j, 3, 3; q^{-1}) \\
&\quad = F_m(j, 1, 1) - q^{-2j-1}F_m(-j-1, 1, 1) \\
&\quad = [F_{m-1}(j, 1, 1) + q^m F_{m-1}(j, 2, 1) + q^m F_{m-1}(j, 0, 1)] \\
&\qquad - q^{-2j-1}[F_{m-1}(-j-1, 1, 1) + q^m F_{m-1}(-j-1, 2, 1) \\
&\qquad + q^m F_{m-1}(-j-1, 0, 1)] \qquad [\text{by } (3.8)] \\
&\quad = X_{m-1}^*(\rho_j, 3, 3; q^{-1}) + q^m X_{m-1}^*(j, 2, 3; q^{-1}) + q^m X_{m-1}^*(j, 1, 3; q^{-1})
\end{aligned}
$$

which is (3.6).

Next $b = 3$, $c = 2$:

$$
\begin{aligned}
&X_m^*(\rho_j, 3, 2; q^{-1}) \\
&\quad = F_m(j, 1, 2) - q^{-2j-1}F_m(-j-1, 1, 2) \\
&\quad = [F_{m-1}(j, 2, 1) + q^m F_{m-1}(j, 1, 1) + q^{2m} F_{m-1}(j, 0, 1)] \\
&\qquad - q^{-2j-1}[F_{m-1}(-j-1, 2, 1) + q^m F_{m-1}(-j-1, 1, 1) \\
&\qquad + q^{2m} F_{m-1}(-j-1, 0, 1)] \qquad [\text{by } (3.9_{\text{b}})] \\
&\quad = X_m^*(\rho_j, 2, 3; q^{-1}) + q^m X_{m-1}^*(\rho_j, 3, 3; q^{-1}) + q^{2m} X_{m-1}^*(\rho_j, 1, 3; q^{-1})
\end{aligned}
$$

For $b = 3$, $c = 1$:

$$
\begin{aligned}
&X_m^*(\rho_j, 3, 1, q^{-1}) \\
&\quad = F_m(j, 1, 0) - q^{-2j-1}F_m(-j-1, 1, 0) \\
&\quad = [q^m F_{m-1}(j, 1, 1) + q^{2m} F_{m-1}(j, 2, 1) + F_{m-1}(j, 0, 1)] \\
&\qquad - q^{-2j-1}[q^m F_{m-1}(-j-1, 1, 1) + q^{2m} F_{m-1}(-j-1, 2, 1) \\
&\qquad + F_{m-1}(-j-1, 0, 1)] \qquad [\text{by } (3.9_{\text{a}})] \\
&\quad = q^m X_{m-1}^*(\rho_j, 3, 3; q^{-1}) + q^{2m} X_{m-1}^*(\rho_j, 2, 3; q^{-1}) \\
&\qquad + X_{m-1}^*(\rho_j, 1, 3; q^{-1})
\end{aligned}
$$

For $b = 2$, $c = 3$:

$$
\begin{aligned}
&X_m^*(\rho_j, 2, 3; q^{-1}) \\
&\quad = F_m(j, 2, 1) - q^{-2j-1}F_m(-j-1, 2, 1) \\
&\quad = [q^m F_{m-1}(j, 2, 2) + q^{2m} F_{m-1}(j, 3, 2) + F_{m-1}(j, 1, 2)] \\
&\qquad - q^{-2j-1}[q^m F_{m-1}(-j-1, 2, 2) + q^{2m} F_{m-1}(-j-1, 3, 2) \\
&\qquad + F_{m-1}(-j-1, 1, 2)] \qquad [\text{by } (3.9_{\text{a}})]
\end{aligned}
$$

$$\begin{aligned}
&= q^m X_m^*(\rho_j, 2, 2; q) + X_m^*(\rho_j, 3, 2; q^{-1}) \\
&\quad + q^{2m}[F_{m-1}(j, 3, 2)_0 - q^{-2j-1}F_{m-1}(-j-1, 3, 2)_0] \\
&= q^m X_m^*(\rho_j, 2, 2; q^{-1}) + X_m^*(\rho_j, 3, 2; q^{-1}) \\
&\quad + q^{2m}[q^{6-2j}F_{m-1}(-j-1, -4, -1)_0 \\
&\quad - q^{-2j-1}F_{m-1}(-j-1, 3, 2)_0] \qquad [\text{by } (3.11)] \\
&= q^m X_m^*(\rho_j, 2, 2; q^{-1}) + X_m^*(\rho_j, 3, 2; q^{-1}) \qquad [\text{by } (3.12)]
\end{aligned}$$

For $b=2,\ c=2$:

$$\begin{aligned}
&X_m^*(\rho_j, 2, 2; q^{-1}) \\
&\quad = F_m(j, 2, 2) - q^{-2j-1}F_m(-j-1, 2, 2) \\
&\quad = [F_{m-1}(j, 2, 2)_1 + q^m F_{m-1}(j, 3, 2)_0 + q^m F_{m-1}(j, 1, 2)_0] \\
&\quad\quad - q^{-2j-1}[F_{m-1}(-j-1, 2, 2)_1 + q^m F_{m-1}(-j-1, 3, 2)_0 \\
&\quad\quad + q^m F_{m-1}(-j-1, 1, 2)_0] \\
&\quad = X_{m-1}^*(\rho_j, 1, 1; q^{-1}) + q^m X_{m-1}^*(\rho_j, 1, 2; q^{-1}) \\
&\quad\quad [\text{as in the case } b=2,\ c=3]
\end{aligned}$$

For $b=1,\ c=3$:

$$\begin{aligned}
&X_m^*(\rho_j, 1, 3; q^{-1}) \\
&\quad = F_m(j, 0, 1) - q^{-2j-1}F_m(-j-1, 0, 1) \\
&\quad = [F_{m-1}(j, 1, 0)_0 + q^m F_{m-1}(j, 0, 0)_1 + q^{2m}F_{m-1}(j, -1, 0)_0] \\
&\quad\quad - q^{-2j-1}[F_{m-1}(-j-1, 1, 0)_0 + q^m F_{m-1}(-j-1, 0, 0)_1 \\
&\quad\quad + q^{2m}F_{m-1}(-j-1, -1, 0)_0] \\
&\quad = X_m^*(\rho_j, 3, 1; q^{-1}) + q^m[F_{m-1}(j, 0, 0)_1 \\
&\quad\quad + q^m F_{m-1}(j, -1, 0)_0 - qF_{m-1}(j, -1, 1)_1 \\
&\quad\quad - q^{m-1}F_{m-1}(j, 0, 1)_0] \\
&\quad = X_m^*(\rho_j, 3, 1; q^{-1}) \qquad [\text{by } (3.10_b]
\end{aligned}$$

Hence $X_m^*(a, b, c; q^{-1})$ satisfies (3.5) and (3.6). Therefore (3.4) is valid. We may now easily conclude with limits of $X_m(a, b, c; q^{-1})$, which are equivalent to (3.16a)–(3.16d) of II.

Recall

$$Q(q)=\prod_{j=1}^{\infty}(1-q^j) \tag{3.13}$$

and

$$R(q)=\prod_{j=1}^{\infty}(1-q^{2j-1})=\prod_{j=1}^{\infty}\frac{1}{1+q^j} \tag{3.14}$$

Then we may rewrite (2.53) and (2.54) as

$$\lim_{\substack{m\to\infty\\ m-A\equiv\lambda\ (\mathrm{mod}\ 2)}} T_0(m,A,q)=\frac{1}{2}\left[R(-q)+(-1)^{\lambda}R(q)\right] \tag{3.15}$$

Hence

$$\begin{aligned}
&\lim_{\substack{m-\infty\\ m-k+j\equiv\lambda\ (\mathrm{mod}\ 2)}} F_m(j,k,l)_0\\
&-\sum_{\substack{\mu=-\infty\\ \mu\ \mathrm{even}}}^{\infty} q^{35\mu^2+(7k+7l-10j-5)\mu+(j-k)(j-l)}\\
&\quad\times\frac{1}{2Q(q^2)}\left[R(-q)+(-1)^{\lambda}R(q)\right]\\
&\quad+\sum_{\substack{\mu=-\infty\\ \mu\ \mathrm{odd}}}^{\infty} q^{35\mu^2+(7k+7l-10j-5)\mu+(j-k)(j-l)}\\
&\quad\times\frac{1}{2Q(q^2)}\left[R(-q)-(-1)^{\lambda}R(q)\right]\\
&=\frac{R(-q)}{2Q(q^2)}\{35,7k+7l-10j-5,(j-k)(j-l);q^2\}\\
&\quad+\frac{(-1)^{\lambda}R(-q)}{2Q(q^2)}\{35,7k+7l-10j-5,(j-k)(j-l);q^2\}_-\\
&\quad\text{[in the notation of (2.13) and (2.14) in II]}
\end{aligned} \tag{3.16}$$

Also, by (2.51),

$$\lim_{m\to\infty} F_m(j,k,l)_1=\frac{1}{Q(q^2)\,R(q^2)}\{35,7k+7l-10j-5,(j-k)(j-l);q^2\} \tag{3.17}$$

Applying these limits to (3.4), we find that

$$\begin{aligned}\lim_{m\to\infty} & X_m(\rho_j, \rho_k, \rho_k; q^{-1})\\ &= \frac{1}{Q(q^2)\,R(q^2)}[\{35, 14k-10j-5, (j-k)^2; q^2\}\\ &\quad -q^{-2j-1}\{35, 14k+10j+5, (j+k+1)^2; q^2\}]\end{aligned}\tag{3.18}$$

and for $l \neq k$

$$\begin{aligned}\lim_{\substack{m\to\infty\\ m-k+j\equiv\lambda(\mathrm{mod}\ 2)}} & X_m(\rho_j, \rho_k, \rho_l; q^{-1})\\ &= \frac{R(-q)}{2Q(q^2)}[\{35, 7k+7l-10j-5, (j-k)(j-l); q^2\}\\ &\quad -\{35, 7k+7l+10j+5, (j+k+1)(j+l+1); q^2\}]\\ &\quad +\frac{(-1)^\lambda R(q)}{2Q(q^2)}[\{35, 7k+7l-10j-5, (j-k)(j-l); q^2\}_-\\ &\quad -\{35, 7k+7l+10j+5, (j+k+1)(j+l+1); q^2\}_-]\end{aligned}\tag{3.19}$$

Inspection shows that Eqs. (3.18) and (3.19) provide the large-m results for $X_m(a, b, c; q^{-1})$ discussed in the section on regime III in II.

4. REGIME I

In this section we consider the polynomials of (2.6) in II:

$$X_m(a, b, c; q) = \sum_{\sigma_2} \cdots \sum_{\sigma_m} q^{\sum_{j=1}^m jH(\sigma_j, \sigma_{j+1}, \sigma_{j+2})}\tag{4.1}$$

under the conditions $0 \leqslant \sigma_j$, $0 \leqslant \sigma_j + \sigma_{j+1} \leqslant 2$, $1 \leqslant j \leqslant m+1$, $\sigma_1 = 3-a$, $\sigma_{m+1} = 3-b$, $\sigma_{m+2} = 3-c$, and

$$H(a, b, c) = b\tag{4.2}$$

i.e., condition (2.5b) of II. Subject to the above conditions, we see that $4-b \leqslant c \leqslant 3$, and otherwise c does not affect the polynomials. To emphasize the independence from c, we write

$$Y_m(a, b; q) = X_m(a, b, c; q) = \sum_{\sigma_2} \cdots \sum_{\sigma_m} q^{\sum j\sigma_{j+1}}\tag{4.3}$$

Inspection of (4.1) and the related summation conditions yields the following set of defining recurrences and initial conditions (with $1 \leqslant a, b \leqslant 3$):

$$Y_m(a, b; q) = q^{m(3-b)} \sum_{j=4-b}^{3} Y_{m-1}(a, j; q) \tag{4.4}$$

$$Y_0(a, b; q) = \begin{cases} 1 & \text{if} \quad a = b \\ 0 & \text{otherwise} \end{cases} \tag{4.5}$$

Our object is to represent the $Y_m(a, b; q)$ in terms of q-trinomial coefficients. We define for ε, $\psi = 0$ or 1

$$\begin{aligned} \mathfrak{F}_m(j, k; \psi, \varepsilon; q) &= \mathfrak{F}_m(j, k; \psi, \varepsilon) \\ &= \sum_{\mu=-\infty}^{\infty} q^{14\mu^2 + (-4j+5-7\psi)\mu} \binom{m; 7\mu - j + k - \varepsilon; q}{7\mu - j + k}_2 \end{aligned} \tag{4.6}$$

The required representation is

$$\begin{aligned} Y_m(\rho_j, \rho_k; q) = {} & q^{m(1-\delta(1,k)) + \delta(0,k)j} \mathfrak{F}_m(j, k; \delta(0, k), \delta(2, k)) \\ & - q^{(2j+1)(1-\delta(0,k))} \mathfrak{F}_m(-j-1, k; \delta(0, k), \delta(2, k)) \end{aligned} \tag{4.7}$$

As in Section 3, we obtain recurrences for the $\mathfrak{F}_m(j, k; 4, \varepsilon)$ from q-trinomial recurrences.

From (2.28)

$$\begin{aligned} \mathfrak{F}_m(j, k; \psi, \varepsilon) = {} & \mathfrak{F}_{m-1}(j, k; \psi, \varepsilon) + q^{m-\varepsilon-1} \mathfrak{F}_{m-1}(j, k+1; \psi, \varepsilon+1) \\ & + q^{m+j-k} \mathfrak{F}_{m-1}(j, k-1; \psi+1, \varepsilon) \end{aligned} \tag{4.8}$$

From (2.29)

$$\begin{aligned} \mathfrak{F}_m(j, k; \psi, \varepsilon) = {} & \mathfrak{F}_{m-1}(j, k; \psi, \varepsilon) + q^{m+j-k} \mathfrak{F}_{m-1}(j, k-1; \psi+1, \varepsilon+1) \\ & + q^{m-j+k-\varepsilon} \mathfrak{F}_{m-1}(j, k+1; \psi-1, \varepsilon) \end{aligned} \tag{4.9}$$

From (2.25)

$$\begin{aligned} \mathfrak{F}_m(j, k; \psi, 1) = {} & q^{m-1} \mathfrak{F}_{m-1}(j, k; \psi, 1) + q^{k-j} \mathfrak{F}_{m-1}(j, k+1; \psi-1, 0) \\ & + \mathfrak{F}_{m-1}(j, k-1; \psi, 0) \end{aligned} \tag{4.10}$$

Replacing μ by $-\mu$ in (4.6) with $\varepsilon = 0$ or 1, we have

$$\mathfrak{F}_m(j, k; \psi, \varepsilon) = q^{\varepsilon(k-j)} \mathfrak{F}_m(-j-1, -k-1; 2+\varepsilon-\psi, \varepsilon) \tag{4.11}$$

Replacing μ by $-\mu-1$ in (4.6), we obtain

$$\mathfrak{F}_m(j, k; -1, 0) = q^{4j+2}\mathfrak{F}_m(-j-1, k; -1, 0) \tag{4.12}$$

Finally, from (2.27)

$$\begin{aligned}&\mathfrak{F}_m(j, k; \psi, 0) + q^m\mathfrak{F}_m(j, k+1; \psi, 1)\\ &\quad -\mathfrak{F}_m(j, k+1; \psi, 0) - q^{m+j-k}\mathfrak{F}_m(j, k; \psi+1, 1) = 0\end{aligned} \tag{4.13}$$

We let $y_m(\rho_j, \rho_k; q)$ denote the right-hand side of (4.7), and we now verify (4.4) and (4.5) for $y_m(\rho_j, \rho_k; q)$:

$$\begin{aligned}y_m(\rho_j, 3; q) &= \mathfrak{F}_m(j, 1; 0, 0) - q^{2j+1}\mathfrak{F}_m(-j-1, 1; 0, 0)\\ &= \mathfrak{F}_{m-1}(j, 1; 0, 0) + q^{m-1}\mathfrak{F}_{m-1}(j, 2; 0, 1)\\ &\quad + q^{m+j-1}\mathfrak{F}_{m-1}(j, 0; 1, 0)\\ &\quad - q^{2j+1}[\mathfrak{F}_{m-1}(-j-1, 1; 0, 0) + q^{m-1}\mathfrak{F}_{m-1}(-j-1, 2; 0, 1)\\ &\quad + q^{m-j-2}\mathfrak{F}_{m-1}(-j-1, 0; 1, 0)] \qquad [\text{by } (4.8)]\\ &= y_{m-1}(\rho_j, 3; q) + y_{m-1}(\rho_j, 2; q) + y_{m-1}(\rho_j, 1; q)\end{aligned} \tag{4.14}$$

which is (4.4) for $b=3$.

$$\begin{aligned}y_m(\rho_j, 2; q) &= q^m[\mathfrak{F}_m(j, 2; 0, 1) - q^{2j+1}\mathfrak{F}_m(-j-1, 2; 0, 1)]\\ &= q^m\{q^{m-1}\mathfrak{F}_{m-1}(j, 2; 0, 1) + q^{2-j}\mathfrak{F}_{m-1}(j, 3; -1, 0)\\ &\quad + \mathfrak{F}_{m-1}(j, 1; 0, 0) - q^{2j+1}[q^{m-1}\mathfrak{F}_{m-1}(-j-1, 2; 0, 1)\\ &\quad + q^{3+j}\mathfrak{F}_{m-1}(-j-1, 3; -1, 0) + \mathfrak{F}_{m-1}(-j-1, 1; 0, 0)]\}\\ &\quad [\text{by } (4.10)]\\ &= q^m y_{m-1}(\rho_j, 2; q) + q^{m+2-j}[\mathfrak{F}_{m-1}(j, 3; -1, 0)\\ &\quad - q^{4j+2}\mathfrak{F}_{m-1}(-j-1, 3; -1, 0)] + q^m y_{m-1}(\rho_j, 3; q)\\ &= q^m[y_{m-1}(\rho_j, 2; q) + y_{m-1}(\rho_j, 3; q)] \qquad [\text{by } (4.12)]\end{aligned} \tag{4.15}$$

which is (4.4) for $b=2$.

$$\begin{aligned}y_m(\rho_j, 1; q) &= q^{m+j}[\mathfrak{F}_m(j, 0; 1, 0) - \mathfrak{F}_m(-j-1, 0; 1, 0)]\\ &= q^{m+j}\{\mathfrak{F}_{m-1}(j, 0; 1, 0) + q^{m+j}\mathfrak{F}_{m-1}(j, -1; 2, 1)\\ &\quad + q^{m-j}\mathfrak{F}_{m-1}(j, 1; 0, 0) - [\mathfrak{F}_{m-1}(-j-1, 0; 1, 0)\\ &\quad + q^{m-j-1}\mathfrak{F}_{m-1}(-j-1, -1; 2, 1)\\ &\quad + q^{m+j+1}\mathfrak{F}_{m-1}(-j-1, 1; 0, 0)]\}\end{aligned}$$

$$
\begin{aligned}
&= q^{2m} y_{m-1}(\rho_j, 3; q) + q^{m+j}[\mathfrak{F}_{m-1}(-j-1, -1; 1, 0) \\
&\quad + q^{m-1}\mathfrak{F}_{m-1}(-j-1, 0; 1, 1) - \mathfrak{F}_{m-1}(-j-1, 0; 1, 0) \\
&\quad - q^{m-j-1}\mathfrak{F}_{m-1}(-j-1, -1; 2, 1)] \\
&= q^{2m} y_{m-1}(\rho_j, 3; q) \qquad [\text{by } (4.13)]
\end{aligned} \tag{4.16}
$$

which is (4.4) for $b = 1$. Thus, (4.4) is established for the $y_m(\rho_j, \rho_k; q)$.

As for (4.5) for $y_0(\rho_j, \rho_k; q)$, we see that this follows immediately from the fact that $\mathfrak{F}_0(j, k; \psi, \varepsilon) = \delta(j, k)$.

Thus, (4.7) is established.

It is an easy consequence of (4.4) that $q^{m(b-3)} Y_m(a, b; q)$ does not depend on b in the limit. Hence,

$$
\begin{aligned}
&\lim_{m\to\infty} q^{m(b-3)} Y_m(\rho_j, b; q) \\
&\quad = \lim_{m\to\infty} Y_m(\rho_j, 3; q) \\
&\quad = \frac{1}{Q(q)}\left(\sum_{\mu=-\infty}^{\infty} q^{14\mu^2-(4j-5)\mu} - q^{2j+1}\sum_{\mu=-\infty}^{\infty} q^{14\mu^2+(4j+9)\mu}\right) \\
&\quad \qquad [\text{by } (4.7) \text{ and } (2.48)] \\
&\quad = \prod_{\substack{n=1 \\ n\not\equiv 0, \pm\rho_j \,(\mathrm{mod}\ 7)}}^{\infty} (1-q^n)^{-1}
\end{aligned} \tag{4.17}
$$

where the last line follows by Jacobi's triple product identity.

5. REGIME II

As was pointed out in II, we regain for regime II in our new model the same limits as in our 8VSOS model.[6] We prove this is the case with a series of propositions.

Proposition 5.1. Let $d_m(j, k, l)$ denote the degree of the polynomial $X_m(\rho_j, \rho_k, \rho_l; q^{-1})$. Then, for $-2 \leqslant \alpha \leqslant 2$,

$$
d_{5m+\alpha}(j, k, l) = 15m^2 + [6\alpha + kl + (l-1)(4l-3k-3)]\, m + d_\alpha(j, k, l) \tag{5.1}
$$

Proof. This is merely a trivial but tedious mathematical induction using the recurrence (3.6). We omit the details. ∎

Proposition 5.2. For $j=0$, 1, or 2

$$q^{m^2-j}X_m(\rho_j, 1, 3; q)$$
$$= \sum_{\mu=-\infty}^{\infty} q^{14\mu^2-(4j+2)\mu}[t_0(m, 7\mu-\alpha, q)-t_0(m, 7\mu-\alpha-1, q)] \tag{5.2}$$

Proof. By (3.4) and (3.1)

$$X_m(\rho_j, 1, 3; q^{-1}) = F_m(j, 0, 1) - q^{-2j-1}F_m(-j-1, 0, 1)$$
$$= \sum_{\mu=-\infty}^{\infty} q^{35\mu^2+(2-10j)\mu+j(j-1)}T_0(m, 7\mu-j, q)$$
$$- \sum_{\mu=-\infty}^{\infty} q^{35\mu^2-(12+10j)\mu+j^2+j+1}T_0(m, 7\mu-j-1, q)$$

where we have replaced μ by $-\mu$ in the second sum. Consequently, replacing q by q^{-1}, we obtain

$$q^{m^2-j}X_m(\rho_j, 1, 3; q)$$
$$= q^{m^2-j}\left\{ \sum_{\mu=-\infty}^{\infty} q^{-35\mu^2-(2-10j)\mu-j^2+j+(7\mu-j)^2-m^2}t_0(m, 7\mu-j, q)\right.$$
$$\left. - \sum_{\mu=-\infty}^{\infty} q^{-35\mu^2+(10j+12)\mu-j^2-j-1+(7\mu-j-1)^2-m^2}t_0(m, 7\mu-j, q)\right\}$$

[by (2.33_0)]

$$= \sum_{\mu=-\infty}^{\infty} q^{14\mu^2-(4j+2)\mu}[t_0(m, 7\mu-j, q)-t_0(m, 7\mu-j-1, q)] \tag{5.3}$$

as desired. ▮

Now we must recall some polynomial definitions (Ref. 6, §2.6):

$$x_m(a, b, c) = q^{(m+a-b)/4}[f_m(a, b, c) - f_m(-a, b, c)] \tag{5.4}$$

$$f_m(a, b, c) = \sum_{\lambda=-\infty}^{\infty} q^{7\lambda^2-a\lambda+(b+1-c)(14\lambda+b-a)/4}\begin{bmatrix} m \\ \frac{1}{2}(m+a-b)-7\lambda \end{bmatrix} \tag{5.5}$$

$$x_m(a, 1, 2) = q^{(m-1)(m+6)/20}\hat{x}_m(a, 1, 2) \tag{5.6}$$

Proposition 5.3. For $\alpha=0, \pm 1, \pm 2$,

$$\lim_{\substack{m\to\infty \\ m\equiv\alpha \pmod 5}} q^{\binom{m+1}{2}-\binom{j}{2}-(2m-1)(m+3)/10}x_m(\rho_j, 1, 3; q^{1/2})$$
$$= \lim_{\substack{m\to\infty \\ m\equiv\alpha \pmod 5}} \hat{x}_{2m}(2j+1, 1, 3) \tag{5.7}$$

Proof. By Proposition 5.2 and (2.34_0) with $j=0, 1,$ or 2

$$
\begin{aligned}
&q^{(m^2-j)/2}X_m(\rho_j, 1, 3; q^{1/2})\\
&= \sum_{\mu=-\infty}^{\infty} q^{7\mu^2-(2j+1)\mu}\left\{\binom{m; 7\mu-\alpha; q}{7\mu-\alpha}_2 - \binom{m; 7\mu-\alpha-1; q}{7\mu-\alpha-1}_2\right\}\\
&= \sum_{h=0}^{m} (-1)^h q^{mh-\binom{h+1}{2}} \begin{bmatrix} m \\ h \end{bmatrix}\\
&\quad\times\left\{\sum_{\mu=-\infty}^{\infty} q^{7\mu^2-(2j+1)\mu}\left(\begin{bmatrix} 2m-2h \\ m-h+7\mu-j \end{bmatrix} - \begin{bmatrix} 2m-2h \\ m-h+7\mu-j-1 \end{bmatrix}\right)\right.\\
&\quad[\text{by } (2.35)]\\
&= \sum_{h=0}^{m} (-1)^h q^{mh-\binom{h+1}{2}} \begin{bmatrix} m \\ h \end{bmatrix} q^{-(m-h+j)/2} x_{2m-2h}(2j+1, 1, 2)\\
&\quad[\text{by } (5.4)]\\
&- \sum_{h=0}^{m} (-1)^h q^{mh-\binom{h+1}{2}} \begin{bmatrix} m \\ h \end{bmatrix} q^{-(m-h+j)/2+(2m-2h-1)(m-h+3)/10}\\
&\quad\times \hat{x}_{2m-2h}(2j+1, 1, 2) \qquad [\text{by } (5.6)]
\end{aligned}
\tag{5.8}
$$

Consequently,

$$
\begin{aligned}
&q^{(m^2-j)/2+(m+j)/2-(2m-1)(m+3)/10}X_m(\rho_j, 1, 3; q^{1/2})\\
&\quad= \sum_{h=0}^{m} (-1)^h q^{3h(2m-h)/10-7h/10} \begin{bmatrix} m \\ h \end{bmatrix} \hat{x}_{2m-2h}(2j+1, 1, 2)
\end{aligned}
\tag{5.9}
$$

Now in (5.9) we note two things: (1) $\hat{x}_{2m-2h}(2j+1, 1, 2)$ converges to a nonzero limit as m passes to infinity in an arithmetic progression of difference 5 (Ref. 6, §2.6), and (2) the exponent on q in the sum is at least $(3m-5)/5$ for $1 \leqslant h \leqslant m$. Therefore,

$$
\begin{aligned}
&\lim_{\substack{m\to\infty\\ m\equiv\alpha \pmod 5}} q^{\binom{m+1}{2}-(2m-1)(m+3)/10}X_m(\rho_j, 1, 3; q^{1/2})\\
&\qquad= \lim_{\substack{m\to\infty\\ m\equiv\alpha \pmod 5}} \hat{x}_{2m}(2j+1, 1, 2) \quad \blacksquare
\end{aligned}
\tag{5.10}
$$

Finally, we note that in order to know the limiting values of the various $X_m(\rho_j, \rho_k, \rho_l; q^{1/2})$, we need only know those of $X_m(\rho_j, 1, 3; q^{1/2})$. This is because by (3.6) with $b=3$ and $c=1$ we have $X_m(\rho_j, 3, 1; q^{1/2})$; then by (3.6) with $b=2$ and $c=3$ we have $X_m(\rho_j, 3, 2; q^{1/2})$; then, by (3.6) with

$b=c=3$ we have $X_m(\rho_j, 3, 3; q^{1/2})$; then, by (3.6) with $b=3$ and $c=2$ we have $X_m(\rho_j, 3, 2; q^{1/2})$, and finally by (3.6) with $b=c=2$ we have $X_m(\rho_j, 2, 2; q^{1/2})$. These observations are made merely by examining each instance of (3.6) and using Proposition 5.1 to determine which term actually contributes the highest power of q.

Consequently, Proposition 5.3 establishes results equivalent to (3.4)–(3.7) of II.

6. REGIME IV

Now we must consider polynomials reciprocal to those of regime I, i.e., we need to consider $\mathfrak{F}_m(j, k, \psi, \varepsilon; q^{-1})$ from (4.6). By (2.60) and (2.61) with $\lambda=0$ or 1

$$\lim_{\substack{m\to\infty\\ m\equiv\lambda\ (\mathrm{mod}\ 2)}} q^{(m^2-\lambda)/2}\mathfrak{F}_m(j, k; \psi, 0; q^{-1})$$

$$= \frac{R(-q^{1/2})}{2Q(q)}\{21, -6j+10-14\psi+14k, (j-k)^2; q\} + \frac{R(q^{1/2})(-1)^{j-k-\lambda}}{2Q(q)}\{21, -6j+10-14\psi+14k, (j-k)^2; q\}_- \tag{6.1}$$

By (2.51), (2.33), and (2.34),

$$\lim_{m\to\infty} q^{\binom{m}{2}}\mathfrak{F}_m(j, k; \psi, 1; q^{-1})$$

$$= \frac{q^{\binom{k-j}{2}}}{Q(q)\,R(q)}\{21, -6j+3-14\psi+14k, (j-k)^2+(j-k); q\} \tag{6.2}$$

We merely substitute these limits into the limiting case of (4.7) in order to obtain the limits for regime IV.

First we note the reciprocal polynomials for those in regime I are defined by

$$y_{2m}(a, b; q) \equiv q^{2m^2+(3-b)m+\min(0,a-b)}Y_{2m}(a, b; q^{-1}) \tag{6.3}$$

$$y_{2m+1}(a, b; q) \equiv q^{2m^2-(b-1)m+\min(0,a+b-4)}Y_{2m-1}(a, b; q^{-1}) \tag{6.4}$$

The limits are

$$\lim_{m\to\infty} y_m(\rho_j, 2; q)$$

$$= \frac{1}{Q(q)\,R(q)}(\{21, 11-6j, 0; q\} - \{21, 17+6j, 2j^2+2; q\}) \tag{6.5}$$

$$\begin{aligned}
&\lim_{m\to\infty} y_{2m}(\rho_j, 3; q) \\
&\quad = \lim_{m\to\infty} y_{2m+1}(\rho_j, 1; q) \\
&\quad = \frac{R(-q^{1/2})}{2Q(q)} [\{21, -4+6j, (j-1)(3-2j); q\} \\
&\qquad - \{21, 32-6j, -2j^2+j+9; q\}] \\
&\qquad - \frac{(-1)^j R(q^{1/2})}{2Q(q)} [\{21, -4+6j, (j-1)(3-2j); q\}_- \\
&\qquad - \{21, 32-6j, -2j^2+j+9; q\}_-]
\end{aligned} \tag{6.6}$$

and

$$\begin{aligned}
&\lim_{m\to\infty} y_{2m}(\rho_j, 1; q) \\
&\quad = \lim_{m\to\infty} y_{2m+1}(\rho_j, 3; q) \\
&\quad = \frac{R(-q^{1/2})}{2Q(q)} [\{21, -4+6j, j(j-2); q\} \\
&\qquad - \{21, 10+6j, j^2+1; q\}] \\
&\qquad + \frac{(-1)^j R(q^{1/2})}{2Q(q)} [\{21, -4+6j, j(j-2); q\}_- \\
&\qquad + \{21, 10+6j, j^2+1; q\}_-]
\end{aligned} \tag{6.7}$$

These results are equivalent to (3.20a) and (3.20b).

7. AN ALTERNATIVE APPROACH TO REGIMES I AND IV

As we remarked in II, we have an alternative treatment of regimes I and IV, which yields different expressions for the limits in (6.5)–(6.7). It appears at this time that the uniformity provided by our q-trinomial coefficients will extend to the models with more than two particles per site. However, our alternative method may turn out to have other applications, so we describe it briefly.

We start with a two-variable generating function for polynomials $D_{r,s}(a;q)$:

$$\sum_{r,s\geqslant 0} D_{r,s}(a;q)\,x^r y^s$$

$$\equiv f_a(x,y)$$

$$=\sum_{i,j\geqslant 0}\frac{q^{(i+j)^2+i^2+(3-a)i+\delta(1,a)j}x^{4i+2j}y^{2i+2j}}{(x;q)_{i+1}\,(y;q)_{j+1}} \tag{7.1}$$

It is then easy to find simple functional equations for $f_a(x,y)$ that directly imply

$$D_{r,s}(a;q)-D_{r-1,s}(a;q)=q^{r+1-a}D_{r-4,s-2}(a;q) \tag{7.2}$$

provided r is odd or $r>s$; and

$$D_{r,s}(a;q)-D_{r,s-1}(a;q)=q^{s-1+\delta(1,a)}D_{r-2,s-2}(a;q) \tag{7.3}$$

provided s is odd or $s>r/2$.

Using these recurrences as well as (4.4) and (4.5), we find for $m\geqslant 1$

$$Y_m(a,b;q)=q^{(3-b)m}D_{2m+6-b,m-\delta(1,a)-\delta(1,b)}(a;q) \tag{7.4}$$

The $D_{r,s}(a;q)$ can also in certain instances be identified with certain polynomials whose limiting behavior is easily discerned:

$$\begin{aligned}\Delta_{K,i}(a,b;q)=&\sum_{\mu=-\infty} q^{\mu(2K\mu-K+2i)}\begin{bmatrix}a+b\\ a-K\mu\end{bmatrix}\\ &-\sum_{\mu=-\infty}^{\infty} q^{(2\mu-1)(K\mu-i)}\begin{bmatrix}a+b\\ a-K\mu+i\end{bmatrix}\end{aligned} \tag{7.5}$$

These polynomials appear in Ref. 18 [p. 56, Eq. (2.10) corrected] as $D_{K,i}(0;a,b;q)$ and in Ref. 19 as $\delta_{K,i}(a,b;1,1)$. Again recurrences of the $\Delta_{K,i}(a,b;q)$ plus (7.2) and (7.3) allow us to prove for $-1\leqslant a-b\leqslant 2$

$$D_{a+b-[|a-b-1/2|],a+b}(3,q)=\Delta_{7,3}(a,b;q) \tag{7.6}$$

$$D_{2b-\lambda(3\lambda-1)/2,a+b}(\alpha;q)+\delta(-1,a-b)\,q^{b^2+(2-\alpha)m}=\Delta_{7,\alpha}(a-\alpha+3,b;q) \tag{7.7}$$

where $\alpha=1$ or 2 and $\lambda=a-b$.

It is then possible to relate the $\Delta_{7,\alpha}(a, b; q)$ to the $Y_m(a, b; q)$ using the following easily established identities for the $D_{r,s}(a; q)$:

$$D_{M+R,M}(a;q)=\sum_{0\leqslant 2j\leqslant R} q^{j(M+3-a)}\begin{bmatrix} R-j \\ j \end{bmatrix} D_{M-2j,M-2j}(a;q) + \sum_{1\leqslant 2j\leqslant R+1} q^{j(M+2-a)}\begin{bmatrix} R-j \\ j-1 \end{bmatrix} D_{M-2j-1,M-2j}(a;q) \quad (7.8)$$

and for $a=1$ or 2

$$D_{M+R,M}(a;q)=\sum_{0\leqslant 2j\leqslant R} q^{(j+1)(M+1-a)}\begin{bmatrix} R-j \\ j \end{bmatrix} \Big\{ D_{m-2j-4,M-2j-2}(a;q) + \frac{1}{2}[1+(-1)^M]\, q^{\left(\frac{M-2j-2}{2}\right)^2+\delta(1,a)\left(\frac{M-2j-2}{2}\right)}\Big\} + \sum_{0\leqslant 2j\leqslant R+1} q^{j(M+2-a)}\begin{bmatrix} R+1-j \\ j \end{bmatrix} D_{M-2j-1,M-2j}(a;q) \quad (7.9)$$

This concludes the full description of our alternative approach. We next provide a prototypical example of how the above actually provides useful representations of the $Y_m(a, b; q)$:

$$Y_{2m}(3,3;q)=\sum_{j=0}^{m} q^{2mj}\begin{bmatrix} 2m-j \\ j \end{bmatrix}\Delta_{7,3}(m-j,m-j;q) + \sum_{j=1}^{m-1} q^{(2m-1)j}\begin{bmatrix} 2m-j \\ j-1 \end{bmatrix}\Delta_{7,3}(m-j+1,m-j-1;q) \quad (7.10)$$

The regime IV result then proceeds as follows:

$$\begin{aligned}\lim_{m\to\infty} & y_{2m}(3,3;q)\\ &= \lim_{m\to\infty} q^{2m^2} Y_{2m}(3,3;q^{-1})\\ &= \lim_{m\to\infty} q^{2m^2}\Big\{\sum_{j=0}^{m} q^{-2j(m-j)}\begin{bmatrix} m+j \\ 2j \end{bmatrix} q^{-2j(m-j)}\Delta_{7,3}(j,j;q^{-1})\\ &\quad + \sum_{j=0}^{m-1} q^{-(2m-1)(m-j)}\begin{bmatrix} m+j \\ 2j+1 \end{bmatrix} q^{-(2j+1)(m-j-1)}\Delta_{7,3}(j+1,j-1;q^{-1})\end{aligned} \quad (7.11)$$

We now replace $\Delta_{7,3}(j+1, j-1; q^{-1})$ by $\Delta_{7,3}(j+2, j-1; q^{-1})$, since these polynomials turn out to be identical. Hence

$$\begin{aligned}
&\lim_{m\to\infty} y_{2m}(3, 3; q) \\
&= \sum_{j=0}^{\infty} q^{2j^2} \left\{ q^{-j^2} \sum_{\mu=-\infty}^{\infty} \frac{q^{35\mu^2+\mu}}{(q)_{j-7\mu}(q)_{j+7\mu}} \right. \\
&\quad \left. - q^{-j^2+6} \sum_{\mu=-\infty}^{\infty} \frac{q^{35\mu^2-29\mu}}{(q)_{j-7\mu+3}\,(q)_{j+7\mu-3}} \right\} \\
&\quad + \sum_{j=0}^{\infty} q^{2j^2+2j+1} \left\{ q^{-(j+2)(j-1)} \sum_{\mu=-\infty}^{\infty} \frac{q^{35\mu^2-20\mu}}{(q)_{j+2-7\mu}\,(q)_{j+7\mu-1}} \right. \\
&\quad \left. - q^{-j^2-j+17} \sum_{\mu=-\infty}^{\infty} \frac{q^{35\mu^2-50\mu}}{(q)_{j+5-7\mu}\,(q)_{j+7\mu-4}} \right\} \\
&= \sum_{\mu=-\infty}^{\infty} q^{35\mu^2+\mu} \sum_{j=0}^{\infty} \frac{q^{j^2}}{(q)_{j-7\mu}\,(q)_{j+7\mu}} \\
&\quad - \sum_{\mu=-\infty}^{\infty} q^{35\mu^2-29\mu+6} \sum_{j=0}^{\infty} \frac{q^{j^2}}{(q)_{j-7\mu+3}\,(q)_{j+7\mu-3}} \\
&\quad + \sum_{\mu=-\infty}^{\infty} q^{35\mu^2-20\mu+3} \sum_{j=0}^{\infty} \frac{q^{j^2+j}}{(q)_{j+2-7\mu}\,(q)_{j+7\mu-1}} \\
&\quad - \sum_{\mu=-\infty}^{\infty} q^{35\mu^2-50\mu+18} \sum_{j=0}^{\infty} \frac{q^{j^2+j}}{(q)_{j+5-7\mu}(q)_{j+7\mu-4}}
\end{aligned} \tag{7.12}$$

The series with index j are all summable by (2.57):

$$\sum_{j=0}^{\infty} \frac{q^{j^2+(B-A)j}}{(q)_{j-A}\,(q)_{j+B}} = \frac{q^{AB}}{(q)_\infty} = \frac{q^{AB}}{Q(q)} \tag{7.13}$$

Hence

$$\begin{aligned}
&\lim_{m\to\infty} y_{2m}(3, 3; q) \\
&= \frac{1}{Q(q)} \left\{ \sum_{\mu=-\infty}^{\infty} q^{84\mu^2+\mu} + \sum_{\mu=-\infty}^{\infty} q^{84\mu^2-41\mu+5} \right. \\
&\quad \left. - \sum_{\mu=-\infty}^{\infty} q^{84\mu^2-71\mu+15} - \sum_{\mu=-\infty}^{\infty} q^{84\mu^2-113\mu+38} \right\}
\end{aligned} \tag{7.14}$$

As is obvious, (7.14) is quite different from the $j=1$ case of (6.6). Of course, all the regime IV limits may be done in this manner, and we find

$$\begin{aligned}
&\lim_{m\to\infty} y_m(j, 2; q) \\
&\quad = \frac{1}{Q(q)}[\{84, 28-12j, 0; q^2\} + \{84, 56-12j, 7-2j; q^2\} \\
&\qquad - \{84, 28+12j, 4j; q^2\} - \{84, 56+12j, 7+6j; q^2\}]
\end{aligned} \tag{7.15}$$

$$\begin{aligned}
&\lim_{m\to\infty} y_{2m+\delta(2,j)}(j, 3; q) \\
&\quad = \lim_{m\to\infty} y_{2m+\delta(2,j)-1}(j, 1; q) \\
&\quad = \frac{1}{Q(q)}[\{84, 35-12j, 0; q^2\} + \{84, 77-12j, 14-3j; q^2\} \\
&\qquad - \{84, 35+12j, 5j; q^2\} - \{84, 91-12j, 21-4j; q^2\}]
\end{aligned} \tag{7.16}$$

$$\begin{aligned}
&\lim_{m\to\infty} y_{2m+\delta(2,j)}(j, 1; q) \\
&\quad = \lim_{m\to\infty} y_{2m+\delta(2,j)-1}(j, 3; q) \\
&\quad = \frac{1}{Q(q)}[\{84, 12j-7, (j-1)(j-2); q^2\} + \{84, 49-12j, (3-j)^2; q^2\} \\
&\qquad - \{84, 12j+7, (j-1)^2+1; q^2\} - \{84, 49+12j, -4j^2+26j-21; q\}]
\end{aligned} \tag{7.17}$$

As we remarked in II, we have a direct proof of the equivalence of (6.5)–(6.7) with (7.15)–(7.17). The proof is somewhat intricate; the primary result at work is (see Ref. 20, p. 921 for notation)

$$\theta_2\left(u-\frac{\pi}{3}, q\right) - \theta_2\left(u+\frac{\pi}{3}, q\right) = \frac{\sqrt{3}\,\theta_1(u, q^3)\,\theta_3(u, q^3)\,\theta_4(u, q^3)}{q^{1/4}Q^2(q^3)} \tag{7.18}$$

We have not seen this result before; it may be verified by applying Liouville's theorem to the left-hand side divided by the right-hand side.

8. CONCLUSION

The next objective is to extend our work from the $n=3$ case to arbitrary n. The mathematical tools for this project clearly appear to be q-analogs of n-polynomial coefficients, i.e., q-analogs of the coefficients in

$$(1+x+x^2+\cdots+x^{n-1})^N \tag{8.1}$$

We also mention that the mapping ρ_j defined by (3.3) should extend as follows:

$$\rho_j = \begin{cases} 2j+1, & 0 \leqslant 2j < n \\ 2n-2j, & n \leqslant 2j < 2n \end{cases} \tag{8.2}$$

and so

$$\rho^{-1}(j) = \begin{cases} (j-1)/2, & j \text{ odd} \\ n-j/2, & j \text{ even} \end{cases} \tag{8.3}$$

This notation allows us to write (2.5a) of II as

$$H(n-a, n-b, n-c) = -|\rho^{-1}(a) - \rho^{-1}(c)| \tag{8.4}$$

and this appears to be the appropriate way to generalize succinctly regime III.

ACKNOWLEDGMENTS

The work of G.E.A. was partially supported by NSF grant DMS-86503324.

REFERENCES

1. R. J. Baxter and G. E. Andrews, *J. Stat. Phys.* **44**:249 (1986).
2. G. E. Andrews and R. J. Baxter, *J. Stat. Phys* **44**:713 (1986).
3. B. Gordon, *Am. J. Math.* **83**:393 (1961).
4. G. E. Andrews, *The Theory of Partitions* (Addison-Wesley, Reading, Massachusetts, 1976).
5. R. J. Baxter, *Exactly Solved Models in Statistical Mechanics* (Academic, London, 1982).
6. G. E. Andrews, R. J. Baxter, and P. J. Forrester, *J. Stat. Phys.* **35**:193 (1984).
7. R. J. Baxter, *J. Phys. A.* **13**:L61 (1980); *J. Stat. Phys.* **26**:427 (1981).
8. A. Kuniba, Y. Akutsu, and M. Wadati, *J. Phys. Soc. Japan* **55**:1092 (1986).
9. A. Kuniba, Y. Akutsu, and M. Wadati, *Phys. Lett. A* **116**:382 (1986).
10. A. Kuniba, Y. Akutsu, and M. Wadati, *Phys. Lett. A.* **117**:358 (1986).
11. A. Kuniba, Y. Akutsu, and M. Wadati, *J. Phys. Soc. Japan* **55**:2605 (1986).
12. E. Date, M. Jimbo, T. Miwa, and M. Okado, *Lett. Math. Phys.* **12** (1986).
13. L. Comtet, *Advanced Combinatorics* (Reidel, Dordrecht, 1974).
14. W. N. Bailey, *Generalized Hypergeometric Series* (Cambridge, London, 1935; reprinted: Hafner, New York, 1964).
15. D. B. Sears, *Proc. Lond. Math. Soc. (2)* **53**:158 (1951).
16. G. E. Andrews, *Duke Math. J.* **33**:575 (1986).
17. L. J. Slater, *Proc. Lond. Math. Soc. (2)* **54**:147 (1952).
18. G. E. Andrews, in *Enumeration and Design* (Academic, Toronto, 1984).
19. G. E. Andrews, R. J. Baxter, D. M. Bressoud, W. H. Burge, P. J. Forrester, and G. Viennot, *Eur. J. Comb.*, to appear.
20. I. S. Gradshteyn and I. M. Ryzhik, *Table of Integrals, Series* and *Products* (Academic, New York, 1965).
21. E. Date, M. Jimbo, A. Kuniba, T. Miwa, and M. Okado, Exactly solvable SOS models: Local height probabilities and theta function identities, preprint.

2.9 The Pfaff Trilogy

In my year at IBM Yorktown Heights, working with the SCRATCHPAD group, I happened empirically upon the following assertion:

$$
{}_5F_4\left(\begin{matrix}-2i-1, x+2i+2, x-z+\frac{1}{2}, x+i+1, z+i+1; 1\\ \frac{x}{2}+\frac{1}{2}, \frac{x}{2}+1, 2z+2i+2, 2x-2z+1\end{matrix}\right)=0.
$$

Finally, in [184], I came up with an intricate recurrence method to prove this result. Upon reflection, I realized that the method of proof must have been used by Pfaff in his 1797 proof of the summation which was subsequently known by the name "Pfaff–Saalschütz." Paper [185], included here, gives a full account of the Pfaff method.

The question of a machine proof of the above identity gave rise to a lively dispute between Doron Zeilberger and myself which led to [186].

2.9.1 Pfaff's Method (II): Diverse applications

ELSEVIER Journal of Computational and Applied Mathematics 68 (1996) 15–23

JOURNAL OF COMPUTATIONAL AND APPLIED MATHEMATICS

Pfaff's method (II): diverse applications

George E. Andrews[1]

Department of Mathematics, The Pennsylvania State University, University Park, PA 16802, United States

Received 25 October 1994; revised 12 April 1995

Abstract

In 1797, Pfaff summed a ${}_3F_2$ hypergeometric series. His result was ignored until Saalschutz rediscovered it in 1890. This paper is devoted to a detailed account of the wide applicability of Pfaff's method.

1. Introduction

In the first paper of this series, we proved the identity

$$ {}_5F_4\left(\begin{matrix} -2n-1, x+2n+2, x-z+\frac{1}{2}, x+n+1, z+n+1; 1 \\ 1+\frac{1}{2}x,\ \frac{1}{2}+\frac{1}{2}x, 2z+2n+2, 2x-2z+1 \end{matrix}\right) = 0, \tag{1.1} $$

where

$$ {}_{r+1}F_r\left(\begin{matrix} a_0, a_1, \dots, a_r; t \\ b_1, \dots, b_r \end{matrix}\right) = \sum_{n=0}^{\infty} \frac{(a_0, a_1, \dots, a_r)_j\, t^j}{(1, b_1, \dots, b_r)_j}, \tag{1.2} $$

with

$$ (A_1, A_2, \dots, A_N)_M = \prod_{i=0}^{N} \prod_{j=0}^{M-1} (A_i + j). \tag{1.3} $$

Our method of proof was inspired by Pfaff's 1797 proof [18, p. 51] of what has become known as Saalschutz's summation [19] (cf. [9, p. 9, Section 2.2]).

---

[1] Partially supported by National Science Foundation Grant DMS 8702695-04.

In this paper, we shall consider a number of hypergeometric identities using Pfaff's method. The point to be emphasized is that Pfaff's method is often effective in proving collections of results (cf. Sections 4–6). Thus, in practice (most particularly so in [4]) we use the computer to determine the existence of natural identities "close to" one of particular interest. Pfaff's method is then applied to establish the conjectures produced from the computer data.

In Section 2, we recreate Pfaff's original proof that

$$ {}_3F_2\left(\begin{matrix} -n, a, b; 1 \\ c, 1+a+b-n-c \end{matrix}\right) = \frac{(c-a, c-b)_n}{(c, c-a-b)_n}. \tag{1.4}$$

Section 3 treats the terminating Kummer identity. Bailey's ${}_4F_3$ summation is proved in Section 4, and Dougall's summation in Section 5.

These last two sections really give the flavor of Pfaff's method. In Section 4 we are forced to prove simultaneously a lesser known companion summation [5, p. 2, Eq. (2.3)]. In Section 5, we not only prove Dougall's theorem [12] but must simultaneously prove Lakin's [16] nearly forgotten ${}_7F_6$ summation.

In Section 6, we discuss the relationship of [4] to this survery, and we briefly illustrate the application of Pfaff's method to q-analogs in Section 7.

We conclude with an examination of some open questions.

2. Pfaff's method

In 1797, Pfaff gave the simplest proof of (1.4) imaginable [18, p. 51]. Namely, let

$$ S_n(a, b, c) = {}_3F_2\left(\begin{matrix} -n, a, b; 1 \\ c, 1+a+b-c-n \end{matrix}\right). \tag{2.1}$$

Then

$$\begin{aligned} S_n(a, b, c) &- S_{n-1}(a, b, c) \\ &= \sum_{j=0}^{n} \frac{(a, b)_j(-n+1)_{j-1}}{(1, c)_j(1+a+b-c-n)_{j+1}}((-n)(1+a+b-c-n+j) \\ &\qquad\qquad -(-n+j)(1+a+b-c-n)) \\ &= -(1+a+b-c) \sum_{j=0}^{n} \frac{(a)_j(b)_j(-n+1)_{j-1}j}{j!(c)_j(1+a+b-c-n)_{j+1}} \\ &= \frac{-(1+a+b-c)ab}{c(1+a+b-c-n)(2+a+b-c-n)} \sum_{j=0}^{n-1} \frac{(-n+1, a+1, b+1)_j}{(1, c+1, 3+a+b-c-n)_j} \\ &= \frac{-(1+a+b-c)ab\, S_{n-1}(a+1, b+1, c+1)}{c(1+a+b-c-n)(2+a+b-c-n)}. \end{aligned} \tag{2.2}$$

By mathematical induction, we see that the recurrence (2.2) together with the initial value $S_0(a,b,c) = 1$ uniquely determine $S_n(a,b,c)$. If we now define

$$\sigma_n(a,b,c) = \frac{(c-a)_n(c-b)_n}{(c)_n(c-a-b)_n}, \tag{2.3}$$

we see that $\sigma_0(a,b,c) = 1$ and

$$\begin{aligned}
&\sigma_n(a,b,c) - \sigma_{n-1}(a,b,c)\\
&= \frac{(c-a)_{n-1}(c-b)_{n-1}}{(c)_n(c-a-b)_n}((c-a+n-1)(c-b+n-1)-(c+n-1)(c-a-b+n-1))\\
&= \frac{ab(c-a)_{n-1}(c-b)_{n-1}}{(c)_n(c-a-b)_n}\\
&= \frac{(1+a+b-c)ab\,\sigma_{n-1}(a+1,b+1,c+1)}{c(1+a+b-c-n)(2+a+b-c-n)}.
\end{aligned} \tag{2.4}$$

Therefore, because S_n and σ_n satisfy the same defining recurrence and initial conditions, we see that for $n \geqslant 0$,

$$S_n(a,b,c) = \sigma_n(a,b,c); \tag{2.5}$$

i.e., (1.4), the Pfaff–Saalschutz theorem, is proved.

3. The terminating Kummer theorem

Our object here is to provide a further easily comprehended example of Pfaff's method. We shall establish [9, p. 9, Section 2.3, $b = -n$]

$${}_2F_1 = \begin{pmatrix} a, & -n; & -1 \\ & 1+a+n & \end{pmatrix} = \frac{(a+1)_n}{(1+\frac{1}{2}a)_n}. \tag{3.1}$$

Let $K_n(a)$ denote the left-hand side of (3.1). Then $K_0(a) = 1$, and

$$\begin{aligned}
&K_n(a) - K_{n-1}(a)\\
&= \sum_{j=0}^{n} \frac{(a)_j}{j!}\frac{(-n+1)_{j-1}}{(a+n)_{j+1}}((-n)(a+n)-(-n+j)(a+n+j))\\
&= -\sum_{j\geqslant 0} \frac{(a)_j(-n+1)_{j-1}j(a+j)}{j!(a+n)_{j+1}}\\
&= -\sum_{j=0}^{n-1} \frac{(a)_{j+2}(-n+1)_j}{j!(a+n)_{j+2}}\\
&= -\frac{a(a+1)}{(a+n)(a+n+1)}K_{n-1}(a+2).
\end{aligned} \tag{3.2}$$

On the other hand, if $k_n(a)$ denotes the right-hand side of (3.1), then $k_0(a) = 1$ and

$$k_n(a) - k_{n-1}(a) = \frac{(a+1)_{n-1}}{(1+\frac{1}{2}a)_n}((a+n) - (\tfrac{1}{2}a+n))$$

$$= \frac{(a+1)_{n-1}}{2(1+\frac{1}{2}a)_n} = \frac{-a(a+1)(a+2)}{(a+n)(a+n+1)} \frac{(a+3)_{n-1}}{(a+2)(2+\frac{1}{2}a)_{n-1}}$$

$$= \frac{-a(a+1)}{(a+n)(a+n+1)} k_{n-1}(a+2). \tag{3.3}$$

As before, this shows that for $n \geqslant 0$,

$$K_n(a) = k_n(a), \tag{3.4}$$

i.e., (3.1) is proved.

4. Bailey's theorem

In 1929, Bailey [8, p. 512, (c)] proved the following summation:

$${}_4F_3\left(\begin{matrix}\frac{1}{2}a, \frac{1}{2}a+\frac{1}{2}, b+n, -n; 1\\ \frac{1}{2}b, \frac{1}{2}(b+1), a+1\end{matrix}\right) = \frac{(b-a)_n}{(b)_n}. \tag{4.1}$$

In a recent work on certain addition theorems for determinants [5, p. 2, Eq. (2.3)], it was necessary to prove

$${}_4F_3\left(\begin{matrix}\frac{1}{2}a, \frac{1}{2}a+\frac{1}{2}, b+n-1, -n; 1\\ \frac{1}{2}b, \frac{1}{2}(b+1), a\end{matrix}\right) = \frac{(b-a)_n}{(b+2n-1)(b)_{n-1}}. \tag{4.2}$$

In order to prove either of these formulas using Pfaff's method, we must prove both at once. Let $B_n(a,b)$ denote the left-hand side of (4.1) and $C_n(a,b)$ denote the left-hand side of (4.2). Then clearly $B_0(a,b) = C_0(a,b) = 1$, and

$$B_n(a,b) - C_n(a,b)$$

$$= \sum_{j=0}^{n} \frac{(\frac{1}{2}a, \frac{1}{2}(a+1), -n)_j (b+n)_{j-1}}{(1, \frac{1}{2}b, \frac{1}{2}(b+1))_j (a)_{j+1}} ((b+n+j-1)a - (b+n-1)(a+j))$$

$$= \frac{-n(a-b-n+1)}{b(b+1)} C_{n-1}(a+2, b+2), \tag{4.3}$$

while

$$C_n(a,b) - B_{n-1}(a,b)$$

$$= \sum_{j=0}^{n} \frac{(\frac{1}{2}a, \frac{1}{2}(a+1), b+n-1)_j (-n+1)_{j-1}}{(1, \frac{1}{2}b, \frac{1}{2}(b+1))_j (a)_{j+1}} ((-n)(a+j) - (-n+j)a)$$

$$= \frac{-(b+n-1)(a+n)}{b(b+1)} C_{n-1}(a+2, b+2). \tag{4.4}$$

Now (4.3) and (4.4) together with the initial conditions $B_0(a,b) = C_0(a,b) = 1$ uniquely define both $B_n(a,b)$ and $C_n(a,b)$. It is then a standard exercise in college algebra to verify that the right-hand sides of (4.1) and (4.2) satisfy the same recurrences and initial conditions. Thus, (4.1) and (4.2) are proved.

5. The theorems of Dougall and Lakin

In 1907, Dougall [12] proved the following famous and useful result:

$$
{}_7F_6\left(\begin{matrix} 2a, a+1, a+b, a+c, a+d, a+e, a+f; 1 \\ a, 1+a-b, 1+a-c, 1+a-d, 1+a-e, 1+a-f \end{matrix}\right)
= \frac{(2a+1, 1-c-d, 1-b-d, 1-b-c)_m}{(1-a-b-c-d, 1+a-b, 1+a-c, 1+a-d)_m}, \tag{5.1}
$$

where

$$
b+c+d+e = m+1 \quad \text{and} \quad a+f = -m. \tag{5.2}
$$

In 1950, Burchnall and Lakin [10] gave an operator-theoretic proof of (5.1). Their method applied subsequently by Lakin [16] yielded also

$$
{}_7F_6\left(\begin{matrix} 2a, a+1, a+b, a+c, a+d, a+e, a+f; 1 \\ a, 1+a-b, 1+a-c, 1+a-d, 1+a-e, 1+a-f \end{matrix}\right)
= \frac{a_3(1-b-c, 1-b-d, 1-c-d)_{m-1}(1+2a)_m}{(1+a-b, 1+a-c, 1+a-d, -a-b-c-d)_m}, \tag{5.3}
$$

where

$$
b+c+d+e = m \quad \text{and} \quad a+f = -m, \tag{5.4}
$$

with

$$
\begin{aligned} a_3 &= a_3(a,b,c,d,e,f) \\ &= \text{the third elementary symmetric function of } a, b, c, d, e \text{ and } f. \end{aligned} \tag{5.5}
$$

While Dougall's theorem is famous, Lakin's theorem was ignored until 1989. Then Askey in a study of variants of Clausen's formula [7] provided two new proofs of (5.3); he also applied it to provide a brand new Clausen-type formula for the square of a special ${}_2F_1$.

In order to apply Pfaff's method, we define

$$
\begin{aligned} L_m(x) &= L_m(x; a,b,c,d) \\ &= {}_7F_6\left(\begin{matrix} 2a, a+1, a+b, a+c, a+d, a+m+1+x-b-c-d, -m; 1 \\ a, 1+a-b, 1+a-c, 1+a-d, a-m-x+b+c+d, 1+2a+m \end{matrix}\right), \end{aligned} \tag{5.6}
$$

and we note that $L_n(0)$ is the left-hand side of (5.1), $L_n(-1)$ the left-hand side of (5.3).

Again term by term differencing produces three-term functional equations:

$$\begin{aligned}
L_m(x) &= L_m(x-1) \\
&= \sum_{j\geqslant 0} \frac{(2a, a+1, a+b, a+c, a+d, -m)_j}{(1, a, 1+a-b, 1+a-c, 1+a-d, 1+2a+m)_j} \\
&\quad \times \frac{(a+m+1+x-b-c-d)_{j-1}}{(a-m-x+b+c+d)_{j+1}} \\
&\quad \times ((a+m+x-b-c-d+j)(a-m-x+b+c+d+j) \\
&\qquad -(a+m+x-b-c-d)(a-m-x+b+c+d)) \\
&= \sum_{j\geqslant 0} \frac{(2a, a+1, a+b, a+c, a+d, -m)_j}{(1, a, 1+a-b, 1+a-c, 1+a-d, 1+2a+m)_j} \\
&\quad \times \frac{(a+m+1+x-b-c-d)_{j-1}\, j(j+2a)}{(a-m-x+b+c+d)_{j+1}} \\
&= \frac{-(2a+1)(2a+2)(a+b)(a+c)(a+d)mL_{m-1}(x; a+1, b, c, d)}{(1+a-b)(1+a-c)(1+a-d)(1+2a+m)(a-m-x+b+c+d)_2},
\end{aligned} \tag{5.7}$$

where the last line is obtained by shifting the index j to $j+1$ in the penultimate line.

$$\begin{aligned}
&L_m(x) - L_{m-1}(x+1) \\
&= \sum_{j\geqslant 0} \frac{(2a, a+1, a+b, a+c, a+d, a+m+1+x-b-c-d)_j}{(1, a, 1+a-b, 1+a-c, 1+a-d, a-m-x+b+c+d)_j} \\
&\quad \times \frac{(-m+1)_{j-1}}{(2a+m)_{j+1}} ((-m)(2a+m) - (-m+j)(2a+m+j)) \\
&= -\sum_{j\geqslant 0} \frac{(2a, a+1, a+b, a+c, a+d, a+m+1+x-b-c-d)_j (-m+1)_{j-1} j(j+2a)}{(1, a, 1+a-b, 1+a-c, 1+a-d, a-m-x+b+c+d)_j (2a+m)_{j+1}} \\
&= \frac{-(2a+1)(2a+2)(a+b)(a+c)(a+d)(a+m+1+x-b-c-d)L_{m-1}(x+1; a+1, b, c, d)}{(1+a-b)(1+a-c)(1+a-d)(a-m-x+b+c+d)(2a+m)(2a+m+1)},
\end{aligned} \tag{5.8}$$

where the last line is obtained by shifting the index j to $j+1$ in the penultimate line.

Specializing (5.7) to $x=0$ and (5.8) to $x=-1$, we find

$$\begin{aligned}
&L_m(0) - L_m(-1) \\
&= \frac{-(2a+1)(2a+2)(a+b)(a+c)(a+d)m\, L_{m-1}(0; a+1, b, c, d)}{(1+a-b)(1+a-c)(1+a-d)(1+2a+m)(a-m+b+c+d)(1+a-m+b+c+d)}
\end{aligned} \tag{5.9}$$

and

$$L_m(-1) - L_{m-1}(0)$$
$$= \frac{-(2a+1)(2a+2)(a+b)(a+c)(a+d)(m+a-b-c-d)L_{m-1}(0; a+1, b, c, d)}{(1+a-b)(1+a-c)(1+a-d)(1+a-m+b+c+d)(2a+m)(2a+m+1)}. \tag{5.10}$$

In the manner of the previous section, we see that the initial values $L_0(-1) = L_0(0) = 1$ and the recurrences (5.9) and (5.10) uniquely determine both $L_m(0)$ and $L_m(-1)$.

Finally, we note that straightforward college algebra reveals that the right-hand sides of (5.1) and (5.3) satisfy the initial conditions and (5.9) and (5.10).

Consequently, both Dougall's theorem (5.1) and Lakin's theorem (5.3) follow by mathematical induction.

6. The ${}_5F_4$'s of [4]

The results of Sections 2–5 are all known previously (some more than others). The true power of Pfaff's method becomes abundantly clear in [4].

It is quite possible that the list of 20 identities proved there simultaneously is unnecessarily long. On the other hand, it may be, in some quite natural sense, too short. Note that each of identities (5.13), (5.14), (5.15), (5.20) and (5.21) of [4] require elimination of five instances of $H(n, m; x, z,; a_1, a_2, a_3)$ not included in our list of 20. If these five had been included, the list would have been five longer, and some of the related identities would not have had entirely linear factors.

In any event, however long the list, it appears that when the amount of effort required to apply Pfaff doubles, triples or n-tuples, the number of results proved simultaneously increases correspondingly.

It should be remarked that (4.2) of [4] has been independently proved by Dennis Stanton and Shalosh B. Ekhad (a.k.a. Doron Zeilberger). We shall provide a more extensive discussion of these proofs in [6]. Apart from (4.2) we know of no proofs of the remaining 19 identities in [4, Section 4] except the one given there.

7. q-analogs

Often Pfaff's method works for q-analogs in the exact same way that it works for ordinary hypergeometric series. In this brief section, we provide a sketch sufficient to convince the reader of this parallel.

For example, let us consider the q-analog of Pfaff's original theorem (due to Jackson [14])

$$S_n(a,b,c;q) = \sum_{j=0}^{n} \frac{(q^{-n}, a, b; q)_j q^j}{(q, c, q^{1-n}ab/c; q)_j}, \tag{7.1}$$

where

$$(A_1, A_2, \ldots, A_r; q)_j = \prod_{i=1}^{r} \prod_{h=0}^{j-1} (1 - A_i q^h). \tag{7.2}$$

Then term-by-term differencing yields

$$S_n(a,b,c;q) - S_{n-1}(a,b,c;q) = \frac{-q^{1-n}(1-a)(1-b)(1-abq/c)S_{n-1}(aq,bq,cq;q)}{(1-c)(1-(abq^{1-n}/c))(1-(abq^{2-n}/c))}. \tag{7.3}$$

As in Section 2, it is routine to show that

$$\frac{(c/a, c/b; q)_n}{(c, (c/ab); q)_n}$$

is equal to 1 when $n = 0$ and satisfies (7.3). Consequently,

$$\sum_{j=0}^{n} \frac{(q^{-n}, a, b; q)_j q^j}{(q, c, q^{1-n}ab/c; q)_j} = \frac{(c/a, c/b; q)_n}{(c, (c/ab); q)_n}. \tag{7.4}$$

The q-analog of the work in Section 3 establishes the terminating form of the q-analog of Kummer's theorem due to Daum [11] (cf. [1]):

$$\sum_{j=0}^{n} \frac{(a, q^{-n}; q)_j}{(q, aq^{n+1}; q)_j} (-q^{1+n})^j = \frac{(-q; q)_n (aq; q^2)_n}{(aq^{n+1}; q)_n}. \tag{7.5}$$

The q-analog of Section 4 establishes

$$\sum_{j \geqslant 0} \frac{(a;q)_{2j}(b^2 q^{2n}, q^{-2n}; q^2)_j q^{2j}}{(b;q)_{2j}(a^2 q^2, q^2; q^2)_j} = \frac{a^n(-q, b/a; q)_n}{(-aq, b; q)_n} \tag{7.6}$$

and

$$\sum_{j \geqslant 0} \frac{(a;q)_{2j}(b^2 q^{2n-2}, q^{-2n}; q^2)_j q^{2j}}{(b;q)_{2j}(a^2, q^2; q^2)_j} = \frac{a^n(-q, b/a; q)_n}{(-a;q)_n (b;q)_{n-1}(1 - bq^{2n-1})}. \tag{7.7}$$

Identity (7.7) was originally proved in [2; p. 14, Eq. (4.3)]; I do not recall seeing (7.6) before.

Jackson [15] first presented a q-analog of Dougall's theorem (5.1), and Lakin himself proved the q-analog of (5.3) [16, p. 231, Section 3]. These two results may then be proved by a step-by-step q-analog of Section 5.

Section 6 (indeed all of [4]) is still a mystery for a q-analog. Pfaff's method is only useful once one has in hand the various conjectured identities. I have no conjectured q-analogs of the results in [4].

8. Conclusion

We shall in Part III of this series contrast Pfaff's method with the WZ-method [20]. There are certainly a number of open problems concerning Pfaff's method.

Pfaff's method seems most effective for balanced and well-poised hypergeometric series (and q-series). Its effectiveness for balanced series is perhaps explained by Theorem 1 of [4, Section 2]. The results in Sections 3 and 5 of this paper are examples for well-poised series and clearly indicate that there is a well-poised analog of Theorem 1 of [4, Section 2]. It is reasonable to ask whether there are applications of Paff's method that fall outside these two categories.

Finally, we note the following identity:

$$ {}_6F_5\left(\begin{matrix} -2n-1, x+2n+1, x-z+\frac{1}{2}, z+n+\frac{1}{2}, x+n+\frac{1}{2}, \frac{2}{3}(x+n+2)\,;\,1 \\ 2z+2n+1, 1+\frac{1}{2}x, \frac{1}{2}(x+1), 2x-2z+1, \frac{1}{3}(2x+2n+1) \end{matrix}\right) = 0. \tag{8.1} $$

This identity was given as a conjecture at the June 1994 Ann Arbor Conference on Algebra and Combinatorics. Dennis Stanton observed that (8.1) is actually a specialization of formulas he found jointly with Ira Gessel [13]. Also Shalosh B. Ekhad (a.k.a. Doron Zeilberger) proved (8.1) using the WZ method [20]. Identity (8.1) has significant applications. In [6], Stanton and I will show how (1.1) and (8.1) can be used to provide a new and reasonably succinct proof of the main result in [3, Theorem 8, p. 196]. We also hope to consider q-analogs.

For Pfaff's method to apply to (8.1) it would be necessary to find a cluster of results like those of [4, Section 4]. So far such a cluster has not been found.

References

[1] G.E. Andrews, On the q-analog of Kummer's theorem and applications, *Duke Math. J.* **40** (1973) 525–528.

[2] G.E. Andrews, Connection coefficient problems and partitions, *Proc. Symp. Pure Math.* **34** (1979) 1–24.

[3] G.E. Andrews, Plane partitions III: the weak Macdonald conjecture, *Invent. Math.* **53** (1979) 193–225.

[4] G.E. Andrews, Pfaff's method I: the Mills–Robbins–Rumsey determinant, *Discrete Math.*, to appear.

[5] G.E. Andrews and W.H. Burge, Determinant identities, *Pacific J. Math.* **158** (1993) 1–14.

[6] G.E. Andrews and D.W. Stanton, Determinants in plane partitions enumeration, to appear.

[7] R. Askey, Variants of Clausen's formula for the square of special ${}_2F_1$, in: *Number Theory and Related Topics*, Ramanujan Colloquium, Bombay (Oxford Univ. Press, Oxford, 1989) 1–12.

[8] W.N. Bailey, Some identities involving generalized hypergeometric series, *Proc. London Math. Soc. Ser.* (2) **29** (1929) 503–516.

[9] W.N. Bailey, *Generalized Hypergeometric Series* (Cambridge Univ. Press, London and New York, 1935) (Reprinted: Hafner, New York, 1964).

[10] J.L. Burchnall and A. Lakin, The theorems of Saalschutz and Dougall, *Quart. J. Math. Oxford* (2) **1** (1950) 161–164.

[11] J.A. Daum, The basic analog of Kummer's theorem, *Bull. Amer. Math. Soc.* **48** (1942) 711–713.

[12] J. Dougall, On Vandermonde's theorem and some more general expansions, *Proc. Edin. Math. Soc.* **25** (1907) 114–132.

[13] I. Gessel and D. Stanton, Strange evaluations of hypergeometric series, *SIAM J. Math. Anal.* **13** (1982) 295–308.

[14] F.H. Jackson, Transformations of q-series, *Messenger Math.* **39** (1910) 145–153.

[15] F.H. Jackson, Summation of q-hypergeometric series, *Messenger Math.* **50** (1921) 101–112.

[16] A. Lakin, A hypergeometric identity related to Dougall's theorem, *J. London Math. Soc.* **27** (1952) 229–234.

[17] W.H. Mills, D.P. Robbins and H. Rumsey, Enumeration of a symmetry class of plane partitions, *Discrete Math.* **67** (1987) 43–55.

[18] J.F. Pfaff, Observationes analyticae ad L. Euler Institutiones Calculi Integralis, Vol. IV, Supplem. II et IV, Historia de 1793, *Nova Acta Acad. Sci. Petropolitanae* **11** (1797) 38–57.

[19] L. Saalschultz, Eine Summationsformel, *Z. Math. Phys.* **35** (1890) 186–188.

[20] H.S. Wilf and D. Zeilberger, Rational functions certify combinatorial identities, *J. Amer. Math. Soc.* **3** (1990) 147–158.

2.10 MacMahon's "Partition Analysis"

MacMahon's "Partition Analysis" is a method pioneered by MacMahon and fully recounted by him in Section VIII of Volume 2 of *Combinatory Analysis* [MacM16]. When I was preparing MacMahon's *Collected Papers*, I found this topic somewhat perplexing. It was clearly a method he had carefully developed to treat his conjectures on plane partitions. However, after 96 pages of theory MacMahon tells us that he is unable to prove his conjectures using Partition Analysis.

Clearly, MacMahon believed in this method even though he could prove nothing of substantial interest in using it. So, in editing Chapter 10 of [Book9], I recorded in the back of my mind that someday, perhaps, this method would prove useful.

In the Fall of 1995, I visited the University of Melbourne and met Mireille Bousquet-Mélou there. She told me of a miraculous theorem that she and Kimmo Eriksson had proved: "The Lecture Hall Partition Theorem" [Bou-E97].

The number of partitions of N of the form $b_1 + \cdots + b_n$ satisfying

$$\frac{b_n}{n} \geqq \frac{b_{n-1}}{n-1} \geqq \cdots \geqq \frac{b_2}{2} \geqq \frac{b_1}{1}$$

equals the number of partitions of N into odd parts each $\leqq 2N-1$.

This problem cried out for MacMahon's Partition Analysis, and I was able to provide such a treatment in [193], although with much more difficulty than I had anticipated. I followed this up with an elementary account of, and introduction to, Partition Analysis [206]: the one paper from the series included here.

Given that Partition Analysis is an algorithm for producing partition generating functions, I was able to convince Peter Paule and Axel Riese to join an effort to automate this algorithm. The result was the Omega Package [207]. Subsequently, there have been 10 joint Partition Analysis papers with Paule, Riese and others: [208], [212], [228], [229], [230], [232], [249], [250], [254], [257]. Perhaps most pleasing of all was [254] (described in leisurely terms in [257]) which retrieved MacMahon's original project of treating plane partitions with Partition Analysis. I am especially indebted to Peter Paule and his collegues at RISC for making all this possible and for contributing many striking insights. Indeed, I have more joint papers with Paule than with anyone else.

There have been improvements and alternative implementations of MacMahon's Partition Analysis by Xin [Xin04], deLera *et al.* [DeL-HTY04], Corteel *et al.* [Cor-LS07], and others.

2.10.1 MacMahon's Partition Analysis: II. Fundamental theorems

Annals of Combinatorics 4 (2000) 327–338

MacMahon's Partition Analysis: II Fundamental Theorems

George E. Andrews*

Department of Mathematics, The Pennsylvania State University, University Park, Pennsylvania 16802, USA
andrews@math.psu.edu

Received April 21, 1999

AMS Subject Classification: 05A17, 05A19, 05A30, 11P81

Abstract. We continue the study of the method outlined by MacMahon in Section VIII of [10]. The long range object is to show the relevance of MacMahon's ideas in current partition-theoretic research. In this paper we present a number of theorems which MacMahon overlooked. For example, the number of partitions of n with non-negative first and second differences between parts equals the number of partitions of n into triangular numbers.

Keywords: Partition Analysis, partitions, plane partitions

1. Introduction

In [2], the history and basic ideas of MacMahon's Partition Analysis are presented. The object in that paper was to show that MacMahon's Partition Analysis was a powerful enough tool to provide an alternative proof of the Lecture Hall Partition Theorem [5].

In this paper, our object is to reveal a myriad of striking results which one might have expected MacMahon to discover. As one reads Section VIII of [10], it becomes fairly clear that MacMahon would very much like to prove some very substantial results with Partition Analysis. However, such achievements eluded him [10, p. 187]. Consequently, it may well be that he failed to notice striking, albeit elementary, results that were well within his reach.

In Section 2, we shall collate those formulas from Partition Analysis that will be used subsequently; we shall not present a "tutorial" as was done in Section 2 of [2]. The interested reader can consult the latter for a more leisurely treatment. Alternatively, the implementation of Partition Analysis in Mathematica is the subject of [3], and the reader knowledgeable in computer algebra may enjoy that approach.

Section 3 is devoted to the Jordan–Walch–Wisner theorem [7, 8] enumerating $\Delta(n)$, the number of non-congruent triangles with integer length sides of perimeter n. In Section 4, we consider a natural generalization of their result.

* Partially supported by National Science Foundation Grant DMS-9206993.

Section 5 looks at partitions with higher order difference conditions. MacMahon glanced at this topic [10, p. 95]; however, he seems to have missed the results presented here. For example,

Theorem 1.1. *The number of partitions of n in which both first and second differences of the parts (arranged in non-increasing order) are non-negative equals the number of partitions of n into triangular numbers.*

In Section 6, we examine two-rowed plane partitions and illustrate how Partition Analysis applies to MacMahon's original object.

2. Background

MacMahon's Partitions Analysis consists of applications of the linear operator $\Omega_{\geq}$ which is applied to multiple Laurent series:

$$\underset{\geq}{\Omega} \sum_{n_1,n_2,\ldots,n_r=-\infty}^{\infty} A_{n_1,n_2,\ldots,n_r}\lambda_1^{n_1}\lambda_2^{n_2}\ldots\lambda_r^{n_r} = \sum_{n_1,n_2,\ldots,n_r=0}^{\infty} A_{n_1,n_2,\ldots,n_r}. \tag{2.1}$$

The $A_{n_1,\ldots,n_r}$ will often be functions of parameters independent of $\lambda_1, \lambda_2, \ldots, \lambda_r$.

It is an elementary exercise in the use of geometric series to show that if a and r are non-negative integers, then

$$\underset{\geq}{\Omega} \frac{\lambda^{-a}}{(1-A\lambda)(1-B\lambda^{-r})} = \frac{A^a}{(1-A)(1-A^rB)}. \tag{2.2}$$

Formula (2.2) will suffice for all our work in the next three sections.

3. Integer-Sided Triangles of Perimeter n

This section is devoted to an elementary problem. Let $\Delta(n)$ denote the number of incongruent triangles with integer length sides and perimeter n. Find a formula for $\Delta(n)$.

This has been posed as a problem in books on combinatorices [9, 13]. It has been solved by providing a simple defining recurrence for $\Delta(n)$ [1, 7, 8]; an exact formula for $\Delta(n)$ has been given:

$$\Delta(n) = \left\{\frac{n^3}{12}\right\} - \left\lfloor\frac{n}{4}\right\rfloor \left\lfloor\frac{n+2}{4}\right\rfloor, \tag{3.1}$$

where $\{x\}$ is the nearest integer to x, and $\lfloor x \rfloor$ is the greatest integer not exceeding x. A lengthy account of the problem has been provided by Honsberger [6, pp. 39–47].

The object here is to show that the problem can be solved algorithmically (i.e., automatically after setting it up) using partition analysis.

Clearly we are trying to enumerate all unordered triples of integers a_1, a_2, and a_3 which satisfy the triangle inequalities $a_1 + a_2 > a_3$, $a_1 + a_3 > a_2$ and $a_2 + a_3 > a_1$.

Hence, ordering the a_i in non-increasing order, we see that

$$\begin{aligned}\sum_{n\geq 0}\Delta(n)q^n &= \sum_{a_1\geq a_2\geq a_3\geq 1 a_2+a_3>a_1} q^{a_1+a_2+a_3}\\ &= \underset{\geq}{\Omega}\sum_{a_1,a_2,a_3\geq 0} q^{a_1+a_2+a_3}\lambda_1^{a_1-a_2}\lambda_2^{a_2-a_3}\lambda_3^{a_2+a_3-a_1-1}\\ &= \underset{\geq}{\Omega}\frac{\lambda_3^{-1}}{(1-(\lambda_1 q)/\lambda_3)(1-(\lambda_2\lambda_3 q)/\lambda_1)(1-(\lambda_3 q)/\lambda_2)}\\ &= \underset{\geq}{\Omega}\frac{\lambda_3^{-1}}{(1-q/\lambda_3)(1-\lambda_2 q^2)(1-(\lambda_3 q)/\lambda_2)} \quad \text{(by (2.2) with } r=1,\ a=0)\\ &= \underset{\geq}{\Omega}\frac{\lambda_3^{-1}}{(1-q/\lambda_3)(1-q^2)(1-\lambda_3 q^3)} \quad \text{(by (2.2) with } r=1,\ a=0)\\ &= \frac{q^3}{(1-q^4)(1-q^2)(1-q^3)} \quad \text{(by (2.2) with } r=a=1). \end{aligned} \tag{3.2}$$

Thus once the problem is set up, we find that three applications of (2.2) provide the generating function for $\Delta(n)$ automatically. From this last expression we can easily obtain (3.1) by the standard methods for deriving the coefficients of rational functions. Indeed partial fractions and the binomial series will suffice.

However, examination of the steps in (3.2) reveals that we can actually obtain much more information using MacMahon's method. Indeed we can keep track of *each* of the three summands:

$$\begin{aligned}\sum_{a_1\geq a_2\geq a_3\geq 1, a_2+a_3>a_1} x^{a_1}y^{a_2}z^{a_3} &= \underset{\geq}{\Omega}\sum_{a_1,a_2,a_3\geq 0} x^{a_1}y^{a_2}z^{a_3}\lambda_1^{a_1-a_2}\lambda_2^{a_2-a_3}\lambda_3^{a_2+a_3-a_1-1}\\ &= \underset{\geq}{\Omega}\frac{\lambda_3^{-1}}{(1-(\lambda_1 x)/\lambda_3)(1-((\lambda_2\lambda_3 x)/\lambda_1)y)(1-(\lambda_3/\lambda_2)z)}\\ &= \frac{xyz}{(1-x^2yz)(1-xy)(1-xyz)}\end{aligned}$$

(where the steps here eliminating λ_1, λ_2 and λ_3 are precisely the same as the steps used in (3.2))

$$= \sum_{n_1,n_2,n_3\geq 0} x^{2n_1+n_2+n_3+1}y^{n_1+n_2+n_3+1}z^{n_1+n_3+1}. \tag{3.3}$$

Now we see that this last expression provides us with a complete parametrization of our integer-sided triangles. In other words, given any three non-negative integers n_1, n_2 and n_3, there is the unique corresponding triangle of sides

$$2n_1+n_2+n_3+1, \quad n_1+n_2+n_3+1, \quad n_1+n_3+1,$$

and this mapping is easily seen to be reversible.

There are two lessons to be learnt from this example:

(I) Once MacMahon's method has been applied, it may very well be that the same steps will work with several independent variables instead of just one variable.

(II) The steps of the algorithm may well suggest a natural generalization.

We have already seen how (I) has worked out. Concerning (II), it seems to be most enlightening to weaken the initial problem to degenerate triangles. In other words, we want to find all partitions of n of the form $a_1+a_2+a_3$ wherein $a_1 \geq a_2 \geq a_3 \geq 0$ and $a_2+a_3 \geq a_1$.

Since passing from non-degenerate to degenerate triangles is uniquely done by subtracting one from each side, it is no surprise that

$$\sum_{m\geq 0} \Delta_1(m) g^m = \frac{1}{(1-q^2)(1-q^3)(1-q^4)},$$

where $\Delta_1(m)$ is the number of degenerate, non-negative integer-sided triangles of perimeter m.

Theorem 3.1. *Let $\Delta_m(n)$ denote the number of partitions of n into $2m+1$ non-negative parts*

$$n = a_1 + a_2 + a_3 + \cdots + a_{2m+1},$$

where the parts are listed in non-increasing order and additionally

$$\begin{aligned} a_1 - a_2 - a_3 + a_4 &\leq 0, \\ a_3 - a_4 - a_5 + a_6 &\leq 0, \\ &\vdots \\ a_{2m-3} - a_{2m-2} - a_{2m-1} + a_{2m} &\leq 0, \\ a_{2m-1} - a_{2m} - a_{2m+1} &\leq 0. \end{aligned}$$

Then $\Delta_m(n)$ equals the number of partitions of n into parts that are either $\leq 2m$ and even or of the form $(j+1)(2m+1-j)$ with $0 \leq j \leq m$.

Remark. This is precisely equivalent to (3.4) when $m=1$.

Proof.

$$\begin{aligned} &\sum_{n\geq 0} \Delta_m(n)\, q^n \\ &\quad = \underset{\geq}{\Omega} \sum_{a_1,a_2,\ldots,a_{2m+1}\geq 0} q^{a_1+a_2+\cdots+a_{2m+1}} \lambda_1^{a_1-a_2} \lambda_2^{a_2-a_3} \cdots \lambda_{2m}^{a_{2m}-a_{2m+1}} \\ &\qquad \times \mu_1^{-a_1+a_2+a_3-a_4} \mu_3^{-a_3+a_4+a_5-a_6} \cdots \mu_{2m-1}^{-a_{2m-1}+a_{2m}+a_{2m+1}} \end{aligned}$$

$$= \underset{\geq}{\Omega} \frac{1}{(1-(\lambda_1 q)/\mu_1)(1-(\lambda_2\mu_1 q)/\lambda_1)(1-(\lambda_3\mu_1 q)/(\lambda_2\mu_3))(1-(\lambda_4\mu_3 q)/(\lambda_3\mu_1))}$$

$$\times \frac{1}{(1-(\lambda_5\mu_3 q)/(\lambda_4\mu_5))(1-(\lambda_6\mu_5 q)/(\lambda_5\mu_3))\cdots(1-(\mu_{2m-1}q)/\lambda_{2m})}$$

$$= \underset{\geq}{\Omega} \frac{1}{(1-q/\mu_1)(1-q^2)(1-(\mu_1/\mu_3)q^3)(1-q^4)}$$

$$\times \frac{1}{(1-(\mu_3 q^5)/\mu_5)(1-q^6)\cdots(1-\mu_{2m-1}q^{2m+1})}$$

(where we have applied (2.2) with $r=1, a=0$ exactly $2m$ times eliminating $\lambda_1, \lambda_2, \ldots, \lambda_{2m}$ in succession)

$$= \frac{1}{(1-q^2)(1-q^4)\cdots(1-q^{2m})}$$

$$\times \underset{\geq}{\Omega} \frac{1}{(1-\mu_{2m-1}q^{2m+1})(1-(\mu_{2m-3}/\mu_{2m-1})q^{2m-1})\cdots(1-(\mu_1/\mu_3)q^3)(1-q/\mu_1)}$$

$$= \frac{1}{(1-q^2)(1-q^4)\cdots(1-q^{2m})}$$

$$\times \frac{1}{(1-q^{2m+1})(1-q^{(2m+1)+(2m-1)})\cdots(1-q^{(2m+1)+\cdots+3})(1-q^{(2m+1)+\cdots+1})}$$

$$= \prod_{h=1}^{m} \frac{1}{1-q^{2h}} \prod_{j=0}^{m} \frac{1}{(1-q^{(j+1)(2m+1-j)})},$$

and this is precisely the generating functions for the partitions described in Theorem 3.1. ∎

Note that to reconstruct the discovery of Theorem 3.1, one must start by examining the proof of (3.4) and, from there, observing how one would extend the number of factors in a way that would produce the domino effect exhibited in the proof of Theorem 3.1.

Furthermore it is easy to replace the appearance of q in each factor by $2m+1$ variables $x_1, x_2, \ldots, x_{2m+1}$ just as (3.3) refines (3.2).

4. Partitions With Higher Order Difference Conditions

Once we have seen the results of Section 3, it is natural to consider a variety of partition identities related to further difference conditions.

Theorem 4.1. *Let $p_2(n)$ denote the number of partitions of n of the form $a_1 + a_2 + \cdots + a_s$ (s arbitrary) where the first differences are non-negative, i.e., $a_i - a_{i+1} \geq 0$ for $1 \leq i \leq s-1$, and the second differences are non-negative, i.e., $a_i - 2a_{i+1} + a_{i+2} \geq 0$ for $1 \leq i \leq s-1$ (assuming $a_{s+1} = 0$). Then $p_2(n)$ equals the number of partitions of n into triangular numbers.*

Proof.

$$\sum_{n\geq 0} p_2(n)q^n$$

$$= \underset{\geq}{\Omega} \sum_{a_1,a_2,a_3,\cdots\geq 0} q^{a_1+a_2+a_3+\cdots}\lambda_1^{a_1-a_2}\lambda_2^{a_2-a_3}\lambda_3^{a_3-a_4}\cdots$$

$$\mu_1^{a_1-2a_2+a_3}\mu_2^{a_2-2a_3+a_4}\mu_3^{a_3-2a_4+a_5}\cdots$$

$$= \underset{\geq}{\Omega} \frac{1}{(1-\lambda_1\mu_1 q)\left(1-((\lambda_2\mu_2)/(\lambda_1\mu_1^2))q\right)\left(1-((\lambda_3\mu_1\mu_3)/(\lambda_2\mu_2^2))q\right)\left(1-((\lambda_4\mu_2\mu_4)/(\lambda_3\mu_3^2))q\right)}$$

$$= \underset{\geq}{\Omega} \frac{1}{(1-\mu_1 q)(1-(\mu_2/\mu_1)q^2)(1-(\mu_3/\mu_2)q^3)(1-(\mu_4/\mu_3)q^4)\cdots}$$

(using (2.2) with $r = 1$, $a = 0$ to eliminate the λ's successively)

$$= \frac{1}{(1-q)(1-q^{1+2})(1-q^{1+2+3})(1-q^{1+2+3+4})\cdots}$$

(using (2.2) with $r = 1$, $a = 0$ to eliminate the μ's)

$$= \prod_{i=1}^{\infty} \frac{1}{\left(1-q^{\binom{i+1}{2}}\right)}, \tag{4.1}$$

which is the generating function for partitions into triangular numbers. ∎

Once we have seen Theorem 4.1, Theorem 4.2 is the natural and immediate next step.

Theorem 4.2. *Let $p_r(n)$ denote the number of partitions of n of the form $a_1 + a_2 + \cdots + a_s$ (s arbitrary) for which*

$$\sum_{j=0}^{t} (-1)^j a_{i+j} \binom{t}{j} \geq 0$$

for $1 \leq i \leq s-1$, $1 \leq t \leq r$ with $a_{s+1}, \ldots, a_{s+t-1} = 0$. Then $p_r(n)$ equals the number of partitions of n into integers that are rth order binomial coefficients, i.e., of the form $\binom{h+r-1}{r}$.

Proof. Now instead of two sets of Greek variables as in Theorem 4.1, there are r sets of Greek variables. Now there are r steps in which the sets of variables coding the 1st, 2nd, ..., and rth differences are eliminated successively. The final result follows directly as before once we observe that

$$\binom{A-1}{A} + \binom{A}{A} + \binom{A+1}{A} + \cdots + \binom{A+m}{A} = \binom{A+m+1}{A+1}. \tag{4.2}$$

∎

Now we should note that again we can replace each instance of the variable q by independent variables. Thus the proof of Theorem 4.1 can be modified to show that

$$\sum_{\substack{a_1 \geq a_2 \geq a_3 \geq a_4 \geq \cdots \\ a_i - 2a_{i+1} + a_{i+2} \geq 0}} x_1^{a_1} x_2^{a_2} x_3^{a_3} x_4^{a_4} \cdots = \prod_{i=1}^{\infty} \frac{1}{(1 - x_1 x_2^2 x_3^3 \cdots x_i^i)}. \tag{4.3}$$

From (4.3) one can almost immediately see that Theorem 4.1 can be proved bijectively by a pure conjugation type argument. In other words, we can expand each factor of the infinite product in (4.3) by a geometric series. A typical partition arising from the multiplication of these series together may be schematically represented by the array:

$$\begin{array}{c}
\begin{array}{c} n_i \\ \text{times} \end{array} \left\{ \begin{array}{lllll} i & i-1 & i-2 & \dots & 1 \\ i & i-1 & i-2 & \dots & 1 \\ & & & & \\ & & & & \end{array} \right. \\[2ex]
\begin{array}{c} n_{i-1} \\ \text{times} \end{array} \left\{ \begin{array}{llll} i-1 & i-2 & \dots & 1 \\ i-1 & i-2 & \dots & 1 \\ & & & \\ i-1 & i-2 & \dots & 1 \end{array} \right. \\[2ex]
\begin{array}{c} n_2 \\ \text{times} \end{array} \left\{ \begin{array}{ll} 2 & 1 \\ 2 & 1 \\ \vdots & \vdots \\ 2 & 1 \end{array} \right. \\[2ex]
\begin{array}{c} n_1 \\ \text{times} \end{array} \left\{ \begin{array}{l} 1 \\ 1 \\ 1 \\ 1 \end{array} \right.
\end{array}$$

If we add up the rows we clearly have the general representation of a partition into triangular numbers. If we add up the columns, we clearly have a unique corresponding partition in which the first and second differences are all ≥ 0. This establishes a bijection between the two classes of partitions considered in Theorem 4.1. Precisely the same technique will produce a bijective proof of Theorem 4.2.

It should be noted that Sylvie Corteel has produced an independent bijective proof of Theorems 4.1 and 4.2.

5. Partitions With Mixed Difference Conditions

Once one becomes aware of the discoveries that Partition Analysis yields almost painlessly, it is possible to produce a plethora of results. Here we shall content ourselves with just one more example; however it should be clear that the possibilities are endless.

Theorem 5.1. *Let $p_{\pm}(m,n)$ denote the number of partitions of n of the form $a_1 + a_2 + \cdots + a_m$, wherein $a_i - a_{i+1} \geq 0$ for $1 \leq i \leq m-1$ while $a_i - 2a_{i+1} + a_{i+2} \leq 0$ for $1 \leq i \leq m-1$ (with $a_{m+1} = 0$). Then $p_{\pm}(m,n)$ equals the number of partitions of n of the form $(m-j)(m+j+1)/2$, where $0 \leq j < m$.*

Proof.

$$\sum_{n\geq 0} p_{\pm}(m,n)q^n$$

$$= \underset{\geq}{\Omega} \sum_{a_1,a_2,\dots,a_m\geq 0} q^{a_1+a_2+\cdots+a_m}\lambda_1^{a_1-a_2}\lambda^{a_1-a_2}\lambda^{a_2-a_3}\cdots\lambda_{m-1}^{a_{m-1}-a_m}$$

$$\mu_1^{-a_1+2a_2-a_3}\mu_2^{-a_2+2a_3-a_4}\cdots\mu_{m-1}^{-a_{m-1}+2a_m}$$

$$= \underset{\geq}{\Omega} \frac{1}{(1-(\lambda_1 q)/\mu_1)\left(1-((\lambda_2\mu_1^2)/(\lambda_1\mu_2))q\right)\left(1-(\lambda_3\mu_2^2 q)/(\lambda_2\mu_1\mu_3)\right)}\cdots$$

$$\times \frac{1}{\left(1-(\mu_{m-1}^2 q)/(\mu_{m-2}\lambda_{m-1})\right)}$$

$$= \underset{\geq}{\Omega} \frac{1}{(1-q/\mu_1)\left(1-(\mu_1 q^2/\mu_2)\right)\left(1-(\mu_2 q^3/\mu_3)\right)\cdots(1-\mu_{m-1}q^m)}$$

(by (2.2) with $r=1$, $a=0$ applied successively to $\lambda_1,\lambda_2,\dots,\lambda_{m-1}$)

$$= \underset{\geq}{\Omega} \frac{1}{(1-\mu_{m-1}q^m)\left(1-(\mu_{m-2}/\mu_{m-1})q^{m-1}\right)\cdots\left(1-(\mu_1/\mu_2)q^2\right)(1-q/\mu_2)}$$

(reversing the product)

$$= \frac{1}{(1-q^m)(1-q^{m+(m-1)})(1-q^{m+(m-1)+(m-2)})\cdots(1-q^{m+(m-1)+\cdots+1})}$$

$$= \prod_{j=1}^{m} \frac{1}{(1-q^{j(2m-j+1)/2})}$$

$$= \prod_{j=0}^{m-1} \frac{1}{(1-q^{(m-j)(m+j+1)/2})}, \tag{5.1}$$

and this last expression is the generating function for the type of partitions mentioned at the end of the statement of Theorem 5.1. ∎

6. Two-Rowed Plane Partitions

In his discussion of plane and solid partitions using partition analysis [10, Section VIII, Ch. II], MacMahon begins by considering plane partitions on a square. In other words, he considers

$$n = a_{11}+a_{12}+a_{21}+a_{22},$$

where

$$\begin{array}{ccc} a_{11} & \geq & a_{12} \\ \text{VI} & & \text{VI} \\ a_{21} & \geq & a_{22}. \end{array}$$

He shows [10, Section 430, p. 183] that

$$\sum_{\substack{a_{11} \geq a_{12} \\ \text{VI} \quad \text{VI} \\ a_{21} \geq a_{22}}} x_1^{a_{11}} x_2^{a_{12}} x_3^{a_{21}} x_3^{a_{22}}$$

$$= \underset{\geq}{\Omega} \frac{1}{(1-\lambda_1\lambda_2 x_1)(1-(\lambda_3/\lambda_2)x_2)(1-(\lambda_4/\lambda_1)x_3)(1-\lambda_4/(\lambda_3 x_4))}$$

$$= \frac{1-x_1^2 x_2 x_3}{(1-x_1)(1-x_1x_2)(1-x_1x_3)(1-x_1x_2x_3)(1-x_1x_2x_3x_4)}.$$

Furthermore, he observes that if $x_1 = x_2 = x_3 = x_4 = q$, the resulting generating function is

$$\frac{1}{(1-q)(1-q^2)^2(1-q^3)}.$$

After considering some further questions about plane partitions on a square [10, Sections 431–433, pp. 184–186], MacMahon writes down [10, Section 434, p. 186] the omega operator expression for the full generating function for plane partitions of m rows, ℓ columns and each part not exceeding n. He then makes his fateful statement:

"Our knowledge of the $\Omega_{\geq}$ operation is not sufficient to enable us to establish the final form of result."

As we have noted before [2, Section 1], this comment effectively killed this method. In the subsequent 75 years, only Stanley [12] was able to use partition analysis in a non-trivial manner.

In this section, we shall point out that partition analysis allows immediate treatment of all two-rowed plane partitions.

Theorem 6.1. *The generating function for two-rowed plane partitions of length n is given by*

$$\frac{1}{(1-q)(1-q^2)^2(1-q^3)^2\cdots(1-q^n)^2(1-q^{n+1})}. \tag{6.1}$$

This result will follow easily from the following lemma from partition analysis:

Lemma 6.2. *If*

$$G\begin{pmatrix} a_1, a_2, \ldots, a_n \\ b_1, b_2, \ldots, b_n \end{pmatrix} = \underset{\geq}{\Omega} \left\{ \frac{1}{(1-\mu_1 a_1)(1-\mu_1\mu_2 a_2)\cdots(1-\mu_1\cdots\mu_n a_n)} \right.$$
$$\left. \times \frac{1}{(1-b_1/\mu_1)(1-b_2/(\mu_1\mu_2))\cdots(1-b_n/(\mu_1\ldots\mu_n))} \right\}, \tag{6.2}$$

then

$$G\begin{pmatrix} a_1, a_2, \ldots, a_n \\ b_1, b_1 a_1, \ldots, b_1 a_{n-1} \end{pmatrix}$$
$$= \frac{1}{(1-a_1)(1-a_2)\cdots(1-a_n)(1-b_1a_1)(1-b_1a_2)\cdots(1-b_1a_n)}. \tag{6.3}$$

Proof. Eliminating μ_n from the right-hand side of (6.2) using (2.2) with $r=1$ and $a=0$, we see that

$$G\begin{pmatrix} a_1, a_2, \ldots, a_n \\ b_1, b_2, \ldots, b_n \end{pmatrix}$$

$$= \frac{1}{(1-a_nb_n)} \underset{\geq}{\Omega} \left\{ \frac{1}{(1-\mu_1 a_1)(1-\mu_1\mu_2 a_2)\cdots(1-\mu_1\mu_2\ldots\mu_{n-1}a_{n-1})} \right.$$

$$\left. \times \frac{1}{(1-\mu_1\mu_2\cdots\mu_{n-1}a_n)(1-b_1/\mu_1)(1-b_2/(\mu_1\mu_2))\cdots(1-b_{n-1}/(\mu_1\mu_2\cdots\mu_{n-1}))} \right\}. \tag{6.4}$$

Now we invoke the following partial fraction identity

$$\frac{1}{(1-A\mu_{n-1})(1-xA\mu_{n-1})} = \frac{1}{(1-x)}\left(\frac{1}{1-A\mu_{n-1}} - \frac{x}{1-xA\mu_{n-1}}\right), \tag{6.5}$$

where $A=\mu_1\mu_2\cdots\mu_{n-2}a_{n-1}$ and $x=a_n/a_{n-1}$.

This yields directly

$$G\begin{pmatrix} a_1, a_2, \ldots, a_n \\ b_1, b_2, \ldots, b_n \end{pmatrix} = \frac{1}{(1-a_nb_n)(1-a_n/a_{n-1})}$$

$$\times \left\{ G\begin{pmatrix} a_1, \ldots, a_{n-1} \\ b_1, \ldots, b_{n-1} \end{pmatrix} - \frac{a_n}{a_{n-1}} G\begin{pmatrix} a_1, \ldots, a_{n-2}, a_n \\ b_1, \ldots, b_{n-2}, b_{n-1} \end{pmatrix} \right\}. \tag{6.6}$$

Now define

$$g_n(b_1; a_1, \ldots, a_n) = G\begin{pmatrix} a_1, a_2, \ldots, a_n \\ b_1, b_1a_1, \ldots, b_1a_{n-1} \end{pmatrix}, \tag{6.7}$$

and we see that the previous recurrence is

$$g_n(b_1, a_1, \ldots, a_n) = \frac{1}{(1-b_1a_na_{n-1})(1-a_n/a_{n-1})}(g_{n-1}(b_1; a_1, \ldots, a_{n-1})$$

$$-a_n/a_{n-1}g_{n-1}(b_1; a_1, \ldots, a_{n-2}, a_n)). \tag{6.8}$$

But the right-hand side of (6.3) satisfies exactly this same recurrence in n. Furthermore, when $n=1$

$$g_1(b_1; a_1) = G\begin{pmatrix} a_1 \\ b_1 \end{pmatrix} = \underset{\geq}{\Omega} \frac{1}{(1-\mu a_1)(1-b_1/\mu)} = \frac{1}{(1-a_1)(1-a_1b_1)}.$$

Hence the lemma is proved. ∎

Proof of Theorem 6.1. The generating function for two-rowed plane partitions of length n is clearly

$$\sum_{\substack{a_1 \geq a_2 \geq \cdots \geq a_n \geq 0 \\ \text{VI} \quad \text{VI} \qquad \text{VI} \\ b_1 \geq b_2 \geq \cdots \geq b_n \geq 0}} q^{a_1+a_2+\cdots+a_n+b_1+b_2+\cdots+b_n}$$

$$= \underset{\geq}{\Omega} \frac{1}{(1-\lambda_1\mu_1 q)(1-((\lambda_2\mu_2)/\lambda_1)q)\cdots(1-(\mu_n q)/\lambda_{n-1})}$$

$$\times \frac{1}{(1-(\nu_1/\mu_1)q)(1-(\nu_2/(\nu_1\mu_1))q)\cdots(1-q/(\nu_{n-1}\mu_n))}$$

(and now use (2.2) with $r = 1$ and $a = 0$ to eliminate all the λ's and ν's)

$$= \underset{\geq}{\Omega} \frac{1}{(1-\mu_1 q)(1-\mu_1\mu_2 q^2)\cdots(1-\mu_1\mu_2\cdots\mu_n q^n)}$$

$$\times \frac{1}{(1-q/\mu_1)(1-q^2/(\mu_1\mu_2))\cdots(1-q^n/(\mu_1\mu_2\cdots\mu_n))}$$

$$= G\begin{pmatrix} q, q^2, q^3, \ldots, q^n \\ q, q^2, q^3, \ldots, q^n \end{pmatrix}$$

$$= g_n(q; q, q^2, q^3, \ldots, q^n)$$

$$= \frac{1}{(1-q)(1-q^2)\cdots(1-q^n)(1-q^2)(1-q^3)\cdots(1-q^{n+1})}$$

(by the Lemma)

$$= \frac{1}{(1-q)(1-q^2)^2\cdots(1-q^n)^2(1-q^{n+1})} \tag{6.9}$$

as desired. ∎

7. Conclusion

In this paper, the object has been to reveal the numerous theorems that are not only easily proved using Partition Analysis but are indeed suggested by considering problems in partitions from the Partition Analysis viewpoint. There are many further possibilities for Partition Analysis that will be examined later in this series.

In the next paper [3], Peter Paule, Axel Riese and the author will discuss the fact that Partition Analysis of rational functions is actually an algorithmic process implementable in a computer algebra system. In a further paper, Paule and the author consider its application to the summation of series of hypergeometric type [4]. Further investigations are on-going as well including the use of Partition Analysis in the study of Rogers–Ramanujan-type identities.

References

1. G.E. Andrews, A note on partitions and triangles with integer sides, Amer. Math. Monthly **86** (1979) 477–478.
2. G.E. Andrews, MacMahon's partition analysis: I. The lecture hall partition theorem, In: Mathematical Essays in Honor Gian-Carlo Rota, B. Sagan and R. Stanley, Eds., Birkhäuser, Boston, 1998, pp. 1–22.
3. G.E. Andrews, P. Paule, and A. Riese, MacMahon's partition analysis: III. The omega package, to appear.
4. G.E. Andrews and P. Paule, MacMahon's partition analysis: IV. Hypergeometric multisums, to appear.
5. M. Bousquet-Mélou and K. Eriksson, Lecture hall partitions, The Ramanujan Journal **1** (1997) 101–111.
6. R. Honsberger, Mathematical Gems III, Math. Assoc. of America, Washington, 1985.
7. J.H. Jordan, R. Walsh, and R.J. Wisner, Triangles with integer sides, Notices Amer. Math. Soc. **24** (1977) A–450.
8. J.H. Jordan, R. Walsh, and R.J. Wisner, Triangles with integer sides, Amer. Math. Monthly **86** (1979) 686–689.
9. C.I. Liu, Introduction to Combinatorial Mathematics, McGraw-Hill, New York, 1968.
10. P.A. MacMahon, Combinatory Analysis, Vol. 2, Cambridge University Press, Cambridge, 1916; Chelsea, New York, 1960, reissued
11. P.A. MacMahon, Collected Papers, Vol. 1, G.E. Andrews, Ed., MIT Press, Cambridge, 1978.
12. R.P. Stanley, Linear homogeneous equations and magic labelings of graphs, Duke Math. J. **40** (1973) 607–632.
13. R.P. Stanley, Enumerative Combinatories, Vol. 1, Wadsworth, Monterey, 1986.

Chapter 3

Partition Identities

This chapter relates to approximately 70 of my papers. Of these, I have chosen 18 to include here. They do not split easily into subtopics; hence, this chapter, although long, is not divided into sections as was Chapter 2.

When I took Rademacher's course at Penn in 1961–1962, I became fascinated with the Rogers–Ramanujan identities, partly because Rademacher clearly found them fascinating. I devoted Section 4 of "Some Debts I Owe" to an account of how I was led from Rademacher's lectures to paper [10], the natural generalization of the Rogers–Ramanujan identities, and beyond. Indeed, I discovered this result unaware that Gordon had already found and published it [Gor61].

The method used in [10] proved to be highly adaptable to further problems in partition identities. Indeed, papers [14], [15], [18], and [27] were natural successors to [10]. Alder surveyed much of this work in [Ald69].

At the same time that the extensions of [10] were found, I was studying the work of Heinz Göllnitz [Göl60], [Göl67], and further papers of Gordon [Gor62], [Gor65]. This led to [16], [20], [21], [22], and the q-series paper [23]. A special word should be said about [16]. Given the results in [10], and the Göllnitz–Gordon results [Göl60], [Gor62], [Gor65], one was naturally led to a dead end in the first attempt to generalize these theorems. The power of [10] lies in the fact that the partitions in question are generated by iterating the q-difference equations. By finding the relevant q-difference equtions for [16] first, one was forced into the surprising constraint that odd parts could not be repeated. I regard this as a further instance of: "Try to hear what the mathematics is trying to tell you!" All this culminated in [31], which is included in this chapter. Paper [31] was improved in the AMS Memoir [58]; although [31] is far more readable.

The methods developed in these papers, combined with a better understanding of q-difference equations, and q-series transformations like those of Heine, led to many further partition identities: [9], [11], [12], [13], [18], [24], [26], [28], [29], [33], [34], [36], [37], [39], [40], [44], [46], [48], [53], [56], [91].

The sheer implausibility of partition identities existing at all, combined with the cornucopia of results I was finding, provided an almost intoxicating initial decade in my career.

The next landmark in my work was the 1970–1971 sabbatical year at MIT with Gian-Carlo Rota. In preparing for that year, I spent a long time studying Rota's papers [Gold-R69], [Mul-R70]. I had hoped to carry my work on partitions over to the study of finite vector spaces. I tried really hard to achieve this, and while failing at my main object, I inadvertently came upon the new generalizations of the Rogers–Ramanujan identities related to successive ranks. The main paper on this topic is [50], but I have instead included [49] which contains the main ideas in [50] but is more readable.

The methods from [49] and [50] have been subsequently employed in: [101], [103], [118], [121], [123] especially, [126], [136], [149], [152], [154], [155], [156], [163], [171], [174], [183], [190], [200], [201], [259]. Perhaps most important is the fact that the methods of [49] and [50] led to most of my contributions to the work on statistical mechanics with Baxter (*cf.* Ch. 2). I have also included the paper [115] relating these results to new statistics connected with partitions. This paper, and Chapter 9 of [Book3], provide the background for the Ph.D. theses of Santos [San91] and Sills [Sil03].

Paper [43], included in this chapter, grew out of the theorems in [20], and eventually led Yee [Yee04] and Alfes *et al.* [Alf-JO10] to a complete proof of the Alder conjecture [Ald56]. Paper [80] (whose main idea just appeared one morning while I was shaving) is, in my opinion, the natural interpretation of the generalized analytic Rogers–Ramanujan series appearing in [61].

My year (1975–1976) spent at Wisconsin, as the guest of Richard Askey (described by Askey in [Ask01]), led mostly to q-series and plane partitions work, along with the discovery of Ramanujan's "Lost" Notebook in the spring of 1976. My 1978–1979 sabbatical in Australia led to further work on partitions [91], and further interaction with my first Australian student, Mike Hirschhorn, who had previously drawn my attention to the Lusztig–Macdonald–Wall conjecture [76].

Perhaps the most fortunate aspect of the year in Australia was a colloquium I gave in Canberra on the "Lost" Notebook. Thus, when Baxter consulted the Mathematics Department at the A.N.U. about q-series, they referred him to me, and the fruitful collaboration with Baxter commenced from this chance beginning (*cf.* Ch. 1 of [Book4]).

The work on generalized Frobenius partitions arose from a mistaken interpretation of some of the eigenvalue summations coming from Baxter's physics. By now, I had learned to "listen to the mathematics." So, even though generalized Frobenius partitions were not what the physics required, they turned out to be the very fruitful subject discussed in the AMS *Memoir* [109]. For brevity, I have instead included [106] which is an introduction to this topic.

My second Australian student was Frank Garvan, whose thesis was devoted to a wonderful explication of a couple of mysterious pages from Ramanujan's "Lost" Notebook. This in turn led to two joint discoveries: (1) The Mock Theta Conjectures (stated in [138] and subsequently proved first by Hickerson, then Bringmann, Ono, and Rhoades [Bri-OR08], and Folsom [Fol08]), and (2) Dyson's crank. Dyson [Dys44] had made a number of fascinating conjectures in 1944. The paper [139], included here, provides the relevant background and reveals the crank. The story surrounding this discovery has been told by Askey [Ask01].

Benson wrote to me in the mid-1980s concerning research in group theory. In particular, he had conjectured but could not prove a partition identity that was needed to complete [Ben88]. It turned out that it was a theorem of Schur [Schu26]. Benson thanked me warmly in his paper, and this alerted other group theorists (notably Olsson and Bessenrodt) that they might contact me in case of similar problems. Papers [158], [176], and [189] were the result. Paper [176] arose from unlikely events. In my memoir on the general Rogers–Ramanujan theorem [58], I put a conjectured partition identity in the conclusion merely to give some indication of how one might extend my main theorem. It lay untouched for more than 15 years. Then, to my surprise, Bessenrodt wrote to me to ask if it had been proved because she and Olsson needed it in their current research. Paper [176] followed, and [189] refines [176]. I have chosen to include [158] because it is so general and builds on [20]. Bessenrodt subsequently gave a beautiful combinatorial proof [Bes91], [Bes95].

I have been fortunate to have a close relationship with colleagues at the University of Florida, where I received an honorary doctorate in 2002. Both Berkovich and Garvan are there. Krishna Alladi has been instrumental in facilitating my visits to Florida. He and I have worked together on extending the work of Göllnitz (*cf.* [22]) and Capparelli's conjecture (*cf.* [174]). This collaboration resulted in: [177], [178], [192], [195], [224], [239], and [253]. I have chosen [177] as an appropriate representation of our collaboration.

The final destination during my "round the world" sabbatical leave in 1977–1978 was Brighton in the U.K., where I collaborated with the late Richard Lewis. This month-long visit (complemented by a subsequent visit by Lewis to Penn State) led to five joint papers on partitions: [202], [204], [205], [216], and [219]. Of these, my favorite is [204], which I have included here. The conjectures Lewis and I raised have been studied by others, including Kane [Kan04], and Bringmann [Bri09].

In the first decade of the twenty-first century three projects were especially important to me, and I have tried to include representations of each. Paper [252], my inaugural paper as a member of the National Academy of Sciences, arose from correspondence with Dan Romik about his work with Holroyd and Liggett [Hol-LR04]. The fact that partitions with short sequences actually had a nice generating function greatly surprised me, along with the connection to mock theta functions. This work was further developed in [259].

Trying to understand the intricacies of Dyson's rank led to paper [263] that laid the groundwork for a number of papers by others: [Kei07], [Bri08], and [Bri-LO10].

A discussion with Alladi on possible connections between partitions and multiplicative number theory reminded me of a paper by Fockink, Fockink, and Wang [Fok-FW95] which gave an identity between the divisor function and a weighted count of partitions with distinct parts. Paper [209] had already been partly inspired by [Fok-FW95], but this reawakening led to a consideration of the total number of smallest parts in the partitions of n, $\mathrm{spt}(n)$, along with the surprising congruence

$$\mathrm{spt}(13n+6) \equiv 0 \pmod{13}.$$

This is the topic of [266].

Paper [270] is sufficiently recent and lengthy that I have decided not to include it; however, it is worth recording the story of its genesis. I needed to respond to an invitation from Alladi to speak at a group theory conference honoring Chat Ho. This forced me to take a new look at an old theorem of Schur [Schu26]:

The number of partitions of n into distinct odd parts equals the number of partitions of n in which the difference between parts is at least 4, at least 8 between multiples of 4 and no part $\equiv 2 \pmod 4$.

I had never actually written out a proof of this theorem before, although after my early work on [Schu26] in [13], and [200], it was straightforward. The result of this exercise revealed that the generating function for the second type of partitions with largest part $\leqq 4N-1$ is:

$$\sum_{j=0}^{N} t^j q^{j^2} (-q^2;q^2)_j \begin{bmatrix} N \\ j \end{bmatrix}_4,$$

where

$$(A;q)_n = (1-A)(1-Aq)\cdots(1-Aq^{n-1}),$$

and

$$\begin{bmatrix} N \\ j \end{bmatrix}_r = \begin{cases} 0 & \text{if } j<0 \text{ or } j>N \\ \frac{(q^r;q^r)_N}{(q^r;q^r)_j (q^r;q^r)_{N-j}}, & \text{otherwise} \end{cases}.$$

This polynomial, which converges to

$$(-q;q^2)_\infty$$

as $N \to \infty$, set me to thinking about other possible non-standard q-analogs of some instance of the binomial theorem. This suggested:

$$\sum_{j=0}^{N} t^j q^{j^2} (-q;q)_j \begin{bmatrix} N \\ j \end{bmatrix}_2$$

which would lead in the limit to the first Rogers–Ramanujan function. The upshot was the study of parity in partition identities in [270]. Kursungoz [Kur09] has given a superb, purely combinatorial account of most of the theorems in [270], and Kim and Yee [Kim-Y10] have proven most of the conjectures at the end of [270].

The final paper included here, [267], could have been included in Chapter 6 owing to the fact that it is filled with divisibility properties of multicompositions; however, it seemed equally natural to include it here. My student, Matt Katz, has solved many of the mysteries presented at the end of [267].

To keep the commentary for this chapter within bounds, I can give only brief mention to a number of papers. Among these, I have chosen to include [114]: partly because it honors my late friend, Nathan Fine, partly because it provides an application of Dyson's adjoint [Dys44], and partly because it was further developed by Igor Pak in his masterly survey of bijective methods in partition identities [Pak06].

Papers [9], [148], and [149] relate to continuing interest in the contributions of Sylvester [Syl82]. Paper [162] was a surprisingly asymptotic work (joint with Almkvist) for restricted partitions. Paper [221] provided a proof of the Friedman–Joichi–Stanton conjecture at primes and set the stage for Stanton and Prellberg [Pre-S03] to finish the problem.

Bipartitions and multipartitions were of great interest to MacMahon [MacM16], [MacM78], and I have published a number of papers on this topic: [74], [131], [146], [205], [211], [264], and [267].

Paper [106] contains a full proof of the general theorem alluded to at the end of [49]. However, [49] gives a feeling for the sieving process that is the more succinet proof in [106]. See also paper [123].

One final note regarding Issai Schur: P. Bachmann's *Niedere Zahlentheorie* [Bac02] attributes a certain unpublished result to "J. Schur." The initial "J" is also found in [Dic19]. This confusion over Schur's initials led me to refer to "I. J. Schur" in many publications, and this error was subsequently duplicated by other authors. Nonetheless, Schur always published as "I. Schur."

3.1 An Analytical Proof of the Rogers–Ramanujan–Gordon Identities

AN ANALYTIC PROOF OF THE ROGERS-RAMANUJAN-GORDON IDENTITIES.

By George E. Andrews.

1. Introduction. In [2], Gordon gives a combinatorial proof of the following beautiful generalization of the Rogers-Ramanujan identities.

Theorem. *Let a and k be integers with $0 < a \leqq k$. Let $A_{k,a}(n)$ denote the number of partitions of n into parts not of the forms $(2k+1)m$, $(2k+1)m \pm a$; $A_{k,a}(0) = 1$. Let $B_{k,a}(n)$ denote the number of partitions of n of the form*

$$n = b_1 + \cdots + b_s$$

with $b_i \geqq b_{i+1}$, $b_i - b_{i+k-1} \geqq 2$, and with 1 appearing as a summand at most $a-1$ times; $B_{k,a}(0) = 1$. Then

$$A_{k,a}(n) = B_{k,a}(n).$$

Gordon states, however, [2; p. 394] that he has been unable to deduce this theorem from the corresponding identities of Alder [1] which generalize the Rogers-Ramanujan identities analytically.

The object of this paper is to give an analytic proof of Gordon's theorem along the lines of Ramanujan's proof of the original identities [4].

2. Proof of the theorem. We define

$$C_{k,i}(x) = 1 - x^i q^i + \sum_{\mu=1}^{\infty} (-1)^{\mu} x^{k\mu} q^{\frac{1}{2}(2k+1)\mu(\mu+1)-i\mu} (1 - x^i q^{(2\mu+1)i}) \frac{(1-xq)\cdots(1-xq^{\mu})}{(1-q)\cdots(1-q^{\mu})}.$$

Selberg [5; p. 4, equation 3] has proved

$$C_{k,i}(x) = C_{k,i-1}(x) + x^{i-1}q^{i-1}(1-xq)C_{k,k-i+1}(xq). \tag{2.1}$$

If we define

$$Q_{k,i}(x) = C_{k,i}(x) \prod_{j=1}^{\infty} (1-xq^j)^{-1},$$

then

Received July 8, 1965.

(2.2) $$Q_{k,i}(x) = Q_{k,i-1}(x) + x^{i-1}q^{i-1}Q_{k,k-i+1}(xq).$$

We may expand $Q_{k,i}(x)$ as follows

(2.3) $$Q_{k,i}(x) = \sum_{N=-\infty}^{\infty} \sum_{M=-\infty}^{\infty} c_{k,i}(M,N)x^M q^N, \ |x| \leqq 1, |q| < 1.$$

We then easily verify by means of the definition of $Q_{k,i}(x)$ and (2.2) that

(2.4) $$c_{k,0}(M,N) = 0 \text{ for all } k, M, N.$$

(2.5) $$c_{k,i}(M,N) = \begin{cases} 1 \text{ if } M=0 \text{ and } N=0, \ 1 \leqq i \leqq k \\ 0 \text{ if either } M \leqq 0 \text{ or } N \leqq 0 \text{ and not both are } 0 \end{cases}$$

(2.6) $$c_{k,i}(M,N) - c_{k,i-1}(M,N) = c_{k,k-i+1}(M-i+1, N-M), \ 1 \leqq i \leqq k.$$

One easily verifies by methematical induction that the $c_{k,i}(M,N)$ are uniquely determined by (2.4), (2.5), and (2.6).

Let $p_{k,i}(M,N)$ denote the number of partitions of N into M parts of the form $N = b_1 + \cdots + b_M$, with $b_l \geqq b_{l+1}$, $b_l - b_{l+k-1} \geqq 2$, and 1 appearing as a summand at most $i-1$ times. The $p_{k,i}(M,N)$ clearly satisfy (2.4) and (2.5). We now show that they satisfy (2.6).

$p_{k,i}(M,N) - p_{k,i-1}(M,N)$ enumerates the number of partitions of the form given in the previous paragraph with the added condition that 1 appears *exactly* $i-1$ times as a summand. We note that if 1 appears exactly $i-1$ times then 2 can appear at most $k-i$ times. Let us now subtract 1 from every summand of the partition under consideration. Since 1 appeared exactly $i-1$ times we have reduced the number of summands to $M-i+1$. Since we have subtracted M ones from our partition, we are now partitioning $N-M$. Since 2 could appear at most $k-i$ times formerly, now 1 appears at most $k-i$ times. Thus we now have a partition of the form enumerated by $p_{k,k-i+1}(M-i+1, N-M)$. The above procedure clearly establishes a one-to-one correspondence between the partitions enumerated by $p_{k,i}(M,N) - p_{k,i-1}(M,N)$ and the partitions enumerated by $p_{k,k-i+1}(M-i+1, N-M)$. Therefore

$$p_{k,i}(M,N) - p_{k,i-1}(M,N) = p_{k,k-i+1}(M-i+1, N-M).$$

Thus the $p_{k,i}(M,N)$ fulfill (2.4), (2.5), and (2.6). Therefore by the remark following (2.6) we have

(2.7) $$p_{k,i}(M,N) = c_{k,i}(M,N).$$

Thus

$$\sum_{N=0}^{\infty} A_{k,a}(N) q^N = \prod_{n \not\equiv 0, \pm a (\mathrm{mod}\ 2k+1)}^{\infty} (1 - q^n)^{-1}$$

$$= Q_{k,a}(1) = \sum_{N=-\infty}^{\infty} \sum_{M=-\infty}^{\infty} p_{k,a}(M, N) q^N = \sum_{N=0}^{\infty} B_{k,a}(N) q^N,$$

where the second equality follows from Jacobi's identity [3; p. 282]. Therefore $A_{k,a}(N) = B_{k,a}(N)$.

THE PENNSYLVANIA STATE UNIVERSITY,
UNIVERSITY PARK, PENNSYLVANIA.

REFERENCES.

[1] H. L. Alder, "Generalizations of the Rogers-Ramanujan identities," *Pacific Journal of Mathematics*, vol. 4 (1954), pp. 161-168.

[2] B. Gordon, "A combinatorial generalization of the Rogers-Ramanujan identities," *American Journal of Mathematics*, vol. 83 (1961), pp. 393-399.

[3] G. H. Hardy and E. M. Wright, *An introduction to the theory of numbers*, 4th ed., Oxford University Press, Oxford, 1960.

[4] S. Ramanujan, "Proof of certain identities in combinatory analysis," *Proceedings of the Cambridge Philosophical Society*, vol. 19 (1919), pp. 214-216.

[5] A. Selberg, "Uber einige arithmetische Identitäten," *Avhandlinger Norske Akademie*, 1936, no. 8.

3.2 Some New Partition Theorems

Reprinted from JOURNAL OF COMBINATORIAL THEORY Vol. 2, No. 4, June 1967
 Printed in Italy

Some New Partition Theorems

GEORGE E. ANDREWS*

Department of Mathematics, The Pennsylvania State University, University Park, Pennsylvania

Communicated by Gian-Carlo Rota

ABSTRACT

In this paper the following theorem is proved and generalized.

The partitions of any positive integer, n, into parts of the forms $6m + 2$, $6m + 3$, $6m + 4$ are equinumerous with those partitions of n into parts $\geqq 2$ which neither involve sequences nor allow any part to appear more than twice.

1. INTRODUCTION

It appears that P. A. MacMahon was the first to recognize the combinatorial significance of the Rogers-Ramanujan identities [4, Ch. III]. The second of these identities was stated thusly:

The partitions of any positive integer, n, into parts of the forms $5m + 2$, $5m + 3$ are equinumerous with those partitions of n into parts ≥ 2 which involve neither sequences nor repetitions.

In the next chapter of MacMahon's book, the following striking (and little known) theorem which resembles the second Rogers-Ramanujan identity is proved, although not explicitly stated:

The partitions of any positive integer, n, into parts of the forms $6m$, $6m + 2$, $6m + 3$, $6m + 4$ are equinumerous with those partitions of n into parts ≥ 2 which do not involve sequences.

* Partially supported by National Science Foundation Grant NSF 6592.

Thus with $n = 12$, the 16 partitions of the first type are

$$12,\ 10+2,\ 9+3,\ 8+4,\ 8+2+2,\ 6+6,\ 6+4+2,\ 6+3+3,$$
$$6+2+2,\ 4+4+4,\ 4+4+2+2,\ 4+3+3+2,$$
$$4+2+2+2+2,\quad 3+3+3+3,\quad 3+3+2+2+2,$$
$$2+2+2+2+2+2;$$

while the 16 partitions of the second type are

$$12,\ 10+2,\ 9+3,\ 8+4,\ 8+2+2,\ 7+5,\ 6+6,\ 6+4+2,$$
$$6+3+3,\ 6+2+2+2,\ 5+5+2,\ 4+4+4,\ 4+4+2+2,$$
$$4+2+2+2+2,\quad 3+3+3+3,\quad 2+2+2+2+2.$$

The following theorem in a certain sense appears to be intermediate between the second Rogers-Ramanujan identity and MacMahon's theorem.

THEOREM 1. *The partitions of any positive integer, n, into parts of the forms $6m+2$, $6m+3$, $6m+4$ are equinumerous with those partitions of n into parts ≥ 2 which neither involve sequences nor allow any part to appear more than twice.*

Thus again with $n = 12$, the 11 partitions of the first type are

$$10+2,\quad 9+3,\quad 8+4,\quad 8+2+2,\quad 4+4+4,\quad 4+4+2+2,$$
$$4+3+3+2,\ 4+2+2+2+2,\ 3+3+3+3,$$
$$3+3+2+2+2,\ 2+2+2+2+2+2;$$

while the 11 partitions of the second type are

$$12,\ 10+2,\ 9+3,\ 8+4,\ 8+2+2,\ 7+5,\ 6+6,\ 6+4+2,$$
$$6+3+3,\ 5+5+2,\ 4+4+2+2.$$

I have been unable to find the above theorem in the literature. Actually it is a special case of the following theorem with $k = 1$, $a = 0$.

THEOREM 2. *Let $0 \leq a < k$ be integers. Let $A_{k,a}(N)$ denote the number of partitions of N into parts $\not\equiv 0, \pm(2a+1) (\bmod 4k+2)$. Let $B_{k,a}(N)$ denote the number of partitions of N of the form $\sum_{i=1}^{\infty} f_i \cdot i$, where (1) $f_1 \leq 2a$, (2) $[\frac{1}{2}(f_i+1)] + [\frac{1}{2}(f_{i+1}+1)] \leq k$. Then*

$$A_{k,a}(N) = B_{k,a}(N).$$

The object of this paper is to prove these two theorems. As we have observed, Theorem 1 is a special case of Theorem 2, so we shall content ourselves with proving Theorem 2.

2. Proof of Theorem 2

We first discuss the second condition on the partitions enumerated by $B_{k,a}(N)$; it states that, if i appears $2j - 1$ or $2j$ times as a summand, then $i + 1$ appears at most $2(k - j)$ times. Thus we see (with $j = k$) that no part appears more than $2k$ times.

We now proceed by the technique developed in [1] and [2]. If

$$C_{k,i}(x; q) = \sum_{\mu=0}^{\infty} (-1)^{\mu} x^{k\mu} q^{1/2(2k+1)\mu(\mu+1)-i\mu}(1 - x^i q^{(2\mu+1)i}) \times \frac{(1 - xq) \cdots (1 - xq^{\mu})}{(1 - q) \cdots (1 - q^{\mu})}, \tag{1}$$

then [5, p. 4]

$$C_{k,i}(x; q) - C_{k,i-1}(x; q) = x^{i-1}q^{i-1}(1 - xq)C_{k,k-i+1}(xq; q), \tag{2}$$

and [5, p. 4]

$$C_{k,-i}(x; q) = - x^{-i}q^{-i}C_{k,i}(x; q) \tag{3}$$

Consequently, if we define

$$R_{k,i}(x) = C_{k,i+1/2}(x^2; q^2) \prod_{j=1}^{\infty} (1 - xq^j)^{-1},$$

then for $0 \leq i \leq k$, (2) implies

$$R_{k,i}(x) - R_{k,i-1}(x) = x^{2i-1}q^{2i-1}(1 + xq)R_{k,k-i}(xq), \tag{4}$$

and (2) and (3) imply

$$R_{k,0}(x) = R_{k,k}(xq). \tag{5}$$

We many expand $R_{k,i}(x)$ as follows

$$R_{k,i}(x) = \sum_{N=-\infty}^{\infty} \sum_{M=-\infty}^{\infty} c_{k,i}(M, N)x^M q^N, \qquad |x| \leq 1, \ |q| < 1. \tag{6}$$

Then by means of (4), (5), (6), and the definition of $R_{k,i}(x)$ we easily verify that

$$c_{k,i}(M, N) = \begin{cases} 1 & \text{if } M = N = 0, \\ 0 & \text{if either } M \leq 0 \text{ or } N \leq 0, \text{ and } M^2 + N^2 \neq 0, \end{cases} \tag{7}$$

$$c_{k,0}(M, N) = c_{k,k}(M, N - M), \tag{8}$$

$$\begin{aligned} c_{k,i}(M, N) - c_{k,i-1}(M, N) = c_{k,k-i}(M - 2i + 1, N - M) \\ + c_{k,k-i}(M - 2i, N - M), \qquad 0 < i \leq k. \end{aligned} \tag{9}$$

One easily verifies by mathematical induction that the $c_{k,i}(M, N)$ for $0 \leq i \leq k$ are uniquely determined by (7), (8), and (9).

Let $p_{k,i}(M, N)$ denote the number of partitions of N into M parts of the form $N = \Sigma_{j=1}^{\infty} f_j \cdot j$ with $f_1 \leq 2i$ and

$$[\tfrac{1}{2}(f_j + 1)] + [\tfrac{1}{2}(f_{j+1} + 1)] \leq k.$$

We wish to show that $p_{k,i}(M, N)$ satisfies (7), (8), and (9). Now (7) is by definition.

As for (8), let us consider any partition enumerated by $p_{k,0}(M, N)$. Since 1 does not appear, every summand is ≥ 2. Subtracting 1 from every summand, we obtain a partition of $N - M$ into M parts with 1 appearing at most $2k$ times and again

$$[\tfrac{1}{2}(f_j + 1)] + [\tfrac{1}{2}(f_{j+1} + 1)] \leq k.$$

Thus we have a partition of the type enumerated by $p_{k,k}(M, N - M)$. The above procedure establishes a one-to-one correspondence between the partitions enumerated by $p_{k,k}(M, N - M)$ and the partitions enumerated by $p_{k,0}(M, N)$. Hence

$$p_{k,0}(M, N) = p_{k,k}(M, N - M).$$

Finally we treat (9). We note that $p_{k,i}(M, N) - p_{k,i-1}(M, N)$ enumerates the number of partitions of N into M parts of the form $N = \Sigma_{j=1}^{\infty} f_j \cdot j$ with

$$f_1 = 2i - 1 \text{ or } 2i \qquad \text{and} \qquad [\tfrac{1}{2}(f_j + 1)] + [\tfrac{1}{2}(f_{j+1} + 1)] \leq k.$$

In case $f_1 = 2i - 1$, we see that $f_2 \leq 2(k - i)$; subtracting 1 from every summand, we obtain a partition of $N - M$ into $M - 2i + 1$ parts with 1 appearing at most $2(k - i)$ times and

$$[\tfrac{1}{2}(f_j + 1)] + [\tfrac{1}{2}(f_{j+1} + 1)] \leq k.$$

Thus we have a partition of the type enumerated by $p_{k,k-i}(M - 2i + 1, N - M)$. In case $f_1 = 2i$, we see that $f_2 \leq 2(k - i)$; subtracting 1 from every summand, we obtain a partition of $N - M$ into $M - 2i$ parts with 1 appearing at most $2(k - i)$ times and

$$[\tfrac{1}{2}(f_j + 1)] + [\tfrac{1}{2}(f_{j+1} + 1)] \leq k.$$

Thus we have a partition of the type enumerated by $p_{k,k-i}(M - 2i, N - M)$. The above procedure establishes a one-to-one correspondence between the partitions enumerated by

$$p_{k,i}(M, N) - p_{k,i-1}(M, N)$$

and the partitions enumerated by

$$p_{k,k-i}(M - 2i + 1, N - M) + p_{k,k-i}(M - 2i, N - M).$$

Hence

$$p_{k,i}(M, N) - p_{k,i-1}(M, N) = p_{k,k-i}(M - 2i + 1, N - M) + p_{k,k-i}(M - 2i, N - M).$$

Thus by the comment following (9),

$$c_{k,i}(M, N) = p_{k,i}(M, N). \tag{10}$$

Thus for $0 \leq a < k$

$$\begin{aligned}\sum_{N=0}^{\infty} A_{k,a}(N)q^N &= \prod_{\substack{n=1 \\ n \not\equiv 0, \pm(2a+1) \pmod{4k+2}}}^{\infty} (1 - q^n)^{-1} \\ &= R_{k,a}(1) \\ &= \sum_{N=-\infty}^{\infty} \sum_{M=-\infty}^{\infty} p_{k,a}(M, N)q^N \\ &= \sum_{N=0}^{\infty} B_{k,a}(N)q^N,\end{aligned}$$

where the second equation follows from Jacobi's identity [3, p. 283].

Therefore

$$A_{k,a}(N) = B_{k,a}(N).$$

This concludes the proof of Theorem 2.

References

1. G. E. Andrews, An Analytic Proof of the Rogers-Ramanujan-Gordon Identities, *Amer. J. Math.* **88** (1966) 844–846.

2. G. E. Andrews, Partition Theorems Related to the Rogers-Ramanujan Identities, J. Combinatorial Theory **2** (1967), 437–465 (following paper).

3. G. H. Hardy and E. M. Wright, *An Introduction to the Theory of Numbers*, 4-th ed., Oxford University Press, London, 1960.

4. P. A. Mac Mahon, *Combinatory Analysis*, Vol. 2, Cambridge Univ. Press, 1916, reprinted, Dover, New York.

5. A. Segberg, Über einige arithmetische Identitäten, *Avh. Norske Vid.-Akad. Oslo* 1936, No. 8.

3.3 A Generalization of the Classical Partition Theorems

TRANSACTIONS OF THE
AMERICAN MATHEMATICAL SOCIETY
Volume 145, November 1969

A GENERALIZATION OF THE CLASSICAL PARTITION THEOREMS

BY
GEORGE E. ANDREWS(1)

1. **Introduction.** Let $P_d(n)$ denote the number of partitions of n into positive integers of the form $n=b_1+\cdots+b_s$ with $b_i-b_{i+1}\geqq d$ where strict inequality holds if $d|b_i$. $P_1(n)$ appears in the first Rogers-Ramanujan identity [5, p. 291]. $P_2(n)$ appears in the first of the Göllnitz-Gordon partition theorems [2, p. 945], and $P_3(n)$ appears in a partition theorem of I. J. Schur [6, p. 489].

For $d>3$, we have the following result due to H. L. Alder [1, p. 713].

(1.1) Let S be any fixed set of positive integers. Then $P_d(n)$ is not identically equal to the number of partitions of n into parts taken from S if $d>3$.

In a different approach, B. Gordon [4, p. 394] has made the following generalization of the Rogers-Ramanujan identities.

(1.2) Let a and k be integers with $0<a\leqq k$. Let $A_{k,a}(n)$ denote the number of partitions of n into parts $\not\equiv 0, \pm a \pmod{2k+1}$. Let $B_{k,a}(n)$ denote the number of partitions of n of the form $n=b_1+\cdots+b_s$ with $b_i\geqq b_{i+1}$, $b_i-b_{i+k-1}\geqq 2$, and with 1 appearing as a summand at most $a-1$ times. Then $A_{k,a}(n)=B_{k,a}(n)$.

The Göllnitz-Gordon identities have also been generalized in this manner [2, p. 946].

(1.3) Let a and k be integers with $0<a\leqq k$. Let $A^{\#}_{k,a}(n)$ denote the number of partitions of n into parts which are neither $\equiv 2 \pmod 4$ nor $\equiv 0, \pm(2a-1) \pmod{4k}$. Let $B^{\#}_{k,a}(n)$ denote the number of partitions of n of the form $n=b_1+\cdots+b_s$, with $b_i\geqq b_{i+1}$, no odd parts are repeated, $b_i-b_{i+k-1}\geqq 2$ with strict inequality if $2|b_i$ and at most $a-1$ parts not exceeding 2 appear. Then $A^{\#}_{k,a}(n)=B^{\#}_{k,a}(n)$.

This leads us to define the following partition function. Let $B_{\lambda,k,a}(n)$ denote the number of partitions of n of the form $n=b_1+\cdots+b_s$, with $b_i\geqq b_{i+1}$, no parts $\not\equiv 0 \pmod{\lambda+1}$ are repeated, $b_i-b_{i+k-1}\geqq \lambda+1$ with strict inequality if $\lambda+1|b_i$ and at most $a-1$ parts not exceeding $\lambda+1$ appear. We note that $B_{0,k,a}(n)$ appears in (1.2); $B_{1,k,a}(n)$ appears in (1.3), and $B_{d-1,2,2}(n)=P_d(n)$.

QUESTION. *If k is sufficiently large with respect to λ, does $B_{\lambda,k,a}(n)$ appear in a partition theorem of the Rogers-Ramanujan type?*

We answer this question affirmatively with the following result.

THEOREM 1. *Let $0\leqq\lambda\leqq a\leqq k$ be integers, and $2\lambda-1\leqq k$. If λ is even, define*

Received by the editors December 6, 1967.

(1) Partially supported by National Science Foundation Grant GP-6663.

$A_{\lambda,k,a}(n)$ to be the number of partitions of n into parts such that no part $\not\equiv 0 \pmod{\lambda+1}$ may be repeated and no part is $\equiv 0$, $\pm(a-\frac{1}{2}\lambda)(\lambda+1) \pmod{(2k-\lambda+1)(\lambda+1)}$. If λ is odd, define $A_{\lambda,k,a}(n)$ to be the number of partitions of n into parts such that no part $\not\equiv 0 \pmod{\frac{1}{2}(\lambda+1)}$ may be repeated, no part is $\equiv \lambda+1 \pmod{2\lambda+2}$, and no part is $\equiv 0$, $\pm(2a-\lambda)\frac{1}{2}(\lambda+1) \pmod{(2k-\lambda+1)(\lambda+1)}$. Then $A_{\lambda,k,a}(n)=B_{\lambda,k,a}(n)$.

There is numerical evidence to suggest that the condition "$k \geqq 2\lambda-1$" is unnecessary; but, the proof given here strongly relies on this condition. Actually the case $k=\lambda=a=2$ violates this condition; however, Theorem 1 becomes Schur's theorem in this case.

In §2, we shall develop results concerning auxiliary partition functions. In §3, we shall prove Theorem 1; our work will be based on the generalized well-poised basic hypergeometric functions studied in [3]. In §4, we prove a more general theorem than Theorem 1 in which the parameter a is allowed to take values in the interval $(\frac{1}{2}\lambda, k]$. In §5, we discuss possibilities for further work.

2. **Auxiliary partition functions.** Let $\pi(k, \lambda, t, j, \mu; n)$ denote the number of partitions of n of the form $n=\sum_{i=1}^{2\lambda+1} f_i \cdot i$ (here f_i is the number of times the summand i appears) where

(2.1) $f_1+\cdots+f_{\lambda+1}=k-\lambda+j$;

(2.2) $f_{\lambda+1}+\cdots+f_{2\lambda+1}=k-t$;

(2.3) $f_m+\cdots+f_{m+\lambda}\geqq k$ for some m;

(2.4) $f_a>1$ implies $a=\lambda+1$;

(2.5) $f_1+\cdots+f_{2\lambda+1}=k+\mu$.

Throughout our discussion we shall assume k, λ, t, j, and μ are all nonnegative integers with $j\leqq\lambda$, $k\geqq\max(1, 2\lambda-1)$, and $\pi(k, \lambda, t, j, \mu; 0)=0$.

Let $g(k, \lambda, t, j, \mu; q)=\sum_{n=0}^{\infty} \pi(k, \lambda, t, j, \mu; n)q^n$.

LEMMA 2.1.

$$g(k, \lambda, t, j, \mu; q) = q^{(k-t)(\lambda+1)}\sigma_\mu(\lambda)\sigma_{\lambda+\mu+t-j}(\lambda), \tag{2.6}$$

where

$$\begin{aligned}\sigma_i(\lambda) &= q^{i(i+1)/2}\begin{bmatrix}\lambda\\ i\end{bmatrix},\\ &= q^{i(i+1)/2}\frac{(q^\lambda-1)\cdots(q^{\lambda-i+1}-1)}{(q^i-1)\cdots(q-1)}, \qquad 0\leqq i\leqq\lambda,\\ &= 0, \qquad \textit{otherwise};\end{aligned}$$

thus $\sigma_i(\lambda)$ is just the generating function for partitions into exactly i distinct parts with all parts $\leqq\lambda$.

Proof. We proceed by induction on the second argument of g, namely λ. If $\lambda=0$, then since $j\leqq\lambda$, $j=0$. The conditions (2.1)–(2.5) imply $f_1=k+j=k-t$

$=k+\mu\geqq k$; thus $j=t=\mu=0$. Thus the only allowable partition is of k into k "ones". Therefore

$$g(k, 0, t, j, \mu; q) = q^k \quad \text{if } \mu = j = t = 0, \\ = 0 \quad \text{otherwise.}$$

On the other hand, consider the right-hand side of (2.6); call it $G(k, 0, t, j, \mu; q)$. The only way $G(k, 0, t, j, \mu; q)$ will be nonzero is if $\mu=0$, $t=j$, and since $j\leqq\lambda$ we have $t=j=0$. Hence

$$G(k, 0, t, j, \mu; q) = q^k \quad \text{if } \mu = j = t = 0, \\ = 0 \quad \text{otherwise.}$$

Thus the lemma holds for $\lambda=0$.

If $\mu=0$, there are just k parts appearing in the partition by (2.5), and since (2.3) must hold we see that there exists a particular m such that

$$f_m+\cdots+f_{m+\lambda} = k, \qquad f_i = 0 \quad \text{for } i < m \text{ or } i > m+\lambda.$$

If we set $A=f_1+\cdots+f_\lambda$, $B=f_{\lambda+2}+\cdots+f_{2\lambda+1}$, then (2.1), (2.2), and (2.5) imply $A=t+\mu$, $f_{\lambda+1}=k+j-\lambda-t-\mu$, $B=\lambda+\mu-j$; thus in case $\mu=0$, $A=t$, $f_{\lambda+1}=k+j-\lambda-t$, $B=\lambda-j$. Let us now consider any partition of the type described in this paragraph. If we subtract $\lambda+1$ from each of the $\lambda-j$ parts greater than $\lambda+1$, we get a partition with $\lambda+1$ appearing $k+j-\lambda-t$ times with $\lambda+t-j$ remaining parts which are distinct and are all $\leqq\lambda$. Conversely if we have a given partition in which $\lambda+1$ appears $k+j-\lambda-t$ times and there are $\lambda+t-j$ remaining parts which are distinct and are $\leqq\lambda$, we define m to be the integer such that $f_1+\cdots+f_{m-1}=\lambda-j$, and we add $\lambda+1$ to each part $<m$; this gives us precisely the type of partition described at the beginning of this paragraph. Hence by the definition of $\sigma_j(\lambda)$,

$$\begin{aligned} g(k, \lambda, t, j, 0; q) &= q^{(k+j-\lambda-t)(\lambda+1)+(\lambda-j)(\lambda+1)}\sigma_{\lambda+t-j}(\lambda) \\ &= q^{(k-t)(\lambda+1)}\sigma_{\lambda+t-j}(\lambda) \\ &= G(k, \lambda, t, j, 0; q), \end{aligned}$$

and (2.6) is verified in this case.

We now proceed by induction on λ (>0), and we may also assume $\mu>0$. Our first task is to verify

$$\begin{aligned} \pi(k, \lambda, t, j, \mu; n) = {} & \pi(k, \lambda-1, t, j-1, \mu; n+j-k-2\mu-\lambda) \\ & +\pi(k, \lambda-1, t, j, \mu; n+j-k-2\mu-\lambda) \\ & +\pi(k, \lambda-1, t, j-2, \mu-1; n+j-k-2\mu-\lambda) \\ & +\pi(k, \lambda-1, t, j-1, \mu-1; n+j-k-2\mu-\lambda). \end{aligned} \tag{2.7}$$

To verify (2.7) we proceed as follows. Let us consider a partition of n of the type enumerated by $\pi(k, \lambda, t, j, \mu; n)$. Let us subtract 1 from every summand $\leqq\lambda+1$, and let us subtract 2 from each summand $>\lambda+1$. The number being partitioned now is $n-(k+j-\lambda)-2(\lambda+\mu-j)=n+j-k-2\mu-\lambda$. Four cases are to be considered depending on: (1) $f_1=f_{\lambda+2}=0$; (2) $f_1=0$, $f_{\lambda+2}=1$; (3) $f_1=1$, $f_{\lambda+2}=0$; (4) $f_1=f_{\lambda+2}=1$.

We let f_i' denote the number of appearances of i in the transformed partition. Thus for example in case (1), conditions (2.1)–(2.5) now read

$(2.1)_1$ $f_1'+\cdots+f_\lambda'=k-(\lambda-1)+j-1$;

$(2.2)_1$ $f_\lambda'+\cdots+f_{2\lambda-1}'=k-t$;

$(2.3)_1$ $f_m'+\cdots+f_{m+\lambda-1}'\geqq k$ for some m;

$(2.4)_1$ $f_a'>1$ implies $a=\lambda$;

$(2.5)_1$ $f_1'+\cdots+f_{2\lambda-1}'=k+\mu$.

Thus we have a partition of the type enumerated by

$$\pi(k,\lambda-1,t,j-1,\mu;n+j-k-2\mu-\lambda).$$

Indeed the above procedure establishes a one-to-one correspondence between those partitions enumerated by $\pi(k,\lambda,t,j,\mu;n)$ which fall in case (1) and the partitions enumerated by $\pi(k,\lambda-1,t,j-1,\mu;n+j-k-2\mu-\lambda)$. Similarly in the other three cases we get the other three partition functions appearing on the right-hand side of equation (2.7). Thus (2.7) is established.

(2.7) implies

$$\begin{aligned}g(k,\lambda,t,j,\mu;q) = q^{k+\lambda+2\mu-j}(&g(k,\lambda-1,t,j-1,\mu;q)+g(k,\lambda-1,t,j,\mu;q)\\&+g(k,\lambda-1,t,j-2,\mu-1;q)\\&+g(k,\lambda-1,t,j-1,\mu-1;q)).\end{aligned}\tag{2.8}$$

Let us now examine the right-hand side of (2.6) which we have called $G(k,\lambda,t,j,\mu;q)$. Now

$$\sigma_j(\lambda) = q^j\sigma_j(\lambda-1)+q^j\sigma_{j-1}(\lambda-1);\tag{2.9}$$

this is merely a restatement of the well-known formula for Gaussian polynomials

$$\begin{bmatrix}\lambda\\ j\end{bmatrix} = \begin{bmatrix}\lambda-1\\ j-1\end{bmatrix}+q^j\begin{bmatrix}\lambda-1\\ j\end{bmatrix}.\tag{2.10}$$

Consequently

$$\begin{aligned}G(k,\lambda,t,j,\mu;q) &= q^{(k-t)(\lambda+1)}\sigma_\mu(\lambda)\sigma_{\lambda+\mu+t-j}(\lambda)\\&= q^{k-t+\mu+\lambda+\mu+t-j}q^{(k-t)\lambda}(\sigma_\mu(\lambda-1)+\sigma_{\mu-1}(\lambda-1))\\&\quad\times(\sigma_{\lambda+\mu+t-j}(\lambda-1)+\sigma_{\lambda+\mu+t-j-1}(\lambda-1))\\&= q^{k+\lambda+2\mu-j}(G(k,\lambda-1,t,j-1,\mu;q)+G(k,\lambda-1,t,j,\mu;q)\\&\qquad+G(k,\lambda-1,t,j-2,\mu-1;q)\\&\qquad+G(k,\lambda-1,t,j-1,\mu-1;q)).\end{aligned}\tag{2.11}$$

Now we know that $g(k,\lambda,t,j,\mu;q)$ and $G(k,\lambda,t,j,\mu;q)$ are identical if either λ or μ is zero. Thus since they both satisfy the recurrence (2.8), we have by induction on λ that they are equal in general. Thus Lemma 2.1 is established.

LEMMA 2.2. *Let $\pi_1(k, \lambda, t, j; m; n)$ denote the number of partitions of n into m parts of the form $n=\sum_{i=1}^{2\lambda+1} f_i \cdot i$, where*

(2.12) $f_1+\cdots+f_{\lambda+1} \leqq k-\lambda+j$;

(2.13) $f_{\lambda+1}+\cdots+f_{2\lambda+1}=k-t$;

(2.14) $f_v+\cdots+f_{v+\lambda} \geqq k$ *for some* v;

(2.15) $f_a>1$ *implies* $a=\lambda+1$.

Let

$$g_1(k, \lambda, t, j; x; q) = \sum_{m=1}^{\infty} \sum_{n=1}^{\infty} x^m q^n \pi_1(k, \lambda, t, j; m; n).$$

Then

$$g_1(k, \lambda, t, j; x; q) = x^k q^{(k-t)(\lambda+1)} \sum_{\mu=0}^{\lambda} x^{\mu} \sigma_{\mu}(\lambda) \sum_{r=0}^{j-\mu-t} \sigma_{\lambda-r}(\lambda).$$

Proof. By the definitions of π and π_1, it is clear that

$$\pi_1(k, \lambda, t, j; k+\mu; n) - \pi_1(k, \lambda, t, j-1; k+\mu; n) = \pi(k, \lambda, t, j, \mu; n).$$

Hence

$$\begin{aligned} g_1(k, \lambda, t, j; x; q) - g_1(k, \lambda, t, j-1; x; q) &= \sum_{\mu=0}^{\infty} x^{k+\mu} g(k, \lambda, t, j, \mu; q) \\ &= \sum_{\mu=0}^{\infty} x^{k+\mu} q^{(k-t)(\lambda+1)} \sigma_{\mu}(\lambda) \sigma_{\lambda+\mu+t-j}(\lambda). \end{aligned}$$

Therefore

$$\begin{aligned} g_1(k, \lambda, t, j; x; q) &= \sum_{r=\lambda-k}^{j} \sum_{\mu=0}^{\infty} x^{k+\mu} q^{(k-t)(\lambda+1)} \sigma_{\mu}(\lambda) \sigma_{\lambda+\mu+t-r}(\lambda) \\ &= x^k q^{(k-t)(\lambda+1)} \sum_{\mu=0}^{\lambda} x^{\mu} \sigma_{\mu}(\lambda) \sum_{r=\mu+t}^{j} \sigma_{\lambda+\mu+t-r}(\lambda) \\ &= x^k q^{(k-t)(\lambda+1)} \sum_{\mu=0}^{\lambda} x^{\mu} \sigma_{\mu}(\lambda) \sum_{r=0}^{j-\mu-t} \sigma_{\lambda-r}(\lambda). \end{aligned}$$

To conclude this section we prove a result concerning the multiplication of generating functions.

LEMMA 2.3. *Let $p_1(A; c; m, n)$ denote the number of partitions of n into m parts all $<c$ where the summands are subject to a set of conditions, A. Let $p_2(B; c; m, n)$ denote the number of partitions of n into m parts all $\geqq c$ where the summands are subject to a set of conditions, B. Let*

$$f_1(x, q) = \sum_{m=0}^{\infty} \sum_{n=0}^{\infty} p_1(A; c; m, n) x^m q^n,$$

$$f_2(x, q) = \sum_{m=0}^{\infty} \sum_{n=0}^{\infty} p_2(B; c; m, n) x^m q^n,$$

where we assume that $f_i(x, q)$ is absolutely convergent for $|x|<b_i$, $|q|<1$.

Then for $|x| < \min(b_1, b_2)$, $|q| < 1$, *if we let*

$$f_1(x,q)f_2(x,q) = \sum_{m=0}^{\infty}\sum_{n=0}^{\infty} p_3(A, B; c; m, n)x^m q^n,$$

then $p_3(A, B; c; m, n)$ *is the number of partitions of n into m parts such that the summands* $<c$ *are subject to condition A and the summands* $\geqq c$ *are subject to condition B.*

Proof The result follows directly from

$$p_3(A, B; c; m, n) = \sum_{m_1+m_2=m;\, n_1+n_2=n} p_1(A; c; m_1, n_1)p_2(B; c; m_2, n_2).$$

3. **Proof of Theorem 1.** We begin by studing certain q-series considered in [3]. As in §2, we require that $k \geqq \max(1, 2\lambda-1)$. We define

$$J_{\lambda,k,i}(x) = J_{\lambda,k,i}(x;q) = J_{k,i}(-q, -q^2, \ldots, -q^\lambda; x; q^{\lambda+1}); \tag{3.1}$$

$$H_{\lambda,k,i}(x) = H_{\lambda,k,i}(x;q) = H_{k,i}(-q, -q^2, \ldots, -q^\lambda; xq^{\lambda+1}; q^{\lambda+1}), \tag{3.2}$$

where (as in [3])

$$\begin{aligned} &J_{k,i}(a_1, \ldots, a_\lambda; x; q) \\ &= \sum_{n=0}^{\infty} (-1)^{n(\lambda+1)} x^{kn} (a_1 a_2 \cdots a_\lambda)^{-n} q^{(2k-\lambda+1)n^2/2+(\lambda+1)n/2+(k-i)n} \\ &\cdot \prod_{r=1}^{n}(1-q^r)^{-1} \prod_{u=n+1}^{\infty}(1-xq^u)^{-1} \prod_{s=1}^{\lambda}\left(\prod_{v=0}^{n-1}(1-a_s q^v)\prod_{w=n+1}^{\infty}\left(1-\frac{xq^w}{a_s}\right)\right) \\ &\cdot\left\{1+\frac{(-1)^{\lambda+1}x^i q^{(2n+1)i-\lambda n}(1-a_1q^n)(1-a_2q^n)\cdots(1-a_\lambda q^n)}{a_1a_2\cdots a_\lambda(1-xq^{n+1}a_1^{-1})(1-xq^{n+1}a_2^{-1})\cdots(1-xq^{n+1}a_\lambda^{-1})}\right\}; \end{aligned} \tag{3.3}$$

$$\begin{aligned} &H_{k,i}(a_1, \ldots, a_\lambda; x; q) \\ &= \sum_{n=0}^{\infty} (-1)^{n(\lambda+1)} x^{kn} (a_1 a_2 \cdots a_\lambda)^{-n} q^{(2k-\lambda+1)n^2/2+(\lambda+1)n/2-in}(1-x^i q^{2ni}) \\ &\cdot \prod_{r=1}^{n}(1-q^r)^{-1} \prod_{u=n}^{\infty}(1-xq^u)^{-1} \prod_{s=1}^{\lambda}\left(\prod_{v=0}^{n-1}(1-a_s q^v)\prod_{w=n+1}^{\infty}\left(1-\frac{xq^w}{a_s}\right)\right). \end{aligned} \tag{3.4}$$

Then from [3, equation (2.1)], we have

$$H_{\lambda,k,i}(x) - H_{\lambda,k,i-1}(x) = (xq^{\lambda+1})^{i-1} J_{\lambda,k,k-i+1}(xq^{\lambda+1}); \tag{3.5}$$

from [3, equation (2.2)], we have

$$J_{\lambda,k,i}(x) = \sum_{j=0}^{\lambda} x^j \sigma_j(\lambda) H_{\lambda,k,i-j}(x); \tag{3.6}$$

from [3, equation (2.3)], we have

$$H_{\lambda,k,-i}(x) = -(xq^{\lambda+1})^{-i} H_{\lambda,k,i}(x), \tag{3.7}$$

and from [3, equation (2.8)], we have

$$H_{\lambda,k,1}(x) = J_{\lambda,k,k}(xq^{\lambda+1}) = J_{\lambda,k,k+1}(xq^{\lambda+1}). \tag{3.8}$$

LEMMA 3.1. *Let $|q|<1$, and let $H^*_{\lambda,k,i}(x)$ $(0 \leqq i \leqq k)$ and $J^*_{\lambda,k,i}(x)$ $(1 \leqq i \leqq k)$ be any collection of functions analytic in x in the neighborhood of zero which satisfy*

$$H^*_{\lambda,k,i}(x) - H^*_{\lambda,k,i-1}(x) = (xq^{\lambda+1})^{i-1} J^*_{\lambda,k,k-i+1}(xq^{\lambda+1}), \qquad 1 \leqq i \leqq k; \tag{3.9}$$

$$H^*_{\lambda,k,0}(x) = 0; \tag{3.10}$$

$$\begin{aligned} J^*_{\lambda,k,i}(x) = & \sum_{j=0}^{\min(i,\lambda)} x^j \sigma_j(\lambda) H^*_{\lambda,k,i-j}(x) \\ & - x^i \sum_{j=\min(i,\lambda)+1}^{\lambda} q^{(\lambda+1)(i-j)} \sigma_j(\lambda) H^*_{\lambda,k,j-i}(x); \end{aligned} \tag{3.11}$$

$$H^*_{\lambda,k,i}(0) = J^*_{\lambda,k,i}(0) = 1, \qquad 1 \leqq i \leqq k. \tag{3.12}$$

Then $H^*_{\lambda,k,i}(x) = H_{\lambda,k,i}(x)$, and $J^*_{\lambda,k,i}(x) = J_{\lambda,k,i}(x)$.

Proof. We need only show that solutions of the above equations are unique since (3.3)–(3.8) imply that the $H_{\lambda,k,i}(x)$ and $J_{\lambda,k,i}(x)$ fulfill the conditions of the lemma. If we can show that the $H^*_{\lambda,k,i}(x)$ are uniquely determined, then since the $J^*_{\lambda,k,i}(x)$ are defined in terms of the $H^*_{\lambda,k,i}(x)$ by (3.11), the $J^*_{\lambda,k,i}(x)$ will be uniquely determined also.

From (3.9) and (3.11) we have

$$\begin{aligned} & H^*_{\lambda,k,i}(x) - H^*_{\lambda,k,i-1}(x) \\ & = (xq^{\lambda+1})^{i-1} \sum_{j=0}^{\min(k-i+1,\lambda)} x^j q^{j(\lambda+1)} \sigma_j(\lambda) H^*_{\lambda,k,k-i+1-j}(xq^{\lambda+1}) \\ & \quad - (xq^{\lambda+1})^{k-i+1} \sum_{j=\min(k-i+1,\lambda)+1}^{\lambda} q^{(\lambda+1)(k-i+1-j)} \sigma_j(\lambda) H^*_{\lambda,k,j-k+i-1}(xq^{\lambda+1}). \end{aligned} \tag{3.13}$$

Hence if $H^*_{\lambda,k,i}(x) = \sum_{n=0}^{\infty} \eta_n(i) x^n$, then by (3.10) and (3.12)

$$\begin{aligned} \eta_0(i) &= 1, \quad 1 \leqq i \leqq k, \\ &= 0, \quad i = 0. \end{aligned}$$

From (3.13),

$$\begin{aligned} \eta_n(1) - \eta_n(k) q^{n(\lambda+1)} &= \text{polynomial in } \eta_j\text{'s with } j < n, \\ \eta_n(2) - \eta_n(1) &= \text{polynomial in } \eta_j\text{'s with } j < n, \\ &\vdots \\ \eta_n(k) - \eta_n(k-1) &= \text{polynomial in } \eta_j\text{'s with } j < n. \end{aligned}$$

We consider the above equations as a system of k equations in the k unknowns $\eta_n(1), \eta_n(2), \ldots, \eta_n(k)$. The determinant of the system is $1-q^{n(\lambda+1)} \neq 0$ since $|q|<1$. Hence the $\eta_n(i)$ are uniquely determined by the η_j's with $j<n$. Hence by mathematical induction the $\eta_n(i)$ are unique, and therefore the $H^*_{\lambda,k,i}(x)$ are unique. Thus Lemma 3.1 is proved.

LEMMA 3.2. *Suppose* $|q|<1$, $k \geqq \max(1, 2\lambda-1)$. *Let* $H^{\#}_{\lambda,k,i}(x)$, $0 \leqq i \leqq k$ *and* $J^{\#}_{\lambda,k,i}(x)$, $1 \leqq i \leqq k$ *be any collection of functions analytic in x in the neighborhood of zero which satisfy*

$$H^{\#}_{\lambda,k,i}(x) - H^{\#}_{\lambda,k,i-1}(x) = (xq^{\lambda+1})^{i-1} J^{\#}_{\lambda,k,k-i+1}(xq^{\lambda+1}), \qquad 1 \leqq i \leqq k; \tag{3.14}$$

$$H^{\#}_{\lambda,k,0}(x) = 0; \tag{3.15}$$

$$J^{\#}_{\lambda,k,i}(x) = \sum_{j=0}^{\min(i,\lambda)} x^j \sigma_j(\lambda) H^{\#}_{\lambda,k,i-j}(x), \qquad 1 \leqq i \leqq k-\lambda+1; \tag{3.16}$$

$$\begin{aligned} J^{\#}_{\lambda,k,k-\lambda+1+j}(x) &= \sum_{r=0}^{\lambda} x^r \sigma_r(\lambda) H^{\#}_{\lambda,k,k-\lambda+1+j-r}(x) \\ &\quad - \sum_{r=0}^{j} g_1(k, \lambda, r, j; x; q) H^{\#}_{\lambda,k,r}(xq^{\lambda+1}), \qquad 1 \leqq j < \lambda, \end{aligned} \tag{3.17}$$

where $g_1(k, \lambda, r, j; x; q)$ *is defined in Lemma* 2.2.

$$H^{\#}_{\lambda,k,i}(0) = 1, \quad 1 \leqq i \leqq k; \qquad J^{\#}_{\lambda,k,i}(0) = 1, \quad 1 \leqq i \leqq k. \tag{3.18}$$

Then if we define

$$\begin{aligned} H^*_{\lambda,k,i}(x) &= H^{\#}_{\lambda,k,i}(x), \qquad 0 \leqq i \leqq k-\lambda+1, \\ &= H^{\#}_{\lambda,k,i}(x) - \sum_{r=1}^{j} H^{\#}_{\lambda,k,r}(xq^{\lambda+1}) x^k q^{(\lambda+1)(k-r)} \sum_{m=0}^{j-r} \sigma_{\lambda-m}(\lambda), \\ &\qquad i = k-\lambda+1+j,\ 1 \leqq j < \lambda; \end{aligned} \tag{3.19}$$

$$J^*_{\lambda,k,i}(x) = x^{-k+i}(H^*_{\lambda,k,k-i+1}(xq^{-\lambda-1}) - H^*_{\lambda,k,k-i}(xq^{-\lambda-1})), \tag{3.20}$$

then $H^*_{\lambda,k,i}(x)$ *and* $J^*_{\lambda,k,i}(x)$ *fulfill the equations of Lemma* 3.1 *and for* $\lambda \leqq i \leqq k$, $J^*_{\lambda,k,i}(x) = J^{\#}_{\lambda,k,i}(x)$.

Proof. Obviously (3.9) and (3.10) are fulfilled from (3.20) and (3.15). Clearly for $\lambda \leqq i \leqq k$,

$$\begin{aligned} J^*_{\lambda,k,i}(x) &= x^{-k+i}(H^*_{\lambda,k,k-i+1}(xq^{-\lambda-1}) - H^*_{\lambda,k,k-i}(xq^{-\lambda-1})) \\ &= x^{-k+i}(H^{\#}_{\lambda,k,k-i+1}(xq^{-\lambda-1}) - H^{\#}_{\lambda,k,k-i}(xq^{-\lambda-1})) \\ &= J^{\#}_{\lambda,k,i}(x). \end{aligned}$$

Now if $\lambda \leqq i \leqq k-\lambda+1$, (3.19) and (3.16) show that (3.11) is valid. If $i=k-\lambda+1+j$, $1 \leqq j < \lambda$, then

$$\begin{aligned}
J^*_{\lambda,k,k-\lambda+1+j}(x) &= J^\#_{\lambda,k,k-\lambda+1+j}(x) \\
&= \sum_{r=0}^{j-1} x^r\sigma_r(\lambda)H^\#_{\lambda,k,k-\lambda+1+j-r}(x)+\sum_{r=j}^{\lambda} x^r\sigma_r(\lambda)H^\#_{\lambda,k,k-\lambda+1+j-r}(x) \\
&\quad -\sum_{r=0}^{j} g_1(k,\lambda,r,j;x;q)H^\#_{\lambda,k,r}(xq^{\lambda+1}) \\
&= \sum_{r=0}^{j-1} x^r\sigma_r(\lambda)\Bigg(H^*_{\lambda,k,k-\lambda+1+j-r}(x)+\sum_{v=1}^{j-r} H^*_{\lambda,k,v}(xq^{\lambda+1})(xq^{\lambda+1})^k q^{-v(\lambda+1)} \\
&\qquad\qquad \sum_{w=0}^{j-r-v} \sigma_{\lambda-w}(\lambda)\Bigg) \\
&\quad +\sum_{r=j}^{\lambda} x^r\sigma_r(\lambda)H^*_{\lambda,k,k-\lambda+1+j-r}(x) \\
&\quad -\sum_{r=0}^{j} g_1(k,\lambda,r,j;x;q)H^*_{\lambda,k,r}(xq^{\lambda+1}) \\
&= \sum_{r=0}^{\lambda} x^r\sigma_r(\lambda)H^*_{\lambda,k,k-\lambda+1+j-r}(x) \\
&\quad +\sum_{v=1}^{j} H^*_{\lambda,k,r}(xq^{\lambda+1})x^k q^{(k-v)(\lambda+1)} \sum_{r=0}^{j-v} x^r\sigma_r(\lambda) \sum_{w=0}^{j-r-v} \sigma_{\lambda-w}(\lambda) \\
&\quad -\sum_{r=0}^{j} g_1(k,\lambda,r,j;x;q)H^*_{\lambda,k,r}(xq^{\lambda+1}) \\
&= \sum_{r=0}^{\lambda} x^r\sigma_r(\lambda)H^*_{\lambda,k,k-\lambda+1+j-r}(x),
\end{aligned}$$

where the last equation follows by Lemma 2.2; thus again we have (3.11).

Finally, if $1 \leqq i < \lambda$, then

$$\begin{aligned}
J^*_{\lambda,k,i}(x) &= x^{-k+i}\Bigg(H^\#_{\lambda,k,k-i+1}(xq^{-\lambda-1})-\sum_{v=1}^{\lambda-i} H^\#_{\lambda,k,v}(x)x^k q^{-v(\lambda+1)} \sum_{w=0}^{\lambda-i-v} \sigma_{\lambda-w}(\lambda) \\
&\qquad -H^\#_{\lambda,k,k-i}(xq^{-\lambda-1})+\sum_{v=1}^{\lambda-i-1} H^\#_{\lambda,k,v}(x)x^k q^{-v(\lambda+1)} \sum_{w=0}^{\lambda-i-1-v} \sigma_{\lambda-w}(\lambda)\Bigg) \\
&= J^\#_{\lambda,k,i}(x)-x^i\sum_{v=1}^{\lambda-i} H^\#_{\lambda,k,v}(x)q^{-v(\lambda+1)}\sigma_{i+v}(\lambda) \\
&= \sum_{j=0}^{i} x^j\sigma_j(\lambda)H^\#_{\lambda,k,i-j}(x)-x^i\sum_{v=i+1}^{\lambda} H^\#_{\lambda,k,v-i}(x)q^{(i-v)(\lambda+1)}\sigma_v(\lambda) \\
&= \sum_{j=0}^{i} x^j\sigma_j(\lambda)H^*_{\lambda,k,i-j}(x)-x^i\sum_{v=i+1}^{\lambda} H^*_{\lambda,k,v-i}(x)q^{(i-v)(\lambda+1)}\sigma_v(\lambda),
\end{aligned}$$

which again is (3.11).

Now from (3.19) it is clear that $H^*_{\lambda,k,i}(0)=1$, $1\leqq i\leqq k$, and since $J^*_{\lambda,k,i}(x)$ satisfies (3.11), $J^*_{\lambda,k,i}(0)=1$, $1\leqq i\leqq k$. The analytic nature of these functions is now obvious from (3.19) and (3.11). Thus we have Lemma 3.2.

COROLLARY 3.3. *The set of functions described in the hypothesis of Lemma* 3.2 *is unique and exists.*

Proof. By Lemma 3.1, the $H^*_{\lambda,k,i}(x)$ and $J^*_{\lambda,k,i}(x)$ defined in Lemma 3.2 are unique, and thus by (3.19), (3.14), and (3.16) so must $H^{\#}_{\lambda,k,i}(x)$ and $J^{\#}_{\lambda,k,i}(x)$ be. For existence take $H^{\#}_{\lambda,k,i}(x)=H_{\lambda,k,i}(x)$, $0\leqq i\leqq k-\lambda+1$,

$$H^{\#}_{\lambda,k,i}(x) = H_{\lambda,k,i}(x)+\sum_{r=1}^{j} H_{\lambda,k,r}(xq^{\lambda+1})x^kq^{(\lambda+1)(k-r)}\sum_{m=0}^{j-r}\sigma_{\lambda-m}(\lambda),$$

$$i = k-\lambda+1+j,\ 1 \leqq j < \lambda,$$

$$J^{\#}_{\lambda,k,i}(x) = J_{\lambda,k,i}(x), \qquad \lambda \leqq i \leqq k,$$

$$J^{\#}_{\lambda,k,i}(x) = \sum_{j=0}^{i} x^j\sigma_j(\lambda)H_{\lambda,k,i-j}(x), \qquad 1 \leqq i < \lambda.$$

Thus we have the corollary.

We are now in a position to establish our main result.

Proof of Theorem 1. Let $P_{\lambda,k,a}(m, n)$, $1\leqq a\leqq k$, denote the number of partitions of n into m parts of the form $n=\sum_{i=1}^{\infty} f_i\cdot i$, where $f_i+\cdots+f_{i+\lambda}\leqq k-1$, $f_{i(\lambda+1)}+\cdots+f_{(i+1)(\lambda+1)}\leqq k-1$, $f_1+\cdots+f_{\lambda+1}\leqq a-1$, $f_b>1$ then $\lambda+1|b$ (equivalently, $n=b_1+\cdots+b_m$, $b_i\geqq b_{i+1}$, $b_i-b_{i+k-1}\geqq\lambda+1$ with strict inequality if $\lambda+1|b_i$, the total number of parts $\leqq\lambda+1$ does not exceed $a-1$, and only multiples of $\lambda+1$ may be repeated).

Let $Q_{\lambda,k,a}(m, n)$ denote the number of partitions described above in which $f_1=f_2=\cdots=f_\lambda=0$ (i.e., the smallest part is $\geqq\lambda+1$).

Let

$$J'_{\lambda,k,a}(x) = 1+\sum_{m=1}^{\infty}\sum_{n=1}^{\infty} P_{\lambda,k,a}(m, n)x^mq^n, \qquad 1 \leqq a \leqq k;$$

$$H'_{\lambda,k,a}(x) = 1+\sum_{m=1}^{\infty}\sum_{n=1}^{\infty} Q_{\lambda,k,a}(m, n)x^mq^n, \qquad 1 \leqq a \leqq k;$$

$$H'_{\lambda,k,0}(x) = 0.$$

By comparison with the generating function for the number of ordinary partitions of n into m parts, we see that we have convergence for $|q|<1$, $|x|<|q|^{-1}$; thus, for given $|q|<1$, the above are analytic in a circle of radius >1 around $x=0$.

Also $H'_{\lambda,k,a}(0)=J'_{\lambda,k,a}(0)=1$, $1\leqq a\leqq k$.

Our next object is to show that these functions satisfy (3.14)–(3.17).

Beginning with (3.14), we consider $Q_{\lambda,k,i}(m, n)-Q_{\lambda,k,i-1}(m, n)$; this expression enumerates those partitions considered by $Q_{\lambda,k,i}(m, n)$ in which $\lambda+1$ appears exactly $i-1$ times. Let us now subtract $\lambda+1$ from each part in such a partition; the number being partitioned is reduced to $n-m(\lambda+1)$, and the number of parts appearing is now $m-i+1$. Since we had originally

$$f_{\lambda+2}+\cdots+f_{2\lambda+2} = k-1-f_{\lambda+1} = (k-i+1)-1,$$

we have after subtraction

$$f_1+\cdots+f_{\lambda+1} = (k-i+1)-1.$$

Otherwise the conditions on the summands are just shifted but unaltered. Thus we are now considering a partition of the type enumerated by

$$P_{\lambda,k,k-i+1}(m-i+1, n-m(\lambda+1)).$$

The above procedure establishes a one-to-one correspondence between the partitions enumerated by $Q_{\lambda,k,i}(m, n)-Q_{\lambda,k,i-1}(m, n)$ and those enumerated by $P_{\lambda,k,k-i+1}(m-i+1, n-m(\lambda+1))$. Hence

$$Q_{\lambda,k,i}(m, n)-Q_{\lambda,k,i-1}(m, n) = P_{\lambda,k,k-i+1}(m-i+1, n-m(\lambda+1)).$$

In terms of generating functions, this implies

$$H'_{\lambda,k,i}(x)-H'_{\lambda,k,i-1}(x) = (xq^{\lambda+1})^{i-1}J'_{\lambda,k,k-i+1}(xq^{\lambda+1}),$$

which is (3.14).

(3.15) is by definition.

We now treat (3.16). I claim that for $1 \leqq i \leqq k-\lambda+1$

$$x^j\sigma_j(\lambda)H'_{\lambda,k,i-j}(x), \qquad 0 \leqq j \leqq \min(i, \lambda), \tag{3.21}$$

is the generating function for the number of partitions of the type enumerated by $P_{\lambda,k,i}(m, n)$ in which exactly j parts $\leqq \lambda$ appear.

By Lemma 2.3, (3.21) is the generating function for partitions in which there are j distinct parts $\leqq \lambda$ and the parts $\geqq \lambda+1$ satisfy the desired conditions. Is it possible then that any of the conditions on the partitions enumerated by $P_{\lambda,k,i}(m, n)$ could be violated? This is not possible because for $1 \leqq v \leqq \lambda$

$$f_v+\cdots+f_\lambda \leqq j; \qquad f_{\lambda+1} \leqq i-j-1; \qquad f_{\lambda+2}+\cdots+f_{\lambda+v} \leqq v-1;$$

hence

$$f_v+\cdots+f_{v+\lambda} \leqq j+i-j-1+v-1 = i+v-2.$$

If $v=1$, this is $f_1+\cdots+f_{\lambda+1} \leqq i-1$ as desired, and if $1<v\leqq\lambda$, this is $f_v+\cdots+f_{v+\lambda} \leqq i+v-2 \leqq k-\lambda+1+\lambda-2 = k-1$ as desired.

Conversely it is easily seen that any partition of the type enumerated by $P_{\lambda,k,i}(m, n)$ with exactly j parts $\leqq \lambda$ is of the type considered by (3.21).

Thus clearly

$$\begin{aligned} J'_{\lambda,k,i}(x) &= \sum_{j=0}^{i} x^j \sigma_j(\lambda) H'_{\lambda,k,i-j}(x) \\ &= \sum_{j=0}^{\min(i,\lambda)} x^j \sigma_j(\lambda) H'_{\lambda,k,i-j}(x), \qquad 1 \leqq i \leqq k-\lambda+1, \end{aligned}$$

which is (3.16).

For (3.17) we note that the above procedure must be modified since now in considering (3.21) with $i>k-\lambda+1$ it is indeed possible that there exists v with $1<v\leqq\lambda$, such that $f_v+\cdots+f_{v+\lambda}\geqq k$.

Indeed by the preceding argument for $1\leqq j<\lambda$

$$\sum_{r=0}^{\lambda} x^r \sigma_r(\lambda) H'_{\lambda,k,k-\lambda+1+j-r}(x) - J'_{\lambda,k,k-\lambda+1+j}(x) \tag{3.22}$$

is the generating function for partitions of the type enumerated by

$$P_{\lambda,k,k-\lambda+1+j}(m, n)$$

except for the fact that for some v, $1<v\leqq\lambda$, $f_v+\cdots+f_{v+\lambda}\geqq k$.

On the other hand, let us consider

$$g_1(k, \lambda, r, j; x; q) H'_{\lambda,k,r}(xq^{\lambda+1}), \qquad 0 \leqq r < \lambda. \tag{3.23}$$

I claim that this is the generating function for partitions of the type described in the preceding paragraph in which $f_{\lambda+1}+\cdots+f_{2\lambda+1}=k-r$.

By Lemma 2.3, Lemma 2.2 (and the fact that $H'_{\lambda,k,r}(xq^{\lambda+1})$ is the generating function for partitions of the type enumerated by $P_{\lambda,k,r}(m, n)$ with the added restriction that $f_1=f_2=\cdots=f_{2\lambda+1}=0$, $f_{2\lambda+2}\leqq r-1$), we see that (3.23) is the generating function for partitions of n into m parts of the form $n=\sum_{i=1}^{\infty} f_i \cdot i$, where

$$f_1+\cdots+f_{\lambda+1} \leqq k-\lambda+j; \tag{3.24}$$

$$f_{\lambda+1}+\cdots+f_{2\lambda+1} = k-r; \tag{3.25}$$

$$f_v+\cdots+f_{v+\lambda} \geqq k \quad \text{for some } v, \qquad 1 < v \leqq \lambda; \tag{3.26}$$

$$f_a > 1 \quad \text{implies } \lambda+1|a; \tag{3.27}$$

$$f_m+\cdots+f_{m+\lambda} \leqq k-1, \quad \text{for } m \geqq 2\lambda+2; \tag{3.28}$$

$$f_{m(\lambda+1)}+\cdots+f_{(m+1)(\lambda+1)} \leqq k-1, \qquad m \geqq 2; \tag{3.29}$$

$$f_{2\lambda+2} \leqq r-1. \tag{3.30}$$

We shall establish the claim made for (3.23) if we can show that the following two conditions hold for the partitions considered.

$$f_m+\cdots+f_{m+\lambda} \leqq k-1, \quad \text{for } \lambda+1 < m < 2\lambda+2; \tag{3.31}$$

$$f_{\lambda+1}+\cdots+f_{2\lambda+2} \leqq k-1. \tag{3.32}$$

Now (3.32) is clear from (3.25) and (3.30). As for (3.31) we have

$$f_m+\cdots+f_{2\lambda+1}+f_{2\lambda+3}+\cdots+f_{m+\lambda} \leqq \lambda; \quad f_{2\lambda+2} \leqq r-1;$$

therefore

$$f_m+\cdots+f_{m+\lambda} \leqq \lambda+r-1 \leqq \lambda+\lambda-2 = 2\lambda-2 \leqq k-1.$$

Thus (3.23) is the desired generating function.

Consequently for $1 \leqq j < \lambda$,

$$\begin{aligned}\sum_{r=0}^{\lambda} x^r\sigma_r(\lambda)H'_{\lambda,k,k-\lambda+1+j-r}(x)-J'_{\lambda,k,k-\lambda+1+j}(x) &= \sum_{r=0}^{\lambda-1} g_1(k,\lambda,r,j;x;q)H'_{\lambda,k,r}(xq^{\lambda+1}) \\ &= \sum_{r=0}^{j} g_1(k,\lambda,r,j;x;q)H'_{\lambda,k,r}(xq^{\lambda+1}),\end{aligned}$$

which is (3.17); the second to last equation is summed to $\lambda-1$ only because (3.25) and (3.24) contradict (3.26) if $r \geqq \lambda$.

Thus by Corollary 3.3 and its proof,

$$J'_{\lambda,k,i}(x) = J_{\lambda,k,i}(x), \qquad \lambda \leqq i \leqq k. \tag{3.33}$$

Finally

$$\begin{aligned}1+\sum_{n=1}^{\infty} B_{\lambda,k,i}(n)q^n &= 1+\sum_{n=1}^{\infty}\sum_{m=1}^{\infty} P_{\lambda,k,i}(m,n)q^n \\ &= J_{\lambda,k,i}(1) \\ &= \prod_{j=1}^{\infty}(1+q^j)(1+q^{j(\lambda+1)})^{-1}(1-q^{j(\lambda+1)})^{-1}(1-q^{(2k-\lambda+1)(\lambda+1)j-(i-\frac{1}{2}\lambda)(\lambda+1)}) \\ &\quad \cdot(1-q^{(2k-\lambda+1)(\lambda+1)(j-1)+(i-\frac{1}{2}\lambda)(\lambda+1)})(1-q^{(2k-\lambda+1)(\lambda+1)j}) \\ &= \begin{cases} \displaystyle\prod_{\substack{j=1 \\ j\not\equiv 0 \,(\mathrm{mod}\ \lambda+1)}}^{\infty}(1+q^j) \prod_{\substack{s=0 \\ s\equiv 0\,(\mathrm{mod}\ \lambda+1) \\ s\not\equiv 0,\,\pm(i-\frac{1}{2}\lambda)(\lambda+1)(\mathrm{mod}\ (2k-\lambda+1)(\lambda+1))}}^{\infty}(1-q^s)^{-1}, & \lambda \text{ even} \\ \displaystyle\prod_{\substack{j=1 \\ j\not\equiv 0\,(\mathrm{mod}\ \frac{1}{2}(\lambda+1))}}^{\infty}(1+q^j) \prod_{\substack{s=0 \\ s\equiv 0,\frac{1}{2}(\lambda+1),\frac{3}{2}(\lambda+1)(\mathrm{mod}\ 2\lambda+2) \\ s\not\equiv 0,\,\pm(2i-\lambda)\frac{1}{2}(\lambda+1)(\mathrm{mod}\ (2k-\lambda+1)(\lambda+1))}}^{\infty}(1-q^s)^{-1}, & \lambda \text{ odd} \end{cases} \\ &= 1+\sum_{n=1}^{\infty} A_{\lambda,k,i}(n)q^n,\end{aligned}$$

where the third to last equation follows from [3, equation (2.9)]. Hence $A_{\lambda,k,i}(n) = B_{\lambda,k,i}(n)$ and Theorem 1 is established.

4. **Extension of Theorem 1.** We extend our definition of $B_{\lambda,k,a}(n)$ as follows. Let $B^*_{\lambda,k,a}(n)$ denote those partitions of n enumerated by $B_{\lambda,k,a}(n)$ such that for any integer j in the interval $[1, \frac{1}{2}(\lambda+1)]$, there must be at most $a-j$ parts of the partition in the interval $[j, \lambda-j+1]$ (in terms of frequency notation the new condition is $f_j+\cdots+f_{\lambda-j+1}\leqq a-j$, $1\leqq j\leqq\frac{1}{2}(\lambda+1)$). We define $P^*_{\lambda,k,a}(m, n)$ similarly.

We remark that for $a\geqq\lambda$, $B^*_{\lambda,k,a}(n)=B_{\lambda,k,a}(n)$. This is because the new conditions are automatically fulfilled for $a\geqq\lambda$, since the number of summands in the interval $[j, \lambda-j+1]$ is $\leqq\min(a-1, \lambda-2j+2)\leqq\min(a-1, a-2j+2)\leqq a-j$.

In Theorem 1 we note that $A_{\lambda,k,a}(n)$ is well defined for $\frac{1}{2}\lambda<a\leqq k$. We now have the following extension of Theorem 1.

THEOREM 2. *$A_{\lambda,k,a}(n)=B^*_{\lambda,k,a}(n)$ for $\frac{1}{2}\lambda<a\leqq k$, $k\geqq 2\lambda-1$.*

This theorem will be easy to deduce from Theorem 1 after we digress briefly for a study of a new auxiliary partition function.

Let $\varphi(i, j, \lambda; n)$, $\frac{1}{2}\lambda<i\leqq\lambda$, $i\leqq j\leqq\lambda$, be the number of partitions of n into $2i-j$ distinct parts of the form $n=\sum_{e=1}^{\lambda} f_e\cdot e$ where *at least one* of the following inequalities holds

$$(4.1)_b \qquad f_b+\cdots+f_{\lambda-b+1} > i-b, \qquad 1\leqq b\leqq\tfrac{1}{2}(\lambda+1).$$

Let $\psi(i, j, \lambda; q)=\sum_{n=0}^{\infty}\varphi(i, j, \lambda; n)q^n$.

LEMMA 4.1. *For $\frac{1}{2}\lambda<i\leqq\lambda$, $i\leqq j\leqq\lambda$,*

$$(4.2) \qquad \psi(i, j, \lambda; q) = q^{(i-j)(\lambda+1)}\sigma_j(\lambda).$$

Proof. Now if $i=j$, then $2i-j=i$. Therefore $f_1+\cdots+f_\lambda=i>i-1$. Hence $\psi(i, i, \lambda; q)$ is just the generating function for partitions into i distinct parts all $\leqq\lambda$. Thus

$$(4.3) \qquad \psi(i, i, \lambda; q) = \sigma_i(\lambda).$$

We now wish to show

$$(4.4) \qquad \psi(i+1, j, \lambda; q) = q^{\lambda+1}\psi(i, j, \lambda; q);$$

(4.4) together with (4.3) imply (4.2) by mathematical induction.

We proceed as follows. For a given partition, π, of the type enumerated by $\varphi(i, j, \lambda; n)$, define

$$(4.5) \qquad m(\pi) = \max_{1\leqq b\leqq\frac{1}{2}(\lambda+1)} f_b+\cdots+f_{\lambda-b+1}-i+b;$$

$$(4.6) \qquad U(\pi) = \min\{b \mid f_b+\cdots+f_{\lambda-b+1}-i+b = m(\pi)\};$$

$$(4.7) \qquad L(\pi) = \max\{b \mid f_b+\cdots+f_{\lambda-b+1}-i+b = m(\pi)\}.$$

We now note that

$$(4.8) \qquad f_{U(\pi)-1} = f_{\lambda-U(\pi)+2} = 0; \qquad f_{L(\pi)} = f_{\lambda-L(\pi)+1} = 1;$$

(4.8) is correct because the falsity of the first assertion would contradict the minimality of $U(\pi)$, and the falsity of the second assertion would contradict the maximality of $L(\pi)$ (note that $U(\pi)>1$ for $j>i$ because $i-1\geqq 2i-j=f_1+\cdots+f_\lambda$ in contrast with the fact that $(4.1)_b$ is valid for some b).

We now prove (4.4). If π is a partition enumerated by $\varphi(i,j,\lambda;n)$ delete from π the summands $L(\pi)$ and $\lambda-L(\pi)+1$, and call the new partition π'. π' is a partition of $n-\lambda-1$ with $2(i-1)-j$ parts.

$$U(\pi') = L(\pi)+1; \tag{4.9}$$

this is because if $b<L(\pi)+1$, then in π' we have

$$f'_b+\cdots+f'_{\lambda-b+1}-(i-1)+b = f_b+\cdots+f_{\lambda-b+1}-2-i+1+b \leqq m(\pi)-1,$$

while for $b\geqq L(\pi)+1$

$$f'_b+\cdots+f'_{\lambda-b+1}-(i-1)+b = f_b+\cdots+f_{\lambda-b+1}-i+1+b \leqq m(\pi),$$

and

$$\begin{aligned} f'_{L(\pi)+1}+\cdots+f'_{\lambda-L(\pi)}-(i-1)+L(\pi)+1 &\\ = f_{L(\pi)}+\cdots+f_{\lambda-L(\pi)+1}-2-i+1+L(\pi)+1 &\\ = m(\pi). & \end{aligned}$$

Now since originally $m(\pi)>0$, we see that $m(\pi)=m(\pi')>0$, and therefore π' is a partition of the type enumerated by $\varphi(i-1,j,\lambda;n-\lambda-1)$.

Conversely if π' is a partition of the type enumerated by $\varphi(i-1,j,\lambda;n-\lambda-1)$ then introduce the summands $U(\pi')-1$, and $\lambda-U(\pi')+2$ and call the new partition π''. π'' is a partition of n into $2i-j$ parts.

$$L(\pi'') = U(\pi')-1; \tag{4.10}$$

this is because if $b>U(\pi')-1$

$$\begin{aligned} f''_b+\cdots+f''_{\lambda-b+1}-i+b &= f'_b+\cdots+f'_{\lambda-b+1}-(i-1)+b-1 \\ &\leqq m(\pi')-1, \end{aligned}$$

while for $b\leqq U(\pi')-1$

$$\begin{aligned} f''_b+\cdots+f''_{\lambda-b+1}-i+b &= f'_b+\cdots+f'_{\lambda-b+1}-(i-1)+b+2-1 \\ &\leqq m(\pi')-1+2-1 = m(\pi'), \end{aligned}$$

and

$$\begin{aligned} f''_{U(\pi')-1}+\cdots+f''_{\lambda-U(\pi')+2}-i+U(\pi')-1 &\\ = f'_{U(\pi')}+\cdots+f'_{\lambda-U(\pi')+1}+2-(i-1)+U(\pi')-2 &\\ = m(\pi'). & \end{aligned}$$

Thus we see that the above procedure establishes a one-to-one correspondence between the partitions enumerated by $\varphi(i,j,\lambda;n)$ and those enumerated by $\varphi(i-1,j,\lambda;n-\lambda-1)$. Therefore

$$\varphi(i+1,j,\lambda;n) = \varphi(i,j,\lambda;n-\lambda-1). \tag{4.11}$$

We deduce (4.4) immediately from (4.11). Thus we have established Lemma 4.1.

Proof of Theorem 2. We have already noted that Theorem 2 is just Theorem 1 if $a \geqq \lambda$. We thus need only treat $\frac{1}{2}\lambda < a < \lambda$. Under these conditions we let $J''_{\lambda,k,a}(x)$ be the generating function for $P^*_{\lambda,k,a}(m, n)$. Then as in the proof of Theorem 1,

$$(4.12)\quad \begin{aligned} J''_{\lambda,k,a}(x) &= J'_{\lambda,k,a}(x) - \sum_{j=a}^{\lambda} x^{2a-j}\psi(a, j, \lambda; q)H_{\lambda,k,j-a}(x) \\ &= \sum_{j=0}^{a} x^j\sigma_j(\lambda)H_{\lambda,k,a-j}(x) - \sum_{j=a+1}^{\lambda} x^{2a-j}\psi(a, j, \lambda; q)H_{\lambda,k,j-a}(x) \end{aligned}$$

for the first term on the right-hand side (from the proof of Theorem 1) generates the partitions enumerated by $P_{\lambda,k,a}(m, n)$ and the second term, by the definition of ψ and Lemma 2.3, just removes those partitions enumerated by $P_{\lambda,k,a}(m, n)$ which are not enumerated by $P^*_{\lambda,k,a}(m, n)$.

Now on the other hand from (3.6) and (3.7)

$$(4.13)\quad J_{\lambda,k,a}(x) = \sum_{j=0}^{a} x^j\sigma_j(\lambda)H_{\lambda,k,a-j}(x) - x^a \sum_{j=a+1}^{\lambda} q^{(a-j)(\lambda+1)}\sigma_j(\lambda)H_{\lambda,k,j-a}(x).$$

Consequently by Lemma 4.1

$$(4.14)\quad J''_{\lambda,k,a}(1) = J_{\lambda,k,a}(1).$$

The remainder of the proof is now identical with that of Theorem 1.

5. **Conclusion.** Let us define $K(\lambda)$ to be that integer such that for $k \geqq K(\lambda)$ Theorem 2 (and consequently Theorem 1) holds.

CONJECTURE. $K(\lambda) = \lambda$.

It is fairly easy to show that $K(2) = 2$, and indeed numerical evidence indicates the truth of $K(\lambda) = \lambda$ for $\lambda = 3, 4, 5$. The main problem in proving the conjecture lies in the fact that the q-difference equations become much more involved for smaller values of k and the related partition identities become even more cumbersome.

A further interesting question seems to be whether or not even smaller values of k may be taken provided new restrictions are put on the partitions enumerated by $B_{\lambda,k,a}(n)$. As an example, Schur has proved [6, p. 495] that

$$A_{3,2,2}(n) = B^0_{3,2,2}(n)$$

where $B^0_{3,2,2}(n)$ denotes the number of partitions enumerated by $B_{3,2,2}(n)$ with the added restriction that no parts are $\equiv 2 \pmod 4$.

REFERENCES

1. H. L. Alder, *The nonexistence of certain identities in the theory of partitions and compositions*, Bull. Amer. Math. Soc. **54** (1948), 712–722.

2. G. E. Andrews, *A generalization of the Göllnitz-Gordon partition theorems*, Proc. Amer. Math. Soc. **18** (1967), 945–952.

3. G. E. Andrews, *q-difference equations for certain well-poised basic hypergeometric series*, Quart. J. Math. Oxford Ser. (2) **19** (1968), 433–447.

4. B. Gordon, *A combinatorial generalization of the Rogers-Ramanujan identities*, Amer. J. Math. **83** (1961), 393–399.

5. G. H. Hardy and E. M. Wright, *An introduction to the theory of numbers*, 4th ed., Oxford Univ. Press, Oxford, 1960.

6. I. J. Schur, *Zur additiven Zahlentheorie*, S.-B. Deutsch. Akad. Wiss. Berlin Math.-Mat. Kl. (1926), 488–495.

The Pennsylvania State University,
University Park, Pennsylvania

3.4 Sieves for Theorems of Euler, Rogers, and Ramanujan

SIEVES FOR THEOREMS OF EULER, ROGERS, AND RAMANUJAN*

George E. Andrews, Massachusetts Institute of Technology
and Pennsylvania State University

1. Introduction. The theory of sieves has become an important and sophisticated tool in the theory of numbers during the past 52 years. Many important and deep results have been proved related to such unanswered questions as the Twin Primes Problem and Goldbach's Conjecture. For the most part sieves have been employed in problems concerning primes. The object of this paper is to introduce a sieve related to the theory of partitions.

The sieve method may be thought of as breaking into two parts. First the principle of inclusion-exclusion (or some similar combinatorial counting process) is applied to obtain an expression for a particular arithmetic function. Then the new expression is treated number-theoretically to yield information about the original function. The most elementary example of this is the simple sieve estimate for $\pi(x)$, the number of primes not exceeding x. If $p_1, p_2, \ldots, p_r$ denote the first r primes, $A(x,r)$ denotes the number of integers not exceeding x that are not divisible by any of $p_1, p_2, \ldots, p_r$, and $[x]$ denotes the largest integer not exceeding x, then by the principle of inclusion-exclusion [LeVeque; 8; p. 98]

$$A(x,r) = [x] - \sum_{i=1}^{r} \left[\frac{x}{p_i}\right] + \sum_{1 \le i < j \le r} \left[\frac{x}{p_i p_j}\right] - \cdots + (-1)^r \left[\frac{x}{p_1 \cdots p_r}\right].$$

The estimate $\pi(x) = O(x/\log\log x)$ is then obtained by observing that $\pi(x) \le A(x,r) + r$ and estimating $A(x,r)$ by carefully comparing it with

*This research was partially supported by National Science Foundation Grant GP-9660.

2

$$x - \sum_{i=1}^{r} \frac{x}{p_i} + \sum_{1 \le i < j \le r} \frac{x}{p_i p_j} - \cdots + (-1)^r \frac{x}{p_1 \cdots p_r}$$

$$= x \prod_{i=1}^{r} \left(1 - \frac{1}{p_i}\right).$$

We propose to examine a new technique in the theory of partitions that is at least a cousin of the sieves of prime number theory. This technique involves an "inclusion-exclusion" type enumeration of certain partitions; then one makes a careful analysis of the expressions obtained.

We propose to give sieve-theoretic proofs of the following three theorems from additive number theory

$$1 = \frac{1}{(q)_\infty} - \frac{q}{(q)_\infty} - \frac{q^2}{(q)_\infty} + \frac{q^5}{(q)_\infty} + \frac{q^7}{(q)_\infty} - \cdots, \tag{1.1}$$

$$\sum_{n=0}^{\infty} \frac{q^{n^2}}{(q)_n} = \frac{1}{(q)_\infty} - \frac{q^2}{(q)_\infty} - \frac{q^3}{(q)_\infty} + \frac{q^9}{(q)_\infty} + \frac{q^{11}}{(q)_\infty} - \cdots \tag{1.2}$$

$$= \frac{1}{(q)_\infty} + \sum_{n=1}^{\infty} (-1)^n \left\{\frac{q^{\frac{1}{2}n(5n-1)}}{(q)_\infty} + \frac{q^{\frac{1}{2}n(5n+1)}}{(q)_\infty}\right\},$$

$$\sum_{n=0}^{\infty} \frac{q^{n^2+n}}{(q)_n} = \frac{1}{(q)_\infty} - \frac{q}{(q)_\infty} - \frac{q^4}{(q)_\infty} + \frac{q^7}{(q)_\infty} + \frac{q^{13}}{(q)_\infty} - \cdots \tag{1.3}$$

$$= \frac{1}{(q)_\infty} + \sum_{n=1}^{\infty} (-1)^n \left\{\frac{q^{\frac{1}{2}n(5n-3)}}{(q)_\infty} + \frac{q^{\frac{1}{2}n(5n+3)}}{(q)_\infty}\right\},$$

where

$$(a;q)_n = (a)_n = (1 - a)(1 - aq)\cdots(1 - aq^{n-1})$$

and

$$(a;q)_\infty = (a)_\infty = \lim_{n\to\infty} (a;q)_n.$$

The identity (1.1) is essentially Euler's Pentagonal Number Theorem [Hardy and Wright; 7; p. 284] while (1.2) and (1.3) are known as the Rogers-Ramanujan identities [Hardy and Wright; 7; p. 290].

The sieves that we shall apply act on the set of all partitions of an integer, each partition being classified according to its "successive ranks" (see Section 2). The sieve related to (1.1) sieves out all partitions except the empty partition, whence the 1 on the left-hand side. The sieves for (1.2) and (1.3) remove all partitions except those related in an obvious way to the left-hand series.

In Sections 3 and 4, we shall treat (1.1) in detail. Section 5 will be devoted to an outline of the proofs of (1.2), (1.3), and a description of other results.

2. Successive ranks. We shall consider partitions of an integer n into positive integral parts, where the order of the parts is irrelevant. To facilitate matters we shall represent each partition with the parts in decreasing order of magnitude. We shall also be interested in the graphical representation of partitions. For example, the partition π of 15 given by $5 + 4 + 4 + 2$ is represented graphically as follows.

```
• • • • •
• • • •
• • • •
• •
```

The number of nodes in the j-th row of the graph equals the j-th part of π. To obtain what is called the conjugate partition $C\pi$, we read the nodes vertically; in this instance $C\pi$ is $4 + 4 + 3 + 3 + 1$.

In 1944, F. J. Dyson [5] introduced an important parameter of a partition called its rank, the largest part minus the number of parts. Thus the rank of the partition $5 + 4 + 4 + 2$ is $5 - 4 = 1$. Dyson conjectured and Atkin and Swinnerton-Dyer [3] proved the following result connecting the rank with the Ramanujan congruences:

Let $K(r,a;n)$ denote the number of partitions of n in which the rank is congruent to a modulo r; then for

$0 \le a \le 4$, $K(5,a;5n+4) = \frac{1}{5}\,p(5m+4)$ and for $0 \le a \le 6$, $K(7,a;7n+5) = \frac{1}{7}\,p(7n+5)$.

From these results, we see directly that

$$p(5n+4) \equiv 0 \pmod{5},$$

and

$$p(7n+5) \equiv 0 \pmod{7}.$$

In a later paper, Atkin [4] introduced what he termed the successive ranks. He was motivated to study these further ranks in an unsuccessful search for a "rank-theoretic" interpretation of the further Ramanujan congruence:

$$p(11n+6) \equiv 0 \pmod{11}.$$

<u>Definition</u> 1. If π denotes a partition, then $d_r(\pi)$, the r-<u>rank</u> of π, is defined as the difference between the r-th part of π and the r-th part of the conjugate of π; furthermore, the r-rank is defined <u>only</u> if both the r-th row and r-th column of π have a node in common.

Computing the successive ranks of a partition is done most easily by means of the graphical representation. Thus for the partition π: $15 = 5 + 4 + 4 + 2$,

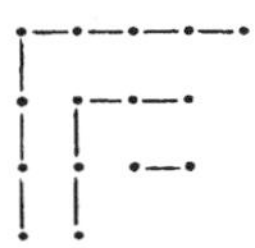

the ranks are $d_1(\pi) = 1$, $d_2(\pi) = 0$, $d_3(\pi) = 1$. The computation is done by subtracting the number of vertical nodes in the r-th indicated right angle from the number of horizontal nodes. We see from the graphical representation that if the Durfee square of π (see [Hardy and Wright; 7; p. 281]) is of side s, then there are exactly s successive ranks for π.

Our interest lies in the "oscillation" of the successive ranks between positive and nonpositive values.

Definition 2. Let h be the largest integer for which there exists a sequence $j_1 < j_2 < \cdots < j_h$ such that

$$d_{j_1}(\pi) > 0, \quad d_{j_2}(\pi) \le 0, \quad d_{j_3}(\pi) > 0, \quad d_{j_4}(\pi) \le 0,$$

and so on (more precisely $(-1)^i(\tfrac{1}{2} - d_{j_i}(\pi)) > 0$ for $1 \le i \le h$). We define h to be the positive oscillation of π.

Definition 3. Let k be the largest integer for which there exists a sequence $j_1 < j_2 < \cdots < j_k$ such that

$$d_{j_1}(\pi) \le 0, \quad d_{j_2}(\pi) > 0, \quad d_{j_3}(\pi) \le 0, \quad d_{j_4}(\pi) > 0,$$

and so on (more precisely $(-1)^i(d_{j_i}(\pi) - \tfrac{1}{2}) > 0$ for $1 \le i \le k$). We define k to be the negative oscillation of π.

For example, let us consider the partition π given by

$$112 = 17 + 15 + 10 + 10 + 10 + 10 + 8 + 8 + 7 + 5 + 5 + 4 + 3.$$

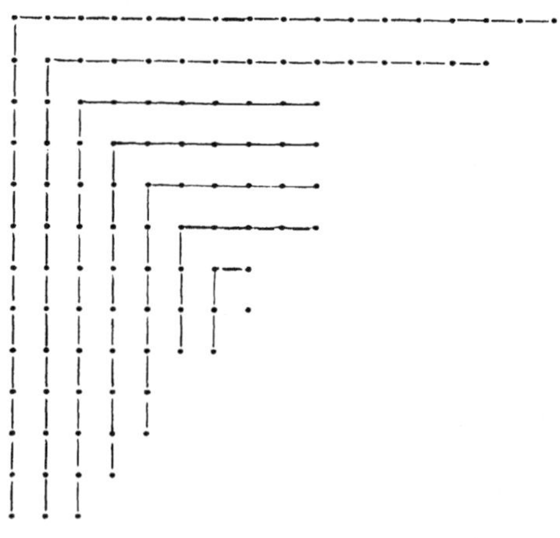

$d_1(\pi) = 4$, $d_2(\pi) = 2$, $d_3(\pi) = -3$, $d_4(\pi) = -2$, $d_5(\pi) = -1$, $d_6(\pi) = 1$, $d_7(\pi) = -1$, $d_8(\pi) = 0$. Consequently the positive oscillation of π is 4 (e.g., choose $j_1 = 1$, $j_2 = 3$, $j_3 = 6$, $j_4 = 7$), and the negative oscillation of π is 3 (e.g., choose $j_1 = 3$, $j_2 = 6$, $j_3 = 7$).

6

Proposition 1. The positive oscillation of π is 1 larger than the negative oscillation if $d_1(\pi) > 0$ and is 1 less if $d_1(\pi) \le 0$.

Proof. Suppose that h denotes the positive oscillation of π and k the negative oscillation. If $d_1(\pi) > 0$, let $j_1 < j_2 < \cdots < j_k$ be the sequence described in Definition 3. Then the sequence $j'_1 < j'_2 < \cdots < j'_k < j'_{k+1}$ given by $j'_1 = 1$, $j'_i = j_{i-1}$ for $2 \le i \le k+1$ is of the type described in Definition 2; consequently $h \ge k+1$. However it is obvious that $k \ge h-1$ because in the maximal sequence $j^*_1 < j^*_2 < \cdots < j^*_h$ defining h (see Definition 2) we see that $j''_1 < j''_2 < \cdots < j''_{h-1}$ (where $j''_i = j^*_{i+1}$) is of the type described in Definition 3. Therefore if $d_1(\pi) > 0$, then $h = k+1$.

If $d_1(\pi) \le 0$, a mirror image of the above argument shows that $k = h+1$. □

In the next section we shall develop a sieve related to successive ranks.

3. Generating functions for partitions with prescribed oscillation. We shall study certain functions related to the positive and negative oscillation of a partition.

Definition 4. Let $p(a,b;\mu;N)$ (resp., $m(a,b;\mu;N)$) denote the number of partitions of N with at most b parts, with largest part at most a and with positive (resp., negative) oscillation at least μ.

To these partition functions we associate the related generating functions:

$$p(a,b;\mu;q) = \sum_{N \ge 0} p(a,b;\mu;N)q^N, \tag{3.1}$$

$$M(a,b;\mu;q) = \sum_{N \ge 0} m(a,b;\mu;N)q^N. \tag{3.2}$$

<u>Lemma</u> 1. <u>The following recurrence relations hold for</u> $\mu \geq 1$:

$$m(a,b;\mu;N) - m(a-1,b;\mu;N) - m(a,b-1;\mu;N) + m(a-1,b-1;\mu;N)$$

$$(3.3) \qquad = \begin{cases} m(a-1,b-1;\mu;N-a-b+1), & \text{if } a > b, \\ p(a-1,b-1;\mu-1;N-a-b+1), & \text{if } a \leq b; \end{cases}$$

$$p(a,b;\mu;N) - p(a-1,b;\mu;N) - p(a,b-1;\mu;N) + p(a-1,b-1;\mu;N)$$

$$(3.4) \qquad = \begin{cases} m(a-1,b-1;\mu-1;N-a-b+1), & \text{if } a > b, \\ p(a-1,b-1,\mu;N-a-b+1), & \text{if } a \leq b. \end{cases}$$

<u>Proof</u>. Let us start by examining the left side of (3.3). The expression $m(a,b;\mu;N) - m(a-1,b;\mu;N)$ denotes the number of partitions of N with at most b parts, with negative oscillation at least μ, and with largest part <u>equal</u> to a. Consequently, the expression $\{m(a,b;\mu;N) - m(a-1,b;\mu;N)\} - \{m(a,b-1;\mu;N) - m(a-1,b-1;\mu;N)\}$ (which is the left-hand side of (3.3)) denotes the number of partitions of N with exactly b parts, with negative osicllation at least μ, and with largest part <u>equal</u> to a. We now transform such partitions by deleting the largest part and subtracting 1 from each of the remaining parts; in this way, we are left with a partition of N-a-b+1 into at most b-1 parts with largest part at most a-1. Furthermore if $a > b$, then $d_1(\pi) > 0$; hence the removal of this outer angle of nodes from the Ferrars graph has no effect on the negative oscillation which is therefore still equal to μ. Thus our transformed partition is of the type enumerated by $m(a-1,b-1;\mu;N-a-b+1)$. Since the above procedure is clearly reversible, we see that

$$m(a,b;\mu;N) - m(a-1,b;\mu;N) - m(a,b-1;\mu;N) + m(a-1,b-1;\mu;N)$$
$$= m(a-1,b-1;\mu;N-a-b+1),$$

and so the top half of (3.3) is verified.

If $a \leq b$, then $d_1(\pi) \leq 0$, and removal of the outer angle of nodes from the Ferrars graph may indeed affect the negative oscillation. In fact, the resulting partition is now characterized by the fact that it has positive oscillation of at least $\mu-1$. The transformed partitions in this instance are of the type enumerated by

$p(a-1,b-1;\mu-1;N-a-b+1)$. Hence the lower half of (3.3) follows.

The proof of (3.4) is a mirror image of the proof for (3.3). □

The results in Lemma 1 may now be translated directly into functional equations for the related generating functions. Hence, for $\mu \geq 1$,

$$M(a,b;\mu;q) - M(a-1,b;\mu;q) - M(a,b-1;\mu;q) + M(a-1,b-1;\mu;q)$$

$$= q^{a+b-1} \begin{cases} M(a-1,b-1;\mu;q), & \text{if } a > b, \\ P(a-1,b-1;\mu-1;q), & \text{if } a \leq b; \end{cases} \tag{3.5}$$

$$P(a,b;\mu;q) - P(a-1,b;\mu;q) - P(a,b-1;\mu;q) + P(a-1,b-1;\mu;q)$$

$$= q^{a+b-1} \begin{cases} M(a-1,b-1;\mu-1;q), & \text{if } a > b, \\ P(a-1,b-1;\mu;q), & \text{if } a \leq b. \end{cases} \tag{3.6}$$

<u>Lemma</u> 2. <u>The following relations hold</u>:

$$P(a,b;0;q) = M(a,b;0;q) = \binom{a+b}{a}_q ; \tag{3.7}$$

<u>for</u> $\mu \geq 1$.

$$P(0,b;\mu;q) = P(a,0;\mu;q) = M(0,b;\mu;q) = M(a,0;\mu;q) = 0, \tag{3.8}$$

where

$$\binom{N}{M}_q = \begin{cases} \dfrac{(q)_N}{(q)_M \, (q)_{N-M}}, & \text{for } 0 \leq M \leq N, \\ 0, & \text{otherwise.} \end{cases}$$

<u>Proof</u>. The requirement that the negative (or positive) oscillation be at least 0 is clearly met by any partition. Therefore $P(a,b;0;q)$ and $M(a,b;0;q)$ are both the generating function for partitions with largest part at most a and at most b parts. It is well-known [Sylvester; 10; p. 269] that this generating function is

$$\binom{a+b}{a}_q .$$

Thus we have (3.7).

The requirement that either the largest part a (or the number of parts b) be at most 0 is not met by any partition save the empty partition, and the empty partition cannot have negative (or positive) oscillation $\mu \geq 1$. Hence (3.8) is established. □

Our next result might be described as the second part of the sieve. It will be the essential result after the "inclusion-exclusion" type procedure has been used.

Theorem 1. If $b = a$ or $a-1$, then

$$M(a,b;2\mu;q) = q^{\mu(6\mu+1)}\binom{a+b}{b-3\mu}_q, \tag{3.9}$$

$$M(a,b;2\mu-1;q) = q^{(2\mu-1)(3\mu-2)}\binom{a+b}{b-3\mu+2}_q, \tag{3.10}$$

$$P(a,b;2\mu;q) = q^{\mu(6\mu-1)}\binom{a+b}{b+3\mu}_q, \tag{3.11}$$

$$P(a,b;2\mu-1;q) = q^{(2\mu-1)(3\mu-1)}\binom{a+b}{a+1-3\mu}_q . \tag{3.12}$$

Proof. We first note that equations (3.5), (3.6), (3.7), and (3.8) uniquely define $M(a,b;\mu;q)$ and $P(a,b;\mu;q)$ for each nonnegative integral a, b, and μ. We now consider the following functions:

$$M^*(a,b;2\mu;q) = q^{\mu(6\mu+1)}\binom{a+b}{b-3\mu}_q, \quad \text{for } a \geq b, \quad \mu \geq 1; \tag{3.13}$$

$$M^*(a,b;2\mu-1;q) = q^{(2\mu-1)(3\mu-2)}\binom{a+b}{b-3\mu+2}_q, \quad \text{for } a \geq b, \quad \mu \geq 1; \tag{3.14}$$

10

$$(3.15)\qquad P^*(a,b;2\mu;q) = q^{\mu(6\mu-1)}\binom{a+b}{b+3\mu}_q, \qquad \text{for } b \ge a-1, \quad \mu \ge 1;$$

$$(3.16)\qquad P^*(a,b;2\mu-1;q) = q^{(2\mu-1)(3\mu-1)}\binom{a+b}{a+1-3\mu}_q, \qquad \text{for } b \ge a-1, \quad \mu \ge 1;$$

$$(3.17)\qquad M^*(a-1,a+r;2\mu;q) = q^{6\mu^2+\mu}\,\frac{1-q^{(r+2)(a-3\mu)}}{1-q^{a-3\mu}}\binom{2a-2}{a-1-3\mu}_q$$

$$+\sum_{j=0}^{r-1} q^{6\mu^2-2\mu+a+1+j}\,\frac{1-q^{(r-j)(a-3\mu)}}{1-q^{a-3\mu}}\binom{2a-2+j}{a-2-3\mu}_q,$$

for $a \ge 1$, $r \ge -1$, $\mu \ge 1$;

$$(3.18)\qquad M^*(a-1,a+r;2\mu-1;q) = q^{(2\mu-1)(3\mu-2)}\,\frac{1-q^{(r+2)(a-3\mu+2)}}{1-q^{a-3\mu+2}}\binom{2a-2}{a+3\mu-3}_q$$

$$+\sum_{j=0}^{r-1} q^{6\mu^2-10\mu+a+5+j}\,\frac{1-q^{(r-j)(a-3\mu+2)}}{1-q^{a-3\mu+2}}\binom{2a-2+j}{a+3\mu-2+j}_q,$$

for $a \ge 1$, $r \ge -1$, $\mu \ge 1$;

$$(3.19)\qquad P^*(b+r,b-1;2\mu;q) = q^{6\mu^2-\mu}\,\frac{1-q^{(r+1)(b-3\mu+1)}}{1-q^{b-3\mu+1}}\binom{2b-1}{b+3\mu-1}_q$$

$$+\sum_{j=0}^{r-2} q^{6\mu^2-4\mu+2+b+j}\,\frac{1-q^{(r-j-1)(b-3\mu+1)}}{1-q^{b-3\mu+1}}\binom{2b-1+j}{b+3\mu+j}_q,$$

for $b \ge 1$, $r \ge 0$, $\mu \ge 1$;

$$(3.20)\qquad P^*(b+r,b-1;2\mu-1;q) = q^{(2\mu-1)(3\mu-1)}\,\frac{1-q^{(r+1)(b-3\mu+2)}}{1-q^{b-3\mu+2}}\binom{2b-1}{b+1-3\mu}_q$$

$$+\sum_{j=0}^{r-2} q^{(2\mu-1)(3\mu-1)+b-3\mu+3+j}\,\frac{1-q^{(r-1-j)(b-3\mu+2)}}{1-q^{b-3\mu+2}}\binom{2b-1+j}{b-3\mu}_q,$$

for $b \ge 1$, $r \ge 0$, $\mu \ge 1$;

$$(3.21)\qquad P^*(a,b;0;q) = M^*(a,b;0;q) = \binom{a+b}{a}_q.$$

We observe immediately that $P^*(a,b;\mu;q)$ and $M^*(a,b;\mu;q)$ satisfy (3.7) and (3.8). It is a matter of elementary (but lengthy) algebraic manipulation using the identities

$$\binom{N}{M}_q = \binom{N-1}{M-1}_q + q^M \binom{N-1}{M}_q, \tag{3.22}$$

$$\binom{N}{M}_q = \binom{N-1}{M}_q + q^{N-M} \binom{N-1}{M-1}, \tag{3.23}$$

and

$$\binom{N}{M}_q = \binom{N}{N-M}_q, \tag{3.24}$$

to prove that $P^*(a,b;\mu;q)$ and $M^*(a,b;\mu;q)$ also satisfy (3.5) and (3.6).

Recalling the first sentence of this proof, we thus see that for each nonnegative integral a, b, and μ,

$$P^*(a,b;\mu;q) = P(a,b;\mu;q), \tag{3.25}$$

and

$$M^*(a,b;\mu;q) = M(a,b;\mu;q). \tag{3.26}$$

The equations (3.9)-(3.12) now follow from (3.13) - (3.16), (3.25), and (3.26).

Definition 5. Let $p(\mu;N)$ (resp., $m(\mu;N)$) denote the number of partitions of N with positive (resp., negative) oscillation at least μ. We also let $P(\mu;q)$ and $M(\mu;q)$ denote the respective generating functions.

Corollary. For each $\mu \geq 0$,

$$M(2\mu;q) = \frac{q^{\mu(6\mu+1)}}{(q)_\infty}, \tag{3.27}$$

$$M(2\mu-1;q) = \frac{q^{(2\mu-1)(3\mu-2)}}{(q)_\infty}, \tag{3.28}$$

(3.29) $$P(2\mu;q) = \frac{q^{\mu(6\mu-1)}}{(q)_\infty},$$

(3.30) $$P(2\mu-1;q) = \frac{q^{(2\mu-1)(3\mu-1)}}{(q)_\infty}.$$

Proof. These results follow directly from the fact that for $|q| < 1$

$$|P(\mu;q) - P(a,a;\mu;q)| \le \sum_{n=a+1}^{\infty} p(n)|q|^n \to 0 \quad \text{as} \quad a \to \infty,$$

$$|M(\mu;q) - M(a,a;\mu;q)| \le \sum_{n=a+1}^{\infty} p(n)|q|^n \to 0 \quad \text{as} \quad a \to \infty,$$

and

$$\lim_{a\to\infty} q^X \binom{2a+Y}{a-Z}_q = \frac{q^X}{(q)_\infty}.$$

4. A sieve for Euler's Pentagonal Number Theorem (1.1). The following theorem may be thought of as the "inclusion-exclusion" aspect of the sieve we are constructing.

Theorem 2. For $n > 0$,

(4.1) $$p(0;n) + \sum_{\mu=1}^{\infty} (-1)^\mu m(\mu;n) + \sum_{\mu=1}^{\infty} (-1)^\mu p(\mu;n) = 0.$$

Proof. The above sum is a weighted count of partitions of n. First if π is a partition of n, $d_1(\pi) > 0$ and the positive oscillation of π is r, then by Proposition 1 the negative oscillation of π is $r-1$. Hence π is counted once by each of $p(0,n)$, $m(1;n)$, ..., $m(r-1,n)$, $p(1,n)$, ..., $p(r,n)$ and in the above sum the weight of the count is

$$1 + \sum_{\mu=1}^{r-1} (-1)^\mu + \sum_{\mu=1}^{r} (-1)^\mu = 1 + \sum_{\mu=1}^{r-1} (-1)^\mu - \sum_{\mu=0}^{r-1} (-1)^\mu = 1 - 1 = 0.$$

If $d_1(\pi) \le 0$, and the positive oscillation of π is r, then by Proposition 1 the negative oscillation of π is $r+1$. Hence the weighted count in this case is

$$1 + \sum_{\mu=1}^{r+1} (-1)^{\mu} + \sum_{\mu=1}^{r} (-1)^{\mu} = 1 - \sum_{\mu=0}^{r} (-1)^{\mu} + \sum_{\mu=1}^{r} (-1)^{\mu} = 1 - 1 = 0.$$

Thus each partition of n is counted with weight 0; that is,

$$p(0,n) + \sum_{\mu=1}^{\infty} (-1)^{\mu} m(\mu,n) + \sum_{\mu=1}^{\infty} (-1)^{\mu} p(\mu,n) = 0$$

for $n > 0$.

Theorem 3. (Euler's Pentagonal Number Theorem (1.1)).

$$1 = \frac{1}{(q)_\infty} + \sum_{n=1}^{\infty} (-1)^n q^{\frac{1}{2}n(3n-1)} \left\{\frac{1 + q^n}{(q)_\infty}\right\}.$$

Proof. Multiply equation (4.1) by q^n, sum for $1 \le n < \infty$ and then add 1 to both sides of the equation.

$$1 = P(0;q) + \sum_{\mu=1}^{\infty} (-1)^{\mu} M(\mu;q) + \sum_{\mu=1}^{\infty} (-1)^{\mu} P(\mu;q)$$

$$= \frac{1}{(q)_\infty} - \sum_{\mu=1}^{\infty} M(2\mu-1;q) + \sum_{\mu=1}^{\infty} M(2\mu;q)$$

$$- \sum_{\mu=1}^{\infty} P(2\mu-1;q) + \sum_{\mu=1}^{\infty} P(2\mu;q)$$

$$= \frac{1}{(q)_\infty} - \sum_{\mu=1}^{\infty} \frac{q^{(2\mu-1)(3\mu-2)}}{(q)_\infty} + \sum_{\mu=1}^{\infty} \frac{q^{\mu(6\mu+1)}}{(q)_\infty}$$

$$- \sum_{\mu=1}^{\infty} \frac{q^{(2\mu-1)(3\mu-1)}}{(q)_\infty} + \sum_{\mu=1}^{\infty} \frac{q^{\mu(6\mu-1)}}{(q)_\infty}$$

$$= \frac{1}{(q)_\infty} - \sum_{\mu=1}^{\infty} \frac{q^{\frac{1}{2}(2\mu-1)(3(2\mu-1)-1)}(1+q^{2\mu-1})}{(q)_\infty}$$

$$+ \sum_{\mu=1}^{\infty} \frac{q^{\frac{1}{2}2\mu(3(2\mu)-1)}(1+q^{2\mu})}{(q)_\infty}$$

$$= \frac{1}{(q)_\infty} + \sum_{n=1}^{\infty} (-1)^n q^{\frac{1}{2}n(3n-1)} \left\{\frac{1+q^n}{(q)_\infty}\right\}. \quad \square$$

If we had used Theorem 1 directly for our sieve instead of the Corollary, we would have been able to obtain the following identity due to Schur [9].

$$\sum_{\lambda=-\infty}^{\infty} (-1)^\lambda q^{\frac{1}{2}\lambda(3\lambda+1)} \binom{n}{[\frac{1}{2}(n-3\lambda)]}_q = 1, \tag{4.2}$$

where $[x]$ denotes the largest integer not exceeding x. The case $a = b$ in Theorem 1 yields (4.2) with $n = 2a$, and $a = b-1$ in Theorem 1 yields (4.2) with $n = 2b-1$.

5. <u>Sieve for the Rogers-Ramanujan identity (1.2) and (1.3)</u>. The technique developed in Sections 4 and 5 can be extended to provide an infinite family of sieves. We shall consider the extension that yields a proof of (1.2) and (1.3).

<u>Definition</u> 6. Let h be the largest integer for which there exists a sequence $j_1 < j_2 < \cdots < j_h$ such that

$$d_{j_1}(\pi) > 2k-a-1, \quad d_{j_2}(\pi) \le -(a-1), \quad d_{j_3}(\pi) > 2k-a-1, \quad d_{j_4}(\pi) \le -(a-1),$$

and so on. We define h to be the (k,a)-<u>positive</u> <u>oscillation</u> of π.

<u>Definition</u> 7. Let g be the largest integer for which there exists a sequence $j_1 < j_2 < \cdots < j_g$ such that

$$d_{j_1}(\pi) \le -(a-1), \quad d_{j_2}(\pi) > 2k-a-1, \quad d_{j_3}(\pi) \le -(a-1), \quad d_{j_4}(\pi) > 2k-a-1,$$

and so on. We define g to be the (k,a)-<u>negative</u> <u>oscillation</u> of π.

We remark that Definitions 2 and 3 correspond to the case $k = a = 1$.

Definition 8. Let $P_{k,a}(\mu;N)$ (resp., $M_{k,a}(\mu;N)$) denote the number of partitions of N with (k,a)-positive (resp., (k,a)-negative) oscillation at least μ.

We also let $P_{k,a}(\mu;q)$ and $M_{k,a}(\mu;q)$ denote the respective generating functions. Suitably modifying the procedures of Sections 3 and 4, one is able to prove the following identities.

$$M_{k,a}(2\mu;q) = \frac{q^{\mu\{(2k+1)2\mu + (2k-2a+1)\}}}{(q)_\infty}, \tag{5.1}$$

$$M_{k,a}(2\mu-1;q) = \frac{q^{(2\mu-1)\{(2k+1)\mu - (2k-a+1)\}}}{(q)_\infty}, \tag{5.2}$$

$$P_{k,a}(2\mu;q) = \frac{q^{\mu\{(2k+1)2\mu - (2k-2a+1)\}}}{(q)_\infty}, \tag{5.3}$$

$$P_{k,a}(2\mu-1;q) = \frac{q^{(2\mu-1)\{(2k+1)\mu-a\}}}{(q)_\infty}. \tag{5.4}$$

Theorem 2 now may be extended to the following result.

Theorem 4. Let $Q_{k,a}(n)$ denote the number of partitions π of n such that $-(a-2) \le r_i(\pi) \le 2k-a-1$ for each of the successive ranks of π. Then

$$Q_{k,a}(n) = P_{k,a}(0;n) + \sum_{\mu=1}^{\infty} (-1)^\mu M_{k,a}(\mu;n) + \sum_{\mu=1}^{\infty} (-1)^\mu P_{k,a}(\mu;n). \tag{5.5}$$

The expression on the right-hand side sieves out all partitions that have at least one rank outside the interval $[-(a-2), 2k-a-1]$.

Theorem 3 now extends to the following series identity

$$\sum_{n=0}^{\infty} Q_{k,a}(n)q^n$$

$$= \frac{1}{(q)_\infty}\left\{1 + \sum_{n=1}^{\infty} (-1)^n q^{\frac{1}{2}n\{(2k+1)n-(2k-2a+1)\}}(1+q^{(2k-2a+1)n})\right\}$$

$$(5.6) \qquad = \frac{(q^{2k+1};q^{2k+1})_\infty (q^a;q^{2k+1})_\infty (q^{2k+1-a};q^{2k+1})_\infty}{(q)_\infty}$$

$$= \prod_{\substack{n=1 \\ n \not\equiv 0,\, \pm a \pmod{2k+1}}}^{\infty} (1-q^n)^{-1},$$

where the penultimate equation follows by Jacobi's triple product identity [Hardy and Wright; 7; p. 282].

Identity (5.6) may now be used to deduce directly the following partition theorem.

<u>Theorem</u> 5. <u>Let</u> $A_{k,a}(n)$ <u>denote the number of partitions of</u> n <u>into parts not congruent to</u> 0, a, <u>or</u> $-a$ <u>modulo</u> $2k+1$. <u>Then</u> $A_{k,k}(n) = Q_{k,a}(n)$.

The following theorem shows how Theorem 5 reduces to the first Rogers-Ramanujan identity (1.2) when $k = 2$, $a = 2$.

<u>Theorem</u> 6. <u>Let</u> $B(n)$ <u>denote the number of partitions of</u> n <u>in which the difference between any two parts is at least</u> 2. <u>Then</u> $B(n) = Q_{2,2}(n)$.

<u>Proof</u>. Let us take a partition π of the type enumerated by $B(n)$, say $n = b_1 + b_2 + \cdots + b_s$, where $b_i - b_{i+1} \geq 2$. We form a graphical representation of π as follows. The i-th part of π is represented by a right angle of nodes: if $b_i = 2m+1$, then we form

17

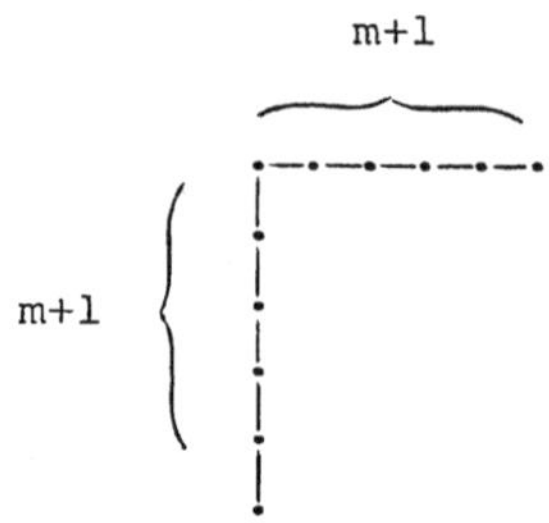

,

if $b_i = 2s$, then we form

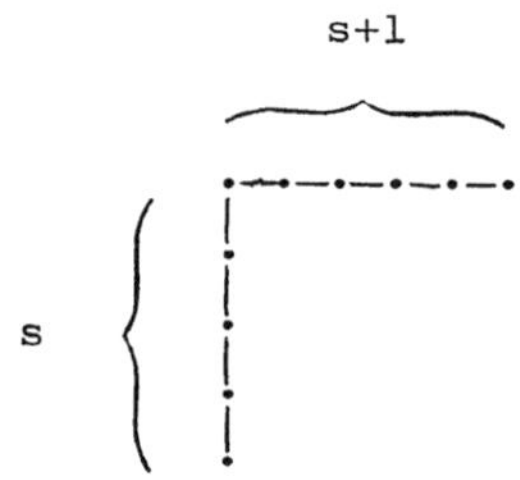

.

Since $b_i - b_{i+1} \geq 2$, we may collect these angles as the successive angles of a Ferrars graph in which the successive ranks are either 0 or 1, that is, a partition of the type enumerated by $Q_{2,2}(n)$. The above transformation is clearly reversible, and thus we have a one-to-one correspondence between the partitions enumerated by $B(n)$ and those enumerated by $Q_{2,2}(n)$. Hence $B(n) = Q_{2,2}(n)$. □

The following is an example of the above transformation for $n = 9$.

9 →  → 5 + 1 + 1 + 1 + 1

8 + 1 →  → 5 + 2 + 1 + 1

7 + 2 →  → 4 + 3 + 1 + 1

6 + 3 →  → 4 + 3 + 2

5 + 3 + 1 →  → 3 + 3 + 3.

As is well-known [Hardy-Wright; 7; p. 290-291] (and easily shown),

$$\sum_{n=0}^{\infty} B(n)q^n = \sum_{n=0}^{\infty} \frac{q^{n^2}}{(q)_n}. \tag{5.7}$$

Hence

$$\sum_{n=0}^{\infty} \frac{q^{n^2}}{(q)_n} = \sum_{n=0}^{\infty} B(n)q^n = \sum_{n=0}^{\infty} Q_{2,2}(n)q^n$$

$$= \frac{1}{(q)_\infty}\left(1 + \sum_{n=1}^{\infty} (-1)^n q^{\frac{1}{2}n(5n-1)}(1+q^{2n})\right),$$

and this proves (1.2).

Finally the following theorem shows how Theorem 5 reduces to the second Rogers-Ramanujan identity (1.3) when $k = 2$, $a = 1$.

Theorem 7. *Let* $C(n)$ *denote the number of partitions of* n *in which* 1 *does not appear and the difference between any two parts is at least* 2. *Then* $C(n) = Q_{2,1}(n)$.

Proof. The proof here is very much like the proof of Theorem 6, so we shall omit most of the details. The only change is that if $b_1 + b_2 + \cdots + b_s$ is a partition of the type enumerated by $C(n)$ and b_i (= 2m+1) is an odd part of that partition, then b_i is

represented by the angle of nodes

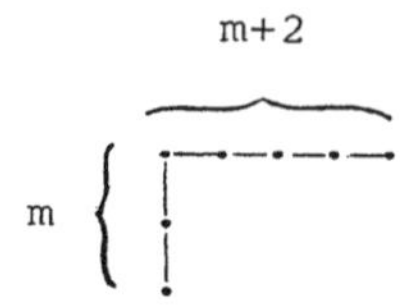

which has rank 2. The fact that $b_s > 1$ now fits in perfectly with the fact that 1 is the only positive odd number that cannot be represented by the angle of nodes of the type given above. □

We remark that the techniques we have described for the Rogers-Ramanujan identities are sufficient to prove the main theorem of [Andrews; 2], a polynomial identity that implies (1.2) and (1.3).

It is interesting to note that there is a third partition function related to Theorem 5. Namely if $B_{k,a}(n)$ denotes the number of partitions of n of the form $n = b_1 + b_2 + \cdots + b_s$ where $b_i \geq b_{i+1}$, $b_i - b_{i+k-1} \geq 2$, and $b_{s-a+1} > 1$, then B. Gordon has shown (see [Gordon; 6] and [Andrews; 1]) that

$$B_{k,a}(n) = A_{k,a}(n).$$

An independent proof that $B_{k,a}(n) = Q_{k,a}(n)$ would certainly illuminate this type of theorem.

It is hoped that the ideas presented here will stimulate research in two directions. First, partition theorems related to the successive ranks of Atkin are clearly of interest and should be studied. Second, it would be nice to find further sieves that yield other partition theorems.

REFERENCES

1. G. E. Andrews, An analytic proof of the Rogers-Ramanujan-Gordon identities, Amer. J. Math. 88 (1966), 844-846.

2. G. E. Andrews, A polynomial identity which implies the Rogers-Ramanujan identities, Scripta Math. 23 (1970), 297-305.

3. A. O. L. Atkin and H. P. F. Swinnerton-Dyer, Some properties of partitions, Proc. London Math. Soc. (3) 4 (1954), 84-106.

4. A. O. L. Atkin, A note on ranks and conjugacy of partitions, Quart. J. Math. Oxford Ser. (2) 17 (1966), 335-338.

5. F. J. Dyson, Some guesses in the theory of partitions, Eureka 8 (1944), 10-15.

6. B. Gordon, A combinatorial generalization of the Rogers-Ramanujan identities, Amer. J. Math. 83 (1961), 393-399.

7. G. H. Hardy and E. M. Wright, An Introduction to the Theory of Numbers, Fourth Edition, Oxford University Press, Oxford, 1960.

8. W. J. LeVeque, Topics in Number Theory, Vol. 1, Addison-Wesley, Reading, 1956.

9. I. J. Schur, Ein Beitag zur additiven Zahlentheorie, Sitzungsber. Akad. Wissensch., Berlin, Phys.-Math. Kl., 1917, 302-321.

10. J. J. Sylvester, A constructive theory of partitions, arranged in three acts, an interact, and an exodion, Amer. J. Math. 5 (1882), 251-330.

3.5 Combinatorics and Ramanujan's "Lost" Notebook

Combinatorics and Ramanujan's "Lost" Notebook

George E. Andrews
The Pennsylvania State University
University Park, Pennsylvania, 16802, U.S.A.

1. Introduction.

L. J. Rogers' paper (Rogers; 1894) which contains the Rogers-Ramanujan identities together with their proof was ignored for 20 years before Ramanujan came across it while leafing through old volumes of the Proceedings of the London Mathematical Society. In the interim, Ramanujan had discovered the Rogers-Ramanujan identities empirically, and they were making the rounds as major unsolved problems (cf. Hardy; 1940, p. 91). This is undoubtedly one of the very few times that a set of significant unsolved problems was solved 20 years before it was posed.

The most obvious reason Rogers' paper lay buried is that it is page after page of q-series identities with the Rogers-Ramanujan identities sneaking past in mild disguise on page 10 of this tour de force.

As more discoveries were made, the subject became even less readable. The Rev. F. H. Jackson was one of the early pioneer q-series researchers. His papers also read much like Rogers'. It is not surprising to read in Jackson's obituary (Chaundy (1962)); "Once (with a whimisical smile one imagines) he [Jackson] recounted the occasion of his quarrel with our Society [the L.M.S.]: he had read a paper when someone remarked: 'Surely, Mr. President, we have heard all this before.' He strode from the room and never darkened our pages again." As it turned out this critical remark was directed at what was, in fact, Jackson's most valuable paper. Again the result was one equation among many which were indistinguishable to the outsider.

These incidents point up one of the main difficulties in presenting results on q-series to a wide audience. While there have been interesting interactions of this subject with physics (Baxter; 1980, 1982 or Andrews et al.; 1984), transcendental number theory (Richmond and Szekeres; 1981), group theory (Lusztig; 1977 or Andrews; 1977, 1984), and additive number theory (Andrews; 1972), nonetheless a paper like Slater's compendium of Rogers-Ramanujan type identities (Slater; 1952) leaves the impression that it is impossible to have any idea of what is really going on.

With this background we turn to Ramanujan's "Lost" Notebook (cf. Andrews; 1979 or Rankin; 1982). This document contains over 600 unproved results of which at least two thirds are q-series identities. Again the superficial sameness of these results leaves one daunted.

Having thus criticized some very fine mathematicians, I hope I will be forgiven if I fail to provide the Olympian overview which you are probably expecting. I do hope to suggest an approach to increasing understanding. In particular I want to describe some means by which one might hope to gain insight about a series like

$$\sum_{n=0}^{\infty} \frac{(-1)^n q^{n^2}(1-q)(1-q^3)\dots(1-q^{2n-1})}{(1+q^2)^2(1+q^4)^2\dots(1+q^{2n})^2} \tag{1.1}$$

for example. I choose this example because it appears prominently in Ramanujan's "Lost" Notebook, and I devoted a lengthy paper to a study of it and several related series (Andrews; 1981a). Indeed I gave the first 36 coefficients of its Maclaurin series expansion (Andrews; 1981a, pp. 44-45). As is clear from my comments in Section 5 of that paper, I hadn't the least notion of any reasonable combinatorial significance of this series. At least the methods I describe herein will easily yield a straightforward combinatorial interpretation of (1.1) (cf. Section 7).

Section 2 of this paper will provide a general setting for series of this type. In Section 3, we shall describe some means to fit various q-series into this framework. The remainder of the paper considers applications of these ideas to some of the more incomprehensible q-series. Among the topics covered are all the classes of mock theta functions given in Ramanujan's last letter to Hardy (Ramanujan; 1927). Also in Sections 6 and 7 we examine two classes of functions that also appear in the "Lost" Notebook. We then show how to derive the Rogers-Ramanujan type identities given in Andrews (1981b) for Regime II of Baxter's hard hexagon model.

2. The Combinatorial Setting.

Part of the astonishing nature of Ramanujan's genius lies in his ability to find important formulas and intricate relationships without possessing either a related general theory or even the results in question in full generality (Hardy; 1940, p. 14).

The value of generality in this subject is immensely important. The only proofs of the Rogers-Ramanujan identities with no generality are Schur's incredibly brilliant and intricate combinatorial arguments (Schur; 1917) and the extension of Schur's treatment to a bijective proof (Garsia & Milne; 1981). In contrast, Watson (1929) gives a proof in such great generality that he only needs to base his arguments on the fact that two polynomials of degree n agreeing at $n+1$ values must be identical.

The introduction of some generality is the key here. In effect we shall extend each function of the one variable q to a function of two variables.

Definition 1. A <u>Ramanujan statistic</u> is an ordered pair $(\rho,\mathcal{S})$ where $\mathcal{S}$ is a subset of the set of all partitions of all nonnegative integers and ρ is a nonnegative arithmetic function on $\mathcal{S}$ with the condition that $\rho^{-1}(n)$ is a finite set for each nonnegative integer n.

Definition 2. The RS-polynomial for the Ramanujan statistic $(\rho,\mathcal{S})$ is

$$(2.1)\quad p_n(\rho,\mathcal{S}|q) = \sum_{\substack{\pi\in\mathcal{S}\\ \rho(\pi)\leq n}} q^{\sigma(\pi)},$$

where $\sigma(\pi)$ is the integer partitioned by π.

Andrews: Combinatorics and Ramanujan's "Lost" Notebook 4

Definition 3. The RS-generating function for the Ramanujan statistic (ρ,S) is

(2.2) $$F(\rho,S|q,t) = \sum_{n=0}^{\infty} p_n(\rho,S|q)t^n.$$

Definition 4. Let $P(S;n)$ denote the number of partitions of n that lie in S.

Definition 5. Let $D(\rho,S;n)$ denote the number of partitions $\pi(\in S)$ of n for which $\rho(\pi)$ is even minus the number of those for which $\rho(\pi)$ is odd.

Lemma 1. For $|q| < 1$,

(2.3) $$\sum_{n=0}^{\infty} P(S;n)q^n = \lim_{t\to 1^-} (1-t)F(\rho,S|q,t),$$

(2.4) $$\sum_{n=0}^{\infty} D(\rho,S;n)q^n = \lim_{t\to -1^+} (1-t)F(\rho,S|q,t).$$

Proof. We note that by Definition 1,

(2.5) $$\sum_{n\geq 0} P(S;n)q^n = \lim_{n\to\infty} p_n(\rho,S|q)$$

$$= \lim_{t\to 1^-} (1-t)F(\rho,S|q,t),$$

by Abel's lemma (Andrews, 1971, p. 190).

(2.6) $$\sum_{n\geq 0} D(\rho,S;n)q^n = p_0(\rho,S|q) + \sum_{n=1}^{\infty} (-1)^n(p_n(\rho,S|q)-p_{n-1}(\rho,S|q))$$

$$= \lim_{t\to -1^+} (1-t)F(\rho,S|q,t).$$

The convergence in each instance is guaranteed by the fact that the $p_n(\rho,S|q)$ are polynomials with nonnegative coefficients and for each j the coefficient of q^j in $p_n(\rho,S|q)$ is bounded by the total number of partitions of j and is fixed for $n > n_0(j)$ [Definition 1].□

In our applications we shall find that (2.3) and (2.4) are often our starting points. It also happens that $F(\rho,S|q^2,q)$ or $F(\rho,S|q,q)$ arise. I have not been able to interpret nicely these functions in general; however it is possible to do so in many specific applications. Of course, if we define for example

$$\sum_{n=0}^{\infty} T_n(\rho,S;n)q^n = F(\rho,S|q^2,q), \tag{2.7}$$

then clearly $T_n(\rho,S;n)$ is the number of solutions of $\rho(\pi)+2\sigma(\pi)\leqslant n$, so this at least provides some general combinatorial significance for $F(\rho,S|q^2,q)$. $F(\rho,S|q,q)$ can be similarly interpreted.

3. Determination of Ramanujan Statistics.

We are assuming that the Ramanujan statistic is not provided for us ab initio. Instead we start with series like

$$\sum_{n=0}^{\infty} \frac{(A_1;q^r)_n(A_2;q^r)_n\cdots(A_s;q^r)_n q^{ra\binom{n}{2}} z^n}{(q^r;q^r)_n(B_1;q^r)_n\cdots(B_j;q^r)_n} \tag{3.1}$$

or

$$\sum_{n=0}^{\infty} \frac{(-A_1;q^r)_n(-A_2;q^r)_n\cdots(-A_s;q^r)_n(-1)^{an}z^n q^{ra\binom{n}{2}}}{(-q^r;q^r)_n(-B_1;q^r)_n\cdots(-B_j;q^r)_n} \tag{3.2}$$

or

$$\sum_{n=0}^{\infty} \frac{(A_1q^r;q^{2r})_n(A_2q^r;q^{2r})_n\cdots(A_sq^r;q^{2r})_n q^{ran^2} z^n}{(q^r;q^{2r})_{n+1}(B_1q^r;q^{2r})_n\cdots(B_jq^r;q^{2r})_n} \tag{3.3}$$

where

$$(a;q)_n \equiv (a)_n = (1-a)(1-aq)\cdots(1-aq^{n-1}). \tag{3.4}$$

Often we find in the literature and especially in Ramanujan's work instances of (3.1)-(3.3) intertwined in various identities.

If we define

$$F\begin{pmatrix} A_1,\ldots,A_s;r,a;q,z;t \\ B_1,\ldots,B_j \end{pmatrix} \equiv F(q,t) \tag{3.5}$$

$$= \sum_{n=0}^{\infty} \frac{(A_1t;q^r)_n(A_2t;q^r)_n\cdots(A_st;q^r)_n z^n t^{an} q^{ra\binom{n}{2}}}{(t;q^r)_{n+1}(B_1t;q^r)_n\cdots(B_jt;q^r)_n},$$

then we note that $\lim_{t\to 1^-}(1-t)F(q,t)$ is (3.1), $\lim_{t\to -1^+}(1-t)F(q,t)$ is (3.2) and $F(q^2,q^r)$ is (3.3). Thus our $F(q,t)$ fits in with the form of the results given in Lemma 1.

Lemma 2. Let $F(q,t)$ be given by (3.5) with a a positive integer. Then the coefficient of t^n in the expansion of $F(q,t)$ is a polynomial in the A_i, the B_i, z and q. Furthermore $F(q,t)$ satisfies the following simple non-homogeneous q-difference equation:

$$(3.6)\quad (1-t)(1-B_1t)\cdots(1-B_jt)F(q,t)$$
$$= (1-B_1t)(1-B_2t)\cdots(1-B_jt)$$
$$+(1-A_1t)\cdots(1-A_st)t^a z\, F(q,tq^r).$$

Proof. The polynomial nature of the coefficients follows immediately from the two classical formulas (Andrews; 1976, p. 36)

$$(3.7)\quad \frac{1}{(A;q)_n} = \sum_{m=0}^{\infty} \begin{bmatrix} n+m-1 \\ m \end{bmatrix} A^m,$$

$$(3.8)\quad (A;q)_n = \sum_{m=0}^{n} \begin{bmatrix} n \\ m \end{bmatrix} (-A)^m q^{\binom{m}{2}},$$

where $\begin{bmatrix} r \\ s \end{bmatrix}$ is the q-binomial coefficient or Gaussian polynomial given by

$$(3.9)\quad \begin{bmatrix} r \\ s \end{bmatrix}_q = \begin{bmatrix} r \\ s \end{bmatrix} = \begin{cases} \dfrac{(1-q^r)(1-q^{r-1})\cdots(1-q^{r-s+1})}{(1-q^s)(1-q^{s-1})\cdots(1-q)}, & r,\ s \geqq 0 \\ 0 \quad \text{otherwise.} & \end{cases}$$

To see (3.6) we observe that

$$(3.10)\quad F(q,t) = \frac{1}{1-t}$$

$$+ \sum_{n=1}^{\infty} \frac{(A_1t;q^r)_n(A_2t;q^r)_n\cdots(A_st;q^r)_n t^{an} z^n q^{ar\binom{n}{2}}}{(t;q^r)_{n+1}(B_1t;q^r)_n\cdots(B_jt;q^r)_n}$$

$$= \frac{1}{1-t} + \frac{(1-A_1t)\ldots(1-A_st)t^az}{(1-t)(1-B_1t)\ldots(1-B_jt)} F(q,tq^r),$$

where the last line follows by replacing n by $n+1$ in the sum. Identity (3.6) is merely (3.10) with denominators cleared. □

The steps in applications are now clear

1. Determine $F(q,t)$ from given instances of (3.1)-(3.3).

2. Examine $\lim_{t\to 1^-}(1-t)F(q,t)$ in order, if possible, to identify it as the generating function for partitions lying in some set S. (Often more than one S may turn up).

3. Study the polynomial coefficients for $F(q,t)$ described in Lemma 2. The object is to identify them as RS-polynomials.

Unfortunately I have no general wisdom about Step 3. In practice however there seem to be two possible occurrences: 1) It may be possible to find a simple representation for $F(q,t)$ which transparently yields the relevant Ramanujan statistic. 2) There may be known families of polynomial generating functions related to the set of partitions S found in Step 2; if so it may be possible to identify ρ from these polynomials.

4. The Third Order Mock Theta Function.

We shall now consider two of the principal third order mock theta functions (Watson: 1936, p. 62):

$$f(q) = 1 + \sum_{n=1}^{\infty} \frac{q^{n^2}}{(-q)_n^2}, \tag{4.1}$$

and

$$\omega(q) = \sum_{n=0}^{\infty} \frac{q^{2n^2+2n}}{(q;q^2)_{n+1}^2}. \tag{4.2}$$

Following the guidance of Section 3, we define

Andrews: Combinatorics and Ramanujan's "Lost" Notebook 8

(4.3) $$M\theta_3(q,t) = \sum_{n=0}^{\infty} \frac{t^{2n}q^{n^2}}{(t)_{n+1}(tq)_n} = F\left(_q;1,2;q,q;t\right),$$

and we note

(4.4) $$f(q) = \lim_{t\to -1^+}(1-t)M\theta_3(q,t),$$

(4.5) $$\omega(q) = M\theta_3(q^2,q)/(1-q).$$

To find our relevant set of partitions we observe that

(4.6) $$\lim_{t\to 1^-} (1-t)M\theta_3(q,t)$$
$$= \sum_{n=0}^{\infty} \frac{q^{n^2}}{(q)_n^2} = \frac{1}{(q)_\infty} = \sum_{n=0}^{\infty} p(n)q^n.$$

This last set of identities is often attributed to Euler (Andrews; 1976, p. 21); however, I believe that this q-series first arose in Jacobi's work (Jacobi: 1829, §64). We now see from (4.6) that $\mathcal{S}=\mathcal{P}$, the set of all partitions of all nonnegative integers.

To determine ρ we apply a transformation to $M\theta_3(q,t)$. Namely

(4.7) $$M\theta_3(q,t) = \frac{1}{1-t} \lim_{x\to 0} \sum_{n=0}^{\infty} \frac{(q)_n(x^{-1})_n(x^{-1})_n t^{2n}q^n x^{2n}}{(q)_n(tq)_n(tq)_n}$$
$$= \sum_{n=0}^{\infty} \frac{t^n}{(tq)_n} .$$

(by Sears; 1951, p. 174, eq. (10.1) with $p=q$, $a=b=x^{-1}$, $e=f=tq$, $c=q$, and $x\to 0$). This last expression is so easy to interpret that we see immediately that the coefficient of $t^M q^N$ in (4.7) is the number of partitions of N with the largest part plus the number of parts $\leq M$. Hence in this instance

$$\rho(\pi) = gn(\pi),$$

where $gn(\pi)$ denotes the largest part plus the number of parts (the choice of notation refers to the gnomon of the Ferrers graph of π).

Consequently we can now easily identify both $f(q)$ and $(1-q)\omega(q)$ as generating functions of partition functions:

Theorem 1. Let $f(q)$ and $\omega(q)$ be given by (4.1) and (4.2). Then

$$f(q) = \sum_{n=0}^{\infty} D(gn,P;n)q^n, \tag{4.8}$$

and

$$(1-q)\omega(q) = \sum_{n=0}^{\infty} \Omega(n)q^n, \tag{4.9}$$

where $D(gn,P;n)$ is defined in Definition 5 and in this instance is the total number of partitions π of n with $gn(\pi)$ even minus the number with $gn(\pi)$ odd. The partition function $\Omega(n)$ is the number of partitions of n into odd parts where the largest part is at most one more than twice the number of ones.

Remark. The result on $f(q)$ is implicit in a forthcoming book by N. J. Fine (1985).

Proof. Equation (4.8) follows immediately from (2.4), (4.4) and our comments following (4.7). From (4.5), we see that

$$\begin{aligned}(1-q)\omega(q) &= M\Theta_3(q^2,q)\\ &= \sum_{n=0}^{\infty} \frac{q^n}{(q^3;q^2)_n} \qquad \text{(by(4.7))}\\ &= \sum_{n=0}^{\infty} \Omega(n)q^n. \end{aligned}$$ □

Corollary. For $n > 0$,

$$p_n(gn,P|q) = \sum_{j=0}^{n-1} \begin{bmatrix} n-1 \\ j \end{bmatrix} q^j. \tag{4.10}$$

Remark. Equation (4.10) identifies the RS-polynomial in this instance as a special case of the Rogers-Szegö polynomials (Andrews; 1976, p. 49).

Proof. By (4.7) and (2.2)

$$\sum_{n=0}^{\infty} p_n(gn,P|q)t^n = M\Theta_3(q,t) \tag{4.11}$$

$$= \sum_{m=0}^{\infty} \frac{t^m}{(tq)_m}$$

$$= 1 + \sum_{m=1}^{\infty} t^m \sum_{j=0}^{\infty} \begin{bmatrix} m+j-1 \\ j \end{bmatrix} t^j q^j \qquad \text{(by (3.7))}$$

$$= 1 + \sum_{n=1}^{\infty} t^n \sum_{j=0}^{n-1} \begin{bmatrix} n-1 \\ j \end{bmatrix} q^j .$$

Comparing coefficients of t^n in the extremes of (4.11), we obtain (4.10). □

Most of the other third order mock theta functions can be treated in a similar manner. However for the most part the combinatorial interpretations of them are straight forward and are covered well in N. J. Fine's soon to be published book (Fine; 1985).

5. The Fifth and Seventh Order Mock Theta Functions.

We lump together the fifth order mock theta functions with the seventh order functions. These families first arose in Ramanujan's last letter to Hardy (see Ramanujan; 1927, pp. 354-355). G. N. Watson (1937) proved most of the assertions about the fifth order mock theta functions. There are alternative combinatorial interpretations available for the fifth order mock theta functions. We shall concentrate on the one which serves as a prototype for the seventh order mock theta functions and other applications. To this end we define an arithmetic function $\rho_{\lambda;a,b}$ on all partitions π by

(5.1) $$\rho_{\lambda;a,b}(\pi) = \max(\lambda\cdot\ell(\pi) - a, \lambda\cdot \#(\pi) - b),$$

where $\ell(\pi)$ is the largest part of π and $\#(\pi)$ is the number of parts of π.

The two fifth order mock theta functions of most interest to us here are (G. N. Watson; 1937, p. 277)

(5.2) $$f_0(q) = \sum_{n=0}^{\infty} \frac{q^{n^2}}{(-q)_n}$$

and

$$(5.3)\qquad F_0(q) = \sum_{n=0}^{\infty} \frac{q^{2n^2}}{(q;q^2)_n} .$$

Recalling the first of the famous Rogers-Ramanujan identities (Andrews; 1976, Ch. 7)

$$(5.4)\qquad \sum_{n=0}^{\infty} \frac{q^{n^2}}{(q)_n} = \prod_{n=0}^{\infty} \frac{1}{(1-q^{5n+1})(1-q^{5n+4})} ,$$

we are led to consider

$$(5.5)\qquad M\theta_5(q,t) = \sum_{n=0}^{\infty} \frac{t^{2n}q^{n^2}}{(t)_{n+1}} = \mathcal{F}(\;;1,2;q,q;t).$$

Note that $\lim_{t\to 1^-} (1-t)M\theta_5(q,t)$ is the left side of (5.4), while

$$(5.6)\qquad \lim_{t\to -1^+}(1-t)M\theta_5(q,t) = f_0(q),$$

and

$$(5.7)\qquad M\theta_5(q^2,q) = q^{-1}(F_0(q)-1).$$

Next we observe that

$$M\theta_5(q,t) = \sum_{n=0}^{\infty} t^{2n}q^{n^2} \sum_{m=0}^{\infty} \begin{bmatrix} m+n \\ m \end{bmatrix} t^m \qquad \text{(by (3.7))}$$

$$(5.8)\qquad = \sum_{N=0}^{\infty} t^N \sum_{0\leq 2n\leq N} \begin{bmatrix} N-n \\ n \end{bmatrix} q^{n^2}$$

$$(5.9)\qquad = \sum_{N=0}^{\infty} t^N \sum_{\lambda=-\infty}^{\infty} (-1)^{\lambda} q^{\lambda(5\lambda+1)/2} \begin{bmatrix} N \\ \left[\frac{N-5\lambda}{2}\right] \end{bmatrix}, \qquad \text{(Andrews; 1976, p. 157, Ex. 4).}$$

Thus we have available two useful representations for our RS-polynomial. If we choose (5.8), then we require S to be $\mathcal{D}_2$ the set of all partitions in each of which the parts differ by at least 2, and for ρ we choose $\ell(\pi)$, the largest part of π. With these choices it is a simple exercise (Andrews; 1976, p. 157, Ex. 4) to see that

$$M\theta_5(q,t) = F(\ell,\ \mathcal{D}_2|q,t). \tag{5.10}$$

In addition it is immediate that $F_0(q)$ is the generating function for partitions into odd parts wherein every odd number up to and including the largest part appears at least twice.

Equation (5.9) yields an interpretation of broader interest.

Definition 6. Let Π be a partition whose Ferrers graph (Andrews; 1976, §1.3) has a node in the i-th row and j-th column; we call this node the (i,j)-th node. We define the <u>hook difference</u> at the (i,j)-th node to be the number of nodes in the i-th row of Π minus the number of nodes in the j-th column of Π.

Definition 7. We say that the (i,j)-th node of Π lies on diagonal c if $i-j = c$.

Definition 8. Let α and β be positive integers. Let $p_{K,i}(N,M;\alpha,\beta;n)$ denote the number of partitions of n into at most M parts each $\leqq N$ such that the hook differences on diagonal $1-\beta$ are $\geqq -i+\beta+1$ and on diagonal $\alpha-1$ are $\leqq K-i-\alpha-1$.

Definition 9. Let $\mathcal{S}_{K,i}(\alpha,\beta)$ denote the set of all partitions for which the hook differences on diagonal $1-\beta$ are $\geqq -i+\beta+1$ and on diagonal $\alpha-1$ are $\leqq K-i-\alpha-1$.

The related generating function is defined to be

$$D_{K,i}(N,M;\alpha,\beta;q) = \sum_{n\geq 0} p_{K,i}(N,M;\alpha,\beta;n)q^n. \tag{5.11}$$

The reason for these last few definitions is that the polynomials in (5.11) have a representation which specializes to the polynomials appearing in (5.9). Namely, by Andrews et al. (1985),

$$M\theta_5(q,t) = \sum_{n=0}^{\infty} t^n D_{5,2}\left(\left[\frac{n+1}{2}\right], \left[\frac{n}{2}\right];1,1;q\right). \tag{5.12}$$

$$= \sum_{n=0}^{\infty} t^n p_n(\rho_{2;1,0}, S_{5,2}(1,1)|q).$$

To add clarity to this example we restate (2.4) in this specific case:

$$(5.13) \qquad f_0(q) = \sum_{n\geq 0} D(\rho_{2;1,0}, S_{5,2}(1,1);n)q^n,$$

where $D(\rho_{2;1,0}, S_{5,2}(1,1);n)$ in this instance may be most simply interpreted as the excess of the number of partitions of n which have hook difference 0 or 1 on diagonal 0 and largest part plus number of parts even over the same type of partitions of n with largest part plus number of parts odd. For example, we note

$$f_0(q) = 1 + q - q^2 + q^3 - q^6 + q^7 - q^8 + q^9 \ldots,$$

and indeed $D(\rho_{2;1,0}, S_{5,2}(1,1);9) = 1$ since the admissable partitions with largest part plus number of parts even are 5+1+1+1+1, 4+3+1+1, 3+3+3 while those with odd sum of largest part plus number of parts are 5+2+1+1, 4+3+2.

The other fifth order mock theta functions can be treated in a similar manner. We shall not do this however preferring instead to consider more mysterious q-series.

Having made this elaborate account for the fifth order functions, we are prepared to treat the seventh order mock theta functions:

$$(5.14) \qquad F_0(q) = \sum_{n=0}^{\infty} \frac{q^{n^2}}{(q^{n+1})_n} = \sum_{n=0}^{\infty} \frac{q^{n^2}(q)_n}{(q)_{2n}},$$

$$(5.15) \qquad F_1(q) = \sum_{n=0}^{\infty} \frac{q^{n^2+2n}}{(q^{n+1})_{n+1}} = \sum_{n=0}^{\infty} \frac{q^{n^2+2n}(q)_n}{(q)_{2n+1}},$$

$$(5.16) \qquad F_2(q) = \sum_{n=0}^{\infty} \frac{q^{n^2+n}}{(q^{n+1})_{n+1}} = \sum_{n=0}^{\infty} \frac{q^{n^2+n}(q)_n}{(q)_{2n+1}}.$$

Recently it has been shown by Andrews (1984) that both the fifth and seventh order mock theta functions are related to certain series that resemble some of Hecke's series involving indefinite quadratic forms

(Hecke; 1959). Other than this there is nothing of substance in the literature on the seventh order mock theta functions.

In order to get some starting point for these functions, we combine some known results:

$$(5.17)\qquad G_0(q) \equiv \sum_{n=0}^{\infty} \frac{q^{n^2}(-q)_n}{(q)_{2n}}$$

$$= \prod_{\substack{n=1\\ n\not\equiv 0,\ \pm 6 (\mathrm{mod}\ 14)}}^{\infty} (1-q^n)^{-1}, \quad \text{(Slater; 1952, eq. (61))}$$

$$= \sum_{n=0}^{\infty} P(S_{7,3}(2,2);n)q^n, \quad \text{(Andrews et. al.; 1985, Th. 3)}$$

$$(5.18)\qquad G_1(q) \equiv \sum_{n=0}^{\infty} \frac{q^{n^2+2n}(-q)_n}{(q)_{2n+1}}$$

$$= \prod_{\substack{n=1\\ n\not\equiv 0,\pm 2 (\mathrm{mod}\ 14)}}^{\infty} (1-q^n)^{1}, \quad \text{(Slater; 1952, eq. (59))}$$

$$= \sum_{n=0}^{\infty} P(S_{7,1}(2,2);n)q^n, \quad \text{(Andrews et al.; 1985, Th. 3)}$$

$$(5.19)\qquad G_2(q) \equiv \sum_{n=0}^{\infty} \frac{q^{n^2+n}(-q)_n}{(q)_{2n+1}}$$

$$= \prod_{\substack{n=1\\ n\not\equiv 0,\pm 4 (\mathrm{mod}\ 14)}}^{\infty} (1-q^n)^{-1}, \quad \text{(Slater; 1952, eq. (60))}$$

$$= \sum_{n=0}^{\infty} P(S_{7,2}(2,2);n)q^n, \quad \text{(Andrews et al., 1985, Th. 3).}$$

Comparing (5.17)-(5.19) with (5.14)-(5.16) respectively, we see immediately the obvious route for applying Lemma 1. Namely we take

$$(5.20)\qquad g_0(q,t) = \sum_{n=0}^{\infty} \frac{t^{2n}q^{n^2}}{(t)_{n+1}(t^2q;q^2)_n} = F\left(q^{1/2},-q^{1/2};1,2;q,q;t\right),$$

$$(5.21)\quad g_1(q,t) = \sum_{n=0}^{\infty} \frac{t^{2n}q^{n^2+2n}}{(t)_{n+1}(t^2q;q^2)_{n+1}} = \frac{1}{(1-t^2q)} F\left(q^{3/2},-q^{3/2};1,2;q,q^3;t\right),$$

$$(5.22)\quad g_2(g,t) = \sum_{n=0}^{\infty} \frac{t^{2n}q^{n^2+n}}{(t)_{n+1}(t^2q;q^2)_{n+1}} = \frac{1}{(1-t^2q)} F\left(q^{3/2},-q^{3/2};1,2;q,q^2;t\right).$$

With this choice of $g_i(q,t)$ we see that $(1-t)g_i(q,t)$ evaluated at $t=1$ yields $G_i(q)$ and at $t=-1$ yields $F_i(q)$. All that remains is to determine whether the polynomial coefficients in the $g_i(q,t)$ are indeed RS-polynomials. We carry out the details only in the case $i=0$.

If $g_0(q,t) = \Sigma_{n\geq 0}\, \alpha_n t^n$, then by (3.6), the α_n are defined by $\alpha_0=1$, $\alpha_1=1$, $\alpha_2=1+q$, and for $n\geq 3$

$$(5.23)\qquad \alpha_n = \alpha_{n-1} + (q+q^{n-1})\alpha_{n-2} - q\,\alpha_{n-3}.$$

Now it follows directly from the recurrences for $D_{7,3}(N,M;\alpha,\beta;q)$ given in the proof of Theorem 1 of Andrews et al. (1985) that $D_{7,3}(\left[\frac{n+1}{2}\right], \left[\frac{n}{2}\right]; 2,2;q)$ also satisfies the above inital conditions and recurrence. Hence

$$(5.24)\quad g_0(q,t) = \sum_{n=0}^{\infty} D_{7,3}(\left[\tfrac{n+1}{2}\right], \left[\tfrac{n}{2}\right];2,2;q)t^n.$$

Thus the relevant ρ is as before $\rho_{2;1,0}$ and the Ramanujan statistic is $(\rho_{2;1,0}, S_{7,3}(2,2))$.

In exactly the same way

$$(5.25)\qquad g_1(q,t) = \sum_{n=0}^{\infty} D_{7,1}(\left[\tfrac{n+5}{2}\right], \left[\tfrac{n}{2}\right];2,2;q)t^n;$$

the relevant ρ is now $\rho_{2;5,0}$ and the Ramanujan statistic is $(\rho_{2;5,0}, S_{7,1}(2,2))$.

Finally

$$(5.26)\quad g_2(q,t) = \sum_{n=0}^{\infty} D_{7,2}(\left[\tfrac{n+3}{2}\right],\left[\tfrac{n}{2}\right];2,2;q)t^n;$$

the relevant ρ is now $\rho_{2;3,0}$ and the Ramanujan statistic is $(\rho_{2;3,0}, S_{7,2}(2,2))$.

6. The Sixth Order Mock Theta Functions.

There are two functions that appear in many identities in Ramanujan's "Lost" Notebook (c.f. Andrews; 1979). For reasons that wil become clear we choose to call these sixth order mock theta functions. These functions are

$$(6.1)\qquad \phi_6(q) = \sum_{n=0}^{\infty} \frac{(-1)^n q^{n^2} (q;q^2)_n}{(-q)_{2n}},$$

and

$$(6.2)\qquad \psi_6(q) = \sum_{n=0}^{\infty} \frac{(-1)^n q^{n^2+2n} (q;q^2)_n}{(-q)_{2n+1}}.$$

As with the seventh order mock theta functions, we must refer to the literature to obtain a starting point of study.

$$(6.3)\qquad R_\phi(q) = \sum_{n=0}^{\infty} \frac{q^{n^2} (-q;q^2)_n}{(q)_{2n}}$$

$$= \prod_{n=1}^{\infty} \frac{(1-q^{12n})(1-q^{12n-6})^2}{(1-q^n)}$$

(while this result does not explicitly appear in the literature it is easily derived by setting $a = -f = 1$ in (iv) page 6 of Bailey (1949) and then substituting the resulting α_n and β_n in (3.1) of the same paper with $\rho_1 \to \infty$, $\rho_2 \to \infty$)

$$= \sum_{n=0}^{\infty} P(S_{6,3}(2,2);n)q^n$$

$$(6.4)\qquad R_\psi(q) = \sum_{n=0}^{\infty} \frac{q^{n^2+2n} (-q;q^2)_n}{(q)_{2n+1}}$$

$$= \prod_{\substack{n=1 \\ n\not\equiv 0,\pm 2 \pmod{12}}}^{\infty} (1-q^n)^{-1}$$

(Slater; 1952, p. 157, eq. (50))

$$= \sum_{n=0}^{\infty} P(S_{6,1}(2,2);n)q^n$$ (Andrews et al.; 1985, Th. 3).

As with the seventh order mock theta functions, we can compare (6.3) with (6.1) and (6.4) with (6.2). In this way we are led to consider

$$\text{(6.5)} \qquad r_\phi(q,t) = \sum_{n=0}^{\infty} \frac{t^n q^{n^2}(-tq;q^2)_n}{(t;q^2)_{n+1}(tq;q^2)_n} = F\begin{pmatrix}-q;2,1;q,q;t\\ q\end{pmatrix}$$

$$\text{(6.6)} \qquad r_\psi(q,t) = \sum_{n=0}^{\infty} \frac{t^n q^{n^2+2n}(-tq;q^2)_n}{(t;q^2)_{n+1}(tq;q^2)_{n+1}} = \frac{1}{1-tq}\, F\begin{pmatrix}-q;2,1;q,q^3;t\\ q^3\end{pmatrix}.$$

Note that the right side of (6.5) multiplied by $(1-t)$ reduces to (6.3) for $t=1$ and to (6.1) for $t = -1$, while the right side of (6.6) multiplied by $(1-t)$ reduces to (6.4) for $t=1$ and to (6.2) for $t = -1$.

In the manner that we treated $g_0(q,t)$ we can also treat $r_\phi(q,t)$ and $r_\psi(q,t)$ so we omit the details.

Theorem 2. $r_\phi(q,t)$ is the RS-generating function for the Ramanujan statistic $(\rho_{1;1,0}, S_{6,3}(2,2))$. $r_\psi(q,t)$ is the RS-generating function for the Ramanujan statistic $(\rho_{1;2,0}, S_{6,1}(2,2))$.

7. A Further Example From the "Lost" Notebook.

As noted in Section 1, I devoted a lengthy paper (Andrews; 1981a) to the study of a number of results in Ramanujan's "Lost" Notebook. All these identities were provable by a q-analog of the classical partial fractions decomposition (Andrews; 1981a, Lemma 2, p. 21). No combinatorial interpretation of these functions was given; however, the first thirty-six Maclaurin series coefficients of $\alpha(q)$, $\beta(q)$ and $\mu(q)$ were listed where

$$\text{(7.1)} \qquad \alpha(q) = \sum_{n=0}^{\infty} \frac{q^{(n+1)^2}(-q;q^2)_n}{(q;q^2)^2_{n+1}}$$

(Andrews (1981); p. 35, eq. (3.27)),

$$(7.2)\qquad \beta(q) = \sum_{n=0}^{\infty} \frac{q^{n^2+n}(-q^2;q^2)_n}{(q;q^2)_{n+1}^2}$$

(Andrews; 1981a, p. 35, eq. (1.8) and eq. (4.3) both corrected),

$$(7.3)\qquad \mu(q) = \sum_{n=0}^{\infty} \frac{(-1)^n q^{n^2}(q;q^2)_n}{(-q^2;q^2)_n^2}$$

(Andrews; 1981a; p. 35, eq. (1.7)).

To get started here, we observe that

$$(7.4)\qquad \sum_{n=0}^{\infty} \frac{q^{n^2}(-q;q^2)_n}{(q^2;q^2)_n^2} = \prod_{n=0}^{\infty} \frac{(1+q^{2n+1})}{(1-q^{2n+2})}$$

(Andrews; 1981a, p. 20, eq. (3.12)).

Thus we define

$$(7.5)\qquad T(q,t) = \sum_{n=0}^{\infty} \frac{t^n q^{n^2}(-tq;q^2)_n}{(t;q^2)_{n+1}(tq^2;q^2)_n} = F\begin{pmatrix}-q;2,1;q,q;t\\ q^2\end{pmatrix}.$$

As always $(1-t)T(q,t)$ reduces to the left hand side of (7.4) at $t=1$ and to (7.3) at -1. Furthermore $(1-q)\beta(q) = T(q,q)$. While $\alpha(q)$ does not specialize directly from (7.5), we shall nonetheless find that it fits in well with our methods; indeed the lead to our treatment of $T(q,t)$ is given by the fact that

$$(7.6)\qquad \alpha(q) = \sum_{n=0}^{\infty} \frac{q^{n+1}(-q^2;q^2)_n}{(q;q^2)_{n+1}}.$$

The proof of (7.6) given by Andrews (1981a) pp. 28-29 is easily adapted to show that

$$(7.7)\qquad T(q,t) = \sum_{n=0}^{\infty} \frac{(-q;q^2)_n t^n}{(tq^2;q^2)_n}.$$

Theorem 3. $T(q,t)$ is the RS-generating function for the Ramanujan statistic $(\gamma,\mathcal{O})$ where $\gamma(\pi) = [\frac{1}{2}(\ell(\pi)+2\epsilon(\pi)+1]$ ($\epsilon(\pi)$ is the number of even parts of π) and $\mathcal{O}$ is the set of all partitions that have no repeated odd parts.

Proof. This is immediate from (7.7). □

Since only β and μ are specializations of (7.7), we have yet to consider α. However from (7.6) we see immediately that $\alpha(q)$ is the generating function for partitions with no repeated even parts and with every part less than twice the number of ones.

8. Regime II of the Hard Hexagon Model.

In Andrews (1981b), I stated six Rogers-Ramanujan identities related to Regime II in Baxter's (1980), (1982) solution of the hard hexagon model. These six identities were obtained from two polynomial identities:

$$(8.1)\qquad \sum_{n,r\geqq 0} q^{n(3n+1)/2+r}\begin{bmatrix}N-2n-2r\\ n\end{bmatrix}_q\begin{bmatrix}r+n\\ r\end{bmatrix}_{q^2} = \sum_{\lambda=-\infty}^{\infty}(-1)^\lambda q^{5\lambda^2+\lambda}\begin{bmatrix}N\\ \left[\frac{N-5\lambda}{2}\right]\end{bmatrix}_q ,$$

$$(8.2)\qquad \sum_{n,r\geqq 0} q^{3n(n+1)/2+r}\begin{bmatrix}N-2n-2r-1\\ n\end{bmatrix}_q\begin{bmatrix}r+n\\ r\end{bmatrix}_{q^2} = \sum_{\lambda=-\infty}^{\infty}(-1)^\lambda q^{5\lambda^2-3\lambda}\begin{bmatrix}N\\ \left[\frac{N-5\lambda}{2}\right]+1\end{bmatrix}_q .$$

The six identities in question are derived from (8.1) and (8.2) by replacing N by $3N+a$ $(a=0, 1)$, then replaicng q by q^{-1} and then multiplying by q raised to the degree of the polynomial being considered. That the righthand sides of the resulting identities converge to the appropriate theta functions has been shown in Andrews et al. (1984), Appendix B and Theorem 2.6.12. The convergence of the lefthand sides to the Rogers-Ramanujan type series is simple algebraic manipulation. The question is then: Where did (8.1) and (8.2) come from?

We may answer this by noting that the generating function identitites for the polynomials in (8.1) and (8.2) are respectively:

$$\sum_{n=0}^{\infty} \frac{t^{3n} q^{n(3n+1)/2}}{(t)_{n+1}(t^2q;q^2)_{n+1}} \quad \sum_{N=0}^{\infty} t^N \sum_{\lambda=-\infty}^{\infty} (-1)^{\lambda} q^{\lambda(5\lambda+1)} \begin{bmatrix} N \\ \left[\frac{N-5\lambda}{2}\right] \end{bmatrix}, \tag{8.3}$$

and

$$\sum_{n=0}^{\infty} \frac{t^{3n} q^{n(3n+3)/2}}{(t)_{n+1}(t^2q;q^2)_{n+1}} = \sum_{N=0}^{\infty} t^N \sum_{\lambda=-\infty}^{\infty} (-1)^{\lambda} q^{\lambda(5\lambda-3)} \begin{bmatrix} N \\ \left[\frac{N-5\lambda}{2}\right]+1 \end{bmatrix}. \tag{8.4}$$

Once we have written down (8.3) and (8.4), we see immediately that we are dealing with RS-generating functions. In fact, we may prove the following result in the exact same way that we treated the seventh order mock theta functions in Section 5.

Theorem 4. The function given in (8.3) is the RS-generating function for the Ramanujan statistic $(\rho_{2;1,0}, S_{5,2}(2,2))$. The function given in (8.4) is the RS-generating function for the Ramanujan statistic $(\rho_{2;3,0}, S_{5,1}(2,2))$.

Finally we observe that the application of (2.3) to (8.3) yields in this instance

$$\sum_{n=0}^{\infty} \frac{q^{n(3n+1)/2}}{(q)_n(q;q^2)_{n+1}} = \prod_{\substack{n=1 \\ n\neq 0,\pm 4 \pmod{10}}}^{\infty} (1-q^n)^{-1}. \tag{8.5}$$

Identity (8.5) does not appear in the literature; instead we find

$$\sum_{n=0}^{\infty} \frac{q^{n(3n-1)/2}}{(q)_n(q;q^2)_n} = \prod_{\substack{n=1 \\ n\neq 0,\pm 4 \pmod{10}}}^{\infty} (1-q^n)^{-1}, \tag{8.6}$$

(Slater; 1952, eq. (46)).

However (8.5) and (8.6) are equivalent because

$$\sum_{n=0}^{\infty} \frac{q^{n(3n+1)/2}}{(q)_n(q;q^2)_{n+1}} - \sum_{n=0}^{\infty} \frac{q^{n(3n-1)/2}}{(q)_n(q;q^2)_n} \tag{8.7}$$

$$= \sum_{n=0}^{\infty} \frac{q^{n(3n-1)/2}(q^n-(1-q^{2n+1}))}{(q)_n(q;q^2)_{n+1}}$$

$$= \sum_{n=0}^{\infty} \frac{q^{n(3n-1)/2}(-(1-q^n)(1-q^{2n+1})+q^{3n+1})}{(q)_n(q;q^2)_{n+1}}$$

$$= -\sum_{n=1}^{\infty} \frac{q^{n(3n-1)/2}}{(q)_{n-1}(q;q^2)_n} + \sum_{n=0}^{\infty} \frac{q^{n(3n-1)/2+3n+1}}{(q)_n(q;q^2)_{n+1}}$$

$$= 0$$

since the first sum becomes identical with the second if we replace n by $n+1$.

9. Conclusion.

Ramanujan statistics are obviously related to many q-series. As we have seen many of these series previously had no known combinatorial interpretation. Obviously much remains to be done in the study of Ramanujan statistics. Our two methods are clearly ad hoc. Furthermore we would like some extension of our results of Sections 2 and 3 that would allow us to consider all the $D_{K,i}(N,M;\alpha,\beta;q)$ polynomials.

REFERENCES

Andrews, G. E. (1971). Number Theory. Philadelphia, Pa.: W. B. Saunders (reprinted in 1984 by Hindustan Publ. of New Delhi).

Andrews, G. E. (1972). Partition identities. Advances in Math., 9, 10-51.

Andrews, G. E. (1976). The Theory of Partitions. Reading, Mass.: Addison-Wesley.

Andrews, G. E. (1977). Partitions, q-series and the Lusztig-Macdonald-Wall conjectures, Invent. Math., 41(1977), 91-102.

Andrews, G. E. (1979). An introduction to Ramanujan's "lost" notebook. Amer. Math. Monthly, 86, 89-108.

Andrews, G. E. (1981a). Mordell integrals and Ramanujan's "lost" notebook. In Analytic Number Theory, Lecture Notes in Mathematics, No. 899, ed. M. I. Knopp, pp. 10-48. Berlin: Springer-Verlag.

Andrews, G. E. (1981b). The hard-hexagon model and Rogers-Ramanujan type identities. Proc. Nat. Acad. Sci. U.S.A., 78, 5290-5292.

Andrews: Combinatorics and Ramanujan's "Lost" Notebook 22

Andrews, G. E. (1984). On the Wall polynomials and the L-M-W conjectures, J. Austral. Math. Soc. (Series A), 37, 17-26.

Andrews, G. E. (1985). The fifth and seventh order mock theta functions. (to appear).

Andrews, G. E. & Baxter, R. J. & Forrester, P. J. (1984). Eight-vertex SOS model and generalized Rogers-Ramanujan-type identities. J. Stat. Phys., 35, 193-266.

Andrews, G. E. & Baxter, R. J. & Bressoud, D. M. & Burge, W. & Forrester, P. J. & Viennot, G. (1985). Partitions with prescribed hook differences. (to appear).

Bailey, W. N. (1949). Identities of the Rogers-Ramanujan type. Proc. London Math. Soc. (2), 50, 1-10.

Baxter, R. J. (1980). Hard hexagons: exact solution. J. Phys. A: Math. Gen., 13, L61-L70.

Baxter, R. J. (1982). Exactly Solved Models in Statistical Mechanics. London: Academic Press.

Chaundy, T. W. (1962). Frank Hilton Jackson. J. London Math. Soc., 37, 126-128.

Fine, N. J. (1985). Some Basic Hypergeometric Series and Applications. (to appear).

Garsia, A. & Milne, S. (1981). A Rogers-Ramanujan bijection. J. Comb. Th. (A), 31, 289-339.

Hardy, G. H. (1940). Ramanujan. Cambridge: Cambridge University Press. (Reprinted by Chelsea, New York).

Hecke, E. (1959). Über einen Zusammenhang zwischen elliptischen Modulfunktionen und indefiniten quadratischen Formen. In Mathematische Werke, pp. 418-427. Göttingen: Vandenhoeck and Ruprecht.

Jacobi, C. G. J. (1829). Fundamenta Nova Theoriae Funktionum Ellipticarum. Regiomonti: Fratrum Bornträger. (Reprinted in Gesammelte Werke, Vol. 1, pp. 49-239. Berlin: Reimer).

Lusztig, G. (1977). Irreducible representation of finite classical groups, Invent. Math., 43, 125-177.

Ramanujan, S. (1927). Collected Papers. Cambridge: Cambridge University Press. (Reprinted by Chelsea, New York).

Rankin, R. A. (1982). Ramanujan's manuscripts and notebooks. Bull. London Math. Soc., 14: 81-97.

Richmond, L. B. & Szekeres, G. (1981). Some formulas related to dilogarithms, the zeta-function and the Andrews-Gordon identities. J. Austral. Math. Soc., 31, 362-373.

Rogers, L. J. (1894). Second memoir on the expansion of certain infinite products. Proc. London Math. Soc., 25, 318-343.

Schur, I. (1917). Ein Beitrag zur additiven Zahlentheorie und zur Theorie der Kettenbrüche, Sitzungsber. Akad. Wissensch., Berlin, Phys. - Math. Kl., 302-321.

Sears, D. B. (1951). On the transformation theory of basic hypergeometric functions. Proc. London Math. Soc. (2), 53, 158-180.

Slater, L. J. (1952). Further identities of the Rogers-Ramanujan type. Proc. London Math. Soc. (2), 54, 147-167.

Watson, G. N. (1929). A new proof of the Rogers-Ramanujan identities. J. London Math. Soc., 4, 4-9.

Watson, G. N. (1936). The final problem: an account of the mock theta functions. J. London Math. Soc., 11, 55-80.

Watson, G. N. (1937). The mock theta functions (2). Proc. London Math. Soc. (2), 42, 272-304.

3.5.1 Erratum

In the preceding article, it is stated, "Identity (8.5) is not in the literature." In fact, (8.5) had appeared at least twice in the literature before its appearance here: on page 330 (2), line 1 of Rogers [Rog17], and on page 158, Eq. (62) of Slater [Sla52].

3.6 On a Partition Problem of H.L. Alder

PACIFIC JOURNAL OF MATHEMATICS
Vol. 36, No. 2, 1971

ON A PARTITION PROBLEM OF H. L. ALDER

GEORGE E. ANDREWS

We study $\Delta_d(n) = q_d(n) - Q_d(n)$, where $q_d(n)$ is the number of partitions of n into parts differing by at least d, and $Q_d(n)$ is the number of partitions of n into parts congruent to 1 or $d+2 \pmod{d+3}$. We prove that $\Delta_d(n) \to +\infty$ with n for $d \geqq 4$, and that $\Delta_d(n) \geqq 0$ for all n if $d = 2^s - 1, s \geqq 4$.

In 1956, H. L. Alder proposed the following problem [1].

"Let $q_d(n)$ = the number of partitions of n into parts differing by at least d; let $Q_d(n)$ = the number of partitions of n into parts congruent to 1 or $d + 2 \pmod{d+3}$; let $\Delta_d(n) = q_d(n) - Q_d(n)$. It is known that $\Delta_1(n) = 0$ for all positive n (Euler's identity), $\Delta_2(n) = 0$ for all positive n (one of the Rogers-Ramanujan identities), $\Delta_3(n) \geqq 0$ for all positive n (from Schur's theorem which states $\Delta_3(n)$ = the number of those partitions of n into parts differing by at least 3 which contain at least one pair of consecutive multiples of 3). (a) Is $\Delta_d(n) \geqq 0$ for all positive d and n? (b) If (a) is true, can $\Delta_d(n)$ be characterized as the number of a certain type of restricted partitions of n as is the case for $d = 3$?"

This problem was again mentioned in [2; p. 743] as still being open. A recent general result on partitions with difference conditions [3] allows us to give some partial answers to Alder's problem.

First we derive a partition theorem which is somewhat analogous to the type of result asked for by Alder.

THEOREM 1. *Let ν be the largest integer such that $2^{\nu+1} - 1 \leqq d$. Let $\mathscr{L}_d(n)$ denote the number of partitions of n into distinct parts $\equiv 1, 2, 4, \cdots,$ or $2^\nu \pmod{d}$. Then*

$$q_d(n) \geqq \mathscr{L}_d(n) .$$

We may utilize some asymptotic formulae of Meinardus [4], [5] to prove

THEOREM 2. *For any $d \geqq 4$, $\lim_{n\to\infty} \Delta_d(n) = +\infty$*

Finally, Theorem 1 may be utilized to prove a result which settles Alder's problem in an infinite number of cases

THEOREM 4. *If $d = 2^s - 1$ and $s = 1, 2,$ or $\geqq 4$, then $\Delta_d(n) \geqq 0$ for all n.*

The proof of Theorem 4 relies on the following result which is of independent interest.

THEOREM 3. *Let* $S = \{a_i\}_{i=1}^{\infty}$ *and* $T = \{b_i\}_{i=1}^{\infty}$ *be two strictly increasing sequences of positive integers such that* $b_1 = 1$ *and* $a_i \geqq b_i$ *for all* i. *Let* $\rho(S; n)$ (resp. $\rho(T; n)$) *denote the number of partitions of* n *into parts taken from* S (resp. T). *Then*

$$\rho(T; n) \geqq \rho(S; n)$$

for all n.

2. Proof of Theorem 1. In Theorem 1 of [3] set $N = d$, $a(1) = 1$, $a(2) = 2, \cdots, a(\nu + 1) = 2^{\nu}$. Thus in the notation of [3], $D(A_N; n)$ becomes $\mathscr{L}_d(n)$. Now $D(A_N; n) = E(A'_N; n)$ where the latter partition function is the number of partitions of n:

$$n = b_1 + b_2 + \cdots + b_s,$$

$$b_i \equiv 1, 2, 3, 4, \cdots, 2^{\nu+1} - 1 \pmod d$$

with

$$b_i - b_{i+1} \geqq dw(\beta_d(b_{i+1})) + v(\beta_d(b_{i+1})) - \beta_d(b_{i+1}) .$$

Here $\beta_d(m)$ is the least positive residue of $m \bmod d$, $w(m)$ is the number of powers of 2 in the binary representation of m and $v(m)$ is the least power of 2 in the binary representation of m. Consequently if $b_{i+1} \equiv 2^j \pmod d$, $0 \leqq j \leqq \nu$,

$$dw(\beta_d(b_{i+1})) + v(\beta_d(b_{i+1})) - \beta_d(b_{i+1}) = d \cdot 1 + 2^j - 2^j = d .$$

If $b_{i+1} \not\equiv 2^j \pmod d$ $0 \leqq j \leqq \nu$, then

$$\begin{aligned} &dw(\beta_d(b_{i+1})) + v(\beta_d(b_{i+1})) - \beta_d(b_{i+1}) \\ &\qquad \geqq 2 \cdot d + 1 - (2^{\nu+1} - 1) \geqq 2d + 1 - d = d + 1 . \end{aligned}$$

Thus the difference condition is always $b_i - b_{i+1} \geqq d$ or stronger. Therefore $E(A'_N; n) \leqq q_d(n)$ and Theorem 1 follows.

3. Proof of Theorem 2. Meinardus has proved a general theorem on asymptotic formulae for partitions with repetitions [4]. Following the notation of Meinardus [4; pp. 388–389], we see that to treat $Q_d(n)$, we must have his

$$a_n = \begin{cases} 1 \text{ if } n \equiv 1, d + 2 \pmod{d + 3} \\ 0 \text{ otherwise} . \end{cases}$$

Under these circumstances, Meinardus's $D(s)$ satisfies

$$D(s) = (d + 3)^{-s}\left(\zeta\left(s, \frac{1}{d + 3}\right) + \zeta\left(s, \frac{d + 2}{d + 3}\right)\right) ,$$

where $\zeta(s, a) = \sum_{n=1}^{\infty}(n + a)^{-s}$, the Hurwitz zeta function [6; Ch. XIII], α, the abscissa of convergence for $D(s)$ is 1, and A, the residue at $s = 1$ is $2/d + 3$.

$$g(\tau) = \frac{e^{-\tau} + e^{-(d+2)\tau}}{1 - e^{-(d+3)\tau}} .$$

One may now easily verify that Meinardus's analytic conditions on $D(s)$ and $g(\tau)$ are fulfilled, thus

$$\log Q_d(n) \sim 2\pi\sqrt{\frac{n}{3d + 9}} . \tag{3.1}$$

In [5], Meinardus has derived the asymptotic formula

$$\log q_d(n) \sim 2\sqrt{A_d n} , \tag{3.2}$$

where

$$A_d = \frac{d}{2}\log^2 \alpha_d + \sum_{r=1}^{\infty}\frac{(\alpha_d)^{rd}}{r^2} ,$$

and α_d is real >0, $\alpha_d^d + \alpha_d - 1 = 0$.

If we put $\alpha_d = e^{-\lambda_d}$, so that $e^{-d\lambda_d} + e^{-\lambda_d} = 1$, then

$$\begin{aligned} A_d &= \frac{d}{2}\lambda_d^2 + \sum_{r=1}^{\infty}\frac{\alpha_d^{d\cdot r}}{r^2} > \frac{d}{2}\lambda_d^2 + \alpha_d^d \\ &= \frac{d}{2}\lambda_d^2 + 1 - e^{-\lambda_d} > \frac{d}{2}\lambda_d^2 + \lambda_d - \frac{1}{2}\lambda_d^2 \\ &= \frac{d-1}{2}\lambda_d^2 + \lambda_d . \end{aligned}$$

Now the following table shows that

$$A_d > \pi^2/(3d + 9) \text{ for } 4 \leqq d \leqq 14$$

TABLE 1.

d	$\lambda_d >$	$\frac{d-1}{2}\lambda_d^2 >$	$A_d >$	$\frac{\pi^2}{3d+9} <$
4	0.32	0.153	0.473	0.471
5	0.28	0.15	0.43	0.42
6	0.25	0.15	0.40	0.37
7	0.22	0.14	0.36	0.33
8	0.20	0.14	0.34	0.30
9	0.19	0.14	0.33	0.28
10	0.18	0.14	0.32	0.26
11	0.16	0.12	0.28	0.24
12	0.15	0.12	0.27	0.22
13	0.15	0.13	0.28	0.21
14	0.14	0.12	0.26	0.20

For $d \geqq 15$, we have

$$e^{-d(2/d)} + e^{-2/d} > e^{-2} + 1 - 2/d > 1 ,$$

Hence, $\lambda_d > 2/d$ and

$$A_d > \frac{d-1}{2}\left(\frac{2}{d}\right)^2 + 2/d = \frac{1}{d}(4 - 2/d) > \frac{10}{3d} > \frac{\pi^2}{3d+9} .$$

Thus for all $d \geqq 4$,

$$A_d > \frac{\pi^2}{3d+9} .$$

Hence comparing (3.1) with (3.2) we find

$$\lim_{n\to\infty} (\log q_d(n) - \log Q_d(n)) = +\infty .$$

Thus $\lim_{n\to\infty} \Delta_d(n) = \lim_{n\to\infty} q_d(n)(1 - Q_d(n)/q_d(n)) = +\infty$. and we have Theorem 2.

I would like to thank the referee for aid in simplifying and extending Theorem 2.

4. Proof of Theorem 3. Let us define $S_i = \{a_1, a_2, \cdots a_i\}$ and $T_i = \{b_1, b_2, \cdots, b_i\}$. We shall proceed to prove by induction on i that $\rho(T_i; n) \geqq \rho(S_i; n)$; this will establish Theorem 3 for if we choose I such that $a_I > n$, $b_I > n$, then $\rho(T; n) = \rho(T_I; n) \geqq \rho(S_I; n) = \rho(S; n)$.

First we remark that $\rho(T_i; n)$ is a nondecreasing function of n; this is because $1 = b_1 \in T_i$ and thus every partition of $n-1$ into parts taken from T_i may be transformed into a partition of n merely by adjoining a 1.

Now $\rho(T_1; n) = 1$ for all n since $T_1 = \{1\}$. Since $S_1 = \{a_1\}$

$$\rho(S_1; n) = \begin{cases} 1 \text{ if } a_1 \mid n \\ 0 \text{ otherwise} . \end{cases}$$

Hence

$$\rho(T_1; n) \geqq \rho(S_1; n) .$$

Now assume that $\rho(T_{i-1}; n) \geqq \rho(S_{i-1}; n)$ for all n. Hence if we define $\rho(T_i; 0) = \rho(S_i; 0) = 1$,

$$\begin{aligned}
&\sum_{n=0}^{\infty} (\rho(T_i; n) - \rho(S_i; n))q^n \\
&= \prod_{j=1}^{i} \frac{1}{1-q^{b_j}} - \prod_{j=1}^{i} \frac{1}{1-q^{a_j}} \\
&= \left(\prod_{j=1}^{i-1} \frac{1}{1-q^{b_j}}\right)\left(\frac{1}{1-q^{a_i}} + \frac{q^{b_i} - q^{a_i}}{(1-q^{a_i})(1-q^{b_i})}\right) - \prod_{j=1}^{i} \frac{1}{1-q^{a_j}}
\end{aligned}$$

$$= \frac{1}{1-q^{a_i}}\left(\prod_{j=1}^{i-1}\frac{1}{1-q^{b_j}} - \prod_{j=1}^{i-1}\frac{1}{1-q^{a_j}}\right) + \frac{q^{b_i}-q^{a_i}}{(1-q^{a_i})}\prod_{j=1}^{i}\frac{1}{1-q^{b_j}}$$

$$= \frac{1}{1-q^{a_i}}\left(\sum_{n=0}^{\infty}(\rho(T_{i-1};n) - \rho(S_{i-1};n))q^n\right.$$

$$\left. + \sum_{n=0}^{\infty}(\rho(T_i;n-b_i) - \rho(T_i;n-a_i))q^n\right).$$

Now the coefficients of these two infinite series are nonnegative: the first by the induction hypothesis, and the second by the fact that $\rho(T_i;n)$ is a nondecreasing sequence. Since $(1-q^{a_i})^{-1} = \sum_{j=0}^{\infty} q^{ja_i}$, we see that all coefficients in the power series expansion of our last expression must be nonnegative. Hence

$$\rho(T_i;n) \geqq \rho(S_i;n),$$

and Theorem 3 is proved.

5. Proof of Theorem 4. Since $d = 2^s - 1$, we see that the ν of Theorem 1 is just $s-1$. Now

$$\sum_{n=0}^{\infty}\mathscr{L}_d(n)q^n = \prod_{j=0}^{\infty}(1+q^{dj+1})(1+q^{dj+2})\cdots(1+q^{dj+2^{\nu}})$$

$$= \prod_{j=0}^{\infty}\frac{1}{(1-q^{2dj+1})(1-q^{2dj+d+2})(1-q^{2dj+d+4})\cdots(1-q^{2dj+d+2^{\nu}})}.$$

Thus $\mathscr{L}_d(n) = \rho(T;n)$ where $T = \{m \mid m \equiv 1, d+2, d+4, \cdots, \text{ or } d+2^{s-1} \pmod{2d}\}$. Clearly, $1 \in T$. We now show that for $s \geqq 4$ the ith element of T (arranged in increasing magnitude) is no larger than the ith element of S where $S = \{m \mid m \equiv 1, d+2 \pmod{d+3}\}$. Since $s \geqq 4$, the first four elements of T are

$$1, d+2, d+4, d+8 \; (2d+5 > d+8 \text{ since } d \geqq 15).$$

Thus the first four elements of T are less than or equal the first four elements of S respectively. In general the $(4m+1)-st$ element of T is $\leqq 2dm+1$ while the $(4m+1)-st$ element of S is $2m(d+3)+1$; for $2 \leqq j \leqq 4$ the $(4m+j)-$th element of T is $\leqq 2dm+d+2^{j-1}$ while the $(4m+j)-$ element of S is $\geqq 2m(d+3)+d+2$ and for $2 \leqq j \leqq 4$, $m \geqq 1$, $2dm+d+2^{j-1} \leqq 2dm+d+8 \leqq 2dm+d+6+2 \leqq 2m(d+3)+d+2$. Hence, the conditions of Theorem 3 are met, and therefore

$$q_d(n) \geqq \mathscr{L}_d(n) = \rho(T;n) \geqq \rho(S;n) = Q_d(n).$$

Thus Theorem 4 is established.

6. Conclusion. By modification of the results in [3], it appears possible to apply the techniques of §4 to prove that $\Delta_d(n) \geqq 0$ for any $d \geqq 15$ which is a difference of powers of 2; however, since this approach does not yield a complete answer to Alder's problem it seems hardly worth undertaking.

Lengthier versions of the following table indicate that Alder's problem may be extended as follows.

Conjecture. $\Delta_d(n) > 0$ for $n \geqq d + 6$ if $d \geqq 8$.

n	$\Delta_3(n)$	$\Delta_4(n)$	$\Delta_5(n)$	$\Delta_6(n)$	$\Delta_7(n)$	$\Delta_8(n)$
1	0	0	0	0	0	0
2	0	0	0	0	0	0
3	0	0	0	0	0	0
4	0	0	0	0	0	0
5	0	0	0	0	0	0
6	0	0	0	0	0	0
7	0	0	0	0	0	0
8	0	0	0	0	0	0
9	1	0	0	0	0	0
10	0	1	0	0	0	0
11	0	1	1	0	0	0
12	0	1	1	1	0	0
13	0	0	2	1	1	0
14	0	0	1	2	1	1
15	1	0	1	2	2	1
16	1	0	0	2	2	2
17	1	1	0	1	3	2
18	1	2	0	1	2	3
19	1	2	1	0	2	3
20	1	2	2	0	1	3
21	2	2	3	1	1	2
22	2	2	3	2	0	2
23	2	3	3	3	1	1
24	2	4	3	4	2	1

References

1. H. L. Alder, *Research problem no.* 4, Bull. Amer. Math. Soc. **62** (1956), 76.
2. ———, *Partition identities—from Euler to the present*, Amer. Math. Monthly **76** (1969), 733–746.
3. G. E. Andrews, *A general theorem on partitions with difference conditions*, Amer. J. Math. **91** (1969), 18–24.
4. G. Meinardus, *Asymptotische Aussagen über Partitionen*, Math. Zeit. **59** (1954), 388–398.
5. ———, *Über Partitionen mit Differenzenbedingungen*, Math. Zeit. **61** (1954), 289–302.
6. E. T. Whittaker and G. N. Watson, *Modern Analysis*, 4th Ed., Cambridge University Press, Cambridge, 1927.

Received April 1, 1970. Partially supported by NSF Grant GP-9660.

THE PENNSYLVANIA STATE UNIVERSITY
UNIVERSITY PARK, PENNSYLVANIA

3.7 Partitions and Durfee Dissection

PARTITIONS AND DURFEE DISSECTION.

By George E. Andrews*.

1. Introduction. The Rogers-Ramanujan identities may be stated analytically as follows:

$$\sum_{n=0}^{\infty} \frac{q^{n^2}}{(q)_n} = \prod_{n=0}^{\infty} (1-q^{5n+1})^{-1}(1-q^{5n+4})^{-1}, \tag{1.1}$$

$$\sum_{n=0} \frac{q^{n^2+n}}{(q)_n} = \prod_{n=0}^{\infty} (1-q^{5n+2})^{-1}(1-q^{5n+3})^{-1}, \tag{1.2}$$

where $(a)_0 = 1$ and $(a)_n = (1-a)(1-aq)\cdots(1-aq^{n-1})$. These identities together with their history are discussed at length in [9, Chapter 7].

Recently the identities (1.1) and (1.2) were generalized [6; 8; 9, p. 111] to

$$\sum_{n_1=0}^{\infty} \cdots \sum_{n_{k-1}=0}^{\infty} \frac{q^{N_1^2+N_2^2+\cdots+N_{k-1}^2+N_a+\cdots+N_{k-1}}}{(q)_{n_1}(q)_{n_2}\cdots(q)_{n_{k-1}}}$$

$$= \prod_{\substack{n=1 \\ n\not\equiv 0, \pm a \ (\mathrm{mod}\ 2k+1)}}^{\infty} (1-q^n)^{-1}, \tag{1.3}$$

where $N_j = n_j + n_{j+1} + \cdots + n_{k-1}$ with $1 \leqslant a \leqslant k$. The identity (1.3) reduces to (1.1) when $k = a = 2$ and to (1.2) when $k = a+1 = 2$.

Now the right side of (1.3) is clearly the generating function for $A_{k,a}(n)$, the number of partitions of n into parts $\not\equiv 0, \pm a \pmod{2k+1}$. B. Gordon [12] (see also [1], [3], [9, Chapter 7, Section 3]) has proved that $A_{k,a}(n) = B_{k,a}(n)$, the number of partitions of n of the form $b_1 + b_2 + \cdots + b_s$ where $b_j \geqslant b_{j+1}$, $b_j - b_{j+k-1} \geqslant 2$, and at most $a-1$ of the b_j equal 1.

In [11], R. Askey and I asked for a combinatorial interpretation of the left side of (1.3) which would show directly that it was the generating function for

Manuscript received June 10, 1977.

*Partially supported by National Science Foundation Grant No. MSP 74-07282.

American Journal of Mathematics, Vol. 101, No. 3, pp. 735–742

$B_{k,a}(n)$. It turns out that there is a very natural interpretation for the left side of (1.3), especially when $a=k$; however, the interpretation is completely different from any given previously. To give the new interpretation we must recall the definition of the Durfee square. For example, let us consider the partition $9+7+5+4+4+3+1+1$; its graphical representation is given by

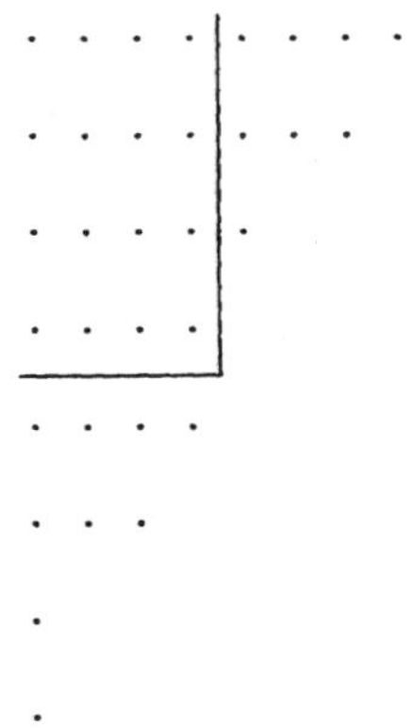

and the "Durfee square" or largest square of dots is the 4×4 square indicated. The Durfee square concept has been extended to other rectangles [2]; however, here we wish to discuss "successive Durfee squares." This concept is almost self-explanatory. Once the Durfee square is determined, it splits the given partition into 3 parts: (1) the square itself, (2) a smaller partition to the right of the square, and (3) a smaller partition below the square. We say that the smaller partition to the right of the square is "attached" or "associated" with the square. One can now determine a "second Durfee square" in the smaller partition below the original Durfee square. Clearly third, fourth, etc. Durfee squares can be determined as long as the lower portion of the partition is not exhausted. Thus the partition $9+7+5+4+4+3+1+1$ has four successive Durfee squares:

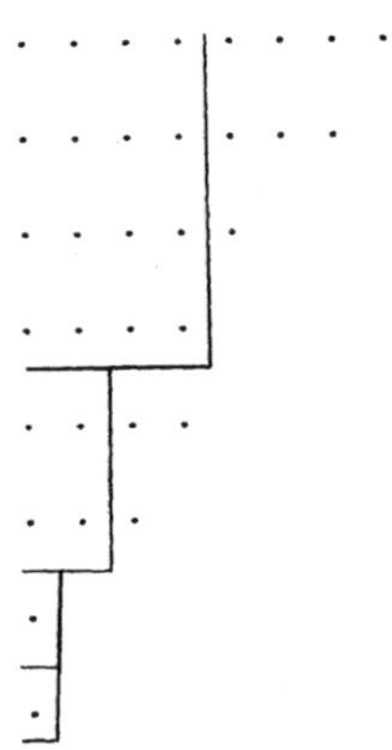

The nicest result we obtain on successive Durfee squares is the following:

THEOREM 1. *The number of partitions of n with at most* $k-1$ *successive Durfee squares equals the number of partitions of n into parts* $\not\equiv 0, \pm k \pmod{2k+1}$.

We remark that Theorem 1 directly implies the first Rogers-Ramanujan identity in the case $k=2$. To see this most easily, we note that the generating function for partitions with at most one Durfee square is

$$\sum_{n=0}^{\infty} \frac{q^{n+n+\cdots+n}}{(q)_n} = \sum_{n=0}^{\infty} \frac{q^{n^2}}{(q)_n},$$

while the generating function for partitions with parts $\not\equiv 0, \pm 2 \pmod 5$ is

$$\prod_{n=0}^{\infty} \frac{1}{(1-q^{5n+1})(1-q^{5n+4})},$$

and (1.1) follows from the assertion that these two functions are identical.

Now, of course, the identity (1.3) contains the two parameters a and k. In Theorem 2 we interpret all the cases of (1.3). To do this we must introduce the concept of a "Durfee dissection," and we do this in the next section.

2. Durfee Dissections. The idea of successive Durfee squares introduced in Section 1 is adequate to state Theorem 1. To treat the identity (1.3) in full generality we must define the "(k,a)-Durfee dissection" of a partition.

We begin by determining the first $a-1$ successive Durfee squares of the partition. Next we determine $k-a$ maximal rectangles (one below the other) for which the horizontal side has one less dot than the vertical side. For example the (5,3)-Durfee dissection of $9+7+6+6+5+5+5+4+4+4+4+4+3+3+2+2+1+1+1$ is (see following page).

We shall say a partition is "(k,a)-admissible" if it has no parts below the last rectangle in its (k,a)-Durfee dissection and furthermore the lower edge of each of the final $k-a$ rectangles of the (k,a)-Durfee dissection is actually a part of the partition (i.e., there are no dots of the Ferrers graph of the partition to the right and on the same row).

Finally we note that "(k,k)-admissible" means only that the partition has at most $k-1$ successive Durfee squares. We are now prepared to interpret (1.3).

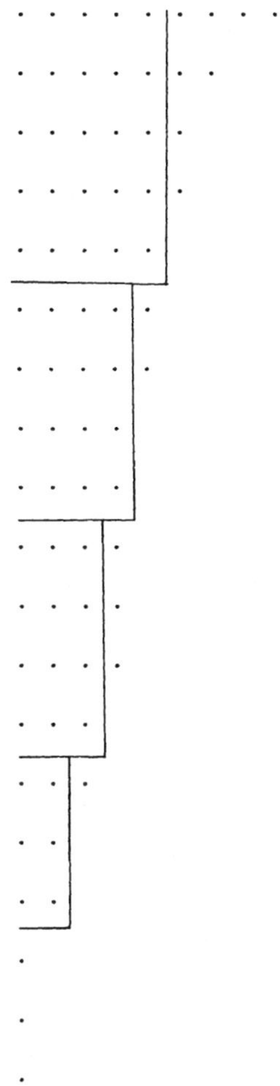

3. The Main Theorems.

THEOREM 2. *For $1 \leqslant a \leqslant k$ the number of (k,a)-admissible partitions of n equals the number of partitions of n into parts* $\not\equiv 0, \pm a \pmod{2k+1}$.

Proof. As is well known, the right hand side of (1.3) is the generating function for $A_{k,a}(n)$ the number of partitions of n into parts $\not\equiv 0, \pm a$ (mod $2k+1$) (see [9, Chapter 7, Section 3]). Thus to prove Theorem 2 we deduce it from (1.3) by showing that the left hand side is the generating function for $\Delta_{k,a}(n)$, the number of (k,a)-admissible partitions of n.

Let us recall that the Gaussian polynomial

$$\begin{bmatrix} n+m \\ m \end{bmatrix} = \frac{(q)_{n+m}}{(q)_n(q)_m}$$

is the generating function for partitions with at most m parts each $\leqslant n$ [9, p. 33]. Now

$$\sum_{n_1 \geqslant 0} \cdots \sum_{n_{k-1} \geqslant 0} \frac{q^{N_1^2+N_2^2+\cdots+N_{k-1}^2+N_a+N_{a+1}+\cdots+N_{k-1}}}{(q)_{n_1}(q)_{n_2}\cdots(q)_{n_{k-1}}}$$

$$= \sum_{n_1 \geqslant 0} \cdots \sum_{n_{k-1} \geqslant 0} \frac{q^{(n_1+n_2+\cdots+n_{k-1})^2}}{(q)_{n_1+n_2+\cdots+n_{k-1}}}$$

$$\times \frac{q^{(n_2+n_3+\cdots+n_{k-1})^2}(q)_{n_1+n_2+\cdots+n_{k-1}}}{(q)_{n_1}(q)_{n_2+n_3+\cdots+n_{k-1}}}$$

$$\times \frac{q^{(n_3+n_4+\cdots+n_{k-1})^2}(q)_{n_2+n_3+\cdots+n_{k-1}}}{(q)_{n_2}(q)_{n_3+n_4+\cdots+n_{k-1}}}$$

$$\times \cdots \times \frac{q^{(n_{a-1}+\cdots+n_{k-1})^2}(q)_{n_{a-2}+\cdots+n_{k-1}}}{(q)_{n_{a-2}}(q)_{n_{a-1}+\cdots+n_{k-1}}}$$

$$\times \frac{q^{(n_a+\cdots+n_{k-1})(n_a+\cdots+n_{k-1}+1)}(q)_{n_{a-1}+\cdots+n_{k-1}}}{(q)_{n_{a-1}}(q)_{n_a+\cdots+n_{k-1}}}$$

$$\times \cdots \times \frac{q^{n_{k-1}(n_{k-1}+1)}(q)_{n_{k-2}+n_{k-1}}}{(q)_{n_{k-2}}(q)_{n_{k-1}}}$$

$$= \sum_{n_1 \geqslant 0} \cdots \sum_{n_{k-1} \geqslant 0} \frac{q^{N_1^2}}{(q)_{N_1}} q^{N_2^2} \begin{bmatrix} N_1 \\ n_1 \end{bmatrix}$$

$$\times q^{N_3^2} \begin{bmatrix} N_2 \\ n_2 \end{bmatrix} \cdots q^{N_{a-1}^2} \begin{bmatrix} N_{a-2} \\ n_{a-2} \end{bmatrix}$$

$$\times q^{N_a(N_a+1)} \begin{bmatrix} N_{a-1} \\ n_{a-1} \end{bmatrix} \cdots q^{N_{k-1}(N_{k-1}+1)} \begin{bmatrix} N_{k-2} \\ n_{k-2} \end{bmatrix}.$$

Now $q^{N_1^2}/(q)_{N_1}$ is clearly the generating function for that portion of the partition attached to a first Durfee square of side N_1. To see this, work with the conjugate of the partition consisting of the dots of the Ferrers graph that are attached to the first Durfee square. The graphical representation of the conjugate of a partition is obtained by reflecting in the main diagonal, the graph of the original partion. For $2 \leqslant j \leqslant a-1$

$$q^{N_j^2} \begin{bmatrix} N_{j-1} \\ n_{j-1} \end{bmatrix}$$

generates those parts of the partition associated with a jth Durfee square of side

N_j. Note that the conjugate of the partition attached to the jth Durfee square has $\leqslant n_{j-1}$ parts each $\leqslant N_{j-1}-n_{j-1}$. For if not, we could enlarge either the $(j-1)$st or the jth maximal Durfee square respectively. Finally, for $a \leqslant j \leqslant k-1$,

$$q^{N_j^2}\begin{bmatrix} N_{j-1} \\ n_{j-1} \end{bmatrix}_{q^{N_j}}$$

generates those parts of the partition associated with a $N_j \times (N_j+1)$ Durfee rectangle where the lower edge of the rectangle is a part of the partition. This last fact follows because $\begin{bmatrix} N_{j-1} \\ n_{j-1} \end{bmatrix}$ enumerates only those partitions attached to the jth Durfee rectangle whose conjugate partition has parts all $\leqslant N_{j-1}-n_{j-1}=N_j$. Thus from this analysis we see that all the (k,a)-admissible partitions are generated by the left side of (1.3), since as $n_1,\ldots,n_{k-1}$ independently range over all nonnegative $(k-1)$-tuples of integers, we see that $N_1,\ldots,N_{k-1}$ range over all nonnegative $(k-1)$-tuples subject to $N_1 \geqslant N_2 \geqslant \cdots \geqslant N_{k-1}$, which is precisely what is required to produce the squares and rectangles in the (k,a)-Durfee dissection. Thus Theorem 2 is established.

Theorem 1 now follows immediately from Theorem 2 and the final remark in Section 2.

4. Generalizations of the Theorems of Euler and Cauchy. In a recent paper [10], we discussed and proved analytically the following identity:

$$\sum_{n_1 \geqslant 0} \cdots \sum_{n_{k-1} \geqslant 0} \frac{q^{N_1^2+N_2^2+\cdots+N_{k-1}^2} z^{N_1+\cdots+N_{k-1}}}{(q)_{n_1}(q)_{n_2}\cdots(q)_{n_{k-1}}(zq)_{n_{k-1}}} = \frac{1}{(zq)_\infty}, \tag{4.1}$$

where $N_j = n_j + n_{j+1} + \cdots + n_{k-1}$, and $(zq)_\infty = \lim_{n\to\infty}(zq)_n$. When $k=2$ this reduces to Cauchy's generalization of an identity of Euler (see [9, p. 20]). The coefficient of $z^n q^n$ on the right hand side of (4.1) is well known to be the number of partitions of n into m parts [9, p. 16]. From our analysis in Section 3 we see that

$$\frac{q^{N_1^2+N_2^2+\cdots+N_{k-1}^2} z^{N_1+\cdots+N_{k-1}}}{(q)_{n_1}(q)_{n_2}\cdots(q)_{n_{k-1}}} \tag{4.2}$$

is the generating function for partitions with at most $k-1$ successive Durfee squares (whose sides are $N_1, N_2, \ldots, N_{k-1}$). The factor $(zq)_{n_{k-1}}^{-1}$ generates partitions whose parts are all $\leqslant N_{k-1} = n_{k-1}$ [9, p. 16]. Hence the product of the expression (4.2) with $(zq)_{n_{k-1}}^{-1}$ generates all partitions with the first $k-1$ successive Durfee squares having edges $N_1, N_2, \ldots, N_{k-1}$. Note that in this case we split the Ferrers graph of a partition up into two parts: (1) the first $N_1 + \cdots + N_{k-1}$ rows, and (2) the remaining rows (each having $\leqslant N_{k-1} = N_{k-1}$ dots). Finally we sum over all nonnegative $(k-1)$-tuples $n_1, \ldots, n_{k-1}$, and we obtain the generating function for all partitions. Hence the left side of (4.1) equals the right side.

5. Conclusion. It is my belief that the type of analysis of partitions described here may lead to other interesting and new partition identities. Obviously the Durfee square generalizations treated in [2] could now be utilized in Durfee dissections. Thus, for example, (4.1) could be extended in the way that identities (2.11) and (2.12) of [2] extend the case $k=2$ of (4.1). Also the various special cases of the multiple series extension of the q-analog of Whipple's theorem [8, Theorem 4] suggest that similar analyses could be made of the series-product identities related to the general Rogers-Ramanujan theorem of [5].

Finally we again draw attention to the fact that there are now known to be four different generalizations of the Rogers-Ramanujan identities. Besides Gordon's result stated in Section 1 and our Theorem 2, there is an interpretation involving the successive ranks [4] (see also [9, Chapter 9]), and one involving the Alder polynomials [7]. In the case $k=2$ it is a simple matter to pass from any one of these interpretations to another. When $k>2$, D. Bressoud has recently found how to interpret each term in the series of (1.3) in terms of the partitions enumerated by $B_{k,a}(n)$; he has thus answered the question Askey and I posed in [11].

PENNSYLVANIA STATE UNIVERSITY.

REFERENCES.

[1] G. E. Andrews, An analytic proof of the Rogers-Ramanujan-Gordon identities, *Amer. J. Math.* **88** (1966), pp. 844–846.

[2] ———, Generalizations of the Durfee square, *J. London Math. Soc.* (2) **3** (1971), pp. 563–570.

[3] ———, Partition identities, *Advances in Math.* **9** (1972), pp. 10–51.

[4] ———, Sieves in the theory of partitions, *Amer. J. Math.* **94** (1972), pp. 1214–1230.

[5] ———, On the general Rogers-Ramanujan theorem, *Mem. Amer. Math. Soc.*, No. 152, 1974.

[6] ———, An analytic generalization of the Rogers-Ramanujan identities for odd moduli, *Proc. Nat. Acad. Sci. U.S.A.* **71** (1974), pp. 4082–4085.

[7] ———, On the Alder polynomials and a new generalization of the Rogers-Ramanujan identities, *Trans. Amer. Math. Soc.* **204** (1975), pp. 40–64.

[8] ———, Problems and prospects for basic hypergeometric functions, in *The Theory and Applications of Special Functions* (R. Askey, Ed.), Academic, New York, 1975, pp. 191–224.

[9] ———, *The Theory of Partitions*, Encyclopedia of Mathematics and Its Applications, Vol. 2, Addison-Wesley, Reading, 1977.

[10] ———, Multiple q-series identities, to appear.

[11] G. E. Andrews and R. Askey, Enumeration of partitions: the role of Eulerian series and q-orthogonal polynomials, in *Higher Combinatorics* (M. Aigner, Ed.), Reidel, Dordrecht, 1977.

[12] B. Gordon, A combinatorial generalization of the Rogers-Ramanujan identities, *Amer. J. Math.* **83** (1961), pp. 393–399.

3.8 Use and Extension of Frobenius' Representation of Partitions

Use and Extension of Frobenius' Representation of Partitions

George E. Andrews(1)

The Pennsylvania State University
University Park, Pennsylvania 16802

1. Introduction.

G. Frobenius [11] was the first to study equi-length two-rowed arrays of nonnegative integers

$$\begin{pmatrix} a_1 a_2,,,a_r \\ b_1 b_2,,,b_r \end{pmatrix} \tag{1.1}$$

where $a_1>a_2>\ldots>a_r\geq 0$, $b_1>b_2>\ldots>b_r\geq 0$. The number n being partitioned is given by $n=r+\sum a_i+\sum b_i$. Frobenius' object was to use these symbols to represent ordinary partitions of integers. The way these symbols represent partitions is best explained by an example. The ordinary partitions 7 + 7 + 7 + 4 + 4 + 3 + 1 (a partition of 35) has as its Ferrers graph

$$\begin{array}{l} \cdot\ \cdot\ \cdot\ \cdot\ \cdot\ \cdot\ \cdot \\ \cdot\ \cdot\ \cdot\ \cdot\ \cdot\ \cdot\ \cdot \\ \cdot\ \cdot\ \cdot\ \cdot\ \cdot\ \cdot\ \cdot \\ \cdot\ \cdot\ \cdot\ \cdot \\ \cdot\ \cdot\ \cdot\ \cdot \\ \cdot\ \cdot\ \cdot \\ \cdot \end{array} \tag{1.2}$$

Frobenius deletes the main diagonal of nodes; reads those to the right in horizontal rows and those to the left in vertical columns:

(1) Partially supported by National Science Foundation Grant MCS 8201733.

ISBN 0-12-379120-0

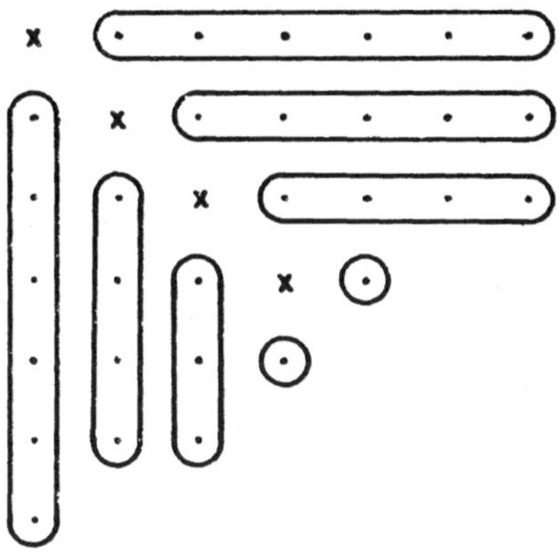

Thus the Frobenius symbol for 7 + 7 + 7 + 4 + 4 + 3 + 1 is

$$\begin{pmatrix} 6\ 5\ 4\ 0 \\ 6\ 4\ 3\ 1 \end{pmatrix},$$

and we note $33=4+(6+5+4+0)+(6+4+3+1)$.

There are two purposes in this paper: First, we wish to show how useful Frobenius symbols are in work on ordinary partition theory; to this end we shall provide a new and conceptually simpler proof of the Successive Ranks Theorem. Second, we shall describe recent studies which alter the conditions on the integers appearing in (1.1). This latter topic is extensively studied in [5] and we shall only scratch the surface in Sections 3,4 and 5.

2. The Successive Ranks Theorem.

In order to understand the Successive Ranks Theorem, we must look to the Frobenius representations of partitions.

DEFINITION 1. If $\begin{pmatrix} a_1 a_2 \ldots a_j \\ b_1 b_2 \ldots b_j \end{pmatrix}$ is the Frobenius representation of the partition π, then the ith successive rank $r_i(\pi)$ is defined to be $a_i - b_i$. (Note that $r_i(\pi)$ is only defined for $1 \leq i \leq j$).

THEOREM 1. (The Successive Ranks Theorem). Let $Q_{K,a}(n)$ denote the number of partitions π of n whose successive ranks all lie in $[-a+2, K-2-a]$. Let $A_{K,a}(n)$ denote the number of partitions of n into parts $\not\equiv 0, \pm a \pmod{K}$. Then for $1 \leq a < K/2$, $Q_{K,a}(n) = A_{K,a}(n)$ for all n.

REMARK 1. This result was proved originally for K odd in [1], [2] and [3; Ch. 9]. Bressoud [7] found a combinatorial explanation of the sieving process used in the previous work, and he observed that the restriction to K odd was unnecessary. Burge [8], [9] has found elegant relations between partitions enumerated by $Q_{K,a}(n)$ and other classes of partitions and permutations.

REMARK 2. This proof will eschew the use of sieves in favor of simple recurrences. The actual sieves will be buried in the Definitions 2, 4 and 6 of certain polynomials.

PROOF. We consider three families of polynomials:

DEFINITION 2. $\Delta_{K,i}(0;a,b;q)$, defined for $-i+1\le a-b\le K-i-1$, $a\ge 0$, $b\ge 0$, is the generating function for all partitions π whose successive ranks lie in $[-i+2,K-1-i]$ and whose first column in the Frobenius symbol for π is $\begin{pmatrix} a_1 \\ b_1 \end{pmatrix}$ with $a_1<a$ and $b_1<b$.

DEFINITION 3. A "lower Frobenius partitions" π is an array of the form

$$\begin{pmatrix} a_1a_2\dots a_r \\ b_0b_1b_2\dots b_r \end{pmatrix} \tag{2.1}$$

where $a_1>a_2>\dots>a_r\ge 0$, $b_0>b_1>\dots b_r\ge 0$ and the number being partitioned is $n=r+\sum_{i=1}^{r}a_i+\sum_{i=0}^{r}b_i$.

DEFINITION 4. $\Delta_{K,i}(1;a,b;q)$, defined for $-i+2\le a-b\le K-i$, $a\ge 0$, $b\ge 0$, is the generating function for all lower Frobenius partitions π (see (2.1)) wherein $-i+2\le a_h-b_h\le K-1-i$ for $1\le h\le r$, $a_1\le a-2$, and $a-K+i+1\le b_0\le b$.

DEFINITION 5. An "upper Frobenius partition" π is an array of the form

$$\begin{pmatrix} a_0a_1\dots a_r \\ b_1\dots b_r \end{pmatrix} \tag{2.2}$$

where $a_0>a_1>\dots>a_r\ge 0$, $b_1>b_2>\dots>b_r\ge 0$ and the number being partitioned is $n=r+\sum_{i=0}^{r}a_i+\sum_{i=1}^{r}b_i$.

DEFINITION 6. $\Delta_{K,i}(-1;a,b;q)$, defined for $-i\le a-b\le K-i-2$, $a\ge 0$, $b\ge 0$, is the generating function for all upper Frobenius partitions π (see (2.2)) wherein $-i+2\le a_h-b_h\le K-1-i$ for $1\le h\le r$, $b_1\le b-2$, and $b-i+1 \le a_0 \le a$.

The object of this deluge of definitions is to provide three families of generating functions (actually polynomials) for which we can find an intertwined system of defining recurrences and initial conditions.

The appropriate initial conditions and recurrences are the following (throughout $a \ge 0$, $b \ge 0$, $1 \le i < K/2$)

$$\Delta_{K,i}(-1;a,b;q)=0 \text{ if } a-b=-i, \tag{2.3}$$

$$\Delta_{K,i}(1;a,b;q)=0 \text{ if } a-b=K-i, \tag{2.4}$$

$$\Delta_{K,i}(c;0,0;q)=1,\ c=-1,0,1, \tag{2.5}$$

$$\Delta_{K,i}(0;a,b;q)=\begin{cases}\Delta_{K,i}(0;a-1,b;q), & a-b=K-i-1\\ \Delta_{K,i}(0;a-1,b;q)+q^a\Delta_{K,i}(1;a,b-1;q), & \\ \quad -i+2 \le a-b \le K-i-2, &\end{cases} \tag{2.6}$$

$$\Delta_{K,i}(-1;a,b;q)=\begin{cases}q^a\Delta_{K,i}(0;a,b-1;q), & a-b=-i+1\\ \Delta_{K,i}(-1;a-1,b;q)+q^a\Delta_{K,i}(0;a,b-1;q), & \\ \quad -i+2 \le a-b \le K-i-2, &\end{cases} \tag{2.7}$$

$$\Delta_{K,i}(0;a,b,q)=\begin{cases}\Delta_{K,i}(0;a,b-1;q), & a-b=-i+1\\ \Delta_{K,i}(0;a,b-1;q)+q^b\Delta_{K,i}(-1;a-1,b;q) & \\ \quad -i+2 \le a-b \le K-i-2, &\end{cases} \tag{2.8}$$

$$\Delta_{K,i}(1;a,b;q)=\begin{cases}q^b\Delta_{K,i}(0;a-1,b;q), & a-b=K-i-1\\ \Delta(1;a,b-1;q)+q^b\Delta_{K,i}(0;a-1,b;q),. & \\ \quad -i+2 \le a-b \le K-i-2 &\end{cases} \tag{2.9}$$

The proofs of the initial conditions are almost immediate. To see (2.3) we merely observe that if $a-b=-i$ then the condition $b-i+1 \le a_0 \le a(=b-i)$ in Definition 6 is impossible. As for (2.4) we note that if $a-b=K-i$ then the condition $a-K+i+1 \le b_0 \le b(=a-K+1)$ is impossible. For the finally initial condition (2.5), we note that when $a=b=0$ the only admissable partition is the empty partition of 0, and this partition satisfies vacuously all the conditions imposed in the various generating functions.

The proofs of (2.6)-(2.9) are all similar; indeed the proofs of (2.8) and (2.9) are the mirror images of (2.6) and (2.7) respectively. Therefore in the interest of brevity we include only the proofs of (2.6) and (2.7).

To see (2.6) let us examine the Frobenius representation of

partitions enumerated by $\Delta_{K,i}(0;a,b;q)$, say

$$\begin{pmatrix} a_1a_2 \ldots a_r \\ b_1b_2 \ldots b_r \end{pmatrix}$$

We form two disjoint classes of these partitions. Class 1 contains those for which $a_1<a-1$; Class 2 contains those for which $a_1=a-1$. Clearly the partitions in Class 1 are enumerated by $\Delta_{K,i}(0;a-1,b;q)$. Now Class 2 is empty if $a-b=K-i-1$ because since $a_1=a-1$, $b_1\le b-1$ therefore $K-i-2\ge a_1-b_1\ge a-b=K-i-1$ which is impossible. Hence the top line of (2.6) is correct. If instead $-i+2\le a-b\le K-i-2$ then delete $a_1=a-1$ from the Frobenius representation of each partition in Class 2, and view the result as a lower Frobenius partition. The number being partitioned has been reduced by $a_1+1=a$. Furthermore the new b_0 (the old b_1) satisfies $b_0\le b-1$ and $(a-1)-b_0\le K-i-2$ or $b_0\ge a-K+i+1$. The new a_1 satisfies $a_1<a-1$ or $a_1\le a-2$. Hence these transformed partitions are precisely those enumerated by $q^a\Delta_{K,i}(1;a,b-1;q)$, where the q^a factor takes account of the deletion of $1+(a-1)$ from the number originally partitioned. Hence (2.6) is completely established.

To establish (2.7) we, as above, separate the upper Frobenius partitions enumerated by $\Delta_{K,i}(-1;a,b;q)$ into two classes. Class 1 consists of those upper Frobenius partitions with $a_0<a$; Class 2 consists of those with $a_0=a$. Now if $a-b=-i+1$, then Class 1 is empty because then $a_0<a=b-i+1$ contradicting the conditions on a_0 in Definition 5. For $-i+2\le a-b\le K-i-2$ we see that the elements of Class 1 are enumerated by $\Delta_{K,i}(-1;a,b;q)$. As for Class 2, we transform each upper Frobenius partition by deleting $a_0=a$ (thus reducing each number being partitioned by an amount a). The resulting partitions are clearly ordinary Frobenius representations of partitions; furthermore the new a_1 satisfies $a_1<a_0=a$, or $a_1\le a-1$ and the new b_1 satisfies $b_1\le b-2=(b-1)-1$. Hence the partitions in Class 2 are enumerated by $q^a\Delta_{K,i}(0;a,b-1;q)$ (where the q^a accounts for the transformation deleting $a_0=a$). Hence (2.7) is verified.

The verification of (2.8) and (2.9) is precisely like that above with the roles of upper and lower Frobenius partitions being reversed.

One may easily prove by induction on $N=a+b$ that (2.3) - (2.9) uniquely define the $\Delta_{K,i}(c;a,b;q)$ for $c=-1,0,1$ with $a-b$ in the appropriate intervals. We now exhibit seemingly new families of polynomials that also satisfy (2.3) - (2.9). Namely let

$$D_{K,i}(c;a,b;q) = \sum_{\mu=-\infty}^{\infty} q^{\mu(2K\mu-(K-2i))} \begin{bmatrix} a+b \\ a-K\mu \end{bmatrix} q^{-cK\mu}$$

$$- \sum_{\mu=-\infty}^{\infty} q^{\mu(2K\mu-(K-2i))} \begin{bmatrix} a+b \\ a-K\mu+i \end{bmatrix} q^{c(-K\mu+i)}, \qquad (2.10)$$

where

$$\begin{bmatrix} A \\ B \end{bmatrix} = \begin{cases} \dfrac{(1-q^A)(1-q^{A-1})\dots(1-q^{A-B+1})}{(1-q^B)(1-q^{B-1})\dots(1-q)}, & B \geq 0 \\ 0, \quad B<0 \end{cases} \qquad (2.11)$$

The expressions in (2.11) are called Gaussian polynomials or q-binomial coefficients. They satisfy the two following recurrences [3;p.35, eqs. (3.3.4), (3.3.3)]:

$$\begin{bmatrix} A \\ B \end{bmatrix} = \begin{bmatrix} A-1 \\ B-1 \end{bmatrix} + q^B \begin{bmatrix} A-1 \\ B \end{bmatrix}, \qquad (2.12)$$

and

$$\begin{bmatrix} A \\ B \end{bmatrix} = \begin{bmatrix} A-1 \\ B \end{bmatrix} + q^{A-B} \begin{bmatrix} A-1 \\ B-1 \end{bmatrix}. \qquad (2.13)$$

Also there is the obvious symmetry relation [3; p.35, eq. (3.3.2)]:

$$\begin{bmatrix} A \\ B \end{bmatrix} = \begin{bmatrix} A \\ A-B \end{bmatrix}. \qquad (2.14)$$

If we apply (2.12) to each term in (2.10), we find that

$$D_{K,i}(c;a,b;q) = D_{K,i}(c;a-1,b;q) + q^a D_{K,i}(c+1;a,b-1;q), \qquad (2.15)$$

and if we apply (2.13) to each term in (2.10), it follows that

$$D_{K,i}(c;a,b;q) = D_{K,i}(c;a,b-1;q) + q^b D_{K,i}(c-1;a-1,b;q). \qquad (2.16)$$

Furthermore for $a=b=0$ it follows directly from (2.10) that

$$D_{K,i}(c;0,0;q) = \begin{bmatrix} 0 \\ 0 \end{bmatrix} = 1, \quad 1 \leq i < K/2. \qquad (2.17)$$

Also if $a-b=-i$

$$D_{K,i}(-1;a,b;q) \qquad (2.18)$$

$$= \sum_{\mu=-\infty}^{\infty} q^{\mu(2K\mu-(K-2i))} \begin{bmatrix} a+b \\ a-K\mu \end{bmatrix} q^{K\mu}$$

$$- \sum_{\mu=-\infty}^{\infty} q^{(2\mu-1)(K\mu-i)} \begin{bmatrix} a+b \\ b-K\mu \end{bmatrix} q^{K\mu-i}$$

$$= \sum_{\mu=-\infty}^{\infty} q^{2K\mu^2+2i\mu} \begin{bmatrix} a+b \\ a-K\mu \end{bmatrix}$$

$$- \sum_{\mu=-\infty}^{\infty} q^{(-2\mu-1)(-K\mu-i)} \begin{bmatrix} a+b \\ b+K\mu \end{bmatrix} q^{-K\mu-i}$$

(replacing μ by $-\mu$, in the second sum)

$$= \sum_{\mu=-\infty}^{\infty} q^{2K\mu^2+2i\mu} \begin{bmatrix} a+b \\ a-K\mu \end{bmatrix}$$

$$- \sum_{\mu=-\infty}^{\infty} q^{2K\mu^2+2i\mu} \begin{bmatrix} a+b \\ a-K\mu \end{bmatrix} \quad \text{(by (2.14))}$$

$$=0.$$

Finally, if $a-b=K-i$, then

$$D_{K,i}(1;a,b;q) \tag{2.19}$$

$$= \sum_{\mu=-\infty}^{\infty} q^{\mu(2K\mu-(K-2i))} \begin{bmatrix} a+b \\ a-K\mu \end{bmatrix} q^{-K\mu}$$

$$- \sum_{\mu=-\infty}^{\infty} q^{(2\mu-1)(K\mu-i)} \begin{bmatrix} a+b \\ b+K-K\mu \end{bmatrix} q^{-K\mu+i}$$

$$= \sum_{\mu=-\infty}^{\infty} q^{\mu(2K\mu-(K-2i))} \begin{bmatrix} a+b \\ a-K\mu \end{bmatrix} q^{-K\mu}$$

$$- \sum_{\mu=-\infty}^{\infty} q^{(-2\mu+1)(-K\mu+K-i)} \begin{bmatrix} a+b \\ a+K\mu \end{bmatrix} q^{K\mu-K+i}$$

(replacing μ by $-\mu+1$ in the second sum)

$$= \sum_{\mu=-\infty}^{\infty} q^{2K\mu^2-2K\mu+2i\mu} \begin{bmatrix} a+b \\ a-K\mu \end{bmatrix}$$

$$- \sum_{\mu=-\infty}^{\infty} q^{2K\mu^2-2K\mu+2i\mu} \begin{bmatrix} a+b \\ a-K\mu \end{bmatrix} \quad \text{(by (2.14))}$$

$$=0.$$

It is now a simple matter to check that the $D_{K,i}(c;a,b;q)$ must satisfy all of (2.3) - (2.9). In fact (2.3) is implied by (2.18); (2.4) follows from (2.19); (2.5) from (2.17); (2.6) from (2.15) with $c=0$ plus (2.19) when $a-b=K-i-1$; (2.7) from (2.15) with $c=-1$ plus (2.18) when $a-b=-i+1$; (2.8) from (2.16) with $c=0$ plus (2.18) when $a-b=i+1$; (2.9) from (2.16) with $c=1$ plus (2.19) when $a-b=K-i-1$.

Hence for $c=-1,0,1$, $1\leq i<K/2$, $a\geq 0$, $b\geq 0$

$$D_{K,i}(c;a,b;q)=\Delta_{K,i}(c;a,b;q) \tag{2.20}$$

provided $-i+1+c\leq a-b\leq K-i-1+c$.

To conclude the proof of Theorem 1 we need only observe that [3; p. 153]:

$$\lim_{a\to\infty} \begin{bmatrix} 2a \\ a-K\mu \end{bmatrix} = \prod_{n=1}^{\infty} \frac{1}{1-q^n}. \tag{2.21}$$

Hence

$$\sum_{n\geq 0} Q_{K,i}(n)q^n = \lim_{a\to\infty} \Delta_{K,i}(0;a,a;q) \tag{2.22}$$

$$= \lim_{a\to\infty} D_{K,i}(0;a,a;q)$$

$$= \prod_{n=1}^{\infty}(1-q^n)^{-1} \left(\sum_{\mu=-\infty}^{\infty} \{q^{\mu(2K\mu-K+2i)} - q^{(2\mu-1)(K\mu-i)}\} \right)$$

$$= \prod_{n=1}^{\infty}(1-q^n)^{-1} \sum_{m=-\infty}^{\infty} (-1)^m q^{K(m^2-m)/2+im}$$

$$= \prod_{n=1}^{\infty}(1-q^n)^{-1} \prod_{s=1}^{\infty}(1-q^{Ks})(1-q^{i+(s-1)K})(1-q^{-i+Ks})$$

by [3;p. 22, Cor. 2.9])

$$= \prod_{\substack{n=1 \\ n \not\equiv 0, \pm i \pmod K}}^{\infty} (1-q^n)^{-1}$$

$$= \sum_{n=0}^{\infty} A_{K,i}(n) q^n.$$

Comparing coefficients in the extremes of (2.22) we see that $Q_{K,i}(n) = A_{K,i}(n)$ for all n. Thus Theorem 1 is established.

3. Generalized Frobenius Partitions.

Having observed the usefulness of the Frobenius Symbol (1.1) in providing a "sieveless" proof of the Successive Ranks Theorem, we shall now briefly sketch some of the things that happen when we alter the conditions on the nonnegative integers appearing in (1.1). In [5] we restrict our attention to two families of Generalized Frobenius Partitions or F-partitions.

Let $\phi_k(n)$ denote the number of F-partitions where no integer is repeated more than k times in each row. Thus $\phi_1(n) = p(n)$ the number of ordinary partitions of n.

For example, the F-partitions enumerated by $\phi_2(4)$ are

$$\begin{pmatrix}3\\0\end{pmatrix} \begin{pmatrix}0\\3\end{pmatrix} \begin{pmatrix}2\\1\end{pmatrix} \begin{pmatrix}1\\2\end{pmatrix} \begin{pmatrix}20\\00\end{pmatrix} \begin{pmatrix}00\\20\end{pmatrix} \begin{pmatrix}11\\00\end{pmatrix} \begin{pmatrix}00\\11\end{pmatrix} \begin{pmatrix}10\\10\end{pmatrix}$$

Actually it turns out that [5; sec. 5]

$$\sum_{n \geq 0} \phi_1(n) q^n = \prod_{n=1}^{\infty} \frac{1}{1-q^n}, \tag{3.1}$$

$$\sum_{n \geq 0} \phi_2(n) q^n \tag{3.2}$$

$$= \prod_{n=1}^{\infty} \frac{1}{(1-q^n)(1-q^{12n-10})(1-q^{12n-9})(1-q^{12n-3})(1-q^{12n-2})}$$

$$\sum_{n \geq 0} \phi_3(n) q^n \tag{3.3}$$

$$= \prod_{n=1}^{\infty} \frac{1-q^{12n-6}}{(1-q^{6n-5})(1-q^{6n-4})^2(1-q^{6n-3})^3(1-q^{6n-2})^2(1-q^{6n-5})(1-q^{12n})},$$

and more generally

$$\sum_{n\geq 0}\phi_k(n)q^n=\frac{\sum\limits_{m_1,\ldots,m_{k-1}=-\infty}^{\infty}\zeta^{(k-1)m_1+(k-2)m_2+\ldots+m_{k-1}}q^{Q(m_1,m_2,\ldots,m_{k-1})}}{\prod\limits_{n=1}^{\infty}(1-q^n)^k}, \tag{3.4}$$

where $\zeta=\exp(2\pi i/(k+1))$, and

$$Q(m_1,\ldots,m_{k-1})=\sum_{j=1}^{k-1}m_j^2+\sum_{1\leq i<j\leq k-1}m_im_j. \tag{3.5}$$

Thus we see arising numerous interconnections of these F-partitions with modular forms and multi-dimensional theta series.

Instead of allowing repeats of the same integer, we can allow k copies of each integer (say k colors) where now we order the integers by $0_1<0_2<\ldots<0_k<1_1<1_2<\ldots<1_k<2_1<2_2<\ldots <2_k<3_1<3_2<\ldots <3_k<4_1<\ldots$. We let $c\phi_k(n)$ denote the number of F-partitions of n with k colors and strict decrease in each row. For example, the 20 F-partitions enumerated by $\phi_2(3)$ are

$$\begin{pmatrix}2_1\\0_1\end{pmatrix}\begin{pmatrix}2_2\\0_1\end{pmatrix}\begin{pmatrix}2_1\\0_2\end{pmatrix}\begin{pmatrix}2_2\\0_2\end{pmatrix}\begin{pmatrix}1_1\\1_1\end{pmatrix}\begin{pmatrix}1_2\\1_1\end{pmatrix}\begin{pmatrix}1_1\\1_2\end{pmatrix}\begin{pmatrix}1_2\\1_2\end{pmatrix}$$

$$\begin{pmatrix}0_1\\2_1\end{pmatrix}\begin{pmatrix}0_1\\2_2\end{pmatrix}\begin{pmatrix}0_2\\2_1\end{pmatrix}\begin{pmatrix}0_2\\2_2\end{pmatrix}\begin{pmatrix}1_10_1\\0_20_1\end{pmatrix}\begin{pmatrix}1_20_1\\0_20_1\end{pmatrix}\begin{pmatrix}1_10_2\\0_20_1\end{pmatrix}\begin{pmatrix}1_20_2\\0_20_1\end{pmatrix}$$

$$\begin{pmatrix}0_20_1\\1_10_1\end{pmatrix}\begin{pmatrix}0_20_1\\1_20_1\end{pmatrix}\begin{pmatrix}0_20_1\\1_10_2\end{pmatrix}\begin{pmatrix}0_20_1\\1_20_2\end{pmatrix}.$$

Again nice relations with modular forms and multiple theta series arise:

$$\sum_{n\geq 0}c\phi_1(n)q^n=\prod_{n=1}^{\infty}\frac{1}{1-q^n}, \tag{3.6}$$

$$\sum_{n\geq 0}c\phi_2(n)q^n=\prod_{n=1}^{\infty}\frac{(1-q^{4n-2})}{(1-q^{2n-1})^4(1-q^{4n})} \tag{3.7}$$

$$\sum_{n\geq 0}c\phi_3(n)q^n=\prod_{n=1}^{\infty}\frac{(1-q^{12})(1-q^{12n-6})^3}{(1-q^{6n-5})^5(1-q^{6n-1})^4(1-q^{4n})^2(1-q^{6n-3})^6} \tag{3.8}$$

$$+4q\prod_{n=1}^{\infty}\frac{(1-q^{12n})(1-q^{4n})}{(1-q^{12n-6})(1-q^{4n-2})(1-q^n)^3}$$

and more generally

$$\sum_{n\geq 0}\phi_k(n)q^n=\frac{\sum\limits_{m_1,\ldots,m_{k-1}=-\infty}^{\infty} q^{Q(m_1,m_2,\ldots,m_{k-1})}}{\prod\limits_{n=1}^{\infty}(1-q^n)^k}. \tag{3.9}$$

These results and many others are extensively explored in [5]. The main theoretical starting point for such results is as follows:

Let

$$f_A(z,q)=\sum_{\pi\in A} z^{\#(\pi)}q^{\sum\pi} \tag{3.10}$$

where A is a set of ordinary partitions, $\#(\pi)$ denotes the number of parts of π and $\sum\pi$ denotes the sum of all the parts of π (i.e. $\sum\pi$ is the number being partitioned).

GENERAL PRINCIPLE. Let $\phi_{A,B}(n)$ denote the number of F-partitions of n where the top row, considered as an ordinary partition, is in A and the bottom row is in B. Then $\sum_{n\geq 0}\phi_{A,B}(n)q^n$ is the constant term relative to z in

$$f_A(zq,q)f_B(z^{-1},q). \tag{3.11}$$

The historical precedent for this principle lies in the many combinatorial proofs of Jacobi's triple product identity which turn out to be essentially an application of the General Principle (cf. [5; Sec. 3]).

Numerous arithmetic, analytic and combinatorial properties of F-paritions are explored in [5] beginning with the General Principle. For example

$$\phi_2(5n+3)\equiv c\phi_2(5n+3)\equiv 0 \pmod 5, \tag{3.12}$$

and whenever k is prime

$$c\phi_k(n)\equiv\begin{cases}0 \pmod{k^2} & \text{if } k\nmid n\\ p(n/k) \pmod{k^2} & \text{if } k\mid n\end{cases} \tag{3.13}$$

The congruences in (3.12) lead to natural conjectures on ranks of F-partitions in analogy with F. J. Dyson's conjectures concerning ranks of partitions and Ramanujan's congruences [10]. Dyson's conjectures were proved by Atkin and Swinnerton-Dyer [6].

4. Euler's Theorem.

We close this eclectic introduction to the many uses and extensions of F-partitions by given an "apparently" new proof of Euler's theorem [3; p.5]:

THEOREM 2 (Euler). The number of partitions of n into distinct parts equals the number of partitions of n into odd parts.

PROOF. Let π be a partition of n with odd parts, say π is

$$\underbrace{(2s_1+1)(2s_1+1)\ldots(2s_1+1)}_{\nu_{s_1}\text{ terms}}\ldots\underbrace{3\,3\ldots3}_{\nu_1\text{ terms}}\,\underbrace{1\,1\,1\,1\ldots1}_{\nu_0\text{ terms}}. \tag{4.1}$$

We can split π into two partitions π_1 and π_2 where π_1 is

$$\underbrace{(s_1+1)(s_1+1)\ldots(s_1+1)}_{\nu_{s_1}\text{ terms}}\ldots\underbrace{2\,2.\ldots2}_{\nu_1\text{ terms}}\,\underbrace{1\,1\,1\,1\ldots1}_{\nu_0\text{ terms}} \tag{4.2}$$

and π_2 is

$$\underbrace{s_1s_1\ldots s_1}_{\nu_{s_1}\text{ terms}}\ldots\underbrace{11..1}_{\nu_1\text{ terms}}. \tag{4.3}$$

Thus in terms of the Ferrers graphs of π_1 and π_2 we see that π_2 is gotten from π_1 by deleting the left hand column of nodes in the Ferrers graph of π_1. Furthermore π_1 and π_2 are uniquely determined by π and uniquely determine π.

Now if the Frobenius representation of π_2 is

$$\begin{pmatrix} a_1a_2\ldots a_r \\ b_1b_2\ldots b_r \end{pmatrix}$$

then that of π_1 is clearly

$$\begin{pmatrix} a_1+1,a_2+1,\ldots,a_r+1,0 \\ b_0,b_1-1,\ldots,b_{r-1}-1,b_r-1 \end{pmatrix}$$

where the last column of π_1 is deleted if $b_r=0$, and b_0+1 is the total number of rows in the Ferrers graph of π_1.

We now form a partition $\overline{\pi}$ by the sequence of parts

$$(a_1+1+b_0+1),\ (a_1+b_1+1),\ (a_2+b_1+1),\ (a_2+b_2+1),\ (a_3+b_2+1)$$
$$,\ldots,(a_r+b_r+1),\ (b_r)$$

and $\overline{\pi}$ is clearly a partition of n into an odd number (namely $2r+1$) of distinct nonnegative parts. Deleting a zero part if it occurs we see that this reversible mapping between $\overline{\pi}$ and π provide a bijection between the partitions of n with odd parts and the partitions of n with distinct parts.

As an example of our bijection let us consider π given by

$$13,9,9,9,9,7,7,7,3,3,3,3,3,1$$

a partition of 86. Thus π_1 is

$$7,5,5,5,5,4,4,4,2,2,2,2,2,1$$

and π_2 is

$$6,4,4,4,4,3,3,3,1,1,1,1,1.$$

Hence the Frobenius representation of π_1 is

$$\begin{pmatrix} 6 & 3 & 2 & 1 & 0 \\ 13 & 11 & 5 & 4 & 0 \end{pmatrix}$$

and for π_2

$$\begin{pmatrix} 5 & 2 & 1 & 0 \\ 12 & 6 & 5 & 1 \end{pmatrix}$$

Therefore $\overline{\pi}$ is

$$20,18,15,9,8,7,6,2,1,$$

a partition of 86 into distinct parts.

Now to let the cat out of the bag. The above procedure is merely an F-partition version of Sylvester's "fish-hook" proof of Euler's theorem [13; p.39], [12; p.13]. To see this, rather than provide a general explanation of Sylvester's method we carry through his construction with the above example.

First write this partition π with odd parts as a sequence of symmetric right angles of nodes centered on the main diagonal in the fourth quadrant:

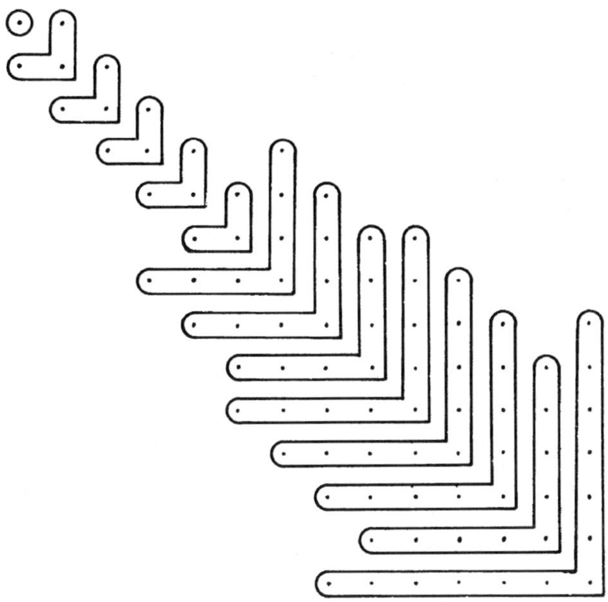

Sylvester now regroups the nodes through use of 45° angles ("fish-hooks"), as follows

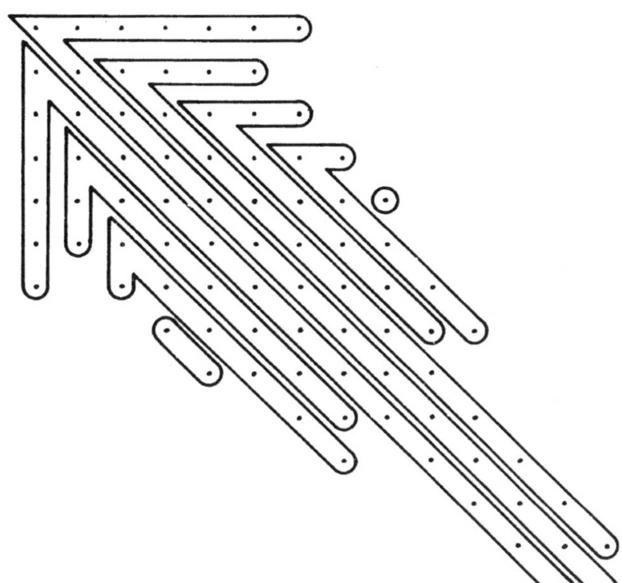

Collecting "fish-hooks" we obtain the partition

$$20,18,15,9,8,7,6,2,1,$$

which is precisely the correspondence provided above.

One may well ask why provide a cumbersome F-partition translation of Sylvester's elegant proof. The only reason lies in the fact that the reversal of Sylvester's algorithm is quite cumbersome [13; p.40] while the F-partition algorithm clearly exhibits the process from back to front as well as front to back.

5. Conclusion

The reader interested in F-partitions should consult [5]. The material in Section 2 was found in an effort to simplify the proof of the Successive Ranks Theorem. Polynomials like $D_{k,i}(c;a,b;q)$ and their combinatorial and analytic generalizations have turned out to be important in the mathematics surrounding Baxter's solution of the hard hexagon model [4].

References

[1] G.E. Andrews, Sieves for theorems of Euler, Rogers and Ramanujan, from The Theory of Arithmetic Functions (A.A. Gioia and D.L. Goldsmith, eds.), Lecture Notes in Math., No. 251, pp. 1-20, Springer, Berlin, 1971.

[2] G.E. Andrews, Sieves in the theory of partitions, Amer. J. Math., 94 (1972), 1214-1230.

[3] G.E. Andrews, The Theory of Partitions, Encyclopedia of Mathematics and Its Applications, Vol. 2, Addison-Wesley, Reading, 1976.

[4] G.E. Andrews, The hard-hexagon model and Rogers-Ramanujan type identities, Proc. Nat. Acad. Sci. (U.S.A.), 78(1981), 5290-5292.

[5] G.E. Andrews, Generalized Frobenius partitions, (to appear).

[6] A.O.L. Atkin and H.P.F. Swinnerton-Dyer, Some properties of partitions, Proc. London Math. Soc. (3), 4(1954), 84-106.

[7] D. Bressoud, Extension of the partition sieve, J. Number Th., 12(1980), 87-100.

[8] W. Burge, A correspondence between partitions related to generalizations of the Rogers-Ramanujan identities, Discrete Math., 34(1981), 9-15.

[9] W. Burge, A three way correspondence between partitions, European J. Math., (to appear).

[10] F.J. Dyson, Some guesses in the theory of partitions, Eureka (Cambridge), 8(1944), 10-15.

[11] G. Frobenius, Über die Charaktere der symmetrischen Gruppe, Sitzber. Pruess. Akad. Berlin, 1900, 516-534.

[12] P.A. MacMahon, Combinatory Analysis, Volume 2, Cambridge University Press, London, 1916 (Reprinted: Chelsea, New York, 1960).

[13] J.J. Sylvester, A constructive theory of partitions in three acts, an interact and an exodion, Amer. J. Math., 5(1982), 251-330 (or pp. 1-83 of the Coll. Math. Papers of J.J. Sylvester, Vol. 4, Cambridge University Press, London, 1912; reprinted, Chelsea, New York, (1974).

3.9 Dyson's Crank of a Partition (with F.G. Garvan)

BULLETIN (New Series) OF THE
AMERICAN MATHEMATICAL SOCIETY
Volume 18, Number 2, April 1988

DYSON'S CRANK OF A PARTITION

GEORGE E. ANDREWS AND F. G. GARVAN

1. Introduction. In [**3**], F. J. Dyson defined the rank of a partition as the largest part minus the number of parts. He let $N(m,t,n)$ denote the number of partitions of n of rank congruent to m modulo t, and he conjectured

$$N(m,5,5n+4) = \tfrac{1}{5}p(5n+4), \qquad 0 \le m \le 4; \tag{1.1}$$

$$N(m,7,7n+5) = \tfrac{1}{7}p(7n+5), \qquad 0 \le m \le 6, \tag{1.2}$$

where $p(n)$ is the total number of partitions of n [**1**, Chapter 1]. These conjectures were subsequently proved by Atkin and Swinnerton-Dyer [**2**].

Dyson [**3**] went on to observe that the rank did not separate the partition of $11n+6$ into 11 equal classes even though Ramanujan's congruence

$$p(11n+6) \equiv 0 \pmod{11} \tag{1.3}$$

holds. He was thus led to conjecture the existence of some other partition statistic (which he called the crank); this unknown statistic should provide a combinatorial interpretation of $\frac{1}{11}p(11n+6)$ in the same way that (1.1) and (1.2) treat the primes 5 and 7.

In [**4**, **5**], one of us was able to find a crank relative to vector partitions as follows:

For a partition π, let $\#(\pi)$ be the number of parts of π and $\sigma(\pi)$ be the sum of the parts of π (or the number π is partitioning) with the convention $\#(\phi) = \sigma(\phi) = 0$ for the empty partition ϕ, of 0. Let

$$V = \{(\pi_1,\pi_2,\pi_3) \mid \pi_1 \text{ is a partition into distinct parts}, \\ \pi_2,\pi_3 \text{ are unrestricted partitions}\}.$$

We shall call the elements of V *vector partitions*. For $\vec{\pi} = (\pi_1,\pi_2,\pi_3)$ in V we define the sum of parts, s, a weight, ω, and a crank, r, by

$$s(\vec{\pi}) = \sigma(\pi_1) + \sigma(\pi_2) + \sigma(\pi_3), \tag{1.4}$$

$$\omega(\vec{\pi}) = (-1)^{\#(\pi_1)}, \tag{1.5}$$

$$r(\vec{\pi}) = \#(\pi_2) - \#(\pi_3). \tag{1.6}$$

We say $\vec{\pi}$ is a vector partition of n if $s(\vec{\pi}) = n$. For example, if

$$\vec{\pi} = (5+3+2, 2+2+1, 2+1+1)$$

Received by the editors August 13, 1987.
1980 *Mathematics Subject Classification* (1985 *Revision*). Primary 11P76.
First author partially supported by National Science Foundation Grant DMS 8503324.

then $s(\vec{\pi}) = 19$, $\omega(\vec{\pi}) = -1$, $r(\vec{\pi}) = 0$ and $\vec{\pi}$ is a vector partition of 19. The number of vector partitions of n (counted according to the weight ω) with crank m is denoted by $N_V(m,n)$, so that

$$N_V(m,n) = \sum_{\substack{\vec{\pi}\in V \\ s(\vec{\pi})=n \\ r(\vec{\pi})=m}} \omega(\vec{\pi}). \tag{1.7}$$

The number of vector partitions of n (counted according to the weight ω) with crank congruent to k modulo t is denoted by $N_V(k,t,n)$, so that

$$N_V(k,t,n) = \sum_{m=-\infty}^{\infty} N_V(mt+k,n) = \sum_{\substack{\vec{\pi}\in V \\ s(\vec{\pi})=n \\ r(\vec{\pi})\equiv k \pmod t}} \omega(\vec{\pi}). \tag{1.8}$$

By considering the transformation that interchanges π_2 and π_3 we have

$$N_V(m,n) = N_V(-m,n) \tag{1.9}$$

so that

$$N_V(t-m,t,n) = N_V(m,t,n). \tag{1.10}$$

We have the following generating function for $N_V(m,n)$:

$$\sum_{m=-\infty}^{\infty}\sum_{n=0}^{\infty} N_V(m,n)z^m q^n = \prod_{n=1}^{\infty} \frac{(1-q^n)}{(1-zq^n)(1-z^{-1}q^n)}. \tag{1.11}$$

By putting $z=1$ in (1.11) we find

$$\sum_{m=-\infty}^{\infty} N_V(m,n) = p(n). \tag{1.12}$$

VECTOR-CRANK THEOREM (GARVAN [**4**, **5**]).

(1.13)

$$N_V(0,5,5n+4) = N_V(1,5,5n+4) = \cdots = N_V(4,5,5n+4) = \frac{p(5n+4)}{5},$$

(1.14)

$$N_V(0,7,7n+5) = N_V(1,7,7n+5) = \cdots = N_V(6,7,7n+5) = \frac{p(7n+5)}{7},$$

$$N_V(0,11,11n+6) = \cdots = N_V(10,11,11n+6) = \frac{p(11n+6)}{11}. \tag{1.15}$$

The above still leaves open the question of whether there is a crank for ordinary partitions. The answer is "yes" when the crank is defined as follows:

DEFINITION. For a partition π, let $l(\pi)$ denote the largest part of π, $\omega(\pi)$ denote the number of ones in π, and $\mu(\pi)$ denote the number of parts of π larger than $\omega(\pi)$. The crank $c(\pi)$ is given by

$$c(\pi) = \begin{cases} l(\pi) & \text{if } \omega(\pi) = 0, \\ \mu(\pi) - \omega(\pi) & \text{if } \omega(\pi) > 0. \end{cases}$$

Our main result is the following.

THEOREM 1. *The number of partitions π of n with $c(\pi) = m$ is $N_V(m, n)$ for all $n > 1$.*

Obviously, in light of the Vector-Crank Theorem, we see that Theorem 1 supplies the crank asked for by Dyson.

2. PROOF OF THEOREM 1. We shall require the standard notation of q-series:

$$(A; q)_n = (A)_n = \prod_{j=0}^{\infty} \frac{(1 - Aq^j)}{(1 - Aq^{j+n})} \tag{2.1}$$

($= (1 - A)(1 - Aq) \cdots (1 - Aq^{n-1})$ when n is a positive integer), and

$$(A; q)_\infty = (A)_\infty = \prod_{j=0}^{\infty} (1 - Aq^j). \tag{2.2}$$

We now transform (1.11):

$$\begin{aligned} \sum_{m=-\infty}^{\infty} \sum_{n=0}^{\infty} N_V(m, n) z^m q^n &= \frac{(1-q)}{(zq)_\infty} \frac{(q^2; q)_\infty}{(q/z)_\infty} \\ &= \frac{(1-q)}{(zq)_\infty} \sum_{j=0}^{\infty} \frac{(zq)_j (q/z)^j}{(q)_j} \quad \text{(by [1, p. 17])} \\ &= \frac{(1-q)}{(zq)_\infty} + \sum_{j=1}^{\infty} \frac{q^j z^{-j}}{(q^2; q)_{j-1} (zq^{j+1})_\infty}. \end{aligned} \tag{2.3}$$

As was noted in (1.12), when we set $z = 1$ the series on the left of (2.3) reduces to the generating function for $p(n)$. For $j > 0$, the jth term in the sum on the right is

$$\frac{z^{-j} q^{\overbrace{1+1+\cdots+1}^{j \text{ times}}}}{(1 - q^2)(1 - q^3) \cdots (1 - q^j)(1 - zq^{j+1})(1 - zq^{j+2}) \cdots}.$$

The standard techniques of partition theory [1, Chapter 1] show that this expression generates partitions with $\omega(\pi) = j$ and the exponent on z is clearly $\mu(\pi) - \omega(\pi)$, i.e. $c(\pi)$, since $j > 0$.

Thus we must interpret

$$\frac{(1 - q)}{(1 - zq)(1 - zq^2)(1 - zq^3) \cdots}$$

as the generating function for partitions without ones. By considering conjugate partitions, we note that

$$\frac{1}{(1 - zq)(1 - zq^2)(1 - zq^3) \cdots}$$

generates all partitions with the exponent on z counting the largest part, and for integers larger than 1

$$\frac{q}{(1-zq)(1-zq^2)(1-zq^3)\cdots}$$

generates partitions with at least one 1 appearing again with the exponent on z counting the largest part. Note that this interpretation fails for 1 because this is the unique instance in which introducing a 1 into the partitions of $n-1$ alters the largest part. Hence

$$\frac{1-q}{(zq)_\infty}$$

counts (for $n > 1$) the number of partitions with no ones and with the exponent on z being the largest part of the partition $l(\pi) = c(\pi)$. Thus in the double series expansion of

$$\frac{(1-q)}{(zq)_\infty} + \sum_{j=1}^{\infty} \frac{q^j z^{-j}}{(q^2;q)_{j-1}(zq^{j+1})_\infty}$$

we see that the coefficient of $z^m q^n$ $(n > 1)$ is the number of ordinary partitions of n in which $c(\pi) = m$. Therefore by (2.3), we have Theorem 1. □

3. Conclusion. We can't resist exhibiting $c(\pi)$ for the first instance of (1.3).

partitions of 6	$l(\pi)$	$\omega(\pi)$	$\mu(\pi)$	$c(\pi)$
6	6	0	1	6
5 + 1	5	1	1	0
4 + 2	4	0	2	4
4 + 1 + 1	4	2	1	− 1
3 + 3	3	0	2	3
3 + 2 + 1	3	1	2	1
3 + 1 + 1 + 1	3	3	0	− 3
2 + 2 + 2	2	0	3	2
2 + 2 + 1 + 1	2	2	0	− 2
2 + 1 + 1 + 1 + 1	2	4	0	− 4
1 + 1 + 1 + 1 + 1 + 1	1	6	0	− 6

As Theorem 1 together with (1.15) predicts, $c(\pi)$ provides eleven different residue classes modulo 11.

References

1. G. E. Andrews, *The theory of partitions*, Encyclopedia of Mathematics and Its Applications, Vol. 2 (G.-C. Rota, ed.), Addison-Wesley, Reading, 1976 (Reprinted: Cambridge Univ. Press, London and New York, 1984).

2. A. O. L. Atkin and P. Swinnerton-Dyer, *Some properties of partitions*, Proc. London Math. Soc. (3) **4** (1954), 84–106.

3. F. J. Dyson, *Some guesses in the theory of partitions*, Eureka (Cambridge) **8** (1944), 10–15.

4. F. G. Garvan, *Generalizations of Dyson's rank*, Ph. D. thesis, Pennsylvania State Univ., 1986.

5. ——, *New combinatorial interpretations of Ramanujan's partition congruences* mod 5, 7 *and* 11, Trans. Amer. Math. Soc. (to appear).

DEPARTMENT OF MATHEMATICS, PENNSYLVANIA STATE UNIVERSITY, UNIVERSITY PARK, PENNSYLVANIA 16802

DEPARTMENT OF MATHEMATICS, UNIVERSITY OF WISCONSIN—MADISON, MADISON, WISCONSIN 53706

Current address (F. G. Garvan): Institute for Mathematics and its Applications, University of Minnesota, Minneapolis, Minnesota 55455

3.10 Partition Identities with an Application to Group Representation Theory (with J.B. Olsson)

J. reine angew. Math. **413** (1991), 198—212

Journal für die reine und angewandte Mathematik

Partition identities with an application to group representation theory

By *George E. Andrews*[1]) at Pennsylvania and *Jorn B. Olsson*[2]) at Copenhagen

1. Introduction

Our object in this paper is to prove a general theorem on partitions with difference conditions.

Theorem 2. *Let $A=\{a_1, a_2, \dots, a_r\}$ be a set of r distinct positive integers arranged in increasing order, and let N be an integer larger than a_r. Let $P_1(A; N, n)$ denote the number of partitions of n into distinct parts each of which is congruent to some a_i* modulo N. *Let $P_2(A; N, n)$ denote the number of partitions of n into parts each of which is congruent to 0 or to some a_i* modulo N, *in addition only parts divisible by N may be repeated, the smallest part is $<N$, the difference between successive parts is at most N and strictly less than N if either part is divisible by N. Then for each $n \geqq 0$,*

$$P_1(A; N, n) = P_2(A; N, n). \tag{1. 1}$$

A special case of this result gives further evidence for a conjecture in the modular representation theory of finite symmetric groups. A full description of this is given in Section 5. In Section 2, we shall provide the necessary generating function background. In Section 3 we shall prove Theorem 2. In Section 4 we relate Theorem 2 to previous work in the theory of partitions.

2. Preliminaries

We shall consider a set $\mathscr{A}=\{\alpha_1, \alpha_2, \dots, \alpha_r\}$ of indeterminates (complex variables). We code the 2^r words with distinct letters taken from $\mathscr{A}$ by means of binary representation as follows:

$$\kappa(\alpha_{i_1}, \alpha_{i_2}, \dots, \alpha_{i_j}) = 2^{i_1-1} + 2^{i_2-1} + \dots + 2^{i_j-1}, \tag{2. 1}$$

$$\kappa(0) = 2^r, \quad (0 \text{ is the empty word}).$$

---

[1]) Partially supported by National Science Foundation Grant DMS-870269.

[2]) Partially supported by the Danish Natural Science Foundation.

Since binary representation is unique, κ has an inverse which we call ω given by

$$\omega(2^{i_1-1}+2^{i_2-1}+\ldots+2^{i_r-1})=\alpha_{i_1}\alpha_{i_2}\cdots\alpha_{i_r}, \qquad \omega(2^r)=0. \tag{2.2}$$

Our coding provides a linear order on the 2^r subsets of $\mathscr{A}$ which corresponds to lexicographic ordering from the right by subscripts. For example if $r=3$, $\kappa(0)=8$, $\kappa(\alpha_1)=1$, $\kappa(\alpha_2)=2$, $\kappa(\alpha_1\alpha_2)=3$, $\kappa(\alpha_3)=4$, $\kappa(\alpha_1\alpha_3)=5$, $\kappa(\alpha_2\alpha_3)=6$, $\kappa(\alpha_1\alpha_2\alpha_3)=7$.

In addition we need two further functions. We let $v(j)=2^s$ where $s+1$ is the least subscript in the word $\omega(j)$, and $\beta(j)$ denotes the number of letters from $\mathscr{A}$ in $\omega(j)$.

We now define a system of 2^r q-difference equations for functions

$$f_j(\alpha_1, \alpha_2, \ldots, \alpha_r; x; q)=f_j(x)$$

as follows:

$$f_j(x)-f_{j+1}(x)=x\omega(j)f_{v(j)}(xq^{\beta(j)}), \qquad 1\leqq j\leqq 2^r-1, \tag{2.3}$$

$$f_{2_r}(x)=f_1(xq). \tag{2.4}$$

Furthermore we stipulate that these are analytic functions of x inside $|x|<|q|^{-1}$ (where throughout $|q|<1$), and $f_j(0)=1$. We shall employ the standard notation

$$(B;q)_n=(B)_n=(1-B)(1-Bq)\cdots(1-Bq^{n-1}), \tag{2.5}$$

$$(B;q)_\infty=(B)_\infty=\prod_{j=0}^{\infty}(1-Bq^j). \tag{2.6}$$

Theorem 1.

$$f_1(x)=(x)_\infty\sum_{n\geqq 0}\frac{(-\alpha_1)_n(-\alpha_2)_n\cdots(-\alpha_r)_n x^n}{(q)_n}. \tag{2.7}$$

Initial remarks. The actual proof of this theorem obscures what is going on due to cumbersome notation. So to make clear what is happening we illustrate the case $r=3$. The 8 relevant q-difference equations are

$$f_1(x)-f_2(x)=x\alpha_1 f_1(xq), \tag{2.8}$$

$$f_2(x)-f_3(x)=x\alpha_2 f_2(xq), \tag{2.9}$$

$$f_3(x)-f_4(x)=x\alpha_1\alpha_2 f_1(xq^2), \tag{2.10}$$

$$f_4(x)-f_5(x)=x\alpha_3 f_4(xq), \tag{2.11}$$

$$f_5(x)-f_6(x)=x\alpha_1\alpha_3 f_1(xq^2), \tag{2.12}$$

$$f_6(x)-f_7(x)=x\alpha_2\alpha_3 f_2(xq^2), \tag{2.13}$$

$$f_7(x)-f_8(x)=x\alpha_1\alpha_2\alpha_3 f_1(xq^3), \tag{2.14}$$

$$f_8(x)=f_1(xq). \tag{2.15}$$

From (2. 8) we see that

$$f_2(x)=f_1(x)-x\alpha_1 f_1(xq). \tag{2. 16}$$

Eliminating $f_2(xq)$ from the right-hand side of (2. 9) using (2. 16) and then adding (2. 8), (2. 9) and (2. 10), we obtain

$$f_4(x)=f_1(x)-x(\alpha_1+\alpha_2)f_1(xq)-x\alpha_1\alpha_2(1-xq)f_1(xq^2). \tag{2. 17}$$

Next we eliminate $f_2(xq)$, $f_2(xq^2)$ and $f_4(xq)$ from the right-hand sides of (2. 8)—(2. 15), and add all of (2. 8)—(2.15) together. The result is

$$\begin{aligned} f_1(x)=(1+x(\alpha_1+\alpha_2+\alpha_3))f_1(xq) \\ +x(\alpha_1\alpha_2+\alpha_1\alpha_3+\alpha_2\alpha_3)(1-xq)f_1(xq^2) \\ +x\alpha_1\alpha_2\alpha_3(1-xq)(1-xq^2)f_1(xq^3). \end{aligned} \tag{2. 18}$$

We now define

$$G(x)=f_1(x)/(x)_\infty, \tag{2. 19}$$

and we divide (2. 18) by $(xq)_\infty$ (letting σ_i denote the ith elementary symmetric function of the a_i). This yields

$$(1-x)G(x)=(1+x\sigma_1)G(xq)+x\sigma_2 G(xq^2)+x\sigma_3 G(xq^3). \tag{2. 20}$$

We now expand $G(x)$ in a power series

$$G(x)=\sum_{n=0}^{\infty}\gamma_n x^n. \tag{2. 21}$$

Since $G(0)=1$, we have $\gamma_0=1$, and for $n>0$ we obtain

$$\gamma_n-\gamma_{n-1}=q^n\gamma_n+(\sigma_1 q^{n-1}+\sigma_2 q^{2(n-1)}+\sigma_3 q^{3(n-1)})\gamma_{n-1} \tag{2. 22}$$

by comparing coefficients of x^n on both sides of (2. 20). Hence

$$\begin{aligned} \gamma_n &= \frac{(1+\sigma_1 q^{n-1}+\sigma_2 q^{2(n-1)}+\sigma_3 q^{3(n-1)})}{(1-q^n)}\gamma_{n-1} \\ &= \frac{(1+\alpha_1 q^{n-1})(1+\alpha_2 q^{n-1})(1+\alpha_3 q^{n-1})}{(1-q^n)}\gamma_{n-1}. \end{aligned} \tag{2. 23}$$

Iterating (2. 23), we find that

$$\gamma_n = \frac{(-\alpha)_n(-\alpha_2)_n(-\alpha_3)_n}{(q)_n},$$

and

$$\begin{aligned}(2.\,24)\qquad f_1(x) &= (x)_\infty\, G(x)\\ &= (x)_\infty \sum_{n=0}^{\infty} \frac{(-\alpha_1)_n(-\alpha_2)_n(-\alpha_3)_n x^n}{(q)_n}.\end{aligned}$$

The full proof consists of merely systematizing in general our creation of (2. 18).

Proof of Theorem 1. First we add together equations (2. 3) for $1 \leqq j \leqq 2^k - 1$. Thus

$$(2.25)\qquad f_1(x) - f_{2^k}(x) = \sum_{j=1}^{2^k-1} x\,\omega(j)\, f_{\nu(j)}(x q^{\beta(j)}).$$

Next we add together equations (2. 3) for $2^k \leqq j \leqq 2^{k+1} - 1$, here we obtain

$$(2.\,26)\qquad f_{2^k}(x) - f_{2^{k+1}}(x) = \sum_{j=2^k}^{2^{k+1}-1} x\,\omega(j)\, f_{\nu(j)}(x q^{\beta(j)});$$

however for j in the interval $[2^k, 2^{k+1})$, $j = 2^k + j'$, with j' in $[0, 2^k)$, so

$$\begin{aligned}(2.\,27)\qquad & f_{2^k}(x) - f_{2^{k+1}}(x)\\ &= x\alpha_{k+1} f_{2^k}(xq) + \sum_{j'=1}^{2^k-1} x\alpha_{k+1}\,\omega(j')\, f_{\nu(j')}(x q^{1+\beta(j')})\\ &= x\alpha_{k+1} f_{2^k}(xq) + \alpha_{k+1} q^{-1}(f_1(xq) - f_{2^k}(xq)) \qquad \text{(by (A. 25))}\\ &= \alpha_{k+1} q^{-1} f_1(xq) - \alpha_{k+1} q^{-1}(1 - xq) f_{2^k}(xq).\end{aligned}$$

We now are in position to prove a centrally important identity: for $k \leqq r$

$$(2.\,28)\qquad f_1(x) = f_{2^k}(x) + \sum_{j=1}^{k} x\,\sigma_j(\alpha_1, \alpha_2, \ldots, \alpha_k)\,(xq)_{j-1} f_1(xq^j).$$

When $k=0$ this trivializes to $f_1(x) = f_1(x)$. We now assume (2. 28) is true up to a particular $k < r$. Then

$$\begin{aligned}(2.\,29)\qquad & f_1(x) - f_{2^{k+1}}(x)\\ &= (f_1(x) - f_{2^k}(x)) + (f_{2^k}(x) - f_{2^{k+1}}(x))\\ &= \sum_{j=1}^{k} x\,\sigma_j(\alpha_1, \alpha_2, \ldots, \alpha_k)\,(xq)_{j-1} f_1(xq^j) + \alpha_{k+1} q^{-1} f_1(xq) - \alpha_{k+1} q^{-1}(1 - xq) f_{2^k}(xq)\end{aligned}$$

(by the induction hypothesis and (2. 27))

$$= \sum_{j=1}^{k} x\sigma_j(\alpha_1, \alpha_2, \ldots, \alpha_k)(xq)_{j-1} f_1(xq^j) + \alpha_{k+1} q^{-1}(1-xq) f_1(xq) + x\alpha_{k+1} f_1(xq)$$

$$- \alpha_{k+1} q^{-1}(1-xq) f_{2^k}(xq)$$

$$= \sum_{j=1}^{k} x\sigma_j(\alpha_1, \alpha_2, \ldots, \alpha_k)(xq)_{j-1} f_1(xq^j)$$
$$+ \alpha_{k+1} q^{-1}(1-xq) \sum_{j=1}^{k} xq\sigma_j(\alpha_1, \alpha_2, \ldots, \alpha_k)(xq^2)_{j-1} f_1(xq^{j+1}) + x\alpha_{k+1} f_1(xq)$$

(by the induction hypothesis)

$$= \sum_{j=1}^{k} x\sigma_j(\alpha_1, \ldots, \alpha_k)(xq)_{j-1} f_1(xq^j) + \sum_{j=2}^{k+1} x\alpha_{k+1}\sigma_{j-1}(\alpha_1, \alpha_2, \ldots, \alpha_k)(xq)_{j-1} f_1(xq^j)$$

$$+ x\alpha_{k+1} f_1(xq)$$

$$= \sum_{j=1}^{k+1} x(\sigma_j(\alpha_1, \ldots, \alpha_k) + \alpha_{k+1}\sigma_{j-1}(\alpha_1, \ldots, \alpha_k))(xq)_{j-1} f_1(xq^j)$$

$$= \sum_{j=1}^{k+1} x\sigma_j(\alpha_1, \ldots, \alpha_{k+1})(xq)_{j-1} f_1(xq^j).$$

We thus have (2. 28) established by mathematical induction.

We now set $k = r$ in (2. 28) and eliminate $f_{2^r}(x)$ using (2. 4).

Therefore

(2. 30) $$f_1(x) = f_1(xq) + \sum_{j=1}^{r} x\sigma_j(xq)_{j-1} f_1(xq^j).$$

This is precisely (2. 18) when $r = 3$, and the remainder of our proof is exactly like steps (2. 19)—(2. 24). The only change is that now

(2. 31) $$\gamma_n = \frac{(1 + \sigma_1 q^{n-1} + \sigma_2 q^{2(n-1)} + \ldots + \sigma_r q^{r(n-1)})}{(1-q^n)} \gamma_{n-1}$$

$$= \frac{(1+\alpha_1 q^{n-1})(1+\alpha_2 q^{n-1}) \cdots (1+\alpha_r q^{n-1})}{(1-q^n)} \gamma_{n-1}$$

$$\vdots$$

$$= \frac{(-\alpha_1)_n(-\alpha_2)_n \cdots (-\alpha_r)_n}{(q)_n}.$$

Consequently

$$f_1(x)=(x)_\infty G(x) \tag{2. 32}$$

$$=(x)_\infty \sum_{n=0}^{\infty} \frac{(-\alpha_1)_n(-\alpha_2)_n\cdots(-\alpha_r)_n x^n}{(q)_n},$$

which is (2. 7) as desired. □

Now it appears that $f_1(x)$ has a singularity at $x=1$; however we see immediately from (2. 30) and (2. 32) that in fact $f_1(x)$ is analytic for $|x|<|q|^{-1}$. This observation is important because in Section 3 we need to consider $f_1(1)$ and to evaluate it as $\lim_{x\to 1^-} f_1(x)$.

3. The main result

Our object now is to prove Theorem 2 which is stated in the introduction. We begin by noting that by standard methods [3], Ch. 1:

$$\sum_{n\geqq 0} P_1(A;N;n)q^n=\prod_{n=0}^{\infty}(1+q^{Nn+a_1})(1+q^{Nn+a_2})\cdots(1+q^{Nn+a_r}) \tag{3. 1}$$

$$=\lim_{x\to 1^-} f_1(q^{a_1},q^{a_2},\ldots,q^{a_r};x;q^N)$$

(by Abel's lemma [2], p. 190, applied to Thm. 1)

$$=f_1(q^{a_1},q^{a_2},\ldots,q^{a_r};1;q^N)$$

by the observation at the end of Section 2. Consequently we must relate this last expression to the partitions connected with $P_2(A;N;n)$ which we shall call P_2-type partitions.

Among the P_2-type partitions we shall now define 2^r classes. The largest part of any P_2-type partition, π, lies in some interval, say $(\lambda N, \lambda N+N]$. Let us consider the set of parts of π lying in $(\lambda N, \lambda N+N)$; the residues modulo N of these parts form a subset of A (possibly empty, in which case the largest part is $\lambda N+N$); if this subset is empty we adjoin to it the element 0, and we write $A^*=A\cup\{0\}$. Call this latter subset of A^* the indicator of π, say $i(\pi)$, and write $\sum(i(\pi))$ for the sum of the elements of $i(\pi)$.

As is clearly indicated in (3. 1) the set of variables $\mathscr{A}=\{\alpha_1,\ldots,\alpha_r\}$ will here be $\{q^{a_1},q^{a_2},\ldots,q^{a_r}\}$. Clearly the word $\alpha_{i_1}\alpha_{i_2}\cdots\alpha_{i_j}$ corresponds naturally to the sum $a_{i_1}+a_{i_2}+\ldots+a_{i_j}$ since $\alpha_{i_1}\alpha_{i_2}\cdots\alpha_{i_j}=q^{a_{i_1}+a_{i_2}+\ldots+a_{i_j}}$. If $i(\pi)=\{a_{i_1},a_{i_2},\ldots,a_{i_r}\}$, we shall write

$$\kappa(i(\pi))=2^{i_1-1}+2^{i_2-1}+\ldots+2^{i_j-1},$$

$$\kappa(0)=2^r,$$

$$\omega(2^{i_1-1}+2^{i_2-1}+\ldots+2^{i_j-1})=\{a_{i_1},a_{i_2},\ldots,a_{i_r}\},$$

$$\omega(2^r)=0.$$

In the same vein, $\nu(j)=2^s$ where $s+1$ is the least subscript on the a_i's appearing in $\omega(j)$, and $\beta(j)$ will denote the number of a_i's in $\omega(j)$. We now define the ith class of P_2-type partitions to be those P_2-type partitions for which $\kappa(i(\pi))\geqq i$.

We also define $P_2(A;\,N;\,i;\,n,\,m)$ to be the number of P_2-type partitions π of n in the ith class with $m=\left\lceil\dfrac{\text{largest part of }\pi}{N}\right\rceil$ ($\lceil x\rceil$ is the smallest integer $\geqq x$).

Next we wish to establish two recurrences for the $P_2(A;\,N;\,i;\,n,\,m): 1\leqq i\leqq 2^r-1$,

$$(3.\ 2)\qquad P_2(A;\,N;\,i;\,n,\,m)-P_2(A;\,N;\,i+1;\,n,\,m)$$

$$=P_2(A;\,N;\,\nu(i);\,n-\textstyle\sum(i(\pi))-N\beta(i)\,(m-1),\,m-1),$$

and

$$(3.\ 3)\qquad P_2(A;\,N;\,2^r;\,n,\,m)=P_2(A;\,N;\,1;\,n-Nm,\,m).$$

To see (3. 2), we observe that its left-hand side enumerates precisely those P_2-tpye partitions π of n with $m=\left\lceil\dfrac{\text{largest part of }\pi}{N}\right\rceil$ and $\kappa(i(\pi))=i$. This means there are $j=\beta(i)$ parts of π of the form $(m-1)N+a_{i_1}$, $(m-1)N+a_{i_2},\ldots,(m-1)N+a_{i_j}$ (in increasing magnitude) where $\{a_{i_1},a_{i_2},\ldots,a_{i_j}\}=i(\pi)$. Note for later that $i_1=\nu(i)$. The sum of these j parts added together is $(m-1)N\beta(i)+\sum(i(\pi))$. Let us delete all these parts from π to form a new partition π' of $n-(m-1)N\beta(i)-\sum(i(\pi))$. Due to the deletions and the limitation on successive differences, we see that $m-1=\left\lceil\dfrac{\text{largest part of }\pi'}{N}\right\rceil$. Again due to the condition on P_2-type partitions concerning differences between successive parts, we see that admissible $i(\pi')$ are precisely those subsets of A that have $\kappa(i(\pi'))\geqq\nu(i)$ (recall $2^{i_1-1}=\nu(i)$). Hence the partition π' is precisely of the type enumerated by $P_2(A;\,N;\,\nu(i);\,n-N\beta(i)\,(m-1)-\kappa(i(\pi)),\,m-1)$. Clearly the correspondence between π and π' is reversible. Therefore a bijection is established between the two classes of partitions which in turn proves (3. 2).

We now turn to (3. 3). In this instance the left-hand side of (3. 3) enumerates those P_2-type partitions π of n with largest part equal to Nm. If we delete one appearance of Nm to form π', we see by the difference conditions on successive parts for P_2-type partitions that still $m=\left\lceil\dfrac{\text{largest part of }\pi'}{N}\right\rceil$. Furthermore the parts of π' in $(N(m-1),\,Nm]$ are now only subject to the constraint $\kappa(i(\pi'))\geqq 1$. Hence the partitions π' are enumerated by $P_2(A;\,N;\,1;\,n-Nm,\,m)$. Since the mapping of π to π' is clearly reversible, we see that this bijection established (3. 3).

Identities (3. 2) and (3. 3) may now be directly translated into generating function identities. Let us define

$$(3.4)\qquad \mathscr{F}_i(A; N; x, q) = \sum_{n, m \geq 0} P_2(A; N; i; n, m) q^n x^m.$$

By comparison with $\prod_{n\geq 1} (1-xq^n)^{-1}$ we see that we have $\mathscr{F}_i$ analytic in x for $|x| < |q|^{-1}$, and furthermore $\mathscr{F}_i(A; N; 0, q) = 1$. From (3. 2) we find

$$(3.5)\qquad \mathscr{F}_i(A; N; x, q) - \mathscr{F}_{i+1}(A; N; x, q) = xq^{\kappa(i(\pi))} \mathscr{F}_{\nu(i)}(A; N; xq; N\beta(i)q),$$

and from (3. 3) we find

$$(3.6)\qquad \mathscr{F}_{2^r}(A; N; x, q) = \mathscr{F}_1(A; N; xq^N; q).$$

Comparison of (3. 5) and (3. 6) with (2. 3) and (2. 4) shows that the $\mathscr{F}_i(A; N; x, q)$ precisely fulfill the boundary conditions and q-difference equations satisfied by the $f_i(q^{a_1}, q^{a_2}, \ldots, q^{a_r}; x; q^N)$. Therefore by Theorem 1

$$(3.7)\qquad \mathscr{F}_1(A; N; x; q) = (x; q^N)_\infty \sum_{n\geq 0} \frac{(-q^{a_1}; q^N)_n \cdots (-q^{a_r}; q^N)_n x^n}{(q^N; q^N)_n}.$$

Clearly

$$(3.8)\qquad \sum_{n\geq 0} P_2(A; N; n) q^n = \mathscr{F}_1(A; N; 1; q)$$

$$= \lim_{x\to 1^-} (x; q^N)_\infty \sum_{n\geq 0} \frac{(-q^{a_1}; q^N)_n \cdots (-q^{a_r}; q^N)_n x^n}{(q^N; q^N)_n}$$

$$= (-q^{a_1}; q^N)_\infty (-q^{a_2}; q^N)_\infty \cdots (-q^{a_r}; q^N)_\infty$$

(by Abel's lemma [2], p. 190)

$$= \sum_{n\geq 0} P_1(A; N; n) q^n$$

(by (3. 1)).

Comparing coefficients of q^n in the extremes of (3. 8) yields Theorem 2. □

4. Application to other parition identities

In Theorem 2, $A=\{a_1, \ldots, a_r\}$ is an arbitrary set of positive integers arranged in increasing order for which $a_r < N$. In this section we require additionally

$$\sum_{i=1}^{k-1} a_i < a_k \quad (k \leqq r), \tag{4. 1}$$

$$N \geqq a_1 + a_2 + \ldots + a_r, \tag{4. 2}$$

and all of the 2^r subsets of A must have distinct sums. Let A' be the set of $2^r - 1$ positive sums arising from the nonempty subsets of A. Let A'_N denote the set of all positive integers that are congruent to some element of A' modulo N. Let $\varrho_N(m)$ denote the least positive residue of m modulo N. For $m \in A'$, let $b(m)$ be the number of terms appearing in the sum of distinct elements of A making up m and let $v(m)$ denote the least a_i in this sum.

In [1], the main result may be restated as follows:

Theorem 3. *Let $E(A'_N; n)$ denote the number of partitions of n into parts taken from $A'_N: n = c_1 + c_2 + \ldots + c_s$, $c_i \geqq c_{i+1}$,*

$$c_i - c_{i+1} \geqq N b(\varrho_N(c_{i+1})) + v(\varrho_N(c_{i+1})) - \varrho_N(c_{i+1}).$$

Then $E(A'; n) = P_1(A; N, n)$.

Remark. While this theorem is less general than Theorem 2, it is, as will be clear subsequently, essentially equivalent to it under the more stringent conditions imposed at the beginning of this section.

Sketch of Proof. The acual proof in [1] of this result is very closed related to the work done in proving Theorem 2. To see the relationship we shall show why $P_2(A; N; n)$ and $E(A'_N; n)$ must be identical.

We require MacMahon's modification of the Ferrers graph called "modular partitions" [7]. For our purposes the modulus N will be used. Each part of a partition is represented by a row of N's with the last entry in the row the least positive residue modulo N of that part. Thus if $N=7$, the modular representation of $23+23+14+12+3+2$ is

```
7772
7772
77
75
3
2
```

Thus for the N of our theorem, the modular representation of a general partition of the type enumerated by $P_2(N; A; n)$ is

$$\begin{array}{l}
NNN.\,.\,.\,.\,.\,.\,NNa_{i_j}\\
NNN.\,.\,.\,.\,.\,.\,NNa_{i_{j-1}}\\
\vdots\\
NNN.\,.\,.\,.\,.\,.\,NNa_{i_1}\\
NNN.\,.\,.\,.\,.\,.\,NN\\
\vdots\\
NNN.\,.\,.\,.\,.\,.\,NN\\
NNN.\,.\,.\,.\,.\,.\,Na_{h_k}\\
NNN.\,.\,.\,.\,.\,.\,Na_{h_{k-1}}\\
\vdots\\
NNN.\,.\,.\,.\,.\,.\,Na_{h_1}\\
NNN.\,.\,.\,.\,.\,.\,N\\
\vdots
\end{array}$$

The partitions under consideration are now mapped bijectively onto the partitions enumerated by $E(A'; N; n)$ by conjugation, i.e. adding up columns instead of rows. To see this we note that due to the difference conditions imposed on summands of P_2-type partitions there must be a subset of A making up the bottom of each column. Thus the conjugate parts are all taken from A'_N. In our example, the smallest part in the conjugate is $c_s = a_{i_j} + \ldots + a_{i_1}$ and the next part is $c_{s-1} = lN + a_{h_k} + \ldots + a_{h_1}$ where $l \geqq a_{i_1}$. Hence

$$\begin{aligned}
c_{s-1} - c_s &= lN + (a_{h_k} + \ldots + a_{h_1}) - (a_{i_j} + \ldots + a_{i_1})\\
&\geqq N\,b(\varrho_N(c_s)) + (a_{h_k} - a_{i_1}) - (a_{i_2} + \ldots + a_{i_j})\\
&\geqq N\,b(\varrho_N(c_s)) + a_{i_1} - (a_{i_1} + \ldots + a_{i_j})\\
&= N\,b(\varrho_N(c_s)) + v(\varrho_N(c_s)) - \varrho_N(c_s).
\end{aligned}$$

Furthermore this minimum is assumed when $k=1$ and $a_{h_1}=a_{i_1}$. Thus this difference between parts is precisely the one required by $E(A'_N;\, n)$, and exactly the same reasoning applies to any adjacent columns in the conjugate.

Therefore $P_2(A;\, N;\, n)=E(A'_n;\, n)$, and Theorem 3 now follows from Theorem 2. □

5. On the Mullineux conjecture

We review a few facts from the representation theory of the finite symmetric groups S_n leading to a description of the so-called "Mullineux conjecture". A general reference is [5] and [6]. When F is any field FS_n denotes the group algebra of the symmetric group S_n over F. An F-representation of S_n is then an FS_n-module of finite F-dimension. We are interested in the irreducible F-representations of S_n, i.e. the simple FS_n-modules. When $\operatorname{char} F \neq 2$ there are 2 non isomorphic simple FS_n-modules of dimension 1, the trivial module $D_{(n)}$ and the sign module $D^0_{(n)}$. (When $\operatorname{char} F=2$ these modules coincide.) In any case, when D is a simple FS_n-module we put

$$D^0 = D \otimes_{FS_n} D^0_{(n)}$$

and call D^0 the conjugate module to D. D^0 is of course also simple. We let $\operatorname{Par}(n)$ denote the set of partitions of n and when p is a prime, $\operatorname{Par}_p(n)$ denotes the set of p-regular partitions of n, i.e. the set of partitions containing no p equal parts. Moreover, $\operatorname{Irr}_F(S_n)$ denotes the set of (isomorphism classes) of simple FS_n-modules.

Consider first the "ordinary" case, when $\operatorname{char} F=0$. Then there is a canonical bijection $\delta: \lambda \mapsto D_\lambda$ between $\operatorname{Par}(n)$ and $\operatorname{Irr}_F(S_n)$.

Using the above we get a bijection $\lambda \mapsto \lambda^0$ on $\operatorname{Par}(n)$, defined by $\lambda^0=\delta^{-1}(D^0_\lambda)$. It is well-known (see [6], 2. 1. 8) that λ^0 is the usual conjugate (dual) partition to λ. To understand Mullineux's "p-version" of the conjugation map, consider the "Frobenius symbol" of a partition λ as described in [4], sec. 2 (or in [11], sec. 1)

$$F(\lambda)=\begin{pmatrix} a_1 & a_2 \dots a_r \\ b_1 & b_2 \dots b_r \end{pmatrix}$$

where $a_1>a_2>\dots>a_r \geqq 0$, $b_1>b_2>\dots>b_r \geqq 0$ and $r+\sum a_i+\sum b_i=n$. Then

$$F(\lambda^0)=\begin{pmatrix} b_1 & b_2 \dots b_r \\ a_1 & a_2 \dots a_r \end{pmatrix}$$

and the diagonal hooklengths of the partitions λ and λ^0 are $h_1, h_2, \dots, h_r$, where $h_i=a_i+b_i+1$. Thus h_1 is the length of the rim of the Ferrer's graph (Young diagram) of the partition $\lambda_1=\lambda$, h_2 is the rim length of the partition λ_2 obtained from λ_1 by removing the rim nodes, etc. We let c_i be the number of parts of λ_i and put

$$G(\lambda)=\begin{pmatrix} h_1 & h_2 \dots h_r \\ c_1 & c_2 \dots c_r \end{pmatrix}.$$

If $F(\lambda)$ is as above, then $c_i = b_i + 1$, $h_i = a_i + b_i + 1$. Moreover

$$G(\lambda^0) = \begin{pmatrix} h_1 & h_2 \dots h_r \\ d_1 & d_2 \dots d_r \end{pmatrix}$$

where $d_i = a_i + 1 = h_i + 1 - c_i$.

The entries in $G(\lambda)$ satisfy

$$(5.\ 1) \qquad \begin{cases} h_i - h_{i+1} > c_i - c_{i+1} > 0 \quad \text{for} \quad i = 1, 2, \dots, r-1, \\ h_r \geqq c_r \geqq 1, \quad h_1 + \dots + h_r = n. \end{cases}$$

Example. $\lambda = (6^2, 5^2, 2)$. In the diagram below the rim nodes of λ_i are replaced by i:

$$\begin{matrix} 4 & 4 & 3 & 2 & 2 & 1 \\ 3 & 3 & 3 & 2 & 1 & 1 \\ 2 & 2 & 2 & 2 & 1 & \\ 2 & 1 & 1 & 1 & 1 & \\ 1 & 1 & & & & \end{matrix}$$

Thus

$$G(\lambda) = \begin{pmatrix} 10 & 8 & 4 & 2 \\ 5 & 4 & 2 & 1 \end{pmatrix}, \qquad G(\lambda^0) = \begin{pmatrix} 10 & 8 & 4 & 2 \\ 6 & 5 & 3 & 2 \end{pmatrix}.$$

Let us note that in the general case

$$\lambda = \lambda^0 \Leftrightarrow F(\lambda) = F(\lambda^0) \Leftrightarrow G(\lambda) = G(\lambda^0)$$

$$\Leftrightarrow G(\lambda) = \begin{pmatrix} 2b_1 + 1 & 2b_2 + 1 \dots 2b_r + 1 \\ b_1 + 1 & b_2 + 1 \dots b_r + 1 \end{pmatrix}$$

for some $b_1 > b_2 > \dots > b_r \geqq 0$. So this is the well-known (trivial) result that the number of self-conjugate partitions of n equals the number of partitions of n into odd distinct parts.

Consider now the "modular" case, when $\operatorname{char} F = p > 0$. Then there is a canonical bijection $\delta_p : \lambda \mapsto D_\lambda$ between $\operatorname{Par}_p(n)$ and $\operatorname{Irr}_F(S_n)$. As above we get a bijection $\lambda \mapsto \lambda^P$ on $\operatorname{Par}_p(n)$ defined by $\lambda^P = \delta_p^{-1}(D_\lambda^0)$. It is of course natural to ask for the relation between the p-regular paratitions λ and λ^P. The answer to this question is not known and indeed seems out of reach at the present time. However, Mullineux in [9] defined what may be seen as a p-analogue of the conjugation map on $\operatorname{Par}_p(n)$. We denote this map by $M(\lambda \mapsto \lambda^M)$. Then, of course, the Mullineux conjecture states:

$$\text{For all } \lambda \in \operatorname{Par}_p(n) : \lambda^P = \lambda^M.$$

We proceed to give a description of the map M. In analogy with $G(\lambda)$ above we characterize $\lambda \in \mathrm{Par}_p(n)$ by a symbol

$$G_p(\lambda)=\begin{pmatrix} h_1' & h_2' \dots h_r' \\ c_1' & c_2' \dots c_r' \end{pmatrix}.$$

We have to replace the rim of λ by the p-rim, which is defined as follows:

The p-rim is a part of the rim of λ, which is composed of p-segments, which all except possibly the last contain p points. The first p-segment consists of the first p points of the rim of λ starting with the longest row. (If the rim contains at most p points it is the entire rim.) The next p-segment is obtained similarly starting in the row next below that which contains the end of the previous p-segment. This process is continued until the final row is reached. Then h_1' is the number of nodes in the p-rim of $\lambda=\lambda_1'$. Removing the p-rim of λ_1' we obtain a new p-regular partition λ_2' and let h_2' be the p-rim length of λ_2', etc. We let c_i' be the number of parts of λ_i' and put

$$G_p(\lambda)=\begin{pmatrix} h_1' & h_2' \dots h_r' \\ c_1' & c_2' \dots c_r' \end{pmatrix}.$$

This is the Mullineux symbol of λ. It is easy to recover λ from its Mullineux symbol. Moreover, by [9] the entries of $G_p(\lambda)$ satisfy

$$(5.2)\qquad \begin{cases} (1) & 0 \leqq c_i' - c_{i+1}' \leqq p, \\ (2) & c_i' - c_{i+1}' + \varepsilon_{i+1} \leqq h_i' - h_{i+1}' < c_i' - c_{i+1}' + \varepsilon_{i+1} + p, \\ (3) & c_i' = c_{i+1}' \quad \Rightarrow \quad p \mid h_i', \\ (4) & c_i' - c_{i+1}' = p \quad \Rightarrow \quad p \nmid h_i', \\ (5) & 1 \leqq c_i' \leqq p, \quad 0 \leqq h_r' - c_r' < p, \quad \sum h_i' = n \end{cases}$$

where $\varepsilon_i = 1$ if $p \nmid h_i'$ and $\varepsilon_i = 0$ if $p \mid h_i'$.

If $G_p(\lambda)$ is as above, then λ^M is defined as the p-regular partition having

$$G_p(\lambda^M)=\begin{pmatrix} h_1' & h_2' \dots h_r' \\ d_1' & d_2' \dots d_r' \end{pmatrix}$$

as Mullineux symbol, where $c_i' + d_i' = h_i' + \varepsilon_i$.

Example. $p=3$, $\lambda=(6^2, 5^2, 2)$. In the following diagram the 3-rim nodes of λ_i' are replaced by i

```
4 4 3 2 2 1
4 3 3 2 1 1
4 3 2 2 1
3 3 2 1 1
1 1
```

so

$$G_3(\lambda)=\begin{pmatrix}8 & 6 & 6 & 4\\ 5 & 4 & 4 & 3\end{pmatrix}.$$

We get

$$G_3(\lambda^M)=\begin{pmatrix}8 & 6 & 6 & 4\\ 4 & 2 & 2 & 2\end{pmatrix}$$

which gives the diagram

$$\begin{matrix}4 & 4 & 4 & 3 & 3 & 3 & 2 & 2 & 2 & 1 & 1 & 1\\ 4 & 3 & 3 & 3 & 2 & 2 & 2 & 1 & 1 & 1 & & \\ 1 & & & & & & & & & & & \\ 1 & & & & & & & & & & & \end{matrix}$$

Thus $\lambda^M=(12, 10, 1, 1)$.

It is easy to describe the Mullineux symbols of the p-regular partitions fixed by the map M. Such a symbol is characterized by its first row $h'_1, h'_2, \ldots, h'_r$ and then the conditions of (5. 2) may be transformed into conditions on the h'_i. We get

Proposition 1. *The number of fixpoints of M is equal to the number of partitions $(h'_1, \ldots, h'_r)$ of n satisfying*

(1) $2|h'_i \Leftrightarrow p|h'_i$;

(2) $0\leqq h'_i-h'_{i+1}\leqq 2p$;

(3) *If* $h'_i=h'_{i+1}$ *then* h'_i *is even*;

(4) *If* $h'_i-h'_{i+1}=2p$ *then* h'_i *is odd*;

(5) $h'_r<2p$.

We are also interested in the number of fixpoints for the map P on $\mathrm{Par}_p(n)$. Using [12], Formula (2. 1)* we get

Proposition 2. *The number of fixpoints of P is equal to the number of partitions $(h'_i, \ldots, h'_r)$ of n satisfying*

(1) $2\nmid h'_i$ *and* $p\nmid h'_i$ *for all* i;

(2) $h'_i\neq h'_{i+1}$ *for all* i.

Now the number of fixpoints for M is $P_2(A; N, n)$ and the number of fixpoints for P is $P_1(A; N, n)$ where $N=2p$ and

$$A=\{a\,|\,1\leqq a<2p,\ a \text{ odd},\ a\neq p\}.$$

Thus, by Theorem 2

Proposition 3. *The maps M and P have the same number of fixpoints.*

Of course Proposition 3 yields further evidence for the validity of the Mullineux conjecture in addition to previous evidence given in [9]—[10] and [8].

References

[1] *G. E. Andrews*, A general theorem on partitions with difference conditions, Amer. J. Math. **91** (1969), 18—24.

[2] *G. E. Andrews*, Number Theory, Philadelphia 1971. (Reissued: New Delhi 1984.)

[3] *G. E. Andrews*, The Theory of Partitions, Vol. 2 of the Encyclopedia of Mathematics and Its Applications, Reading 1976. (Reissued: New York and London 1985.)

[4] *G. E. Andrews*, Generalized Frobenius parititions, Mem. Amer. Math. Soc. **49** (1984), No. 301, iv.—44 pp.

[5] *G. James*, The representation theory of the symmetric groups, Proc. Symp. Pure Math. **47** (1987), 111—126.

[6] *G. James* and *A. Kerber*, The Representation Theory of the Symmetric Group, Vol. 16 of the Encyclopedia of Mathematics and Its Applications, Reading 1981. (Reissued: New York and London 1985).

[7] *P. A. MacMahon*, The theory of modular partitions, Proc. Cambridge Phil. Soc. **21** (1923), 197—204. (Reprinted in The Coll. Papers of P. A. MacMahon, Vol. I, Cambridge 1978.)

[8] *S. Martin*, On the ordinary quiver of the principal block of certain symmetric groups, Quart. J. Math., Oxford second series, **40** (1989), 209—223.

[9] *G. Mullineux*, Bijections of p-regular partitions and p-modular irreducibles of the symmetric group, J. London Math. Soc. (2) **20** (1979), 60—66.

[10] *G. Mullineux*, On the p-cores of p-regular diagrams, J. London Math. Soc. (2) **20** (1979), 222—226.

[11] *J. B. Olsson*, Frobenius symbols for partitions and degrees of spin characters, Math. Scand. **61** (1987), 223—247.

[12] *J. B. Olsson*, The number of modular characters in certain blocks, in preparation.

Department of Mathematics, 218 McAllister Building, Pennsylvania State University, University Park, PA 16802, USA

Københavns Universitets Matematiske Institut, Universitetsparken 5, 2100 København Ø. Danmark

Eingegangen 24. Januar 1990

3.11 Generalizations and Refinements of a Partition Theorem of Göllnitz (with K. Alladi and B. Gordon)

J. reine angew. Math. **460** (1995), 165—188

Journal für die reine und angewandte Mathematik

Generalizations and refinements of a partition theorem of Göllnitz

By *Krishnaswami Alladi*[1]) at Gainesville, *George E. Andrews*[2]) at University Park, and *Basil Gordon* at Los Angeles

§ 1. Introduction

One of the deepest results in the theory of partitions is a theorem of Göllnitz [17]:

Theorem G. *Let $A(n)$ denote the number of partitions of n into parts $\equiv 2, 5, 11 \pmod{12}$.*

Let $B(n)$ denote the number of partitions of n into distinct parts $\equiv 2, 4, 5 \pmod 6$.

Let $C(n)$ denote the number of partitions of n in the form $m_1 + m_2 + \cdots + m_s$, no part $= 1$ or 3, and such that $m_\ell - m_{\ell+1} \geqq 6$ with strict inequality if $m_\ell \equiv 6, 7$ or $9 \pmod 6$.

Then $A(n) = B(n) = C(n)$.

In Theorem G, and throughout, we adopt the convention that $j \equiv \mu \pmod M$ means that $j = \mu + \ell M$ with $j \geqq \mu$ and $j > 0$.

The equality $A(n) = B(n)$ is trivial because

$$(1.1)\qquad \sum_{n=0}^{\infty} A(n)q^n = \prod_{m=0}^{\infty} \frac{1}{(1-q^{12m+2})(1-q^{12m+5})(1-q^{12m+11})}$$
$$= \prod_{m=0}^{\infty} (1+q^{6m+2})(1+q^{6m+4})(1+q^{6m+5}) = \sum_{n=0}^{\infty} B(n)q^n .$$

Thus, the real challenge is the equality $B(n) = C(n)$. Göllnitz' proof [17] of this is very complicated and the details are forbidding; but he succeeded in obtaining the refinement

$$(1.2)\qquad B(n; v) = C(n; v) .$$

---

[1]) Work done at The Pennsylvania State University on sabbatical leave from Florida during 1992–93.

[2]) Research supported in part by NSF grant DMS 8702695-04.

In (1.2), $B(n; v)$ and $C(n; v)$ denote the number of partitions counted by $B(n)$ and $C(n)$ respectively, with the additional restriction that there are precisely v parts and the convention that parts $\equiv 6, 7$ and $9 \pmod 6$ are counted twice. Such refinements are not possible with $A(n)$ and so we will not be concerned with $A(n)$ in this paper.

Göllnitz also obtained two generalizations of Theorem G ([17], Sätze 4.8 and 4.10). In his Satz 4.8, the modulus 6 is replaced by $M + 4$ with $M \geqq 2$, and the residues $2, 4, 5, 6, 7$ and $9 \pmod 6$ are replaced by $M, M + 2, M + 3, 2M + 2, 2M + 3$ and $2M + 5 \pmod{M + 4}$. The difference conditions in Satz 4.8 are that $m_\ell - m_{\ell+1} \geqq M + 4$, with strict inequality if $m_\ell \equiv 2M + 2, 2M + 3$ or $2M + 5 \pmod{M + 4}$. In his Satz 4.10 also, the modulus 6 is replaced by $M + 4$ with $M \geqq 2$, but the residues 2, 4, 5, 6, 7 and 9 are replaced by $2, M + 2, M + 3, M + 4, M + 5$ and $2M + 5 \pmod{M + 4}$. Both these results are established in [17] in the refined form given by (1.2).

Our purpose here is to obtain an entirely new proof of the following result from which Theorem G, (1.2), and Sätze 4.8 and 4.10 of [17] fall out as special cases.

Theorem 1. *Let $M \geqq 6$ and r_1, r_2, r_3 residues satisfying the following inequalities*:

$$(1.3) \qquad 0 < r_1 < r_2 < r_3 < M \leqq r_1 + r_2 \quad \text{and} \quad r_1 + M < r_2 + r_3 .$$

Let $B(n; v)$ denote the number of partitions of n into v distinct parts $\equiv r_1, r_2$ or $r_3 \pmod M$.

Let $C(n; v)$ denote the number of partitions of n into v distinct parts $m_1 > m_2 > m_3 \dots$ such that

(i) *each part m_ℓ is $\equiv r_1, r_2, r_3, r_1 + r_2, r_1 + r_3$ or $r_2 + r_3 \pmod M$*,

(ii) *$m_\ell - m_{\ell+1} \geqq M$ with strict inequality if $m_\ell \equiv r_1 + r_2, r_1 + r_3$ or $r_2 + r_3 \pmod M$*,

(iii) *the parts $\equiv r_1 + r_2, r_1 + r_3$ and $r_2 + r_3 \pmod M$ are counted twice.*

Then $B(n; v) = C(n; v)$.

It is to be noted that in Theorem 1 we have almost total freedom in choosing the three residues $r_1, r_2, r_3 \pmod M$, whereas in Göllnitz' Sätzen 4.8 and 4.10, the residue classes are prescribed. Actually the restrictions (1.3) in Theorem 1 can be removed by introducing the concept of *levels* as in Alladi and Gordon [1]; but then the difference condition (ii) for $C(n; v)$ will be slightly more complicated.

Theorem 1 is proved in §2–§4 by means of a new technique called *the method of weighted words*, which was introduced recently by Alladi and Gordon [1], [2] to obtain generalizations, refinements and companions of Schur's celebrated 1926 partition theorem [18]. Indeed the method and results of this paper are substantial extensions of those in [1]. Andrews gave two proofs of Theorem G (and (1.2)), both of them simpler than Göllnitz' original proof. Andrews' first proof [7] was by the method of generating functions; his second proof (see [12], §10) made use of computer algebra to simplify the calculations. In [12], Andrews asks for a proof which lends more insight into the equality (1.2). We believe that the proof given here does provide this insight.

From the point of view of the method of weighted words, Theorem 1 is seen as emerging out of the following *key identity*:

$$(1.4)\quad \sum_{i,j,k} a^i b^j c^k \sum_{\substack{s=\alpha+\beta+\gamma+\delta+\varepsilon+\phi \\ i=\alpha+\delta+\varepsilon,\, j=\beta+\delta+\phi,\, k=\gamma+\varepsilon+\phi}} \frac{q^{T_s+T_\delta+T_\varepsilon+T_\phi-1}(1-q^\alpha(1-q^\phi))}{(q)_\alpha(q)_\beta(q)_\gamma(q)_\delta(q)_\varepsilon(q)_\phi}$$

$$= \sum_{i,j,k} \frac{a^i b^j c^k q^{T_i+T_j+T_k}}{(q)_i(q)_j(q)_k} = (-aq)_\infty(-bq)_\infty(-cq)_\infty .$$

In (1.4), $T_m = m(m+1)/2$ is the m-th triangular number. We have also made use of the standard notation

$$(z)_n = (z;q)_n = \prod_{j=0}^{n-1}(1-zq^j),$$

and

$$(z)_\infty = \prod_{j=0}^{\infty}(1-zq^j), \quad \text{for } |q|<1 .$$

Theorem 1 is a consequence of (1.4) under the substitutions

$$(1.5)\quad \text{(dilation) } q \mapsto q^M \text{ and (translations) } a \mapsto xq^{r_1-M},\ b \mapsto xq^{r_2-M},\ c \mapsto xq^{r_3-M},$$

if we compare the coefficients of $x^\nu q^n$ on both sides. While it is clear that the right hand side of (1.4) corresponds to the generating function of $B(n)$, it is not obvious (and indeed quite difficult to show) that the left side of (1.4) corresponds to the generating function of $C(n)$; this is established in § 3. (See also the explanation following Theorem 2 in § 2.)

We prove (1.4) in § 4. The depth of (1.4) becomes plain when we see that we not only require Watson's q-analog of Whipple's theorem but also the ${}_6\psi_6$ summation of Bailey; this constitutes substantially more than is required to prove either Schur's theorem [4] or the Rogers-Ramanujan identities [12].

Since there are three free parameters in (1.4), it is possible to refine Theorem 1 even further (see § 2, Theorem 2), by keeping track of the number of parts in each of the residue classes r_1, r_2 and r_3 (mod M). Another advantage with this method is that (1.4) deals with the *base case* without any dilations; it therefore offers the possibility of considering different dilations and translations.

Theorem G can be considered to be the partition result which is closest to, and in a certain sense an extension of Schur's famous partition theorem [18] which is

Theorem S. *Let $T(n)$ denote the number of partitions of n into parts* $\equiv 1, 5 \pmod 6$.

Let $S(n)$ denote the number of partitions of n into distinct parts $\equiv 1, 2 \pmod 3$.

Let $S_1(n)$ denote the number of partitions of n into parts $m_1 + m_2 + \cdots$ such that $m_\ell - m_{\ell+1} \geqq 3$ with strict inequality if $m_\ell \equiv 3 \pmod 3$.

Then $T(n) = S(n) = S_1(n)$.

As in (1.1), the equality $T(n) = S(n)$ is trivial because

$$\sum_{n=0}^{\infty} T(n)q^n = \prod_{m=0}^{\infty} \frac{1}{(1-q^{6m+1})(1-q^{6m+5})} \tag{1.6}$$

$$= \prod_{m=0}^{\infty} (1+q^{3m+1})(1+q^{3m+2}) = \sum_{n=0}^{\infty} S(n)q^n.$$

So, the more interesting equality is $S(n) = S_1(n)$. Using the method of weighted words, Alladi and Gordon [1] view Theorem S as emerging out of the *key identity*

$$\sum_{i,j} a^i b^j \sum_m \frac{q^{T_{i+j-m}+T_m}}{(q)_{i-m}(q)_{j-m}(q)_m} = \sum_{i,j} \frac{a^i b^j q^{T_i+T_j}}{(q)_i (q)_j} = (-aq)_\infty(-bq)_\infty. \tag{1.7}$$

More precisely, under the substitutions $q \mapsto q^3$, $a \mapsto q^{-2}$, $b \mapsto q^{-1}$, the product in (1.7) yields the product on the right in (1.6) which is the generating function of $S(n)$; the sum on the left of (1.7) under these substitutions turns out to be the generating function of $S_1(n)$.

Yet another advantage of the method of this paper is that (1.7) falls out as a special case of (1.4) by taking $c = 0$. This makes $\gamma = \varepsilon = \phi = 0$, $\delta = m$, $\alpha = i - m$, $\beta = j - m$ and $s = \alpha + \beta + \delta = i + j - m$. Thus Theorems G and S can both be considered as particular cases of (1.4).

The method of weighted words developed here considers integers occurring in six colors, of which three colors a, b, c are primary, and three colors ab, ac and bc are secondary (see § 2 for a description of the method). In order to discuss partitions, a specific lexicographic ordering of these colored integers is chosen. For Schur's theorem, Alladi and Gordon [1] use two primary colors a and b, and one secondary color ab. The scheme considered here is an extension because by setting $c = 0$, we see that c, ac and bc drop out and only a, b and ab survive.

Alladi and Gordon [2] consider $S_1(n)$ as one of six companion partition functions. These six functions are obtained via the six permutations of the colors a, b and ab, which yield six orderings of the colored integers. Similarly, here also, one may consider $6! = 720$ lexicographic orderings of the colored integers corresponding to the different permutations of the colors a, b, c, ab, ac and bc. These yield 720 companion partition functions $C_j(n)$, $j = 1, 2, \ldots, 720$, all equal, one of which is $C(n) = C_1(n)$ (see § 5). The difference conditions defining these $C_j(n)$, are, in general more complicated than those defining $C(n)$, but there are a few that are as nice as $C(n)$; one interesting example is presented in § 5. There are natural bijections between the partitions counted by $C_j(n)$ and there is also a *conservation law* for the difference conditions defining these $C_j(n)$ (see § 5).

Finally, in § 6, we discuss some related problems.

Given a partition π, we let $\sigma(\pi)$ denote the sum of its parts and $\nu(\pi)$, the number of parts of π. Often we will be counting parts of a certain type and this will be indicated by a subscript; for instance, $v_a(\pi)$ denotes the number of a-parts of π. The q-multinomial coefficients of order t are given by

$$(1.8)\qquad \begin{bmatrix} n_1+n_2+\cdots+n_t \\ n_1, n_2, \ldots, n_t \end{bmatrix} = \begin{bmatrix} n_1+n_2+\cdots+n_t \\ n_1, n_2, \ldots, n_t \end{bmatrix}_q = \frac{(q)_{n_1+n_2+\cdots+n_t}}{(q)_{n_1}(q)_{n_2}\cdots(q)_{n_t}}.$$

The q-multinomial coefficients of order 6 will play a crucial role in connection with *local generating functions* related to the partitions counted by $C(n)$ (see § 3).

§ 2. The method of weighted words

Let the integer 1 occur in three primary colors red, blue and yellow, and let integers $j \geqq 2$ occur in six colors, of which red, blue and yellow are primary, and purple, orange and green are secondary. We introduce the symbols a_j, b_j, c_j, d_j, e_j and f_j to denote the integer j in colors red, blue, yellow, purple, orange and green respectively. We will often refer to a_j as an a-part, b_j as a b-part, and so on. The letters a, b, c, d, e, f are free parameters and their powers will keep track of the number of a-parts, b-parts, ..., f-parts in partitions counted by $C(n)$. However, in making the transition from $C(n)$ to $B(n)$ we need to take $d = ab$, $e = ac$ and $f = bc$. That is why the subscripts of d, e and f are integers $j \geqq 2$ since they are obtained by combining two parts in primary colors.

The weight of a symbol is its subscript. The gap between symbols is the difference in absolute value of their weights. The gap is therefore a non-negative integer without color. For example, the gap between a_5 and e_3 is 2.

Next, in order to discuss partitions using these symbols, we need a lexicographic ordering. We first consider the following lexicographic ordering which we call *Scheme* 1:

$$(2.1)\qquad a_1 \prec b_1 \prec c_1 \prec d_2 \prec e_2 \prec a_2 \prec f_2 \prec b_2 \prec c_2 \prec d_3 \prec e_3 \prec a_3 \prec \cdots .$$

Under the transformations

$$(2.2)\qquad \text{(dilation) } q \mapsto q^6 \text{ and (translations) } a \mapsto aq^{-4},\ b \mapsto bq^{-2},\ c \mapsto cq^{-1},$$

$$\text{and (combinations) } d = ab,\ e = ac,\ f = bc,$$

the symbols become

$$(2.3)\qquad a_j = 6j-4,\quad b_j = 6j-2,\quad c_j = 6j-1,$$

$$d_j = 6j-6,\quad e_j = 6j-5,\quad f_j = 6j-3,$$

and so the lexicographic ordering in (2.1) becomes

$$(2.4)\qquad 2 < 4 < 5 < 6 < 7 < 8 < 9 < 10 < 11 < 12 < 13 < \cdots,$$

the standard ordering among the integers. Note that the integers 1 and 3 are absent in (2.4). This is because $e_1 = ac_1$ and $f_1 = bc_1$ are absent in (2.1). For the sake of completeness (and for reasons that will become clear in § 5) we consider the full set of symbols arranged in the following order:

$$(2.5)\qquad \underline{d}_1 \prec \underline{e}_1 \prec a_1 \prec \underline{f}_1 \prec b_1 \prec c_1 \prec d_2 \prec e_2 \prec a_2 \prec f_2 \prec b_2 \prec c_2 \prec \cdots .$$

The symbols d_1, e_1 and f_1 are underlined here to indicate that they are absent in (2.1) which is the list we will be using. We refer to the transformations in (2.2) as *standard transformations.*

By a partition π we mean a sum involving the symbols which are arranged in decreasing order according to (2.1). Instead of thinking of π as a sum, we may consider it as a word with symbols in decreasing order. By $\nu(\pi)$ we mean the number of parts of π, and if we only count parts of a certain color, we indicate this with a subscript. For example, if

$$\pi = f_4 + a_4 + a_4 + d_4 + c_3 + b_3 + b_3 + f_3 + e_3 + b_2 + a_2 + c_1 + b_1 + a_1 + c_1 ,$$

then $\sigma(\pi) = 39$, $\nu(\pi) = 15$, $\nu_a(\pi) = 4$, $\nu_b(\pi) = 4$ and so on.

Observe that for symbols with the same weight, the colors occur in the following order:

$$(2.6)\qquad d \prec e \prec a \prec f \prec b \prec c .$$

Given two colors, we use (2.6) to determine which of the two is of lower order. For example, between e and b, e is of lower order and b of higher order.

Now we consider partitions $\pi = m_1 + m_2 + \cdots$ using symbols in Scheme 1 such that the gap between symbols is $\geqq 1$, with the added restriction that the gap between consecutive symbols m_ℓ and $m_{\ell+1}$ is $\geqq 2$ if

(2.7) m_ℓ is of lower order and $m_{\ell+1}$ is of higher order,

or if m_ℓ and $m_{\ell+1}$ are of the same secondary color.

We will refer to π as a Type 1 partition. Under the standard transformations, Type 1 partitions become the partitions counted by $C(n)$ is Theorem G. More generally, under the transformations (1.5), Type 1 partitions become the partitions counted by $C(n)$ in Theorem 1.

The principal partition result of this paper is Theorem 2 which generalizes and refines Theorem 1, which in itself is a generalization of Satz 4.8, (4.10) and (1.2):

Theorem 2. *Let $B(n; i, j, k)$ denote the number of vector partitions $\pi' = (\pi_1; \pi_2; \pi_3)$ of n such that π_1, π_2 and π_3 have distinct a-parts, b-parts and c-parts respectively and also $\nu(\pi_1) = i$, $\nu(\pi_2) = j$, and $\nu(\pi_3) = k$.*

Let $C(n;\alpha,\beta,\gamma,\delta,\varepsilon,\phi)$ *denote the number of Type* 1 *partitions* π *of* n *such that* $\nu_a(\pi)=\alpha$, $\nu_b(\pi)=\beta,\ldots,\nu_f(\pi)=\phi$. *Then*

$$B(n;i,j,k)=\sum_{\alpha+\delta+\varepsilon=i,\,\beta+\delta+\phi=j,\,\delta+\varepsilon+\phi=k} C(n;\alpha,\beta,\gamma,\delta,\varepsilon,\phi)\,.$$

Theorem 2 is the combinatorial version of the *key identity* (1.4). Theorem 1 follows from Theorem 2 (and (1.4)) under the substitutions given in (1.5). Under these transformations, the *a*-, *b*- and *c*- parts in Theorem 2 correspond to the parts $\equiv r_1$, r_2 and $r_3 \pmod M$ respectively in Theorem 1. Also, the *d*-, *e*- and *f*- parts in Theorem 2 correspond to the parts $\equiv r_1+r_2$, r_1+r_3 and $r_2+r_3 \pmod M$ in Theorem 1 because $d=ab$, $e=ac$ and $d=ef$. The inequalities (1.3) of Theorem 1 are a direct consequence of the lexicographic ordering (2.1).

The combinatorial explanation for counting the *d*-, *e*- and *f*-parts twice in Theorem 2 is as follows: Given a partition counted by $B(n;i,j,k)$, take δ of the *a*-parts and δ of the *b*-parts to form the *ab*-parts. Similarly, ε of the *a*-parts and ε of the *c*-parts combine to form the *ac*-parts. Finally, ϕ of the *b*-parts and ϕ of the *c*-parts combine to form the *bc*-parts. Thus, in partitions counted by the function C, we have

$$\alpha=i-\delta-\varepsilon,\quad \beta=j-\delta-\phi,\quad \gamma=k-\varepsilon-\phi\,. \tag{2.8}$$

Therefore

$$\alpha+\beta+\gamma+2\delta+2\varepsilon+2\phi=i+j+k=\nu\,.$$

This also explains the necessity of making the choices $d=ab$, $e=ac$ and $f=bc$ in (1.4). Finally, the number of parts in partitions counted by $C(n;\alpha,\beta,\gamma,\delta,\varepsilon,\phi)$ is

$$s=\alpha+\beta+\gamma+\delta+\varepsilon+\phi\,. \tag{2.9}$$

Throughout, we assume (2.8) and (2.9).

Actually, a stronger form of Theorem 1 follows from Theorem 2 if we employ the transformations

$$q\mapsto q^M,\quad a\mapsto aq^{r_1-M},\quad b\mapsto bq^{r_2-M},\quad c\mapsto cq^{r_3-M}$$

which is more general than (1.5). Then the equality $B(n,\nu)=C(n;\nu)$ in Theorem 1 can be strengthened to:

the number of parts which are $\equiv r_i \pmod M$ *counted by* $B(n)$

equals the number of parts $\equiv r_i$, r_i+r_j *or* r_i+r_k *counted by* $C(n)$,

where i, j, k is any permutation of the integers 1, 2, 3. It is to be noted that in this refinement, the parts of secondary color are counted only once.

We now proceed to the proof of Theorem 2.

Clearly the generating function of $B(n; i, j, k)$ is

$$\sum_n B(n; i, j, k) q^n = \frac{q^{T_i + T_j + T_k}}{(q)_i (q)_j (q)_k} \tag{2.10}$$

which is the summand on the right hand side of (1.4). In order to prove Theorem 2, we need to do two things. Firstly, we need to show that

Lemma 1.

$$\sum_n C(n; \alpha, \beta, \gamma, \delta, \varepsilon, \phi) q^n = \frac{q^{T_s + T_\delta + T_\varepsilon + T_\phi - 1}(1 - q^\alpha(1 - q^\phi))}{(q)_\alpha (q)_\beta (q)_\gamma (q)_\delta (q)_\varepsilon (q)_\phi}.$$

Lemma 1 is proved in the next section. Secondly, we need to prove

Lemma 2 (Key identity).

$$\sum_{\alpha+\delta+\varepsilon=i,\, \beta+\delta+\phi=j,\, \gamma+\varepsilon+\phi=k} \frac{q^{T_s + T_\delta T_\varepsilon + T_\phi - 1}(1 - q^\alpha(1 - q^\phi))}{(q)_\alpha (q)_\beta (q)_\gamma (q)_\delta (q)_\varepsilon (q)_\phi} = \frac{q^{T_i + T_j + T_k}}{(q)_i (q)_j (q)_k}.$$

This is clearly equivalent to the key identity (1.4). Lemma 2 is proved in §4.

Consider the generating function for all Type 1 partitions where the numbers α, β, γ, δ, ε and ϕ are specified and the order in which the colors occur is also given. Then this generating function can be computed in the following manner: First write down the minimal partition π corresponding to the given order of colors. Then the required generating function is

$$q^{\sigma(\pi)}/(q)_{v(\pi)}. \tag{2.11}$$

The expression in (2.11) is the required generating function because Type 1 partitions are given by lower bound gap conditions. More precisely, the partitions in question are obtained by adding to π all (ordinary) partitions ϱ with $v(\varrho) \leqq v(\pi)$. For example, if the colors occur in the order

$$a\,d\,d\,e\,c\,b\,f,$$

then the minimal Type 1 partition is

$$\pi = a_{11} + d_{10} + d_8 + e_6 + c_4 + b_3 + f_2.$$

So $\sigma(\pi) = 44$ and $v(\pi) = 7$. Hence the required generating function is

$$q^{44}/(q)\tau.$$

Having made this observation, let us rewrite the expression on the right in Lemma 1 in the form

$$G = G(\alpha, \beta, \gamma, \delta, \varepsilon, \phi) = \frac{q^{T_s + T_\delta + T_\varepsilon + T_\phi - 1}(1 - q^\alpha(1 - q^\phi))}{(q)_s} \begin{bmatrix} s \\ \alpha, \beta, \gamma, \delta, \varepsilon, \phi \end{bmatrix}. \tag{2.12}$$

Since s is the number of parts in every partition counted by G, it follows from (2.11) that the generating function for *all* minimal partitions π having

$$v_a(\pi) = \alpha,\ v_b(\pi) = \beta, \ldots, v_f(\pi) = \phi$$

is

$$H = H(\alpha, \beta, \gamma, \delta, \varepsilon, \phi) = q^{T_s + T_\delta + T_\varepsilon + T_{\phi-1}}(1 - q^\alpha(1 - q^\phi)) \begin{bmatrix} s \\ \alpha, \beta, \gamma, \delta, \varepsilon, \phi \end{bmatrix}. \tag{2.13}$$

In order to prove (2.13), we will consider in § 3 various *local generating functions*, namely, generating functions $H_a(\alpha, \beta, \ldots, \phi)$, $H_b(\alpha, \beta, \ldots, \phi), \ldots, H_f(\alpha, \beta, \ldots, \phi)$, for all minimal partitions counted by H, but with smallest part $a_1, b_1, \ldots, f_2$ respectively. Clearly

$$H_a + H_b + H_c + H_d + H_e + H_f = H. \tag{2.14}$$

In the next section, explicit formulae for these local generating functions will be established (see Lemma 3) by induction on s. From these formulae and (2.14), (2.13) will follow, thereby completing the proof of Lemma 1.

§ 3. Local generating functions

The aim of this section is to prove

Lemma 3. *Let* $s = \alpha + \beta + \gamma + \delta + \varepsilon + \phi$ *and* $\sigma = T_s + T_\delta + T_\varepsilon + T_{\phi-1}$. *Then*

(i) $$H_a(\alpha, \beta, \gamma, \delta, \varepsilon, \phi) = q^\sigma \begin{bmatrix} s-1 \\ \alpha-1, \beta, \gamma, \delta, \varepsilon, \phi \end{bmatrix},$$

(ii) $$H_b(\alpha, \beta, \gamma, \delta, \varepsilon, \phi) = q^{\sigma+\alpha+\phi} \begin{bmatrix} s-1 \\ \alpha, \beta-1, \gamma, \delta, \varepsilon, \phi \end{bmatrix},$$

(iii) $$H_c(\alpha, \beta, \gamma, \delta, \varepsilon, \phi) = q^{\sigma+\alpha+\phi+\beta} \begin{bmatrix} s-1 \\ \alpha, \beta, \gamma-1, \delta, \varepsilon, \phi \end{bmatrix},$$

(iv) $$H_d(\alpha, \beta, \gamma, \delta, \varepsilon, \phi) = q^{\sigma+\alpha+\phi+\beta+\gamma} \begin{bmatrix} s-1 \\ \alpha, \beta, \gamma, \delta-1, \varepsilon, \phi \end{bmatrix},$$

(v) $$H_e(\alpha, \beta, \gamma, \delta, \varepsilon, \phi) = q^{\sigma+\alpha+\phi+\beta+\gamma+\delta} \begin{bmatrix} s-1 \\ \alpha, \beta, \gamma, \delta, \varepsilon-1, \phi \end{bmatrix},$$

(vi) $$H_f(\alpha, \beta, \gamma, \delta, \varepsilon, \phi) = q^{\sigma+s+\alpha} \begin{bmatrix} s-1 \\ \alpha, \beta, \gamma, \delta, \varepsilon, \phi-1 \end{bmatrix}.$$

Before proving Lemma 3 we first check that when combined with (2.14), it yields (2.13).

For this we use the following well-known recurrence for q-multinomial coefficients [11], Ch. 3,

$$(3.1)\quad \begin{bmatrix} n_1+n_2+\cdots+n_t \\ n_1, n_2, \ldots, n_t \end{bmatrix} = \begin{bmatrix} n_1+n_2+\cdots+n_t-1 \\ n_1-1, n_2, n_3, \ldots, n_t \end{bmatrix} + q^{n_1} \begin{bmatrix} n_1+n_2+\cdots+n_t-1 \\ n_1, n_2-1, n_3, \ldots, n_t \end{bmatrix}$$

$$+q^{n_1+n_2} \begin{bmatrix} n_1+n_2+\cdots+n_t-1 \\ n_1, n_2, n_3-1, n_4, \ldots, n_t \end{bmatrix} + \cdots + q^{n_1+\cdots+n_{t-1}} \begin{bmatrix} n_1+n_2+\cdots+n_t-1 \\ n_1, n_2, \ldots, n_{t-1}, n_t-1 \end{bmatrix}.$$

More precisely, rewrite the formula (vi) for H_f in the form

$$(3.2)\quad H_f = q^{\sigma+\alpha} \begin{bmatrix} s-1 \\ \alpha, \beta, \gamma, \delta, \varepsilon, \phi-1 \end{bmatrix} - q^{\sigma+\alpha}(1-q^s) \begin{bmatrix} s-1 \\ \alpha, \beta, \gamma, \delta, \varepsilon, \phi-1 \end{bmatrix}.$$

Then with $t = 6$,

$$(3.3)\qquad n_1 = \alpha,\quad n_2 = \phi,\quad n_3 = \beta,\quad n_4 = \gamma,\quad n_5 = \delta,\quad n_6 = \varepsilon$$

in (3.1) we see that

$$H_a + H_b + H_c + H_d + H_e + H_f = q^\sigma \begin{bmatrix} s \\ \alpha, \beta, \gamma, \delta, \varepsilon, \phi \end{bmatrix} - q^{\sigma+\alpha}(1-q^s) \begin{bmatrix} s-1 \\ \alpha, \beta, \gamma, \delta, \varepsilon, \phi-1 \end{bmatrix}$$

$$= q^\sigma \begin{bmatrix} s \\ \alpha, \beta, \gamma, \delta, \varepsilon, \phi \end{bmatrix} - q^{\sigma+\alpha}(1-q^\phi) \begin{bmatrix} s \\ \alpha, \beta, \gamma, \delta, \varepsilon, \phi \end{bmatrix}$$

$$= q^\sigma(1-q^\alpha(1-q^\phi)) \begin{bmatrix} s \\ \alpha, \beta, \gamma, \delta, \varepsilon, \phi \end{bmatrix}.$$

Thus (2.13) follows from (3.4) and (2.14).

Now we prove Lemma 3 by induction on s.

Proof of Lemma 3. The formulae (i)–(vi) are clearly valid when $s = 0$ or 1. So let $s \geqq 2$. To prove Lemma 3 (i) observe that the partitions counted by H_a can be divided into six cases.

Case 1.1. Partitions ending with $a_2 + a_1$.

In this case, if a_1 is deleted, then the remaining partition have $s-1$ parts and are obtained by adding 1 to each part of minimal partitions counted by $H_a(\alpha-1, \beta, \gamma, \delta, \varepsilon, \phi)$. Hence by the induction hypothesis the generating function for Case 1.1 is

$$q^1 \cdot q^{T_{s-1}+T_\delta+T_\varepsilon+T_{\phi-1}+s-1} \begin{bmatrix} s-2 \\ \alpha-2, \beta, \gamma, \delta, \varepsilon, \phi \end{bmatrix} = q^\sigma \begin{bmatrix} s-2 \\ \alpha-2, \beta, \gamma, \delta, \varepsilon, \phi \end{bmatrix}.$$

Case 1.2. Partitions ending with $f_2 + a_1$.

In this case, if a_1 is deleted, we have a partition counted by $H_f(\alpha-1,\beta,\gamma,\delta,\varepsilon,\phi)$. Hence the generating function is

$$q^1\cdot q^{T_{s-1}+T_\delta+T_\varepsilon+T_{\phi-1}+s-1+\alpha-1}\begin{bmatrix} s-2 \\ \alpha-1,\beta,\gamma,\delta,\varepsilon,\phi-1\end{bmatrix}=q^{\sigma+\alpha-1}\begin{bmatrix} s-2 \\ \alpha-1,\beta,\gamma,\delta,\varepsilon,\phi-1\end{bmatrix}.$$

Case 1.3. Partitions ending with b_2+a_1.

For such partitions, by deleting a_1, we are left with a partition counted by

$$H_b(\alpha-1,\beta,\gamma,\delta,\varepsilon,\phi)$$

with 1 added to each part. Thus the generating function is

$$q^1\cdot q^{T_{s-1}+T_\delta+T_\varepsilon+T_{\phi-1}+s-1+\alpha-1+\phi}\begin{bmatrix} s-2 \\ \alpha-1,\beta-1,\gamma,\delta,\varepsilon,\phi\end{bmatrix}$$
$$=q^{\sigma+\alpha-1+\phi}\begin{bmatrix} s-2 \\ \alpha-1,\beta-1,\gamma,\delta,\varepsilon,\phi\end{bmatrix}.$$

Case 1.4. Partitions ending with c_2+a_1.

For such partitions, by deleting a_1, we are left with a partition counted by

$$H_c(\alpha-1,\beta,\gamma,\delta,\varepsilon,\phi)$$

with 1 added to each part. Thus the generating function is

$$q^1\cdot q^{T_{s-1}+T_\delta+T_\varepsilon+T_{\phi-1}+s-1+\alpha-1+\phi+\beta}\begin{bmatrix} s-2 \\ \alpha-1,\beta,\gamma-1,\delta,\varepsilon,\phi\end{bmatrix}$$
$$=q^{\sigma+\alpha-1+\phi+\beta}\begin{bmatrix} s-2 \\ \alpha-1,\beta,\gamma-1,\delta,\varepsilon,\phi\end{bmatrix}.$$

Case 1.5. Partitions ending with d_3+a_1.

Here by deleting a_1, we are left with partitions counted by $H_d(\alpha-1,\beta,\gamma,\delta,\varepsilon,\phi)$ with 1 added to each part. This yields the generating function

$$q^1\cdot q^{T_{s-1}+T_\delta+T_\varepsilon+T_{\phi-1}+s-1+\alpha-1+\phi+\beta+\gamma}\begin{bmatrix} s-2 \\ \alpha-1,\beta,\gamma,\delta-1,\delta,\varepsilon,\phi\end{bmatrix}$$
$$=q^{\sigma+\alpha-1+\phi+\beta+\gamma}\begin{bmatrix} s-2 \\ \alpha-1,\beta,\delta-1,\delta,\varepsilon,\phi\end{bmatrix}.$$

Case 1.6. Partitions ending with e_3+a_1.

In this case by deleting a_1, we are left with partitions counted by $H_e(\alpha-1, \beta, \gamma, \delta, \varepsilon, \phi)$ with 1 added to each part. So the generating function is

$$q^1 \cdot q^{T_{s-1}+T_\delta+T_\varepsilon+T_{\phi-1}+s-1+\alpha-1+\phi+\beta+\gamma+\delta} \begin{bmatrix} s-2 \\ \alpha-1, \beta, \gamma, \delta, \varepsilon-1, \phi \end{bmatrix}$$

$$= q^{\sigma+\alpha-1+\phi+\beta+\gamma+\delta} \begin{bmatrix} s-2 \\ \alpha-1, \beta, \gamma, \delta, \varepsilon-1, \phi \end{bmatrix}.$$

The sum of the generating functions in Cases 1.1–1.6 can be evaluated by applying (3.1) with n_i as in (3.3). This gives the formula for H_a in Lemma 3 (i).

Rules of cycles. Observe that while computing H_a using (3.1), the values $n_1, n_2, \ldots, n_t$ in (3.3) are determined by starting with the color a in (2.6) and moving through the list of colors cyclically to the right. The procedure is the same for H_b and H_c, where we would start with b and c respectively and more to the right cyclically in (2.6). Thus the generating functions for Cases 2.1–2.6 and 3.1–3.6 are determined by the choices

Lemma 3. (ii) $n_1 = \beta,\ n_2 = \gamma,\ n_3 = \delta,\ n_4 = \varepsilon,\ n_5 = \alpha,\ n_6 = \phi$.

Lemma 3. (iii) $n_1 = \gamma,\ n_2 = \delta,\ n_3 = \varepsilon,\ n_4 = \alpha,\ n_5 = \phi,\ n_6 = \beta$.

We skip the details of the proofs of parts (ii) and (iii).

The rule of cycles for parts (iv), (v) and (vi) is slightly different. For secondary colors d, e and f, start with the color to its right in (2.6) and complete the cycle. Thus the choices of n_i will be

Lemma 3. (iv) $n_1 = \varepsilon,\ n_2 = \alpha,\ n_3 = \phi,\ n_4 = \beta,\ n_5 = \gamma,\ n_6 = \delta$.

Lemma 3. (v) $n_1 = \alpha,\ n_2 = \phi,\ n_3 = \beta,\ n_4 = \gamma,\ n_5 = \delta,\ n_6 = \varepsilon$.

Lemma 3. (vi) $n_1 = \beta,\ n_2 = \gamma,\ n_3 = \delta,\ n_4 = \varepsilon,\ n_5 = \alpha,\ n_6 = \phi$.

We now give the proof only for part (vi) (which is the most interesting of the above three cases) and skip the details for parts (iv) and (v) which are similar and simpler.

Case 6.1. Partitions ending with $b_3 + f_2$.

Delete f_2 to see that we have partitions counted by $H_b(\alpha, \beta, \gamma, \delta, \varepsilon, \phi-1)$ with 2 added to each part. So the generating function is

$$q^2 \cdot q^{T_{s-1}+T_\delta+T_\varepsilon+T_{\phi-2}+2(s-1)+\alpha+\phi-1} \begin{bmatrix} s-2 \\ \alpha, \beta-1, \gamma, \delta, \varepsilon, \phi-1 \end{bmatrix}$$

$$= q^{\sigma+s+\alpha} \begin{bmatrix} s-2 \\ \alpha, \beta-1, \gamma, \delta, \varepsilon, \phi-1 \end{bmatrix}.$$

Case 6.2. Partitions ending with $c_3 + f_2$.

Delete f_2 to get partitions counted by $H_c(\alpha, \beta, \gamma, \delta, \varepsilon, \phi - 1)$ with 2 added to each part. This gives the generating function

$$q^2 \cdot q^{T_{s-1} + T_\delta + T_\varepsilon + T_\phi - 2 + 2(s-1) + \alpha + \phi - 1 + \beta} \begin{bmatrix} s-2 \\ \alpha, \beta, \gamma - 1, \delta, \varepsilon, \phi - 1 \end{bmatrix}$$

$$= q^{\sigma + s + \alpha + \beta} \begin{bmatrix} s-2 \\ \alpha, \beta, \gamma - 1, \delta, \varepsilon, \phi - 1 \end{bmatrix}.$$

Case 6.3. Partitions ending with $d_4 + f_2$.

Here by deleting f_2 we are left with partitions counted by $H_d(\alpha, \beta, \gamma, \delta, \varepsilon, \phi - 1)$ with 2 added to each part. This yields the generating function

$$q^2 \cdot q^{T_{s-1} + T_\delta + T_\varepsilon + T_\phi - 2 + 2(s-1) + \alpha + \phi - 1 + \beta + \gamma} \begin{bmatrix} s-2 \\ \alpha, \beta, \gamma, \delta - 1, \varepsilon, \phi - 1 \end{bmatrix}$$

$$= q^{\sigma + s + \alpha + \beta + \gamma} \begin{bmatrix} s-2 \\ \alpha, \beta, \gamma, \delta - 1, \varepsilon, \phi - 1 \end{bmatrix}.$$

Case 6.4. Partitions ending with $e_4 + f_2$.

Delete f_2 to get partitions counted by $H_e(\alpha, \beta, \gamma, \delta, \varepsilon, \phi - 1)$ with 2 added to each part. So the generating function is

$$q^2 \cdot q^{T_{s-1} + T_\delta + T_\varepsilon + T_\phi - 2 + 2(s-1) + \alpha + \phi - 1 + \beta + \gamma + \delta} \begin{bmatrix} s-2 \\ \alpha, \beta, \gamma, \delta, \varepsilon - 1, \phi - 1 \end{bmatrix}$$

$$= q^{\sigma + s + \alpha + \beta + \gamma + \delta} \begin{bmatrix} s-2 \\ \alpha, \beta, \gamma, \delta, \varepsilon - 1, \phi - 1 \end{bmatrix}.$$

Case 6.5. Partitions ending with $a_4 + f_2$.

In this case by deleting f_2 we are left with partitions counted by $H_a(\alpha, \beta, \gamma, \delta, \varepsilon, \phi - 1)$ with 3 added to each part. So the generating function is

$$q^2 \cdot q^{T_{s-1} + T_\delta + T_\varepsilon + T_\phi - 2 + 3(s-1)} \begin{bmatrix} s-2 \\ \alpha - 1, \beta, \gamma, \delta, \varepsilon, \phi - 1 \end{bmatrix}$$

$$= q^{\sigma + s + \alpha + \beta + \gamma + \delta + \varepsilon} \begin{bmatrix} s-2 \\ \alpha - 1, \beta, \gamma, \delta, \varepsilon, \phi - 1 \end{bmatrix}.$$

Case 6.6. Partitions ending with $f_4 + f_2$.

In this final case, we see by deleting f_2 that we have partitions counted by

$$H_f(\alpha, \beta, \gamma, \delta, \varepsilon, \phi - 1)$$

with 2 added to each part. Hence the generating function is

$$q^2 \cdot q^{T_{s-1}+T_\delta+T_\varepsilon+T_{\phi-2}+2(s-1)+s-1+\alpha}\begin{bmatrix} s-2 \\ \alpha,\beta,\gamma,\delta,\varepsilon,\phi-2 \end{bmatrix}$$

$$= q^{\sigma+s+\alpha+\beta+\gamma+\delta+\varepsilon+\alpha}\begin{bmatrix} s-2 \\ \alpha,\beta,\gamma,\delta,\varepsilon,\phi-2 \end{bmatrix}.$$

The generating functions for Cases 6.1–6.6 can be added by applying (3.1) with the choices given in (3.5). This proves Lemma 3 (vi) and thus completes the proof of Lemma 1.

§ 4. Proof of the key identity

First we standardize notation. For non-negative integers n, the expression

$$(a)_n = (1-a)(1-aq)\cdots(1-aq^{n-1})$$

can be rewritten in the form

$$(4.1) \qquad (a)_n = \frac{(a)_\infty}{(aq^n)_\infty}.$$

Since the expression on the right in (4.1) makes sense for all real values of n (except when the denominator is zero), we may use (4.1) to define $(a)_n$ for all n. In particular if $n > 0$ is an integer, (4.1) yields the relation

$$(4.2) \quad (a)_{-n} = \frac{1}{(1-aq^{-n})(1-aq^{-n+1})\cdots(1-aq^{-1})} = \frac{(-1)^n a^{-n} q^{\binom{n+1}{2}}}{(q/a)_n}.$$

Next, let

$$(4.3) \quad (A_1, A_2, \ldots, A_r; q)_n = (A_1, A_2, \ldots, A_r)_n = (A_1)_n(A_2)_n\cdots(A_r)_n.$$

Then from (4.2) and (4.3) we get for integers $n \geqq 0$

$$(4.4) \quad \frac{(A_1, A_2, \ldots, A_r)_{-n}}{(B_1, B_2, \ldots, B_r)_{-n}} = \frac{(q/B_1, q/B_2, \ldots, q/B_r)_n}{(q/A_1, q/A_2, \ldots, q/A_r)_n}\left(\frac{B_1 B_2 \cdots B_r}{A_1 A_2 \cdots A_r}\right)^n.$$

As a first step toward the proof of Lemma 2 we establish

Lemma 4. *Let $F(\omega) = F(\omega; a, b, c, q)$ be defined by*

$$(4.5) \qquad F(\omega)$$

$$= \left(1+\frac{abc}{\omega^2}\right)\sum_{n=-\infty}^{-1}\frac{(1+\omega^{-1}q^{2n})(-a^{-1},-b^{-1},c^{-1})_n}{\left(\frac{a}{\omega},\frac{b}{\omega},\frac{c}{\omega}\right)_{n+1}}\left(\frac{abc}{\omega^2}\right)^n q^{\binom{n+1}{2}}.$$

Then $F(\omega)$ satisfies the functional equation

$$F(\omega) = -F\left(-\frac{abc}{\omega}\right).$$

Proof. Rewrite the expression in (4.5) in the form

$$F(\omega) = \frac{\left(1+\frac{abc}{\omega^2}\right)}{\left(1-\frac{a}{\omega}\right)\left(1-\frac{b}{\omega}\right)\left(1-\frac{c}{\omega}\right)} \sum_{n=-\infty}^{-1} \frac{(1+\omega^{-1}q^{2n})(-a^{-1},-b^{-1},-c^{-1})_n}{\left(\frac{aq}{\omega},\frac{bq}{\omega},\frac{cq}{\omega}\right)_n}\left(\frac{abc}{\omega^2}\right)^n q^{\binom{-n}{2}}$$

$$= \frac{\left(1+\frac{abc}{\omega^2}\right)}{\left(1-\frac{a}{\omega}\right)\left(1-\frac{b}{\omega}\right)\left(1-\frac{c}{\omega}\right)} \sum_{n=1}^{\infty} \frac{(1+\omega^{-1}q^{-2n})\left(\frac{\omega}{a},\frac{\omega}{b},\frac{\omega}{c}\right)_n}{(-aq,-bq,-cq)_n}\left(\frac{abc}{\omega}\right)^n q^{\binom{n}{2}+3n}(-1)^n$$

$$= \frac{-\left(1+\frac{\omega^2}{abc}\right)}{(1+aq)(1+bq)(1+cq)} \sum_{n=1}^{\infty} \frac{(1+\omega q^2)\left(\frac{\omega q}{a},\frac{\omega q}{b},\frac{\omega q}{c}\right)_{n-1}}{(-aq^2,-bq^2,-cq^2)_{n-1}}\left(\frac{abc}{\omega}\right)^n q^{\binom{n+1}{2}}(-1)^n$$

using (4.4). Hence

$$-(1+aq)(1+bq)(1+cq)F(\omega) \tag{4.6}$$

$$= \left(1+\frac{\omega^2}{abc}\right)\sum_{n=0}^{\infty} \frac{(1+\omega q^{2n+2})\left(\frac{\omega q}{a},\frac{\omega q}{b},\frac{\omega q}{c}\right)_n}{(-aq^2,-bq^2,-cq^2)_n}\left(\frac{abc}{\omega}\right)^{n+1} q^{\binom{n+2}{2}}(-1)^{n+1}$$

$$= -\left(\frac{abc}{\omega}+\omega\right)(1+\omega q^2)q \lim_{N\to\infty} {}_8\phi_7\left(\begin{matrix} u, q\sqrt{u}, -q\sqrt{u}, \frac{\omega q}{a}, q, \frac{\omega q}{b}, \frac{\omega q}{c}, q^{-N}; \frac{abc}{\omega}q^{N+2} \\ \sqrt{u}, -\sqrt{u}, -aq^2, u, -bq^2, -cq^2, uq^{N+1} \end{matrix}\right),$$

where $u = -\omega q^2$, and we have made use of the ${}_r\phi_s$ notation as in Gasper and Rahman [16], p. 4, eqn. (1.2.22). At this stage we make use of Watson's q-analogue of Whipple's theorem (see [16], p. 242, eqn. (III.17)) to rewrite (4.6) in the form

$$= -\left(\frac{abc}{\omega}+\omega\right)(1+\omega q^2)q\frac{\left(-\omega q^3, \frac{-bcq}{\omega}\right)_\infty}{(-bq^2,-cq^2)_\infty}\,{}_3\phi_2\left(\begin{matrix} -aq, \frac{\omega q}{b}, \frac{\omega q}{c}; \frac{-bcq}{\omega} \\ -\omega q^2, -aq^2 \end{matrix}\right). \tag{4.7}$$

Next we use [16], p. 241, eqn. (III.9) to rewrite the expression in (4.7) in the form

$$= -\left(\frac{abc}{\omega}+\omega\right)q\frac{\left(-\omega q^2, -\frac{bcq}{\omega}\right)_\infty}{(-bq^2,-cq^2)_\infty}\cdot\frac{\left(q,\frac{abcq^2}{\omega}\right)_\infty}{\left(-aq^2,\frac{-bcq}{\omega}\right)_\infty}\,{}_3\phi_2\left(\begin{matrix} -aq, -bq, -cq; q \\ -\omega q^2, \frac{abcq^2}{\omega} \end{matrix}\right).$$

Therefore

$$(4.8)\qquad F(\omega)=\frac{q\left(\dfrac{abc}{\omega}+\omega\right)\left(-\omega q^2, q, \dfrac{abcq^2}{\omega}\right)_\infty}{(-aq,-bq,-cq)_\infty}\,{}_3\phi_2\left(\begin{matrix}-aq,-bq,-cq;q\\ -\omega q^2, \dfrac{abcq^2}{\omega}\end{matrix}\right).$$

Notice that replacing ω by $-abc/\omega$ merely changes the sign of the right hand side of (4.8), and this is precisely the assertion of Lemma 4.

To state the next result and for later use, we need the following notation:

$$[\omega^m]\,P(\omega)=\text{coefficient of } \omega^m \text{ in the Laurent series expansion of } P.$$

So we have

Corollary. $[\omega^0]\,F(\omega)=0.$

Proof. By Lemma 4, $[\omega^0]\,F(\omega)$ must be equal to its negative. Hence it is zero.

Now we are ready for the

Proof of the key identity. We begin by rewriting (1.4) more uniformly as

$$\sum_{\substack{\alpha,\beta,\gamma,\delta,\varepsilon,\phi\geqq 0\\ s=\alpha+\beta+\gamma+\delta+\varepsilon+\phi}}\frac{q^{\binom{s+1}{2}+\binom{\delta+1}{2}+\binom{\varepsilon+1}{2}+\binom{\phi+1}{2}}(q^{-\phi}-q^{\alpha-\phi}+q^{\alpha})a^{\alpha+\delta+\varepsilon}b^{\beta+\delta+\phi}c^{\gamma+\varepsilon+\phi}}{(q)_\alpha(q)_\beta(q)_\gamma(q)_\delta(q)_\varepsilon(q)_\phi}$$

$$(4.9)\qquad =(-aq)_\infty(-bq)_\infty(-cq)_\infty\,.$$

We can rid ourselves of the condition on s in (4.9) by writing it in the equivalent form

$$(4.10)\qquad [\omega^0]\sum_{s}\sum_{\alpha,\beta,\gamma,\delta,\varepsilon,\phi\geqq 0}$$

$$\frac{\omega^{s-\alpha-\beta-\gamma-\delta-\varepsilon-\phi}q^{\binom{s+1}{2}+\binom{\delta+1}{2}+\binom{\varepsilon+1}{2}+\binom{\phi+1}{2}}(q^{-\phi}-q^{\alpha-\phi}+q^{\alpha})a^{\alpha+\delta+\varepsilon}b^{\beta+\delta+\phi}c^{\gamma+\varepsilon+\phi}}{(q)_\alpha(q)_\beta(q)_\gamma(q)_\delta(q)_\varepsilon(q)_\phi}$$

$$=(-aq)_\infty(-bq)_\infty(cq)_\infty\,.$$

It is to be noted that in (4.10) all seven sums over the variables s, α, β, γ, δ, ε and ϕ can be evaluated to be products. To deal with the sum over s, use Jacobi's triple product identity

$$(4.11)\qquad \sum_{s=-\infty}^{\infty}\omega^s q^{\binom{s+1}{2}}=(q)_\infty(-\omega q)_\infty(-\omega^{-1})_\infty\,.$$

To deal with sums over δ, ε and ϕ, use Euler's identity

$$\sum_{n=0}^{\infty} \frac{z^n q^{\binom{n+1}{2}}}{(q)_n} = (-zq)_\infty . \tag{4.12}$$

And for sums over α, β and γ, use another of Euler's identities, namely,

$$\sum_{n=0}^{\infty} \frac{z^n q^n}{(q)_n} = \frac{1}{(zq)_\infty} . \tag{4.13}$$

So, using (4.11), (4.12) and (4.13), (4.10) can be rewritten in the form

$$1 = [\omega^0] \frac{\left(q, -\omega q, -\omega^{-1}, \frac{-abq}{\omega}, \frac{-acq}{\omega}\right)_\infty}{\left(\frac{b}{\omega}, \frac{c}{\omega}, -aq, -bq, -cq\right)_\infty} \left\{ \frac{\left(\frac{-bc}{\omega}\right)_\infty}{\left(\frac{a}{\omega}\right)_\infty} - \frac{\left(\frac{-bc}{\omega}\right)_\infty}{\left(\frac{aq}{\omega}\right)_\infty} + \frac{\left(\frac{-bcq}{\omega}\right)_\infty}{\left(\frac{aq}{\omega}\right)_\infty} \right\}$$

$$= [\omega^0] \frac{\left(q, -\omega q, -\omega^{-1}, \frac{-abq}{\omega}, \frac{-acq}{\omega}, \frac{-bcq}{\omega}\right)_\infty}{\left(\frac{a}{\omega}, \frac{b}{\omega}, \frac{c}{\omega}, -aq, -bq, -cq\right)_\infty} \left(1 + \frac{abc}{\omega^2}\right). \tag{4.14}$$

Now we use Bailey's ${}_6\psi_6$ summation [16], p. 239, eqn. (II, 33) to rewrite (4.14) in the form

$$1 = [\omega^0] \left(1 + \frac{abc}{\omega^2}\right) \times \sum_{n=-\infty}^{\infty} \frac{(1 + \omega^{-1} q^{2n})(-a^{-1}, -b^{-1}, -c^{-1})_n}{\left(\frac{a}{\omega}, \frac{b}{\omega}, \frac{c}{\omega}\right)_{n+1}} \left(\frac{abc}{\omega^2}\right)^n q^{\binom{n+1}{2}} . \tag{4.15}$$

Using (4.5) the sum in (4.15) can be decomposed, and so it can be put in the form

$$1 = [\omega^0] \left\{ \left(1 + \frac{abc}{\omega^2}\right) \sum_{n=1}^{\infty} \frac{(1 + \omega^{-1} q^{2n})(-a^{-1}, -b^{-1}, -c^{-1})_n}{\left(\frac{a}{\omega}, \frac{b}{\omega}, \frac{c}{\omega}\right)_{n+1}} \left(\frac{abc}{\omega^2}\right)^n q^{\binom{n+1}{2}} + \frac{\left(1 + \frac{abc}{\omega^2}\right)(1 + \omega^{-1})}{\left(1 - \frac{a}{\omega}\right)\left(1 - \frac{b}{\omega}\right)\left(1 - \frac{c}{\omega}\right)} + F(\omega) \right\} . \tag{4.16}$$

Note that the sum on the right of (4.16) contains only negative powers of ω. Clearly

$$[\omega^0]\frac{\left(1+\dfrac{abc}{\omega^2}\right)(1+\omega^{-1})}{\left(1-\dfrac{a}{\omega}\right)\left(1-\dfrac{b}{\omega}\right)\left(1-\dfrac{c}{\omega}\right)}=1\,.$$

Finally, we know that $[\omega^0]\ F(\omega)=0$, by the corollary. Hence (4.16) is true, and this completes the proof of the key identity.

§ 5. Companions

In 1971, using a computer search, Andrews [8] found the following companion to Schur's theorem:

Theorem A. *Let $S_2(n)$ denote the number of partitions of n in the form $m_1+m_2+\dots$, such that $m_\ell-m_{\ell+1}\geqq 3,2$ or 5 if $m_\ell\equiv 1,2$ or 3 (mod 3). Then*

$$S_1(n)=S_2(n)\,.$$

Andrews proved Theorem A using generating functions in a manner similar to his proof [4] of Theorem S, but the exact relationship between Theorems S and A remained unclear. By considering the six lexicographic orderings of the colored integers generated by the six permutations of the symbols

$$a_1,\, b_1,\quad \text{and}\quad ab_2\,,$$

Alladi and Gordon [2] showed that $S_1(n)$ and $S_2(n)$ are only two of six partition functions $S_\mu(n)$, $\mu=1,2,\dots,6$, all equal. That is

$$S(n)=S_\mu(n),\quad \mu=1,2,\dots,6\,. \tag{5.1}$$

The gap conditions defining $S_\mu(n)$ for $\mu\geqq 3$ are slightly more complicated and so did not show up in Andrews' computer search. It was also shown in [2] that the functions $S_\mu(n)$ correspond to the six ways in which the q-multinomial coefficient $\begin{bmatrix} i+j+k \\ i,j,k \end{bmatrix}$ can be expanded in terms of $\begin{bmatrix} i+j+k-1 \\ i-1,j,k \end{bmatrix}$, $\begin{bmatrix} i+j+k-1 \\ i,j-1,k \end{bmatrix}$ and $\begin{bmatrix} i+j+k-1 \\ i,j,k-1 \end{bmatrix}$.

Our purpose now is to discuss companions to the function $C(n)$ (hereinafter denoted by $C_1(n)$) obtained by considering lexicographic orderings other than Scheme 1. Before introducing these other orderings, we make two observations about Type 1 partitions.

Firstly, the gap conditions (2.7) defining Type 1 partitions can be embodied in the form of the matrix given below.

Type 1 gap matrix

	d	e	a	f	b	c
d	2	1	1	1	1	1
e	2	2	1	1	1	1
a	2	2	1	1	1	1
f	2	2	2	2	1	1
b	2	2	2	2	1	1
c	2	2	2	2	2	1

This matrix is to be read row-wise and the entries in this matrix represent the minimum permissible gaps between consecutive symbols in Type 1 partitions. For example, the entries in the fourth row are to be interpreted as follows. If an f-part has weight ω, then the next larger part, if it is a d-part has weight $\geqq \omega + 2$, if it is an e-part has weight $\geqq \omega + 2, \ldots$, if it is a c-part has weight $\geqq \omega + 1$.

Secondly, let $x_m = x_m^{(1)}$ denote the symbol occupying position m in the full list (2.5). That is $x_0^{(1)} = \underline{d}_1$, $x_1^{(1)} = \underline{e}_1$, $x_2^{(1)} = a_1$, and so on. We set $x_0^{(1)} = \underline{d}_1$ because under the standard transformations d_1 becomes the integer 0. With this notation, the gap conditions (2.7) defining Type 1 partitions can be recast in the following equivalent form: Type 1 partitions are those of the form $x_{m_1} + x_{m_2} + \cdots$, where

$$(5.2) \qquad m_\ell - m_{\ell+1} \geqq 6, \text{ with strict inequality if } x_{m_\ell} \text{ is of secondary order}.$$

From now on, we will refer to (5.2) as the *standard gap conditions*.

In considering other lexicographic orderings we omit the symbol $\underline{d}_1$ because $\underline{d}_1 = 0$ under the standard transformations. We therefore choose any ordering of the symbols

$$(5.3) \qquad \underline{e}_1 \quad a_1 \quad \underline{f}_1 \quad b_1 \quad c_1 \quad d_2 .$$

For instance, Scheme 1 is generated by the basic ordering

Scheme 1.

$$(5.4) \qquad \underline{e}_1 \prec a_1 \prec \underline{f}_1 \prec b_1 \prec c_1 \prec d_2 .$$

We then get the full list of symbols in (2.5) by increasing the weights in (5.4) by 1 in succession. Similarly, we may consider another ordering of the symbols, namely,

Scheme 2.

$$(5.5) \qquad \underline{e}_1 \prec a_1 \prec \underline{f}_1 \prec b_1 \prec d_2 \prec c_1 .$$

Then the full list of symbols generated by (5.5) is

$$(5.6) \quad \underline{e}_1 \prec a_1 \prec \underline{f}_1 \prec b_1 \prec d_2 \prec c_1 \prec e_2 \prec a_2 \prec f_2 \prec b_2 \prec d_3 \prec c_2 \prec \cdots .$$

As in (2.5), the symbols $\underline{e}_1$ and $\underline{f}_1$ are underlined in (5.6) to indicate that they will never appear in the partitions we will be considering presently.

Let $x_m^{(2)}$ denote the symbol occupying position m in (5.6). That is

$$x_1^{(2)} = \underline{e}_1, \quad x_2^{(2)} = a_1, \quad x_3^{(2)} = \underline{f}_1,$$

and so on. We do not have $x_0^{(2)}$ because we dropped $\underline{d}_1$ from the list. We now define Type 2 partitions to be those of the form $x_{m_1} + x_{m_2} + \cdots$, where $x_{m_\ell} = x_{m_\ell}^{(2)}$ satisfy the standard gap conditions (5.2). Now let $C_2(n; \alpha, \beta, \gamma, \delta, \varepsilon, \phi)$ denote the number of Type 2 partitions π of n such that $v_a(\pi) = \alpha$, $v_b(\pi) = \beta, \ldots, v_f(\pi) = \phi$. We then have the equality

$$C_1(n; \alpha, \beta, \gamma, \delta, \varepsilon, \phi) = C_2(n; \alpha, \beta, \gamma, \delta, \varepsilon, \phi). \tag{5.7}$$

A bijective (combinatorial) proof of (5.7) can be given as follows:

Step 1. Consider a partition counted by $C_1(n; \alpha, \beta, \gamma, \delta, \varepsilon, \phi)$. Write the parts in descending order in the form of a column.

Step 2. Subtract 0 from the bottom element of the column, 1 from the element above that, 2 from the next one above, and so on, and display the new values and the ones subtracted in two adjacent columns $\mathscr{C}_1 | \mathscr{C}_2$. The element of $\mathscr{C}_2$ have no color, while those of $\mathscr{C}_1$ retain the color of the parts from which they were derived.

Step 3. Observe that the standard gap conditions ensure that the elements of $\mathscr{C}_1$ from the bottom upwards are in increasing order in Scheme 1. The symbols of primary color alone may repeat in $\mathscr{C}_1$. Now rearrange the elements of $\mathscr{C}_1$ from the bottom upwards to be of increasing order according to Scheme 2 to get a rearranged column $\mathscr{C}_1^R$. Display the elements of $\mathscr{C}_1^R$ and $\mathscr{C}_2$ in two adjacent columns.

Step 4. Add the corresponding elements of $\mathscr{C}_1^R$ and $\mathscr{C}_2$ to get a partition counted by $C_2(n; \alpha, \beta, \gamma, \delta, \varepsilon, \phi)$. The parts of this Type 2 partition retain the colors of the symbols in $\mathscr{C}_1^R$ from which they were obtained.

The constructions in each of these steps are reversible and so this completes the bijective proof of (5.7).

More generally, one may consider all $6! = 720$ orderings of the symbols in (5.3) and the full list of symbols generated by these orderings. Thus we get 720 schemes. Now, let $x_m^{(\mu)}$, for $m \geqq 1$, denote the symbol occupying position m in scheme μ. We define a Type μ partition to be one of the form $x_{m_1} + x_{m_2} + \ldots$, where $x_{m_\ell} = x_{m_\ell}^{(\mu)}$ satisfy the standard gap conditions (5.2). Next, let $C_\mu(n; \alpha, \beta, \gamma, \delta, \varepsilon, \phi)$ denote the number of Type μ partitions π of n with $v_a(\pi) = \alpha$, $v_b(\pi) = \beta, \ldots, v_f(\pi) = \phi$. Then extending (5.7) we have

Theorem 3. *For* $\mu = 2, 3, \ldots, 720$ *we have*

$$C_1(n; \alpha, \beta, \gamma, \delta, \varepsilon, \phi) = C_\mu(n; \alpha, \beta, \gamma, \delta, \varepsilon, \phi).$$

A bijective proof of Theorem 3 can be given in a manner similar to the proof of (5.7) given above. We do not repeat the details.

As in the case of Type 1 partitions, the gap conditions defining Type μ partitions can be put in matrix form. For instance, the matrix for Type 2 partitions is:

Type 2 gap matrix

	d	e	a	f	b	c
d	2	1	1	1	1	0
e	2	2	1	1	1	1
a	2	2	1	1	1	1
f	2	2	2	2	1	1
b	2	2	2	2	1	1
c	3	2	2	2	2	1

There are only two changes in this matrix from the matrix for Type 1, namely, the 2 at the bottom of the first column has now increased to 3, and consequently, the 1 at the right end of the first row has now been replaced by 0. This is just an instance of a *conservation law for gaps* which holds for Type μ partitions. In order to describe this, let $\psi_\mu(x, y)$ denote the entry in the position determined by the row corresponding to x and the column corresponding to y of the Type μ gap matrix. Then, from the standard gap conditions it is obvious that for all μ

$$\psi_\mu(d, d) = \psi_2(e, e) = \psi_\mu(f, f) = 2 \tag{5.8}$$

and

$$\psi_\mu(a, a) = \psi_\mu(b, b) = \psi_\mu(c, c) = 1\,. \tag{5.9}$$

In addition, there is a *gap conservation law*, namely

$$\psi_\mu(x, y) + \psi_\mu(y, x) = 3 \quad \text{for } x \neq y\,. \tag{5.10}$$

Under the standard transformations, the full list of symbols in Scheme 2 yields the following ordering of the positive integers:

$$\underline{1} \prec 2 \prec \underline{3} \prec 4 \prec 6 \prec 5 \prec 7 \prec 8 \prec 9 \prec 10 \prec 12 \prec 11 \prec \cdots\,. \tag{5.11}$$

The ordering in (5.11) leads to Theorem 4 below, which is an analogue of Theorem A.

Theorem 4. *Let $C_2(n)$ denote the number of partitions of n in the form $m_1 + m_2 + \cdots$, such that no $m_\ell = 1$ or 3, and $m_\ell - m_{\ell+1} \geqq 7, 6, 7, 6, 5$ or 8 if $m_\ell \equiv 7, 2, 9, 4, 5$ or 6 (mod 6). Then*

$$C_1(n) = C_2(n)\,.$$

Remarks. The inequalities defining most of the other companions $C_\mu(n)$ are generally more complicated than the ones for $C_1(n)$ and $C_2(n)$, but there may be a few which are as nice as these two; it may be worthwhile to determine them all.

What we have produced are bijections $\mathscr{F}_{\mu,\omega}$ between partitions of Type μ and those of Type ω, for $\mu, \omega = 1, 2, \ldots, 720$; that is, bijections between partitions counted by $C_\mu(n)$ and $C_\omega(n)$. But it is a different story altogether to produce a bijection between partitions counted by $C_\mu(n)$ and $B(n)$, and indeed no such bijection is known at present. This appears to be extremely difficult.

As was pointed out earlier in this section, it was shown in [2] that the six companions $S_\mu(n)$ to Schur, correspond to the six ways in which the q-multinomial coefficient $\begin{bmatrix} i+j+k \\ i, j, k \end{bmatrix}$ can be expanded. We note here that there are $6! = 720$ ways to expand the q-multinomial coefficient

$$\begin{bmatrix} \alpha+\beta+\gamma+\delta+\varepsilon+\phi \\ \alpha, \beta, \gamma, \delta, \varepsilon, \phi \end{bmatrix}.$$

Although the multinomial coefficients of order $t = 6$ played a crucial role in the determination of the local generating functions in § 3, it is unclear how the 720 expansions of these q-multinomial coefficients correspond to the 720 companions $C_\mu(n)$. One reason for the difficulty is the presence of the factor

$$1 - q^\alpha(1 - q^\phi)$$

on the lefthand side of the key identity (1.4). We intend to explore the connections between the companions $C_\mu(n)$ and the expansions of the q-multinomial coefficients of order $t = 6$, subsequently.

§ 6. Related problems

In approaching Schur's theorem via the method of weighted words, Alladi and Gordon [1], [2] considered integers in two primary colors a, b and one secondary color ab. Here, for Göllnitz theorem, we dealt with integers in three primary colors a, b, c and three secondary colors ab, ac and bc. The principal reason for the substantial increase in the depth and difficulty of the problem considered here is that the ternary color abc is dropped; that is, we are not dealing with the full non-empty alphabet of colors that can be generated using a, b, c, but only with a proper subset.

Andrews [5], [6] has obtained general partition theorems extending Schur's theorem by choosing a set $a_1, a_2, \ldots, a_r$ of distinct residues $(\operatorname{mod} M)$, with $M \geqq 2^r - 1$, and by considering all possible residue classes which are obtained as non-empty sums

$$\sum \varepsilon_i a_i, \quad \varepsilon_i = 0 \text{ or } 1 .$$

The main reason he was able to obtain such extensions was that he was dealing with the complete set of $2^r - 1$ residues generated by $a_1, a_2, \ldots, a_r$.

One way of extending Göllnitz's theorem would be to consider four primary colors a, b, c, d. We then have a choice of dropping either $abcd$, or may be even some of the secondary or ternary colors. At the moment we do not know which of the choices (if any) would be connected with the expansion of

$$(-aq)_\infty(-bq)_\infty(-cq)_\infty(-dq)_\infty\,.$$

An attempt at this question might give some clues about the general situation with r primary colors $a_1, a_2, \ldots, a_r$; once again, the emphasis is that we should not deal with the full alphabet of colors generated by $a_1, a_2, \ldots, a_r$. Such a study was attempted computationally in 1971 by Andrews [8], p. 384–385; however, the effort was limited in scope by the amount of computer power and memory available at that time. Also, a theoretical analytic approach to this problem was outlined in [10], § 6, p. 218–221; but that project also came up dry. Again the computing facilities available then to back up the study were quite limited by today's standards. In light of the fact that many problems from that era (see for example [9], p. 84, Conjecture 2) and of this general type have now fallen to the power of investigations using modern computer algebra systems [13], it is not unreasonable to consider these questions now.

In connection with the extensions of Schur's theorem obtained in [5] and [6], Andrews and Olsson [14] and more recently Andrews, Bessenrodt and Olsson [13],have considered partitions with upper bound conditions on differences rather than lower bound gap conditions. This, from a combinatorial point of view, has been shown to be more natural.

Another related problem is a partition conjecture of Capparelli [15] which arose in his study of vertex operators in Lie Algebras. Using the method of weighted words we recently obtained generalizations and refinements of Capparelli's conjecture [3]. More precisely, we viewed Capparelli's conjecture as emerging out of an expansion of the product

$$(-aq^2; q^2)_\infty(-bq^2; q^2)_\infty(-q)_\infty\,.$$

For Göllnitz's theorem we dealt with the product

$$(-aq)_\infty(-bq)_\infty(-cq)_\infty\,.$$

Although these products look similar, it is unclear how exactly these two problems are related. It might be worthwhile to explore connections between Capparelli's conjecture and Göllnitz's theorem.

Acknowledgements. The first author would like to thank the Mathematics Department at The Pennsylvania State University for all the hospitality extended during his sabbatical visit in 1992–93 when this work was completed. We also thank the referee for his comments and suggestions.

References

[1] *K. Alladi* and *B. Gordon*, Generalizations of Schur's partition theorem, Manuscr. Math. **79** (1993), 113–126.

[2] *K. Alladi* and *B. Gordon*, Schur's partition theorem, companions, refinements and generalizations, Trans. Amer. Math. Soc., to appear.

[3] *K. Alladi, G. E. Andrews* and *B. Gordon*, Refinements and generalizations of Capparelli's conjecture on partitions, J. Algebra, to appear.

[4] *G.E. Andrews*, On Schur's second partition theorem, Glasgow Math. J. **9** (1967), 127–132.
[5] *G.E. Andrews*, A new generalization of Schur's secound partition theorem, Acta Arith. **14** (1968), 429–434.
[6] *G.E. Andrews*, A general partition theorem with difference conditions, Amer. J. Math. **91** (1969), 18–24.
[7] *G.E. Andrews*, On a partition theorem of Göllnitz and related formulae, J. reine angew. Math. **236** (1969), 37–42.
[8] *G.E. Andrews*, The use of computers in search of identities of the Rogers-Ramanujan type, in: Computers in Number Theory (A.O.L. Atkin and B.J. Birch Eds.), Academic Press (1971), 377–387.
[9] *G.E. Andrews*, On the general Rogers-Ramanujan theorem, Amer. Math. Soc. Mem. **152** (1974).
[10] *G.E. Andrews*, Problems and prospects for basic hyper-geometric functions, in: Theory and Applications of Special Functions (R. Askey Ed.), Academic Press (1975), 191–224.
[11] *G.E. Andrews*, The theory of partitions, Encyclopedia of Math. **2**, Addison-Wesley, Reading 1976.
[12] *G.E. Andrews*, *q*-series: Their development and applications in Analysis Number Theory, Combinatorics, Physics and Computer Algebra, NSF-CBMS Lectures **66**, Amer. Math. Soc., Providence 1985.
[13] *G.E. Andrews*, *C. Bessenrodt* and *J.B. Olsson*, Partition identities and labels for some modular characters, Trans. Amer. Math. Soc. **344** (1994), 597–615.
[14] *G.E. Andrews* and *J.B. Olsson*, Partition identities with an application to group representation theory, J. reine angew. Math. **413** (1991), 198–212.
[15] *S. Capparelli*, On some representations of twisted affine Lie Algebras and combinatorial identities, J. Algebra **154** (1993), 335–355.
[16] *G. Gasper* and *M. Rahman*, Basic hyper-geometric series, Encyclopedia of Mathematics and its Applications **35**, Cambridge 1990.
[17] *H. Göllnitz*, Partitionen mit Differenzenbedingungen, J. reine angew. Math. **225** (1967), 154–190.
[18] *I. Schur*, Zur Additiven Zahlentheorie, Gesammelte Abhandlungen **2**, Springer, Berlin (1973), 43–50.

University of Florida, Gainesville, FL, 32611

Pennsylvania State University, University Park, PA, 16802

University of California, Los Angeles, CA, 90024

Eingegangen 5. November 1993, in revidierter Fassung 8. September 1994

3.12 An Algebraic Identity of Jackson and its Implications for Partitions (with R. Lewis)

ELSEVIER

Discrete Mathematics 232 (2001) 77–83

DISCRETE MATHEMATICS

www.elsevier.com/locate/disc

Note

An algebraic identity of F.H. Jackson and its implications for partitions

George E. Andrews[a], Richard P. Lewis[b,*]

[a]Department of Mathematics, 218 McAllister Building, Pennsylvania State University, PA 16802, USA
[b]School of Mathematical Sciences, The University of Sussex, Brighton BN1 9QH, UK

Received 29 January 2000; revised 15 May 2000; accepted 16 June 2000

Abstract

An identity of F.H. Jackson is used to derive new partition generating functions and related identities.

1. Introduction

In the study of the theory of partitions, we often consider infinite product generating functions. This is perhaps epitomised in the study of partition ideals of order 1 (cf. [2, pp. 124–128]). In every previous instance, the factors in the infinite product arise from either the finite geometric series

$$1+q^n+\cdots+q^{(r-1)n}=\frac{1-q^{rn}}{1-q^n} \tag{1.1}$$

or the infinite geometric series

$$1+q^n+q^{2n}+\cdots=\frac{1}{1-q^n}. \tag{1.2}$$

Thus in every case, we see the familiar factors $1-q^{an}$ arising in the numerator and denominator. Multiplying instances of these two identities together yields every generating function for partitions in which different parts appear independently although the number of appearances of each individual part may be restricted (i.e. partition ideals of order 1).

* Corresponding author.
E-mail address: r.p.lewis@sussex.ac.uk (R.P. Lewis).

In this way, we are led quickly to the products of Euler [2, pp. 4 and 5]:

$$\prod_{n=1}^{\infty}\frac{1}{1-q^n}=\sum_{n=0}^{\infty}p(n)q^n, \tag{1.3}$$

$$\sum_{n=0}^{\infty}\mathcal{O}(n)q^n=\prod_{n=1}^{\infty}\frac{1}{1-q^{2n-1}}$$

$$=\prod_{n=1}^{\infty}\frac{1-q^{2n}}{1-q^n}$$

$$=\prod_{n=1}^{\infty}(1+q^n)=\sum_{n=0}^{\infty}\mathcal{D}(n)q^n, \tag{1.4}$$

where $p(n)$ is the total number of partitions of n, $\mathcal{O}(n)$ is the number of partitions of n into odd parts and $\mathcal{D}(n)$ is the number of partitions of n into distinct parts.

In the same vein, we have the infinite product portions of the Rogers–Ramanujan identities [2, Sections 7.1.6 and 7.1.7]

$$G(q)=\prod_{n=1}^{\infty}\frac{1}{(1-q^{5n-4})(1-q^{5n-1})}, \tag{1.5}$$

$$H(q)=\prod_{n=1}^{\infty}\frac{1}{(1-q^{5n-3})(1-q^{5n-2})}. \tag{1.6}$$

In this note, we bring to light identities more recondite than (1.1) and (1.2) which give rise to interesting and useful results in the theory of partitions.

We begin with a surprising but little known identity due to Jackson [4]:

$$1-\frac{a(1-b)(1-c)(1-d)(1-a^2bcd)}{(1-ab)(1-ac)(1-ad)(1-abcd)}=\frac{(1-a)(1-abc)(1-abd)(1-acd)}{(1-ab)(1-ac)(1-ad)(1-abcd)}. \tag{1.7}$$

In fact, as Jackson pointed out, this identity is the instance $n=q=1$ of his q-analogue of Dougall's theorem [4].

If we differentiate both sides of (1.7) w.r.t. a and then set $a=1$, we find

$$1+\frac{b}{1-b}+\frac{c}{1-c}+\frac{d}{1-d}-\frac{bcd}{1-bcd}=\frac{(1-bc)(1-bd)(1-cd)}{(1-b)(1-c)(1-d)(1-bcd)}. \tag{1.8}$$

Setting $d=0$ in (1.8) gives the simple

$$1+\sum_{n=1}^{\infty}(b^n+c^n)=1+\frac{b}{1-b}+\frac{c}{1-c}=\frac{1-bc}{(1-b)(1-c)} \tag{1.9}$$

and, setting $d = bc$ in (1.8), we find

$$1 + \sum_{n=1}^{\infty} (b^n + c^n + (bc)^{2n-1}) = 1 + \frac{b}{1-b} + \frac{c}{1-c} + \frac{bc}{1-b^2c^2}$$
$$= \frac{(1-b^2c)(1-bc^2)}{(1-b)(1-c)(1-b^2c^2)}. \tag{1.10}$$

In Section 2, we will use (1.9) and (1.10) to obtain a number of infinite product generating functions for partitions. Subsequently, in Section 3, we shall exhibit some identities for these newly defined partition functions.

2. Generating functions

We shall be considering partitions whose parts lie in prescribed arithmetic progressions mod k. In each example, we shall ask that the parts differ by a certain amount.

Definition 1. Suppose $0 < a < b < k$. Let $W_1(a,b;k;n)$ denote the number of partitions of n in which the parts are congruent to a or $b \bmod k$ and such that, for any j, $kj + a$ and $kj + b$ are not both parts.

Definition 2. Suppose $0 < 2a < 2b < k$. Let $W_2(a,b;k;n)$ denote the number of partitions of n in which the parts are congruent to $\pm a$ or $\pm b \bmod k$ and such that, for $r \geqslant 1$, the parts $kr + a$ and $kr - a$ do not both appear and, for $s \geqslant 0$, the parts $ks + b$ and $ks + k - b$ do not both appear.

Definition 3. Suppose $0 < a < b < a + b \leqslant k$. Let $W_3(a,b;k;n)$ denote the number of partitions of n in which the parts are congruent to a or $b \bmod k$ and in which, for $s \geqslant 0$, if parts $ks+a$ and $ks+b$ both appear, they each appear the same, odd number of times.

Definition 4. For $i = 1,2,3$,

$$\mathscr{W}_i(a,b;k;q) := \sum_{n=0}^{\infty} W_i(a,b;k;n)q^n. \tag{2.1}$$

We have, by (1.9),

$$\mathscr{W}_1(a,b;k;q) = \prod_{n=0}^{\infty} \left(1 + \sum_{m=1}^{\infty} q^{(kn+a)m} + \sum_{m=1}^{\infty} q^{(kn+b)m}\right)$$
$$= \prod_{n=0}^{\infty} \frac{(1-q^{2kn+a+b})}{(1-q^{kn+a})(1-q^{kn+b})} \tag{2.2}$$

and (1.9) also gives

$$\begin{aligned}\mathscr{W}_2(a,b;k;q) &= \sum_{n=0}^{\infty} q^{an} \prod_{n=0}^{\infty}\left(1+\sum_{m=1}^{\infty} q^{(kn+k+a)m}+\sum_{m=1}^{\infty} q^{(kn+k-a)m}\right)\\ &\quad\times \prod_{n=0}^{\infty}\left(1+\sum_{m=1}^{\infty} q^{(kn+b)m}+\sum_{m=1}^{\infty} q^{(kn+k-b)m}\right)\\ &= \frac{1}{1-q^a}\prod_{n=0}^{\infty}\frac{(1-q^{2kn+2k})}{(1-q^{kn+k+a})(1-q^{kn+k-a})}\\ &\quad\times\prod_{n=0}^{\infty}\frac{(1-q^{2kn+k})}{(1-q^{kn+b})(1-q^{kn+k-b})}\\ &= \prod_{n=0}^{\infty}\frac{1-q^{kn+k}}{(1-q^{kn+a})(1-q^{kn+k-a})(1-q^{kn+b})(1-q^{kn+k-b})}. \end{aligned} \tag{2.3}$$

Finally, (1.10) gives

$$\begin{aligned}\mathscr{W}_3(a,b;k;q) &= \prod_{n=0}^{\infty}\left(1+\sum_{m=1}^{\infty} q^{(kn+a)m}+\sum_{m=1}^{\infty} q^{(kn+b)m}+\sum_{m=1}^{\infty} q^{(2kn+a+b)(2m-1)}\right)\\ &= \prod_{n=0}^{\infty}\frac{(1-q^{3kn+2a+b})(1-q^{3kn+a+2b})}{(1-q^{kn+a})(1-q^{kn+b})(1-q^{4kn+2a+2b})}. \end{aligned} \tag{2.4}$$

Note that the infinite products appearing in (2.2)–(2.4) are not a priori power series with nonnegative coefficients. However, since each is a partition generating function, this must in fact be the case.

3. Identities

Theorem 1. *The number of partitions of n into non-multiples of* 3 *in which no two parts differ by* 1 *equals the number of partitions of n in which no part appears more than twice and no two parts differ by* 1.

Proof. The first class of partitions is enumerated by $W_1(1,2;3;n)$. The second class of partitions is enumerated by the partition function $B_{1,1}(n)$ from [1] and we know from this reference that

$$\sum_{n=0}^{\infty} B_{1,1}(n)q^n = \prod_{n=0}^{\infty}\frac{(1-q^{6n+3})^2(1-q^{6n+6})}{(1-q^{n+1})}. \tag{3.1}$$

Now by (2.2)

$$\sum_{n=0}^{\infty} W_1(1,2;3;n)q^n = \mathscr{W}_1(1,2;3;q) = \prod_{n=0}^{\infty} \frac{(1-q^{6n+3})}{(1-q^{3n+1})(1-q^{3n+2})}$$

$$= \prod_{n=0}^{\infty} \frac{(1-q^{6n+3})(1-q^{3n+3})}{(1-q^{n+1})}$$

$$= \prod_{n=0}^{\infty} \frac{(1-q^{6n+3})^2(1-q^{6n+6})}{(1-q^{n+1})} \tag{3.2}$$

and, comparing (3.1) and (3.2) we see that

$$W_1(1,2;3;n) = B_{1,1}(n)$$

for every $n \in \mathbb{N}$ and so we have proved Theorem 1. □

To illustrate Theorem 1, we note that the five partitions enumerated by $W(1,2;3;7)$ are 7, $5+2$, $5+1+1$, $4+1+1+1$ and $1+1+1+1+1$ while the five partitions enumerated by $B_{1,1}(7)$ are 7, $6+1$, $5+2$, $5+1+1$ and $3+3+1$.

Corollary. The number of partitions of n into non-multiples of 3 in which at least two parts differ by 1 equals the number of partitions of n in which no part appears more than twice and at least two parts differ by 1.

Proof. This follows from Theorem 1 and Glaisher's Theorem [2, p. 6], a special case of which asserts that the number of partitions of n into non-multiples of 3 equals the number of partitions of n in which no part appears more than twice. □

Theorem 2. *Suppose* $0 < a < b < k = a+b$. *Then* $W_3(a,b;k;n)$ *equals the number of partitions into parts congruent to* $a, 3a+2b, b$ *or* $3b+2a$ *modulo* $3k$ *or congruent to* $2k$ *mod* $4k$.

Proof. Bearing in mind that $a+b=k$, we see that, by (2.4),

$$\sum_{n=0}^{\infty} W_3(a,b;k;n)q^n$$

$$= \prod_{n=0}^{\infty} \frac{(1-q^{3kn+2a+b})(1-q^{3kn+a+2b})}{(1-q^{kn+a})(1-q^{kn+b})(1-q^{4kn+2a+2b})}$$

$$=\prod_{n=0}^{\infty}\frac{(1-q^{3kn+2a+b})}{(1-q^{3kn+a})(1-q^{3kn+2a+b})(1-q^{3kn+3a+2b})}$$
$$\times\prod_{n=0}^{\infty}\frac{(1-q^{3kn+a+2b})}{(1-q^{3kn+b})(1-q^{3kn+a+2b})(1-q^{3kn+2a+3b})}\prod_{n=0}^{\infty}\frac{1}{(1-q^{4kn+2k})}$$
$$=\prod_{n=0}^{\infty}\frac{1}{(1-q^{3kn+a})(1-q^{3kn+3a+2b})(1-q^{3kn+b})(1-q^{3kn+2a-3b})(1-q^{4kn+2k})},$$

and this last infinite product is the generating function for the partitions of the type described in the theorem. □

Theorem 3. *$W_1(1,3;4;n)$ equals the number of partitions of n into odd parts in which no part appears more than three times.*

Proof.

$$\sum_{n=0}^{\infty}W_1(1,3;4;n)q^n=\prod_{n=0}^{\infty}\frac{(1-q^{8n+4})}{(1-q^{4n+1})(1-q^{4n+3})}$$
$$=\prod_{n=0}^{\infty}\frac{(1-q^{8n+4})}{(1-q^{2n+1})}$$
$$=\prod_{n=0}^{\infty}(1+q^{2n+1}+q^{2(2n+1)}+q^{3(2n+1)})$$

and this last infinite product is the generating function for the partitions described in the theorem. □

We end this section with a result of a different type.

Theorem 4. *For $0<a<b<a+b<k$,*

$$\mathscr{W}_2(a,b;k;q)=\frac{\sum_{n=-\infty}^{\infty}q^{an}/1-q^{kn+b}}{\sum_{n=-\infty}^{\infty}(-1)^n q^{kn(n+1)/2-(a+b)n}}$$

Proof. By (2.3)

$$\mathscr{W}_2(a,b;k;n)=\frac{1}{\prod_{n=0}^{\infty}(1-q^{kn+a+b})(1-q^{kn+k-a-b})(1-q^{kn+k})}$$
$$\times\prod_{n=0}^{\infty}\frac{(1-q^{kn+a+b})(1-q^{kn+k-a-b})(1-q^{kn+k})^2}{(1-q^{kn+a})(1-q^{kn+k-a})(1-q^{kn+b})(1-q^{kn+k-b})}$$
$$=\frac{\sum_{n=-\infty}^{\infty}\frac{q^{an}}{1-q^{kn+b}}}{\sum_{n=-\infty}^{\infty}(-1)^n q^{kn(n+1)/2-(a+b)n}}$$

which follows by applying Jacobi's Triple Product identity [2; 2.2.10] to the first factor and Ramanujan's ${}_1\Psi_1$ summation ([3, (C.1), p. 115], with q^a for t, q^k for q, q^{b+k} for b and q^b for a) to the second. □

4. Conclusion

Our object has not been to provide an exhaustive account of these new partition functions. Indeed our main goal has been to reveal that infinite product generating functions can arise in subtler ways than had previously been encountered.

It would be of great interest if bijective proofs could be found for Theorems 1–3. Theorems 2 and 3 are the most likely to yield to this approach, since their proofs involve only the manipulation of infinite products; a bijective proof of Theorem 1 would probably require a bijective proof of the expression given for $\sum_{n=0}^{\infty} B_{1,1}(n)q^n$.

We remark that the classes of partitions related to Definitions 1 and 3 can be extended to a very general setting. Namely, suppose $S_1=\{a_n\}_{n=1}^{\infty}$ and $S_2=\{b_n\}_{n=1}^{\infty}$ are disjoint sets of integers. Define $W_1(S_1,S_2;n)$ to be the number of partitions of N into elements of $S_1 \cup S_2$ which do not contain as parts both a_j and b_j, for any j and define $W_3(S_1,S_2;n)$ to be the number of partitions of n into elements of $S_1 \cup S_2$ in which, if a_j and b_j do both appear as parts, they appear the same odd number of times. Then

$$\sum_{n=0}^{\infty} W_1(S_1,S_2;n)q^n = \prod_{j=1}^{\infty} \frac{(1-q^{a_j+b_j})}{(1-q^{a_j})(1-q^{b_j})}$$

and

$$\sum_{n=0}^{\infty} W_3(S_1,S_2;n)q^n = \prod_{j=1}^{\infty} \frac{(1-q^{2a_j+b_j})(1-q^{2b_j+a_j})}{(1-q^{a_j})(1-q^{b_j})(1-q^{2a_j+2b_j})}.$$

We choose not to begin with these more general partition functions because all the results of interest that we found related to arithmetic progressions.

We thank Robin Chapman for pointing out some inaccuracies in our original formulations of Definitions 2 and 3. Robin has also shown us an elegant bijective proof of Theorem 3.

References

[1] G.E. Andrews, Some new partition theorems II, J. Combin. Theory 7 (1969) 262–263.
[2] G.E. Andrews, The theory of partitions, in: G.-C. Rota (Ed.), Encyclopedia of Mathematics and Its Applications, Vol. 2, Addison-Wesley, Reading, MA, 1976 (reissued: CUP, London, 1985).
[3] G.E. Andrews, q-series: their development and application ..., CBMS Regional Conference Lecture Series, 66, American Mathematical Society Providence, RI, 1986.
[4] F.H. Jackson, Summation of q-hypergeometric series, Messenger Math. 50 (1921) 101–112.

3.13 Partitions with Short Sequences and Mock Theta Functions

Partitions with short sequences and mock theta functions

George E. Andrews*

Department of Mathematics, Pennsylvania State University, University Park, PA 16802

This contribution is part of the special series of Inaugural Articles by members of the National Academy of Sciences elected on April 29, 2003.

Contributed by George E. Andrews, January 10, 2005

P. A. MacMahon was the first to examine integer partitions in which consecutive integers were not allowed as parts. Such partitions may be described as having sequences of length 1. Recently it was shown that partitions containing no sequences of consecutive integers of length *k* are of interest in seemingly unrelated problems concerning threshold growth models. The object now is to develop the subject intrinsically to both provide deeper understanding of the theory and application of partitions and reveal the surprising role of Ramanujan's mock theta functions.

1. Preliminaries

In MacMahon's monumental work on combinatorial analysis (1), we find the first study of partitions without sequences. For example, there are eight partitions of 7 without sequences: $7, 6 + 1, 5 + 2, 5 + 1 + 1, 4 + 1 + 1 + 1, 3 + 3 + 1, 3 + 1 + 1 + 1 + 1$, and $1 + 1 + 1 + 1 + 1 + 1 + 1$. MacMahon was able to obtain the generating function for such partitions as a q series. He concluded his chapter on the subject by showing that the number of partitions of n without sequences and without ones equals the number of partitions of n into parts not congruent to 1 or 5 modulo 6.

Subsequently, in the 20th century, there were a few additional papers (2–4) devoted to furthering the study of such partitions.

Recently, however, Holroyd *et al.* (5) made an ingenious study of several definite integrals with widespread applications. The main integral in question is

$$-\int_0^1 \frac{\log f(x)}{x}\,dx = \frac{\pi^2}{3ab}, \qquad \textbf{[1.1]}$$

where $0 < a < b$ and $f = f_{a,b}$ is the unique decreasing function mapping the unit interval into itself and satisfying

$$[f(x)]^a - [f(x)]^b = x^a - x^b, \quad 0 \leqq x \leqq 1. \qquad \textbf{[1.2]}$$

This result has wide application. Holroyd *et al.* (5) prove an asymptotic theorem in probability.

Let $0 < s < 1$, and let $C_1, C_2, \ldots$ be independent events with probabilities

$$\mathbf{P}_s(C_n) = 1 - (1 - s)^n$$

under a probability measure $\mathbf{P}_s$. (We can think of C_n as the event that at least one occurs of a further set of n independent events, each of probability s.) Let k be a positive integer, and let A_k be the event

$$A_k = \bigcap_{i=1}^{\infty} (C_i \cup C_{i+1} \cup \cdots \cup C_{i+k-1})$$

that there is no sequence of k consecutive C_i values that do not occur.

Theorem 1 (5). *For every positive integer k,*

$$-\log \mathbf{P}_s(A_k) \sim \frac{\pi^2}{3k(k+1)}\frac{1}{s} \quad \text{as } s \to 0.$$

Holroyd *et al.* apply this theorem directly to a central probability question regarding threshold growth models. They also consider the application of Eq. **1.1** to partitions with short sequences and speculate that it might be possible to base many of their discoveries on the combinatorial aspects of the theory of partitions.

Indeed, in ref. 5, the more general partition function $p_k(n)$ is considered; $p_k(n)$ is the number of partitions of n that do not contain any sequence of consecutive integers of length k. Thus, from our previous calculation we see that $p_2(7) = 8$. Holroyd *et al.* also study the related generating function:

See accompanying Biography on page 4663.

*E-mail: andrews@math.psu.edu.

$$G_k(q) = \sum_{n=0}^{\infty} p_k(n) q^n. \quad [1.3]$$

Using the close connection of $G_k(q)$ to the probability function $\mathbf{P}_s(A_k)$, they then prove that

$$\log G_k(q) \sim \frac{\pi^2}{6}\left(1 - \frac{2}{k(k+1)}\right)\frac{1}{1-q} \quad \text{as } q \uparrow 1. \quad [1.4]$$

From this result they deduce the asymptotic behavior of $\log p_k(n)$ as $n \to \infty$.

In the final section of ref. 5, the authors show for the case $k = 2$ how to deduce their results from MacMahon's theorem. Section 4 of ref. 5 concludes with the observation that if Eq. **1.4** could be partition-theoretically proved for general k, then it would allow a reversal of their arguments and by analytic continuation provide an independent proof of Eq. **1.1**.

We shall prove the following results:

Theorem 2. *For* $|q| < 1$,

$$G_k(q) = \frac{1}{(q;q)_\infty} \sum_{s,r \geqq 0} \frac{(-1)^s q^{\binom{k+1}{2}(s+r)^2 + (k+1)\binom{r+1}{2}}}{(q^k;q^k)_s (q^{k+1};q^{k+1})_r}, \quad [1.5]$$

where

$$(A;q)_n = (1-A)(1-Aq)\cdots(1-Aq^{n-1}), \quad [1.6]$$

and

$$(A;q)_\infty = \lim_{n\to\infty} (A;q)_n. \quad [1.7]$$

To make our second result clear, we introduce further notation:

$$P(v,k;q) := (q^{k(k+1)};q^{k(k+1)})_\infty \frac{(q^{\binom{k+1}{2}+kv};q^{k(k+1)})_\infty (q^{\binom{k+1}{2}-kv};q^{k(k+1)})_\infty}{(q;q)_\infty (q^k;q^k)_\infty}, \quad [1.8]$$

$$S(v,k;q) := (q^v;q^{k+1})_\infty \sum_{j \geqq 0} \frac{(q^{k+1-v};q^{k+1})_{kj+v-1} q^{k(k+1)j+kv}}{(q^k;q^k)_{(k+1)j+v}}. \quad [1.9]$$

Theorem 3. *For* $|q| < 1$,

$$G_k(q) = (q^{k+1};q^{k+1})_\infty P(0,k;q) - \sum_{v=1}^{\lfloor \frac{k}{2} \rfloor} P(v,k;q)(S(v,k;q) - S(k+1-v,k;q)). \quad [1.10]$$

Now the results in *Theorems 1* and *2* are true for all k. In light of the fact that $P(v, k; q)$ is a quotient of three theta functions, we can easily obtain asymptotic results on it of great accuracy as $q \uparrow 1$. On the other hand, we do not yet have enough information about $S(v, k; q)$ to handle its asymptotics as $q \uparrow 1$ in general. However, we shall provide a full treatment when $k = 2$ with the hope that it may point the way to understanding $G_k(q)$ fully as $q \uparrow 1$.

It is surprising that the key to the $k = 2$ case lies in one of Ramanujan's mock theta functions (6, 7):

$$\chi(q) := 1 + \sum_{n=1}^{\infty} \frac{q^{n^2}}{\prod_{j=1}^{n}(1 - q^j + q^{2j})}.$$

Theorem 4. *For* $|q| < 1$,

$$G_2(q) = \frac{(-q^3;q^3)_\infty}{(q^2;q^2)_\infty}\chi(q).$$

From here we are led directly to the full asymptotics of $P_s(A_2)$. Namely:

Theorem 5. $\mathbf{P}_s(A_2) \sim \sqrt{\pi/2}\, s^{-1/2} e^{-\pi^2/18s}$ *as* $s \downarrow 0$.

Note that this is a much stronger result than the $k = 2$ instance of theorem 2 of ref. 5.

We hope that these initial successes will lead to an effort to understand the many implications of *Theorems 1* and *2*. In particular, it would be very appealing to obtain the asymptotics of $\mathbf{P}_s(A_k)$ not just for $k = 2$.

Numerical calculations of the series arising in *Theorem 2* lead us to the following:

Conjecture. *For each $k \geq 2$, there is a positive constant C_k such that*

$$\mathbf{P}_s(A_k) \sim C_k s^{-\frac{1}{2}} e^{-\frac{\pi^2}{3k(k+1)s}} \quad as \quad s \downarrow 0.$$

2. Proof of *Theorem 1*

To accomplish our argument, we must refine $P_k(n)$ to $P_k(m, n)$, the number of partitions of n into m parts containing no sequences of consecutive integers of length k or longer. We then define

$$G_k(x, q) = \sum_{m,n \geqq 0} p_k(m, n) x^m q^n, \tag{2.1}$$

and we note that

$$G_k(1, q) = G_k(q). \tag{2.2}$$

We begin by noting that $G_k(0, q) = 1$, and

$$G_k(x, q) = G_k(xq, q) + \sum_{j=1}^{k-1} \frac{x^j q^{\binom{j+1}{2}} G_k(xq^{j+1}, q)}{(xq; q)_j}. \tag{2.3}$$

The proof of the functional Eq. **2.3** is quite straight forward. We split the partitions enumerated by $G_k(x)$ into k classes, where the jth class contains those partitions in which 1 occurs in a maximal sequence of precisely j consecutive integers. If $j = 0$ (i.e., 1 is not a part), then the relevant partitions are clearly generated by $G_k(xq, q)$. If $1 \leqq j \leqq k - 1$, the relevant generating function is easily seen to be

$$\frac{x^j q^{1+2+3+\cdots+j}}{(1 - xq)(1 - xq^2)\cdots(1 - xq^j)} G_k(xq^{j+1}, q), \tag{2.4}$$

and adding up the generating functions for these k exhaustive classes, we obtain Eq. **2.3**.

To prove Eq. **1.5** and consequently *Theorem 1*, we shall prove more generally

$$G_k(x, q) = \frac{1}{(xq; q)_\infty} \sum_{r,s \geqq 0} \frac{(-1)^s x^{ks+(k+1)r} q^{\binom{k+1}{2}(r+s)^2 + (k+1)\binom{r+1}{2}}}{(q^k; q^k)_s (q^{k+1}; q^{k+1})_r} := g_k(x, q). \tag{2.5}$$

Once Eq. **2.5** is proved, we may deduce *Theorem 1* by setting $x = 1$ therein and applying Eq. **2.2**.

To prove Eq. **2.5**, we proceed by denoting the right side of Eq. **2.5** by $g_k(x, q)$; we need only show that $g_k(0, q) = 1$ (which is obvious by inspection) and that $g_k(x, q)$ satisfies the functional Eq. **2.3**, which together with $g_k(0, q) = 1$ uniquely defines $G_k(x, q)$.

Apart from the right side of Eq. **2.5**, there are two other useful representations of $g_k(x, q)$. First,

$$\sum_{r,s \geqq 0} \frac{x^{k(s-1)+(k+1)r}(-1)^{s-1} q^{\binom{k+1}{2}(r+s-1)^2 + (k+1)\binom{r}{2}} (1 - q^{ks+(k+1)r})}{(x; q)_\infty (q^k; q^k)_s (q^{k+1}; q^{k+1})_r}$$

$$= \sum_{r,s \geqq 0} \frac{x^{ks+1+(k+1)r}(-1)^{s-1} q^{\binom{k+1}{2}(r+s)^2 + (k+1)\binom{r+1}{2}}}{(z; q)_\infty (q^k; q^k)_s (q^{k+1}; q^{k+1})_r} + \sum_{r,s \geqq 0} \frac{x^{ks+(k+1)r}(-1)^s q^{\binom{k+1}{2}(r+s)^2 + (k+1)\binom{r+1}{2}}}{(x; q)_\infty (q^k; q^k)_s (q^{k+1}; q^{k+1})_r}$$

{because $1 - qks+(k+1)r = (1 - q(k+1)r) + q(k+1)r(1 - qks)$}

$$= \frac{(1 - x)}{(x; q)_\infty} \sum_{r,s \geqq 0} \frac{x^{ks+(k+1)r}(-1)^s q^{\binom{k+1}{2}(r+s)^2 + (k+1)\binom{r+1}{2}}}{(q^k; q^k)_s (q^{k+1}; q^{k+1})_r} = g_k(x, q), \tag{2.6}$$

by Eq. **2.5**.

Second, we may start with the top line of Eq. **2.6**, and noting $1 - q^{ks+(k+1)r} = 1 - q^{ks} + q^{ks}(1 - q^{(k+1)r})$, we see that also

$$g_k(x, q) = \sum_{r,s \geqq 0} \frac{x^{ks+(k+1)r}(-1)^s q^{\binom{k+1}{2}(r+s)^2 + (k+1)\binom{r}{2}}}{(x; q)_\infty (q^k; q^k)_s (q^{k+1}; q^{k+1})_r} + \sum_{r,s \geqq 0} \frac{x^{ks+1+(k+1)r}(-1)^{s+1} q^{\binom{k+1}{2}(r+s)^2 + (k+1)\binom{r+1}{2} + ks}}{(x; q)_\infty (q^k; q^k)_s (q^{k+1}; q^{k+1})_r}. \tag{2.7}$$

We take the representation of $g_k(x, q)$ given by Eq. **2.6** and insert it in the right side of the required functional equation

$$\sum_{j=0}^{k-1} \frac{x^j q^{\binom{j+1}{2}} g_k(xq^{j+1}, q)}{(xq; q)_j} = \left(\sum_{j=0}^{k-1} x^j q^{\binom{j+1}{2}} \sum_{r,s \geqq 0} \frac{(-1)^s x^{ks+(k+1)r} q^{(j+1)(ks+(k+1)r) + \binom{k+1}{2}(r+s)^2 + (k+1)\binom{r}{2}}}{(xq; q)_\infty (q^k; q^k)_s (q^{k+1}; q^{k+1})_r} - \sum_{j=0}^{k-1} x^j q^{\binom{j+1}{2}} \right.$$

$$\times \sum_{r,s\geqq 0} \frac{(-1)^s x^{ks+1+(k+1)r} q^{(j+1)(ks+1+(k+1)r)+\binom{k+1}{2}(r+s)^2+(k+1)\binom{r+1}{2}+ks}}{(xq;q)_\infty (q^k;q^k)_s (q^{k+1};q^{k+1})_r} \Bigg)$$

$$= \Bigg(\sum_{j=0}^{k-1} x^j q^{\binom{j+1}{2}} \sum_{r,s\geqq 0} \frac{(-1)^s x^{ks+(k+1)r} q^{(j+1)(ks+(k+1)r)+\binom{k+1}{2}(r+s)^2+(k+1)\binom{r}{2}}}{(xq;q)_\infty (q^k;q^k)_s (q^{k+1};q^{k+1})_r} - \sum_{j=1}^{k} x^j q^{\binom{j}{2}}$$

$$\times \sum_{r,s\geqq 0} \frac{(-1)^s x^{ks+(k+1)r} q^{j(ks+1+(k+1)r)+\binom{k+1}{2}(r+s)^2+(k+1)\binom{r+1}{2}+ks}}{(xq;q)_\infty (q^k;q^k)_s (q^{k+1};q^{k+1})_r} \Bigg)$$

$$= \Bigg(\sum_{r,s\geqq 0} \frac{(-1)^s x^{ks+(k+1)r} q^{(ks+(k+1)r)+\binom{k+1}{2}(r+s)^2+(k+1)\binom{r}{2}}}{(xq;q)_\infty (q^k;q^k)_s (q^{k+1};q^{k+1})_r} - \sum_{r,s\geqq 0} \frac{(-1)^s x^{k(s+1)+(k+1)r}}{(xq;q)_\infty (q^k;q^k)_s (q^{k+1};q^{k+1})_r}$$

$$\times q^{k(ks+1+(k+1)r)+\binom{k}{2}+\binom{k+1}{2}(r+s)^2+(k+1)\binom{r+1}{2}+ks} \Bigg)$$

{because all terms have cancelled except that $j = 0$ in the first sum and $j = k$ in the second}

$$= \Bigg(\sum_{r,s\geqq 0} \frac{(-1)^s x^{ks+(k+1)r} q^{ks+(k+1)r+\binom{k+1}{2}(r+s)^2+(k+1)\binom{r}{2}}}{(xq;q)_\infty (q^k;q^k)_s (q^{k+1};q^{k+1})_r}$$

$$+ \sum_{r,s\geqq 0} \frac{(-1)^s (1-q^{ks}) x^{ks+(k+1)r}}{(xq;q)_\infty (q^k;q^k)_s (q^{k+1};q^{k+1})_r} q^{k(k(s-1)+1+(k+1)r)+\binom{k}{2}+\binom{k+1}{2}(r+s-1)^2+(k+1)\binom{r+1}{2}+k(s-1)} \Bigg)$$

(where s has been shifted to $s - 1$ in the second sum)

$$= \frac{1}{(xq;q)_\infty} \sum_{r,s\geqq 0} \frac{(-1)^s x^{ks+(k+1)r} q^{\binom{k+1}{2}(r+s)^2+(k+1)\binom{r+1}{2}}}{(q^k;q^k)_s (q^{k+1};q^{k+1})_r}$$

$$= g_k(x,q),$$

by Eq. **2.5**, and thus we obtain the desired defining functional equation.

Hence,

$$g_k(x,q) = G_k(x,q),$$

and thus *Theorem 1* follows.

3. Proof of *Theorem 2*

The object in this second result is to obtain a representation of $G_k(q)$ that will yield the asymptotics promised in *Theorem 3*. To this end, we require several results from the literature. First is Euler's identity (see p. 490, corollary 10.2.2.b, in ref. 8),

$$\sum_{n=0}^{\infty} \frac{(-1)^n q^{\binom{n}{2}} x^n}{(q;q)_n} = (x;q)_\infty; \qquad \textbf{[3.1]}$$

next is Jacobi's triple product identity (see p. 497, theorem 1.4.1., in ref. 8; with $q \to q^{2a}$, $x = q^{a+h}$),

$$\theta(a,h;q) := \sum_{s=-\infty}^{\infty} (-1)^s q^{as^2+hs} = (q^{2a};q^{2a})_\infty (q^{a+h};q^{2a})_\infty (q^{a-h};q^{2a})_\infty . \qquad \textbf{[3.2]}$$

We proceed with the proof of *Theorem 2* starting from Eq. **1.5**. Hence,

$$G_k(q) = \frac{1}{(q;q)_\infty} \sum_{s,r\geqq 0} \frac{(-1)^s q^{\binom{k+1}{2}(s+r)^2+(k+1)\binom{r+1}{x}}}{(q^k;q^k)_s (q^{k+1};q^{k+1})_r} = \sum_{r=0}^{\infty} \sum_{s=-\infty}^{\infty} \frac{(-1)^s q^{\binom{k+1}{2}(s+r)^2+(k+1)\binom{r+1}{2}} (q^{k(s+1)};q^k)_\infty}{(q;q)_\infty (q^k;q^k)_\infty (q^{k+1};q^{k+1})_r}$$

$$= \sum_{r=0}^{\infty} \sum_{s=-\infty}^{\infty} \frac{(-1)^{s-r} q^{\binom{k+1}{2}s^2+(k+1)\binom{r+1}{2}} (q^{k(s-r+1)};q^k)_\infty}{(q;q)_\infty (q^k;q^k)_\infty (q^{k+1};q^{k+1})_r} = \sum_{r=0}^{\infty} \sum_{s=-\infty}^{\infty} \sum_{j=0}^{\infty} \frac{(-1)^{s-r+j} q^{\binom{k+1}{2}s^2+(k+1)\binom{r+1}{2}+\binom{j+1}{2}k+jk(s-r)}}{(q;q)_\infty (q^k;q^k)_\infty (q^{k+1};q^{k+1})_r (q^k;q^k)_j}$$

[3.3]

(by Eq. **3.1**)

$$= \sum_{r=0}^{\infty}\sum_{j=0}^{\infty} \frac{\theta\left(\frac{k+1}{2}, j; q^k\right)(-1)^{r+j} q^{(k+1)\binom{r+1}{2}+k\binom{j+1}{2}-jkr}}{(q;q)_\infty (q^k;q^k)_\infty (q^{k+1};q^{k+1})_r (q^k;q^k)_j}$$

(by Eq. **3.2**)

$$= \sum_{j=0}^{\infty} \frac{\theta\left(\frac{k+1}{2}, j; q^k\right)(-1)^j q^{k\binom{j+1}{2}}}{(q;q)_\infty (q^k;q^k)_\infty (q^k;q^k)_j} (q^{k+1-jk}; q^{k+1})_\infty,$$

by Eq. **3.1**.

Now we note several things about this last expression. First, if j is a positive multiple of $(k+1)$, then the infinite product in the summand vanishes. Next, we observe that $\theta(a, -h; q) = \theta(a, h; q)$; also,

$$\theta(a, h+2a; q) = \sum_{s=-\infty}^{\infty} (-1)^s q^{a(s+1)^2 + h(s+1) - a - h} = -q^{-a-h}\theta(a, h; q), \tag{3.4}$$

and thus,

$$\theta(a, a; q) = -\theta(a, -a; q) = -\theta(a, a; q) = 0.$$

This last identity means that if $j \equiv k + 1/2 (\text{mod } k+1)$, the sum in Eq. **3.4** also vanishes. Finally, we note that the $j = 0$ term in Eq. **3.3** is in fact $(q^{k+1}; q^{k+1})_\infty P(0, k; q)$ by Eqs. **1.8** and **3.2**.

Hence, separating the terms of the sum in Eq. **3.2** according to the residue of j modulo $k+1$, we see that

$$G_k(q) = (q^{k+1}; q^{k+1})_\infty P(0, k; q) + \sum_{v=1}^{\lfloor \frac{k}{2} \rfloor} \left\{ \sum_{j=0}^{\infty} \frac{\theta\left(\frac{k+1}{2}, (k+1)j + v; q^k\right)(-1)^{(k+1)j+v}(q^{k+1-((k+1)j+v)k}; q^{k+1})_\infty}{(q;q)_\infty (q^k;q^k)_\infty (q^k;q^k)_{(k+1)j+v}} q^{k\binom{(k+1)j+v+1}{2}} \right.$$

$$\left. + \sum_{j=0}^{\infty} \frac{\theta\left(\frac{k+1}{2}, (k+1)j + k + 1 - v; q^k\right)(-1)^{(k+1)j+k+1-v}(q^{k+1-((k+1)j+k+1-v)k}; q^{k+1})_\infty}{(q;q)_\infty (q^k;q^k)_\infty (q^k;q^k)_{(k+1)j+k+1-v}} q^{k\binom{(k+1)j+k+1-v+1}{2}} \right\}.$$

This last expression simplifies term by term into the formulation in *Theorem 2*; in the simplification one must invoke Eq. **3.2** and the fact that

$$(Aq^{-N}; q)_\infty = (-1)^N q^{-\binom{N+1}{2}} A^N (q/A; q)_N (A; q)_\infty. \tag{3.5}$$

4. Proof of *Theorem 3*

It is surprising that half the proof of *Theorem 3* relies on an identity of MacMahon (1), and half relies on an identity of Fine (7). Thus, we shall refer to the result as the MacMahon–Fine identity. We begin with MacMahon, (ref. 1, p. 52) who proved

$$G_2(q) = 1 + \sum_{n=1}^{\infty} \frac{q^n (q^6; q^6)_{n-1}}{(1-q^n)(q^2;q^2)_{n-1}(q^3;q^3)_{n-1}}, \tag{4.1}$$

which may be algebraically simplified term by term to

$$G_2(q) = 1 + \sum_{n=1}^{\infty} \frac{q^n \prod_{j=1}^{n-1}(1 - q^j + q^{2j})}{(q;q)_n}. \tag{4.2}$$

Next, in his study of Ramanujan's mock theta functions, Fine (7) showed that

$$\chi(q) = \frac{(q^2 q^2)_\infty}{(-q^3; q^3)_\infty} \left(1 + \sum_{n=1}^{\infty} \frac{q^n \prod_{j=1}^{n-1}(1 - q^j + q^{2j})}{(q;q)_n} \right). \tag{4.3}$$

Combining Eqs. **4.2** and **4.3**, we obtain the assertion of *Theorem 3*.

5. Proof of *Theorem 4*

Here we rely principally on one of Ramanujan's mock theta function identifies (6), which we write as

$$\chi(q) = \frac{1}{4}f(q) + \frac{3}{4}\frac{(q^3; q^3)_\infty^4}{(q)_\infty(q^6; q^6)_\infty^2}, \quad [5.1]$$

where

$$f(q) = \sum_{n=0}^{\infty} \frac{q^{n^2}}{(-q; q)_n^2}. \quad [5.2]$$

In light of the fact that $q/(1 + q)^2$ is increasing for $0 \leqq q \leqq 1$, one may easily deduce that

$$f(q) \to \sum_{n=0}^{\infty} \left(\frac{1}{4}\right)^n = \frac{4}{3} \quad \text{as} \quad q \uparrow 1. \quad [5.3]$$

Finally, we require (see theorem 1.12.8 in ref. 8)

$$\frac{1}{(e^{-s}; e^{-s})_\infty} \sim \frac{1}{\sqrt{2\pi}} s^{\frac{1}{2}} e^{\frac{\pi^2}{6s}} \quad \text{as} \quad s \downarrow 0. \quad [5.4]$$

Combining these results with *Theorem 3*, we see that as $s \downarrow 0$ and $q = e^{-s}$,

$$G_2(q) = \frac{(-q^3; q^3)}{(q^2; q^2)_\infty}\chi(q) = \frac{(q^6; q^6)_\infty}{(q^2; q^2)_\infty(q^3; q^3)_\infty}\left(\frac{1}{4}f(q) + \frac{3}{4}\frac{(q^3; q^3)_\infty^4}{(q)_\infty(q^6; q^6)_\infty^2}\right) \sim \frac{1}{6\pi} s^{\frac{1}{2}} e^{\frac{\pi^2}{9s}} + \frac{1}{2} e^{\frac{\pi^2}{9s}} \sim \frac{1}{2} e^{\frac{\pi^2}{9s}}. \quad [5.5]$$

Finally, then, from the identity of Holroyd *et al.* (5),

$$P_s(A_2) = \frac{G_2(q)}{G(q)} \sim \frac{1}{2} e^{\frac{\pi^2}{9s}} \sqrt{2\pi s}^{-\frac{1}{2}} e^{\frac{-\pi^2}{6s}} \sim \sqrt{\frac{\pi}{2}} s^{-\frac{1}{2}} e^{\frac{-\pi^2}{18s}} \quad \text{as} \quad s \downarrow 0, \quad [5.6]$$

which is the assertion in *Theorem 4*.

6. Conclusions

Several themes have been brought together in this article. First, the partition function, $G_k(q)$, of Holroyd *et al.* (5) has been represented in explicit q series. In *Theorem 1*, $G_k(q)$ is represented as an infinite product times a double series, and *Theorem 2*, $G_k(q)$ is represented by less than k single-fold series multiplied by infinite products.

Empirical studies of the functions appearing on the right side in *Theorem 2* suggest that $(q^{k+1}; q^{k+1})_\infty P(0, k; q)$ is the dominant term. This observation is what led to *Conjecture*.

We hope that *Theorem 3* provides a hint of further discoveries for $G_k(q)$ in general. The fact that $G_2(q)$ is an infinite product (in fact, a modular form) times the mock theta function, $\chi(q)$, suggests that the analytic nature of $G_k(q)$ for $k > 2$ may be very interesting. Indeed, additional discoveries may well lead to the extension of *Theorem 4* to results such as those stated in *Conjecture*.

The striking applications of such results pioneered by Holroyd *et al.* (5) make clear the potential significance of additional discoveries.

1. MacMahon, P. A. (1916) *Combinatory Analysis* (Cambridge Univ. Press, Cambridge, U.K.), Vol. II, pp. 49–58.
2. Andrews, G. E. (1967) *J. Comb. Theory* **2,** 431–436.
3. Andrews, G. E. (1967) *J. Comb. Theory* **3,** 100–101.
4. Andrews, G. E. & Lewis, R. P. (2001) *Discrete Math.* **232,** 77–83.
5. Holroyd, A. E., Liggett, T. M. & Romik, D. (2004) *Trans. Am. Math. Soc.* **356,** 3349–3368.
6. Ramanujan, S. (1927) *Collected Papers* (Cambridge Univ. Press, Cambridge, U.K.), p. 354.
7. Fine, N. J. (1988) *Basic Hypergeometric Series and Applications* (Am. Math. Soc., Providence, RI), p. 57.
8. Andrews, G. E., Askey, R. A. & Roy, R. (1999) *Special Functions* (Cambridge Univ. Press, Cambridge, U.K.).

3.13.1 Errata

There is a problem with the numbering of the theorems in the original article as published in the Proceedings of the National Academy of Sciences USA. Apparently the copy eidtor changed the numbering of the theorems without adjusting any of the internal references. To correct the problem, the following changes would need to be made:

— Page 556, line -8: "Theorem 1" should be relabelled as either "Theorem 0" or simply "Theorem."
— Page 557, line 9: "Theorem 2" should be relabelled as "Theorem 1."
— Page 557, line 18: "Theorem 3" should be relabelled as "Theorem 2."
— Page 557, line -8: "Theorem 4" should be relabelled as "Theorem 3."
— Page 557, line -5: "Theorem 5" should be relabelled as "Theorem 4."

No internal references to these theorem numbers should be changed.

3.14 Partitions, Durfee Symbols, and the Atkin–Garvan Moments of Ranks

Invent. math. 169, 37–73 (2007)
DOI: 10.1007/s00222-007-0043-4

Partitions, Durfee symbols, and the Atkin–Garvan moments of ranks

George E. Andrews *

Pennsylvania State University, University Park, PA 16802, USA
(e-mail: andrews@math.psu.edu)

Oblatum 8-XI-2006 & 30-I-2007
Published online: 7 March 2007 –

Abstract. Atkin and Garvan introduced the moments of ranks of partitions in their work connecting ranks and cranks. Here we consider a combinatorial interpretation of these moments. This requires the introduction of a new representation for partitions, the Durfee symbol, and subsequent refinements. This in turn leads us to a variety of new congruences for our 'marked' Durfee symbols much in the spirit of Dyson's original conjectures on the ranks of partitions.

1. Introduction

In [16], F.J. Dyson, defined the rank of a partition as the largest part minus the number of parts. He conjectured that this partition statistic provided a combinatorial explanation of the famous congruences of Ramanujan. Namely

$$p(5n+4) \equiv 0 \pmod{5} \tag{1.1}$$

and

$$p(7n+5) \equiv 0 \pmod{7}, \tag{1.2}$$

where $p(n)$ is the number of partitions of n.

If we denote the number of partitions of n with rank m by $N(m, n)$ and the number of partitions of n with rank $\equiv m \pmod{Q}$ by $N(m, Q, n)$, then Dyson [16] (cf. [11, pp. 85–87]) showed that

$$\sum_{n=0}^{\infty} N(m,n)q^n = \frac{1}{(q;q)_\infty} \sum_{n=1}^{\infty} (-1)^n q^{n(3n-1)/2+mn}(1-q^n), \tag{1.3}$$

* Partially supported by National Science Foundation Grant DMS 0457003.

Mathematics Subject Classification (2000): 05A17, 05A19, 11P83

and

$$(1.4)\quad \sum_{n=0}^{\infty} N(m,Q,n)q^n = \frac{1}{(q;q)_\infty} \sum_{\substack{n=-\infty \\ n\neq 0}}^{\infty} (-1)^n q^{n(3n+1)/2} \frac{q^{mn}+q^{n(Q-m)}}{1-q^{Qn}},$$

where

$$(1.5)\qquad (A;q)_m = \prod_{j=0}^{m-1}(1-A\,q^j),$$

and

$$(1.6)\qquad (A;q)_\infty = \lim_{m\to\infty}(A;q)_m,$$

These results follow easily from the identities:

$$R_1(z;q) := 1 + \sum_{m=-\infty}^{\infty}\sum_{n=1}^{\infty} N(m,n)z^m q^n$$

$$(1.7)\qquad = \sum_{n=0}^{\infty} \frac{q^{n^2}}{(zq;q)_n(z^{-1}q;q)_n}$$

$$(1.8)\qquad = \frac{1}{(q;q)_\infty}\left(1+\sum_{n=1}^{\infty}(-1)^n q^{n(3n+1)/2}\frac{(1+q^n)(1-z)(1-z^{-1})}{(1-zq^n)(1-z^{-1}q^n)}\right)$$

$$(1.9)\qquad = \frac{1}{(q;q)_\infty}\left(1+\sum_{n=1}^{\infty}(-1)^n q^{n(3n+1)/2}\left(\frac{1-z}{1-zq^n}+\frac{1-z^{-1}}{1-z^{-1}q^n}\right)\right)$$

$$(1.10)\qquad = \frac{(1-z)}{(q;q)_\infty}\sum_{n=-\infty}^{\infty}\frac{(-1)^n q^{n(3n+1)/2}}{1-zq^n}.$$

The first equality follows by direct combinatorial argument wherein partitions are classified by their Durfee square [22, Ch. 18, §19.7]. The second equality was given by G.N. Watson [32, p. 64], and the last lines follow by partial fractions (cf. Garvan [19, Lemma 7.9]).

Dyson [16] conjectured that for $0 \leqq a \leqq 4$, $N(a,5,5n+4) = p(5n+4)/5$ and for $0 \leq a \leqq 6$, $N(a,7,7n+5) = p(7n+5)/7$.

Using the above generating functions, Atkin and Swinnerton-Dyer [11] were able to prove all of Dyson's conjectures about the rank.

Recently Bringmann and Ono [13] have greatly extended the study of Dyson's rank relating it to the theory of Maass forms.

Dyson's original paper [16] has been a continuing source of inspiration over the years. Partly because of his "crank" conjecture. The rank did **not** explain Ramanujan's third congruence.

$$(1.11)\qquad p(11n+6) \equiv 0 \pmod{11}.$$

In Dyson's words,

> "One is thus led irresistably to the conclusion that there must be some analogue modulo 11 ....
>
> I hold in fact:
>
> That there exists an arithmetical coefficient similar to, but more recondite than, the rank of a partition; I shall call this hypothetical coefficient the "crank" of the partion, and denote by $M(m, q, n)$ the number of partitions of n whose crank is congruent to m modulo q:
>
> that $M(m, q, n) = M(q - m, q, n)$;
>
> that
>
> $$\begin{aligned} M(0, 11, 11n + 6) &= M(1, 11, 11n + 6) = M(2, 11, 11n + 6) \\ &= M(3, 11, 11n + 6) = M(4, 11, 11n + 6); \end{aligned}$$
>
> that numerous other relations exist analogous to (12)–(19), and in particular
>
> $$\begin{aligned} M(1, 11, 11n + 1) &= M(2, 11, 11n + 1) = M(3, 11, 11n + 1) \\ &= M(4, 11, 11n + 1); \end{aligned}$$
>
> that $M(m, 11, n)$ has a generating function not completely different in form from (24);
>
> that the values of the differences such as $M(0, 11, n) - M(4, 11, n)$ are always extremely small compared with $p(n)$.
>
> Whether these guesses are warranted by the evidence, I leave to the reader to decide. Whatever the final verdict of posterity may be, I believe the "crank" is unique among arithmetical functions in having been named before it was discovered. May it be preserved from the ignominious fate of the planet Vulcan!"

The discovery of the vector crank by F. Garvan [19] eventually led to the discovery of the crank [9]. Most recently K. Mahlburg [25] has revealed that the crank explains in a grand and glorious manner the infinite panoply of partition congruences discovered and proved by K. Ono et al. [2], [26], [27].

In a paper relating cranks and ranks [10], Atkin and Garvan found that the moments of ranks and cranks were important in the study of further partition congruences. In particular, they define the kth moment of the rank by

$$N_k(n) = \sum_{m=-\infty}^{\infty} m^k N(m, n). \tag{1.12}$$

In light of the fact that $N(m, n) = N(-m, n)$, we see that the odd order moments are all zero.

It turns out that there is a rich combinatorial and enumerative structure associated with the moments of ranks. Actually what is required is a symmetrized k^{th} moment function

$$\eta_k(n) = \sum_{m=-\infty}^{\infty} \binom{m + \lfloor \frac{k-1}{2} \rfloor}{k} N(m,n). \tag{1.13}$$

In Sect. 2, we shall present the necessary background for consideration of the symmetrized k^{th} moment function. Section 3 is devoted to the definition of new combinatorial representations of partitions; we shall call these objects **Durfee symbols**. Subsequently we shall refine these symbols to k-marked Durfee symbols. Each of the latter will have k Dyson-like ranks associated with it. When $k = 1$, the objects in question reduce to classical partitions, and the new rank is precisely Dyson's original rank, i.e. the largest part minus the number of parts.

The next section of the paper is devoted to congruence theorems for the $\eta_k(n)$ which have combinatorial explanations. Special corollaries of our results include

$$N_2(n) \equiv \eta_2(n) \equiv 0 \pmod 5 \quad \text{if } n \equiv \pm 1 \pmod 5 \tag{1.14}$$

$$N_2(n) \equiv \eta_2(n) \equiv 0 \pmod 7 \quad \text{if } n \equiv 1, 5 \pmod 7 \tag{1.15}$$

$$N_4(n) \equiv \eta_4(n) \equiv 0 \pmod 7 \quad \text{if } n \equiv 1, 5 \pmod 7. \tag{1.16}$$

These congruences may, of course, be deduced easily from the work of Atkin and Swinnerton-Dyer [11]. For example

$$\begin{aligned} N_2(5n+4) &= \sum_{m=-\infty}^{\infty} m^2 N(m, 5n+4) \\ &\equiv N(1,5,5n+4) - N(2,5,5n+4) \\ &\quad - N(3,5,5n+4) + N(4,5,5n+4) \\ &\equiv 0 \pmod 5. \end{aligned}$$

The others can be proved similarly. Our object in Sect. 7 will be the provision of a Dyson-like combinatorial explanation of such congruences.

The remainder of the paper considers "odd Durfee symbols". These resemble ordinary Durfee symbols with a modified Durfee square and with odd numbers as entries.

2. Background

Here we collect several results that will be relevant in subsequent sections.

Theorem 1. *$\eta_k(n) = 0$ if k is odd.*

Proof. Let $k = 2\nu + 1$

$$\begin{aligned}\eta_{2\nu+1}(n) &= \sum_{m=-\infty}^{\infty} \binom{m+\nu}{2\nu+1} N(m,n) \\ &= \sum_{m=-\infty}^{\infty} \binom{-m+\nu}{2\nu+1} N(-m,n) \\ &= (-1)^{2\nu+1} \sum_{m=-\infty}^{\infty} \binom{m+\nu}{2\nu+1} N(m,n) \\ &\quad \text{(by the symmetry of binomial coefficients and [16, p. 11, Eq. (10)])} \\ &= -\eta_{2\nu+1}(n).\end{aligned}$$

□

Theorem 2.

$$\sum_{n=1}^{\infty} \eta_{2\nu}(n) q^n = \frac{1}{(q;q)_\infty} \sum_{n=1}^{\infty} \frac{(1+q^n)(-1)^{n-1} q^{n(3n-1)/2+\nu n}}{(1-q^n)^{2\nu}} \tag{2.1}$$

$$= \frac{1}{(q;q)_\infty} \sum_{\substack{n=-\infty \\ n\neq 0}}^{\infty} \frac{(-1)^{n-1} q^{n(3n-1)/2+\nu n}}{(1-q^n)^{2\nu}}. \tag{2.2}$$

Proof. If $\nu > 0$, then

$$\begin{aligned}\sum_{n=1}^{\infty} \eta_{2\nu}(n) q^n &= \frac{1}{(2\nu)!} \left(\frac{d^{2\nu}}{dz^{2\nu}} z^{\nu-1} R_1(z;q) \right) \Big|_{z=1} \\ &= \frac{1}{(2\nu)!} \sum_{j=0}^{\nu-1} \binom{2\nu}{j} (\nu-1)(\nu-2)\cdots(\nu-j) R_1^{(2\nu-j)}(1;q)\end{aligned}$$

(by Leibniz's rule).

Now

$$R_1'(z;q) = \frac{-1}{(q;q)_\infty} \sum_{\substack{n=-\infty \\ n\neq 0}}^{\infty} \frac{(-1)^n q^{n(3n+1)/2}(1-q^n)}{(1-zq^n)^2},$$

and consequently for $j \geqq 1$

$$R_1^{(j)}(z;q) = \frac{-j!}{(q;q)_\infty} \sum_{\substack{n=-\infty \\ n\neq 0}}^{\infty} \frac{(-1)^n q^{n(3n-1)/2+jn}(1-q^n)}{(1-zq^n)^{j+1}}.$$

Therefore

$$\sum_{n=1}^{\infty}\eta_{2\nu}(n)q^n$$

$$= \frac{1}{(q;q)_\infty}\sum_{j=0}^{\nu-1}\binom{\nu-1}{j}\sum_{\substack{n=-\infty\\ n\neq 0}}^{\infty}\frac{(-1)^{n-1}q^{n(3n-1)/2+(2\nu-j)n}(1-q^n)}{(1-q^n)^{2\nu-j+1}}$$

$$= \frac{1}{(q;q)_\infty}\sum_{\substack{n=-\infty\\ n\neq 0}}^{\infty}\frac{(-1)^{n-1}q^{n(3n-1)/2+2\nu n}}{(1-q^n)^{2\nu}}\left(1+\frac{q^{-n}}{(1-q^n)^{-1}}\right)^{\nu-1}$$

$$= \frac{1}{(q;q)_\infty}\sum_{\substack{n=-\infty\\ n\neq 0}}^{\infty}\frac{(-1)^{n-1}q^{n(3n+1)/2+\nu n}}{(1-q^n)^{2\nu}}$$

$$= \frac{1}{(q;q)_\infty}\left(\sum_{n=1}^{\infty}\frac{(-1)^{n-1}q^{n(3n+1)/2+\nu n}}{(1-q^n)^{2\nu}}+\sum_{n=1}^{\infty}\frac{(-1)^{-n-1}q^{n(3n-1)/2-\nu n}}{(1-q^{-n})^{2\nu}}\right)$$

$$= \frac{1}{(q;q)_\infty}\sum_{n=1}^{\infty}\frac{(-1)^{n-1}q^{n(3n-1)/2+\nu n}(1+q^n)}{(1-q^n)^{2\nu}}.$$

If $\nu = 0$, then by (1.13) $\eta_0(n) = p(n)$ for $n > 0$, while $\eta_0(0) = 0$. Hence starting with the right-hand side of (2.2) with $\nu = 0$

$$\frac{1}{(q;q)_\infty}\sum_{\substack{n=-\infty\\ n\neq 0}}^{\infty}(-1)^{n-1}q^{n(3n-1)/2} = \frac{1}{(q;q)_\infty}(-(q;q)_\infty+1)$$

(by [5, p. 11, Corollary 1.7])

$$= \frac{1}{(q;q)_\infty}-1 = \sum_{n=1}^{\infty}p(n)q^n = \sum_{n=1}^{\infty}\eta_0(n)q^n,$$

as desired. □

Our next theorem treats

(2.3) $R_k(x_1,x_2,\ldots,x_k;q)$

$$= \sum_{\substack{m_1>0\\ m_2,m_3,\ldots,m_k\geqq 0}}\frac{q^{(m_1+m_2+\cdots+m_k)^2+(m_1+\cdots+m_{k-1})+(m_1+\cdots+m_{k-2})+\cdots+m_1}}{(x_1q;q)_{m_1}\left(\frac{q}{x_1};q\right)_{m_1}\left(x_2q^{m_1};q\right)_{m_2+1}\left(\frac{q^{m_1}}{x_2};q\right)_{m_2+1}}$$

$$\times\frac{1}{\left(x_3q^{m_1+m_2};q\right)_{m_3+1}\left(\frac{q^{m_1+m_2}}{x_3};q\right)_{m_3+1}}$$

$$\times\frac{1}{\cdots\left(x_kq^{m_1+\cdots+m_{k-1}};q\right)_{m_k+1}\left(\frac{q^{m_1+\cdots+m_{k-1}}}{x_k};q\right)_{m_k+1}}.$$

Theorem 3. *For $k \geqq 2$,*

$$R_k(x_1,x_2,\dots,x_k;q) = \frac{1}{(q;q)_\infty} \sum_{n=1}^{\infty} \frac{(-1)^{n-1}(1+q^n)(1-q^n)^2 q^{3n(n-1)/2+kn}}{\prod_{j=1}^{k} \left(1-x_j q^n\right)\left(1-\frac{q^n}{x_j}\right)}.$$

Proof. We recall the k-fold generalization of Watson's q-analog of Whipple's theorem [5, p. 199, Theorem 4]:

For $k \geqq 1$, a nonnegative integer,

(2.4)

$$ {}_{2k+4}\phi_{2k+3}\left[\begin{matrix} a, q\sqrt{a}, -q\sqrt{a}, b_1, c_1, b_2, c_2, \\ \dots, b_k, c_k, q^{-N}; q, \frac{a^k q^{k+N}}{b_1\dots b_k c_1 \dots c_k} \\ \sqrt{a}, -\sqrt{a}, aq/b_1, aq/c_1, aq/b_2, aq/c_2, \\ \dots, aq/b_k, aq/c_k, aq^{N+1} \end{matrix}\right]$$

$$= \frac{(aq;q)_N (aq/b_k c_k;q)_N}{(aq/b_k;q)_N (aq/c_k;q)_N}$$

$$\times \sum_{m_1,\dots,m_{k-1}\geqq 0} \frac{(aq/b_1c_1;q)_{m_1}(aq/b_2c_2;q)_{m_2}\cdots(aq/b_{k-1}c_{k-1};q)_{m_{k-1}}}{(q;q)_{m_1}(q;q)_{m_2}\cdots(q;q)_{m_{k-1}}}$$

$$\times \frac{(b_2;q)_{m_1}(c_2;q)_{m_1}(b_3;q)_{m_1+m_2}(c_3;q)_{m_1+m_2}}{(aq/b_1;q)_{m_1}(aq/c_1;q)_{m_1}(aq/b_2;q)_{m_1+m_2}(aq/c_2;q)_{m_1+m_2}}$$

$$\times \frac{\cdots (b_k;q)_{m_1+\cdots+m_{k-1}}}{\cdots (aq/b_{k-1};q)_{m_1+\cdots+m_{k-1}}}$$

$$\times \frac{(c_k;q)_{m_1+\cdots+m_{k-1}}}{(aq/c_{k-1};q)_{m_1+\cdots+m_{k-1}}} \cdot \frac{(q^{-N};q)_{m_1+m_2+\cdots+m_{k-1}}}{\left(b_k c_k q^{-N}/a;q\right)_{m_1+m_2+\cdots+m_{k-1}}}$$

$$\times \frac{(aq)^{m_{k-2}+2m_{k-3}+\cdots+(k-2)m_1} q^{m_1+m_2+\cdots+m_{k-1}}}{(b_2c_2)^{m_1}(b_3c_3)^{m_1+m_2}\cdots(b_{k-1}c_{k-1})^{m_1+m_2+\cdots+m_{k-2}}}$$

here (cf. [21, p. 4])

$$(2.5)\quad {}_{r+1}\phi_r\left(\begin{matrix} a_0, a_1, \dots, a_r; q, z \\ b_1, \dots, b_r \end{matrix}\right) = \sum_{n=0}^{\infty} \frac{(a_0;q)_n(a_1;q)_n\cdots(a_r;q)_n z^n}{(q;q)_n(b_1;q)_n\cdots(b_r;q)_n}.$$

In (2.4), replace k by $k+1$, for $1 \leqq j \leqq k$ set $b_j = c_j^{-1} = x_j$ set $a = 1$ and let b_{k+1}, c_{k+1} and $N \to \infty$. After simplification, we find that

$$1+\sum_{n=1}^{\infty} \frac{\prod_{j=1}^{k}(1-x_j)(1-x_j^{-1})}{\prod_{j=1}^{k}(1-q^n x_j)(1-\frac{q^n}{x_j})}(-1)^n q^{n(3n-1)/2+kn}(1+q^n) \tag{2.6}$$

$$= (q;q)_\infty \sum_{m_1,\dots,m_k \geqq 0} \prod_{j=2}^{k}(1-x_j)(1-x_j^{-1})$$

$$\times \frac{1}{(x_1 q;q)_{m_1}\left(\frac{q}{x_1};q\right)_{m_1}\left(x_2 q^{m_1};q\right)_{m_2+1}\left(\frac{q^{m_1}}{x_2};q\right)_{m_2+1}}$$

$$\times \frac{1}{\cdots\left(x_k q^{m_1+\cdots+m_{k-1}};q\right)_{m_k+1}\left(\frac{q^{m_1+\cdots+m_{k-1}}}{x_k};q\right)_{m_k+1}}$$

$$\times q^{(m_1+\cdots+m_k)^2+m_1+(m_1+m_2)+(m_1+m_2+m_3)+\cdots+(m_1+m_2+\cdots+m_{k-1})}.$$

Now let us examine the term $m_1 = 0$ on the right-hand side of (2.6). Upon inspection and simplification, we observe that it is precisely what we would obtain if, on the right-hand side, we replaced k by $k-1$ and replaced each x_i by x_{i+1}. Hence we see that

$$1+\sum_{n=1}^{\infty} \frac{\prod_{j=1}^{k}(1-x_j)(1-x_j^{-1})(-1)^n q^{n(3n-1)/2+kn}(1+q^n)}{\prod_{j=1}^{k}(1-q^n x_j)(1-\frac{q^n}{x_j})}$$

$$-\left(1+\sum_{n=1}^{\infty} \frac{\prod_{j=2}^{k}(1-x_j)(1-x_j^{-1})(-1)^n q^{n(3n-1)/2+(k-1)n}(1+q^n)}{\prod_{j=2}^{k}(1-q^n x_j)(1-\frac{q^n}{x_j})}\right) \tag{2.7}$$

$$= (q;q)_\infty \sum_{\substack{m_1>0\\ m_2,\dots,m_k \geqq 0}} \prod_{j=2}^{k}(1-x_j)(1-x_j^{-1})$$

$$\times \frac{1}{(x_1 q;q)_{m_1}\left(\frac{q}{x_1};q\right)_{m_1}\left(x_2 q^{m_1};q\right)_{m_2+1}\left(\frac{q^{m_1}}{x_2};q\right)_{m_2+1}}$$

$$\times \frac{1}{\cdots\left(x_k q^{m_1+\cdots+m_{k-1}};q\right)_{m_k+1}\left(\frac{q^{m_1+\cdots+m_{k-1}}}{x_k};q\right)_{m_k+1}}$$

$$\times q^{(m_1+\cdots+m_k)^2+m_1+(m_1+m_2)+(m_1+m_2+m_3)+\cdots+(m_1+m_2+\cdots+m_{k-1})}.$$

Now we may simplify the left-hand side of (2.7); indeed, it reduces to

$$\sum_{n=1}^{\infty} \frac{\prod_{j=2}^{k}(1-x_j)\left(1-x_j^{-1}\right)(-1)^n q^{n(3n-1)/2+(k-1)n}}{\prod_{j=1}^{k}\left(1-q^n x_j\right)\left(1-\frac{q^n}{x_j}\right)}(1+q^n)$$
$$\times\left((1-x_1)\left(1-x_1^{-1}\right)q^n-\left(1-q^n x_1\right)\left(1-\frac{q^n}{x_1}\right)\right)$$
$$=\prod_{j=2}^{k}(1-x_j)\left(1-x_j^{-1}\right)$$
$$\times\sum_{n=1}^{\infty}\frac{(-1)^{n-1}(1+q^n)(1-q^n)^2 q^{n(3n-1)/2+(k-1)n}}{\prod_{j=1}^{k}\left(1-q^n x_j\right)\left(1-\frac{q^n}{x_j}\right)}$$

Substituting this last expression for the left-hand side of (2.7) and dividing both sides of the result by $\prod_{j=2}^{k}(1-x_j)(1-x_j^{-1})(q;q)_\infty$ we obtain the desired result. □

Corollary 4. *$R_k(x_1, x_2, \ldots, x_k; q)$ is a symmetric function of $x_1, x_2, \ldots, x_k$.*

Proof. Immediate from inspection of Theorem 3. □

Corollary 5. *For $\nu \geqq 1$,*

$$R_{\nu+1}(1, 1, \ldots, 1; q) = \sum_{n=1}^{\infty} \eta_{2\nu}(n) q^n.$$

Proof. By Theorem 3,

$$R_{\nu+1}(1,1,\ldots,1;q) = \frac{1}{(q;q)_\infty}\sum_{n=1}^{\infty}\frac{(-1)^{n-1}(1+q^n)q^{n(3n-1)/2+\nu n}}{(1-q^n)^{2\nu}}$$
$$= \sum_{n=1}^{\infty}\eta_{2\nu}(n)q^n,$$

by Theorem 2. □

Corollary 6. *For $k \geqq 2$,*

$$R_k(x_1, x_2, \ldots, x_k; q)$$
$$= \frac{1}{2(q;q)_\infty}\sum_{n=-\infty}^{\infty}\frac{(-1)^{n-1}(1+q^n)(1-q^n)^2 q^{3n(n-1)/2+kn}}{\prod_{j=1}^{k}\left(1-x_j q^n\right)\left(1-\frac{q^n}{x_j}\right)}.$$

Proof. In the expansion given in Theorem 3, we see that the summand would equal 0 if $n = 0$ and that it is unaltered if n is replaced by $-n$. □

Theorem 7.

$$R_k(x_1, \dots, x_k; q) = \sum_{i=1}^{k} \frac{R_1(x_i; q)}{\prod\limits_{\substack{j=1 \\ j\neq i}}^{k} (x_i - x_j)\left(1 - \frac{1}{x_i x_j}\right)}.$$

Remark. While $R_k(x_1, \dots, x_k; q)$ is well-defined if one sets $x_i = x_j$, we see that on the right-hand side we have term-by-term singularities. In applications where such an identification might be useful, one must modify the partial fraction decomposition employed.

Proof. First we note a partial fraction identity

$$\frac{(1+w)(1-w)^2 w^{k-2}}{\prod\limits_{j=1}^{k} (1 - x_j w)\left(1 - \frac{w}{x_j}\right)}$$

$$= \sum_{i=1}^{k} \frac{(x_i - 1)}{\prod\limits_{\substack{j=1 \\ j\neq i}}^{k} (x_i - x_j)\left(1 - \frac{1}{x_i x_j}\right)} \left(\frac{1}{1 - x_i w} - \frac{x_i^{-1}}{1 - \frac{w}{x_i}}\right).$$

Therefore

$$R_k(x_1, \dots, x_k; q)$$

$$= \frac{1}{2(q;q)_\infty} \sum_{n=-\infty}^{\infty} (-1)^{n-1} q^{n(3n+1)/2} \sum_{i=1}^{k} \frac{(x_i - 1)}{\prod\limits_{\substack{j=1 \\ j\neq i}}^{k} (x_i - x_j)\left(1 - \frac{1}{x_i x_j}\right)}$$

$$\times \left(\frac{1}{1 - x_i q^n} - \frac{x_i^{-1}}{1 - \frac{q^n}{x_i}}\right)$$

$$= \frac{1}{(q;q)_\infty} \sum_{n=-\infty}^{\infty} (-1)^{n-1} q^{n(3n+1)/2} \sum_{i=1}^{k} \frac{(x_i - 1)}{(1 - x_i q^n) \prod\limits_{\substack{i=1 \\ i\neq j}}^{k} (x_i - x_j)\left(1 - \frac{1}{x_i x_j}\right)}$$

$$= \sum_{i=1}^{k} \frac{R_1(x_i; q)}{\prod\limits_{\substack{j=1 \\ j\neq i}}^{k} (x_i - x_j)\left(1 - \frac{1}{x_i x_j}\right)}.$$

□

Corollary 8.

$$R_2(z, z^2; q) = \frac{z^2}{(1-z)(1-z^3)}\big(R_1(z^2; q) - R_1(z; q)\big).$$

Proof. Take $k = 2$, $x_1 = z$, $x_2 = z^2$ in Theorem 7. □

Corollary 9.

$$\begin{aligned} R_3(z, z^2, z^3; q) = {} & \frac{R_1(z; q)}{(z-z^2)(z-z^3)(1-z^{-3})(1-z^{-4})} \\ & + \frac{R_1(z^2; q)}{(z^2-z)(z^2-z^3)(1-z^{-3})(1-z^{-5})} \\ & + \frac{R_1(z^3; q)}{(z^3-z)(z^3-z^2)(1-z^{-4})(1-z^{-5})}. \end{aligned}$$

Proof. Take $k = 3$, $x_1 = z$, $x_2 = z^2$ and $x_3 = z^3$ in Theorem 7. □

3. The Durfee symbol

There are numerous ways of representing a partition of an integer. For example, the partition $5 + 5 + 4 + 4 + 3 + 1$ of 22, has a Ferrers graph [6, §1.3]

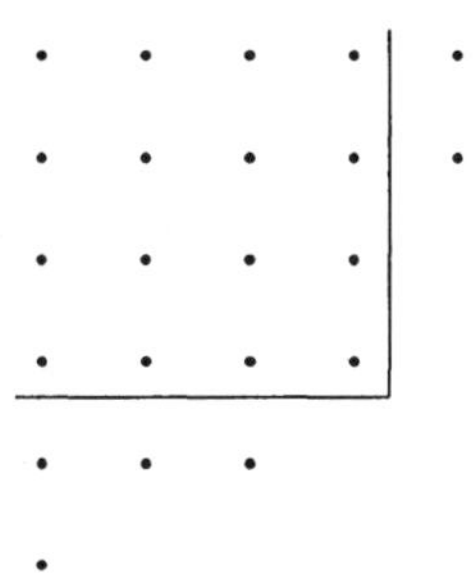

The largest square of nodes in the Ferrers graph is called the Durfee square [6, p. 28].

There is also the Frobenius symbol

$$\begin{pmatrix} 4 & 3 & 1 & 0 \\ 5 & 3 & 2 & 0 \end{pmatrix}$$

where the top line indicates the rows to the right of the main diagonal in the Ferrers graph and the bottom line indicates the columns below, viz.

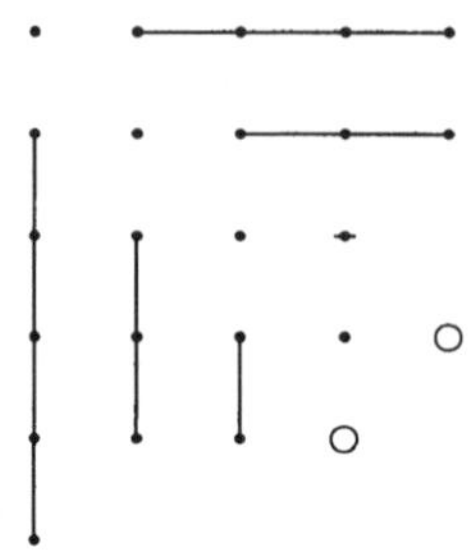

Note that the numbers being partitioned (i.e. 22) is equal to the sum of both rows plus the number of nodes on the main diagonal (i.e. 8 + 10 + 4).

We know from Jacobi's identity [6, p. 28]

$$\begin{aligned}
(3.1)\qquad \sum_{n=0}^{\infty} p(n)q^n &= \frac{1}{(q;q)_\infty} \\
&= \sum_{n=0}^{\infty} \frac{q^{n^2}}{(q;q)_n^2} \\
&= \sum_{n=0}^{\infty} q^n \cdot \frac{q^{\frac{n(n-1)}{2}}}{(q;q)_n} \cdot \frac{q^{\frac{n(n-1)}{2}}}{(q;q)_n}
\end{aligned}$$

that the Frobenius symbols are, in fact, naturally generated from the final entry in (3.1). However examination of the penultimate series in (3.1) suggests a new symbol for representing partitions. Namely

$$\begin{pmatrix} 2 & \\ 3 & 1 \end{pmatrix}_4,$$

where the top row consists of the columns to the right of the Durfee square; the bottom row consists of the rows below the Durfee square

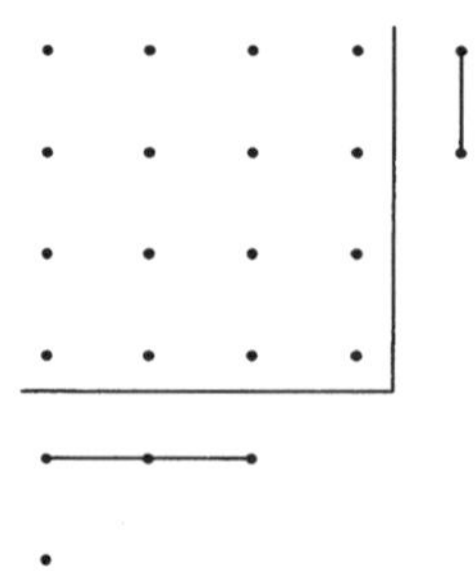

and the subscript 4 denotes the side length of the Durfee square. In this case, the number being partitioned is equal to the sum of the rows of the symbol plus the number of nodes in the Durfee square (i.e. $2 + (3 + 1) + 4^2 = 22$).

Sylvester's analysis of (3.1) combinatorially introduced the name **Durfee square**, and consequently we naturally name our new symbol the **Durfee symbol**.

For example, when $n = 5$

Partitions	Ferrers graph	Frobenius symbol	Durfee symbol
5	· · · · ·	$\begin{pmatrix} 4 \\ 0 \end{pmatrix}$	$\begin{pmatrix} 1 & 1 & 1 & 1 \\ & & & \end{pmatrix}_1$
$4+1$	· · · · ·	$\begin{pmatrix} 3 \\ 1 \end{pmatrix}$	$\begin{pmatrix} 1 & 1 & 1 \\ 1 & & \end{pmatrix}_1$
$3+2$	· · · · ·	$\begin{pmatrix} 2 & 0 \\ 1 & 0 \end{pmatrix}$	$\begin{pmatrix} 1 & \\ & \end{pmatrix}_2$
$3+1+1$	· · · · ·	$\begin{pmatrix} 2 \\ 2 \end{pmatrix}$	$\begin{pmatrix} 1 & 1 \\ 1 & 1 \end{pmatrix}_1$
$2+2+1$	· · · · ·	$\begin{pmatrix} 1 & 0 \\ 2 & 0 \end{pmatrix}$	$\begin{pmatrix} & \\ 1 & \end{pmatrix}_2$
$2+1+1+1$	· · · · ·	$\begin{pmatrix} 1 \\ 3 \end{pmatrix}$	$\begin{pmatrix} 1 & & \\ 1 & 1 & 1 \end{pmatrix}_1$
$1+1+1+1+1$	· · · · ·	$\begin{pmatrix} 0 \\ 4 \end{pmatrix}$	$\begin{pmatrix} & & & \\ 1 & 1 & 1 & 1 \end{pmatrix}_1$

Note that the rank of a partition is given by the number of entries in the top row of the Durfee symbol minus the number of entries in the bottom row.

4. The definition of marked Durfee symbols and their ranks

There have been many partition studies in which several copies of the integers have been used as parts of the partitions, e.g. [1], [3], [4]. To define "k-marked" Durfee symbols, we shall require k copies of the integers designated by $\{1_1, 2_1, 3_1, \dots\}$, $\{1_2, 2_2, 3_2, \dots\}, \dots, \{1_k, 2_k, 3_k, \dots\}$. We now form Durfee symbols as in Sect. 3, but now we use the k copies of the integers for parts in both rows. In addition, we require that: (1) the sequence of parts and the sequence of subscripts in each row be non-increasing; (2) each of the subscripts $1, 2, \dots, k-1$ occur at least once in the top row, and finally (3) if $M_1, N_2, \dots, V_{k-2}, W_{k-1}$ are the largest parts with their respective subscripts in the top row, then all parts in the bottom row with subscript 1 lie in $[1, M]$, with subscript 2 lie in $[M, N], \dots$, with subscript $k-1$ lie in $[V, W]$ and with subscript k lie in $[W, S]$ where S is the side of the Durfee square.

As an example, we exhibit the 21 2-marked Durfee symbols associated with the partitions of 5.

$$\begin{pmatrix} 1_1 & 1_1 & 1_1 & 1_1 \\ & & & \end{pmatrix}_1 \quad \begin{pmatrix} 1_2 & 1_1 & 1_1 \\ 1_2 & & \end{pmatrix}_1 \quad \begin{pmatrix} 1_2 & 1_1 \\ 1_2 & 1_1 \end{pmatrix}_1$$

$$\begin{pmatrix} 1_2 & 1_1 & 1_1 & 1_1 \\ & & & \end{pmatrix}_1 \quad \begin{pmatrix} 1_2 & 1_2 & 1_1 \\ 1_1 & & \end{pmatrix}_1 \quad \begin{pmatrix} 1_1 & 1_1 \\ 1_2 & 1_2 \end{pmatrix}_1$$

$$\begin{pmatrix} 1_2 & 1_2 & 1_1 & 1_1 \\ & & & \end{pmatrix}_1 \quad \begin{pmatrix} 1_2 & 1_2 & 1_1 \\ 1_2 & & \end{pmatrix}_1 \quad \begin{pmatrix} 1_2 & 1_1 \\ 1_2 & 1_2 \end{pmatrix}_1$$

$$\begin{pmatrix} 1_2 & 1_2 & 1_2 & 1_1 \\ & & & \end{pmatrix}_1 \quad \begin{pmatrix} 1_1 \\ \ \end{pmatrix}_2 \quad \begin{pmatrix} 1_1 & & \\ 1_1 & 1_1 & 1_1 \end{pmatrix}_1$$

$$\begin{pmatrix} 1_1 & 1_1 & 1_1 \\ 1_1 & & \end{pmatrix}_1 \quad \begin{pmatrix} 1_1 & 1_1 \\ 1_1 & 1_1 \end{pmatrix}_1 \quad \begin{pmatrix} 1_1 & & \\ 1_2 & 1_1 & 1_1 \end{pmatrix}_1$$

$$\begin{pmatrix} 1_1 & 1_1 & 1_1 \\ 1_2 & & \end{pmatrix}_1 \quad \begin{pmatrix} 1_2 & 1_1 \\ 1_1 & 1_1 \end{pmatrix}_1 \quad \begin{pmatrix} 1_1 & & \\ 1_2 & 1_2 & 1_1 \end{pmatrix}_1$$

$$\begin{pmatrix} 1_2 & 1_1 & 1_1 \\ 1_1 & & \end{pmatrix}_1 \quad \begin{pmatrix} 1_1 & 1_1 \\ 1_2 & 1_1 \end{pmatrix}_1 \quad \begin{pmatrix} 1_1 & & \\ 1_2 & 1_2 & 1_2 \end{pmatrix}_1$$

Definition. $\mathcal{D}_k(n)$ denotes the number of k-marked Durfee symbols arising from partitions of n.

In the next section, we shall show that $\mathcal{D}_{k+1}(n) = \eta_{2k}(n)$. This will form the basis of our subsequent investigations.

To conclude this section we introduce k parameters associated with each k-marked Durfee symbol.

Definition. Suppose δ is a k-marked Durfee symbol. We denote the number of entries in the top (resp. bottom) row of δ with subscript i by $\tau_i(\delta)$

(resp. $\beta_i(\delta)$). We define $\rho_i(\delta)$, **the ith rank of δ**, by

$$\rho_i(\delta) = \begin{cases} \tau_i(\delta) - \beta_i(\delta) - 1 & \text{for } 1 \leqq i < k \\ \tau_i(\delta) - \beta_i(\delta) & \text{for } i = k. \end{cases} \tag{4.1}$$

Definition. $\mathcal{D}_k(m_1, m_2, \ldots, m_k; n)$ denotes the number of k-marked Durfee symbols arising from partitions of n with ith rank equal to m_i.

Note that if $k = 1$, then $\rho_1(\delta)$ is just Dyson's rank and $\mathcal{D}_1(m_1; n) = N(m, n)$.

5. Generating functions (1)

Theorem 10. *For $k \geqq 1$*

$$\sum_{n_1, n_2, \ldots, n_k = -\infty}^{\infty} \sum_{n \geq 0} \mathcal{D}_k(n_1, n_2, \ldots, n_k; n) x_1^{n_1} x_2^{n_2} \cdots x_k^{n_k} q^n$$
$$= R_k(x_1, x_2, \ldots, x_k; q).$$

Proof. If $k = 1$, this is just (1.7); so in the following we assume $k \geqq 2$. Let us consider a general Durfee symbol wherein the largest entry with subscript 1 on the top row is m_1, the largest entry on the top row with subscript 2 is $m_1 + m_2, \ldots$, the largest entry on the top row with subscript $k - 1$ is $m_1 + m_2 + \cdots + m_{k-1}$. Finally the underlying Durfee square has side $m_1 + m_2 + \cdots + m_k$.

Then parts with subscript 1 on both rows are generated by

$$\frac{q^{m_1}}{(x_1 q; q)_{m_1} \left(\frac{q}{x_1}; q\right)_{m_1}}.$$

Parts with subscript 2 on both rows are generated by

$$\frac{q^{m_1 + m_2}}{\left(x_2 q^{m_1}; q\right)_{m_2+1} \left(\frac{q^{m_1}}{x_2}; q\right)_{m_2+1}}.$$

Parts with subscript $k - 1$ on both rows are generated by

$$\frac{q^{m_1 + \cdots + m_{k-1}}}{\left(x_{k-1} q^{m_1 + \cdots + m_{k-2}}; q\right)_{m_{k-1}+1} \left(q^{m_1 + \cdots + m_{k-2}} / x_{k-1}; q\right)_{m_{k-1}+1}}.$$

In light of the fact that there are not necessarily any parts with subscript k, we see that these parts are generated by

$$\frac{1}{\left(x_k q^{m_1 + \cdots + m_{k-1}}; q\right)_{m_k+1} \left(q^{m_1 + \cdots + m_{k-1}} / x_k; q\right)_{m_k+1}},$$

and finally the Durfee square is generated by

$$q^{(m_1+m_2+\cdots+m_k)^2}.$$

Multiplying all these expressions together and summing over all m_i's with $m_1 > 0$, $m_i \geqq 0$ for $2 \leqq i \leqq k$, we see that the required generating function is $R_k(x_1, \ldots, x_k; q)$ by (2.3). □

Corollary 11. *For* $k \geqq 1$

$$\sum_{n_1,\ldots,n_k=-\infty}^{\infty} \sum_{n=0}^{\infty} \mathcal{D}_k(n_1, n_2, \ldots, n_k; n) x_1^{n_1} x_2^{n_2} \cdots x_k^{n_k} q^n$$
$$= \frac{1}{(q;q)_\infty} \sum_{n=1}^{\infty} \frac{(-1)^{n-1}(1+q^n)(1-q^n)^2 q^{3n(n-1)/2+kn}}{\prod_{j=1}^{k} \left(1 - x_j q^n\right)\left(1 - \frac{q^n}{x_j}\right)}.$$

Proof. This follows immediately by combining Theorem 10 with Theorem 3. □

Corollary 12. $\mathcal{D}_k(n_1, n_2, \ldots, n_k; n)$ *is symmetric in* $n_1, n_2, \ldots, n_k$.

Proof. The generating function as represented in Corollary 11 is clearly symmetric in $x_1, x_2, \ldots, x_k$. □

Corollary 13. *For* $k \geqq 1$,

$$\mathcal{D}_{k+1}(n) = \eta_{2k}(n).$$

Remark. In words, the number of $(k + 1)$-marked Durfee symbols related to partitions of n equals the symmetrized $(2k)$-th moment function at n.

Proof. By Theorem 10,

$$\sum_{n \geqq 0} \mathcal{D}_{k+1}(n) q^n = R_{k+1}(1, 1, \ldots 1; q) = \sum_{n=1}^{\infty} \eta_{2k}(n) q^n,$$

by Corollary 5. □

6. Self-conjugate k-marked Durfee symbols

Sylvester and his student Durfee [29], [30] were the first to study self-conjugate partitions. This led to the analysis of the largest square within the

Ferrers graph (named the Durfee square by Sylvester), indeed their analysis of

$$\sum_{n=0}^{\infty} \frac{q^{n^2}}{(q^2;q^2)_n} = (-q;q^2)_\infty \tag{6.1}$$

led to the now well-known result [6, p. 14, Ex. 8] that the number of self-conjugate partitions of n equals the number of partitions of n into distinct odd parts.

Definition. A self-conjugate k-marked Durfee symbol is one in which the two rows are identical, and $SC_k(n)$ denotes the number of self-conjugate k-Durfee symbols related to n. Finally,

$$\mathcal{SC}_k(q) = \sum_{n \geqq 0} SC_k(n) q^n.$$

Theorem 14. *The number of self-conjugate k-marked Durfee symbols representing n equals the number of partitions of n into distinct unmarked odds and with* $(k-1)$ *different (albeit possibly repeated)* $(k-1)$*-marked evens each* $\leqq$ *twice the number of odd parts.*

Proof. Let $\mathcal{O}_k(n)$ denote the cardinality of the set of partitions mentioned in the theorem. It is straightforward to provide a combinatorial proof of this result patterned on the Sylvester–Durfee argument [29]. However, to prepare for subsequent results we provide an analytic proof.

By following exactly the argument in Theorem 10, we see that

(6.2)

$$\begin{aligned}
\mathcal{SC}_k(q) &= \sum_{\substack{m_1>0 \\ m_2,\dots,m_k \geqq 0}} \frac{q^{(m_1+m_2+\cdots+m_k)^2+2(m_1+\cdots+m_{k-1})+2(m_1+\cdots+m_{k-1})+\cdots+2m_1}}{(q^2;q^2)_{m_1}(q^{2m_1};q^2)_{m_2+1}(q^{2(m_1+m_2)};q^2)_{m_3+1}} \\
&\qquad \times \frac{1}{\cdots(q^{2(m_1+\cdots+m_{k-1})};q^2)_{m_k+1}} \\
&= \sum_{M_k \geqq 1} \frac{q^{M_k^2}}{(q^2;q^2)_{M_k}} \\
&\qquad \times \sum_{1 \leqq M_1 \leqq M_2 \leqq \cdots \leqq M_k} \frac{q^{2(M_1+M_2+\cdots+M_{k-1})}}{(1-q^{2M_1})(1-q^{2M_2})\cdots(1-q^{2M_{k-1}})} \\
&= \sum_{n \geqq 0} \mathcal{O}_k(n) q^n,
\end{aligned}$$

where in the penultimate line we have set $M_i = m_1 + m_2 + \cdots + m_i$, and to obtain the final line we have invoked the Sylvester–Durfee observation that $q^{m^2}/(q^2;q^2)_m$ is the generating function for partitions with exactly m distinct odd parts. □

Theorem 15. *For $k \geqq 2$,*

$$\frac{2\mathcal{S}C_k(-q)}{(q;q^2)_\infty} = -\sum_{m=1}^{\infty} \frac{q^{m(m-1)/2+(k-1)m}(1+(-1)^m)}{(q;q)_m(1-q^m)^{k-1}}.$$

Proof. To accomplish this result we must invoke Dilcher's wonderful identity [15, p. 91, Theorem 4]

$$\text{(6.3)}\qquad \sum_{m=1}^{n} \frac{(-1)^{m-1}q^{m(m+1)/2+(k-1)m}(q;q)_n}{(1-q^m)^k(q;q)_m(q;q)_{n-m}}$$
$$= \sum_{j_1=1}^{n} \frac{q^{j_1}}{1-q^{j_1}} \sum_{j_2=1}^{j_1} \frac{q^{j_2}}{1-q^{j_2}} \cdots \sum_{j_k=1}^{j_{k-1}} \frac{q^{j_k}}{1-q^{j_k}}$$

In (6.3), we replace k by $k-1$, n by M_k and q by q^2. The result is the inner sum in the penultimate line of the proof of Theorem 14.

Hence

$$\begin{aligned}
\mathcal{S}C_k(q) &= \sum_{M_k \geqq 1} \frac{q^{M_k^2}}{(q^2;q^2)_{M_k}} \sum_{m=1}^{M_k} \frac{(-1)^{m-1}q^{m^2+m+2(k-2)m}(q^2;q^2)_{M_k}}{(1-q^{2m})^{k-1}(q^2;q^2)_m(q^2;q^2)_{M_k-m}} \\
&= \sum_{m=1}^{\infty} \frac{(-1)^{m-1}q^{m^2+m+2km-4m}}{(1-q^{2m})^{k-1}(q^2;q^2)_m} \sum_{M_k=0}^{\infty} \frac{q^{(M_k+m)^2}}{(q^2;q^2)_{M_k}} \\
&= \sum_{m=1}^{\infty} \frac{(-1)^{m-1}q^{2m^2+(2k-3)m}}{(1-q^{2m})^{k-1}(q^2;q^2)_m}(-q^{2m+1};q^2)_\infty \\
&= (-q;q^2)_\infty \sum_{m=1}^{\infty} \frac{(-1)^{m-1}q^{2m^2+(2k-3)m}}{(q^2;q^2)_m(-q;q^2)_m(1-q^{2m})^{k-1}}.
\end{aligned}$$

Therefore

$$\begin{aligned}
\frac{\mathcal{S}C_k(-q)}{(q;q^2)_\infty} &= -\sum_{m=1}^{\infty} \frac{q^{2m^2+(2k-3)m}}{(q;q)_{2m}(1-q^{2m})^{k-1}} \\
&= -\sum_{m=1}^{\infty} \frac{q^{m(m-1)/2+(k-1)m}}{(q;q)_m(1-q^m)^{k-1}} \frac{1}{2}(1+(-1)^m),
\end{aligned}$$

and this assertion is clearly equivalent to the desired identity. □

Corollary 16.

$$\frac{2\mathcal{S}C_2(-q)}{(q;q^2)_\infty} = -\sum_{m=1}^{\infty} \frac{q^{m(m+1)/2}}{(q;q)_m(1-q^m)} + \sum_{m=1}^{\infty} \frac{q^m}{1-q^m}.$$

Proof. In light of Theorem 15, we need only show that

$$\sum_{m=1}^{\infty} \frac{(-1)^{m-1} q^{m(m+1)/2}}{(q;q)_m(1-q^m)} = \sum_{m=1}^{\infty} \frac{q^m}{1-q^m}.$$

and this result was proved by both Kluyver [23] and Uchimura [31] (as noted by Dilcher [15]). □

In fact, it is known ([7], cf. [8], [15]) that

$$\sum_{m=1}^{\infty} \frac{(-1)^{m-1} q^{m(m+1)/2}}{(q;q)_m(1-q^m)^{k-1}}$$

is representable as an algebraic expression of divisor sums; however, the expression is sufficiently complicated that we have only exhibited the $k = 2$ case.

7. The full rank and congruences for $\mathcal{D}_k(n)$

In this section, we shall establish the modulus 5 and 7 congruences given in (1.14), (1.15) and (1.16). Indeed we shall prove much more including combinatorial, rank-theoretic explanations of a number of congruences.

Definition. For each k-marked Durfee symbol δ, we define the **full rank** $FR(\delta)$ by

$$FR(\delta) = \rho_1(\delta) + 2\rho_2(\delta) + 3\rho_3(\delta) + \cdots + k\rho_k(\delta),$$

and we let $NF_k(m,n)$ denote the number of k-marked Durfee symbols related to n with full rank m. In addition, $NF_k(m,Q,n)$ denotes the number of k-marked Durfee symbols related to n with full rank $\equiv m \pmod{Q}$.

Theorem 17. *For $n \geqq 0$ and $a = 2, 3$ or 5*

$$\begin{aligned} NF_2(1,5,5n+a) &= NF_2(2,5,5n+a) = NF_2(3,5,5n+a) \\ &= NF_2(4,5,5n+a), \end{aligned}$$

and for $a = 1$ or 4

$$\begin{aligned} NF_2(0,5,5n+a) &= NF_2(1,5,5n+a) = NF_2(2,5,5n+a) \\ &= NF_2(3,5,5n+a) = NF_2(4,5,5n+a). \end{aligned}$$

Proof. Clearly by Corollary 8

$$\sum_{n\geqq 1}\sum_{m=-\infty}^{\infty} NF_2(m,n)z^m q^n = R_2(z,z^2;q)$$

$$= \frac{1}{(z-z^2)(1-z^{-3})}\big(R_1(z;q)-R_1(z^2;q)\big).$$

Building on the work of Atkin and Swinnerton-Dyer [11], Garvan [19] established the following formula from Ramaujan's Lost Notebook [28, p. 20]:

$$\begin{aligned}(7.1)\quad R_1(\zeta;q) &= A(q^5)+(\zeta+\zeta^{-1}-2)\phi(q^5)\\ &\quad + qB(q^5)+(\zeta+\zeta^{-1})q^2C(q^5)\\ &\quad -(\zeta+\zeta^{-1})q^3\Big(D(q^5)-(\zeta^2+\zeta^{-2}-2)\frac{\psi(q^5)}{q^5}\Big),\end{aligned}$$

where ζ is any primitive fifth root of unity. Garvan, following Ramanujan, provided and proved explicit beautiful representations for each of $A(q)$, $B(q)$, $C(q)$, $D(q)$, $\phi(q)$ and $\psi(q)/q$. For our purposes in this proof we need only know that each of these six functions is an analytic function inside $|q|<1$ with integer coefficients in their power series. We use Corollary 8 here which is valid term by term as long as $|q|<|z|<|q|^{-1}$ and z is not set equal to a cube root of unity.

Therefore,

$$\sum_{n\geqq 1}\sum_{m=-\infty}^{\infty} NF_2(m,n)\zeta^m q^n = R_2(\zeta,\zeta^2;q)$$

$$= \frac{1}{(\zeta-\zeta^2)(1-\zeta^{-3})}\big(R_1(\zeta;q)-R_1(\zeta^2;q)\big)$$

$$= \phi(q^5)+q^2C(q^5)-q^3\Big(D(q^5)+2\frac{\psi(q^5)}{q^5}\Big).$$

Note first that there are no powers of q on the right-hand side that are congruent to either 1 or 4 modulo 5, and second that if q is real, the right-hand side is real.

Collecting terms on the left-hand side that make up the coefficient of q^{5n+a} $(1\leqq a\leqq 5)$, we deduce that

$$\sum_{j=0}^{4} NF_2(j,5,5n+a)\zeta^j = \begin{cases}0 & \text{if } a=1 \text{ or } 4\\ \text{some real number} & \text{if } a=2,3, \text{ or } 5.\end{cases}$$

But ζ is one of the roots of the irreducible polynomial $1+z+z^2+z^3+z^4=0$. Hence the only way the above sum can be true is if the assertions in the Theorem are true; otherwise we would have produced a new irreducible polynomial with ζ as a root, an impossibility. □

There appear to be many more such results. We shall content ourselves with two of the most appealing.

Theorem 18. *For $n \geqq 0$ and $0 \leqq i, j \leqq 6$,*

$$NF_2(i, 7, 7n+5) = NF_2(j, 7, 7n+5),$$

and

$$NF_2(i, 7, 7n+1) = NF_2(j, 7, 7n+1).$$

Proof. We note that if ζ is any primitive seventh root of unity, then

$$\begin{aligned}
&\sum_{j=0}^{6} \zeta^j NF_2(j, 7, 7n+5)\\
&\quad = \text{coeff. of } q^{7n+5} \text{ in } R_2(\zeta, \zeta^2; q)\\
&\quad = \text{coeff. of } q^{7n+5} \text{ in } \frac{1}{(\zeta-\zeta^2)(1-\zeta^{-3})}\big(R_1(\zeta; q) - R_1(\zeta^2; q)\big)\\
&\quad = \frac{1}{(\zeta-\zeta^2)(1-\zeta^{-3})}\left(\sum_{j=0}^{6}(\zeta^j - \zeta^{2j})N(j, 7, 7n+5)\right)\\
&\quad = 0,
\end{aligned}$$

where the last line follows from the fact that the $N(j, 7, 7n+5)$ are identical (by [11, Eq. (2.10) and the equation preceeding (2.2)]) and ζ and ζ^2 are both primitive seventh roots of unity. Finally we see that the $NF_2(j, 7, 7n+5)$ must all be identical otherwise we would contradict the fact that the minimal polynomial for ζ is $1 + z + z^2 + z^3 + z^4 + z^5 + z^6$.

Next

$$\begin{aligned}
&\sum_{j=0}^{6} \zeta^j NF_2(j, 7, 7n+1)\\
&\quad = \text{coeff. of } q^{7n+1} \text{ in } R_2(\zeta, \zeta^2; q)\\
&\quad = \text{coeff. of } q^{7n+1} \text{ in } \frac{1}{(\zeta-\zeta^2)(1-\zeta^{-3})}\big(R_1(\zeta; q) - R_1(\zeta^2; q)\big)\\
&\quad = \frac{1}{(\zeta-\zeta^2)(1-\zeta^{-3})}\sum_{j=0}^{6}(\zeta^j - \zeta^{2j})N(j, 7, 7n+1)\\
&\quad = \frac{1}{(\zeta-\zeta^2)(1-\zeta^{-3})}\sum_{j=1}^{6}(\zeta^j - \zeta^{2j})N(j, 7, 7n+1)
\end{aligned}$$

$$= \frac{N(1,7,7n+1)}{(\zeta-\zeta^2)(1-\zeta^{-3})}\sum_{j=1}^{6}(\zeta^j-\zeta^{2j})$$

(by [11, Eq. (2.6) and the equation preceeding (2.2)])

$$= \frac{N(1,7,7n+1)}{(\zeta-\zeta^2)(1-\zeta^{-3})}\sum_{j=0}^{6}(\zeta^j-\zeta^{2j}) = 0,$$

and by the same argument used for $NF_2(5,7,m+5)$ we see that the $NF_2(j,7,7n+1)$ must all be identical. □

Next we provide a similar result for $k = 3$.

Theorem 19. *For* $n \geqq 0$, $0 \leqq i, j \leqq 6$, $a = 0, 1, 5$

$$NF_3(i,7,7n+a) = NF_3(j,7,7n+a).$$

Proof. For simplicity of notation we define ζ to be any primitive 7th root of unity, and we let

$$P(\zeta) = \frac{1}{\zeta^3(1-\zeta)(1-\zeta^2)(1-\zeta^3)(1-\zeta^4)}.$$

If we now take $k = 3$, and $z = \zeta$ in Theorem 7 we find that

$$\begin{aligned} &R_3(\zeta,\zeta^2,\zeta^3;q) \\ &\quad = P(\zeta)\big(\zeta R_1(\zeta;q) - (1+\zeta+\zeta^2)R_1(\zeta^2;q) + (1+\zeta^2)R_1(\zeta^3;q)\big). \end{aligned}$$

Consequently

$$\begin{aligned} &\sum_{j=0}^{6}\zeta^j NF_3(j,7,7n+a) \\ &\quad = \text{coeff. of } q^{7n+a} \text{ in } R_3(\zeta,\zeta^2,\zeta^3;q) \\ &\quad = P(\zeta)\sum_{j=0}^{6}(\zeta^{j+1}-\zeta^{2j}-\zeta^{2j+1}-\zeta^{2j+2}+\zeta^{3j}+\zeta^{3j+2})N(j,7,7n+a). \end{aligned}$$

Now if $a = 5$, then as in Theorem 18, we note that all the $N(j,7,7n+5)$ are identical and immediately the sums all are zero, which leads to the conclusion that the $NF_3(j,7,7n+5)$ are all identical.

If $a = 1$, then things are a bit trickier but essentially the same as in the proof of Theorem 18. First we note that

$$\zeta - 1 - \zeta - \zeta^2 + 1 + \zeta^2 = 0; \tag{7.2}$$

so

$$\begin{aligned}
&\sum_{j=0}^{6} \zeta^j NF_3(j,7,7n+1)\\
&\quad = \text{coeff. of } q^{7n+1} \text{ in } R_3(\zeta,\zeta^2,\zeta^3;q)\\
&\quad = P(\zeta)\sum_{j=0}^{6}(\zeta^{j+1}-\zeta^{2j}-\zeta^{2j+1}-\zeta^{2j+2}+\zeta^{3j}+\zeta^{3j+2})N(j,7,7n+1)\\
&\quad = P(\zeta)\sum_{j=1}^{6}(\zeta^{j+1}-\zeta^{2j}-\zeta^{2j+1}-\zeta^{2j+2}+\zeta^{3j}+\zeta^{3j+2})N(j,7,7n+1)\\
&\quad = P(\zeta)N(1,7,7n+1)\sum_{j=1}^{6}(\zeta^{j+1}-\zeta^{2j}-\zeta^{2j+1}-\zeta^{2j+2}+\zeta^{3j}+\zeta^{3j+2})\\
&\quad = P(\zeta)N(1,7,7n+1)\sum_{j=0}^{6}(\zeta^{j+1}-\zeta^{2j}-\zeta^{2j+1}-\zeta^{2j+2}+\zeta^{3j}+\zeta^{3j+2})\\
&\quad = 0,
\end{aligned}$$

and as before all the $NF_3(j,7,7n+1)$ are identical.

The most complicated case is $a = 0$.

$$\begin{aligned}
&\sum_{j=0}^{6} \zeta^j NF_3(j,7,7n)\\
&\quad = \text{coeff. of } q^{7n} \text{ in } R_3(\zeta,\zeta^2,\zeta^3;q)\\
&\quad = P(\zeta)\sum_{j=0}^{6}(\zeta^{j+1}-\zeta^{2j}-\zeta^{2j+1}-\zeta^{2j+2}+\zeta^{3j}+\zeta^{3j+2})N(j,7,7n)\\
&\quad = P(\zeta)\Big((\zeta^5-\zeta^4)(N(1,7,7n)-N(6,7,7n))\\
&\qquad\qquad + N(2,7,7n)\sum_{j=2}^{5}(\zeta^{j+1}-\zeta^{2j}-\zeta^{2j+1}-\zeta^{2j+2}+\zeta^{3j}+\zeta^{3j+2})\Big)\\
&\qquad \text{(by (7.2) and [11, p. 85, Eq. (2.5) and the equation preceeding (2.2)])}\\
&\quad = 0
\end{aligned}$$

because $N(1,7,7n) = N(6,7,7n)$ by [11, p. 85] and it is purely a matter of calculation to determine that the remaining $\sum_{j=2}^{5}$ vanishes. Now, as before the third and final assertion of our theorem follows. □

Corollary 20. *The congruences (1.14), (1.15) and (1.16) are valid.*

Proof. As noted earlier $N_2(n) = 2\eta_2(n)$. If $n \equiv \pm 1 \pmod 5$, then by Theorem 17

$$\eta_2(n) = 5NF_2(0, 5; n);$$

if $n \equiv 1$ or $5 \pmod 7$, then by Theorem 18

$$\eta_2(n) = 7NF_2(0, 7; n).$$

Finally $N_4(n) = 24\eta_4(n) + 2\eta_2(n)$, and

$$\eta_4(n) = 7NF_3(0, 7; n)$$

if $n \equiv 1$ or $5 \pmod 7$. □

8. $\omega(q)$ and odd Durfee symbols

In [32], G.N. Watson defined one of the third order mock theta functions, $\omega(q)$, by

$$\begin{aligned} q\omega(q) &:= \sum_{n=0}^{\infty} \frac{q^{2n(n+1)+1}}{(q;q^2)_{n+1}^2} \\ &= q + 2q^2 + 3q^3 + 4q^4 + 6q^5 + 8q^6 \\ &\quad + 10q^7 + 14q^8 + \cdots . \end{aligned} \tag{8.1}$$

For our purposes, the most convenient partition-theoretic interpretation of $\omega(q)$ is given by considering partitions wherein at least all but one instance of the largest part is one of a pair of consecutive non-negative integers. For example, there are 8 such partitions of 6: 6, $5+(1+0)$, $4+(1+0)+(1+0)$, $3+(2+1)$, $3+(1+0)+(1+0)+(1+0)$, $2+(2+1)+(1+0)$, $2+(1+0)+(1+0)+(1+0)+(1+0)$, $1+(1+0)+(1+0)+(1+0)+(1+0)+(1+0)$.

The fact that $q\omega(q)$ is the generating function for such partitions follows immediately from inspection of Fine's formula for $\omega(q)$ [17, Eq. (26.84), p. 61]

$$\begin{aligned} q\omega(q) &= \sum_{n=0}^{\infty} \frac{q^{n+1}}{(q;q^2)_{n+1}} \\ &= \sum_{n=1}^{\infty} \frac{q^n}{(1-q^{1+0})(1-q^{2+1})(1-q^{3+2})\cdots(1-q^{n+(n-1)})}. \end{aligned} \tag{8.2}$$

It should be mentioned that S. Garthwaite [18] has recently completed a study of the coefficients of $\omega(q)$ that is analogous to the work of Bringmann and Ono [13] for $f(q) = R_1(-1; q)$.

Our study turns out to be most interested in how (8.1) can be seen as naturally generating such partitions. The way to see best what is happening is to invoke P.A. MacMahon's modular partitions [24] with modulus 2 where the dots in the Ferrers graph are replaced by 2's and the graph is bordered by 1's as follows. Let us consider $8 + (4+3) + (4+3) + (3+2) + (2+1) + (2+1) + (1+0) + (1+0) + (1+0)$

$$\begin{array}{ccc|cccccc}
1 & 1 & 1 & 1 & 1 & 1 & 1 & 1 & 1 \\
1 & 2 & 2 & 2 & & & & & \\
1 & 2 & 2 & 2 & & & & & \\ \hline
1 & 2 & 2 & & & & & & \\
1 & 2 & & & & & & & \\
1 & 2 & & & & & & & \\
1 & & & & & & & & \\
1 & & & & & & & & \\
1 & & & & & & & &
\end{array}$$

The top row of 1's represents 8; the second row is $(4+3)$ as is the third row; the fourth row is $(3+2)$ followed by two rows each representing $(2+1)$ and the final three singleton rows representing $(1+0)$.

As before, we isolate now the Durfee square of 2's indicated by the dotted lines. In general, a Durfee square of size n will contribute $2n^2+2n+1$ to the partition (where the $2n+1$ comes from the bordering 1's), the rows below the Durfee square yield partitions using consecutive integers the largest of which is $((n+1)+n)$. Equivalently the rows below the Durfee square can be interpreted as partitions with odd parts each $\leqq 2n+1$. Finally if we read the columns to the right of the Durfee square we also see an arbitrary partition with odd parts each $\leqq 2n+1$. Consequently the generating function for these modified 2 modular partitions with a Durfee square of side n is given by

$$\frac{q^{2n^2+2n+1}}{(q;q^2)_{n+1}^2},$$

and summing over all n we arrive at (8.1) for the generating function.

Now it becomes quite easy to define a modified Durfee symbol.

Definition. Each partition into pairs of consecutive integers (excepting one instance of the largest part) has associated with it the **odd Durfee symbol**

$$\begin{pmatrix} a_1 & a_2 & \cdots & a_i \\ b_1 & b_2 & \cdots & b_j \end{pmatrix}_n$$

wherein the a's and b's are all odd numbers each $\leqq 2n+1$ and the number being partitioned is $2n^2+2n+1+\sum_{h+1}^{i} a_h + \sum_{k=1}^{j} b_k$.

Thus in the example we considered originally, the related odd Durfee symbol is

$$\begin{pmatrix} 5 & 1 & 1 & 1 & 1 & \\ 5 & 3 & 3 & 1 & 1 & 1 \end{pmatrix}_2$$

Following Dyson's lead as described in the introduction, we now define the **odd rank** of an odd Durfee symbol to be the number of entries in the top row minus the number of entries in the bottom row.

In analogy with ordinary Durfee symbols, we let $N^0(m,n)$ denote the number of partitions related to an odd Durfee symbol with odd rank m, and similar to the ordinary rank

$$R_1^0(z;q) := \sum_{n\geqq 1} \sum_{m=-\infty}^{\infty} N^0(m,n) z^m q^n \tag{8.3}$$

$$= \sum_{n\geqq 0} \frac{q^{2n(n+1)+1}}{(zq;q^2)_{n+1}(z^{-1}q;q^2)_{n+1}}. \tag{8.4}$$

Watson's first identity on page 66 of [32] is equivalent to the following

$$R_1^0(z;q) = \frac{1}{(q^2;q^2)_\infty} \sum_{n=-\infty}^{\infty} \frac{(-1)^n q^{3n^2+3n+1}}{1-zq^{2n+1}}. \tag{8.5}$$

We shall not devote a separate section to **k-marked odd Durfee symbols** because their definition is **almost** identical to that of the ordinary k-marked Durfee symbols. The only change is that: (1) all the entries in the symbol are odd numbers, (2) the subscripting rules are precisely the same, and instead of adding n^2 to the sum of the entries (as is done for ordinary Durfee symbols) when the Durfee square is of side n, we **now** add $2n^2+2n+1$ because of the modified Durfee square configuration

$$\begin{array}{ccccc} 1 & 1 & 1 & \cdots & 1 \\ 1 & 2 & 2 & \cdots & 2 \\ 1 & 2 & 2 & \cdots & 2 \\ \vdots & \vdots & \vdots & & \\ 1 & 2 & 2 & \cdots & 2 \\ n & & & & \end{array}$$

We shall maintain the conventions of the prior notation and shall add a superscript "0" to indicate that we are treating the related object connected with odd k-marked Durfee symbols.

9. Moments of odd ranks

The study of the moments of odd ranks is more straightforward than the results found in Sect. 2.

Definition.

$$\eta_k^0(n) = \sum_{m=-\infty}^{\infty} \binom{m + \lfloor \frac{k}{2} \rfloor}{k} N^0(m, n).$$

The fact that $N^0(m, n) = N^0(-m, n)$ (just interchange the rows of the symbol) means that

$$\eta_{2m+1}^0(n) = 0,$$

and the proof is exactly the same as that of Theorem 1.

Theorem 21.

$$\sum_{n=1}^{\infty} \eta_{2\nu}^0(n) q^n = \frac{1}{(q^2; q^2)_\infty} \sum_{n=-\infty}^{\infty} \frac{(-1)^n q^{3n^2+(2\nu+3)n+\nu+1}}{(1 - q^{2n+1})^{2\nu+1}}$$

Proof. Clearly

$$\begin{aligned}
&\sum_{n=1}^{\infty} \eta_{2\nu}^0(n) q^n \\
&= \frac{1}{(2\nu)!} \left(\frac{d^{2\nu}}{dz^{2\nu}} z^\nu R^0(z; q) \right) \Bigg|_{z=1} \\
&= \frac{1}{(2\nu)!} \sum_{j=0}^{2\nu} \binom{2\nu}{j} \nu(\nu - 1) \cdots (\nu - j + 1) R^{0^{(2\nu-j)}}(1; q) \\
&\hspace{20em} \text{(by Leibniz's rule)} \\
&= \sum_{j=0}^{\nu} \binom{\nu}{j} \frac{1}{(q^2; q^2)_\infty} \sum_{n=-\infty}^{\infty} \frac{(-1)^n q^{3n^2+3n+1+(2\nu-j)(2n+1)}}{(1 - q^{2n+1})^{2\nu-j+1}} \\
&= \frac{1}{(q^2; q^2)_\infty} \sum_{n=-\infty}^{\infty} \frac{(-1)^n q^{3n^2+3n+1+2\nu(2n+1)}}{(1 - q^{2n+1})^{2\nu+1}} \left(1 + \frac{q^{-(2n+1)}}{(1 - q^{2n+1})^{-1}} \right)^\nu \\
&= \frac{1}{(q^2; q^2)_\infty} \sum_{n=-\infty}^{\infty} \frac{(-1)^n q^{3n^2+(2\nu+3)n+\nu+1}}{(1 - q^{2n+1})^{2\nu+1}}.
\end{aligned}$$

□

Following the progression of results in Sect. 2, we define

(9.1)

$$
\begin{aligned}
&R_k^0(x_1, x_2, \dots, x_k; q) \\
&= \sum_{m_1, m_2, \dots, m_k \geqq 0} \frac{q^{2(m_1+m_2+\cdots+m_k)^2+2m_1+2(m_1+m_2)+\cdots+2(m_1+m_2+\cdots+m_k)+k}}{\left(x_1 q; q^2\right)_{m_1+1}\left(\frac{q}{x_1}; q^2\right)_{m_1+1}} \\
&\quad \times \frac{1}{\left(x_2 q^{2m_1+1}; q^2\right)_{m_2+1}\left(\frac{q^{2m_1+1}}{x_2}; q^2\right)_{m_2+1}} \\
&\quad \times \frac{1}{\cdots\left(x_k q^{2(m_1+\cdots+m_{k-1})+1}; q^2\right)_{m_k+1}\left(\frac{q^{2(m_1+\cdots+m_{k-1})+1}}{x_k}; q^2\right)_{m_k+1}}
\end{aligned}
$$

Theorem 22.

$$
R_k^0(x_1, x_2, \dots, x_k; q) = \frac{1}{(q^2; q^2)_\infty} \sum_{n \geqq 0} \frac{(1-q^{4n+2})(-1)^n q^{3n^2+(2k+1)n+k}}{\prod_{i=1}^{k}\left(1 - x_i q^{2n+1}\right)\left(1 - \frac{q^{2n+1}}{x_i}\right)}.
$$

Proof. In (2.4), we replace q by q^2, then set $a = q^2$, $b_i = x_i q$, $c_i = q/x_i$. Finally we divide both sides by $q^{-k}(q^4; q^2)_\infty \prod_{i=1}^{k}(1 - x_i q)(1 - q/x_i)$ and simplify. □

Corollary 23. *$R_k^0(x_1, \dots, x_k; q)$ is a symmetric function of $x_1, x_2, \dots, x_k$.*

Proof. Immediate from inspection of Theorem 22. □

Corollary 24.

$$
R_{\nu+1}^0(1, 1, \dots, 1; q) = \sum_{n=1}^{\infty} \eta_{2\nu}^0(n) q^n.
$$

Proof. By Theorem 22,

$$
\begin{aligned}
R_{\nu+1}^0(1, 1, \dots, 1; q) &= \frac{1}{(q^2; q^2)_\infty} \sum_{n=0}^{\infty} \frac{(1+q^{2n+1})(-1)^n q^{3n^2+(2\nu+3)n+\nu+1}}{(1-q^{2n+1})^{2\nu+1}} \\
&= \frac{1}{(q^2; q^2)_\infty} \sum_{n=-\infty}^{\infty} \frac{(-1)^n q^{3n^2+(2\nu+3)n+\nu+1}}{(1-q^{2n+1})^{2\nu+1}} \\
&= \sum_{n=1}^{\infty} \eta_{2\nu}^0(n) q^n,
\end{aligned}
$$

by Theorem 21. □

Theorem 25.

$$R_k^0(x_1, x_2, \dots, x_k) = \sum_{i=1}^{k} \frac{R_1^0(x_i; q)}{\prod\limits_{\substack{j=1 \\ j\neq i}}^{k} (x_i - x_j)\left(1 - \frac{1}{x_i x_j}\right)}.$$

Proof. First we note that $k = 1$ is a tautology. So we now assume $k \geqq 2$. Following the line of proof of Theorem 7, we note a new partial fraction identity

(9.2)

$$\frac{(1-w^2)w^{k-1}}{\prod\limits_{j=1}^{k}(1-x_j w)\left(1-\frac{w}{x_j}\right)} = \sum_{i=1}^{k} \frac{1}{\prod\limits_{\substack{j=1 \\ j\neq i}}^{k} (x_i - x_j)\left(1 - \frac{1}{x_i x_j}\right)} \left(\frac{1}{1-wx_i} + \frac{1}{1-\frac{w}{x_i}}\right).$$

If we set $w = 1$ in this identity, we see that

$$\sum_{i=1}^{k} \frac{1}{\prod\limits_{\substack{j=1 \\ j\neq i}}^{k} (x_i - x_j)\left(1 - \frac{1}{x_i x_j}\right)} = 0. \tag{9.3}$$

Combining these observations with Theorem 22, we see that

$$\begin{aligned}
&R_k^0(x_1, x_2, \dots, x_k) \\
&= \frac{1}{(q^2;q^2)_\infty} \sum_{n\geqq 0} \frac{(1-q^{4n+2})(-1)^n q^{3n^2+3n+1+(2n+1)(k-1)}}{\prod\limits_{j=1}^{k}\left(1-x_i q^{2n+1}\right)\left(1-\frac{q^{2n+1}}{x_i}\right)} \\
&= \frac{1}{(q^2;q^2)_\infty} \sum_{n\geqq 0} (-1)^n q^{3n^2+3n+1} \\
&\quad \times \sum_{i=1}^{k} \frac{1}{\prod\limits_{\substack{j=1 \\ j\neq i}}^{k} (x_i - x_j)\left(1 - \frac{1}{x_i x_j}\right)} \left(\frac{1}{1-q^{2n+1}x_i} + \frac{1}{1-\frac{q^{2n+1}}{x_i}}\right) \\
&\qquad\qquad\qquad\qquad\qquad\qquad\qquad\qquad\qquad\qquad \text{(by (9.2))} \\
&= \frac{1}{(q^2;q^2)_\infty} \sum_{n\geqq 0} (-1)^n q^{3n^2+3n+1} \\
&\quad \times \sum_{i=1}^{k} \frac{1}{\prod\limits_{\substack{j=1 \\ j\neq i}}^{k} (x_i - x_j)\left(1 - \frac{1}{x_i x_j}\right)} \left(1 + \frac{1-q^{4n+2}}{\left(1-x_i q^{2n+1}\right)\left(1-\frac{q^{2n+1}}{x_i}\right)}\right)
\end{aligned}$$

$$= \frac{1}{(q^2;q^2)_\infty} \sum_{n \geqq 0} (-1)^n q^{3n^2+3n+1}(1-q^{4n+2})$$

$$\times \sum_{i=1}^{k} \frac{1}{\prod\limits_{\substack{j=1\\ j\neq i}}^{k} (x_i - x_j)\left(1-\frac{1}{x_i x_j}\right)} \frac{1}{\left(1-x_i q^{2n+1}\right)\left(1-\frac{q^{2n+1}}{x_i}\right)} \qquad \text{(by (9.3))}$$

$$= \sum_{i=1}^{k} \frac{1}{\prod\limits_{\substack{j=1\\ j\neq i}}^{k} (x_i - x_j)\left(1-\frac{1}{x_i x_j}\right)} R_1^0(x_i;q).$$

□

10. Generating functions (2)

Theorem 26. *For $k \geqq 1$,*

$$\sum_{n_1,n_2,\ldots,n_k=-\infty}^{\infty} \sum_{n \geqq 0} \mathcal{D}_k^0(n_1,n_2,\ldots,n_k;n) x_1^{n} x_2^{n_2} \cdots x_k^{n_k} q^n$$
$$= R_k^0(x_1,x_2,\ldots,x_k;q).$$

Proof. The proof here is identical with that of Theorem 10 except for the fact that all parts are odd and the Durfee square given

$$(m_1+m_2+\cdots+m_k)^2$$

is now replaced by the modified square containing

$$2(m_1+m_2+\cdots+m_k)^2 + 2(m_1+m_2+\cdots+m_k)+1$$

nodes. The rest follows from inspection of (9.1). □

Corollary 27. *For $k \geqq 1$*

$$\sum_{n_1,\ldots,n_k=-\infty}^{\infty} \sum_{n=0} \mathcal{D}_k^0(n_1,n_2,\ldots,n_k;n) x_1^{n_1} x_2^{n_2} \cdots x_k^{n_k} q^n$$
$$= \frac{1}{(q^2;q^2)_\infty} \sum_{n \geqq 0} \frac{(1-q^{4n+2})(-1)^n q^{3n^2+(2k+1)n+k}}{\prod\limits_{i=1}^{k}\left(1-x_i q^{2n+1}\right)\left(1-\frac{q^{2n+1}}{x_i}\right)}.$$

Proof. This is immediate by comparing Theorems 26 and 22. □

Corollary 28. *$\mathcal{D}_k^0(n_1,n_2,\ldots,n_k;n)$ is symmetric in $n_1,n_2,\ldots,n_k$.*

Proof. The generating function as represented in Corollary 27 is clearly symmetric in $x_1, x_2, \ldots, x_k$. □

Corollary 29. *For* $k \geqq 0$,

$$\mathcal{D}^0_{k+1}(n) = \eta^0_{2k}(n)$$

Remark. In words, the number of $(k+1)$-marked odd Durfee symbols related to n equals the symmetrized $(2k)$-th moment function at n for odd Durfee symbols.

Proof. By Theorem 26 and Corollary 24

$$\sum_{n \geqq 0} \mathcal{D}^0_{k+1}(n) q^n = R^0_{k+1}(1, 1, \ldots, 1; q) = \sum_{n=1}^{\infty} \eta^0_{2k}(n) q^n.$$

□

11. Self-conjugate odd Durfee symbols

This section is much shorter than Sect. 6 owing to the fact that we have been unable to obtain a result for odd summands analogous to Dilcher's formula (6.3).

The definition of self-conjugate k-marked Durfee symbols carries over exactly to k-marked odd Durfee symbols: **the two rows in the symbol are identical.**

Furthermore in analogy with (6.2) we see that

(11.1)

$$\begin{aligned}
&\mathcal{SC}^0_k(q) \\
&= \sum_{m_1, m_2, \ldots, m_k \geqq 0} \frac{q^{2(m_1+m_2+\cdots+m_k)^2+2(m_1+\cdots+m_k)+4(m_1+\cdots+m_{k-1})+\cdots+4m_1+2k-1}}{(q^2;q^4)_{m_1+1}(q^{4m_1+2};q^4)_{m_2+1}\cdots(q^{4(m_1+\cdots+m_{k-1})+2};q^4)_{m_k+1}} \\
&= \sum_{M_k \geqq 0} \frac{q^{2M_k^2+2M_k+1}}{(q^2;q^4)_{M_k+1}} \\
&\quad \times \sum_{0 \leqq M_1 \leqq M_2 \leqq \cdots \leqq M_{k-1}} \frac{q^{4(M_1+M_2+\cdots+M_{k-1})+2(k-1)}}{(1-q^{4M_1+2})(1-q^{4M_2+2})\cdots(1-q^{4M_{k-1}+2})}
\end{aligned}$$

It is possible to deduce from (11.1) a theorem analogous to Theorem 13; however, the statement is somewhat complicated. It is of interest to note the following which follows directly from inspection of (11.1).

Lemma 30. $\mathcal{SC}^0_k(q)$ *is an odd function of* q.

Also one final remark for $k = 1$.

$$\mathcal{S}C_1^0(q) = \sum_{m_1 \geqq 0} \frac{q^{2m_1^2+2m_1+1}}{(q^2;q^4)_{m_1+1}} = q\upsilon(q^2),$$

where $\upsilon(q)$ is another of the third order mock theta functions [32, p. 62].

12. Parity of moments of odd ranks

In light of the fact that

$$R_1^0(1;q) = q\omega(q)$$

is a mock theta function rather than a modular form, we might expect fewer congruence theorems. In this brief section we shall comment on the parity of the $\mathcal{D}_k^0(n)$.

We begin by noting a theorem easily deduced from observations of N.J. Fine [17, pp. 61–62]

Theorem 31. *$\mathcal{D}_1^0(n)$ is even except when $n = 6j^2 \pm 4j + 1$.*

Proof. By Theorem 26,

$$\begin{aligned}\sum_{n=0}^{\infty} \mathcal{D}_1^0(n)q^n &= R_1^0(1;q) \\ &= q\omega(q) && \text{(by (8.4))} \\ &\equiv q\sum_{n\geqq 0} \frac{(-1)^n q^n}{(q;q^2)_{n+1}} \pmod 2 && \text{(by [17, Eq. (26.92)])} \\ &= q\left(1 + \sum_{j\geqq 1}(-1)^j q^{6j^2}(q^{4j} - q^{-4j})\right) && \text{(by [17, Eqs. (26.96) and (26.97)])}\end{aligned}$$

and the result follows immediately. □

We note that for $n < 300$, $\mathcal{D}_2^0(n)$ is odd for $n = 2, 8, 10, 16, 18, 24, 34, 50, 56, 58, 64, 66, 80, 82, 88, 90, 98, 104, 106, 120, 128, 130, 144, 154, 160, 168, 170, 186, 192, 200, 202, 216, 218, 226, 234, 240, 242, 248, 258, 266, 274, 280, 282, 288, 296$. If parity were distributed randomly we would expect something on the order of 150 odd values instead of 45.

We note that for $n < 300$, $\mathcal{D}_3^0(n)$ is odd for $n = 3, 5, 7, 21, 31, 39, 53, 63, 67, 69, 71, 85, 95, 101, 117, 131, 133, 135, 149, 181, 199, 223, 231, 245, 255, 263, 293$. If parity were distributed randomly we would expect something on the order of 150 odd values instead of 27.

At least part of the discrepancies suggested in these calculations can be explained by the following

Theorem 32. *For each $k \geqq 1$, if $n \equiv k-1 \pmod 2$, then $\mathcal{D}_k^0(n)$ is even.*

Proof. First we note that $\binom{2k}{2j+1}$ is always even because

$$(1+x)^{2k} = (1+2x+x^2)^k \equiv (1+x^2)^k \pmod 2.$$

The assertion in our theorem is equivalent to the assertion that

$$\frac{1}{2}\left(R_k^0(1,\dots,1;q) - (-1)^k R_k^0(1,\dots,1;-q)\right)$$

has even coefficients, and by Theorem 22

$$\begin{aligned}
&\frac{1}{2}\left(R_k^0(1,\dots,1;q) - (-1)^k R_k^0(1,\dots,1;-q)\right) \\
&= \frac{1}{2(q^2;q^2)_\infty} \sum_{n\geqq 0} (1-q^{4n+2})(-1)^n q^{3n^2+(2k+1)n+k} \\
&\qquad \times \left(\frac{1}{(1-q^{2n+1})^{2k}} - \frac{1}{(1+q^{2n+1})^{2k}}\right) \\
&= \frac{1}{2(q^2;q^2)_\infty} \sum_{n\geqq 0} \frac{(-1)^n q^{3n^2+(2k+1)n+k}}{(1-q^{4n+2})^{2k-1}} \\
&\qquad \times \left((1+q^{2n+1})^{2k} - (1-q^{2n+1})^{2k}\right) \\
&= \frac{1}{(q^2;q^2)_\infty} \sum_{n\geqq 0} \frac{(-1)^n q^{3n^2+(2k+1)n+k}}{(1-q^{4n+2})^{2k-1}} \sum_{j=0}^{k} \binom{2k}{2j+1} q^{(2n+1)(2j+1)} \\
&\equiv 0 \pmod 2,
\end{aligned}$$

which proves our theorem. □

Seemingly more is true than is contained in this result.

Conjecture A. $\mathcal{D}_2^0(n)$ is even if $n \equiv 4$ or $6 \pmod 8$.

Conjecture B. $\mathcal{D}_3^0(n)$ is even if $n \equiv 1, 9, 11$ or $13 \pmod{16}$.

13. Mock theta functions and multiple series

In light of Theorems 7 and 25, we may find numerous multiple series equal to linear combinations of Ramanujan's mock theta functions. We shall restrict our considerations to a few typical examples. Recall the assertions of Ramanujan and Watson [32, p. 62]:

$$f(q) := \sum_{n=0}^{\infty} \frac{q^{n^2}}{(-q;q)_n^2} = R_1(-1;q) \tag{13.1}$$

$$\phi(q) := \sum_{n=0}^{\infty} \frac{q^{n^2}}{(-q^2;q^2)_n} = R_1(i;q) \tag{13.2}$$

$$q\omega(q) := \sum_{n=0}^{\infty} \frac{q^{2n^2+2n+1}}{(q;q^2)_{n+1}^2} = R_1^0(1;q) \tag{13.3}$$

$$qv(q^2) := \sum_{n=0}^{\infty} \frac{q^{2n^2+2n+1}}{(-q^2;q^4)_{n+1}} = R_1^0(i;q); \tag{13.4}$$

also there is the result of Jacobi immediately deduced from (1.7) and (1.8)

$$\frac{1}{(q;q)_\infty} = R_1(1;q). \tag{13.5}$$

Hence by Theorem 7 with $k = 2$,

$$R_2(-1,1;q) = \frac{1}{4}\left(\frac{1}{(q;q)_\infty} - f(q)\right) \tag{13.6}$$

$$R_2(i,-1;q) = \frac{1}{2}(\phi(q) - f(q)). \tag{13.7}$$

By Theorem 25, with $k = 2$,

$$R_2^0(-1,1;q) = \frac{1}{4}(q\omega(q) - q\omega(-q)) \tag{13.8}$$

$$R_2^0(i,-1;q) = \frac{1}{2}(qv(q^2) - q\omega(-q)). \tag{13.9}$$

By Theorem 7, with $k = 3$,

$$R_3(1,-1,i;q) = \frac{1}{8}\left(\frac{1}{(q;q)_\infty} + f(q) - 2\phi(q)\right), \tag{13.10}$$

and by Theorem 25, with $k = 3$

$$R_3^0(1,-1,i;q) = \frac{1}{8}(q\omega(q) + q\omega(-q) - qv(q^2)). \tag{13.11}$$

While we have written these results succinctly, we remark that in (13.6)–(13.9) the left-hand sides are all double series, and in (13.10) and (13.11) the left-hand sides are triple series. For emphasis we exhibit (13.10)

$$\sum_{\substack{m_1>0 \\ m_2,m_3\geqq 0}} \frac{q^{(m_1+m_2+m_3)^2+2m_1+m_2}}{(q;q)_{m_1}^2(-q^{m_1};q)_{m_2+1}^2(-q^{2(m_1+m_2)};q^2)_{m_3+1}} = \frac{1}{8(q;q)_\infty} + \frac{1}{8}\sum_{n=0}^{\infty}\frac{q^{n^2}}{(-q;q)_n^2} - \frac{1}{4}\sum_{n=0}^{\infty}\frac{q^{n^2}}{(-q^2;q^2)_n}. \tag{13.12}$$

14. Open problems

There exist a variety of serious questions that naturally arise from this study. They fall into 3 basic groups: Combinatorial, asymptotic and congruential.

1. Prove Theorem 7 combinatorially.
2. Prove Corollary 11 combinatorially.
3. Prove Corollary 12 combinatorially.
4. Prove Corollary 13 combinatorially.
5. Prove Theorem 15 combinatorially.
6. Prove Theorem 25 combinatorially.
7. Prove Corollary 27 combinatorially.
8. Prove Corollary 28 combinatorially.
9. Prove Corollary 29 combinatorially.
10. Find a simpler representation for $\mathcal{SC}^0(q)$ analogous to the one given for $\mathcal{SC}(q)$ in Theorem 15.
11. In light of Bringmann's asymptotics for the rank partition functions [12], and the recent breakthroughs by Bringmann and Ono on the exact formulas for the coefficients in the power series of mock theta functions [13], provide similar results for $\mathcal{D}_k(n)$ and $\mathcal{D}_k^0(n)$.
12. Prove that $\eta_4(n) \equiv 0 \pmod 5$ if $n \equiv 24 \pmod{25}$.
13. Develop a full combinatorial theory of congruences for the $NF_k(m, Q, n)$ analogous to the achievements for the ordinary rank and crank chronicled by Mahlburg [25] and Bringmann and Ono [14].
14. F. Garvan [20] successfully utilized (2.4) in a very different generalization of Dyson's rank. Are there other "ranks" that may lie hidden in specializations of (2.4).

References

1. Agarwal, A.K., Andrews, G.E.: Rogers–Ramanujan identities for partitions with "N copies of N". J. Comb. Theory, Ser. A, **45**, 40–49 (1987)
2. Ahlgren, S., Ono, K.: Addition and counting: The arithmetic of partitions. Notices Am. Math. Soc. **48**, 978–984 (2001)

3. Alladi, K.: The method of weighted words and applications to partitions. In: David, S. (ed.) Number Theory, pp. 1–36. Cambridge University Press, Cambridge (1995)
4. Alladi, K., Andrews, G.E., Berkovich, A.: A new four parameter q-series identity and its partition implications. Invent. Math. **153**, 231–260 (2003)
5. Andrews, G.E.: Problems and prospects for basic hypergeometric functions. In: Askey, R. (ed.) Theory and Appl. of Special Functions, pp. 191–224. Academic Press, New York (1975)
6. Andrews, G.E.: The theory of partitions. In: Rota, G.-C. (ed.) Encycl. of Math. and Its Appl. Addison-Wesley, Reading, (1976) (Reissued: Cambridge University Press, New York, 1998)
7. Andrews, G.E., Crippa, D., Simon, K.: q-Series arising from the study of random graphs. SIAM J. Discrete Math. **10**, 41–56 (1997)
8. Andrews, G.E., Freitas, P.: Extension of Abel's Lemma with q-series implications. Ramanujan J. **10**, 137–152 (2005)
9. Andrews, G.E., Garvan, F.: Dyson's crank of a partition. Bull. Am. Math. Soc. **18**, 167–171 (1988)
10. Atkin, A.O.L., Garvan, F.: Relations between the ranks and cranks of partitions. Ramanujan J. **7**, 343–366 (2003)
11. Atkin A.O.L., Swinnerton-Dyer, H.P.F.: Some properties of partitions. Proc. Lond. Math. Soc., III Ser. **4**, 84–106 (1954)
12. Bringmann, K.: Asymptotics for rank partition functions. (to appear)
13. Bringmann, K., Ono, K.: The $f(q)$ mock theta function conjecture and partition ranks. Invent. Math. **165**, 243–266 (2006)
14. Bringmann, K., Ono, K.: Dyson's ranks and Maass forms. Submitted
15. Dilcher, K.: Some q-series identities related to divisor functions. Discrete Math. **145**, 98–110 (1995)
16. Dyson, F.: Some guesses in the theory of partitions. Eureka **8**, 10–15 (1944) (Reprinted: Selected Papers, Am. Math. Soc., Providence, pp. 51–56 (1996))
17. Fine, N.J.: Basic Hypergeometric Series and Applications. Am. Math. Soc., Providence (1988)
18. Garthwaite, S.: The coefficients of the $\omega(q)$ mock theta function. Proc. Am. Math. Soc. (to appear)
19. Garvan, F.G.: New combinatorial interpretations of Ramanujan's partition congruences. Trans. Am. Math. Soc. **305**, 47–77 (1988)
20. Garvan, F.G.: Generalizations of Dyson's rank and non-Rogers–Ramanujan partitions. Manuscr. Math. **84**, 343–359 (1994)
21. Gasper, G., Rahman, M.: Basic Hypergeometric Series. Cambridge University Press, Cambridge (1990)
22. Hardy, G.H., Wright, E.M.: An Introduction to the Theory of Numbers, 5th edn. Clarendon Press, Oxford (1979)
23. Kluyver, J.C.: Vraagstuk XXXVII. (Solution by van Veen, S.C.). Wiskundige Opgaven, 92–93 (1919)
24. MacMahon, P.A.: The theory of modular partitions. Proc. Camb. Phil. Soc. **21**, 197–204 (1923) (Reprinted: Coll. Papers, vol. 1, M.I.T. Press, Cambridge, pp. 1090–1097 (1978))
25. Mahlburg, K.: Partition congruences and the Andrews–Garvan–Dyson crank. Proc. Nat. Acad. Sci. **102**, 15373–15376 (2005)
26. Ono, K.: Distribution of the partition function modulo m. Ann. Math. **151**, 293–307 (2000)
27. Ono, K.: Arithmetic of the partition function. In: Bustoz, J., Suslov, S. (eds.) Proc. NATO Adv. Study Inst. on Special Functions, Special Functions, pp. 243–253. Kluwer, New York (2000)
28. Ramanujan, S.: The Lost Notebook and Other Unpublished Papers. Intro. by G.E. Andrews. Narosa, New Delhi (1997)

29. Sylvester, J.J.: Note on the paper of Mr.Durfee. Johns Hopkins Univ. Circulars **2**, pp. 23, 24; 42, 43 (1883) (Reprinted: Coll. Papers, vol. 3, pp. 660–662, Chelsea, New York (1973))
30. Sylvester, J.J.: A constructive theory of partitions .... Am. J. Math. **5**, 251–330 (1882), **6**, 334–336 (1884) (Reprinted: Coll. Papers, vol. 4, pp. 1–83, Chelsea, New York (1973))
31. Uchimura, K.: An identity for the divisor generating function arising from sorting theory. J. Comb. Theory, Ser. A **31**, 131–135 (1981)
32. Watson, G.N.: The final problem. J. Lond. Math. Soc. **11**, 55–80 (1936)

Note: The main definitions on Durfee symbols are greatly improved and made much clearer in [Bou-K11].

3.15 The Number of Smallest Parts in the Partitions of n

J. reine angew. Math. **624** (2008), 133—142
DOI 10.1515/CRELLE.2008.083

Journal für die reine und angewandte Mathematik

Berlin · New York 2008

The number of smallest parts in the partitions of n

By *George E. Andrews* at University Park

Abstract. We denote by spt(n) the total number of appearances of the smallest part in each integer partition of n. We shall relate spt(n) to the Atkin-Garvan moments of ranks, and we shall prove that $5 \mid \mathrm{spt}(5n+4)$, $7 \mid \mathrm{spt}(7n+5)$ and $13 \mid \mathrm{spt}(13n+6)$.

1. Introduction

In the study of the partitions of integers, it has often been of interest to consider weighted counts of partitions. In a series of three papers [1], [2], [3], K. Alladi really initiated the study of weighted counts of partitions including [2], Sec. 3, attaching certain simple weights to the Rogers-Ramanujan partitions (i.e. partitions with parts differing by at least 2) so that the weighted count was actually $p(n)$, the number of partitions of n.

In [10], Fokkink, Fokkink and Wang proved that a weighted sum over partitions of n into distinct parts equals the number of divisors of n.

Theorem 1 ([10]). *If $\mathscr{D}_n$ denotes the set of partitions π of n into distinct parts, then*

$$-\sum_{\pi \in \mathscr{D}_n} (-1)^{\#(\pi)} \sigma(\pi) = d(n), \tag{1.1}$$

where $\#(\pi)$ is the number of parts of π, $\sigma(\pi)$ is the smallest part of π and $d(n)$ is the number of divisors of n.

Their proof is a lovely combination of combinatorial analysis applied to a remarkable sequence of polynomials. However, it is instructive to note that the result is effectively a corollary of the differentiation of the q-analog of Gauss's theorem [4], p. 20, Cor. 2.4. Namely, if FFW(n) denotes the left-hand side of (1.1), then standard partition-theoretic arguments (cf. [4], Ch. 1) show that

---

Partially supported by National Science Foundation Grant DMS 0457003.

$$\sum_{n\geq 0} \mathrm{FFW}(n)q^n = \sum_{n=1}^{\infty} \frac{(-1)^{n-1}q^{n(n+1)/2}}{(q;q)_n(1-q^n)} \quad \text{(by consideration of conjugate partitions)}$$

$$= \left[\frac{d}{dz}\sum_{n\geq 0}\frac{(z;q)_n(-1)^n q^{n(n+1)/2}}{(q;q)_n(zq;q)_n}\right]_{z=1}$$

$$= \frac{d}{dz}\left[\frac{(q;q)_\infty}{(zq;q)_\infty}\right]_{z=1} \quad \text{(by [4], p. 20, Cor. 2.4 with } a=z,\ b=q/\tau,\ c=zq,\ \tau\to 0)$$

$$= \left[\frac{(q;q)_\infty}{(zq;q)_\infty}\sum_{j=1}^{\infty}\frac{q^j}{1-zq^j}\right]_{z=1}$$

$$= \sum_{j=1}^{\infty}\frac{q^j}{1-q^j} = \sum_{n=1}^{\infty} d(n)q^n.$$

We have used the standard notation $(A;q)_n = (1-A)(1-Aq)\dots(1-Aq^{n-1})$.

Our purpose in presenting this argument in the introduction is to highlight the role of differentiation in proving theorems on weighted counts of partitions.

As mentioned in the abstract, this paper is devoted to the study of $\mathrm{spt}(n)$, the total number of appearances of the smallest parts in all of the partitions of n. In light of the nice theorem about $\mathrm{FFW}(n)$, it is not unreasonable that $\mathrm{spt}(n)$ be interesting as well.

As is well-known [4], Ch. 10, $p(n)$ has a number of interesting congruence properties, namely

$$p(5n+4) \equiv 0 \pmod 5, \tag{1.2}$$

$$p(7n+5) \equiv 0 \pmod 7, \tag{1.3}$$

$$p(11n+6) \equiv 0 \pmod{11}. \tag{1.4}$$

Similar results hold for $\mathrm{spt}(n)$.

Theorem 2.

$$\mathrm{spt}(5n+4) \equiv 0 \pmod 5, \tag{1.5}$$

$$\mathrm{spt}(7n+5) \equiv 0 \pmod 7, \tag{1.6}$$

$$\mathrm{spt}(13n+6) \equiv 0 \pmod{13}. \tag{1.7}$$

The appearance of 13 in this result is completely unexpected.

In Section 2, we shall collect the necessary background results. Section 3 will be devoted to proving

Theorem 3.

$$\mathrm{spt}(n) = np(n) - \frac{1}{2}N_2(n),$$

where $N_2(n)$ is the second Atkin-Garvan moment (see eq. (2.13) *for the definition).*

Section 4 is devoted to a proof of Theorem 2.

I must give special thanks to Frank Garvan who not only pointed out O'Brien's thesis [12] to me but also supplied me with a copy.

2. Background

In this section, we collect necessary facts from analysis, q-series and the theory of partitions.

We begin with three minor observations concerning any function $f(z)$ that is at least twice differentiable at $z = 1$.

$$-\frac{1}{2}\left[\frac{d^2}{dz^2}(1-z)(1-z^{-1})f(z)\right]_{z=1} = f(1). \tag{2.1}$$

This follows immediately once one computes the second derivative of the expression in question.

We also require Watson's celebrated q-analog of Whipple's theorem [11], p. 43, eq. (2.5.1):

$$\begin{aligned}
{}_8\phi_7&\left(\begin{matrix} a, q\sqrt{a}, -q\sqrt{a}, b, c, d, e, q^{-N}; q, \dfrac{a^2q^{2+N}}{bcde} \\ \sqrt{a}, -\sqrt{a}, \dfrac{aq}{b}, \dfrac{aq}{c}, \dfrac{aq}{d}, \dfrac{aq}{e}, aq^{N+1} \end{matrix}\right) \\
&= \frac{(aq;q)_N\left(\dfrac{aq}{de};q\right)_N}{\left(\dfrac{aq}{d};q\right)_N\left(\dfrac{aq}{e};q\right)_N}\,{}_4\phi_3\left(\begin{matrix} \dfrac{aq}{bc}, d, e, q^{-N}; q, q \\ \dfrac{aq}{b}, \dfrac{aq}{c}, \dfrac{deq^{-N}}{a} \end{matrix}\right),
\end{aligned} \tag{2.2}$$

where

$$ {}_{r+1}\phi_r\left(\begin{matrix} a_0, a_1, \dots, a_r; q, t \\ b_1, \dots, b_r \end{matrix}\right) = \sum_{n=0}^{\infty} \frac{(a_0,q)_n(a_1;q)_n \cdots (a_r;q)_n t^n}{(q;q)_n(b_1;q)_n \cdots (b_r;q)_n}. \tag{2.3}$$

Next we have another differentiation identity:

$$-\frac{1}{2}\left[\frac{d^2}{dz^2}(zq;q)_\infty(z^{-1}q;q)_\infty\right]_{z=1}=(q;q)_\infty^2\sum_{n=1}^{\infty}\frac{nq^n}{1-q^n}. \tag{2.4}$$

This must be well-known, but we were unable to find a reference. The proof of (2.4) depends on the Jacobi triple product identity [4], p. 21, Th. 2.8:

$$\begin{aligned}
&-\frac{1}{2}\left[\frac{d^2}{dz^2}(zq;q)_\infty(z^{-1}q;q)_\infty\right]_{z=1}\\
&=-\frac{1}{2(q;q)_\infty}\left[\frac{d^2}{dz^2}\frac{\sum_{n=-\infty}^{\infty}(-z)^n q^{n(n-1)/2}}{1-z}\right]_{z=1}\\
&=-\frac{1}{2(q;q)_\infty}\left[\frac{d^2}{dz^2}\sum_{n=0}^{\infty}(-z)^{-n}q^{n(n+1)/2}\left(\frac{1-z^{2n+1}}{1-z}\right)\right]_{z=1}\\
&=-\frac{1}{2(q;q)_\infty}\left[\frac{d^2}{dz^2}\sum_{n=0}^{\infty}(-1)^n q^{n(n+1)/2}\sum_{j=0}^{2n}z^{-n+j}\right]_{z=1}\\
&=-\frac{1}{2(q;q)_\infty}\sum_{n=0}^{\infty}(-1)^n q^{n(n+1)/2}\sum_{j=0}^{2n}(-n+j)(-n+j-1)\\
&=-\frac{1}{2(q;q)_\infty}\sum_{n=0}^{\infty}(-1)^n q^{n(n+1)/2}\frac{1}{3}n(n+1)(2n+1)\\
&=-\frac{q}{3(q;q)_\infty}\sum_{n=0}^{\infty}(-1)^n(2n+1)\frac{n(n+1)}{2}q^{n(n+1)/2-1}\\
&=-\frac{q}{3(q;q)_\infty}\frac{d}{dq}\sum_{n=0}^{\infty}(-1)^n(2n+1)q^{n(n+1)/2}\\
&=-\frac{q}{3(q;q)_\infty}\frac{d}{dq}(q;q)_\infty^3 \quad \text{(by [4], p. 176, ex. 7)}\\
&=(q;q)_\infty^2\sum_{n=1}^{\infty}\frac{nq^n}{1-q^n},
\end{aligned}$$

where the last line follows by logarithmic differentiation. The referee has noted that (2.4) can also be proved by twice using logarithmic differentiation on the original product.

To conclude this section we collect some results concerning the ranks of partitions, a topic begun by Dyson with his celebrated conjectures. We shall follow closely the notation as developed in [7], [8] and [12].

The rank of a partition is defined to be the largest part minus the number of parts. Thus the partition of 18 given by $5+4+4+2+1+1+1$ has rank $5-7=-2$.

We define $N(m,n)$ to be the number of partitions of n with rank m, and $N(m,Q,n)$ to be the number of partitions of n with rank $\equiv m \pmod{Q}$. Next

$$r_{a,b}(d) = \sum_{n=0}^{\infty}(N(a,Q,Qn+d) - N(b,Q,Qn+d))q^{Qn}. \tag{2.5}$$

The only case we shall be considering for $r_{a,b}(d)$ in detail is where $Q = 13$. In this instance, we additionally define

$$S_1(d) = r_{0,1}(d) - 6r_{5,6}(d), \tag{2.6}$$

$$S_2(d) = r_{1,2}(d) - 5r_{5,6}(d), \tag{2.7}$$

$$S_3(d) = r_{2,3}(d) - 4r_{5,6}(d), \tag{2.8}$$

$$S_4(d) = r_{3,4}(d) - 3r_{5,6}(d), \tag{2.9}$$

$$S_5(d) = r_{4,5}(d) - 2r_{5,6}(d). \tag{2.10}$$

With these definitions (remembering $Q = 13$), we have [12], p. 44:

$$S_1(6) + 2S_2(6) - 5S_5(6) \equiv 0 \pmod{13}, \tag{2.11}$$

and

$$S_2(6) + 5S_3(6) + 3S_4(6) + 3S_5(6) \equiv 0 \pmod{13}. \tag{2.12}$$

In [6], Atkin and Garvan introduce the moments of ranks

$$N_j(n) = \sum_{m=-\infty}^{\infty} m^j N(m,n). \tag{2.13}$$

Conjugation of partitions maps a partition with rank m to one of n with rank $-m$. Consequently,

$$N_j(n) = 0 \quad \text{if } j \text{ is odd}. \tag{2.14}$$

In [5], eqs. (1.3) and (1.10), it is shown that

$$R_1(z;q) := 1 + \sum_{m=-\infty}^{\infty}\sum_{n=1}^{\infty} N(m,n)z^m q^n \tag{2.15}$$

$$= \sum_{n=0}^{\infty} \frac{q^{n^2}}{(zq;q)_n(z^{-1}q;q)_n} \tag{2.16}$$

$$= \frac{(1-z)}{(q;q)_\infty}\sum_{n=-\infty}^{\infty}\frac{(-1)^n q^{n(3n+1)/2}}{1-zq^n}. \tag{2.17}$$

3. The two main theorems

Theorem 4.

$$\sum_{n\geqq 1}\operatorname{spt}(n)q^n = \frac{1}{(q;q)_\infty}\sum_{n=1}^{\infty}\frac{nq^n}{1-q^n} + \frac{1}{(q;q)_\infty}\sum_{n=1}^{\infty}\frac{(-1)^n q^{n(3n+1)/2}(1+q^n)}{(1-q^n)^2}. \tag{3.1}$$

We restate Theorem 3 from the introduction.

Theorem 3.

$$\text{spt}(n) = np(n) - \frac{1}{2}N_2(n). \tag{3.2}$$

From these results we will deduce the congruences presented in the introduction.

Proof of Theorem 4.

$$\sum_{n\geq 1} \text{spt}(n)q^n$$

$$= \sum_{n=1}^{\infty}(q^n + 2q^{2n} + 3q^{3n} + \cdots)\prod_{j=n+1}^{\infty}(1 + q^j + q^{2j} + q^{3j} + \cdots)$$

$$= \sum_{n=1}^{\infty}\frac{q^n}{(1-q^n)^2(q^{n+1};q)_\infty}$$

$$= -\frac{1}{2(q;q)_\infty}\left[\frac{d^2}{dz^2}\sum_{n=0}^{\infty}\frac{(z;q)_n(z^{-1};q)_n q^n}{(q;q)_n}\right]_{z=1} \quad \text{(by (2.1))}$$

$$= -\frac{1}{2(q;q)_\infty}\left[\frac{d^2}{dz^2}\frac{(zq;q)_\infty(z^{-1}q;q)_\infty}{(q;q)_\infty^2}\left(1 + \sum_{n=1}^{\infty}\frac{(-1)^n q^{n(3n+1)/2}(1+q^n)(z;q)_n(z^{-1};q)_n}{\left(\frac{q}{z};q\right)_n(zq;q)_n}\right)\right]_{z=1}$$

(by (2.2) with $a \to 1$, $d = e^{-1} = z$, $b, c, N \to \infty$)

$$= -\frac{1}{2(q;q)_\infty^3}\left[\frac{d^2}{dz^2}(zq;q)_\infty(z^{-1}q;q)_\infty\right]_{z=1} + \frac{1}{(q;q)_\infty}\sum_{n=1}^{\infty}\frac{(-1)^n q^{n(3n+1)/2}(1+q^n)}{(1-q^n)^2} \quad \text{(by (2.1))}$$

$$= \frac{1}{(q;q)_\infty}\sum_{n=1}^{\infty}\frac{nq^n}{1-q^n} + \frac{1}{(q;q)_\infty}\sum_{n=1}^{\infty}\frac{(-1)^n q^{n(3n+1)/2}(1+q^n)}{(1-q^n)^2} \quad \text{(by (2.4))}.$$

Proof of Theorem 3. The object here is to show that the first expression on the right-hand side of (3.1) is the generating function for $np(n)$ and the second is the generating function for $\frac{1}{2}N_2(n)$.

Recalling ([4], Ch. 1)

$$\sum_{n=0}^{\infty}p(n)q^n = \frac{1}{(q;q)_\infty},$$

we see that

$$\sum_{n=0}^{\infty}np(n)q^n = q\frac{d}{dq}\frac{1}{(q;q)_\infty} = \frac{1}{(q;q)_\infty}\sum_{n=1}^{\infty}\frac{nq^n}{1-q^n}, \tag{3.3}$$

by logarithmic differentiation.

Next

$$\begin{aligned}
(3.4)\qquad \sum_{n\geqq 0}\frac{1}{2}N_2(n)q^n &= \sum_{n=0}^{\infty}\frac{N_2(n)-N_1(n)}{2}q^n \quad (\text{because } N_1(n)=0 \text{ by } (2.14))\\
&= \frac{1}{2}\left[\frac{d^2}{dz^2}R(z;q)\right]_{z=1} \quad (\text{by } (2.15))\\
&= \frac{1}{2(q;q)_\infty}\left[\frac{d^2}{dz^2}\sum_{n=-\infty}^{\infty}\frac{(-1)^n q^{n(3n+1)/2}(1-z)}{1-zq^n}\right]_{z=1} \quad (\text{by } (2.17))\\
&= \frac{-1}{2(q;q)_\infty}\left[\frac{d}{dz}\sum_{\substack{n=-\infty\\ n\neq 0}}^{\infty}\frac{(-1)^n q^{n(3n+1)/2}(1-q^n)}{(1-zq^n)^2}\right]_{z=1}\\
&= \frac{-1}{(q;q)_\infty}\sum_{\substack{n=-\infty\\ n\neq 0}}^{\infty}\frac{(-1)^n q^{n(3n+1)/2+n}}{(1-q^n)^2}\\
&= \frac{-1}{(q;q)_\infty}\left(\sum_{n=1}^{\infty}\frac{(-1)^{-n}q^{n(3n-1)/2-n}}{(1-q^{-n})^2}+\sum_{n=1}^{\infty}\frac{(-1)^n q^{n(3n+1)/2+n}}{(1-q^n)^2}\right)\\
&= \frac{-1}{(q;q)_\infty}\sum_{n=1}^{\infty}\frac{(-1)^n q^{n(3n+1)/2}(1+q^n)}{(1-q^n)^2}.
\end{aligned}$$

Finally substituting (3.3) and (3.4) into (3.1) and comparing coefficients of q^n, we derive (3.2).

4. The congruences

The object in this section is to prove Theorem 2 which contains three congruences. The first two are fairly direct consequences of the results of Atkin and Swinnerton-Dyer [8].

$$\begin{aligned}
\mathrm{spt}(5n+4) &= (5n+4)p(5n+4)-\frac{1}{2}N_2(5n+4) \quad (\text{by Th. } 3)\\
&\equiv -\frac{1}{2}N_2(5n+4) \pmod 5 \quad (\text{by } (1.2))\\
&= -\frac{1}{2}\sum_{m=-\infty}^{\infty} m^2 N(m,5n+4)\\
&\equiv -\frac{1}{2}(N(1,5,5n+4)-N(2,5,5n+4)\\
&\qquad -N(3,5,5n+4)+N(4,5,5n+4)) \pmod 5\\
&\equiv 0 \pmod 5 \quad (\text{by [8], p. 85, eq. } (2.4)),
\end{aligned}$$

$$\begin{aligned}\mathrm{spt}(7n+5) &= (7n+5)p(7n+5) - \frac{1}{2}N_2(7n+5)\\ &\equiv -\frac{1}{2}N_2(7n+5) \pmod 7 \quad (\text{by } (1.3))\\ &= -\frac{1}{2}\sum_{m=-\infty}^{\infty} m^2 N(m, 7n+5)\\ &\equiv -\big(N(1,7,7n+5) + 4N(2,7,7n+5) + 2N(3,7,7n+5)\big)\\ &\equiv 0 \pmod 7 \quad (\text{by [8], p. 85, eq. (2.10)}).\end{aligned}$$

The final congruence is a bit trickier and requires J. N. O'Brien's deep achievements concerning the modulus 13.

$$\begin{aligned}\mathrm{spt}(13n+6) &= (13n+6)p(13n+6) - \frac{1}{2}N_2(13n+6)\\ &\equiv 6p(13n+6) - \frac{1}{2}\sum_{j=0}^{12}(12+j^2-12)N(j,13,13n+6) \pmod{13}\\ &\equiv \sum_{j=0}^{12} 6(j^2-12)N(j,13,13n+6) \pmod{13}\\ &\equiv 6N(0,13,13n+6) + 12N(1,13,13n+6)\\ &\quad + 4N(2,13,13n+6) + 8N(3,13,13n+6)\\ &\quad + 11N(4,13,13n+6) + 0\cdot N(5,13,13n+6)\\ &\quad + N(6,13,13n+6) + N(7,13,13n+6)\\ &\quad + 0\cdot N(8,13,13n+6) + 11N(9,13,13n+6)\\ &\quad + 8N(10,13,13n+6) + 4N(11,13,13n+6)\\ &\quad + 12N(12,13,13n+6) \pmod{13}\\ &\equiv 2\big(3N(0,13,13n+6) - N(1,13,13n+6)\\ &\qquad + 4N(2,13,13n+6) - 5N(3,13,13n+6)\\ &\qquad - 2N(4,13,13n+6) + N(6,13,13n+6)\big) \pmod{13}\\ &\quad (\text{since } N(m,13,13+6) = N(13-m,13,13n+6)).\end{aligned}$$

This last congruence can be cast equivalently in terms of generating functions:

$$3r_{0,1}(6) + 2r_{1,2}(6) + 6r_{2,3}(6) + r_{3,4}(6) - r_{4,5}(6) - r_{5,6}(6) \equiv 0 \pmod{13}.$$

Using (2.6)–(2.10), we may rewrite this last congruence using the $S_i(d)$, namely

$$
\begin{aligned}
3\big(S_1(6) + 6r_{5,6}(6)\big) &+ 2\big(S_2(6) + 5r_{5,6}(6)\big) \\
&+ 6\big(S_3(6) + 4r_{5,6}(6)\big) + S_4(6) + 3r_{5,6}(6) \\
&- S_5(6) - 2r_{5,6}(6) - r_{5,6}(6) \equiv 0 \pmod{13}.
\end{aligned}
$$

In this last congruence we see that the coefficient of $r_{5,6}(6)$ is $52 = 4 \cdot 13$; so all we need to prove is

$$
3S_1(6) + 2S_2(6) + 6S_3(6) + S_4(6) - S_5(6) \equiv 0 \pmod{13}. \tag{4.1}
$$

But

$$
\begin{aligned}
&3S_1(6) + 2S_2(6) + 6S_3(6) + S_4(6) - S_5(6) \\
&\quad \equiv 3\big(S_1(6) + 2S_2(6) - 5S_5(6)\big) - 4\big(S_2(6) + 5S_3(6) + 3S_4(6) + 3S_5(6)\big) \\
&\quad \equiv 0 \pmod{13},
\end{aligned}
$$

by (2.11) and (2.12).

This concludes the proof of Theorem 3. □

5. Conclusion

The connection between spt(n) and the second order moment of Atkin and Garvan [6] in Theorem 3 suggests that there should be much more arithmetic information about spt(n) still to be unearthed.

In addition the connection of $N_2(n)/2$ to the enumeration of 2-marked Durfee symbols in [5] suggests the fact that there are also serious problems concerning combinatorial mappings that should be investigated.

References

[1] *K. Alladi*, Weighted partition identities and applications, from Analytic Number Theory, Vol. 1, B. C. Berndt, H. G. Diamond and A. J. Hildebrand, eds., Birkhäuser, Boston 1996.

[2] *K. Alladi*, Partition identities involving gaps and weights, Trans. Amer. Math. Soc. **349** (1997), 5001–5019.

[3] *K. Alladi*, Partition identities involving gaps and weights, Ramanujan J. **2** (1998), 21–37.

[4] *G. E. Andrews*, The Theory of Partitions, Encycl. Math. Appl. **2**, Addison-Wesley, Reading 1976; reprinted: Cambridge University Press, 1998.

[5] *G. E. Andrews*, Partitions, Durfee Symbols and the Atkin-Garvan Moments of Ranks, Invent. Math., to appear.

[6] *A. O. L. Atkin* and *F. G. Garvan*, Relations between the ranks and cranks of partitions, Ramanujan J. **7** (2003), 343–366.

[7] *A. O. Atkin* and *S. M. Hussain*, Some properties of partitions (2), Trans. Amer. Math. Soc. **89** (1958), 184–200.

[8] *A. O. L. Atkin* and *H. P. F. Swinnerton-Dyer*, Some propertities of partitions, Proc. London Math. Soc. (3) **4** (1954), 84–106.

[9] *F. J. Dyson*, Some guesses in the theory of partitions, Eureka **8** (1944), 10–15.

[10] *R. Fokkink, W. Fokkink* and *B. Wang*, A relation between partitions and the number of divisors, Amer. Math. Monthly **102** (1995), 345–347.

[11] *G. Gasper* and *M. Rahman*, Basic Hypergeometric Series, Encylc. Math. Appl. **35**, Cambridge University Press, Cambridge 1990.

[12] *J. N. O'Brien*, Some Properties of Partitions, With Special Reference to Primes Other than 5, 7 and 11, Ph. D. thesis, Durham University, 1965.

The Pennsylvania State University, University Park, PA 16802, USA
e-mail: andrews@math.psu.edu

Eingegangen 22. Juni 2007, in revidierter Fassung 31. Juli 2007

3.16 On a Partition Theorem of N.J. Fine

J. Nat. Acad. Math.
Vol. 1 (1933), 105–107.

ON A PARTITION THEOREM OF N. J. FINE

By George E. Andrews

A generalization of Euler's partition theorem due to N. J. Fine is given a combinational proof utilizing Dyson's concept of the adjoint of a partition.

1. Introduction

The concept of the rank of a partition is originally due to Dyson [1; p. 10]; it is the largest part of the partition less the number of parts.

Let $D_r(n)$ denote the number of partitions of n into positive distinct parts, the rank of each partition being r.

Let $U_{2r+1}(n)$ denote the number of partitions of n into positive odd parts, the maximum part being $2r+1$.

Fine [3; p. 617] proved the following generalization of Euler's partition theorem.

Theorem 1. *For all* $n \geqslant 1$, $r \geqslant 0$, *we have*

$$U_{2r+1}(n) = D_{2r+1}(n) + D_{2r}(n).$$

Fine's proof of this result [4] is based on an elegant and rather difficult analytic identity, namely

$$\sum_{m=1}^{\infty} \alpha^m q^m \prod_{j=0}^{m-1} (1+\alpha^{-1} q^j) - \sum_{m=1}^{\infty} (-1)^m \alpha^m q^m \prod_{j=0}^{m-1} (1+\alpha^{-1} q^j)$$

$$= 2 \sum_{m=0}^{\infty} (\alpha q)^{2m+1} \Big/ \prod_{j=0}^{m} (1-q^{2j+1}).$$

The object of this paper is to prove Theorem 1 by purely combinatorial means utilizing a new property of partitions due to Dyson [2; p. 58] called the adjoint. First we restate Theorem 1 in the form in which we shall prove it.

Let $\triangle_r(n)$ denote the number of partitions of n into non-negative distinct parts, the rank of each partition being r. Then clearly

Received August 17, 1983

$$\Delta_r(n) = D_{r+1}(n) + D_r(n)$$

since the introduction of zero into a partition with positive parts lowers the rank by 1 and does not alter the sum.

Thus the following is equivalent to Fine's theorem

Theorem 2. *For all* n $\geqslant$ 1, r $\geqslant$ 0, *we have*

$$U_{2r+1}(n) = \Delta_{2r}(n).$$

2. Proof of Theorem 2.

First we note that

$$U_{2r+1}(1) = \begin{cases} 1, & r = 0 \\ 0, & r > 0 \end{cases},$$

and

$$\Delta_{2r}(1) = \begin{cases} 1, & r = 0 \\ 0, & r > 0 \end{cases}$$

Consequently Theorem 2 is valid for n = 1.

Now we note that

$$U_{2r+1}(n) = \sum_{j=0}^{r} U_{2j+1}(n - 2r - 1),$$

where the $U_{2r+1}(n)$ is defined to be zero for negative n. The truth of this formula follows from the fact that if we delete the a part of size 2r + 1 from a partition of n with largest part equal to 2r + 1, we obtain a partition of n − 2r − 1 with largest part $\leqslant$ 2r + 1; conversely all partitions of the type counted by $U_{2r+1}(n)$ may be formed by introducing 2r + 1 into a partition of n − 2r − 1 into odd parts with largest part $\leqslant$ 2r + 1.

We now consider $\Delta_{2r}(n)$. We recall Dyson's definition of the adjoint of P, a partition into non-negative parts. From P we derive a partition P_1 with strictly positive parts by adding 1 to each part in P. Let P'_1 be the conjugate of P_1. Then P′, the adjoint of P, is derived from P'_1 by subtracting one from each part.

Now for each partition, P, counted by $\Delta_{2r}(n)$, let us form the conjugate of the adjoint of P, say P*. P* is a partition with strictly positive distinct parts. If λ is the largest part of P and ν is the number of parts of P, then $2r = \lambda - \nu$

and furthermore P^* is a partition of $n + \nu - (\lambda + 1) = n - (\lambda - \nu) - 1 = n - 2r - 1$. Also we note that the rank of P^* is $\leqslant 2r + 1$ since the largest part of P^* is $\leqslant \lambda$ and the number of parts of P^* is $= \nu - 1$. We have thus established a one-to-one correspondence between the partitions of n with rank 2r and non-negative distinct parts and partitions of $n - 2r - 1$ with rank $\leqslant 2r + 1$ and positive distinct parts. Hence

$$\triangle_{2r}(n) = \sum_{j=0}^{2r+1} D_j(n - 2r - 1)$$

$$= \sum_{j=0}^{r} \triangle_{2j}(n - 2r - 1).$$

Thus both $\triangle$ and U satisfy the same recurrence and since they are identical for all r with $n = 1$, we have Theorem 2 by mathematical induction.

Department of Mathematics
Pennsylvania State University
University Park, Pennsylvania 16802

REFERENCES

[1] F. J. Dyson, Some guesses in the theory of partitions, Eureka, 8 (1944), 10–15.

[2] F. J. Dyson, A new symmetry of partitions, J. Comb. Theory, 7 (1969), 56–61.

[3] N. J. Fine, Some new results on partitions, Proc. Nat. Acad. Sci, 34 (1948), 616–618.

[4] N. J. Fine, Scme Basic Hypergeometric Series and Applications, unpnblished monograph.

3.17 The Friedman–Joichi–Stanton Monotonicity Conjecture at Primes

DIMACS Series in Discrete Mathematics
And Theoretical Computer Science
Volume **64**, 2004

The Friedman-Joichi-Stanton monotonicity conjecture at primes

George E. Andrews

ABSTRACT. The object is to prove that if n is an odd prime, then the number of partitions of N into parts taken from $(n, n+1, \ldots, 2n-1)$ is weakly increasing for $N \geq 1$.

1. Introduction.

In [2] (recently summarized in [6]) Friedman, Joichi and Stanton posed the following:

Conjecture. *Suppose that $n \geq 3$ is an odd integer. If*

$$\frac{1}{(1-q^n)(1-q^{n+1})\ldots(1-q^{2n-1})} = \sum_{m=0}^{\infty} a_m q^m,$$

then a_m is weakly increasing for $m \geq 1$.

From this conjecture, they are able to characterize all sets of positive integers P such that

$$\prod_{n \in P} \frac{1}{1-q^n}$$

has a power series expansion with weakly increasing coefficients past the constant term.

The object here is to prove the following:

Theorem 3. *Suppose that $n \geq 3$ is a prime. If*

$$\frac{1}{(1-q^n)(1-q^{n+1})\ldots(1-q^{2n-1})} = \sum_{m=0}^{\infty} a_m q^m, \tag{1.1}$$

then a_m is weakly increasing for $m \geq 1$.

The proof is divided into two major parts. In Section 2, we shall derive a new identity to represent the product in (1.1). In Section 3, we shall discuss a new set of polynomials that arise in Section 2. In Section 4, we will tie these two results

2000 *Mathematics Subject Classification.* 11P81, 05A17, 33D15.
Key words and phrases. partitions, primes, monotonicity.
Partially supported by National Science Foundation Grant DMS-020004.

together to prove Theorem 1. We conclude with comments and a look at open problems.

2. The identity.

To make clear the significance of the identity to be proved in this section, we list its first two instances

$$\frac{(1-q)}{(1-q^3)(1-q^4)(1-q^5)} = \left(1-q+\frac{q^3}{1-q^5}\right)\frac{1}{(1-q^{12})}, \tag{2.1}$$

and

$$\begin{aligned}&\frac{(1-q)}{(1-q^5)(1-q^6)(1-q^7)(1-q^8)(1-q^9)} \\ &= \left(1-q+\frac{q^5}{1-q^9}+\frac{q^{12}}{(1-q^6)(1-q^8)(1-q^9)}\right)\frac{1}{(1-q^{35})}.\end{aligned} \tag{2.2}$$

The truth of Theorem 3 in the cases $n = 3$ or 5 is readily deduced from (2.1) and (2.2) respectively.

We shall use the following notation:

$$(A;q)_n = (1-A)(1-Aq)\dots(1-Aq^{n-1}), \tag{2.3}$$

$${}_{r+1}\Theta_r\left(\begin{array}{c} a_0, a_1, \dots, a_r; q, t \\ b_1, \dots, b_r \end{array}\right) = \sum_{n=0}^{\infty} \frac{(a_0;q)_n(a_1;q)_n\dots(a_r;q)_n t^n}{(q;q)_n(b_1;q)_n\dots(b_r;q)_n}, \tag{2.4}$$

$$\left[\begin{array}{c} N \\ M \end{array}\right] = \frac{(q;q)_N}{(q;q)_M(q;q)_{N-M}}, \tag{2.5}$$

$$A(n,j) = \frac{(q^{n-j+1};q)_{j-1}}{(q^2;q)_{j-1}} = \frac{(1-q)}{(1-q^n)}\left[\begin{array}{c} n \\ j \end{array}\right]. \tag{2.6}$$

Theorem 1. *For $n \geq 3$ and odd,*

$$\begin{aligned}&\frac{(1-q)}{(1-q^n)(1-q^{n+1})\dots(1-q^{2n-1})} \\ &= \frac{1}{(1-q^{n(3n-1)/2})}\left(1-q+\sum_{j=0}^{\frac{n-3}{2}}\frac{q^{(j+1)(n+j)}A(n-j-1,j+1)}{(q^{n+1};q)_j(q^{2n-j-1};q)_{j+1}}\right)\end{aligned} \tag{2.7}$$

Proof. Multiplying both sides of (2.7) by $(1-q^{n(3n-1)/2})(q^{n+1};q)_{n-1}/(1-q)$, we see that (2.7) is equivalent to the assertion that

$$\frac{1-q^{n(3n-1)/2}}{1-q^n} = \sum_{j=0}^{\frac{n-1}{2}} \frac{q^{j(n+j-1)}(q^{n+j};q)_{n-2j}(q^{n-2j};q)_j}{(q;q)_j(1-q^{n-2j})}. \tag{2.8}$$

Now the expression on the right equals

$$
\begin{aligned}
&\frac{(q;q)_{2n-1}}{(q;q)_n}\ {}_4\Theta_3\left(\begin{matrix} q^{-\frac{n}{2}}, -q^{-\frac{n}{2}}, q^{-\frac{n}{2}+\frac{1}{2}}, -q^{-\frac{n}{2}+\frac{1}{2}}; q, q \\ q^n, q^{-2n+1}, q^{-n+1} \end{matrix}\right) \\
= \ &\frac{(q;q)_{2n-1}(q^{-\frac{n}{2}};q)_{(\frac{n-1}{2})}(-q^{n+1};q)_{(\frac{n-1}{q})}(-q^{n+\frac{1}{2}};q)_{(\frac{n-1}{2})}}{(q;q)_n(q^{1-n};q)_{(\frac{n-1}{q})}(q^n;q)_{(\frac{n-1}{2})}(q^{(3n+1)/2};q)_{(\frac{n-1}{2})}} \\
&\times {}_4\Theta_3\left(\begin{matrix} q^{-(\frac{n-1}{2})}, q^{1-\frac{n}{2}}, q^{\frac{3n}{2}}, q^{\frac{3n}{2}+\frac{1}{2}}; q, q \\ -q^{n+1}, -q^{n+\frac{1}{2}}, q^{\frac{3}{2}} \end{matrix}\right) \\
&\text{(by [3; eq. (III.16), p. 242])} \\
= \ &\frac{(q;q)_{2n-1}(q^{-\frac{n}{2}};q)_{(\frac{n-1}{2})}(-q^{n+1};q)_{(\frac{n-1}{2})}(-q^{n+\frac{1}{2}};q)_{(\frac{n-1}{2})}}{(q;q)_n(q^{1-n};q)_{(\frac{n-1}{2})}(q^n;q)_{(\frac{n-1}{2})}(q^{(3n+1)/2};q)_{(\frac{n-1}{2})}} \\
&\times \frac{(1+q^n)(1-q^{\frac{1}{2}})q^{-\frac{1}{2}}}{(1-q^{(3n-1)/2})(1-q^{-\frac{n}{2}})} \sum_{j\geq 0} \frac{(q^{-\frac{n}{2}};q^{\frac{1}{2}})_{2j+1}(q^{(3n-1)/2};q^{\frac{1}{2}})_{2j+1}q^{j+\frac{1}{2}}}{(q^{\frac{1}{2}};q^{\frac{1}{2}})_{2j+1}(-q^n;q^{\frac{1}{2}})_{2j+1}} \\
= \ &\frac{(q;q)_{2n-1}(q^{-\frac{n}{2}};q)_{(\frac{n-1}{2})}(-q^{n+1};q)_{(\frac{n-1}{2})}(-q^{n+\frac{1}{2}};q)_{(\frac{n-1}{2})}}{(q;q)_n(q^{1-n};q)_{(\frac{n-1}{2})}(q^n;q)_{(\frac{n-1}{2})}(q^{(3n+1)/2};q)_{(\frac{n-1}{2})}} \\
&\times \frac{(1+q^n)(1-q^{\frac{1}{2}})q^{-\frac{1}{2}}}{(1-q^{(3n-1)/2})(1-q^{-\frac{n}{2}})} \sum_{j\geq 0} \frac{(q^{-\frac{n}{2}};q^{\frac{1}{2}})_j(q^{(3n-1)/2};q^{\frac{1}{2}})_j q^{j/2}}{(q^{\frac{1}{2}};q^{\frac{1}{2}})_j(-q^n;q^{\frac{1}{2}})_j} \left(\frac{1-(-1)^j}{2}\right) \\
= \ &\frac{(q;q)_{2n-1}(q^{-\frac{n}{2}};q)_{(\frac{n-1}{2})}(-q^{n+1};q)_{(\frac{n-1}{2})}(-q^{n+\frac{1}{2}};q)_{(\frac{n-1}{2})}}{2(q;q)_n(q^{1-n};q)_{(\frac{n-1}{2})}(q^n;q)_{(\frac{n-1}{2})}(q^{(3n+1)/2};q)_{(\frac{n-1}{2})}} \\
&\times \frac{(1+q^n)(1-q^{\frac{1}{2}})q^{-\frac{1}{2}}}{(1-q^{(3n-1)/2})(1-q^{-\frac{1}{2}})} \left(\frac{q^{(3n-1)n/2}(-q^{(1-n)/2};q^{\frac{1}{2}})_n}{(-q^n;q)_n} - \frac{(-q^{(1-n)/2};q^{\frac{1}{2}})_n}{(-q^n;q)_n}\right) \\
&\text{(by [3; eq.(II.6), p.236] applied to the first} \\
&\text{sum and [3; eq. (II.7), p.236] applied to the second)} \\
= \ &\frac{(1-q^{n(3n-1)/2})}{(1-q^n)},
\end{aligned}
$$

where the last line follows from a routine but lengthy cancellation of most of the products. Having obtained the desired expression for our sum, we have proved Theorem 1. □

3. The $A(n, j)$.

Some of the $A(n, j)$ are polynomials. For example

$$A(8,3) = 1 + q^2 + q^3 + q^4 + q^5 + q^6 + q^8. \tag{3.1}$$

Some are not; for example

$$A(8,6) = \frac{1 + q^2 + q^3 + q^4 + q^5 + q^6}{1+q}. \tag{3.2}$$

Theorem 2. *If n and j are positive integers with $\gcd(n, j) = 1$, then $A(n, j)$ is a reciprocal polynomial of degree $(j-1)(n-j-1)$ with non-negative coefficients. If $\gcd(n, j) > 1$, then $A(n, j)$ is not a polynomial.*

Proof. We recall from (2.6) that

$$A(n,j) = \frac{(1-q^{n-j+1})(1-q^{n-j+2})\dots(1-q^{n-1})}{(1-q^2)(1-q^3)\dots(1-q^j)}. \tag{3.3}$$

Let us first determine whether $A(n, j)$ is a polynomial or not. We note that, as a rational function of q, the only possible poles of $A(n, j)$ occur for primitive k-th roots of unity, ρ, with $2 \leq k \leq j$. Thus ρ is a root of the denominator of order $\lfloor \frac{j}{k} \rfloor$ and a root of the numerator of order $\lfloor \frac{n-1}{k} \rfloor - \lfloor \frac{n-j}{k} \rfloor$.

We write $n = k\nu + r$, $0 \le r < k$, and $j = k\mu + s$, $0 \le s < k$; so

$$\begin{aligned} & \left(\left\lfloor \tfrac{n-1}{k} \right\rfloor - \left\lfloor \tfrac{n-j}{k} \right\rfloor\right) - \left\lfloor \tfrac{j}{k} \right\rfloor \\ = & \left\lfloor \frac{r-1}{k} \right\rfloor - \left\lfloor \frac{r-s}{k} \right\rfloor - \left\lfloor \tfrac{s}{k} \right\rfloor \\ = & \left\lfloor \frac{r-1}{k} \right\rfloor - \left\lfloor \frac{r-s}{k} \right\rfloor \end{aligned}$$

because $0 \le s < k$.

Case 1. $\gcd(n, j) = 1$. Consequently it is impossible that both r and s are 0 because if they were this would mean $k(> 1)$ is a common divisor of n and j. So if $r > 0$, then

$$\left\lfloor \frac{r-1}{k} \right\rfloor - \left\lfloor \frac{r-s}{k} \right\rfloor \ge - \left\lfloor \frac{r-s}{k} \right\rfloor = 0 \text{ or } 1.$$

If $r = 0$, then $s > 0$; so

$$\left\lfloor \frac{r-1}{k} \right\rfloor - \left\lfloor \frac{r-s}{k} \right\rfloor = -1 - \left\lfloor \frac{-s}{k} \right\rfloor = -1 - (-1) = 0.$$

In any event, we see that ρ is at least as often a zero of the numerator as it is of the denominator. Hence in this case $A(n, j)$ has no poles, and so it is a polynomial.

Case 2. $\gcd(n, j) = k > 1$. Then $n = k\nu$ and $j = k\mu$ (with $r = s = 0$). In this case, if ρ is a primitive k^{th} root of unity, then ρ is a root of the denominator of order μ, and a root of the numerator of order

$$\left\lfloor \frac{k\nu - 1}{k} \right\rfloor - \left\lfloor \frac{k\nu - k\mu}{k} \right\rfloor = \mu - 1.$$

Thus $A(n, j)$ has a pole at ρ and so cannot be a polynomial.

Next we note that when $\gcd(n, j) = 1$ (i.e. when $A(n, j)$ is a polynomial), then degree of $A(n, j)$ is

$$\binom{n}{2} - \binom{j+1}{2} - \binom{n-j+1}{2} + 1 = (j-1)(n-j-1).$$

Now if in $A(n, j)$ we replace q by q^{-1} and multiply by $q^{(j-1)(n-j-1)}$, we obtain

$$\begin{aligned} & q^{(j-1)(n-j-1)} \tfrac{(1-q^{-n+j-1})(1-q^{-n+j-2})\dots(1-q^{-n+1})}{(1-q^{-2})(1-q^{-3})\dots(1-q^{-j})} \\ = & q^{(j-1)(n-j-1)-(j-1)n+\binom{j}{2}\binom{j+1}{2}-1} \tfrac{(q^{n-j+1};q)_{j-1}}{(q^2;q)_{j-1}} \\ = & A(n, j). \end{aligned}$$

There, if $\gcd(n, j) = 1$, then $A(n, j)$ is a reciprocal polynomial.

Finally, we recall the second line of (2.6)

$$A(n, j) = \frac{(1-q)}{(1-q^n)} \begin{bmatrix} n \\ j \end{bmatrix}. \tag{3.4}$$

The Gassian polynomial $\begin{bmatrix} n \\ j \end{bmatrix}$ has been proved to have weakly increasing coefficients up to $q^{\lfloor (j(n-j)+1)/2 \rfloor}$ (cf. [1; p. 48], [4], [5; p. 76], [7]). In other words, $(1-q)\begin{bmatrix} n \\ j \end{bmatrix}$ has non-negative coefficients up through $q^{\lfloor (j(n-j)+1)/2 \rfloor}$. Consequently as a rational function expanded in a power series around $q = 0$, $A(n, j)$ has non-negative coefficients up through $q^{\lfloor (j(n-j)+1)/2 \rfloor}$. But $A(n, j)$ is a reciprocal polynomial of degree $(j-1)(n-j-1)$. Hence if it has non-negative coefficients up through $q^{\lfloor (j-1)(n-j-1)/2 \rfloor}$, then all its coefficients are non-negative. Since

$$\lfloor (j-1)(n-j-1)/2 \rfloor \leq \lfloor (j(n-j)+1)/2 \rfloor,$$

we know that all the coefficients of $A(n,j)$ are non-negative whenever $\gcd(n,j)=1$. □

4. Proof of Theorem 3.

We suppose n is an odd prime. We note that for any j with $0 \leq j \leq \frac{n-3}{2}$, $\gcd(n-j-1, j+1) = 1$ because if those two numbers had a common factor $d > 1$, then d would divide $(n-j-1)+(j+1) = n$, a prime which is impossible because $d \leq j+1 \leq \frac{n-1}{2}$. Hence when we examine the entries in (2.7) we see that every summand in the sum of j is a rational function with non-negative coefficients by Theorem 2. Of course, the coefficient of q on the right-hand side of (2.7) is -1. However if we can show that $q^{n(3n-1)/2+1}$ appears with a positive coefficient in the term $j = (n-3)/2$, then we will have proved that $-q$ is the only term with a negative coefficient on the right hand side of (2.7) and Theorem 3 will be proved.

Case 1. $n = 4\nu+1$. In this case, the $j = (n-3)/2$ term is

$$\frac{q^{12\nu^2}}{(1-q^{4\nu+2})(1-q^{4\nu+3})\dots(1-q^{6\nu})(1-q^{6\nu+2})(1-q^{6\nu+3})\dots(1-q^{8\nu+1})} \tag{4.1}$$

Now

$$\begin{aligned} & 1+n(3n-1)/2 \\ = \ & 1+(n+(n+1)+(n+2)+\dots+(2n-1)) \\ = \ & 1+(n+(5\nu+2)+(5\nu-1)+\dots+7\nu)+(n+1)+(n+2)+\dots+(5\nu+1) \\ & \qquad +(7\nu+1)+\dots+(8\nu+1) \\ = \ & 12\nu^2+2\times(n+2)+(n+3)+\dots+(5\nu+1)+(7\nu+1)+\dots+(8\nu+1). \end{aligned}$$

This last expression clearly arises as a partition generated by (4.1). Hence the coefficient of $q^{n(3n-1)/2+1}$ is positive in (4.1).

Case 2. $n = 4\nu+3$ with $\nu > 2$. In this case, the $j = (n-3)/2$ term is

$$\frac{q^{3(2\nu+1)^2}}{(1-q^{4\nu+4})(1-q^{4\nu+5})\dots(1-q^{6\nu+3})(1-q^{6\nu+5})(1-q^{6\nu+6})\dots(1-q^{8\nu+5})} \tag{4.2}$$

Again

$$\begin{aligned} & 1+n(3n-1)/2 \\ = \ & 1+(n+(n+1)+(n+2)+\dots+(2n-1) \\ = \ & 1+((5\nu+3)+(5\nu+4)+\dots+(7\nu+3)) \\ & +((4\nu+3)+(4\nu+4)+\dots+(5\nu+2)) \\ & +((7\nu+4)+(7\nu+5)+\dots+(8\nu+5)) \\ = \ & 3(2\nu+1)^2+2\times(4\nu+4)+(4\nu+5)+\dots+(5\nu+2) \\ & +(7\nu+4)+(7\nu+5)+\dots+(8\nu+5). \end{aligned}$$

This last expression clearly arises as a partition generated by (4.2). Hence the coefficient of $q^{n(3n-1)/2+1}$ is positive in (4.2).

Case 3. $n = 3$. In this case $j = 0$, and we see immediately that since $13 = 3+2$, the coefficient of $q^{n(3n-1)/2+1}$ in

$$\frac{q^3}{(1-q^5)}$$

is positive.

Case 4. $n = 7$. In this case $j = 1$, and we see directly that since $71 = 27 + 8 + 11 + 12 + 13$ the coefficient of $q^{n(3n-1)/2+1}$ in

$$\frac{27}{(1-q^8)(1-q^9)(1-q^11)(1-q^12)(1-q^13)}$$

is positive.

Therefore, we may rewrite the right hand side of (2.7) as

$$= -q + \frac{1}{(1-q^{n(3n-1)/2})}\left\{1 - q^{n(3n-1)/2+1} + \sum_{j=0}^{\frac{n-3}{2}} \frac{q^{(j+1)(n+j)}A(n-j-1,j+1)}{(q^{n+1};q)_j(q^{2n-j-1};q)_{j+1}}\right\},$$

and we have just proved that the expression inside the brackeets has non-negative coefficients. Therefore every power of q in the expansion of

$$\frac{(1-q)}{(1-q^n)(1-q^{n+1})\dots(1-q^{2n-1})}$$

beyond q^1 has non-negative coefficients. Thus Theorem 3 is proved. □

5. Conclusion.

One would hope that it might be possible to prove the full Friedman-Joichi-Stanton conjecture from Theorem 1. This is tempting but not immediate. When n is composite some of the terms in the sum in (2.7) have negative coefficients.

Conjecture. For each k with $0 \le k \le (n-1)/2$ and n odd, the coefficients in

$$\sum_{j=0}^{k} \frac{q^{(j+1)(n+j)}A(n-j-1,j+1)}{(q^{n+1};q)_j(q^{2n-j-1};q)_{j+1}}$$

are non-negative.

The conjecture is clearly true for $k = 0$ and 1, because $A(n-1,1)$ and $A(n-2,2)$ are both polynomials with nonnegative coefficients since n is odd. Obviously the Friedman-Joichi-Stanton conjecture should follow with minimal effort from this conjecture.

Dennis Stanton has observed that one may alternatively deduce Theorem 1 from [3; eq. (III.14), p.242] under the substitution $a = q^n$, $b = 1$, $z = q^{2n}$. Further study of this identity might lead to new positivity results. He has also noted that one can extend the results in this paper to strict montonicity of the coefficients past q^{3n+4} for primes ≥ 7; furthermore this observation is required for the application in [2].

Finally it is of interest to learn more about the $A(n,j)$ especially when n and j are relatively prime. Since $A(\infty,j)$ is the generating function for partitions with no 1's and all parts $\le j$, we would expect that $A(n,j)$ is the generating function for some finite subsets of the partitions with no 1's and all parts $\le j$. Furthermore something must appear in the constraints defining these partitions that makes clear the role of $\gcd(n,j) = 1$.

References

[1] G.E.Andrews, *The Theory of Partitions*, Encycl. of Math. and Its Applications, Vol. 2, G.-C. Rota Ed., Addison-Wesley, Reading, 1976. (Reissued: Cambridge University Press, Cambridge, 1998).

[2] J.Friedman, J.T.Joichi, and D.Stanton, *More monotonicity theorems for partitions*, Exp. Math. 4 (1994), 31-37.

[3] G.Gasper and M.Rahman, *Basic Hypergeometric Series*, Encycl. of Math. and Its Applications, Vol. 35, G.-C. Rota Ed., Cambridge University Press, Cambridge, 1990.

[4] R.Proctor, *Solutions of two difficult combinatorial problems with linear algebra*, Amer. Math. Monthly 89(1982), 721-734.

[5] I.Schur, *Vorlesungen über Invariantentheorie*, Springer, Berlin, 1968.

[6] D.Stanton, *Open positivity conjectures for integer partitioins*, preprint.

[7] J.J.Sylvester, *Proof of the hitherto undemonstrated fundamental theorem of invariants*, Philosophical Mag., 5 (1878), 178-188. (Reprinted: Coll. Papers of J.J.Sylvester, Vol. 3, 117-126, Chelsea, New York, 1973).

Department of Mathematics, The Pennsylvania State University, University Park, PA 16802

E-mail address: `andrews@math.psu.edu`

3.18 The Theory of Compositions, IV: Multicompositions

The Mathematics Student,
Special Centenary Volume (2007), 25-31

THE THEORY OF COMPOSITIONS IV: MULTICOMPOSITIONS

GEORGE E. ANDREWS

(Received: 31-08-2007)

ABSTRACT. The theory of ordered partitions, or compositions, originated with P. A. MacMahon. In this paper, we explore compositions wherein several copies of the integers are used as summands.

1. Introduction

Compositions are ordered partitions of integers. For example, there are eight compositions of 4: 4, $3+1$, $1+3$, $2+2$, $2+1+1$, $1+2+1$, $1+1+2$, $1+1+1+1$. In his initial study of compositions ([4, p. 151], [5, p. 620]), P. A. MacMahon noted that there are 2^{n-1} compositions of n.

Our object in this paper is to consider compositions wherein we use k copies of the positive integers as summands: $\{1_1, 1_2, \ldots, 1_k, 2_1, 2_2, \ldots, 2_k, 3_1, 3_2, \ldots, 3_k, \ldots.\}$. We shall refer to these new compositions as multicompositions (or k-compositions when using exactly k copies of the integers). We shall add one restriction on the summands, namely, *THE LAST SUBSCRIPT IN THE COMPOSITION MUST BE 1.* If we do not throw this in then we will have k sets of k-compositions which will be identical except for the last subscript. For example, in the case of bi-partitions of 3, our restriction allows 3_1, 2_1+1_1, 2_2+1_1, 1_1+2_1, 1_2+2_1, $1_1+1_1+1_1$, $1_2+1_1+1_1$, $1_1+1_2+1_1$, $1_2+1_2+1_1$ Whereas allowing any final subscript would add nine additional bi-compositions of 3, namely 3_2, 2_1+1_2, 2_2+1_2, 1_1+2_2, 1_2+2_2, $1_1+1_1+1_2$, $1_2+1_1+1_2$, $1_1+1_2+1_2$, $1_2+1_2+1_2$. In the interest of keeping the hand calculation of the relevant sums within reason, we have added this restriction.

In 1964, H. Gould [3, p. 251] studied compositions of n with relatively prime summands. We shall call this number the Gould function, $g_1(n)$.

2000 AMS Classification Numbers: 11P81, 11P83
Key words and phrases: compositions, multicompositions, Gould function.
Partially supported by National Science Foundation Grant No. DMS 0457003.

Thus $g_1(4) = 6$ because the compositions enumerated are $3+1$, $1+3$, $2+1+1$, $1+2+1$, $1+1+2$ and $1+1+1+1$. Clearly as is implied in Gould's paper

$$2^{n-1} = \sum_{d|n} g_1(d). \tag{1.1}$$

Recently E. Deutsch, in a problem in the American Mathematical Monthly [2], noted that $3|g_1(n)$ if $n \geq 3$, which means that since $g(1) = g(2) = 1$ the only time $g(n)$ is prime is when $n = 3$. Actually $g_1(n)$ is highly composite as the following table reveals.

n	$g_1(n)$	$g_1(n)$ factored	n	$g_1(n)$	$g_1(n)$ factored
1	1	1	11	1023	$3 \cdot 11 \cdot 31$
2	1	1	12	2010	$2 \cdot 3 \cdot 5 \cdot 67$
3	3	3	13	4095	$3^2 \cdot 5 \cdot 7 \cdot 13$
4	6	$2 \cdot 3$	14	8127	$3^3 \cdot 7 \cdot 43$
5	15	$3 \cdot 5$	15	16365	$3 \cdot 5 \cdot 1091$
6	27	3^3	16	32640	$2^7 \cdot 3 \cdot 5 \cdot 17$
7	63	$3^2 7$	17	65535	$3 \cdot 5 \cdot 17 \cdot 257$
8	120	$2^3 35$	18	130788	$2^2 \cdot 3^3 \cdot 7 \cdot 173$
9	252	$2^2 3^2 7$	19	262143	$3^3 \cdot 7 \cdot 19 \cdot 73$
10	495	$3^2 511$	20	523770	$2 \cdot 3 \cdot 5 \cdot 13 \cdot 17 \cdot 19$

Obviously there is much more divisibility going on here than just divisibility by 3. Is there a more general factorization theorem for $g_1(n)$ than that given by Deutsch's problem?

Of course, we may naturally generalize Gould's function to $g_k(n)$ the number of k-compositions of n wherein the summands are relatively prime (here we ignore subscripts, so the summands in $4_2 + 2_1 + 2_2$ are viewed as having 2 as a common divisor).

Here is a table of the first twenty values of $g_2(n)$:

n	$g_2(n)$	$g_2(n)$ factored	n	$g_2(n)$	$g_2(n)$ factored
1	1	1	11	59048	$2^3 \cdot 11^2 \cdot 61$
2	2	2	12	176880	$2^4 \cdot 3 \cdot 5 \cdot 11 \cdot 67$
3	8	2^3	13	531440	$2^4 \cdot 5 \cdot 7 \cdot 13 \cdot 73$
4	24	$2^3 \cdot 3$	14	1593592	$2^3 \cdot 7 \cdot 11 \cdot 13 \cdot 199$
5	80	$2^4 \cdot 5$	15	4782880	$2^5 \cdot 5 \cdot 167 \cdot 179$
6	232	$2^3 \cdot 29$	16	14346720	$2^5 \cdot 3^7 \cdot 5 \cdot 41$
7	728	$2^3 \cdot 7 \cdot 13$	17	43046720	$2^6 \cdot 5 \cdot 17 \cdot 41 \cdot 193$
8	2160	$2^4 \cdot 3^3 \cdot 5$	18	129133368	$2^3 \cdot 3^2 \cdot 7 \cdot 13 \cdot 19709$
9	6552	$2^3 \cdot 3^2 \cdot 7 \cdot 13$	19	387420488	$2^3 \cdot 7 \cdot 13 \cdot 19 \cdot 37 \cdot 757$
10	19600	$2^4 \cdot 5^2 \cdot 7^2$	20	1162241760	$2^5 \cdot 3 \cdot 5 \cdot 41 \cdot 73 \cdot 809$

The object of this paper will be to prove the following result. We use the notation $\phi(n)$ to denote the number of positive integers $\leq n$ and relatively prime to n.

Theorem 1. *If the prime factorization of n is $p_1^{\alpha_1}p_2^{\alpha_2}\dots p_r^{\alpha_r}$, then the number*

$$\ell.c.m\left(k(k+2),\ (k+1)^{\phi(p_1^{\alpha_1})}-1,\ (k+1)^{\phi(p_2^{\alpha_2})}-1,\dots,(k+1)^{\phi(p_r^{\alpha_r})}-1\right)$$

divides $g_k(n)$ provided $n \geqq 3$.

For example, if $n = 18 = 3^2 2$, $k = 2$, then

$$\operatorname{lcm}\left(2\cdot 4, 3^{\phi(2)}-1, 3^{\phi(3)}-1\right) = \operatorname{lcm}(8,2,728) = 728 = 2^3\cdot 7\cdot 13.$$

The next section will be devoted to a proof of this theorem. We will conclude with some observations and open problems.

2. Proof of Theorem 1

In order to prove Theorem 1, it is sufficient to prove that each of the entries in the $\ell.c.m$ (= least common multiple) expression divides $g_k(n)$. First we require some preliminary results.

Lemma 2. *Suppose $\{a_n\}$ and $\{b_n\}$ are sequences of integers,*

$$\sum_{n=m}^{\infty} b_n q^n = \sum_{n=m}^{\infty} \frac{a_n q^n}{1-q^n}, \tag{2.1}$$

and $j|b_n$ for $n \geqq m$. Then $j|a_n$ for $n \geqq m$.

Proof. We proceed by induction. We note that

$$\sum_{n=m}^{\infty} \frac{a_n q^n}{1-q^n} = \sum_{n=m}^{\infty}\sum_{s=1}^{\infty} a_n q^{ns}. \tag{2.2}$$

Hence the coefficient of q^m on the right-hand side is a_m; therefore $a_m = b_m$, so $j|a_m$.

Now assume that $j|a_n$ for $m \leqq n < N$. So

$$\begin{aligned}\sum_{n=N}^{\infty} \frac{a_n q^n}{1-q^n} &= \sum_{n=m}^{\infty} b_n q^n - \sum_{n=m}^{N-1} \frac{a_n q^n}{1-q^n}\\ &= \sum_{n=m}^{\infty} b_n q^n - \sum_{n=m}^{N-1}\sum_{s=1}^{\infty} a_n q^{ns}.\end{aligned} \tag{2.3}$$

All the coefficients on the right-hand side of (2.3) are divisible by j, and in particular the coefficient of q^N is so divisible. But on the left-hand side we see that the coefficient of q^N is just a_N. Hence $j|a_N$, and the result follows by mathematical induction. □

Lemma 3. *The total number of k-compositions of n is $(k+1)^{n-1}$.*

Proof. We modify MacMahon's proof of the case $k=1$. Namely, we can geometrically represent the k-compositions as follows: choose n_1 numbers among $\{1, 2, \ldots, n-1\}$ to be labelled "1"; choose n_2 numbers among those remaining to be labelled "2", etc. up through k. The number of possible choices is

$$\binom{n-1}{n_1, n_2, \ldots, n_k, n-1-n_1-n_2-\cdots-n_k} = \frac{(n-1)!}{n_1! n_2! \cdots n_k! (n-1-n_1-n_2-\cdots-n_k)!}$$

These labelled points define integer length segments on the unit line,

and these segments make up the k-composition with the stipulation that the label of the right end determines the subscript of the part (n is labelled "1").

For example. If $k=3$, $n=10$, $n_1=3$, $n_2=2$, $n_3=1$, then one possible choice is

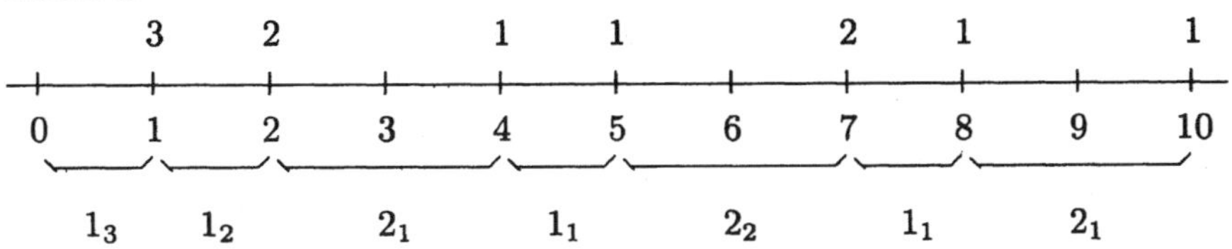

yielding the 3-composition of 10: $1_3 + 1_2 + 2_1 + 1_1 + 2_2 + 1_1 + 2_1$. Thus the number of k-compositions of n with n_1 parts with subscript 1 (excluding the final part), n_2 parts with subscript 2, etc. is

$$\binom{n-1}{n_1, n_2, \ldots, n_k, n-1-n_1-n_2-\cdots-n_k}$$

and so the total number of k-compositions of n is

$$\sum_{n_1, n_2, \ldots, n_k \geqq 0} \binom{n-1}{n_1, n_2, \ldots, n_k, n-1-n_1-n_2-\cdots-n_k}$$
$$= (1 + \underbrace{1+1+1+\cdots+1}_{k \text{ terms}})^{n-1}$$
$$= (k+1)^{n-1}.$$

We are now in a position to prove Theorem 1. First we shall prove that $k(k+2)$ divides $g_k(n)$ for $n \geqq 3$.

We see immediately that

$$(k+1)^{n-1} = \sum_{d|n} g_k(d),$$

by classifying the k-compositions of n according to the greatest common divisor of their parts.

We may translate this into generating function form as follows:

$$\begin{aligned}\frac{q}{1-(k+1)q} &= \sum_{n=1}^{\infty}(k+1)^{n-1}q^n = \sum_{n=1}^{\infty}\sum_{d\cdot e=n} g_k(d)q^n = \sum_{d=1}^{\infty}\sum_{e=1}^{\infty} g_k(d)q^{ed} \\ &= \sum_{d=1}^{\infty}\frac{g_k(d)q^d}{1-q^d}.\end{aligned}$$

Now $g_k(1) = 1$ because $g_k(1)$ only counts 1_1, and $g_k(2) = k$ because $g_k(2)$ counts $1_1 + 1_1$, $1_2 + 1_1, \ldots, 1_k + 1_1$. Hence

$$\begin{aligned}\sum_{d=3}^{\infty}\frac{g_k(d)q^d}{1-q^d)} &= \frac{q}{1-(k+1)q} - \frac{q}{1-q} - \frac{kq^2}{1-q^2} \\ &= \frac{k(k+2)q^3}{(1-q)(1-q^2)(1-(k+1)q)}\end{aligned}$$

Therefore by Lemma 1, $k(k+2)$ divides $g_k(n)$ for $n \geqq 3$.

Next we must consider each of the primes p_i which occur in the prime factorizations of $n = p_1^{\alpha_1} \cdots p_r^{\alpha_r}$.

So let

$$n = p^{\alpha}m,$$

where $(p,m) = 1$.

Then by Möbius inversion [1, p. 87]

$$\begin{aligned}g_k(n) &= \sum_{d|n}\mu(d)(k+1)^{\frac{n}{d}-1} \\ &= \sum_{d|m}\mu(d)(k+1)^{p^{\alpha}\frac{m}{d}-1} + \sum_{d|m}\mu(pd)(k+1)^{p^{\alpha-1}\frac{m}{d}-1} \\ &\qquad (\text{because } \mu(d) = 0 \text{ if } p^2|d) \\ &= \sum_{d|m}\mu(d)(k+1)^{p^{\alpha-1}\frac{m}{d}-1}\left((k+1)^{\frac{m}{d}(p^{\alpha}-p^{\alpha-1})} - 1\right).\end{aligned}$$

Now each term in this last expression is clearly divisible by

$$(k+1)^{p^{\alpha}-p^{\alpha-1}} - 1 = (k+1)^{\phi(p^{\alpha})} - 1$$

This concludes the proof of Theorem 1 because we have now shown that each of the factors in the ℓ.c.m. divides $g_k(n)$. □

Corollary 4. *If all the prime factors of n are relatively prime to $k+1$, then $n | g_k(n)$.*

Proof. Let $n = p_1^{\alpha_1} p_2^{\alpha_2} \cdots p_r^{\alpha_r}$. Since $(p_i, k+1) = 1$, we know by Euler's Theorem [1, p. 62] that

$$(k+1)^{\phi(p_i^{\alpha_i})} \equiv 1 \pmod{p_i^{\alpha_i}}.$$

Hence for each i,

$$p_i^{\alpha_i} \left| \left\{ (k+1)^{\phi(p_i^{\alpha_i})} - 1 \right\} \right| g_k(n);$$

therefore, $n = p_1^{\alpha_1} \cdots p_r^{\alpha_r} \Big| g_k(n)$. □

Corollary 5. *If n is odd, then $n | g_1(n)$.*

Proof. This follows from Corollary 4 with $k = 1$. □

3. Conclusion

While Theorem 1 explains a lot about why $g_k(n)$ has many small prime factors, it clearly doesn't explain everything. For example,

$$g_3(36) = 2^{10} \cdot 3^2 \cdot 5 \cdot 7 \cdot 13 \cdot 17 \cdot 67 \cdot 241 \cdot 1{,}025{,}663{,}893.$$

whereas

$$\begin{aligned} \ell cm(3 \cdot 5, 4^{\phi(9)} - 1,\ 4^{\phi(4)} - 1) &= \ell cm(3 \cdot 5,\ 4^6 - 1,\ 4^2 - 1) \\ &= 3^2 \cdot 5 \cdot 7 \cdot 13. \end{aligned}$$

This leaves $2^{10} \cdot 17$ unexplained.

So a natural project is the following:

Find other theorems like Theorem 1 that account for the other small prime factors of $g_k(n)$.

Also one may view $g_k(n)$ as a polynomial in k. In that case, Theorem 1 is still valid and asserts the divisibility of polynomials. So when $n = 36$

$$\begin{aligned} g_k(36) &= k(k+1)^5(k+2)(k^2+k+1)(k^2+2k+2) \\ &\quad \times (k^2+3k+3)(k^4+4k^3+5k^2+2k+1) \\ &\quad \times (k^{18} + 18k^{17} + \cdots + 48620\,k^9 + \cdots + 24k + 1). \end{aligned}$$

On the other hand

$$\begin{aligned} &\ell.c.m.(k(k+2), (k+1)^{\phi(2^2)} - 1,\ (k+1)^{\phi(3^2)} - 1) \\ &= k(k+2)(k^2+k+1)(k^2+3k+3), \end{aligned}$$

and this leaves unexplained the factors

$$\begin{aligned}&(k+1)^5(k^2+2k+2)(k^4+4k^3+5k^2+2k+1)\\&=(k+1)^5((k+1)^2+1)((k+1)^4-(k+1)^2+1).\end{aligned}$$

We note that each of the irreducible polynomial factors $p(k)$ of our ℓ.c.m. is such that $p(k-1)$ is a cyclotomic polynomial. So our final project is to account for all the other factors or $g_k(n)$ that are instances of cyclotomic polynomials evaluated at $k+1$.

References

[1] G. E. Andrews, *Number Theory*, Dover, New York, 1994.

[2] E. Deutsch, *Problem 11161 solution*, Amer. Math. Monthly, 114 (2007), 363.

[3] H. W. Gould, *Binomial coefficients, the bracket function, and compositions with relatively prime summands*, Fibonacci Quart., 2 (1964), 241–260.

[4] P. A. MacMahon, *Combinatory Analysis*, Vol. 1, Cambridge University Press, London, 1916. (Reprinted: Chelse, New York, 1960).

[5] P. A. MacMahon, *Collected Papers*, Vol. 1, G. E. Andrews, ed., MIT Press, Cambridge, 1978.

George E. Andrews
Department of Mathematics,
306 McAllister Bldg., The Pennsylvania State University,
University Park, PA 16802
E-mail address: andrews@math.psu.edu

Chapter 4

Plane Partitions

I have chosen to include four of my sixteen papers on plane partitions. Four papers related to plane partitions: [230], [249], [254], and [257] are applications of MacMahon's Partition Analysis; a topic dealt with at length in Chapter 2. Paper [89] examines a few corollaries of [84], and paper [95] is effectively an announcement of [84]. Paper [41] is a brief disproof of a plane partitions conjecture of Guinand.

My first real research carried out on plane partitions was my successful assault on the MacMahon conjecture for plane partitions [79]. Instead of including [79], I have chosen [72], which is the case $q = 1$ of [79]. Paper [72] avoids the complications associated with arbitrary q that arise in [79].

Paper [75] (which was written after [79]) proved the equivalence of two famous plane partitions conjectures: the MacMahon conjecture and the Bender–Knuth conjecture [Ben-K72]. Richard Stanley [Sta71] had remarked that Basil Gordon had a proof of the Bender–Knuth conjecture. Gordon eventually published his proof [Gor83]. Thus [75], together with Gordon's paper [Gor83], provide an alternative proof of the MacMahon conjecture.

Paper [84] has probably been of wider interest than any of my other papers on plane partitions. It is in this paper that descending plane partitions are introduced, and these objects are shown to be enumerated by the sequence:

$$1, 2, 7, 42, 429, \ldots, \prod_{i=0}^{n-1} \frac{(3i+1)!}{(n+i)!}, \ldots.$$

Subsequent developments surrounding this sequence are well chronicled by Robbins [Rob91], Zeilberger [Zei94], and in Bressoud's prizewinning book, *Proofs and Confirmations* [Bre99].

I have included the paper [135] because it best illustrates the method used throughout the six papers in my "Plane Partitions" series. It settles one of the Mills–Robbins–Rumsey conjectures.

The fifth paper in the "Plane Partitions" series [166] proves the conjecture of Macdonald on totally symmetric self-complementary plane partitions (TSSCPP). It builds on path-breaking work by John Stembridge [Ste90] that reduced the problem to the evaluation of a Pfaffian. I have chosen to include [173] instead of [166] because

[173] contains a much simpler (actually readable) proof of the TSSCPP conjecture. The efforts on TSSCPP were all done at IBM-Yorktown Heights when I was a guest of the SCRATCHPAD group, and [173] is joint with Bill Burge.

Paper [187] is joint with Dennis Stanton. It both provides an alternative proof of the identity proved in [184], and settles a conjecture from the same paper.

In 1995, John Stembridge [Ste95] proved the totally symmetric plane partitions conjecture (TSPP). Then, in 2003, a conference was convened to honor my friend David Robbins who was terminally ill. Thus, I wanted to present a paper that would be related to Robbins' interests. The first thing I thought of was an approach I had taken to the TSPP conjecture when I was at IBM. My efforts had fallen short, but I had reduced the problem to the evaluation of some extremely complicated hypergeometric identities. My friends at RISC, Carsten Schneider and Peter Paule, had developed the automated methods of series summation to the point that they were able to rescue my stranded project. The result was [248].

4.1 MacMahon's Conjecture on Symmetric Plane Partitions

Reprinted from

Proc. Natl. Acad. Sci. USA
Vol. 74, No. 2, pp. 426–429, February 1977
Mathematics

MacMahon's conjecture on symmetric plane partitions

(combinatorial identities/hypergeometric series)

GEORGE E. ANDREWS*

Mathematics Research Center, 610 Walnut Street, Madison, Wisconsin 53706

Communicated by Donald E. Knuth, December 3, 1976

ABSTRACT In 1898, P. A. MacMahon conjectured that the generating function for symmetric plane partitions with at most s rows and each part at most j has an especially simple closed form. An outline of the proof of this conjecture is given here.

1. Introduction

In 1898, P. A. MacMahon (1) presented his first study of symmetric higher dimensional partitions. One of the most interesting parts of his paper is a conjecture concerning the generating function for $M(j, s; n)$ the number of representations of n of the form

$$n = \sum_{i,k \geqq 1} n_{ik}, \qquad [1.1]$$

in which the n_{ik} are nonnegative integers such that (i) $n_{ik} \geqq n_{i,k+1}$, (ii) $n_{ik} \geqq n_{i+1,k}$, (iii) $n_{ik} = n_{ki}$, (iv) $n_{ik} = 0$ if $i > s$, (v) $n_{11} \leqq j$. In the standard terminology of plane partitions [see Stanley (2)], $M(j, s; n)$ is the number of symmetric plane partitions with at most s rows and with each part at most j. MacMahon (ref. 1, p. 153) conjectures

$$\sum_{n \geqq 0} M(j, s; n) q^n = \prod_{i=1}^{s} \left\{ \frac{(1 - q^{j+2i-1})}{(1 - q^{2i-1})} \prod_{h=i+1}^{s} \frac{[1 - q^{2(j+i+h-1)}]}{[1 - q^{2(i+h-1)}]} \right\}. \qquad [1.2]$$

He demonstrates the truth of this conjecture in a few instances and remarks: "The proof of this formula, the truth of which seems unquestionable, is much to be desired."

Subsequently, in his monumental treatise *Combinatory Analysis*, MacMahon (ref. 3; pp. 262–271) again discusses this conjecture at length, and again he asserts that "The result has not been rigorously established." At the conclusion of his discussion (ref. 3; pp. 270–271), MacMahon carefully examines the symmetry in the product in [**1.2**] and states that "This property of the enumerating function is of great beauty and mathematical elegance."

In the late 1960's Gordon (4, 5) proved MacMahon's conjecture when $s = \infty$. Gordon (ref. 5; p. 158) observes that Sylvester's mapping of self-conjugate partitions into partitions with distinct odd parts may be directly extended to plane partitions to show that $M(j, s; n)$ is also the number of plane partitions of n with strict decrease along rows where each part is odd and at most $2s - 1$ and there are at most j rows.

Also in the late 1960's, Bender and Knuth (6) developed a very powerful combinatorial method for treating many problems in plane partitions. They also extended much of the work of Gordon (4, 5) and of Gordon and Houten (7, 8), and they showed (ref. 6; p. 50) that if

$$g_j(q) = \sum_{n \geqq 0} M(j, m; n) q^n, \qquad [1.3]$$

then

$$g_{2n}(q) = \det(C_{i-j} + C_{i+j-1})_{n \times n}, \qquad [1.4]$$

and

$$g_{2n+1}(q) = \left\{ \prod_{i=1}^{m} (1 + q^{2i-1}) \right\} \det(C_{i-j} - C_{i+j})_{n \times n}, \qquad [1.5]$$

in which

$$C_k = q^{k^2} \binom{2m}{m+k}_2, \qquad [1.6]$$

and $\binom{N}{M}_r$ is the Gaussian polynomial (or q-binomial coefficient) defined by

$$\binom{N}{M}_r = \begin{cases} \dfrac{(1 - q^{Nr})[1 - q^{(N-1)r}] \ldots [1 - q^{(N-M+1)r}]}{(1 - q^{Mr})[1 - q^{(M-1)r}] \ldots (1 - q^r)} & , 0 < M \leqq N \\ & , M = 0 \\ 1 & , M < 0 \\ 0 & M > N. \end{cases} \qquad [1.7]$$

However, Bender and Knuth (ref. 6; p. 50) go on to state that "We have not been able to simplify these determinants any further, even for the limiting case as $q \to 1$ . . . But the known results, and calculations for small j give overwhelming empirical evidence that the answer has a simple form."

The object of this paper is to sketch the proof of MacMahon's conjecture (a complete account will appear in the *Advances in Mathematics*). In Section 2, we shall give a complete proof of MacMahon's conjecture in the case $q = 1$. In Section 3, we shall indicate how to extend the work in Section 2 to prove the conjecture in general.

2. The MacMahon conjecture for $q = 1$

We begin by observing that Eq. **1.2** reduces to

$$\sum_{n \geqq 0} M(j, s; n) = \prod_{i=1}^{s} \left\{ \frac{(j + 2i - 1)}{(2i - 1)} \prod_{h=i+1}^{s} \frac{(j + i + h - 1)}{(i + h - 1)} \right\}, \qquad [2.1]$$

when $q \to 1$. Furthermore the C_k defined by Eq. **1.6** becomes the binomial coefficient $\binom{2m}{m+k}$. We now define ten $n \times n$ matrices:

$$\alpha_n = \left[\binom{2m}{m+i-j} + \binom{2m}{m+i+j-1}\right]_{n \times n}, \qquad [2.2]$$

$$\alpha'_n = \left[\binom{2m}{m+i-j} - \binom{2m}{m+i+j}\right]_{n \times n}, \qquad [2.3]$$

$$\delta_n = \left[\binom{2j-1}{j-i}\right]_{n \times n}, \qquad [2.4]$$

* Present address: Department of Mathematics, Pennsylvania State University, University Park, Pa. 16802.

$$\delta'_n = \left[\binom{2j}{j-i} \frac{i}{j} \right]_{n \times n}, \quad [2.5]$$

$$\beta_n = \left[\binom{2m+2j-1}{m+i+j-1}\right]_{n \times n}, \quad [2.6]$$

$$\beta'_n = \left[\binom{2m+2j}{m+i+j} \frac{i}{m+j} \right]_{n \times n}, \quad [2.7]$$

$$\epsilon_n = \left[(-1)^{i+j} \frac{(2m+2j-1)!(m+i+j-2)!(m+i-1)!}{(2m+2i-1)!(m+2j-2)!(m+j-1)!} \binom{j-1}{i-1} \right]_{n \times n} \quad [2.8]$$

$$\epsilon'_n = \left[(-1)^{i+j} \frac{(2m+2j)!(m+i+j-1)!(m+i)!}{(2m+2i)!(m+2j-1)!(m+j-1)!} \binom{j-1}{i-1} \right]_{n \times n}, \quad [2.9]$$

$$\zeta_n = \left[\frac{(-1)^{j+1(2m+2j-1)!m!(1-i)}j - 1^{(2-i-j)}j - 1}{(m+2j-2)!(m+i)!(m+j-i)!(1-i-m-j)_{j-1}} \right]_{n \times n}, \quad [2.10]$$

$$\zeta'_n = \left[\frac{(-1)^j(2m+2j)!m!(-i)_j(1-i-j)_{j-1}(m+j)}{(m+2j-1)!(m+i+1)!(m+j-1)!(-i-m-j)_{j-1}} \right]_{n \times n}, \quad [2.11]$$

in which $(a)_n = a(a+1)\ldots(a+n-1)$.
Given the results of Bender and Knuth described in Section 1, we see that we must evaluate $g_{2n}(1) = \det \alpha_n$ and $g_{2n+1}(1) = 2^m \det \alpha'_n$ To perform these evaluations we prove two lemmas that provide the necessary relations among out 10 matrices.

LEMMA 1. *For each* $n \geqq 1$: $\alpha_n\delta_n = \beta_n$, $\alpha'_n\delta'_n = \beta'_n, \beta_n\epsilon_n = \zeta_n$, $\beta'_n\epsilon'_n = \zeta'_n$.

Proof: The first two identities follow directly from the classical and elementary Chu–Vandermonde summation (Eq. 3, p. 8 of ref. 9)

$$\sum_{j=-\infty}^{\infty} \binom{A}{B+j}\binom{C}{D-j} = \binom{A+C}{B+D}, \quad [2.12]$$

which is equivalent to the simple terminating hypergeometric series summation

$${}_2F_1\left[{a,-n;1 \atop b}\right] = \sum_{j=0}^{n} \frac{(a)_j(-n)_j}{j!(b)_j} = \frac{(b-a)_n}{(b)_n}, \text{ [ref. (10), p. 3].}$$

$$\begin{aligned}
\alpha_n \cdot \delta_n &= \left[\sum_{l=1}^{n} \left\{\binom{2m}{m+i-l} + \binom{2m}{m+i+l-1}\right\} \binom{2j-1}{j-l} \right]_{n \times n} \\
&= \left[\sum_{l=1}^{n} \binom{2m}{m+i-l}\binom{2j-1}{j-l} + \sum_{l=0}^{-n+1} \binom{2m}{m+i-l}\binom{2j-1}{j+l-1} \right]_{n \times n} \\
&= \left[\sum_{l=-n+1}^{n} \binom{2m}{m+i-l}\binom{2j-1}{j+l-1} \right]_{n \times n} \\
&= \left[\binom{2m+2j-1}{m+i+j-1}\right]_{n \times n} \qquad \text{(by Eq. 2.12)} \\
&= \beta_n.
\end{aligned}$$

$$\begin{aligned}
\alpha'_n\delta'_n &= \left[\sum_{l=1}^{n} \left\{\binom{2m}{m+i-l} - \binom{2m}{m+i+l}\right\} \binom{2j}{j-l} \frac{l}{j} \right]_{n \times n} \\
&= \left[\frac{1}{j} \sum_{l=1}^{n} \binom{2m}{m+i-l}\binom{2j}{j-l} l + \frac{1}{j} \sum_{l=-1}^{-n} \binom{2m}{m+i-l}\binom{2j}{j+l} l \right]_{n \times n} \\
&= \left[\frac{1}{j} \sum_{l=-n}^{n} \binom{2m}{m+i-l}\binom{2j}{j-l} l \right]_{n \times n} \\
&= \left[\frac{1}{j} \sum_{l=0}^{2j} \binom{2m}{m+i-j+l}\binom{2j}{l}(j-l) \right]_{n \times n} \\
&= \left[\binom{2m+2j}{m+j-i} - \frac{1}{j} \sum_{l=0}^{2j} \binom{2m}{m+i-j+l}\binom{2j}{l} l \right]_{n \times n} \\
&= \left[\binom{2m+2j}{m+j-i} - 2 \sum_{l=0}^{2j} \binom{2m}{m+i-j+l}\binom{2j-1}{l-1} \right]_{n \times n} \\
&= \left[\binom{2m+2j}{m+j-i} - 2\binom{2m+2j-1}{m+j-i-1}\right]_{n \times n} \\
&= \left[\binom{2m+2j}{m+i+j} \frac{i}{m+j} \right]_{n \times n} \\
&= \beta'_n.
\end{aligned}$$

The final two identities of Lemma 1 rely on the Pfaff–Saalschütz summation (§2.2, p. 9 of ref. 10):

$$\begin{aligned}
&{}_3F_2\left[{a,b,-n;1 \atop c,1+a+b-c-n}\right] \\
&= \sum_{j=0}^{n} \frac{(a)_j(b)_j(-n)_j}{j!(c)_j(1+a+b-c-n)!} \\
&= \frac{(c-a)_n(c-b)_n}{(c)_n(c-a-b)_n}.
\end{aligned} \quad [2.13]$$

Hence

$$\begin{aligned}
&\beta_n\epsilon_n \\
&= \left[\sum_{l=1}^{n} \binom{2m+2l-1}{m+i+l-1}(-1)^{l+j} \frac{(2m+2j-1)!\,(m+l+j-2)!\,(m+l-1)!}{(2m+2l-1)!\,(m+2j-2)!(m+j-1)!} \binom{j-1}{l-1} \right]_{n \times n} \\
&= \left[(-1)^{j+1} \frac{(2m+2j-1)!m!!}{(m+2j-2)!(m+i)!(m-i+1)!} \sum_{l=0}^{n-1} \frac{(-j+1)_l(m+j)_l(m+1)_l}{l!(m+i+1)_l(m-i+2)_l} \right]_{n \times n} \\
&= \left[(-1)^{j+1} \frac{(2m+2j-1)!m!}{(m+2j-2)!(m+i)!(m-i+1)!} \frac{(1-i)_{j-1}(2-i-j)_{j-1}}{(m-i+2)_{j-1}(1-i-m-j)_{j-1}} \right]_{n \times n} \qquad \text{(by Eq. 2.13)} \\
&= \left[(-1)^{j+1} \frac{(2m+2j-1)!m!(1-i)_{j-1}(2-i-j)_{j-1}}{(m+2j-2)!(m+i)!(m-i+j)!(1-i-m-j)_{j-1}} \right]_{n \times n} \\
&= \zeta n.
\end{aligned}$$

Finally

$$\begin{aligned}
\beta_n{}'\epsilon_n{}' &= \left[\sum_{l=1}^{n} \binom{2m+2l}{m+i+l} \frac{i}{m+l} (-1)^{l+j} \frac{(2m+2j)!(m+l+j-1)!(m+l)!}{(2m+2l)!(m+2j-1)!(m+j-1)!} \binom{j-1}{l-1}\right]_{n\times n} \\
&= \left[(-1)^{j+1} \frac{(2m+2j)!m!(m+j)}{(m+2j-1)!(m+i+1)!(m-i-1)!} \sum_{l=0}^{n-1} \frac{(-j+1)_l(m+j+1)_l(m+1)_l}{l!(m+i+2)_l(m-i+2)_l}\right]_{n\times n} \\
&= \left[i(-1)^{j+1} \frac{(2m+2j)!m!(m+j)}{(m+2j-1)!(m+i+1)!(m-i+1)!} \frac{(1-i)_{j-1}(1-i-j)_{j-1}}{(m-i+2)_{j-1}(-i-m-j)_{j-1}}\right]_{n\times n} \qquad \text{(by Eq. 2.13)} \\
&= \left[\frac{(-1)^j(2m+2j)!m!(-i)_j(1-i-j)_{j-1}(m+j)}{(m+2j-1)!(m+i+1)!(m+j-i)!(-i-m-j)_{j-1}}\right]_{n\times n} \\
&= \zeta'_n.
\end{aligned}$$

This concludes the proof of *Lemma 1*.

Now six of the above matrices are triangular, and so we evaluate their determinants easily in the next lemma.

LEMMA 2. *For $n \geqq 1$: $\det \delta_n = \det \delta'_n = \det \epsilon_n = 1$, $\det \epsilon'_n = (m+1)_n$,*

$$\det \zeta_n = \prod_{i=1}^{m} \left\{\frac{(2n+2i-1)}{(2i-1)} \prod_{h=i+1}^{m} \frac{(2n+i+h-1)}{(i+h-1)}\right\}, \qquad [2.14]$$

$$2^m \det \zeta'_n/(m+1)_n = \prod_{i=1}^{m} \left\{\frac{(2n+2i)}{(2i-1)} \prod_{h=i+1}^{m} \times \frac{(2n+i+h)}{(i+h-1)}\right\}. \qquad [2.15]$$

Proof: From Eqs. **2.4**, **2.5**, and **2.7** we see that δ_n, δ'_n, and ϵ_n (respectively) are upper triangular with ones on the main diagonal; therefore their determinants are all 1. The matrix ϵ_n is upper triangular with $m + i$ as the i-th entry on the main diagonal; consequently $\det \epsilon_n = \Pi_{i=1}^{n} (m+i) = (m+1)_n$. Finally, we note that ζ_n and ζ'_n are both lower triangular. Hence the determinant of each is obtained by multiplying together the entries on the main diagonal. The resulting products can then be readily rearranged to produce identities **2.14** and **2.15**.

THEOREM 1. (*MacMahon's conjecture for $q = 1$*).

$$\sum_{N\geqq 0} M(n, m; N) = \prod_{i=1}^{m} \left\{\frac{(n+2i-1)}{(2i-1)} \prod_{h=i+1}^{m} \times \frac{(n+i+h-1)}{(i+h-1)}\right\}. \qquad [2.16]$$

Proof: By Eq. **1.4**,

$$\begin{aligned}
\sum_{N\geqq 0} M(2n, m; N) &= g_{2n}(1) \\
&= \det \alpha_n && \text{(by Eq. 2.2)} \\
&= \det \beta_n/\det \delta_n && \text{(by Lemma 1)} \\
&= \det \beta_n && \text{(by Lemma 2)} \\
&= \det \zeta_n/\det \epsilon_n && \text{(by Lemma 1)} \\
&= \det \zeta_n && \text{(by Lemma 2)} \\
&= \prod_{i=1}^{m} \left\{\frac{(2n+2i-1)}{(2i-1)} \times \prod_{h=i+1}^{m} \frac{(2n+i+h-1)}{(i+h-1)}\right\} && \text{(by Eq. 2.14).}
\end{aligned}$$

Hence Eq. **2.16** holds when n is even.

When $n = 1$, $\Sigma_{N\geqq 0} M(1, m; N)$ is just 2^m because it is the total number of partitions with distinct odd parts each at most $2m - 1$. On the other hand,

$$\begin{aligned}
\prod_{i=1}^{m} \left\{\frac{2i}{2i-1} \prod_{h=i+1}^{m} \frac{i+h}{i+h-1}\right\} &= \prod_{i=1}^{m} \left\{\frac{2i}{2i-1} \frac{i+m}{2i}\right\} \\
&= \frac{(m+1)_m}{1 \cdot 3 \ldots (2m-1)} \\
&= \frac{(2m)! 2^m \cdot m!}{m!(2m)!} = 2^m.
\end{aligned}$$

Hence Eq. **2.16** is valid when $n = 1$.

When n is odd and larger than 1, we see by Eq. **1.5** that

$$\begin{aligned}
\sum_{N\geqq 0} M(2n+1, m; N) &= g_{2n+1}(1) \\
&= 2^m \det \alpha'_n && \text{(by Eq. 2.3)} \\
&= 2^m \det \beta'_n/\det \delta'_n && \text{(by Lemma 1)} \\
&= 2^m \det \beta'_n && \text{(by Lemma 2)} \\
&= 2^m \det \zeta'_n/\det \epsilon'_n && \text{(by Lemma 1)} \\
&= 2^m \det \zeta'_n/(m+1)_n && \text{(by Lemma 2)} \\
&= \prod_{i=1}^{m} \left\{\frac{(2n+2i)}{(2i-1)} \times \prod_{h=i+1}^{m} \frac{(2n+i+h)}{(i+h-1)}\right\} && \text{(by Eq. 2.15).}
\end{aligned}$$

Hence Eq. **2.16** holds also when n is odd, and this completes the proof of MacMahon's conjecture when $q = 1$.

3. The MacMahon conjecture for arbitrary q

The proof of the full conjecture closely resembles the proof in the case $q = 1$. The matrices α_n, α'_n, β_n, β'_n are generalized as follows:

$$\alpha_n = \left[q^{(i-j)^2} \binom{2m}{m+i-j}_2 + q^{(i+j-1)^2} \binom{2m}{m+i+j-1}_2\right]_{n\times n}, \qquad [3.1]$$

$$\alpha'_n = \left[q^{(i-j)^2} \binom{2m}{m+i-j}_2 - q^{(i+j)^2} \binom{2m}{m+i+j}\right]_{n\times n}, \qquad [3.2]$$

$$\beta_n = \left[q^{(i-j)^2} \binom{2m+2j-1}{m+i+j-1}_2 \frac{[1+q^{(2i-1)(2j-1)}]}{[1+q^{(2m+2j-1)(2j-1)}]}\right]_{n\times n}, \qquad [3.3]$$

$$\beta'_n = \left[q^{(i-j)^2} \binom{2m+2j}{m+i+j}_2 \frac{(1-q^{4ij})}{[1-q^{4j(m+j)}]}\right]_{n\times n}. \qquad [3.4]$$

As in *Lemma 1* above, there exist upper triangular δ_n and δ'_n with ones on the main diagonal such that $\alpha_n \delta_n = \beta_n$ and $\alpha'_n \delta'_n = \beta'_n$. The proof uses the q-analog of the Chu–Vandermonde summation (p. 97 of ref. 11). Unfortunately, the q-analogs of $\beta_n \epsilon_n = \zeta_n$ and $\beta'_n \epsilon'_n = \zeta'_n$ are not so easy. To accomplish the necessary transformation, we must introduce two further matrices:

$$\gamma_n = \left[q^{(i-1)^2-2m(j-1)-(j-1)^2} \binom{2m+2j-1}{m+i+j-1}_2 \times \frac{(1+q^{2i-1})}{(1+q^{2m+2j-1})} \right]_{n\times n}, \quad [3.5]$$

and

$$\gamma'_n = \left[q^{(i-1)^2-2m(j-1)-(j-1)^2} \binom{2m+2j}{m+i+j}_2 \frac{(1-q^{4i})}{(1-q^{4m+4j})} \right]_{n\times n}. \quad [3.6]$$

Then it is possible to define upper-triangular ϵ_n and ϵ'_n so that $\gamma_n \epsilon_n = \zeta_n$ and $\gamma'_n \epsilon'_n = \zeta'_n$ where the ζ_n and ζ'_n are lower triangular; the proof of these identities requires the q-analog of the Pfaff–Saalschutz summation (§8.4, Eq. 1, p. 68 of ref. 10). The real stumbling block in the proof of the full conjecture lies in showing that $\beta_n^{-1} \cdot \gamma_n$ and $\beta'_n{}^{-1} \cdot \gamma'_n$ are upper triangular with ones on the main diagonal. The proof of this is indirect, and four new complicated basic hypergeometric series identities are required.

Note Added in Proof. Ian Macdonald of Queen Mary College informs me that he had obtained a proof of MacMahon's conjecture from group representation theory.

This work was sponsored in part by the United States Army under Contract no. DAAG29-75-C-0024 and the National Science Foundation under Grant MSP 74-07282.

1. MacMahon, P. A., (1898) "Partitions of numbers whose graphs possess symmetry," *Trans. Cambridge Phil. Soc.* **17**, 149–170.
2. Stanley, R. P. (1971) "Theory and applications of plane partitions II," *Stud. Appl. Math.* **50**, 259–279.
3. MacMahon, P. A. (1916) *Combinatory Analysis* (Cambridge University Press, Cambridge, Vol. 2 [Reprinted: Chelsea, New York, 1960]).
4. Gordon, B. (1962) "Two new representations of the partition function," *Proc. Am. Math. Soc.* **13**, 869–873.
5. Gordon, B. (1971) "Notes on plane partitions V," *J. Combinatorial Th.* **B-11**, 157–168.
6. Bender, E. A. & Knuth, D. E. (1972) "Enumeration of plane partitions," *J. Combinatorial Th.* **13**, 40–54.
7. Gordon, B. & Houten, L. (1968) "Notes on plane partitions I," *J. Combinatorial Th.* **4**, 72–80.
8. Gordon, B. & Houten, L. (1968) "Notes on plane partitions II," *J. Combinatorial Th.*, **4**, 81–99.
9. Riordan, J. (1968) *Combinatorial Identities* (John Wiley, New York).
10. Bailey, W. N. (1935) *Generalized Hypergeometric Series* (Cambridge Mathematical Tract no. 32, Cambridge University Press, Cambridge [Reprinted: Hafner, New York, 1964]).
11. Slater, L. J. (1966) *Generalized Hypergeometric Functions* (Cambridge University Press, Cambridge).

4.2 Plane Partitions, III: The Weak Macdonald Conjecture

Inventiones math. 53, 193–225 (1979)

Plane Partitions (III): The Weak Macdonald Conjecture

George E. Andrews*

The Pennsylvania State University, University Park, Pennsylvania 16802, U.S.A.

Dedicated to the memory of Alfred Young and F.J.W. Whipple

1. Introduction

A plane partition π of n is an array of positive integers a_{ij} such that $\sum a_{ij}=n$ (more briefly $|\pi|=n$) where $a_{ij}\geqq\max(a_{i+1,j}, a_{i,j+1})$. Generally we represent a plane partition by the array

$$\begin{matrix} a_{11} & a_{12} & a_{13} & \cdots & & a_{1r} \\ a_{21} & a_{22} & a_{23} & \cdots & a_{2s} & \\ \vdots & & & & & \\ a_{j1} & \cdots & \cdots & \cdots & a_{jw}. & \end{matrix}$$

Thus the plane partitions of 3 are

$$\begin{matrix} 3, & 21, & 2, & 111, & 11, & 1 \\ & & 1 & & 1 & 1 \\ & & & & & 1. \end{matrix}$$

There are some well-known theorems and open questions related to plane partitions in which the number of rows and columns and the size of the parts is restricted. I.G. Macdonald [10] has devised a notation that allows a uniform consideration of these questions. First one considers the "Ferrers graph" $D(\pi)$ of a plane partition π; this is the set of integer points (i, j, k) in the first octant that satisfy $1\leqq k\leqq a_{ij}$. Next define the height of $p=(i, j, k)$ to be $ht(p)=i+j+k-2$. Defining

$$\mathscr{B}_{l,m,n}=[1, l]\times[1, m]\times[1, n]\subset\mathbb{Z}^3, \tag{1.1}$$

* Partially supported by National Science Foundation Grant MCS 75-19162

Macdonald notes [10] that MacMahon's [12; p. 293] main result on plane partitions asserts that

$$\sum_{D(\pi)\subseteq\mathscr{B}_{l,m,n}} q^{|\pi|}=\prod_{p\in\mathscr{B}_{l,m,n}}\frac{1-q^{1+ht(p)}}{1-q^{ht(p)}}. \tag{1.2}$$

Furthermore, MacMahon [11] made a conjecture about symmetric plane partitions. Let G_2 be the two element subgroup of the symmetric group on three letters that contains the identity and the transposition $(i, j, k)\to(j, i, k)$.

Macdonald [10] points out that MacMahon's conjecture reduces to

$$\sum_{\substack{D(\pi)\subseteq\mathscr{B}_{l,l,n}\\ D(\pi)\text{ is }G_2\text{ invariant}}} q^{|\pi|}=\prod_{\xi\in\mathscr{B}_{l,l,n}/G_2}\frac{(1-q^{|\xi|+ht(\xi)})}{(1-q^{ht(\xi)})}, \tag{1.3}$$

where $|\xi|=\text{Card}(\xi)$, $ht(\xi)=\sum_{p\in\xi} ht(p)$. Macdonald (unpublished) and I [1], [4] have independently proved this conjecture.

Next Macdonald [10] considers G_3, the three element group of cyclic permutations of (i, j, k), and he conjectures

$$\sum_{\substack{D(\pi)\subseteq\mathscr{B}_{m,m,m}\\ D(\pi)\text{ is }G_3\text{ invariant}}} q^{|\pi|}=\prod_{\xi\in\mathscr{B}_{m,m,m}/G_3}\frac{(1-q^{|\xi|+ht(\xi)})}{(1-q^{ht(\xi)})}. \tag{1.4}$$

While the formulae (1.2), (1.3) and (1.4) beautifully illustrate the parallel nature of these assertions, one becomes aware very quickly that an immense number of cancellations occur in each of these products. Hence we choose to reformulate (1.4) less elegantly but more tractably as follows:

Macdonald's Conjecture. Let $M(m, n)$ denote the number of plane partitions π such that $|\pi|=n$, $D(\pi)$ is invariant under G_3 and $D(\pi)\subseteq\mathscr{B}_{m,m,m}$. Then

$$\sum_{n\geq 0} M(m, n)\, q^n=\prod_{i=1}^{n}\left\{\frac{(1-q^{3i-1})}{(1-q^{3i-2})}\prod_{j=i}^{m}\frac{(1-q^{3(m+i+j-1)})}{(1-q^{3(2i+j-1)})}\right\}. \tag{1.5}$$

If we let $m\to\infty$, then we obtain

Limiting Macdonald Conjecture.

$$\sum_{n\geq 0} M(\infty, n)\, q^n=\prod_{n=1}^{\infty}\frac{(1-q^{3n-1})}{(1-q^{3n-2})(1-q^{3n})^{\left[\frac{n+1}{3}\right]}}. \tag{1.6}$$

The above conjectures are still open. Our main object in this paper is to prove the following:

Weak Macdonald Conjecture (Theorem 9 below). The total number of plane partitions π such that $D(\pi)\subseteq\mathscr{B}_{m,m,m}$ and $D(\pi)$ is G_3 invariant is

$$\prod_{i=1}^{m}\left\{\frac{3i-1}{3i-2}\prod_{j=i}^{m}\frac{m+i+j-1}{2i+j-1}\right\}$$

(i.e. (1.5) holds at $q=1$).

In our treatment of the Weak Macdonald Conjecture we are led to consider what we term descending plane partitions. To define these objects we must first consider shifted plane partitions, a topic treated extensively in [9] (see also [14] and [15]). A shifted plane partition ψ of n is an array of positive integers a_{ij} *defined only for* $j \geqq i$ such that $\sum a_{ij} = n$ (more briefly $|\psi| = n$), $a_{ij} \geqq a_{i,j+1}$, and for $j > i$, $a_{ij} \geqq a_{i+1,j}$. Thus the seven shifted plane partitions of 4 are

$$\begin{array}{lllllll} 4, & 31, & 22, & 211, & 21, & 1111, & 111 \\ & & & & \;\;1 & & \;\;1\,. \end{array}$$

We shall call shifted plane partitions "strict" if there is strict decrease down each column. Finally we define a *descending plane partition* to be a strict shifted plane partition such that the first entry in each row does not exceed the number of parts in the preceding row but is larger than the number of parts in its own row. Thus

$$\begin{array}{cccc} 5 & 4 & 4 & 3 \\ & 3 & 3 & \\ & & 2 & \end{array}$$

is a descending plane partition of 24.

Descending Plane Partitions Conjecture. Let $De(m, n)$ denote the number of descending plane partitions of n in which each part is $\leqq m$. Then

$$\sum_{n \geqq 0} De(m, n)\, q^n = \prod_{i=1}^{m} \prod_{j=i}^{m} \frac{(1 - q^{m+i+j-1})}{(1 - q^{2i+j-1})}. \tag{1.7}$$

Limiting Descending Plane Partitions Conjecture.

$$\sum_{n \geqq 0} De(\infty, n)\, q^n = \prod_{n=1}^{\infty} \frac{1}{(1 - q^n)^{\left[\frac{n+1}{3}\right]}}. \tag{1.8}$$

These two conjectures are also open. However as a parallel consequence of our solution to the weak Macdonald conjecture we are able to prove also the following assertion:

Weak Descending Plane Partitions Conjecture (Theorem 10 below). The total number of descending plane partitions whose parts do not exceed m is $\prod_{i=1}^{m} \prod_{j=i}^{m} \frac{m+i+j-1}{2i+j-1}$.

In Sect. 2 we present a combinatorial study of shifted plane partitions with the object of showing their relationship to the conjectures we have described. The three main results to be proved there are:

Theorem 1. *Let* $\delta(a_1, \ldots, a_r; \lambda_1, \ldots, \lambda_r; n)$ denote the number of strict shifted r-rowed plane partitions of n with a_i the largest part in the i-th row and λ_i the number of parts in the i-th row. Then for $a_1 > a_2 > \ldots > a_r > 0$, $\lambda_1 > \lambda_2 > \ldots > \lambda_r > 0$,

$$\sum_{n\geq 0} \delta(a_1, \ldots, a_r; \lambda_1, \ldots, \lambda_r; n)\, q^n = q^{a_1+\ldots+a_r+\lambda_1+\ldots+\lambda_r-r} \det\left(\begin{bmatrix} a_j+\lambda_i-2 \\ \lambda_i-1 \end{bmatrix}\right), \quad (1.9)$$

where $\begin{bmatrix} A \\ B \end{bmatrix}$ is the Gaussian polynomial defined by

$$\begin{bmatrix} A \\ B \end{bmatrix} = \begin{cases} \dfrac{(1-q^A)(1-q^{A-1})\ldots(1-q^{A-B+1})}{(1-q^B)(1-q^{B-1})\ldots(1-q)}, & B\geq 0 \\ 0, & B<0. \end{cases} \quad (1.10)$$

Theorem 3. *For* $m\geq 2$

$$\sum_{n\geq 0} De(m, n)\, q^n = \det\left(\delta_{ij} + q^{j+2}\begin{bmatrix} i+j+2 \\ i \end{bmatrix}\right)_{0\leq i,\, j\leq m-2}, \quad (1.11)$$

where $\delta_{ij}=0$ *if* $i\neq j$ *and* $\delta_{ii}=1$.

Theorem 4

$$\sum_{n\geq 0} M(m, n)\, q^n = \det\left(\delta_{ij} + q^{3j+1}\begin{bmatrix} i+j \\ i \end{bmatrix}_{q^3}\right)_{0\leq i,\, j\leq m-1} \quad (1.12)$$

where $\begin{bmatrix} A \\ B \end{bmatrix}_{q^3}$ *is the Gaussian polynomial with* q *replaced by* q^3.

Obviously then the Macdonald conjecture is reduced to evaluating the determinant in Theorem 4, while the descending plane partitions conjecture is reduced to evaluating the determinant in Theorem 3. Unfortunately we have not been able to evaluate these determinants except when $q=1$. In Sect. 3 we prove a hypergeometric series identity utilizing the work of Whipple [16, 17]. Then in Sect. 4, we prove the following result on determinants of binomial coefficients:

Theorem 8

$$\det\left(\delta_{ij} + \binom{\mu+i+j}{i}\right)_{0\leq i,\, j\leq m-1} = \prod_{j=0}^{m-1} \Delta_j(\mu), \quad (1.13)$$

where

$$\Delta_{2j}(\mu) = \begin{cases} \dfrac{(\mu+2j+2)_j \left(\frac{\mu}{2}+2j+\frac{3}{2}\right)_{j-1}}{(j)_j \left(\frac{\mu}{2}+j+\frac{3}{2}\right)_{j-1}}, & j>0 \\ 2, & j=0 \end{cases} \quad (1.14)$$

and

$$\Delta_{2j-1}(\mu) = \frac{(\mu+2j)_{j-1} \left(\frac{\mu}{2}+2j+\frac{1}{2}\right)_j}{(j)_j \left(\frac{\mu}{2}+j+\frac{1}{2}\right)_{j-1}}, \quad j>0, \quad (1.15)$$

with $(A)_j = A(A+1)\ldots(A+j-1)$.

It is then a straight forward exercise to show that the weak Macdonald conjecture (Theorem 9) follows from the case $\mu=0$ of Theorem 8, while the weak descending plane partitions conjecture follows from the case $\mu=2$.

One might expect that the proof of Theorem 8 would be easy especially in light of the fact that if the δ_{ij} is removed then it is easy to prove that

$$\det\left(\binom{\mu+i+j}{i}\right)_{0\leqq i,\, j\leqq m-1}=1, \qquad [13;\text{ p. } 682].$$

However, it will become abundantly obvious in Sects. 3 and 4 that we have not found any simple proof of Theorem 8.

We conclude with an examination of the problems involved in proving the two main conjectures.

Since the results of this paper were first announced at Alfred Young Day in Waterloo on June 2, 1978, it is certainly fitting to include him in the dedication. Since the crucial results for the determinant evaluations (Sect. 3) rely heavily on the little known but highly significant work of Whipple [16, 17], we have also included him in the dedication.

2. Shifted and Descending Plane Partitions

The cornerstone of all our work is Theorem 1 (stated in Sect. 1) which allows us to represent various generating functions as determinants. Our proof resembles that of Carlitz [8] for his derivation of (1.2).

Before we treat Theorem 1 we must make some conventions which vary from the ordinary theory of partitions. Namely, we shall admit into consideration two different partitions of zero: (i) the "empty" partition of zero (i.e. that partition that has no parts) will be said to have no parts and largest part equal to zero; (ii) the "non-empty" partition of zero is that partition which has one part and that part is zero. These rather strange conventions greatly simplify the recurrences that we must treat.

Proof of Theorem 1. When $r=1$, $\delta(a_1;\lambda_1;n)$ is the number of partitions of n into exactly λ_1 parts with largest part equal to a_1. Hence by Theorem 3.1 of [3; p. 33] we find

$$\sum_{n\geqq 0}\delta(a_1;\lambda_1;n)\,q^n$$

$$=\begin{cases} q^{a_1+\lambda_1-1}\begin{bmatrix}(a_1-1)+(\lambda_1-1)\\ \lambda_1-1\end{bmatrix}, & \text{if } a_1\geqq 0,\ \lambda_1\geqq 0,\ \textit{not both}=0\\ 1, & a_1=\lambda_1=0.\end{cases}$$

$$=\begin{cases} q^{a_1+\lambda_1-1}\begin{bmatrix}a_1+\lambda_1-2\\ \lambda_1-1\end{bmatrix}, & \text{if } a_1\geqq 0,\ \lambda_1\geqq 0,\ \text{not both}=0\\ 1, & \text{if } a_1=\lambda_1=0.\end{cases} \tag{2.1}$$

We next define

$$G(a_1,\ldots,a_r;\lambda_1,\ldots,\lambda_r;q)=\sum_{n\geqq 0}\delta(a_1,\ldots,a_r;\lambda_1,\ldots,\lambda_r;n)\,q^n. \tag{2.2}$$

Instead of proving (1.9) for $a_1 > a_2 > \ldots > a_r \geqq 1$, $\lambda_1 > \ldots > \lambda_r \geqq 1$, we prove more generally that for $a_1 > \ldots > a_r \geqq 0$, $\lambda_1 > \ldots > \lambda_r \geqq 0$,

$$G(a_1, \ldots, a_r; \lambda_1, \ldots, \lambda_r; q) = \det(G(a_i; \lambda_j; q))_{1 \leqq i, j \leqq r}. \tag{2.3}$$

We proceed by a double induction first on r then on λ_r. If $r=1$, (2.3) is trivial and (2.1) provides the Gaussian polynomial representation of $G(a_i; \lambda_1; q)$.

Assume that (2.3) is valid up to but not including a fixed r. We now start a second induction on λ_r. If $\lambda_r = 0$, we see that

$$G(a_1, \ldots, a_r; \lambda_1, \ldots, \lambda_{r-1}, 0; q) = \begin{cases} G(a_1, \ldots, a_{r-1}; \lambda_1, \ldots, \lambda_{r-1}; q), & \text{if } a_r = 0 \\ 0, & \text{if } a_r > 0. \end{cases} \tag{2.4}$$

However expanding the determinant in (2.3) along the right hand column we see that with $\lambda_r = 0$:

$$\det(G(a_i; \lambda_j; q))_{1 \leqq i, j \leqq r} = G(a_r; 0; q) \det(G(a_i; \lambda_j; q))_{1 \leqq i, j \leqq r}, \tag{2.5}$$

and from (2.1) we see that (2.5) is the same assertion for the right side of (2.3) as (2.4) is for the left side. Consequently by the induction hypothesis on r, (2.3) holds for $\lambda_r = 0$. We now assume $\lambda_r > 0$; consequently a_r must be > 0. Hence for $\lambda_j \geqq 0$, we have (assuming $a_{r+1} = -1$)

$$\sum_{b_i = a_{i+1}+1}^{a_i} G(b_i; \lambda_j - 1; q) = q^{-a_i} G(a_i; \lambda_j; q) - q^{-a_{i+1}} G(a_{i+1}; \lambda_j; q); \tag{2.6}$$

this identity is the assertion for $\lambda_j > 1$ that (by [3; p. 37, Eq. (3.3.9))

$$\sum_{b_i = a_{i+1}+1}^{a_i} q^{b_i + \lambda_j - 2} \begin{bmatrix} b_i + \lambda_j - 3 \\ \lambda_j - 2 \end{bmatrix} = q^{\lambda_j - 1} \left(\begin{bmatrix} a_i + \lambda_j - 2 \\ \lambda_{j+1} - 1 \end{bmatrix} - \begin{bmatrix} a_{i+1} + \lambda_j - 2 \\ \lambda_j - 1 \end{bmatrix} \right), \tag{2.7}$$

and for $\lambda_j = 1$ (and therefore $j = r$) that $0 = q^{-a_i} \cdot q^{a_i} - q^{-a_{i+1}} \cdot q^{a_{i+1}}$ if $i < r$, and $1 = q^{-a_r} q^{a_r}$.

By directing attention to what remains if we remove each of the first parts in each row of a strict shifted plane partition, we see that (where $a_{r+1} = -1$)

$$G(a_1, \ldots, a_r; \lambda_1, \ldots, \lambda_r; q) = q^{a_1 + \ldots + a_r} \sum_{\substack{a_1 \geqq b_1 > a_2 \\ a_2 \geqq b_2 > a_3 \\ \vdots \\ a_r \geqq b_r > a_r + 1}} G(b_1, \ldots, b_r; \lambda_1 - 1, \ldots, \lambda_r - 1; q). \tag{2.8}$$

However it is true that the right side of (2.3) also satisfies this recurrence, a fact which may be seen as follows:

$$q^{a_1+\dots+a_r} \sum_{\substack{a_1 \geqq b_1 > a_2 \\ a_2 \geqq b_2 > a_3 \\ \vdots \\ a_r \geqq b_r > a_{r+1}}} \det(G(b_i;\lambda_j-1;q))$$

$$= q^{a_1+\dots+a_r} \det(q^{-a_i} G(a_i;\lambda_j;q) - q^{-a_{i+1}} G(a_{i+1};\lambda_j;q))$$

(applying (2.6) to each row)

$$= q^{a_i+\dots+a_r} \det(q^{-a_i} G(a_i;\lambda_j;q))$$

(noting $G(a_{r+1};\lambda_j;q)=0$ since $\lambda_j \geqq \lambda_r > 0$, and starting from the bottom adding each row to the one above it)

$$= \det(G(a_i;\lambda_j;q)). \tag{2.9}$$

Therefore by the fact that each side of (2.3) satisfies the recurrence (2.8) we see that therefore (2.3) holds for every λ_r by the induction on λ_r and therefore for every r by the induction on r. □

The following theorem provides a further application of Carlitz's technique of [8]. It allows us to treat several families of plane partition problems.

Theorem 2. *For $d \leqq 2$, and fixed $a_1 > a_2 > \dots > a_r \geqq 0$*

$$\sum_{a_1+d>\lambda_1 \geqq a_2+d > \dots \geqq a_r+d > \lambda_r \geqq 1} \det(G(a_i;\lambda_j;q)) = q^{a_1+\dots+a_r} \det\left(\begin{bmatrix} a_i+a_j-2+d \\ a_j \end{bmatrix}\right) \tag{2.10}$$

where the λ_i are the summation indices.

Proof. In parallel with (2.6), we now require (assuming $a_{r+1} = -d+1$)

$$\sum_{\lambda_j=a_{j+1}+d}^{a_j+d-1} G(a_i;\lambda_j;q) = q^{a_i}\left(\begin{bmatrix} a_i+a_j+d-2 \\ a_i \end{bmatrix} - \begin{bmatrix} a_i+a_{j+1}+d-2 \\ a_i \end{bmatrix}\right); \tag{2.11}$$

this result is equivalent to

$$\sum_{\lambda_j=a_{j+1}+d}^{a_j+d-1} q^{a_i+\lambda_j-1} \begin{bmatrix} a_i+\lambda_j-2 \\ \lambda_j-1 \end{bmatrix} = q^{a_i}\left(\begin{bmatrix} a_i+a_j+d-2 \\ a_i \end{bmatrix} - \begin{bmatrix} a_i+a_{j+1}+d-2 \\ i \end{bmatrix}\right). \tag{2.12}$$

Thus applying (2.11) to each column independently in (2.10), we find that the left side of (2.10) equals

$$q^{a_1+\dots+a_r} \det\left(\begin{bmatrix} a_i+a_j-2+d \\ a_i \end{bmatrix} - \begin{bmatrix} a_i+a_{j+1}-2+d \\ a_i \end{bmatrix}\right)$$

$$= q^{a_1+\dots+a_r} \det\left(\begin{bmatrix} a_i+a_j-2+d \\ a_j \end{bmatrix}\right),$$

since $\begin{bmatrix} a_i+a_{r+1}-2+d \\ a_i \end{bmatrix} = \begin{bmatrix} a_i-1 \\ a_i \end{bmatrix} = 0$ the second determinant is obtained from the first by adding each column to the one on its left starting from the right. □

We are now prepared to prove a theorem more general than Theorem 3 (stated in the Introduction). We let $De(d; m, n)$ denote the number of strict shifted plane partitions of n with largest part $\leqq m+1-d$ such that the largest part on each row is $\leqq$ the number of parts on the previous row less d at the same time being larger than the number of parts on the same row less d. Note that $De(0; m, n) = De(m+1, n)$.

Theorem 3′. *For $d \leqq 2$ and integral,*

$$\sum_{n \geqq 0} De(d; m, n)\, q^n = \det\left(\delta_{ij} + q^{j-d+2} \begin{bmatrix} i+j+2-d \\ i \end{bmatrix}\right)_{0 \leqq i, j \leqq m-1}. \tag{2.13}$$

Proof. From (2.3) and Theorem 2, we see that

$$\sum_{n \geqq 0} De(d; m, n)\, q^n = \sum_{\substack{1+m-d \geqq a_1 > a_2 > \ldots > a_r \geqq 2-d \\ r \text{ arbitrary}}} \det\left(q^{a_j} \begin{bmatrix} a_i + a_j - 2 + d \\ a_j \end{bmatrix}\right).$$

But this is merely the formula for the expansion along the main diagonal of

$$\det\left(\delta_{ij} + q^j \begin{bmatrix} i+j-2+d \\ j \end{bmatrix}\right)_{2-d \leqq i, j \leqq 1+m-d}$$
$$= \det\left(\delta_{ij} + q^{j-d+2} \begin{bmatrix} i+j-d+2 \\ i \end{bmatrix}\right)_{0 \leqq i, j \leqq m-1}.$$

We next utilize Theorem 2 to prove Theorem 4 (stated in the Introduction).

Proof of Theorem 4. We begin with an arbitrary plane partition π that is invariant under G_3. We form a new semi-graphical representation for π as follows:

(1) Associate a 1 with each point (i, i, i) in the positive 1st octant that lies in $D(\pi)$.

(2) Associate a 3 with each point (i, j, k) in the positive first octant that lies in $D(\pi)$ for which $k < \min(j, i)$ or $k = j < i$.

This process clearly associates with each π invariant under G_3 a set of two dimensional arrays (for each $k \geqq 1$). For example the arrays

$k=1$						$k=2$	
1	3	3	3	3		1	3
	3	3	3				
	3	3					

correspond to

```
5 4 3 1 1
3 3 2 1
3 2 1
2
1
```

Notice that the 1's and 3's are associated with corresponding representatives of orbits of size 1 and 3 respectively. Consequently the sum total of the 1's and 3's in all the layers equals $|\pi|$. Now we add up the entries in each row of each layer and write the results as a shifted plane partition. Thus the above example yields

$$\begin{matrix} 13 & 9 & 6 \\ & 4 & . \end{matrix}$$

note that the first entries in each row are of the form $3a_{ii}+1=3a_i+1$, while the remaining entries are of the form $3a_{ij}$. Furthermore in order that the shifted plane partition π' come from a plane partition π invariant under G_3 it is clearly necessary and sufficient that

$$a_1>\lambda_1-2\geqq a_2>\lambda_2-2\geqq \ldots \geqq a_r>\lambda_r-2\geqq -1,$$

where λ_i is the number of parts on the i-th row of π'. Note that $a_r=0$ is permissable since it corresponds to the r-th layer having the single entry 1. Applying (2.3) and Theorem 2, we thus see that

$$\begin{aligned}
&\sum_{n\geqq 0} M(m,n)\,q^n \\
&= \sum_{m+1\geqq a_1+2>\lambda_1\geqq a_2+2>\ldots>a_r+2>\lambda_r\geqq 1} q^r\, G(a_1,\ldots,a_r;\lambda_1,\ldots,\lambda_r;q^3) \\
&= \sum_{m+1\geqq a_1+2>\lambda_1\geqq a_2+2>\ldots\geqq a_r+2>\lambda_r\geqq 1} q^r \det(G(a_i;\lambda_j;q^3)) \\
&= \sum_{m-1\geqq a_1>a_2>\ldots>a_r\geqq 0} \det\left(q^{3a_i+1}\begin{bmatrix} a_i+a_j \\ a_j \end{bmatrix}_{q^3}\right) \\
&= \det\left(\delta_{ij}+q^{3j+1}\begin{bmatrix} i+j \\ i \end{bmatrix}_{q^3}\right)_{0\leqq i,j\leqq m-1}.
\end{aligned}$$

3. Hypergeometric Series

In order to evaluate the determinant in Theorem 8, we require the following hypergeometric series identity which follows from some fundamental transformations due to F.J.W. Whipple [16, 17].

Theorem 5. *Let*

$$M(i,j,m;a,\omega)=\frac{\left(\frac{m}{2}+2j\right)_{a+\omega-1}}{\left(\frac{m}{2}+j+1\right)_{a+\omega-1}} \times\sum_{s\geqq 0}\binom{m+i+j+a-s-1}{i-j+a+s-1}\frac{(j-s)_{2s}(-m-3j-a-\omega+1)_s 4^{-s}}{s!\left(-\frac{m}{2}-2j-a-\omega+2\right)_s\left(-j-\frac{m}{2}+\frac{1}{2}\right)_s}, \tag{3.1}$$

and

$$\Lambda(i,j,m;a,\omega)=\frac{(2j-a-i)!}{(2j+\omega-i-1)!}(-1)^{i+a}\sum_{s\geqq 0}\binom{j-s-1}{i-j+a+s-1}$$
$$\times\frac{(j-s)_{2s+a+\omega-1}}{s!\,(m+4j-2)(m+4j-4)\dots(m+4j-2s)}$$
$$\times\frac{(-m-3j-a-\omega+1)_s}{(m+2j-1)(m+2j-3)\dots(m+2j-2s+1)}. \tag{3.2}$$

Then for a,ω,i *and* j *integers with* $a+\omega\geqq 1$, $2j-a\geqq i$,

$$M(i,j,m;\,a,\omega)=\Lambda(i,j,m;\,a,\omega). \tag{3.3}$$

Remark. For a restatement of Theorem 5 as a purely hypergeometric identitiy see Sect. 5, Eq. (5.1).

Proof. We shall require formulas involving the generalized hypergeometric function:

$${}_{s+1}F_s\left[\begin{matrix}a_0,a_1,\dots,a_s;\,t\\ b_1,\dots,b_s\end{matrix}\right]=\sum_{n\geqq 0}\frac{(a_0)_n(a_1)_n\dots(a_s)_n t^n}{n!\,(b_1)_n\dots(b_s)_n}, \tag{3.4}$$

where

$$(A)_n=A(A+1)\dots(A+n-1)=\frac{\Gamma(A+n)}{\Gamma(A)},$$

[7; p. 8].

The two formulas we need are as follows (M and N nonnegative integers):

$${}_4F_3\left[\begin{matrix}A,B,C,-M;\,1\\ R,S,A+B+C-R-S+1-M\end{matrix}\right]$$
$$=\frac{(R+S-A-B)_M(S-C)_M}{(R+S-A-B-C)_M(S)_M}\times{}_4F_3\left[\begin{matrix}C,R-B,R-A,-M;\,1\\ R,R+S-A-B,\,C-S+1-M\end{matrix}\right],$$
$$[16;\text{ Eq. (10.11)}], \tag{3.5}$$

$${}_4F_3\left[\begin{matrix}-N,B,C,D;\,1\\ 1-N-B,1-N-C,W\end{matrix}\right]$$
$$=\frac{(W-D)_N}{(W)_N}\,{}_5F_4\left[\begin{matrix}D,1-N-B-C,-\frac{N}{2},\frac{1}{2}-\frac{N}{2},1-N-W;\,1\\ 1-N-B,1-N-C,\frac{1}{2}(1+D-W-N),1+\frac{1}{2}(D-W-N)\end{matrix}\right]$$
$$[17;\text{ Eq. (6.6)}],\ [7;\text{ p. 33, Eq. (1)}]. \tag{3.6}$$

Identity (3.5) is Whipple's relation between two Saalschutzian or balanced ${}_4F_3$'s; the hypergeometric series given by (3.4) is called Saalschutzian or balanced if $1+a_0+a_1+\dots+a_s=b_1+b_2+\dots+b_s$. It should be noted in passing that (3.5) is a little known generalization of the celebrated Pfaff-Saalschutz summation [7; p. 9] which follows from (3.5) if we set $R=A$. Identity (3.6) is a transformation of a nearly poised ${}_4F_3$ series of the second kind into a balanced or Saalschutzian ${}_5F_4$; the hypergeometric series given by (3.4) is called "nearly poised of the second kind" if $1+a_0=a_1+b_1=a_2+b_2=\ldots=a_{s-1}+b_{s-1}$.

$$M(i,j,m;\,a,\omega)=\frac{\left(\frac{m}{2}+2j\right)_{a+\omega-1}}{\left(\frac{m}{2}+j+1\right)_{a+\omega-1}}$$

$$\times\sum_{s\geqq 0}\binom{m+i+j+a-s-1}{i-j+a+s-1}\frac{(j-s)_{2s}(-m-3j-a-\omega+1)_s\cdot 4^{-s}}{s!\left(-\frac{m}{2}-2j-a-\omega+2\right)_s\left(-j-\frac{m}{2}+\frac{1}{2}\right)_s}$$

$$=\frac{\left(\frac{m}{2}+2j\right)_{a+\omega-1}}{\left(\frac{m}{2}+j+1\right)_{a+\omega-1}}\sum_{s\geqq 0}\frac{(m+i+j+a-1)!\,(j-s)_{2s}(-m-3j-a-\omega+1)_s\,2^{-s}}{(i-j+a-1)!\,(m+2j-2s)!\left(-\frac{m}{2}-2j-a-\omega+2\right)_s}$$

$$\times\frac{1}{(-m-i-j-a+1)_s(i-j+a)_s(m+2j-1)(m+2j-3)\ldots(m+2j-2s+1)}$$

$$=\frac{\left(\frac{m}{2}+2j\right)_{a+\omega-1}}{\left(\frac{m}{2}+j+1\right)_{a+\omega-1}}\binom{m+i+j+a-1}{i-j+a-1}$$

$$\times{}_4F_3\left[\begin{matrix}-j+1,j,\;-m-3j-a-\omega+1,\;-\frac{m}{2}-j;\;1\\[6pt] i-j+a,\;-\frac{m}{2}-2j-a-\omega+2,\;-m-i-j-a+1\end{matrix}\right]$$

$$=\binom{m+i+j+a-1}{i-j+a-1}\frac{\left(\frac{m}{2}+2j\right)_{a+\omega-1}}{\left(\frac{m}{2}+j+1\right)_{a+\omega-1}}$$

$$\times\frac{\left(i-j+\frac{m}{2}+a+1\right)_{j-1}(-j-a-\omega+2)_{j-1}}{(i+m+a+1)_{j-1}\left(-\frac{m}{2}-2j-a-\omega+2\right)_{j-1}}$$

$$\times{}_4F_3\left[\begin{matrix}i-2j+a,\;-j+1,\;-j-\frac{m}{2},\;m+i+2j+2a+\omega-1;\;1\\[6pt] i-j+a,\;i-j+\frac{m}{a}+a+1,\;a+\omega\end{matrix}\right]$$

$\Big($by (3.5) with $A=j$, $B=-m-3j-a-\omega+1$,

$C=-j-\frac{m}{2}, R=i-j+a, S=-\frac{m}{2}-2j-a-\omega+2\Big)$

$$=\binom{m+i+j+a-1}{i-j+a-1}\frac{\left(\frac{m}{2}+2j\right)_{a+\omega-1}}{\left(\frac{m}{2}+j+1\right)_{a+\omega-1}}$$

$$\times\frac{\left(i-j+\frac{m}{2}+a+1\right)_{j-1}(-j-a-\omega+2)_{j-1}}{(i+m+a+1)_{j-1}\left(-\frac{m}{2}-2j-a-\omega+2\right)_{j-1}}\frac{(-m-i-2j-a+1)_{2j-i-a}}{(a+\omega)_{2j-i-a}}$$

$$\times{}_4F_3\left[\begin{matrix}m+i+2j+2a+\omega-1,\ -j+\frac{i+a}{2},\ -j+\frac{i+a+1}{2},\ 1+i-2j-\omega;\ 1\\ i-j+a,\ i-j+\frac{m}{2}+a+1,\ 1+\frac{1}{2}(m+2i+2a-1)\end{matrix}\right]$$

$\Big($by (3.6) with $N=2j-i-a,\ B=-j+1,\ C=-j-\frac{m}{2}$,

$D=m+i+2j+2a+\omega-1,\ W=a+\omega\Big)$

$$=\binom{m+i+j+a-1}{i-j+a-1}\frac{\left(\frac{m}{2}+2j\right)_{a+\omega-1}\left(i-j+\frac{m}{2}+a+1\right)_{j-1}(-j-a-\omega+2)_{j-1}}{\left(\frac{m}{2}+j+1\right)_{a+\omega-1}(i+m+a+1)_{j-1}\left(-\frac{m}{2}-2j-a-\omega+2\right)_{j-1}}$$

$$\times\frac{(-m-i-2j-a+1)_{2j-i-a}\,\Gamma\left(-\frac{m}{2}-j-\frac{i+a}{2}+\frac{1}{2}\right)\Gamma\left(i-j+\frac{m}{2}+a+1\right)}{(a+\omega)_{2j-i-a}\,\Gamma\left(\frac{i+a+1+m}{2}\right)\Gamma\left(\frac{i+a+2+m}{2}\right)}$$

$$\times\frac{\Gamma\left(-\frac{m}{2}-j-\frac{i+a}{2}+1\right)\Gamma\left(\frac{m}{2}+j+\frac{1}{2}\right)}{\Gamma\left(1-\frac{m}{2}-2j\right)\Gamma\left(-\frac{m}{2}-i-a+\frac{1}{2}\right)}$$

$$\times{}_4F_3\left[\begin{matrix}-j+\frac{i+a}{2},\ -j+\frac{i+a+1}{2},\ j+a+\omega-1,\ -m-3j-a-\omega+1;1\\ i-j+a,\ -\frac{m}{2}-2j+1,\ -\frac{m}{2}-j+\frac{1}{2}.\end{matrix}\right]$$

$\Big($by (3.5) once we have noted that it is symmetric in C and $-M$ so that we need only require one of them to be a nonpositive integer; we therefore may take A

$=1+i-2j-\omega,\ B=m+i+2j+2a+\omega-1,\ C=-j+\frac{i+a+1}{2},\ -M=-j+\frac{i+a}{2},$
$R=i-j+a,\ S=i-j+\frac{m}{2}+a+1\Big).$

The ${}_4F_3$ series that we now have is basically what we want for $\Lambda(i,j,m;a,\omega)$. First we must simplify the massive expression in front of the ${}_4F_3$. In our treatment we shall assume that m is not an integer; the case when m is an integer (i.e. the interesting case for us) will then be obtained by continuity. We shall utilize (3.5) as well as Legendre's duplication formula [18; p. 240]:

$$\Gamma(2x)=\frac{2^{2x-1}}{\sqrt{\pi}}\,\Gamma(x)\,\Gamma(x+\tfrac{1}{2}), \tag{3.7}$$

and the functional equation [18; p. 239]:

$$\Gamma(x)\,\Gamma(1-x)=\frac{\pi}{\sin\pi x}. \tag{3.8}$$

$$\binom{m+i+j+a-1}{i-j+a-1}\frac{\left(\frac{m}{2}+2j\right)_{a+\omega-1}\left(i-j+\frac{m}{2}+a+1\right)_{j-1}(-j-a-\omega+2)_{j-1}}{\left(\frac{m}{2}+j+1\right)_{a+\omega-1}(i+m+a+1)_{j-1}\left(-\frac{m}{2}-2j-a-\omega+2\right)_{j-1}}$$

$$\times\frac{(-m-i-2j-a+1)_{2j-i-a}\,\Gamma\left(-\frac{m}{2}-j-\frac{i+a}{2}+\frac{1}{2}\right)\Gamma\left(i-j+\frac{m}{2}+a+1\right)}{(a+\omega)_{2j-i-a}\,\Gamma\left(\frac{i+a+1+m}{2}\right)\Gamma\left(\frac{i+a+2+m}{2}\right)}$$

$$\times\frac{\Gamma\left(-\frac{m}{2}-j-\frac{i+a}{2}+1\right)\Gamma\left(\frac{m}{2}+j+\frac{1}{2}\right)}{\Gamma\left(1-\frac{m}{2}-2j\right)\Gamma\left(-\frac{m}{2}-i-a+\frac{1}{2}\right)}$$

$$=\binom{m+i+j+a-1}{i-j+a-1}\frac{\left(i-j+\frac{m}{2}+a+1\right)_{j-1}(a+\omega)_{j-1}}{(i+m+a+1)_{j-1}\left(\frac{m}{2}+j+a+\omega\right)_{j-1}}$$

$$\times\frac{(-1)^{i+a}(m+2i+2a)_{2j-i-a}(a+\omega-1)!}{(2j-i+\omega-1)!}$$

$$\times\frac{\left(\frac{m}{2}+2j\right)_{a+\omega-1}\Gamma\left(-\frac{m}{2}-j-\frac{i+a}{2}+\frac{1}{2}\right)}{\left(\frac{m}{2}+j+1\right)_{a+\omega-1}\Gamma\left(\frac{i+a+1+m}{2}\right)\Gamma\left(\frac{i+a+2+m}{2}\right)}$$

$$\times\frac{\Gamma\left(i-j+\frac{m}{2}+a+1\right)\Gamma\left(-\frac{m}{2}-j-\frac{i+a}{2}+1\right)\Gamma\left(\frac{m}{2}+j+\frac{1}{2}\right)}{\Gamma\left(1-\frac{m}{2}-2j\right)\Gamma\left(-\frac{m}{2}-i-a+\frac{1}{2}\right)}$$

$$=\frac{(-1)^{i+a}\Gamma(m+i+a+1)\left(i-j+\frac{m}{2}+a+1\right)_{j-1}}{(i-j+a-1)!\,\Gamma(m+2j+1)\left(\frac{m}{2}+j+a+\omega\right)_{j-1}}$$

$$\times\frac{(a+\omega+j-2)!\,(m+2i+2a)_{2j-i-a}}{(2j-i+\omega-1)!}$$

$$\times\frac{\left(\frac{m}{2}+2j\right)_{a+\omega-1}\Gamma\left(-\frac{m}{2}-j-\frac{i+a}{2}+\frac{1}{2}\right)\Gamma\left(i-j+\frac{m}{2}+a+1\right)}{\left(\frac{m}{2}+j+1\right)_{a+\omega-1}\Gamma\left(\frac{i+a+1+m}{2}\right)\Gamma\left(\frac{i+a+2+m}{2}\right)}$$

$$\times\frac{\Gamma\left(-\frac{m}{2}-j-\frac{i+a}{2}+1\right)\Gamma\left(\frac{m}{2}+j+\frac{1}{2}\right)}{\Gamma\left(1-\frac{m}{2}-2j\right)\Gamma\left(-\frac{m}{2}-i-a+\frac{1}{2}\right)}$$

$$=(-1)^{i+a}\binom{j+a+\omega-2}{i-j+a-1}\frac{\Gamma(i+a+m+1)}{\Gamma(m+2j+1)}$$

$$\times\frac{\left(i-j+\frac{m}{2}+a+1\right)_{j-1}(m+2i+2a)_{2j-i-a}}{\left(\frac{m}{2}+j+a+\omega\right)_{j-1}}$$

$$\times\frac{\left(\frac{m}{2}+2j\right)_{a+\omega-1}\Gamma\left(-\frac{m}{2}-j-\frac{i+a}{2}+\frac{1}{2}\right)\Gamma\left(i-j+\frac{m}{2}+a+1\right)}{\left(\frac{m}{2}+j+1\right)_{a+\omega-1}\Gamma\left(\frac{i+a+1+m}{2}\right)\Gamma\left(\frac{i+a+2+m}{2}\right)}$$

$$\times\frac{\Gamma\left(-\frac{m}{2}-j-\frac{i+a}{2}+1\right)\Gamma\left(\frac{m}{2}+j+\frac{1}{2}\right)}{\Gamma\left(1-\frac{m}{2}-2j\right)\Gamma\left(-\frac{m}{2}-i-a+\frac{1}{2}\right)}$$

$$=(-1)^{i+a}\binom{j+a+\omega-2}{i-j+a-1}\frac{\Gamma(m+i+a+1)\Gamma\left(i+\frac{m}{2}+a\right)\Gamma\left(\frac{m}{2}+j+a+\omega\right)}{\Gamma(m+2j+1)\Gamma\left(\frac{m}{2}+2j+a+\omega-1\right)\Gamma(m+2i+2a)}$$

$$\times\frac{\left(\frac{m}{2}+2j\right)_{a+\omega-1}\Gamma(m+2j+i+a)\Gamma\left(-\frac{m}{2}-j-\frac{i+a}{2}+\frac{1}{2}\right)}{\left(\frac{m}{2}+j+1\right)_{a+\omega-1}\Gamma\left(\frac{i+a+1+m}{2}\right)\Gamma\left(\frac{i+a+2+m}{2}\right)}$$

$$\times\frac{\Gamma\left(-\frac{m}{2}-j-\frac{i+a}{2}+1\right)\Gamma\left(\frac{m}{2}+j+\frac{1}{2}\right)}{\Gamma\left(1-\frac{m}{2}-2j\right)\Gamma\left(-\frac{m}{2}-i-a+\frac{1}{2}\right)}$$

$$=(-1)^{i+a}\binom{j+a+\omega-2}{i-j+a-1}2^{2(i+a+m+j)-1}\Gamma(-m-2j-i-a+1)$$

$$\times\frac{\left(\frac{m}{2}+2j\right)_{a+\omega-1}\Gamma\left(i+\frac{m}{2}+a\right)\Gamma\left(\frac{m}{2}+j+a+\omega\right)}{\left(\frac{m}{2}+j+1\right)_{a+\omega-1}\Gamma(m+2j+1)\Gamma\left(\frac{m}{2}+2j+a+\omega-1\right)}$$

$$\times\frac{\Gamma(m+2j+i+a)\Gamma\left(\frac{m}{2}+j+\frac{1}{2}\right)}{\Gamma(m+2i+a)\Gamma\left(1-\frac{m}{2}-2j\right)\Gamma\left(-\frac{m}{2}-i-a+\frac{1}{2}\right)}$$

$$=(-1)^{i+a}\binom{j+a+\omega-2}{i-j+a-1}\frac{2\pi\sin\pi\left(i+\frac{m}{2}+a+\frac{1}{2}\right)}{\sin\pi(m+2j+i+a)}$$

$$\times\frac{\left(\frac{m}{2}+2j\right)_{a+\omega-1}\Gamma\left(\frac{m}{2}+j+a+\omega\right)}{\left(\frac{m}{2}+j+1\right)_{a+\omega-1}\Gamma\left(\frac{m}{2}+j+1\right)\Gamma\left(\frac{m}{2}+2j+a+\omega-1\right)\Gamma\left(1-\frac{m}{2}-2j\right)}$$

$$=(-1)^{i+a}\binom{j+a+\omega-2}{i-j+a-1}\frac{2\sin\pi\left(i+\frac{m}{2}+a+\frac{1}{2}\right)\sin\frac{\pi m}{2}}{\sin\pi(m+i+a)}$$

$$=(-1)^{i+a}\binom{j+a+\omega-2}{i-j+a-1}\frac{2(-1)^{i+a}\cos\frac{\pi m}{2}\sin\frac{\pi m}{2}}{(-1)^{i+a}\sin\pi m}$$

$$=(-1)^{i+a}\binom{j+a+\omega-2}{i-j+a-1}\qquad(\text{since }\sin 2\theta=2\sin\theta\cos\theta).$$

Hence we have

$$M(i,j,m;a,\omega)=(-1)^{i+a}\binom{j+a+\omega-2}{i-j+a-1}$$

$$\times{}_4F_3\left[\begin{matrix}-j+\frac{i+a}{2},\ -j+\frac{i+a+1}{2},\ j+a+\omega-1,\ -m-3j-a-\omega+1;\ 1\\ i-j+a,\ -\frac{m}{2}-2j+1,\ -\frac{m}{2}-j+\frac{1}{2}\end{matrix}\right]$$

$$=(-1)^{i+a}\sum_{s\geqq 0}\frac{(j+a+\omega+s-2)!(-2j+i+a)_{2s}}{s!(i-j+a+s-1)!(2j-i+\omega-1)!}$$

$$\times\frac{4^{-s}(-m-3j-a-\omega+1)_s}{\left(-\frac{m}{2}-2j+1\right)_s\left(-\frac{m}{2}-j+\frac{1}{2}\right)_s}=\frac{(-1)^{i+a}(2j-i-a)!}{(2j-i+\omega-1)!}$$

$$\times \sum_{s \geqq 0} \binom{j-s-1}{i-j+a+s-1} \frac{4^{-s}(-m-3j-a-\omega+1)_s}{s!\left(-\frac{m}{2}-2j+1\right)_s \left(-\frac{m}{2}-j+\frac{1}{2}\right)_s}$$

$$\times \frac{(j+a+\omega+s-2)!}{(j-s-1)!}$$

$$= \frac{(-1)^{i+a}(2j-i-a)!}{(2j-i+\omega-1)!} \sum_{s \geqq 0} \binom{j-s-1}{i-j+a+s-1} \frac{(-m-3j-a-\omega+1)_s}{s!}$$

$$\times \frac{(j-s)_{2s+a+\omega-1}}{(m+4j-2)(m+4j-4)\dots(m+4j-2s)(m+2j-1)(m+2j-3)\dots(m+2j-2s+1)}$$

$= A(i, j, m; a, \omega)$, as desired.

In actual fact our above treatment would appear to be valid only provided $i-2j+a$ is a nonpositive integer and $i-j+a$ is any complex (or real) number other than a nonpositive integer. The result then follows for $i-j+a$ nonpositive integral by continuity. Also we had to assume m nonintegral to allow $\sin \pi m$ in a denominator; however again the excluded values on m are admissible by continuity. □

Corollary 5a

$$A(i, j, m; a, \omega) = \begin{cases} \binom{j+a+\omega-2}{j-1} & \text{for } i = 2j-a \\ 0 & \text{for } i > 2j-a. \end{cases}$$

Proof. In (3.2) we see that if $i=2j-a$ then only the first term does not vanish while if $i>2j-a$ then all terms vanish. □

Corollary 5b. *For j a positive integer,*

$$M(2j+\omega, j, m; a, \omega)$$
$$= \binom{m+2j+a+\omega-1}{a+\omega-1} \frac{(m+2j+a+\omega)_j \left(\frac{m}{2}+2j\right)_{a+\omega-1} \left(2j+\frac{m}{2}+a+\omega\right)_{j-1}}{(j+a+\omega-1)_j \left(\frac{m}{2}+j+1\right)_{a+\omega-1} \left(\frac{m}{2}+j+a+\omega\right)_{j-1}}.$$

Proof. After the first transformation applied to $M(i, j, m; a, \omega)$ in the proof of Theorem 5 we find that

$$M(2j+\omega, j, m; a, \omega)$$
$$= \binom{m+3j+a+\omega-1}{j+a+\omega-1} \frac{\left(\frac{m}{2}+2j\right)_{a+\omega-1} \left(j+\frac{m}{2}+a+\omega+1\right)_{j-1}}{\left(\frac{m}{2}+j+1\right)_{a+\omega-1} (2j+m+a+\omega+1)_{j-1}}$$

$$\times \frac{(-j-a-\omega+2)_{j-1}}{\left(-\frac{m}{2}-2j-a-\omega+2\right)_{j-1}} {}_3F_2\left[\begin{matrix} -j+1, -j-\frac{m}{2}, m+4j+2a+2\omega-1; 1 \\ j+a+\omega, j+\frac{m}{2}+a+\omega+1 \end{matrix}\right]$$

$$=\binom{m+3j+a+\omega-1}{j+a+\omega-1}\frac{\left(\frac{m}{2}+2j\right)_{a+\omega-1}\left(j+\frac{m}{2}+a+\omega+1\right)_{j-1}}{\left(\frac{m}{2}+j+1\right)_{a+\omega-1}(2j+m+a+\omega+1)_{j-1}}$$

$$\times\frac{(a+\omega)_{j-1}\left(2j+\frac{m}{2}+a+\omega\right)_{j-1}(m+2j+a+\omega+1)_{j-1}}{\left(\frac{m}{2}+j+a+\omega\right)_{j-1}(j+a+\omega)_{j-1}\left(\frac{m}{2}+j+a+\omega+1\right)_{j-1}}\qquad\text{(by [7; p. 9])}$$

$$=\frac{(m+2j+a+\omega-1)!\,(m+2j+a+\omega)_j}{(a+\omega-1)!\,(j+a+\omega-1)(m+2j)!}$$

$$\times\frac{\left(\frac{m}{2}+2j\right)_{a+\omega-1}\left(2j+\frac{m}{2}+a+\omega\right)_{j-1}}{\left(\frac{m}{2}+j+1\right)_{a+\omega-1}\left(\frac{m}{2}+j+a+\omega\right)_{j-1}(j+a+\omega)_{j-1}}$$

$$=\binom{m+2j+a+\omega-1}{a+\omega-1}\frac{(m+2j+a+\omega)_j\left(\frac{m}{2}+2j\right)_{a+\omega-1}\left(2j+\frac{m}{2}+a+\omega\right)_{j-1}}{(j+a+\omega-1)_j\left(\frac{m}{2}+j+1\right)_{a+\omega-1}\left(\frac{m}{2}+j+a+\omega\right)_{j-1}},$$

as desired. □

While Theorem 5 and its corollaries play the central role in the next section, it turns out that the following relation among three balanced ${}_4F_s$'s is crucial in order to conclude the proof of Theorem 7.

Theorem 6. *For* $0\leqq i\leqq 2j-2$ *with* $i-2j+2$ *integral,*

$${}_4F_3\left[\begin{matrix}i-2j+1,\,1-j,\,-j-\frac{\mu}{2}+\frac{1}{2},\,i+\mu+2j;\,1\\ i-j+1,\,i-j+\frac{\mu}{2}+\frac{3}{2},\,1\end{matrix}\right]+\frac{(2i+\mu+1)\left(\frac{\mu}{2}+j-\frac{1}{2}\right)j(j-1)}{\left(i-j+\frac{\mu}{2}+\frac{3}{2}\right)(i-j+1)(i-j+2)}$$

$$\times{}_4F_3\left[\begin{matrix}i-2j+2,\,2-j,\,-j-\frac{\mu}{2}+\frac{3}{2},\,i+\mu+2j+1;\,1\\ i-j+3,\,i-j+\frac{\mu}{2}+\frac{5}{2},\,2\end{matrix}\right]$$

$$+\frac{\left(\frac{\mu}{2}+i+\frac{1}{2}\right)(i+1)(2j-i-2)j}{\left(i-j+\frac{\mu}{2}+\frac{3}{2}\right)(i-j+1)(i-j+2)}$$

$$\times{}_4F_3\left[\begin{matrix}i-2j+3,\,1-j,\,-j-\frac{\mu}{2}+\frac{3}{2},\,i+\mu+2j+1;\,1\\ i-j+3,\,i-j+\frac{\mu}{2}+\frac{5}{2},\,2\end{matrix}\right]=0. \tag{3.9}$$

Proof. If we denote the index of summation in each of the above ${}_4F_3$'s by s and if we shift s to $s-1$ in the second ${}_4F_3$, then we find that the expression L on the left of (3.9) is

$$L=\sum_{s\geqq 0}\frac{(i-2j+1)_s(1-j)_s\left(-j-\frac{\mu}{2}+\frac{1}{2}\right)_s(i+\mu+2j)_s}{s!\,(i-j+1)_s\left(i-j+\frac{\mu}{2}+\frac{3}{2}\right)_s s!}$$

$$-(2i+\mu+1)\sum_{s\geqq 0}\frac{s\left(-\frac{\mu}{2}-j+\frac{1}{2}\right)_s(i+2j+\mu+1)_{s-1}(-j)_{s+1}(i-2j+2)_{s-1}}{s!\,(i-j+1)_{s+1}\left(i-j+\frac{\mu}{2}+\frac{3}{2}\right)_s s!}$$

$$+\left(\frac{\mu}{2}+i+\frac{1}{2}\right)(i+1)\sum_{s\geqq 0}\frac{(i-2j+2)_{s+1}(-j)_{s+1}\left(-j-\frac{\mu}{2}+\frac{3}{2}\right)_s(i+\mu+2j+1)_s}{s!\,(i-j+1)_{s+2}\left(i-j+\frac{\mu}{2}+\frac{3}{2}\right)_{s+1}(s+1)!}.$$

Since $(2i+\mu+1)=(i+\mu+2j)+(i-2j+1)$, we combine the first sum with the $(i+\mu+2j)$ portion of the second sum:

$$\sum_{s\geqq 0}\frac{(i-2j+2)_{s-1}(1-j)_s\left(-\frac{\mu}{2}-j+\frac{1}{2}\right)_s(i+\mu+2j)_s}{s!\,(i-j+1)_{s+1}\left(i-j+\frac{\mu}{2}+\frac{3}{2}\right)_s s!}$$
$$\times\{(i-2j+1)(i-j+s+1)-(-j)\,s\},$$

and since the expression inside { } is just $(s+i-2j+1)(i-j+1)$ we see that this last expression equals

$$\sum_{s\geqq 0}\frac{(i-2j+2)_s(1-j)_s\left(-\frac{\mu}{2}-j+\frac{1}{2}\right)_s(i+\mu+2j)_s}{s!\,(i-j+2)_s\left(i-j+\frac{\mu}{2}+\frac{3}{2}\right)_s s!}.$$

Therefore the expression on the left of (3.9) may be written as follows:

$$L=\sum_{s\geqq 0}\frac{(i-2j+2)_s(1-j)_s\left(-\frac{\mu}{2}-j+\frac{1}{2}\right)_s(i+\mu+2j)_s}{s!\,(i-j+2)_s\left(i-j+\frac{\mu}{2}+\frac{3}{2}\right)_s s!}$$

$$-\sum_{s\geqq 0}\frac{s\left(-\frac{\mu}{2}-j+\frac{1}{2}\right)_s(i+2j+\mu+1)_{s-1}(-j)_{s+1}(i-2j+1)_s}{s!\left(i-j+\frac{\mu}{2}+\frac{3}{2}\right)_s(i-j+1)_{s+1}\,s!}$$

$$+\tfrac{1}{2}(i+1)(\mu+2i+1)$$

$$\times \sum_{s\geqq 0} \frac{(i-2j+2)_{s+1}(-j)_{s+1}\left(-j-\frac{\mu}{2}+\frac{3}{2}\right)_s (i+\mu+2j+1)_s}{s!\,(i-j+1)_{s+2}\left(i-j+\frac{\mu}{2}+\frac{3}{2}\right)_{s+1}(s+1)!}.$$

Let us now shift the index of summation in the second sum from s to $s+1$ and then combine it with the third sum. Hence

$$L=\sum_{s\geqq 0} \frac{(i-2j+2)_s(1-j)_s\left(-\frac{\mu}{2}-j+\frac{1}{2}\right)_s(i+\mu+2j)_s}{s!\,(i-j+2)_s\left(i-j+\frac{\mu}{2}+\frac{3}{2}\right)_s s!}$$

$$+\sum_{s\geqq 0} \frac{\left(-j-\frac{\mu}{2}+\frac{3}{2}\right)_s(i+\mu+2j+1)_s(-j)_{s+1}(i-2j+2)_s}{s!\left(i-j+\frac{\mu}{2}+\frac{3}{2}\right)_{s+1}(i-j+1)_{s+2}(s+1)!}$$

$$\times\left[-\left(-\frac{\mu}{2}-j+\frac{1}{2}\right)(-j+s+1)(i-2j+1)+(i+1)\right.$$

$$\left.\times\left(\frac{\mu}{2}+i+\frac{1}{2}\right)(i-2j+s+2)\right].$$

The expression inside "[ ]" factors into

$$(i+1-j)\left\{(1+s-j)(\mu+2j+i)+(i+1)\left(\frac{\mu}{2}+i+\frac{1}{2}\right)\right\}.$$

Hence we conclude that L (the expression on the left side of (3.9)) is given by

$$L=\sum_{s\geqq 0} \frac{(i-2j+2)_s(1-j)_s\left(-\frac{\mu}{2}-j+\frac{1}{2}\right)_s(i+\mu+2j)_s}{s!\,(i-j+2)_s\left(i-j+\frac{\mu}{2}+\frac{3}{2}\right)_s s!}$$

$$+\sum_{s\geqq 0} \frac{(i-2j+2)_s(-j)_{s+1}\left(-\frac{\mu}{2}-j+\frac{3}{2}\right)_s(i+\mu+2j+1)_s}{s!\left(i-j+\frac{\mu}{2}+\frac{3}{2}\right)_{s+1}(i-j+2)_{s+1}(s+1)!}$$

$$\times\left\{(1+s-j)(\mu+2j+i)+(i+1)\left(\frac{\mu}{2}+i+\frac{1}{2}\right)\right\}. \tag{3.10}$$

Equation (3.10) now provides the starting point for the final assault. We propose to prove that if for $0\leqq n\leqq 2j-2-i$

$$L_n = \sum_{s=2j-2-i-n}^{2j-i-2} \frac{(i-2j+2)_s(1-j)_s\left(-\frac{\mu}{2}-j+\frac{1}{2}\right)_s(i+\mu+2j)_s}{s!\,(i-j+2)_s\left(i-j+\frac{\mu}{2}+\frac{3}{2}\right)_s s!}$$

$$+\sum_{s=2j-2-i-n}^{2j-i-2} \frac{(i-2j+2)_s(-j)_{s+1}\left(-\frac{\mu}{2}-j+\frac{3}{2}\right)_s(i+\mu+2j+1)_s}{s!\left(i-j+\frac{\mu}{2}+\frac{3}{2}\right)_{s+1}(i-j+2)_{s+1}(s+1)!}$$

$$\times\left\{(1+s-j)(\mu+2j+i)+(i+1)\left(\frac{\mu}{2}+i+\frac{1}{2}\right)\right\},$$

then

$$L_n = \frac{-(-1)^i(i+\mu+2j+1)_{2j-i-3-n}(i-2j+2)_{n+1}}{n!\,(2j-i-2-n)!\,(i-j+2)_n} \times\frac{(1-j)_n\left(-\frac{\mu}{2}-j+\frac{3}{2}\right)_n\left(\frac{\mu}{2}+3j-n-\frac{3}{2}\right)}{\left(i-j+\frac{\mu}{2}+\frac{3}{2}\right)_{n+1}}. \tag{3.11}$$

If we can prove (3.11) for $0 \leqq n \leqq 2j-i-2$ this will imply the main result since $L = L_{2j-i-2}$, and by (3.11) we see that $L_{2j-i-2}=0$.

We proceed to treat (3.11) by induction on n. For $n=0$,

$$L_0 = \frac{(i-2j+2)_{2j-i-2}(1-j)_{2j-i-2}\left(-\frac{\mu}{2}-j+\frac{1}{2}\right)_{2j-i-2}(i+\mu+2j)_{2j-i-2}}{(2j-i-2)!\,(i-j+2)_{2j-i-2}\left(i-j+\frac{\mu}{2}+\frac{3}{2}\right)_{2j-i-2}(2j-i-2)!}$$

$$+\frac{(i-2j+2)_{2j-i-2}(-j)_{2j-i-1}\left(-\frac{\mu}{2}-j+\frac{3}{2}\right)_{2j-i-2}(i+\mu+2j+1)_{2j-i-2}}{(2j-i-2)!\left(i-j+\frac{\mu}{2}+\frac{3}{2}\right)_{2j-i-1}(i-j+2)_{2j-i-1}(2j-i-1)!}$$

$$\times\left\{(j-i-1)(\mu+2j+i)+(i+1)\left(\frac{\mu}{2}+i+\frac{1}{2}\right)\right\}$$

$$=\frac{(-1)^i\left(\frac{\mu}{2}+j-\frac{1}{2}\right)(i+\mu+2j)_{2j-i-2}}{(2j-i-2)!\left(i-j+\frac{\mu}{2}+\frac{3}{2}\right)}$$

$$-\frac{(-1)^i(i+\mu+2j+1)_{2j-i-2}}{(2j-i-1)!\left(\frac{\mu}{2}+j-\frac{1}{2}\right)}\left\{(\mu+2j-1)\left(j-\frac{i}{2}-\frac{1}{2}\right)\right\}$$

$$=\frac{-(-1)^i(i+\mu+2j+1)_{2j-i-3}}{(2j-i-2)!\left(i-j+\frac{\mu}{2}+\frac{3}{2}\right)}$$

$$\times\left[-(i+\mu+2j)\left(\frac{\mu}{2}+j-\frac{1}{2}\right)+(\mu+4j-2)\left(i-j+\frac{\mu}{2}+\frac{3}{2}\right)\right]$$

$$=\frac{-(-1)^i(i+\mu+2j+1)_{2j-i-3}(i-2j+2)\left(\frac{\mu}{2}+3j-\frac{3}{2}\right)}{(2j-i-2)!\left(i-j+\frac{\mu}{2}+\frac{3}{2}\right)},$$

which is (3.11) in the case $n=0$.

We now assume that (3.11) is proved for all nonnegative integers up to but not including a specific n in the interval $(0, 2j-i-2]$. Hence by the induction hypothesis

$$L_n=\frac{(i-2j+2)_{2j-2-i-n}(1-j)_{2j-2-i-n}}{(2j-2-i-n)!\,(i-j+2)_{2j-2-i-n}}$$

$$\times\frac{\left(-\frac{\mu}{2}-j+\frac{1}{2}\right)_{2j-2-i-n}(1+\mu+2j)_{2j-2-i-n}}{\left(i-j+\frac{\mu}{2}+\frac{3}{2}\right)_{2j-2-i-n}(2j-2-i-n)!}$$

$$+\frac{(i-2j+2)_{2j-2-i-n}(-j)_{2j-1-i-n}}{(2j-2-i-n)!\,(i-j+2)_{2j-1-i-n}}$$

$$\times\frac{\left(-\frac{\mu}{2}-j+\frac{3}{2}\right)_{2j-2-i-n}(i+\mu+2j+1)_{2j-2-i-n}}{\left(i-j+\frac{\mu}{2}+\frac{3}{2}\right)_{2j-1-i-n}(2j-i-n-1)!}$$

$$\times\left\{(j-i-n-1)(\mu+2j+i)+(i+1)\left(\frac{\mu}{2}+i+\frac{1}{2}\right)\right\}$$

$$-\frac{(-1)^i(i+\mu+2j+1)_{2j-i-2-n}(i-2j+2)_n}{(n-1)!\,(2j-i-1-n)!\,(i-j+2)_{n-1}}$$

$$\times\frac{(1-j)_{n-1}\left(-\frac{\mu}{2}-j+\frac{3}{2}\right)_{n-1}\left(\frac{\mu}{2}+3j-n-\frac{1}{2}\right)}{\left(i-j+\frac{\mu}{2}+\frac{3}{2}\right)_n}$$

$$=\frac{-(-1)^i(i+\mu+2j)_{2j-2-i-n}(1-j)_n\left(-\frac{\mu}{2}-j+\frac{1}{2}\right)_{n+1}(i-2j+2)_n}{n!\,(2j-2-i-n)!\,(i-j+2)_n\left(i-j+\frac{\mu}{2}+\frac{3}{2}\right)_{n+1}}$$

$$+\frac{(-1)^i(i+\mu+2j+1)_{2j-2-i-n}(-j)_n\left(-\frac{\mu}{2}-j+\frac{3}{2}\right)_{n-1}(i-2j+2)_n}{n!\,(2j-i-n-1)!\,(i-j+2)_n\left(i-j+\frac{\mu}{2}+\frac{3}{2}\right)_n}$$

$$\times\left\{(j-i-n-1)(\mu+2j+i)+(i+1)\left(\frac{\mu}{2}+i+\frac{1}{2}\right)\right\}$$

$$-\frac{(-1)^i(i+\mu+2j+1)_{2j-i-2-n}(i-2j+2)_n}{(n-1)!\,(2j-1-i-n)(i-j+2)_{n-1}}$$

$$\times\frac{(1-j)_{n-1}\left(-\frac{\mu}{2}-j+\frac{3}{2}\right)_{n-1}\left(\frac{\mu}{2}+3j-n-\frac{1}{2}\right)}{\left(i-j+\frac{\mu}{2}+\frac{3}{2}\right)_n}$$

$$=\frac{-(-1)^i(i+\mu+2j)_{2j-2-i-n}(1-j)_n\left(-\frac{\mu}{2}-j+\frac{1}{2}\right)_{n+1}(i-2j+2)_n}{n!\,(2j-2-i-n)!\,(i-j+2)_n\left(i-j+\frac{\mu}{2}+\frac{3}{2}\right)_{n+1}}$$

$$-\frac{(-1)^i(i+\mu+2j+1)_{2j-2-i-n}(-j)_n\left(-\frac{\mu}{2}-j+\frac{3}{2}\right)_{n-1}(i-2j+2)_{n-1}}{n!\,(2j-i-n-2)!\,(i-j+2)_n\left(i-j+\frac{\mu}{2}+\frac{3}{2}\right)_n}$$

$$\times\left\{(j-i-n-1)(\mu+2j+i)+(i+1)\left(\frac{\mu}{2}+i+\frac{1}{2}\right)\right\}$$

$$+\frac{(-1)^i(i+\mu+2j+1)_{2j-i-2-n}(i-2j+2)_{n-1}}{(n-1)!\,(2j-i-n-2)!\,(i-j+2)_{n-1}}$$

$$\times\frac{(1-j)_{n-1}\left(-\frac{\mu}{2}-j+\frac{3}{2}\right)_{n-1}\left(\frac{\mu}{2}+3j-n-\frac{1}{2}\right)}{\left(i-j+\frac{\mu}{2}+\frac{3}{2}\right)_n}$$

$$=\frac{-(-1)^i(i+\mu+2j+1)_{2j-i-3-n}}{n!\,(2j-i-n-2)!\,(i-j+2)_n}$$

$$\times\frac{(i-2j+2)_{n-1}(1-j)_{n-1}\left(-\frac{\mu}{2}-j+\frac{3}{2}\right)_{n-1}}{\left(i-j+\frac{\mu}{2}+\frac{3}{2}\right)_{n+1}}$$

$$\times\left[(i+\mu+2j)(n-j)\left(-\frac{\mu}{2}-j+\frac{1}{2}\right)(i-2j+n+1)\left(-\frac{\mu}{2}-j+n+\frac{1}{2}\right)\right.$$

$$+(\mu+4j-n-2)(-j)\left(i-j+\frac{\mu}{2}+\frac{3}{2}+n\right)$$

$$\times\left\{(j-i-n-1)(\mu+2j+i)+(i+1)\left(\frac{\mu}{2}+i+\frac{1}{2}\right)\right\}$$

$$\left.-(\mu+4j-n-2)\left(\frac{\mu}{2}+3j-n-\frac{1}{2}\right)n(i-j+n+1)\left(i-j+\frac{\mu}{2}+\frac{3}{2}+n\right)\right].$$

Now we must simplify the expression inside the square brackets. When we multiply it out, we find that there are 752 separate terms initially. After cancellation we find that

$$
\begin{aligned}
L_n = {} & \frac{(-1)^i (i+\mu+2j+1)_{2j-i-3-n} (i-2j+2)_{n-1} (1-j)_{n-1} \left(-\frac{\mu}{2}-j+\frac{3}{2}\right)_{n-1}}{n!\,(2j-i-n-2)!\,(i-j+2)_n \left(i-j+\frac{\mu}{2}+\frac{3}{2}\right)_{n+1}} \\
& \times [\tfrac{3}{2}j-\tfrac{3}{2}n+\tfrac{9}{4}ji+\tfrac{39}{2}jin-15jin\mu+\tfrac{3}{2}jin\mu^2-3ji\mu \\
& +33jin^2-10jin^2\mu+14jin^3+\tfrac{3}{4}ji\mu^2+\tfrac{67}{4}jn-15jn\mu \\
& +\tfrac{9}{4}jn\mu^2-2j\mu+\tfrac{3}{4}ji^2+5ji^2n-3ji^2n\mu-ji^2\mu+5ji^2n^2 \\
& +\tfrac{1}{4}ji^2\mu^2+\tfrac{163}{4}jn^2-20jn^2\mu+\tfrac{5}{4}jn^2\mu^2+34jn^3-7jn^3\mu \\
& +9jn^4+\tfrac{1}{2}j\mu^2-\tfrac{9}{4}in+3in\mu-\tfrac{3}{4}in\mu^2-\tfrac{15}{2}in^2+5in^2\mu \\
& -\tfrac{1}{2}in^2\mu^2-7in^3+2in^3\mu-2in^4+2n\mu-\tfrac{1}{2}n\mu^2-12j^2i-47j^2in \\
& +16j^2in\mu+10j^2i\mu-34j^2in^2-j^2i\mu^2-59j^2n+32j^2n\mu-2j^2n\mu^2 \\
& +10j^2\mu-3j^2i^2-7j^2i^2n+2j^2i^2\mu-82j^2n^2+18j^2n^2\mu-31j^2n^3-\tfrac{3}{2}j^2\mu^2 \\
& +21j^3i+34j^3in-8j^3i\mu+83j^3n-20j^3n\mu-16j^3\mu+3j^3i^2+51j^3n^2 \\
& +j^3\mu^2-12j^4i-40j^4n+8j^4\mu-\tfrac{3}{4}i^2n+i^2n\mu-\tfrac{1}{4}i^2n\mu^2-2i^2n^2 \\
& +i^2n^2\mu-i^2n^3+5n^2\mu-\tfrac{3}{4}n^2\mu^2+4n^3\mu-\tfrac{1}{4}n^3\mu^2+n^4\mu-\tfrac{21}{2}j^2 \\
& +27j^3-30j^4+12j^5-\tfrac{25}{4}n^2-\tfrac{35}{4}n^3-5n^4-n^5] \\
= {} & \frac{(-1)^i (i+\mu+2j+1)_{2j-i-3-n} (i-2j+2)_{n-1} (1-j)_{n-1} \left(-\frac{\mu}{2}-j+\frac{3}{2}\right)_{n-1}}{n!\,(2j-i-n-2)!\,(i-j+2)_n \left(i-j+\frac{\mu}{2}+\frac{3}{2}\right)_{n+1}} \\
& \times \left[(i-2j+n+1)(i-2j+n+2)(n-j)\left(-\frac{\mu}{2}-j+n+\frac{1}{2}\right)\left(\frac{\mu}{2}+3j-n-\frac{3}{2}\right)\right] \\
= {} & \frac{-(-1)^i (i+\mu+2j+1)_{2j-i-3-n} (i-2j+2)_{n+1}}{n!\,(2j-i-n-2)!\,(i-j+2)_n} \\
& \times \frac{(1-j)_n \left(-\frac{\mu}{2}-j+\frac{3}{2}\right)_n \left(\frac{\mu}{2}+3j-n-\frac{3}{2}\right)}{\left(i-j+\frac{\mu}{2}+\frac{3}{2}\right)_{n+1}},
\end{aligned}
\tag{3.12}
$$

and so (3.11) is proved by mathematical induction. As we remarked earlier (3.11) implies (3.9) the assertion of our theorem. □

4. Proof of the Weak Conjectures

We begin with a matrix identity that will immediately imply Theorem 8.

Theorem 7. *The following matrix identity holds* (i *and* j *lie in* $[0, m-1]$)

$$\left(\delta_{ij} + \binom{\mu+i+j}{i}\right) \times (e_{ij}) = (f_{ij}), \tag{4.1}$$

where

$$e_{ij}=0 \quad \text{for } i>j \tag{4.2}$$

$$e_{ii}=1, \tag{4.3}$$

and more generally for $2j\geqq i$

$$e_{i,2j}=\sum_{s\geqq 0}(-1)^i\binom{j-s}{2j-2s-i}\frac{(j-s)_{2s}(-\mu-3j-1)_s 4^{-s}}{s!\left(-\frac{\mu}{2}-j\right)_s\left(-\frac{\mu}{2}-2j+\frac{1}{2}\right)_s}, \tag{4.4}$$

and for $2j-1\geqq i$

$$e_{i,2j-1}=\frac{(-1)^{i-1}}{2}\sum_{s\geqq 0}\left(\binom{j-s}{2j-1-i-2s}+\binom{j-s-1}{2j-1-i-2s}\right)\frac{(j-s)_{2s}(-\mu-3j+1)_s 4^{-s}}{s!\left(-\frac{\mu}{2}-j+1\right)_s\left(-\frac{\mu}{2}-2j+\frac{3}{2}\right)_s}\times\frac{\mu+3j-1-3s}{\mu+3j-1}; \tag{4.5}$$

furthermore

$$f_{ij}=0 \quad \text{for } j>i, \tag{4.6}$$

and

$$f_{ii}=\Delta_i(\mu), \tag{4.7}$$

where $\Delta_i(\mu)$ *is defined in the statement of Theorem* 8 *in Section* 1.

Proof. Since there is no assertion about the f_{ij} for $j<i$, we need only show that the assumption of (4.2), (4.3), (4.4) and (4.5) implies (4.6) and (4.7). We note that (4.2) and (4.3) are implicit in (4.4) and (4.5).

We begin by treating $f_{i,2j}$:

$$\begin{aligned}
f_{i,2j}&=\sum_{l\geqq 0}\left\{\delta_i+\binom{\mu+i+l}{i}\right\}e_{l,2j}\\
&=\sum_{l\geqq 0}\left\{\delta_i+\binom{\mu+i+l}{i}\right\}\\
&\quad\times\sum_{s\geqq 0}(-1)^l\binom{j-s}{2j-2s-l}\frac{(j-s)_{2s}(-\mu-3j-1)_s 4^{-s}}{s!\left(-\frac{\mu}{2}-j\right)_s\left(-\frac{\mu}{2}-2j+\frac{1}{2}\right)_s}\\
&=\sum_{s\geqq 0}\frac{(j-s)_{2s}(-\mu-3j-1)_s 4^{-s}}{s!\left(-\frac{\mu}{2}-j\right)_s\left(-\frac{\mu}{2}-2j+\frac{1}{2}\right)_s}\sum_{l\geqq 0}(-1)^l\left\{\binom{\mu+i+l}{i}+\delta_{il}\right\}\binom{j-s}{2j-2s-l}\\
&=\sum_{s\geqq 0}\frac{(j-s)_{2s}(-\mu-3j-1)_s 4^{-s}}{s!\left(-\frac{\mu}{2}-j\right)_s\left(-\frac{\mu}{2}-2j+\frac{1}{2}\right)_s}\left\{\binom{\mu+i+j-s}{i-j+s}+(-1)^i\binom{j-s}{2j-2s-i}\right\}
\end{aligned}$$

(by the Chu-Vandermonde summation [7; p. 3])

$$
\begin{aligned}
&=(M(i+1,j,\mu+1;0,1)-M(i,j,\mu+1;0,1))\\
&\quad+(\Lambda(i,j,\mu+1;0,1)-\Lambda(i+1,j,\mu+1;0,1))\\
&=(M(i+1,j,\mu+1;0,1)-\Lambda(i+1,j,\mu+1;0,1))\\
&\quad-(M(i,j,\mu+1;0,1)-\Lambda(i,j,\mu+1;0,1))\\
&=\begin{cases}0 & \text{for } 0\leqq i\leqq 2j-1 \text{ by Theorem 5}\\ M(2j+1,j,\mu+1;0,1) & \text{for } i=2j \text{ by Theorem 5 and Corollary 5a}\end{cases}\\
&=\begin{cases}0 \quad \text{for } 0\leqq i\leqq 2j-1\\ 2 \quad \text{for } i=j=0\\ \dfrac{(\mu+2j+2)_j\left(\dfrac{\mu}{2}+2j+\dfrac{3}{2}\right)_{j-1}}{(j)_j\left(\dfrac{\mu}{2}+j+\dfrac{3}{2}\right)_{j-1}} & \text{if } i=2j>0\end{cases}\\
&=\begin{cases}0 & \text{for } 0\leqq i\leqq 2j-1\\ \Delta_{2j}(\mu) & \text{for } i=2j,\end{cases}
\end{aligned}
\tag{4.8}
$$

where $\Delta_{2j}(\mu)$ is defined by (1.14).

Next we treat $f_{i,\,2j-1}$:

$$
\begin{aligned}
f_{i,\,2j-1}&=\sum_{l\geqq 0}\left\{\delta_{il}+\binom{\mu+i+l}{i}\right\}e_{l,\,2j-1}\\
&=\tfrac{1}{2}\sum_{l\geqq 0}\left\{\binom{m+i+l}{i}+\delta_{il}\right\}\sum_{s\geqq 0}(-1)^{l-1}\left(\binom{j-s}{2j-1-l-2s}+\binom{j-s-1}{2j-1-l-2s}\right)\\
&\quad\times\frac{(j-s)_{2s}(-\mu-3j+1)_s4^{-s}}{s!\left(-\dfrac{\mu}{2}-j+1\right)_s\left(-\dfrac{\mu}{2}-2j+\dfrac{3}{2}\right)_s}\,\frac{\mu+3j-1-3s}{\mu+3j-1}\\
&=-\tfrac{1}{2}\sum_{s\geqq 0}\frac{(j-s)_{2s}(-\mu-3j+1)_s4^{-s}}{s!\left(-\dfrac{\mu}{2}-j+1\right)_s\left(-\dfrac{\mu}{2}-2j+\dfrac{3}{2}\right)_s}\,\frac{\mu+3j-1-3s}{\mu+3j-1}\\
&\quad\times\sum_{l\geqq 0}(-1)^l\left\{\binom{m+i+l}{i}+\delta_{il}\right\}\left\{\binom{j-s}{2j-1-l-2s}+\binom{j-s-1}{2j-1-l-2s}\right\}\\
&=-\tfrac{1}{2}\sum_{s\geqq 0}\left\{(-1)^i\binom{j-s}{2j-2s-i-1}+(-1)^i\binom{j-s-1}{2j-2s-i-1}\right.\\
&\quad\left.-\binom{\mu+i+j-s-1}{i-j+s}-\binom{\mu+i+j-s}{i-j+s+1}\right\}\\
&\quad\times\frac{(j-s)_{2s}(-\mu-3j+1)_s4^{-s}(-\mu-3j+1+3s)}{s!\left(-\dfrac{\mu}{2}-j+1\right)_s\left(-\dfrac{\mu}{2}-2j+\dfrac{3}{2}\right)_s(-\mu-3j+1)}\\
&=-\tfrac{1}{2}\sum_{s\geqq 0}\left\{(-1)^i\binom{j-s}{2j-2s-i-1}+(-1)^i\binom{j-s-1}{2j-2s-i-1}\right.\\
&\quad\left.-\binom{\mu+i+j-s-1}{i-j+s}-\binom{\mu+i+j-s}{i-j+s+1}\right\}
\end{aligned}
$$

$$\times \frac{(j-s)_{2s}(-\mu-3j+1)_s 4^{-s}}{s!\left(-\frac{\mu}{2}-j+1\right)_s\left(-\frac{\mu}{2}-2j+\frac{3}{2}\right)_s}$$

$$-\tfrac{3}{2}\sum_{s\geq 0}\left\{(-1)^i\binom{j-s-1}{2j-2s-i-3}+(-1)^i\binom{j-s-2}{2j-2s-i-3}\right.$$

$$\left.-\binom{\mu+i+j-s-2}{i-j+s+1}-\binom{\mu+i+j-s-1}{i-j+s+2}\right\}$$

$$\times \frac{(j-s-1)_{2s+2}(-\mu-3j+2)_s 4^{-s-1}}{s!\left(-\frac{\mu}{2}-j+1\right)_{s+1}\left(-\frac{\mu}{2}-2j+\frac{3}{2}\right)_{s+1}} \tag{4.9}$$

(here the single sum has been split into two by $(-\mu-3j+1+3s)=(-\mu-3j+1)+(3s)$)

$$=-\tfrac{1}{2}\sum_{s\geq 0}\left\{(-1)^i\binom{j-s}{2j-2s-i-1}+(-1)^i\binom{j-s-1}{2j-2s-i-1}\right.$$

$$\left.-\binom{\mu+i+j-s-1}{i-j+s}-\binom{\mu+i+j-s}{i-j+s+1}\right\}$$

$$\times \frac{(j-s)_{2s}(-\mu-3j+1)_s 4^{-s}}{s!\left(-\frac{\mu}{2}-j+1\right)_s\left(-\frac{\mu}{2}-2j+\frac{3}{2}\right)_s}$$

$$-\tfrac{3}{2}\sum_{s\geq 0}\left\{(-1)^i\binom{j-s-1}{2j-2s-i-3}+(-1)^i\binom{j-s-2}{2j-2s-i-3}\right\}$$

$$\times \frac{(j-s-1)_{2s+2}(-\mu-3j+2)_s 4^{-s-1}}{s!\left(-\frac{\mu}{2}-j+1\right)_{s+1}\left(-\frac{\mu}{2}-2j+\frac{3}{2}\right)_{s+1}}$$

$$+\tfrac{3}{2}\sum_{s\geq 0}\left\{\binom{\mu+i+j-s-2}{i-j+s+1}+\binom{\mu+i+j-s-1}{i-j+s+2}\right\}$$

$$\times \frac{(j-s-1)_{2s+2}(-\mu-3j+2)_s 4^{-s-1}}{s!\left(-\frac{\mu}{2}-j+1\right)_{s+1}\left(-\frac{\mu}{2}-2j+\frac{3}{2}\right)_{s+1}}$$

$$=-S_1-S_2+S_3 \text{ say.}$$

Now

$$\begin{aligned}S_1=&\tfrac{1}{2}(-(\Lambda(i+1,j,\mu-1;0,1)-\Lambda(i+2,j,\mu-1;0,1))\\&-\Lambda(i+1,j,\mu-1;0,1)-M(i+1,j,\mu-1;0,1)\\&-M(i+2,j,\mu-1;0,1))\\=&\begin{cases}-\tfrac{3}{2}\Lambda(i+1,j,\mu-1;0,1) & \text{for } 0\leqq i\leqq 2j-2\\ -\tfrac{3}{2}\Lambda(i+1,j,\mu-1;0,1)-\tfrac{1}{2}M(2j+1,j,\mu-1;0,1) & \text{for } i=2j-1\end{cases}\\=&\begin{cases}-\tfrac{3}{2}\Lambda(i+1,j,\mu-1;0,1) & \text{for } 0\leqq i\leqq 2j-2\\ -\tfrac{3}{2}-\tfrac{1}{2}M(2j+1,j,\mu-1;0,1) & \text{for } i=2j-1 \text{ (by Corollary 5a).}\end{cases}\end{aligned}$$

As for S_2, we see that

$$S_2 = +\tfrac{3}{2}(i+1)(-1)^i \times \sum_{s\geqq 0} \binom{j-s-1}{2j-2s-i-3} \frac{(j-s)_{2s+1}(-\mu-3j+2)_s 4^{-s-1}}{s!\left(-\frac{\mu}{2}-j+1\right)_{s+1}\left(-\frac{\mu}{2}-2j+\frac{3}{2}\right)_{s+1}}$$

$$= -\tfrac{3}{2}(i+1)\frac{(2j-i-2)\,\Lambda(i,j,\mu-3;\,3,\,-1)}{4\left(-\frac{\mu}{2}-j+1\right)\left(-\frac{\mu}{2}-2j+\frac{3}{2}\right)}.$$

Finally we treat S_3:

$$S_3 = \tfrac{3}{2}(\mu+2i+1)\sum_{s\geqq 0}\frac{(\mu+i+j-s-2)!\,(j-s-1)_{2s+2}}{(i-j+s+2)!\,(\mu+2j-2s-3)!\,s!} \times \frac{(-\mu-3j+2)_s 4^{-s-1}}{\left(-\frac{\mu}{2}-j+1\right)_{s+1}\left(-\frac{\mu}{2}-2j+\frac{3}{2}\right)_{s+1}}$$

$$= \frac{3(\mu+2i+1)\,2^{-1}(\mu+i+j-2)!\,j(-j+1)}{2\left(-\frac{\mu}{2}-2j+\frac{3}{2}\right)(i-j+2)!\,(\mu+2j-2)!} \times \sum_{s\geqq 0}\frac{(-j+2)_s(j+1)_s\left(-\frac{\mu}{2}-j+\frac{3}{2}\right)_s(-\mu-3j+2)_s}{s!\left(-\frac{\mu}{2}-2j+\frac{5}{2}\right)_s(i-j+3)_s(-\mu-i-j+2)_s}$$

$$= \frac{3(\mu+2i+1)(\mu+i+j-2)!\,j(j-1)}{2(\mu+4j-3)(i-j+2)!\,(\mu+2j-2)!} \times {}_4F_3\left[\begin{matrix}-j+2, j+1, -\frac{\mu}{2}-j+\frac{3}{2}, -\mu-3j+2;\ 1\\ -\frac{\mu}{2}-2j+\frac{5}{2}, i-j+3, -\mu-i-j+2\end{matrix}\right]$$

$$= \frac{3(\mu+2i+1)(\mu+i+j-2)!\,j(j-1)}{2(\mu+4j-3)(i-j+2)!\,(\mu+2j-2)!}\,\frac{\left(i-j+\frac{\mu}{2}+\frac{5}{2}\right)_{j-2}(-j+1)_{j-2}}{(\mu+i+1)_{j-2}\left(-\frac{\mu}{2}-2j+\frac{5}{2}\right)_{j-2}} \times {}_4F_3\left[\begin{matrix}-\frac{\mu}{2}-j+\frac{3}{2}, i+2j+\mu+1, -j+2, i-2j+2;\ 1\\ i-j+\frac{\mu}{2}+\frac{5}{2}, i-j+3, 2\end{matrix}\right].$$

Now when $i=2j-1$ we see that $S_2=0$. Consequently

$$f_{2j-1,\,2j-1}=\tfrac{3}{2}+\tfrac{1}{2}M(2j+1,j,\mu-1;0,1)$$

$$+\tfrac{3}{2}\sum_{s\geqq 0}\frac{\left(-\frac{\mu}{2}-j+\frac{1}{2}\right)_{s+1}(\mu+4j-1)_{s+1}(-j+1)_{s+1}}{(s+1)!\left(j+\frac{\mu}{2}+\frac{1}{2}\right)_{s+1}(j+1)_{s+1}}$$

$$=\tfrac{1}{2}M(2j+1,j,\mu-1;0,1)$$

$$+\tfrac{3}{2}\,{}_3F_2\left[\begin{matrix}-j+1,\ \mu+4j-1,\ -\frac{\mu}{2}-j+\frac{1}{2};\ 1\\ j+\frac{\mu}{2}+\frac{1}{2},\ j+1\end{matrix}\right]$$

$$=\frac{1}{2}\frac{(\mu+2j)_j\left(2j+\frac{\mu}{2}+\frac{1}{2}\right)_{j-1}}{(j)_j\left(\frac{\mu}{2}+j+\frac{1}{2}\right)_{j-1}}+\frac{3}{2}\frac{(\mu+2j)_{j-1}\left(-\frac{\mu}{2}-3j+\frac{3}{2}\right)_{j-1}}{\left(j+\frac{\mu}{2}+\frac{1}{2}\right)_{j-1}(-2j+1)_{j-1}}$$

$$=\frac{1}{2}\frac{(\mu+2j)_{j-1}\left(\frac{\mu}{2}+2j+\frac{1}{2}\right)_{j-1}}{(j)_j\left(\frac{\mu}{2}+j+\frac{1}{2}\right)_{j-1}}(\mu+3j-1+3j)$$

$$=\frac{(\mu+2j)_{j-1}\left(\frac{\mu}{2}+2j+\frac{1}{2}\right)_j}{(j)_j\left(\frac{\mu}{2}+j+\frac{1}{2}\right)_{j-1}}=\Delta_{2j-1}(\mu)$$

by (1.15). Thus we have now established (4.7) for all $i\geqq 0$.

Finally for $i<2j-1$, we see that

$$f_{i,\,2j-1}=-S_1-S_2+S_3$$

$$=\tfrac{3}{2}\Lambda(i+1,j,\mu-1;0,1)$$

$$+\frac{3}{2}\frac{(i+1)(2j-i-2)\,\Lambda(i,j,\mu-3;3,-1)}{4\left(-\frac{\mu}{2}-j+1\right)\left(-\frac{\mu}{2}-2j+\frac{3}{2}\right)}$$

$$+\frac{3(\mu+2i+1)(\mu+i+j-2)!\,j(j-1)\left(i-j+\frac{\mu}{2}+\frac{5}{2}\right)_{j-2}(1-j)_{j-2}}{(\mu+4j-3)(i-j+2)!\,(\mu+2j-2)!\,(\mu+i+1)_{j-2}\left(-\frac{\mu}{2}-2j+\frac{5}{2}\right)_{j-2}}$$

$$\times\,{}_4F_3\left[\begin{matrix}-\frac{\mu}{2}-j+\frac{3}{2},\ i+2j+\mu+1,\ -j+2,\ i-2j+2;\ 1\\ i-j+\frac{\mu}{2}+\frac{5}{2},\ i-j+3,\ 2\end{matrix}\right].$$

For the two Λ-functions we use the second ${}_4F_3$ expression that arises in the string of equations equating $M(i,j,m;a,\omega)$ with $\Lambda(i,j,m;a,\omega)$ in the proof of Theorem 5. Hence

$$
\begin{aligned}
f_{i,2j-1} &= \frac{3(\mu+i)!\,(1-j)_{j-1}\left(i-j+\frac{\mu}{2}+\frac{3}{2}\right)_{j-1}}{2(i-j)!\,(\mu+2j-1)!\left(-\frac{\mu}{2}-2j+\frac{3}{2}\right)_{j-1}} \\
&\quad \times \left\{ {}_4F_3\left[\begin{matrix} i-2j+1,\ 1-j,\ -j-\frac{\mu}{2}+\frac{1}{2},\ i+\mu+2j;\ 1 \\ i-j+1,\ i-j+\frac{\mu}{2}+\frac{3}{2},\ 1 \end{matrix}\right]\right. \\
&\quad + \frac{(2i+\mu+1)\left(\frac{\mu}{2}+j-\frac{1}{2}\right)j(j-1)}{\left(i-j+\frac{\mu}{2}+\frac{3}{2}\right)(i-j+1)(i-j+2)} \\
&\quad \times {}_4F_3\left[\begin{matrix} i-2j+2,\ 2-j,\ -j-\frac{\mu}{2}+\frac{3}{2},\ i+\mu+2j+1;\ 1 \\ i-j+3,\ i-j+\frac{\mu}{2}+\frac{5}{2},\ 2 \end{matrix}\right] \\
&\quad + \frac{\left(\frac{\mu}{2}+i+\frac{1}{2}\right)(i+1)(2j-i-2)j}{\left(i-j+\frac{\mu}{2}+\frac{3}{2}\right)(i-j+1)(i-j+2)} \\
&\quad \left. \times {}_4F_3\left[\begin{matrix} i-2j+3,\ 1-j,\ -j-\frac{\mu}{2}+\frac{3}{2},\ i+\mu+2j+1;\ 1 \\ i-j+3,\ i-j+\frac{\mu}{2}+\frac{5}{2},\ 2 \end{matrix}\right]\right\} \\
&= 0, \quad \text{by Theorem 6.}
\end{aligned}
$$

It appears at first glance that we can invoke Theorem 6 for i and j integers only for $i \geqq j$. However the above expression is indentically zero provided $i-2j+2$ is a nonpositive integer and $i-j$ is any complex (or real) number other than a negative integer. The conclusion for all integral i with $0 \leqq i < 2j-1$ then follows by continuity.

As a direct corollary of Theorem 7, we immediately deduce:

Theorem 8.

$$
\det\left(\delta_{ij}+\binom{\mu+i+j}{i}\right)_{0\leqq i,\,j\leqq m-1} = \prod_{j=0}^{m-1} \Delta_j(\mu),
$$

where

$$\Delta_{2j}(\mu)=\begin{cases}\dfrac{(\mu+2j+2)_j\left(\frac{\mu}{2}+2j+\frac{3}{2}\right)_{j-1}}{(j)_j\left(\frac{\mu}{2}+j+\frac{3}{2}\right)_{j-1}}, & j>0\\ 2, & j=0,\end{cases} \tag{4.10}$$

and

$$\Delta_{2j-1}(\mu)=\frac{(\mu+2j)_{j-1}\left(\frac{\mu}{2}+2j+\frac{1}{2}\right)_j}{(j)_j\left(\frac{\mu}{2}+j+\frac{1}{2}\right)_{j-1}}, \quad j>0 \tag{4.11}$$

Proof. By Theorem 7, we see that

$$\begin{aligned}\det\left(\delta_{ij}+\binom{\mu+i+j}{i}\right)_{0\leqq i,\,j\leqq m-1}&=\frac{\det(f_{ij})}{\det(e_{ij})}\\&=f_{00}f_{11}\cdots f_{m-1,m-1}\\&=\prod_{j=0}^{m-1}\Delta_j(\mu).\quad\square\end{aligned}$$

As a consequence of Theorem 8, we can now prove the weak forms of both conjectures given in the introduction.

Theorem 9 (The Weak Macdonald Conjecture). *The total number of plane partitions π such that $D(\pi)\subseteq\mathscr{B}_{m,m,m}$ and $D(\pi)$ is G_3 invariant is*

$$\prod_{i=1}^{m}\left\{\frac{3i-1}{3i-2}\prod_{j=i}^{m}\frac{m+i+j-1}{2i+j-1}\right\}.$$

Proof. We let γ_m denote the product given in Theorem 9. Then by Theorem 4 with $q=1$ and Theorem 8, we see that we must prove that

$$\gamma_1=\Delta_0(0), \tag{4.12}$$

$$\frac{\gamma_m}{\gamma_{m-1}}=\Delta_{m-1}(0). \tag{4.13}$$

Now (4.12) is clear since $\gamma_1=\frac{2}{1}\cdot\frac{2}{2}=2=\Delta_0(0)$. As for (4.13), we see that

$$\begin{aligned}\frac{\gamma_m}{\gamma_{m-1}}&=\frac{3m-1}{3m-2}\times\prod_{i=1}^{m}\frac{2m+i-1}{2i+m-1}\times\prod_{i=1}^{m-1}\frac{2m+i-2}{m+2i-2}=\frac{3m-1}{3m-2}\,\frac{(2m)_m(2m-1)_{m-1}}{(m)_{2m-2}(3m-1)}\\&=\frac{3m-1}{3m-2}\times\frac{(2m)_{m-1}(2m-1)_{m-1}}{(m)_{2m-2}}=\frac{(3m-1)}{(3m-2)}\times\frac{(2m)_{m-1}}{(m)_{m-1}},\end{aligned}$$

and by (1.15)

$$\begin{aligned}\Delta_{2j-1}(0)&=\frac{(2j)_{j-1}(2j+\frac{1}{2})_j}{(j)_j(j+\frac{1}{2})_{j-1}}=\frac{(3j-\frac{1}{2})(4j)_{2j-1}}{(3j-1)(2j)_{2j-1}}\\&=\frac{6j-1}{6j-2}\,\frac{(4j)_{2j-1}}{(2j)_{2j-1}}=\frac{\gamma_{2j}}{\gamma_{2j-1}},\end{aligned}$$

while by (1.14)

$$\Delta_{2j}(0)=\frac{(2j+2)_j(2j+\frac{3}{2})_{j-1}}{(j)_j(j+\frac{3}{2})_{j-1}}=\frac{3j+1}{3j+\frac{1}{2}}\times\frac{(2j+2)_{j-1}(2j+\frac{3}{2})}{(j)_j(j+\frac{3}{2})_{j-1}}$$
$$=\frac{(3j+1)}{(3j+\frac{1}{2})}\times\frac{(2j+1)_j(2j+\frac{3}{2})_j}{(j+1)_j(j+\frac{1}{2})_j}=\frac{(6j+2)}{(6j+1)}\times\frac{(4j+2)_{2j}}{(2j+1)_{2j}}=\frac{\gamma_{2j+1}}{\gamma_{2j}}.$$

Thus Theorem 9 is established. □

Theorem 10. *The total number of descending plane partitions whose parts do not exceed m is* $\prod_{i=1}^{m}\prod_{j=1}^{m}\frac{m+i+j-1}{2i+j-1}$.

Proof. We let δ_m denote the product given in Theorem 10. Then by Theorem 3′ with $d=0$, $q=1$ and Theorem 8, we see that we must establish

$$\delta_2=\Delta_0(2), \tag{4.14}$$

$$\frac{\delta_m}{\delta_{m-1}}=\Delta_{m-2}(2). \tag{4.15}$$

Now (4.14) is clear since $\delta_2=\frac{3}{2}\cdot\frac{4}{3}\cdot\frac{5}{5}=2=\Delta_0(2)$. As for (4.15), we see that

$$\frac{\delta_m}{\delta_{m-1}}=\frac{3m-2}{3m-1}\cdot\frac{\gamma_m}{\gamma_{m-1}}=\frac{(2m)_{m-1}}{(m)_{m-1}},$$

and by (1.15)

$$\Delta_{2j-1}(2)=\frac{(2j+2)_{j-1}(2j+\frac{3}{2})_j}{(j)_j(j+\frac{3}{2})_{j-1}}=\frac{(2j+1)_j(2j+\frac{3}{2})_j}{(j+1)_j(j+\frac{1}{2})_j}$$
$$=\frac{(4j+2)_{2j}}{(2j+1)_{2j}}=\frac{\delta_{2j+1}}{\delta_{2j}},$$

while by (1.14)

$$\Delta_{2j}(2)=\frac{(2j+4)_j(2j+\frac{5}{2})_{j-1}}{(j)_j(j+\frac{5}{2})_{j-1}}$$
$$=\frac{(2j+3)_{j+1}(2j+\frac{5}{2})_{j-1}}{(j+1)_j(j+\frac{3}{2})_j}=\frac{(2j+3)_{j+1}(2j+\frac{3}{2})_j}{(j+1)_j(j+\frac{3}{2})_{j+1}}$$
$$=\frac{(2j+2)_{j+1}(2j+\frac{5}{2})_j}{(j+1)_{j+1}(j+\frac{3}{2})}\times\frac{(2j+1)(3j+3)}{(2j+2)(3j+\frac{3}{2})}$$
$$=\frac{(4j+2)_{2j+1}}{(2j+2)_{2j+1}}=\frac{\delta_{2j+2}}{\delta_{2j+1}}.$$

Thus Theorem 10 is established. □

Clearly we can apply Theorem 8 to the case $q=1$ of Theorem 3′; in [5] we discuss this application in more detail.

5. Conclusion

Obviously the final assault on Macdonald's conjecture is the first order of business now. One is tempted to hope that the q-analogs of the results on hypergeometric series utilized in this paper will be adequate to treat both Macdonald's conjecture and the descending plane partitions conjecture. However just as the full q-series proof of MacMahon's conjecture [4] was significantly more complicated than just the case $q=1$ [1], so too is it clear upon inspection that the full conjectures studied here are not obtained purely by direct substitution of q-analogs of Theorems 5 and 6. A discussion of the evidence for these conjectures and related problems occurs in [5].

In [2] the equivalence of the Bender-Knuth conjecture with MacMahon's conjecture was proved using only elementary row operations on determinants. So far similar efforts to prove the equivalence of Macdonald's conjecture with the descending plane partitions conjecture have gone astray.

The assertion in Theorem 5 is the following hypergeometric series identity:

$$\begin{aligned}&\frac{\left(\frac{m}{2}+2j\right)_{a+\omega-1}}{\left(\frac{m}{2}+j+1\right)_{a+\omega-1}}\binom{m+i+j+a-1}{i-j+a-1}\\ &\quad\times{}_4F_3\left[\begin{matrix}1-j, j, -m-3j-a-\omega+1, -\frac{m}{2}-j; 1\\ i-j+a, -\frac{m}{2}-2j-a-\omega+2, -m-i-j-a+1\end{matrix}\right]\\ &=(-1)^{i+a}\binom{j+a+\omega-2}{i-j+a-1}\\ &\quad\times{}_4F_3\left[\begin{matrix}-j+\frac{i+a}{2}, -j+\frac{i+a+1}{2}, j+a+\omega-1, -m-3j-a-\omega+1; 1\\ i-j+a, -\frac{m}{2}-2j+1, -\frac{m}{2}-j+\frac{1}{2}\end{matrix}\right].\end{aligned} \tag{5.1}$$

Since the crucial step of the proof of this identity is obtained from applying (3.6) when the series on the left is not only nearly poised of the second kind but also balanced, it would appear that such series have a significant amount of structure beyond that of merely balanced ${}_4F_3$'s. Recent work of J. Wilson [19] and [20] and R. Askey and J. Wilson [6] has made clear the importance of balanced ${}_4F_3$'s.

Theorem 6 is essentially a contiguous relation among three balanced ${}_4F_3$'s. Consequently it should be derivable from the ${}_4F_3$ contiguous relations described by J. Wilson [19]. Indeed, R. Askey has outlined how the derivation should proceed from the work in Chapter 4 of Wilson's thesis [20]. Equation (3.11) essentially asserts that the tails of the three series in Theorem 6 are indeed summable to the right side of (3.11). Thus we may deduce from (3.11) and (3.9) that the sum of three truncated balanced ${}_4F_3$'s of this type is summable. Askey also suggests that Wilson's work can be used to explain this phenomenon.

Finally we mention that the "massive" computation in (3.12) for L_n which was essential to the proof of Theorem 6 was done by the FORMAC Utility Program due to H.D. Noble of the Pennsylvania State University Computation Center.

References

1. Andrews, G.E.: MacMahon's conjecture on symmetric plane partitions. Proc. Nat. Acad. Sci. USA, **74**, 426–429 (1977)
2. Andrews, G.E.: Plane partitions (II): the equivalence of the Bender-Knuth and MacMahon conjectures. Pac. J. Math., **72**, 283–291 (1977)
3. Andrews, G.E.: The Theory of Partitions, Addision-Wesley, Reading, 1976
4. Andrews, G.E.: Plane partitions (I): the MacMahon conjecture, Advances in Math. (to appear)
5. Andrews, G.E.: On Macdonald's conjecture and descending plane partitions, Proceedings of the Alfred Young Day Conference (to appear)
6. Askey, R., Wilson, J.: A set of orthogonal polynomials that generalize the Racah coefficients or 6-j symbols, S.I.A.M. J. Math. Anal., (to appear)
7. Bailey, W.N.: Generalized Hypergeometric Series. London and New York: Cambridge University Press 1935
8. Carlitz, L.: Rectangular arrays and plane partitions, Acta Arith., **13**, 22–47 (1967)
9. Gansner, E.: Matrix correspondences and the enumeration of plane partitions, Ph.D. Thesis, M.I.T., 1978
10. Macdonald, I.G.: Lecture on plane partitions, Oberwolfach Conference on Combinatorics and Special Functions, May 1977
11. MacMahon, P.A.: Partitions of numbers whose graphs possess symmetry. Trans. Cambridge Phil. Soc., **17**, 149–170 (1898–99)
12. MacMahon, P.A.: Combinatory Analysis, Vol. 2. London and New York: Cambridge University Press 1916
13. Muir, T.: A Treatise on the Theory of Determinants. London: Longmans 1933
14. Stanley, R.P.: Theory and applications of plane partitions I, Studies in Appl. Math., **50**, 167–188 (1971)
15. Stanley, R.P.: Theory and applications of plane partitions II, Studies in Appl. Math., **50**, 259–279 (1971)
16. Whipple, F.J.W.: Well-poised series and other generalized hypergeometric series, Proc. London Math. Soc. (2), **25**, 525–544 (1926)
17. Whipple, F.J.W.: Some transformations of generalized hypergeometric series, Proc. London Math. Soc. (2), **26**, 257–272 (1927)
18. Whittaker, E.T., Watson, G.N.: A Course of Modern Analysis, 4th ed., London and New York: Cambridge University Press 1958
19. Wilson, J.: Three-term contiguous relations and some new orthogonal polynomials, from Pade and Rational Approximation (Saff and Varga, eds.). New York-London: Academic Press, 1977
20. Wilson, J.: Ph.D. Thesis, University of Wisconsin, 1978

Received October 3, 1978

4.3 Plane Partitions, IV: A Conjecture of Mills–Robbins–Rumsey

Aequationes Mathematicae **33** (1987) 230–250
University of Waterloo

Plane partitions IV: A conjecture of Mills–Robbins–Rumsey

GEORGE E. ANDREWS*

Dedicated to the memory of my mother

Abstract. In this paper we consider a problem posed by W. Mills. D. Robbins and H. Rumsey for a certain plane partition generating function $Z_n(x, m)$. The special case $Z_n(1, m)$ is the generating function that arose in the weak Macdonald conjecture. Mills–Robbins–Rumsey conjectured that $Z_n(2, m)$ also possesses a nice finite product representation. Their conjecture is proved as Theorem 1.

The method of proof resembles that of the evaluation of $Z_n(1, m)$ given previously. Many results for the ${}_3F_2$ hypergeometric function are required including Whipple's theorem, the Pfaff–Saalschütz summation and contiguous relations.

In passing we note that our Lemma 2 provides a new and simpler representation of $Z_n(2, m)$ as a determinant:

$$Z_n(2, m) = \det\left(\delta_{ij} + \sum_{i=0}^{i}\binom{m+j+t}{t}\binom{m+i}{m+t}\right)_{0 \leqq i, j \leqq n-1}$$

Conceivably this new representation may provide new interpretations of the combinatorial significance of $Z_n(2, m)$. In the final analysis, one would like a combinatorial explanation of $Z_n(2, m)$ that would provide an algorithmic proof of the Mills–Robbins–Rumsey conjecture.

1. Introduction.

In [5], W. H. Mills, D. P. Robbins and H. Rumsey enumerate classes of cyclically symmetric plane partitions. These are plane partitions whose Ferrers graph is invariant under cyclic permutation of the coordinate axes. In their work, they consider $Z_n(x, m)$ the generating function for the following set $Z_n(m)$ of plane partitions. Namely those arrays of positive integers

---

AMS (1980) subject classification: Primary 05A19, 10A45, 33A30.

Manuscript received September 30, 1986.

* *Partially supported by National Science Foundation Grant DMS-8503324.*

$$\begin{matrix} a_{11}\ a_{12}\ a_{13} \ldots a_{1,\lambda_1} \\ a_{22}\ a_{23} \ldots a_{2,\lambda_2} \\ a_{33} \\ \cdot \\ \cdot \\ \cdot \\ a_{rr} \ldots a_{r,\lambda_r} \end{matrix}$$

wherein there is strict decrease along columns, nonincrease along rows, $\lambda_i \leqq n$, and $a_{ii} = \lambda_i - i + 1 + 2m$. Mills–Robbins–Rumsey [5] describe the relationship of these partitions to descending plane partitions and alternating sign matrices. The generating function $Z_n(x, m)$ can be represented as an $n \times n$ determinant [5]:

$$Z_n(x, m) = \det\left(\delta_{ij} + \sum_{t,k=0}^{n-1} \binom{j+m}{t}\binom{k}{t}\binom{i-k+m-1}{i-k} x^{k-t}\right)_{0 \leqq i,j \leqq n-1}. \tag{1.1}$$

It is also pointed out in [5] that the case $x = 1$ reduces to the determinant evaluation done in [2] for the weak Macdonald conjecture. The authors then note that there also appears to be a nice evaluation when $x = 2$. In fact, they make the following:

Conjecture (henceforward the M–R–R conjecture). Let v be a positive integer. Then

$$\frac{Z_{2v}(2, m)}{Z_{2v-1}(2, m)} = 2^v \prod_{k=1}^{v} \frac{m + 2v + 2k - 1}{v + k}, \tag{1.2}$$

and

$$\frac{Z_{2v+1}(2, m)}{Z_{2v}(2, m)} = 2^{v+1} \prod_{k=1}^{v} \frac{m + 2v + 2k - 1}{v + k}. \tag{1.3}$$

The object of this paper is to prove the M–R–R conjecture. The proof is similar in broad outline to that of the weak Macdonald conjecture given in [2]. In Section 2 we describe the empirical background and motivation for our proof. In Section 3 we prepare the necessary hypergeometric machinery. Section 4 provides the full proof of the conjecture, and Section 5 examines some of the implications of this work.

It should be noted that this conjecture is Conjecture 6 in R. P. Stanley's baker's dozen of conjectures on plane partitions [7].

2. SCRATCHPAD and the M–R–R conjecture.

The proof of the M–R–R conjecture is based on the same simple row-reduction principle as the weak Macdonald conjecture. Namely, if $Z_n(2, m) = \det(A_{ij})_{0 \leqq i,j \leqq n-1}$, then we wish to determine e_{ij} and f_{ij} such that $e_{ij} = 0$ if $j < i$, $e_{ii} = 1$, $f_{ij} = 0$ if $i < j$, and as $n \times n$ matrices

$$(A_{ij})(e_{ij}) = (f_{ij}).$$

Clearly from these conditions

$$\begin{aligned} Z_n(2, m) &= \det(A_{ij})_{0 \leqq i,j \leqq n-1} \\ &= (\det(A_{ij})_{0 \leqq i,j \leqq n-1}) \cdot (\det(e_{ij})_{0 \leqq i,j \leqq n-1}) \\ &= (\det(f_{ij})_{0 \leqq i,j \leqq n-1}) \\ &= \prod_{i=0}^{n-1} f_{ii}. \end{aligned} \tag{2.1}$$

Hence we proceed empirically. In order to make $f_{ij} = 0$ for $i < j$, we must have the following j equations valid:

$$\sum_{l=0}^{j-1} A_{il} e_{lj} + A_{ij} = 0, \qquad i = 0, 1 \ldots, j-1. \tag{2.2}$$

Provided the M–R–R conjecture is valid, the above system is nonsingular and the e_{lj} are uniquely determined. We can then theoretically at least work them out. However from (1.1) we note that for $0 \leqq i, l \leqq n-1$

$$A_{il} = \delta_{il} + \sum_{t,k=0}^{n-1} \binom{l+m}{t}\binom{k}{t}\binom{i-k+m-1}{i-k} 2^{k-t}. \tag{2.3}$$

Thus computation of the e_{ij} is quite cumbersome by hand. However one can ask IBM's symbolic algebra package SCRATCHPAD to solve systems of equations like (2.2) in order to find the e_{ij} as rational functions of m. The first few of these are

$$e_{0,1} = -\frac{1}{2}$$

$$(e_{0,2}, e_{1,2}) = (0, -1)$$

$$(e_{0,3}, e_{1,3}, e_{2,3}) = \left(\frac{1}{2m+6}, \frac{m+1}{2m+6}, -\frac{3}{2}\right)$$

$$(e_{0,4}, \ldots, e_{3,4}) = \left(0, \frac{1}{m+3}, \frac{m+2}{m+3}, -2\right)$$

$$(e_{0,5}, \ldots, e_{4,5}) = \left(\frac{-3}{2m^2+16m+30}, \frac{-3m-3}{2m^2+16m+30}, \frac{-m+4}{2m+10}, \frac{2m+7}{m+5}, -\frac{5}{2}\right).$$

It is a simple matter to simplify these rational functions as follows:

$$e_{0,1} = -\frac{1}{2}$$

$$(e_{0,2}, e_{1,2}) = (0, -1)$$

$$(e_{0,3}, e_{1,3}, e_{2,3}) = \left(\frac{1}{2(m+3)}, \frac{1}{2} - \frac{1}{m+3}, -\frac{3}{2}\right)$$

$$(e_{0,4}, e_{1,4}, e_{2,4}, e_{3,4}) = \left(0, \frac{1}{m+3}, 1 - \frac{1}{m+3}, -2\right)$$

$$(e_{0,5}, \ldots, e_{4,5}) = \left(\frac{-3}{2(m+3)(m+5)}, -\frac{3}{2(m+5)} + \frac{3}{(m+3)(m+5)}, -\frac{1}{2} + \frac{1}{2(m+5)}, 2 - \frac{3}{m+5}, -\frac{5}{2}\right)$$

Comparison of these values and the remaining e_{ij} for $j \leqq 9$ with their counterparts in [2; eqs. (4.4) and (4.5)], yields the following conjectural representations of the e_{ij}:

$$e_{i,2j} = \sum_{j \geqq s \geqq 0} (-1)^j \binom{j-s}{2j-2s-i} \frac{4^{-s}(j-s)_{2s}}{s!\left(-\frac{m}{2} - j + \frac{1}{2}\right)_s}, \tag{2.4}$$

and

$$e_{i,2j-1} = \frac{(-1)^j}{2} \sum_{j \geqq s \geqq 0} \left(\binom{j-s}{2j-2s-i-1} + \binom{j-s-1}{2j-2s-i-1}\right) \frac{4^{-s}(j-s)_{2s}}{s!\left(-\frac{m}{2} - j + \frac{1}{2}\right)_s}. \tag{2.5}$$

Having empirically determined the e_{ij}, we still have the problem of proving that they fit (2.2) in general, and we must evaluate the f_{ii} for all i. In order to do this, we require a fair amount of machinery from the theory of hypergeometric series.

3. Hypergeometric series.

We must collect and develop a number of results for hypergeometric functions. The series we wish to consider is

$$ {}_{s+1}F_s\left(\begin{matrix} a_0, a_1, \ldots, a_s; t \\ b_1, \ldots, b_s \end{matrix}\right) = \sum_{n=0}^{\infty} \frac{(a_0)_n (a_1)_n \ldots (a_s)_n t^n}{n!(b_1)_n \ldots (b_s)_n}, \tag{3.1} $$

where [3; p.1]

$$ (A)_n = \Gamma(A+n)/\Gamma(A) \quad (= A(A+1)\ldots(A+n-1) \text{ for } n \text{ a nonnegative integer}). \tag{3.2} $$

We shall only consider ${}_{s+1}F_s$ in cases where at least one of the a_i is a nonpositive integer; consequently the series we shall consider are all finite.

First we collect the necessary known results:

Whipple's theorem:

$$ {}_3F_2\left[\begin{matrix} a, 1-a, \frac{e+f-1}{2}; 1 \\ e, f \end{matrix}\right] = \frac{\pi\Gamma(e)\Gamma(f)}{2^{e+f-2}\Gamma\left(\frac{a+e}{2}\right)\Gamma\left(\frac{a+f}{2}\right)\Gamma\left(\frac{1-a+e}{2}\right)\Gamma\left(\frac{1-a+f}{2}\right)} \tag{3.3} $$

[3; p.16, §3.4]

$$ {}_3F_2\left(\begin{matrix} \alpha_1, \alpha_2, \alpha_3; 1 \\ \beta_1, \beta_2 \end{matrix}\right) = \frac{\Gamma(\beta_2)\Gamma(\beta_1+\beta_2-\alpha_1-\alpha_2-\alpha_3)}{\Gamma(\beta_2-\alpha_3)\Gamma(\beta_1+\beta_2-\alpha_1-\alpha_2)} {}_3F_2\left(\begin{matrix} \beta_1-\alpha_1, \beta_1-\alpha_2, \alpha_3; 1 \\ \beta_1, \beta_1+\beta_2-\alpha_1-\alpha_2 \end{matrix}\right) \tag{3.4} $$

[3; p.98, ex.7];

for one of b and n a nonnegative integer,

$$ {}_3F_2\left(\begin{matrix} -n, -b, c; 1 \\ e, f \end{matrix}\right) = \frac{(e+b)_n}{(e)_n} {}_3F_2\left(\begin{matrix} -n, -b, f-c; 1 \\ 1-n-b-e, f \end{matrix}\right) \tag{3.5$_a$} $$

$$ = \frac{\Gamma(e+b+n)\Gamma(e)}{\Gamma(e+b)\Gamma(e+n)} {}_3F_2\left(\begin{matrix} -n, -b, f-c; 1 \\ 1-n-b-e, f \end{matrix}\right) \tag{3.5$_b$} $$

$$= \frac{\Gamma(1-e-\mathrm{b})\Gamma(1-e-n)}{\Gamma(1-e)\Gamma(1-e-b-n)} {}_3F_2\left(\begin{matrix}-n, -b, f-c; 1\\ 1-n-b-e, f\end{matrix}\right)$$

[6; p.16, eq. (3.8)]; $(3.5)_c$

Pfaff-Saalschütz summation [3; p.9, §2.2, eq. (1)]

$$ {}_3F_2\left(\begin{matrix}a, b, -n; 1\\ c, d\end{matrix}\right) = \frac{(c-a)_n(c-b)_n}{(c)_n(c-a-b)_n}, \tag{3.6}$$

provided n is a nonnegative integer and $1+a+b-n=c+d$;

Chu-Vandermonde summation [3; p.3, last equation in §1.3]

$$ {}_2F_1\left(\begin{matrix}-n, b; 1\\ c\end{matrix}\right) = \frac{(c-b)_n}{(c)_n} \tag{3.7}$$

where n is a nonnegative integer.

$$ {}_2F_1\left[\begin{matrix}a, b; \frac{1}{2}\\ c\end{matrix}\right] = 2^a {}_2F_1\left(\begin{matrix}a, c-b; -1\\ c\end{matrix}\right) \tag{3.8}$$

[3; p.11, second equation]

Next we require an immediate consequence of some of the three term ${}_3F_2$ contiguous relations; we refer the reader to J. Wilson's article [9] for a complete account of such relations.

$$\begin{aligned}(1-d)(d+e-a-b-c-1)\,{}_3F_2\left(\begin{matrix}a, b, c; 1\\ d-1, e\end{matrix}\right) &+ ((a-d)(e-1) \\ + (d-1)(d+e-a-b-c-1)+bc)\,{}_3F_2\left(\begin{matrix}a, b, c; 1\\ d, e\end{matrix}\right) & \\ + (d-a)(e-1)\,{}_3F_2\left(\begin{matrix}a-1, b, c; 1\\ d, e-1\end{matrix}\right) &= 0\end{aligned} \tag{3.9}$$

(interchange d and e in equation (17) of [9] and replace a by $a-1$ therein; multiply the resulting equation by $(a-d)$ and add it to equation (20) of [9]).

In addition to the above results, we shall, at times, require some of the identities for the Γ-function:

$$\Gamma(z)\Gamma(1-z)=\frac{\pi}{\sin \pi z} \text{ [8; p.239]}, \tag{3.10}$$

$$\Gamma(2z)=\frac{2^{2z-1}\Gamma(z)\Gamma(z+\frac{1}{2})}{\sqrt{\pi}} \text{ [8; p.240]}, \tag{3.11}$$

and we note that $\frac{1}{\Gamma(z)}$ is analytic in the entire z-plane with zeros at the nonpositive integers [8; p.236].

Finally we need a hypergeometric identity which does not appear in the literature. It plays a key role and is the counterpart of Theorem 5 in [2].

LEMMA 1. *If M and N are nonnegative integers with $M \geqq N/2$, then*

$$\begin{aligned}
&{}_3F_2\begin{pmatrix}-N/2,\ -N/2+1/2,\ A;\ 1\\ 1/2-M,\ B\end{pmatrix}\\
&=\frac{2^{2B-2}\Gamma(B)\Gamma(1+A-B)\Gamma\left(B+M-\frac{1}{2}\right)\Gamma(N+1)(-1)^{N+M}}{\Gamma(1+A+M-B-N)\Gamma\left(M+\frac{1}{2}\right)\Gamma(1-M+N)\Gamma(2B+N-1)}\\
&\quad\times {}_3F_2\begin{bmatrix}\frac{1}{2}-M+\frac{N}{2},\ -M+\frac{N}{2},\ 1+A-B;\ 1\\ \frac{3}{2}-M-B,\ 1-M+N\end{bmatrix}
\end{aligned} \tag{3.12}$$

Proof. By (3.4) with $\alpha_1=-N/2, \alpha_2=-N/2+1/2, \alpha_3=A, \beta_1=\frac{1}{2}-M, \beta_2=B$,

$$\begin{aligned}
&{}_3F_2\begin{bmatrix}-\frac{N}{2},\ -\frac{N}{2}+\frac{1}{2},\ A;\ 1\\ \frac{1}{2}-M,\ B\end{bmatrix}\\
&=\frac{\Gamma(B)\Gamma(B+N-A-M)}{\Gamma(B-A)\Gamma(B+N-M)}\ {}_3F_2\begin{bmatrix}\frac{1}{2}-M+\frac{N}{2},\ -M+\frac{N}{2},\ A;\ 1\\ \frac{1}{2}-M,\ B+N-\mathrm{M}\end{bmatrix}
\end{aligned} \tag{3.13}$$

$$= \frac{\Gamma(B)\Gamma(B+N-A-M)\Gamma\left(1+\frac{N}{2}\right)\Gamma\left(\frac{1}{2}+\frac{N}{2}\right)}{\Gamma(B-A)\Gamma(B+N-M)\Gamma\left(\frac{1}{2}+M\right)\Gamma(1-M+N)}$$

$$\times\ {}_3F_2\left[\begin{matrix}\frac{1}{2}-M+\frac{N}{2}, -M+\frac{N}{2}, B+N-M-A; 1\\ 1-M+N, B+N-M\end{matrix}\right]$$

(by $(3.5)_c$ with $n = M - \frac{N}{2} - \frac{1}{2}, b = M - \frac{N}{2}, c = A,$

$$e = \frac{1}{2} - M, f = B + N - M)$$

$$= \frac{\Gamma(B)\Gamma(B+N-A-M)\Gamma\left(1+\frac{N}{2}\right)\Gamma\left(\frac{1}{2}+\frac{N}{2}\right)\Gamma\left(B+M-\frac{1}{2}\right)}{\Gamma(B-A)\Gamma\left(\frac{1}{2}+M\right)\Gamma(1-M+N)\Gamma\left(B+\frac{N}{2}-\frac{1}{2}\right)\Gamma\left(B+\frac{N}{2}\right)}$$

$$\times\ {}_3F_2\left[\begin{matrix}\frac{1}{2}-M+\frac{N}{2}, -M+\frac{N}{2}, 1-B+A;\\ \frac{3}{2}-M-B, 1-M+N\end{matrix}\right]$$

(by $(3.5)_b$ with $n = M - \frac{N}{2} - \frac{1}{2}, b = M - \frac{N}{2},$

$$c = B + N - M - A, e = B + N - M, f = 1 - M + N).$$

This last expression contains the ${}_3F_2$ given in (3.12), and direct applications of (3.10) and (3.11) to the gamma functions quotient reduces it to the one appearing in (3.12). □

We conclude this section with a few brief remarks about the passage back and forth between binomial coefficient sums and hypergeometric series. The standard identities used are

$$\binom{n}{j} = \frac{(-1)^j(-n)_j}{j!}, \tag{3.14}$$

$$\binom{m+j}{j} = \frac{(m+1)_j}{j!}. \tag{3.15}$$

A detailed account of this relationship is presented in [1; §5].

4. The main theorem.

We are now in a position to evaluate the determinant in (1.1) when $x = 2$. Recall that for $0 \leqq i, j \leqq n - 1$ we let A_{ij} be the (i, j)-th entry in (1.1) when $x = 2$.

LEMMA 2.

$$A_{ij} = \delta_{i,j} + \sum_{t=0}^{i} \binom{m + j + t}{t} \binom{i + m}{t + m}. \tag{4.1}$$

Proof.

$$\begin{aligned} A_{ij} &= \delta_{ij} + \sum_{k=0}^{i} \sum_{t=0}^{k} \binom{j + m}{t} \binom{k}{t} \binom{i - k + m - 1}{i - k} 2^{k-t} \\ &= \delta_{ij} + \sum_{k=0}^{i} \binom{i - k + m - 1}{i - k} 2^k {}_2F_1 \left(\begin{matrix} -j - m, -k; \frac{1}{2} \\ 1 \end{matrix} \right) \\ &= \delta_{ij} + \sum_{k=0}^{i} \binom{i - k + m - 1}{i - k} {}_2F_1 \left(\begin{matrix} -k, 1 + j + m; -1 \\ 1 \end{matrix} \right) \\ & \qquad \text{(by (3.8))} \\ &= \delta_{ij} + \sum_{k=0}^{i} \sum_{t=0}^{k} \binom{i - k + m - 1}{i - k} \binom{k}{t} \binom{m + j + t}{t} \\ &= \delta_{ij} + \sum_{t=0}^{i} \sum_{k \geqq 0} \binom{i - k - t + m - 1}{i - k - t} \binom{k + t}{t} \binom{m + j + t}{t} \\ &= \delta_{ij} + \sum_{t=0}^{i} \binom{m + j + t}{t} \binom{i - t + m - 1}{i - t} {}_2F_1 \left(\begin{matrix} t + 1, -i + t; 1 \\ -i + t - m + 1 \end{matrix} \right) \\ &= \delta_{ij} + \sum_{t=0}^{j} \binom{m + j + t}{t} \binom{i - t + m - 1}{i - t} \frac{(-i - m)_{i-t}}{(-i + t - m + 1)_{i-t}} \qquad (4.2) \\ & \qquad \text{(by (3.7))} \\ &= \delta_{ij} + \sum_{t=0}^{j} \binom{m + j + t}{t} \binom{i + m}{m + t}. \end{aligned}$$

□

We are now prepared for the main result.

THEOREM 1. *The M-R-R conjecture is valid,* (*i.e.* (1.2) *and* (1.3) *are true*).

Proof. We proceed along the lines described in Section 2. We take A_{ij} from (4.1) in Lemma 1 and e_{ij} from (2.4) and (2.5). We then let

$$\mu_{2j-1}(m) = 2^j \prod_{k=1}^{j} \frac{m+2j+2k-1}{j+k}, \tag{4.3}$$

and

$$\mu_{2j}(m) = 2^{j+1} \prod_{k=1}^{j} \frac{m+2j+2k-1}{j+k}. \tag{4.4}$$

We define f_{ij} by the matrix multiplication

$$(A_{ij})(e_{ij}) = (f_{ij}),$$

and Theorem 1 reduces to proving that $f_{ij} = 0$ for $i < j$ and $f_{jj} = \mu_j(m)$.

We see immediately from (4.3) and (4.4) that even numbered and odd numbered columns must be treated differently. We begin with the even numbered case:

$$\begin{aligned} f_{i,2j} &= \sum_{l=0}^{2j} A_{i,l} e_{l,2j} \\ &= \sum_{l \geqq 0} \left(\delta_{il} + \sum_{t=0}^{i} \binom{m+l+t}{t} \binom{m+i}{i-t} \right) e_{l,2j} \\ &= e_{i,2j} + \sum_{t=0}^{i} \binom{m+i}{i-t} \sum_{s \geqq 0} \frac{4^{-s}(j-s)_{2s}}{s! \left(-\frac{m}{2} - j + \frac{1}{2} \right)_s} \\ &\quad \times \sum_{l \geqq 0} (-1)^l \binom{m+l+t}{t} \binom{j-s}{2j-2s-l} \\ &= e_{i,2j} + \sum_{t=0}^{i} \binom{m+i}{i-t} \sum_{s \geqq 0} \frac{4^{-s}(j-s)_{2s}}{s! \left(-\frac{m}{2} - j + \frac{1}{2} \right)_s} \binom{m+t+j-s}{t-j+s} \\ &\qquad \text{(by (3.7))} \\ &= e_{i,2j} + T_1(i,j,m). \end{aligned} \tag{4.5}$$

Now,

$$T_1(i,j,m) = \sum_{t=0}^{i} \binom{m+i}{i-t} \binom{m+t+j}{t-j} {}_3F_2 \left(\begin{matrix} j,\ 1-j,\ -\frac{m}{2} - j;\ 1 \\ t-j+1,\ -m-t-j \end{matrix} \right)$$

$$= \sum_{t=0}^{i} \binom{m+i}{i-t}\binom{m+t+j}{t-j} \times \frac{\pi\Gamma(t-j+1)\Gamma(-m-t-j)2^{m+2j+1}}{\Gamma\left(\frac{t+1}{2}\right)\Gamma\left(1-j+\frac{t}{2}\right)\Gamma\left(-\left(\frac{m+t}{2}\right)\right)\Gamma\left(-j-\frac{(m+t-1)}{2}\right)}$$

(by (3.3))

$$= (-1)^{j-1}\sum_{t=0}^{i}\binom{m+i}{i-t}\binom{m+t+1}{t-2j+1}\frac{\left(\frac{m}{2}+\frac{t}{2}+\frac{3}{2}\right)_{j-1}}{\left(-\frac{t}{2}+\frac{1}{2}\right)_{j-1}} \tag{4.6}$$

(by (3.2), (3.10) and (3.11)).

In this last expression all terms with t even and $<2j$ are zero; hence the only case with a nonzero even term occurs for $i = t = 2j$. Hence noting that the term $i = t = 2j$ is $\frac{1}{2}\mu_{2j}(m)$, we obtain

$$T_1(i,j,m) = \binom{m+i}{i-1}\frac{(-1)^{j-1}\left(\frac{1}{2}\right)_{j-1}}{\left(\frac{m}{2}+\frac{3}{2}\right)_{j-1}}\,{}_3F_2\left[\begin{matrix}-\frac{i}{2}+\frac{1}{2}, -\frac{i}{2}+1, \frac{m}{2}+j+1; 1\\ \frac{m}{2}+1, \frac{3}{2}-j\end{matrix}\right] + \left[\begin{matrix}0 & \text{if} & i<2j\\ \frac{1}{2}\mu_{2j}(m) & \text{if} & i=2j\end{matrix}\right].$$

$$= T_2(i,j,m) + \left[\begin{matrix}0 & \text{if} & i<2j\\ \frac{1}{2}\mu_{2j}(m) & \text{if} & i=2j\end{matrix}\right]. \tag{4.7}$$

Now

$$\frac{\left(\frac{1}{2}-j\right)_s(1-j)_s}{\left(\frac{3}{2}-j\right)_s} = \frac{\left(\frac{1}{2}-j\right)(1-j)_s}{\frac{1}{2}-j+s}$$

$$= \frac{\left(\frac{1}{2}-j\right)(-j)_s}{(-j)} - \frac{1}{2}\frac{\left(\frac{1}{2}-j\right)(-j)_s}{\left(\frac{1}{2}-j+s\right)(-j)}; \tag{4.8}$$

therefore

$$
\begin{aligned}
&(-1)^j T_2(2j, j, m) \\
&= -\binom{m+2j}{2j-1} \frac{\left(\frac{1}{2}\right)_{j-1}}{\left(\frac{m}{2}+\frac{3}{2}\right)_{j-1}} \left[\frac{\left(\frac{1}{2}-j\right)}{(-j)} {}_2F_1\left[\begin{matrix} -j, \frac{m}{2}+j+1; 1 \\ \frac{m}{2}+1 \end{matrix}\right] \right. \\
&\quad \left. + \frac{1}{2j} {}_3F_2\left[\begin{matrix} \frac{1}{2}-j, -j, \frac{m}{2}+j+1; 1 \\ \frac{3}{2}-j, \frac{m}{2}+1 \end{matrix}\right]\right] \\
&= \binom{m+2j}{2j-1}\left[\frac{\left(\frac{1}{2}\right)_j (-j)_j}{j\left(\frac{m}{2}+1\right)_j \left(\frac{m}{2}+\frac{3}{2}\right)_j} - \frac{(-j)_j \left(\frac{m}{2}+\frac{1}{2}+j\right)_j}{\left(\frac{m}{2}+\frac{3}{2}\right)_{j-1}\left(\frac{m}{2}+1\right)_j}\right] \qquad (4.9)
\end{aligned}
$$

(by (3.7) applied to the ${}_2F_1$ and (3.6) applied to the ${}_3F_2$)

$$
\begin{aligned}
&= -(-1)^j + \frac{(-1)^j 2^{j-1} j!(m+2j+1)(m+2j+3)\ldots(m+4j-1)}{(2j-1)!} \\
&= -(-1)^j + (-1)^j \frac{1}{2}\mu_{2j}(m).
\end{aligned}
$$

Collecting (4.5), (4.6), (4.7) and (4.9) in the case $i = 2j$, we find

$$
\begin{aligned}
f_{2j,2j} &= 1 + T_1(2j, j, m) \\
&= 1 + \frac{1}{2}\mu_{2j}(m) + T_2(2j, j, m) \\
&= 1 + \frac{1}{2}\mu_{2j}(m) + \left(-1 + \frac{1}{2}\mu_{2j}(m)\right) \\
&= \mu_{2j}(m), \qquad (4.10)
\end{aligned}
$$

as desired.

Now we assume $0 \leqq i < 2j$; so by (4.5), (4.6) and (4.7)

$$
\begin{aligned}
f_{i,2j} &= e_{i,2j} + T_1(i, j, m) \\
&= e_{i,2j} + T_2(i, j, m)
\end{aligned}
$$

$$= e_{i,2j} + \binom{m+i}{i-1}\frac{(-1)^{j-1}\left(\frac{1}{2}\right)_{j-1}}{\left(\frac{m}{2}+\frac{3}{2}\right)_{j-1}}{}_3F_2\left[\begin{matrix}-\frac{i}{2}+\frac{1}{2}, -\frac{i}{2}+1, \frac{m}{2}+j+1; 1\\ \frac{m}{2}+1, \frac{3}{2}-j\end{matrix}\right]$$

$$= e_{i,2j} + \frac{(-1)^{i-1}(m+i)}{(m+2j-1)}\binom{j}{i-j}{}_3F_2\left[\begin{matrix}\frac{i}{2}-j+\frac{1}{2}, \frac{i}{2}-j+1, j+1; 1\\ 1+i-j, \frac{3}{2}-j-\frac{m}{2}\end{matrix}\right]$$

(by Lemma 1 with $N = i-1$, $A = \frac{m}{2}+j-1$, $M = j-1$, $B = \frac{m}{2}+1$)

$$(4.11)$$

$$= e_{i,2j} + \frac{(-1)^{i}(m+i)}{2(2j-i)}\sum_{s\geqq 0}\binom{j-s-1}{2j-2s-i-1}\frac{(j-s)_{2s+1}4^{-s}}{s!\left(\frac{1}{2}-j-\frac{m}{2}\right)_{s+1}}$$

$$= e_{i,2j} + \frac{(-1)^{j}}{2(2j-i)}\sum_{s\geqq 0}\binom{j-s-1}{2j-2s-i-1}\frac{(j-s)_{2s+1}4^{-s}}{s!\left(\frac{1}{2}-j-\frac{m}{2}\right)_{s+1}}$$
$$\times \{(m+2j-1-2s)-(2j-2s-i-1)\}.$$

We now form two sums out of the one by splitting the expression inside the curly brackets as indicated; we then shift s to $s-1$ in the second arising sum and recombine. Hence

$$f_{i,2j} = e_{i,2j} - \frac{(-1)^{j}}{2(2j-i)}\sum_{s\geqq 0}\frac{(j-s-1)!(j-s)_{2s}4^{-s}}{(2j-2s-i)!(s-j+i)!\left(\frac{1}{2}-j-\frac{m}{2}\right)_{s}s!}$$
$$\times (2(j+s)(2j-2s-i)+4s(s-j+i))$$

$$= e_{i,2j} - \frac{(-1)^{j}}{2(2j-i)}\sum_{s\geqq 0}\frac{(j-s-1)!(j-s)_{2s}4^{-s}(2j-i)(2j-2s)}{(2j-2s-i)!(s-j+i)!\left(\frac{1}{2}-j-\frac{m}{2}\right)_{s}s!} \qquad (4.12)$$

$$= e_{i,2j} - e_{i,2j}$$
$$= 0,$$

as desired, and this completes our work for even numbered columns.

We now proceed with the odd numbered columns. First of all, we rewrite (2.5) in hypergeometric series notation:

$$e_{i,2j-1} = \frac{(-1)^{i+1}(i+1)}{2j}\binom{j}{i-j+1}{}_3F_2\left[\begin{matrix}\frac{i}{2}-j+\frac{1}{2}, \frac{i}{2}-j+1, j; 1\\ i-j+2, -\frac{m}{2}-j+\frac{1}{2}\end{matrix}\right] \tag{4.13}$$

Now

$$\begin{aligned} f_{i,2j-1} &= \sum_{l=0}^{2j-1} A_{i,l}e_{l,2j-1} \\ &= \sum_{l=0}^{2j-1}\left(\delta_{il} + \sum_{t=0}^{i}\binom{m+l+t}{t}\binom{m+i}{i-t}\right)e_{l,2j-1} \\ &= e_{i,2j-1} + \sum_{t=0}^{i}\binom{m+i}{i-t}\frac{1}{2}\sum_{s\geqq 0}\frac{4^{-s}(j-s)_{2s}}{s!\left(-\frac{m}{2}-j+\frac{1}{2}\right)_s} \\ &\quad\times\sum_{l=0}^{2j-1}(-1)^{l-1}\left(\binom{j-s}{2j-2s-l-1}+\binom{j-s-1}{2j-2s-l-1}\right)\binom{m+l+t}{t} \\ &= e_{i,2j-1} + \frac{1}{2}\sum_{t=0}^{i}\binom{m+i}{i-t}\sum_{s\geqq 0}\frac{4^{-s}(j-s)_{2s}}{s!\left(-\frac{m}{2}-j+\frac{1}{2}\right)_s} \\ &\quad\times\left(\binom{m+j-s-1+t}{t-j+s}+\binom{m+j-s+t}{t-j+s+1}\right) \end{aligned}$$

(by (3.7))

$$\begin{aligned} &= e_{i,2j-1} + \frac{1}{2}\sum_{t=0}^{i}\binom{m+i}{i-t}\frac{(m+2t+1)(m+j+t-1)!}{(t-j+1)!(m+2j-1)!} \\ &\quad\times{}_3F_2\left[\begin{matrix}j, 1-j, -\frac{m}{2}-j+1; 1\\ t-j+2, -m-j-t+1\end{matrix}\right] \\ &= e_{i,2j-1} + \frac{1}{2}\sum_{t=0}^{i}\binom{m+i}{i-t}\frac{(m+2t+1)}{m+j+t}\binom{m+j+t}{t-j+1} \end{aligned}$$

$$\times \frac{(t-2j+3)_{j-1}\left(1-j-\frac{t}{2}-\frac{m}{2}\right)_{j-1}}{\left(2-j+\frac{t}{2}\right)_{j-1}(-m-t-j+1)_{j-1}}$$

(by (3.3))

$$= e_{i,2j-1} + \frac{(-1)^{j-1}}{2}\sum_{t=0}^{i}\frac{(m+2t+1)}{m+t+1}\binom{m+i}{i-t} \times \binom{m+t+1}{t-2j+2}\frac{\left(\frac{t}{2}+\frac{m}{2}+1\right)_{j-1}}{\left(-\frac{t}{2}\right)_{j-1}}. \tag{4.14}$$

In this last sum all terms with t odd and $<2j-1$ are zero; hence the only case with a nonzero odd term occurs for $i = t = 2j-1$. Now noting that the term $i = t = 2j-1$ is $\frac{1}{2}\mu_{2j-1}(m)$, we obtain

$$f_{i,2j-1} = e_{i,2j-1} + \frac{(-1)^{j-1}}{2}\sum_{t\geqq 0}\frac{m+4t+1}{m+2t+1}\binom{m+i}{i-2t} \times \binom{m+2t+1}{2t-2j+2}\frac{\left(t+\frac{m}{2}+1\right)_{j-1}}{(-t)_{j-1}} + \begin{bmatrix} 0 & \text{if} & i<2j-1 \\ \frac{1}{2}\mu_{2j-1}(m) & \text{if} & i=2j-1 \end{bmatrix}$$

$$= e_{i,2j-1} + S(i,j,m) + \begin{bmatrix} 0 & \text{if} & i<2j-1 \\ \frac{1}{2}\mu_{2j-1}(m) & \text{if} & i=2j-1 \end{bmatrix}, \tag{4.15}$$

and in hypergeometric terms

$$S(i,j,m) = \frac{1}{2}\frac{\left(\frac{m}{2}+1\right)_{j-1}\left(\frac{3}{2}-j\right)_{j-1}(m+i)!}{2^{2-2j}(m+2j-1)!\,i!}$$

$$\times\left\{(m+1)\,{}_3F_2\left[\begin{matrix}-\frac{i}{2},\ -\frac{i}{2}+\frac{1}{2},\ \frac{m}{2}+j;\ 1\\ \frac{3}{2}-j,\ \frac{m}{2}+1\end{matrix}\right]\right.$$

$$\left.+\frac{4\left(-\frac{i}{2}\right)\left(-\frac{i}{2}+\frac{1}{2}\right)\left(\frac{m}{2}+j\right)}{\left(\frac{3}{2}-j\right)\left(\frac{m}{2}+1\right)}\,{}_3F_2\left[\begin{matrix}1-\frac{i}{2},\ \frac{3}{2}-\frac{i}{2},\ \frac{m}{2}+j+1;\ 1\\ \frac{5}{2}-j,\ \frac{m}{2}+2\end{matrix}\right]\right\} \tag{4.16}$$

We next treat $S(2j-1, j, m)$:

$$S(2j-1,j,m)=\frac{2^{2j-3}\left(\frac{m}{2}+1\right)_{j-1}\left(\frac{3}{2}-j\right)_{j-1}}{(2j-1)!}$$

$$\times\left\{(m+1)\,{}_3F_2\left[\begin{matrix}\frac{1}{2}-j,\ 1-j,\ \frac{m}{2}+j;\ 1\\ \frac{3}{2}-j,\ \frac{m}{2}+1\end{matrix}\right]\right.$$

$$\left.+\frac{4\left(j-\frac{1}{2}\right)(j-1)\left(\frac{m}{2}+j\right)}{\left(\frac{3}{2}-j\right)\left(\frac{m}{2}+1\right)}\,{}_3F_2\left[\begin{matrix}\frac{3}{2}-j,\ 2-j,\ \frac{m}{2}+j+1;\ 1\\ \frac{5}{2}-j,\ \frac{m}{2}+2\end{matrix}\right]\right\}$$

$$=\frac{2^{2j-3}\left(\frac{m}{2}+1\right)_{j-1}\left(\frac{3}{2}-j\right)_{j-1}}{(2j-1)!}$$

$$\times\left\{\frac{(m+1)(j-1)!\left(\frac{3}{2}-2j-\frac{m}{2}\right)_{j-1}}{\left(\frac{3}{2}-j\right)_{j-1}\left(1-\frac{m}{2}-j\right)_{j-1}}\right.$$

$$+\frac{4\left(j-\frac{1}{2}\right)(j-1)\left(\frac{m}{2}+j\right)(1-j)_{j-2}}{\left(\frac{3}{2}-j\right)\left(\frac{m}{2}+j\right)\left(\frac{m}{2}+2\right)_{j-2}}$$

$$\left.\times\ {}_3F_2\left[\begin{matrix}2-j,\ \frac{m}{2}+j+1,\ 1;\ 1\\ 2,\ \frac{5}{2}-j\end{matrix}\right]\right\} \tag{4.17}$$

(by (3.6) applied to the first ${}_3F_2$ and $(3.5)_a$ applied to the second ${}_3F_2$ with $n = j - 2$,

$b = -\frac{m}{2} - j - 1$, $c = \frac{3}{2} - j$, $e = \frac{m}{2} + 2$, $f = \frac{5}{2} - j$)

Now

$$
{}_3F_2\left[\begin{matrix} 2-j, \frac{m}{2}+j+1, 1; 1 \\ 2, \frac{5}{2}-j \end{matrix}\right]
= \frac{\left(\frac{3}{2}-j\right)}{(1-j)\left(\frac{m}{2}+j\right)}\left[{}_2F_1\left[\begin{matrix} 1-j, \frac{m}{2}+j; 1 \\ \frac{3}{2}-j \end{matrix}\right]-1\right]
= \frac{\left(\frac{3}{2}-j\right)}{(1-j)\left(\frac{m}{2}+j\right)}\left[\frac{\left(\frac{3}{2}-2j-\frac{m}{2}\right)_{j-1}}{\left(\frac{3}{2}-j\right)_{j-1}}-1\right].
$$

We now substitute this expression back into (4.17), and we find

$$
\begin{aligned}
S(2j-1, j, m) &= \frac{(-1)^{j-1}(j-1)!\left(\frac{3}{2}-2j-\frac{m}{2}\right)_{j-1}(m+1)}{2^{3-2j}(2j-1)!} \\
&\quad - \frac{(-1)^{j-1}(j-1)!\left(\frac{3}{2}-2j-\frac{m}{2}\right)_{j-1}4\left(\frac{1}{2}-j\right)}{2^{3-2j}(2j-1)!} - 1 \\
&= \frac{2^{j-2}(j-1)!(m+2j+1)(m+2j+3)\ldots(m+4j-1)}{(2j-1)!} - 1 \\
&= \frac{1}{2}\mu_{2j-1}(m) - 1. \qquad (4.18)
\end{aligned}
$$

Hence by (4.15) and (4.18)

$$
f_{2j-1,2j-1} = 1 + S(2j-1, j, m) + \frac{1}{2}\mu_{2j-1}(m)
$$

$$= 1 + \left(\frac{1}{2}\mu_{2j-1}(m) - 1\right) + \frac{1}{2}\mu_{2j-1}(m)$$

$$= \mu_{2j-1}(m), \tag{4.19}$$

as desired.

Now we assume $0 \leqq i < 2j - 1$, so by (4.13), (4.15) and (4.16) we see that

$$f_{i,2j-1} = e_{i,2j-1} + S(i, j, m)$$

$$= \frac{(-1)^{i+1}(i+1)}{2j}\binom{j}{i-j+1}{}_3F_2\left[\begin{matrix}\frac{i}{2}-j+\frac{1}{2}, \frac{i}{2}-j+1, j; 1\\ i-j+2, -\frac{m}{2}-j+\frac{1}{2}\end{matrix}\right]$$

$$+\frac{1}{2}\;\frac{\left(\frac{m}{2}+1\right)_{j-1}\left(\frac{3}{2}-j\right)_{j-1}(m+i)}{2^{2-2j}(m+2j-1)!i!}$$

$$\times\left[(m+1)\,{}_3F_2\left[\begin{matrix}-\frac{i}{2}, -\frac{i}{2}+\frac{1}{2}, \frac{m}{2}+j; 1\\ \frac{3}{2}-j, \frac{m}{2}+1\end{matrix}\right]\right.$$

$$\left.+\frac{4\left(-\frac{i}{2}\right)\left(-\frac{i}{2}+\frac{1}{2}\right)\left(\frac{m}{2}+j\right)}{\left(\frac{3}{2}-j\right)\left(\frac{m}{2}+1\right)}{}_3F_2\left[\begin{matrix}1-\frac{i}{2}, \frac{3}{2}-\frac{i}{2}, \frac{m}{2}+j+1; 1\\ \frac{5}{2}-j, \frac{m}{2}+2\end{matrix}\right]\right]$$

$$= \frac{(-1)^{i+1}(i+1)}{2j}\binom{j}{i-j+1}{}_3F_2\left[\begin{matrix}\frac{i}{2}-j+\frac{1}{2}, \frac{i}{2}-j+1, j; 1\\ i-j+2, -\frac{m}{2}-j+\frac{1}{2}\end{matrix}\right]$$

$$\times (m+1)\frac{2^m\Gamma\left(\frac{m}{2}+1\right)\Gamma\left(\frac{m}{2}+j-\frac{1}{2}\right)\Gamma(j)\Gamma(i+1)(-1)^{i+j-1}}{\Gamma(2j-i-1)\Gamma\left(j-\frac{1}{2}\right)\Gamma(2-j+i)\Gamma(m+1+i)}$$

$$\times {}_3F_2\left[\begin{matrix}\frac{3}{2}-j+\frac{i}{2}, 1-j+\frac{i}{2}, j; 1\\ \frac{3}{2}-j-\frac{m}{2}, 2-j+i\end{matrix}\right]$$

$$+\frac{i(i-1)\left(\frac{m}{2}+j\right)2^{m+2}\Gamma(j)\Gamma\left(\frac{m}{2}+2\right)\Gamma\left(\frac{m}{2}+j-\frac{1}{2}\right)\Gamma(i-1)(-1)^{i+j}}{\left(\frac{3}{2}-j\right)\left(\frac{m}{2}+1\right)\Gamma(2j-i)\Gamma\left(j-\frac{3}{2}\right)\Gamma(1-j+i)\Gamma(m+i+1)}$$

$$\times {}_3F_2\left[\begin{matrix}\frac{3}{2}-j+\frac{i}{2}, 1-j+\frac{i}{2}, j; 1\\ \frac{3}{2}-j-\frac{m}{2}, 1-j+i\end{matrix}\right]$$

(by Lemma 1 applied to the first ${}_3F_2$ inside the curly brackets with $N=i, A=\frac{m}{2}+j, M=j-1, B=\frac{m}{2}+1$ and applied to the second ${}_3F_2$ inside the curly brackets with $N=i-2, A=\frac{m}{2}+j+1, M=j-2, B=\frac{m}{2}+2$)

$$=\frac{(-1)^{i+1}}{2j}\binom{j}{i-j+1}\left\{(i+1){}_3F_2\left[\begin{matrix}\frac{i}{2}-j+\frac{1}{2}, \frac{i}{2}-j+1, j; 1\\ i-j+2, -\frac{m}{2}-j+\frac{1}{2}\end{matrix}\right]\right.$$

$$+\frac{(m+1)(2j-i-1)}{2\left(-\frac{m}{2}-j+\frac{1}{2}\right)}{}_3F_2\left[\begin{matrix}\frac{3}{2}-j+\frac{1}{2}, 1-j+\frac{i}{2}, j; 1\\ \frac{3}{2}-j-\frac{m}{2}, 2-j+i\end{matrix}\right]$$

$$\left.+\frac{(2m+4j)(1-j+i)}{2\left(-\frac{m}{2}-j+\frac{1}{2}\right)}{}_3F_2\left[\begin{matrix}\frac{3}{2}-j+\frac{i}{2}, 1-j+\frac{i}{2}, j; 1\\ \frac{3}{2}-j-\frac{m}{2}, 1-j+i\end{matrix}\right]\right\} \tag{4.20}$$

(where the ${}_3F_2$'s are unaltered and the coefficients have been simplified by means of (3.2), (3.10) and (3.11))

$$=0 \qquad \left(\text{by (3.9) with } a=\frac{3}{2}-j+\frac{i}{2}, b=1-j+\frac{i}{2}, c=j, d=2-j+i, e=\frac{3}{2}-j-\frac{m}{2}\right).$$

Thus we have completely proved that $f_{ij}=0$ for $i<j$ and $f_{jj}=\mu_j(m)$. Consequently

$$\begin{aligned} Z_n(2, m) &= \det(A_{ij})_{0 \leqq i,j \leqq n-1} \\ &= \det(A_{ij})\det(e_{ij}) \\ &= \det(f_{ij}) = \prod_{j=0}^{n-1} f_{jj} = \prod_{j=0}^{n-1} \mu_j(m), \end{aligned}$$

and Theorem 1 is proved. □

At several places in our development there arose expressions in which a gamma function with a nonpositive integer argument occurred in a numerator or such an argument occurred as a denominator parameter in a hypergeometric series. Such singularities were apparent and not real; each time such a singularity arose there was a compensating zero to cancel it at least. Rather than lengthen our account substantially by remarking on this phenomenon each time it arose, we chose to deal with it here at the end.

5. Conclusion.

As in the case of the weak Macdonald conjecture [2], we again have a very complicated evaluation of the determinant under consideration. It would be very nice to have a simpler proof of Theorem 1. Is there a nice q-analog of Theorem 1 for special values of m especially $m = 0$ or 2? We recall that for $m = 0$ or 2 in Theorem 8 of [2] there are striking q-analogs (the Macdonald conjecture and the descending plane partitions conjecture), and these have been beautifully proved by Mills, Robbins and Rumsey [4]. Is there a reasonable combinatorial explanation of Lemma 2? Since all the terms in the sum in (4.1) are positive for m a positive integer, we might expect this to be a reasonable starting point for further investigation of Theorem 1 and possible q-analogs of $Z_n(2, m)$.

REFERENCES

[1] Andrews, G. E., *Applications of basic hypergeometric functions.* SIAM Rev. *16* (1974), 441–484.

[2] Andrews, G. E., *Plane partitions (III): the weak Macdonald conjecture.* Invent. Math. *53* (1979), 193–225.

[3] Bailey, W. N., *Generalized Hypergeometric Series.* Cambridge University Press, London and New York, 1935 (Reprinted: Hafner, New York, 1964).

[4] Mills, W. H., Robbins, D. P., and Rumsey Jr., H., *Proof of the Macdonald conjecture.* Invent. Math. *66* (1982), 73–87.

[5] Mills, W. H., Robbins, D. P., and Rumsey, Jr., H., *Enumeration of a symmetry class of plane partitions.* Working Paper No. 822, April 1985, IDA, Princeton.

[6] Sears, D. B., *On the transformation theory of hypergeometric functions.* Proc. London Math. Soc. (**2**), *52* (1950), 14–35.

[7] Stanley, R. P., *A baker's dozen of conjectures concerning plane partitions.* In: *Combinatoire énumérative,* (G. Labelle and P. Lerous eds.), Lecture Notes in Math. 1234. Springer, Berlin–New York, 1986, pp. 285–293.

[8] Whittaker, E. T. and Watson G. N., *A Course of Modern Analysis.* 4th ed., Cambridge University Press, London and New York, 1927.

[9] Wilson, J. A., *Three-term contiguous relations and some new orthogonal polynomials.* In: *Padé and Rational Approximation* (E. B. Saff and R. S. Varga, Eds.), Academic Press, New York, 1977, pp. 227–232.

Department of Mathematics,
Pennsylvania State University,
University Park, PA 16802
U.S.A.

4.4 Determinant Identities (with W.H. Burge)

PACIFIC JOURNAL OF MATHEMATICS
Vol. 158, No. 1, 1993

DETERMINANT IDENTITIES

GEORGE E. ANDREWS AND WILLIAM H. BURGE

A number of determinants are evaluated in closed form including

$$\det\left(\binom{i+j+x}{2i-j}+\binom{i+j+y}{2i-j}\right)_{0\le i,j\le n-1}.$$

1. Introduction. In one of their series of papers on plane partitions and related questions, Mills, Robbins and Rumsey [**9**; p. 53] prove the following determinant formula.

$$m_n(x)=\det\left(\binom{i+j+x}{2i-j}\right)_{0\le i,j\le n-1}=\frac{1}{2^n}\prod_{k=0}^{n-1}\Delta_{2k}(2x), \tag{1.1}$$

where $\Delta_0(u)=2$ and for $j>0$

$$\Delta_{2j}(u)=\frac{(u+2j+2)_j(\frac{1}{2}u+2j+\frac{3}{2})_{j-1}}{(j)_j(\frac{1}{2}u+j+\frac{3}{2})_{j-1}} \tag{1.2}$$

with

$$(A)_j=A\cdot(A+1)\cdots(A+j-1). \tag{1.3}$$

Our object here is primarily to prove the following generalization of (1.1).

THEOREM 1. *Let*

$$M_n(x,y)=\det\left(\binom{i+j+x}{2i-j}+\binom{i+j+y}{2i-j}\right)_{0\le i,j\le n-1}; \tag{1.4}$$

$$N_n(x,y)=\det\left(\frac{2}{x+1-y}\left\{\binom{i+j+x+1}{2i-j+1}-\binom{i+j+y}{2i-j+1}\right\}\right)_{0\le i,j\le n-1}. \tag{1.5}$$

Then

$$M_n(x,y)=N_n(x,y)=\prod_{k=0}^{n-1}\Delta_{2k}(x+y). \tag{1.6}$$

Sections 2 and 3 will be devoted to the proof of this result. In §§4 and 5 we shall show how our work leads to two alternative proofs of the T.S.S.C.P.P. conjecture [**2**], and we shall mention a related application of Ishikawa [**6**].

2. Bailey's balanced $_4F_3$ summation. In this section, we consider summation formulas for hypergeometric series [**4**; p. 8], [**10**, p. 41]:

$$ {}_{n+1}F_n\left[\begin{matrix} a_0, a_1, \dots, a_n; t \\ b_1, \dots, b_n \end{matrix}\right] = \sum_{j=0}^{\infty} \frac{(a_0)_j(a_1)_j\cdots(a_n)_j t^n}{j!(b_1)_j\cdots(b_n)_j}. \tag{2.1} $$

The formula of Bailey [**3**; p. 512, (c)] [**10**; p. 245, (III.20)] alluded to in the title of this section is

$$ {}_4F_3\left[\begin{matrix} \frac{a}{2}, \frac{a+1}{2}, b+n, -n; 1 \\ \frac{b}{2}, \frac{b+1}{2}, a+1 \end{matrix}\right] = \frac{(b-a)_n}{(b)_n}, \tag{2.2} $$

and a closely related companion is

$$ {}_4F_3\left[\begin{matrix} \frac{a}{2}, \frac{a+1}{2}, b+n, -n; 1 \\ \frac{b+1}{2}, \frac{b+2}{2}, a \end{matrix}\right] = \frac{(b-a+1)_n}{(b+1)_{n-1}(b+2n)}. \tag{2.3} $$

Also useful for our work is a transformation due to F.J.W. Whipple [**14**; p. 537, eq. (10.1)]. If one of z and n is a nonnegative integer, then

$$ \begin{aligned} {}_4F_3&\left[\begin{matrix} a, b, -z, -n; 1 \\ u, v, w \end{matrix}\right] \\ &= \frac{\Gamma(v+z+n)\Gamma(w+z+n)\Gamma(v)\Gamma(w)}{\Gamma(v+z)\Gamma(v+n)\Gamma(w+n)\Gamma(w+z)} \\ &\times {}_4F_3\left[\begin{matrix} u-a, u-b, -z, -n; 1 \\ 1-v-z-n, 1-w-z-n, u \end{matrix}\right]. \end{aligned} \tag{2.4} $$

We note in passing that (2.4) specialized to $a=\alpha/2$, $b=(\alpha+1)/2$, $z=-\beta-\nu$, $n=\nu$, $v=(\beta+1)/2$, $w=(\beta+2)/2$, $u=\alpha$ yields

$$ \begin{aligned} {}_4F_3&\left[\begin{matrix} \frac{\alpha}{2}, \frac{\alpha+1}{2}, \beta+\nu, -\nu; 1 \\ \frac{\beta+1}{2}, \frac{\beta+2}{2}, \alpha \end{matrix}\right] \\ &= \frac{\beta}{\beta+2\nu} {}_4F_3\left[\begin{matrix} \frac{\alpha-1}{2}, \frac{\alpha}{2}, \beta+\nu, -\nu; 1 \\ \frac{\beta+1}{2}, \frac{\beta}{2}, 1+(\alpha-1) \end{matrix}\right] \\ &= \frac{\beta}{\beta+2\nu}\,\frac{(\beta-\alpha+1)_\nu}{(\beta)_\nu} \qquad \text{(by (2.2))}. \end{aligned} \tag{2.5} $$

Thus (2.3) is an immediate consequence of (2.2) and (2.4).

LEMMA 1. *Let α be a positive integer, then*

$$ {}_4F_3\left[\begin{matrix} -\frac{\alpha}{2}, -\frac{\alpha-1}{2}, b+z, -z; 1 \\ \frac{b}{2}, \frac{b+1}{2}, 1-\alpha \end{matrix}\right] = \frac{(b+z)_\alpha}{(b)_\alpha} + \frac{(-z)_\alpha}{(b)_\alpha}. \tag{2.6} $$

Proof. We begin by considering (2.2) when $a = -\alpha$ a negative integer. The index of summation j runs from 0 to n (when $j > n$ all terms $= 0$). Furthermore, the terms with $\alpha/2 < j < \alpha$ are all identically zero. For $n \geq j \geq \alpha$ there are cancelling zeros in numerator and denominator.

Thus by (2.2)

$$\begin{aligned}\frac{(b+\alpha)_n}{(b)_n} &= \sum_{0\leq 2j\leq\alpha} \frac{\left(-\frac{\alpha}{2}\right)_j \left(\frac{1-\alpha}{2}\right)_j (b+n)_j(-n)_j}{j!\left(\frac{b}{2}\right)_j\left(\frac{b+1}{2}\right)_j(1-\alpha)_j} \\ &\quad + \sum_{j=\alpha}^{n} \frac{(-\alpha)(-\alpha+1)\cdots(-1)(1)(2)\cdots(-\alpha+2j-1)(b+n)_j(-n)_j}{j!(b)_{2j}(1-\alpha)(2-\alpha)\cdots(-1)\cdot(1)\cdot(2)\cdots(j-\alpha)} \\ &= S_1 + S_2.\end{aligned}$$

Now

$$\begin{aligned} S_2 &= \sum_{j=0}^{n-\alpha} \frac{(-1)^\alpha \alpha!(\alpha+2j-1)!(b+n)_{j+\alpha}(-n)_{j+\alpha}}{(j+\alpha)!(b)_{2j+2\alpha}(-1)^{\alpha-1}(\alpha-1)!j!} \\ &= -\frac{(b+n)_\alpha(-n)_\alpha}{(b)_{2\alpha}} \sum_{j=0}^{n-\alpha} \frac{(\alpha)_{2j}(b+n+\alpha)_j(-n+\alpha)_j}{j!(b+2\alpha)_{2j}(\alpha+1)_j} \\ &= -\frac{(b+n)_\alpha(-n)_\alpha}{(b)_{2\alpha}}\, {}_4F_3\left(\begin{matrix}\frac{\alpha}{2}, \frac{\alpha+1}{2}, b+2\alpha+(n-\alpha), -(n-\alpha); 1 \\ \frac{b+2\alpha}{2}, \frac{b+2\alpha+1}{2}, 1+\alpha\end{matrix}\right) \\ &= -\frac{(b+n)_\alpha(-n)_\alpha}{(b)_{2\alpha}} \frac{(b+\alpha)_{n-\alpha}}{(b+2\alpha)_{n-\alpha}} \qquad \text{(by (2.2))} \\ &= -\frac{(-n)_\alpha}{(b)_\alpha}.\end{aligned}$$

Hence

$$\begin{aligned}\sum_{0\leq 2j\leq\alpha} &\frac{\left(-\frac{\alpha}{2}\right)_j\left(\frac{1-\alpha}{2}\right)_j(b+n)_j(-n)_j}{j!\left(\frac{b}{2}\right)_j\left(\frac{b+1}{2}\right)_j(1-\alpha)_j} \\ &= \frac{(b+\alpha)_n}{(b)_n} + \frac{(-n)_\alpha}{(b)_\alpha} = \frac{(b+n)_\alpha}{(b)_\alpha} + \frac{(-n)_\alpha}{(b)_\alpha},\end{aligned} \tag{2.7}$$

which is precisely (2.6) when z is any positive integer, n. However, both sides of (2.6) are polynomials in z of degree at most α, and since they agree for all positive integral z we see that (2.6) holds for all real z. □

LEMMA 2. *Let α be a nonnegative integer, then*

$$(2.8)\qquad {}_4F_3\left[\begin{matrix}-\frac{\alpha}{2}, \frac{-\alpha+1}{2}, b+z, -z ; 1\\ \frac{b+1}{2}, \frac{b+2}{2}, -\alpha\end{matrix}\right] = \frac{(b+z)_{\alpha+1}}{(b+2z)(b+1)_\alpha} - \frac{(-z)_{\alpha+1}}{(b+1)_\alpha(b+2z)}.$$

Proof. In parallel with Lemma 1, we begin by considering (2.3) with $a=-\alpha$ a negative integer. If j is the index of summation in the ${}_4F_3$, then the nonzero terms of the sum occur for $0 \le j \le \alpha/2$ and $\alpha < j \le n$. If we call the two resulting sums T_1 and T_2, then

$$(2.9)\qquad \frac{(b+\alpha+1)_n}{(b+1)_{n-1}(b+2n)} = T_1 + T_2.$$

Now

$$\begin{aligned}
T_2 &= \sum_{j=\alpha+1}^{n} \frac{(-\alpha)_\alpha(1)_{2j-\alpha-1}(b+n)_j(-n)_j}{j!(b+1)_{2j}(-\alpha)_\alpha(1)_{j-\alpha-1}}\\
&= \sum_{j=0}^{n-\alpha-1} \frac{(1)_{2j+\alpha+1}(b+n)_{j+\alpha+1}(-n)_{j+\alpha+1}}{(j+\alpha+1)!(b+1)_{2j+2\alpha+2}j!}\\
&= \frac{(\alpha+1)!(b+n)_{\alpha+1}(-n)_{\alpha+1}}{(\alpha+1)!(b+1)_{2\alpha+2}}\\
&\quad\times {}_4F_3\left[\begin{matrix}\frac{\alpha}{2}+1, \frac{\alpha}{2}+\frac{3}{2}, b+n+\alpha+1, -n+\alpha+1 ; 1\\ \frac{b+2\alpha+3}{2}, \frac{b+2\alpha+4}{2}, \alpha+2\end{matrix}\right]\\
&= \frac{(b+1)_{\alpha+n}(-n)_{\alpha+1}}{(b+1)_{n-1}(b+1)_{2\alpha+2}} \frac{(b+\alpha+1)_{n-\alpha-1}}{(b+2\alpha+3)_{n-\alpha-2}(b+2n)} \qquad \text{(by (2.3))}\\
&= \frac{(-n)_{\alpha+1}}{(b+1)_\alpha(b+2n)}.
\end{aligned}$$

Hence

$$\begin{aligned}
(2.10)\qquad {}_4F_3&\left[\begin{matrix}-\frac{\alpha}{2}, \frac{-\alpha+1}{2}, b+n, -n ; 1\\ \frac{b+1}{2}, \frac{b+2}{2}, -\alpha\end{matrix}\right]\\
&= T_1 = \frac{(b+\alpha+1)_n}{(b+1)_{n-1}(b+2n)} - \frac{(-n)_{\alpha+1}}{(b+1)_\alpha(b+2n)}\\
&= \frac{(b+n)_{\alpha+1}}{(b+1)_\alpha(b+2n)} - \frac{(-n)_{\alpha+1}}{(b+1)_\alpha(b+2n)},
\end{aligned}$$

which is (2.8) when z is any positive integer. Since (2.8) is an identity of rational functions in z, the result in full generality follows immediately. □

3. The main theorem. Our proof of Theorem 1 relies on the following binomial coefficient summations.

LEMMA 3. *For integers* $i, j \geq 0$

$$(3.1)\quad 2\sum_{k=0}^{i} \frac{(y-x)}{(y-x+i-k)}\binom{y-x+i-k}{2i-2k}\binom{k+j+x+y}{2k-j} = \binom{i+j+2x}{2i-j}+\binom{i+j+2y}{2i-j}.$$

$$(3.2)\quad \sum_{k=0}^{i} \frac{(2x-2y+1)}{(y-x+i-k)}\binom{y-x+i-k}{2i-2k+1}\binom{k+j+x+y}{2k-j} = \binom{i+j+2x+1}{2i-j+1}-\binom{i+j+2y}{2i-j+1}.$$

Proof. We note that the only nonzero terms on the left-hand side of (3.1) occur for $i \geq k \geq j/2$. Consequently, if $j > 2i$ then both sides of (3.1) are zero. Hence we may assume $2i \geq j$.

Therefore

$$\begin{aligned}
(3.3)\quad & 2\sum_{k=0}^{i} \frac{(y-x)}{(y-x+i-k)}\binom{y-x+i-k}{2i-2k}\binom{k+j+x+y}{2k-j} \\
&= 2\sum_{k=0}^{i} \frac{(y-x)}{(y-x+k)}\binom{y-x+k}{2k}\binom{i+j-k+x+y}{2i-j-2k} \\
&= 2\binom{i+j+x+y}{2i-j}\sum_{k=0}^{j} \frac{(y-x)_k(-y+x)_k(-2i+j)_{2k}}{(2k)!(-i-j-x-y)_k(2j-i+x+y+1)_k} \\
&= 2\binom{i+j+x+y}{2i-j}\,{}_4F_3\left[\begin{matrix} y-x,\ -y+x,\ -i+\frac{j}{2},\ -i+\frac{j+1}{2};\ 1 \\ \frac{1}{2},\ -i-j-x-y,\ 2j-i+x+y+1 \end{matrix}\right] \\
&= 2\binom{i+j+x+y}{2i-j} \\
&\quad\times \frac{\Gamma(i+j+x+y+1/2)\Gamma(2i-j)\Gamma(2j-i+x+y+1)\Gamma(1/2)}{\Gamma(3j/2+x+y+1)\Gamma(1/2+i-j/2)\Gamma(3j/2+x+y+1/2)\Gamma(i-j/2)} \\
&\quad\times {}_4F_3\left[\begin{matrix} -i-j-2y,\ -i-j-2x,\ -i+\frac{j}{2},\ -i+\frac{j+1}{2};\ 1 \\ -j-i-x-y+\frac{1}{2},\ -i-j-x-y,\ 1-2i+j \end{matrix}\right] \quad \text{(by (2.4))} \\
&= \binom{i+j+x+y}{2i-j}\frac{\Gamma(i+j+x+y+1/2)\Gamma(2j-i+x+y+1)2^{2i+2j+2x+2y}}{\Gamma(1/2)\Gamma(3j+2x+2y+1)} \\
&\quad\times\left(\frac{(-i-j-2x)_{2i-j}}{(-2i-2j-2x-2y)_{2i-j}}+\frac{(-i-j-2y)_{2i-j}}{(-2i-2j-2x-2y)_{2i-j}}\right) \\
&\qquad \text{(by Lemma 1 and Gauss's duplication formula [5; p. 5, eq. (15)])} \\
&= \binom{i+j+2x}{2i-j}+\binom{i+j+2y}{2i-j}
\end{aligned}$$

as desired for (3.1).

For (3.2) again we obtain zero equals zero unless $2i \geq j$. Hence

$$
\begin{aligned}
(3.4)\quad \sum_{k=0}^{i} & \frac{(2x-2y+1)}{(y-x+i-k)} \binom{y-x+i-k}{2i-2k+1} \binom{k+j+x+y}{2k-j} \\
&= \sum_{k=0}^{i} \frac{(2x-2y+1)}{(y-x+k)} \binom{y-x+k}{2k+1} \binom{i-k+j+x+y}{2i-j-2k} \\
&= (2x-2y+1) \binom{i+j+x+y}{2i-j} \\
&\quad \times {}_4F_3 \left[\begin{matrix} y-x,\ 1+x-y,\ -i+\frac{j}{2},\ -i+\frac{j}{2}+\frac{1}{2};\ 1 \\ \frac{3}{2},\ -i-j-x-y,\ x+y+2j-i+1 \end{matrix} \right] \\
&= (2x-2y+1) \binom{i+j+x+y}{2i-j} \\
&\quad \times \frac{\Gamma(2i-j+1)\Gamma(i+j+x+y+1/2)\Gamma(3/2)\Gamma(2j-i+x+y+1)}{\Gamma(i-\frac{j}{2}+\frac{3}{2})\Gamma(i-\frac{j}{2}+1)\Gamma(\frac{3j}{2}+x+y+1)\Gamma(\frac{3j}{2}+x+y+\frac{1}{2})} \\
&\quad \times {}_4F_3 \left[\begin{matrix} -i-j-2y,\ -i-j-2x-1,\ -i+\frac{j}{2},\ -i+\frac{j}{2}+\frac{1}{2};\ 1 \\ -i-j-x-y,\ -2i+j,\ -i-j-x-y+\frac{1}{2} \end{matrix} \right] \quad \text{(by (2.4))} \\
&= (2x-2y+1) \binom{i+j+x+y}{2i-j} \\
&\quad \times \frac{\Gamma(i+j+x+y+\frac{1}{2}) 2^{2i+2j+2x+2y} \Gamma(2j-i+x+y+1)}{\Gamma(\frac{1}{2})\Gamma(3j+2x+2y+1)(2i-j+1)} \\
&\quad \times \left(\frac{(-i-j-2x-1)_{2i-j+1}}{(2y-2x-1)(-2i-2j-2x-2y)_{2i-j}} \right. \\
&\qquad \left. - \frac{(-i-j-2y)_{2i-j+1}}{(2y-2x-1)(-2i-2j-2x-2y)_{2i-j}} \right) \\
&= \binom{i+j+2x+1}{2i-j+1} - \binom{i+j+2y}{2i-j+1}. \qquad \square
\end{aligned}
$$

Proof of Theorem 1. We define five matrices

$$
(3.5) \qquad \mathbf{M}_n(x) = \left(\binom{i+j+x}{2i-j} \right)_{0 \leq i,\, j \leq n-1},
$$

$$
(3.6) \qquad \mu_n(x, y) = \left(\binom{i+j+x}{2i-j} + \binom{i+j+y}{2i-j} \right)_{0 \leq i,\, j \leq n-1},
$$

$$
(3.7) \qquad \nu_n(x, y) = \left(\frac{2}{x+1-y} \left(\binom{i+j+x+1}{2i-j+1} - \binom{i+j+y}{2i-j+1} \right) \right)_{0 \leq i,\, j \leq n-1},
$$

$$
(3.8) \qquad \tau_n(x, y) = \left(\frac{2(y-x)}{(y-x+i-j)} \binom{y-x+i-j}{2i-2j} \right)_{0 \leq i,\, j \leq n-1},
$$

(3.9) $$\sigma_n(x, y) = \left(\frac{2}{(y-x+i-j)}\binom{y-x+i-j}{2i-2j+1}\right)_{0\le i,j\le n-1}.$$

Clearly

(3.10) $$m_n(x) = \det(\mathbf{M}_n(x)),$$

(3.11) $$M_n(x, y) = \det(\mu_n(x, y)),$$

(3.12) $$N_n(x, y) = \det(\nu_n(x, y)),$$

(3.13) $$\det(\tau_n(x, y)) = 2^n,$$

and

(3.14) $$\det(\sigma_n(x, y)) = 2^n.$$

Equations (3.10)–(3.12) are restatements of (1.1), (1.4) and (1.5), while (3.13) and (3.14) are obvious since each matrix in question is lower triangular.

Now

(3.15) $$\begin{aligned}&\tau_n(x, y)\mathbf{M}_n(x+y)\\ &= \left(\sum_{k=0}^{i}\frac{2(y-x)}{(y-x+i-k)}\binom{y-x+i-k}{2i-2k}\binom{k+j+x+y}{2k-j}\right)_{0\le i,j\le n-1}\\ &= \left(\binom{i+j+2x}{2i-j}+\binom{i+j+2y}{2i-j}\right)_{0\le i,j\le n-1} \qquad \text{(by Lemma 3, eq. (3.1))}\\ &= \mu_n(2x, 2y).\end{aligned}$$

Hence by (3.15) and (1.1)

$$\begin{aligned}M_n(x, y) &= \det(\mu_n(x, y))\\ &= \det\left(\tau_n\left(\frac{x}{2}, \frac{y}{2}\right)\mathbf{M}_n\left(\frac{x+y}{2}\right)\right)\\ &= 2^n\cdot\det\left(\mathbf{M}_n\left(\frac{x+y}{2}\right)\right)\\ &= \prod_{k=0}^{n-1}\Delta_{2k}(x+y),\end{aligned}$$

which proves the first part of (1.6).

Similarly

(3.16) $\sigma_n(x, y)\mathbf{M}_n(x+y)$

$$= \left(\sum_{k=0}^{i} \frac{2}{(y-x+i-k)} \binom{y-x+i-k}{2i-2k+1} \binom{k+j+x+y}{2k-j} \right)_{0\le i,j\le n-1}$$

$$= \left(\frac{2}{(2x-2y+1)} \left(\binom{i+j+2x+1}{2i-j+1} - \binom{i+j+2y}{2i-j+1} \right) \right)_{0\le i,j\le n-1}$$

(by Lemma 3, eq. (3.2))

$$= \nu_n(2x, 2y).$$

Therefore by (3.16) and (1.1)

$$\begin{aligned} N_n(x, y) &= \det(\nu_n(x, y)) \\ &= \det\left(\sigma_n\left(\frac{x}{2}, \frac{y}{2}\right)\mathbf{M}_n\left(\frac{x+y}{2}\right)\right) \\ &= 2^n \cdot \det\left(\mathbf{M}_n\left(\frac{x+y}{2}\right)\right) \\ &= \prod_{k=0}^{n-1} \Delta_{2k}(x+y). \end{aligned}$$

□

4. Applications to the T.S.S.C.P.P. conjecture. We shall consider in this section a few instances of the results we have obtained . We begin with a rather odd determinant for generalized harmonic numbers.

Corollary 1. *Let* $H_n(x) = \sum_{j=0}^{n} \frac{1}{x+j}$, *then*

(4.1) $$\det\left(\binom{i+j+x}{2i-j+1}(H_{i+j}(x) - H_{2j-i-1}(x))\right)_{0\le i,j\le n-1} = \frac{1}{2^n}\prod_{k=0}^{n-1} \Delta_{2k}(2x-1).$$

Proof. From (1.5) and (1.6) with x replaced by $x-1$ we find

(4.2) $$\det\left(\frac{1}{x-y}\left(\binom{i+j+x}{2i-j+1} - \binom{i+j+y}{2i-j+1}\right)\right)_{0\le i,j\le n-1} = 2^{-n}\prod_{k=0}^{n-1} \Delta_{2k}(x+y-1).$$

Now let $y \to x$, and we obtain

(4.3) $$\det\left(\frac{d}{dx}\binom{i+j+x}{2i-j+1}\right)_{0\le i,j\le n-1} = 2^{-n}\prod_{k=0}^{n-1} \Delta_{2k}(2x-1).$$

Identity (4.1) is merely (4.3) after the differentiation has been completed. □

In [**8**], Mills, Robbins and Rumsey define "*a totally symmetric plane partition* of size n (to be) a plane partition whose three-dimensional Ferrers graph is contained in the box

$$X_n = [1\,,\, n] \times [1\,,\, n] \times [1\,,\, n]$$

and which is mapped to itself under all permtutations of the coordinate axes. The complement of the Ferrers graph of such a plane partition (that is, the set of lattice points in the box X_n that do not belong to the Ferrers graph) is again totally symmetric when viewed from the vantage point of the vertex $(n+1\,,\, n+1\,,\, n+1)$. A totally symmetric plane partition is self complementary if it is congruent (in the geometrical sense) to its complement. This cannot occur unless $n = 2m$ is even".

If we define A_n by the recurrence (4.6), then Mills et al. [**8**] conjecture that A_n is the total number of TSSCPP's in X_{2n}.

In [**12**], J. Stembridge essentially proved that the TSSCPP conjecture reduces to the following result (the details of Stembridge's result and its equivalence to the following are provided in [**2**; Sec. 2]). It should be noted that our proof of Corollary 2 in the final analysis relies on (1.1) and thus is quite different from the proof of the T.S.S.C.P.P. Conjecture given in [**2**].

COROLLARY 2.

$$\det(a_{ij})_{0\le i\,,\, j\le n-1} = A_n^2\,, \tag{4.4}$$

where

$$a_{ij} = \begin{cases} 1 & \text{if } i=j=0\,, \\ 0 & \text{if } i=j>0\,, \\ \displaystyle\sum_{s=2i-j+1}^{2j-i} \binom{i+j}{s} & \text{if } i<j\,, \\ -a_{ji} & \text{if } i>j\,, \end{cases} \tag{4.5}$$

and

$$A_n = \prod_{i=0}^{n-1} \frac{(3i+1)!}{(n+i)!} = \frac{(3n-2)!(n-1)!}{(2n-1)!(2n-2)!} A_{n-1}. \tag{4.6}$$

Proof. We define several new matrices:

$$w(n) = \left(\binom{i+j+1}{2j-i} + \binom{i+j}{2j-i-1}\right)_{0\le i,j\le n-1}, \tag{4.7}$$

$$u(n) = (\delta_{ij} - 2\delta_{i,j+1})_{0\le i,j\le n-1}, \tag{4.8}$$

$$v(n) = \left(\binom{i+j+2}{2j-i+1}\frac{(3i+1)(3j+1)(3j-3i)}{(i+j)(i+j+1)(i+j+2)}\right)_{0\le i,j\le n-1} \tag{4.9}$$

(where we define the $(0,0)$-th entry of $v(n)$ to be 1),

$$\begin{aligned} u_1(n) &= u(n) + (2\delta_{(i-1)^2+j,0})_{0\le i,j\le n-1} \\ &= \begin{pmatrix} 1 & 0 & 0 & 0 & \cdots \\ 0 & 1 & 0 & 0 & \cdots \\ 0 & -2 & 1 & 0 & \cdots \\ 0 & 0 & -2 & 1 & \cdots \end{pmatrix}_{n\times n}, \end{aligned} \tag{4.10}$$

$$st(n) = (a_{ij})_{0\le i,j\le n-1}. \tag{4.11}$$

The matrices u and u_1 are introduced to perform certain simple row and column operations. In particular, elementary algebra reveals that

$$u(n)w(n) = v(n) \tag{4.12}$$

and

$$u_1(n)st(n)(u_1(n))^T = v(n). \tag{4.13}$$

Finally, if we expand $M_{n+1}(-2,-1)$ along the top row we find

$$\begin{aligned} M_{n+1}(-2,-1) &= 2\det\left(\binom{i+j}{2i-j+1} + \binom{i+j+1}{2i-j+1}\right)_{0\le i,j\le n-1} \\ &= 2\det\left(\binom{i+j}{2j-i-1} + \binom{i+j+1}{2j-i}\right)_{0\le i,j\le n-1} \\ &\qquad \left(\text{since } \binom{A}{B} = \binom{A}{A-B}\right) \\ &= 2\det(w(n)). \end{aligned} \tag{4.14}$$

Consequently, since the determinants of u and u_1 are each 1, it follows from (1.6), (4.12), (4.13) and (4.14) that

$$\begin{aligned}(4.15)\qquad \det(a_{ij})_{0\le i,j\le n-1} &= \det(st(n)) = \det(v(n))\\ &= \det(w(n)) = \frac{1}{2}M_{n+1}(-2,-1)\\ &= \prod_{k=1}^{n}\Delta_{2k}(-3) = A_n^2\end{aligned}$$

because setting $u=-3$ in (1.2) we find

$$(4.16)\qquad \Delta_{2k}(-3) = \left(\frac{(3k-2)!(k-1)!}{(2k-2)!(2k-1)!}\right)^2. \qquad \square$$

5. A related identity and another proof of the T.S.S.C.P.P. conjecture. We have not found any related identities for two variables that are genuinely different from (1.6). However, there is one with one variable that merits mention. To this end we need the $\Delta_j(n)$ with odd subscript [**1**; p. 196]:

$$(5.1)\qquad \Delta_{2j-1}(\mu) = \frac{(\mu+2j)_{j-1}(\frac{1}{2}\mu+2j+\frac{1}{2})_j}{(j)_j(\frac{1}{2}\mu+j+\frac{1}{2})_{j-1}}.$$

The next formula is implicit in the work of Mills-Robbins-Rumsey [**9**]. However, they do not state it so we record it here: Let

$$(5.2)\qquad P_n(\mu) = \left(\binom{i+j+\mu}{2i-j} + 2\binom{i+j+\mu+2}{2i-j+1}\right)_{0\le i,j\le n-1}.$$

In the notation of [**9**; p. 50], the determinant of $P_n(\mu)$ is $R_n(1,\mu)$. This is easily seen by setting $x=1$ in their definition of $R_n(x,\mu)$ [**9**; p. 50] and applying the Chu-Vandermonde summation. Furthermore, from their Theorems 5 and 7, it is easy to see that

$$\begin{aligned}(5.3)\qquad \det(P_n(\mu)) &= \frac{\det\left(\delta_{ij} + \binom{i+j+2\mu}{i}\right)_{0\le i,j\le 2n-1}}{\det\left(\binom{i+j+\mu}{2i-j}\right)_{0\le i,j\le n-1}}\\ &= \frac{2^n\prod_{k=0}^{2n-1}\Delta_k(2\mu)}{\prod_{k=0}^{n-1}\Delta_{2k}(2\mu)} = 2^n\prod_{k=1}^{n}\Delta_{2k-1}(2\mu).\end{aligned}$$

Our final result gives the determinant for

$$(5.4)\qquad W_n(x) = \left(\binom{i+j+x+1}{2i-j} + \binom{i+j+x}{2i-j-1}\right)_{1\le i,j\le n}.$$

THEOREM 2.

$$\det(W_n(x)) = \prod_{k=1}^{n} \Delta_{2k-1}(2x+3). \tag{5.5}$$

Proof. We require an auxiliary matrix

$$\mathcal{S}_n(x) = \left(\frac{(-1/2)_{i-j}(-1)^{i-j}}{2^{2i-2j-1}(i-j)!}\right)_{0\leq i,j\leq n-1}. \tag{5.6}$$

Consequently,

$$\mathcal{S}_n(x)\cdot W_n(x) = P_n(x+3/2). \tag{5.7}$$

This is easily seen if we introduce

$$f(i,j,X) = \sum_{k=0}^{i} \frac{(-1/2)_{i-k}(-1)^{i-k}}{2^{2i-2k-1}(i-k)!}\binom{k+j+X}{2k-j}, \tag{5.8}$$

and note that as an immediate corollary of the Pfaff-Saalschultz summation [**4**; p. 9]:

$$f(i,j,X) = \binom{X+i+j-1/2}{2i-j} + \binom{X+i+j+1/2}{2i-j}. \tag{5.9}$$

Thus

$$\begin{aligned}
&\mathcal{S}_n(x)W_n(x)\\
&= \left(\sum_{k\geq 0}\frac{(-1/2)_{i-k}(-1)^{i-k}}{2^{2i-2k-1}(i-k!}\left\{\binom{k+j+x+3}{2k-j+1}\right.\right.\\
&\qquad\qquad \left.\left.+\binom{k+j+x+2}{2k-j}\right\}\right)_{0\leq i,j\leq n-1}\\
&= (f(i,j-1,x+4)+f(i,j,x+2))_{0\leq i,j\leq n-1}\\
&= \left(\binom{x+i+j+5/2}{2i-j+1}+\binom{x+i+j+7/2}{2i-j+1}\right.\\
&\qquad \left.+\binom{x+i+j+3/2}{2i-j}+\binom{x+i+j+5/2}{2i-j}\right)_{0\leq i,j\leq n-1}\\
&= \left(2\binom{x+i+j+7/2}{2i-j+1}+\binom{x+i+j+3/2}{2i-j}\right)_{0\leq i,j\leq n-1}\\
&= P_n(x+3/2),
\end{aligned}$$

as asserted in (5.7). Consequently since $\det(\mathcal{S}_n(x)) = 2^n$, we see from (5.7) and (5.3) that

$$\det(W_n(x)) = \prod_{k=1}^{n} \Delta_{2k-1}(2x+3)$$

as asserted in Theorem 2. □

We note that Corollary 2 is also derivable from Theorem 2. This is because from (4.7), (5.4) and (5.5)

$$\det(w(n)) = \det(W_{n-1}(0)) = \prod_{k=1}^{n-1} \Delta_{2k-1}(3) = A_n^2,$$

since

$$\Delta_{2k-1}(3) = \frac{(3+2k)_{k-1}(2+2k)_k}{(k)_k(2+k)_{k-1}} = \left(\frac{(3k+1)!k!}{(2k)!(2k+1)!}\right)^2.$$

We also remark that other special values for $W_n(x)$ can be derived from Theorem 2 besides $\det(W_n(0)) = A_{n+1}^2$. Namely

$$\det(W_n(-1)) = H_{2n+1},$$
$$\det(W_n(-2)) = H_{2n},$$
$$\det(W_n(-3)) = A_{n-1}A_n,$$

where the sequence H_n is defined by $H_0 = 1$,

$$\frac{H_{2n+1}}{H_{2n}} = \binom{3n}{n} \Big/ \binom{2n}{n}, \quad \frac{H_{2n}}{H_{2n-1}} = \frac{4}{3}\frac{\binom{3n}{n}}{\binom{2n}{n}}.$$

The sequences A_n and H_n occur in a number of unsolved problems of Mills-Robbins-Rumsey (cf. **[11]** for a survey of the problems).

6. Conclusion. The problem of enumerating symmetry classes of plane partitions is considered extensively in **[12]**. Indeed the identity (1.1) which we rely on heavily throughout our work was used by Mills et al. to treat plane partitions in a different symmetry class **[9]**. G. Kuperberg **[7]** has recently prepared an appealing survey of this topic.

Also recently M. Ishikawa **[6]** has found a nice plane partition theoretic interpreation of $\det(v(n)) = A_n^2$ (see (4.9) and (4.15)).

Acknowledgment. Every stage of this work and each theorem and lemma was found empirically using the symbolic algebra package AXIOM. While our proofs do not rely on the computer, each of our discoveries would have been impossible without the flexibility and power of AXIOM.

References

[1] G. E. Andrews, *Plane partitions* III: *the weak Macdonald conjecture*, Invent. Math., **53** (1979), 193–225.

[2] ——, *Plane partitions* V: *the T.S.S.C.P.P. conjecture*, J. Combin. Theory Ser. A, (to appear).

[3] W. N. Bailey, *Some identities involving generalized hypergeometric series*, Proc. London Math. Soc. Ser. 2, **29** (1929), 503–516.

[4] ——, *Generalized Hypergeometric Series* (Reprinted: Hafner, New York, 1964), Cambridge University Press, London and New York, 1935.

[5] A. Erdelyi et al., *Higher Transcendental Functions*, vol. 1, McGraw-Hill, New York, 1953.

[6] M. Ishikawa, *A remark on totally symmetric self-complementary plane partition*, (to appear).

[7] G. Kuperberg, *Symmetries of plane partitions and the permanent-determinant method*, (to appear).

[8] W. H. Mills, D. P. Robbins and H. Rumsey, *Self-complementary, totally symmetric plane partitions*, J. Combin. Theory Ser. A, **42** (1986), 277–292.

[9] ——, *Enumeration of a symmetry class of plane partitions*, Discrete Math., **67** (1987), 43–55.

[10] L. J. Slater, *Generalized Hypergeometric Functions*, Cambridge University Press, Cambridge, 1966.

[11] R. P. Stanley, *A baker's dozen of conjectures concerning plane partitions*, from Combinatoire énumérative (G. Labelle and P. Leroux, eds.), Lecture Notes in Math. No. 1234, Springer, Berlin/New York, 1986, pp. 285–293.

[12] ——, *Symmetries of plane partitions*, J. Combin. Theory Ser. A, **3** (1986), 103–113.

[13] J. Stembridge, *Nonintersecting paths, Pfaffians, and plane partitions*, Advances in Math., **83** (1990), 96–131.

[14] F. J. W. Whipple, *Well-poised series and other generalized hypergeometric series*, Proc. London Math. Soc. Ser. 2, **25** (1926), 525–544.

Received May 13, 1991 and in revised form March 16, 1992. Partially supported by NFS Grant DMS 8702695-03 and the IBM Thomas J. Watson Research Center.

PENNSYLVANIA STATE UNIVERSITY
UNIVERSITY PARK, PA 16802

AND

IBM THOMAS J. WATSON RESEARCH CENTER
YORKTOWN HEIGHTS, NY 10598

Chapter 5

Combinatorics, Fibonacci Numbers, and Computers

I have written at least a dozen papers that are more or less pure combinatorics. Of these I have chosen only three for inclusion in this chapter. This may seem somewhat inadequate for two reasons: (1) I have always considered myself a "combinatory analyst" (from the name of MacMahon's classic two-volume work [MacM16]), and (2) I was the editor of MacMahon's *Collected Papers* for the MIT Press [MacM78]. However, I suppose I feel that the material in Chapters 2 and 3 are my central work in combinatory analysis.

I have included [134] on Catalan numbers because of the utility of viewing Catalan identities in a hypergeometric context, and paper [163] because it is a natural paper to follow [134], owing to the interpretation of q-Catalan numbers. Paper [272] provides a combinatorial interpretation of the Legendre–Stirling numbers which appears to have extensive applications.

I designated a sequence of four minor papers "Identities in Combinatorics": [63], [66], [113], and [119]. Of these, I would note the natural interpretation of the Pfaff–Saalschütz identity in [63], and the integration of harmonic numbers into hypergeometric identities via differentiation in [119].

One of the many pleasant aspects of my career was a ten-year adjunct appointment at the University of Waterloo (1982–1992), which was followed up by an honorary docorate from Waterloo in 2004. This resulted in a number of joint papers with David Jackson and Ian Goulden: [122], [143], [160], [175], and [240]. For the most part, these papers are combinatorial studies to which I was able to provide some application of hypergeometric functions.

Papers [215] and [217] are properly papers in algebra where I could provide some hypergeometric applications.

I have included [238] of my three papers (also [19], and [54]) on Fibonacci numbers. In my youthful exuberance, I was very excited by [19]. It did eventually lead to the partition results on sieves in partitions: [49], and [50]. However, its actual appearance was rather dismaying. I never received galley proofs, and the typesetter had replaced "Eisenstein" with "Einstein" throughout. However, the ignominy of this was mitigated by the fact the typesetter also replaced my middle initial "E." with "H."; George H. Andrews was for many years Chair of Mathematics at Oberlin

College. I chose [238] for inclusion because of the lessons it teaches about the search for q-analogs.

I have written at least half a dozen papers describing my use of computers to search for and explain theorems in partitions and q-series. I started using computers in the mid-1960s, programming in FORTRAN. Papers [38] and [69] provide an account of some of this work from the 1960's and 1970's. My work with Rodney Baxter (described at length in Chapter 2) required many computer explorations. I have included the most interesting: [141], but [110], [111], and [168] provide accounts of related explorations.

Paper [172] with Doron Zeilberger, and his computer Shalosh B. Ekhad, led to the first of my public debates with Zeilberger over the role of computers now and in the future. An account of this controversy is given in [Essay1].

Paper [180] was the beginning of my lengthy collaboration with Peter Paule.

5.1 Catalan Numbers, q-Catalan Numbers and Hypergeometric Series

Reprinted from Journal of Combinatorial Theory, Series A
Vol. 44, No. 2, March 1987

Catalan Numbers, q-Catalan Numbers and Hypergeometric Series

GEORGE E. ANDREWS

Department of Mathematics, The Pennsylvania State University, University Park, Pennsylvania 16802

Communicated by the Managing Editors

Received April 3, 1986

Catalan numbers are examined in the context of hypergeometric series. We are thus able to produce new and simple q-analogs related to the theory of partitions.

1. INTRODUCTION

In a recent paper, Fürlinger and Hofbauer [4] present a survey and unification of the q-analogs of Catalan numbers. In the introduction they point out that the Catalan numbers C_n most often arise from combinatorial problems as solutions of the recurrence

$$C_n = \sum_{k=1}^{n-1} C_k C_{n-k}, \qquad C_1 = 1. \tag{1.1}$$

There are generally two ways of solving this recurrence: (1) generating functions and (2) Lagrange inversion. The closed form of the solution is

$$C_n = \frac{1}{n}\binom{2n-2}{n-1}. \tag{1.2}$$

The authors note that q-analogs of Lagrange inversion have been the source of interesting q-analogs of the Catalan numbers. This leaves open the question of whether there are other approaches to Catalan numbers that lead to interesting q-analogs.

In particular it should be pointed out that no q-Catalan number considered to date has both a simple representation as a finite product like (1.2) and fulfills an exact q-analog of (1.1).

We shall consider the problem from the hypergeometric series viewpoint. Surprisingly we find that if $C_1(q) = 2q/(1+q)$ and

$$\frac{(1+q^n)}{2} C_n(q) = \sum_{k=1}^{n-1} C_k(q)\, C_{n-k}(q)\, q^k, \tag{1.3}$$

then

$$C_n(q) = \frac{(1-q)}{(1-q^n)} \begin{bmatrix} 2n-2 \\ n-1 \end{bmatrix} \frac{(2q)^{2n-1}}{(-q;q)_n(-q;q)_{n-1}}, \tag{1.4}$$

where

$$(A;q)_n = (1-A)(1-Aq)\cdots(1-Aq^{n-1}), \tag{1.5}$$

and

$$\begin{bmatrix} A \\ B \end{bmatrix} = \frac{(q;q)_A}{(q;q)_B(q;q)_{A-B}}. \tag{1.6}$$

In Section 2 we examine the classical Catalan recurrence as a hypergeometric series identity, the Chu–Vandermonde summation. This approach provides a q-Catalan number $\mathscr{C}_n(\lambda, q)$ which has not arisen previously, and $-2^{2n-1}\mathscr{C}_n(-1, q) = C_n(q)$ is the special case given by (1.3) and (1.4). In Section 3, we examine the $\mathscr{C}_n(\lambda, q)$ as certain partition generating functions, and in Section 4 we provide a combinatorial proof of the related Catalan recurrence. We conclude with some open questions and some examination of how $C_n(q)$ could have been missed.

2. Catalan Numbers and Hypergeometric Series

We begin by noting from (1.2) that

$$C_n = \frac{1}{n}\binom{2n-2}{n-1} = \frac{1}{2(2n-1)}\binom{2n}{n}, \qquad n > 0. \tag{2.1}$$

Hence we may naturally take

$$C_0 = -\tfrac{1}{2}. \tag{2.2}$$

This allows us to define the Catalan numbers more succinctly by

$$C_0 = -\tfrac{1}{2}, \qquad C_1 = 1, \tag{2.3}$$

and

$$\sum_{k=0}^{n} C_k C_{n-k} = 0 \qquad \text{for} \quad n > 1. \tag{2.4}$$

Now immediately from (2.1) we see that

$$C_n = \frac{4^{n-1}(\frac{1}{2})_{n-1}}{n!} = \frac{-2^{2n-1}(-\frac{1}{2})_n}{n!}, \tag{2.5}$$

where

$$(a)_n = a(a+1)\cdots(a+n-1). \tag{2.6}$$

Hence

$$\begin{aligned}\sum_{k=0}^{n} C_k C_{n-k} &= \sum_{k=0}^{n} \frac{2^{2k-1}(-\frac{1}{2})_k}{k!} \frac{2^{2(n-k)-1}(-\frac{1}{2})_{n-k}}{(n-k)!} \\ &= \frac{2^{2n-2}(-\frac{1}{2})_n}{n!} {}_2F_1\left[\begin{matrix} -n, -\frac{1}{2}; 1 \\ -n+\frac{3}{2} \end{matrix}\right] \qquad \text{(by [3, p. 8])} \\ &= \frac{2^{2n-2}(-\frac{1}{2})_n(-n+2)_n}{n!\,(-n+\frac{3}{2})_n} \qquad \text{(by [3, p. 3], the Chu–Vandermonde sum)} \\ &= 0 \qquad \text{for} \quad n > 1. \end{aligned} \tag{2.7}$$

Thus (2.4) (or equivalently (1.1)) has been established by a simple application of the Chu–Vandermonde summation.

As was pointed out in [1, Sect. 5], a useful approach to q-analogs is to look for the appropriate q-analog of an underlying hypergeometric series identity. Since q-analogs of the Chu–Vandermonde sum are well known, it is clear that this approach merits consideration.

3. Partitions with Parts Restricted by Parity

This section is devoted to the partition theoretic aspects of

$$\mathscr{C}_n(\lambda, q) = \frac{q^{2n}(-\lambda q^{-1}; q^2)_n}{(q^2; q^2)_n}. \tag{3.1}$$

THEOREM 1. *Let P_0 (resp. P_1) denote the set of all partitions of positive integers in which the even (resp. odd) parts are not repeated and in the case of P_0 each part is >1. Let $\Sigma(\pi)$ denote the number partitioned by π; $\#(\pi)$*

denote the total number of parts of π, and $\#_1(\pi)$ denote the number of odd parts of π. Then

$$\mathscr{C}_n(\lambda, q) = \sum_{\substack{\pi \in P_1 \\ \#(\pi) = n}} \lambda^{\#_1(\pi)} q^{\Sigma(\pi)}, \tag{3.2}$$

and

$$\lambda^n q^n \mathscr{C}_n(\lambda^{-1}, q) = \sum_{\substack{\pi \in P_0 \\ \#(\pi) = n}} \lambda^{\#_1(\pi)} q^{\Sigma(\pi)}. \tag{3.3}$$

Proof. We see using the standard elementary techniques [2, Chap. 1] that since

$$\mathscr{C}_n(\lambda, q) = \frac{q^{2n}(1+\lambda q^{-1})(1+\lambda q)(1+\lambda q^3)\cdots(1+\lambda q^{2n-3})}{(1-q^2)(1-q^4)(1-q^6)\cdots(1-q^{2n})}, \tag{3.4}$$

$\mathscr{C}_n(\lambda, q)$ may be immediately interpreted as the generating function for partitions in which the largest part is $2n$, even parts may be repeated and odd parts (enumerated by the exponent on λ) are distinct and lie in the interval $[-1, 2n-3]$. We exhibit such partitions using a slight variation on P. A. MacMahon's modular partitions [6]. Each even part $2v$ is represented by a row of v twos; each odd part $2v-1$ is represented by a row of v twos with a single -1 at the end. Thus $6+6+5+4+4+4+(-1)$ becomes

$$\begin{array}{rrr} 2 & 2 & 2 \\ 2 & 2 & 2 \\ 2 & 2 & -1 \\ 2 & 2 & \\ 2 & 2 & \\ 2 & 2 & \\ -1 & & \end{array}$$

Now we take these partitions generated in (3.4) and read parts by columns instead of rows. There are now exactly n parts. Since the top row is n twos, we see that all parts are $\geqslant 1$. The number of odd parts is the same as before and they are still distinct. This is precisely the assertion of (3.2).

To establish (3.3) we follow up on the above construction by adding a top row of ones, and again we read off parts by columns. The parity of all parts has been reversed; however, the multiplication by λ^n and the replacement of λ by λ^{-1} as the first argument of $\mathscr{C}_n$ means that the exponent on λ still counts the number of odd parts in the partition. The only change is that now the even parts are distinct (since previously the odd parts were distinct), and all parts are $\geqslant 2$. ∎

4. The q-Catalan Recurrence

Following the lead from Section 2, we may easily obtain the following q-Catalan recurrence:

THEOREM 2. *For* $n > 1$,

$$\sum_{j=0}^{n} \mathscr{C}_j(\lambda^{-1}, q)\, \mathscr{C}_{n-j}(\lambda, q)(-\lambda q)^j = 0. \tag{4.1}$$

First Proof.

$$\begin{aligned}
&\sum_{j=0}^{n} \mathscr{C}_j(\lambda^{-1}, q)\, \mathscr{C}_{n-j}(\lambda, q)(-\lambda q)^j \\
&\quad = \sum_{j=0}^{n} \frac{q^{2j}(-\lambda^{-1}q^{-1}; q^2)_j}{(q^2; q^2)_j} \frac{q^{2n-2j}(-\lambda q^{-1}; q^2)_{n-j}}{(q^2; q^2)_{n-j}} (-\lambda q)^j \\
&\quad = \frac{q^{2n}(-\lambda q^{-1}; q^2)_n}{(q^2; q^2)_n} \sum_{j=0}^{n} \frac{(q^{-2n}; q^2)_j(-\lambda^{-1}q^{-1}; q^2)_j}{(q^2; q^2)_j(-\lambda^{-1}q^{-2n+3}; q^2)_j} q^{4j} \\
&\quad = \frac{q^{2n}(-\lambda q^{-1}; q^2)_n (q^{-2n+4}; q^2)_n}{(q^2; q^2)_n(-\lambda^{-1}q^{-2n+3}; q^2)_n} \qquad \text{(by [7, p. 97, Eq. (3.3.2.6)])} \\
&\quad = 0, \qquad \text{for} \quad n > 1. \quad \blacksquare
\end{aligned}$$

Of course, the above follows our prescription in the introduction for finding q-analogs. However, once such results are found we may naturally ask for an alternative combinatorial treatment.

Second Proof. Here we follow the lead of Garsia and Milne [5] by constructing a weight preserving, sign reversing involution on a class of pairs of partitions.

We define (using the notation of Theorem 1) $\mathscr{P}_n(l, m)$ as the set of all pairs of partitions (π, ψ), where $\pi \in P_0$, $\psi \in P_1$, $\Sigma(\pi) + \Sigma(\psi) = m$, $\#(\pi) + \#(\psi) = n$, $\#_1(\pi) + \#_1(\psi) = l$. From Theorem 1 we see that

$$\sum_{j=0}^{n} \mathscr{C}_j(\lambda^{-1}, q)\, \mathscr{C}_{n-j}(\lambda, q)(-\lambda q)^j = \sum_{\substack{l \geq 0 \\ m \geq 0}} \lambda^l q^m \sum_{(\pi, \psi) \in \mathscr{P}_n(l,m)} (-1)^{\#(\pi)}.$$

To conclude this proof we need only provide an involution on $\mathscr{P}_n(l, m)$ that maps the partitions (π, ψ) with $\#(\pi)$ even onto those with $\#(\pi)$ odd.

The involution works as follows: identify the smallest part $r > 1$ appearing in either π or ψ and then move it to the other partition in such a way that the resulting pair (π', ψ') is still in $\mathscr{P}_n(l, m)$. Obviously if r appears in only one of π and ψ then deleting one appearance of it where it is and inserting it once in the other partition is clearly legitimate. Suppose r is odd and appears in both π and ψ. Then we cannot add another r to ψ without violating the condition $\psi' \in P_1$. Hence if r is odd and appears in both π and ψ then we must delete it from ψ and add a further appearance of it to π. Similarly if r is even and appears in both π and ψ, then we must delete it from π and add it to ψ. The above construction shows that our mapping is indeed the required weight preserving, sign reversing involution. Hence

$$\sum_{(\pi,\psi)\in \mathscr{P}_n(l,m)} (-1)^{\#(\pi)} = 0,$$

and again we have (4.1).

Note that our argument fails in the cases $n=0$ or 1 for in these cases above the smallest part $r>1$ does not exist. ∎

THEOREM 3. *Let* $C_1(q) = 2q/(1+q)$ *and*

$$\frac{(1+q^n)}{2} C_n(q) = \sum_{k=1}^{n-1} C_k(q)\, C_{n-k}(q)\, q^k, \tag{4.2}$$

then

$$C_n(q) = \frac{(1-q)}{(1-q^n)} \begin{bmatrix} 2n-2 \\ n-1 \end{bmatrix} \frac{(2q)^{2n-1}}{(-q;q)_n(-q;q)_{n-1}}. \tag{4.3}$$

Proof. We note that the expression in (4.3) is $-2^{2n-1}\mathscr{C}_n(-1,q)$. This is $-\frac{1}{2}$ when $n=0$ and $2q/(1+q)$ when $n=1$. Consequently (4.2) is merely (4.1) multiplied by 2^{2n-2} with $\lambda = -1$. ∎

5. CONCLUSION

An important lesson to be drawn from this paper is that each approach to q-analogues has its own merits. Perhaps the reason that the results in this paper had been previously overlooked is that the Lagrange inversion technique provided insight for q-analog constructions, and generating functions appeared "ad hoc." The correct interpretation of this state of affairs is not that generating functions are inadequate and inferior but rather that they should be studied in the context of a well-developed and sophisticated framework such as (in this instance) hypergeometric series.

Finally we may ask for a natural interpretation of $C_n(q)$ that yields $C_n(1) = C_n$. This is a somewhat value request since we, of course, have a combinatorial proof of (4.1); however, the specialization of (4.1) to obtain Theorem 3 leaves us with the hope that there may be a more transparent explanation.

References

1. G. E. Andrews, Applications of basic hypergeometric functions, *SIAM Rev.* **16** (1974), 441–484.
2. G. E. Andrews, "The Theory of Partitions, Encyclopedia of Mathematics and Its Applications," Vol. 2 (G. C. Rota, Ed.), Addison–Wesley, Reading, MA, 1976. (Reprinted: Cambridge Univ. Press, London/New York, 1984.)
3. W. N. Bailey, "Generalized Hypergeometric Series," Cambridge Mathematical Tract No. 32, Cambridge Univ. Press, 1935. (Reprinted: Hafner, New York, 1964.)
4. J. Fürlinger and J. Hofbauer, q-Catalan numbers, *J. Combin. Theory Ser. A* **40** (1985), 248–264.
5. A. M. Garsia and S. C. Milne, A Rogers–Ramanujan bijection, *J. Combin. Theory Ser. A* **31** (1981), 289–339.
6. P. A. MacMahon, The theory of modular partitions, *Proc. Cambridge Philos. Soc.* **21** (1923), 197–204. (Reprinted: Collected Papers, Vol. 1, pp. 1090–1097, M.I.T. Press, Cambridge, MA, 1978.)
7. L. J. Slater, "Generalized Hypergeometric Functions," Cambridge Univ. Press, London/New York, 1966.

5.2 Fibonacci Numbers and the Rogers–Ramanujan Identities

FEATURED ARTICLE

Edited by Andrew Granville

From time to time *The Fibonacci Quarterly* will publish invited papers by well known mathematicians and scientists describing how the authors use Fibonacci numbers, or similar recurrence sequences, in their research and teaching. We are pleased to present here the first paper in the series. The author is the distinguished scholar and teacher George Andrews, who is currently the Evan Pugh Professor of Mathematics at Penn State University. In 2003 Professor Andrews was elected to the National Academy of Sciences, which is one of the highest honors that can be accorded a scientist or engineer.

FIBONACCI NUMBERS AND THE ROGERS-RAMANUJAN IDENTITIES

George E. Andrews[1]
The Pennsylvania State University
University Park, PA 16802
andrews@math.psu.edu

ABSTRACT

In my research on the theory of partitions and related questions, I have often used recurrent integer sequences such as the Fibonacci numbers to obtain hints about the behavior of certain partition generating functions. In this article, I will illustrate this process by studying the relationship between the Fibonacci numbers and a sequence of polynomials used by Schur in his second proof of the Rogers-Ramanujan identities. I will try to make clear how generating function proofs of Fibonacci identities lead to analogous results for Schur's polynomials. We shall find that this approach helps explain why the generalization of Fibonacci formulas to Schur's polynomial is sometimes rather intricate.

1. INTRODUCTION

Everyone reading these words knows the Fibonacci sequence:

$$F_n = \begin{cases} 0 & \text{if } n = 0 \\ 1 & \text{if } n = 1 \\ F_{n-1} + F_{n-2} & \text{if } n > 1. \end{cases} \tag{1.1.1}$$

Less well-known is a sequence of polynomials first considered by I. Schur [12]:

$$S_n(q) = S_n = \begin{cases} 0 & \text{if } n = 0 \\ 1 & \text{if } n = 1 \\ S_{n-1} + q^{n-2} S_{n-2} & \text{if } n > 1. \end{cases} \tag{1.1.2}$$

Of course, it is immediately obvious that

$$S_n(1) = F_n. \tag{1.1.3}$$

[1] Partially supported by National Science Foundation Grant DMS9206993.

FIBONACCI NUMBERS AND THE ROGERS-RAMANUJAN IDENTITIES

It is, however, less than obvious that

$$S_{n+1}(q) = \sum_{0\le 2j\le n} q^{j^2} \begin{bmatrix} n-j \\ j \end{bmatrix}$$

$$= \sum_{j=-\infty}^{\infty} (-1)^j q^{j(5j+1)/2} \begin{bmatrix} n \\ \lfloor \frac{n-5j}{2} \rfloor \end{bmatrix} \tag{1.1.4}$$

where

$$\begin{bmatrix} A \\ B \end{bmatrix} = \begin{cases} 0 & \text{if } B<0 \text{ or } B>A \\ \frac{(1-q^A)(1-q^{A-1})\cdots(1-q^{A-B+1})}{(1-q^B)(1-q^{B-1})\cdots(1-q)} & \text{otherwise} \end{cases} \tag{1.1.5}$$

and $\lfloor x \rfloor$ denotes the greatest integer not exceeding x. Consult [2] for a full proof of (1.4).

The rational functions of q given in (1.5) are in fact polynomials. They are sometimes called Gaussian polynomials and sometimes q-binomial coefficients.

If we allow $n \to \infty$ in formula (1.4) then it is easily shown that what results is

$$1 + \sum_{j=1}^{\infty} \frac{q^{j^2}}{(1-q)(1-q^2)\cdots(1-q^j)} = \frac{\sum_{j=-\infty}^{\infty}(-1)^j q^{j(5j+1)/2}}{\prod_{n=1}^{\infty}(1-q^n)}$$

$$= \prod_{n=0}^{\infty} \frac{1}{(1-q^{5n+1})(1-q^{5n+4})}, \text{ (by [4: p. 113])} \tag{1.1.6}$$

the first of the celebrated Rogers-Ramanujan identities.

Schur also considered a sequence slightly different from (1.2):

$$\mathcal{T}_n(q) = \mathcal{T}_n = \begin{cases} 0 & \text{if } n=0 \\ 1 & \text{if } n=1 \\ \mathcal{T}_{n-1} + q^{n-1}\mathcal{T}_{n-2} & \text{if } n>1. \end{cases} \tag{1.1.7}$$

As before in (1.3), now

$$\mathcal{T}_n(1) = F_n. \tag{1.1.8}$$

In analogy with (1.5)

$$\mathcal{T}_{n+1}(q) = \sum_{0\le 2j\le n} q^{j^2+j} \begin{bmatrix} n-j \\ j \end{bmatrix}$$

$$= \sum_{j=-\infty}^{\infty} (-1)^j q^{j(5j-3)/2} \begin{bmatrix} n+j \\ \lfloor \frac{n+1-5j}{2} \rfloor + 1 \end{bmatrix}. \tag{1.1.9}$$

Allowing $n \to \infty$ in (1.9), we obtain the second Rogers-Ramanujan identity

$$1+\sum_{j=1}^{\infty} \frac{q^{j^2+j}}{(1-q)(1-q^2)\cdots(1-q^j)} = \frac{\sum_{j=-\infty}^{\infty}(-1)^j q^{j(5j-3)/2}}{\prod_{n=1}^{\infty}(1-q^n)} = \prod_{n=0}^{\infty} \frac{1}{(1-q^{5n+2})(1-q^{5n+3})}, \text{ (by [4: p. 113]).} \tag{1.1.10}$$

Now it is already obvious in light of (1.3) and (1.8) that Schur's polynomial sequences are polynomial- or q-analogs of the Fibonacci numbers. Indeed, we see immediately from (1.4) and (1.9) that

$$\begin{aligned} F_{n+1} &= \sum_{0\le 2j\le n} \binom{n-j}{j} \\ &= \sum_{j=-\infty}^{\infty} (-1)^j \binom{n}{\lfloor \frac{n-5j}{2} \rfloor + 1} \\ &= \sum_{j=-\infty}^{\infty} (-1)^j \binom{n+1}{\lfloor \frac{n+1-5j}{2} \rfloor + 1}. \end{aligned} \tag{1.1.11}$$

The first line of (1.11) is quite familiar [14; p. 155]; however the next two lines are much less well known. In fact, they form the basis of the extensive study of generalized Fibonacci numbers in [1] (cf. [10], [13]).

Equation (1.11) suggests that interesting information about F_n may be obtained from studying $\mathcal{S}_n$ and $\mathcal{T}_n$. Our object here is the reverse. Can we systematize a study of the generating function for F_n so that we can deduce formulas for $\mathcal{S}_n$ and $\mathcal{T}_n$ from generating function proofs of identities for F_n?

2. GENERATING FUNCTION PROOFS FOR FIBONACCI IDENTITIES

As is well-known, [14; p. 221]

$$\sum_{n=0}^{\infty} F_n x^n = \frac{x}{1-x-x^2}. \tag{2.2.1}$$

Perhaps less widely appreciated is the fact that the vast majority of Fibonacci number identities are easily verified using (2.1). To make this point briefly, we have chosen a few results that are usually proved by other means (cf. [12; Sec 10.14], [14; Chs. 5, 8 and 12]). However, we should note that the generating function method is effectively explored by Koshy ([14; Chs. 18 and 19]).

$$F_0+F_1+\cdots+F_n = F_{n+2}-1 \quad \text{[14; Ch. pp. 228-229];} \tag{2.2.2}$$

$$F_{n+m} = F_{m-1}F_n + F_m F_{n+1} \quad [14; \text{p. } 88, \#6; \text{p. } 363, \text{eq. } (32.3)]; \tag{2.2.3}$$

$$F_n = 2^{1-n} \sum_{0 \le 2j+1 \le n} \binom{n}{2j+1} 5^j \quad [12; \text{eq. } (10.14.11), \text{p. } 150]. \tag{2.2.4}$$

Proving these results via generating functions will not be as illuminating as some other proofs nor as elegant. However, the ideas arising from an awareness of the power of this method will serve us well in Section 5.

To treat each identity we consider the generating functions for the sequences defined by each side. If these generating functions are identical the identities are proved.

Proof of (2.2) following [14; pp. 228-229]:

$$\begin{aligned}
\sum_{n=0}^{\infty}(F_0 + F_1 + \cdots + F_n)x^n &= \frac{1}{1-x}\sum_{n=0}^{\infty} F_n x^n \\
&= \frac{x}{(1-x)(1-x-x^2)} \\
&= \frac{1+x}{1-x-x^2} - \frac{1}{1-x} \\
&= x^{-2}\left(\frac{x}{1-x-x^2} - x\right) - \frac{1}{1-x} \\
&= x^{-2}\sum_{n=2}^{\infty} F_n x^n - \frac{1}{1-x} \\
&= \sum_{n=0}^{\infty}(F_{n+2} - 1)x^n,
\end{aligned} \tag{2.2.5}$$

and (2.2) is proved by comparing the coefficients of x^n in the extremes of (2.5).

Proof of (2.3). Here we follow the treatment due to R.T. Hansen [11] (cf. [12; p. 231]). We fix m, then

$$\begin{aligned}
(1-x-x^2)\sum_{n=0}^{\infty} F_{n+m}x^n &= F_m + F_{m+1}x + \sum_{n=2}^{\infty} F_{n+m}x^n \\
&\quad - F_m x - \sum_{n=1}^{\infty} F_{n+m}x^{n+1} - \sum_{n=0}^{\infty} F_{n+m}x^{n+2} \\
&\quad + F_m + xF_{m-1} + \sum_{n=0}^{\infty}(F_{n+m+2} - F_{n+m+1} - F_{n+m})x^{n+2} \\
&= F_m + xF_{m-1}. \quad \text{(where } n \text{ has been replaced by } n+2 \text{ in the first sum and } n+1 \text{ in the second)}
\end{aligned} \tag{2.2.6}$$

Hence

$$\begin{aligned}\sum_{n=0}^{\infty} F_{n+m}x^n &= \frac{F_m + xF_{m-1}}{(1-x-x^2)} \\ &= \sum_{n=0}^{\infty} F_m F_{n+1} x^n + \sum_{n=0}^{\infty} F_{m-1}F_n x^n.\end{aligned} \tag{2.2.7}$$

Consequently, (2.3) follows by comparing coefficients of x^n on each side of (2.7).

Finally we consider (2.4):

$$\begin{aligned}\sum_{n=0}^{\infty} 2^{1-n} \sum_{0\le 2j+1\le n} \binom{n}{2j+1} 5^j x^n &= \sum_{n,j\ge 0} 2^{1-n-2j-1}\binom{n+2j+1}{2j+1} 5^j x^{n+2j+1} \\ &\quad \text{(where } n \text{ has been shifted to } n+2j+1) \\ &= \sum_{j\ge 0} \frac{5^j x^{2j+1}}{4^j}\left(1-\frac{x}{2}\right)^{-2j-2} \\ &= \frac{1}{(1-\frac{x}{2})^2}\frac{x}{1-\frac{5x^2}{4(1-\frac{x}{2})^2}} \\ &= \frac{x}{(1-\frac{x}{2})^2-\frac{5x^2}{4}} \\ &= \frac{x}{1-x-x^2} = \sum_{n=0}^{\infty} F_n x^n\end{aligned} \tag{2.2.8}$$

and (2.4) is proved.

As we stated at the beginning of this section, the object of this section was *not* to provide the most elegant of illuminating proofs of identities (2.2)-(2.4) (indeed, not even novel). Rather we have taken a uniform generating function approach in the hope that in Sections 3-5 we can use the work in this section as guidance to point us to correct analogs of (2.2)-(2.4) for Schur's polynomials. We should note that if we can write down a generating function for both sides of a general 'Fibonacci number identity' then we have just one identity involving power series to verify (rather than infinitely many numerical identities), which in each of these cases turns out to be an identity involving rational functions.

3. SCHUR'S POLYNOMIALS, A BEGINNING

In order to find analogs of (2.1) for $\mathcal{S}_n$ and $\mathcal{T}_n$, we need the q-analog of the binomial series:

$$\sum_{j=0}^{\infty} \begin{bmatrix} n+j \\ j \end{bmatrix} x^j = \frac{1}{(x;q)_{n+1}}, \quad [4; \text{p. } 36, \text{eq. } (3.3.7)] \tag{3.3.1}$$

where

$$(A;q)_n = (1-A)(1-Aq)\cdots(1-Aq^{n-1}). \tag{3.3.2}$$

Throughout this section $|x| < 1$, $|q| < 1$.

We may now obtain an analog of (2.1) for $\mathcal{S}_n(q)$.

Theorem 1: $\sum_{n=0}^{\infty} \mathcal{S}_n(q)x^n = \frac{1}{1-x-x^2\eta}x$, *where* η *is an operator on functions of* x *defined by*

$$\eta f(x) = f(xq)$$

Remark: *For a full account of the algebra of operators, I recommend [6]. As a readily comprehensible example, we consider*

$$\frac{1}{1-x\eta}f(x) = \sum_{n=0}^{\infty}(x\eta)^n f(x)$$

$$= \sum_{n=0}^{\infty}\underbrace{(x\eta)(x\eta)\ldots(x\eta)}_{n \text{ times}} f(x)$$

$$= \sum_{n=0}^{\infty} x^n q^{0+1+2+\cdots+(n-1)} f(xq^n)$$

$$= \sum_{n=0}^{\infty} x^n q^{n(n-1)/2} f(xq^n).$$

The algebra of operators (see [6] again) tells us that the inverse operator is $1 - x\eta$, *and we can easily convince ourselves of this as follows*

$$(1-x\eta)\sum_{n=0}^{\infty} x^n q^{n(n-1)/2} f(xq^n)$$

$$= \sum_{n=0}^{\infty} x^n q^{n(n-1)/2} f(xq^n) - \sum_{n=0}^{\infty} x(xq)^n q^{n(n-1)/2} f(xq^{n+1})$$

$$= \sum_{n=0}^{\infty} x^n q^{n(n-1)/2} f(xq^n) - \sum_{m=1}^{\infty} x^m q^{m(m-1)/2} f(xq^m)$$

(where we have replaced $n+1$ by m in the second sum)

$$= f(x).$$

Proof: Let us denote by $\sigma(x)$ the expression on the left side of the equation in Theorem 1. Hence equivalently, we are to prove

$$(1 - x - x^2\eta)\sigma(x) = x.$$

Now,

$$x+\sum_{n=2}^{\infty}\mathcal{S}_n(q)x^n-\sum_{n=0}^{\infty}\mathcal{S}_n(q)x^{n+1}-\sum_{n=0}^{\infty}\mathcal{S}_n(q)x^{n+2}q^n$$

$$=x+\sum_{n=2}^{\infty}(\mathcal{S}_n(q)-\mathcal{S}_{n-1}(q)-q^{n-2}\mathcal{S}_{n-2}(q))x^n$$

$$=x, \quad \text{(where } n \text{ is replaced by } n-1 \text{ in the second sum and } n-2 \text{ in the third)}$$

by (1.2). □

Lemma 1:

$$(x+x^2\eta)^n x = x^{n+1}\sum_{j=0}^{n} x^j q^{j^2}\begin{bmatrix} n \\ j \end{bmatrix}.$$

Proof: For $n=0$, this asserts $x=x$. Now assume the result for a particular n; then

$$(x+x^2\eta)^{n+1}x=(x+x^2\eta)x^{n+1}\sum_{j=0}^{n}x^jq^{j^2}\begin{bmatrix} n \\ j \end{bmatrix}$$

$$=x^{n+2}\sum_{j=0}^{n}x^jq^{j^2}\begin{bmatrix} n \\ j \end{bmatrix}+x^{n+3}\sum_{j=0}^{n}x^jq^{j^2+n+1+j}\begin{bmatrix} n \\ j \end{bmatrix}$$

$$=x^{n+2}\left(\sum_{j\geq 0}x^jq^{j^2}\left(\begin{bmatrix} n \\ j \end{bmatrix}+q^{n+1-j}\begin{bmatrix} n \\ j-1 \end{bmatrix}\right)\right)$$

$$=x^{n+2}\sum_{j\geq 0}x^jq^{j^2}\begin{bmatrix} n+1 \\ j \end{bmatrix} \quad \text{(by [4; eq. (3.3.3), p. 35])},$$

and the result follows by mathematical induction. □

Corollary 1:

$$\sum_{n\geq 0}\mathcal{S}_n(q)x^n=\sum_{j\geq 0}\frac{x^{2j+1}q^{j^2}}{(x;q)_{j+1}}.$$

Proof:

$$\sum_{n\geq 0} \mathcal{S}_n(q)x^n = \sum_{j=0}^{\infty} \frac{1}{1-x-x^2\eta}x$$

$$= \sum_{n=0}^{\infty}(x+x^2\eta)^n x$$

$$= \sum_{n=0}^{\infty} x^{n+1} \sum_{j=0}^{n} x^j q^{j^2} \begin{bmatrix} n \\ j \end{bmatrix}$$

$$= \sum_{n,j\geq 0} x^{n+2j+1} q^{j^2} \begin{bmatrix} n+j \\ j \end{bmatrix}$$

(where n has been replaced by $n+j$)

$$= \sum_{j\geq 0} \frac{x^{2j+1}q^{j^2}}{(x;q)_{j+1}}$$

(by (3.1)). □

Corollary 2:

$$\mathcal{S}_n(q) = \sum_{j\geq 0} q^{j^2} \begin{bmatrix} n-j-1 \\ j \end{bmatrix}.$$

Proof: Extract the coefficient of x^n in the penultimate line in the proof of Corollary 1.1. □

Using precisely this technique, we can prove the following analog of (2.2).

Theorem 2:

$$\mathcal{S}_{n+2}(q) - 1 = \sum_{j=0}^{n} q^j \mathcal{S}_j(q).$$

Proof:

$$\begin{aligned}
\sum_{n\geq 0}(\mathcal{S}_{n+2}(q)-1)x^n &= x^{-2}\left(\sum_{n=0}^{\infty}\mathcal{S}_n(q)x^n - x\right) - \frac{1}{1-x} \\
&= x^{-2}\frac{1}{1-x-x^2\eta}x - \frac{x^{-1}}{1-x} \\
&= \frac{1}{1-x}(-x^{-2}(1-x-x^2\eta)+x^{-2}-x^{-1})\frac{1}{1-x-x^2\eta}x \\
&= \frac{1}{1-x}\eta\frac{1}{1-x-x^2\eta}x \\
&= \frac{1}{1-x}\sum_{j\geq 0}\mathcal{S}_j(q)(xq)^j \\
&= \sum_{n=0}^{\infty}\left(\sum_{j=0}^{n}q^j\mathcal{S}_j(q)\right)x^n,
\end{aligned} \tag{3.3.3}$$

and extracting the coefficients of x^n in the above extremes we obtain Theorem 3. □

Note that if we let $q \to 1$ in Corollary 2 it reduces directly to the first line of (1.11), and if we let $q \to 1$ in Theorem 2 it reduces to (2.2).

4. EXTENDING (2.3) IS HARDER

If we try to follow the proof of (2.3) directly, we find troublesome powers of q (which vanish when $q = 1$) making a direct generalization of (2.3) impossible. In order for us to overcome this difficulty we must, in fact, generalize $\mathcal{S}_n(q)$:

$$\mathcal{S}_n(t,q) = \begin{cases} 0 & \text{if } n=0 \\ 1 & \text{if } n=1 \\ \mathcal{S}_{n-1}(t,q)+tq^{n-2}\mathcal{S}_{n-2}(t,q) & \text{for } n>1. \end{cases} \tag{4.4.1}$$

Now for $m \geq 0$

$$
\begin{aligned}
(1 - x - tx^2q^m\eta)&\sum_{n=0}^{\infty} \mathcal{S}_{n+m}(t,q)x^n \\
&= \mathcal{S}_m(t,q) + x\mathcal{S}_{m+1}(t,q) + \sum_{n=2}^{\infty} \mathcal{S}_{n+m}(t,q)x^n \\
&\quad - x\mathcal{S}_m(t,q) - \sum_{n=1}^{\infty} \mathcal{S}_{n+m}(t,q)x^{n+1} \\
&\quad - \sum_{n=0}^{\infty} \mathcal{S}_{n+m}(t,q)tq^{m+n}x^{n+2} \qquad (4.4.2)\\
&= \mathcal{S}_m(t,q) + x(\mathcal{S}_{m+1}(t,q) - \mathcal{S}_m(t,q)) \\
&\quad + \sum_{n=2}^{\infty}(\mathcal{S}_{n+m}(t,q) - \mathcal{S}_{n+m-1}(t,q) - tq^{m+n-2}\mathcal{S}_{n+m-2}(t,q))x^n
\end{aligned}
$$

(where n has been replaced by $n-1$ in the second sum and $n-2$ in the third)

$$
= \begin{cases} x & \text{if } m = 0 \\ \mathcal{S}_m(t,q) + xtq^{m-1}\mathcal{S}_{m-1}(t,q) & \text{if } m > 0. \end{cases}
$$

Thus

$$
\sum_{n=0}^{\infty} \mathcal{S}_n(t,q)x^n = \frac{1}{1 - x - tx^2\eta}x, \qquad (4.4.3)
$$

and

$$
\sum_{n=0}^{\infty} \mathcal{S}_{n+1}(t/q,q)x^n = \frac{1}{1 - x - tx^2\eta}. \qquad (4.4.4)
$$

So by (4.2) with t replaced by t/q^m, and $m > 0$,

$$\sum_{n=0}^{\infty} \mathcal{S}_{n+m}(tq^{-m})x^n = \frac{1}{1-x-tx^2\eta}(\mathcal{S}_m(tq^{-m},q) + xtq^{-1}\mathcal{S}_{m-1}(tq^{-m},q))$$

$$= \mathcal{S}_m(tq^{-m},q)\sum_{n=0}^{\infty}\mathcal{S}_{n+1}(t/q,q)x^n \tag{4.4.5}$$

$$+ tq^{-1}\mathcal{S}_{m-1}(tq^{-m},q)\sum_{n=0}^{\infty}\mathcal{S}_n(t,q)x^n.$$

We now replace t by tq^m on both sides of (4.5) and compare coefficients of x^n to obtain:

Theorem 3: *For* $n \geq 0,\ m > 0$,

$$\begin{aligned}\mathcal{S}_{n+m}(t,q) &= \mathcal{S}_m(t,q)\mathcal{S}_{n+1}(tq^{m-1},q)\\ &+ tq^{m-1}\mathcal{S}_{m-1}(t,q)\mathcal{S}_n(tq^m,q).\end{aligned} \tag{4.4.6}$$

Consequently (4.6) is the desired generalization of (2.3). Note that the new variable t was an essential addition. If it had not been included, it would not have been possible to transform the factor

$$(1-x-tx^2q^m\eta)$$

into

$$(1-x-tx^2\eta),$$

which formed the essential step in obtaining (4.6).

Results essentially equivalent to Theorem 3 were first proved by MacMahon [15; pp. 43-46]. Related investigations were given by Andrews, Knopfmacher and Paule [5], Berkovich and Paule [7], and Garrett [9].

5. EXTENDING (2.4) IS BIZARRE

We have chosen to obtain a q-analog of (2.4) precisely because there is currently no q-analog of (2.4) in the literature. By using the operator generating function technique we will see precisely how complications arise as we uncover the q-analog.

To this end we require some definitions and another lemma. In the definitions we employ the idea of a partition and its conjugate. A partition is simply a representation of an integer as a sum of positive integers. For example, the partitions of 4 are 4, $3+1$, $2+2$, $2+1+1$ and $1+1+1+1$. Each partition has a Ferrers graph which is left justified array of rows of dots where in the jth row there are λ_j dots where λ_j is the jth part of the partition. Thus the Ferrers graph of $5+5+3+2+1+1$ is

The conjugate partition is obtained by reading columns instead of rows. So in this instance the conjugate to $5+5+3+2+2+1$ is $6+4+3+2+2$.

Definition 5.5.1: By $C_+(n)$ we denote the set of the conjugates of all the partitions with distinct parts each $\leq n$.

So $C_+(3) = \{3+2+1, 2+2+1, 2+1+1, 2+1, 1+1+1, 1+1, 1, \phi\}$ because these are the conjugates of $3+2+1, 3+2, 3+1, 2+1, 3, 2, 1, \phi$, the partitions with distinct parts each ≤ 3.

A moment's reflection reveals that $C_+(n)$ consists of all partitions with at most n parts in which each integer not exceeding the largest part appears at least once.

Definition 5.5.2: By $C_0(n)$ we denote the set of partitions in $C_+(n)$ to which have been added 0's so that there are exactly $n+1$ parts in each partition.

Thus $C_0(3) = \{3+2+1+0, 2+2+1+0, 2+1+1+0, 2+1+0+0, 1+1+1+0, 1+1+0+0, 1+0+0+0, 0+0+0+0\}$.

Lemma 2:

$$\eta^r(f(x)(A+\eta))^m g(x)$$

$$= \sum_{\substack{\pi \in C_0(m) \\ \pi=\{\pi_1,\pi_2,\cdots,\pi_{m+1}\} \\ \pi_0 \geq \pi_1 \geq \cdots \geq \pi_m = 0}} A^{m-\pi_0} g(xq^{\pi_0+r}) f(xq^{\pi_1+r}) \cdots f(xq^{\pi_m+r})$$

Proof: When $m = 0$, the Lemma asserts $g(xq^r) = q(xq^r)$. Now assume that the result is true for a particular m. Then

$$\begin{aligned}
&\eta^r(f(x)(A+\eta))^{m+1}g(x) \\
&= \eta^r f(x)(A+\eta)(f(x)(A+\eta))^m g(x) \\
&= f(xq^r)(A\eta^r + \eta^{r+1})(f(x)(A+\eta))^m g(x) \\
&= f(xq^r) \sum_{\substack{\pi \in C_0(m) \\ \pi=\{\pi_0,\pi_1,\cdots,\pi_m\} \\ \pi_0 \geq \cdots \geq \pi_m = 0}} A^{m+1-\pi_0} g(xq^{\pi_0+r}) f(xq^{\pi_1+r}) \cdots f(xq^{\pi_m+r}) \\
&+ f(xq^r) \sum_{\substack{\pi \in C_0(m) \\ \pi=\{\pi_0,\pi_1,\cdots,\pi_m\} \\ \pi_0 \geq \pi_1 \geq \cdots \geq \pi_m = 0}} A^{(m+1)-(\pi_0+1)} g(xq^{\pi_0+1+r}) f(xq^{\pi_1+1+r}) \cdots f(xq^{\pi_m+1+r}) \\
&= \sum_{\substack{\pi \in C_0(m+1) \\ \pi=\{\pi_0,\pi_1,\cdots,\pi_{m+1}\} \\ \pi_0 \geq \pi_1 \geq \cdots \geq \pi_{m+1} = 0}} A^{m+1-\pi_0} g(xq^{\pi_0+r}) f(xq^{\pi_1+r}) \cdots f(xq^{\pi_{m+1}+r}),
\end{aligned}$$

where the first sum in the penultimate line covers those elements of $C_0(m+1)$ where $\pi_m = 0$ and the second sum treats those elements where $\pi_m > 0$ in which case $\pi_m = 1$ (see comment following Definition 5.1).

The Lemma follows by mathematical induction. □

FIBONACCI NUMBERS AND THE ROGERS-RAMANUJAN IDENTITIES

Before we state Theorem 4, we require some further observations about $C_0(n)$. We note that $C_0(n)$ consists of sequences

$$\pi_0 \geq \pi_1 \geq \cdots \geq \pi_n = 0$$

where $\pi_{i-1} = \pi_i$ or $\pi_i + 1$. So if we write

$$\pi_{i-1} = \pi_i + b_i \quad (b_i = 0 \text{ or } 1, \text{ for } 1 \leq i \leq n),$$

then clearly

$$\pi_{i-1} = b_i + b_{i+1} + \cdots + b_n.$$

We are now prepared to establish a very unlikely q-analog of (2.4).

Theorem 4: *For $n > 0$*

$$\mathcal{S}_n(q) = \frac{n}{2^{n-1}} + \frac{1}{2^{n-1}} \sum_{\substack{1\leq R_1<\cdots<R_m\leq n-1 \\ R_{j+1}\geq R_j+2 \text{ for } 1\leq j\leq m-1}} R_1(R_2-R_1-1)(R_3-R_2-1)\cdots(R_m-R_{m-1}-1) \times \prod_{1\leq i\leq m}(1+4q^{R_i}).$$

Remark: *It is not difficult to see that as $q \to 1$ this result converges to (2.4). Clearly the inner product becomes 5^m, and for $m > 1$ it is an exercise in mathematical induction to show that for $m > 0$*

$$\sum_{\substack{1\leq R_1<\cdots<R_m\leq n-1 \\ R_{j+1}\geq R_j+2 \text{ for } 1\leq j\leq m-1}} R_1(R_2-R_1-1)(R_3-R_2-1)\cdots(R_m-R_{m-1}-1)(n-1-R_m) = \binom{n}{2m+1}.$$

Indeed, to prove this by mathematical induction on m, it is easiest to prove it simultaneously with

$$\sum_{\substack{1\leq R_1<\cdots<R_m\leq n-1 \\ R_{j+1}\geq R_j+2 \text{ for } 1\leq j\leq m-1}} R_1(R_2-R_1-1)(R_3-R_2-1)\cdots(R_m-R_{m-1}-1) = \binom{n}{2m}.$$

Proof: Returning to the proof of Theorem 1, we recall

$$(1-x-x^2\eta)\sigma(x) = x.$$

FIBONACCI NUMBERS AND THE ROGERS-RAMANUJAN IDENTITIES

Hence

$$\left(\left(1-\frac{x}{2}\right)^2 - x^2\left(\frac{1}{4}+\eta\right)\right)\sigma(x) = x$$

or

$$\left(1-\left(\frac{x}{1-\frac{x}{2}}\right)^2\left(\frac{1}{4}+\eta\right)\right)\sigma(x) = \frac{x}{(1-\frac{x}{2})^2}.$$

Consequently

$$\sum_{n\geq 0} \mathcal{S}_n(q)x^n = \frac{1}{\left(1-\left(\frac{x}{1-\frac{x}{2}}\right)^2\left(\frac{1}{4}+\eta\right)\right)}\frac{x}{(1-\frac{x}{2})^2},$$

which is precisely the q-analog of what was done in the proof of (2.4).

Hence

$$\sum_{n\geq 0} \mathcal{S}_n(q)x^n$$

$$= \sum_{m=0}^{\infty}\left(\frac{x^2}{(1-\frac{x}{2})^2}\left(\frac{1}{4}+\eta\right)\right)^m \frac{x}{(1-\frac{x}{2})^2}$$

$$= \frac{x}{(1-\frac{x}{2})^2} + \sum_{m=1}^{\infty} \sum_{\substack{\pi\in C_0(m) \\ \pi=\{\pi_1,\cdots,\pi_{m+1}\} \\ \pi_1\geq\cdots\geq\pi_{m+1}=0}} \left(\frac{1}{4}\right)^{m-\pi_1} \frac{xq^{\pi_1}}{(1-\frac{xq^{\pi_1}}{2})^2} \frac{x^{2m}q^{2(\pi_2+\cdots+\pi_{m+1})}}{(1-\frac{xq^{\pi_1}}{2})^2\cdots\left(1-\frac{xq^{\pi_{m+1}}}{2}\right)^2}$$

$$= \sum_{n=0}^{\infty}\frac{nx^n}{2^{n-1}} + \sum_{m=1}^{\infty} \sum_{\substack{\pi\in C_0(m) \\ \pi=\{\pi_1,\cdots,\pi_{m+1}\} \\ \pi_1\geq\cdots\geq\pi_{m+1}=0}} \left(\frac{1}{4}\right)^{m-\pi_1} x^{2m+1}q^{\pi_1+2(\pi_2+\cdots+\pi_{m+1})}$$

$$\times \sum_{r_1,r_2,\cdots,r_{m+1}\geq 0} (r_1+1)\cdots(r_{m+1}+1)\left(\frac{x}{2}\right)^{r_1+r_2+\cdots+r_{m+1}} q^{r_1\pi_1+\cdots+r_{m+1}\pi_{m+1}}.$$

Therefore

$$\mathcal{S}_n(q) = \frac{n}{2^{n-1}} + \sum_{m=1}^{\lfloor\frac{n-1}{2}\rfloor} \sum_{\substack{r_1,\cdots,r_{m+1}\geq 0\\ r_1+r_2+\cdots+r_{m+1}\leq n-2m-1}} 2^{1-n}(r_1+1)(r_2+1)\cdots(r_{m+1}+1)$$

$$\sum_{\substack{\pi\in C_0(m)\pi=\{\pi_1,\cdots,\pi_{m+1}\}\\ \pi_1\geq\cdots\geq\pi_{m+1}=0}} 4^{\pi_1} q^{(r_1+1)\pi_1+(r_2+2)\pi_2+\cdots+(r_{m+1}+2)\pi_{m+1}}$$

$$= \frac{n}{2^{n-1}} + \sum_{m=1}^{\lfloor\frac{n-1}{2}\rfloor} \sum_{\substack{r_1,\cdots,r_m\geq 0\\ r_1+\cdots+r_m\leq n-2m}} 2^{1-n}(r_1+1)(r_2+1)\cdots(r_m+1)$$

$$\times (n-2m-r_1-\cdots-r_m)$$

$$\times \sum_{\substack{\pi\in C_0(m)\\ \pi=\{\pi_1,\cdots,\pi_{m+1}\}\\ \pi_1\geq\cdots\geq\pi_{m+1}=0}} 4^{\pi_1} q^{(r_1+1)\pi_1+(r_2+2)\pi_2+(r_3+2)\pi_3+\cdots+(r_m+1)\pi_m}.$$

Now given the alternative way of describing the partitions in $C_0(n)$ presented just prior to the statement of Theorem 4, we see that the interior sum above may be rewritten as

$$\sum_{0\leq b_1,\cdots,b_m\leq 1} 4^{b_1+b_2+\cdots+b_m} q^{(r_1+1)b_1+(r_1+r_2+3)b_2+\cdots+(r_1+r_2+\cdots+r_m+2m-1)b_m}.$$

So we now change variables by

$$R_1 = r_1+1$$

$$R_{j+1} = R_j + r_{j+1} + 2$$

and the interior sums now become

$$\sum_{\substack{1\leq R_1<\cdots<R_m\leq n-1\\ R_{j+1}\geq R_j+2 \text{ for } 1\leq j\leq m-1}} R_1(R_2-R_1-1)(R_3-R_2-1)\cdots(R_m-R_{m-1}-1)(n-1-R_m),$$

and so we conclude

$$\sum_{n\geq 0}\mathcal{S}_n(q)x^n = \frac{n}{2^{n-1}}$$

$$+ \sum_{\substack{1\leq R_1<\cdots<R_m\leq n-1\\ R_{j+1}\geq R_j+2 \text{ for } 1\leq j\leq m-1}} R_1(R_2-R_1-1)\cdots(R_m-R_{m-1}-1)(n-1-R_m)$$

$$\times \prod_{i=1}^{m}(1+q^{4R_i}),$$

which is the desired result. □

6. CONCLUSION

I hope that Theorem 1 (the extension of (2.1)), Corollary 1 (the extension of the first line of (1.11)), Theorem 2 (the extension of (2.2)), Theorem 3 (the extension of (2.3)), and Theorem 4 (the very unlikely extension of (2.4)) have shown the richness and the power of using operator methods to study Schur's polynomials as extensions (or q-analogs) of Fibonacci numbers.

There are other identities which do not seem to lend themselves to our operator methods. For example, it is not too difficult to generalize

$$\sum_{j=0}^{n} \binom{n}{j} F_{n-j} = F_{2n} \tag{6.6.1}$$

to

$$\sum_{j=0}^{n} \begin{bmatrix} n \\ j \end{bmatrix} t^j q^{j(n-1)} \mathcal{S}_{n-j}(t,q) = \mathcal{S}_{2n}(t,q). \tag{6.6.2}$$

Indeed the combinatorics of (6.2) were discussed (at least when $t = 1$) in [3; p. 144]. However, it would be of interest to prove (6.2) using the operator methods presented here. The method of Church and Bicknell [8] (cf. [14; pp. 232-234]) would seem to suggest a natural method to emulate, but, so far, this path has failed for me.

In furtherance of such a project, I would note that one may prove in the same way Corollary 2 is proved that

$$\mathcal{S}_n(t,q) = \sum_{0 \le 2j \le n-1} q^{j^2} t^j \begin{bmatrix} n-1-j \\ j \end{bmatrix}. \tag{6.6.3}$$

¿From (6.3) one may prove (6.2) by comparing coefficients of t^j on both sides and invoking the q-analog of the Chu-Vandermonde summation [4; eq. (3.3.10), p. 37].

ACKNOWLEDGMENT

Finally I wish to express my thanks to Andrew Granville, who went to great pains to edit this paper carefully. His recommendations made the paper much more readable, and even I like it much more than the original version.

REFERENCES

[1] G.E. Andrews. "Some Formulae for the Fibonacci Sequence with Generalizations." *The Fibonacci Quarterly* **7** (1969): 113-130.

[2] G.E. Andrews. "A Polynomial Identity Which Implies the Rogers-Ramanujan Identities." *Scripta Math.* **28** (1970): 297-305.

[3] G.E. Andrews. "Combinatorial Analysis and Fibonacci Numbers." *The Fibonacci Quarterly* **12** (1974): 141-146.

[4] G.E. Andrews. *The Theory of Partitions, Encycl. Math. and Its Appl.* Volume 2. G.-C. Rota ed., Addison-Wesley, Reading, 1976 (Reissued: Cambridge University Press, Cambridge 1998).

[5] G.E. Andrews, A. Knopfmacher and P. Paule. "An Infinite Family of Engel Expansions of Rogers-Ramanujan Type." *Adv. in Appl. Math.* **25** (2000): 273-280.

[6] L. Berg. *Introduction to the Operational Calculus.* North Holland, Amsterdam, 1967.

[7] A. Berkovich and P. Paule. "Variants of the Andrews-Gordon Identities." To appear.

[8] C.A. Church and M. Bicknell. "Exponential Generating Functions for Fibonacci Identities." *The Fibonacci Quarterly* **11** (1973): 275-281.

[9] K.C. Garrett. "Lattice Paths and Generalized Rogers-Ramanujan Type Identities." Ph.D. thesis, University of Minnesota, 2001.

[10] H. Gupta. "The Andrews Formula for Fibonacci Numbers." *The Fibonacci Quarterly* **16** (1978): 552-555.

[11] R.T. Hansen. "Generating Identities for Fibonacci and Lucas Triples." *The Fibonacci Quarterly* **10** (1972): 571-578.

[12] G.H. Hardy and E.M. Wright. *An Introduction to the Theory of Numbers.* 5th Edition, Oxford University Press, Oxford, 1995.

[13] M.D. Hirschhorn. "The Andrews Formula for Fibonacci Numbers." *The Fibonacci Quarterly* **19** (1981): 1-2.

[14] T. Koshy. *Fibonacci and Lucas Numbers With Applications.* John Wiley, New York, 2001.

[15] P.A. MacMahon. *Combinatory Analysis.* Volume II. Cambridge University Press, Cambridge, 1918. (Reissued: Chelsea (AMS), New York, 1984.)

[16] I. Schur. "Ein Beitrag zur Additiven Zahlentheorie." Sitzungsber., Akad. Wissensch. Berlin, Phys.-Math. Klasse, (1917), pp. 302-321.

AMS Classification Numbers: 05A15, 05A30, 05A40

✠✠✠

5.3 SCRATCHPAD Explorations for Elliptic Theta Functions (with R.J. Baxter)

SCRATCHPAD Explorations for Elliptic Theta Functions

GEORGE E. ANDREWS [(1)] The Pennsylvania State University, University Park, Pennsylvania

R. J. BAXTER The Australian National University, Canberra, Australia

1. Introduction. In a recent series of papers [5], [6], [9] we have examined a two-particle generalization of hard hexagon model in statistical mechanics. Our object in this paper is to examine the known algorithms that failed us in our research, the lucky accidents that provided our insight for obtaining our solutions and the developments in exploratory algorithms we think might systematize problems like this.

The explicit problem from [5], [6] and [9] that we wish to consider is easily stated. Let $X_m(a,b,c;q)$ be a three parameter sequence of polynomials in q where $1 \leq a,b,c \leq 3$, $b + c \geq 4$ (a,b,c integers) with

---

[(1)] Partially supported by National Science Foundation Grant MCS8201733.

(1.1) $$X_0(a,b,c;q) = \begin{cases} 1 & \text{for } a = b \\ 0 & \text{otherwise} \end{cases}.$$

(1.2) $$X_m(a,b,c;q) = \sum_{h=4-b}^{3} q^{\frac{1}{2}|\nu(h)-\nu(c)|m} X_{m-1}(a,h,b;q)\ ,$$

where $\nu(1) = 0$, $\nu(2) = 2$ and $\nu(3) = 1$. What we want is $X_\infty(a,b,c;q)$ as a modular form of some sort. However initially we have little to go on except our faith arising from the history of exactly solved models that there is a positive answer to our quest.

In Section 2 we briefly consider classical algorithms that were useful in the past. Section 3 is devoted to the lucky accidents that led to the right answers. Section 4 provides a brief discussion of open problems concerning exploratory algorithms in computer algebra.

The symbolic algebra package used throughout this work was IBM's SCRATCHPAD.

2. Classical Algorithms for Solving the Hard Hexagon Model.

We refer the reader to [8; Ch. 14] for a complete treatment of the Hard Hexagon Model which was originally solved by one of us (RJB) in 1979 [7]. In this section we shall concentrate on that aspect of it concerning a limit of polynomial sequences. Instead of the sequence described by (1.1) and (1.2), we consider

(2.1) $$D_0(q) = D_1(q) = 1\ ,$$

(2.2) $$D_n(q) = D_{n-1}(q) + q^{n-1}D_{n-2}\ ,\ n \geq 2.$$

Again we would like to find $D_\infty(q)$ as essentially a modular form.

We note that the coefficients b_n in

$$D_\infty(q) = \sum_{n=0}^{\infty} b_n q^n . \tag{2.3}$$

are easily computed using (2.1) and (2.2) (indeed $D_\infty(q)$ and $D_n(q)$ are identical in coefficients of q^m for $m < n - 1$).

Once a large number of b_n's have been computed we may insert them into Euler's infinite product representation algorithm (EIPRA) [2; p. 98], [3; p. 104]:

E.I.P.R.A. If

$$f(q) \equiv 1 + \sum_{n=1}^{\infty} b_n q^n = \prod_{n=1}^{\infty} (1-q^n)^{-a_n} . \tag{2.4}$$

where the a_n and b_n are integers, then

$$nb_n = \sum_{j=1}^{n} D_j b_{n-j} . \tag{2.5}$$

where

$$D_j = \sum_{d|j} da_d . \tag{2.6}$$

Furthermore if either sequence a_n or b_n is given, the other is uniquely determined by (2.5) and (2.6).

Using EIPRA, we find that for as many a_n as we care to compute (RJB computed a few hundred in 1979) the sequence is periodic modulo 5 and the first five terms are $a_1 = 1$, $a_2 = a_3 = 0$, $a_4 = 1$, $a_5 = 0$. Thus EIPRA yields the following conjecture (which is in fact equivalent to the first Rogers-Ramanujan identity):

$$D_\infty(q) = \prod_{n=1}^{\infty} \frac{1}{(1-q^{5n-4})(1-q^{5n-1})} . \tag{2.7}$$

Next we ask: how can we prove (2.7)? Instead of following the original method, we try to approximate $D_n(q)$ by Gaussian polynomials. The Gaussian polynomials $\begin{bmatrix} n \\ m \end{bmatrix}$ are defined by

$$\begin{bmatrix} n \\ m \end{bmatrix} = \begin{bmatrix} n \\ m \end{bmatrix}_q = \begin{cases} \dfrac{(1-q^n)(1-q^{n-1}...(1-q^{n-m+1})}{(1-q^m)(1-q^{m-1})...(1-q)} , & n \geq 0, \ m \geq 0 \\ 0, \ n \geq 0, \ m < 0. \end{cases} \tag{2.8}$$

They are simply encoded in SCRATCHPAD directly from the definition (2.8).

So we encode the $D_n(q)$ from (2.2) and the $\begin{bmatrix} n \\ m \end{bmatrix}$ from (2.8). We first ask for several $D_n(q)$:

(2.9) $D_0(Q) : 1$

(2.10) $D_1(Q) : Q + 1$

(2.11) $D_2(Q) : Q^2 + Q + 1$

(2.12) $D_3(Q) : Q^4 + Q^3 + Q^2 + Q + 1$

(2.13) $D_4(Q) : Q^6 + Q^5 + 2Q^4 + Q^3 + Q^2 + Q + 1$

(2.14) $D_5(Q) : Q^9 + Q^8 + Q^7 + 2Q^6 + 2Q^5 + 2Q^4 + Q^3 + Q^2 + Q + 1$

(2.15) $D_6(Q)$

$$Q^{12} + Q^{11} + 2Q^{10} + 2Q^9 + 2Q^8 + 2Q^7\ 3Q^6 + 2Q^5 + 2Q^4 + Q^3$$
$$+$$
$$Q^2 + Q + 1$$

For purposes of illustration we restrict our attention to the $D_n(q)$ with odd subscripts; those with even subscripts are treated similarly.

We note that $D_{2n-1}(q)$ appears to be of degree n^2. Since $\begin{bmatrix} 2n \\ n \end{bmatrix}$ is also, we consider $D_{2n-1}(q) - \begin{bmatrix} 2n \\ n \end{bmatrix}$ and ask SCRATCHPAD for these polynomials with $n = 1,2,3,4$ and 5: The results are

(2.16) 0

(2.17) $(-Q^2)$

(2.18) $- Q^7 - Q^6 - Q^5 - Q^4 - 2Q^3 - Q^2$

(2.19)
$$- Q^{14} - Q^{13} - 2Q^{12} - 2Q^{11} - 4Q^{10} - 4Q^9 - 5Q^8 - 4Q^7 - 4Q^6 - 3Q^5$$
$$+$$
$$- 3Q^4 - 2Q^3 - Q^2$$

(2.20)
$$- Q^{23} - Q^{22} - 2Q^{21} - 3Q^{20} - 5Q^{19} - 6Q^{18} - 9Q^{17} - 10Q^{18} - 12Q^{15}$$
$$+$$
$$- 13Q^{14} - 14Q^{13} - 14Q^{12} - 14Q^{11} - 13Q^{10} - 11Q^9 - 10Q^8 - 8Q^7$$
$$+$$
$$- 6Q^6 - 5Q^5 - 3Q^4 - 2Q^3 - Q^2$$

For each $n \geq 2$ it appears that the resulting polynomial is q^2 times polynomials of degree $n^2 - 4 = (n-2)(n+2)$. This suggests that for our next approximation to $D_{2n-1}(q)$ we use $\begin{bmatrix} 2n \\ n \end{bmatrix} - q^2\begin{bmatrix} 2n \\ n+2 \end{bmatrix}$.

We now ask SCRATCHPAD to give us $D_{2n-1}(q) - \begin{bmatrix} 2n \\ n \end{bmatrix} + q^2\begin{bmatrix} 2n \\ n+2 \end{bmatrix}$ and the results for $n = 1,2,3,4$ and 5 are:

(2.21) $\quad 0$

(2.22) $\quad 0$

(2.23) $\quad (-Q^3)$

(2.24) $\quad - Q^{10} - Q^9 - Q^8 - Q^7 - Q^6 - Q^5 - Q^4 - Q^3$

(2.25) $\quad - Q^{19} - Q^{18} - 2Q^{17} - 2Q^{16} - 3Q^{15} - 3Q^{14} - 4Q^{13} - 4Q^{12} - 4Q^{11}$

$+$

$- 4Q^{10} - 3Q^9 - 3Q^8 - 3Q^7 - 2Q^6 - 2Q^5 - Q^4 - Q^3$

In exactly the same manner as above, we see that

$$\begin{bmatrix} 2n \\ n \end{bmatrix} - q^2\begin{bmatrix} 2n \\ n+2 \end{bmatrix} - q^3\begin{bmatrix} 2n \\ n+3 \end{bmatrix}$$

is a still better approximation to $D_{2n-1}(q)$.

It should be clear by now that with a symbolic algebra package like SCRATCHPAD this process can be carried on indefinitely, and the inevitable conclusion is the following conjecture (found and proved by Schur [11]):

$$(2.26)\quad D_{2n-1}(q) = \sum_{\lambda=-\infty}^{\infty} q^{\lambda(10\lambda+1)}\begin{bmatrix}2n\\n-5\lambda\end{bmatrix} - \sum_{\lambda=-\infty}^{\infty} q^{(2\lambda-1)(5\lambda-2)}\begin{bmatrix}2n\\n-5\lambda+2\end{bmatrix} .$$

Once you have conjectured (2.26) and the counterpart for $D_{2n}(q)$ the proof is quite straightforward (see e.g. [1]).

Indeed the general result proved by Schur [11] for all n is

$$(2.27)\quad D_n(q) = \sum_{\lambda=-\infty}^{\infty} (-1)^{\lambda} q^{\lambda(5\lambda+1)/2}\begin{bmatrix} n+1 \\ \left[\frac{n+1-5\lambda}{2}\right]\end{bmatrix} ,$$

where $\begin{bmatrix}a\\b\end{bmatrix}$ is the Gaussian polynomial and $[a]$ is the greatest integer in a.

The above procedure of approximating polynomials by Gaussian polynomials of the same degree and then iterating we shall call Schur's Gaussian Polynomial Approximation Algorithm (SGPAA).

3. The limiting behavior of $X_m(3,3,3;q)$.

We return now to the problem posed in the introduction. We want $X_\infty(a,b,c;q)$ expressed as a modular form, presumably some nice linear combination of infinite products and theta functions. In the following we restrict our consideration to $X_\infty(3,3,3;q)$.

We begin following the lead of Section 2. Consider

$$(3.1)\quad X_\infty(3,3,3;q) = \prod_{n=1}^{\infty} (1-q^n)^{-a_n} .$$

The first thirty-six a_n are as follows:

Table 1.

n	a_n	n	a_n	n	a_n
1	2	13	2	25	2
2	1	14	1	26	4
3	1	15	0	27	-3
4	1	16	1	28	1
5	2	17	3	29	7
6	0	18	0	30	-5
7	2	19	2	31	2
8	1	20	2	32	9
9	1	21	-1	33	-8
10	1	22	1	34	0
11	2	23	4	35	13
12	0	24	-2	36	-12

Clearly Table 1 suggests that EIPRA unaided will not tell us the nice answer we want. Let us try an alternative; this new approach was successful in treating the more complicated cases of Regime II in the original hard hexagon model [8; p. 437, eq. (14.5.22a), (14.5.22b)]. Namely if we multiply both sides of (2.7) by $\prod_{n\geq 1}(1-q^n)$ and apply Jacobi's Triple Product [2; p. 22, Cor. 2.9], we obtain

(3.2) $$\prod_{n\geq 1}(1-q^n) \cdot D_\infty(q) = \sum_{\lambda=-\infty}^{\infty}(-1)^\lambda q^{\lambda(5\lambda+1)/2}$$

$$= 1 - q^2 - q^3 + q^9 + q^{11} - q^{21} - q^{24} + \cdots\cdots$$

Once one contemplates (3.2) it is easy to deduce the rule of formation of the exponents in the series. So we consider

(3.3) $$\prod_{n\geq 1}(1-q^n)X_\infty(3,3,3;q) = \sum_{n\geq 0} c_n q^n .$$

The first twenty c_n are as follows:

n	c_n	n	c_n
0	1	10	3
1	1	11	3
2	1	12	3
3	1	13	3
4	1	14	4
5	2	15	4
6	1	16	4
7	2	17	5
8	2	18	6
9	2	19	6

Again we face unrevealing data. Undaunted we return to Table 1. While we see eventual trouble, we nonetheless note that the a_n looked initially like they were going to be periodic with period 6. Indeed for fourteen terms of the power series we have the following approximation:

(3.4) $\quad X_\infty(3,3,3;q)$

$$\approx \prod_{n=1}^{\infty} \frac{(1-q^{6n})}{(1-q^n)(1-q^{6n-1})(1-q^{6n-5})}$$

$$= \prod_{n=1}^{\infty} \frac{(1-q^{6n-3})(1-q^{6n})}{(1-q^n)(1-q^{2n-1})}$$

$$= \prod_{n=1}^{\infty} \frac{(1+q^n)(1-q^{3n})}{(1-q^n)}$$

$$= \frac{\prod_{n=1}^{\infty}(1-q^{3n})}{1-2q+2q^4-2q^9+2q^{16}-2q^{25}+\cdots\cdot} \ .$$

$$= \frac{1-q^3-q^6+q^{15}+q^{21}-q^{36}-q^{45}+\cdots\cdot}{1-2q+2q^4-2q^9+2q^{16}-2q^{25}+\cdots\cdot} \ .$$

Thus we see that this last expression would become a power series with very few nonzero terms if we were to multiply by $(1 + 2\sum_{n\geq 1}(-1)^n q^{n^2})$. Hence, in desperation we consider

(3.5) $$(1 + 2\sum_{n\geq 1}(-1)^n q^{n^2})\ X_\infty(3,3,3;q)\ .$$

$$= 1 - q^3 - q^6 + q^{17} + q^{18}\cdots\cdot$$

$$= \sum_{\lambda=-\infty}^{\infty} q^{\lambda(35\lambda+1)/2} - q^3\sum_{\lambda=-\infty}^{\infty} q^{\lambda(35\lambda-29\lambda)/2}\ .$$

valid at least for the first 200 terms!!!

The device of multiplication by $(1 + 2 \sum_{n \geq 1} (-1)^n q^{n^2})$ works well for $X_m(a,b,c;q)$ provided $b = c$; however when $b \neq c$ things remain mysterious. Hence we hope that Schur's Gaussian Polynomial Approximation Algorithm (SGPAA) may be appliable. The following table gives the first few $X_m(3,3,3;q)$:

Table 3

n	$X_n(3,3,3;q)$
0	1
1	1
2	$1 + 2q$
3	$1 + 2q + 2q^2 + q^3$
4	$1 + 2q + 4q^2 + 3q^3 + 3q^4 + q^5$
5	$1 + 2q + 4q^2 + 5q^3 + 6q^4 + 6q^5 + 4q^6 + 2q^7 + q^8$

In analogy with the work in Section 2, we want families of polynomials $p_i(m,\lambda,q)$ $(i=1$ or $2)$ such that

$$X_m(3,3,3;q) = \sum_{\lambda=-\infty}^{\infty} q^{\lambda(35\lambda+1)/2} p_1(m,\lambda,q) \tag{3.6}$$

$$- \sum_{\lambda=-\infty}^{\infty} q^{3+\lambda(35\lambda-29)/2} p_2(m,\lambda,q) ,$$

where

(3.7)
$$\lim_{m\to\infty} p_1(m,\lambda,q) = \frac{1}{1-2q+2q^4-2q^9+2q^{16}\ldots} .$$

When we try the SGPAA directly (i.e. taking the $p_1(m,\lambda,q)$ to be Gaussian polynomials) nothing sensible emerges (as we well expect).

Again we have to resort to ad hoc methods. Examining the very useful representation of $D_n(q)$ given by (2.27), we might hope that

(3.8)
$$g_n(q) = \sum_{\lambda=-\infty}^{\infty} (-1)^{\lambda} q^{\lambda(7\lambda+1)/2} \begin{bmatrix} N \\ \left[\frac{N-7\lambda}{2}\right] \end{bmatrix}$$

would somehow apply to our work here. A close study of $g_N(q)$ yields the following:

(3.9)
$$q^{N^2} g_{2N}(q^{-1}) = \sum_{\lambda=-\infty}^{\infty} q^{35\lambda^2+\lambda} \begin{bmatrix} 2N \\ N-7\lambda \end{bmatrix}$$
$$- \sum_{\lambda=-\infty}^{\infty} q^{35\lambda^2-29\lambda+6} \begin{bmatrix} 2N \\ N-7\lambda+3 \end{bmatrix} .$$

Let us, therefore, define

(3.10)
$$G(q) = \lim_{N\to\infty} q^{N^2} g_{2N}(q^{-1})$$
$$= \frac{1}{\prod_{n=1}^{\infty}(1-q^n)} \left\{ \sum_{\lambda=-\infty}^{\infty} q^{35\lambda^2+\lambda} - \sum_{\lambda=-\infty}^{\infty} q^{35\lambda^2-29\lambda+6} \right\} .$$

Thus our conjectured formula (3.5) may be rewritten as

$$(3.11)\qquad X_\infty(3,3,3;q^2) = \left\{\prod_{n=1}^{\infty}\frac{1}{1+q^{2n-1}}\right\}G(q)$$

Now both $G(q)$ and $X_\infty(3,3,3;q^2)$ are limits of polynomial sequences. Hence we wish to formulate a polynomial version of (3.11) that converges to (3.11) in the limit. A few hours of dodging and weaving with SCRATCHPAD led to the following conjecture which was easily checked for $m \leq 7$:

$$(3.12)\qquad X_m(3,3,3;,q^2) = \sum_{j=0}^{m}(-q)^j\begin{bmatrix}m\\j\end{bmatrix}_{q^2} q^{(m-j)^2}g_{2(m-j)}(q^{-1}).$$

Now (3.12) is certainly a major step forward. If we can prove it we have (3.11) immediately by [2; p. 153, limit equation]. However (3.12) is much different from the representation (3.6) we were hoping for. Thus we would like to rewrite (3.12) so that it resembles (3.6). This is easily accomplished by using (3.8) for the $q^{(m-j)^2}g_{2(m-j)}(q^{-1})$. Consequently (3.12) may be rewritten as

$$(3.13)\qquad X_m(3,3,3,q^2)$$

$$= \sum_{\lambda=-\infty}^{\infty} q^{35\lambda^2+\lambda}\left[\sum_{i=0}^{m}(-q)^i\begin{bmatrix}m\\i\end{bmatrix}_{q^2}\begin{bmatrix}2m-2i\\m-i-7\lambda\end{bmatrix}\right]$$

$$- q^6\sum_{\lambda=-\infty}^{\infty} q^{35\lambda^2-29\lambda}\left[\sum_{i=0}^{m}(-q)^i\begin{bmatrix}m\\i\end{bmatrix}_{q^2}\begin{bmatrix}2m-2i\\m-i-7\lambda+3\end{bmatrix}\right] .$$

Let us consider the first intereior sum in the case $\lambda = 0$, $q = 1$; it is thus just a sequence of integers say T_m , whose values are:

m	T_m	m	T_m
0	1	4	19
1	1	5	51
2	3	6	141
3	7	7	393

A check of Neal Sloane's book on integer sequences [12] reveals that T_m is the sequence 1070, namely the central term in the expansion of $(1 + x + x^2)^m$.

This last observation is the crucial one in unravelling the behavior of $X_\infty(a,b,c;q)$ in general. The polynomials we have been searching for are q-analogs of trinomial coefficients (i.e. the coefficients in $(1 + x + x^2)^n$). The literature on the trinomial coefficients themselves is sparse to say the least [10; p. 78, p. 163]. q-Trinomial coefficients have never arisen before to our knowledge. Indeed all the polynomials $X_m(a,b,c;q^2)$ have expansions like (3.6) wherein the $p_i(m,\lambda,q)$ are replaced by various q-analogs of trinomial coefficients.

In [6] we have developed an extensive account of the q-trinomial coefficients. It turns out that they have numerous surprising and subtle properties. For example, let us examine the polynomial from the first sum in (3.13) with $\lambda = 0$:

$$T_m(q) = \sum_{i=0}^{m} (-q)^i \begin{bmatrix} m \\ i \end{bmatrix}_{q^2} \begin{bmatrix} 2m-2i \\ m-i \end{bmatrix} . \tag{3.14}$$

we obtain easily using SCRATCHPAD that

m	T_m
0	1
1	1
2	$1 + 2q^2$
3	$1 + 2q^2 + 2q^4 + 2q^6$
4	$1 + 2q^2 + 4q^4 + 4q^6 + 4q^8 + 2q^{10} + 2q^{12}$
5	$1 + 2q^2 + 4q^4 + 6q^6 + 8q^8 + 8q^{10} + 8q^{12} + 6q^{14} + 4q^{16} + 2q^{18} + 2q^{20}$

Clearly $T_m(q)$ is an even function of q (i.e. a polynomial in q^2); however, this is not at all obvious in the definition (3.14). The evenness of $T_m(-q)$ turns out to be important, and we had a hard time proving it. We refer the reader to [6; Section 2.4, eq. (2.4.2)] for details.

4. Open Problems. In this paper we have examined the explorations via SCRATCHPAD that led to our complete solution of a generalization of the hard hexagon model. In Section 2 we examined two very useful algorithms: EIPRA and SGPAA. If instead of EIPRA, we had had an algorithm to search for

$$f(q) = 1 + \sum_{n=1}^{\infty} b_n q^n \tag{4.1}$$

$$= \prod_{n=1}^{\infty} (1-q^n)^{-a_n} \pm q^d \prod_{n=1}^{\infty} (1-q^n)^{-c_n} ,$$

then we could have guessed (3.5) without difficulty. To determine useful a_n, d, and c_n in (4.1) given b_n is a tall order since there is no longer uniqueness of the representation. If some sensible algorithm can be developed for (4.1), we can obviously ask whether such a thing can be done for larger linear combinations of infinite products. The number-theoretic implications of this problem are discussed in [4].

In Section 3, our discovery of q-trinomial coefficients suggests clearly that SGPAA can be extended to approximations using q-trinomial coefficients. Indeed we are now reasonably certain of what the useful q-analogs of the coefficients in $(1 + x + x^2 + \ldots + x^k)^n$ are. Whether these will prove useful in approximation algorithms like SGPAA remains to be seen.

References

1. G. E. Andrews, A polynomial identity which implies the Rogers-Ramanujan identities, Scripta Math., 28 (1970), 297-305.

2. G. E. Andrews, The Theory of Partitions, Encyclopedia of Mathematics and Its Applications, Vol. 2., Addison-Wesley, Reading, 1976. (Reissued: Cambridge University Press, London and New York, 1984).

3. G. E. Andrews, q-Series: Their Development and Application in Analysis, Number Theory, Combinatorics, Physics, and Computer Algebra, CBMS Regional Conference Series, No. 66, American Math. Soc., Providence, 1986.

4. G. E. Andrews, Further problems on partitions, Amer. Math. Monthly, (to appear).

5. G. E. Andrews and R. J. Baxter, Lattice gas generalization of the hard hexagon model: II. The local densities as elliptic functions, J. Stat. Phys., 44 (1986), 713-728.

6. G. E. Andrews and R. J. Baxter, Lattice gas generalization of the hard hexagon model: III. q-trinomial coefficients (to appear).

7. R. J. Baxter, Hard hexagons: exact solution, J. Phys., A13 (1980), L61-L70.

8. R. J. Baxter, Exactly Solved Models in Statistical Mechanics, Academic Press, London and New York, 1982.

9. R. J. Baxter and G. E. Andrews, Lattice gas generalization of the hard hexagon model: I. star-triangle relation and local densities, J. Stat. Phys., 44 (1986), 249-271.

10. L. Comtet, Advaned Combinatorics, Reidel, Dordrecht and Boston, 1974.

11. I. Schur, Ein Beitrag zur additiven Zahlentheorie und zur Theorie der Kettenbrüche, S.-B. Preuss. Akad. Wiss. Phys.-Math. Kl., 1917, pp. 302-321. (Reprinted in I. Schur: Gesammelte Abhandlungen, Vol. 2., pp. 117-136, Springer, Berlin, 1973).

12. N. J. A. Sloane, A Handbook of Integer Sequences, Academic Press, New York, 1973.

5.3.1 *Erratum*

At the bottom of page 18 of the preceding article, the Schur polynomials $D_n(q)$ are defined with initial conditions:

$$D_0(q) = D_1(q) = 1, \tag{2.1}$$

but on page 20, the indices are inadvertently shifted down by one:

$$D_1(q) = q + 1, \qquad D_2(q) = q^2 + q + 1,$$

etc.

Chapter 6

Number Theory

This chapter was hard to put together. On the one hand, while considering myself a "combinatory analyst," I started out as a number theorist. After all, Rademacher was a great number theorist, and I was his last student. However, most of what I do has a combinatorial, analytic flavor.

This section is really a miscellany of results that are in some way related to classical questions in number theory.

I was drawn to Penn State by the presence of Nathan Fine and Sarvadaman Chowla. Chowla was the grand old man of number theory at Penn State. The only course he taught was known locally as "Chowla's Seminar." This course was devoted to Chowla's interest of the moment, as is clear from the fragmentary notes I took in 1964. One day, Chowla mentioned Bombieri's paper [Bom63] developing an analog of the Selberg elementary proof of the Prime Number Theorem for algebraic curves. I had been participating in a related seminar on algebraic function fields, and this set me on the path to [30] (included here).

During my sabbatical year at MIT (1970–1971), Gian-Carlo Rota asked me to edit the *Collected Papers* of MacMahon [MacM78]. I took the task very seriously: writing to everyone who, according the Science Citation Index, had ever cited MacMahon. When I was researching [68] I thought I could learn something by consulting Neal Sloane's *Handbook of Integer Sequences* [Slo73]. I was greatly excited to find the sequence listed, but disappointment set in when I found Sloane citing MacMahon's prime numbers of measurement. The appearance of [68] generated a long letter from Paul Erdős to me, and drew my attention to the related work of Porubský [Por77].

Paper [127] is included here only because Gauss's "Eureka" theorem is such a classic of number theory. The methods all belong in Chapter 1.1.1, and it is merely a simplification of the proof in [116]. Paper [194] arose when Ranjan Roy and I were preparing Chapter 11 of [Book5] *Special Functions*: a book jointly authored with Askey and Roy. Congruence properties of partitions have also figured in [41], [42], [109], and [263]. Paper [196] had its origins in discussions with Lou Solomon about congruences for compositions, a project that may be completed someday.

My 1997–1998 sabbatical was a trip around the world. One stop was the University of Witwatersrand in Johannesburg, where Arnold Knopfmacher suggested the topic for [218]. The next stop was RISC in Linz, where Burkhardt Zimmermann supplied a number of additional theorems.

There are three other papers that fit in number theory. Paper [25] is an elementary application of partition ideas to basis representation, and its method is included in Chapter 1 of [Book1]. Paper [86] was formulated while listening to a lecture by Jim Jordan [Jor-WW79]; proving that, asymptotically, there are $n^2/48$ integer sided triangles of perimeter n as $n \to \infty$. Paper [153], joint with Kitaoka, applies the methods of q-series to algebraic number theory proving a conjecture in [Kit88].

6.1 A Note on the Bombieri–Selberg Formula for Algebraic Curves

PORTUGALIAE MATHEMATICA
Vol. 27 — Fasc. 2 — 1968

A NOTE ON THE BOMBIERI-SELBERG FORMULA FOR ALGEBRAIC CURVES *

BY GEORGE E. ANDREWS (1)
Pennsylvania State University, University Park, Pa. — U. S. A.

1. Introduction. Let Γ be an absolutely irreducible and nonsingular curve of genus $g > 0$ over the Galois field GF (q). Let N_m denote the number of points of Γ in GF (q^m) (including points at infinity). A. Weil [5] has proved

$$|N_m - (q^m + 1)| \leqq 2 g q^{\frac{1}{2} m}. \tag{1.1}$$

From this inequality one can deduce the Riemann hypothesis for the ζ-function related to Γ [4; p. 432].

E. Bombieri [1] has proved by purely elementary means that

$$m N_m + \sum_{i=0}^{m} N_i N_{m-i} = 2 m q^m + 0 (q^m), \tag{1.2}$$

which is an analog of the formula used by A. Selberg to prove the «Prime Number Theorem».

From (1. 2), Bombieri [1; eq. (26)] then deduced

$$\sum_{i=1}^{m} q^i N_{m-i} = m q^m + 0 (q^m) \tag{1.3}$$

by means of a Tauberian theorem of P. Erdos [3]. Utilizing (1. 2) and (1. 3), Bombieri then deduced

$$N_m \sim q^m, \quad \text{as} \quad m \to \infty. \tag{1.4}$$

* Received June, 1967.
(1) Partially supported by National Science Foundation grant GP-5593.

In this paper we shall show that a relatively simple derivation of (1. 2) may be given by means of generating functions and partial fraction decompositions. While such techniques are not strictly elementary, they are equivalent to elementary techniques. We also show that (1. 3) can be proved in the same manner without the use of a TAUBERIAN theorem. As a direct consequence of (1. 3), we deduce

$$\sum_{i=1}^{m} F_i q^{-i} = \log m + 0(1), \tag{1.5}$$

where F_i is the number of prime divisors of Γ of degree i.

2. Proof of (1. 2). We shall need some results from the elementary theory of algebraic functions. We shall use the symbol $\mathfrak{a}$ to denote integral divisors of Γ, and the symbol $\mathfrak{p}$ to denote prime divisors of Γ. First we have

$$\zeta(s,\Gamma) = \sum_{\mathfrak{a}} \frac{1}{(N\,\mathfrak{a})^s}, \quad \operatorname{Re} s > 1, \quad [4;\ \text{p. } 424],$$

where $\mathfrak{a}$ runs over all integral divisors of Γ.

$$(q-1)\zeta(s,\Gamma) = F(u) + R(u) \tag{2.1}$$

where $u = q^{-s}$, $F(u)$ is a polynomial of degree $\leq 2g-2$, and,

$$R(u) = \frac{h\,q^{1-g}(u\,q)^{2g-1}}{1-uq} - \frac{h}{1-u}, \tag{2.2}$$

with h being the class number of the related algebraic function field.

If we let $Z(u) = \zeta(s,\Gamma)$, then for $|u| < q^{-1}$

$$Z(u) = \sum_{m=0}^{\infty} E_m u^m, \tag{2.3}$$

$$E_m = \sum_{d(\mathfrak{a})=m} 1, \tag{2.4}$$

and for $m > 2g-2$

$$\mathrm{E}_m = h\,\frac{q^{m-g+1}-1}{q-1} \tag{2.5}$$ [2; p. 302].

Also for $|u| < q^{-1}$,

$$\frac{1}{Z(u)} = \sum_{m=0}^{\infty} \mu_m u^m \tag{2.6}$$

where

$$\mu_m = \sum_{d(\mathfrak{a})=m} \mu(\mathfrak{a}), \tag{2.7}$$

$$\mu(\mathfrak{a}) = \begin{cases} 0 & \text{if } \mathfrak{p}^2 \mid \mathfrak{a} \\ 1 & \text{if } \mathfrak{a} = \text{unit divisor} \\ (-1)^k & \text{if } \mathfrak{a} = \mathfrak{p}_1 \cdots \mathfrak{p}_k, \text{ all } \mathfrak{p}_i \text{ district.} \end{cases} \tag{2.8}$$

Since $|\mu(\mathfrak{a})| \leqq 1$,

$$|\mu_m| \leqq \mathrm{E}_m. \tag{2.9}$$

Next if

$$f(u) = \sum_{m=1}^{\infty} \mathrm{N}_m u^m, \ |u| < q^{-1}, \tag{2.10}$$

then

$$f(u) = \frac{u Z'(u)}{Z(u)}, \tag{2.11}$$ [4; p. 433].

It is important to note that $Z(u)$ is a rational function of u which has simple poles at $u = 1$, $1/q$ and is regular elsewhere. Finally, when the «0» notation is used it will be with respect to $m \to \infty$; the constant involved may depend on q.

Lemma. *If*

$$q > 1, \varphi(u) = \sum_{m=0}^{\infty} C_m u^m, \quad \Psi(u) = \sum_{m=0}^{\infty} d_m u^m,$$

$$\varphi(u)\Psi(u) = \sum_{m=0}^{\infty} e_m u^m, \quad C_m = 0(q^m), \quad d_m = 0(m^j),$$

j *a fixed integer, then* $e_m = 0(q^m)$.

PROOF.

$$|e_m| = \left| \sum_{s=0}^{m} C_{m-s} d_s \right| \leq K \sum_{s=0}^{m} s^j q^{m-s} \leq K q^m \sum_{s=0}^{m} s^j q^{-s}$$

$$\leq K q^m \sum_{s=0}^{\infty} s^j q^{-s} = K' q^m .$$

Proof of (1.2). First we define

$$Z^*(u) = \sum_{m=0}^{\infty} m^2 E_m u^m .$$

Then

$$Z^*(u) = \sum_{m=0}^{\infty} (m(m-1) + m) E_m u^m = u^2 Z''(u) + u Z'(u) . \tag{2.12}$$

On the other hand

$$Z^*(u) = u \frac{d}{du} (u Z'(u)) = u \frac{d}{du} (Z(u) f(u)) = u Z'(u) f(u) + u Z(u) f'(u) . \tag{2.13}$$

From (2.12) we note that $Z^*(u)$ has a pole of third order at $u = 1, q^{-1}$. Consequently there exists C_1 such that $Z^*(u) - C_1 Z(u) Z'(u)$ has a pole of second order at $u = 1/q$. There exists C_2 such that

$$Z^*(u) - C_1 Z(u) Z'(u) - C_2 Z^2(u)$$

has a pole of first order at $u = q^{-1}$. There exists C_3 such that

$$Z^*(u) - C_1 Z(u) Z'(u) - C_2 Z^2(u) - C_3 Z(u)$$

has no pole at $u = q^{-1}$ and therefore has as its only possible pole $u = 1$ with order ≤ 3. Thus there exist C_4, C_5, C_6, and a polynomial $b(u)$ such that

$$(2.14)\qquad Z^*(u) = C_1 Z(u) Z'(u) + C_2 Z^2(u) + C_3 Z(u) + C_4(1-u)^{-3} + C_5(1-u)^{-2} + C_6(1-u)^{-1} + b(u).$$

To evaluate C_1, we multiply (2.14) by $(1-uq)^3$ and let $u \to q^{-1}$. Utilizing (2.2) we obtain

$$C_1 = 2(q-1) q^g h^{-1}.$$

If we divide (2.14) by $Z(u)$, then by (2.13)

$$u f'(u) + f^2(u) = C_1 Z'(u) + C_2 Z(u) + C_3 + \frac{C_4}{(1-u)^3 Z(u)} + \frac{C_5}{(1-u)^2 Z(u)} + \frac{C_6}{(1-u)^2 Z(u)} + \frac{b(u)}{Z(u)}.$$

If we compare coefficients of u^m on both sides of (2.15) we find

$$(2.16)\qquad m N_m + \sum_{i=1}^{m} N_i N_{m-i} = C_1 m E_{m-1} + 0(q^m)$$
$$= 2(q-1) h^{-1} q^g m h q^{m-g} (q-1)^{-1} + 0(q^m)$$
$$= 2 m q^m + 0(q^m).$$

In our estimates we have utilized (2.5), (2.9) and the Lemma.

3. Proof of (1.3). Just as with $Z^*(u)$, we make a similar decomposition of $u(1-uq)^{-1} Z'(u)$ and obtain

$$(3.1)\qquad u(1-uq)^{-1} Z'(u) = d_1 Z(u) Z'(u) + d_2 Z^2(u) + d_3 Z(u) + \frac{d_4}{(1-u)^2} + \frac{d_5}{(1-u)^2} + \frac{d_6}{(1-u)} + C(u).$$

Multiplying both sides by $(1-uq)^3$ and letting $u \to 1/q$, we find

$$d_1 = (q-1) h^{-1} q^g.$$

If we divide (2.14) by $Z(u)$, then

$$(3.2)\qquad (1-uq)^{-1}f(u)=d_1 Z'(u)+d_2 Z(u)+d_3+\frac{d_4}{(1-u)^2 Z(u)}$$
$$+\frac{d_5}{(1-u)^2 Z(u)}+\frac{d_6}{(1-u)Z(u)}+\frac{C(u)}{Z(u)}.$$

Comparing coefficients of u^m on both sides of (2.15) we find

$$(3.3)\qquad \sum_{i=0}^{m} q^i N_{m-i}=d_1 m E_{m-1}+0(q^m)$$
$$=(q-1)h^{-1}q^g m h q^{m-g}(q-1)^{-1}+0(q^m)$$
$$=m q^m+0(q^m).$$

Corollary. $N_m=0(q^m)$.

Proof.
$$N_m q^{-m}=\sum_{i=0}^{m} N_i q^{-i}-\sum_{i=0}^{m-1} N_i q^{-i}$$
$$=m+0(1)-(m-1)+0(1)$$
$$=0(1).$$

4. Proof of (1.5). We now note that since

$$N_m=\sum_{d(\mathfrak{p})\mid m} d(\mathfrak{p}) \qquad [4;\ p.\ 432],$$

we may rewrite this formula as

$$N_m=\sum_{d\mid m} d F_d .$$

Hence by the ordinary Mobius inversion formula,

$$m F_m=\sum_{d\mid m}\mu\left(\frac{m}{d}\right)N_d .$$

Thus by the Corollary of (1.3),

$$m F_m=N_m+0(m q^{m/2}).$$

Therefore

$$\sum_{j=1}^{m}\frac{F_j}{q^j}=\sum_{j=1}^{m}(j^{-1}N_j q^{-j}+0(q^{-j/2}))=\sum_{j=1}^{m}j^{-1}N_j q^{-j}+0(1)$$

$$=\sum_{j=1}^{m-1}\left(\sum_{k=1}^{j}q^{-k}N_k\right)(j^{-1}-(j+1)^{-1})+m^{-1}\sum_{k=1}^{m}q^{-k}N_k+0(1)$$

$$=\sum_{j=1}^{m-1}(j+0(1))j^{-1}(j+1)^{-1}+m^{-1}(m+0(1))+0(1)$$

$$=\sum_{j=1}^{m-1}(j+1)^{-1}+0(1)=\log m+0(1).$$

REFERENCES

[1] E. Bombieri, *Sull' analogo della formula di Selberg nei corpi di funzioni,* Atti Accad. Naz. Lincei Rend. Cl. Sci. Fis. Mat. Natur, (8) **35** (1963), 252-257.

[2] M. Eichler, *Introduction to the Theory of Algebraic Numbers and Functions,* Academic Press, New York, 1966.

[3] P. Erdös, *On a tauberian theorem connected with the new proof of the prime number theorem,* J. Indian Math. Soc., **13** (1949), 133-147.

[4] I. S. Gal, *Lectures on Number Theory,* Jones Letter Service, Minneapolis, 1961.

[5] A. Weil, *Sur les Courbes Algébrique et les variétés qui s'en déduisent,* Actualités Sci. Ind. No. 1041 (1948).

6.2 MacMahon's Prime Numbers of Measurement

Reprinted from the AMERICAN MATHEMATICAL MONTHLY
Vol. 82, No. 9, November 1975
pp. 922–923

MAC MAHON'S PRIME NUMBERS OF MEASUREMENT

GEORGE E. ANDREWS

Let us define a sequence of integers m_n by $m_1 = 1$, and for $n > 1$ let m_n be the least positive integer that is not the sum of consecutive terms of $m_1, m_2, \cdots, m_{n-1}$. Thus $m_2 = 2$; since $3 = m_2 + m_1$, $m_3 = 4$; $m_4 = 5$; since $6 = m_3 + m_2$, $7 = m_3 + m_2 + m_1$, $m_5 = 8$, etc.

P. A. MacMahon [1] called these numbers "segmented numbers" and also defined $M_n = m_1 + m_2 + \cdots + m_n$ as the "prime numbers of measurement." MacMahon posed the problem of placing division marks on an infinite ruler so that every integral distance can be measured between some two division points on the ruler; if in addition the marks are put on the scale sequentially so that successively larger distances may be measured, then the marks must be placed at distances M_n from the end.

The first few values of m_n and M_n are given below.

n	1	2	3	4	5	6	7	8	9	10	11	12	13	14
m_n	1	2	4	5	8	10	14	15	16	21	22	25	26	28
M_n	1	3	7	12	20	30	44	59	75	96	118	143	169	197

N. J. A. Sloane in [2] lists these sequences as # 363, and # 1044 respectively.

Apparently no results of any real depth are known concerning m_n. MacMahon asserts that: "It would be interesting to determine how either set of numbers behaves at great distances from the origin."

Conjecture: $\lim_{n\to\infty} \dfrac{n \log n}{m_n \operatorname{loglog} n} = 1.$

The following table indicates the behavior of the above ratio:

n	500	1000	1500	2000	2500	3000	3500	4000	4500	5000
m_n	1477	3165	4963	6853	8802	10725	12704	14624	16521	18524
$\dfrac{n \log}{m_n \operatorname{loglog} n}$	1.1516	1.1293	1.1109	1.0937	1.0802	1.0766	1.0709	1.0724	1.0759	1.0732

While this is certainly not conclusive evidence for the conjecture, it is not atypical behavior. For example, we know that $p_n/n \log n$ tends to 1 as $n \to \infty$ (p_n is the nth prime); however, $p_{5000}/5000 \log 5000 \approx 1.14$. The truth of our conjecture would directly imply $M_n \sim \frac{1}{2} n^2 \log n / \log\log n$.

Problems of less difficulty than the conjecture also suggest themselves:

PROBLEM 1. Prove: $\lim_{n\to\infty} n^{-\Delta} m_n = 0$ for some $\Delta < 2$.

PROBLEM 2. Prove: $\lim_{n\to\infty} \dfrac{m_n}{n} = +\infty$.

PROBLEM 3. Prove: $m_n < p_n$ for every n, where p_n is the nth prime.

It is not difficult to show that for $n \geqq 1$, $n \leqq m_n \leqq \binom{n}{2} + 1$. The left hand inequality is obvious. To see the right hand inequality, we note that since there are only

$$n + (n-1) + (n-2) + \cdots + 2 + 1 = \binom{n+1}{2}$$

possible sums of consecutive members of $m_1, m_2, \cdots, m_n$, we must have m_{n+1} among the first $\binom{n+1}{2} + 1$ integers.

Added in proof. N.J.A. Sloane and Victor Meally have pointed out to me that the sequence $M_n + 1$ appears as sequence No. 416 in [**2**]. This observation supplies two further relevant references: Problem E1910 by R.L. Graham, this MONTHLY, 75 (1968) 80; *Elementary Theory of Numbers* by W. Sierpiński, Warszawa, 1964, pp. 411–412.

Partially supported by NSF Grant 23774.

References

1. P. A. MacMahon, The prime numbers of measurement on a scale, Proc. Cambridge Phil. Soc., 21 (1923) 651–654.

2. N. J. A. Sloane, A Handbook of Integer Sequences, Academic Press, New York, 1973.

DEPARTMENT OF MATHEMATICS, PENNSYLVANIA STATE UNIVERSITY, UNIVERSITY PARK, PA 16802.

6.3 EΥΡΗΚΑ! Num $= \Delta + \Delta + \Delta$

Reprinted from Journal of Number Theory
Vol. 23, No. 3, July 1986

EΥΡΗΚΑ! num $= \Delta + \Delta + \Delta$

GEORGE E. ANDREWS*

Department of Mathematics, Pennsylvania State University, University Park, Pennsylvania 16802

Communicated by H. Zassenhaus

Received May 16, 1984; revised February 26, 1985

A formula for the generating function for the number of representations of n as a sum of three triangular numbers is given. From this formula Gauss's theorem that every natural number is a sum of three triangular numbers follows immediately.

1. INTRODUCTION

The title of this paper is, of course, the famous quotation from Gauss's diary dated July 10, 1796 [5, p. 497], where he notes with Archimedean glee that every natural number is the sum of three triangular numbers. Chapter VII of Volume II of Dickson's History of the Theory of Numbers is devoted to representations of a number by three squares, and it is a well-known and simple exercise to show that the representations of $8n+3$ as a sum of three squares may be bijectively mapped onto the representations of n as a sum of three triangular numbers. While Legendre (in 1798) [11, p. 393] was the first to publish a proof that any natural number not of the form $4^a(8n+7)$ is a sum of three squares, the theorem on three triangular numbers is generally attributed to Gauss due to the famous entry in his diary.

There are famous identities due to Jacobi [9] which provide generating function proofs for the results on representations by two, four, six, and eight squares. For example, if $r_s(n)$ denotes the number of representations of n as a sum of s squares, then

$$\sum_{n=0} (-1)^n r_2(n) q^n = \left(\sum_{n=-\infty}^{\infty} (-1)^n q^{n^2} \right)^2 = \left(\prod_{n=1}^{\infty} \frac{(1-q^n)}{(1+q^n)} \right)^2$$
$$= 1 + 4 \sum_{n=1}^{\infty} \frac{(-1)^n q^{2n-1}}{1-q^{2n-1}}, \tag{1.1}$$

* Partially supported by National Science Foundation Grant MCS 8201733.

$$\sum_{n=0}^{\infty} (-1)^n r_4(n) q^n = \left(\sum_{n=-\infty}^{\infty} (-1)^n q^{n^2} \right)^4 = \left(\prod_{n=1}^{\infty} \frac{(1-q^n)}{(1+q^n)} \right)^4$$

$$= 1 + 8 \sum_{\substack{m=1 \\ 4 \nmid m}}^{\infty} \frac{mq^m}{1-q^m}, \tag{1.2}$$

and similar formulas were given by Jacobi for six and eight squares [9, Sects. 40–42].

In a lengthy study of the fifth and seventh order mock theta functions [3] which was based on the theory of Bailey chains [2], we derived incidentally

$$\left(\sum_{n=0}^{\infty} q^{\binom{n+1}{2}} \right)^3 = \sum_{n=0}^{\infty} \sum_{j=0}^{2n} \frac{q^{2n^2+2n-\binom{j+1}{2}}(1+q^{2n+1})}{(1-q^{2n+1})}. \tag{1.3}$$

From (1.3) the theorem in our title is easy.

The object of this paper is to provide a completely self-contained proof of (1.3) and to examine briefly its consequences. In Section 2 we prove five lemmas, each of which is either explicitly or implicitly in the literature. Section 3 is devoted to the derivation of (1.3) from these lemmas. Finally in Section 4 we examine the consequences of this result briefly.

2. The Lemmas

We shall require five results. To make our exposition reasonably short we shall utilize the standard notation

$$(a; q)_\infty = (a)_\infty = \prod_{n=0}^{\infty} (1 - aq^n), \tag{2.1}$$

$$(a; q)_n = (a)_n = (a; q)_\infty / (aq^n; q)_\infty$$
$$(= (1-a)(1-aq) \cdots (1-aq^{n-1})$$
$$\text{when } n \text{ is a nonnegative integer}). \tag{2.2}$$

Throughout our work convergence questions will be trivial provided we always assume $|q| < 1$.

Lemma 1 (F. H. Jackson [8]).

$$\sum_{n=0}^{\infty} \frac{(a)_n (1-aq^{2n})(y)_n q^{n^2} a^n y^{-n}}{(q)_n (1-a)(aq/y)_n} = \frac{(aq)_\infty}{(aq/y)_\infty}. \tag{2.3}$$

Proof. Let $f(a)$ denote the left-hand side of (2.3). Then

$$\begin{aligned} f(a) &= \sum_{n=0}^{\infty} \frac{(a)_n((1-q^n)+(q^n-aq^{2n}))(y)_n\, q^{n^2}a^n y^{-n}}{(q)_n(1-a)(aq/y)_n} \\ &= \sum_{n=1}^{\infty} \frac{(a)_n(y)_n\, q^{n^2}a^n y^{-n}}{(q)_{n-1}(1-a)(aq/y)_n} + \sum_{n=0}^{\infty} \frac{(a)_{n+1}(y)_n q^{n^2+n}a^n y^{-n}}{(q)_n(1-a)(aq/y)_n} \\ &= \sum_{n=0}^{\infty} \frac{(a)_{n+1}(y)_n\, q^{n^2+n}a^n y^{-n}}{(q)_n(1-a)(aq/y)_{n+1}} \left((1-yq^n)\, q^{n+1}ay^{-1} + \left(1-\frac{aq^{n+1}}{y}\right)\right) \\ &\qquad \text{(where we shifted } n \text{ to } n+1 \text{ in the first sum)} \\ &= \frac{(1-aq)}{(1-(aq/y))} \sum_{n=0}^{\infty} \frac{(aq)_n(1-aq^{2n+1})(y)_n\, q^{n^2+n}a^n y^{-n}}{(q)_n(1-aq)(aq^2/y)_n} \\ &= \frac{(1-aq)}{(1-(aq/y))} f(aq). \end{aligned}$$

Iteration of this functional equation shows that

$$f(a) = \frac{(aq)_N}{(aq/y)_N} f(aq^N), \tag{2.4}$$

and since $f(0)=1$, we see that (2.3) follows by letting $N \to \infty$ in (2.4). ∎

LEMMA 2 (Cauchy [4, Eqs. (15) and (16)]).

$$\sum_{n=0}^{\infty} \frac{(a)_n\, t^n}{(q)_n} = \frac{(at)_\infty}{(t)_\infty}; \tag{2.5}$$

$$\sum_{n=0}^{m} \frac{(q)_m(-z)^n\, q^{\binom{n}{2}}}{(q)_n(q)_{m-n}} = (z)_m. \tag{2.6}$$

Proof. Let $g(t)$ denote the right-hand side of (2.5). Then clearly $(1-t)\, g(t) = (1-at)\; g(tq)$. Since $g(t)$ is analytic around $t=0$ with $g(0)=1$, we substitute its power series expansion $\sum_{n \geqslant 0} A_n t^n$ into our functional equation and compare coefficients of t^n. Hence

$$A_n - A_{n-1} = q^n A_n - aq^{n-1}A_{n-1},$$

or

$$A_n = \frac{(1-aq^{n-1})}{(1-q^n)} A_{n-1},$$

and n iterations together with the fact that $A_0 = g(0) = 1$ imply that $A_n = (a)_n/(q)_n$. Hence the power series expansion on the left-hand side of (2.5) is established.

If in (2.5) we let $a = q^{-m}$, and $t = zq^m$, then (2.6) follows after algebraic simplification. ∎

LEMMA 3 (E. Heine [6, p. 107, Eq. (6)].

$$\sum_{n=0}^{\infty} \frac{(a)_n(b)_n(c/ab)^n}{(q)_n(c)_n} = \frac{(c/a)_\infty(c/b)_\infty}{(c)_\infty(c/ab)_\infty}. \tag{2.7}$$

Proof.

$$\begin{aligned}\sum_{n=0}^{\infty} \frac{(a)_n(b)_n(c/ab)^n}{(q)_n(c)_n} &= \frac{(b)_\infty}{(c)_\infty} \sum_{n=0}^{\infty} \frac{(a)_n(c/ab)^n(cq^n)_\infty}{(q)_n(bq^n)_\infty} \\ &= \frac{(b)_\infty}{(c)_\infty} \sum_{n=0}^{\infty} \sum_{m=0}^{\infty} \frac{(a)_n(c/ab)^n\, b^m q^{nm}(c/b)_m}{(q)_m} \quad \text{(by (2.5))} \\ &= \frac{(b)_\infty}{(c)_\infty} \sum_{m=0}^{\infty} \frac{(c/b)_m\, b^m((c/b)\, q^m)_\infty}{(q)_m(cq^m/ab)_\infty} \\ &= \frac{(b)_\infty(c/b)_\infty}{(c)_\infty(c/ab)_\infty} \sum_{m=0}^{\infty} \frac{(c/ab)_m\, b^m}{(q)_m} \\ &= \frac{(b)_\infty(c/b)_\infty(c/a)_\infty}{(c)_\infty(c/ab)_\infty(b)_\infty},\end{aligned}$$

and now (2.7) now follows by cancellation of $(b)_\infty$ from numerator and denominator in the last expression. ∎

LEMMA 4 (Inverted Bailey transform [2]). *If* $\alpha_0 = \beta_0 = 1$, *and for* $n \geqslant 1$,

$$\alpha_n = (1 - aq^{2n}) \sum_{j=0}^{n} \frac{(aq)_{n+j-1}(-1)^{n-j}\, q^{\binom{n-j}{2}}\beta_j}{(q)_{n-j}}, \tag{2.8}$$

then

$$\sum_{n=0}^{\infty} (y)_n(-1)^n\, a^n y^{-n} q^{\binom{n+1}{2}}\beta_n = \frac{(aq/y)_\infty}{(aq)_\infty} \sum_{n=0}^{\infty} \frac{(y)_n(-1)^n\, a^n y^{-n} q^{\binom{n+1}{2}}\alpha_n}{(aq/y)_n}. \tag{2.9}$$

Proof. We substitute the expression for α_n given in (2.8) into the right-hand side of (2.9). Thus the right-hand side of (2.9) is

$$\frac{(aq/y)_\infty}{(aq)_\infty} \sum_{n=0}^{\infty} \frac{(y)_n(-1)^n a^n y^{-n} q^{\binom{n+1}{2}}(1-aq^{2n})}{(aq/y)_n}$$

$$\times \sum_{j=0}^{n} \frac{(aq)_{n+j-1}(-1)^{n-j} q^{\binom{n-j}{2}} \beta_j}{(q)_{n-j}}$$

$$= \frac{(aq/y)_{\infty'}}{(aq)_\infty} \sum_{j=0}^{\infty} \sum_{n=0}^{\infty}$$

$$\times \frac{(y)_{n+j}(-1)^j a^{n+j} y^{-n-j} q^{\binom{n+j+1}{2}+\binom{n}{2}}(1-aq^{2n+2j})(aq)_{n+2j-1}\beta_j}{(aq/y)_{n+j}(q)_n}$$

$$= \frac{(aq/y)_\infty}{(aq)_\infty} \sum_{j=0}^{\infty} \frac{(y)_j(-1)^j a^j y^{-j} q^{\binom{j+1}{2}} \beta_j (aq)_{2j-1}}{(aq/y)_j}$$

$$\times \sum_{n=0}^{\infty} \frac{(aq^{2j})_n(1-aq^{2j+2n})(yq^j)_n q^{n^2+nj} a^n y^{-n}}{(q)_n(aq^{j+1}/y)_n}$$

$$= \frac{(aq/y)_\infty \cdot}{(aq)_\infty} \sum_{j=0}^{\infty} \frac{(y)_j(-1)^j a^j y^{-j} q^{\binom{j+1}{2}} \beta_j (aq)_{2j-1}}{(aq/y)_j} \cdot \frac{(aq^{2j})_\infty}{(aq^{j+1}/y)_\infty} \quad \text{(by (2.3))}$$

$$= \sum_{j=0}^{\infty} (y)_j(-1)^j a^j y^{-j} q^{\binom{j+1}{2}} \beta_j. \quad \blacksquare$$

LEMMA 5 (D. Shanks [2]).

$$\frac{(q^2;q^2)_n}{(q;q^2)_n} \sum_{s=0}^{n} \frac{(q;q^2)_s q^{-n(2s+1)}}{(q^2;q^2)_s} = \sum_{s=0}^{2n} q^{-\binom{s+1}{2}}. \tag{2.10}$$

Proof. Let S_n denote the left-hand side of (2.10), and let T_n denote the same expression except that the sum on s is up to $n-1$ instead of n. Then clearly

$$S_n - T_n = q^{-\binom{2n+1}{2}}. \tag{2.11}$$

Furthermore

$$T_n - S_{n-1} = \frac{(q^2;q^2)_{n-1}}{(q;q^2)_n} \sum_{s=0}^{n-1} \frac{(q;q^2)_s q^{-n(2s+1)}\{(1-q^{2n}) - q^{2s+1}(1-q^{2n-1})\}}{(q^2;q^2)_s}$$

$$= \frac{(q^2;q^2)_{n-1}}{(q;q^2)_n} \left(\sum_{s=0}^{n-1} \frac{(q;q^2)_{s+1} q^{-n(2s+1)}}{(q^2;q^2)_s} - \sum_{s=1}^{n-1} \frac{(q;q^2)_s q^{-n(2s-1)}}{(q^2;q^2)_{s-1}} \right)$$

(where we have grouped the expression in { } as $(1-q^{2s+1}) - q^{2n}(1-q^{2s})$)

$$=\frac{(q^2;q^2)_{n-1}(q;q^2)_n\,q^{-\binom{2n}{2}}}{(q;q^2)_n(q^2;q^2)_{n-1}}$$

(since the second sum cancels every term in the first save $s=n-1$)

$$=q^{-\binom{2n}{2}}. \tag{2.12}$$

Thus combining (2.11) and (2.12), we see that $S_n - S_{n-1} = q^{-\binom{2n+1}{2}} + q^{-\binom{2n}{2}}$ and since $S_0 = 1$ we see that Lemma 5 follows. ∎

3. Proof of (1.3)

Suppose in all expressions in Lemma 4 we replace q by q^2, and then set $a = q^2$, $y = q$, and let

$$\beta_j - \frac{(-1)^j (q;q^2)_j\,q^{-j^2-j}}{(q^2;q^2)_j(q^3;q^2)_j}.$$

With these specializations, the left-hand side of (2.9) becomes

$$\sum_{n=0}^{\infty}\frac{(q;q^2)_n\,q^n(q;q^2)_n}{(q^2;q^2)_n(q^3;q^2)_n}=\frac{(q^2;q^2)_\infty^2}{(q^3;q^2)_\infty(q;q^2)_\infty}\quad\text{(by Lemma 3).} \tag{3.1}$$

To evaluate the right-hand side of (2.9) we must first consider (2.8),

$$\begin{aligned}
\alpha_n &= \frac{(-1)^n(1-q^{4n+2})}{(1-q^2)}\sum_{j=0}^{n}\frac{(q^2;q^2)_{n+j}(q;q^2)_j\,q^{n^2-n-2nj}}{(q^2;q^2)_{n-j}(q^2;q^2)_j(q^3;q^2)_j}\\
&= \frac{(-1)^n q^{n^2-n}(1-q^{4n+2})(q^2;q^2)_n}{(1-q^2)}\sum_{j=0}^{n}\frac{(q;q^2)_j\,q^{-2nj}(q^{2n+2};q^2)_j}{(q^2;q^2)_{n-j}(q^2;q^2)_j(q^3;q^2)_j}\\
&= \frac{(-1)^n q^{n^2-n}(1-q^{4n+2})(q^2;q^2)_n}{(1-q^2)}\sum_{j=0}^{n}\frac{(q;q^2)_j\,q^{-2nj}}{(q^2;q^2)_{n-j}(q^3;q^2)_j}\\
&\quad\times\sum_{h=0}^{j}\frac{(-1)^h q^{2nh+h^2+h}}{(q^2;q^2)_{j-h}(q^2;q^2)_h}\qquad\text{(by (2.6))}\\
&= \frac{(-1)^n q^{n^2-n}(1-q^{4n+2})}{(1-q^2)}\\
&\quad\times\sum_{j=0}^{n}\sum_{h=0}^{j}\frac{(q^{-2n};q^2)_j(-1)^{h+j}q^{2nh+h^2+h-j^2+j}(q;q^2)_j}{(q^3;q^2)_j(q^2;q^2)_{j-h}(q^2;q^2)_h}
\end{aligned}$$

$$= \frac{(-1)^n q^{n^2-n}(1-q^{4n+2})}{(1-q^2)}$$

$$\times \sum_{\substack{h=0\\h+j\leqslant n}}^{n} \sum_{j=0}^{n} \frac{(q^{-2n};q^2)_{j+h}(-1)^j q^{2nh-2jh-j^2+j+2h}(q;q^2)_{j+h}}{(q^3;q^2)_{j+h}(q^2;q^2)_j(q^2;q^2)_h}$$

$$= \frac{(-1)^n q^{n^2-n}(1-q^{4n+2})}{(1-q^2)} \sum_{j=0}^{n} \frac{(q^{-2n};q^2)_j(q;q^2)_j q^{-j^2+j}(-1)^j}{(q^2;q^2)_j(q^3;q^2)_j}$$

$$\times \sum_{h=0}^{n-1} \frac{(q^{-2n+2j};q^2)_h(q^{2j+1};q^2)_h q^{2h(n-j+1)}}{(q^2;q^2)_h(q^{2j+3};q^2)_h}$$

$$= \frac{(-1)^n q^{n^2-n}(1-q^{4n+2})}{(1-q^2)} \sum_{j=0}^{n} \frac{(q^{-2n};q^2)_j(q;q^2)_j q^{-j^2+j}(-1)^j}{(q^2;q^2)_j(q^3;q^2)_j}$$

$$\times \frac{(q^2;q^2)_{n-j}}{(q^{2j+3};q^2)_{n-j}} \qquad \text{(by Lemma 3)}$$

$$= \frac{(-1)^n q^{n^2-n}(1-q^{4n+2})}{(1-q^2)} \frac{(q^2;q^2)_n}{(q^3;q^2)_n} \sum_{j=0}^{n} \frac{(q;q^2)_j q^{-2nj}}{(q^2;q^2)_j}$$

$$= \frac{(-1)^n q^{n^2}(1+q^{2n+1})}{(1+q)} \sum_{j=0}^{2n} q^{-\binom{j+1}{2}} \tag{3.2}$$

by Lemma 5.

Thus with the specializations given at the beginning of this section, we may now evaluate the right-hand side of (2.9) as

$$\frac{(q^3;q^2)_\infty}{(q^4;q^2)_\infty} \sum_{n=0}^{\infty} \frac{(1-q)\,q^{n^2+2n}(-1)^n}{(1-q^{2n+1})} \cdot (-1)^n q^{n^2} \frac{(1+q^{2n+1})}{(1+q)} \sum_{j=0}^{2n} q^{-\binom{j+1}{2}}$$

$$= \frac{(1-q)(q;q^2)_\infty}{(q^2;q^2)_\infty} \sum_{n=0}^{\infty} \sum_{j=0}^{2n} \frac{q^{2n^2+2n-\binom{j+1}{2}}(1+q^{2n+1})}{(1-q^{2n+1})}. \tag{3.3}$$

Equating (3.1) with (3.3) and recalling that

$$\frac{(q^2;q^2)_\infty}{(q;q^2)_\infty} = \sum_{r=0}^{\infty} q^{\binom{r+1}{2}} \quad \text{(replace } q \text{ by } q^{-1} \text{ in Lemma 5 and let } n\to\infty\text{)},$$

we obtain (1.3).

4. Conclusion

As we mentioned in the Introduction, Gauss's theorem on triangular numbers follows immediately from (1.3). To see this just replace j by $2n-j$,

$$\left(\sum_{n=0}^{\infty} q^{\binom{n+1}{2}}\right)^3 = \sum_{n=0}^{\infty} q^n \sum_{j=0}^{2n} \frac{q^{j(4n+1-j)/2}(1+q^{2n+1})}{(1-q^{2n+1})}, \tag{4.1}$$

and clearly all terms on the right-hand side of (4.1) have positive coefficients when expanded in powers of q and the nth term produces a positive contribution to the coefficient of q^n. In [3], we also show that

$$\left(\sum_{n=-\infty}^{\infty} (-1)^n q^{n^2}\right)^3 = 1 + 4 \sum_{n=1}^{\infty} \frac{(-1)^n q^n}{1+q^n} - 2 \sum_{\substack{n=1 \\ |j|<n}}^{\infty} \frac{q^{n^2-j^2}(1-q^n)(-1)^j}{1+q^n}. \tag{4.2}$$

Of course the coefficients in this representation are no longer obviously positive, and as we know, infinitely many integers are not the sum of three squares (namely, all those of the form $4^a(8b+7)$).

Also from the classical results of Kronecker [10] and Gauss we can use (1.3) and (4.2) to provide computational formulas for the class numbers of binary quadratic forms. However, there already exist reasonable formulae for the generating functions for the class numbers of binary quadratic forms due to Humbert [7, pp. 349–350] (see also Mordell [12] and Watson [13]). Consequently it is doubtful that these new formulas greatly improve the efficiency of computation of class numbers.

References

1. G. E. Andrews, "The Theory of Partitions" (G.-C. Rota, Ed.), Encyclopedia of Math. and Its Applications, Vol. 2, Addison–Wesley, Reading, Mass., 1976.
2. G. E. Andrews, Multiple series Rogers–Ramanujan type identities, *Pacific J. Math.* **114** (1984), 267–283.
3. G. E. Andrews, The fifth and seventh order mock theta functions, *Trans. Amer. Math. Soc.* **293** (1986), 113–134.
4. A. Cauchy, Mémoire sur les fonctions dont plusieurs valeurs sont liées entre elles par une équation lineaire, et sur diverses transformations de produits composes d'un nombre indéfini de facteurs, *C. R. Paris* **17** (1983), 523; Œuvres (1), VIII, 42–50.
5. C. F. Gauss, "Werke," Bd. 10, p. 497, Teubner, Leipzig, 1917.
6. E. Heine, "Handbuch der Kugelfunctionen," Reimer, Berlin, 1878.
7. G. Humbert, Formules relatives aux nombres de classes des formes quadratiques binaires et positives, *J. Math. Pures Appl.* (6) **3** (1907), 337–449.

8. F. H. JACKSON, Summation of q-hypergeometric series, *Messenger of Math.* **50** (1921), 101–112.
9. C. G. J. JACOBI, Fundamenta nova theoriae functionum ellipticarum, Regiomonti fratrum Bornträger, reprinted in "Gesammelte Werke," Vol. 1, pp. 49–239; Reimer, Berlin, 1881.
10. L. KRONECKER, Über die Anzahl der verschiedenen Classen quadratischer Formen von negativer Determinante, *J. Reine Angew. Math.* **57** (1860), 248–255.
11. A. M. LEGENDRE, "Theorie des Nombres," Vol. I., 3rd ed., reprinted, Blanchard, Paris, 1955.
12. L. J. MORDELL, Note on class relation formulae, *Messenger of Math.* **45** (1916), 76–80.
13. D. SHANKS, Two theorems of Gauss, *Pacific J. Math.* **8** (1958), 609–612.
14. G. N. WATSON, Generating functions of class numbers, *Compositio Math.* **1** (1934), 39–68.

6.4 Ramanujan's Method in q-Series Congruences (with R. Roy)

Electronic J. Combin., (Wilf Festschrift) 4(2):R2, 7 pp. (1997)

RAMANUJAN'S METHOD IN q-SERIES CONGRUENCES

BY

GEORGE E. ANDREWS[(1)] AND RANJAN ROY

Written in honor of Herb Wilf's 65th birthday

ABSTRACT. We show that the method developed by Ramanujan to prove $5|p(5n+4)$ and $7|p(7n+5)$ may, in fact, be extended to a wide variety of q-series and products including some with free parameters.

1. Introduction.

Ramanujan [11] is the discoverer of the surprising fact that the partition function, $p(n)$, satisfies numerous congruences. Among the infinite family of such congruences, the two simplest examples are

$$p(5n+4) \equiv 0 \pmod 5 \tag{1.1}$$

and

$$p(7n+5) \equiv 0 \pmod 7. \tag{1.2}$$

Ramanujan used an ingenious and elementary argument to prove these congruences which relied on Jacobi's famous formula [10; last eqn. p.5]:

$$(q;q)_\infty^3 = \prod_{n=1}^{\infty}(1-q^n)^3 = \sum_{j=0}^{\infty}(-1)^j(2j+1)q^{j(j+1)/2}, \tag{1.3}$$

where

$$(A)_N = (A;q)_N = \prod_{j=0}^{N-1}(1-Aq^j). \tag{1.4}$$

(1) Partially supported by National Science Foundation Grant DMS-8702695-04

A rather more general result of this nature was proved in [3; p. 27, Th. 10.1] to account for certain congruences connected with generalized Frobenius partitions.

Indeed J. M. Gandhi [7], [8], [9], J. Ewell [5], L. Winquist [12] and many others (cf., Gupta [10; Sec. 6.3]) have proved partition function congruences based on this idea. In all these theorems, the underlying generating functions were either modular forms or simple linear combinations thereof.

The point of this paper is to show that Ramanujan's original method is applicable to an infinite number of congruence theorems including many non-modular functions defined by q-series.

Our main result is:

Theorem 1. *Suppose p is a prime > 3, and $0 < a < p$ and b are integers. Also, $-a$ must be a quadratic nonresidue mod p. Suppose $\{\alpha_n\}_{n=-\infty}^{\infty} = \{\alpha_n(z_1, z_2, \dots, z_j)\}$ is a doubly infinite sequence of Laurent polynomials over $\mathbb{Z}$ with variables $z_1, \dots, z_j$ independent of q. Then there is an integer c such that the coefficient of $z_1^{m_1} z_2^{m_2} \cdots z_j^{m_j}$ q^{pN} in*

$$\frac{q^c \sum_{n=-\infty}^{\infty} \alpha_n q^{a\binom{n}{2}+bn}}{(q;q)_\infty^{p-3}} \tag{1.5}$$

is divisible by p. For each integer m, we shall denote by $\overline{m}$ the multiplicative inverse of m mod p. The integer $c = c_p(a, b)$ may be chosen as the least nonnegative integer congruent to $\bar{8}(a(2b\bar{a} - 1)^2 + 1)$ mod p.

In Section 2, we shall prove this result. In Section 3, we examine the implications of Theorem 1 for a variety of modular forms. In Section 4, we collect a number of congruences for the coefficients in several q-series.

2. The Proof of Theorem 1.

With the various hypotheses of the theorem, we note that

$$(2.1) \qquad \frac{q^c \sum\limits_{n=-\infty}^{\infty} \alpha_n q^{a\binom{n}{2}+bn}}{(q;q)_\infty^{p-3}}$$

$$= \frac{q^c \sum\limits_{n=-\infty}^{\infty} \sum\limits_{j=0}^{\infty} (-1)^j (2j+1) \alpha_n q^{a\binom{n}{2}+bn+j(j+1)/2}}{(q;q)_\infty^p}$$

$$\equiv \frac{q^c \sum\limits_{n=-\infty}^{\infty} \sum\limits_{j=0}^{\infty} (-1)^j (2j+1) \alpha_n q^{a\binom{n}{2}+bn+j(j+1)/2}}{(q^p;q^p)_\infty} \pmod{p}.$$

We see that in this last expression the denominator is a function of q^p. Let us now examine the exponent of q in the numerator; for ease of computation we multiply by 8:

$$(2.2) \qquad 8\left(c + a\binom{n}{2} + bn + j(j+1)/2\right)$$

$$= 8c + a(4n^2 - 4n) + 8bn + 4j^2 + 4j$$

$$\equiv a(2n + 2b\bar{a} - 1)^2 + (2j+1)^2 \pmod{p}$$

Now we observe (by the definition of c) that if $j \equiv (p-1)/2 \bmod p$ (i.e. $(2j+1) \equiv 0 \pmod{p}$), then the last expression above is congruent to 0 mod p precisely when

$$n \equiv (1 - 2b\bar{a})\bar{2} \equiv \frac{p+1}{2} - b\bar{a} \pmod{p}.$$

If $j \not\equiv \frac{p-1}{2} \pmod{p}$, then the last expression in (2.2) can never be congruent to zero mod p because by the conditions on a

$$-a(2n + 2b\bar{a} - 1)^2$$

is either 0 or a quadratic nonresidue mod p and so cannot be congruent to a quadratic residue (i.e. $(2j+1)^2$) mod p.

Hence the coefficients of q^{pN} in (2.1) will all be linear combinations over $p\mathbb{Z}$ of various α_n (which are Laurent polynomials in several variables over $\mathbb{Z}$). □

3. Modular Forms.

Ramanujan, Ewell, Gandhi (and probably many others) have proved instances of Theorem 1 (as mentioned in Section 1).

Congruence (1.1) follows from Theorem 1 with $p = 5, a = 3, b = 1, c_5(3,1) = 1$ and $\alpha_m = (-1)^m$. Congruence (1.2) follows from Theorem 1 with $p = 7, a = b = 1, c_7(1,1) = 2$ and $\alpha_m = (-1)^m(2m+1)$ if $m \geqq 0, \alpha_m = 0$ if $m < 0$.

Gandhi's Theorem IV in [7] corresponds to $\alpha_m = \delta_{m,0}$, while Theorem 2 in [8] corresponds to $a = b = 1$ and $\alpha_m = (-1)^m(2m+1)$ if $m \geqq 0$, $\alpha_m = 0$ if $m < 0$. Finally, Theorem 4 in [8] corresponds to $a = 3, b = 1$ and $\alpha_m = (-1)^m$.

Theorem 10.1 of [3] is the case $p = 5, a = 2, b = 1, c_5(2,1) = 2$; in that result the α_m were assumed to be 0 if $m < 0$ and to be integers otherwise.

The generality of Theorem 1 allows for a variety of other modular forms. To illustrate, we consider

$$\sum_{n=0}^{\infty} V_n q^n = \frac{\sum\limits_{n=0}^{\infty} p(n) q^n}{\sum\limits_{n=0}^{\infty} (-1)^n r_2(n) q^n},$$

where $r_2(n)$ is the number of representations of n as a sum of two squares. We note that

$$\begin{aligned}\sum_{n=0}^{\infty} V_n q^n &= \frac{1}{(q)_\infty \left(\sum\limits_{n=-\infty}^{\infty} (-1)^n q^{n^2} \right)^2} \\ &= \frac{(-q)_\infty^2}{(q)_\infty^3} \\ &= \frac{(q^2;q^2)_\infty}{(q)_\infty^4 (q;q^2)_\infty} \\ &= \frac{\sum\limits_{m=0}^{\infty} q^{m(m+1)/2}}{(q)_\infty^4}.\end{aligned}$$

Now by Theorem 1 with $p = 7, a = b = 1, c_7(1,1) = 2, \alpha_m = 1$ if $m \geq 0$, and 0 if $m < 0$, we see that

$$V_{7m+5} \equiv 0 \pmod{7}.$$

4. q-Series.

Of course, our point here is not to extend slightly Ramanujan's basic idea to a few more modular forms. Rather we hope to illustrate its applicability to q-series.

Theorem 2. *For any prime $p \equiv 3 \pmod 4$ with $4c \equiv 1 \pmod p$, the coefficient of $z^m q^{pn-c}$ in*

$$\frac{(zq)_\infty}{(q)_\infty^{p-4}} \sum_{n=0}^{\infty} \frac{q^n}{(q)_n (zq)_n}$$

is divisible by p.

Proof. By Heine's transformation [1; Cor. 2.3, p. 19]

$$\sum_{n=0}^{\infty} \frac{q^n}{(q)_n (zq)_n} = \frac{1}{(q)_\infty (zq)_\infty} \sum_{n=0}^{\infty} (-1)^n q^{n(n+1)/2} z^n.$$

Now apply Theorem 1 with $a = b = 1$ and $c \equiv \overline{4} \pmod p$. □

Theorem 3. *For any prime $p \equiv 5$ or $7 \pmod 8$ with $8c \equiv 1 \pmod p$, the coefficient of $z^m q^{pn-c}$ in*

$$\frac{1}{(q)_\infty^{p-3}} \sum_{n=0}^{\infty} \frac{(z)_{n+1} z^n}{(-zq)_n}$$

is divisible by p.

Proof. By the Rogers-Fine identity [6; p. 15, eqn. (14.31)]

$$\sum_{n=0}^{\infty} \frac{(z)_{n+1} z^n}{(-zq)_n} = 1 + 2 \sum_{n \geq 1} (-z^2)^n q^{n^2}.$$

Now apply Theorem 1 with $a = 2, b = 1$ noting that for $p \equiv 5 \pmod 8$ 2 is a non-quadratic residue, and for $p \equiv 7 \pmod 8$, 2 is a quadratic residue. Also we must have $c \equiv \overline{8} \pmod p$. □

Theorem 4. *For any prime $p \equiv 5$ or $11 \pmod{12}$, the coefficient of $q^{pn-(p+1)/2}$ in*

$$\frac{1}{(q)_\infty^{p-4}} \sum_{n=0}^{\infty} \begin{bmatrix} 2n \\ n \end{bmatrix} q^n$$

(where $\begin{bmatrix} A \\ B \end{bmatrix} = (q)_A/((q)_B(q)_{A-B})$ *is the q-binomial coefficient) is divisible by p.*

Proof. By Lemma 3 of [2; p. 159],

$$\frac{1}{(q)_\infty}\sum_{n=0}^{\infty}(-1)^n q^{3n(n+1)/2} = \sum_{n=0}^{\infty}\begin{bmatrix} 2n \\ n \end{bmatrix} q^n.$$

Now apply Theorem 1 with $a = b = 3$ noting that for $p \equiv 5 \pmod{12}$, 3 is a non-quadric residue and for $p \equiv 11 \pmod{12}$, 3 is a quadratic residue.

Theorem 5. *For any prime* $p \equiv 5$ *or* $11 \pmod{12}$ *and* $6c \equiv 1 \pmod{p}$, *the coefficient of* q^{pn-c} *in*

$$\frac{1}{(q)_\infty^{p-4}}\sum_{n=1}^{\infty}\frac{q^{n^2}}{(1-q^n)\begin{bmatrix} 2n \\ n \end{bmatrix}}$$

is divisible by p.

Proof. By (5.1) of [4; p. 272]

$$\sum_{n=1}^{\infty}\frac{q^{n^2}}{(1-q^n)\begin{bmatrix} 2n \\ n \end{bmatrix}}$$

$$= \sum_{n=1}^{\infty}\frac{q^{n^2}(q)_{n-1}}{(q^{n+1})_n} = \frac{1}{(q)_\infty}\sum_{n=-\infty}^{\infty}(-1)^{n-1} n\, q^{n(3n-1)/2}.$$

Now apply Theorem 1 with $a = 3, b = 1$. As in Theorem 4, p must be $\equiv 5$ or 11 mod 12, and now $6c = 1 \pmod{p}$. □

5. Conclusion.

There are undoubtedly significant extensions of the ideas we have presented here. We note that Winquist's proof that

$$p(11n+6) \equiv 0 \pmod{11}$$

does not fit into Theorem 1. However now that the application of (1.3) to Theorem 1 has been made, it should be possible to find a variety of congruences for the coefficients of other q-series. In addition, the function given in Theorem 1 may be multiplied by any function $f_1(q) \in \mathbb{Z}[[q]]$ for which

$$f_1(q) \equiv f_2(q^p) \pmod{p}.$$

Finally, it may well be asked what happens when a is a quadratic residue mod p and $p \equiv 1 \pmod 4$ or a is a quadratic non-residue and $p \equiv 3 \pmod 4$. It is not difficult to show that our analysis produces indices $j \not\equiv (p-1)/2 \pmod p$ which make the exponent on q congruent to zero mod p irrespective of $c_p(a,b)$. Consequently one would have to invoke special conditions on the α_m to produce coefficients that are multiples of p.

References

1. G.E. Andrews, The Theory of Partitions, Encyclopedia of Math. and Its Applications, Vol. 2, Addison-Wesley, Reading, 1976 (Reissued: Cambridge University Press, Cambridge, 1985).
2. G.E. Andrews, *Ramanujan's "lost" notebook: I. partial θ-functions*, Adv. in Math. **41** (1981), 137–172.
3. G.E. Andrews, Generalized Frobenius partitions, Memoirs of the Amer. Math. Soc., 49(1984), No. 301, iv+, 44 pp.
4. G.E. Andrews, *Bailey chains and generalized Lambert series: I. Four identities of Ramanujan*, Illinois J. Math. **36** (1992), 251–274.
5. J.A. Ewell, *Completion of a Gaussian derivation*, Proc. Amer. Math. Soc. **84** (1982), 311–314.
6. N.J. Fine, Basic Hypergeometric Series and Applications, Math. Surveys No. 27, Amer. Math. Soc., Providence, 1988.
7. J.M. Gandhi, *Congruences for $p_r(n)$ and Ramanujan's τ function*, Amer. Math. Monthly **70** (1963), 265–274.
8. J.M. Gandhi, *Generalization of Ramanujan's congruences* $p(5m+4) \equiv 0 \pmod 5$ *and* $p(7m+5) \equiv 0 \pmod 7$, Monatsch. Math. **69** (1965), 389–392.
9. J.M. Gandhi, *Some congruences for k-line partitions of a number*, Amer. Math. Monthly **74** (1967), 179–181.
10. H. Gupta, *Partitions – a survey*, Journal of Res. of Nat. Bur. Standards - B Math. Sciences **74B** (1970), 1–29.
11. S. Ramanujan, Some properties of $p(n)$, the number of partitions of n, Proc. Cambridge Phil. Soc., 19(1919), 207–210 (Reprinted: Coll. Papers., Cambridge Univ. Press, Cambridge, 1927, Reissued: Chelsea, New York, 1962).
12. L. Winquist, *An elementary proof of* $p(11m+6) \equiv 0 \pmod{11}$, J. Combinatorial Theory **6** (1969), 56–59.

The Pennsylvania State University
University Park, PA 16802
email: andrews@math.psu.edu
and
Beloit College
Beloit, Wisconsin
email: royr@beloit.edu

6.5 q-Analogs of the Binomial Coefficient Congruences of Babbage, Wolstenholme, and Glaisher

ELSEVIER
Discrete Mathematics 204 (1999) 15–25
DISCRETE MATHEMATICS
www.elsevier.com/locate/disc

q-Analogs of the binomial coefficient congruences of Babbage, Wolstenholme and Glaisher [1]

George E. Andrews *

Department of Mathematics, Pennsylvania State University, 410 McAllister Bldg., University Park, PA 16802, USA

Received 5 August 1997; revised 15 May 1998; accepted 3 August 1998

Dedicated to Henry Gould on his 70th birthday

Abstract

The object of this paper is to provide q-analogs of several classical binomial coefficient congruences due to Babbage, Wolstenholme and Glaisher. The q-analogs turn out to be succinct and unexpected. © 1999 Elsevier Science B.V. All rights reserved

1. Introduction

In 1819, inspired by Wilson's Theorem which asserts that if p is a prime

$$(p-1)! \equiv -1 \pmod{p}, \tag{1.1}$$

Charles Babbage [2] proved for odd primes p that

$$\binom{2p-1}{p-1} \equiv 1 \pmod{p^2}. \tag{1.2}$$

This is apparently the only paper on number theory written by Babbage, the famous pioneer of computing machines. His proof is a nice application of the Chu-Vandermonde summation.

In 1862, the Reverend Wolstenholme [21] improved Babbage's theorem by proving for primes $p > 3$

$$\binom{2p-1}{p-1} \equiv 1 \pmod{p^3}. \tag{1.3}$$

Gow [13] presented a biographical note on Wolstenholme.

* E-mail: andrews@math.psu.edu.
[1] Partially supported by National Science Foundation Grant DMS-9501101

PII: S0012-365X(98)00364-1

Finally, Glaisher in a series of wonderful papers [8–11] proved a large number of results of this nature, for example [11, p. 110] if p is a prime > 3, then

$$\binom{mp+p-1}{p-1} \equiv 1 \,(\mathrm{mod}\, p^3), \tag{1.4}$$

or as he stated the result

$$(mp+1)(mp+2)\dots(mp+p-1) \equiv (p-1)! \quad (\mathrm{mod}\, p^3). \tag{1.5}$$

From (1.5), Glaisher noted that a variety of congruences for binomial coefficients could be obtained.

In recent times, a number of papers [4–7, 14, 17, 18, Problem 1.6(d)], have appeared which contribute to our understanding of such congruences. For the most part these papers take as their starting point Lucas's congruence for binomial coefficients [15]. Two papers that I know of consider q-binomial coefficients. In [4], Clark proves some congruences for q-binomial coefficients that are closely related to Theorem 3 in our Section 3. Sved [19] considers q-binomial coefficient congruences mod p when q is also an integer.

In this paper, we shall prove the q-analog of (1.5), namely

Theorem 1. *If p is an odd prime and $m \geqslant 1$, then*

$$\frac{(q^{mp+1};q)_{p-1} - q^{mp(p-1)/2}(q;q)_{p-1}}{(1-q^{(m+1)p})(1-q^{mp})} \equiv (p^2-1)p/24 \,(\mathrm{mod}\,[p]), \tag{1.6}$$

where

$$(A;q)_n = (1-A)(1-Aq)\dots(1-Aq^{n-1}), \tag{1.7}$$

and

$$[n] = [n]_q = (1-q^n)/(1-q). \tag{1.8}$$

In Section 2 we shall prove Theorem 1. In Section 3 we shall prove a number of q-binomial coefficient congruences. In Section 4 we shall confirm that (1.6) implies Wolstenholme's congruence (1.3). In Section 5, we consider q-analogs of Wolstenholme's harmonic series congruences, and we consider a natural conjecture in Section 6.

While Henry Gould does not appear to have considered Wolstenholme's Theorem directly, he has dealt explicitly [12, p. 317, eq. (2)] with generalizations of identities used by Wolstenholme [21] in proving his congruences.

2. Proof of Theorem 1

Throughout the paper we will tacitly use the fact that when p is a prime $[p](=1+q+\dots+q^{p-1})$ is an irreducible polynomial in $\mathbb{Q}$ (q) (cf. [20, p. 12]). Furthermore, this means that $\mathbb{Q}(q)/[p]$ is a field and consequently rational functions $r(q)/s(q)$ are well defined modulo $[p]$ provided $s(q)$ is prime to $[p]$.

To prove Theorem 1, we require the following facts:

$$q^p \equiv 1 \pmod{[p]}. \tag{2.1}$$

This follows immediately from (1.8).

$$\sum_{j=0}^{N} \begin{bmatrix} N \\ j \end{bmatrix} (-1)^j q^{\binom{j}{2}} z^j = (z;q)_N, \tag{2.2}$$

where

$$\begin{bmatrix} N \\ j \end{bmatrix} = \begin{bmatrix} N \\ j \end{bmatrix}_q = \begin{cases} 0 & \text{if } j<0 \text{ or } j>N, \\ \dfrac{(q^{N-j+1};q)_j}{(q;q)_j} & \text{if } 0 \leqslant j \leqslant N. \end{cases}$$

This is the q-binomial theorem [1, p. 36].

Finally,

$$q^{\binom{j+1}{2}} \begin{bmatrix} p-1 \\ j \end{bmatrix} \equiv (-1)^j \pmod{[p]}. \tag{2.3}$$

To see that (2.3) follows from (2.1), we note

$$\begin{aligned} q^{\binom{j+1}{2}} \begin{bmatrix} p-1 \\ j \end{bmatrix} &= q^{1+2+\cdots+j} \frac{(1-q^{p-1})(1-q^{p-2})\cdots(1-q^{p-j})}{(1-q)(1-q^2)\cdots(1-q^j)} \\ &= \frac{(q-q^p)(q^2-q^p)\cdots(q^j-q^p)}{(1-q)(1-q^2)\cdots(1-q^j)} \\ &\equiv \frac{(q-1)(q^2-1)\cdots(q^j-1)}{(1-q)(1-q^2)\cdots(1-q^j)} \pmod{[p]} \quad (\text{by } (2.1)) \\ &= (-1)^j. \end{aligned}$$

We are now ready to prove Theorem 1.

$$\begin{aligned} &\frac{(q^{mp+1};q)_{p-1} - q^{mp(p-1)/2}(q;q)_{p-1}}{(1-q^{(m+1)p})(1-q^{mp})} \\ &= (1-q^{(m+1)p})^{-1}(1-q^{mp})^{-1} \sum_{j=0}^{p-1} \begin{bmatrix} p-1 \\ j \end{bmatrix} (-1)^j q^{\binom{j+1}{2}} (q^{mpj} - q^{mp(p-1)/2}) \\ &= (1-q^{(m+1)p})^{-1}(1-q^{mp})^{-1} \\ &\quad \times \sum_{j=0}^{(p-1)/2} \begin{bmatrix} p-1 \\ j \end{bmatrix} (-1)^j \left(q^{\binom{j+1}{2}+mpj} - q^{\binom{j+1}{2}+mp(p-1)/2} \right. \\ &\quad \left. + q^{\binom{p-j}{2}+mp(p-1-j)} - q^{\binom{p-j}{2}+mp(p-1)/2} \right) \end{aligned}$$

(by grouping the jth term with the $(p-1-j)$th term)

$$= (1-q^{(m+1)p})^{-1}(1-q^{mp})^{-1}$$
$$\times \sum_{j=0}^{(p-1)/2} \begin{bmatrix} p-1 \\ j \end{bmatrix} (-1)^j q^{\binom{j+1}{2}} \left(q^{mpj} - q^{mp(p-1)/2} \right.$$
$$\left. + q^{\binom{p}{2} - pj + mp(p-1-j)} - q^{\binom{p}{2} - pj + mp(p-1)/2} \right)$$
$$= (1-q^{(m+1)p})^{-1}(1-q^{mp})^{-1}$$
$$\times \sum_{j=0}^{(p-1)/2} \begin{bmatrix} p-1 \\ j \end{bmatrix} (-1)^j q^{\binom{j+1}{2}} (q^{mpj}(1-q^{mp((p-1)/2-j)})$$
$$-q^{\binom{p}{2} - pj + mp(p-1)/2}(1-q^{mp((p-1)/2-j)}))$$
$$= (1-q^{(m+1)p})^{-1}(1-q^{mp})^{-1}$$
$$\times \sum_{j-0}^{(p-1)/2} \begin{bmatrix} p-1 \\ j \end{bmatrix} (-1)^j q^{\binom{j+1}{2} + mpj}(1-q^{mp((p-1)/2-j)})$$
$$\times \left(1 - q^{(m+1)\binom{p}{2} - (m+1)pj}\right)$$
$$= \sum_{j=0}^{(p-1)/2} q^{\binom{j+1}{2}} \begin{bmatrix} p-1 \\ j \end{bmatrix} (-1)^j q^{mpj}$$
$$\frac{(1-q^{mp((p-1)/2-j)})}{1-q^{mp}} \cdot \frac{(1-q^{(m+1)p((p-1)/2-j)})}{(1-q^{(m+1)p})}$$
$$\equiv \sum_{j=0}^{(p-1)/2} ((p-1)/2-j)^2$$

(mod $[p]$) (by (2.3), (2.1) and the finite geometric series)

$$= \sum_{j=0}^{(p-1)/2} j^2$$
$$= \frac{1}{6}((p-1)/2)((p+1)/2)p$$
$$= (p^2-1)p/24. \qquad \square$$

Corollary 1. $(1-q^{mp})(1-q^{(m+1)p})$ *divides*

$$(q^{mp+1};q)_{p-1} - q^{mp(p-1)/2}(q;q)_{p-1}. \tag{2.4}$$

Proof. This is immediately seen from the line of the proof in Theorem 1 just preceding the invocation of the congruence modulo $[p]$ because of all the terms in the sum are polynomials. □

3. q-Binomial congruences

The most natural consequences of the results in Section 2 are actually direct q-generalizations of Babbage's Theorem (1.1).

Theorem 2. *Let p be an odd prime and $h > 1$. Then*

$$\begin{bmatrix} hp-1 \\ p-1 \end{bmatrix} \equiv q^{(h-1)p(p-1)/2} \pmod{[p]_{q^h}[p]_{q^{h-1}}}; \tag{3.1}$$

$$\begin{bmatrix} hp-1 \\ p-1 \end{bmatrix} \equiv q^{(h-1)p(p-1)/2} \pmod{[p]^2}. \tag{3.2}$$

Proof. We see immediately by Corollary 1 that the product of the polynomials $[p]_{q^h}$ and $[p]_{q^{h-1}}$ divides

$$(q^{(h-1)p+1};q)_{p-1} - q^{(h-1)p(p-1)/2}(q;q)_{p-1}\,.$$

Consequently,

$$(q^{(h-1)p+1};q)_{p-1} \equiv q^{(h-1)p(p-1)/2}(q;q)_{p-1} \pmod{[p]_{q^h}[p]_{q^{h-1}}}$$

Finally, since $(q;q)_{p-1}$ is prime to $[p]_{q^h}[p]_{q^{h-1}}$ (because the latter's roots are primitive pth power roots of unity), it follows on division by $(q;q)_{p-1}$ that

$$\begin{bmatrix} hp-1 \\ p-1 \end{bmatrix} \equiv q^{(h-1)p(p-1)/2} \pmod{[p]_{q^h}[p]_{q^{h-1}}}.$$

Congruence (3.2) follows since $[p]$ is a factor of $[p]_{q^h}$ for each $h \geqslant 1$. □

Theorem 3. *Let p be an odd prime and $h > j \geqslant 1$. Then*

$$\begin{bmatrix} hp-1 \\ jp-1 \end{bmatrix}_q \equiv q^{(h-j)jp(p-1)/2} \begin{bmatrix} h-1 \\ j-1 \end{bmatrix}_{q^p} \pmod{[p]^2}. \tag{3.3}$$

Proof. First, we note that each polynomial of the form

$$\left(q^{mp+1};q\right)_{p-1} \tag{3.4}$$

is prime to $[p]$ because no primitive pth root of unity is a root of the polynomial in (3.4).

Therefore,

$$\begin{aligned}
&\begin{bmatrix} hp-1 \\ jp-1 \end{bmatrix} \\
&= \frac{(q^{(h-1)p+1};q)_{p-1}(1-q^{(h-1)p})(q^{(h-2)p+1};q)_{p-1}(1-q^{(h-2)p})\cdots}{(q^{(j-1)p+1};q)_{p-1}(1-q^{(j-1)p})(q^{(j-2)p+1};q)_{p-1}(1-q^{(j-2)p})\cdots} \\
&\quad\times\frac{\cdots(q^{(h-j+1)p+1};q)_{p-1}(1-q^{(h-j+1)p})(q^{(h-j)p+1};q)_{p-1}}{\cdots(q^{p+1};q)(1-q^p)(q;q)_{p-1}} \\
&\equiv \begin{bmatrix} h-1 \\ j-1 \end{bmatrix}_{q^p} \frac{q^{(h-1)\binom{p}{2}+(h-2)\binom{p}{2}+\cdots+(h-j)\binom{p}{2}}}{q^{(j-1)\binom{p}{2}+(j-2)\binom{p}{2}+\cdots+1\binom{p}{2}+0\binom{p}{2}}} \\
&\quad (\mathrm{mod}[p]^2) \quad (\text{by}(3.2)) \\
&\equiv \begin{bmatrix} h-1 \\ j-1 \end{bmatrix}_{q^p} q^{(h-j)jp(p-1)/2} \quad (\mathrm{mod}[p]^2) \qquad \square
\end{aligned}$$

4. Wolstenholme's binomial congruence

To prove (1.3) with $p>3$ Glaisher [11] only needs to invoke (1.5) with $m=1$. Hence to show that we really have a q-generalization of Wolstenholme's theorem in Theorem 1 we need only show that (1.6) implies (1.5). Now (1.6) is equivalent to the assertion that there exists a polynomial $F(q)$ with integer coefficients such that

$$\frac{(q^{mp+1};q)_{p-1} - q^{mp(p-1)/2}(q;q)_{p-1}}{(1-q^{(m+1)p})(1-q^{mp})} = \frac{(p^2-1)p}{24} + F(q)[p]. \tag{4.1}$$

So divide both sides of (4.1) by $(1-q)^{p-3}$. If we do that and let $q \to 1$, the left hand side reduces to

$$\frac{(mp+1)(mp+2)\cdots(mp+p-1)-(p-1)!}{(m+1)mp^2}. \tag{4.2}$$

Therefore, the expression (4.2) is an integer (by Corollary 1) which is equal to

$$\lim_{q\to 1}\left(\frac{(p^2-1)p}{24} + F(q)[p]\right)/(1-q)^{p-3}. \tag{4.3}$$

Therefore 1 must be a $(p-3)$rd-order zero of the numerator of (4.3), and the evaluation of this expression will follow from application of L'Hopital's rule $p-3$ times (remember $p>3$). Hence by Leibnitz's rule for differentiation, the expression in (4.3) is

$$\sum_{j=0}^{p-3}\binom{p-3}{j}F^{(j)}(1)\left(\frac{\mathrm{d}^{p-3-j}}{\mathrm{d}q^{p-3-j}}[p]\right)_{q=1}/(p-3)! \tag{4.4}$$

But

$$\left(\frac{\mathrm{d}^N}{\mathrm{d}q^N}[p]\right)_{q=1} = \left(\frac{d^N}{dq^N}(1+q+q^2+\cdots+q^{p-1})\right)_{q=1} \tag{4.5}$$

$$= N!\sum_{j=0}^{p-1}\frac{j(j-1)\cdots(j-N+1)}{N!}$$

$$= N!\sum_{j=0}^{p-1}\binom{j}{N}$$

$$= N!\binom{p}{N+1},$$

which is divisible by p if $N \leqslant p-2$. Therefore the expression in (4.4) is an integer divisible by p. Hence equating (4.4) and (4.2) we see that

$$(mp+1)(mp+2)\cdots(mp+p-1)-(p-1)!$$

is divisible by p^3. Consequently (1.5) is proved, and thus Wolstenholme's congruence follows from Theorem 1.

5. Wolstenholme's harmonic series congruence

In order to prove (1.3), Wolstenholme [21] proves auxiliary congruences. Indeed one of these is much more famous than any of the congruences previously cited, namely for each prime $p > 3$

$$H_{p-1} := \sum_{j=1}^{p-1}\frac{1}{j} \equiv 0 \pmod{p^2}. \tag{5.1}$$

Using Theorem 1, we can prove a related congruence for the q-harmonic series:

Theorem 4. *For odd primes p,*

$$H_{p-1}(q) := \sum_{j=1}^{p-1}\frac{1}{[j]} \equiv \frac{p-1}{2}(1-q) \pmod{[p]}, \tag{5.2}$$

and

$$\sum_{j=1}^{p-1}\frac{q^j}{[j]} \equiv \frac{p-1}{2}(q-1) \pmod{[p]}. \tag{5.3}$$

Remark. This result implies (5.1) with p^2 replaced by p upon setting $q=1$.

Proof. We rewrite the expression in (2.4) and note that Corollary 1 implies

$$\prod_{j=1}^{p-1}([mp]+q^{mp}[j]) - q^{m\binom{p}{2}}\prod_{j=1}^{p-1}[j] \equiv 0 \pmod{[p]^2}.$$

If we multiply out the first expression and delete those terms divisible by $[p]^2$, we find upon dividing by $[1][2]\cdots[p-1]$ that

$$[mp]q^{mp(p-2)}H_{p-1}(q) + q^{mp(p-1)} - q^{m\binom{p}{2}} \equiv 0 \pmod{[p]^2}.$$

Consequently, invoking (2.1), we see that

$$mH_{p-1}(q) - m\binom{p-1}{2}(1-q) \equiv 0 \pmod{[p]};$$

so

$$H_{p-1}(q) \equiv \frac{p-1}{2}(1-q) \pmod{[p]},$$

as desired.

Of course, once such a congruence has been found, it is easy to verify it independently. Namely,

$$\begin{aligned} &H_{p-1}(q) - \frac{p-1}{2}(1-q) \\ &= (1-q)\sum_{j=1}^{(p-1)/2}\left(\frac{1}{1-q^j} + \frac{1}{1-q^{p-j}} - 1\right) \\ &= (1-q)\sum_{j=1}^{(p-1)/2}\frac{1-q^p}{(1-q^j)(1-q^{p-j})} = \sum_{j=1}^{(p-1)/2}\frac{[p]}{[j][p-j]} \equiv 0 \pmod{[p]}. \end{aligned}$$

Finally, since

$$H_{p-1}(q) = \sum_{j=1}^{p-1}\frac{1}{[j]} = \sum_{j=1}^{p-1}\frac{1-q^j+q^j}{[j]} = (p-1)(1-q) + \sum_{j=1}^{p-1}\frac{q^j}{[j]},$$

we see that

$$\sum_{j=1}^{p-1}\frac{q^j}{[j]} \equiv \frac{p-1}{2}(q-1) \pmod{[p]}. \qquad \square$$

6. Conclusion

It should be noted that Babbage's original proof of (1.2) can be directly extended to prove the q-analog. Namely,

$$\begin{bmatrix} 2p-1 \\ p-1 \end{bmatrix} = \frac{1}{1+q^p}\begin{bmatrix} 2p \\ p \end{bmatrix} \tag{6.1}$$

$$= \frac{1}{1+q^p}\sum_{j=0}^{p} q^{j^2}\begin{bmatrix} p \\ j \end{bmatrix}^2 \quad \text{(by ([1, p. 37, Eq. (3.3.10)]))}$$

$$\equiv \frac{1+q^{p^2}}{1+q^p} \pmod{[p]^2}$$

$$\equiv q^{p(p-1)/2} \pmod{[p]^2},$$

The last line follows from the fact that

$$(1+q^{p^2}) - q^{\binom{p}{2}}(1+q^p) = (1-q^{\binom{p}{2}})(1-q^{\binom{p+1}{2}}).$$

Note that (6.1) is an immediate corollary of Theorem 2.

One would like a perfect q-analog of (1.4) and (1.5); however, the empirical studies I made which led to Theorem 1 do not yield any instances of congruences modulo $[p]^3$.

We note immediately from Theorem 1 that for $a \geqslant 1$ and p an odd prime

$$H(a,p;q) = \frac{(q^{ap+1};q)_{p-1} - q^{ap(p-1)/2}(q;q)_{p-1}}{(1-q)^{p-3}(1-q^{(a+1)p})(1-q^{ap})} \tag{6.2}$$

is a polynomial. Furthermore, a little algebra reveals that it is a reciprocal polynomial.

A unimodal polynomial

$$p(q) = \sum_{j=0}^{N} a_j q^j$$

is one for which $a_0 \leq a_1 \leqslant a_2 \leqslant \ldots \leqslant a_M \geqslant a_{M+1} \geqslant \ldots \geqslant a_N$ for some M with $0 \leqslant M \leqslant N$.

Conjecture. For $a \geqslant 1$ and p an odd prime, $H(a,p;q)$ is a unimodal polynomial with positive coefficients.

This conjecture should be much easier than K. O'Hara's proof that the q-binomial coefficients (defined in (2.2)) are unimodal. From (6.2) we see directly that

$$H(a,p;q) = \frac{\prod_{j=1}^{p-1}(1+q+\cdots+q^{j-1})\left(\begin{bmatrix} (a+1)p-1 \\ p-1 \end{bmatrix} - q^{a\binom{p}{2}}\right)}{[ap][(a+1)p]} \tag{6.3}$$

The numerator is the product of unimodal and 'almost' unimodal polynomials (for $p = 7$, $a = 1$, $\begin{bmatrix} 13 \\ 6 \end{bmatrix} - q^{21}$ is not unimodal), and the denominator (which we know divides the numerator) is the product of two rather 'flat' unimodal polynomials.

Finally, it should be pointed out that the famous congruence (5.1) as well as

$$\sum_{j=1}^{p-1} \frac{1}{j^2} \equiv 0 \pmod{p} \tag{6.4}$$

are fairly easily deduced by expanding the left-hand side of (1.5) as a polynomial in m. Thus Section 4 reveals that both (5.1) and (6.4) follow from Theorem 1.

Acknowledgements

I wish to thank Lou Solomon who recently drew my attention to this fascinating branch of number theory and who has provided me numerous insights on this subject. Doug Bowman pointed me to most of the recent work, and Ken Stolarsky pointed out [13] to me.

References

[1] G.E. Andrews, in: G.-C. Rota (Ed.), The Theorem of Partitions, Encyclopedia of Mathematics and Its Applications, vol. 2, Addison-Wesley, Reading, 1976. (Reissued: Cambridge University Press, Cambridge, 1984).

[2] C. Babbage, Demonstration of a theorem relating to prime numbers, Edinburgh Philosophical J. 1 (1819) 46–49.

[3] D.F. Bailey, Two p^3 variations of Lucas' theorem, J. Number Theory 35 (1990) 208–215.

[4] W.E. Clark, q-Analog of a binomial coefficient congruence, Internat. J. Math. and Math. Sci. 18 (1995) 197–200.

[5] K.S. Davis and W.A. Webb, Lucas' theorem for prime powers, European J. Combin. 11 (1990) 229–233.

[6] K.S. Davis and W.A. Webb, A binomial coefficient congruence modulo prime powers, J. Number Th. 43 (1993) 20–23.

[7] N.J. Fine, Binomial coefficients modulo a prime, Amer. Math. Monthly 54 (1947) 489–592.

[8] J.W.L. Glaisher, On the residue of a binomial-theorem coefficient with respect to a prime modulus, Quart. J. Math. Oxford Series 30 (1899) 150–156.

[9] J.W.L. Glaisher, On the residue with respect to p^{n+1} of a binomial-theorem coefficient divisible by p^n, Quart. J. Math. Oxford Series 30 (1899) 349–360.

[10] J.W.L. Glaisher, Congruences relating to the sums of products of the first n numbers and to other sums of products, Quart. J. Math. Oxford Series 31 (1900) 1–35.

[11] J.W.L. Glaisher, Residues of binomial-theorem coefficients with respect to p^3, Quart. J. Math., Oxford Series 31 (1900) 110–124.

[12] H.W. Gould, Some relations involving the finite harmonic series, Math. Mag. 34 (1961) 317–321.

[13] R. Gow, Joseph Wolstenholme, Leslie Stephen 'To the Lighthouse', Irish Math. Soc. Bull. 34 (1995) 40–46.

[14] A. Granville, Binomial coefficients (mod p^q), Proc. Organic Math Workshop at http://www.cecm.sfu.ca/projects/OMP, 1996.

[15] E. Lucas, Sur les congruences des nombres eulériens et des coefficients différentiels des fonctions trigonométriques, suivant un module premier, Bull. Soc. Math. de France 6 (1878) 49–54.

[16] R.J. McIntosh, On the converse of Wolstenholme's theorem, preprint.
[17] D. Singmaster, Divisibility of binomial and multinomial coefficients by primes and prime powers, in 'A Collection of Manuscripts Related to the Fibonacci Sequence', Fibonacci Association, Santa Clara, 1980, pp. 98–113.
[18] R.P. Stanley, 'Enumerative Combinatorics', Wadsworth & Brooks/Cole, 1986.
[19] M. Sved, Divisibility — with visibility, Math. Intelligencer 10 (1988) 56–64.
[20] L. Washington, Introduction to Cyclotomic Fields, Springer, New York, 1982.
[21] J. Wolstenholme, On certain properties of prime numbers, Quart. J. Math., Oxford Ser. 5 (1862) 35–39.

6.6 On the Number of Distinct Multinomial Coefficients (with A. Knopfmacher and B. Zimmermann)

ELSEVIER

Available online at www.sciencedirect.com
SCIENCE DIRECT

Journal of Number Theory 118 (2006) 15–30

JOURNAL OF Number Theory

www.elsevier.com/locate/jnt

On the number of distinct multinomial coefficients

George E. Andrews [a,*,1], Arnold Knopfmacher [b,2], Burkhard Zimmermann [c,3]

[a] *Mathematics Department, 410 McAllister Building, The Pennsylvania State University, University Park, PA 16802, USA*

[b] *The John Knopfmacher Centre for Applicable Analysis and Number Theory, University of the Witwatersrand, Johannesburg, South Africa*

[c] *Research Institute for Symbolic Computation, Johannes Kepler Universität Linz, A-4040 Linz, Austria*

Received 10 November 2003

Available online 17 November 2005

Communicated by David Goss

Abstract

We study $M(n)$, the number of distinct values taken by multinomial coefficients with upper entry n, and some closely related sequences. We show that both $p_{\mathbb{P}}(n)/M(n)$ and $M(n)/p(n)$ tend to zero as n goes to infinity, where $p_{\mathbb{P}}(n)$ is the number of partitions of n into primes and $p(n)$ is the total number of partitions of n. To use methods from commutative algebra, we encode partitions and multinomial coefficients as monomials.

Keywords: Factorials; Binomial coefficients; Combinatorial functions; Partitions of integers; Polynomial ideals; Gröbner bases

* Corresponding author.

E-mail addresses: andrews@math.psu.edu (G.E. Andrews), arnoldk@cam.wits.ac.za (A. Knopfmacher), zimmermann@risc.uni-linz.ac.at (B. Zimmermann).

[1] Partially supported by National Science Foundation Grant DMS-0200047.

[2] Partially supported by The John Knopfmacher Center for Applicable Analysis and Number Theory of the University of the Witwatersrand.

[3] Supported by SFB grant F1305 of the Austrian FWF.

doi:10.1016/j.jnt.2005.08.012

1. Introduction

The classical multinomial expansion is given by

$$(x_1+x_2+\cdots+x_k)^n=\sum\binom{n}{i_1,i_2,\ldots,i_k}x_1^{i_1}x_2^{i_2}\cdots x_k^{i_k}, \tag{1}$$

where the sum runs over all $(i_1,i_2,\ldots,i_k)$ such that $i_1+i_2+\cdots+i_k=n$ and $i_1,i_2,\ldots,i_k\geqslant 0$. Multinomial coefficients are defined by

$$\binom{n}{i_1,i_2,\ldots,i_k}:=\frac{n!}{i_1!i_2!\cdots i_k!}. \tag{2}$$

It is natural to ask about $M_k(n)$, the number of different values of

$$\binom{n}{i_1,i_2,\ldots,i_k}, \tag{3}$$

where $i_1+i_2+\cdots+i_k=n$. Obviously if the $i_1,i_2,\ldots,i_k$ are merely permuted, then the value of $\binom{n}{i_1,i_2,\ldots,i_k}$ is unchanged. However identical values do not necessarily arise only by permuting the $i_1,i_2,\ldots,i_k$. For example,

$$\binom{7}{3,2,2}=\binom{7}{4,1,1,1} \tag{4}$$

and

$$\binom{236}{64,55,55,52,7,3}=\binom{236}{62,56,54,51,13}. \tag{5}$$

We note that if $k\geqslant n$, then $M_k(n)=M_n(n)$, and we define $M(n):=M_n(n)$ to be the total number of distinct multinomial coefficients with upper entry n.

Since permuting its lower indices leaves the value of a multinomial coefficient unchanged it is immediately clear that

$$M_k(n)\leqslant p_k(n) \tag{6}$$

and

$$M(n)\leqslant p(n), \tag{7}$$

where $p_k(n)$ is the number of partitions of n into at most k parts, and $p(n)$ is the total number of partitions of n, respectively [2]. Observing that the binomial coefficients $\binom{n}{k,n-k}$ are strictly increasing for $0\leqslant k\leqslant \frac{n}{2}$, we deduce that, in fact,

$$M_2(n)=p_2(n). \tag{8}$$

However the inequality (6) seems to be stronger for large k. Indeed (Theorem 8),

$$\lim_{n\to\infty} \frac{M(n)}{p(n)} = 0. \tag{9}$$

Bounding $M(n)$ from below we will prove (Theorem 1) that

$$M(n) \geqslant p_{\mathbb{P}}(n), \tag{10}$$

where $p_{\mathbb{P}}(n)$ is the number of partitions of n into parts belonging to the set of primes $\mathbb{P}$. Indeed (Theorem 12),

$$\lim_{n\to\infty} \frac{p_{\mathbb{P}}(n)}{M(n)} = 0. \tag{11}$$

It is natural to generalize the problem from $M(n)$ to $M_S(n)$, the number of different multinomial coefficients with upper entry n whose lower entries belong to a given set S of natural numbers. Let

$$\mathcal{M}_S(q) := \sum_n M_S(n) q^n \quad \text{and} \quad \mathcal{P}_S(q) := \sum_n p_S(n) q^n, \tag{12}$$

where $p_S(n)$ is the number of partitions of n into elements from S. Define $[s] := \{1, 2, \ldots, s\}$. Results of numerical calculations such as

$$\mathcal{M}_{[4]}(q)/\mathcal{P}_{[4]}(q) = 1 - q^7 + O\left(q^{100}\right) \tag{13}$$

and

$$\mathcal{M}_{[7]}(q)/\mathcal{P}_{[7]}(q) = 1 - q^7 - q^8 - q^{10} + q^{12} + q^{13} + O\left(q^{100}\right) \tag{14}$$

suggest that $\mathcal{M}_S(q)/\mathcal{P}_S(q)$ is a polynomial for any finite S. This is indeed true (Theorem 5) and leads to an algorithm for computing a closed form for the sequence $M_S(n)$ for a given finite set S (Section 4).

Partitions and multinomial coefficients can be written as monomials in a natural way: For instance, the monomial $q_4 q_1^3$ represents the partition $4 + 1 + 1 + 1$, and $x_7 x_5 x_3 x_2$ represents the multinomial coefficient $\binom{7}{4,1,1,1}$ whose factorization into primes is $7 \cdot 5 \cdot 3 \cdot 2$. This encoding serves as a link between our counting problem and Hilbert functions (Section 3). Sections 4, 5 and 6 are based on that link.

We call a pair of partitions of n that yield the same multinomial coefficient but have no common parts an *irreducible pair*. For example, the partitions $4+1+1+1$ and $3+2+2$ form an irreducible pair according to Eq. (4). In Section 7, we study $i(n)$ the total number of irreducible pairs of partitions of n, and we prove (Theorem 13) that $i(n) > \frac{n}{56} - 1$.

Related work: Erdös and Niven [9] prove that the number of multinomial coefficients $\binom{n}{i_1, i_2, \ldots, i_k}$ less than a fixed natural number x is $(1+\sqrt{2})\sqrt{x} + o(\sqrt{x})$, once the trivial case $\binom{n}{n,1} = n$ is excluded.

2. A lower bound for $M(n)$

We relate $M(n)$ to $p_{\mathbb{P}}(n)$ whose asymptotics is known by a theorem of Kerawala [11]:

$$\log p_{\mathbb{P}}(n) \sim \frac{2\pi}{\sqrt{3}}\sqrt{\frac{n}{\log n}}. \tag{15}$$

Theorem 1. $M(n) \geqslant p_{\mathbb{P}}(n)$.

Theorem 1 is implied by the following lemma:

Lemma 2. *Any two distinct partitions of the same natural number n into primes yield different multinomial coefficients.*

Proof of Lemma 2. It suffices to show that if

$$p_1!p_2!\cdots p_r! = q_1!q_2!\cdots q_s!, \tag{16}$$

where $p_1 \leqslant p_2 \leqslant \cdots \leqslant p_r$ and $q_1 \leqslant q_2 \leqslant \cdots \leqslant q_s$ are all primes, then $r = s$ and $p_i = q_i$ for $i = 1, \ldots, s$. We proceed by mathematical induction on r.

If $r = 1$, then q_s must equal p_1 because if $q_s < p_1$ then p_1 divides the left side of the above equation but not the right side. If $q_s > p_1$ then q_s divides the right side but not the left. Hence $q_s = p_1$, and dividing both sides by $p_1!$ we see that there can be no other q_i. Hence $s = 1$ and $q_1 = p_1$.

Assume now that our result holds up to but not including a particular r. As in the case $r = 1$, we must have $q_s = p_r$. Cancel $p_r!$ from both sides and apply the induction hypothesis to conclude that $s - 1 = r - 1$ and $p_i = q_i$ for $i = 1, \ldots, s - 1$. Hence the lemma follows by mathematical induction. □

Some values of $p_{\mathbb{P}}(n)$ and $M(n)$ are listed in Table 1. We will refine Theorem 1 in Section 6.

Table 1

n	$p_{\mathbb{P}}(n)$	$L_{[4]}(n)$	$p^*(n)$	$M(n)$	$p^\#(n)$	$U_{\mathbb{N}^+,[4]}(n)$	$p(n)$
0	1	1	1	1	1	1	1
10	5	30	36	36	39	39	42
20	26	232	357	366	445	526	627
30	98	1102	2064	2131	2875	4349	5604
40	302	4020	8853	9292	13 549	27 195	37 338
50	819	12 405	31 639	33 799	52 321	140 965	204 226
60	2018	34 016	99 245	107 726	175 426	636 536	966 467
70	4624	85 333	281 307	310 226	527 909	2 582 469	4 087 968

3. The algebraic setting

Encoding partitions and multinomial coefficients as monomials allows us to apply constructive methods from commutative algebra to the problem of counting multinomial coefficients. Let us assume that $S \subseteq \mathbb{N}$ throughout the paper. We will see that $M_S(n)$ finds a natural interpretation as the Hilbert function of a certain graded ring (Lemma 4). In the case of finite S, it can be computed by the method of Gröbner bases [1,5,6,8].

We represent the partition $\lambda_0 + \lambda_1 + \cdots + \lambda_i$ of n by the monomial $q_{\lambda_0} q_{\lambda_1} \cdots q_{\lambda_i}$ whose degree is n if we define the degree of variables suitably by $\deg q_j := j$. For convenience, we will use the notions "partition of n" and "monomial of degree n" interchangeably.

Let k be a field of characteristic zero. We abbreviate the ring $k[q_i\colon i \in S]$ of polynomials in the variables q_i for $i \in S$ over k by $k[S]$. Define the degree of monomials by $\deg q_i := i$, and let $k[S]_n$ denote the subspace of all homogeneous polynomials of degree n. In other words, $k[S]_n$ is the k-vector space whose basis are the partitions of n into parts S. Note that $k[S]$ is graded by $k[S] = \bigoplus_n k[S]_n$. For instance,

$$k\big[\{1,3,\ldots\}\big]_4 = k \cdot q_3 q_1 \oplus k \cdot q_1^4 \tag{17}$$

corresponding to the partitions $3+1$ and $1+1+1+1$ of 4 into odd parts.

The multinomial coefficients with upper entry n into parts belonging to S are the numbers $n!/\prod_j j!^{a_j}$ where $\prod_j q_j^{a_j}$ ranges over the monomials in $k[S]_n$. Since the numerator $n!$ of these fractions is fixed, it suffices to count the set of all denominators:

$$M_S(n) = \left|\left\{\prod_j j!^{a_j}\colon \prod_j q_j^{a_j} \in k[S]_n\right\}\right|. \tag{18}$$

To count the values taken by $\prod_j j!^{a_j}$, we look at their factorization into primes. Let $h(q_j)$ be the factorization of $j!$ into primes, written as a monomial in $k[x] := k[x_p\colon p \text{ prime}]$, multiplied by q^j. For example, $h(q_5) = q^5 x_2^3 x_3 x_5$ corresponding to $5! = 2^3 \cdot 3 \cdot 5$. An elementary counting argument [10] shows that the prime p occurs in the factorization of $j!$ with exponent $\sum_{l=1}^{\infty} \lfloor j/p^l \rfloor$, where $\lfloor x \rfloor$ denotes the largest integer that does not exceed the real number x. Therefore,

$$h(q_j) = q^j \prod_{p \text{ prime}} x_p^{\sum_{l=1}^{\infty} \lfloor j/p^l \rfloor}. \tag{19}$$

Since factorization into primes is unique, (18) can be written as

$$M_S(n) = \left|\left\{\prod_j h(q_j)^{a_j}\colon \prod_j q_j^{a_j} \in k[S]_n\right\}\right|. \tag{20}$$

Extending h to a k-algebra homomorphism $k[S] \to k[x,q]$ allows us to reformulate (20) as

Lemma 3.

$$M_S(n) = \dim_k / h\big(k[S]_n\big). \tag{21}$$

Example. Since there are 10 partitions of 7 into parts 1, 2, 3 and 4, the dimension of

$$k\big[\{1,2,3,4\}\big]_7 = kq_4q_3 \oplus kq_4q_2q_1 \oplus kq_4q_1^3 \oplus kq_3q_2^2 \oplus \cdots \oplus kq_1^7 \tag{22}$$

over k is 10. However, the dimension of its image

$$h\big(k\big[\{1,2,3,4\}\big]_7\big) = kq^7x_2^4x_3^2 \oplus kq^7x_2^4x_3 \oplus kq^7x_2^3x_3 \oplus \cdots \oplus kq^7 \tag{23}$$

under h is only 9 and so $M_{[4]}(7) = 9$. The defect is due to $h(q_4q_1^3) = h(q_3q_2^2)$ which is nothing but a restatement of (4).

To use Lemma 3 for effective computation (in the case of finite S), we express $\dim_k h(k[S]_n)$ as the value (at n) of the Hilbert function of a certain elimination ideal. This method is taken from [1]; the result in our case is Lemma 4 below.

First we make the map h degree-preserving (graded) by defining $\deg q := 1$ and $\deg x_p := 0$ in the ring $k[x_p\colon\ p \text{ prime}][q]$. (This is why we introduced the extra factor of q^j in the defining Eq. (19) of h.) Second, note that

$$h\big(k[S]_n\big) \cong k[S]_n / \big(k[S]_n \cap \ker h\big) \tag{24}$$

as k-vector spaces, since h is a k-linear map on $k[S]_n$. In particular, dimensions agree. Therefore,

$$M_S(n) = \dim_k k[S]_n / \big(k[S]_n \cap \ker h\big). \tag{25}$$

Recall that the (projective) Hilbert function H_R of a graded k-algebra $R = \bigoplus_n R_n$ is defined by $H_R(n) := \dim_k R_n$. Thus (25) relates M_S to the Hilbert function of $k[S]/\ker h$:

$$M_S(n) = H_{k[S]/\ker h}(n). \tag{26}$$

By [1, Theorem 2.4.2], $\ker h$ can be computed by elimination:

$$\ker h = I \cap k[S], \tag{27}$$

where the ideal I of $k[S][q][x_p\colon\ p \text{ prime}]$ is defined by

$$I := \big\langle q_j - h(q_j)\colon\ j \in S\big\rangle. \tag{28}$$

Summarizing this section, we have proved the following lemma:

Lemma 4. *Let $k[S]$ be graded by $\deg q_i := i$. Define a k-algebra homomorphism from $k[S]$ to $k[q,x]$ by*

$$h(q_j) := q^j \prod_p x_{p^{\sum_{l=1}^{\infty} \lfloor j/p^l \rfloor}}. \tag{29}$$

Let the ideal I of $k[S,q,x]$ be defined by

$$I := \langle q_j - h(q_j)\colon\ j \in S \rangle \tag{30}$$

and let

$$J := I \cap k[S]. \tag{31}$$

Then M_S is the (projective) Hilbert function of the k-algebra $k[S]/J$:

$$M_S(n) = H_{k[S]/J}(n). \tag{32}$$

Example. If $S = [4]$, then $I = \langle q_1 - q,\ q_2 - q^2 x_2,\ q_3 - q^3 x_2 x_3,\ q_4 - q^4 x_2^3 x_3 \rangle$ and $J = \langle q_4 q_1^3 - q_3 q_2^2 \rangle$. For $M_{[4]}(n)$, see (40).

4. Explicit answers

Let S be a given finite set throughout this section. Lemma 4 allows to compute a closed form for the sequence $M_S(n)$ by well-known methods from computational commutative algebra. For the sake of completeness, let us briefly review them:

(1) Fix a term order $\preccurlyeq$ on $k[S,q,x]$ that allows the elimination of the variable q and the variables x_p in step 2 below. Compute a Gröbner basis F for the (toric) ideal $I = \langle q_j - h(q_j)\colon\ j \in S \rangle$ with respect to this term order using Buchberger's algorithm [5,6].
(2) Let $G := F \cap k[S]$. By the elimination property of Gröbner bases with respect to a suitable elimination order $\preccurlyeq$, the set G is a Gröbner basis for the elimination ideal $J = I \cap k[S]$.
(3) Let $L := I_{\preccurlyeq}(G)$ be the set of leading terms of polynomials in G.
(4) Compute $\mathcal{M}_S(q)$ using

$$\mathcal{M}_S(q) = \mathcal{H}_{k[S]/J}(q) = \mathcal{H}_{k[S]/I_{\preccurlyeq}(J)}(q) = \mathcal{H}_{k[S]/\langle L \rangle}(q). \tag{33}$$

The first equality holds by Lemma 4. The second equality is an identity of Macaulay [12]. Since G is a Gröbner basis, its initial terms L generate the initial term ideal of $\langle G \rangle$ with respect to $\preccurlyeq$, which explains the third equation sign. A naive method for computing the Hilbert–Poincaré series of $k[S]/\langle L\rangle$ is to apply the inclusion–exclusion relation

$$\mathcal{H}_{k[S]/\langle\{t\}\cup L\rangle}(q)=\mathcal{H}_{k[S]/\langle L\rangle}(q)-q^{\deg t}\mathcal{H}_{k[S]/\langle L\rangle:t}(q), \tag{34}$$

recursively until the base case

$$\mathcal{H}_{k[S]/\langle\rangle}(q)=\mathcal{H}_{k[S]}(q)=\frac{1}{\prod_{j\in S}(1-q_j)} \tag{35}$$

is reached. For better (faster) algorithms, see [4].

(5) Extract a closed form expression for $H_{k[S]/\langle L\rangle}(n)$ from its generating function $\mathcal{H}_{k[S]/\langle L\rangle}(q)$. (Use partial fraction decomposition and the binomial series). It is the desired answer $M_S(n)$.

One of the authors computed (1)–(4) for several finite S using a few different computer algebra systems. It turned out that CoCoA [7] was fastest for that purpose.

Theorem 5. *Let S be a finite subset of the positive natural numbers. Then*

(1) *$\mathcal{M}_S(q)$ can be written as*

$$\mathcal{M}_S(q)=\frac{f_S(q)}{\prod_{j\in S}(1-q^j)}, \tag{36}$$

where $f_S(q)$ is a polynomial with integer coefficients.

(2) *There exists n_0 such that $M_S(n)$ can be written as a quasipolynomial* [13] *for $n\geqslant n_0$. Moreover, it suffices to use periods which are divisors of elements of S.*

Proof. Relations (34) and (35) prove the first statement. The second statement follows from the first easily. □

Let us follow the algorithm in the case $S=[4]$, which is the simplest nontrivial case. We have $I=\langle q_1-q, g_2-q^2x_2, q_3-q^3x_2x_3, q_4-q^4x_2^3x_3\rangle$. To eliminate the variables x_3, x_2, and q we choose the lexical term order where $x_3\succ x_2\succ q\succ q_4\succ q_3\succ q_2\succ q_1$. The corresponding reduced Gröbner basis of I is

$$\begin{aligned}F=\{&q_1^3q_4-q_2^2q_3,\ q-q_1,\ q_1^2x_2-q_2,\ q_2q_3x_2-q_1q_4,\ q_1q_3x_2^2-q_4,\ q_1q_2x_3-q_3,\\ &q_2^2x_3-q_1q_3x_2,\ q_1^2q_4x_3-q_3^2x_2,\ q_2q_4x_3-q_3^2x_2^2,\ q_1q_4^2x_3-q_3^3x_2^3,\ q_4^3x_3-q_3^4x_2^5\}.\end{aligned}$$

By the elimination property of Gröbner bases $G:=F\cap k[q_1,q_2,q_3,q_4]=\{q_1^3q_4-q_2^2q_3\}$ is a Gröbner basis for the elimination ideal $J=I\cap k[q_1,q_2,q_3,q_4]$. Collecting leading terms of G gives $L = \{q_1^3 q_4\}$. Since G is a Gröbner basis of J we know that $I_{\preccurlyeq}(J) = \langle q_1^3 q_4\rangle$. The Hilbert–Poincaré series of $k[q_1, q_2, q_3, q_4]/\langle q_1^3 q_4\rangle$ gives

$$\mathcal{M}_{[4]}(q) = \frac{1-q^7}{(1-q)(1-q^2)(1-q^3)(1-q^4)}. \tag{37}$$

It is clear that we may replace any occurrence of the partition $4+1+1+1$ in a multinomial coefficient by $3+2+2$ without changing the value of the multinomial coefficient. Therefore, there are at most as many multinomial coefficients as there are partitions avoiding $4+1+1+1$. Equation (38) states that this upper bound gives in fact the exact number in the case of $S = \{1, 2, 3, 4\}$.

Note that all denominators in the partial fraction decomposition

$$\begin{aligned}\mathcal{M}_{[4]}(q) = {} & -\frac{7}{24}\frac{1}{(q-1)^3} - \frac{77}{288}\frac{1}{(q-1)} + \frac{1}{16}\frac{1}{(q+1)^2} + \frac{1}{32}\frac{1}{(q+1)} \\ & + \frac{1}{9}\frac{(q+2)}{(q^2+q+1)} + \frac{1}{8}\frac{(q+1)}{(q^2+1)}\end{aligned} \tag{38}$$

of (37) are powers of cyclotomic polynomials $C_j(q)$, where j divides an element of $S = \{1, 2, 3, 4\}$. We rewrite this as

$$\begin{aligned}\mathcal{M}_{[4]}(q) = {} & \frac{7}{24}\frac{1}{(1-q)^3} + \frac{77}{288}\frac{1}{(1-q)} + \frac{1}{16}\frac{(1-q)^2}{(1-q^2)} + \frac{1}{32}\frac{(1-q)}{(1-q^2)} \\ & + \frac{1}{9}\frac{(2-q-q^2)}{(1-q^3)} + \frac{1}{8}\frac{(1+q-q^2-q^3)}{(1-q^4)},\end{aligned} \tag{39}$$

in order to use the binomial series $(1-z)^{-a-1} = \sum_{n=0}^{\infty}\binom{a+n}{a}z^n$. The result is

$$\begin{aligned}M_{[4]}(n) = {} & \frac{7}{48}n^2 + \left(\frac{1}{16}[1,-1](n) + \frac{7}{16}\right)n + \frac{1}{8}[1,1,-1,-1](n) \\ & + \frac{1}{9}[2,-1,-1](n) + \frac{3}{32}[1,-1](n) + \frac{161}{288},\end{aligned} \tag{40}$$

where $[a_0, a_1, \ldots, a_m](n) := a_j$ for $n \equiv j\,(m)$. Similar computations show that

$$\mathcal{M}_{[5]}(q) = \frac{1-q^7}{(1-q)(1-q^2)\cdots(1-q^5)}, \tag{41}$$

$$\mathcal{M}_{[6]}(q) = \frac{1-q^7-q^8-q^{10}+q^{12}+q^{13}}{(1-q)(1-q^2)\cdots(1-q^6)}, \tag{42}$$

and

$$\mathcal{M}_{[7]}(q) = \frac{1-q^7-q^8-q^{10}+q^{12}+q^{13}}{(1-q)(1-q^2)\cdots(1-q^7)}. \tag{43}$$

It is no coincidence that the numerators of (42) and (43) agree (Theorem 11).

5. Upper bounds

Trivially, $M(n) \leqslant p(n)$. Our goal is to find sharper upper bounds.

Lemma 6. *Assume $S' \subseteq S$.*

Let $\tilde{I}$ be the ideal of $k[S,q,x]$ generated by the set of polynomials $\{q_j - h(q_j)\colon\ j \in S'\}$. Let $\tilde{J}$ be the ideal generated by $\tilde{I} \cap k[S']$ in the ring $k[S]$. Let $U_{S,S'}(n) := H_{k[S]/\tilde{J}}(n)$. Then

(1) $M_S(n) \leqslant U_{S,S'}(n)$.
(2) *We have*

$$\sum_n U_{S,S'}(n) q^n = \frac{f_{S'}(q)}{\prod_{j\in S}(1-q^j)}, \tag{44}$$

where $f_{S'}(q)$ is defined by

$$\sum_n M_{S'}(n) q^n = \frac{f_{S'}(q)}{\prod_{j\in S'}(1-q^j)}. \tag{45}$$

Proof. We prove the first statement. Let I be the ideal of $k[S,q,x]$ generated by the set of polynomials $\{q_j - h(q_j)\colon\ j \in S\}$ and let $J = I \cap k[S]$. Since $\tilde{J}$ is a k-vector subspace of J we have

$$\dim_k k[S]_n \cap J \geqslant \dim_k k[S]_n \cap \tilde{J} \tag{46}$$

and therefore

$$\dim_k \big(k[S]/J\big)_n \leqslant \dim_k \big(k[S]/\tilde{J}\big)_n \tag{47}$$

i.e.

$$M_S(n) \leqslant U_{S,S'}(n). \tag{48}$$

To prove the second statement, let I' be the ideal generated by $\{q_j - h(q_j)\colon\ j \in S'\}$ in the ring $k[S',q,x]$ and let $J' := I' \cap k[S']$. Since the ideals $\tilde{J}$ and J' are generated by the same set of polynomials (albeit in different rings), the Hilbert functions $U_{S,S'}(n) = H_{k[S]/\tilde{J}}(n)$ and $M'_S(n) = H_{k[S']/J'}(n)$ correspond in the way claimed by (44) and (45). $\square$

To get upper bounds for $M(n)$, we use the preceding lemma in the special case $S = \mathbb{N}$ getting:

Theorem 7. *For any S' we have*

$$M(n) \leqslant [q^n]\frac{f_{S'}(q)}{\prod_{j=1}^{\infty}(1-q^j)} \tag{49}$$

(where $[q^n]\mathcal{A}(q)$ denotes the coefficient of q^n in the power series expansion of $\mathcal{A}(q)$). For instance, the cases $S' = [4]$ and $S' = [6]$ yield the bounds

$$M(n) \leqslant p(n) - p(n-7) \tag{50}$$

and

$$M(n) \leqslant p(n) - p(n-7) - p(n-8) - p(n-10) + p(n-12) + p(n-13). \tag{51}$$

Note that a direct proof of $M(n) \leqslant p(n) - p(n-7)$ could be given by exploiting the equivalence of the partitions $4+1+1+1$ and $3+2+2$ in the sense of Eq. (4).

The bound $M(n) \leqslant p(n) - p(n-7)$ is good enough to imply:

Theorem 8. $M(n) = o(p(n))$, *i.e.* $\lim_{n\to\infty} M(n)/p(n) = 0$.

Proof. Due to the monotonicity of $p(n)$ and the fact that the unit circle is the natural boundary for

$$\sum_{n=0}^{\infty} p(n)q^n = \prod_{n=1}^{\infty} \frac{1}{1-q^n}, \tag{52}$$

we see that

$$\lim_{n\to\infty} \frac{p(n-7)}{p(n)} = 1. \tag{53}$$

Hence

$$0 \leqslant \lim_{n\to\infty} \frac{M(n)}{p(n)} \leqslant \lim_{n\to\infty} \frac{p(n)-p(n-7)}{p(n)} = 1 - 1 = 0, \tag{54}$$

which proves Theorem 8. □

6. Lower bounds

Recall that $M(n) \geqslant p_{\mathbb{P}}(n)$ (Theorem 1). The numbers given in Table 1 suggest that $M(n)$ grows much faster than $p_{\mathbb{P}}(n)$. We will prove that this is indeed the case: $\lim_{n\to\infty} p_{\mathbb{P}}(n)/M(n) = 0$ (Theorem 12) and we will give better lower bounds for $M(n)$.

Let us write $S < P$ if each element of S is less than each element of P. We need the following generalization of Lemma 2:

Lemma 9. *Assume $S < P$ where P is a set of primes. Let s and s' be any two power products in $k[S]$ and let p and p' be distinct power products in $k[P]$. Then $h(sp) \neq h(s'p')$.*

In the case $S = \emptyset$, Lemma 9 states that distinct partitions p and p' into primes yield different multinomial coefficients: $h(p) \neq h(p')$. Lemma 9 can be proved by the same induction argument as Lemma 2.

Lemma 10. *Assume $S < P$ where P is a set of primes. Define h on $k[S \cup P]$ by* (19). *Then* $\ker h$ *is generated, as an ideal of* $k[S \cup P]$, *by* $\ker h \cap k[S]$.

Proof. Let $f \in \ker h$. Since $k[S \cup P] = k[S][P]$, we can f as a finite sum $f = \sum_s \sum_p c_{s,p} sp$ indexed by power products s and p from $k[S]$ and $k[P]$, respectively, with coefficients $c_{s,p} \in k$. As $f \in \ker h$, $\sum_s \sum_p c_{s,p} h(sp) = 0$. By Lemma 9, this implies $\sum_s c_{s,p} h(sp) = 0$ for arbitrary but fixed p. Canceling $h(p)$ from this equation shows that $h(f_p) = 0$ where $f_p := \sum_s c_{s,p} s$. In this way we succeed in writing f as $f = \sum_p f_p p$ where each f_p is in $\ker h \cap k[S]$. □

As an immediate consequence of Lemma 10 we get:

Theorem 11. *Assume $S < P$ where P is a set of primes. Then*

$$\mathcal{M}_{S \cup P}(q) = \mathcal{M}_S(q) / \prod_{j \in P} (1 - q^j). \tag{55}$$

As a first application of Theorem 11, we count multinomial coefficients with lower entries which are either prime or equal to 1:

$$\mathcal{M}_{\{1\} \cup \mathbb{P}}(q) = \frac{1}{(1-q) \prod_{j \in \mathbb{P}} (1 - q^j)}, \tag{56}$$

which allows for improving Theorem 1:

Theorem 12. *We have*

$$\lim_{n \to \infty} p_{\mathbb{P}}(n) / M_{\{1\} \cup \mathbb{P}}(n) = 0 \tag{57}$$

and therefore $\lim_{n \to \infty} p_{\mathbb{P}}(n) / M(n) = 0$.

Proof. Let $A(n) := M_{\{1\} \cup \mathbb{P}}(n)$. Due to the monotonicity of $A(n)$ and the fact that the unit circle is the natural boundary for we see that

$$\lim_{n \to \infty} A(n-1)/A(n) = 1. \tag{58}$$

By (56),

$$p_{\mathbb{P}}(n) = A(n) - A(n-1). \tag{59}$$

Therefore,

$$0 \leqslant \lim_{n\to\infty} \frac{p_{\mathbb{P}}(n)}{A(n)} \leqslant \lim_{n\to\infty} \frac{A(n)-A(n-1)}{A(n)} = 1-1=0, \tag{60}$$

which proves Theorem 12. □

Let $L_S(n) := M_{S\cup\mathbb{P}}(n)$; clearly, $L_S(n)$ is a lower bound for $M(n)$. Theorem 11 allows us deduce

$$\mathcal{L}_{[4]}(q) = \mathcal{L}_{[5]}(q) = \frac{1-q^7}{\prod_{j\in[4]\cup\mathbb{P}}(1-q^j)} \tag{61}$$

and

$$\mathcal{L}_{[6]}(q) = \mathcal{L}_{[7]}(q) = \frac{1-q^7-q^8-q^{10}+q^{12}+q^{13}}{\prod_{j\in[6]\cup\mathbb{P}}(1-q^j)} \tag{62}$$

from the Eqs. (37)–(43); some values of $L_{[4]}(n)$ are listed in Table 1.

7. The irreducible pairs

An *irreducible pair* is a pair of partitions of n that yield the same multinomial coefficient but have no parts in common. For example,

$$(4,1,1,1) \quad \text{and} \quad (3,2,2) \tag{63}$$

is an irreducible pair.

It turns out that there are infinitely many irreducible pairs of partitions. The following is a partial list: Generalizing (63) we see that

$$(2^m, \underbrace{1,1,\ldots,1}_{2m-1}) \quad \text{and} \quad (2^m-1, \underbrace{2,2,\ldots,2}_{m}) \tag{64}$$

form an irreducible pair of partitions of 2^m+2m-1. More generally, for any integers $a \geqslant 2$ and $m \geqslant 1$ the partitions

$$(a^m, \underbrace{a-1,a-1,\ldots,a-1}_{m}, \underbrace{1,1,\ldots,1}_{m-1}) \quad \text{and} \quad (a^m-1, \underbrace{a,a,\ldots,a}_{m}) \tag{65}$$

form an irreducible pair of partitions of a^m+am-1.

The pair

$$(6,1,1) \quad \text{and} \quad (5,3) \tag{66}$$

can be generalized to irreducible pairs

$$(j!, \underbrace{1, \ldots, 1}_{(j-1)}) \quad \text{and} \quad (j! - 1, j) \tag{67}$$

of partitions of $(j! + j - 1)$ for $j \geqslant 3$.

From any two irreducible pairs we can get a third one by combining them in a natural way. For instance, combining a copies of (66) with b copies of (63) gives the pair (69) which is used in the proof below.

The above examples show that $i(n)$ is positive infinitely often. Indeed we have:

Theorem 13. $i(n) \geqslant \frac{n}{56} - 1$.

Proof. For each pair of non-negative integers a and b satisfying

$$8a + 7b = n, \tag{68}$$

we see that

$$(\underbrace{6, \ldots, 6}_{a}, \underbrace{4, \ldots, 4}_{b}, \underbrace{1, \ldots, 1}_{2a+3b}) \quad \text{and} \quad (\underbrace{5, \ldots, 5}_{a}, \underbrace{3, \ldots, 3}_{a+b}, \underbrace{2, \ldots, 2}_{2b}) \tag{69}$$

forms a new irreducible pair of partitions of n. Consequently $i(n)$ is at least as large as the number of non-negative solutions of the linear Diophantine equation (68).

Now the segment of the line $8a + 7b = n$ in the first quadrant is of length $n\sqrt{113}/56$. Furthermore from the full solution of the linear Diophantine equation we note that the integral solutions of (3.7) are points on this spaced a distance $\sqrt{113}$ apart. Hence in the first quadrant there must be at least

$$\left\lfloor \frac{n\sqrt{113}/56}{\sqrt{113}} \right\rfloor = \left\lfloor \frac{n}{56} \right\rfloor > \frac{n}{56} - 1 \tag{70}$$

such points. Therefore

$$i(n) > \frac{n}{56} - 1. \quad \square \tag{71}$$

Theorem 13 shows that $i(n) > 0$ for all $n \geqslant 56$. Direct computation shows that $i(n) > 0$ for all $n > 7$ with the exception of $n = 9, 11$, and 12.

8. Further problems

Clearly we have only scratched the surface concerning the order of magnitude of $M_k(n)$, $M(n)$, and $i(n)$. We have computed tables of the functions, and based on that evidence we make the following conjectures.

Conjecture 14. *$M(n) \geqslant p^*(n)$ for $n \geqslant 0$, where $p^*(n)$ is the total number of partitions of n into parts that are either $\leqslant 6$ or multiples of 3 or both.*

Conjecture 15. *There exists a positive constant C so that*

$$\lim_{n\to\infty} \frac{\log M(n)}{\sqrt{n}} = C. \tag{72}$$

If C exists and if Conjecture 14 is true, then [3, Theorem 6.2, p. 89]

$$\frac{\pi}{3}\sqrt{2} \leqslant C \leqslant \pi\sqrt{\frac{2}{3}}. \tag{73}$$

Conjecture 16. *Let C_k be the infimum of the quotients $M_k(n)/p_k(n)$ where n ranges over the natural numbers. Then $C_k > 0$ for all natural numbers k. Moreover, C_k is a strictly decreasing function of k for $k \geqslant 3$ and $C_k \to 0$ as $k \to \infty$.*

Conjecture 17. *$M(n) \leqslant p^{\#}(n)$ for $n \geqslant 0$ where $p^{\#}(n)$ is the total number of partitions of n into parts that are either $\leqslant 7$ or multiples of 3 or both.*

Conjecture 17 together with Conjecture 14 allows us to replace Conjecture 15 with

Conjecture 18.

$$\lim_{n\to\infty} \frac{\log M(n)}{\sqrt{n}} = \frac{\pi}{3}\sqrt{2}. \tag{74}$$

Acknowledgments

We thank Anna Bigatti for expert advice on CoCoA [7] and Bogdan Matasaru for rewriting a program for computing $M(n)$ in the programming language C.

References

[1] W.E. Adams, P. Loustaunau, An Introduction to Gröbner Bases, Grad. Stud. Math., vol. 3, Amer. Math. Soc., Providence, 1994.
[2] G.E. Andrews, Number Theory, Saunders, Philadelphia, 1971; reprinted by Dover, New York, 1994.
[3] G.E. Andrews, The Theory of Partitions, Encyclopaedia Math. Appl., vol. 2, Addison–Wesley, Reading, 1976; reprinted by Cambridge Univ. Press, Cambridge, 1985.
[4] A.M. Bigatti, Computation of Hilbert–Poincaré series, J. Pure Appl. Algebra 119 (3) (1997) 237–253.
[5] B. Buchberger, An algorithmic criterion for the solvability of algebraic systems of equations, Aequationes Math. 4 (3) (1970) 374–383.
[6] B. Buchberger, Gröbner bases: An algorithmic method in polynomial ideal theory, in: N.K. Bose (Ed.), Multidimensional Systems Theory, Reidel, Dordrecht, 1985, pp. 184–232.
[7] A. Capani, G. Niesi, L. Robbiano, CoCoA, a system for doing computations in commutative algebra, 2000, available via anonymous ftp from ftp://cocoa.dima.unige.it.

[8] D. Cox, J. Little, D. O'Shea, Ideals, Varieties, and Algorithms: An Introduction to Computational Algebraic Geometry and Commutative Algebra, second ed., Undergrad. Texts Math., Springer-Verlag, New York, 1997.

[9] P. Erdös, I. Niven, The number of multinomial coefficients, Amer. Math. Monthly 61 (1) (1954) 37–39.

[10] R.L. Graham, D.E. Knuth, O. Patashnik, Concrete Mathematics: A Foundation for Computer Science, second ed., Addison–Wesley, Amsterdam, 1994.

[11] S.M. Kerawala, On the asymptotic values of $\ln p_A(n)$ and $\ln p_A^{(d)}(n)$ with A as the set of primes, J. Natur. Sci. Math. 9 (1969) 209–216.

[12] F.S. Macaulay, Some properties of enumeration in the theory of modular systems, Proc. London Math. Soc. 26 (1927) 531–555.

[13] R.P. Stanley, Combinatorics and Commutative Algebra, second ed., Progr. Math., vol. 41, Birkhäuser, Boston, 1996.

Chapter 7

Surveys

At various points in my career I have been urged to survey some particular aspect of partitions and q-series. I have chosen six surveys to include in this chapter.

Three ([51], [57], and [227]) are on partitions. The two early ones provide a rather full account of the progress of work on partition identities up to the mid-1970s. Paper [227] draws attention to a number of open questions, some of which have been explored recently (for example, [Bri-O06], and [Bri-OR08]).

Three papers ([59], [70], and [191]) are devoted to hypergeometric and q-hypergeometric series. Paper [59] in Section 3 provides a full account of how to deduce classical theorems on sums of squares from q-hypergeometric identities, and Section 5 provides a nearly algorithmic recipe for producing q-analogs of classical identities. Paper [70] provided the full ${}_{2k+4}\phi_{2k+3}$ generalization of Watson's q-analog of Whipple's theorem [Wat29]. This result has turned out to have a variety of applications, both within q-series and partitions [256], [270], and outside [Krat-R07].

The questions raised at the end of Section 3 in [59], and in Problems 2, 3, and 4 in [70], were part of the inspiration for Steve Milne in his long and grand quest to develop a theory of well-poised hypergeometric series connected to the classical groups (for example, [Miln80], [Miln85]). This eventually led to his monograph on sums of squares [Miln02]. I had the honor to write the introduction for the latter.

Section 5 of [70] inspired two marvelous papers by Zeilberger: first [Zei-B85], his proof (joint with Bressoud) of the q-Dyson conjecture [70]; second, a q-analog of MacMahon's master theorem [Gar-LZ06].

Of the surveys not included, a few comments are in order. Paper [24] is only partly a survey, most of which is included in [51]. Paper [150], surveying mock theta functions, is now rather dated, owing to the immense accomplishments of Ken Ono and his collaborators (*cf.* [Bri-OR08]). Paper [169] is actually only a summary of the work in [154], [158], and [165]. Of these papers, [154], and [158] are included in Chapter 2, and [165] is recounted by Berndt in [Bern85-98]. Paper [213] provides an account of very early influences in my work. [247] and [253] are commentaries on two people who have greatly influenced my work: Basil Gordon and Ramanujan.

7.1 Partition Identities

Reprinted from ADVANCES IN MATHEMATICS
Vol. 9, No. 1, August 1972

Printed in Belgium

Partition Identities*

GEORGE E. ANDREWS

Department of Mathematics, Massachusetts Institute of Technology and Pennsylvania State University

1. INTRODUCTION

Intuitively a partition of a positive integer n may be thought of as an unordered representation of n as a sum of other positive integers. Thus $3 + 2 + 2$, $2 + 3 + 2$, $2 + 2 + 3$ represent the same partition of 7. For convenience we usually consider that form of the partition in which the parts are arranged in nonincreasing order of magnitude.

We may also completely describe a partition by giving a list that tells the number of appearances of the various parts. Thus we shall sometimes denote a partition Π by $\{f_i\}_{i=1}^{\infty}$ (or merely $\{f_i\}$) where the f_i are nonnegative integers of which only finitely many are nonzero. The number partitioned is given by

$$n = \sum_{i=1}^{\infty} f_i i.$$

In the above example $f_1 = 0$, $f_2 = 2$, $f_3 = 1$, and $f_n = 0$ for $n > 3$.

The first problem in the theory of partitions is the study of the *unrestricted partition function* $p(n)$, the total number of partitions of n. Many important results are known about $p(n)$; in particular, accurate asymptotic formulae have been found [67, Chap. 6], and infinite families of congruences are known for special sequences of n [67, Chap. 8]. The elementary aspects of the theory of partitions are given in detail in [66, Chap. 19], and the reader is referred there for such things as a proof of the following formula for the generating function related to $p(n)$.

$$1 + \sum_{n=1}^{\infty} p(n)\, q^n = \prod_{n=1}^{\infty} \frac{1}{1 - q^n}.$$

* This research was partially supported by National Science Foundation Grant 6P-9660

PARTITION IDENTITIES 11

In this paper we shall be concerned with what might be called *restricted partition functions*, that is, partition functions that enumerate only those partitions of n that satisfy certain given conditions.

For example, we consider the number $Q(n)$ of partitions of n into odd parts, and the number $D(n)$ of partitions of n into distinct parts. In 1748, L. Euler [54, Chap. 16] presented and proved the first known theorem asserting the identity of two apparently different partition functions. Since

$$\sum_{n=0}^{\infty} D(n)\, q^n = \prod_{n=1}^{\infty} (1 + q^n) = \prod_{n=1}^{\infty} \frac{(1 - q^{2n})}{(1 - q^n)}$$

$$= \frac{\prod_{n=1}^{\infty} (1 - q^{2n})}{\prod_{n=1}^{\infty} (1 - q^n)} = \prod_{n=1}^{\infty} \frac{1}{1 - q^{2n-1}} = \sum_{n=0}^{\infty} Q(n)\, q^n,$$

we see that $D(n) = Q(n)$ for each integer n. We may state this result as follows:

(1.1) *The number of partitions of n in which the summands are odd equals the number of partitions of n in which the summands are distinct.*

Theorems such as this are generally referred to as partition identities. The fact that such theorems may be quite difficult is not at all obvious from the simple proof of Euler's theorem. Indeed it was not until the early 1900's that substantially deeper partition identities were discovered. In 1917, P. A. MacMahon [71, Chap. 3] describes the following two theorems which he states are unproven but have been verified for $n \leqslant 89$.

(1.2) *The number of partitions of n in which the minimal difference between summands is at least 2 equals the number of partitions of n into parts of the forms $5m + 1$ and $5m + 4$.*

(1.3) *The number of partitions of n in which the minimal difference between summands is at least 2 and 1 does not appear equals the number of partitions of n into parts of the forms $5m + 2$ and $5m + 3$.*

These theorems are known as the Rogers–Ramanujan identities. In 1894, Rogers [73] proved the following series-product identities:

$$1 + \sum_{n=1}^{\infty} \frac{q^{n^2}}{(1 - q)(1 - q^2) \cdots (1 - q^n)} = \prod_{n=0}^{\infty} \frac{1}{(1 - q^{5n+1})(1 - q^{5n+4})}, \tag{1.4}$$

$$1 + \sum_{n=1}^{\infty} \frac{q^{n^2+n}}{(1-q)(1-q^2)\cdots(1-q^n)} = \prod_{n=0}^{\infty} \frac{1}{(1-q^{5n+2})(1-q^{5n+3})} . \tag{1.5}$$

From these analytic identities it is possible to deduce (1.2) and (1.3) (see [71, Chap. 3]); however, Rogers did not give the combinatorial interpretation and his work was soon forgotten. Around 1913, Ramanujan rediscovered (1.4) and (1.5) but could not find a proof. He communicated these identities to G. H. Hardy who also was baffled by them. It was only in 1917 when Ramanujan was looking through some old volumes of the Proceedings of the London Mathematical Society that he ran onto Rogers's paper containing proofs of (1.4) and (1.5). After that several further proofs of these results were found by Rogers and Ramanujan [74, 75]. Also it should be mentioned that Schur [77] discovered these theorems independently in 1917.

In 1926, Schur [78] discovered the following theorem of a similar type.

(1.6) *The number of partitions of n in which the minimal difference between summands is at least* 3 *and at least* 6 *between summands that are both multiples of* 3 *equals the number of partitions of n into parts of the forms* $6m + 1$ *and* $6m + 5$.

Certain nonexistence theorems for identities like (1.1)–(1.3), and (1.6) were proved by Lehmer and Alder in the late 1940's [2, 70].

It was not until 1960 that two further identities of this nature were discovered by Göllnitz [59] (they were discovered independently by B. Gordon [62] in 1965 and have been called the Göllnitz–Gordon identities).

(1.7) *The number of partitions of n in which the minimal difference between summands is at least* 2 *and at least* 4 *between even summands. equals the number of partitions of n into parts of the forms* $8m + 1$, $8m + 4$, *and* $8m + 7$.

(1.8) *The number of partitions of n in which the minimal difference between summands is at least* 2, *at least* 4 *between even summands, and all summands are greater than* 2 *equals the number of partitions of n into parts of the forms* $8m + 3$, $8m + 4$, *and* $8m + 5$.

Then in 1961, Gordon [61] made a significant breakthrough in this study by proving a theorem for an infinite number of identities; the Rogers–Ramanujan identities (1.2) and (1.3) correspond to the cases $k = 2$, $a = 1, 2$.

(1.9) *Let $B_{k,a}(n)$ denote the number of partitions of n of the form $n = b_1 + b_2 + \cdots + b_s$ where $b_i \geqslant b_{i+1}$, $b_i - b_{i+k-1} \geqslant 2$ and 1 appears as a summand at most $a - 1$ times; let $A_{k,a}(n)$ denote the number of partitions of n into parts that are not congruent to $0, \pm a (\text{mod } 2k + 1)$. Then $A_{k,a}(n) = B_{k,a}(n)$ for each n.*

A great deal of subsequent work has related to theorems similar to (1.8). The following seems to be the most general result of this nature known [20].

(1.10) *Let $B_{\lambda,k,a}(n)$ denote the number of partitions of n of the form $n = b_1 + b_2 + \cdots + b_s$, where $b_i \geqslant b_{i+1}$, only parts divisible by $\lambda + 1$ may be repeated, $b_i - b_{i+k-1} \geqslant \lambda + 1$ with strict inequality if $(\lambda + 1) \mid b_i$, and the total number of appearances of summands in the set $\{1, 2, \ldots, \lambda + 1\}$ is at most $a - 1$. If λ is even, let $A_{\lambda,k,a}(n)$ denote the number of partitions of n into parts such that no part $\not\equiv 0 (\text{mod } \lambda + 1)$ may be repeated and no part is congruent to $0, \pm(a - \frac{1}{2}\lambda)(\lambda + 1) \ (\text{mod}(2k - \lambda + 1)(\lambda + 1))$; if λ is odd, let $A_{\lambda,k,a}(n)$ denote the number of partitions of n into parts such that no part $\not\equiv 0 (\text{mod } \frac{1}{2}(\lambda + 1))$ may be repeated, no part is congruent to $\lambda + 1 (\text{mod } 2\lambda + 2)$, and no part is congruent to*

$$0, \ \pm(2a - \lambda)\, \tfrac{1}{2}(\lambda + 1)(\text{mod}(2k - \lambda + 1)(\lambda + 1)).$$

Then provided $k \geqslant 2\lambda - 1$, and $k \geqslant a \geqslant \lambda$, we have that $A_{\lambda,k,a}(n) = B_{\lambda,k,a}(n)$.

The condition $k \geqslant 2\lambda - 1$ is thought to be unnecessary although the only known proof of (1.10) relies heavily on this inequality.

Identity (1.10) includes many of the previous identities as special cases. In particular (1.2) is the case $\lambda = 0$, $k = 2 = a$; (1.3) is the case $\lambda = 0$, $k = 2$, $a = 1$; (1.6) is essentially equivalent to the case $\lambda = 2$, $k = 2 = a$; (1.7) is the case $\lambda = 1$, $k = 2 = a$; (1.8) is the case $\lambda = 1$, $k = 2$, $a = 1$, and (1.10) is the case $\lambda = 0$. The case $\lambda = 1$ yields a previously discovered identity generalizing the Göllnitz–Gordon identities [7]. The case $\lambda = 2$ yields a generalization of Schur's identity (1.6), and we shall study this generalization in Section 4. We chose this example because it has not been proved independently before; also it illustrates some of the complexities that arise in the full proof of (1.10) but is not nearly as troublesome as the general result.

We observe that the partition function $B_{2,k,a}(n)$ (or in general $B_{\lambda,k,a}(n)$) is related to partitions in which there are different conditions on the differences $b_i - b_{i+k-1}$ in the partitions $n = b_1 + b_2 + \cdots + b_s$. There are, however, at least two general theorems on partitions [17, 18] in which the conditions are on adjacent differences, that is, $b_i - b_{i+1}$. To state these theorems we require some further terminology.

Let $A = \{a_1, a_2, ..., a_r\}$ be a set of positive integers listed in increasing order of magnitude, and suppose that $\sum_{j=1}^{i-1} a_j < a_i$ for each i (this guarantees that the various sums $a_{i_1} + \cdots + a_{i_r}$ are all distinct). We let A' denote the $2^r - 1$ different sums of distinct elements of A, and we write $A' = \{\alpha_1, \alpha_2, ..., \alpha_{2^r-1}\}$ where the α_i are listed in increasing order of magnitude (and so $\alpha_{2^i} = a_{i+1}$). The letter N denotes a positive integer not smaller than $a_1 + a_2 + \cdots + a_r$. We let A_N denote the set of all positive integers congruent to some a_i modulo N; $-A_N$ denotes the set of all positive integers congruent to some $-a_i$ modulo N; A_N' denotes the set of all positive integers congruent to some α_i modulo N, and $-A_N'$ denotes the set of all positive integers congruent to some $-\alpha_i$ modulo N. If $m \in A'$, then m is uniquely representable as $a_{i_1} + \cdots + a_{i_j}$ with $a_{i_1} > \cdots > a_{i_j}$; we let $\omega(m) = j$, and $\nu(m) = a_{i_j}$. By $\beta_N(s)$, we denote the least positive residue of s modulo N.

(1.11) *Let $\mathscr{D}(A_N : n)$ denote the number of partitions of n into distinct parts taken from A_N. Let $\mathscr{E}(A_N' : n)$ denote the number of partitions of n of the form, $n = b_1 + \cdots + b_s$, where $b_i \in A_N'$ and $b_i - b_{i+1} \geqslant N\omega(\beta_N(b_{i+1})) + \nu(\beta_N(b_{i+1})) - \beta_N(b_{i+1})$, for each i. Then for each n, $\mathscr{D}(A_N : n) = \mathscr{E}(A_N' : n)$.*

(1.12) *Let $\mathscr{F}(-A_N : n)$ denote the number of partitions of n into distinct parts taken from $-A_N$. Let $\mathscr{J}(-A_N' : n)$ denote the number of partitions of n of the form, $n = b_1 + \cdots + b_s$, where $b_i \in -A_N'$ and*

$$b_i - b_{i+1} \geqslant N \cdot \omega(\beta_N(-b_i)) + \nu(\beta_N(-b_i)) - \beta_N(-b_i),$$

for each i, and furthermore $b_s \geqslant N \cdot (\omega(\beta_N(-b_s)) - 1)$. Then

$$\mathscr{F}(-A_N : n) = \mathscr{J}(-A_N' : n).$$

If $N = 3$, $a_1 = 1$, $a_2 = 2$, then both (1.11) and (1.12) reduce to a result essentially equivalent to Schur's Theorem (1.6). In Section 5, we shall treat (1.11) in the case $N = 7$, $a_1 = 1$, $a_2 = 2$, $a_3 = 4$. Other specializations of these results may be found in [17] and [18]. In [33], (1.11) is used to solve partially a problem posed by Alder [3].

As the number of results for partition identities has increased tremendously recently, it becomes important to construct a general theory for such results so that the relationships between the various identities may be better perceived. Also a general theory allows one to pose the general goals and important problems of the subject in reasonably concise and comprehensible terms. In Section 2, we undertake to formulate the foundations of such a general theory. Our starting point lies in the observation that all the identities we have here discussed concern conditions on the frequency of appearance of parts in partitions according to various rules.

Section 3 is devoted to an extensive study of the simplest type of partition identities—those essentially similar to Euler's identity (1.1). Sections 4–6 provide some examples of the various techniques used in proving partition theorems. In Section 7, we return to the general theory begun in Sections 2 and 3. Finally Section 8 concludes by discussing possible generalizations and extensions of the theory of partition identities.

2. The Central Problems

We are going to couch the study of partition identities in the terminology of lattice theory. Lattice theory provides a viewpoint which allows us to unify most of the known results on partition identities, and it provides an especially simple statement of the fundamental problems.

Lemma 2.1. *Let $\mathscr{S}$ denote the set of all sequences $\{f_i\}_{i=1}^{\infty}$ (more briefly $\{f_i\}$) where each f_i is a nonnegative integer and where only finitely many f_i are nonzero. Then $\mathscr{S}$ forms a distributive lattice under the partial ordering*

$$\{f_i\} \leqslant \{g_i\} \text{ provided } f_i \leqslant g_i \text{ for each } i.$$

Proof. Observing that

$$\{f_i\} \cap \{g_i\} = \{\min(f_i, g_i)\},$$

and

$$\{f_i\} \cup \{g_i\} = \{\max(f_i, g_i)\},$$

we may easily verify that $\mathscr{S}$ is a distributive lattice [76; Section 9; 44, Chap. 1, Section 6].

LEMMA 2.2. *If N denotes the set of nonnegative integers, and if*

$$\sigma : \mathscr{S} \to N$$

where $\sigma(\{f_i\}_{i=1}^{\infty}) = \sum_{i=1}^{\infty} f_i \cdot i$, *then* σ *is a positive valuation on* $\mathscr{S}$.

Proof. If $\Pi = \{f_i\}_{i=1}^{\infty}$ and $\Pi^1 = \{g_i\}_{i=1}^{\infty}$

$$\sigma(\Pi \cap \Pi^1) + \sigma(\Pi \cup \Pi^1) = \sum_{i=1}^{\infty} (\min(f_i, g_i)i + \max(f_i, g_i)i)$$

$$= \sum_{i=1}^{\infty} f_i i + \sum_{i=1}^{\infty} g_i i = \sigma(\Pi) + \sigma(\Pi^1).$$

If $\Pi > \Pi^1$, then there exists j such that $f_j > g_j$. Hence

$$\sigma(\Pi) = \sum_{i=1}^{\infty} f_i \cdot i \geqslant \sum_{\substack{i=1 \\ i \neq j}}^{\infty} g_i \cdot i + (g_j + 1)j = \sigma(\Pi^1) + j.$$

First let us observe how these concepts relate to the classical study of partitions. The sequences $\{f_i\}$ making up $\mathscr{S}$ may be thought of as defining a partition where f_i gives the number of times the part i appears. The function σ merely maps each sequence $\{f_i\}$ onto the number being partitioned.

Recall now that a *semi-ideal* J in a lattice L is a subset of L such that if $a \in J$, $x \in L$, and $x \leqslant a$, then $x \in J$ [44, p. 56].

DEFINITION 1. A semi-ideal in the lattice $\mathscr{S}$ is called a *partition ideal.*

The concept of a partition ideal has evolved from that of a "partition condition" as described in [28]. Partition ideals are less general than partition conditions; however, we shall see that they describe most all known partition identities.

DEFINITION 2. If C is a partition ideal in $\mathscr{S}$, we say that $p(C; n)$ is the *C-partition function* if for each n, $p(C; n)$ denotes the cardinality of the set $\{\Pi \mid \Pi \in C, \sigma(\Pi) = n\}$.

DEFINITION 3. We say that two partition ideals C_1 and C_2 of $\mathscr{S}$ are *partition-theoretically equivalent* (more briefly PT-equivalent) if for each nonnegative integer

$$p(C_1\,;n) = p(C_2\,;n),$$

and we shall write $C_1 \sim^{\mathrm{PT}} C_2$.

DEFINITION 4. We say that two partition ideals C_1 and C_2 of S are *asymptotically equivalent* (more briefly A-equivalent) if $p(C_2\,;n) \neq 0$ for $n > n_0$ and

$$\lim_{n\to\infty} [p(C_1\,;n)/p(C_2\,;n)] = 1,$$

and we write $C_1 \sim^{\mathrm{A}} C_2$.

DEFINITION 5. We say that two partition ideals C_1 and C_2 of S are *weakly asymptotically equivalent* (more briefly WA-equivalent) if $p(C_2\,;n) \neq 0$ for $n > n_0$ and

$$\lim_{n\to\infty} \frac{\log p(C_1\,;n)}{\log p(C_2\,;n)} = 1,$$

and we write $C_1 \sim^{\mathrm{WA}} C_2$.

It is obvious that $\sim^{\mathrm{PT}}$, $\sim^{\mathrm{A}}$, and $\sim^{\mathrm{WA}}$ are equivalence relations in the standard sense.

Before considering some examples to justify this terminology, let us present the fundamental classification problem in the theory of partition identities.

First Problem

Fully describe the equivalence classes in $\mathscr{S}$ under the equivalence $\sim^{\mathrm{PT}}$.

Of interest also is the same problem with $\sim^{\mathrm{PT}}$ replaced by $\sim^{\mathrm{A}}$ or $\sim^{\mathrm{WA}}$. In Sections 3 and 7 we shall prove some general theorems concerning the complete problem. For the remainder of this section we shall endeavor to justify calling this the fundamental classification problem.

The intuitive description of a partition ideal in $\mathscr{S}$ is just this: if a given partition Π satisfies a certain specified condition then all partitions formed by deleting one or many summands from Π also satisfy the same specified condition. This situation is clearly in evidence in all the theorems described in the introduction as we shall see.

DEFINITION 6. Let $C(a_1, a_2, \ldots, a_r; m)$ denote $\{\Pi \mid \Pi = \{f_i\} \in \mathscr{S}$ and if $f_i \neq 0$, then $i \equiv a_1, a_2, \ldots,$ or $a_r (\text{mod } m)\}$. Let $C_d(a_1, a_2, \ldots, a_r; m)$ denote $\{\Pi \mid \Pi = \{f_i\} \in \mathscr{S}, 0 \leqslant f_i \leqslant 1$, and if $f_i \neq 0$, then $i \equiv a_1, a_2, \ldots,$ or $a_r (\text{mod } m)\}$.

It is obvious that both $C(a_1, a_2, \ldots, a_r; m)$ and $C_d(a_1, a_2, \ldots, a_r; m)$ are partition ideals in $\mathscr{S}$. Furthermore, $p(C(a_1, a_2, \ldots, a_r; m); n)$ is the number of partitions of n into parts congruent to $a_1, a_2, \ldots, a_r (\text{mod } m)$ while $p(C_d(a_1, a_2, \ldots, a_r; m); n)$ is the number of such partitions in which the parts are distinct.

DEFINITION 7. *Let* $D(r; b_1, b_2, \ldots, b_m; m)$ denote $\{\Pi \mid \Pi = \{f_i\} \in \mathscr{S}$, $f_j = 0$ when $j < r$, $0 \leqslant f_j \leqslant 1$, and if $f_j = 1$ with $j \equiv k (\text{mod } m)$, then $f_{j+1} = f_{j+2} = \cdots = f_{j+b_k-1} = 0\}$.

Again it is clear that $D(b_1, b_2, \ldots, b_m; m)$ is an order ideal in $\mathscr{S}$. Furthermore, $p(D(b_1, b_2, \ldots, b_m; m); n)$ is the number of partitions of n of the form $n = C_1 + C_2 + \cdots + C_s$ where $C_i - C_{i+1} \geqslant b_h$ if $C_{i+1} \equiv h (\text{mod } m)$. With these definitions we may briefly state (1.1), (1.2), (1.3), (1.6), (1.7) and (1.8) in the following equivalences, respectively:

$$C(1; 2) \overset{PT}{\sim} C_d(1; 1) \tag{2.1}$$

$$C(1, 4; 5) \overset{PT}{\sim} D(1; 2; 1) \tag{2.2}$$

$$C(2, 3; 5) \overset{PT}{\sim} D(2; 2; 1) \tag{2.3}$$

$$C(1, 5; 6) \overset{PT}{\sim} D(1; 3, 3, 4; 3) \tag{2.4}$$

$$C(1, 4, 7; 8) \overset{PT}{\sim} D(1; 2, 3; 2) \tag{2.5}$$

$$C(3, 4, 5; 8) \overset{PT}{\sim} D(3; 2, 3; 2). \tag{2.6}$$

Göllnitz [60] has also proved the following results

$$C_d(1, 2, 4; 4) \overset{PT}{\sim} C(1, 5, 6; 8) \overset{PT}{\sim} D(1; 3, 2; 2) \tag{2.7}$$

$$C_d(2, 3, 4; 4) \overset{PT}{\sim} C(2, 3, 7; 8) \overset{PT}{\sim} D(2; 3, 2; 2). \tag{2.8}$$

We point out that (1.9)–(1.12) as well as the theorems in [8–11], and [50] can all be interpreted as equivalences of partition ideals modulo $\sim^{PT}$.

From these comments it is clear that several equivalence classes modulo $\sim^{PT}$ contain more than one element, and at least two contain at least three elements. Before we begin we introduce two more concepts that will be useful in our future discussions.

Definition 8. We say that C, a partition ideal, has *order k* provided k is the least integer such that whenever $\{f_i\} \notin C$, then there exists m such that the sequence $\{f_i'\}$ defined by

$$f_i' = \begin{cases} f_i, & i = m, m+1,\ldots, m+k-1, \\ 0, & \text{otherwise}, \end{cases}$$

is also not in C.

Intuitively, C being of order k makes explicit the idea that summands of Π at least k units apart cannot affect whether or not $\Pi \in C$.

Proposition 1. *The partition ideals*

$$C(a_1,\ldots, a_r; m) \quad \textit{and} \quad C_d(a_1,\ldots, a_r; m)$$

each have order 1. *The partition ideal* $D(r; b_1, b_2,\ldots, b_m; m)$ *has order* $\max(b_1, b_2,\ldots, b_m)$.

Proof. If $\{f_i\} \notin C(a_1, a_2,\ldots, a_r; m)$ then there exists j such that $f_j \neq 0$ and $j \not\equiv a_1, a_2,\ldots,$ or $a_r (\text{mod } m)$. Thus the sequence $\{f_i'\}$ defined by

$$f_i' = \begin{cases} f_j, & \text{if } i = j, \\ 0, & \text{otherwise} \end{cases}$$

is not in $C(a_1, a_2,\ldots, a_r; m)$. The same reasoning shows that $C_d(a_1, a_2,\ldots, a_r; m)$ has order 1.

Finally we consider $D(r; b_1, b_2,\ldots, b_m; m)$, and we let

$$M = \max(b_1, b_2,\ldots, b_m).$$

If $\{f_i\} \notin D(r; b_1, b_2,\ldots, b_m; m)$, then there exist j and k such that $f_j = f_k = 1$ and $0 < k - j < b_r$ where $r \equiv j (\text{mod } m)$. Then the sequence $\{f_i'\}$ defined by

$$f_i' = \begin{cases} f_i, & i = j, j+1,\ldots, k,\ldots, j+M-1, \\ 0, & \text{otherwise} \end{cases}$$

is not in $D(r; b_1, b_2, ..., b_m; m)$. To see that M satisfies the minimality condition we let $b_t = M$ and consider the sequence $\{f''_i\}_{i=1}^{\infty}$ defined by

$$f''_i = \begin{cases} 1, & i = mM + t, (m+1)M + t, \\ 0, & \text{otherwise.} \end{cases}$$

Clearly $\{f''_i\}$ is in $D(r; b_1, b_2, ..., b_m; m)$. ∎

The special role of partition ideals of order 1 will become obvious in succeeding sections. In Section 3, we shall prove that the generating function for C-partition functions where C is of order 1 are infinite products of a very simple sort. Actually as we shall see most of the known results on partition identities concern equivalences of two partition ideals where at least one has order 1. This has certainly been the case with (2.1)–(2.8).

Second Problem

Fully describe those equivalence classes in $\mathscr{S}$ modulo $\sim^{\mathrm{PT}}$ that contain a partition ideal of order 1.

3. Partition Ideals of Order 1

Theorem 1. *Let C be a partition ideal of order* 1. *Let γ_l be the set of $\{f_i\} \in C$ such that $f_i = 0$ for all $i \neq l$. Let*

$$d_l = \sup_{\{f_i\} \in \gamma_l} f_l .$$

Then for $|q| < 1$

$$\sum_{n=0}^{\infty} p(C; n)\, q^n = \frac{\prod_{l=1, d_l<\infty}^{\infty} (1 - q^{l(d_l+1)})}{\prod_{l=1}^{\infty} (1 - q^l)} .$$

Proof. We first must prove that $\Pi = \{f_i\} \in C$ if and only if $f_i \leqslant d_i$ for each i. Suppose that this condition holds for $\Pi = \{f_i\}$. If $\Pi \notin C$, then there must exist $\Pi' = \{f_i'\}_{i=1}^{\infty} \notin C$, where all $f_i' = 0$ except for a single j and $f_j' = f_j$. Now by the fact that d_j is a supremum of a set of nonnegative integers we see that d_j may be $+\infty$; however, there must exist $\{f''_i\} \in \gamma_j$ such that $f_j' = f_j \leqslant f''_j$ since $f_j \leqslant d_j$. Hence

$$\Pi' = \{f_i'\} \leqslant \{f''_i\} \in \gamma_j \subset C,$$

and since C is a partition ideal of $\mathcal{S}$, we see that $\Pi' \in C$, a contradiction. Hence $\Pi \in C$.

On the other hand, if $\Pi = \{f_i\} \in C$, then $\Pi_j' = \{f_i'\}$ defined by

$$f_i' = \begin{cases} f_j, & \text{if } i = j, \\ 0, & \text{if } i \neq j, \end{cases}$$

is in γ_j for each j. Hence

$$f_j = f_j' \leqslant \sup_{\{g_i\} \in \gamma_j} g_j = d_j,$$

and the prescribed condition holds for each j.

Thus $p(C; n)$ is the number of partitions of n in which each j appears at most d_j times. Therefore,

$$\begin{aligned}\sum_{n=0}^{\infty} p(C; n)\, q^n &= \prod_{\substack{j=1 \\ d_j \neq \infty}}^{\infty} (1 + q^j + q^{2j} + \cdots + q^{d_j j}) \prod_{\substack{j=1 \\ d_j = \infty}}^{\infty} (1 + q^j + q^{2j} + \cdots) \\ &= \prod_{\substack{j=1 \\ d_j \neq \infty}}^{\infty} \frac{(1 - q^{j(d_j+1)})}{(1 - q^j)} \prod_{\substack{j=1 \\ d_j = \infty}}^{\infty} \frac{1}{(1 - q^j)} \\ &= \frac{\prod_{j=1, d_j < \infty}^{\infty} (1 - q^{j(d_j+1)})}{\prod_{j=1}^{\infty} (1 - q^j)}. \qquad \blacksquare\end{aligned}$$

Theorem 1 is the reason that infinite products play such an important role in the study of partition identities. Since most known identities involve a partition ideal of order 1, Theorem 1 shows that the associated identity for the generating functions will involve an infinite product.

Since partition ideals of order 1 have such nice associated generating functions we shall be able to prove some rather general theorems concerning these ideals.

THEOREM 2. *The ideals in $\mathcal{S}$ are the partition ideals of order* 1.

Proof. We recall that an ideal in a lattice is a semi-ideal closed under union [44, p. 25].

Suppose C is of order 1. Then we know that there exist integers d_i (d_i may also equal $+\infty$) such that $\{f_i\} \in C$ if and only if $f_i \leqslant d_i$ for each i (see the proof of Theorem 1). Hence if $\{f_i\}$ and $\{f_i'\}$ are both in C, then for each i

$$\max(f_i, f_i') \leqslant d_i;$$

therefore

$$\{\max(f_i, f_i')\} = \{f_i\} \cup \{f_i'\} \in C.$$

Hence C is an ideal in $\mathscr{S}$.

Conversely, suppose that C is an ideal $\mathscr{S}$. Define d_j by

$$d_j = \sup_{\{f_i\} \in C} f_j ;$$

we note that d_j may be $+\infty$.

Suppose now that $\{f_i\} \in \mathscr{S}$; obviously if $f_i > d_i$ for some i then $\{f_i\} \notin C$. Suppose that $f_{i_0} \leqslant d_{i_0}$ for each i_0. Then by the definition of d_i, there must exist for each i_0 a $\{g_i(i_0)\} \in C$ such that $f_{i_0} \leqslant g_{i_0}(i_0)$. Now define

$$h_i(i_0) = \begin{cases} f_i, & \text{if } i = i_0, \\ 0, & \text{otherwise.} \end{cases}$$

Since C is an ideal and since $h_i(i_0) \leqslant g_i(i_0)$ for each i, we see that $\{h_i(i_0)\} \leqslant \{g_i(i_0)\}$ and therefore $\{h_i(i_0)\} \in C$.

Now by the definition of elements of $\mathscr{S}$ we know that there exists N such that $f_i = 0$ for $i > N$. Consequently,

$$\{f_i\} = \bigcup_{i_0=1}^{N} \{h_i(i_0)\},$$

and this is in C, since C is an ideal. Thus $\{f_i\} \in C$ if and only if $f_i \leqslant d_i$ for each i. Thus C is a partition ideal of order 1. ∎

THEOREM 3. *Suppose C is a partition ideal of order 1 and let the d_j be the associated suprema as in Theorem 1. Let C' be a second partition ideal of order 1 and let d_j' be the associated suprema. Then $C \sim^{\mathrm{PT}} C'$ if and only if the two sequences $\{j(d_j+1)\}_{d_j<\infty}$ and $\{j(d_j'+1)\}_{d_j'<\infty}$ are merely reorderings of one another.*

Proof. If the above two sequences are identical except for ordering, then

$$\sum_{n=0}^{\infty} p(C; n)\, q^n = \frac{\prod_{j=1, d_j<\infty}^{\infty} (1 - q^{j(d_j+1)})}{\prod_{j=1}^{\infty} (1 - q^j)}$$

$$= \frac{\prod_{j=1, d_j'<\infty}^{\infty} (1 - q^{j(d_j'+1)})}{\prod_{j=1}^{\infty} (1 - q^j)} = \sum_{n=0}^{\infty} p(C'; n)\, q^n.$$

Comparing coefficients of q^n in the above sums, we see that $p(C; n) = p(C'; n)$ for each n. Thus $C \sim^{\mathrm{PT}} C'$.

If, however, the two sequences are not identical, then there exists a least integer h that appears a different number of times in each sequence. Without loss of generality, we may assume that h appears r times in the first sequence, s times in the second sequence and that $r > s$. Consequently,

$$\prod_{\substack{j=1 \\ d_j<\infty \\ j(d_j+1)\geqslant h}}^{\infty} (1 - q^{j(d_j+1)}) \neq \prod_{\substack{j=1 \\ d_j'<\infty \\ j(d_j'+1)\geqslant h}}^{\infty} (1 - q^{j(d_j+1)})$$

since on the left hand side the coefficient of q^h arises from $(1 - q^h)^r = 1 - rq^h + \cdots$ and is therefore $-r$, while on the right hand side the coefficient of q^h arises from $(1 - q^h)^s = 1 - sq^h + \cdots$ and is therefore $-s$ which is not $-r$. From the minimality of h, we see directly that

$$\prod_{\substack{d_j<\infty \\ j(d_j+1)<h}} (1 - q^{j(d_j+1)}) = \prod_{\substack{d_j'<\infty \\ j(d_j'+1)<h}} (1 - q^{j(d_j'+1)}).$$

Therefore combining this equality with our previous inequality, we see that

$$\sum_{n=0}^{\infty} p(C; n)\, q^n = \frac{\prod_{j=1, d_j<\infty}^{\infty} (1 - q^{j(d_j+1)})}{\prod_{j=1}^{\infty} (1 - q^j)}$$

$$\neq \frac{\prod_{j=1, d_j'<\infty}^{\infty} (1 - q^{j(d_j'+1)})}{\prod_{j=1}^{\infty} (1 - q^j)} = \sum_{n=0}^{\infty} p(C'; n)\, q^n.$$

Examining the extremes of the above inequality, we see that $p(C; n) \neq p(C'; n)$ for some n. Consequently $C \not\sim^{\mathrm{PT}} C'$. ∎

Theorem 3 is quite useful in the study of what are known as "Euler pairs" (see [23, 85]). To discuss fully Euler pairs we shall require some further notation.

DEFINITION 9. If T is a set of nonnegative integers, then

$$C(T) = \{\{f_i\} \mid \{f_i\} \in \mathscr{S} \text{ and if } f_j > 0 \text{ then } j \in T\},$$
$$C_d(T) = \{\{f_i\} \mid \{f_i\} \in \mathscr{S}\ \ 0 \leqslant f_j \leqslant 1, \text{ and if } f_j > 0 \text{ then } j \in T\}.$$

We remark that $C(T)$ and $C_d(T)$ are partition ideals in $\mathscr{S}$ of order 1. Furthermore, $p(C(T); n)$ is the number of partitions of n into parts taken from the set T, and $p(C_d(T); n)$ is the number of partitions of n into distinct parts taken from the set T.

DEFINITION 10 (see [23]). We say that two sets of positive integers T_1 and T_2 form an *Euler pair* (T_1, T_2) if

$$C_d(T_1) \overset{PT}{\sim} C(T_2).$$

The central theorem in [23] characterizes the set of all Euler pairs. In [85], Subbarao extends this work by defining Euler pairs of higher order and giving a much simpler proof of his more general result. As we shall see Theorem 3 also allows us to give a simple proof of the characterization of Euler pairs.

THEOREM 4. *(S_1, S_2) is an Euler pair if and only if $2S_1 \subseteq S_1$ and $S_2 = S_1 - 2S_1$.*

Proof. We shall apply Theorem 2 to $C_d(S_1)$ and $C(S_2)$. The associated sequence for $C_d(S_1)$ is $\{n \mid n = 2k, k \in S_1\}$ together with $\{n \mid n \notin S_1\}$; the associated sequence for $C(S_2)$ is $\{n \mid n \notin S_2\}$.

We first observe that if $2S_1 \not\subseteq S_1$ then the first associated sequence involves at least one repetition; this means that the two sequences are unequal since the second involves no repetitions.

If $2S_1 \subseteq S_1$, then for the two sequences to be equal we see that it is necessary and sufficient that S_2' the complement of S_2 in N be $2S_1 \cup S_1'$. Hence $S_2 = S_1 - 2S_1$. ∎

This result on Euler pairs allows us to exhibit an infinite number of equivalence classes modulo $\sim^{\mathrm{PT}}$ that contain an infinity of partition ideals. Our approach is suggested by some results of Guy [63].

THEOREM 5. *Let $\mathscr{C}$ be an equivalence class modulo $\sim^{\mathrm{PT}}$ that contains $C_d(S_1)$ and $C(S_2)$ where (S_1, S_2) is an Euler pair. If S_2 is an infinite set, then $\mathscr{C}$ is infinite.*

Proof. Suppose

$$S_2 = \{\sigma_1, \sigma_2, \sigma_3, \ldots\},$$

where the σ_i are positive integers arranged in increasing size. Define the order ideal C_j by $C_j = \{\{f_i\} \mid$ if $i = 2^s\sigma_k$ for some $s \geqslant 0$ and $k < j$ then $0 \leqslant f_i \leqslant 1$, otherwise $f_i = 0$ unless $i = \sigma_r$ for $r \geqslant j$ in which case the f_i is unrestricted$\}$.

First we comment on whether C_j is well-defined. There is clearly a possible conflict if $2^s\sigma_k = \sigma_r$ for some $s \geqslant 0$, $k < j$, $r \geqslant j$. However, this is impossible for since $\sigma_k \in S_2 \subset S_1$, $2\sigma_k \in S_1$, $2^2\sigma_k \in S_1$,..., $2^s\sigma_k \in S_1$ because $2S_1 \subseteqq S_1$ by Theorem 3. Now $s > 0$ since $k < j \leqslant r$. Therefore $2^s\sigma_k \in 2S_1$. Hence $2^s\sigma_k \notin S_2 = S_1 - 2S_1$ while $\sigma_r \in S_2$. Thus C_j is always well-defined.

Clearly the C_j are partiiion ideals of order 1 and each is different from the others. To conclude our proof, we show that

$$C_1 \overset{PT}{\sim} C_2 \overset{PT}{\sim} C_3 \overset{PT}{\sim} C_4 \overset{PT}{\sim} \cdots$$

$$\sum_{n=0}^{\infty} p(C_j\,;n)\,q^n = \frac{\prod_{i=2^s\sigma_k, s\geqslant 0, k<j}(1+q^i)}{\prod_{l=j}^{\infty}(1-q^{\sigma_l})}$$

$$= \frac{1}{(1-q^{\sigma_j})} \cdot \frac{\prod_{i=2^s\sigma_k, s\geqslant 0, k<j}(1+q^i)}{\prod_{l=j+1}^{\infty}(1-q^{\sigma_l})}$$

$$= \frac{\prod_{t=0}^{\infty}(1+q^{2^t\sigma_j})\prod_{i=2^s\sigma_k, s\geqslant 0, k<j}(1+q^i)}{\prod_{l=j+1}^{\infty}(1-q^{\sigma_l})}$$

$$= \frac{\prod_{i=2^s\sigma_k, s\geqslant 0, k<j+1}(1+q^i)}{\prod_{l=j+1}^{\infty}(1-q^{\sigma_l})} = \sum_{n=0}^{\infty} p(C_{j+1}\,;n)\,q^n.$$

Hence comparing coefficients of q^n in the two above sums, we see that $p(C_j\,;n) = p(C_{j+1}\,;n)$ for each n. Therefore $C_j \sim^{\text{PT}} C_{j+1}$ for each j. Since $C_1 = C(S_2)$, we see that $\mathscr{C}$ contains an infinite number of elements. ∎

It is not difficult to deduce from Theorem 5 that there exist infinitely many such $\mathscr{C}$ with infinitely many elements. Indeed if S_2 is any set of odd positive integers and if S_1 is the set of all positive integers whose largest odd factor lies in S_2, then clearly by Theorem 3, (S_1, S_2) is an Euler pair. Furthermore if $S_2 \neq S_2{}^*$, then $C(S_2) \not\sim^{\text{PT}} C(S_2{}^*)$ (for a proof of this see the Lemma in [23]). Thus each different infinite set of odd integers S_2 gives rise to a different Euler pair and a new equivalence class containing the partition ideals associated with that Euler pair.

The results in this section have described some of the partition ideals of order 1 and have given some idea of the nature of some of the equivalence classes modulo $\sim^{\text{PT}}$. As the Introduction and equivalences (2.1)–(2.8) indicate, most of the more important theorems concerning partition identities arise form equivalences of the sort $C_1 \sim^{\mathrm{PT}} C_2$ where C_1 is of order 1 and C_2 is of higher order. In the next three sections we shall examine some of the techniques used in proving these deeper results. Later in Section 7, we shall return to the study of the structure of the equivalence classes of $\mathscr{S}$.

4. Well-Poised Theorems

We refer to the family of partition identities given in (1.10) as well-poised since the generating function for the case $a = k$ is an infinite product times the well-poised basic hypergeometric series:

$$
{}_{\lambda+3}\Phi_{\lambda+2}\left[\begin{matrix} x, q^{\lambda+1}\sqrt{x}, -q^{\lambda+1}\sqrt{x}, -q, -q^2,\ldots, -q^{\lambda}; q^{\lambda+1}, -x^k q^{\frac{1}{2}(\lambda+1)(2k-\lambda+1)n+\frac{1}{2}(\lambda-1)} \\ -\sqrt{x}, -\sqrt{x}, -xq^{\lambda}, -xq^{\lambda-1},\ldots, -xq \end{matrix}\right] \tag{4.1}
$$

(for more details concerning well-poised basic, hypergeometric series, see [81, 21]). We have chosen to here present a complete proof of (1.10) in the case $\lambda = 2$. This choice allows us to observe some of the difficulties that arise in the proof of (1.10) without embroiling us in the intricate details that accompany proof of the full result [20]. Also Theorem 6 has independent interest since it is a generalization of Schur's Theorem (1.6).

We shall study the following functions:

$$
C_{k,a}(x) = \sum_{n=0}^{\infty} \frac{(-1)^n x^{kn} q^{\frac{3}{2}(2k-1)n^2+\frac{3}{2}n-3an}(1 - x^a q^{6na})(x, q^3)_n(-q; q^3)_n(-q^2; q^3)_n}{(q^3; q^3)_n(1-x)(-xq; q^3)_n(-xq^2; q^3)_n}, \tag{4.2}
$$

$$
\begin{aligned}
D_{k,a}(x) = \sum_{n=0}^{\infty} & \frac{(-1)^n x^{kn} q^{\frac{3}{2}(2k-1)n^2+\frac{3}{2}n+3(k-a)n}(xq^3; q^3)_n(-q; q^3)_n(-q^2; q^3)_n}{(q^3; q^3)_n(-xq; q^3)_n(-xq^2; q^3)_n} \\
& \times \left\{1 - \frac{x^a q^{(6n+3)(a-1)}(1 + q^{3n+1})(1 + q^{3n+2})}{(1 + xq^{3n+1})(1 + xq^{3n+2})}\right\},
\end{aligned} \tag{4.3}
$$

where $|q| < 1$, $x \neq q^{-3n-1}, q^{-3n-2}$ for any integer $n \geqslant 0$, and

$$
(a; q)_n = (a)_n = (1-a)(1-aq)\cdots(1-aq^{n-1}),
$$
$$
(a; q)_\infty = (a)_\infty = \lim_{n\to\infty} (a; q)_n .
$$

$$H_{k,a}(x) = \frac{(-xq;\, q^3)_\infty(-xq^2;\, q^3)_\infty}{(xq^3;\, q^3)_\infty} C_{k,a}(x), \tag{4.4}$$

$$J_{k,a}(x) = \frac{(-xq;\, q^2)_\infty(-xq^2;\, q^3)_\infty}{(xq^3;\, q^3)_\infty} D_{k,a}(x). \tag{4.5}$$

From these four equations, we see directly that

$$C_{k,0}(x) = H_{k,0}(x) = 0, \tag{4.6}$$

and

$$H_{k,a}(0) = J_{k,a}(0) = 0 \qquad \text{for} \quad 1 \leqslant a \leqslant k. \tag{4.7}$$

We now must prove some lemmas.

LEMMA 4.1. $H_{k,a}(x) - H_{k,a-1}(x) = x^{a-1} J_{k,k-a+1}(x)$.

Proof. We shall prove the equivalent identity

$$C_{k,a}(x) - C_{k,a-1}(x) = x^{a-1} D_{k,k-a+1}(x). \tag{4.8}$$

we see that

$$C_{k,a}(x) - C_{k,a-1}(x) = \sum_{n=0}^{\infty} \frac{(-1)^n\, x^{kn} q^{\frac{3}{2}(2k-1)n^2+\frac{3}{2}n}(x;\, q^3)_n(-q;\, q^3)_n(-q^2;\, q^3)_n}{(q^3;\, q^3)_n(1-x)(-xq;\, q^3)_n(-xq^2;\, q^3)_n}$$
$$\times \{q^{-3an}(1-q^{3n}) + x^{a-1} q^{3n(a-1)}(1-xq^{3n})\}$$

We split this sum into two separate parts one for each of the main terms inside the curly brackets, and we replace n by $n+1$ in the first sum. Hence we see that

$$C_{k,a}(x) - C_{k,a-1}(x)$$
$$= -\sum_{n=0}^{\infty} \frac{(-1)^n\, x^{kn+k} q^{\frac{3}{2}(2k-1)n^2+3(2k-a+\frac{1}{2})n+3(k-a)}(x;\, q^3)_{n+1}(-q;\, q^3)_{n+1}(-q^2;\, q^3)_{n+1}}{(q^3;\, q^3)_n(1-x)(-xq;\, q^3)_{n+1}(-xq^2;\, q^3)_{n+1}}$$
$$+ x^{a-1} \sum_{n=0}^{\infty} \frac{(-1)^n\, x^{kn} q^{\frac{3}{2}(2k-1)n^2+\frac{3}{2}n+3n(a-1)}(x;\, q^3)_{n+1}(-q;\, q^3)_n(-q^2;\, q^3)_n}{(q^3;\, q^3)_n(1-x)(-xq;\, q^3)_n(-xq^2;\, q^3)_n}$$
$$= x^{a-1} \sum_{n=0}^{\infty} \frac{(-1)^n\, x^{kn} q^{\frac{3}{2}(2k-1)n^2+\frac{3}{2}n+3n(a-1)}(xq^3;\, q^3)_n(-q;\, q^3)_n(-q^2;\, q^3)_n}{(q^3;\, q^3)_n(-xq;\, q^3)_n(-xq^2;\, q^3)_n}$$
$$\times \left\{1 - \frac{x^{k-a+1} q^{6kn-6an+3(k-a)}(1+q^{3n+1})(1+q^{3n+2})}{(1+xq^{3n+1})(1+xq^{3n+2})}\right\}$$
$$= x^{a-1} D_{k,k-a+1}(x).$$

LEMMA 4.2.

$$J_{k,a} = H_{k,a}(xq^3) + (xq + xq^2) H_{k,a-1}(xq^3) + x^2q^3H_{k,a-2}(xq^3).$$

Proof. We shall prove the equivalent identity:

$$D_{k,a}(x) = \frac{(1 - xq^3)}{(1 + xq)(1 + xq^2)} \times \{C_{k,a}(xq^3) + (xq + xq^2) C_{k,a-1}(xq^3) + x^2q^3C_{k,a-2}(xq^3)\}$$

$$\begin{aligned} D_{k,a}(x) &= \sum_{n=0}^{\infty} \frac{(-1)^n x^{kn}q^{\frac{1}{2}(2k-1)n^2+\frac{1}{2}n+3(k-a)n}(xq^3; q^3)_n(-q; q^3)_n(-q^2; q^3)_n}{(q^3; q^3)_n(-xq; q^3)_n(-xq^2; q^3)_n} \\ &\quad \times \left\{\frac{(1 + xq^{3n+1})(1 + xq^{3n+2}) - x^aq^{(6n+3)(a-1)}(1 + q^{3n+1})(1 + q^{3n+2})}{(1 + xq^{3n+1})(1 + xq^{3n+2})}\right\} \\ &= \frac{(1 - xq^3)}{(1 + xq)(1 + xq^2)} \\ &\quad \times \sum_{n=0}^{\infty} \frac{(-1)^n(xq^3)^{kn} q^{\frac{3}{2}(2k-1)n^2+\frac{3}{2}n}(xq^3; q^3)_n(-q; q^3)_n(-q^2; q^3)_n}{(1 - xq^3)(q^3; q^3)_n(-xq^4; q^3)_n(-xq^5; q^3)_n} \\ &\quad \times \{q^{-3an}(1 - (xq^3)^a q^{6an}) + x(q + q^2) q^{-3(a-1)n}(1 - (xq^3)^{a-1} q^{6n(a-1)}) \\ &\qquad + x^2q^3q^{-3(a-2)n}(1 - (xq^3)^{a-2} q^{6n(a-2)})\} \\ &= \frac{(1 - xq^3)}{(1 + xq)(1 + xq^2)} \\ &\quad \times \{C_{k,a}(xq^3) + x(q + q^2) C_{k,a-1}(xq^3) + x^2q^3C_{k,a-2}(xq^3)\}. \qquad \blacksquare \end{aligned} \tag{4.9}$$

LEMMA 4.3. $H_{k,1}(x) = J_{k,k}(x)$.

Proof. Set $a = 1$ in Lemma 4.1 and recall that $H_{k,0}(x)$ is identically 0. ∎

LEMMA 4.4. $J_{k,2}(x) = (1 + xq + xq^2) J_{k,k}(xq^3) + xq^3J_{k,k-1}(xq^3)$.

Proof. By Lemmas 4.1–4.3, we have that

$$\begin{aligned} J_{k,2}(x) &= H_{k,2}(xq^3) + x(q + q^2) H_{k,1}(xq^3) \\ &= H_{k,1}(xq^3) + xq^3J_{k,k-1}(xq^3) + x(q + q^2) J_{k,k}(xq^3) \\ &= (1 + xq + xq^2) J_{k,k}(xq^3) + xq^3J_{k,k-1}(xq^3). \qquad \blacksquare \end{aligned}$$

LEMMA 4.5.

$$J_{k,a}(x) - J_{k,a-1}(x)$$
$$= x^{a-1}q^{3a-6}(q^3 J_{k,k-a+1}(xq^3) + (q^2 + q)\, J_{k,k-a+2}(xq^3) + J_{k,k-a+3}(xq^3)).$$

Proof. By Lemmas 4.1, and 4.2, we see that

$$\begin{aligned} J_{k,a}(x) - J_{k,a-1}(x) &= (H_{k,a}(xq^3) - H_{k,a-1}(xq^3)) + x(q + q^2)(H_{k,a-1}(xq^3) - H_{k,a-2}(xq^3)) \\ &\quad + x^2q^3(H_{k,a-2}(xq^3) - H_{k,a-3}(xq^3)) \\ &= x^{a-1}q^{3a-6}(q^3 J_{k,k-a+1}(xq^3) + (q + q^2)\, J_{k,k-a+2}(xq^3) + J_{k,k-a+3}(xq^3)). \quad \blacksquare \end{aligned}$$

LEMMA 4.6. $H_{k,-a}(x) = -x^{-a}H_{k,a}(x)$.

Proof. This follows directly from (4.2), (4.4) and the fact that

$$q^{3an}(1 - x^{-a}q^{-6na}) = -x^{-a}q^{-3an}(1 - x^a q^{6an}). \quad \blacksquare$$

LEMMA 4.7. $J_{k,1}(x) = (1 - x)\, J_{k,k}(xq^3)$.

Proof. By Lemmas 4.2, 4.3, and 4.6,

$$\begin{aligned} J_{k,1}(x) &= H_{k,1}(xq^3) + x^2q^3H_{k,-1}(xq^3) \\ &= J_{k,k}(xq^3) - xH_{k,1}(xq^3) \\ &= (1 - x)\, J_{k,k}(xq^3). \quad \blacksquare \end{aligned}$$

LEMMA 4.8.

$$J_{k,k}(x) - J_{k,k-1}(x)$$
$$= x^{k-1}q^{3k-6}((1 - xq^3)\, q^3 J_{k,k}(xq^6) + (q + q^2)\, J_{k,2}(xq^3) + J_{k,3}(xq^3)).$$

Proof. Set $a = k$ in Lemma 4.5, and apply Lemma 4.7 to $J_{k,1}(xq^3)$. ∎

We are now prepared to prove the following result.

THEOREM 6. *Let $A_{2,k,a}(n)$ denote the number of partitions of n in which only multiples of 3 may be repeated and no part is $\equiv 0, \pm(3a - 3)$ (mod $6k - 3$). Let $B_{2,k,a}(n)$ denote the number of partitions of n of the* *form* $n = b_1 + b_2 + \cdots + b_s$, *where* $b_i \geqslant b_{i+1}$, *only parts divisible by* 3 *may be repeated,* $b_i - b_{i+k-1} \geqslant 3$ *with strict inequality if* $3 \mid b_i$, *and the total number of appearances of the parts* 1, 2, *and* 3 *is at most* $a - 1$. *Let* $A^*_{2,k,a}(n)$ *denote the number of partitions of* n *into parts that are* $\equiv 0, 1, 3, 5 \pmod 6$ *but* $\not\equiv 0, \pm(3a-3) \pmod{6k-3}$. *Then for* $2 \leqslant a \leqslant k$ *and* $k \geqslant 3$, *we have*

$$A_{2,k,a}(n) = B_{2,k,a}(n) = A^*_{2,k,a}(n).$$

Remark. The case $k = a = \lambda = 2$ reduces to Schur's theorem (1.6). The restriction $k \geqslant 3$ is unnecessary, but when $k = 2$ the details are somewhat different [15].

Proof. Let us consider the following expansion of $J_{k,a}(x)$.

$$J_{k,a}(x) = \sum_{N=-\infty}^{\infty} \sum_{M=-\infty}^{\infty} j_{k,a}(M, N)\, x^M q^N,$$

where the variables x and q are subject to the conditions listed after (4.2).

We shall now substitute this series into the identities we have obtained in the Lemmas. This will allow us to compute certain recursion relations for the $j_{k,a}(M, N)$.

By (4.2), (4.4), and (4.7), we see that

$$\begin{aligned} j_{k,a}(M, N) &= 1 \qquad \text{if } M = N = 0 \\ &= 0 \qquad \text{if either } M \leqslant 0 \text{ or } N \leqslant 0, \text{ but } M^2 + N^2 \neq 0 \end{aligned} \tag{4.10}$$

By Lemma 4.4,

$$\begin{aligned} j_{k,2}(M, N) &= j_{k,k}(M, N - 3M) + j_{k,k}(M - 1, N - 3M + 2) \\ &\quad + j_{k,k}(M - 1, N - 3M + 1) + j_{k,k-1}(M - 1, N - 3M). \end{aligned} \tag{4.11}$$

By Lemma 4.5 (with $3 \leqslant a \leqslant k - 1$),

$$\begin{aligned} &j_{k,a}(M, N) - j_{k,a-1}(M, N) \\ &\quad = j_{k,k-a+1}(M - a + 1, N - 3M) + j_{k,k-a+2}(M - a + 1, N - 3M + 1) \\ &\qquad + j_{k,k-a+2}(M - a + 1, N - 3M + 2) \\ &\qquad + j_{k,k-a+3}(M - a - 1, N - 3M + 3). \end{aligned} \tag{4.12}$$

By Lemma 4.8,

$$\begin{aligned} j_{k,k}(M, N) - j_{k,k-1}(M, N) \\ &= j_{k,k}(M-k+1, N-6M+3k-3) \qquad (4.13) \\ &\quad - j_{k,k}(M-k, N-6M+3k) + j_{k,2}(M-k+1, N-3M+1) \\ &\quad + j_{k,2}(M-k+1, N-3M+2) + j_{k,2}(M-k+1, N-3M+3). \end{aligned}$$

We now observe that Eqs. (4.10)–(4.13) uniquely define $j_{k,a}(M, N)$ for all M and N with $2 \leqslant a \leqslant k$: Eq. (4.10) provides the initial conditions; then one proceeds by a triple mathematical induction first on N, then M, then a (for $2 \leqslant a \leqslant k$) to verify the uniqueness of $j_{k,a}(M, N)$.

Let $P_{k,a}(M, N)$ denote the number of partitions of N of the type enumerated by $B_{2,k,a}(N)$ with the added condition that there are exactly M parts. We shall show that $P_{k,a}(M, N)$ also satisfies (4.10)–(4.13).

The fact that $P_{k,a}(M, N)$ satisfies (4.10) is obvious; note that $P_{k,a}(0, 0) = 1$ since the empty partition of 0 is counted here.

Next we prove that (4.12) is valid with $P_{k,a}(M, N)$ replacing $j_{k,a}(M,N)$. First we observe that $P_{k,a}(M, N) - P_{k,a-1}(M, N)$ counts those partitions of the type enumerated by $P_{k,a}(M, N)$ with the condition that the total number of appearances of the summands 1, 2, and 3 is *exactly* $a - 1$. We split these partitions into 4 disjoint classes: (1) those in which neither 1 nor 2 appear; (2) those in which 1 appears but 2 does not; (3) those in which 2 appears but 1 does not, and (4) those in which both 1 and 2 appear. We now transform all our partitions by deleting all summands that are smaller than 4 and subtracting 3 from each of the other summands.

The number of parts of each partition in each class is now reduced to $M - a + 1$. The number being partitioned is reduced to $N - 3M$ in the first class, $N - 3(M-1) - 1 = N - 3M + 2$ in the second class $N - 3(M-1) - 2 = N - 3M + 1$ in the third class, and $N - 3(M-2) - 3 = N - 3M + 3$ in the fourth class.

Since originally for partitions in the first class, $f_3 = a - 1$, and $f_3 + f_4 + f_5 + f_6 \leqslant k - 1$, we see that $f_4 + f_5 + f_6 \leqslant (k - a + 1) - 1$; consequently the transformed partitions from the first class are of the type enumerated by $P_{k,k-a+1}(M - a + 1, N - 3M)$. Reversal of the above argument shows that there are exactly

$$P_{k,k-a+1}(M - a + 1, N - 3M)$$

elements of the first class. Exactly similar arguments show that the other three classes produce the corresponding partition functions appearing in (4.12).

The argument to prove (4.11) is quite similar except that now the four classes are: (1) 3 appears while 2 and 1 do not; (2) 1 appears while 2 and 3 do not; (3) 2 appears while 1 and 3 do not, and (4) none of 1, 2, or 3 appears. One now proceeds in the manner used to verify (4.12).

Equation (4.13) seems unnecessary; it should be a special case of (4.12). Unfortunately, the function $J_{k,1}(x)$ is *not* the generating function for $P_{k,1}(M, N)$ and so there is no hope of proving that it is. Consequently to avoid the appearance of $J_{k,1}(x)$ and $P_{k,1}(M, N)$, we must treat (4.13).

Equation (4.13) is treated at the outset exactly as is (4.12). Indeed the entire left-hand side and the last three terms on the right side are produced exactly as before. In the case where $f_3 = k - 1$, we subtract 6 from all summands larger than 3 and delete all summands that equal 3. This produces partitions of the type enumerated by

$$P_{k,k}(M - a + 1, N - 6M - 3k + 3).$$

Examining (4.13), we notice that there is still a term

$$-P_{k,k}(M - k, N - 6M + 3k)$$

left over. This arises from the fact that by including the complete third term on the right-side of (4.13) we counted too much. Namely, if $f_1 = 0$, $f_2 = 1$, and $f_3 = k - 2$, then $f_2 + f_3 + f_4 \leqslant k - 1$ implies $f_4 = 0$ a condition separate from and *not* implied by $f_4 + f_5 + f_6 \leqslant 1$. Thus we must subtract off the number of those partitions for which $f_1 = 0, f_2 = 1, f_3 = k - 2, f_4 = 1, f_5 = 0, f_6 = 0$ but which otherwise are partitions of the type enumerated by $P_{k,k}(M, N)$. To count such partitions, we delete all summands $\leqslant 6$ and subtract 6 from each of the remaining summands. In this way we obtain just those partitions enumerated by $P_{k,k}(M - k, N - 6M + 3k)$, and this accounts for the remaining term on the right hand side of (4.13).

Thus we see that $P_{k,a}(M, N)$ satisfies (4.10)–(4.13); therefore since the $j_{k,a}(M, N)$ are the unique solutions of these equations, we have proved that

$$P_{k,a}(M, N) = j_{k,a}(M, N) \tag{4.14}$$

for each M and N with $2 \leqslant a \leqslant k$.

Equation (4.14) is the crucial matter in the proof. The rest is fairly simple. We observe that

$$B_{2,k,a}(N) = \sum_{M=0}^{\infty} P_{k,a}(M, N).$$

$$\begin{aligned}
\sum_{N=0}^{\infty} B_{2,k,a}(N)\, q^N &= \sum_{N=0}^{\infty} \sum_{M=0}^{\infty} P_{k,a}(M, N)\, q^N \\
&= J_{k,a}(1) \\
&= \frac{(-q; q^3)_\infty(-q^2; q^3)_\infty}{(q^2; q^3)_\infty} \sum_{n=-\infty}^{\infty} (-1)^n\, q^{\frac{3}{2}(2k-1)n^2+\frac{3}{2}n+3(k-a)n} \\
&\quad \times (1 - q^{(6n+3)(a-1)}) \\
&= \frac{(-q; q^3)_\infty(-q^2; q^3)_\infty}{(q^3; q^3)_\infty} \sum_{n=-\infty}^{\infty} (-1)^n\, q^{\frac{3}{2}(2k-1)n^2+\frac{3}{2}n+3(k-a)n} \\
&= \frac{(-q; q^3)_\infty(-q^2; q^3)_\infty}{(q^3; q^3)_\infty} (q^{6k-3}; q^{6k-3})_\infty (q^{6k-3-3(a-1)}; q^{6k-3})_\infty \\
&\quad \times (q^{3a-3}; q^{6k-3})_\infty \\
&= \sum_{N=0}^{\infty} A_{2,k,a}(N)\, q^N,
\end{aligned}$$

where the penultimate expression follows by Jacobi's identity [66, p. 282]. Thus comparing coefficients of q^N we see that $B_{2,k,a}(N) = A_{2,k,a}(N)$. To obtain the result for $A^*_{2,k,a}(n)$ we merely substitute $(q; q^6)^{-1}\,(q^5; q^6)_\infty^{-1}$ for $(-q; q^3)_\infty\,(-q^2; q^3)_\infty$ in the penultimate expression above. Thus Theorem 6 is proved. ∎

Several closing comments are in order. First of all we see that this result provides an example of two partition ideals that are equivalent modulo $\sim^{PT}$ even though the partition ideals related to $A_{2,k,a}(n)$ and to $A^*_{2,k,a}(n)$ have order 1 while the partition ideal related to $B_{2,k,a}(n)$ has order 4.

With regard to the general Eq. (1.10), we see easily that the partition ideal related to $A_{\lambda,k,a}(n)$ has order 1 while the one related to $B_{\lambda,k,a}(n)$ has order $\lambda + 2$. Thus we see clearly that order is *not* a property of equivalence classes in general.

Finally it should be remarked that the snag of proving (4.13) for $P_{k,a}(M, N)$ becomes worse and worse for larger and larger λ. This accounts for the complex special partition functions studied in detail in Sections 2 and 4 of [20].

5. A Theorem for the Modulus 7

The result of Section 4 (a special case of (1.10)) relied on a combinatorial interpretation of the q-difference equations associated with certain well-poised basic hypergeometric series; also the proof utilized the fact that these series reduced to an instance of the Jacobi triple product identity when $x = 1$.

Here we are concerned with a special case of (1.12). The general proofs of (1.11) and (1.12) both rely on Abel's lemma concerning the limiting value of $\sum a_n x^n$ as $x \to 1^-$; however, (1.11) also uses q-difference Eqs. [17] while (1.12) is concerned with what might be termed q-recurrent sequences [18]. We have chosen for our study a special case of (1.11) which illustrates fairly clearly the techniques related to the general proofs of (1.11) and (1.12).

Theorem 7.

$$C(1, 9, 11; 14) \overset{PT}{\sim} C_d(1, 2, 4; 7) \overset{PT}{\sim} D(1; 7, 7, 12, 7, 10, 10, 15; 7).$$

Remark. To simplify notation we shall write D for $D(1; 7, 7, 12, 7, 10, 10, 15; 7)$.

Proof. The equivalence

$$C(1, 9, 11; 14) \overset{PT}{\sim} C_d(1, 2, 4; 7)$$

follows directly from our theorem on Euler pairs, Theorem 4.

To prove the second equivalence we start by defining $p_a(m, n)$ to be the cardinality of the following set:

$$\left\{\{f_i\} \mid \{f_i\} \in D, \sum_{i=1}^{\infty} f_i i = n, \sum_{i=1}^{\infty} f_i = m, \text{ and } f_i = 0 \text{ if } i \leqslant a\right\}.$$

We must now establish the following identities:

$$p_0(m, n) = p_1(m, n) + p_0(m - 1, n - 7m + 6) \tag{5.1}$$
$$p_1(m, n) = p_2(m, n) + p_1(m - 1, n - 7m + 5) \tag{5.2}$$
$$p_2(m, n) = p_3(m, n) + p_0(m - 1, n - 14m + 11) \tag{5.3}$$
$$p_3(m, n) = p_4(m, n) + p_3(m - 1, n - 7m + 3) \tag{5.4}$$
$$p_4(m, n) = p_5(m, n) + p_0(m - 1, n - 14m + 9) \tag{5.5}$$
$$p_5(m, n) = p_6(m, n) + p_1(m - 1, n - 14m + 8) \tag{5.6}$$
$$p_6(m, n) = p_7(m, n) + p_0(m - 1, n - 21m + 14) \tag{5.7}$$
$$p_7(m, n) = p_0(m, n - 7m). \tag{5.8}$$

The proofs of these eight identities all resemble one another. We therefore choose (5.7) to present in detail. First we see that $p_6(m, n) - p_7(m, n)$ is *exactly* the number of partitions π of n with m parts such that $\pi \in D$ and 7 is a summand of π. By the requirements on D we see that all other summands besides the 7 must be at least as large as 22. We now transform these partitions by deleting the 7 and subtracting 21 from every other part. This leaves us with a partition of the type enumerated by $p_0(m - 1, n - 21m + 14)$. Clearly the above process is reversible, and so we have established a on-to-one correspondence between the partitions enumerated by $p_6(m, n) - p_7(m, n)$ and those enumerated by $p_0(m - 1, n - 21m + 14)$. Thus (5.7) is established.

We now define

$$f_a(x) = \sum_{m=0}^{\infty} \sum_{n=0}^{\infty} p_a(m, n)\, x^n q^n.$$

Then directly from our eight identities, we deduce that

$$f_0(x) = f_1(x) + xq^1 f_0(xq^7), \tag{5.9}$$
$$f_1(x) = f_2(x) + xq^2 f_1(xq^7), \tag{5.10}$$
$$f_2(x) = f_3(x) + xq^3 f_0(xq^{14}), \tag{5.11}$$
$$f_3(x) = f_4(x) + xq^4 f_3(xq^7), \tag{5.12}$$
$$f_4(x) = f_5(x) + xq^5 f_0(xq^{14}), \tag{5.13}$$
$$f_5(x) = f_6(x) + xq^6 f_1(xq^{14}), \tag{5.14}$$
$$f_6(x) = f_7(x) + xq^7 f_0(xq^{21}), \tag{5.15}$$
$$f_7(x) = f_0(xq^7). \tag{5.16}$$

We wish now to solve these equations to obtain an identity in $f_0(x)$, $f_0(xq^7)$, $f_0(xq^{14})$,... . To do this we note that adding all these equations produces an identity $f_0(x) = E_1$ where E_1 involves only f_0, f_1, and f_3 (with various arguments). Adding the first three equations produces an identity $f_3(x) = E_2$, where E_2 involves only f_0 and f_1. Furthermore (5.9) may be transformed to produce an identity

$$f_1(x) = f_0(x) - xqf_0(xq^7) = E_3,$$

an expression involving only f_0. Thus we may use E_3 for substitutions into E_2 to produce an identity $f_3(x) = E_4$ where E_4 involves only f_0. Finally we may use E_3 and E_4 for substitutions into E_1 to produce an identity involving only f_0 with various arguments. The final result is

$$f_0(x) = (1 + xq + xq^2 + xq^4) f_0(xq^7) + x(q^3 + q^5 + q^6)(1 - xq^7) f_0(xq^{14}) + xq^7(1 - xq^7)(1 - xq^{14}) f_0(xq^{21}). \quad (5.17)$$

If we define $G(x) = f_0(x)_7(x; q^7)_\infty$, then by dividing (5.17) by $(xq^7; q^7)_\infty$ we obtain

$$(1 - x) G(x) = (1 + xq + xq^2 + xq^4) G(xq^7) + x(q^3q^5 + q^6) G(xq^{14}) + xq^7G(xq^{21}). \quad (5.18)$$

We now consider the MacLaurin series for $G(x)$, namely

$$G(x) = \sum_{n=0}^{\infty} \gamma_n x^n,$$

and we substitute into (5.18). Thus from the coefficient of x^n after the substitution we see that

$$\gamma_n - \gamma_{n-1} = q^{7n}\gamma_n + q^{7n-6}\gamma_{n-1} + q^{7n-5}\gamma_{n-1} + q^{7n-3}\gamma_{n-1} + q^{14n-11}\gamma_{n-1} + q^{14-1}\gamma_{n-1} + q^{14n-1}\gamma_{n-1} + q^{21n-14}\gamma_{n-1}. \quad (5.19)$$

Therefore

$$(1 - q^{7n}) \gamma_n = (1 + q^{7n-6})(1 + q^{7n-5})(1 + q^{7n-3}) \gamma_{n-1}. \quad (5.20)$$

Since $\gamma_0 = G(0) = f_0(0) = 1$, we may easily solve the recurrence (5.20) by iteration.

Therefore,

$$\gamma_n = (-q; q^7)_n(-q^2; q^7)_n(-q^4; q^7)_n/(q^7; q^7)_n \tag{5.21}$$

Hence

$$f_0(x) = (x; q^7)_\infty\, G(x) = (x; q^7)_\infty \sum_{n=0}^{\infty} \frac{(-q; q^7)_n(-q^2; q^7)_n(-q^4; q^7)_n\, x^n}{(q^7; q^7)_n}.$$

Letting $x \to 1^-$ we deduce using Abel's lemma (or the more general Appell Comparison theorem [52, p. 101]) that

$$\begin{aligned}\sum_{n=0}^{\infty} p_0(D; n)\, q^n &= f_0(1)\\ &\lim_{x\to 1^-} f_0(X)\\ &= (q^7; q^7)_\infty \lim_{x\to 1^-} (1-x) \sum_{n=0}^{\infty} \frac{(-q; q^7)_n(-q^2; q^7)_n(-q^4; q^7)_n\, x^n}{(q^7; q^7)_n}\\ &= (-q; q^7)_\infty(-q^2; q^7)_\infty(-q^4; q^7)_\infty\\ &= \sum_{n=0}^{\infty} p(C_d(1, 2, 4; 7); n)\, q^n. \end{aligned} \tag{5.22}$$

Thus $D \sim^{\mathrm{PT}} C_d(1, 2, 4; 7)$, and Theorem 7 is established. ∎

6. Elementary Techniques

In Sections 4 and 5, we have studied the important interrelationship between partition identities and q-difference equations. The results obtained were always very general yielding families of identities rather than just single equivalences. In this section we shall discuss how more elementary techniques may be utilized. We choose, for our example, the identity of Göllnitz and Gordon, namely, (1.7) [or equivalently (2.5)]. It is especially appropriate for our purposes since the proof will involve both work with infinite series and a study of the graphical representation of partitions.

Theorem 8. *$p(D(1; 2, 3, 2); n)$ equals the number of partitions of n of the form $\sum_{i=1}^{s} a_i + \sum_{i=1}^{t} b_i$ where a_i and b_i are odd, $a_i > a_{i+1}$, $b_i > b_{i+1}$, and $a_i \leqslant 2t - 1$.*

Proof. Let us consider the second type of partitions first. As an example, we take the partition of 48 in which the a-sum is $9 + 7 + 1$ and the b-sum is $11 + 9 + 7 + 3 + 1$. The b-sum is now represented as an ordinary Ferrars graph [66, Ch. 19]

```
· · · · · · · · · · ·
· · · · · · · · ·
· · · · · · ·
· · ·
·
```

The a-sum is represented in a modified graphical form, namely, each part $2j + 1$ is represented by two vertical columns of $j + 1$ and j nodes respectively. Thus we have

```
 9    7    1
xx   xx   x
xx   xx
xx   xx
xx   x
x
```

These columns are now inserted in the obvious places in the Ferrars graph for the b-sum (there is only one place each pair of columns will fit except for the 1 which is put on the far right):

```
· x x · · x x · · · · · · · · x
· x x · · x x · · · · · ·
· x x · · x x · · · ·
· x x · · x
· x
```

This merged graph (read rows horizontally again) now represents $16 + 13 + 11 + 6 + 2$.

In general, the above procedure establishes a one-to-one correspondence between the two types of partitions mentioned in our theorem. This is because the b-sum has only distinct odd parts; then the a-sum is joined into the Ferrars graph of the b sum in such a way that the distance between parts is not decreased and if an even part arises (which occurs due to the lower left hand node in the representation of an a_i) then it is at least 3 units larger than the next smallest part. Conversely, given a partition of the type enumerated by $p(D(1; 2, 3; 2); n)$ we form the corresponding partition by removing for the a-sum those pairs of columns the lower left hand node of which is the right-hand-most node of an even part. ▌

COROLLARY.

$$\sum_{n=0}^{\infty} p(D(1, 2, 3; 2); n)\, q^n = \sum_{n=0}^{\infty} \frac{q^{n^2}(-q; q^2)_n}{(q^2; q^2)_n}.$$

Proof. By Theorem 8, we know that $p(D(1; 2, 3; 2); n)$ equals the number of partitions of n of the form $\sum_{i=1}^{s} a_i + \sum_{i=1}^{t} b_i$ where a_i and b_i are odd, $a_i > a_{i+1}$, $b_i > b_{i+1}$, and $a_i \leqslant 2t - 1$. For a given t, the generating function for the b-sum is

$$\frac{q^{1+3+\cdots+(2t-1)}}{(1-q^2)(1-q^4)\cdots(1-q^{2t})} = \frac{q^{t^2}}{(q^2; q^2)_t}$$

while the generating function for the a-sum is

$$(1+q)(1+q^3)\cdots(1+q^{2t-1}) = (-q; q^2)_t .$$

Consequently for given t the generating function for the two sums taken together is

$$q^{t^2}(-q; q^2)_t/(q^2; q^2)_t .$$

Summing over all t, we obtain the desired result. ▌

We shall now consider this generating function (actually we replace q by $-q$ for convenience) in the following theorem.

THEOREM 9.

$$\sum_{n=0}^{\infty} \frac{(-1)^n q^{n^2}(q; q^2)_n}{(q^2; q^2)_n} = (-q; q^8)_\infty^{-1}(q^4; q^8)_\infty^{-1}(-q^7; q^8)_\infty^{-1}.$$

Proof. For the proof of this identity we shall need the two elementary identities due to Euler [66, Chap. 19]:

$$\sum_{n=0}^{\infty} \frac{Z^n}{(q)_n} = (Z)_\infty^{-1}, \tag{6.1}$$

and

$$\sum_{n=0}^{\infty} \frac{Z^n q^{\frac{1}{2}n(n-1)}}{(q)_n} = (-Z)_\infty , \tag{6.2}$$

as well as the famous triple product identity due to Jacobi [66, Chap. 19]:

$$\sum_{n=-\infty}^{\infty} q^{n^2} Z^n = (q^2; q^2)_\infty (-Zq; q^2)_\infty (-Z^{-1}q; q^2)_\infty . \tag{6.3}$$

Although it was long believed that (6.3) was a much deeper identity than (6.1) and (6.2), it has been shown [6] that (6.3) follows rather easily from (6.1) and (6.2). Thus our Theorem is deducible from (6.1) and (6.2) above.

$$\sum_{n=0}^{\infty} \frac{(-1)^n q^{n^2}(q; q^2)_n}{(q^2; q^2)_n}$$

$$= (q; q^2)_\infty \sum_{n=0}^{\infty} \frac{(-1)^n q^{n^2}}{(q^2; q^2)_n (q^{2n+1}; q^2)_\infty}$$

$$= (q; q^2)_\infty \sum_{n=0}^{\infty} \frac{(-1)^n q^{n^2}}{(q^2; q^2)_n} \sum_{m=0}^{\infty} \frac{q^{2mn+m}}{(q^2; q^2)_m} \qquad \text{(by (6.1))}$$

$$= (q; q^2)_\infty \sum_{m=0}^{\infty} \frac{q^m}{(q^2; q^2)_m} \sum_{n=0}^{\infty} \frac{(-1)^n q^{n^2+2mn}}{(q^2; q^2)_n}$$

$$= (q; q^2)_\infty \sum_{m=0}^{\infty} \frac{q^m}{(q^2; q^2)_m} (q^{2m+1}; q^2)_\infty \qquad \text{(by (6.2))}$$

$$= (q; q^2)_\infty^2 \sum_{m=0}^{\infty} \frac{q^m}{(q^2; q^2)_m (q; q^2)_m}$$

$$= (q; q^2)_\infty^2 \sum_{m=0}^{\infty} \frac{q^m}{(q; q)_{2m}}$$

$$= \frac{1}{2} (q; q^2)_\infty^2 \sum_{m=0}^{\infty} \frac{q^{m/2}(1 + (-1)^m)}{(q; q)_m}$$

$$= \frac{1}{2} (q; q^2)_\infty^2 ((q^{1/2})_\infty^{-1} + (-q^{1/2})_\infty^{-1}) \qquad \text{(by 6.1))}$$

$$= \frac{1}{2}(q;q^2)_\infty(q^{1/2})_\infty(-q^{1/2})_\infty((q^{1/2})_\infty^{-1} + (-q^{1/2})_\infty^{-1})$$

$$= \frac{1}{2}(q;q^2)_\infty((-q^{1/2})_\infty + (q^{1/2})_\infty)$$

$$= \frac{1}{2}\frac{(q;q^2)_\infty}{(q^2;q^2)_\infty}((q^2;q^2)_\infty(-q^{1/2};q^2)_\infty(-q^{3/2};q^2)_\infty$$

$$+ (q^2;q^2)_\infty(q^{1/2};q^2)_\infty(q^{3/2};q^2)_\infty)$$

$$= \frac{1}{2}\frac{(q;q^2)_\infty}{(q^2;q^2)_\infty}\left(\sum_{n=-\infty}^{\infty} q^{n^2+n/2} + \sum_{n=-\infty}^{\infty}(-1)^n q^{n^2+n/2}\right) \qquad \text{(by (6.3))}$$

$$= \frac{(q;q^2)_\infty}{(q^2;q^2)_\infty}\sum_{n=-\infty}^{\infty} q^{4n^2+n}$$

$$= \frac{(q;q^2)_\infty}{(q^2;q^2)_\infty}(q^8;q^8)_\infty(-q^3;q^8)_\infty(-q^5;q^8)_\infty \qquad \text{(by (6.3))}$$

$$= \frac{(q^2;q^4)_\infty}{(-q;-q)_\infty}(q^8;q^8)_\infty(-q^3;q^8)_\infty(-q^5;q^8)_\infty$$

$$= \frac{(q^2;q^8)_\infty(q^6;q^8)_\infty(q^8;q^8)_\infty(-q^3;q^8)_\infty(-q^5;q^8)_\infty}{(-q;-q)_\infty}$$

$$= (-q;q^8)_\infty^{-1}(q^4;q^8)_\infty^{-1}(-q^7;q^8)_\infty^{-1}. \qquad \blacksquare$$

Corollary. $C(1, 4, 7; 8) \sim^{\mathrm{PT}} D(1; 2, 3; 2)$.

Proof.

$$\sum_{n=0}^{\infty} p(D(1;2,3;2);n)\,q^n$$

$$= \sum_{n=0}^{\infty}\frac{q^{n^2}(-q;q^2)_n}{(q^2;q^2)_n} \qquad \text{(by the Corollary to Theorem 8)}$$

$$= \prod_{n=0}^{\infty}\frac{1}{(1-q^{8n+1})(1-q^{8n+4})(1-q^{8n+7})} \qquad \text{(by Theorem 9)}$$

$$= \sum_{n=0}^{\infty} p(C(1,4,7;8);n)\,q^n.$$

Hence comparing coefficients of q^n in the extreme sums above, we see that

$$p(D(1; 2, 3; 2); n) = p(C(1, 4, 7; 8); n)$$

for each n. Hence our result follows. ∎

It should be emphasized that the techniques used in this chapter are merely a sample of many that are used in this subject. For further analytic theorems like Theorem 9, we refer the reader to [87, 12–14, 31, 1, 79–81]. The elementary study of partitions via graphical and other arithmetical techniques has been extensively studied. There is an excellent introduction to this aspect in [66; Chap. 19], and in [28] it is shown that many theorems can be proved purely using such means.

7. Partition Ideals of Higher Order

In Sections 2 and 3 we considered general questions on the structure of the lattice $\mathscr{S}$ relative to $\sim^{\mathrm{PT}}$. In Section 3, we proved that there were infinitely many equivalence classes modulo $\sim^{\mathrm{PT}}$ that each contained infinitely many partition ideals of order 1. In later sections, we saw that the deeper aspects of this theory concern theorems of the form $C_1 \sim^{\mathrm{PT}} C_2$ where C_1 is of order 1 and C_2 is of higher order. A natural question arises now, namely: are there equivalence classes modulo $\sim^{\mathrm{PT}}$ that each contain infinitely many partition ideals of order greater than 1? We can settle this question with the following theorem although it still leaves many questions unanswered.

Theorem 10. *Suppose $r \geqslant 5$. Then there exists an equivalence class modulo $\sim^{\mathrm{PT}}$ that contains infinitely many partition ideals of order r.*

Proof. Let $k = r - 2$ so that $k \geqslant 3$.

Define the partition ideal D_j by letting it be those $\{f_i\}$ associated with partitions π of the form $b_1 + b_2 + \cdots + b_s$ where 1) if $b_i \leqslant jk$, then $b_i - b_{i+1} \geqslant k$ and if $k \mid b_i, b_i - b_{i+1} > k$; 2) if $b_i > jk$ then $b_i - b_{i+1} \geqslant k$ if $b_i \not\equiv 0, -1 (\mathrm{mod}\, k)$, when $b_i = kl - 1$, $b_i - b_{i+1} \geqslant k - 1$, when $b_i = kl$, $b_i - b_{i+1} \geqslant k + 2$. We also consider the case $j = \infty$, where now the parts of the partitions are subject always to condition 1. We shall prove that

$$D_0 \overset{PT}{\sim} D_1 \overset{PT}{\sim} D_2 \overset{PT}{\sim} D_3 \overset{PT}{\sim} \cdots.$$

To prove this we define $p_m(D_j\,; n)$ as the number of partitions of the type enumerated by $p(D_j\,; n)$ with the additional restriction that all summands do not exceed m. We also consider the related generating functions

$$f_m(j; q) = \sum_{n=0}^{\infty} p_m(D_j\,; n)\, q^n.$$

It is clear that $f_m(j; q)$ is a polynomial in q, and that

$$\lim_{m\to\infty} f_m(j; q) = \sum_{n=0}^{\infty} p(D_j\,; n)\, q^n.$$

The crux of the proof now lies in establishing the following identities.

$$f_m(j; q) = f_m(\infty; q) \qquad \text{if} \quad k \nmid (m+1) \quad \text{or} \quad m \leqslant jk, \tag{7.1}$$

$$f_{lk-1}(j; q) = f_{lk-1}(\infty; q) + q^{2k(l-1)+k-1} f_{(l-2)k-1}(\infty; q) \qquad \text{for} \quad l > j. \tag{7.2}$$

We prove this by observing that the difference conditions used in defining D_∞ and D_j imply the following recursive relations.

$$f_m(\infty; q) = f_{m-1}(\infty; q) + q^m f_{m-k}(\infty; q), \qquad \text{if} \quad k \nmid m; \tag{7.3}$$

$$f_{lk}(\infty; q) = f_{lk-1}(\infty; q) + q^{lk} f_{lk-k-1}(\infty; q); \tag{7.4}$$

$$f_m(j; q) = f_{m-1}(j; q) + q^m f_{m-k}(j; q), \qquad \text{if} \quad k \nmid m \quad \text{for} \quad m \leqslant jk$$
$$\text{and for} \quad m > jk \quad \text{if} \quad k \nmid m \quad \text{or} \quad m-1; \tag{7.5}$$

$$f_{lk}(j; q) = f_{lk-1}(j; q) + q^{lk} f_{lk-k-1}(j; q) \qquad \text{for} \quad l \leqslant j; \tag{7.6}$$

$$f_{lk-1}(j; q) = f_{lk-2}(j; q) + q^{lk-1} f_{lk-k}(j; q), \qquad \text{if} \quad l > j; \tag{7.7}$$

$$f_{lk}(j; q) = f_{lk-1}(j; q) + q^{lk} f_{lk-k-2}(j; q), \qquad \text{if} \quad l > j. \tag{7.8}$$

By (7.3)–(7.5) and (7.7) and the fact that $f_0(\infty; q) = f_0(j; q) = 1$, $f_m(0; q) = f_m(j; q) = 0$ for $m < 0$, we obtain using mathematical induction that (7.1) is valid for all $m \leqslant jk$. We now assume that the theorem has been verified for all $m \leqslant rk$, where $r \geqslant j$. Then for $1 \leqslant i \leqslant k-2$, we have (using a second finite induction on i)

$$\begin{aligned} f_{rk+i}(j; q) &= f_{rk+i-1}(j; q) + q^{rk+i} f_{rk-k+i}(j; q) \\ &= f_{rk+i-1}(\infty; q) + q^{rk+i} f_{rk-k+i}(\infty; q) \\ &= f_{rk+i}(\infty; q). \end{aligned}$$

$$\begin{aligned}
f_{rk+k-1}(j; q) &= f_{rk+k-2}(j; q) + q^{rk+k-1} f_{rk}(j; q) \\
&= f_{rk+k-2}(\infty; q) + q^{rk+k-1} f_{rk}(\infty; q) \\
&= f_{rk+k-2}(\infty; q) + q^{rk+k-1} f_{rk-1}(\infty; q) \\
&\quad + q^{rk+k-1}(f_{rk}(\infty; q) - f_{rk-1}(\infty; q)) \\
&= f_{rk+k-2}(\infty; q) + q^{rk+k-1} f_{rk-1}(\infty; q) + q^{rk+k-1} f_{rk-k-1}(\infty; q)\, q^{rk} \\
&= f_{rk+k-1}(\infty; q) + q^{2rk+k-1} f_{rk-k-1}(\infty; q),
\end{aligned}$$

$$\begin{aligned}
f_{rk+k}(j; q) &= f_{rk+k-1}(j; q) + q^{rk+k} f_{rk-2}(j; q) \\
&= f_{rk+k-1}(\infty; q) + q^{2rk+k-1} f_{rk-k-1}(\infty; q) + q^{rk+k} f_{rk-2}(\infty; q) \\
&= f_{rk+k-1}(\infty; q) + q^{rk+k} f_{rk-1}(\infty; q) \\
&\quad + q^{rk+k}(f_{rk-2}(\infty; q) - f_{rk-1}(\infty; q) + q^{rk-1} f_{rk-k-1}(\infty; q)) \\
&= f_{rk+k-1}(\infty; q) + q^{rk+k} f_{rk-1}(\infty; q) \\
&= f_{rk+k}(\infty; q).
\end{aligned}$$

Thus, in general, we see that if Eqs. (7.1) and (7.2) are valid for $m \leqslant rk$, then they are valid for $m \leqslant rk + k$. Consequently, they are valid for all m. Clearly then we see that for $|q| < 1$,

$$\lim_{m\to\infty} f_m(j; q) = \lim_{m\to\infty} f_m(\infty; q).$$

Therefore,

$$\sum_{n=0}^{\infty} p(D_j\,; n)\, q^n = \sum_{n=0}^{\infty} p(D_\infty\,; n)\, q^n,$$

and so for each n

$$p(D_j\,; n) = p(D_\infty\,; n).$$

Therefore $D_j \sim^{\mathrm{PT}} D_\infty$. Since the order of D_j is $k + 2 = r$ for each $j > 0$, Theorem 10 is established. ▮

The genesis of Theorem 10 lies in the discovery [25] that for $k = 3$,

$$D_\infty \overset{PT}{\sim} D_0 = D(1; 2, 3, 4; 3) \overset{PT}{\sim} C_d(1, 2; 3) \overset{PT}{\sim} C(1, 5; 6),$$

an extension of Schur's theorem.

The proof of Theorem 10, may easily be modified to prove that there exist infinitely many equivalence classes modulo $\sim^{\mathrm{PT}}$ that each contain an infinite number of partition ideals of order r provided $r \geqslant 7$. The modification is made by requiring that in $b_1 + b_2 + \cdots + b_s$, $b_i - b_{i+1} \geqslant k + 1$ if $b_i = hk + 2$ where h is fixed and $k = r - 2$. This change produces a new equivalence class for each h but does not affect the essential relationships involving $f_m(\infty; q)$ and $f_m(j; q)$ for $m \equiv 0, -1 (\mathrm{mod}\, k)$. It remains an open question whether or not there exist any equivalence classes with an infinite number of elements of order r for $r = 2, 3, 4$. Much deeper (and more interesting) is the question of whether there exists an equivalence class with an infinite number of orders.

8. Conclusion

It must be pointed out here that not all partition functions of interest in the theory of partition identities are C-partition functions for some partition ideal C. For example, from Euler's famous pentagonal number theorem

$$\sum_{n=-\infty}^{\infty} (-1)^n q^{\frac{1}{2}n(3n-1)} = (q)_\infty .$$

Legendre deduced that

$$p_e(n) - p_0(n) = \begin{cases} (-1)^\lambda & \text{if } n = \frac{1}{2}\lambda(3\lambda \pm 1), \\ 0, & \text{otherwise,} \end{cases}$$

where $p_e(n)$ (resp. $p_0(n)$) is the number of partitions of n into an even (resp. odd) number of distinct parts. Thus

$$p_e(n) + p_0(n) = p(D(1; 1; 1); n);$$

however $p_e(n)$ counts only a certain restricted subset of $D(1; 1; 1) = D_1$, namely, those $\{f_i\} \in D_1$ for which $\sum_{i=1}^{\infty} f_i$ is even. Similarly $p_0(n)$ counts those elements of D_1 for which $\sum_{i=_k}^{\infty} f_i$ is odd. Thus the study of equivalence classes of $\mathscr{S}$ modulo $\sim^{\mathrm{PT}}$ is not all that is of interest in partition identities. The internal structure of certain partition ideals is also of interest in certain cases (see also [34]).

In this vein, we also remark that a sieve method has recently been introduced into the theory of partition identities [38]. This technique allows new proofs of Euler's pentagonal number theorem, the Rogers-Ramanujan identities, and certain new partition identities related to the

$A_{k,a}(n)$ of (1.10) for each n. The sieve acts on partitions classified according to successive ranks [40], and so the relationship with the partition ideals in $\mathcal{S}$ is not clear at this time; however, one suspects some canonical mapping between such partitions and partitions of the type enumerated by the $B_{k,a}(n)$ of (1.10).

As is clear from the theorems presented in this paper, the theory of partition identities is intimately related to basic hypergeometric series (through q-difference equations) and to techniques for transforming Ferrars graphs. Recent work of Goldman and Rota [57, 58], and Knuth [68] also indicates that the combinatorial theory of finite vector spaces may play an important role in future developments.

With regard to the lattice-theoretic side of partition identities, we note that if L is a lattice and σ is a positive, integer-valued valuation on L such that the sets

$$\{x \in L \mid \sigma(x) = n\}$$

are all finite, then we may study "generalized partition identities" in L by defining an equivalence on the semi-ideals of L just as $\sim^{\mathrm{PT}}$ was defined on $\mathcal{S}$. For example, if L is the lattice of subspaces of an n-dimensional vector space over the finite field $GF(q)$, then the dimension function $d(S)$ is a positive, integer-valued valuation on L. Furthermore, if $p_L(r)$ denotes the cardinality of $\{S \in L \mid d(S) = r\}$, then it is known that

$$p_L(r) \equiv \begin{bmatrix} n \\ r \end{bmatrix} = (q)_n/(q)_r(q)_{n-r} \qquad [57].$$

Thus the generating function for $p_L(r)$ is

$$\sum_{r=0}^{n} p_L(r)\, x^r = \sum_{r=0}^{n} \begin{bmatrix} n \\ r \end{bmatrix} x^r = H_n(x),$$

the q-Hermite polynomial [48]. More interesting questions arise when the lattice L is infinite.

At the conclusion of his survey article [4], Alder asks about the discovery of families of partition identities. The identities given in (1.9)–(1.12) as well as those discussed in [8 and 9] are all examples of families of identities. If $B_{\lambda,k,a}$ denotes the partition ideal of which $B_{\lambda,k,a}(n)$ is the associated partition function, then we see that

$$B_{\lambda,k,a} \subseteq B_{\lambda',k',a'},$$

whenever $\lambda = \lambda'$, $k' \geqslant k$, $a' \geqslant a$. Thus $B_{\lambda,k,a}$ may be viewed as an increasing function of k and a. General results related to chains of partition ideals would be of interest.

We close with some open problems for which even partial answers would be important.

1. For a given partition ideal C of order k determine an asymptotic formula for $p(C; n)$ that depends only on some parameters related to C, for example the numbers

$$d_{r,s} = \sup_{\{f_i\} \in C} \{f_r + f_{r+1} + \cdots + f_{r+s}\},$$

where $0 \leqslant s \leqslant k - 1$, $\infty > r \geqslant 1$.

2. Suppose C is a partition ideal of order k. Suppose also there exist integers r and m such that for $\{f_i\} \in C$ with $f_i = 0$ for $i < r$ then $\{f_{i-m}\}$ and $\{f_{i+m}\}$ are both in C. Does

$$f(x) - \sum_{\{f_i\} \in C} x^{\Sigma f_i} q^{\Sigma f_i i}$$

satisfy a linear q-difference equation of finite order whose coefficients are polynomials in x and q?

3. Can $p(C_1, n) \neq p(C_2, n)$ for only finitely many n?

Subbarao has posed the latter question for partition functions involving Euler pairs of various orders.

References

The following references are intended to supplement the bibliographies of [2, 22, 28, and 89]. The reader is referred to [51, Chap. 3] for older references. Reference [82] is included for reference to work on higher dimensional partitions where a fairly complete bibliography of that subject is given; Ref. [67] provides a good introduction to the analytic number theory aspects of partitions.

1. R. P. Agarwal, Certain basic hypergeometric identities associated with mock theta functions, *Quart. J. Math.* **20** (1969), 121–128.
2. H. L. Alder, The nonexistence of certain identities in the theory of partitions and compositions, *Bull. Amer. Math. Soc.* **54** (1948), 712–722.
3. H. L. Alder, Research problems, *Bull. Amer. Math. Soc.* **62** (1956), 76.
4. H. L. Alder, Partition identities—from Euler to the present, *Amer. Math. Monthly* **76** (1969), 733–746.
5. H. L. Alder, Proof of Andrews' conjecture on partition identities, *Proc. Amer. Math. Soc.* **22** (1969), 688–689.

6. G. E. ANDREWS, A simple proof of Jacobi's triple product identity, *Proc. Amer. Math. Soc.* **16** (1965), 333–334.
7. G. E. ANDREWS, A generalization of the Göllnitz–Gordon partition theorems, *Proc. Amer. Math. Soc.* **18** (1967), 945–952.
8. G. E. ANDREWS, Partition theorems related to the Rogers–Ramanujan identities, *J. Combinatorial Theory* **2** (1967), 422–430.
9. G. E. ANDREWS, Some new partition theorems, *J. Combinatorial Theory* **2** (1967), 431–436.
10. G. E. ANDREWS, A generalization of a partition theorem of MacMahon, *J. Combinatorial Theory* **2** (1967), 100–101.
11. G. E. ANDREWS, On Schur's second partition theorem, *Glasgow Math. J.* **8** (1967), 127–132.
12. G. E. ANDREWS, On basic hypergeometric series, mock-theta functions, and partitions I, *Quart. J. Math.* **17** (1966), 64–80.
13. G. E. ANDREWS, On basic hypergeometric series, mock-theta functions, and partitions II, *Quart. J. Math.* **17** (1966), 132–143.
14. G. E. ANDREWS, q-Identities of Auluck, Carlitz, and Rogers, *Duke Math. J.* **33** (1966), 575–582.
15. G. E. ANDREWS, On partition functions related to Schur's second partition theorem, *Proc. Amer. Math. Soc.* **19** (1968), 441–444.
16. G. E. ANDREWS, Some formulae for the Fibonacci sequence with generalizations, *Fibonacci Quart.* **7** (1969), 113–130.
17. G. E. ANDREWS, A general theorem on partitions with difference conditions, *Amer. J. Math.* **91** (1969), 18–24.
18. G. E. ANDREWS, A new generalization of Schur's second partition theorem, *Acta Arithmetica* **14** (1968), 429–434.
19. G. E. ANDREWS, On a partition theorem of Göllnitz and related formulae, *J. Reine Angew. Math.* **236** (1969), 37–42.
20. G. E. ANDREWS, A generalization of the classical partition theorems, *Trans. Amer. Math. Soc.* **145** (1969), 205–221.
21. G. E. ANDREWS, On q-difference equations for certain well-poised basic hypergeometric series, *Quart. J. Math.* **19** (1968), 433–447.
22. G. E. ANDREWS, A polynomial identity which implies the Rogers–Ramanujan identities, *Scripta Math.* **23** (1970), 297–305.
23. G. E. ANDREWS, Two theorems of Euler and a general partition theorem, *Proc. Amer. Math. Soc.* **20** (1969), 499–502.
24. G. E. ANDREWS, Some new partition theorems II, *J. Combinatorial Theory* **7** (1969). 262–263.
25. G. E. ANDREWS, An extension of Schur's second partition theorem, *Notices Amer. Math. Soc.* **17** (1970), 300.
26. G. E. ANDREWS, A new property of partitions with applications to the Rogers–Ramanujan identities, *J. Combinatorial Theory*, **10** (1971), 266–270.
27. G. E. ANDREWS, Note on a partition theorem, *Glasgow Math. J.* **11** (1970), 108–109.
28. G. E. ANDREWS, On a calculus of partition functions, *Pacific J. Math.* **31** (1969), 555–562.
29. G. E. ANDREWS, On a partition problem of J. J. Sylvester, *J. London Math. Soc.* **2** (1970), 571–576.

30. G. E. Andrews, On a transformation of bilateral series with applications, *Proc. Amer. Math. Soc.* **25** (1970), 554–558.
31. G. E. Andrews, On the Rogers–Ramanujan identities and partial q-difference equations, *Illinois J. Math.*, to appear.
32. G. E. Andrews, The use of computers in search of identities of the Rogers–Ramanujan type, Computers in Number Theory, Edited by A. O. L. Atkin and B. J. Birch, Academic Press (1971), 377–387.
33. G. E. Andrews, On a partition problem of H. L. Alder, *Pacific J. Math.* **36** (1971), 279–284.
34. G. E. Andrews, Lambert series, false theta functions, and partitions, *Rocky Mountain J. Math.*, **2** (1972), 49–56.
35. G. E. Andrews, On a theorem of Schur and Gleissberg, *Archiv Math.* (Basel), **22** (1971), 165–167.
36. G. E. Andrews, On identities implying the Rogers–Ramanujan identities, to appear.
37. G. E. Andrews, Two theorems of Gauss and allied identities proved arithmetically, *Pacific J. Math.*, to appear.
38. G. E. Andrews, Sieves for theorems of Euler, Rogers, and Ramanujan, Proceedings of the Kalamazoo Conferences on Arithmetic Functions, Springer-Verlag, Lecture Notes in Math., No. 251, 1–20.
39. G. E. Andrews, M. V. Subbarao, and M. Vidyasagar, A family of combinatorial identities, *Canad. Math. Bull.* **15** (1972), 11–18.
40. A. O. L. Atkin, A note on ranks and conjugacy of partitions, *Quart. J. Math.* **17** (1966), 335–338.
41. F. C. Auluck, On some new types of partitions associated with generalized Ferrars graphs, *Proc. Cambridge Phil. Soc.* **47** (1951), 679–686.
42. D. P. Banerjee, On some identities in theory of partition, *Proc. Nat. Acad. Sci. India Sect. A* **34** (1964), 68–73.
43. E. A. Bender and J. R. Goldman, The enumerative uses of generating functions, *Indiana Math. J.* **20** (1971), 753–765.
44. G. Birkhoff, Lattice Theory, Amer. Math. Soc. Colloq. Publ., No. 25, 1948, revised 1967.
45. G. R. Blakley, Combinatorial remarks on partitions of a multipartite number, *Duke Math. J.* **31** (1964), 335–340.
46. G. R. Blakley, Algebra of formal power series, *Duke Math. J.* **31** (1964), 341–346.
47. G. R. Blakley, Formal solution of nonlinear simultaneous equations: reversion of series in several variables, *Duke Math. J.* **31** (1964), 347–358.
48. L. Carlitz, Some polynomials related to theta functions, *Ann. Math. Pura Appl.* **41** (1956), 359–373.
49. L. Carlitz, A note on the Rogers–Ramanujan identities, *Duke Math. J.* **35** (1968), 839–842.
50. M. P. Chen, A generalization of a partition theorem of Andrews, Studies and Essays Presented to Yu-Why Chen on His Sixtieth Birthday, Math. Res. Center, National Taiwan University, Taipei, 1970.
51. L. E. Dickson, "History of the Theory of Numbers," Vol. 2, Chelsea, New York, 1952.
52. P. Dienes, "The Taylor Series," Dover, New York, 1957.
53. F. J. Dyson, A new symmetry of partitions, *J. Combinatorial Theory* **7** (1969), 56–61.
54. L. Euler, Introductio in Analysin Infinitorum," Lausanne, 1748.

55. N. J. FINE, Some new results on partitions, *Proc. Nat. Acad. Sci.* **34** (1948), 616–618.
56. N. J. FINE, Some Basic Hypergeometric Series and Applications, to appear.
57. J. GOLDMAN AND G.-C. ROTA, The number of subspaces of a vector space, *in* "Recent Progress in Combinatories," (W. Tutte, Ed.), pp. 75–83, Academic Press, New York, 1969.
58. J. GOLDMAN AND G.-C. ROTA, On the foundations of combinatorial theory IV: finite vector spaces and Eulerian generating functions, *Studies Appl. Math.* **49** (1970), 239–258.
59. H. GÖLLNITZ, Einfache Partitionen, (unpublished), Diplomarbeit W. S. (1960), Göttingen.
60. H. GÖLLNITZ, Partitionen mit Differenzenbedingungen, *J. Reine Angew. Math.* **225** (1965), 154–190.
61. B. GORDON, A combinatorial generalization of the Rogers–Ramanujan identities, *Amer. J. Math.* **83** (1961), 393–399.
62. B. GORDON, Some continued fractions of the Rogers–Ramanujan type, *Duke Math. J.* **31** (1965), 741–748.
63. R. GUY, Two theorems on partitions, *Math. Gaz.* **42** (1958), 84–86.
64. R. R. HALL, Three identities connected with the numbers 5 and 7, *Quart. J. Math.* **18** (1967), 113–117.
65. G. H. HARDY, "Ramanujan," Cambridge University Press, Cambridge, 1940.
66. G. H. HARDY AND E. M. WRIGHT, "An Introduction to the Theory of Numbers," 4th ed., Oxford University Press, Oxford, 1960.
67. M. I. KNOPP, "Modular Functions in Analytic Number Theory," Markham, Chicago, 1970.
68. D. E. KNUTH, Subspaces, subsets, and partitions, *J. Combinatorial Th. (A)* **10** (1971), 178–180.
69. H. R. LATTMAN, q-Difference equation proofs for some identities of the Rogers–Ramanujan type, Masters paper, Pennsylvania State Univ., 1969.
70. D. H. LEHMER, Two nonexistence theorems on partitions, *Bull. Amer. Math. Soc.* **52** (1946), 538–544.
71. P. A. MACMAHON, "Combinatory Analysis," Vol. 2, Cambridge University Press, Cambridge, 1916.
72. E. A. PARBERRY, On a partition function related to a theorem of Schur, Ph.D. dissertation, Pennsylvania State Univ., 1969.
73. L. J. ROGERS, Second memoir on the expansion of certain infinite products, Proc. London Math. Soc. **25** (1894), 318–343.
74. L. J. ROGERS, On two theorems of combinatory analysis and some allied identities, *Proc. London Math. Soc.* **16** (1917), 315–336.
75. L. J. ROGERS AND S. RAMANUJAN, Proof of certain identities in combinatory analysis, *Proc. Camb. Phil. Soc.* **19** (1919), 211–214.
76. D. E. RUTHERFORD, "Introduction to Lattice Theory," Hafner, New York, 1965.
77. I. J. SCHUR, Ein Beitrag zur additiven Zahlentheorie, *Sitzungsber. Akad. Wissensch. Berlin Phys.-Math. Klasse* (1917), 302–321.
78. I. J. SCHUR, Zur additiven Zahlentheorie, Sitzungsber, *Akad. Wissensch. Berlin, Phys.-Math. Klasse* (1926), 488–495.
79. L. J. SLATER, A new proof of Rogers's transformations of infinite series, *Proc. London Math. Soc.* **53** (1951), 460–475.

80. L. J. Slater, Further identities of the Rogers–Ramanujan type, *Proc. London Math. Soc.* **54** (1952), 147–167.
81. L. J. Slater, "Generalized Hypergeometric Functions," Cambridge, Univ. Press, Cambridge, 1966.
82. R. P. Stanley, Theory and application of plane partitions (I) and (II), *Studies in Appl. Math.* **50** (1971), 167–188, 259–279.
83. K. B. Stolarsky, Higher partition functions and their relation to finitely generated nilpotent groups, Ph.D. dissertation, Univ. of Wisconsin, 1968.
84. K. B. Stolarsky, A vanishing finite sum associated with Jacobi's triple product identity, *J. Combinatorial Theory* **6** (1969), 392–398.
85. M. V. Subbarao, Partition theorems for Euler pairs, *Proc. Amer. Math. Soc.* **28** (1971), 330–336.
86. M. V. Subbarao, On a partition theorem of MacMahon–Andrews, *Proc. Amer. Math. Soc.* **27** (1971), 449–450.
87. G. N. Watson, The mock theta functions II, *Proc. London Math. Soc.* **42** (1937), 272–304.
88. G. E. Andrews, Generalizations of the Durfee square, *J. London Math. Soc.* **3** (1971), 563–570.
89. H. Gupta, Partitions—a survey, *J. Res. Nat. Bur. Stand. B* **74** (1970), 1–29.

7.2 A General Theory of Identities of the Rogers–Ramanujan Type

BULLETIN OF THE
AMERICAN MATHEMATICAL SOCIETY
Volume 80, Number 6, November 1974

A GENERAL THEORY OF IDENTITIES OF THE ROGERS-RAMANUJAN TYPE

BY GEORGE E. ANDREWS[1]

1. **Introduction.** This paper is a direct successor to [**8**], "Partition identities", henceforward referred to as PI. The lecture presented under the title of the present paper [**9**] can be roughly broken into three parts. Part 1 was a historical survey; Part 2 presented the basic elements of the theory of partition ideals and discussed partition ideals of order 1 since they possess rather elegant properties that are easily developed. Both Parts 1 and 2 essentially appear in §§ 1–3 of PI. Part 3 (of [**9**]) was devoted to a discussion of how the more difficult questions in the theory of partition identities might be attacked. In §2 we shall present a resumè of the elementary definitions and results for partition ideals. In §3 below, we shall present in detail the discussion of the Rogers-Ramanujan identities that made up [**9**, Part 3]. It is now possible to provide a substantial partial answer to the first of the two main questions raised in §3, and this result will be presented as Theorem 4.1 in §4. §5 will outline how the ideas involved in the proof of Theorem 4.1 have extended our knowledge of the general Rogers-Ramanujan theorem [**6**], [**14**]. In §6, we shall briefly survey the analytic results related to the second question raised in §3. We conclude with a look at the open problems in this subject.

2. **Partition ideals.** All of the results in this section are presented (and proved) in PI. We shall, therefore, omit proofs and shall state only those results that are essential to the developments that follow. The lattice-theoretic definitions may be found in [**19**, Chapter 1].

Let $\mathscr{S}$ denote the set of all sequences $\{f_i\}_{i=1}^{\infty}$ (more briefly $\{f_i\}$), where each f_i is a nonnegative integer and where only finitely many f_i are nonzero. Then $\mathscr{S}$ forms a distributive lattice under the partial ordering $\{f_i\}\leqq\{g_i\}$ whenever $f_i\leqq g_i$ for each i. Furthermore the functions $\#$ and $\sigma:\mathscr{S}\rightarrow N$ (where N is the set of nonnegative integers) given by $\#(\{f_i\})=\sum f_i$, $\sigma(\{f_i\})=\sum f_i\cdot i$ are positive valuations on $\mathscr{S}$. We let $\mathbf{0}=\{0, 0, 0, \cdots\}$ denote the constant sequence of zeros.

An expanded version of an invited address delivered before the 77th Summer Meeting of the Society at Dartmouth College, Hanover, New Hampshire on September 1, 1972; received by the editors February 9, 1974.

AMS (MOS) subject classifications (1970). Primary 10A45, 33A30; Secondary 05A15, 05A19.

Key words and phrases. Partitions, partition identities, Rogers-Ramanujan identities, generating functions.

[1] Partially supported by National Science Foundation Grant GP-23774.

In intuitive terms, the sequences $\{f_i\}$ that make up $\mathscr{S}$ may each be thought of as defining a partition of a nonnegative integer where f_i gives the number of times i appears as a part. The function σ maps each sequence $\{f_i\}$ onto the number being partitioned, and $\#(\{f_i\})$ is the number of parts of the partition.

Recall now that a *semi-ideal* J in a lattice L is a subset of L such that whenever $a \in J$, $x \in L$, and $x \leqq a$, then $x \in J$ [**19**, p. 56].

DEFINITION 1. A nonempty semi-ideal in the lattice $\mathscr{S}$ is called a *partition ideal*.

DEFINITION 2. If C is a partition ideal (in $\mathscr{S}$), we say that $p(C; n)$ is the *C-partition function* if for each n, $p(C; n)$ denotes the cardinality of the set $\{\{f_i\} \in C | \sigma(\{f_i\}) = n\}$.

DEFINITION 3. We say two partition ideals C_1 and C_2 are equivalent (or partition-theoretically equivalent, or PT-equivalent) if for each non-negative integer n, $p(C_1; n) = p(C_2; n)$. We shall write $C_1 \sim^{\mathrm{PT}} C_2$.

FUNDAMENTAL PROBLEM. Fully describe the equivalence classes of partition ideals under the equivalence $\sim^{\mathrm{PT}}$.

DEFINITION 4. We say that a partition ideal C has *order* k provided k is the least integer such that whenever $\{f_i\} \notin C$, there exists m such that $\{f'_i\} \notin C$, where

$$f'_i = f_i, \quad i = m, m+1, \cdots, m+k-1, \\ = 0, \quad \text{otherwise.}$$

Intuitively, the assertion that C is of order k makes explicit the idea that if $\pi \in \mathscr{S}$, then summands of π at least k units apart cannot "interact" to affect whether $\pi \in \mathscr{S}$.

SECOND PROBLEM. Fully describe the equivalence classes of partition ideals that contain a partition ideal of order 1 under the equivalence $\sim^{\mathrm{PT}}$.

The partition ideals of order 1 have especially nice properties and are extremely amenable to various analytic and combinatorial techniques. The following results make this assertion abundantly obvious. The first result shows that if C is of order 1 then the generating function for the C-partition function is an infinite product.

THEOREM 2.1 [**8**, P. 20, THEOREM 1]. *Let C be a partition ideal of order* 1. *Let $d_l = \sup_{\{f_i\} \in C} f_l$. Then for $|q| < 1$,*

$$\sum_{n \geqq 0} p(C; n) q^n = \prod_{l=1; d_l < \infty}^{\infty} (1 - q^{l(d_l+1)}) \Big/ \prod_{l=1}^{\infty} (1 - q^l).$$

The ideals of $\mathscr{S}$ are the partition ideals C that are closed under union [**19**, p. 56], where $\{f_i\} \cup \{g_i\} = \{\max(f_i, g_i)\}$.

THEOREM 2.2 [**8**, P. 21, THEOREM 2]. *The ideals in $\mathscr{S}$ are the partition ideals of order* 1.

Finally Theorem 2.1 implies a simple computational device for determining the equivalence of two partition ideals of order 1.

THEOREM 2.3 [**8**, P. 22, THEOREM 3]. *Let C be a partition ideal of order* 1 *and let the d_l be defined as in Theorem* 2.1. *Let C' be a second partition ideal of order* 1 *with associated d'_l. Then $C \sim^{PT} C'$ if and only if the two multisets (i.e. sets with possible multiplicities of elements; see* [**29**, *p.* 22] *for detailed description of multisets) $\{j(d_j+1)\}_{d_j<\infty}$ and $\{j(d'_j+1)\}_{d'_j<\infty}$ are identical.*

Theorem 2.3 provides us with the type of characterization one might possibly hope for in answer to the Fundamental Problem. As was shown in §3 of **PI**, questions restricted to possible equivalence of partition ideals of order 1 are extremely routine in virtue of Theorem 2.3. A theorem that dropped the restriction to order 1 but was similar in character to Theorem 2.3 in which whether or not $C \sim^{PT} C'$ could be answered by an algorithm that utilized certain easily ascertainable parameters of each partition ideal would certainly constitute an adequate answer to the Fundamental Problem.

3. **The Rogers-Ramanujan identities.** Our object now is to analyze one of the classical proofs of the Rogers-Ramanujan identities [**31**] using the terminology and approach of §2. As a result of our analysis we shall pose two questions that appear to hold reasonable promise for attacking the Fundamental Problem (and the Second Problem) described in §2.

THEOREM 3.1. *If $\mathscr{R}=\{\{f_i\}\in\mathscr{S} \mid f_i+f_{i+1}\leqq 1\}$ and $\mathscr{T}=\{\{f_i\}\in\mathscr{S} \mid f_i>0$ implies $i\equiv 1, 4 \pmod 5\}$, then*

$$\mathscr{R}\sim^{PT}\mathscr{T}.$$

REMARK. This is the first of the Rogers-Ramanujan identities written in the language of partition ideals. We point out that the condition "$f_i+f_{i+1}\leqq 1$" that defines $\mathscr{R}$ merely insures that in the partition $\sum_{i=1}^{\infty} f_i \cdot i=\sigma(\{f_i\})$, no part repeats and no consecutive integers appear. Hence $p(\mathscr{R}; n)$ is the number of partitions of n in which the difference between parts is at least 2. On the other hand if $\{f_i\}\in\mathscr{T}$, then $\sum_{i=1}^{\infty} f_i \cdot i$ is a partition in which only parts $\equiv 1, 4 \pmod 5$ may appear.

Thus Theorem 3.1 asserts that the partitions of each integer n in which the difference between parts is at least 2 are equinumerous with the partitions of n into parts $\equiv 1$ or $4 \pmod 5$. This is the standard number-theoretic formulation of the first Rogers-Ramanujan identity.

SKETCH OF PROOF OF THEOREM 3.1. We begin by considering

$$f_{\mathscr{R}}(x;q) \equiv f_{\mathscr{R}}(x) \equiv \sum_{\{f_i\}\in\mathscr{R}} x^{\sum f_i} q^{\sum f_i \cdot i}, \tag{3.1}$$

a double series convergent at least for $|q|<1$, $|x|<|q|^{-1}$. Thus

$$f_{\mathscr{R}}(1) = \sum_{n\geqq 0} p(\mathscr{R};n)q^n, \tag{3.2}$$

and, in fact, the coefficient of $x^m q^n$ in (3.1) is the number of partitions of n belonging to $\mathscr{R}$ that have exactly m parts.

Now

$$\mathscr{R} = \mathscr{R}_1 \cup \mathscr{R}_2,$$

where

$$\mathscr{R}_1 = \{\{f_i\} \in \mathscr{R} \mid f_1 = 0\}, \qquad \mathscr{R}_2 = \{\{f_i\} \in \mathscr{R} \mid f_1 = 1\}.$$

By the definition of $\mathscr{R}$ it is clear that

$$\mathscr{R}_1 = \{\{f_i'\} \mid f_i' = f_{i-1}, i \geqq 2, f_1' = 0 \text{ for some } \{f_i\} \in \mathscr{R}\}$$

$$\mathscr{R}_2 = \{\{f_i''\} \mid f_i'' = f_{i-2}, i \geqq 3, f_1'' = 1, f_2'' = 0 \text{ for some } \{f_i\} \in \mathscr{R}\}.$$

Hence since $\mathscr{R}_1 \cap \mathscr{R}_2 = \varnothing$,

$$\begin{aligned} f_{\mathscr{R}}(x) &= \sum_{\{f_i\}\in\mathscr{R}} x^{\sum f_i} q^{\sum f_i i} = \left(\sum_{\{f_i\}\in\mathscr{R}_1} + \sum_{\{f_i\}\in\mathscr{R}_2}\right) x^{\sum f_i} q^{\sum f_i \cdot i} \\ &= \sum_{\{f_i\}\in\mathscr{R}} \left(x^{\sum f_{i-1}} q^{\sum f_{i-1} i} + x^{1+\sum f_{i-2}} q^{1+\sum f_{i-2} i}\right) \\ &= \sum_{\{f_i\}\in\mathscr{R}} \left(x^{\sum f_i} q^{\sum f_i(i+1)} + xqx^{\sum f_i} q^{\sum f_i(i+2)}\right) \\ &= f_{\mathscr{R}}(xq) + xqf_{\mathscr{R}}(xq^2). \end{aligned}$$

Now it is possible to prove in a completely elementary manner (coefficient comparison) that there is only one function $F(x)$ that is analytic in x around $x=0$ and satisfies

$$F(0) = 1, \tag{3.3}$$

$$F(x) = F(xq) + xqF(xq^2). \tag{3.4}$$

Clearly $f_{\mathscr{R}}(x)$ is such a function. On the other hand, ingenious but elementary arguments involving only the rearrangements of series **[31]** allow one to establish that if

$$\phi(x) = \sum_{n\geqq 0} \frac{(-1)^n x^{2n} q^{n(5n+1)/2}(1 - x^2q^{4n+2})}{(1-q)(1-q^2)\cdots(1-q^n)(1-xq^{n+1})(1-xq^{n+2})\cdots}, \tag{3.5}$$

then $\phi(0)=1$, $\phi(x)$ is analytic in x for $|q|<1$, $|x|<|q|^{-1}$, and $\phi(x)$ satisfies (3.4).

Therefore $\phi(x)=f_{\mathscr{R}}(x)$ for $|q|<1$, $|x|<|q|^{-1}$. Hence

$$\begin{aligned}\sum_{n\geqq 0} p(\mathscr{R}; n)q^n &= f_{\mathscr{R}}(1) = \phi(1)\\ &= \sum_{n\geqq 0}(-1)^n q^{n(5n+1)/2}(1-q^{4n+2})\Big/\prod_{n=1}^{\infty}(1-q^n)\\ &= \prod_{n=0}^{\infty}\frac{1}{(1-q^{5n+1})(1-q^{5n+4})} \quad \text{(by Jacob's identity [7, pp. 186–187])}\\ &= \sum_{n\geqq 0} p(\mathscr{T}; n)q^n \quad \text{(by Theorem 2.1).}\end{aligned} \tag{3.6}$$

Comparing coefficients in the extremes of this string of equations we see that $p(\mathscr{R}; n)=p(\mathscr{T}; n)$ for each n, i.e. $\mathscr{R} \sim^{\mathrm{PT}} \mathscr{T}$. Thus Theorem 3.1 is established. □

The steps of the above proof may be analyzed as follows:

1. Produce a functional equation (namely (3.4) for $f_{\mathscr{R}}(x)$—a linear, second order q-difference equation).

2. Produce some q-series representation of the essentially unique solution of the functional equation.

3. Deduce the theorem from some q-series and product identity like (3.6).

This analysis applies to the vast majority of results that have been obtained in this area (see §§1, 4, and 5 of PI for examples of this technique).

The following two questions then naturally arise when one looks to how widely the above technique may be applied.

Question 1. For what partition ideals C does

$$f_C(x; q) = \sum_{\{f_i\}\in C} x^{\Sigma f_i} q^{\Sigma f_i \cdot i}$$

satisfy a finite linear q-difference equation with polynomial coefficients?

Question 2. What finite linear q-difference equations with polynomial coefficients (such as (3.4)) have solutions that can be represented by "higher-dimensional" q-series?

Question 2 leaves (intentionally) vague the concept of q-series. What one wants are solutions that are as amenable to manipulation and identity proving as the basic hypergeometric series studied by Bailey **[16]** and Slater **[37]**:

$$\sum_{n\geqq 0}\frac{(a_1)_n\cdots(a_r)_n t^n}{(q)_n(b_1)_n\cdots(b_s)_r},$$

where $(A)_n=(1-A)(1-Aq)\cdots(1-Aq^{n-1})$. As we shall see in §6, presumably acceptable solutions would involve many-fold q-series, i.e. higher-dimensional q-series and limiting cases thereof. F. H. Jackson's basic

Appell series **[28]** (see also **[10]**) provide examples of two-dimensional q-series.

4. **Linked partition ideals.** The dissection of $\mathscr{R}$ into $\mathscr{R}_1 \cup \mathscr{R}_2$ utilized in §3 to prove that $f_{\mathscr{R}}(x)$ satisfies (3.4) may be extended to a wide class of partition ideals. We shall now present a series of new concepts which will allow us to define a *linked* partition ideal. We shall see in Theorem 4.1 that for each linked partition ideal C, $f_C(x; q)$ satisfies a finite q-difference equation with polynomial coefficients.

DEFINITION 5. If C is a partition ideal,

$$C^{(j)} = \{\{f_i\} \in C \mid f_1 = f_2 = \cdots = f_j = 0\}.$$

DEFINITION 6. We denote by φ the bijection $\mathscr{S} \to \mathscr{S}^{(1)}$ given by

$$\varphi(\{f_i\}) = \{g_i\} \quad \text{where } g_i = 0 \quad \text{if } i = 1, \qquad = f_{i-1} \quad \text{if } i > 1. \tag{4.1}$$

In intuitive terms, φ adds one to each part of each partition since

$$\sigma(\varphi(\{f_i\})) = \sum_{i=2}^{\infty} f_{i-1} \cdot i = \sum_{i=1}^{\infty} f_i(i+1).$$

DEFINITION 7. We say a partition ideal C has a modulus m if m is an integer such that $\varphi^m C = C^{(m)}$.

If we look back at §3, we see immediately that the least modulus for $\mathscr{R}$ is 1 and $\mathscr{T}$ has least modulus 5.

DEFINITION 8. If C has a modulus, say m, and if the set

$$L_C = \{\{f_i\} \in C \mid f_i = 0 \text{ for } i > m\}$$

is finite, then we say that C is a locally finite partition ideal.

Thus $L_{\mathscr{R}}$ is the two element set consisting of $\mathbf{0} = \{0, 0, 0, \cdots\}$ and $\{1, 0, 0, 0, \cdots\}$, while $L_{\mathscr{T}}$ is infinite and consists of all elements of $\mathscr{S}$ of the form $\{m_1, 0, 0, m_2, 0, 0, 0, 0, 0\}$, where $m_1 \geqq 0$, $m_2 \geqq 0$.

It is clear that if C has a modulus, then for C to be locally finite it is necessary and sufficient that there exists an absolute constant M such that $f_i \leqq M$ for all i whenever $\{f_i\} \in C$.

It is also clear that if m is the least positive modulus for C, then every modulus M is a multiple of m. This is because if $M = km + r$, $0 \leqq r < m$, then

$$C = \varphi^{-M} C^{(M)} = \varphi^{-r} \varphi^{-km} C^{(M)} = \varphi^{-r} C^{(M-km)} = \varphi^{-r} C^{(r)};$$

hence $r = 0$, otherwise the minimality of m is contradicted. Generally when we refer to the modulus of C, we shall mean the least positive modulus unless otherwise indicated.

DEFINITION 9. For each $\pi=\{f_i\}\in\mathscr{S}$, $\pi_{<N}=\{f_{i<N}\}$ shall denote $\{f_1, f_2, f_3, \cdots, f_{N-1}, 0, 0, 0, \cdots\}$, while $\pi_{\leq N}=\{f_{i\leq N}\}$ denotes $\{f_1, f_2, f_3, \cdots, f_N, 0, 0, 0, \cdots\}$.

DEFINITION 10. We write $\{f_i\}\oplus\{g_i\}=\{f_i+g_i\}$.

DEFINITION 11. Suppose C is locally finite of modulus m. We shall say that $\pi=\{f_i\}\in L_C$ is *linked* with *span* $l=l(\pi)$ if (1) l is the least positive integer such that $\{f_i\}\oplus\varphi^{lm}\{g_i\}\in C$ for some $\{g_i\}\in L_C$ with $\{g_i\}\neq\mathbf{0}$, and (2) $\{f_i\}\oplus\{h_i\}\in C$ whenever $\{h_i\}\in C^{(lm)}$ and $\{f_i\}\oplus\{h_{i\leq(l+1)m}\}\in C$. We say C is *linked* if C is locally finite and every element of L_C is linked.

DEFINITION 12. Suppose C is linked. For each $\pi\in L_C$, we define a subset $\mathscr{L}_C(\pi)$ of L_C called the *linking set for* π as those $\pi'\in L_C$ such that $\pi\oplus\varphi^{l(\pi)\cdot m}\pi'\in C$.

Returning to $\mathscr{R}$ we see that $\mathscr{R}$ is linked since $\mathbf{0}$ has span 1 and $\mathscr{L}_{\mathscr{R}}(\mathbf{0})=L_{\mathscr{R}}$ while $\boldsymbol{i}=\{1, 0, 0, 0, \cdots\}$ has span 2 and $\mathscr{L}_{\mathscr{R}}(\boldsymbol{i})=L_{\mathscr{R}}$ also.

THEOREM 4.1. *If C is a linked partition ideal, then $f_C(x;q)$ satisfies a finite homogeneous linear q-difference equation whose coefficients are polynomials in x and q.*

PROOF. We begin by deriving several results about C from the fact that C is linked. First we know that C has at least one modulus, say m, for which we also know that the corresponding L_C is finite. Let κ denote the cardinality of L_C, and we write $L_C=\{\pi_0, \pi_1, \pi_2, \cdots, \pi_{\kappa-1}\}$, where $\pi_0=\mathbf{0}$.

Next we define for each $\pi_i\in L_C$

$$H_i(x)=\sum_{\pi'}{}^{\dagger}\, x^{\#(\pi')}q^{\sigma(\pi')}$$

where $\sum^{\dagger}$ runs over all $\pi'=\{f_i\}\in C$ such that $\{f_{j\leq m}\}=\pi_i$. We may now easily deduce several important relations for these functions. First

$$H_0(x)=\sum_{\pi\in C^{(m)}} x^{\#(\pi)}q^{\sigma(\pi)}=\sum_{\pi\in C} x^{\#(\pi)}q^{\sigma(\pi)+m\cdot\#(\pi)}=f_C(xq^m;q). \tag{4.2}$$

Next for each $\pi_i\in L_C$,

$$\begin{aligned} H_i(x) &= \sum_{\substack{\pi=\pi_i\oplus\varphi^{l(\pi_i)m}\pi^*\\ \pi^*_{\leq m}\in\mathscr{L}_C(\pi_i);\,\pi^*\in C}} x^{\#(\pi)}q^{\sigma(\pi)}\\ &= x^{\#(\pi_i)}q^{\sigma(\pi_i)}\sum_{\substack{\pi=\varphi^{l(\pi_i)m}\pi^*\\ \pi^*_{\leq m}\in\mathscr{L}_C(\pi_i);\,\pi^*\in C}} x^{\#(\pi)}q^{\sigma(\pi)}\\ &= x^{\#(\pi_i)}q^{\sigma(\pi_i)}\sum_{\pi^*_{\leq m}\in\mathscr{L}_C(\pi_i);\,\pi^*\in C} x^{\#(\pi^*)}q^{\sigma(\pi^*)+l(\pi_i)m\#(\pi^*)}\\ &= x^{\#(\pi_i)}q^{\sigma(\pi_i)}\sum_{\pi_j\in\mathscr{L}_C(\pi_i)} H_j(xq^{l(\pi_i)m}). \end{aligned} \tag{4.3$_i$}$$

Now the $(4.3)_i$ constitute a system of κ q-difference equations in the κ functions $H_i(x)$. All that remains (by (4.2)) is to show that this system may be reduced to a single q-difference equation for $H_0(x)$.

Here we shall rely on the anology with linear differential equations. In [**30**, Chapter 5, §5], Murray and Miller provide an algorithm for reducing a system of first order differential equations in $y_1, \cdots, y_n$ to n higher order equations wherein the jth equation involves only y_j. For our purposes the significant result lies in the first part of their Theorem 7 [**30**, p. 129]:

MURRAY-MILLER THEOREM. *Every function y which is a component of the solution of*

$$\dot{y}_j = \sum_{k=1}^{n} p_{jk}(x)y_k + q_j(x), \qquad j = 1, \cdots, n,$$

is a solution of a single linear differential equation of order r where $r \leqq n$.

On examination of the algorithm that is used to prove this theorem, we see that the differentiation operator may be replaced by any linear operator L on the field of functions analytic in some complex domain $\mathscr{D}$ subject to the restriction

$$L(a(x)y) = \alpha_L(x)y + \beta_L(x)L(y),$$

where $\alpha_L(x)$, $\beta_L(x)$ and $a(x)$ all lie in the field of functions that contains the coefficients of the equations considered, and where $\beta_L(x) \equiv 0$ if and only if $a(x) \equiv 0$.

When $L = d/dx$, $\alpha_L(x) = a'(x)$ and $\beta_L(x) = a(x)$. When $L(f(x)) = f(xq)$, then $\alpha_L(x) = 0$ and $\beta_L(x) = a(xq)$.

MODIFIED MURRAY-MILLER THEOREM. *Let $\mathscr{F} = C(x)$ be the field of rational functions of the complex variable x. Let L be a linear operator subject to the above described conditions. Every function y which is part of a solution of*

(4.4)
$$L(y_j) = \sum_{k=1}^{n} p_{jk}(x)y_k + q_j(x), \qquad 1 \leqq j \leqq n, \quad p_{jk}(x) \in \mathscr{F}, \quad q_j(x) \in \mathscr{F}.$$

satisfies an equation of the form

$$\sum_{h=0}^{r} s_h(x)L^h(y) = t(x), \quad \text{where } r \leqq n, \quad s_h(x) \in \mathscr{F}, \quad t(x) \in \mathscr{F}.$$

Furthermore if the $q_j(x)$ are all zero so are the $t(x)$.

The final assertion in this theorem is easily deduced by examination of the algorithm used in the proof.

For our purposes L will denote the operator given by $L(f(x))=f(xq^m)$, $0<|q|<1$. We must now transform the system $(4.3)_i$ into the form (4.4). First we write

$$h_{ij}(x) = H_i(x^{-1}q^{-jm}), \qquad 0 \leqq i \leqq \kappa - 1, \quad 0 \leqq j < \omega_i,$$

where ω_i is the largest value of ω such that $H_i(xq^{\omega m})$ actually appears in the system $(4.3)_i$ (by $(4.3)_0$ we see that $\omega_i \geqq 1$ always). The system $(4.3)_i$ is now equivalent to the following system of $\Omega=\omega_0+\omega_1+\cdots+\omega_{\kappa-1}$ equations in Ω dependent variables h_{ij}.

(4.5)

$$L(h_{i,\omega_i-1}) = h_{i,\omega_i-1}(xq^m) = \sum_{j=0}^{\kappa}\sum_{k=0}^{\omega_i-1} p_{j,k}(i;x)h_{j,k}(x), \qquad 0 \leqq i \leqq \kappa - 1,$$

$$L(h_{ij}) = h_{i,j+1}, \qquad 0 \leqq i \leqq \kappa - 1, \quad 0 \leqq j < \omega_i - 1,$$

where the top system of equations in (4.5) is merely the $(4.3)_i$ with x replaced by $x^{-1}q^{-\omega_i m}$. Since (4.5) is of the form (4.4), we see that Theorem 4.1 follows. Furthermore the order of the implied q-difference equation is $\leqq m\Omega$. □

To illustrate the workings of Theorem 4.1, let us return to $f_{\mathfrak{R}}(x;q)$. In this case $(4.3)_0$ becomes

$$H_0(x) = H_0(xq) + H_1(xq),$$

and $(4.3)_1$ becomes

$$H_1(x) = xq(H_0(xq^2) + H_1(xq^2)).$$

Hence eliminating H_1 from these equations, we find

$$H_0(x) - H_0(xq) = xq^2(H_0(xq^3) + H_0(xq^2) - H_0(xq^3)),$$

or

$$H_0(x) = H_0(xq) + xq^2H_0(xq^2).$$

Hence

$$f_{\mathfrak{R}}(x;q) = f_{\mathfrak{R}}(xq;q) + xqf_{\mathfrak{R}}(xq^2;q).$$

In [2], H. L. Alder defined polynomials $G_{k,n}(q)$ that have subsequently become known as the Alder polynomials. These polynomials may be defined by

$$\sum_{\pi\in\mathscr{B}_{0,k,k}} x^{\#(\pi)}q^{\sigma(\pi)} = \sum_{n\geqq 0} G_{k,n}(q)x^n/(q)_n,$$

where $\mathscr{B}_{0,k,k}=\{\{f_i\}\in\mathscr{S}\,|\,f_i+f_{i+1}\leqq k-1\}$. (The truth of this assertion follows from equations (5.2) and (5.15) in the next section after comparison with Alder's original definition.)

The Alder polynomials have been found to be of number-theoretic significance in **[13]**. We shall now show that any linked partition ideal has an associated family of Alder polynomials:

THEOREM 4.2. *Suppose that C is a linked partition ideal with a modulus m. Then there exists a family of polynomials $A_n(C; m; q)$ (to be called the "Alder polynomials for C relative to m" or more briefly the C-Alder polynomials) such that*

$$f_C(x; q) = \sum_{\pi \in C} x^{\#(\pi)} q^{\sigma(\pi)} = \sum_{n \geqq 0} \frac{A_n(C; m, q) x^n}{(q^m; q^m)_n}.$$

PROOF. It is clear that the $A_n(C, m; q)$ exist as functions of q since $f_C(x; q)$ is analytic in x around $x=0$ for $|q|<1$. The problem is to show that they are polynomials. Referring to the proof of Theorem 4.1, we define

$$H_j(x) = \sum_{n \geqq 1} \frac{a_n(j; q) x^n}{(q^m; q^m)_{n-1}}, \qquad 1 \leqq j \leqq \kappa - 1, \tag{4.6}$$

$$H_0(x) = \sum_{n \geqq 0} \frac{a_n(0; q) x^n q^{mn}}{(q^m; q^m)_n} \tag{4.7}$$

which is admissible since $a_0(0; q)=1$, $a_0(j; q)=0$ for $1 \leqq j \leqq \kappa-1$, and $a_n(j; q)=0$ for $n<0$.

Furthermore when $i=0$ in $(4.3)_i$, we see that $\pi_0=\mathbf{0}$, $\#(\pi_0)=0$, $\sigma(\pi_0)=0$, and every element of L_C is linked to π_0 with span 1. Hence $(4.3)_0$ is just

$$H_0(x) = \sum_{j=0}^{\kappa-1} H_j(xq^m);$$

therefore for $n \geqq 1$,

$$\frac{a_n(0; q) q^{mn}}{(q^m; q^m)_n} = \frac{q^{2mn} a_n(0; q)}{(q^m; q^m)_n} + \sum_{j=1}^{\kappa-1} \frac{q^{mn} a_n(j; q)}{(q^m; q^m)_{n-1}},$$

or

$$a_n(0; q) = \sum_{j=1}^{\kappa-1} a_n(j; q). \tag{4.8}$$

For $i>0$, we see from $(4.3)_i$ that for $n>0$,

$$\frac{a_n(i; q)}{(q^m; q^m)_{n-1}} = q^{\sigma(\pi_i)} \sum_{\pi_j \in \mathscr{L}_C(\pi_i); \pi_j \neq \pi_0} q^{l(\pi_i) m (n - \#(\pi_i))} \frac{a_{n-\#(\pi_i)}(j; q)}{(q^m; q^m)_{n-\#(\pi_i)-1}}$$
$$+ \frac{q^{\sigma(\pi_i) + (l(\pi_i)+1) m (n-\#(\pi_i))} a_{n-\#(\pi_i)}(0; q)}{(q^m; q^m)_{n-\#(\pi_i)}},$$

or

$$a_n(i;q) = q^{\sigma(\pi_i)} \sum_{\pi_j \in \mathscr{L}_C(\pi_i); \pi_j \neq \pi_0} q^{l(\pi_i)m(n-\#(\pi_i))} a_{n-\#(\pi_i)}(j;q) \cdot \prod_{h=0}^{\#(\pi_i)-1} (1-q^{m(n-\#(\pi_i)+h)}) + q^{\sigma(\pi_i)+(l(\pi_i)+1)m(n-\#(\pi_i))} a_{n-\#(\pi_i)}(0;q) \cdot \prod_{h=0}^{\#(\pi_i)} (1-q^{m(n-\#(\pi_i)+h)}). \tag{4.9}$$

It is now an easy matter to prove that the functions $a_n(j;q)$ are all polynomials and are uniquely determined by (4.8), (4.9), and the initial conditions $a_0(0;q)=1$, $a_0(j;q)=0$ for $j>0$, $a_n(j;q)=0$ for $n<0$. The initial conditions establish our assertion for $n=0$. Assume the result for $n<n_0$. Then for $i\neq 0$, we know $\#(\pi_i)>0$; hence the fact that $a_{n_0}(i;q)$ is unique and a polynomial follows from (4.9) (note that if $\#(\pi_i)>n_0$, then $a_{n_0-\#(\pi_i)}(j;q)=0$). For $i=0$, we now use (4.8) to establish the result for $a_{n_0}(0;q)$ and, therefore, to complete the induction. To conclude we note that (by (4.2)),

$$\sum_{n\geqq 0} \frac{A_n(C;m;q)x^m q^m}{(q^m;q^m)_n} = f_C(xq^m;q) = H_0(x) = \sum_{n\geqq 0} \frac{a_n(0;q)x^m q^m}{(q^m;q^m)_n};$$

hence $A_n(C;m;q)=a_n(0;q)$ is a polynomial in q. □

COROLLARY. *The $H_j(x)$ are the unique functions analytic in x at $x=0$ that satisfy the system* $(4.3)_i$ *and the initial conditions $H_0(0)=1$, $H_j(x)=0$, $1\leqq j\leqq \kappa-1$.*

PROOF. The uniqueness of the $a_n(j;q)$ proved in Theorem 4.2 was derived using only the facts described in the Corollary. □

5. **Discussion of linked partition ideals.** The linked partition ideals introduced in §4 appear in almost all of the partition identities of the Rogers-Ramanujan type. As a vehicle for discussion of Theorem 4.1 we state the most general identity known that contains the Rogers-Ramanujan identities as a special case **[14]**.

THEOREM 5.1. *Let $0\leqq\lambda\leqq a\leqq k$ be integers. If λ is even, let $A_{\lambda,k,a}(n)$ denote the number of partitions of n into parts such that no part $\not\equiv 0$* (mod $\lambda+1$) *may be repeated and no part is* $\equiv 0$,

$$\pm(a-\tfrac{1}{2}\lambda)(\lambda+1) \pmod{(2k-\lambda+1)(\lambda+1)}.$$

If λ is odd, let $A_{\lambda,k,a}(n)$ denote the number of partitions of n into parts such that no part $\not\equiv 0$ (mod $\frac{1}{2}(\lambda+1)$) *may be repeated, no part is* $\equiv\lambda+1$

(mod $2\lambda+2$), and no part is $\equiv 0, \pm(2a-\lambda)\frac{1}{2}(\lambda+1) \pmod{(2k-\lambda+1)(\lambda+1)}$. *Let* $B_{\lambda,k,a}(n)$ *denote the number of partitions of n of the form* $b_1+\cdots+b_s$, *with* $b_i \geqq b_{i+1}$, *no parts* $\not\equiv 0 \pmod{\lambda+1}$ *are repeated,* $b_i - b_{i+k-1} \geqq \lambda+1$ *with strict inequality if* $(\lambda+1)|b_i$ *and at most* $a-1$ *parts are* $\leqq \lambda+1$. *Then* $A_{\lambda,k,a}(n) = B_{\lambda,k,a}(n)$ *for each n.*

REMARKS. The above theorem may be easily phrased in terms of partition ideals. Namely, let

$$\mathscr{B}_{\lambda,k,a} = \left\{ \{f_i\} \in \mathscr{S} \,\middle|\, \sum_{j=0}^{\lambda} f_{m+j} \leqq k-1 \text{ and } \sum_{j=0}^{\lambda+1} f_{m(\lambda+1)+j} \leqq k-1 \right.$$

$$\left. \text{for each } m \geqq 1, f_1 + \cdots + f_{\lambda+1} \leqq a-1, \text{ and } f_i > 1 \text{ implies } (\lambda+1) \mid i \right\},$$

$$\mathscr{A}_{2\lambda,k,a} = \{\{f_i\} \in \mathscr{S} \mid f_i > 1 \text{ implies } (2\lambda+1) \mid i; \text{ and } f_i > 0 \text{ implies}$$

$$i \not\equiv 0, \pm(a-\lambda)(2\lambda+1) \pmod{(2k-2\lambda+1)(2\lambda+1)}\},$$

$$\mathscr{A}_{2\lambda+1,k,a} = \{\{f_i\} \in \mathscr{S} \mid f_i > 1 \text{ implies } (\lambda+1) \mid i; f_i > 0 \text{ implies}$$

$$i \not\equiv 2\lambda+2 \pmod{4\lambda+4} \text{ and } i \not\equiv 0,$$

$$\pm(2a-2\lambda-1)(\lambda+1) \pmod{(2k-2\lambda)(2\lambda+2)}\}.$$

Then Theorem 5.1 is equivalent to $\mathscr{A}_{\lambda,k,a} \sim^{\text{PT}} \mathscr{B}_{\lambda,k,a}$ provided $0 \leqq \lambda \leqq a \leqq k$.

The first Rogers-Ramanujan identity is the case $\lambda=0$, $k=a=2$. The second Rogers-Ramanujan identity is the case $\lambda=0$, $k=2$, $a=1$. Theorem 5.1 contains a number of other well-known partition identities as special cases; a weaker form of this theorem which stipulates $k \geqq 2\lambda-1$ is discussed in §1 of PI and proved in **[6]**.

It is easy to show that $\mathscr{B}_{\lambda,k,k}$ is a linked partition ideal with least modulus $\lambda+1$ and that $L_{\mathscr{B}_{\lambda,k,k}}$ is a set consisting of

$$\kappa = k + (k-1)\binom{\lambda}{1} + (k-2)\binom{\lambda}{2} + \cdots + (k-\lambda)\binom{\lambda}{\lambda} = 2^{\lambda-1}(2k-\lambda)$$

elements. Furthermore the span of each element is at most 2 so that in terms of Theorem 4.1 the order of the resulting q-difference equation is at most $(\lambda+1)\cdot 2 \cdot 2^{\lambda-1}(2k-\lambda) = (\lambda+1)2^{\lambda}(2k-\lambda)$. In order to prove Theorem 5.1, much more subtlety is required than is available from a direct application of Theorem 4.1, and in fact a system of q-difference equations is obtained from which it is possible to deduce a single one of order $k(\lambda+1)$ for $f_{\mathscr{B}_{\lambda,k,k}}(x;q)$. The technique of proof relies on cutting the number of auxiliary functions down from $\kappa = 2^{\lambda-1}(2k-\lambda)$ to $2k+1$. In fact, in most interesting special cases, the relevant q-difference equation will have a much lower order than the upper bound given in Theorem 4.1. Below we outline the general proof of Theorem 5.1. The details of this work will appear in **[14]**.

SKETCH OF PROOF OF THEOREM 5.1. We begin with functions that correspond to $\phi(x)$ defined in equation (3.5):

$$\varphi_{\lambda,k,i}(x) = \frac{(-xq)_\infty}{(x^2q^{2\lambda+2};q^{2\lambda+2})_\infty}\sum_{n=0}^{\infty}(-1)^n x^{kn} q^{(\lambda+1)((2k-\lambda+1)n^2+(2k-2i+1)n)/2} \cdot\frac{(1-x^iq^{(2n+1)i(\lambda+1)})(x^2q^{2\lambda+2};q^{2\lambda+2})_n(-q)_{\lambda n+n}}{(q^{2\lambda+2};q^{2\lambda+2})_n(-xq)_{\lambda n+\lambda+n}}; \tag{5.1}$$

$$\psi_{\lambda,k,i}(x) = \frac{(-xq)_\infty}{(x^2q^{2\lambda+2};q^{2\lambda+2})_\infty}\sum_{n=0}^{\infty}(-1)^n x^{kn} q^{(\lambda+1)((2k-\lambda+1)n^2+(2k-2i+1)n)/2} \cdot\frac{(x^2q^{2\lambda+2};q^{2\lambda+2})_n(-q)_{(\lambda+1)n}}{(q^{2\lambda+2};q^{2\lambda+2})_n(-xq)_{\lambda n+n}} \times\left(1-\frac{x^iq^{(\lambda+1)(2n+1)(2i-\lambda)/2}(-q^{\lambda n+n+1})_\lambda}{(-xq^{n\lambda+n+1})_\lambda}\right), \tag{5.2}$$

where $(a;q)_n=(a)_n=(1-a)(1-aq)\cdots(1-aq^{n-1})$, and $(a;q)_\infty=(a)_\infty=\lim_{n\to\infty}(a;q)_n$.

Note that $\varphi_{0,2,2}(x)=\psi_{0,2,2}(x)=\varphi(x)$.

It is possible to prove that these functions satisfy the following relations:

$$\varphi_{\lambda,k,i}(x)-\varphi_{\lambda,k,i-1}(x)=(xq^{\lambda+1})^{i-1}\psi_{\lambda,k,k-i+1}(xq^{\lambda+1}); \tag{5.3}$$

$$\psi_{\lambda,k,i}(x)=\sum_{j=0}^{\lambda}x^jq^{j(j+1)/2}\begin{bmatrix}\lambda\\ j\end{bmatrix}\varphi_{\lambda,k,i-j}(x); \tag{5.4}$$

$$\varphi_{\lambda,k,-i}(x)=-(xq^{\lambda+1})^{-i}\varphi_{\lambda,k,i}(x), \tag{5.5}$$

where

$$\begin{bmatrix}\lambda\\ j\end{bmatrix}=\frac{(q)_\lambda}{(q)_j(q)_{\lambda-j}},\quad 0\leqq j\leqq\lambda, \qquad =0,\quad\text{otherwise.} \tag{5.6}$$

Utilizing Jacobi's triple product identity [7, pp. 169–170], one may easily deduce from (5.2) that

$$\begin{aligned}\psi_{\lambda,k,i}(1)&=(-q)_\infty(q^{2\lambda+2};q^{2\lambda+2})_\infty^{-1}(q^{(i-\lambda/2)(\lambda+1)};q^{(2k-\lambda+1)(\lambda+1)})_\infty\\&\quad\cdot(q^{(2k-\lambda/2-i+1)(\lambda+1)};q^{(2k-\lambda+1)(\lambda+1)})_\infty(q^{(2k-\lambda+1)(\lambda+1)};q^{(2k-\lambda+1)(\lambda+1)})_\infty\\&=\sum_{n=0}^{\infty}A_{\lambda,k,a}(n)q^n.\end{aligned} \tag{5.7}$$

On the other hand, we may define generating functions related to the partition ideals $\mathscr{B}_{\lambda,k,a}$. Namely

$$j_{\lambda,k,a}(x)=\sum_{\{f_i\}\in\mathscr{B}_{\lambda,k,a}}x^{\#(\{f_i\})}q^{\sigma(\{f_i\})}, \tag{5.8}$$

and

$$h_{\lambda,j,a}(x) = \sum_{\{f_i\}\in\mathscr{B}_{\lambda,k,a}} x^{\#(\{f_i\})} q^{\sigma(\{f_i\})} \quad (\text{where } \{f_{i\leqq\lambda}\} = \mathbf{0}). \tag{5.9}$$

It is a straightforward matter to prove that

$$h_{\lambda,k,a}(x) - h_{\lambda,k,a-1}(x) = (xq^{\lambda+1})^{a-1} j_{\lambda,k,k-a+1}(xq^{\lambda+1}), \tag{5.10}$$

a perfect replica of (5.3).

Surprisingly instead of a replica of (5.4), the following relation holds.

$$\begin{aligned} j_{\lambda,k,a}(x) &= \sum_{j=0}^{a} x^j q^{j(j+1)/2} \begin{bmatrix} \lambda \\ j \end{bmatrix} h_{\lambda,k,a-j}(x) \\ &\quad + \sum_{l\geqq 2} (-1)^{l-1} \sum_{r=0}^{k} g(l; k, \lambda, r, a; x; q) h_{\lambda,k,r}(xq^{(l-1)(\lambda+1)}), \end{aligned} \tag{5.11}$$

where

$$\begin{aligned} &g(l; k, \lambda, r, a; x; q) \\ &\quad = \sum_{\mathscr{C}} x^{k(l-1)+\Sigma A-\Sigma D} q^{\Sigma_{b=1}^{l-1}(k-D_b+A_b)b(\lambda+1)+D_b(D_b+1)/2} \\ &\quad\quad \cdot q^{(\Sigma A-\Sigma D)(\Sigma A-\Sigma D+1)/2} \begin{bmatrix} \lambda \\ D_1 \end{bmatrix} \begin{bmatrix} \lambda \\ D_2 \end{bmatrix} \cdots \begin{bmatrix} \lambda \\ D_{l-1} \end{bmatrix} \begin{bmatrix} \lambda \\ \Sigma A - \Sigma D \end{bmatrix}, \end{aligned} \tag{5.12}$$

with $\Sigma A = A_0 + A_1 + \cdots + A_{l+1}$, $\Sigma D = D_1 + \cdots + D_{l-1}$, and $\mathscr{C}$ is the set of those $(2l-1)$-tuples $(A_0, A_1, \cdots, A_{l-1}, D_1, \cdots, D_{l-1})$ that satisfy $A_0 + k - D_1 \leqq i-1$, $A_{l-1} = D_{l-1} - r$, $0 \leqq A_0 \leqq \lambda$, $1 \leqq D_j \leqq k$, $0 \leqq A_j \leqq D_j$, $A_j \leqq D_j + D_{j+1} - k - 1$.

From these q-difference equations (5.8)–(5.11), it is possible to deduce that

$$\varphi_{\lambda,k,i}(x) = h_{\lambda,k,i}(x) + \sum_{l\geqq 2} (-1)^{l-1} \sum_{r=1}^{k} h_{\lambda,k,r}(xq^{(l-1)(\lambda+1)}) \gamma(l; k, \lambda, r, i; x; q),$$

with

$$\begin{aligned} &\gamma(l; k, \lambda, r, i; x; q) \\ &\quad = x^{(l-1)k} \sum_{\mathscr{D}} q^{\Sigma_{b=1}^{l-1}(k-D_b+A_b)b(\lambda+1)+D_b(D_b+1)/2} \begin{bmatrix} \lambda \\ D_1 \end{bmatrix} \begin{bmatrix} \lambda \\ D_2 \end{bmatrix} \cdots \begin{bmatrix} \lambda \\ D_{l-1} \end{bmatrix}, \end{aligned} \tag{5.13}$$

where $\mathscr{D}$ is the set of those $(2l-2)$-tuples $(A_1, \cdots, A_{l-1}, D_1, \cdots, D_{l-1})$ that satisfy $D_2 + \cdots + D_{l-2} - A_1 - \cdots - A_{l-2} \leqq i-k-r-1$, $1 \leqq D_j \leqq k$, $A_{l-1} = D_{l-1} - r$, $0 \leqq A_j \leqq D_j$, $A_j \leqq D_j + D_{j+1} - k - 1$ (when $l=2$, $\mathscr{D}$ is to be defined by $-D_1 \leqq i-k-r-1$, $A_1 = D_1 - r$). It turns out that

$$\gamma(l; k, \lambda, r, i; x; q) \equiv 0 \quad \text{for } 0 \leqq i \leqq k - \lambda + 1. \tag{5.14}$$

Hence for $\lambda \leqq a \leqq k$,

$$\begin{aligned} j_{\lambda,k,a}(x) &= x^{-a+1}(h_{\lambda,k,k-a+1}(xq^{-\lambda-1}) - h_{\lambda,k,k-a}(xq^{-\lambda-1})) \\ &= x^{-a+1}(\varphi_{\lambda,k,k-a+1}(xq^{-\lambda-1}) - \varphi_{\lambda,k,k-a}(xq^{-\lambda-1})) \\ &= \psi_{\lambda,k,a}(x). \end{aligned} \tag{5.15}$$

Consequently,

$$\begin{aligned} \sum_{n\geqq 0} A_{\lambda,k,a}(n)q^n &= j_{\lambda,k,a}(1) = \psi_{\lambda,k,a}(1) \\ &= \sum_{\{f_i\}\in\mathscr{B}_{\lambda,k,a}} q^{\sigma(\{f_i\})} = \sum_{n\geqq 0} B_{\lambda,k,a}(n)q^n. \end{aligned} \tag{5.16}$$

Comparing coefficients of q^n in the extremes of (5.16), we find that $A_{\lambda,k,a}(n)=B_{\lambda,k,a}(n)$ for each n, provided $\lambda \leqq a \leqq n$.

This concludes our sketch of the proof of Theorem 5.1. □

The above work should make clear that the establishment of Theorem 5.1 relies on a lot of detailed results other than those implied by Theorem 4.1. In particular the polynomials $g(l; k, \lambda, r, i; x; q)$ and $\gamma(l; k, \lambda, r, i; x; q)$ undergo extensive consideration in [**14**] before one can represent them as sums of products of Gaussian polynomials $\left[{\lambda \atop D_j}\right]$.

Thus while Theorem 4.1 tells us that all linked partition ideals are amenable to attack via q-difference equations, it tells us nothing about what sorts of analytic representations (like $\varphi_{\lambda,k,a}(x)$ and $\psi_{\lambda,k,a}(x)$) are needed in order to establish new partition identities.

Before closing our discussion of Theorem 5.1, we remark that in [**14**] the result is generalized to include parameters satisfying $\lambda/2 < a \leqq k$, $\lambda \leqq k$; in fact this generalization appears in [**6**] subject to the extra condition $k \geqq 2\lambda - 1$. Also the following conjecture is raised in [**14**].

CONJECTURE. $A_{4,3,3}(n) = B^0_{4,3,3}(n) = p(\mathscr{B}^0; n)$, where

$$\begin{aligned} \mathscr{B}^0 = \{\{f_i\} \in \mathscr{S} \mid\, & f_j + f_{j+1} + \cdots + f_{j+4} \leqq 2, f_{5j} + f_{5j+1} + \cdots + f_{5j+5} \leqq 2, \\ & f_{5j-3} + f_{5j-2} \leqq 1, f_{5j-1} + f_{5j+1} \leqq 1, f_{5j-1} + f_{5j} + f_{5j+5} + f_{5j+6} \leqq 3 \\ & \text{for } j \geqq 1, \text{ and } f_m > 1 \text{ implies } 5 \mid m\}. \end{aligned}$$

This conjecture has been verified for $n \leqq 59$ where $A_{4,3,3}(59) = B^0_{4,3,3}(59) = 2938$.

This problem seems to lie beyond the scope of the techniques introduced in [**14**], and if it is true, its proof should provide valuable insights and techniques for the theory of partition identities.

6. **Representations of solutions of q-difference equations.** Now we turn to Question 2 (stated at the end of §3), an aspect of this study where relatively little is known, and no truly general theorems like Theorem 4.1 are available for guidance.

We begin by remarking that if we expand the $F(x)$ defined by (3.3) and (3.4) into a power series in x, say $\Sigma A_n x^n$, we find directly that $A_0=1$ and $(1-q^n)A_n=q^{2n-1}A_{n-1}$. Hence

$$F(x) = \sum_{n\geqq 0} \frac{q^{n^2}x^n}{(q)_n}. \tag{6.1}$$

Comparing (6.1) with (3.5), we obtain an analytic identity:

$$\sum_{n\geqq 0} \frac{q^{n^2}x^n}{(q)_n} = \sum_{n\geqq 0} \frac{(-1)^n x^{2n} q^{n(5n+1)/2}(1 - x^2q^{4n+2})}{(q)_n(xq^{n+1})_\infty}. \tag{6.2}$$

Thus two very different q-series representations of the solution of (3.4) are available to us.

In the same vein, one can prove that

$$\varphi_{0,3,3}(x) = \sum_{m\geqq 0} \frac{x^{2m}q^{2m^2}(-xq^{2m+1})_\infty}{(q^2;q^2)_m} \qquad \text{[3, Theorem 8, p. 433]}$$

$$\varphi_{0,4,4}(x) = \sum_{m\geqq 0} \frac{x^{3m}q^{3m^2}(x^3q^{6m+3};q^3)_\infty}{(q^3;q^3)_m(xq^{3m+1})_\infty} \qquad \text{[3, Theorem 9, p. 444]},$$

$$\varphi_{0,5,5}(x) = \sum_{j\geqq 0}\sum_{n\geqq 0} \frac{q^{4n^2+12nj+8j^2+j}(-1)^j(xq^{4n+2j+1})_\infty x^{4n+6j}}{(q^4;q^4)_n(q^2;q^2)_j(x^4q^{8n+8j+4};q^4)_\infty} \qquad \text{[15]}.$$

Other series representations of $\varphi_{\lambda,k,i}(x)$ and $\psi_{\lambda,k,i}(x)$ have been found; however in all cases $(2k-\lambda+1)(\lambda+1)\leqq 64$, and in these cases the only prime factors of $(2k-\lambda+1)(\lambda+1)$ are 2, 3, 5, 7 (apart from the case $\lambda=0$, $2k+1=11$ mentioned above and treated in **[15]**). These results are all (with the exception of **[15]**) to be found in the work of W. N. Bailey **[16]**, **[17]**, **[18]** and L. J. Slater **[35]**, **[36]**, **[37]**. To date the methods employed seem of limited scope as Slater's cataloging of known results includes only 130 identities.

Unfortunately most of the classical investigations of q-difference equations were concerned with existence theorems given specified analyticity conditions on the coefficients **[1]**, **[20]**, **[21]**, **[33]**. We, however, have polynomial coefficients and already know that a unique solution exists given our simple boundary conditions. Our concern is not with the existence of a solution but with an appropriate representation of the solution.

Presumably the work of W. Hahn **[25]**, **[26]**, and F. Ryde **[33]**, as well as the extension of Bailey's techniques contained in **[15]** may point the way to progress in this area, but much remains to be done.

7. **Conclusion.** I hope that this paper has given some indication of what has been done and what remains to be done in the theory of identities of the Rogers-Ramanujan type.

Obviously a project that holds promise of rather immediate positive results is that of examining linked partition ideals C with modulus $\leqq M_1$, all spans $\leqq M_2$ and $|L_C| \leqq M_3$. Clearly the number of such C is finite, and this systematic approach promises to reveal C with related q-difference equations of reasonably low order.

More important for the general goals, one hopes to make advances on the analytic questions considered in §6.

Before concluding, we should point out that not all of the known theory of partition identities falls into the theory of linked partition ideals.

In particular, the theorems in [**4**] are not related to linked partition ideals except in the special case of Schur's theorem [**34**]. However examination of these results and comparison with the results in [**5**] suggest that a "dual" theory could be developed wherein q-recurrent sequences would replace q-difference equations.

There have been, however, other theorems developed that are definitely partition identities but do not at all concern an equivalence of partition ideals.

Frobenius [**23**, p. 523] introduced a notation for partitions that is especially useful here. He writes a partition π as

$$\begin{pmatrix} a_1, a_2, \cdots, a_r \\ b_1, b_2, \cdots, b_r \end{pmatrix},$$

where π has Durfee square of side r (see [**27**, p. 281] for discussion of the Durfee square), and where there are a_i nodes to the right of the main diagonal in the ith row of the Ferrars graph representation of π and b_i nodes below the main diagonal in the ith column of the Ferrars graph representation of π. For example, if π is $5+5+4+4+2+1+1$ (in $\mathscr{S}\{2, 1, 0, 2, 2, 0, 0, 0, \cdots\}$), then the Ferrars graph is

$$\begin{array}{lllll} \cdot & \cdot & \cdot & \cdot & \cdot \\ \cdot & \cdot & \cdot & \cdot & \cdot \\ \cdot & \cdot & \cdot & \cdot & \\ \cdot & \cdot & \cdot & \cdot & \\ \cdot & \cdot & & & \\ \cdot & & & & \\ \cdot & & & & \end{array}$$

and the Frobenius representation is

$$\begin{pmatrix} 4, 3, 1, 0 \\ 6, 3, 1, 0 \end{pmatrix}.$$

We now define $F_{\max}(\pi)$ and $F_{\min}(\pi)$ to be $\max(a_i-b_i)$ and $\min(a_i-b_i)$, respectively. The following theorem then holds [**11**], [**12**].

THEOREM 7.1. *Let $Q_{k,a}(n)$ denote the number of partitions π of n such that $-a+2 \leqq F_{\min}(\pi) \leqq F_{\max}(\pi) \leqq 2k-a-1$. Then $Q_{k,a}(n)=A_{0,k,a}(n)$ for all n.*

It is not difficult to show that the Rogers-Ramanujan identities are equivalent to this result in the cases $k=2$, $a=2$, 1. However all aspects of the multiple result

$$Q_{k,a}(n) = A_{0,k,a}(n) = B_{0,k,a}(n)$$

(implied by Theorem 5.1) are mostly unexplored for $k \geqq 3$. One could introduce a new lattice structure on partitions by the partial ordering $\pi_1 \leqq \pi_2$ if all the columns in the Frobenius representation of π_1 appear in the Frobenius representation of π_2. Where such an approach would lead concerning partition identities has not been investigated at all.

Still other partition functions $H_{k,a}(n)$ have been found [**13**] such that $H_{k,a}(n)=A_{0,k,a}(n)$; here a study is made of the Alder polynomials $G_{k,a}(n;q)$ defined by

$$\varphi_{0,k,a}(x) = \sum_{n \geqq 0} \frac{G_{k,a}(n;q)x^n}{(q)_n}.$$

Finally, we mention that W. Connor [**22**] has obtained infinite families of partition identities also unlike any of the previously mentioned results. Connor's most important theorem in the simplest instance is a new set of partition-identities related to two elegant analytic theorems originally due to L. J. Rogers [**32**] but put in their most interesting form by Connor:

$$\sum_{n=0}^{\infty} \frac{q^{n^2+n}}{(q)_{2n}} = \prod_{n=1;\, n \not\equiv \pm 1, \pm 8, \pm 9, 10 \pmod{20}}^{\infty} (1-q^n)^{-1}; \tag{7.1}$$

$$\sum_{n=0}^{\infty} \frac{q^{n^2+n}}{(q)_{2n+1}} = \prod_{n=1;\, n \not\equiv \pm 3, \pm 4, \pm 7, 10 \pmod{20}}^{\infty} (1-q^n)^{-1}. \tag{7.2}$$

Connor's partition identity related to (7.1) is the following:

THEOREM 7.2. *Let $A(n)$ denote the number of partitions of n into parts* $\not\equiv 0, \pm 1, \pm 8, \pm 9, 10 \pmod{20}$. *Let $C(n)$ denote the number of partitions of n into parts where even parts may not be repeated and where odd parts occur only if an even part one unit away occurs. Let $E(n)$ denote the number of partitions of n of the form $b_1+b_2+\cdots+b_s$, where s is even and $b_1 \geqq b_2 > b_3 \geqq b_4 > b_5 \geqq \cdots$. Then $A(n)=C(n)=E(n)$.*

We note that the identity

$$\sum_{n\geqq 0} E(n)q^n = \sum_{n\geqq 0} A(n)q^n$$

is rather directly deduced from (7.1); however, the appearance of $C(n)$ in Theorem 7.2 is a complete surprise. While $A(n)$ is related to a partition ideal, neither $C(n)$ nor $E(n)$ is. Identity (7.2) produces a perfect companion for Theorem 7.2. B. Gordon [**24**] has also proved several interesting theorems of this type.

It might be supposed that we are concluding on a disappointing note. Theorem 4.1 suggests that we may see the beginnings of order amidst chaos, and so the last few paragraphs tend to suggest otherwise. However perhaps a subject seems most vital when (1) general methods are available to begin the attack on numerous problems while (2) other problems and results indicate that many quite unexplored areas remain.

References

(See [**8**] for an extensive bibliography of this subject.)

1. C. R. Adams, *Linear q-difference equations*, Bull. Amer. Math. Soc. **37** (1931), 361–400.

2. H. L. Alder, *Generalizations of the Rogers-Ramanujan identities*, Pacific J. Math. **4** (1954), 161–168. MR **15**, 856.

3. G. E. Andrews, *On q-difference equations for certain well-poised basic hypergeometric series*, Quart. J. Math. Oxford Ser. (2) **19** (1968), 433–447. MR **38** #6112.

4. ———, *A new generalization of Schur's second partition theorem*, Acta Arith. **14** (1967/68), 429–434. MR **37** #4042.

5. ———, *A general theorem on partitions with difference conditions*, Amer. J. Math. **91** (1969), 18–24. MR **39** #4109.

6. ———, *A generalization of the classical partition theorems*, Trans. Amer. Math. Soc. **145** (1969), 205–221. MR **40** #4226.

7. ———, *Number theory*, Saunders, Philadelphia, Pa., 1971. MR **46** #8943.

8. ———, *Partition identities*, Advances in Math. **9** (1972), 10–51. MR **46** #5232.

9. ———, *A general theory of identities of the Rogers-Ramanujan type*, Amer. Math. Soc. Audio Recording No. 75 and Supplementary Manual, Providence, R.I., 1972.

10. ———, *Summations and transformations of basic Appell series*, J. London Math. Soc. (2) **4** (1972), 618–622. MR **46** #5682.

11. ———, *Sieves for theorems of Euler, Rogers, and Ramanujan*, Proc. Western Michigan Conference on Arithmetic Functions, Lecture Notes in Math., no. 251, Springer-Verlag, New York and Berlin, 1972.

12. ———, *Sieves in the theory of partitions*, Amer. J. Math. **94** (1972), 1214–1230.

13. ———, *On the Alder polynomials and a new generalization of the Rogers-Ramanujan identities*, Trans. Amer. Math. Soc. (to appear).

14. ———, *On the general Rogers-Ramanujan theorem* (to appear).

15. ———, *On Rogers-Ramanujan type identities related to the modulus* 11, Proc. London Math. Soc. (to appear).

16. W. N. Bailey, *Generalized hypergeometric series*, Cambridge Math. Tract, no. 32, Cambridge Univ. Press, Cambridge, 1935.

17. ———, *Some identities in combinatory analysis*, Proc. London Math. Soc. (2) **49** (1947), 421–435. MR **9**, 263.

18. ———, *Identities of the Rogers-Ramanujan type*, Proc. London Math. Soc. (2) **50** (1948), 1–10. MR **9**, 585.

19. G. Birkhoff, *Lattice theory*, Amer. Math. Soc. Colloq. Publ. vol. 25, Amer. Math. Soc., Providence, R.I., 1948, revised 1967.

20. G. D. Birkhoff, *The generalized Riemann problem for linear differential equations and the allied problems for linear difference and q-difference equations*, Proc. Amer. Acad. Arts and Sci. **49** (1913), 521–568.

21. R. D. Carmichael, *The general theory of linear q-difference equations*, Amer. J. Math. **34** (1912), 147–168.

22. W. G. Connor, *Partition theorems related to some identities of Rogers and Watson*, Trans. Amer. Math. Soc. (to appear).

23. G. Frobenius, *Über die Charaktere der symmetrischen Gruppe*, Sitzber. Preuss. Akad. Berlin **1900**, 516–534.

24. B. Gordon, *Some continued fractions of the Rogers-Ramanujan type*, Duke Math. J. **32** (1965), 741–748. MR **32** #1477.

25. W. Hahn, *Beiträge zur Theorie der Heineschen Reihen. Die 24 Integrale der hypergeometrischen q-Differenzengleichung. Das q-Analogen der Laplace-Transformation*, Math. Nachr. **2** (1949), 340 379. MR **11**, 720.

26. ———, *Über die höheren Heineschen Reihen und eine einheitliche Theorie der sogenannten speziellen Funktionen*, Math. Nachr. **3** (1950), 257–294. MR **12**, 711.

27. G. H. Hardy and E. M. Wright, *An introduction to the theory of numbers*, 4th ed., Oxford Univ. Press, Oxford, 1960.

28. F. H. Jackson, *On basic double hypergeometric functions*, Quart. J. Math. Oxford Ser. **13** (1942), 69–82. MR **4**, 141.

29. D. E. Knuth, *The art of computer programming*. Vol. 3. *Sorting and searching*, Addison-Wesley, Reading, Mass., 1973.

30. F. J. Murray and K. S. Mller, *Existence theorems for ordinary differential equations*, New York Univ. Press, New York, 1954. MR **16**, 358.

31. S. Ramanujan and L. J. Rogers, *Proof of certain identities in combinatory analysis*, Proc. Cambridge Philos. Soc. **19** (1919), 211–216.

32. L. J. Rogers, *On two theorems of combinatory analysis and some allied identities*, Proc. London Math. Soc. (2) **16** (1917), 315–336.

33. F. Ryde, *A contribution to the theory of linear homogeneous geometric difference equations (q-difference equations)*, Ph.D. Dissertation, Lund, 1921.

34. I. J. Schur, *Zur additiven Zahlentheorie*, Sitzler. Akad. Wiss. Berlin Phys.-Math. Kl. **1926**, 488–495.

35. L. J. Slater, *A new proof of Roger's transformations of infinite series*, Proc. London Math. Soc. (2) **53** (1951), 460–475. MR **13**, 227.

36. ———, *Further identities of the Rogers-Ramanujan type*, Proc. London Math. Soc. (2) **54** (1952), 147–167. MR **14**, 138.

37. ———, *Generalized hypergeometric functions*, Cambridge Univ. Press, Cambridge, 1966. MR **34** #1570.

DEPARTMENT OF MATHEMATICS, PENNSYLVANIA STATE UNIVERSITY, UNIVERSITY PARK, PENNSYLVANIA 16802

7.3 Partitions: At the Interface of q-Series and Modular Forms

THE RAMANUJAN JOURNAL, 7, 385–400, 2003

Partitions: At the Interface of q-Series and Modular Forms

GEORGE E. ANDREWS*
The Pennsylvania State University, University Park, Pennsylvania 16802

In memory of Robert A. Rankin

Received February 10, 2003; Accepted February 20, 2003

Abstract. In this paper we explore five topics from the theory of partitions: (1) the Rademacher conjecture, (2) the Herschel-Cayley-Sylvester formulas, (3) the asymptotic expansions of E.M. Wright, (4) the asymptotics of mock theta function coefficients, (5) modular transformations of q-series.

Key words: partitions, q-series, modular forms

2000 Mathematics Subject Classification: Primary—11P82, 11P83, 11P81, 05A30, 33D15

1. Introduction

The twentieth century found two major currents in the theory of partitions. The first is characterized by the role played by modular forms. Seminal papers in this vein include the Hardy-Ramanujan paper on $p(n)$ [20], the proofs of the Ramanujan congruences by Watson [41] and Atkin [7], and the many recent contributions by Ono [28, 29] to name only a few. It should be noted that Ramanujan himself proved the congruence for general powers of 5 and likely had the corresponding proof for powers of 7 (see [12] for details).

The second has been extensive study of the application of basic hypergeometric series or q-series to partitions. This has its genesis in the work of MacMahon [25] and Schur [37]. In the 1960's a variety of results were discovered; these are described in a survey by Alder [1] (cf. [3]).

Surprisingly perhaps, there has not been a large amount of interplay between these two themes. In this paper, I hope to survey topics that lie at the interface of modular forms and q-series. Some concern important and unjustly neglected conjectures. Each should suggest a number of research possibilities.

Section 2 will be devoted to Rademacher's conjecture [32, p. 302] which concerns his partial fraction decomposition for the generating function of $p(n)$. Section 3 considers alternative representations of the restricted partition function $p(n, m)$. Section 4 looks at a sequence of papers by Wright [42–44]. The third of these is quite unlike any other in the history of partition asymptotics and foreshadows many possibilities. In Section 5, we

*Partially supported by National Science Foundation Grant DMS-9206993.

look again at problems posed by Dragonette [15] for the asymptotics of the coefficients of Ramanujan's mock theta functions. The sixth section considers the efforts by Ehrenpreis [16] and others to develop a modular type transformation theory for q-series. We conclude with brief synthesizing observations.

2. The Rademacher conjecture

Near the end of his book, Topics in Analytic Number Theory [32, p. 300], Rademacher derives the following partial fraction decomposition for the generating function of the partition function $p(n)$.

2.1. Rademacher's partial fraction decomposition

We use the following notation for forward differences

$$\Delta_\alpha f(\alpha) = f(\alpha+1) - f(\alpha),$$

and for nonnegative integers j

$$\Delta_\alpha^j f(\alpha) = \sum_{h=0}^{j} (-1)^h \binom{j}{h} f(\alpha + j - h).$$

Also

$$L_j(y) = \sum_{n=0}^{\infty} \frac{y^n}{n!\,\Gamma(j+n+1)}.$$

Then

$$F(x) = -2\pi\left(\frac{\pi}{12}\right)^{3/2} \sum_{k=1}^{\infty} k^{-5/2} \sum_{\substack{h \bmod k \\ (h,k)=1}} w_{h,k} \times \sum_{j=0}^{\infty} \Delta_\alpha^j \left[L_{3/2}\left(-\frac{\pi^2}{6k^2}(\alpha+1)\right)\right]\left(xe^{-\frac{2\pi i h}{k}} - 1\right)^{-j-1}, \qquad (2.1)$$

where $\alpha = 1/24$, w_{hk} is a root of unity [32, p. 269] and

$$F(x) = \begin{cases} \sum_{n=0}^{\infty} p(n)x^n, & |x| < 1 \\ 0, & |x| > 1. \end{cases}$$

Hardy and Ramanujan had several other results of this sort, with the generating function being equal to 0 outside the unit circle. See their last published paper. In particular, see Theorem 3 on page 319 of Ramanujan's *Collected Papers*. In letters to Hardy written by

Ramanujan from nursing homes, he gives several more examples. Some of these results are recounted and proved by Berndt et al. in [11].

Immediately after proving the above expansion for $F(x)$, Rademacher [29, p. 301] observes that if $|x| < 1$

$$\lim_{N\to\infty} \frac{1}{(1-x)(1-x^2)\dots(1-x^N)} = \frac{1}{\prod_{m=1}^{\infty}(1-x^m)} = \sum_{n=0}^{\infty} p(n)x^n = F(x), \tag{2.3}$$

and if $|x| > 1$

$$\lim_{N\to\infty} \frac{1}{(1-x)(1-x^2)\dots(1-x^N)} = 0 = F(x). \tag{2.4}$$

Of course, one can perform the classical partial fraction decomposition of this finite product. Indeed, we may write

$$\frac{1}{\prod_{m=1}^{N}(1-x^m)} = \sum_{\substack{0\le h<k\le n \\ (h,k)=1}} \sum_{j=1}^{\lfloor N/k \rfloor} \frac{C_{hkj}(N)}{\left(x - e^{\frac{2\pi i h}{k}}\right)^j}. \tag{2.5}$$

As Rademacher notes, there is a natural identification suggested by these two expansions.

Rademacher's Conjecture [32, p. 302].

$$\lim_{N\to\infty} C_{hkj}(N)$$

exists and is given by

$$C_{hkj}(\infty) = -2\pi\left(\frac{\pi}{12}\right)^{3/2} \frac{w_{hk}e^{\frac{2\pi i h j}{k}}}{k^{5/2}} \Delta_{\alpha}^{j-1} L_{3/2}\left(-\frac{\pi^2}{6k^2}(\alpha+1)\right),$$

where $\alpha = 1/24$, and

$$L_{3/2}(-y^2) = -\frac{1}{2\sqrt{\pi}y}\frac{d}{dy}\left(\frac{\sin 2y}{y}\right) = -\frac{1}{2\sqrt{\pi}y^2}\left(2\cos 2y - \frac{\sin 2y}{y}\right).$$

Rademacher goes on to provide a small table of values (reprinted here with one correction):

N	C_{011}	C_{012}	C_{121}
1	-1	0	0
2	$-\frac{1}{4} = -0.25$	$\frac{1}{2} = 0.5$	$\frac{1}{4} = 0.25$
3	$-\frac{17}{72} = -0.23611\ldots$	$\frac{1}{4} = 0.25$	$\frac{1}{8} = 0.125$
4	$-\frac{17}{72} = -0.23611\ldots$	$\frac{59}{288} = 0.204861\ldots$	$\frac{1}{8} = 0.125$
5	$-\frac{20831}{86400} = -0.24101\ldots$	$\frac{3}{16} = 0.1875$	$\frac{13}{128} = 0.1015625$

and he lists explicitly three of the conjectured limits:

$$C_{011}(\infty) = -\frac{6}{25}\left(1 + \frac{2\sqrt{3}}{5\pi}\right) = -0.273339\ldots,$$

$$C_{012}(\infty) = \frac{24}{25 \cdot 49}\left(6 + \frac{109\sqrt{3}}{35\pi}\right) = 0.15119\ldots,$$

$$C_{121}(\infty) = -\frac{\sqrt{6}}{25}\left(\cos\frac{5\pi}{12} - \frac{12}{5\pi}\sin\frac{5\pi}{12}\right) = 0.046941.$$

Of these coefficients he says [32, p. 301]: "No explicit formula for $C_{hkl}(N)$ is known, not even for the simplest case $h = 0$, $k = 1$, $l = 1$, and variable N."

In the hope of stimulating serious consideration of this problem, I will present an explicit formula for $C_{011}(N+1)$ below. The same procedure can obviously be applied to provide a formula for $C_{hkj}(N+1)$ in general. However, owing to the fact that we cannot prove Rademacher's conjecture from our result, we shall not complicate matters by presenting the full formula for $C_{hkj}(N+1)$. It should be pointed out that Rademacher's conjecture lies at the interface of the theory of modular forms and the theory of q-series. Thus progress on this problem may require contributions from two areas that have had less contact in the past than might have been expected or hoped for.

Theorem 1. *With $\rho_j = e^{2\pi i/j}$, and $H_N(x_1, \ldots, x_n)$ the N-th homogeneous symmetric function of $x_1, x_2, \ldots, x_n$, then*

$$C_{011}(N+1) = \frac{-1}{(N+1)!}\sum_{h_1=1}^{1}\sum_{h_2=1}^{2}\cdots\sum_{h_N=1}^{N}\left(\prod_{i=1}^{N}\rho_{i+1}^{-h_i}\right)H_N\left(\frac{\rho_2^{h_1}}{1-\rho_2^{h_1}}, \ldots, \frac{\rho_{N+1}^{h_N}}{1-\rho_{N+1}^{h_N}}\right).$$

Proof: Noting that (2.5) may be presented as

$$\frac{1}{(x;x)_{N+1}} = \frac{C_{01N+1}(N+1)}{(x-1)^{N+1}} + \frac{C_{01N}(N+1)}{(x-1)^{N}} + \cdots + \frac{C_{011}(N+1)}{(x-1)}$$
$$+ \sum_{\substack{0 \le h < k \le N+1 \\ (h,k)=1 \\ k>1}} \sum_{j=1}^{\lfloor n/k \rfloor} \frac{C_{hkj}(N+1)}{(x - e^{2\pi i h/k})^j},$$

we find by multiplying by $(x-1)^{N+1}$, differentiating N times and then setting $x=1$ that

$$N!\,C_{011}(N+1)=\frac{d^N}{dx^N}\frac{(x-1)^{N+1}}{(x;x)_{N+1}}\bigg|_{x=1},$$

where

$$(A;q)_N=(1-A)(1-Aq)\dots(1-Aq^{N-1}).$$

So we clearly need the r-dimensional Leibnitz rule

$$\frac{d^N}{dx^N}f_1\,f_2\dots f_r=\sum_{\substack{n_1,\dots,n_r\ge 0\\ n_1+\dots+n_r=N}}\binom{N}{n_1,n_2,\dots,n_r}f_1^{(n_1)}f_2^{(n_2)}\dots f_r^{(n_r)}.$$

Now recalling that

$$\frac{1}{1-q^M}=\frac{1}{M}\sum_{j=0}^{M-1}\frac{1}{(1-\rho_M^j q)},$$

where $\rho_M=e^{2\pi i/M}$, we easily deduce

$$\frac{1-q}{1-q^M}=\frac{1}{M}\sum_{j=1}^{M-1}\frac{(1-\rho_M^{-j})}{(1-\rho_M^j q)}.$$

Consequently

$$\frac{d^k}{dq^k}\frac{(1-q)}{(1-q^M)}=\frac{k!}{M}\sum_{j=1}^{M-1}\frac{(1-\rho_M^{-j})\rho_M^{jk}}{(1-\rho_M^j q)^{k+1}}.$$

Next we recall the standard representation of the homogeneous symmetric function in terms of monomials:

$$H_N(X_1,X_2,\dots,X_N)=\sum_{\substack{n_1,n_2,\dots,n_N\ge 0\\ n_1+n_2+\dots+n_N=N}}X_1^{n_1}X_2^{n_2}\dots X_N^{n_N}.$$

As a result,

$$\begin{aligned}
&N!\,C_{011}(N+1)\\
&=(-1)^{N+1}\sum_{\substack{n_1,\dots,n_N\ge 0\\ n_1+\dots+n_N=N}}\binom{N}{n_1,n_2,\dots,n_N}\left(\frac{d^{n_1}}{dx^{n_1}}\frac{(1-x)}{(1-x^2)}\right)\cdots\left(\frac{d^{n_N}}{dx^{n_N}}\frac{(1-x)}{(1-x^{N+1})}\right)\bigg|_{x=1}\\
&=(-1)^{N+1}\sum_{\substack{n_1,\dots,n_N\ge 0\\ n_1+\dots+n_N=N}}\binom{N}{n_1,n_2,\dots,n_N}\left(\frac{n_1!}{2}\sum_{j_1=1}^{2-1}\frac{(1-\rho_1^{-j_1})\rho_2^{j_1n_1}}{(1-\rho_2^{j_1}q)^{n_1+1}}\right)
\end{aligned}$$

$$
\cdots\left(\frac{n_N!}{(N+1)}\sum_{j_N=1}^{(N+1)-1}\frac{\left(1-\rho_{N+1}^{-j_N}\right)\rho_{N+1}^{j_N n_N}}{\left(1-\rho_{N+1}^{j_N}\right)^{n_N+1}}\right)\Bigg|_{x=1}
$$

$$
=\frac{(-1)^{N+1}}{(N+1)!}\sum_{j_1=1}^{1}\sum_{j_2=1}^{2}\cdots\sum_{j_N=1}^{N}\frac{\left(1-\rho_2^{-j_1}\right)}{\left(1-\rho_2^{j_1}\right)}\frac{\left(1-\rho_3^{-j_2}\right)}{\left(1-\rho_3^{j_2}\right)}\cdots\frac{\left(1-\rho_{N+1}^{-j_N}\right)}{\left(1-\rho_{N+1}^{j_N}\right)}
$$

$$
\times\sum_{\substack{n_1,\ldots,n_N\geq 0\\ n_1+\cdots+n_N=N}}\left(\frac{\rho_{N+1}^{j_N}}{1-\rho_{N+1}^{j_N}}\right)^{n_N}\cdots\left(\frac{\rho_2^{j_1}}{1-\rho_2^{j_1}}\right)^{n_1}
$$

$$
=\frac{-1}{(N+1)!}\sum_{j_1=1}^{1}\sum_{j_2=1}^{2}\cdots\sum_{j_N=1}^{N}\left(\prod_{i=1}^{N}\rho_{i+1}^{-j_i}\right)H_N\left(\frac{\rho_2^{j_1}}{1-\rho_2 j_1},\ldots,\frac{\rho_{N+1}^{j_N}}{1-\rho_{N+1}^{j_N}}\right),
$$

as desired. □

It may be reasonably objected that Theorem 1 provides little hope of proving Rademacher's Conjecture. At the most it may suggest the value of finding better formulas for $C_{hkj}(N)$.

Ehrenpreis in [16, p. 317] mentions that his student, Jane Friedman, studied computer algorithms for this problem but he states that: "Unfortunately, the computer study proved inconclusive." While this is still true, we can add a few more values to those already computed for $C_{011}(N)$:

N	C_{011}
6	$-\dfrac{85823}{345600} = -0.24833$
7	$-\dfrac{19554517}{76204800} = -0.25660$
8	$-\dfrac{80858443}{304819200} = -0.26527$

3. Formulae for $p(n, m)$

We denote the number of partitions of n into at most m parts by $p(n, m)$. There is an extensive literature concerning formulae for $p(n, m)$, there is a full account of the history up to 1958 given in the Royal Society Table of Partitions [19] which is mostly devoted to extensive tables of $p(n, m)$. Many authors, among them, Paoli [30], DeMorgan [14], Herschel [21], Cayley [13], Sylvester [38], Glaisher [17] and more recently Arkin [6], have all made major contributions to the study of exact formulae for $p(n, m)$.

When one proceeds to examine the actual formulae, one is struck by a substantial difference that arises between $m \leq 4$ and $m > 4$. Namely, we see immediately that

$$
p(n,1) = 1, \tag{3.1}
$$

$$
p(n,2) = \left\lfloor \frac{n+2}{2} \right\rfloor. \tag{3.2}
$$

DeMorgan [14] first proved that

$$p(n,3)=\left\{\frac{(n+3)^2}{12}\right\}, \tag{3.3}$$

and Glösel proved [18, p. 138]:

$$p(n,4)=\left\{\left\lfloor\frac{n+4}{2}\right\rfloor^2\left(3\left\lfloor\frac{n+9}{2}\right\rfloor-\left\lfloor\frac{n+10}{2}\right\rfloor\right)\frac{1}{36}\right\}, \tag{3.4}$$

where $\lfloor x\rfloor$ is the greatest integer $\leq x$, and $\{x\}$ is the nearest integer to x.

Now there are several immediate observations to make about these formulae. First of all, they are all obviously real numbers, indeed positive integers. Furthermore, they are given in one line and are computed with familiar functions. We remark that (3.4) could be improved if one could eliminate some of the instances of the greatest integer function.

For $m > 5$ (Glösel has a lengthy formula for $p(n,5)$ that is of the same type as (3.1)–(3.4)), one runs into the stark fact that one needs representations of periodic sequences. As is well-known [31, p. 76], such sequences are representable by Finite Fourier series. So, for example, an alternative to (3.4) is [26, p. 153]

$$\begin{aligned}p(n,4)&=\frac{1}{144}(n+5)^3-\frac{5}{96}(n+5)+\frac{1}{32}(-1)^n(n+5)\\&\quad+\frac{1}{27}\left(r_3^n+r_3^{-n}-r_3^{n+1}-r_3^{-n-1}\right)\\&\quad+\frac{1}{16}i^n(1+(-1)^n),\end{aligned}$$

where $r_3=e^{2\pi i/3}$.

Now this formula is rather disturbing. A little thought reveals that it is real because it is equal to its complex conjugate. However, its integrality is far from obvious.

One way to avoid this difficulty is to introduce the circulator notation for a periodic sequence of period m [19, p. xvii]

$$(A_0,A_1,\dots,A_{m-1})crm_n=A_i \quad \text{if } n\equiv i \pmod{m}. \tag{3.6}$$

Computationally this approach avoids complex numbers, but it merely introduces notation for periodic sequences.

In this new notation we find [19, p. xxvii, Eq. (6.7)]

$$\begin{aligned}p(n,4)&=\frac{1}{24}\binom{n+6}{3}+\left(\frac{1}{9}-\frac{1}{16}(2,3)cr2_n\right)(n+5)\\&\quad+\frac{1}{9}(1,0,-1)cr3_n+\frac{1}{8}(1,0,-1,0)cr4_n.\end{aligned} \tag{3.7}$$

In this section, we shall make the simple observation that periodic sequences may also be represented by the greatest integer function. From that knowledge it is easy to derive formulas such as

$$p(n,4)=\left\{(n+5)\left(n^2+n+22+18\left\lfloor\frac{n}{2}\right\rfloor\right)\Big/144\right\}, \tag{3.8}$$

and

$$p(n,5)=\left\{(n+8)\left(n^3+n+22n^2+44n+248+180\left\lfloor\frac{n}{2}\right\rfloor\right)\Big/2880\right\}, \tag{3.9}$$

Note that these formulas are much more in the spirit of (3.1)–(3.3) than is (3.5). The formula (3.9) is strikingly simpler than any of the several given in [19]. Furthermore such formulas for $p(n, m)$ continue for larger m. For example,

$$\begin{aligned}p(n,6)=\Bigg\{&(n+11)\Bigg((6n^4+249n^3+2071n^2-4931n+40621)/518400\\&+\left\lfloor\frac{n}{2}\right\rfloor(n+10)/192+\left(\left\lfloor\frac{n+1}{3}\right\rfloor+2\left\lfloor\frac{n}{3}\right\rfloor\right)\Big/54\Bigg)\Bigg\},\end{aligned} \tag{3.10}$$

$$\begin{aligned}p(n,7)=\Bigg\{&(n+14)\Bigg((n^5+70n^4+1785n^3-15365n^2+9702n+277032\big/3628800\\&+\left\lfloor\frac{n}{2}\right\rfloor(n+14)/384+\left\lfloor\frac{n}{3}\right\rfloor\Big/54\Bigg)\Bigg\},\end{aligned} \tag{3.11}$$

$$\begin{aligned}p(n,8)=\Bigg\{&(n+18)\Bigg((n^6+108n^5+4503n^4+79911n^3+522148n^2-202687n\\&+9441216)/203212800+\left\lfloor\frac{n}{2}\right\rfloor(n^2+36+231)\Big/9216\\&+\left(\left\lfloor\frac{n+1}{3}\right\rfloor+2\left\lfloor\frac{n}{3}\right\rfloor\right)\Big/162+\left\lfloor\frac{n}{4}\right\rfloor\Big/64\Bigg)\Bigg\},\end{aligned} \tag{3.12}$$

$$\begin{aligned}p(n,9)=\Bigg\{&(n+22)\Bigg((n^7+158n^6+10034n^5+327352n^4\\&+5419144n^3-32063602n^2+5172096n+564401888)/14631321600\\&+\left\lfloor\frac{n}{2}\right\rfloor(2n^2+91n+728)\Big/36864\\&+\left((n+20)\left\lfloor\frac{n+1}{3}\right\rfloor+2(n+23)\left\lfloor\frac{n}{3}\right\rfloor\right)\Big/2916\\&+\left(\left\lfloor\frac{n}{4}\right\rfloor+\left\lfloor\frac{n+2}{4}\right\rfloor\right)\Big/256\Bigg)\Bigg\}.\end{aligned} \tag{3.13}$$

In order to establish such results, we require the following:

Theorem 2. *For a fixed positive integer m, the periodic sequence in n is given by*

$$\begin{aligned}
&(A_0, A_1, \dots, A_{m-1})crm_n \\
&= A_0\left\lfloor \frac{n+m}{m} \right\rfloor + \sum_{j=1}^{m-1}(A_j - A_{j-1})\left\lfloor \frac{n+m-j}{m} \right\rfloor - A_{m-1}\left\lfloor \frac{n}{m} \right\rfloor \\
&= A_0 + \sum_{j=1}^{m}(A_j - A_{j-1})\left\lfloor \frac{n+m-j}{m} \right\rfloor, \qquad (3.14)
\end{aligned}$$

where we define $A_m = A_0$.

Proof: Let us define $f(n)$ to be the right-hand side of (3.14).

$$\begin{aligned}
f(n+m) - f(n) &= \sum_{j=1}^{m}(A_j - A_{j-1})\left(\left\lfloor \frac{n+2m-j}{m} \right\rfloor - \left\lfloor \frac{n+m-j}{m} \right\rfloor\right) \\
&= \sum_{j=1}^{m}(A_j - A_{j-1}) = A_m - A_0 = 0.
\end{aligned}$$

Hence $f(n)$ has period m. Furthermore for $0 \leq i < m$,

$$\begin{aligned}
f(i) &= A_0 + \sum_{j=1}^{m}(A_j - A_{j-1})\left\lfloor \frac{i+m-j}{m} \right\rfloor \\
&= A_0 + \sum_{j=1}^{i}(A_j - A_{j-1}) \\
&= A_i.
\end{aligned}$$

Therefore $f(n) = (A_0, A_1, \dots, A_{m-1})crm_n$ as desired. □

Corollary.

$$\begin{aligned}
&(A_0, A_1, \dots, A_{m-1})crm_n \\
&= A_0 + (A_1 - A_0)n + \sum_{j=2}^{m}(A_j - A_{j-1} - A_1 + A_0)\left\lfloor \frac{n+m-j}{m} \right\rfloor.
\end{aligned}$$

Proof: This follows immediately from Theorem 1 once we recall that

$$\left\lfloor \frac{n+m-1}{m} \right\rfloor = n - \left\lfloor \frac{n}{m} \right\rfloor - \left\lfloor \frac{n+1}{m} \right\rfloor - \dots - \left\lfloor \frac{n+m-2}{m} \right\rfloor.$$

□

Formulas (3.8)–(3.13) are now easily derived from the literature, in particular, from the formulas [19, pp. xxvii–xxviii, Eq. (6.7)]. Since the process is routine, we illustrate with the cases $m = 4$ and 5.

Note that in (3.7) we may drop the $\frac{1}{9}(1, 0, -1)cr3_n + \frac{1}{8}(1, 0, -1, 0)cr4_n$ because this expression is in absolute value $<\frac{1}{9} + \frac{1}{8} < \frac{1}{2}$. Hence immediately

$$\begin{aligned} p(n, 4) &= \left\{\binom{n+6}{3}\frac{1}{24} + \left(\frac{1}{9} - \frac{1}{16}\left(2 + n - 2\left\lfloor\frac{n}{2}\right\rfloor\right)\right)(n+5)\right\} \\ &= \left\{(n+5)\left(n^2 + n + 22 + 18\left\lfloor\frac{n}{2}\right\rfloor\right)\Big/144\right\}, \end{aligned} \tag{3.16}$$

as desired.

Similarly [6, p. xxvii, Eq. (6.7)]

$$\begin{aligned} p(n, 5) &= \frac{1}{120}\binom{n+9}{4} - \frac{5}{288}\binom{n+8}{2} + \frac{1}{64}(n+8)(1, -1)cr2 \\ &\quad + \frac{1}{9}(1, 0, 0)cr3_n + \frac{1}{64}(0, 1, -8, -7)cr4_n + \frac{1}{5}(1, 0, 0, 0, 0)cr5_n \\ &= \left\{\frac{1}{120}\binom{n+9}{4} - \frac{5}{288}\binom{n+8}{2} + \frac{1}{64}(n+8)\left(1 - 2n + 4\left\lfloor\frac{n}{2}\right\rfloor\right)\right\} \\ &= \left\{(n+8)\left(n^3 + 22n^2 + 44n + 248 + 180\left\lfloor\frac{n}{2}\right\rfloor\right)\Big/2880\right\} \end{aligned} \tag{3.17}$$

as desired.

Formulas (3.10)–(3.13) are handled in exactly this way.

4. Asymptotics of non-modular generating functions

The topic of this section has had extensive development since 1941 when Ingham's seminal paper appeared [22]. Among the authors obtaining general theorems on this topic are Bateman and Erdös [8], Nicolas and Sárközy [27], Richmond [35], Roth and Szekeres [36, 39], and McIntosh [24]. This list is undoubtedly incomplete and is actually only tangential to the real object of this section. Of special interest are Ramanujan's observations as explicated by Berndt [10, p. 269, Entry 7].

Prior to Ingham's work, E.M. Wright wrote three papers on the asymptotics of particular generating functions. The first of these [42] was devoted to the asymptotics of the plane partition function. Its method was essentially the saddlepoint method that forms the foundation of many of the succeeding works mentioned above. Wright's second paper [43] in this series was based on a more elementary than analytic account but was not a surprising development.

The point I wish to make in the section is that Wright's third paper [44] on partitions into powers IS UNIQUE in the history of this subject. Its starting point and fundamental philosophy are different from anything that has come before or since.

Wright's focus in [44] is

$$f(x)=\prod_{\ell=1}^{\infty}\left(1-x^{\ell^k}\right)^{-1}=1+\sum_{n=1}^{\infty}p_k(n)x^n, \tag{4.1}$$

where $p_k(n)$ is the number of partitions of n into k-th powers.

His results are intricate to say the least. However, after stating his central results we shall examine the import of his work.

Write

$$a=\frac{1}{k},\quad b=\frac{1}{k+1},\quad j=j(k)=0\quad (k\text{ even}); \tag{4.2}$$

$$j=j(k)=\frac{(-1)^{(k+1)/2}}{(2\pi)^{k+1}}\Gamma(k+1)\zeta(k+1)\quad (k\text{ odd}). \tag{4.3}$$

Define a generalized Bessel function as follows:

$$\varphi(z)=\sum_{\ell=0}^{\infty}\frac{z^\ell}{\Gamma(\ell+1)\Gamma\left(\ell a-\frac{1}{2}\right)}. \tag{4.4}$$

Wright then produces the following asymptotic formula for $p_k(n)$ which is obtained by examining in detail the singularity of $f(x)$ at $x=1$:

$$p_k(n)=(n+j)^{-\frac{3}{2}}(2\pi)^{-\frac{k}{2}}\varphi(\Gamma(1+a)\zeta(1+a)(n+j)^a)+O\left(e^{(\Delta-\alpha)n^b}\right), \tag{4.5}$$

where

$$\Delta=(k+1)(a\Gamma(1+a)\zeta(1+a))^{1-b},\quad \alpha=\alpha(k)>0, \tag{4.6}$$

and $\zeta(s)$ is the Riemann zeta function.

Now the important point here is that the error term in (4.5) is of *exponentially* lower order of magnitude than the main term.

Wright accomplishes this by proving a complete generalization of the modular transformation formulae for Dedekind's eta function.

Let p, q, h, s and l be integers with $1\le p<q$, $(p,q)=1$, except that, when $q=1$, $p=0$,

$$1\le h\le q,\quad 1\le s\le k,\quad l\ge 0.$$

Define d_h by

$$ph^k\equiv d_h\ (\mathrm{mod}\ q),\ 0\le d_h<q.$$

If $d_h\neq 0$, we write

$$\mu_{h,s}=\frac{d_h}{q}(s\text{ odd}),\quad \mu_{h,s}=\frac{q-d_h}{q}\ (s\text{ even}). \tag{4.7}$$

If $d_h=0$, we take $\mu_{h,s}=1$. Hence always $q\mu_{h,s}\ge 1$.

In connection with any particular values of p and q we write

$$X = xe_q(-p) = e^{-y}, \quad Y = q^k y, \tag{4.8}$$

and we take y real and positive when X is real and $0 < X < 1$. We write also

$$t_s = \left(\frac{2\pi}{Y}\right)^a \exp\left\{a\,\pi\, i\left(s - \frac{1}{2}\right)\right\}, \tag{4.9}$$

where Y^a is that k-th root of Y which is real and positive when Y is real and positive,

$$\begin{aligned} g(h, l, s) &= \exp\left\{2\pi\, i(l + \mu_{h,s})^a t_s\right\} e_q(-h) \\ &= \exp\left\{\frac{(2\pi)^{1+a}(l + \mu_{h,s})^a e^{\frac{1}{2}\pi a i(2s+k-1)}}{q y^a} - \frac{2h\pi\, i}{q}\right\}, \end{aligned} \tag{4.10}$$

and

$$P_{p,q} = \prod_{h=1}^{q}\prod_{s=1}^{k}\prod_{l=0}^{\infty}\{1 - g(h, l, s)\}^{-1}. \tag{4.11}$$

Then we have:

Wright's Transformation Theorem [*44, p. 149, Theorem 4*]. *If $\Re\,(y) > 0$, then $P_{p,q}$ is convergent and*

$$f(x) = f(e^{-y} e_q(p)) = C_{p,q}\, y^{\frac{1}{2}} e^{jy} \exp\left(\frac{\Lambda_{p,q}}{y^a}\right) P_{p,q}, \tag{4.12}$$

where $C_{p,q}$ and $\Lambda_{p,q}$ are explicitly given constants depending on p and q.

In the case $k = 1$, (4.12) reduces to the modular function transformation used by Hardy and Ramanujan [20].

In addition, various instances of $k = 2$ appear in [20] and have been extended by Baxter [9] in his proof of the Doochel Kim conjectures [23].

There are a couple of things to say about Wright's theorems. First, how large is the class of partition generating functions for which there are analogs of Wright's Transformation Theorem? Second, how good is the full analog of the Hardy-Ramanujan-Rademacher formula for $p(n)$ that Wright gives for $p_k(n)$ (Theorem 3 of [44] but not restated here). Wright suggests that each of his main terms has order

$$n^{c_1} e^{c_2 n^{\frac{1}{k+1}}}.$$

He goes on to compare his error term to the famous $O(n^{-\frac{1}{4}})$ that appears in the Hardy-Ramanujan formula for $p(n)$ [20]. "If we attempt to make a similar improvement for $p_k(n)$, we find that we can choose $q_0 = q_k(k, n)$ so that the error term is $O(e^{n^d})$, where $d < b$.

This is not so good as the result for $k = 1$, and, in view of the heavy analysis required for this further step, I am content to prove $[O(e^{\epsilon n^b})]$."

One would hope that the advances on closely related work on modular forms would allow a complete analysis of this error term. Does it tend to 0 as $n \to \infty$? If not, is it bounded by polynomial growth? If not, what then?

5. The power series coefficients of Mock theta functions

In his last letter to Hardy [34, pp. 127–131], Ramanujan first presents his mock theta functions. Hardy [33, p. 354] summarizes the idea of a mock theta function: It is a function, "defined by a q-series convergent for $|q| < 1$, for which we can calculate asymptotic formulae, when q tends to a "rational point $e^{2\pi ir/s}$, of the same degree of precision as those furnished, for the ordinary ϑ-functions, by the theory of linear transformation." Subsequently, starting with G.N. Watson's famous L.M.S. Presidential address [40], there have been a number of studies of mock theta functions. Much of this work was summarized in a survey article [5]. Ramanujan focuses upon one mock theta function in particular:

$$f(q) = 1 + \sum_{n=1}^{\infty} \frac{q^{n^2}}{(1+q)^2(1+q^2)^2 \dots (1+q^n)^2}. \tag{5.1}$$

He asserts that: "The coefficient of q^n in $f(q)$ is

$$\frac{(-1)^{n-1} e^{\pi(\frac{n}{6} - \frac{1}{144})^{\frac{1}{2}}}}{2(n - \frac{1}{24})^{\frac{1}{2}}} + O\left(\frac{e^{\frac{\pi}{2}(\frac{n}{6} - \frac{1}{144})^{\frac{1}{2}}}}{(n - \frac{1}{24})^{\frac{1}{2}}}\right). \tag{5.2}$$

In 1964, building on the work of Watson [40] and Dragonette [15], I proved [2] a theorem equivalent to the assertion that if

$$f(q) = \sum_{n=0}^{\infty} \alpha(n) q^n, \tag{5.3}$$

then

$$\alpha(n) = \sum_{k=1}^{\lfloor n^{\frac{1}{2}} \rfloor} \frac{(-1)^{\lfloor \frac{k+1}{2} \rfloor} A_{2k}(n - k(1 + (-1)^k)/4) \sinh\left(\frac{\pi}{k}\left(\frac{n}{6} - \frac{1}{144}\right)^{\frac{1}{2}}\right)}{k^{\frac{1}{2}}\left(n - \frac{1}{24}\right)^{\frac{1}{2}}} + O(n^{\epsilon}) \tag{5.4}$$

[2, p. 455, Theorem 5.1]. The $A_k(n)$ is the exponential sum that appears in the formula for $p(n)$ [4, p. 70].

However, earlier Dragonette [15, p. 494], [2, p. 456] had suggested that in fact the actual error might indeed be smaller than $\frac{1}{2}$ in absolute value. If so, then the main sum rounded to the nearest integer would give $\alpha(n)$ exactly. Dragonette supported this possibility with calculations at $n = 100$ and 200.

Indeed it is possible that the main sum *converges* when extended to infinity and, if so, may actually equal $\alpha(n)$ [2, p. 456]. The following table provides further evidence:

n	$\alpha(n)$	The main sum in (5.4)	The main sum in (5.4) extended to n terms
100	−18520	−18520.18	−18520.01
200	−2660008	−2660008.01	−2660008.05
300	−128045286	−128045285.83	−128045285.99
400	−3447212602	−3447212601.86	−34472122602.05
500	−63676485905	−63676485905.01	−63676485905.06
600	−897840541970	−897840541969.91	−897840541969.98
700	−10302538222405	−10302538222405.14	−10302538222405.02
800	−100343014357869	−100343014357869.11	−100343014357869.01
900	−854282301584078	−854282301584078.03	−854282301584078.02
1000	−649583638177105	−649583638177105.004	−649583638177105.105

While three of the errors of $\sqrt{n}$ terms are smaller than the corresponding errors for n terms, it is nonetheless the case that the average error in column 2 is 0.0884 and in column 4 it is 0.0355 which offers some support to our conjecture that the series in (5.4) actually converges to $\alpha(n)$.

Indeed, in order to prove convergence of the series, it would suffice to prove that

$$\sum_{k=1}^{\infty} \frac{(-1)^{\lfloor \frac{k+1}{2} \rfloor} A_{2k}(n - k(1 + (-1)^k)/4)}{k^{\frac{3}{2}}}$$

converges. This in turn asks for an analysis of Dirichlet series such as

$$A(n, s) = \sum_{k=1}^{\infty} \frac{(-1)^{\lfloor \frac{k+1}{2} \rfloor} A_{2k}(n - k(1 + (-1)^k)/4)}{k^s} \tag{5.6}$$

which can easily be shown to converge absolutely for Re $s > \frac{3}{2}$.

6. Modular transformations and q-hypergeometric series

Let us state an example of the type of problem that concerns us here.

Problem. Prove directly that if

$$P(q) = 1 + \sum_{n=1}^{\infty} \frac{q^{n^2}}{(1-q)^2(1-q^2)^2 \dots (1-q^n)^2}, \tag{6.1}$$

then for $\Im(\tau) > 0$

$$P(e^{2\pi i\tau}) = e^{\frac{-\pi i}{4}}\tau^{\frac{1}{2}}q^{\frac{\pi i}{12}(\tau+\frac{1}{\tau})}P\big(e^{-\frac{2\pi i}{\tau}}\big). \tag{6.2}$$

The catch in this problem is the single word "directly." Otherwise the problem is merely a reformulation of the transformation of Dedekind's η-function because [4, p. 70]

$$P(q) = \frac{q^{\frac{1}{24}}}{\eta(\tau)} = \prod_{n=1}^{\infty}\frac{1}{1-q^n} \tag{6.3}$$

where $q = e^{2\pi i\tau}$.

None of the methods which have been used to prove (6.3) seems to adapt to a direct treatment of the series in (6.1).

The only work I know of on this topic is by L. Ehrenpreis. His observations are presented in [16]; see especially Section 3 thereof. Of his work, he says: "This method sheds light on the question of why the generating function for Rogers-Ramanujan is an automorphic function and why automorphicity is difficult to prove."

7. Conclusion

The five topics chosen for this paper are tightly related. The problems presented do not fall naturally into the home turf of either modular forms or q-series. Consequently they have received little attention. However, they are calling out for an attack by a joint effort of these two areas.

References

1. H.L. Alder, "Partition identities—from Euler to the present," *Amer. Math. Monthly* **76** (1969), 733–746.
2. G.E. Andrews, "On the theorems of Watson and Dragonette for Ramanujan's mock theta functions," *Amer. J. Math* **88** (1966), 454–490.
3. G.E. Andrews, "Partition identities," *Advances in Math.* **9** (1972), 10–51.
4. G.E. Andrews, *The Theory of Partitions*, Addison-Wesley, Reading 1976; reprinted, Cambridge University Press, Cambridge, 1984, 1998.
5. G.E. Andrews, "Mock theta functions," *Proc. Symp. in Pure Math* **49** (1989), 283–298.
6. J. Arkin, "Researches on partitions," *Duke Math. J.* **38** (1970), 403–409.
7. A.O.L. Atkin, "Proof of a conjecture of Ramanujan," *Glasgow Math. J.* **8** (1967), 14–32.
8. P. Bateman and P. Erdös, "Monotonicity of partition functions," *Mathematika* **3** (1956), 1–14.
9. R.J. Baxter, "A direct proof of Kim's identities," *J. Phys. A: Math. Gen.* **31** (1998), 1105–1108.
10. B.C. Berndt, *Ramanujan's notebooks*, Part IV, Springer Verlag, New York, 1994.
11. B.C. Berndt, P.B. Bialek, and A.J. Yee, "Formulas of Ramanujan for the power series coefficients of certain quotients of eisenstein series," *Inter. Math. Res. Not.* **21** (2002), 1077–1109.
12. B.C. Berndt and K. Ono, "Ramanujan's unpublished manuscript on the partition and tau functions with proofs and commentary," (Andrews Festschrift, D. Foata, and G.-N. Han, eds.), Springer Verlag, Berlin, 2001, pp. 39–110.
13. A. Cayley, "Researches on the partition of numbers," *Phil. Trans. Royal Soc.* **146** (1856), 127–140, (Reprinted: *Coll. Math. Papers*, **2** (1889), 235–249).

14. A. DeMorgan, "On a new form of difference equation," *Cambridge Math. J.* **4** (1843), 87–90.
15. L. Dragonette, "Some asymptotic formulae for the mock theta series of Ramanujan," *Trans. Amer. Math. Soc.* **72** (1952), 474–500.
16. L. Ehrenpreis, "Function theory for Rogers-Ramanujan-like partition identities," *Contemporary Math* **143** (1993), 259–320.
17. J.W.L. Glaisher, "On the number of partitions of a number of partitions of a number into a given number of parts," *Quart. J. Pure and Appl. Math.* **40** (1908), 57–143.
18. K. Glösel, "Über die Zerlegung der ganzen Zahlen," *Monatschefte Math. Phys.* **7** (1896), 133–141.
19. H. Gupta, E.E. Gwyther, and J.C.P. Miller, *Tables of Partitions*, Royal Soc. Math. Tables, Cambridge University Press, Cambridge, 1958, Vol. 4.
20. G.H. Hardy and S. Ramanujan, "Asymptotic formulae in combinatory analysis," *Proc. London Math. Soc.* **17**(2) (1918), 75–115.
21. J.F.W. Herschel, "On circulating functions and on the integration of a class of equations of finite differences into which they enter as coefficients," *Phil. Trans. Royal Soc. London* **108** (1818), 144–168.
22. A.E. Ingham, "A Tauberian theorem for partitions," *Annals of Math.* **42** (1941), 1075–1090.
23. D. Kim, "Asymmetric XXZ chain at the antiferromagnetic transition: Spectra and partition functions," *J. Phys. A: Math. Gen.* **30** (1996), 3817–3836.
24. R. McIntosh, "Some asymptotic formulae for q-hypergeometric series," *J. London Math. Soc.* **51**(2) (1995), 120–136.
25. P.A. MacMahon, *Combinatory Analysis*, Cambridge University Press, London, 1916, Vol. 2. reprinted Chelsea, New York, 1960.
26. E. Netto, *Lehrbuch der Combinatorik* 2nd edn., Teubner, Berlin, 1927; reprinted: Chelsea, New York, 1958.
27. J.L. Nicolas and A. Sárközy, "On the asymptotic behavior of general partition functions," *Ramanujan Journal* **4** (2000), 29–39.
28. K. Ono, "On the parity of the partition function in arithmetic progressions," *J. für die r. und a. Math.* **472** (1996), 1–16.
29. K. Ono, "Distribution of the partition function modulo m," *Annals of Math.* **151** (2000), 293–307.
30. P. Paoli, *Opuscula Analytica*, Liburni, 1780, Opusc. II (Meditations Arith.), Section 1.
31. H. Rademacher, *Lectures on Elementary Number Theory*, Blaisdell, New York, 1964.
32. H. Rademacher, *Topics in Analytic Number Theory*, Springer, New York, 1973.
33. S. Ramanujan, *Collected Papers*, Cambridge University Press, London, 1927; reprinted: A.M.S. Chelsea, 2000 with new preface and extensive commentary by B. Berndt.
34. S. Ramanujan, *The Lost Notebook and Other Unpublished Papers*, Narosa, New Delhi, 1988.
35. L.B. Richmond, "A general asymptotic result for partitions," *Canadian J. Math.* **27** (1975) 1083–1091.
36. K.F. Roth and G. Szekeres, "Some asymptotic formulae in the theory of partitions," *Quant. J. Math. Oxford Series* **5**(2) (1954), 241–259.
37. I. Schur, *Ein Beitrag zur additiven Zahlentheorie und zur Theorie der Kettenbrüche*, S.-B. Preuss. Akad. Wiss., Phys.-Math. Kl., 1926, 488–495.
38. J.J. Sylvester, "On subinvariants, i.e. semi-invariants to binary quantics of an unlimited order. With an excursus on rational fractions and partitions," *Amer. J. Math.* **5** (1882), 79–136.
39. G. Szekeres, "An asymptotic formula in the theory of partitions, II," *Quart. J. Math. Oxford Series* **2**(2) (1951), 85–108.
40. G.N. Watson, "The final problem: An account of the mock theta functions," *J. London Math. Soc.* **11** (1936), 55–80.
41. G.N. Watson, "Ramanujans Vermutung über Zerfälllungsanzahlen," *J. reine und angew. Math.* **179** (1938), 97–128.
42. E.M. Wright, "Asymptotic partition formulae, I. plane partitions," *Quart. J. Math., Oxford Series* **2** (1931), 177–189.
43. E.M. Wright, "Asymptotic partition formulae, II. Weighted partitions," *Proc. London Math. Soc.* **36**(2) (1932), 117–141.
44. E.M. Wright, "Asymptotic partitions formulae, III. Partitions into k-th powers," *Acta Math.* **63** (1934), 143–191.

7.3.1 Errata

Sills and Zeilberger pointed out in [Sil-Z12] that Rademacher's [Rad73] stated values for $C_{011}(\infty), C_{012}(\infty)$, and $C_{121}(\infty)$ contained some errors, which were duplicated on page 388 of the article reproduced in section 7.3. Specifically, the decimal approximation for $C_{011}(\infty)$ should be $-0.29292757\ldots$; the exact value is stated correctly. The correct value for $C_{012}(\infty)$ is $\frac{144}{1225} + \frac{5616}{42875\pi} = 0.1897670688\ldots$. Finally, the stated value of $C_{112}(\infty)$ is exactly one half the correct value.

7.4 Applications of Basic Hypergeometric Functions

SIAM REVIEW
Vol. 16, No. 4, October 1974

APPLICATIONS OF BASIC HYPERGEOMETRIC FUNCTIONS*

GEORGE E. ANDREWS†

Abstract. This paper surveys recent applications of basic hypergeometric functions to partitions, number theory, finite vector spaces, combinatorial identities and physics.

1. Introduction. The theory of basic hypergeometric functions has until recently existed very much in the shadow of ordinary hypergeometric functions. The basic hypergeometric function is

$$ {}_m\phi_n\begin{bmatrix} a_1, \cdots, a_m; q, z \\ b_1, \cdots, b_n \end{bmatrix} = \sum_{j \geqq 0} \frac{(a_1)_j(a_2)_j \cdots (a_m)_j z^j}{(b_1)_j(b_2)_j \cdots (b_n)_j(q)_j}, \tag{1.1} $$

where $(a)_n = (a;q)_n = (1-a)(1-aq)\cdots(1-aq^{n-1})$, and $|z| < 1$, $|q| < 1$, $b_i \neq q^{-n}$ for any nonnegative integer n.

The applications of the ordinary

$$ {}_mF_n\begin{bmatrix} a_1, \cdots, a_m; z \\ b_1, \cdots, b_n \end{bmatrix} = \sum_{j \geqq 0} \frac{z^j}{j!} \prod_{s=0}^{j-1} \frac{(a_1+s)(a_2+s)\cdots(a_m+s)}{(b_1+s)(b_2+s)\cdots(b_n+s)} \tag{1.2} $$

are so widespread that a survey of their applications would hardly be possible. Indeed we might summarize their application with the following paragraph of W. W. Sawyer [50, p. 63]:

> Besides the functions that occur in school work, there are many functions used by engineers or physicists—the Legendre polynomials and the Bessel functions, for example—which are particular cases of the hypergeometric function. In fact there must be many universities today where 95 per cent, if not 100 per cent, of the functions studied by physicists, engineering and even mathematics students are covered by this single symbol $F(a,b;c;x)$ $\left[= {}_2F_1\begin{bmatrix} a, b; x \\ c \end{bmatrix}\right]$.

Surprisingly then, basic hypergeometric functions have been for the most part the province of a few specialists. In fact prior to 1960, most of the known results on these functions had been obtained by E. Heine, C. R. Adams, F. H. Jackson, W. N. Bailey, D. B. Sears, W. Hahn, L. J. Slater and R. P. Agarwal. Perhaps the reason for this general neglect lay in the view that the theory of these functions was merely a generalization of the theory of ordinary hypergeometric functions; furthermore, since there were few applications of basic hypergeometric functions known, there was not widespread interest in investigating basic hypergeometric functions.

This paper is devoted to an exposition of several areas of pure and applied mathematics in which basic hypergeometric functions have assumed significant

* Received by the editors May 11, 1973.

† Department of Mathematics, Pennsylvania State University, University Park, Pennsylvania 16802. This work was supported in part by the National Science Foundation under Grant GP-23774.

importance. We shall develop in detail only those properties of basic hypergeometric functions that are important to our applications. Extensive accounts of the general theory of basic hypergeometric functions can be found in books by L. J. Slater [52] and W. N. Bailey [15], and in the papers of W. Hahn [36], [37].

In § 2, we briefly discuss the application of basic hypergeometric functions to partitions; since there are recent surveys of partition theory that delve deeply into this application [4], [9], we shall restrict ourselves to only a few examples that are related to Heine's fundamental transformation of the ${}_2\phi_1$.

In § 3, we shall show how bilateral basic hypergeometric series have application in number theory (apart from partitions). In particular, we shall present a reasonably elementary proof of W. N. Bailey's summation of the well-poised ${}_6\psi_6$. This summation formula has many interesting applications in number theory; from it we shall derive Jacobi's triple product identity, the quintuple product identity, Ramanujan's congruence $p(5n + 4) \equiv 0 \pmod 5$ and the Jacobi formulas for $r_s(n)$, $s = 2, 4$ and 8, where $r_s(n)$ denotes the number of representations of n as a sum of s squares.

In § 4, we shall discuss the relationship between finite vector spaces and basic hypergeometric functions. In this section, we shall briefly describe the theory of Eulerian differential operators which arose from the combinatorial theory of finite vector spaces and which has interesting applications to basic hypergeometric series.

In § 5. we shall discuss the usefulness of both ordinary and basic hypergeometric functions in the proof of and classification of combinatorial identities such as those discussed in the books of J. Riordan [46] and H. W. Gould [35].

In § 6, we shall describe some of the recent applications of basic hypergeometric functions in physics. We shall relate in detail W. Hahn's application of ${}_m\phi_n$ to mechanics.

Neither this survey nor the bibliography is exhaustive. Each section of the paper provides only a sample of the work done in each area. It is my hope that these samples will be adequate to indicate the breadth of application of basic hypergeometric functions, and the bibliography should at least provide the basic leads to the extensive literature of this area.

2. Application to the theory of partitions. The starting point of this application lies in the following three simple theorems.

THEOREM 2.1. *Let $p_N(m, n)$ denote the number of partitions of n into m parts, each of which does not exceed N. Then*

$$\sum_{n=0}^{\infty} \sum_{m=0}^{\infty} p_N(m, n) z^m q^n = \frac{1}{(zq)_N}, \qquad |z| < |q|^{-1}, \quad |q| < 1.$$

Proof.

$$\begin{aligned} \frac{1}{(zq)_N} &= \prod_{i=1}^{N} \frac{1}{1 - zq^i} \\ &= \prod_{i=1}^{N} \sum_{m_i=0}^{\infty} z^{m_i} q^{m_i \cdot i} \qquad \text{(by geometric series)} \\ &= \sum_{m_1=0}^{\infty} \sum_{m_2=0}^{\infty} \cdots \sum_{m_N=0}^{\infty} z^{m_1 + m_2 + \cdots + m_N} q^{m_1 \cdot 1 + m_2 \cdot 2 + \cdots + m_N \cdot N}. \end{aligned}$$

Hence the term $z^m q^n$ arises in the above series as many times as there are non-negative solutions to the Diophantine equation $n = m_1 \cdot 1 + m_2 \cdot 2 + \cdots + m_N \cdot N$ subject to the restriction $m_1 + m_2 + \cdots + m_N = m$. But this Diophantine equation may be viewed as a partition of n in which 1 appears m_1 times, 2 appears m_2 times, and so on, and where the total number of parts in the partition is just m. Thus the N-fold series given above equals

$$\sum_{n=0}^{\infty} \sum_{m=0}^{\infty} p_N(m, n) z^m q^n.$$

The conditions $|q| < 1$ and $|z| < |q|^{-1}$ guarantee absolute convergence of all N series $\sum z^{m_i} q^{m_i \cdot i}$; consequently our rearrangements of series were admissible. □

THEOREM 2.2. *Let $Q_N(m, n)$ denote the number of partitions of n into m distinct parts, each of which does not exceed N. Then*

$$\sum_{n=0}^{\infty} \sum_{m=0}^{\infty} Q_N(m, n) z^m q^n = (-zq)_N.$$

Sketch of proof.

$$\begin{aligned}(-zq)_N &= \prod_{i=1}^{N} (1 + zq^i) \\ &= \prod_{i=1}^{N} \sum_{m_i=0}^{1} z^{m_i} q^{m_i \cdot i} \\ &= \sum_{m_1=0}^{1} \sum_{m_2=0}^{1} \cdots \sum_{m_N=0}^{1} z^{m_1+m_2+\cdots+m_N} q^{m_1 \cdot 1 + m_2 \cdot 2 + \cdots + m_N \cdot N} \\ &= \sum_{n=0}^{\infty} \sum_{m=0}^{\infty} Q_N(m, n) z^m q^n. \quad \square\end{aligned}$$

Several remarks are in order about extensions of these theorems. First we note that they remain valid if we let $N \to \infty$ (for a justification of this, see [10, Chap. 13]). Therefore, with $p(m, n) = p_\infty(m, n)$ and $Q(m, n) = Q_\infty(m, n)$, we see that

$$\sum_{n=0}^{\infty} \sum_{m=0}^{\infty} p(m, n) z^m q^n = \frac{1}{(zq)_\infty}, \qquad |z| < |q|^{-1}, \quad |q| < 1, \tag{2.1}$$

$$\sum_{n=0}^{\infty} \sum_{m=0}^{\infty} Q(m, n) z^m q^n = (-zq)_\infty, \qquad |q| < 1. \tag{2.2}$$

Also the same arguments may be applied mutatis mutandis to $p_N(S; m, n)$ and $Q_N(S; m, n)$, where S is some set of positive integers and $p_N(S; m, n)$ (resp. $Q_N(S; m, n)$) is the number of partitions of the type enumerated by $p_N(m, n)$ (resp. $Q_N(m, n)$) with the added condition that all the parts of the partition are in S. In this way, we see that

$$\sum_{n=0}^{\infty} \sum_{m=0}^{\infty} P_N(S; m, n) = \prod_{\substack{j \in S \\ j \leq N}} (1 - zq^j)^{-1}, \qquad |z| < |q|^{-1}, \quad |q| < 1, \tag{2.3}$$

$$\sum_{n=0}^{\infty}\sum_{m=0}^{\infty} Q_N(S;m,n) = \prod_{\substack{j\in S\\ j\leqq N}} (1+zq^j), \tag{2.4}$$

$$\sum_{n=0}^{\infty}\sum_{m=0}^{\infty} P(S;m,n) = \prod_{j\in S} (1-zq^j)^{-1}, \qquad |z|<|q|^{-1},\quad |q|<1, \tag{2.5}$$

$$\sum_{n=0}^{\infty}\sum_{m=0}^{\infty} Q(S;m,n) = \prod_{j\in S} (1+zq^j), \qquad |q|<1. \tag{2.6}$$

The next theorem allows the pasting together of different partition generating functions to produce more complex generating functions.

THEOREM 2.3. *Let $P_j(B_j, S_j; m, n)$ ($j = 1$ or 2) denote the number of partitions of n into m parts, where all the parts lie in S_j and each partition is subject to the condition B_j. Let $P_3(B_3, S_3; m, n)$ denote the number of partitions π of n into m parts, where $S_3 = S_1 \cup S_2$ and B_3 is the condition that the two subpartitions of π made up of parts belonging to S_1 (resp. S_2) satisfy condition B_1 (resp. B_2). Finally we let*

$$f_j(z,q) = \sum_{n=0}^{\infty}\sum_{m=0}^{\infty} P_j(B_j, S_j; m, n) z^m q^n,$$

where $|z| < |q|^{-1}$, $|q| < 1$. Then if $S_1 \cap S_2 = \varnothing$,

$$f_3(z,q) = f_1(z,q) f_2(z,q). \tag{2.7}$$

Proof. Absolute convergence of the series to be considered follows by comparison with the series in (2.1).

Suppose now that π is a partition of the type enumerated by $P_3(B_3, S_3; m, n)$. Then since $S_1 \cap S_2 = \varnothing$, π may be written uniquely in the form

$$a_1 + \cdots + a_{m_1} + b_1 + \cdots + b_{m_2}$$

where $n_1 = a_1 + \cdots + a_{m_1}$ is a partition of the type enumerated by $P_1(B_1, S_1; m_1, n_1)$ and $n_2 = b_1 + \cdots + b_{m_2}$ is a partition of the type enumerated by $P_2(B_2, S_2; m_2, n_2)$. Consequently $P_3(B_3, S_3; m, n)$, the total number of such partitions, is given by

$$\sum_{\substack{m_1+m_2=m\\ n_1+n_2=n\\ m_1,m_2,n_1,n_2\geqq 0}} P_1(B_1, S_1; m_1, n_1) P_2(B_2, S_2; m_2, n_2),$$

and this establishes that the coefficients on either side of (2.7) are identical. □

We remark that Theorem 2.3 is an example of a large number of theorems of this nature. The disjointness of S_1 and S_2 is the essential feature that makes the theorem work.

We hope now to illustrate how partitions and basic hypergeometric functions interact. Theorem 2.4 is a basic hypergeometric function identity with a partition-theoretic proof. Theorems 2.5–2.8 are analytic identities that are corollaries of Theorem 2.4. The final three theorems are partition-theoretic consequences of Theorem 2.8.

THEOREM 2.4. *For $|z| < 1$, $|q| < 1$,*

$${}_1\phi_0[a; q, z] = \sum_{n=0}^{\infty} \frac{(a)_n z^n}{(q)_n} = \frac{(az)_\infty}{(z)_\infty}.$$

Proof. We shall prove the equivalent identity:

$$\sum_{n=0}^{\infty} \frac{(-aq;q^2)_n z^n q^{2n}}{(q^2;q^2)_n} = \frac{(-azq^3;q^2)_\infty}{(zq^2;q^2)_\infty}. \tag{2.8}$$

Let $W_1(r, m, n)$ denote the number of partitions of n into m parts, of which exactly r are odd, subject to the condition that no odd parts are repeated and all odd parts are larger than 1. By splitting such partitions into two subpartitions containing odd parts (resp. even parts), we see that (2.5), (2.6) and Theorem 2.3 (slightly modified to account for the extra parameter r) imply

$$\sum_{n=0}^{\infty} \sum_{m=0}^{\infty} \sum_{r=0}^{\infty} W_1(r, m, n) a^r z^m q^n = (-azq^3;q^2)_\infty \cdot \frac{1}{(zq^2;q^2)_\infty} = \frac{(-azq^3;q^2)_\infty}{(zq^2;q^2)_\infty}.$$

Let $W_2(r, m, n)$ denote the number of partitions of n with largest part $2m$, with exactly r parts odd, and with the condition that no odd parts are repeated. Arguments of the types used in Theorems 2.1–2.3 may be easily seen to establish that

$$\sum_{n=0}^{\infty} \sum_{r=0}^{\infty} W_2(r, m, n) a^r q^n = (-aq;q^2)_m \cdot \frac{q^{2m}}{(q^2;q^2)_m} = \frac{(-aq;q^2)_m q^{2m}}{(q^2;q^2)_m}.$$

Therefore

$$\sum_{n=0}^{\infty} \sum_{m=0}^{\infty} \sum_{r=0}^{\infty} W_2(r, m, n) a^r z^m q^n = \sum_{m=0}^{\infty} \frac{(-aq;q^2)_m z^m q^{2m}}{(q^2;q^2)_m}.$$

Thus we see that (2.8) is equivalent to the following partition function identity:

$$W_1(r, m, n) = W_2(r, m, n). \tag{2.9}$$

To establish (2.9), we must utilize a modification of the Ferrars graph of a partition [39, p. 273]. Let π be a typical partition of the type enumerated by $W_2(r, m, n)$. Each even part 2ν of π is to be represented in the graph by two rows of ν nodes each, and each odd part $2\nu - 1$ of π is to be represented by one row of ν nodes and one row of $\nu - 1$ nodes. Thus $8 + 8 + 5$ is represented graphically by

We note that the graph of π contains n nodes, has m columns, has exactly r columns with an odd number of nodes. Furthermore, no two columns which each have an odd number of nodes can have the same number of nodes. Finally, every column must contain at least two nodes since the largest part of π was even. Let $\pi^1 : b_1 + \cdots + b_m$ denote the partition of n in which b_i is the number of nodes in the

ith column of the graphical representation of π. From the above comments we see that π^1 is a partition of the type enumerated by $W_1(r, m, n)$, and indeed since the above procedure is reversible it establishes a bijection between the partitions enumerated by $W_2(r, m, n)$ and those enumerated by $W_1(r, m, n)$. Hence (2.9) is established, and so therefore are (2.8) and the theorem. □

THEOREM 2.5 (Heine [40, (5a), p. 106]). *Subject to obvious convergence conditions,*

$$ {}_2\phi_1\begin{bmatrix} a, b; q, z \\ c \end{bmatrix} = \frac{(b)_\infty (az)_\infty}{(c)_\infty (z)_\infty} {}_2\phi_1\begin{bmatrix} c/b, z; q, b \\ az \end{bmatrix}. $$

Remark. The phrase "subject to obvious convergence conditions" means here $|z| < 1$, $|b| < 1$, $|q| < 1$, $c \neq q^{-n}$, $z \neq q^{-n}$, $az \neq q^{-n}$, for any nonnegative integer n. In general, conditions like $c \neq q^{-n}$ are needed to keep zeros out of denominators, while the conditions $|z| < 1$, $|b| < 1$, $|q| < 1$ guarantee absolute convergence of all the series and products appearing. For the most part, such conditions as these will be tacitly assumed; only in Theorem 3.3 will convergence problems be difficult enough to require mention.

Proof.

$$
\begin{aligned}
{}_2\phi_1\begin{bmatrix} a, b; q, z \\ c \end{bmatrix} &= \sum_{n=0}^{\infty} \frac{(a)_n (b)_n z^n}{(q)_n (c)_n} \\
&= \frac{(b)_\infty}{(c)_\infty} \sum_{n=0}^{\infty} \frac{(a)_n z^n}{(q)_n} \cdot \frac{(cq^n)_\infty}{(bq^n)_\infty} \\
&= \frac{(b)_\infty}{(c)_\infty} \sum_{n=0}^{\infty} \frac{(a)_n z^n}{(q)_n} \sum_{m=0}^{\infty} \frac{(c/b)_m}{(q)_m} b^m q^{nm} \qquad \text{(by Theorem 2.4)} \\
&= \frac{(b)_\infty}{(c)_\infty} \sum_{m=0}^{\infty} \frac{(c/b)_m b^m}{(q)_m} \cdot \frac{(azq^m)_\infty}{(zq^m)_\infty} \qquad \text{(by Theorem 2.4)} \\
&= \frac{(b)_\infty (az)_\infty}{(c)_\infty (z)_\infty} \sum_{m=0}^{\infty} \frac{(c/b)_m (z)_m b^m}{(q)_m (az)_m} \\
&= \frac{(b)_\infty (az)_\infty}{(c)_\infty (z)_\infty} {}_2\phi_1\begin{bmatrix} c/b, z; q, b \\ az \end{bmatrix}. \quad \square
\end{aligned}
$$

Our next result is the q-analogue of Gauss's summation formula [15, p. 2] for ${}_2F_1\begin{bmatrix} a, b; 1 \\ c \end{bmatrix}$.

THEOREM 2.6 (Heine [40, (6), p. 107]).

$$ {}_2\phi_1\begin{bmatrix} a, b; q, c/ab \\ c \end{bmatrix} = \frac{(c/a)_\infty (c/b)_\infty}{(c)_\infty (c/ab)_\infty}. $$

Proof. By Theorem 2.5, we see that

$$
\begin{aligned}
{}_2\phi_1\begin{bmatrix} a, b; q, c/ab \\ c \end{bmatrix} &= \frac{(b)_\infty (c/b)_\infty}{(c)_\infty (c/ab)_\infty} {}_1\phi_0[c/ab; q, b] \\
&= \frac{(b)_\infty (c/b)_\infty (c/a)_\infty}{(c)_\infty (c/ab)_\infty (b)_\infty} \qquad \text{(by Theorem 2.4)} \\
&= \frac{(c/a)_\infty (c/b)_\infty}{(c)_\infty (c/ab)_\infty}. \quad \square
\end{aligned}
$$

The following result is less well known than Theorem 2.6 and was found independently in the 1940's by W. N. Bailey [16] and J. A. Daum [29]; this theorem is the q-analogue of Kummer's summation [52, III. 5, p. 243] of ${}_2F_1\begin{bmatrix} a, b; -1 \\ a+1-b \end{bmatrix}$.

THEOREM 2.7.

$$
{}_2\phi_1\begin{bmatrix} a, b; q, -q/b \\ qa/b \end{bmatrix} = \frac{(aq; q^2)_\infty (-q)_\infty (q^2a/b^2; q^2)_\infty}{(qa/b)_\infty (-q/b)_\infty}.
$$

Proof.

$$
\begin{aligned}
{}_2\phi_1\begin{bmatrix} a, b; q, -q/b \\ qa/b \end{bmatrix} &= {}_2\phi_1\begin{bmatrix} b, a; q, -q/b \\ qa/b \end{bmatrix} \\
&= \frac{(a)_\infty (-q)_\infty}{(qa/b)_\infty (-q/b)_\infty} \sum_{n=0}^{\infty} \frac{(q/b)_n (-q/b)_n a^n}{(q)_n (-q)_n} \qquad \text{(by Theorem 2.5)} \\
&= \frac{(a)_\infty (-q)_\infty}{(qa/b)_\infty (-q/b)_\infty} \sum_{n=0}^{\infty} \frac{(q^2/b^2; q^2)_n a^n}{(q^2; q^2)_n} \\
&= \frac{(a)_\infty (-q)_\infty (aq^2/b^2; q^2)_\infty}{(qa/b)_\infty (-q/b)_\infty (a; q^2)_\infty} \qquad \text{(by Theorem 2.4)} \\
&= \frac{(aq; q^2)_\infty (-q)_\infty (aq^2/b^2; q^2)_\infty}{(qa/b)_\infty (-q/b)_\infty}. \quad \square
\end{aligned}
$$

COROLLARY 2.7.1.

$$
\sum_{n=0}^{\infty} \frac{(a)_n q^{n(n+1)/2}}{(q)_n} = (aq; q^2)_\infty (-q)_\infty.
$$

Proof. The identity in question follows directly from letting $b \to \infty$ in Theorem 2.7. To justify the limit we need only observe that if we replace b by $1/\beta$ in Theorem 2.7, each side of the identity holds for β in a deleted neighborhood of zero. Since both sides are continuous functions of β at zero, both sides must be equal at $\beta = 0$. Alternatively one may justify this process by an appeal to Tannery's theorem [43, p. 371]. $\square$

Our first application to partitions will be to J. J. Sylvester's generalization of Euler's partition theorem [53, p. 293]. Our proof is based on one recently given by V. Ramamani and K. Venkatachaliengar [45].

THEOREM 2.8. *Let $A_k(n)$ denote the number of partitions of n into odd parts (repetitions allowed) with exactly k different parts in each partition. Let $B_k(n)$ denote the number of partitions of n into distinct parts such that exactly k sequences of consecutive integers appear in each partition. Then for each k and n, $A_k(n) = B_k(n)$.*

Remark. Euler's theorem merely asserts $\sum_{k=0}^{\infty} A_k(n) = \sum_{k=0}^{\infty} B_k(n)$.

Proof. First we note that

$$\begin{aligned}\sum_{k=0}^{\infty}\sum_{n=0}^{\infty} A_k(n)a^kq^n &= \prod_{j=1}^{\infty}(1 + aq^{2j-1} + aq^{2(2j-1)} + aq^{3(2j-1)} + \cdots)\\ &= \prod_{j=1}^{\infty}\left(1 + \frac{aq^{2j-1}}{1-q^{2j-1}}\right)\\ &= \prod_{j=1}^{\infty}\frac{(1+(a-1)q^{2j-1})}{(1-q^{2j-1})}.\end{aligned}$$

The reasoning here exactly parallels that used in Theorem 2.1. Now since

$$\prod_{m=1}^{\infty}(1+q^m) = \prod_{m=1}^{\infty}\frac{(1-q^{2m})}{(1-q^m)} = \prod_{j=1}^{\infty}\frac{1}{1-q^{2j-1}},$$

we see that

$$\sum_{k=0}^{\infty}\sum_{n=0}^{\infty} A_k(n)a^kq^n = (-q)_\infty((1-a)q;q^2)_\infty. \tag{2.10}$$

On the other hand, let us consider the graphical representation of a partition of the type enumerated by $B_k(n)$. For example, $10 + 9 + 5 + 4 + 3 + 1$ is a partition of the type enumerated by $B_3(32)$, and it has the following graphical representation:

```
. . . . . . . . . .
. . . . . . . . .
. . . . .
. . . .
. . .
.
```

If we examine the *conjugate* partition (i.e., the partition obtained by reading the graph vertically), we see that the conjugate partition (namely, $6 + 5 + 5 + 4 + 3 + 2 + 2 + 2 + 2 + 1$) has exactly two parts that are repeated. Indeed if π is any partition of the type enumerated by $B_k(n)$ in which 1 is a summand, then π', the conjugate of π, is a partition of n in which the largest part appears only once, all positive integers not exceeding the largest part appear and exactly $k - 1$ of these appear with repetition.

If the condition that 1 appears in π is removed, then π' is a partition of n in which the largest part appears more than once, all positive integers not exceeding the largest part appear and exactly k parts appear with repetition.

Since the union of the two classes of partitions described above is in one-to-one correspondence with the partitions enumerated by $B_k(n)$, we see that

$$\begin{aligned}
\sum_{k=0}^{\infty}\sum_{n=0}^{\infty} B_k(n)a^k q^n &= 1 + \sum_{N=1}^{\infty} aq^N \prod_{j=1}^{N-1}(q^j + aq^{2j} + aq^{3j} + \cdots) \\
&\quad + \sum_{N=1}^{\infty}\left\{\prod_{j=1}^{N-1}(q^j + aq^{2j} + aq^{3j} + \cdots)\right\}(aq^{2N} + aq^{3N} + \cdots) \\
&= 1 + \sum_{N=1}^{\infty}\left\{\prod_{j=1}^{N-1} q^j\left(1 + \frac{aq^j}{1-q^j}\right)\right\}\frac{aq^N}{1-q^N} \\
&= 1 + \sum_{N=1}^{\infty}\frac{aq^{N(N+1)/2}((1-a)q)_{N-1}}{(q)_N} \\
&= 1 + \sum_{N=1}^{\infty}\frac{(1-(1-a))q^{N(N+1)/2}((1-a)q)_{N-1}}{(q)_N} \\
&= \sum_{N=0}^{\infty}\frac{((1-a))_N q^{N(N+1)/2}}{(q)_N}.
\end{aligned} \tag{2.11}$$

Comparing (2.10) and (2.11), we see that Sylvester's theorem follows immediately from the identity

$$\sum_{N=0}^{\infty}\frac{((1-a))_N q^{N(N+1)/2}}{(q)_N} = (-q)_\infty((1-a)q; q^2)_\infty,$$

which is merely Corollary 2.7.1 with a replaced by $1 - a$. □

Next we consider a result due to H. Göllnitz [31, Satz 2.3, p. 166] that may also be deduced from Corollary 2.7.1.

THEOREM 2.9. *Let $G_1(n)$ denote the number of partitions of n into parts, where each part is congruent to one of* 1, 5, *or* 6 (mod 8). *Let $H_1(n)$ denote the number of partitions of n of the form $b_1 + b_2 + \cdots + b_j$, where $b_i \geqq b_{i+1} + 2$ and strict inequality holds if b_i is odd. Then for each n, $G_1(n) = H_1(n)$.*

Proof. Let $h_a(m, n)$ ($a = 1$ or 2) denote the number of partitions of the type enumerated by $H_1(n)$ with the added restrictions that there be exactly m parts and that each part is $\geqq a$.

We shall now prove two elementary partition identities for the $h_a(m, n)$. First,

$$h_1(m, n) = h_2(m, n) + h_2(m-1, n-2m+1). \tag{2.12}$$

To see this identity, we split the partitions enumerated by $h_1(m, n)$ into two classes: (i) those partitions that contain 1 as a summand, (ii) those that do not. Clearly there are $h_2(m, n)$ elements of class (ii). We now transform the partitions in class (i) by deleting the summand 1 and subtracting 2 from all the remaining parts. This produces a partition of $n - 2(m-1) - 1 = n - 2m + 1$ into $m - 1$ parts, each of which is $\geqq 2$ (since originally the second smallest part was $\geqq 4$); furthermore, since the inequalities between the parts are not disturbed, we see that the transformed partition is of the type enumerated by $h_2(m-1, n-2m+1)$. The above transformation clearly establishes a bijection between the partitions in class (i) and those enumerated by $h_2(m-1, n-2m+1)$. Thus identity (2.12) is established.

By considering those partitions of the type enumerated by $h_2(m, n)$ that contain a 2 and those that do not, we may in exactly the same way prove that

$$h_2(m,n) = h_1(m, n-2m) + h_2(m-1, n-2m). \tag{2.13}$$

Let

$$\eta_a(z,q) = \sum_{n=0}^{\infty}\sum_{m=0}^{\infty} h_a(m,n))z^m q^n$$

Then identity (2.12) implies that

$$\begin{aligned}\eta_1(z,q) &= \sum_{n=0}^{\infty}\sum_{m=0}^{\infty} h_1(m,n)z^m q^n\\ &= \sum_{n=0}^{\infty}\sum_{m=0}^{\infty} (h_2(m,n) + h_2(m-1, n-2m+1))z^m q^n\\ &= \eta_2(z,q) + zq\sum_{n=0}^{\infty}\sum_{m=0}^{\infty} h_2(m-1, n-2m+1)(zq^2)^{m-1}q^{n-2m+1}\\ &= \eta_2(z,q) + zq\eta_2(zq^2, q).\end{aligned} \tag{2.14}$$

Similarly, identity (2.13) implies that

$$\eta_2(z,q) = \eta_1(zq^2, q) + zq^2\eta_2(zq^2, q). \tag{2.15}$$

We now substitute the formula for $\eta_1(z,q)$ from (2.14) into (2.15). Hence

$$\begin{aligned}\eta_2(z,q) &= \eta_2(zq^2, q) + zq^3\eta_2(zq^4, q) + zq^2\eta_2(zq^2, q)\\ &= (1 + zq^2)\eta_2(zq^2, q) + zq^3\eta_2(zq^4, q).\end{aligned} \tag{2.16}$$

Therefore, if $\eta_2(z,q) = \sum_{n=0}^{\infty}\lambda_n z^n$, then by comparing coefficients of z^n on each side of (2.16), we see that $\lambda_n = \lambda_n q^{2n} + \lambda_{n-1}q^{2n} + \lambda_{n-1}q^{4n-1}$. Therefore

$$\lambda_n = \frac{q^{2n}(1 + q^{2n-1})\lambda_{n-1}}{(1-q^{2n})}. \tag{2.17}$$

Iterating (2.17) n times and observing that $\lambda_0 = 1$, we see that

$$\lambda_n = \frac{q^{n(n+1)}(-q;q^2)_n}{(q^2;q^2)_n}. \tag{2.18}$$

Therefore

$$\eta_2(z;q) = \sum_{n=0}^{\infty}\lambda_n z^n = \sum_{n=0}^{\infty}\frac{z^n q^{n(n+1)}(-q;q^2)_n}{(q^2;q^2)_n}; \tag{2.19}$$

and by (2.14), we see that

$$\eta_1(z;q) = \sum_{n=0}^{\infty}\frac{z^n q^{n(n+1)}(-q;q^2)_n(1 + zq^{2n+1})}{(q^2;q^2)_n}. \tag{2.20}$$

From (2.20) we may easily deduce our theorem by utilizing Corollary (2.7.1):

$$\begin{aligned}
&\sum_{n=0}^{\infty} H_1(n)q^n \\
&= \sum_{n=0}^{\infty}\left(\sum_{m=0}^{\infty} h_1(m,n)\right) q^n = \eta_1(1,q) \\
&= \sum_{n=0}^{\infty} \frac{q^{n(n+1)}(-q;q^2)_{n+1}}{(q^2;q^2)_n} \\
&= (1+q)\sum_{n=0}^{\infty} \frac{q^{n(n+1)}(-q^3;q^2)_n}{(q^2;q^2)_n} \\
&= (1+q)(-q^2;q^2)_\infty(-q^5;q^4)_\infty \qquad \text{(by Corollary 2.7.1)} \\
&= \prod_{m=1}^{\infty} (1+q^{2m})(1+q^{4m-3}) \\
&= \prod_{m=1}^{\infty} \frac{(1-q^{4m})(1-q^{8m-6})}{(1-q^{2m})(1-q^{4m-3})} \\
&= \prod_{m=1}^{\infty} \frac{(1-q^{8m-4})(1-q^{8m})(1-q^{8m-6})}{(1-q^{8m-6})(1-q^{8m-4})(1-q^{8m-2})(1-q^{8m})(1-q^{8m-7})(1-q^{8m-3})} \\
&= \prod_{m=1}^{\infty} \frac{1}{(-q^{8m-7})(1-q^{8m-3})(1-q^{8m-2})} \\
&= \sum_{n=0}^{\infty} G_1(n)q^n .
\end{aligned}$$

Consequently $H_1(n) = G_1(n)$ for all n. □

It is now a very simple matter to prove a companion theorem for Theorem 2.9 that is also due to Göllnitz [31, Satz 2.4, p. 167].

THEOREM 2.10. *Let $G_2(n)$ denote the number of partitions of n into parts, where each part is congruent to one of* 2, 3, *or* 7 (mod 8). *Let $H_2(n)$ denote the number of partitions of n of the form $b_1 + b_2 + \cdots + b_j$, where $b_i \geqq b_{i+1} + 2$ and strict inequality holds if b_i is odd; in addition, $b_j \geqq 2$. Then for each n, $G_2(n) = H_2(n)$.*

Proof. From (2.19), we see that

$$\begin{aligned}
\sum_{n=0}^{\infty} H_2(n)q^n &= \sum_{n=0}^{\infty}\left(\sum_{m=0}^{\infty} h_2(m,n)\right) q^n = \eta_2(1,q) \\
&= \sum_{n=0}^{\infty} \frac{q^{n(n+1)}(-q;q^2)_n}{(q^2;q^2)_n} \\
&= (-q^2;q^2)_\infty(-q^3;q^4)_\infty \qquad \text{(by Corollary 2.7.1)} \\
&= \prod_{m=1}^{\infty} \frac{1}{(1-q^{8m-6})(1-q^{8m-5})(1-q^{8m-1})} \\
&= \sum_{n=0}^{\infty} G_2(n)q^n .
\end{aligned}$$

Consequently $G_2(n) = H_2(n)$ for all n. □

This concludes our sample of the relationship between partition identities and basic hypergeometric functions. The reader is referred to [4], [9] and [10, Chaps. 12–14] for a more extensive account of this subject.

3. Application to number theory. In this section we shall present an account of bilateral basic hypergeometric functions, i.e.,

$$ {}_m\psi_n\begin{bmatrix} a_1, a_2, \cdots, a_m; q, t \\ b_1, b_2, \cdots, b_n \end{bmatrix} = \sum_{j=-\infty}^{\infty} \frac{(a_1)_j(a_2)_j \cdots (a_m)_j t^j}{(b_1)_j(b_2)_j \cdots (b_n)_j}, $$

where

$$ (a)_j = (a;q)_j = (a)_\infty (aq^j)_\infty^{-1}, $$

or

$$ (a)_{-n} = \left(1 - \frac{a}{q^n}\right)^{-1} \cdots \left(1 - \frac{a}{q}\right)^{-1} = (-a)^{-n} q^{n(n+1)/2} (q/a)_n^{-1}. $$

Thus

$$ {}_m\psi_n\begin{bmatrix} a_1, \cdots, a_m; q, t \\ b_1, \cdots, b_n \end{bmatrix} $$

$$ = \sum_{j=0}^{\infty} \frac{(a_1)_j \cdots (a_m)_j t^j}{(b_1)_j \cdots (b_n)_j} + \sum_{j=1}^{\infty} \frac{\left(\frac{q}{b_1}\right)_j \cdots \left(\frac{q}{b_n}\right)_j}{\left(\frac{q}{a_1}\right)_j \cdots \left(\frac{q}{a_m}\right)_j} \left(\frac{b_1 \cdots b_n}{a_1 \cdots a_m t}\right)^j (-1)^{j(m-n)} q^{j(j+1)(m-n)/2}. $$

Hence we see that to insure convergence we must require $n \leqq m$. Also $b_h \neq q^{-N}$, $a_h \neq q^{N+1}$ for any nonnegative integer N. Finally if $n < m$, we need only in addition require $|t| < 1$; however, if $n = m$, we need also

$$ \left|\frac{b_1 \cdots b_n}{a_1 \cdots a_m}\right| < |t| < 1. $$

Our first object is to provide a proof of the following identity due to W. N. Bailey [14, (4.7), p. 113]:

$$ (3.1) \quad {}_6\psi_6\begin{bmatrix} q\sqrt{a}, -q\sqrt{a}, b, c, d, e & ; q, \frac{a^2q}{bcde} \\ \sqrt{a}, -\sqrt{a}, \frac{qa}{b}, \frac{qa}{c}, \frac{qa}{d}, \frac{qa}{e} & \end{bmatrix} = \prod\begin{bmatrix} aq, \frac{aq}{bc}, \frac{aq}{bd}, \frac{aq}{be}, \frac{aq}{cd}, \frac{aq}{ce}, \frac{aq}{de}, q, \frac{q}{a} \\ \frac{q}{b}, \frac{q}{c}, \frac{q}{d}, \frac{q}{e}, \frac{aq}{b}, \frac{aq}{c}, \frac{aq}{d}, \frac{aq}{e}, \frac{a^2q}{bcde} \end{bmatrix}, $$

where

$$ \prod\begin{bmatrix} \alpha_1, \cdots, \alpha_r \\ \beta_1, \cdots, \beta_s \end{bmatrix} = \frac{(\alpha_1)_\infty \cdots (\alpha_r)_\infty}{(\beta_1)_\infty \cdots (\beta_s)_\infty}. $$

The identity (3.1) is probably the most general summation identity known for bilateral basic hypergeometric series. As we shall see, we shall be able to deduce many important and diverse results in number theory from (3.1).

There are three known proofs of (3.1). Bailey's original proof [14, § 4] relies on ingenious combinations of various transformation formulas he had developed for ordinary and basic hypergeometric series. L. J. Slater [51] uses an analogue of the Barnes-type integral, and A. Lakin [51] combines a q-difference equation technique with Carlson's theorem on entire functions.

We shall provide a more elementary proof; in fact we shall utilize q-difference equations together with the uniqueness of a Laurent series expansion about the origin. For the most part, our proof will rely on simple manipulations of series. In our proof, we shall need the following function:

$$\begin{aligned} K_{\lambda,k,i}(a_0, a_1, \cdots, a_\lambda; z; q) &= K_{\lambda,k,i}((a); z; q) \\ &= \sum_{n=-\infty}^{\infty} (-1)^{n(\lambda+1)} z^{(k+1)n} q^{(2k-\lambda+1)n^2/2 - in + (\lambda+1)n/2} \cdot \\ &\quad \cdot (a_0 a_1 \cdots a_\lambda)^{-n} \\ &\quad \cdot \frac{(a_0)_n (a_1)_n \cdots (a_\lambda)_n (1 - z^i q^{2ni})}{(zq/a_0)_n (zq/a_1)_n \cdots (zq/a_\lambda)_n (1 - z)}. \end{aligned} \tag{3.2}$$

I remark that $K_{3,1,1}(b, c, d, e; z, q)$ is the left-hand side of identity (3.1). In the following theorem we shall utilize a series rearrangement technique originally due to Cayley [27] that has also been effectively utilized in the study of the Rogers–Ramanujan identities [10] and more general basic hypergeometric functions [5].

THEOREM 3.1. *The following identities are valid*:

$$K_{\lambda,k,0}(a_0, a_1, \cdots, a_\lambda; z; q) = 0; \tag{3.3}$$

$$K_{\lambda,k,-i}(a_0, a_1, \cdots, a_\lambda; z; q) = -z^{-i} K_{\lambda,k,i}(a_0, a_1, \cdots, a_\lambda; z; q); \tag{3.4}$$

$$\begin{aligned} &K_{\lambda,k,i}(a_0, a_1, \cdots, a_\lambda; z; q) - \frac{z}{a_j} K_{\lambda,k,i-1}(a_0, a_1, \cdots, a_\lambda; z; q) \\ &= \frac{(1 - a_j) a_j^{-1} z^i (1 - zq)\left(1 - \dfrac{zq}{a_j}\right)}{(1 - z)\left(1 - \dfrac{zq}{a_0}\right) \cdots \left(1 - \dfrac{zq}{a_\lambda}\right)} \\ &\quad \cdot \sum_{r=0}^{\lambda} (-1)^r z^r q^r \sigma_r(j) K_{\lambda,k,k-i-r+1}(a_0, a_1, \cdots, a_j q, \cdots, a_\lambda; zq; q); \end{aligned} \tag{3.5}$$

$$\begin{aligned} &K_{\lambda,k,i}(a_0, a_1, \cdots, a_\lambda; z; q) \\ &= \frac{(-1)^{\lambda+1} z^{k+1} (a_0 a_1 \cdots a_\lambda)^{-1} q^{k-i+1} (1 - zq^2)(1 - a_0)(1 - a_1) \cdots (1 - a_\lambda)}{(1 - z)\left(1 - \dfrac{zq}{a_0}\right)\left(1 - \dfrac{zq}{a_1}\right) \cdots \left(1 - \dfrac{zq}{a_\lambda}\right)} \\ &\quad \cdot K_{\lambda,k,i}(a_0 q, a_1 q, \cdots, a_\lambda q; zq^2; q); \end{aligned} \tag{3.6}$$

where in (3.5), $\sigma_r(j)$ *denotes the r-th elementary symmetric function of* $a_0^{-1}, a_1^{-1}, \cdots, a_{j-1}^{-1}, a_{j+1}^{-1}, \cdots, a_\lambda^{-1}$.

Remark. Identities (3.5) and (3.6) are called *q-difference equations*; it is from these equations that we shall deduce (3.1).

Proof. Equation (3.3) is obvious since $1 - z^0 q^{2n\cdot 0} = 1 - 1 = 0$. Equation (3.4) is almost as obvious since $q^{in}(1 - z^{-i}q^{-2ni}) = -z^{-i}q^{-in}(1 - z^i q^{2ni})$. Next we handle (3.6). We replace n by $n + 1$ in the series in (3.2). Hence

$$
\begin{aligned}
K_{\lambda,k,i}&(a_0, a_1, \cdots, a_\lambda; z; q) \\
&= (-1)^{\lambda+1} z^{k+1} q^{k-i+1} (a_0 a_1 \cdots a_\lambda)^{-1} \\
&\quad \cdot \sum_{n=-\infty}^{\infty} (-1)^{n(\lambda+1)} z^{(k+1)n} q^{(2k-\lambda+1)n^2/2 - in + (\lambda+1)n/2 + (2k-\lambda+1)n} (a_0 a_1 \cdots a_\lambda)^{-n} \\
&\quad \cdot \frac{(a_0)_{n+1}(a_1)_{n+1} \cdots (a_\lambda)_{n+1}(1 - z^i q^{2i} q^{2in})}{\left(\frac{zq}{a_0}\right)_{n+1}\left(\frac{zq}{a_1}\right)_{n+1} \cdots \left(\frac{zq}{a_\lambda}\right)_{n+1}(1 - z)} \\
&= \frac{(-1)^{\lambda+1} z^{k+1} (a_0 a_1 \cdots a_\lambda)^{-1} q^{k-i+1}(1 - zq^2)(1 - a_0)(1 - a_1) \cdots (1 - a_\lambda)}{(1 - z)\left(1 - \frac{zq}{a_0}\right)\left(1 - \frac{zq}{a_1}\right) \cdots \left(1 - \frac{zq}{a_\lambda}\right)} \\
&\quad \cdot \sum_{n=-\infty}^{\infty} (-1)^{n(\lambda+1)} (zq^2)^{(k+1)n} q^{(2k-\lambda+1)n^2/2 - in + (\lambda+1)n/2} \\
&\qquad \cdot ((a_0 q)(a_1 q) \cdots (a_\lambda q))^{-n} \\
&\qquad \cdot \frac{(a_0 q)_n (a_1 q)_n \cdots (a_\lambda q)_n (1 - (zq^2)^i q^{2ni})}{\left(\frac{(zq^2)q}{a_0 q}\right)_n \left(\frac{(zq^2)q}{a_1 q}\right)_n \cdots \left(\frac{(zq^2)q}{a_\lambda q}\right)_n (1 - zq^2)} \\
&= \frac{(-1)^{\lambda+1} z^{k+1} (a_0 a_1 \cdots a_\lambda)^{-1} q^{k-i+1}(1 - zq^2)(1 - a_0)(1 - a_1) \cdots (1 - a_\lambda)}{(1 - z)\left(1 - \frac{zq}{a_0}\right)\left(1 - \frac{zq}{a_1}\right) \cdots \left(1 - \frac{zq}{a_\lambda}\right)} \\
&\qquad \cdot K_{\lambda,k,i}(a_0 q, a_1 q, \cdots, a_\lambda q; zq^2; q).
\end{aligned}
$$

Finally we treat (3.5). Since $K_{\lambda,k,i}(a_0, a_1, \cdots, a_\lambda; z; q)$ is symmetric in the a_j, I shall assume for simplicity of notation that $j = 0$, i.e., $a_j = a_0$. We begin by examining the combination of the two series on the left-hand side of (3.5). In particular, we start by considering only that factor of the nth term that actually involves i, namely,

$$
\begin{aligned}
q^{-in}&(1 - z^i q^{2ni}) - \frac{z}{a_0} q^{-in+n}(1 - z^{i-1} q^{2ni-2n}) \\
&= \left\{ q^{-in}\left(1 - \frac{zq^n}{a_0}\right)\right\} + \left\{\frac{z^i}{a_0} q^{n(i-1)}(1 - a_0 q^n)\right\}.
\end{aligned}
$$

Consequently, we see that the series that results on the left-hand side of (3.5) may be split into two series: the first contains the first expression in curly braces given above, and the second series contains the second curly braced expression. Therefore

$$K_{\lambda,k,i}(a_0, a_1, \cdots, a_\lambda; z; q) - \frac{z}{a_0} K_{\lambda,k,i}(a_0, a_1, \cdots, a_\lambda; z; q)$$

$$= \sum_{n=-\infty}^{\infty} (-1)^{n(\lambda+1)} z^{(k+1)n} q^{(2k-\lambda+1)n^2/2 - in + (\lambda+1)n/2} (a_0 a_1 \cdots a_\lambda)^{-n} \cdot \frac{(a_0)_n (a_1)_n \cdots (a_\lambda)_n}{\left(\frac{zq}{a_0}\right)_{n-1} \left(\frac{zq}{a_1}\right)_n \cdots \left(\frac{zq}{a_\lambda}\right)_n (1-z)}$$

$$+ z^i a_0^{-1} \sum_{n=-\infty}^{\infty} (-1)^{n(\lambda+1)} z^{(k+1)n} q^{(2k-\lambda+1)n^2/2 + n(i-1) + (\lambda+1)n/2} (a_0 a_1 \cdots a_\lambda)^{-n} \cdot \frac{(a_0)_{n+1} (a_1)_n \cdots (a_\lambda)_n}{\left(\frac{zq}{a_0}\right)_n \left(\frac{zq}{a_1}\right)_n \cdots \left(\frac{zq}{a_\lambda}\right)_n (1-z)}.$$

We now replace n by $n + 1$ in the first sum and then combine the two series term by term. Thus

$$K_{\lambda,k,i}((a); z; q) - \frac{z}{a_0} K_{\lambda,k,i-1}((a); z; q)$$

$$= a_0^{-1} z^i \sum_{n=-\infty}^{\infty} (-1)^{n(\lambda+1)} z^{(k+1)n} q^{(2k-\lambda+1)n^2/2 + n(i-1) + (\lambda+1)n/2} (a_0 a_1 \cdots a_\lambda)^{-n}$$

$$\cdot \frac{(a_0)_{n+1} (a_1)_n \cdots (a_\lambda)_n}{(zq/a_0)_n (zq/a_1)_n \cdots (zq/a_\lambda)_n (1-z)} \left\{ 1 + \frac{(-1)^{\lambda+1} z^{k-i+1} q^{(2k-\lambda-2i+2)n+k-i+1} (1-a_1 q^n)(1-a_2 q^n) \cdots (1-a_\lambda q^n)}{a_1 \cdots a_\lambda \left(1 - \frac{zq^{n+1}}{a_1}\right)\left(1 - \frac{zq^{n+1}}{a_2}\right) \cdots \left(1 - \frac{zq^{n+1}}{a_\lambda}\right)} \right\}.$$

We now combine fractions inside the curly brackets, and we make the observation that

$$\prod_{j=1}^{\lambda} \left(1 - \frac{X}{a_j}\right) = \sum_{r=0}^{\lambda} (-1)^r \sigma_r(0) X^r.$$

Consequently, we see that

$$K_{\lambda,k,i}((a); z; q) - \frac{z}{a_0} K_{\lambda,k,i-1}((a); z; q)$$

$$= \frac{a_0^{-1} z^i (1 - zq)(1 - a_0)}{(1-z)\left(1 - \frac{zq}{a_1}\right) \cdots \left(1 - \frac{zq}{a_\lambda}\right)}$$

(*cont.*)

$$\cdot \sum_{n=-\infty}^{\infty} (-1)^{n(\lambda+1)}(zq)^{(k+1)n}q^{(2k-\lambda+1)n^2/2-(k-i+1)n+(\lambda+1)n/2}((a_0q)a_1 \cdots a_\lambda)^{-n}$$

$$\cdot \frac{(a_0q)_n(a_1)_n \cdots (a_\lambda)_n}{(zq^2/a_0q)_n(zq^2/a_1)_n \cdots (zq^2/a_\lambda)_n(1-zq)}$$

$$\cdot \left\{ \sum_{r=0}^{\lambda} (-1)^r\sigma_r(0)(zq)^r q^{nr} - (zq)^{k-i+1}q^{2(k-i+1)n} \sum_{r=0}^{\lambda} (-1)^r\sigma_r(0)q^{-nr} \right\}$$

$$= \frac{a_0^{-1}z^i(1-zq)(1-a_0)}{(1-z)\left(1-\frac{zq}{a_1}\right)\cdots\left(1-\frac{zq}{a_\lambda}\right)} \sum_{r=0}^{\lambda} (-1)^r z^r q^r \sigma_r(0)$$

$$\cdot \sum_{n=-\infty}^{\infty} (-1)^{n(\lambda+1)}(zq)^{(k+1)n}q^{(2k-\lambda+1)n^2/2-(k-i-r+1)n+(\lambda+1)n/2}$$

$$\cdot ((a_0q)a_1 \cdots a_\lambda)^{-n}$$

$$\cdot \frac{(a_0q)_n(a_1)_n \cdots (a_\lambda)_n(1-(zq)^{k-i-r+1}q^{2(k-i-r+1)n})}{\left(\frac{zq^2}{a_0q}\right)_n\left(\frac{zq^2}{a_1}\right)_n \cdots \left(\frac{zq^2}{a_\lambda}\right)_n(1-zq)}$$

$$= \frac{a_0^{-1}z^i(1-zq)(1-a_0)}{(1-z)\left(1-\frac{zq}{a_1}\right)\cdots\left(1-\frac{zq}{a_\lambda}\right)} \sum_{r=0}^{\lambda} (-1)^r z^r q^r \sigma_r(0)$$

$$\cdot K_{\lambda,k,k-i-r+1}(a_0q, a_1, \cdots, a_\lambda; zq; q).$$

Thus we have established (3.5) for $j = 0$, and as we remarked earlier, the general case now follows from the symmetry of the a_j. □

Before we establish (3.1), we must first establish a special case of (3.1) known as the limiting form of Jackson's theorem [52, p. 96, (3.3.1.3)]. Our proof is essentially the one given in [5, Thm. 5].

THEOREM 3.2.

$${}_6\phi_5\begin{bmatrix} z, qz^{1/2}, -qz^{1/2}, a_1, a_2, a_3; q, \dfrac{zq}{a_1a_2a_3} \\ \\ z^{1/2}, -z^{1/2}, \dfrac{zq}{a_1}, \dfrac{zq}{a_2}, \dfrac{zq}{a_3} \end{bmatrix} = \prod \begin{bmatrix} zq, \dfrac{zq}{a_1a_2}, \dfrac{zq}{a_1a_3}, \dfrac{zq}{a_2a_3} \\ \\ \dfrac{zq}{a_1}, \dfrac{zq}{a_2}, \dfrac{zq}{a_3}, \dfrac{zq}{a_1a_2a_3} \end{bmatrix}.$$

Proof. We let

$$g_i(a_0, a_1, a_2, a_3; z) = K_{3,1,i}(a_0, a_1, a_2, a_3; z; q), \tag{3.7}$$

and we let

$$f(z) = g_1(z, a_1, a_2, a_3; z). \tag{3.8}$$

From Theorem 3.1, we directly deduce the following relationships:

$$g_0(a_0, a_1, a_2, a_3; z) = 0, \tag{3.9}$$

$$g_{-i}(a_0, a_1, a_2, a_3; z) = -z^{-i}g_i(a_0, a_1, a_2, a_3; z), \tag{3.10}$$

$$g_i(a_0, a_1, a_2, a_3; z) - \frac{z}{a_j} g_{i-1}(a_0, a_1, a_2, a_3; z)$$

$$(3.11) \quad = \frac{(1 - a_j)a_j^{-1} z^i (1 - zq)\left(1 - \frac{zq}{a_j}\right)}{(1 - z)\left(1 - \frac{zq}{a_0}\right)\left(1 - \frac{zq}{a_1}\right)\left(1 - \frac{zq}{a_2}\right)\left(1 - \frac{zq}{a_3}\right)} \sum_{r=0}^{3} (-1)^r z^r q^r \sigma_r(j)$$

$$\cdot g_{2-i-r}(a'_0, a'_1, a'_2, a'_3; zq),$$

where $a'_h = a_h$ if $h \neq j$ and $a'_j = a_j q$.

Setting $i = 1$ and $j = 0$ in (3.11) and utilizing (3.9) and (3.8), we see that

$$(3.12) \quad g_1(a_0, a_1, a_2, a_3; z) = \frac{(1 - a_0)a_0^{-1} z(1 - zq)}{(1 - z)\left(1 - \frac{zq}{a_1}\right)\left(1 - \frac{zq}{a_2}\right)\left(1 - \frac{zq}{a_3}\right)} \cdot (1 - zq\sigma_2(0) g_1(a_0 q, a_1, a_2, a_3; zq) + zq\sigma_3(0) g_2(a_0 q, a_1, a_2, a_3; zq)).$$

Setting $i = 0$ and $j = 0$ in (3.11) and utilizing (3.9) and (3.8), we see that

$$(3.13) \quad g_1(a_0, a_1, a_2, a_3; z) = \frac{(1 - a_0)(1 - zq)}{(1 - z)\left(1 - \frac{zq}{a_1}\right)\left(1 - \frac{zq}{a_2}\right)\left(1 - \frac{zq}{a_3}\right)} \cdot (g_2(a_0 q, a_1, a_2, a_3; zq) - (zq\sigma_1(0) - z^2 q^2 \sigma_3(0)) g_1(a_0 q, a_1, a_2, a_3; zq)).$$

Eliminating $g_2(a_0 q, a_1, a_2, a_3; zq)$ from (3.11) and (3.12), we find that

$$(3.14) \quad \left(\frac{a_0}{z} - zq\sigma_3(0)\right) g_1(a_0, a_1, a_2, a_3; z) = \frac{(1 - a_0)(1 - zq)}{(1 - z)\left(1 - \frac{zq}{a_1}\right)\left(1 - \frac{zq}{a_2}\right)\left(1 - \frac{zq}{a_3}\right)} \cdot (1 - zq\sigma_2(0) + zq\sigma_3(0)(zq\sigma_1(0) - z^2 q^2 \sigma_3(0))) g_1(a_0 q, a_1, a_2, a_3; zq).$$

Simplifying (3.14), we obtain that

$$g_1(a_0, a_1, a_2, a_3; z)$$

$$(3.15) \quad = \frac{a_0^{-1} z(1 - a_0)(1 - zq)\left(1 - \frac{zq}{a_1 a_2}\right)\left(1 - \frac{zq}{a_1 a_3}\right)\left(1 - \frac{zq}{a_2 a_3}\right)}{(1 - z)\left(1 - \frac{zq}{a_1}\right)\left(1 - \frac{zq}{a_2}\right)\left(1 - \frac{zq}{a_3}\right)\left(1 - \frac{z^2 q}{a_0 a_1 a_2 a_3}\right)}$$

$$\cdot g_1(a_0 q, a_1, a_2, a_3; zq).$$

We now set $a_0 = z$ and utilize (3.8) to deduce that

$$(3.16) \quad f(z) = \frac{(1 - zq)\left(1 - \frac{zq}{a_1 a_2}\right)\left(1 - \frac{zq}{a_1 a_3}\right)\left(1 - \frac{zq}{a_2 a_3}\right) f(zq)}{\left(1 - \frac{zq}{a_1}\right)\left(1 - \frac{zq}{a_2}\right)\left(1 - \frac{zq}{a_3}\right)\left(1 - \frac{zq}{a_1 a_2 a_3}\right)}.$$

Repeated iteration of (3.16) then implies that

$$f(z) = \frac{(zq)_N\left(\frac{zq}{a_1a_2}\right)_N\left(\frac{zq}{a_1a_3}\right)_N\left(\frac{zq}{a_2a_3}\right)_N f(zq^N)}{\left(\frac{zq}{a_1}\right)_N\left(\frac{zq}{a_2}\right)_N\left(\frac{zq}{a_3}\right)_N\left(\frac{zq}{a_1a_2a_3}\right)_N}. \tag{3.17}$$

Identity (3.7) now follows if we let $N \to \infty$ and observe that $\lim_{N\to\infty} f(zq^N) = f(0) = 1$ and

$$f(z) = {}_6\phi_5\left[\begin{matrix} z, qz^{1/2}, -qz^{1/2}, a_1, a_2, a_3; q, \frac{zq}{a_1a_2a_3} \\ z^{1/2}, -z^{1/2}, \frac{zq}{a_1}, \frac{zq}{a_2}, \frac{zq}{a_3} \end{matrix}\right]. \quad \square$$

The following result is a lemma that will be necessary in the proof of Theorem 3.3.

LEMMA 3.1. $g_1(a_0, a_1, a_2, a_3; q^m) = 0$ *for* $m = 1, 2, 3, \cdots$.

Proof. First we see that

$$\lim_{z\to 1}(1-z)g_1(a_0, a_1, a_2, a_3; z)$$

$$= \lim_{z\to 1}\sum_{n=-\infty}^{\infty}(1-zq^{2n})\frac{(a_0)_n(a_1)_n(a_2)_n(a_3)_n\left(\frac{z^2q}{a_0a_1a_2a_3}\right)^n}{\left(\frac{zq}{a_0}\right)_n\left(\frac{zq}{a_1}\right)_n\left(\frac{zq}{a_2}\right)_n\left(\frac{zq}{a_3}\right)_n}$$

$$= \sum_{n=-\infty}^{\infty}\frac{(a_0)_n(a_1)_n(a_2)_n(a_3)_n\left(\frac{q}{a_0a_1a_2a_3}\right)^n}{\left(\frac{q}{a_0}\right)_n\left(\frac{q}{a_1}\right)_n\left(\frac{q}{a_2}\right)_n\left(\frac{q}{a_3}\right)_n} - \sum_{n=-\infty}^{\infty}\frac{(a_0)_n(a_1)_n(a_2)_n(a_3)_n\left(\frac{q^3}{a_0a_1a_2a_3}\right)^n}{\left(\frac{q}{a_0}\right)_n\left(\frac{q}{a_1}\right)_n\left(\frac{q}{a_2}\right)_n\left(\frac{q}{a_3}\right)_n}$$

$$= 0,$$

since the second sum may be transformed into the first by replacing n by $-n$. Consequently if in (3.15) we replace a_0 by a_0q^{-1}, multiply by $(1-z)$ and let $z \to 1$, we see that $g_1(a_0, a_1, a_2, a_3; q) = 0$. In general, in (3.15) we replace a_0 by a_0q^{-1}, z by q^{k-1} and from the truth of the theorem for $m = k-1$, we deduce that $g_1(a_0, a_1, a_2, a_3; q^m) = 0$. $\square$

As we shall see, the theorems on two squares and four squares are implied by Theorem 3.2; its main usefulness to us, however, lies in its application in the following proof of (3.1).

THEOREM 3.3. *Identity* (3.1) *holds.*

Proof. In order to avoid convergence problems, we shall initially assume that $|q| < 1$, $|a_h| \geqq 1$ for $0 \leqq h \leqq 3$, $0 < |z| < |q|^{-1/2}$. These conditions will simplify our work and at the conclusion of the proof we shall show how to weaken them to the obvious minimal conditions required.

We note that $g_1(b, c, d, e; z)$ is the left-hand side of (3.1). We have already translated most of the identities of Theorem 3.1 into identities for $g_i(a_0, a_1, a_2,$

$a_3; z)$, namely (3.18), (3.19) and (3.20). There remains identity (3.6), and when $\lambda = 3$, $k = 1$, we see that (3.6) yields the following result:

$$g_i(a_0, a_1, a_2, a_3; z) = \frac{z^2(a_0a_1a_2a_3)^{-1}q^{2-i}(1 - zq^2)(1 - a_0)(1 - a_1)(1 - a_2)(1 - a_3)}{(1 - z)\left(1 - \frac{zq}{a_0}\right)\left(1 - \frac{zq}{a_1}\right)\left(1 - \frac{zq}{a_2}\right)\left(1 - \frac{zq}{a_3}\right)} \cdot g_i(a_0q, a_1q, a_2q, a_3q; zq^2). \tag{3.18}$$

If we let

$$\mathscr{P}(y; b, c, d; z) = \frac{y^{-1}z(1 - y)(1 - zq)\left(1 - \frac{zq}{bc}\right)\left(1 - \frac{zq}{bd}\right)\left(1 - \frac{zq}{cd}\right)}{(1 - z)\left(1 - \frac{zq}{b}\right)\left(1 - \frac{zq}{c}\right)\left(1 - \frac{zq}{d}\right)\left(1 - \frac{z^2q}{ybcd}\right)}, \tag{3.19}$$

then we may write (3.15) in the following simple form:

$$g_1(a_0, a_1, a_2, a_3; z) = \mathscr{P}(a_0; a_1, a_2, a_3; z)g_1(a_0q, a_1, a_2, a_3; zq). \tag{3.20}$$

We note that the left-hand side of (3.20) is symmetric in a_0, a_1, a_2, a_3, and therefore (3.20) is valid under any permutation of the a_j. Consequently by (3.20), we see that

$$\begin{aligned} g_1(a_0, a_1, a_2, a_3; z) &= \mathscr{P}(a_0; a_1, a_2, a_3; z)\mathscr{P}(a_1; a_0q, a_2, a_3, zq)g_1(a_0q, a_1q, a_2, a_3; zq^2) \\ &= \mathscr{P}(a_0; a_1, a_2, a_3; z)\mathscr{P}(a_1; a_0q, a_2, a_3; zq) \\ &\quad \cdot \mathscr{P}(a_2; a_0q, a_1q, a_3; zq^2)g_1(a_0q, a_1q, a_2q, a_3; zq^3) \\ &= \mathscr{P}(a_0; a_1, a_2, a_3; z)\mathscr{P}(a_1; a_0q, a_2, a_3; zq) \\ &\quad \cdot \mathscr{P}(a_2; a_0q, a_1q, a_3; zq^2)\mathscr{P}(a_3; a_0q, a_1q, a_2q; zq^3) \\ &\quad \cdot g_1(a_0q, a_1q, a_2q, a_3q; zq^4). \end{aligned} \tag{3.21}$$

To simplify the final result in (3.21), we write it in the following form:

$$\begin{aligned} g_1(a_0, a_1, a_2, a_3; z) &= \Lambda(a_0, a_1, a_2, a_3; z)g_1(a_0q, a_1q, a_2q, a_3q; zq^4) \\ &= \frac{(1 - zq^2)\left(1 - \frac{zq^3}{a_0}\right)\left(1 - \frac{zq^3}{a_1}\right)\left(1 - \frac{zq^3}{a_2}\right)\left(1 - \frac{zq^3}{a_3}\right)\Lambda(a_0, a_1, a_2, a_3; z)}{z^2q^5(a_0a_1a_2a_3)^{-1}(1 - zq^4)(1 - a_0)(1 - a_1)(1 - a_2)(1 - a_3)} \\ &\quad \cdot g_1(a_0, a_1, a_2, a_3; zq^2), \end{aligned} \tag{3.22}$$

where the last line follows from (3.18) with z replaced by zq^2. Thus when (3.22) is completely written out, we have

$$g_1(a_0, a_1, a_2, a_3; z) = \frac{z^2q\left(\frac{zq}{a_0a_1}\right)_2\left(\frac{zq}{a_0a_2}\right)_2\left(\frac{zq}{a_0a_3}\right)_2\left(\frac{zq}{a_1a_3}\right)_2\left(\frac{zq}{a_2a_3}\right)_2(1-zq^2)}{\left(\frac{zq}{a_0}\right)_2\left(\frac{zq}{a_1}\right)_2\left(\frac{zq}{a_2}\right)_2\left(\frac{zq}{a_3}\right)_2\left(\frac{z^2q}{a_0a_1a_2a_3}\right)_4(1-z)} \cdot g_1(a_0, a_1, a_2, a_3; zq^2). \tag{3.23}$$

Hence if

$$h(z) = \frac{\left(\frac{zq}{a_0}\right)_\infty\left(\frac{zq}{a_1}\right)_\infty\left(\frac{zq}{a_2}\right)_\infty\left(\frac{zq}{a_3}\right)_\infty\left(\frac{z^2q}{a_0a_1a_2a_3}\right)_\infty}{\left(\frac{zq}{a_0a_1}\right)_\infty\left(\frac{zq}{a_0a_2}\right)_\infty\left(\frac{zq}{a_0a_3}\right)_\infty\left(\frac{zq}{a_1a_2}\right)_\infty\left(\frac{zq}{a_1a_3}\right)_\infty\left(\frac{zq}{a_2a_3}\right)_\infty(zq)_\infty\left(\frac{q}{z}\right)_\infty} \cdot g_1(a_0, a_1, a_2, a_3; z) \tag{3.24}$$

then we deduce directly from (3.23) that

$$h(z) = h(zq^2). \tag{3.25}$$

It is clear from our initial conditions (namely, $|q| < 1$, $|a_j| \geqq 1$, $|z| < |q|^{-1/2}$) that $h(z)$ is analytic in any annulus centered on the origin which lies inside the circle $|z| = |q|^{-1/2}$ except possibly at $z = q^m$ for $m = 1, 2, 3, \cdots$; however, by Lemma 3.1 these are removable singularities. Thus

$$h(z) = \sum_{n=-\infty}^{\infty} A_n z^n, \qquad 0 < |z| < |q|^{-1/2}. \tag{3.26}$$

Substituting (3.26) into (3.25) and comparing coefficients of z^n, we see that

$$A_n = q^{2n} A_n. \tag{3.27}$$

Thus $A_n = 0$ for all $n \neq 0$, and $h(z) = A_0$. Since $h(z)$ is a constant function, we need only know one value of it. We thus obtain from (3.24) that

$$\begin{aligned} h(z) &= h(a_0) \\ &= \frac{(q)_\infty\left(\frac{a_0q}{a_1}\right)_\infty\left(\frac{a_0q}{a_2}\right)_\infty\left(\frac{a_0q}{a_3}\right)_\infty\left(\frac{a_0q}{a_1a_2a_3}\right)_\infty}{\left(\frac{q}{a_1}\right)_\infty\left(\frac{q}{a_2}\right)_\infty\left(\frac{q}{a_3}\right)_\infty\left(\frac{a_0q}{a_1a_2}\right)_\infty\left(\frac{a_0q}{a_1a_3}\right)_\infty\left(\frac{a_0q}{a_2a_3}\right)_\infty(a_0q)_\infty\left(\frac{q}{a_0}\right)_\infty} \\ &\quad \cdot \frac{(a_0q)_\infty\left(\frac{a_0q}{a_1a_2}\right)_\infty\left(\frac{a_0q}{a_1a_3}\right)_\infty\left(\frac{a_0q}{a_2a_3}\right)_\infty}{\left(\frac{a_0q}{a_1}\right)_\infty\left(\frac{a_0q}{a_2}\right)_\infty\left(\frac{a_0q}{a_3}\right)_\infty\left(\frac{a_0q}{a_1a_2a_3}\right)_\infty}, \end{aligned} \tag{3.28}$$

where we have evaluated $g_1(a_0, a_1, a_2, a_3; a_0) = f(a_0)$ from Theorem 3.2. Hence

$$h(z) = \frac{(q)_\infty}{\left(\frac{q}{a_0}\right)_\infty \left(\frac{q}{a_1}\right)_\infty \left(\frac{q}{a_2}\right)_\infty \left(\frac{q}{a_3}\right)_\infty}. \tag{3.29}$$

Comparing the expression in (3.29) for $h(z)$ with that in (3.24) we see that

$$g_1(a_0, a_1, a_2, a_3; z) = \prod \left[\begin{matrix} zq, \frac{zq}{a_0a_1}, \frac{zq}{a_0a_2}, \frac{zq}{a_0a_3}, \frac{zq}{a_1a_2}, \frac{zq}{a_1a_3}, \frac{zq}{a_2a_3}, q, \frac{q}{z} \\ \frac{q}{a_0}, \frac{q}{a_1}, \frac{q}{a_2}, \frac{q}{a_3}, \frac{zq}{a_0}, \frac{zq}{a_1}, \frac{zq}{a_2}, \frac{zq}{a_3}, \frac{z^2q}{a_0a_1a_2a_3} \end{matrix}\right], \tag{3.30}$$

and this identity is easily seen to be (3.1). By analytic continuation, we see that (3.1) is valid in any region containing the one we have treated in which both sides of the identity converge; however the conditions $|q| < 1$, $|z| < |q|^{-1/2}$, $|a_j| \geqq 1$ will be adequate for most of our purposes. □

We shall now utilize (3.1) and Theorem 3.2 (which may be deduced from (3.1) by setting $b = z$) to prove various results important in number theory. We start with Jacobi's triple product identity [10, Thm. 13-8, pp. 169–170].

THEOREM 3.4. *For* $|q| < 1$, $z \neq 0$,

$$\sum_{n=-\infty}^{\infty} q^{n^2}z^n = (q^2; q^2)_\infty(-zq; q^2)_\infty(-z^{-1}q; q^2)_\infty.$$

Proof. Let b, c, d, e all $\to\infty$ in (3.1) (such limit-taking may be justified by an appeal to Tannery's theorem; see [43, p. 371] for details). Then

$$\sum_{n=-\infty}^{\infty} \frac{q^{2n^2-n}a^{2n}(1 - aq^{2n})}{(1 - a)} = (aq)_\infty(q)_\infty\left(\frac{q}{a}\right)_\infty.$$

Replacing q by q^2 and then a by $-zq$ in this result, we see that

$$\sum_{n=-\infty}^{\infty} q^{4n^2}z^{2n}(1 + zq^{4n+1}) = (-zq; q^2)_\infty(q^2; q^2)_\infty(-z^{-1}q; q^2)_\infty. \tag{3.31}$$

Now

$$\begin{aligned} \sum_{n=-\infty}^{\infty} q^{4n^2}z^{2n}(1 + zq^{4n+1}) &= \sum_{n=-\infty}^{\infty} q^{(2n)^2}z^{2n} + \sum_{n=-\infty}^{\infty} q^{(2n+1)^2}z^{2n+1} \\ &= \sum_{n=-\infty}^{\infty} q^{n^2}z^n. \end{aligned} \tag{3.32}$$

We may now combine (3.31) and (3.32) to obtain the desired result. □

COROLLARY 3.4.1. *For* $|q| < 1$,

$$\sum_{n=-\infty}^{\infty} (-1)^n q^{n^2} = \frac{(q)_\infty}{(-q)_\infty}.$$

Proof. Setting $z = -1$ in Theorem 3.4 and recalling from the proof of Theorem 2.8 that $(-q)_\infty = 1/(q; q^2)_\infty$, we see that

$$\sum_{n=-\infty}^{\infty} (-1)^n q^{n^2} = (q^2; q^2)_\infty(q; q^2)_\infty(q; q^2)_\infty = (q)_\infty(q; q^2)_\infty = \frac{(q)_\infty}{(-q)_\infty}. \quad \square$$

THEOREM 3.5. *For $n \geqq 1$, $r_2(n) = 4(d_1(n) - d_3(n))$, where $r_2(n)$ is the number of representations of n as a sum of two squares and $d_a(n)$ is the number of divisors of n congruent to a* (mod 4).

Remark. We say that $n = a^2 + b^2 = c^2 + d^2$ are two different representations of n as a sum of two squares if $a \neq c$ or $b \neq d$. Thus 1 has the four representations $1^2 + 0^2, (-1)^2 + 0, 0^2 + 1^2, 0^2 + (-1)^2$.

Proof. First we note

$$\left(\sum_{-\infty}^{\infty} (-1)^n q^{n^2}\right)^2 = \sum_{n=0}^{\infty} r_2(n)(-q)^n. \tag{3.33}$$

$$\begin{aligned}
1 &+ 4 \sum_{n=1}^{\infty} (-1)^n (d_1(n) - d_3(n)) q^n \\
&= 1 + 4 \sum_{m=0}^{\infty} \sum_{r=1}^{\infty} (-1)^r (q^{(4m+1)r} - q^{(4m+3)r}) \\
&= 1 + 4 \sum_{m=0}^{\infty} \sum_{r=1}^{\infty} (-1)^{m+r} q^{(2m+1)r} \\
&= 1 + 4\left(\sum_{m=0}^{\infty} \sum_{r=m+1}^{\infty} (-1)^{m+r} q^{(2m+1)r} + \sum_{m=0}^{\infty} \sum_{r=1}^{m} (-1)^{m+r} q^{(2m+1)r}\right) \qquad (3.34) \\
&= 1 - 4 \sum_{m=0}^{\infty} \frac{q^{(2m+1)(m+1)}}{1 + q^{2m+1}} + 4 \sum_{r=1}^{\infty} \sum_{m=r}^{\infty} (-1)^{m+r} q^{(2m+1)r} \\
&= 1 - 4 \sum_{m=0}^{\infty} \frac{q^{(2m+1)(m+1)}}{1 + q^{2m+1}} + 4 \sum_{r=1}^{\infty} \frac{q^{(2r+1)r}}{1 + q^{2r}} \\
&= 1 + 4 \sum_{n=1}^{\infty} \frac{(-1)^n q^{n(n+1)/2}}{1 + q^n}.
\end{aligned}$$

Comparing (3.33) with (3.34) and using Corollary 3.4.1, we see that our theorem is equivalent to the following analytic identity:

$$1 + 4 \sum_{n=1}^{\infty} \frac{(-1)^n q^{n(n+1)/2}}{1 + q^n} = \left(\frac{(q)_\infty}{(-q)_\infty}\right)^2. \tag{3.35}$$

In Theorem 3.2 set $a_1 = a_2 = -1$, let $a_3 \to \infty$, and $z \to 1$ (equivalently, in (3.1) set $b = a$, then $c = d = -1$, and let $e \to \infty$, $a \to 1$). This yields

$$1 + \sum_{n=1}^{\infty} \frac{(q)_{n-1}(1 - q^{2n})(-1)_n^2(-1)^n q^{n(n+1)/2}}{(q)_n(-q)_n^2} = \left(\frac{(q)_\infty}{(-q)_\infty}\right)^2. \tag{3.36}$$

Now we may simplify the terms of the series as follows:

$$\frac{(q)_{n-1}(1 - q^{2n})(-1)_n^2(-1)^n q^{n(n+1)/2}}{(q)_n(-q)_n^2} = \frac{4(-1)^n q^{n(n+1)/2}}{1 + q^n}.$$

Therefore (3.36) reduces to (3.35), and our theorem is established. □

THEOREM 3.6. *For $n \geqq 1$, $r_4(n)$ equals 8 times the sum of the divisors of n not divisible by 4.*

Proof. As in the previous theorem,

$$\left(\sum_{-\infty}^{\infty}(-1)^n q^{n^2}\right)^4 = \sum_{n=0}^{\infty} r_4(n)(-q)^n. \tag{3.37}$$

Next we note that if $N = 2^\alpha m$, where m is odd, then

$$\begin{aligned}
\sum_{d|n}(-1)^{d+Nd^{-1}}d &= \sum_{d|m}\sum_{j=0}^{\alpha}(-1)^{2^j d+N2^{-j}d^{-1}}2^j d \\
&= -\sum_{d|m}(-1)^N d + \sum_{d|m}\sum_{j=1}^{\alpha}(-1)^{N2^{-j}d^{-1}}2^j d \\
&= \left(\sum_{d|m} d\right)\left(-(1)^N + \sum_{j=1}^{\alpha-1}2^j - 2^\alpha\right) \\
&= \begin{cases} \sum_{d|m} d & \text{if } \alpha = 0 \text{ (i.e., } N \text{ odd)}, \\ -3\sum_{d|m} d & \text{if } \alpha \geqq 1. \end{cases}
\end{aligned}$$

Recalling that the σ function $\sigma(n) = \sum_{d|N} d$ is multiplicative and that $\sigma(2) = 3$, we see that

$$-\sum_{d|N}(-1)^{d+Nd^{-1}}d = (-1)^N \sum_{\substack{d|N \\ 4\nmid d}} d. \tag{3.38}$$

Therefore by (3.38)

$$\begin{aligned}
1 + 8\sum_{N=1}^{\infty}\left(\sum_{d|N} d\right)(-q)^N &= 1 - 8\sum_{r=1}^{\infty}\sum_{m=1}^{\infty}(-1)^{r+m} r q^{rm} \\
&= 1 + 8\sum_{m=1}^{\infty}\frac{(-1)^m q^m}{(1+q^m)^2}
\end{aligned} \tag{3.39}$$

(here we used $z/(1+z)^2 = -\sum m(-1)^m z^m$). Comparing (3.37) with (3.39) and using Corollary 3.4.1, we see that our theorem is equivalent to the following analytic identity:

$$1 + 8\sum_{m=1}^{\infty}\frac{(-1)^m q^m}{(1+q^m)^2} = \left(\frac{(q)_\infty}{(-q)_\infty}\right)^4. \tag{3.40}$$

In Theorem 3.2 set $a_1 = a_2 = a_3 = -1$ and let $z \to 1$ (equivalently, in (3.1) set $b = a$, $c = d$, $e = -1$, and let $a \to 1$). This yields

$$1 + \sum_{n=1}^{\infty}\frac{(q)_{n-1}(1-q^{2n})(-1)_n^3(-1)^n q^n}{(q)_n(-q)_n^3} = \left(\frac{(q)_\infty}{(-q)_\infty}\right)^4. \tag{3.41}$$

Now we may simplify the terms of this series as follows:

$$\frac{(q)_{n-1}(1-q^{2n})(-1)_n^3(-1)^n q^n}{(q)_n(-q)_n^3} = \frac{8(-1)^n q^n}{(1+q^n)^2}.$$

Therefore (3.41) reduces to (3.40) and our theorem is established. □

To simplify our work on eight squares, we require a minor lemma.

LEMMA 3.2. *For* $N \geqq 1$,

$$\lim_{z\to 1} \frac{(-q)_n^N(-1)_n^N - z^L(-z^{-1})_n^N(-zq)_n^N}{(1-z)} = (-1)_n^N(-q)_n^N\left(L - \frac{N}{2} + \frac{Nq^n}{1+q^n}\right).$$

Proof. Let $Q(z) = z^L(-z^{-1})_n^N(-zq)_n^N$. Then the above limit is just $Q'(1)$. Hence by the product rule for derivatives,

$$Q'(1) = (-1)_n^N(-q)_n^N\left\{\frac{L}{z} - N\sum_{j=0}^{n-1}\frac{q^j z^{-2}}{1+z^{-1}q^j} + N\sum_{j=0}^{n-1}\frac{q^{j+1}}{1+zq^{j+1}}\right\}_{z=1}$$
$$= (-1)_n^N(-q)_n^N\left(L - \frac{N}{2} + \frac{Nq^n}{1+q^n}\right);$$

the simplification in the last equation occurs because the two sums cancel at $z = 1$ except for first and last terms. □

THEOREM 3.7. *For* $n \geqq 1$, $r_8(n) = 16(-1)^n \sum_{d|n}(-1)^d d^3$.

Proof. As before,

$$\left(\sum_{-\infty}^{\infty}(-1)^n q^{n^2}\right)^8 = \sum_{n=0}^{\infty} r_8(n)(-q)^n. \tag{3.42}$$

Utilizing the elementary summation

$$\sum_{m=1}^{\infty}(-1)^m m^3 z^m = \frac{-z+4z^2-z^3}{(1+z)^4},$$

we see that

$$\begin{aligned} 1 + 16\sum_{n=1}^{\infty}\left(\sum_{d|n}(-1)^d d^3\right)q^n &= 1 + 16\sum_{r=1}^{\infty}\sum_{m=1}^{\infty}(-1)^m m^3 q^{mr} \\ &= 1 + 16\sum_{r=1}^{\infty}\frac{(-q^r+4q^{2r}-q^{3r})}{(1+q^r)^4}. \end{aligned} \tag{3.43}$$

Comparing (3.42) with (3.43) and using Corollary 3.4.1, we see that our theorem is equivalent to the following analytic identity:

$$1 + 16\sum_{n=1}^{\infty}\frac{(-q^n+4q^{2n}-q^{3n})}{(1+q^n)^4} = \left(\frac{(q)_\infty}{(-q)_\infty}\right)^8. \tag{3.44}$$

In (3.1) set $b = c = d = e = -1$, and let $z \to 1$. This yields

$$\left(\frac{(q)_\infty}{(-q)_\infty}\right)^8 = \lim_{z\to 1}\sum_{-\infty}^{\infty}\frac{(1-zq^{2n})z^{2n}q^n(-1)_n^4}{(1-z)(-zq)_n^4}$$
$$= \lim_{z\to 1}\left\{\sum_{-\infty}^{\infty}\frac{z^{2n}q^n(-1)_n^4}{(1-z)(-zq)_n^4} - z\sum_{n=-\infty}^{\infty}\frac{z^{-2n}q^{-3n}(-z^{-1})_n^4 z^{4n}q^{2n}}{(1-z)(-q)_n^4}\right\}$$

(where $-n$ has replaced n in the second sum)

$$= \sum_{-\infty}^{\infty}\frac{q^n}{(-q)_n^8}\lim_{z\to 1}\frac{(-1)_n^4(-q)_n^4 - z(-z^{-1})_n^4(-zq)_n^4}{(1-z)}$$

(where $\lim\sum = \sum\lim$ by Tannery's theorem [43, p. 371])

(*cont.*)

(3.45)
$$= \sum_{-\infty}^{\infty} \frac{q^n}{(-q)_n^8}(-1)_n^4(-q)_n^4\left(-1 + \frac{4q^n}{1+q^n}\right) \qquad \text{(by Lemma 3.2)}$$

$$= 1 + 16 \sum_{\substack{n=-\infty \\ n\neq 0}}^{\infty} \frac{q^n(3q^n - 1)}{(1+q^n)^5}$$

$$= 1 + 16 \sum_{n=1}^{\infty} \frac{q^n(3q^n - 1)}{(1+q^n)^5} + 16 \sum_{n=1}^{\infty} \frac{q^{-n}(3q^{-n} - 1)q^{5n}}{(1+q^n)^5}$$

$$= 1 + 16 \sum_{n=1}^{\infty} \frac{(-q^n + 3q^{2n} + 3q^{3n} - q^{4n})}{(1+q^n)^5}$$

$$= 1 + 16 \sum_{n=1}^{\infty} \frac{(-q^n + 4q^{2n} - q^{3n})}{(1+q^n)^4}.$$

Thus (3.45) is identical with (3.44), and so our theorem is proved. □

A similar treatment may be given to $r_6(n)$, where in (3.1) we set $b = c = d = -1$ and let $e \to \infty$, $z \to 1$. Algebraic manipulations of the resulting series are somewhat messy however.

There are a number of other applications of (3.1) to number theory. Our next theorem is originally due to S. Ramanujan. We shall give a proof due to W. N. Bailey [17].

THEOREM 3.8. *For* $n \geqq 0$, $p(5n+4) \equiv 0 \pmod 5$, *where* $p(n)$ *is the number of partitions of* n.

Remark. It might seem that this result belongs in § 2; however, since it is deducible from (3.1) and since it is a congruence, we have included it here.

Proof.

$$\sum_{n=0}^{\infty}\left\{\frac{q^{5n+1}}{(1-q^{5n+1})^2} - \frac{q^{5n+2}}{(1-q^{5n+2})^2} - \frac{q^{5n+3}}{(1-q^{5n+3})^2} + \frac{q^{5n+4}}{(1-q^{5n+4})^2}\right\}$$

$$= \sum_{n=-\infty}^{\infty}\left\{\frac{q^{5n+1}}{(1-q^{5n+1})^2} - \frac{q^{5n+3}}{(1-q^{5n+3})^2}\right\}$$

$$= \sum_{n=-\infty}^{\infty} \frac{q^{5n+1}(1-q^2)(1+q^{5n+2})(1-q^{5n+2})}{(1-q^{5n+1})(1-q^{5n+3})^2}$$

(3.46)
$$= \frac{q(1-q^2)(1-q^4)}{(1-q)^2(1-q^3)^2}\, {}_6\psi_6\left[\begin{matrix} q^7, -q^7, q, q, q^3, q^3; q^5, q^5 \\ q^2, -q^2, q^8, q^8, q^6, q^6 \end{matrix}\right]$$

$$= \frac{q(1-q^2)(1-q^4)}{(1-q)^2(1-q^3)^2} \frac{(q^9;q^5)_\infty(q^7;q^5)_\infty(q^5;q^5)_\infty^4(q^3;q^5)_\infty(q;q^5)_\infty}{(q^4;q^5)_\infty^2(q^2;q^5)_\infty^2(q^8;q^5)_\infty^2(q^6;q^5)_\infty^2}$$

$$= q\frac{(q^5;q^5)_\infty^5}{(q)_\infty}.$$

Now from (2.1) we see that $\sum_{n=0}^{\infty} p(n)q^n = 1/(q)_\infty$. Using this fact and comparing terms in the extreme members of (3.46) that involve powers of q^5, we see that

$$(q^5;q^5)_\infty^5 \sum_{n=0}^{\infty} p(5n+4)q^{5n+5}$$

$$= \text{terms involving } q^5 \text{ in } \sum_{n=1}^{\infty} \frac{(n/5)q^n}{(1-q^n)^2}$$

(where $(n/5) = 0, 1, -1, 1$ for $n \equiv 0, 1, 2, 3, 4 \pmod 5$, respectively)

$$= \text{terms involving } q^5 \text{ in } \sum_{n=1}^{\infty}\sum_{m=1}^{\infty} \left(\frac{n}{5}\right) mq^{mn}$$

$$= \sum_{n=1}^{\infty}\sum_{m=1}^{\infty} \left(\frac{n}{5}\right) 5mq^{5mn}$$

$$= 5\sum_{n=1}^{\infty} \frac{(n/5)q^{5n}}{(1-q^{5n})^2}.$$

Therefore

$$\sum_{n=0}^{\infty} p(5n+4)q^{n+1} = \frac{5\sum_{n=1}^{\infty} \frac{(n/5)q^n}{(1-q^n)^2}}{(q)_\infty^5}. \tag{3.47}$$

Hence comparing coefficients of q^{n+1} on each side of (3.47), we see that $p(5n+4) \equiv 0 \pmod 5$ for each $n \geqq 0$. □

To conclude this section we shall derive the quintuple product identity from (3.1). This identity has been rediscovered many times (see [25] for a brief history), and has been of use in several problems in number theory (see Bailey [18], Gordon [32], Andrews [7]).

THEOREM 3.9. *For* $|q| < 1$, $|z| \neq 0$,

$$\sum_{n=-\infty}^{\infty} (-1)^n q^{n(3n-1)/2} z^{3n}(1+zq^n) = (-qz^{-1})_\infty(-z)_\infty(qz^{-2};q^2)_\infty(qz^2;q^2)_\infty(q)_\infty.$$

Proof. In (3.1) let $b, c, d \to \infty$ and set $e = q^{1/2}a^{1/2}$; then

$$\sum_{-\infty}^{\infty} (1-zq^{2n})a^{3/2n}q^{n(3n-2)/2}(-1)^n$$

$$= (1-a)\frac{(aq)_\infty(q)_\infty(qa^{-1})_\infty}{(q^{1/2}a^{-1/2})_\infty(a^{1/2}q^{1/2})_\infty}$$

$$= (-a^{1/2};q^{1/2})_\infty(a^{1/2};q)_\infty(q)_\infty(-q^{1/2}a^{-1/2};q^{1/2})_\infty(a^{-1/2}q;q)_\infty$$

$$= (-a^{1/2})_\infty(-a^{1/2}q^{1/2})_\infty(a^{1/2})_\infty(q)_\infty(-q^{1/2}a^{-1/2})_\infty(-qa^{-1/2})_\infty(a^{-1/2}q)_\infty$$

$$= (a;q^2)_\infty(-a^{1/2}q^{1/2})_\infty(q)_\infty(-q^{1/2}a^{-1/2})_\infty(a^{-1}q^2;q^2)_\infty.$$

Replacing a by $z^{-2}q$, we see that

$$\sum_{-\infty}^{\infty}(1-z^{-2}q^{2n+1})q^{n(3n+1)/2}z^{-3n}(-1)^n = (qz^{-2};q^2)_\infty(-z^{-1}q)_\infty(q)_\infty(-z)_\infty(qz^2;q^2)_\infty. \tag{3.48}$$

Therefore

$$\sum_{n=-\infty}^{\infty}(-1)^n q^{n(3n-1)/2}z^{3n}(1+zq^n)$$

$$= \sum_{n=-\infty}^{\infty}(-1)^n q^{n(3n+1)/2}z^{-3n} + \sum_{n=-\infty}^{\infty}(-1)^n q^{n(3n+1)/2}z^{3n+1}$$

(n replaced by $-n$ in first sum)

$$= \sum_{n=-\infty}^{\infty}(-1)^n q^{n(3n+1)/2}z^{-3n} - \sum_{n=-\infty}^{\infty}(-1)^n q^{n(3n+1)/2+2n+1}z^{-3n-2}$$

(n replaced by $-n-1$ in second sum)

$$= \sum_{n=-\infty}^{\infty}(1-z^{-2}q^{2n+1})q^{n(3n+1)/2}z^{-3n}(-1)^n$$

$$= (-qz^{-1})_\infty(-z)_\infty(qz^{-2};q^2)_\infty(qz^2;q^2)_\infty(q)_\infty \quad \text{(by (3.48))},$$

and this is precisely the desired result. □

It has been our object in this section merely to scratch the surface of applications of basic hypergeometric series to number theory. Apart from Theorems 3.2 and 3.8, none of the proofs given in this section have appeared before. Therefore the techniques introduced here appear to hold promise for many further number-theoretic investigations. We remark that M. Jackson has found significant number-theoretic application of formulas involving a well-poised ${}_8\psi_8$ (i.e., $K_{5,2,2}(a_0, a_1, a_2, a_3, a_4, a_5; z; q)$), and it should be possible to extend our q-difference equation techniques to prove the following identity of W. N. Bailey [19, (3.2), p. 197]:

$$
{}_2\psi_2\begin{bmatrix} e, f; \dfrac{aq}{ef} \\ \dfrac{aq}{c}, \dfrac{aq}{d} \end{bmatrix} = \prod\begin{bmatrix} \dfrac{q}{c}, \dfrac{q}{d}, \dfrac{aq}{e}, \dfrac{aq}{f} \\ aq, \dfrac{q}{a}, \dfrac{aq}{cd}, \dfrac{aq}{ef} \end{bmatrix} \cdot \sum_{s=-\infty}^{\infty} \frac{(q\sqrt{a})_s(-q\sqrt{a})_s(c)_s(d)_s(e)_s(f)_s}{(\sqrt{a})_s(-\sqrt{a})_s\left(\dfrac{aq}{c}\right)_s\left(\dfrac{aq}{d}\right)_s\left(\dfrac{aq}{e}\right)_s\left(\dfrac{aq}{f}\right)_s}\left(\frac{a^3q}{cdef}\right)^s q^{s^2}. \tag{3.49}
$$

This identity may be used to prove the Rogers–Ramanujan identities (as Bailey remarks), and it should be applicable to various fifth order mock theta function identities [6].

It is quite conceivable that a more extensive study of $K_{\lambda,k,i}(a_0, \cdots, a_\lambda; z; q)$ could lead to new proofs of Mordell's formulas [44] for $r_{2s}(n)$, to new proofs of more of the Ramanujan congruences [13], and to other number-theoretic results related to elliptic functions.

Finally, it should be mentioned that Theorem 2.7 is a special case of Theorem 3.2 (which is in turn the special case of (3.1) when $b = a$); to see this, set $a_0 = z^{1/2}$, $a_1 = -z^{1/2}$ in Theorem 3.2. Thus Theorems 2.8, 2.9 and 2.10 may be seen as corollaries of (3.1) also.

Theorems on sums of squares are often treated as corollaries of identities involving elliptic functions. Indeed, L. Carlitz (Nieuw Arch. Wisk., 3 (1955), pp. 193–196, and Proc. Amer. Math. Soc., 8 (1957), pp. 120–124) has treated the sums of four, six and eight squares in a very elegant manner utilizing certain elliptic function identities. It is quite likely that these more general elliptic function identities are also special cases of (3.1).

4. Application to finite vector spaces. The relationship between basic hypergeometric series and finite vector spaces arises from the following well-known theorem (see, for example, [48, p. 240]).

THEOREM 4.1. *Let V_n be a finite-dimensional vector space of dimension n over $GF(q)$, the finite field of q elements. Then there are exactly*

$$\begin{pmatrix} n \\ k \end{pmatrix}_q = \frac{(q)_n}{(q)_k(q)_{n-k}}$$

subspaces of V_n of dimension k.

Remark. $\begin{pmatrix} n \\ k \end{pmatrix}_q$ is called a *q-binomial coefficient* or a *Gaussian polynomial.*

Proof. First we determine the number of k-tuples of linearly independent vectors $\{v_1, v_2, \cdots, v_k\}$ that exist in V_n. We may select any nonzero vector for v_1, and since there are q^n vectors in V_n, we may select v_1 in $q^n - 1$ ways. The vector v_2 must be selected to lie outside the subspace spanned by v_1, and so v_1 may be selected in $q^n - q$ ways. In general, v_i must lie outside the subspace spanned by $\{v_1, \cdots, v_{i-1}\}$, and so v_i may be chosen in $q^n - q^{i-1}$ ways. Hence the number of k-tuples of linearly independent vectors is

$$(q^n - 1)(q^n - q) \cdots (q^n - q^{k-1}). \tag{4.1}$$

Each such k-tuple spans a k-dimensional subspace; however, two different k-tuples may span the same subspace. In fact the number of k-tuples spanning the same subspace is just the number of linearly independent k-tuples that exist in a k-dimensional space, and by (4.1) this number is

$$(q^k - 1)(q^k - q) \cdots (q^k - q^{k-1}). \tag{4.2}$$

Therefore the number of k-dimensional subspaces of V_n is just

$$\begin{aligned}
&\frac{(q^n - 1)(q^n - q) \cdots (q^n - q^{k-1})}{(q^k - 1)(q^k - q) \cdots (q^k - q^{k-1})} \\
&= \frac{q^{k(k-1)/2}(-1)(1 - q^n)(1 - q^{n-1}) \cdots (1 - q^{n-k+1})}{q^{k(k-1)/2}(-1)^k(1 - q^k)(1 - q^{k-1}) \cdots (1 - q)} \\
&= \frac{(q)_n}{(q)_k(q)_{n-k}} = \begin{pmatrix} n \\ k \end{pmatrix}_q. \qquad \square
\end{aligned}$$

To illustrate how finite vector spaces may be used in the theory of basic hypergeometric functions, we prove the following q-analogue of Vandermonde's theorem [52, (3.3.2.6), p. 97]; E. A. Bender [21] has proved a more general result of this nature using finite vector spaces.

THEOREM 4.2. *For nonnegative integers m, n and h,*

$$\sum_{l=0}^{h} \binom{n}{l}_q \binom{m}{h-l}_q q^{l(m-h+l)} = \binom{m+n}{h}_q .$$

Remark. From the definition of $\binom{n}{l}_q$, we may rewrite the above result as

$${}_2\phi_1\left[\begin{matrix} q^{-h}, q^{-n}; q, q^{n+m+1} \\ q^{m-h+1} \end{matrix}\right] = \frac{(q)_{m-h}(q)_{n+m}}{(q)_m(q)_{n+m-h}},$$

and if we assume $|q| < 1$, then this result follows directly from Theorem 2.6, the q-analogue of Gauss's theorem. In the proof below we assume that q is a prime power.

Proof. Let us consider V_{m+n} (a finite vector space of dimension $m + n$ over $GF(q)$), and let V_m be a chosen subspace of dimension m. We note from Theorem 4.1 that there are $\binom{m+n}{h}$ subspaces of V_{m+n} of dimension h.

We now ask for the number T_l of h-dimensional subspaces of V_{m+n} that intersect V_m in an l-dimensional subspace. To count these spaces, we ask for linearly independent h-tuples $(x_1, x_2, \cdots, x_l, y_1, y_2, \cdots, y_{h-l})$ such that the x's lie in V_m and the y's outside V_m. Just as in Theorem 4.1, the x's may be chosen in $(q^m - 1)(q^m - q) \cdots (q^m - q^{l-1})$ ways, and the y's may be chosen in $(q^{m+n} - q^m) \cdots (q^{m+n} - q^{m+(h-l)-1})$ ways. Thus the number of h-tuples of the desired type is

$$(4.3) \qquad \begin{aligned} &(q^m - 1)(q^m - q) \cdots (q^m - q^{l-1}) \\ &\quad \cdot (q^{m+n} - q^m)(q^{m+n} - q^{m+1}) \cdots (q^{m+n} - q^{m+h-l-1}). \end{aligned}$$

As in Theorem 4.1, several h-tuples may span the same space; indeed exactly those h-tuples lying in the same V_h with the first l entries in V_l and the rest outside V_l are the ones spanning the same space. Consequently to obtain T_l, we must divide (4.3) by $(q^l - 1) \cdots (q^l - q^{l-1})(q^h - q^l) \cdots (q^h - q^{l+(h-l)-1})$. Therefore

$$\begin{aligned} T_l &= \frac{(q^m - 1)(q^m - q) \cdots (q^m - q^{l-1})(q^{m+n} - q^m)(q^{m+n} - q^{m+1}) \cdots (q^{m+n} - q^{m+h-l-1})}{(q^l - 1)(q^l - q) \cdots (q^l - q^{l-1})(q^h - q^l)(q^h - q^{l+1}) \cdots (q^h - q^{h-1})} \\ &= \binom{m}{l}_q \frac{q^{m(h-l)}(q^n - 1)(q^n - q) \cdots (q^n - q^{h-l-1})}{q^{l(h-l)}(q^{h-l} - 1)(q^{h-l} - q) \cdots (q^{h-l} - q^{h-l-1})} \\ &= \binom{m}{l}_q \binom{n}{h-l}_q q^{(m-l)(h-l)}. \end{aligned}$$

Hence

$$\binom{n+m}{h}_q = \sum_{l=0}^{h} T_l = \sum_{l=0}^{h} \binom{m}{l}_q \binom{n}{h-l}_q q^{(m-l)(h-l)} = \sum_{l=0}^{h} \binom{m}{h-l}_q \binom{n}{l}_q q^{(m-h+l)l},$$

where we have replaced l by $h - l$ in the last sum. □

G.-C. Rota and J. Goldman in [47] and [48] have shown how fruitful the relationship between basic hypergeometric functions and finite vector spaces is; at the conclusion of [48] they pose a number of problems that appear to be quite important in this study. In a short note [42] D. Knuth presented canonical mappings between certain sets of partitions, the lattice of subspaces of V_n, and the lattice of subsets of $\{1, 2, \cdots, n\}$; Knuth in this way answered one of the central problems posed by Rota and Goldman [48, p. 258].

A second problem posed by Rota and Goldman concerns the extension of the theory of binomial enumeration [49] to finite vector spaces. This extension was made in [8]. Below we present a brief introduction to this topic.

Our consideration will be devoted to certain linear operators acting on $\mathbf{P} = \mathbf{R}[X]$, the ring of polynomials with real coefficients. If $f(X) \in \mathbf{P}$, we define η, the *Eulerian shift operator*, by

$$\eta^a f(X) = f(q^a X) = f(AX), \tag{4.4}$$

where $A = q^a$.

An *Eulerian differential operator* τ is a linear operator on $\mathbf{P}$ satisfying the following two conditions:

$$q^{-a}\tau\eta^a = \eta^a\tau, \quad \text{and} \quad \tau X^n \not\equiv 0 \quad \text{for each } n > 0. \tag{4.5}$$

An *Eulerian family of polynomials* is a sequence of polynomials $\{p_n(X)\}_{n=0}^{\infty}$ such that $p_0(X) = 1$, $p_n(X)$ is of degree n, and

$$p_n(XY) = \sum_{j \geq 0} \begin{bmatrix} n \\ j \end{bmatrix}_q p_j(X) Y^j p_{n-j}(Y). \tag{4.6}$$

Given an Eulerian differential operator τ, a sequence of polynomials $\{p_n(X)\}_{n=0}^{\infty}$ is the sequence of *Eulerian basic polynomials for* τ if $p_0(X) = 1$, $p_n(1) = 0$ for $n > 0$, and

$$\tau p_n(X) = (1 - q^n) p_{n-1}(X). \tag{4.7}$$

We now present the central theorems related to these concepts. Since the proofs are somewhat lengthy and are readily available in [8], we shall for the most part omit them.

THEOREM 4.3. (a) *If* $\{p_n(X)\}_{n=0}^{\infty}$ *is an Eulerian basic sequence for some Eulerian differential operator, then it is an Eulerian family of polynomials.*

(b) *If* $\{p_n(X)\}_{n=0}^{\infty}$ *is an Eulerian family of polynomials, then it is an Eulerian basic sequence for some Eulerian differential operator.*

Proof. See [8, pp. 351–352].

THEOREM 4.4. *If* $\{p_n(X)\}_{n=0}^{\infty}$ *is an Eulerian family of polynomials and if* C_n *is the leading coefficient of* $p_n(X)$, *then*

$$\sum_{n \geq 0} \frac{p_n(X)t^n}{(q)_n} = \frac{f(Xt)}{f(t)}, \tag{4.8}$$

where

$$f(t) = \sum_{n \geq 0} \frac{C_n t^n}{(q)_n}. \tag{4.9}$$

Proof. See [8, pp. 356–357].

THEOREM 4.5. *Let* $\{p_n(X)\}_{n=0}^{\infty}$ *and* $f(t)$ *be as in Theorem* 4.4. *Then*

$$f(t) = \exp\left\{\sum_{n\geq 0} \frac{p_n'(1)t^n}{(q)_n n}\right\}.$$

We choose a simple example to show how this theory is applied. Let

$$D_q = (1/X)(1 - \eta). \tag{4.10}$$

Then one trivially verifies (4.5) for D_q; hence D_q is an Eulerian differential operator (called the *q-derivative*).

Next we define

$$P_n(X, Y) = (X - Y)(X - Yq)\cdots(X - Yq^{n-1}), \qquad P_0(X, Y) = 1. \tag{4.11}$$

Since

$$\begin{aligned} D_q P_n(X,1) &= \frac{(X-1)(X-q)\cdots(X-q^{n-1}) - (Xq-1)\cdots(Xq-q^{n-1})}{X} \\ &= (X-1)(X-q)\cdots(X-q^{n-2})\ \frac{X - q^{n-1} - q^n(X - 1/q)}{X} \\ &= (1-q^n)P_{n-1}(X,1) \end{aligned}$$

and $P_0(X, 1) = 1$, $P_n(1, 1) = 0$, we see (by Theorem 4.3) that $\{P_n(X, 1)\}_{n=0}^{\infty}$ is the Eulerian family of polynomials associated with D_q.

Therefore, by Theorem 4.4,

$$\sum_{n\geq 0} \frac{P_n(X,1)t^n}{(q)_n} = \frac{e(Xt)}{e(t)}, \tag{4.12}$$

where $e(t) = \sum_{n\geq 0} t^n/(q)_n$.

Since

$$P_n'(1,1) = \lim_{X\to 1} \frac{P_n(X,1)}{X-1} = (q)_{n-1},$$

we see by Theorem 4.5 that

$$\begin{aligned} e(t) &= \exp\left\{\sum_{n\geq 1} \frac{(q)_{n-1}t^n}{(q)_n n}\right\} = \exp\left\{\sum_{n\geq 1} \frac{t^n}{(1-q^n)n}\right\} \\ &= \exp\left\{\sum_{n\geq 1}\sum_{m\geq 0} \frac{t^n q^{mn}}{n}\right\} = \exp\left\{-\sum_{m\geq 0} \log(1 - tq^m)\right\} = (t)_{\infty}^{-1}. \end{aligned} \tag{4.13}$$

Combining (4.13) with (4.12), we see that

$$\sum_{n\geq 0} \frac{P_n(X,1)t^n}{(q)_n} = \frac{(t)_\infty}{(tX)_\infty},$$

or

$$\sum_{n\geq 0} \frac{(X^{-1})_n (tX)^n}{(q)_n} = \frac{(t)_\infty}{(tX)_\infty}, \tag{4.14}$$

a result equivalent to Theorem 2.4.

If we define

$$\gamma = \frac{1}{X}\frac{(1-\eta)}{1-b\eta q^{-1}},$$

then γ is also an Eulerian differential operator, and if $\{g_n(X)\}_{n=0}^{\infty}$ is the related Eulerian family of polynomials, then it turns out that [8, (9.4), p. 365]

$$\sum_{n\geq 0}\frac{g_n(X)t^n}{(q)_n} = \frac{(bXt)_\infty(t)_\infty}{(Xt)_\infty(bt)_\infty}. \tag{4.15}$$

If we expand $g_n(X)$ in terms of $P_n(X, 1)$ so that

$$\sum_{n\geq 0}\frac{g_n(X)t^n}{(q)_n} = \sum_{n\geq 0}\frac{C_n(t)P_n(X,1)}{(q)_n}, \tag{4.16}$$

then it is possible to deduce [8, first Eq. (9.9), p. 367] that

$$C_n(t) = \frac{t^n(b)_n}{(bt)_n}. \tag{4.17}$$

Combining (4.15), (4.16) and (4.17), we see that [8, second Eq. (9.9), p. 367]

$$\sum_{n\geq 0}\frac{(X^{-1})_n(b)_n(tX)^n}{(q)_n(bt)_n} = \frac{(bXt)_\infty(t)_\infty}{(Xt)_\infty(bt)_\infty}, \tag{4.18}$$

a result equivalent to Theorem 2.6.

Other important results in basic hypergeometric functions are deducible from this theory including the Rogers–Ramanujan identities, which are related to the Eulerian differential operator $(1/X)(\eta^{-2} - \eta^{-1})$.

Finally, we note [48, p. 252] that when $X = q^x$, $Y = q^y$ and $x > y$, then $P_n(X, Y)$ is the number of nonsingular linear transformations of V_n into V_x where the image of V_n has only 0 in common with V_y, a fixed subspace of V_x. This is easily seen, for if $v_1, \cdots, v_n$ form a basis for V_n, we need only determine the action of a linear transformation f on this basis. There are clearly $X - Y$ choices for $f(v_1)$. Now $f(v_2)$ must lie outside the space spanned by V_y and $f(v_1)$; hence there are $X - Yq$ choices for $f(v_2)$. Continuing in this manner, we see that the total number of such mappings is $(X - Y)(X - Yq) \cdots (X - Yq^{n-1}) = P_n(X, Y)$. Thus we see the relationship between properties of finite vector spaces and Eulerian families of polynomials.

Another more complicated type of nonsingular linear transformation of a finite vector space is also studied in [8, § 11]; here the Eulerian family is

$$h_n(X) = \sum_{j\geq 0}\binom{n}{j}_q P_j(X,1)P_{n-j}(X,1)U^{n-j},$$

where U is a fixed parameter. The $h_n(X)$ are seen to be closely related to what are called the *q-Hermite polynomials* [24], $\sum_{j\geq 0}\binom{n}{j}_q U^{n-j}$.

We should add that recent work by M. Henle [41] suggests that both the Rota–Mullin theory of binomial enumeration [49] and the theory of Eulerian differential operators [8] may be subsumed in a more general structure with diverse combinatorial applications.

5. Application to combinatorial identities. With the rekindling of interest in combinatorial analysis has come a large amount of work on binomial coefficient summations such as

$$\sum_{k\geq 0}\binom{2n+1}{2p+2k+1}\binom{p+k}{k}=\binom{2n-p}{p}2^{2n-2p} \tag{5.1}$$

(see [30, p. 71], [34]), and

$$\sum_{k\geq 0}\binom{x+y+k}{k}\binom{y}{a-k}\binom{x}{b-k}=\binom{x+a}{b}\binom{y+b}{a} \tag{5.2}$$

(see [33]).

The impression one gains from treatises on the subject (cf. Riordan's book [46] or Gould's table [35]) is that, for the most part, such identities cannot be put into a coherent setting. Riordan's statement is, perhaps, the most pessimistic [46, p. vii]: "The central fact developed is that identities are both inexhaustible and unpredictable; the age-old dream of putting order in this chaos is doomed to failure."

This situation seems quite unfortunate. Indeed, binomial coefficient identities arise in many different areas of mathematics, and it would be quite useful if it were possible for mathematicians generally to be able to prove such identities in short order rather than being diverted into a general study of binomial identities just for the purpose of cracking one particular problem. H. W. Gould [35] has made a significant contribution to mitigating this situation. He has tabulated 555 combinatorial identities, of which over 450 are purely binomial coefficient identities. However, there still remains the difficulty of finding one's own identity among a vast number of others, and as Gould points out [35, p. viii], the situation may be further complicated through changes of variable and introduction of redundant factors.

The object of this section is to emphasize the close relationship between binomial coefficient identities and ordinary hypergeometric series (and simultaneously that between q-binomial coefficient identities and basic hypergeometric series). There are several important reasons for studying these relationships: first, by translating binomial coefficient identities into hypergeometric series identities, we then need only check the 4-page, 32-entry table of L. J. Slater [52, Appendix III] rather than the 450 entries of Gould. Second, we not only avoid the need to fashion ad hoc methods for summing our series, but we have our result completely established since we are quoting a known theorem. Third, the number of known q-binomial coefficient identities is much smaller than the number of known ordinary binomial coefficient identities. The approach used here allows the production of q-analogues of many known combinatorial identities by merely replacing the hypergeometric series translation by its q-analogue.

Theorem 5.1 indicates the procedure for translating binomial coefficient identities into hypergeometric series identities, and Theorem 5.2 illustrates why hypergeometric series identities are not easily disguised. The remaining theorems of this section show how our technique applies to some of the more popular binomial coefficient summations.

We define

$$[a]_n = a(a+1)\cdots(a+n-1). \tag{5.3}$$

The standard notation for this product is $(a)_n$; however, this last symbol has already been utilized extensively in this paper with another meaning.

THEOREM 5.1. *Let A, B, C, D be fixed nonnegative integers with $A > C \geqq 0$. Then*

$$(An + DA + C)! = (DA + C)!A^{An}\left[D + \frac{C+1}{A}\right]_n\left[D + \frac{C+2}{A}\right]_n \cdots \left[D + \frac{C}{A} + 1\right]_n; \tag{5.4}$$

$$(B - An)! = \frac{(-1)^{An}B!}{A^{An}\left[-\frac{B}{A}\right]_n\left[-\frac{B-1}{A}\right]_n \cdots \left[-\frac{B-A+1}{A}\right]_n}, \tag{5.5}$$

provided $B - An \geqq 0$;

$$(q)_{An+DA+C} = (q)_{DA+C}(q^{DA+C+1}; q^A)_n \cdots (q^{DA+C+A}; q^A)_n; \tag{5.6}$$

$$(q)_{B-An} = \frac{q^{-BAn+An(An-1)/2}(-1)^{An}(q)_B}{(q^{-B}; q^A)_n(q^{-B+1}; q^A)_n \cdots (q^{-B+A-1}; q^A)_n}. \tag{5.7}$$

Proof. These results are all obtained by simple algebraic manipulation. First,

$$\begin{aligned}(An + DA + C)! &= (DA + C)![DA + C + 1]_{An}\\ &= (DA + C)!A^{An}\left[D + \frac{C+1}{A}\right]_n\left[D + \frac{C+2}{A}\right]_n \cdots \left[D + \frac{C}{A} + 1\right]_n.\end{aligned}$$

Next,

$$\begin{aligned}(B - An)! &= \frac{B!}{B(B-1)\cdots(B-An+1)} = \frac{(-1)^{An}B!}{[-B]_{An}}\\ &= \frac{(-1)^{An}B!A^{-An}}{\left[-\frac{B}{A}\right]_n\left[-\frac{B-1}{A}\right]_n \cdots \left[-\frac{B-A+1}{A}\right]_n}.\end{aligned}$$

Since (5.6) and (5.7) are proved in an entirely analogous manner, we shall omit the proofs. □

We now need some standard definitions from the theory of ordinary hypergeometric functions.

Consider

$${}_nH_n\left[\begin{matrix}a_1, \cdots, a_n; z\\ b_1, \cdots, b_n\end{matrix}\right] = \sum_{m=-\infty}^{\infty} \frac{[a_1]_m \cdots [a_n]_m}{[b_1]_m \cdots [b_n]_m} z^m, \tag{5.8}$$

where $[\alpha]_{-n} = (-1)^n[1-\alpha]_n^{-1}$. This series reduces to ${}_nF_{n-1}$ (see (1.2)) if $b_1 = 1$.

We say that the series in (5.8) is *Saalschutzian* if $a_1 + a_2 + \cdots + a_n + 2 = b_1 + b_2 + \cdots + b_n$. We say that the series in (5.8) is *well-poised* if $a_1 + b_1 = a_2 + b_2 = \cdots = a_n + b_n$, and we say *nearly-poised* if only $n - 1$ of these equalities hold.

Analogously we may consider (as in § 3)

$$(5.9) \qquad {}_n\psi_n\begin{bmatrix}\alpha_1, \cdots, \alpha_n; q, t\\ \beta_1, \cdots, \beta_n\end{bmatrix} = \sum_{m=-\infty}^{\infty} \frac{(\alpha_1)_m \cdots (\alpha_n)_m t^m}{(\beta_1)_m \cdots (\beta_n)_m}.$$

This series reduces to ${}_n\phi_{n-1}$ (see (1.1)) if $\beta_1 = q$. We say that the series in (5.9) is *Saalschutzian* if $\alpha_1\alpha_2 \cdots \alpha_n q^2 = \beta_1\beta_2 \cdots \beta_n$; *well-poised* if $\alpha_1\beta_1 = \alpha_2\beta_2 = \cdots = \alpha_n\beta_n$, and *nearly-poised* if only $n - 1$ of these equalities hold.

In Slater's table of hypergeometric summations [52, Appendix III] only three identities [52, (III. 23), (III. 24) and (III. 28), p. 245] do not involve either a Saalschutzian, well-poised, or nearly-poised series; furthermore, all of the basic hypergeometric summations in [52, Appendix IV] involve basic hypergeometric series that are either Saalschutzian or well-poised.

THEOREM 5.2. *The properties "Saalschutzian", "well-poised" and "nearly-poised" are invariant under shifting the index of summation or reversing the order of summation.*

Remark. This result provides evidence for why Gould's table of binomial coefficient identities is so much longer than Slater's; namely, altering the index of summation now does not alter the most important properties of the series in question, and redundant factors are easily spotted for they will appear identically in both the numerators and denominators of the given series.

Proof. Replacing m by $-m$ in (5.8), we find

$$(5.10) \qquad \sum_{m=-\infty}^{\infty} \frac{[1 - b_1]_m \cdots [1 - b_n]_m z^{-m}}{[1 - a_1]_m \cdots [1 - a_n]_m};$$

hence $2 + \sum a_i = \sum b_i$ implies $2 + \sum (1 - b_i) = \sum (1 - a_i)$, and $a_1 + b_1 = \cdots = a_n + b_n$ implies $(1 - b_1) + (1 - a_1) = \cdots = (1 - b_n) + (1 - a_n)$. Also if only $n - 1$ of the latter equations hold for $a_i + b_i$, then the corresponding $n - 1$ equations hold for $(1 - a_i) + (1 - b_i)$. Thus the three properties listed in the theorem are preserved when m is replaced by $-m$.

If m is replaced by $m + k$ in (5.8), we obtain

$$(5.11) \qquad \frac{[a_1]_k \cdots [a_n]_k z^k}{[a_1]_k \cdots [a_n]_k} \sum_{m=-\infty}^{\infty} \frac{[a_1 + k]_m \cdots [a_n + k]_m z^m}{[b_1 + k]_m \cdots [b_n + k]_m}.$$

In this case $2 + \sum a_i = \sum b_i$ implies $2 + \sum (a_i + k) = \sum (b_i + k)$, and $a_1 + b_1 = \cdots = a_n + b_n$ implies $(a_1 + k) + (b_1 + k) = \cdots = (a_n + k) + (b_n + k)$. Similarly the nearly-poised condition holds, and thus the theorem is established for ordinary hypergeometric series.

The same approach applied to the basic series (5.9) produces first

$$\sum_{m=-\infty}^{\infty} \frac{\left(\frac{q}{\beta_1}\right)_m \cdots \left(\frac{q}{\beta_n}\right)_m \left(\frac{\beta_1\beta_2 \cdots \beta_n}{\alpha_1\alpha_2 \cdots \alpha_n t}\right)^n}{(q/\alpha_1)_m \cdots (q/\alpha_n)_m},$$

and next

$$\frac{(\alpha_1)_k \cdots (\alpha_n)_k t^k}{(\beta_1) \cdots (\beta_n)_k} \sum_{m=-\infty}^{\infty} \frac{(\alpha_1 q^k)_m \cdots (\alpha_n q^k)_m t^m}{(\beta_1 q^k)_m \cdots (\beta_n q^k)_m}.$$

The invariance of the three properties is again easily checked, and so our theorem is established. □

We shall now illustrate the applicability of our method by proving some identities of widespread combinatorial interest. In each instance, we easily produce the corresponding q-analogue.

THEOREM 5.3.

$$\sum_{k\geq 0} \binom{2n+1}{2p+2k+1}\binom{p+k}{k} = \binom{2n-p}{p} 2^{2n-2p}, \tag{5.12}$$

$$\sum_{k\geq 0} \binom{2n}{2p+2k}\binom{p+k}{k} = \frac{n}{2n-p}\binom{2n-p}{p} 2^{2n-2p}. \tag{5.13}$$

Remark. H. T. Davis [30] attributes these identities to the fictional, arch-criminal James Moriarty of the Sherlock Holmes short story, "The Final Problem". The reference is to Holmes's description of Moriarty in which he says, "At the age of twenty-one he wrote a treatise upon the binomial theorem, which has had a European vogue." Apparently Moriarty had turned to crime even earlier than had been thought previously (except by Gould [34, (15)]); for as we shall see below, these two results are merely disguised forms of Gauss's summation of ${}_2F_1\left[\begin{matrix} a, b; 1 \\ c \end{matrix}\right]$.

Proof. For (5.12), we see that

$$\sum_{k\geq 0} \binom{2n+1}{2p+2k+1}\binom{p+k}{k}$$

$$= \binom{2n+1}{2p+1} \sum_{k\geq 0} \frac{[p-n]_k[p-n+\frac{1}{2}]_k[p+1]_k}{[p+1]_k[p+\frac{3}{2}]_k k!} \qquad \text{(by Theorem 5.1)}$$

$$= \binom{2n+1}{2p+1} {}_2F_1\left[\begin{matrix} p-n, p-n+\frac{1}{2}; 1 \\ p+\frac{3}{2} \end{matrix}\right]$$

$$= \binom{2n+1}{2p+1} \frac{[n+1]_{n-p}}{[p+\frac{3}{2}]_{n-p}} \qquad \text{(by Gauss's theorem [52, III. 3 (or III. 4), p. 243])}$$

$$= \frac{2^{2n-2p}[1]_n[\frac{3}{2}]_2(2n-p)![\frac{3}{2}]_p}{(2n-2p)![1]_p[\frac{3}{2}]_p[1]_n[\frac{3}{2}]_n} \qquad \text{(by Theorem 5.1)}$$

$$= \binom{2n-p}{p} 2^{2n-2p}.$$

For (5.13), we see that

$$\sum_{k\geqq 0}\binom{2n}{2p+2k}\binom{p+k}{k}$$

$$=\binom{2n}{2p}\sum_{k\geqq 0}\frac{[p-n]_k[p-n+\frac{1}{2}]_k[p+1]_k}{[p+\frac{1}{2}]_k[p+1]_k k!} \qquad \text{(by Theorem 5.1)}$$

$$=\binom{2n}{2p}{}_2F_1\left[\begin{matrix}p-n, p-n+\frac{1}{2}; 1\\ p+\frac{1}{2}\end{matrix}\right]$$

$$=\binom{2n}{2p}\frac{[n]_{n-p}}{[p+\frac{1}{2}]_{n-p}} \qquad \text{(by Gauss's theorem [52, III. 3 (or III. 4), p. 243])}$$

$$=\frac{2^{2n-2p}[1]_n[\frac{3}{2}]_{n-1}(2n-p-1)![\frac{1}{2}]_p}{(2n-2p)![1]_p[\frac{3}{2}]_{p-1}(n-1)![\frac{1}{2}]_n}$$

$$=2^{2n-2p}\binom{2n-p}{p}\frac{n}{n-2p}. \quad \square$$

If we consider the q-analogue of the above application of Gauss's theorem, we find the following q-analogues of the Moriarty identities.

THEOREM 5.3.

$$\sum_{k\geqq 0}\binom{2n+1}{2p+2k+1}_q\binom{p+k}{k}_{q^2}q^{k(2p+2k+1)} = \binom{2n-p}{p}_{q^2}(-q)_{2n-2p}. \tag{5.14}$$

$$\sum_{k\geqq 0}\binom{2n}{2p+2k}_q\binom{p+k}{k}_{q^2}q^{k(2p+2k-1)} = \binom{2n-p}{p}_{q^2}(-q)_{2n-2p}\frac{(1-q^{2n})}{(1-q^{4n-2p})}. \tag{5.15}$$

Proof. For (5.14), we see that

$$\sum_{k\geqq 0}\binom{2n+1}{2p+2k+1}_q\binom{p+k}{k}_{q^2}q^{k(2p+2k+1)}$$

$$=\binom{2n+1}{2p+1}_q\sum_{k\geqq 0}\frac{q^{(2n-2p)2k-k(2k-1)}(q^{2p-2n})_{2k}(q^{2p+2};q^2)_k q^{k(2p+2k+1)}}{(q^{2p+2})_{2k}(q^2;q^2)_k}$$

(by Theorem 5.1)

$$=\binom{2n+1}{2p+1}_q\sum_{k\geqq 0}\frac{(q^{2p-2n};q^2)_k(q^{2p-2n+1};q^2)_k(q^{2p+2};q^2)_k q^{4nk-2kp+2k}}{(q^{2p+2};q^2)_k(q^{2p+3};q^2)_k(q^2;q^2)_k}$$

$$=\binom{2n+1}{2p+1}_q{}_2\phi_1\left[\begin{matrix}q^{2p-2n}, q^{2p-2n+1}; q^2, q^{4n-2p+2}\\ q^{2p+3}\end{matrix}\right] \qquad (cont.)$$

$$= \begin{pmatrix} 2n+1 \\ 2p+1 \end{pmatrix}_q \frac{(q^{2n+2};q^2)_{n-p}}{(q^{2p+3};q^2)_{n-p}} \qquad \text{(by the } q\text{-analogue of Gauss's theorem [52, IV. 2 or IV. 3, p. 247], i.e., our Theorem 2.6)}$$

$$= \frac{(q;q^2)_{n+1}(q^2;q^2)_n(q^2;q^2)_{2n-p}(q;q^2)_{p+1}}{(q;q^2)_{p+1}(q^2;q^2)_p(q)_{2n-2p}(q^2;q^2)_n(q;q^2)_{n+1}}$$

$$= \begin{pmatrix} 2n-p \\ p \end{pmatrix}_{q^2} (-q)_{2n-2p}.$$

For (5.14), we see that

$$\sum_{k\geqq 0} \begin{pmatrix} 2h \\ 2p+2k \end{pmatrix}_q \begin{pmatrix} p+k \\ k \end{pmatrix}_{q^2} q^{k(2p+2k-1)}$$

$$= \begin{pmatrix} 2n \\ 2p \end{pmatrix}_q \sum_{k\geqq 0} \frac{q^{(2n-2p)2k-k(2k-1)}(q^{2p-2n})_{2k}(q^{2p+2};q^2)_k q^{k(2p+2k-1)}}{(q^{2p+1})_{2k}(q^2;q^2)_k}$$

$$= \begin{pmatrix} 2n \\ 2p \end{pmatrix}_q \sum_{k\geqq 0} \frac{(q^{2p-2n+1};q^2)_k(q^{2p-2n};q^2)_k q^{(4n-2p)k}}{(q^2;q^2)_k(q^{2p+1};q^2)_k}$$

$$= \begin{pmatrix} 2n \\ 2p \end{pmatrix}_q {}_2\phi_1\left[\begin{matrix} q^{2p-2n+1}, q^{2p-2n}; q^2, q^{4n-2p} \\ q^{2p+1} \end{matrix}\right]$$

$$= \begin{pmatrix} 2n \\ 2p \end{pmatrix}_q \frac{(q^{2n};q^2)_{n-p}}{(q^{2p+1};q^2)_{n-p}} \qquad \text{(by the } q\text{-analogue of Gauss's theorem [52; IV. 2 or IV. 3, p. 247], i.e., our Theorem 2.6)}$$

$$= \frac{(q;q^2)_n(q^2;q^2)_n(q;q^2)_p(q^2;q^2)_{2n-p-1}}{(q;q^2)_p(q^2;q^2)_p(q)_{2n-2p}(q;q^2)_n(q^2;q^2)_{n-1}}$$

$$= \begin{pmatrix} 2n-p \\ p \end{pmatrix}_{q^2} (-q)_{2n-2p} \frac{(1-q^{2n})}{(1-q^{4n-2p})}. \qquad \square$$

We next treat a slightly harder problem, namely the Reed Dawson identities [46, p. 71]; these results show that sometimes it is necessary to reverse summation when treating a ${}_2F_1$ (or ${}_2\phi_1$). In this case (as we shall see), the results are corollaries of Gauss's second summation of a ${}_2F_1$.

THEOREM 5.4

$$\sum_{k\geqq 0} (-1)^k \binom{n}{k} 2^{-k} \binom{2k}{k} = \begin{cases} 2^{-2\nu}\dbinom{2\nu}{\nu} & \textit{if } n = 2\nu, \\ 0 & \textit{if } n \textit{ is odd}. \end{cases}$$

Proof. First we see that

$$\sum_{k\geqq 0} (-1)^k \binom{n}{k} 2^{-k} \binom{2k}{k} = \sum_{k\geqq 0} \frac{[-n]_k[1]_k[\frac{1}{2}]_k 2^k}{k!k!k!} \qquad \text{(by Theorem 5.1)}$$

$$= {}_2F_1\left[\begin{matrix} -n, \frac{1}{2}; 2 \\ 1 \end{matrix}\right].$$

A check of Slater's Appendix III shows no ${}_2F_1$ series with argument 2; however, since a number have argument $\frac{1}{2}$, we should clearly reverse the order of summation (i.e., replace k by $n - k$). Hence

$$\sum_{k \geqq 0} (-1)^k \binom{n}{k} 2^{-k} \binom{2k}{k}$$

$$= \sum_{k \geqq 0} (-1)^{n-k} \binom{n}{k} 2^{-n+k} \binom{2n-2k}{n-k}$$

$$= \sum_{k \geqq 0} (-1)^{n-k} \frac{(-n)_k(-1)^k 2^{-n+k}(-n)_k(-n)_k(2n)!}{k!n!n!2^{2k}(-n)_k(-n+\frac{1}{2})_k} \qquad \text{(by Theorem 5.1)}$$

$$= (-1)^n 2^{-n} \binom{2n}{n} {}_2F_1 \left[\begin{matrix} -n, -n; \frac{1}{2} \\ -n+\frac{1}{2} \end{matrix} \right]$$

$$= (-1)^n 2^{-n} \binom{2n}{n} \frac{\Gamma(\frac{1}{2})\Gamma(-n+\frac{1}{2})}{\Gamma\left(\frac{1}{2} - \frac{n}{2}\right)^2} \qquad \text{(by Gauss's second theorem [52, III. 6, p. 243])}$$

$$= \begin{cases} 0 & \text{if } n \text{ is odd (since } \Gamma(z)^{-1} \text{ has zeros when } z \text{ is a negative integer)}, \\ 2^{-2\nu} \binom{4\nu}{2\nu} \dfrac{[\frac{1}{2} - \nu]_\nu}{[\frac{1}{2} - 2\nu]_\nu} & \text{if } n = 2\nu \text{ since } [a]_n = \Gamma(a+n)/\Gamma(a) \end{cases}$$

$$= \begin{cases} 0 & \text{if } n \text{ is odd}, \\ \dfrac{2^{-2\nu}(4\nu)!(2\nu)!2^{2\nu}(2\nu)!2^\nu[\frac{1}{2}]_\nu}{(2\nu)!(2\nu)!2^\nu \nu!(4\nu)!} & \text{if } n = 2\nu \end{cases}$$

$$= \begin{cases} 0 & \text{if } n \text{ is odd}, \\ 2^{-2\nu} \binom{2\nu}{\nu} & \text{if } n = 2\nu. \quad \square \end{cases}$$

In passing to the q-analogue of the Reed Dawson identities, we encounter the problem of finding the q-analogue of Gauss's second theorem; it is not listed by Slater and has in fact only been discovered recently [11]. Namely,

$$\sum_{k=0}^{\infty} \frac{(a)_k(b)_k q^{k(k+1)/2}}{(q)_k(abq; q^2)_k} = \frac{(-q)_\infty(aq; q^2)_\infty(bq; q^2)_\infty}{(qab; q^2)_\infty}. \tag{5.16}$$

THEOREM 5.5.

$$\sum_{k \geqq 0} (-1)^k \binom{n}{k}_q \frac{1}{(-q)_k} \binom{2k}{k}_q q^{(n-k)(n-k-1)/2} = \begin{cases} 0 & \textit{if } n \textit{ is odd} \\ \dfrac{q^{2\nu^2}}{(-q)_\nu^2} \binom{2\nu}{\nu}_q & \textit{if } n = 2\nu. \end{cases} \tag{5.17}$$

Proof. We begin by replacing k by $n - k$.

Hence

$$\sum_{k\geqq 0}(-1)^{n-k}\begin{bmatrix}n\\k\end{bmatrix}_q \frac{1}{(-q)_{n-k}}\begin{bmatrix}2n-2k\\n-k\end{bmatrix}_q q^{k(k-1)/2}$$

$$= (-1)^n \sum_{k\geqq 0}\frac{(q^{-n})_k q^{kn-k(k-1)/2}(q)_{2n-2k}q^{k(k-1)/2}}{(q)_k(q)_{n-k}(q^2;q^2)_{n-k}} \qquad \text{(by Theorem 5.1)}$$

$$= (-1)^n \sum_{k\geqq 0}\frac{(q^{-n})_k q^{kn}(q;q^2)_{n-k}}{(q)_k(q)_{n-k}}$$

$$= \frac{(-1)^n(q;q^2)_n}{(q)_n}\sum_{k\geqq 0}\frac{(q^{-n})_k(q^{-n})_k q^{2kn-k^2/2+k/2-k2n+k^2}}{(q)_k(q^{-2n+1};q^2)_k}$$

$$= \frac{(-1)^n(q;q^2)_n}{(q)_n}\sum_{k\geqq 0}\frac{(q^{-n})_k(q^{-n})_k q^{k(k+1)/2}}{(q)_k(q^{-2n+1};q^2)_k}$$

$$= \frac{(-1)^n(q;q^2)_n(-q)_\infty(q^{-n+1};q^2)_\infty^2}{(q)_n(q^{-2n+1};q^2)_\infty} \qquad \text{(by (5.16), the } q\text{-analogue of Gauss's second theorem)}$$

$$= \begin{cases} 0 & \text{if } n \text{ is odd, for then } (q^{-n+1};q^2)_\infty = 0,\\ \dfrac{(q;q^2)_{2\nu}(q^{-2\nu+1};q^2)_\nu^2}{(q)_{2\nu}(q^{-4\nu+1};q^2)_{2\nu}} & \text{if } n = 2\nu \end{cases}$$

$$= \begin{cases} 0 & \text{if } n \text{ is odd,}\\ \dfrac{q^{2\nu^2}}{(-q)_\nu^2}\begin{bmatrix}2\nu\\\nu\end{bmatrix} & \text{if } n = 2\nu. \end{cases}$$

Theorems 5.3–5.5 illustrate well how our method works. As theorems, 5.4 and 5.5 are new; it should be clear that this technique is extremely useful in producing q-analogues. As far as ordinary binomial coefficient identities go, we shall have to wait and see how useful this approach is. There is a great deal of evidence to support our belief that the use of hypergeometric series is useful in such problems. For example, of the first 50 entries in Table 3 of Gould's book [35], over 40 may be proved by our method using only Gauss's first and second theorems, Kummer's theorem, Saalschutz's theorem, and Dixon's theorem [52, III.3, III.6, III.5, III.2, III.8, p. 243]. Another example is the Chinese identity

$$\binom{n+p}{p}^2 = \sum_{k\geqq 0}\binom{p}{k}^2\binom{n+2p-k}{2p}; \tag{5.18}$$

Riordan points out [46, p. 16] that many proofs of (5.18) were given (at least 8). As Carlitz pointed out in [23], this is merely a corollary of Saalschutz's theorem [52, III.2, p. 243]. Similarly the binomial identity obtained by Cartier and Foata in their study of graphs on three vertices [26, Chap. 6] and (5.2) above are both a direct corollary of Saalschutz's theorem.

Since the approach described in this section is like an algorithm for summing series of binomial coefficients, we are preparing a computer program [12] that should greatly simplify work on such problems.

6. Application to physics. There have been several applications of basic hypergeometric series to physics in recent years. M. Baker and D. D. Coon [20] have successfully utilized basic hypergeometric functions in a series of papers (in the *Physical Review* D) on particle physics. Also N. G. van Kampen [54] has applied basic hypergeometric functions to fluctuations in an electric circuit consisting of a condenser and a diode.

We have chosen as an illustration W. Hahn's solution to the equation of motion of a weightless elastic cable that carries n point masses, where both the weights and mutual distances of successive masses decrease in geometric progression with ratio q, $0 < q < 1$. The solution is a q-analogue of the Laguerre polynomials; in fact, if the weights and distances are constant (i.e., $q = 1$) the solution reduces to the ordinary Laguerre polynomials. The classical case $q = 1$ is treated in detail by O. Bottema [22].

We consider n-point masses P_i $(1 \leqq i \leqq n)$ on a weightless elastic cable acting only under the force of gravity. We let k_i denote the distance from P_i to P_{i+1}, m_i the mass of P_i, and $u_i(t) = u_i$ the displacement of the point P_i at time t. We denote the tension in the portion of cable from P_i to P_{i+1} by S_i. The force which acts on m_i in the positive u-direction is then given by

$$-\frac{S_{i-1}}{k_{i-1}}(u_i - u_{i-1}) + \frac{S_i}{k_i}(u_{i+1} - u_i) \qquad (1 \leqq i \leqq n, u_0 = u_{n+1} = 0).$$

Consequently the equations of motion are

$$m_i u_i''(t) = -\frac{S_{i-1}}{k_{i-1}}(u_i - u_{i-1}) + \frac{S_i}{k_i}(u_{i+1} - u_i).$$

One now applies the method of Lagrange, making the substitution

$$u_i = y_i \cos(\sqrt{u}\, t + \varphi),$$

which leads to the following set of linear equations for the y_i:

$$\mu m_1 y_1 = -\frac{S_1}{k_1}(y_2 - y_1) \tag{6.1}$$

$$\mu m_i y_i = \frac{S_{i-1}}{k_{i-1}}(y_i - y_{i-1}) - \frac{S_i}{k_i}(y_{i+1} - y_1) \qquad (2 \leqq i \leqq n). \tag{6.2}$$

If we now assume

$$m_i = q^{i-1} m_1, \qquad k_i = q^{i-1} k, \qquad 0 < q < 1, \tag{6.3}$$

and for simplicity

$$m_1 = g^{-1}, \qquad k_1 = k/(1 - q), \tag{6.4}$$

then utilizing the substitution

$$T_i(\mu) = (-1)^i y_{i+1}(\mu) \prod_{r=1}^{i} \frac{S^r}{k^r m^r}, \qquad T_{-1} = 0, \qquad T_0 = 1,$$

we deduce that

$$T_i(\mu) = \left(\mu - \frac{1-q^{i-1}}{kq^{2i-3}} - \frac{1-q^i}{kq^{2i-2}}\right) T_{i-1}(\mu) - \frac{(1-q^{i-1})^2}{k^2q^{4i-5}} T_{i-2}(\mu).$$

It is now a simple matter to check (by comparing coefficients of μ^r) that

$$\begin{aligned} T_i(\mu) &= (-1)^i q^{-i(i-1)/2}(q)_i \, {}_2\phi_1\left[\begin{matrix} q^{-i}, 0; q, q^i k\mu \\ q \end{matrix}\right] \\ &= (-1)^i q^{-i(i-1)/2}(q)_i \sum_{r=0}^{i} (-1)^r \frac{k^r \mu^r q^{r(r-1)/2}}{(q)_r} \begin{pmatrix} i \\ r \end{pmatrix}_q . \end{aligned} \tag{6.5}$$

We note that

$$\lim_{q\to 1} \frac{(-1)^i q^{i(i-1)/2}}{(q)_i} T_i(\mu(1-q)) = \sum_{r=0}^{i} (-1)^r \frac{k^r \mu^r}{r!} \binom{i}{r} = L_i(k\mu),$$

the ith Laguerre polynomial [22, (3), p. 46].

If we now assume that our cable possesses infinitely many point masses subject to (6.3) and (6.4), we shall see that a q-difference equation arises whose solution is $\lim_{i\to\infty} (-1)^i q^{i(i-1)/2} T_i(q)/(q)_i$. The length of our cable is in this case $L = k(1-q)^{-2}$. The rest point is taken to be P_1, and now $S_i = g(m_{i+1} + m_{i+2} + \cdots) = q^i/(1-q)$. Then instead of (6.2) we are led to

$$\mu q^{i-1} y_i = \frac{1}{k}(y_i - y_{i-1}) - \frac{1}{k}(y_{i+1} - y_i). \tag{6.6}$$

If we write $x = q^{i-1}$ and $V(x) = y_{i-1}$, we see that (6.6) becomes

$$V(q^2x) + (k\mu x - 2)V(qx) + V(x) = 0. \tag{6.7}$$

The simplest solution of (6.7) analytic in x at 0 may easily be determined by substituting $V(x) = \sum_{j=0}^{\infty} A_j x^j$ into (6.7). As a result, we find

$$A_j(q^j - 1)^2 = -k\mu q^{j-1} A_{j-1}, \tag{6.8}$$

or

$$V(x) = A_0 \sum_{r=0}^{\infty} \frac{(-1)^r q^{r(r-1)/2} (\mu k x)^r}{(q)_r^2}, \tag{6.9}$$

a q-analogue of the Bessel function

$$J_0(2\sqrt{\mu k x}) = \sum_{r=0}^{\infty} \frac{(-1)^r (\mu k x)^r}{(r!)^2}.$$

By the above simple example, we hope to make clear that many problems related to arithmetic differences have q-analogues in which geometric differences replace arithmetic differences. Just as solutions in the arithmetic case many times are hypergeometric series, so are solutions in the geometric case basic hypergeometric series.

7. Conclusion. As was stated in the Introduction, we have in this paper only briefly sampled several areas of mathematics in which basic hypergeometric functions play an important role. Not only are there many further applications in each of the areas we have mentioned, but there are many unexplored possibilities for applications still open. I would hope that the variety and novelty of the results already found might provide inspiration for further research. The interesting work of R. P. Agarwal and his students [2], [3] on bi-basic hypergeometric series should also yield interesting applications in the future.

The following list of references is by no means exhaustive; in fact, we only list those papers cited in the text. The reader is referred to [1], [2], [4], [9], [15], [28], [35], [36] and [52] for much more extensive bibliographies.

REFERENCES

[1] C. R. ADAMS, *Linear q-difference equations*, Bull. Amer. Math. Soc., 37 (1931), pp. 361–400.

[2] R. P. AGARWAL AND A. VERMA, *Generalized basic hypergeometric series with unconnected bases*, Proc. Cambridge Philos. Soc., 63 (1967), pp. 727–734.

[3] ———, *Generalized basic hypergeometric series with unconnected bases II*, Quart. J. Math. Oxford Ser., 18 (1967), pp. 181–192.

[4] H. L. ALDER, *Partition identities from Euler to the present*, Amer. Math. Monthly, 76 (1969), pp. 733–746.

[5] G. E. ANDREWS, *On q-difference equations for certain well-poised basic hypergeometric series*, Quart. J. Math. Oxford Ser., 19 (1968), pp. 433–447.

[6] ———, *On a transformation of bilateral series with applications*, Proc. Amer. Math. Soc., 25 (1970), pp. 554–558.

[7] ———, *On a partition problem of J. J. Sylvester*, J. London Math. Soc., 2 (1970), pp. 571–576.

[8] ———, *On the foundations of combinatorial theory, V: Eulerian differential operators*, Studies in Appl. Math., 50 (1971), pp. 345–375.

[9] ———, *Partition identities*, Advances in Math., 9 (1972), pp. 10–51.

[10] ———, *Number Theory*, W. B. Saunders, Philadelphia, 1971.

[11] ———, *On the q-analog of Kummer's theorem with applications*, Duke Math. J., 40 (1973), pp. 525–528.

[12] ———, *On a computer program for summing series of binomial coefficients*, in preparation.

[13] A. O. L. ATKIN, *Proof of a conjecture of Ramanujan*, Glasgow Math. J., 8 (1967), pp. 14–32.

[14] W. N. BAILEY, *Series of hypergeometric type which are infinite in both directions*, Quart. J. Math., 7 (1936), pp. 105–115.

[15] ———, *Generalized hypergeometric series*, Cambridge Mathematical Tract No. 32, Cambridge University Press, Cambridge, 1935.

[16] ———, *A note on certain q-identities*, Quart. J. Math., 12 (1941), pp. 173–175.

[17] ———, *A note on two of Ramanujan's formulae*, Quart. J. Math. Oxford Ser., 3 (1952), pp. 29–31.

[18] ———, *On the simplification of some identities of the Rogers–Ramanujan type*, Proc. London Math. Soc., 1 (1951), pp. 217–222.

[19] ———, *On the basic bilateral hypergeometric series* ${}_2\Psi_2$, Quart. J. Math. Oxford Ser., 1 (1950), pp. 194–198.

[20] M. BAKER AND D. COON, *Dual resonance theory with nonlinear trajectories*, Phys. Rev. D, 2 (1970), pp. 2349–2358.

[21] E. A. BENDER, *A generalized q-binomial Vandermonde convolution*, Discrete Math., 1 (1971), pp. 115–119.

[22] O. BOTTEMA, *Die Schwingungen eines zusammengesetzten Pendels*, Jber. Deutsche Math.-Verein., 42 (1932), pp. 42–60.

[23] L. CARLITZ, *On a problem in the history of Chinese mathematics*, Mat. Lapok, 6 (1955), pp. 219–220.

[24] ———, *Some polynomials related to theta functions*, Ann. Mat. Pura Appl., 41 (1956), pp. 359–373.

[25] L. Carlitz and M. V. Subbarao, *A simple proof of the quintuple product identity*, Proc. Amer. Math. Soc., 32 (1972), pp. 42–44.

[26] P. Cartier and D. Foata, *Problèmes combinatoires de commutation et réarrangements*, Lecture Notes in Math., No. 85, Springer, New York, 1969.

[27] A. Cayley, *Note on a partition series*, Amer. J. Math., 6 (1884), pp. 63–64.

[28] T. W. Chaundy, *F. H. Jackson*, J. London Math. Soc., 37 (1962), pp. 126–128; (obituary).

[29] J. A. Daum, *The basic analog of Kummer's theorem*, Bull. Amer. Math. Soc., 48 (1942), pp. 711–713.

[30] H. T. Davis, *The Summation of Series*, Principia Press, San Antonio, Texas, 1962.

[31] H. Göllnitz, *Partitionen unit Differenzenbedingungen*, J. Reine Angew. Math., 225 (1965), pp. 154–190.

[32] B. Gordon, *Some identities in combinatorial analysis*, Quart. J. Math. Oxford Ser., 12 (1961), pp. 285–290.

[33] H. W. Gould, *A new symmetrical combinatorial identity*, J. Combinatorial Theory, 13 (1972), pp. 278–286.

[34] ———, *The case of the strange binomial identities of Professor Moriarty*, Fibonacci Quart., 10 (1962), pp. 381–391, p. 402.

[35] ———, *Combinatorial Identities: A Standardized Set of Tables Listing 500 Binomial Coefficient Summations*, published by the author, 1972.

[36] W. Hahn, *Bertrage zur Theorie der Heineschen Reihen, Die 24 integrale der hypergeometrischen q-Differenzengleichung, Das q-Analogen der Laplace Transformation*, Math. Nachr., 2 (1949), pp. 263–278.

[37] ———, *Über die hoheren Heineschen Rechen und eine einheitliche Theorie der sogennanten speziellen Funktionen*, Ibid., 3 (1950), pp. 257–294.

[38] ———, *Die mechanische Deutung einer geometrischen Differenzgleichung*, Z. Angew. Math. Mech., 33 (1953), pp. 1–3.

[39] G. H. Hardy and E. M. Wright, *An Introduction to the Theory of Numbers*, 4th ed., Oxford University Press, Oxford, 1960.

[40] E. Heine, *Handbuch die Kugelfunctionen; Theorie und Anwendung*, vol. I, Reimer, Berlin, 1878.

[41] M. Henle, *Binomial enumeration on dissects*, to appear.

[42] D. Knuth, *Subspaces, subsets, and partitions*, J. Combinatorial Theory A, 10 (1971), pp. 178–180.

[43] T. M. MacRobert, *Functions of a Complex Variable*, 4th ed., Macmillan, London, 1958.

[44] L. J. Mordell, *On the representation of numbers as the sum of 2r squares*, Quart. J. Math., 48 (1920), pp. 93–104.

[45] V. Ramamani and K. Venkatachaliengar, *On a partition theorem of Sylvester*, Michigan Math. J., 19 (1972), pp. 137–140.

[46] J. Riordan, *Combinatorial Identities*, John Wiley, New York, 1968.

[47] G. C. Rota and J. Goldman, *The number of subspaces of a vector space*, Recent Progress in Combinatorics, W. T. Tutte, ed., Academic Press, New York, 1969, pp. 75–83.

[48] ———, *On the foundations of combinatorial theory, IV: Finite vector spaces and Eulerian generating functions*, Studies in Appl. Math., 49 (1970), pp. 239–258.

[49] G. C. Rota and R. Mullin, *On the foundations of combinatorial theory, III: Theory of binomial enumeration*, Graph Theory and its Applications, B. Harris, ed., Academic Press, New York, 1970, pp. 167–213.

[50] W. W. Sawyer, *Prelude to Mathematics*, Penguin, Baltimore, 1955.

[51] L. J. Slater and A. Lakin, *Two proofs of the* ${}_6\Psi_6$ *summation theorem*, Proc. Edinburgh Math. Soc., 9 (1956), pp. 116–121.

[52] L. J. Slater, *Generalized Hypergeometric Functions*, Cambridge University Press, Cambridge, 1966.

[53] J. J. Sylvester, *A constructive theory of partitions arranged in three acts, an interact, and an exodion*, Amer. J. Math., 5 (1882), pp. 251–330, 6 (1884), pp. 334–336; (or *Collected Mathematical Papers of James Joseph Sylvester*, vol. 4, Cambridge University Press, Cambridge, 1912, pp. 1–83).

[54] N. G. van Kampen, *Exact calculation of the fluctuation spectrum for a nonlinear model system*, J. Mathematical Phys., 2 (1961), pp. 592–601.

7.5 Problems and Prospects for Basic Hypergeometric Functions

Reprinted from:
THEORY AND APPLICATION OF SPECIAL FUNCTIONS

ACADEMIC PRESS, INC.
New York San Francisco London

Problems and Prospects for Basic Hypergeometric Functions

George E. Andrews

1. Introduction.

The ordinary basic hypergeometric function is

$$ {}_m\phi_n\left[\begin{matrix} a_1, \ldots, a_m; q, z \\ b_1, \ldots, b_n \end{matrix}\right] = \sum_{j \geq 0} \frac{(a_1)_j (a_2)_j \ldots (a_m)_j z^j}{(b_1)_j (b_2)_j \ldots (b_n)_j (q)_j} , $$

where $(a)_n = (a;q)_n = (1-a)(1-aq)\ldots(1-aq^{n-1})$, $|z| < 1$, $|q| < 1$. In a recent survey paper, "Applications of basic hypergeometric functions" (Andrews (1974a), henceforward referred to ABHF) the applications of basic hypergeometric functions to partitions, number theory, finite vector spaces, combinatorial identities, and physics were discussed. Since that paper was written, a number of advances have been made (especially in combinatorial identities and the theory of partitions), which suggest that the theory of basic hypergeometric functions can be greatly extended and that the resulting theory will have considerable impact on the subjects that originally suggested the extensions. We might have called this paper "Multiple Basic Hypergeometric Series" since most of the extensions we shall consider relate to multiple series; however, such a title would not really convey the intended spirit of this paper.

In Section 2, we shall describe the recent work in the theory of partitions which suggests the need for an extended theory of basic hypergeometric functions. In Section 3 we shall present an extension of Watson's q-analog of Whipple's theorem (Watson (1929)) and we shall show how this result implies many new analytic identities of the Rogers-

Ramanujan type including some recently discovered ones (Andrews (1974b)); we also describe the recent work of Holman, Biedenharn, and Louck (1975) which suggests alternative analogs of Whipple's theorem (Whipple (1926)). In Section 4, we shall study basic Appell and Lauricella functions in order to provide generalizations of the q-analogs of the Chu-Vandermonde summation and Saalschütz's theorem. In Section 5, we shall study the Dyson conjecture as a higher dimensional analog of Dixon's summation, and we shall inquire about possible q-analogs; also a possible q-analog of MacMahon's Master Theorem (MacMahon (1894), (1915)) will be considered. In Section 6, we examine a partition theorem of Göllnitz (1967) which leads to the consideration of nonterminating Saalschützian series; also q-inversion theorems are briefly discussed in this section.

In each section of the paper we shall begin with the statement of one or more problems. The main body of each section will then be devoted to the "prospects" for solution (s) and sometimes a related problem will be stated. In some instances such as Section 4, we shall be able to describe a number of known results.

2. Partitions identities.

PROBLEM 1. What finite linear homogeneous ordinary q-difference equations with coefficients that are polynomials in x and q have multiple basic hypergeometric series as solutions?

Let us begin by showing why this problem is important. For more details we refer the reader to Andrews (1972a), (1974c). Let S denote the set of all $\pi = (1^{f_1} 2^{f_2} 3^{f_3} \ldots)$, the partitions of non-negative integers. The set S forms a distributive lattice under the partial ordering $(1^{f_1} 2^{f_2} 3^{f_3} \ldots) \leqq (1^{g_1} 2^{g_2} 3^{g_3} \ldots)$, whenever $f_i \leqq g_i$ for all i; the "meet" and "join" operations are defined by

$$(1^{t_1} 2^{f_2} \ldots) \cap (1^{g_1} 2^{g_2} \ldots) = (1^{\min(f_1, f_1)} 2^{\min(f_2, g_2)} \ldots)$$

$$(1^{f_1} 2^{f_2} \ldots) \cup (1^{g_1} 2^{g_2} \ldots) = (1^{\max(f_1, g_1)} 2^{\max(f_2, g_2)} \ldots) .$$

BASIC HYPERGEOMETRIC FUNCTIONS

A subset C of S which has the property that whenever $\pi \in C$ and $\pi' \leqq \pi$ then $\pi' \in C$ is called a partition ideal.

Let C be a partition ideal, and denote by $C^{(m)}$ that subset of C that consists of those $\pi = (1^{f_1} 2^{f_2} 3^{f_3} \ldots)$ with $f_1 = f_2 = \ldots = f_m = 0$, i.e. $C^{(m)}$ consists of those partitions in C that have all their parts $> m$. Define $\varphi: S \to S^{(1)}$ by $\varphi((1^{f_1} 2^{f_2} 3^{f_3} \ldots)) = (1^0 2^{f_1} 3^{f_2} \ldots)$; intuitively φ adds 1 to each part of each partition. We shall say that C has modulus $m > 0$ if m is the least positive integer such that $\varphi^m C = C^{(m)}$; it is not difficult to show that if C has a modulus m then $\varphi^M C = C^{(M)}$ if and only if $m | M$.

We shall say that C satisfies the finiteness property if for each $j \geqq 1$ there exists M_j such that if $\pi = (1^{f_1} 2^{f_2} 3^{f_3} \ldots) \in C$ then $f_j \leqq M_j$ for each j.

We shall add partitions in the obvious manner:

$$(1^{f_1} 2^{f_2} 3^{f_3} \ldots) \oplus (1^{f'_1} 2^{f'_2} 3^{f'_3} \ldots)$$

$$= (1^{f_1+f'_1} 2^{f_2+f'_2} 3^{f_3+f'_3} \ldots).$$

Now if C has a modulus m, then it is easy to see that for each $\pi \in C$, there exists $\pi_0, \pi_1, \pi_2, \ldots \in L_C$ (not necessarily distinct) such that

$$\pi = \pi_0 \oplus \varphi^m \pi_1 \oplus \varphi^{2m} \pi_2 \oplus \varphi^{3m} \pi_3 \oplus \ldots, \tag{2.1}$$

where $L_C = \left\{ \left(1^{f_1} 2^{f_2} 3^{f_3} \ldots\right) \in C \,|\, f_i = 0 \text{ for } i > m \right\}$.

We shall say that C is a <u>linked partition ideal</u> if (1) C has a modulus, (2) C satisfies the finiteness property, (3) for each $\pi' \in L_C$ there exists a positive integer $\ell(\pi')$ (called the <u>span</u> of π') and a subset $L_C(\pi')$ of L_C (called the <u>linking set</u> of π') such that for each $\pi \in C$ the representation (2.1) fulfills the requirement that if $\pi_j = \pi'$, then $\pi_{j+1} = \pi_{j+2} = \ldots = \pi_{j+\ell(\pi')-1} = (1^0 2^0 3^0 \ldots)$ and $\pi_{j+\ell(\pi')} \in L_C(\pi')$.

We define

$$f_C(x;q) = \sum_{\pi \in C} x^{f_1+f_2+f_3+\cdots} q^{f_1\cdot 1+f_2\cdot 2+f_3\cdot 3+\cdots} .$$

Theorem 1. If C is a linked partition ideal, then $f_C(x;q)$ satisfies a finite homogeneous linear q-difference equation whose coefficients are polynomials in x and q.

We refer the reader to Andrews (1974c) for a proof of this theorem. We shall content ourselves with an example of Theorem 1 for which Problem 1 can also be answered affirmatively. As a result we shall be abel to derive a well-known partition identity of I. J. Schur (1926). Let

$$C = \{\pi = (1^{f_1}2^{f_2}3^{f_3}\cdots) \mid f_i+f_{i+1}+f_{i+2} \leqq 1,\ f_{3i}+f_{3i+1}+f_{3i+2}+f_{3i+3} \leqq 1\} ;$$

then

$$L_C = \{\pi_0 = (1^0 2^0 3^0 4^0 5^0 \cdots),\quad \pi_1 = (1^1 2^0 3^0 4^0 5^0 \cdots),$$

$$\pi_2 = (1^0 2^1 3^0 4^0 5^0 \cdots),\quad \pi_3 = (1^0 2^0 3^1 4^0 5^0 \cdots)\} ;$$

$\ell(\pi_j) = 1$ for $j = 0, 1, 2, 3,$ and

$$L_C(\pi_0) = \{\pi_0, \pi_1, \pi_2, \pi_3\},\quad L_C(\pi_1) = \{\pi_0, \pi_1, \pi_2, \pi_3\}$$

$$L_C(\pi_2) = \{\pi_0, \pi_2, \pi_3\},\quad L_C(\pi_3) = \{\pi_0\} .$$

The fact that C is linked with modulus 3 makes it a simple matter to deduce several relations among generating functions related to C. Namely if $\pi_i \in L_C$, let

$$H_i(x) = \Sigma' x^{f_1+f_2+\cdots} q^{f_1\cdot 1+f_2\cdot 2+\cdots} \tag{2.2}$$

BASIC HYPERGEOMETRIC FUNCTIONS

where Σ' is over those $\{f_j\} \in C$ such that $\{f_1, f_2, \ldots, f_m, 0, 0, 0, \ldots\} = \pi_i$, then

$$H_0(x) = H_0(xq^3) + H_1(xq^3) + H_2(xq^3) + H_3(xq^3) \tag{2.3}$$

$$H_1(x) = xq(H_0(xq^3) + H_1(xq^3) + H_2(xq^3) + H_3(xq^3)) \tag{2.4}$$

$$H_2(x) = xq^2(H_0(xq^3) + H_2(xq^3) + H_3(xq^3)) \tag{2.5}$$

$$H_3(x) = xq^3 H_0(xq^3) \ . \tag{2.6}$$

Each of these equations is a special case of the following general identity for linked partition ideals (proved in Andrews (1974c))

$$H_i(x) = x^{\#(\pi_i)} q^{\sigma(\pi_i)} \sum_{\pi_\ell \in L_C(\pi_i)} H_\ell\left(xq^{\ell(\pi_i)}\right) ; \tag{2.7}$$

here $\#(\pi_i) = \sum f_j$, $\sigma(\pi_i) = \sum f_j j$, where $\pi_i = \{f_j\}$.

We may eliminate H_1, H_2 and H_3 from (2.3)-(2.6) and in this way we deduce that

$$H_0(x) = (1 + xq^4 + xq^5) H_0(xq^3) + xq^6 (1 - xq^6) H_0(xq^6) \ .$$

Since clearly

$$H_0(x) = f_C(xq^3;q) \ ,$$

we see that

$$f_C(x;q) = (1 + xq + xq^2) f_C(xq^3;q) + xq^3 (1 - xq^3) f_C(xq^6;q) \ . \tag{2.8}$$

It is not difficult to show (see Andrews (1968a) for details) that given the boundary condition $f_C(0;q) = 1$, the unique solution of (2.8) is

$$(2.9)\qquad f_C(x;q) = \sum_{m\geq 0}\sum_{n\geq 0} \frac{(-q;q^3)_n(-q^2;q^3)_n(-1)^m x^{n+m} q^{3/2\, m(m-1)}}{(q^3;q^3)_m(q^3;q^3)_n} .$$

Thus we have a solution of (2.8) of the type envisaged in Problem 1.

Applying a simple identity of Euler to the outer sum in (2.9), we see that

$$f_C(x;q) = (x;q^3)_\infty \sum_{n\geq 0} \frac{(-q;q^3)_n(-q^2;q^3)_n x^n}{(q^3;q^3)_n} .$$

Therefore if $B(n)$ denotes the number of partitions of n in C (i.e. the number of partitions of n with difference between all parts at least 3 and at least 6 between multiples of 3), then

$$(2.10)\qquad \sum_{n=0}^{\infty} B(n)q^n = f_C(1;q)$$

$$= \lim_{x\to 1^-} (x;q^3)_\infty \sum_{n\geq 0} \frac{(-q;q^3)_n(-q^2;q^3)_n x^n}{(q^3;q^3)_n}$$

$$= (q^3;q^3)_\infty \frac{(-q;q^3)_\infty(-q^2;q^3)_\infty}{(q^3;q^3)_\infty}$$

(by Abel's theorem, Andrews (1971b))

$$= \prod_{\substack{n=1\\ 3\nmid n}}^{\infty} (1+q^n) = \prod_{\substack{n=1\\ 3\nmid n}}^{\infty} \frac{(1-q^{2n})}{(1-q^n)} = \prod_{\substack{n=1\\ 3\nmid n,\\ \text{and } 2\nmid n}}^{\infty} \frac{1}{1-q^n}$$

$$= \prod_{n=0}^{\infty} (1-q^{6n+1})^{-1}(1-q^{6n+5})^{-1} = \sum_{n=0}^{\infty} A(n)q^n ,$$

where $A(n)$ is the number of partitions of n into parts $\equiv 1$ or 5 (modulo 6). Comparing coefficients of q^n in the extremes of (2.10), we deduce

Theorem 2 (Schur (1926)). The partitions of n into parts $\equiv 1$ or 5 (mod 6) are equinumerous with the partitions of n in which all parts differ by at least 3 and all parts that are multiples of 3 differ by at least 6.

The above proof of Schur's theorem illustrates how useful a general solution to Problem 1 would be. We conclude this section with one more example of a known case of Problem 1. A. Selberg (1936) has shown that there exist polynomials $a_j(x, q)$ in x and q such that if

$$f(x) = \sum_{n \geqq 0} \frac{(-1)^n x^{kn} q^{\frac{1}{2}((2k+1)n^2+n)}(1-x^k q^{(2n+1)k})}{(q)_n (xq^{n+1})_\infty}, \tag{2.11}$$

then

$$\sum_{j=0}^{k} a_j(x, q)\, f(xq^j) = 0 . \tag{2.12}$$

Recently (Andrews (1974b)) a multiple series representation of $f(x)$ has been found:

$$f(x) = \sum_{n_1, \ldots, n_{k-1} \geqq 0} \frac{x^{N_1+\ldots+N_{k-1}} q^{N_1^2+N_2^2+\ldots+N_{k-1}^2}}{(q)_{n_1}(q)_{n_2}\cdots(q)_{n_{k-1}}}, \tag{2.13}$$

where $N_j = n_j + n_{j+1} + \ldots + n_{k-1}$.

We may directly deduce from (2.11) and (2.13) that

$$\sum_{n_1, \ldots, n_{k-1} \geqq 0} \frac{q^{N_1^2+\ldots+N_{k-1}^2}}{(q)_{n_1}(q)_{n_2}\cdots(q)_{n_{k-1}}} = \prod_{\substack{n=1 \\ n \not\equiv 0, \pm k \pmod{2k+1}}} (1-q^n)^{-1} \tag{2.14}$$

simply applying Jacobi's triple product identity (Andrews (1971b)) to (2.11) when $x = 1$.

The analytic identity (2.14) (and $k-1$ related identities) are closely related to the following partition theorem of B. Gordon (1961) (see Andrews (1966), (1974b) for a more complete discussion):

Theorem 3. Let $A_{k,a}(n)$ denote the number of partitions of n into parts $\not\equiv 0, \pm a \pmod{2k+1}$. Let $B_{k,a}(n)$ denote the number of partitions of n of the form $n = b_1 + b_2 + \dots + b_s$, $b_j \geq b_{j+1}$, $b_j - b_{j+k-1} \geq 2$, $b_{s-a+1} > 1$. Then $A_{k,a}(n) = B_{k,a}(n)$ for all n provided $1 \leqq a \leqq k$.

The cases $k = 2$, $a = 1, 2$ are the celebrated Rogers-Ramanujan identities (Andrews 1971b)).

2. Identities for Multiple Hypergeometric Series.

PROBLEM 2. What are possible multiple series generalizations of the q-analog of Whipple's theorem?

In a way Problem 2 seems quite specialized in comparison with Problem 1; indeed, it is. However we remark that Watson's (1929) q-analog of Whipple's theorem:

$$(3.1)\qquad {}_8\phi_7\left[\begin{matrix} a, q\sqrt{a}, -q\sqrt{a}, c, d, e, f, q^{-N}; q, \dfrac{a^2q^{N+2}}{cdef} \\ \sqrt{a}, -\sqrt{a}, aq/c, aq/d, aq/e, aq/f, aq^{N+1} \end{matrix}\right]$$

$$= \frac{(a)_N(aq/ef)_N}{(aq/e)_N(aq/f)_N}\ {}_4\phi_3\left[\begin{matrix} aq/cd, e, f, q^{-N}; q, q \\ efq^{-N}/a, aq/c, aq/d \end{matrix}\right]$$

has been utilized to prove numerous partition identities (see Andrews (1971a)) including the Rogers-Ramanujan identities (Watson (1929)).

BASIC HYPERGEOMETRIC FUNCTIONS

As an example of a generalization of (3.1) we prove the following identity. We shall show its relationship to (2.14), and we shall derive several new analytic identities of the Rogers-Ramanujan type.

Theorem 4. For $k \geqq 1$, N a nonnegative integer,

$$
{}_{2k+4}\phi_{2k+3}\left[\begin{matrix} a, q\sqrt{a}, -q\sqrt{a}, b_1, c_1, b_2, c_2, \ldots, b_k, c_k, q^{-N}; q, \dfrac{a^k q^{k+N}}{b_1 \ldots b_k c_1 \ldots c_k} \\ \sqrt{a}, -\sqrt{a}, aq/b_1, aq/c_1, aq/b_2, aq/c_2, \ldots, aq/b_k, aq/c_k, aq^{N+1} \end{matrix}\right]
$$

$$
= \frac{(aq)_N (aq/b_k c_k)_N}{(aq/b_k)_N (aq/c_k)_N} \sum_{m_1, \ldots, m_{k-1} \geqq 0} \frac{(aq/b_1c_1)_{m_1} (aq/b_2c_2)_{m_2} \ldots (aq/b_{k-1}c_{k-1})_{m_{k-1}}}{(q)_{m_1} (q)_{m_2} \ldots (q)_{m_{k-1}}}
$$

$$
\cdot \frac{(b_2)_{m_1} (c_2)_{m_1} (b_3)_{m_1+m_2} (c_3)_{m_1+m_2} \ldots (b_k)_{m_1+\ldots+m_{k-1}}}{(aq/b_1)_{m_1} (aq/c_1)_{m_1} (aq/b_2)_{m_1+m_2} (aq/c_2)_{m_1+m_2} \ldots (aq/b_{k-1})_{m_1+\ldots m_{k-1}}}
$$

$$
\frac{(c_k)_{m_1+\ldots+m_{k-1}}}{(aq/c_{k-1})_{m_1+\ldots+m_{k-1}}} \cdot \frac{(q^{-N})_{m_1+m_2+\ldots+m_{k-1}}}{(b_k c_k q^{-N}/a)_{m_1+m_2+\ldots+m_{k-1}}}
$$

$$
\frac{(aq)^{m_{k-2}+2m_{k-3}+\ldots+(k-2)m_1} q^{m_1+m_2+\ldots+m_{k-1}}}{(b_2c_2)^{m_1} (b_3c_3)^{m_1+m_2} \ldots (b_{k-1}c_{k-1})^{m_1+m_2+\ldots+m_{k-2}}} .
$$

Remark. Since this identity is between two terminating series, we can replace each b_i by q^{b_i}, each c_i by q^{c_i}, a by q^a and then let $q \to 1$; this procedure will provide a multiple series identity that contains Whipple's (1926) theorem in the case $k = 2$. Also we should point out that this result is closely related to some work of V. N. Singh (1957); for a complete discussion of this relationship see Andrews (1974b).

Proof. We proceed by induction on k. When $k = 1$, the theorem reduces to the terminating, limiting case of Jackson's theorem (Slater (1966), p. 247. eq. (IV. 7), $d = q^{-N}$). When $k = 2$, the result is Watson's q-analog of Whipple's theorem (Slater (1966), p. 100, eq. (3.4.1.5)). Let us therefore assume that the theorem is true for each integer smaller than a given $k \geqq 3$.

Now by the q-analog of Saalschutz's theorem (Slater (1966), p. 247, eq. (IV. 4))

$$\frac{a^n q^n (b_1)_n (c_1)_n}{b_1^n c_1^n (aq/b_1)_n (aq/c_1)_n} = \sum_{m_1=0}^{n} \frac{(aq/b_1c_1)_{m_1} (aq^n)_{m_1} (q^{-n})_{m_1} q^{m_1}}{(q)_{m_1} (aq/b_1)_{m_1} (aq/c_1)_{m_1}}. \tag{3.2}$$

Hence

$$\begin{aligned}
&{}_{2k+4}\phi_{2k+3}\left[\begin{matrix} a, q\sqrt{a}, -q\sqrt{a}, b_1, c_1, \ldots, b_k, c_k, q^{-N}; q, \dfrac{a^k q^{k+N}}{b_1 \cdots b_k c_1 \cdots c_k} \\ \sqrt{a}, -\sqrt{a}, aq/b_1, aq/c_1, \ldots, aq/b_k, aq/c_k, aq^{1+N} \end{matrix}\right] \\
&= \sum_{n\geqq 0} \frac{(a)_n (1-aq^{2n})(b_2)_n (c_2)_n \cdots (b_k)_n (c_k)_n}{(q)_n (1-a)(aq/b_2)_n (aq/c_2)_n \cdots (aq/b_k)_n (aq/c_k)_n} \frac{(q^{-N})_n}{(aq^{N+1})_n} \\
&\quad \left(\frac{a^{k-1} q^{k-1+N}}{b_2 \cdots b_k c_2 \cdots c_k}\right)^n \sum_{m_1\geqq 0} \frac{(aq/b_1c_1)_{m_1} (aq^n)_{m_1} (q^{-n})_{m_1} q^{m_1}}{(q)_{m_1} (aq/b_1)_{m_1} (aq/c_1)_{m_1}} \qquad \text{(by (3.2))} \\
&= \sum_{m_1\geqq 0} \frac{(aq/b_1c_1)_{m_1} q^{m_1 + \binom{m_1}{2}} (-1)^{m_1}}{(q)_{m_1} (aq/b_1)_{m_1} (aq/c_1)_{m_1}} \cdot \sum_{n\geqq m_1} \frac{(a)_{n+m_1} (1-aq^{2n})}{(q)_{n-m_1} (1-a)}
\end{aligned}$$

BASIC HYPERGEOMETRIC FUNCTIONS

$$\frac{(b_2)_n(c_2)_n\cdots(b_k)_n(c_k)_n(q^{-N})_n}{(aq/b_2)_n(aq/c_2)_n\cdots(aq/b_k)_n(aq/c_k)_n(aq^{N+1})_n}\left(\frac{a^{k-1}q^{k-1+N-m_1}}{b_2\cdots b_kc_2\cdots c_k}\right)^n$$

$$= \sum_{m_1\geq 0}\frac{(aq/b_1c_1)_{m_1}q^{\binom{m_1+1}{2}}(-1)^{m_1}(b_2)_{m_1}(c_2)_{m_1}\cdots(b_k)_{m_1}(c_k)_{m_1}(q^{-N})_{m_1}}{(q)_{m_1}(aq/b_1)_{m_1}(aq/c_1)_{m_1}(aq/b_2)_{m_1}(aq/c_2)_{m_1}\cdots(aq/b_k)_{m_1}(aq/c_k)_{m_1}}$$

$$\left(\frac{a^{k-1}q^{k-1+N-m_1}}{b_2\cdots b_kc_2\cdots c_k}\right)^{m_1}\frac{(aq)_{2m_1}}{(aq^{1+N})_{m_1}}\sum_{n\geq 0}\frac{(aq^{2m_1})_n(1-aq^{2m_1}q^{2n})(b_2q^{m_1})_n}{(q)_n(1-aq^{2m_1})(aq^{m_1+1}/c_1)_n}$$

$$\frac{(c_2q^{m_1})_n\cdots(b_kq^{m_1})_n(c_kq^{m_1})_n(q^{-N+m_1})_n}{(aq^{m_1+1}/b_2)_n\cdots(aq^{m_1+1}/b_k)_n(aq^{m_1+1}/c_k)_n(aq^{N+m_1+1})_n}\left(\frac{a^{k-1}q^{k-1+N-m_1}}{b_2\cdots b_kc_2\cdots c_k}\right)^n$$

$$= (aq)_N\sum_{m_1\geq 0}\frac{(aq/b_1c_1)_{m_1}q^{\binom{m_1+1}{2}}(-1)^{m_1}(b_2)_{m_1}(c_2)_{m_1}\cdots(b_k)_{m_1}(c_k)_{m_1}(q^{-N})_{m_1}}{(q)_{m_1}(aq/b_1)_{m_1}(aq/c_1)_{m_1}\cdots(aq/b_k)_{m_1}(aq/c_k)_{m_1}}$$

$$\left(\frac{a^{k-1}q^{k-1+N-m_1}}{b_2\cdots b_kc_2\cdots c_k}\right)^{m_1}\frac{1}{(aq^{2m_1+1})_{N-m_1}}\,{}_{2k+2}\phi_{2k+1}\left[\begin{matrix}aq^{2m_1},\ q\sqrt{aq^{2m_1}},\ -q\sqrt{aq^{2m_1}},\\ \sqrt{aq^{2m_1}},\ -\sqrt{aq^{2m_1}},\end{matrix}\right.$$

$$\left.\begin{matrix}b_2q^{m_1},\ c_2q^{m_1},\ \ldots,\ b_kq^{m_1},\ c_kq^{m_1},\ q^{-N+m_1}\\ \frac{aq^{m_1+1}}{b_2},\ \frac{aq^{m_1+1}}{c_2},\ \ldots,\ \frac{aq^{m_1+1}}{b_k},\ \frac{aq^{m_1+1}}{c_k},\ aq^{N+m_1+1}\end{matrix};q,\ \frac{a^{k-1}q^{k-1+N-m_1}}{b_2\cdots b_kc_2\cdots c_k}\right]$$

$$= (aq)_N\sum_{m_1\geq 0}\frac{(aq/b_1c_1)_{m_1}q^{\binom{m_1+1}{2}}(-1)^{m_1}(b_2)_{m_1}(c_2)_{m_1}\cdots(b_k)_{m_1}(c_k)_{m_1}(q^{-N})_{m_1}}{(q)_{m_1}(aq/b_1)_{m_1}(aq/c_1)_{m}\cdots(aq/b_k)_{m_1}(aq/c_k)_{m_1}}$$

$$\left(\frac{a^{k-1}q^{k-1+N-m_1}}{b_2\cdots b_k c_2\cdots c_k}\right)^{m_1} \frac{(aq/b_kc_k)_{N-m_1}}{(\frac{aq^{m_1+1}}{b_k})_{N-m}(\frac{aq^{m_1+1}}{c_k})_{N-m_1}}$$

$$\sum_{m_2,\ldots,m_{k-1}\geq 0} \frac{(aq/b_2c_2)_{m_2}\cdots(aq/b_{k-1}c_{k-1})_{m_{k-1}}}{(q)_{m_2}\cdots(q)_{m_{k-1}}}$$

$$\frac{(b_3q^{m_1})_{m_2}(c_3q^{m_1})_{m_2}\cdots(b_kq^{m_1})_{m_2+\cdots+m_{k-1}}(c_kq^{m_1})_{m_2+\cdots+m_{k-1}}}{(\frac{aq^{m_1+1}}{b_2})_{m_2}(\frac{aq^{m_1+1}}{c_2})_{m_2}\cdots(\frac{aq^{m_1+1}}{b_{k-1}})_{m_2+\cdots+m_{k-1}}(\frac{aq^{m_1+1}}{c_{k-1}})_{m_2+\cdots+m_{k-1}}}$$

$$\frac{(q^{-N+m_1})_{m_2+\cdots+m_{k-1}}(aq)^{m_{k-2}+2m_{k-3}+\cdots+(k-3)m_2}q^{m_2+\cdots+m_{k-1}}}{(b_kc_kq^{-N+m_1}/a)_{m_2+\cdots+m_{k-1}}(b_3c_3)^{m_2}(b_4c_4)^{m_2+m_3}\cdots(b_{k-1}c_{k-1})^{m_2+\cdots+m_{k-2}}}$$

$$= \frac{(aq)_N(aq/b_kc_k)_N}{(\frac{aq}{b_k})_N(\frac{aq}{c_k})_N} \sum_{m_1,m_2,\ldots,m_{k-1}\geq 0} \frac{(aq/b_1c_1)_{m_1}(aq/b_2c_2)_{m_2}\cdots(aq/b_{k-1}c_{k-1})_{m_{k-1}}}{(q)_{m_1}(q)_{m_2}\cdots(q)_{m_{k-1}}}$$

$$\frac{(b_2)_{m_1}(c_2)_{m_1}(b_3)_{m_1+m_2}(c_3)_{m_1+m_2}\cdots(b_k)_{m_1+\cdots+m_{k-1}}}{(aq/b_1)_{m_1}(aq/c_1)_{m_1}(aq/b_2)_{m_1+m_2}(aq/c_2)_{m_1+m_2}\cdots(aq/b_k)_{m_1+\cdots+m_{k-1}}}$$

$$\frac{(c_k)_{m_1+\cdots+m_{k-1}}}{(aq/c_k)_{m_1+\cdots+m_{k-1}}} \cdot \frac{(q^{-N})_{m_1+\cdots+m_{k-1}}\left(\frac{a^{k-1}q^{k-1+N-m_1}}{b_2\cdots b_kc_2\cdots c_k}\right)^{m_1} q^{\binom{m_1+1}{2}}(-1)^{m_1}}{(1-\frac{aq^N}{b_kc_k})\cdots(1-\frac{aq^{N-m_1+1}}{b_kc_k})(\frac{b_kc_kq^{-N+m_1}}{d})_{m_2+\cdots+m_{k-1}}}$$

BASIC HYPERGEOMETRIC FUNCTIONS

$$\cdot \frac{(aq)^{m_{k-2}+2m_{k-3}+\dots+(k-3)m_2}\, q^{m_2+\dots+m_{k-1}}}{(b_3c_3)^{m_2}(b_4c_4)^{m_2+m_3}\cdots(b_{k-1}c_{k-1})^{m_2+\dots+m_{k-2}}}$$

$$= \frac{(aq)_N(aq/b_kc_k)_N}{(aq/b_k)_N(aq/c_k)_N} \sum_{m_1,\dots,m_{k-1}\geqq 0} \frac{(aq/b_1c_1)_{m_1}\cdots(aq/b_{k-1}c_{k-1})_{m_{k-1}}}{(q)_{m_1}(q)_{m_2}\cdots(q)_{m_{k-1}}}$$

$$\frac{(b_2)_{m_1}(c_2)_{m_1}(b_3)_{m_1+m_2}(c_3)_{m_1+m_2}\cdots(b_k)_{m_1+\dots+m_{k-1}}}{(aq/b_1)_{m_1}(aq/c_1)_{m_1}(aq/b_2)_{m_1+m_2}(aq/c_2)_{m_1+m_2}\cdots(aq/b_k)_{m_1+\dots+m_{k-1}}}$$

$$\frac{(c_k)_{m_1+\dots+m_{k-1}}}{(aq/c_k)_{m_1+\dots+m_{k-1}}} \quad \frac{(q^{-N})_{m_1+\dots+m_{k-1}}}{(b_kc_kq^{-N}/a)_{m_1+\dots+m_{k-1}}}$$

$$\frac{(aq)^{m_{k-2}+2m_{k-3}+\dots+(k-2)m_1}\, q^{m_1+\dots+m_{k-1}}}{(b_2c_2)^{m_1}(b_3c_3)^{m_1+m_2}\cdots(b_{k-1}c_{k-1})^{m_1+\dots+m_{k-1}}}$$

which is the desired formula. Hence Theorem 4 is established. □

Corollary 4.1.

$$\sum_{n\geqq 0} \frac{(a)_n(1-aq^{2n})(b)_n(c)_n(-1)^n(aq^k/bc)^n q^{(2k-1)n(n-1)/2}}{(q)_n(1-a)(aq/b)_n(aq/c)_n}$$

$$= \frac{(aq)_\infty(aq/ac)_\infty}{(aq/b)_\infty(aq/c)_\infty} \sum_{m_1,\dots,m_{k-1}\geqq 0} \frac{(b)_{m_1+\dots+m_{k-1}}(c)_{m_1+\dots+m_{k-1}}}{(q)_{m_1}(q)_{m_2}\cdots(q)_{m_{k-1}}}$$

$$(q/bc)^{m_1+m_2+\cdots+m_{k-1}} \cdot a^{(k-1)m_1+(k-2)m_2+\cdots+2m_{k-2}+m_{k-1}}$$

$$q^{m_1^2+(m_1+m_2)^2+\cdots+(m_1+\cdots+m_{k-2})^2} .$$

Proof. Set $b_k = b$, $c_k = c$ in Theorem 4; then let $b_1, c_1, \ldots, b_{k-1}, c_{k-1}$ and N all tend to infinity. An appeal to a multiple series analog of Tannery's theorem (see Slater (1966), p. 104 for the case $k = 2$) justifies taking these limits. □

Corollary 4.2. Equation (2.14) is valid.

Proof. Let b and c tend to ∞ in Corollary 4.1; then set $a = 1$, and utilize the fact that

$$1 + \sum_{n=1}^{\infty} (-1)^n q^{\frac{1}{2}((2k+1)n^2-n)} (1+q^n) = \prod_{\substack{n=1 \\ n\equiv 0, \pm k (\mathrm{mod}\ 2k+1)}}^{\infty} (1-q^n)$$

a specialization of Jacobi's triple product identity. □

As a final corollary we generalize an identity of L. J. Slater (Slater (1952), eq. (36)) thus obtaining the analytic counterpart of the generalization of the Göllnitz-Gordon identities (see Andrews (1967b) in the case $k = a$).

Corollary 4.3.

$$\sum_{m_1,\ldots,m_{k-1}\geq 0} \frac{(-q;q^2)_{M_1} q^{M_1^2+M_2^2+\cdots+M_{k-1}^2}}{(q^2;q^2)_{m_1}(q^2;q^2)_{m_2}\cdots(q^2;q^2)_{m_{k-1}}}$$

$$= \prod_{\substack{n=1 \\ n\not\equiv 2 (\mathrm{mod}\ 4) \\ n\not\equiv 0, \pm(2k-1) (\mathrm{mod}\ 4k)}}^{\infty} (1-q^n)^{-1} ,$$

where $M_j = m_j+m_{j+1}+\cdots+m_{k-1}$.

Proof. In Corollary 4.1 replace q by q^2, then set $b = q$, $a = 1$ and let $c \to \infty$. In the resulting identity, we observe that

$$1 + \sum_{n=1}^{\infty} (-1)^n q^{2kn^2-n}(1+q^{2n}) = \prod_{\substack{n=1 \\ n\equiv 0, \pm(2k-1)(\mathrm{mod}\ 4k)}}^{\infty} (1-q^n) ,$$

another specialization of Jacobi's triple product identity. □

Numerous other q-series identities can be derived from Theorem 4. We have chosen a small sampling to illustrate the possibilities.

In closing this section we wish to stress that Theorem 4 may well not be the only useful multiple series generalization of Watson's q-analog of Whipple's theorem.

In particular Holman, Biedenharn and Louck (1975) have proved a generalization of Whipple's theorem that is definitely not in the same direction as Theorem 4 above. Their result arises from work in physics and relies for its proof on various combinatorial and recurrent series arguments. Significant work must be undertaken before we can determine whether q-analogs exist for their results; past experience suggests that such q-analogs may well exist. If they do exist, their importance in the theory of partitions may well depend on whether they provide a solution to the following question.

PROBLEM 3. Are there multiple series q-analogs of well-poised hypergeometric series that specialize to either cases of the quituple product identity, or Winquist's identity or some other multiple theta series that sums to an infinite product?

The quintuple product identity is proved neatly by Carlitz and Subbarao (1972b):

$$\prod_{n=1}^{\infty} (1-q^n)(1-zq^n)(1-z^{-1}q^{n-1})(1-z^2q^{2n-1})(1-z^{-2}q^{2n-1})$$

$$= \sum_{n=-\infty}^{\infty} q^{(3n^2+n)/2}(z^{3n}-z^{-3n-1}) .$$

Winquist's (1969) identity has also been nicely proved and generalized by Carlitz and Subbarao (1972a),

$$\prod_{n=1}^{\infty} (1-q^n)^2 (1-yq^{n-1})(1-zq^{n-1})(1-yz^{-1}q^{n-1})(1-yzq^{n-1})$$

$$\cdot (1-y^{-1}q^n)(1-z^{-1}q^n)(1-y^{-1}zq^n)(1-y^{-1}z^{-1}q^n)$$

$$= \sum_{i=0}^{\infty} \sum_{j=-\infty}^{\infty} (-1)^{i+j} \Big\{ (y^{-3i}-y^{3i+3})(z^{-3j}-z^{3j+1})$$

$$+ (z^{-3j+1}-z^{3j+2})(z^{3i+2}-z^{-3i-1}) \Big\} q^{\frac{3}{2}i(i+1)+\frac{1}{2}j(3j+1)} .$$

As we have seen in special cases of Theorem 4, the Jacobi triple product identity is the one invoked when the specializations of the well-poised ${}_{2k+4}\phi_{2k+3}$ are made in Corollaries 4.2 and 4.3.

As we have shown in ABHF, p. 466, the quintuple product identity may be deduced by specializing parameters in a bilateral well-poised basic hypergeometric series, and this observation suggests that a more extensive study of the $K_{\lambda,k,i}(a_0, a_1, \ldots, a_\lambda; a; q)$ (see ABHF, p. 453) is in order.

In closing this section, we remark that the work of Andrews (1968b) may be used to show that the ${}_{2k+4}\phi_{2k+3}$ appearing in Theorem 4 is the unique solution of a certain k-th order q-difference equation on the variable a of the type described in Problem 1.

4. Basic Appell and Lauricella Series.

In view of the directions indicated by Sections 2 and 3 it is quite natural to ask if there are theories of multiple basic hypergeometric series identities comparable to results known for the ordinary multiple hypergeometric series, namely the Appell and more generally Lauricella functions. To illustrate the possibilities we pose the following problem to which we shall give some reasonably satisfactory answers.

BASIC HYPERGEOMETRIC FUNCTIONS

PROBLEM 4. Are there multiple q-series analogs of the Chu-Vandermonde summation and Saalschütz's theorem?

We begin by defining the basic Lauricella function which is the q-analog of the fourth ordinary Lauricella function (see Slater (1966), p. 288, eq. (8.6.4))

$$\Phi_D[a;b_1,b_2,\ldots,b_n;c;x_1,\ldots,x_n]$$

$$= \sum_{m_1,\ldots,m_n\geq 0} \frac{(a)_{m_1+\ldots+m_n}(b_1)_{m_1}\cdots(b_n)_{m_n} x_1^{m_1}\cdots x_n^{m_n}}{(q)_{m_1}(q)_{m_2}\cdots(q)_{m_n}(c)_{m_1+m_2+\ldots+m_n}}.$$

Theorem 5 (Andrews (1972), p. 621).

$$\Phi_D[a;b_1,b_2,\ldots,b_n;c;x_1,\ldots,x_n]$$

$$= \frac{(a)_\infty(b_1x_1)_\infty(b_2x_2)_\infty\cdots(b_nx_n)_\infty}{(c)_\infty(x_1)_\infty(x_2)_\infty\cdots(x_n)_\infty}\ {}_{n+1}\phi_n\left[\begin{matrix}c/a, x_1, x_2, \ldots, x_n; q, a\\ b_1x_1, b_2x_2, \ldots, b_nx_n\end{matrix}\right]$$

Proof.

$$\Phi_D[a;b_1,b_2,\ldots,b_n;c;x_1,x_2,\ldots,x_n]$$

$$= \frac{(a)_\infty}{(c)_\infty}\sum_{m_1,\ldots,m_n\geq 0} \frac{(b_1)_{m_1}\cdots(b_n)_{m_n}}{(q)_{m_1}\cdots(q)_{m_n}} x_1^{m_1}\cdots x_n^{m_n}\frac{(cq^{m_1+m_2+\ldots+m_n})_\infty}{(aq^{m_1+m_2+\ldots+m_n})_\infty}$$

$$= \frac{(a)_\infty}{(c)_\infty}\sum_{m_1,\ldots,m_n\geq 0} \frac{(b_1)_{m_1}\cdots(b_n)_{m_n}}{(q)_{m_1}\cdots(q)_{m_n}} x_1^{m_1}\cdots x_n^{m_n}\sum_{r\geq 0}\frac{(c/a)_r}{(q)_r}a^r q^{(m_1+\ldots+m_n)r}$$

(by Slater (1966), p. 92, eq. (3.2.2.11))

$$= \frac{(a)_\infty (b_1x_1)_\infty \cdots (b_nx_n)_\infty}{(c)_\infty (x_1)_\infty \cdots (x_n)_\infty} \sum_{r \geqq 0} \frac{(c/a)_r (x_1)_r \cdots (x_n)_r a^r}{(q)_r (b_1x_1)_r \cdots (b_nx_n)_r}$$

(by Slater (1966), p. 92, eq. (3.2.2.11))

$$= \frac{(a)_\infty (b_1x_1)_\infty \cdots (b_nx_n)_\infty}{(c)_\infty (x_1)_\infty \cdots (x_n)_\infty} {}_{n+1}\phi_n \left[\begin{matrix} c/a, x_1, \ldots, x_n; q, a \\ b_1x_1, \ldots, b_nx_n \end{matrix} \right]. \qquad \square$$

Corollary 5.1.

$$\Phi_D[a; b_1, x_1/x_2, x_2/x_3, \ldots, x_{n-1}/x_n; c; x_1, \ldots, x_n]$$

$$= \frac{(a)_\infty}{(c)_\infty (x_n)_\infty} \sum_{r \geqq 0} \frac{(c/a)_r (x_n)_r (b_1x_1q^r)_\infty a^r}{(q)_r}.$$

Proof. Set $b_2 = x_1/x_2$, $b_3 = x_2/x_3, \ldots, b_n = x_{n-1}/x_n$ in Theorem 5. $\square$

We can now easily deduce an analog of the Chu-Vandermonde summation

Corollary 5.2. If N is a nonnegative integer

$$\Phi_D[a; q^{-N+1}/x_1, x_1/x_2, x_2/x_3, \ldots, x_{n-2}/x_{n-1}, x_{n-1}/q; c; x_1, x_2, \ldots, x_{n-1}, q]$$

$$= \frac{a^N (c/a)_N}{(c)_N}.$$

Remark. When $n = 1$, Corollary 5.2 asserts that

$${}_2\phi_1 \left[\begin{matrix} a, q^{-N}; q, q \\ c \end{matrix} \right] = \frac{a^N (c/a)_N}{(c)_N},$$

which is one of the forms of the q-analog of the Chu-Vandermonde summation [Slater (1966), p. 97, eq. (3.3.2.7)).

Proof. By Corollary 5.1,

$$\Phi_D[a;q^{-N+1}/x_1, x_1/x_2, x_2/x_3, \ldots, x_{n-2}/x_{n-1}, x_{n-1}/q;c;x_1, x_2, \ldots, x_{n-1}, q]$$

$$= \frac{(a)_\infty}{(c)_\infty (q)_\infty} \sum_{r \geqq 0} (c/a)_r (q^{-N+r+1})_\infty \, a^r$$

$$= \frac{(a)_\infty}{(c)_\infty (q)_\infty} \sum_{r \geqq 0} (c/a)_{r+N} (q^{r+1})_\infty \, a^{r+N}$$

$$= \frac{a^N (a)_\infty (c/a)_N}{(c)_\infty} \sum_{r \geqq 0} \frac{(cq^N/a)_r \, a^r}{(q)_r}$$

$$= \frac{a^N (a)_\infty (c/a)_N (cq^N)_\infty}{(c)_\infty (a)_\infty} \qquad \text{(by Slater (1966), p. 92, eq. (3.2.2.11))}$$

$$= \frac{a^N (c/a)_N}{(c)_N} .$$

Corollary 5.3 (Al-Salam (1965), eq. (9)).

$$\Phi_D(\beta\beta'/\alpha;\beta, \beta';\beta\beta';x, y)$$

$$= \frac{(x\beta'/\alpha)_\infty (y\beta/\alpha)_\infty}{(x)_\infty (y)_\infty} \Phi_D[\alpha;\beta, \beta'; \beta\beta';\beta'x/\alpha, \beta y/\alpha] .$$

Proof. Let us recall an identity of Newman Hall for the ${}_3\phi_2$-function (Hall (1936)):

$$(4.1) \qquad {}_3\phi_2\left[\begin{matrix} a, b, c;q, ef/abc \\ e, f \end{matrix}\right] = \frac{(a)_\infty (ef/ab)_\infty (ef/ac)_\infty}{(e)_\infty (f)_\infty (ef/abc)_\infty} {}_3\phi_2\left[\begin{matrix} e/a, f/a, ef/abc;q, a \\ ef/ab, ef/ac \end{matrix}\right] .$$

In (4.1), we set $a = \alpha$, $b = x$, $c = y$, $e = \beta x$, $f = \beta' y$, and then we multiply the resulting identity by

$$\frac{(\beta\beta'/\alpha)_\infty (\beta x)_\infty (\beta' y)_\infty}{(\beta\beta')_\infty (x)_\infty (y)_\infty} .$$

Hence

$$(4.2) \qquad \frac{(\beta\beta'/\alpha)_\infty (\beta x)_\infty (\beta' y)_\infty}{(\beta\beta')_\infty (x)_\infty (y)_\infty} \; {}_3\phi_2 \begin{bmatrix} \alpha, x, y; q, \gamma/\alpha \\ \beta x, \beta' y \end{bmatrix}$$

$$= \frac{(\alpha)_\infty (y\beta\beta'/\alpha)_\infty (x\beta\beta'/\alpha)_\infty}{(\beta\beta')_\infty (x)_\infty (y)_\infty} \; {}_3\phi_2 \begin{bmatrix} \beta x/\alpha, \beta' y/\alpha, \beta\beta'/\alpha; q, \alpha \\ y\beta\beta'/\alpha, x\beta\beta'/\alpha \end{bmatrix}$$

$$= \frac{(x\beta'/\alpha)_\infty (y\beta/\alpha)_\infty}{(x)_\infty (y)_\infty}$$

$$\cdot \frac{(\alpha)_\infty (x\beta\beta'/\alpha)_\infty (y\beta\beta'/\alpha)_\infty}{(\beta\beta')_\infty (x\beta'/\alpha)_\infty (y\beta/\alpha)_\infty} \; {}_3\phi_2 \begin{bmatrix} \beta\beta'/\alpha, \beta x/\alpha, \beta' y/\alpha; q, \alpha \\ x\beta\beta'/\alpha, y\beta\beta'/\alpha \end{bmatrix} .$$

Hence by Theorem 5 applied to (4.2), we see that

$$\Phi_D[\beta\beta'/\alpha; \beta, \beta'; \beta\beta'; x, y] = \frac{(x\beta'/\alpha)_\infty (y\beta/\alpha)_\infty}{(x)_\infty (y)_\infty} \cdot \Phi_D[\alpha; \beta, \beta'; \beta\beta'; \beta' x/\alpha, \beta y/\alpha] ,$$

which is the desired result. □

Next we derive Al-Salam's Saalschützian theorem for double q-series.

Corollary 5.4 (Al-Salam (1965), eq. (7)).

$$\sum_{r,s \geqq 0} \frac{(q^{-n})_r (q^{-m})_s (\alpha)_{r+s} (\beta)_r (\beta')_s}{(q)_r (q)_s (\gamma)_{r+s} (\delta)_r (\delta')_s} q^{r+s} = \frac{(\beta\beta'/\alpha)_{n+m} (\beta)_m (\beta')_n}{(\beta\beta')_{n+m} (\beta/\alpha)_m (\beta'/\alpha)_n}$$

provided $\delta = q^{1-n}\alpha/\beta'$, $\delta' = q^{1-m}\alpha/\beta$, $\gamma = \beta\beta'$.

Proof. We note the simple identity

$$\sum_{n\geq 0} \frac{(A)_n t^n}{(q)_n} = \frac{(At)_\infty}{(t)_\infty} \qquad \text{(Slater (1966), p. 92, eq. (3.2.2.11))} .$$

Hence we may compare coefficients of $x^m y^n$ on both sides of the identity in Corollary 5.3:

$$\frac{(\beta\beta'/\alpha)_{m+n}(\beta)_m(\beta')_n}{(\beta\beta')_{m+n}(q)_n(q)_n} = \sum_{r,s\geq 0} \frac{(\alpha)_{r+s}(\beta)_r(\beta')_s(\beta'/\alpha)_{m-r}(\beta/\alpha)_{n-s}}{(q)_r(q)_s(\beta\beta')_{r+s}(q)_{m-r}(q)_{n-s}}\left(\frac{\beta'}{\alpha}\right)^r\left(\frac{\beta}{\alpha}\right)^s .$$

Therefore

$$\frac{(\beta\beta'/\alpha)_{m+n}(\beta)_m(\beta')_n}{(\beta\beta')_{m+n}(\beta/\alpha)_m(\beta'/\alpha)_n} = \sum_{r,s\geq 0} \frac{(\alpha)_{r+s}(\beta)_r(\beta')_s(q^{-m})_r(q^{-n})_s}{(q)_r(q)_s(\beta\beta')_{r+s}(q^{1-n}\alpha/\beta')_r(q^{1-m}\alpha/\beta)_s} q^{r+s} ,$$

which is the desired result. □

Corollary 5.5 (Carlitz (1963), eq. (9)).

$$\sum_{r,s\geq 0} \frac{[-m]_r[-n]_s[\alpha]_{r+s}[\beta]_r[\beta']_s}{r!s![\gamma]_{r+s}[\delta]_r[\delta']_s} = \frac{[\beta+\beta'-\alpha]_{m+n}[\beta']_m[\beta]_n}{[\beta+\beta']_{m+n}[\beta'-\alpha]_m[\beta-\alpha]_n} ,$$

where $[A]_r = A(A+1)\ldots(A+r-1)$, $\gamma = \beta+\beta'$, $\delta = \alpha-\beta'-m+1$, $\delta' = \alpha-\beta-n+1$.

Proof. Replace α by q^α, β by q^β and γ by q^γ in Corollary 5.4; then let $q \to 1^-$. The limiting process is clearly legitimate since all series terminate. □

Corollary 5.6. If $x_1 = c/ab$, then

$$\Phi_D[a;b, x_1/x_2, x_2/x_3, \ldots, x_{n-1}/x_n;c;x_1, x_2, \ldots, x_n] = \frac{(c/a)_\infty (ax_n)_\infty}{(c)_\infty (x_n)_\infty} .$$

Proof. By Corollary 5.1

$$\Phi_D[a;b,x_1/x_2,x_2/x_3,\ldots,x_{n-1}/x_n;c;x_1,x_2,\ldots,x_n]$$

$$= \frac{(a)_\infty(bx_1)_\infty}{(c)_\infty(x_n)_\infty}\sum_{r=0}\frac{(c/a)_r(x_n)_r a^r}{(q)_r(bx_1)_r}$$

$$= \frac{(a)_\infty(bx_1)_\infty}{(c)_\infty(x_n)_\infty}\sum_{r=0}\frac{(x_n)_r a^r}{(q)_r}$$

$$= \frac{(a)_\infty(bx_1)_\infty(x_n a)_\infty}{(c)_\infty(x_n)_\infty(a)_\infty} = \frac{(bx_1)_\infty(x_n a)_\infty}{(c)_\infty(x_n)_\infty} . \qquad \square$$

The above result is an extension of the q-analog of Gauss's theorem to basic Lauricella functions. We may directly deduce an analog of Gauss's summation:

Corollary 5.7. If $\mathrm{Re}\, c \geqq \mathrm{Re}\, a$, $\mathrm{Re}(a-c+nb) < 0$, $\mathrm{Re}((1-n)(a-c)-nb-nx_2) < 0$ and for $2 \leqq i \leqq n-1$ $\mathrm{Re}(a-c+nx_i-nx_{i+1}) < 0$, then

$$F_D[a;b,c-a-b+x_2,x_2-x_3,\ldots,x_{n-1}-x_n;c;1,1,1,\ldots,1]$$

$$= \frac{\Gamma(c)\Gamma(x_n)}{\Gamma(c-a)\Gamma(a+x_n)} .$$

Remark. When $n = 1$ this is just Gauss's summation. The case $n = 2$ has been treated by Appell and Kampe de Feriet (1926), and in fact, their method can also be easily generalized to prove the above result. B. C. Carlson (1974) gives the terminating form of this result as his Theorem 3.1.

Proof. Replace a by q^a, b by q^b, c by q^c, and x_i by q^{x_i} for $1 \leqq i \leqq n$ in Corollary 5.6 and let $q \to 1^-$. The inequalities prescribed for the parameter guarantee convergence of the Lauricella function since the general term in absolute value is

$$\left|\frac{\Gamma(c)}{\Gamma(a)\Gamma(b)\Gamma(c-a-b+x_2)\Gamma(x_2-x_3)\ldots\Gamma(x_{n-1}-x_n)}\right|$$

$$\cdot\left|\frac{\Gamma(a+m_1+\ldots+m_n)\Gamma(b+m_1)\Gamma(c-a-b+x_2+m_2)\Gamma(x_2-x_3+m_3)\ldots\Gamma(x_{n-1}-x_n+m_n)}{\Gamma(c+m_1+\ldots+m_n)\Gamma(1+m_1)\Gamma(1+m_2)\ldots\Gamma(1+m_n)}\right|$$

$$\sim K(a,b,c,x_1,\ldots,x_n)(m_1+\ldots+m_n)^{\mathrm{Re}(a-c)} m_1^{\mathrm{Re}\,b-1}$$

$$m_2^{\mathrm{Re}(c-a-b+x_2)-1}\ldots m_n^{\mathrm{Re}(x_{n-1}-x_n)-1}$$

$$\leqq n^{\mathrm{Re}(a-c)}K(a,b,c,x_1,\ldots,x_n)m_1^{\frac{1}{n}\mathrm{Re}(a-c)+\mathrm{Re}\,b-1}\, m_2^{\mathrm{Re}((1-\frac{1}{n})(c-a)-b+x_2)-1}$$

$$m_3^{\mathrm{Re}(\frac{1}{n}(a-c)+x_2-x_3)-1}\ldots m_n^{\mathrm{Re}(\frac{1}{n}(a-c)+x_{n-1}-x_n)-1},$$

where the last inequality is just the arithmetic-geometric mean inequality and can be used here since $\mathrm{Re}(a-c) \leqq 0$. The remaining inequalities stated in the hypothesis now imply convergence by the comparison test. □

As we have remarked many of the results of this section were originally derived in different ways. It is hoped that the unification provided by Theorem 5 may be one of a class of general multiple q-series identities useful in dealing with some of our earlier problems.

5. MacMahon's Master Theorem and the Dyson Conjecture.

PROBLEM 5. Are there q-analogs of MacMahon's Master Theorem and the Dyson Conjecture?

First let us recall:

MacMahon's Master Theorem (MacMahon (1894), (1915)). The coefficient of $X_1^{P_1}X_2^{P_2}\ldots X_n^{P_n}$ in

$$\prod_{i=1}^{n} (A_{i1}X_1+A_{i2}X_2+\ldots+A_{in}X_n)^{P_i}$$

is the same as the corresponding coefficient in

$$(\det(\delta_{ij}-A_{ij}X_j))^{-1}$$

where $\delta_{ij} = 1$ if $i = j$ and $\delta_{ij} = 0$ if $i \neq j$.

This theorem has remarkable consequences in the theory of permutations and has been used to provide solutions to generalized Rencontre problems (MacMahon (1915)), the Menage problem (Percus (1971)), and numerous other results. Section 3 of MacMahon's book (MacMahon (1915)) as well as pages 18-31 of J. K. Percus (1971) provide ample evidence of this assertion.

I. J. Good (1962) utilized the Master Theorem to prove the following terminating torm of Dixon's theorem (Dixon (1891), (1903)):

$$(5.1) \qquad \sum_{k=-\infty}^{\infty} (-1)^k \binom{b+c}{b+k}\binom{c+a}{c+k}\binom{a+b}{a+k} = \frac{(a+b+c)!}{a!b!c!}.$$

Good's observation is that the coefficient of $x^{b+c}y^{a+c}z^{a+b}$ in $(y-z)^{b+c}(z-x)^{c+a}(x-y)^{a+b}$ is just the left and side of (5.1); the Master Theorem may then be applied to produce the right hand side (5.1). D. Foata mentioned that the coefficient of $x^{b+c}y^{a+c}z^{a+b}$ in $(y-z)^{b+c}$ $(z-x)^{c+a}(x-y)^{a+b}$ is precisely the same as the constant term in

$$(1-\frac{x}{z})^a(1-\frac{x}{y})^a(1-\frac{y}{x})^b(1-\frac{y}{z})^b(1-\frac{z}{x})^c(1-\frac{z}{y})^c \quad ;$$

thus Foata observes that the above special case of the Master Theorem coincides with the three variable case of the following conjecture of F. J. Dyson (1962):

Dyson's Conjecture (Proved by Gunson (1962), Wilson (1962), and Good (1970)). The constant term in the expansion of

$$\prod_{1 \leq i \neq j \leq n} \left(1 - \frac{x_i}{x_j}\right)^{a_i}$$

is $(a_1+a_2+\ldots+a_n)!/(a_1!a_2!\ldots a_n!)$.

Proof. Good (1970) presents the most elegant proof of Dyson's conjecture; so we include it. Define $F_n(a_1, \ldots, a_n) = \prod_{1 \leq i \neq j \leq n} (1 - \frac{x_i}{x_j})^{a_i}$ and let $c_n(a_1, \ldots, a_n)$ be the constant term in $F_n(a_1, \ldots, a_n)$. Clearly $c_n(0, 0, \ldots, 0) = 1$. Furthermore if $a_h = 0$, then $c_n(a_1, \ldots, a_n) = c_{n-1}(a_1, \ldots, a_{h-1}, a_{h+1}, \ldots, a_n)$ since $a_h = 0$ implies that only nonpositive powers of x_h appear in $F_n(a_1, \ldots, a_n)$ and consequently the constant term in this case coincides with that in $F_{n-1}(a_1, \ldots, a_{h-1}, a_{h+1}, \ldots, a_n)$.

Finally we note that

$$(5.2) \qquad 1 = \sum_{j=1}^{n} \prod_{\substack{i=1 \\ i \neq j}}^{n} \frac{(x-x_j)}{(x_i-x_j)} ,$$

since the expression on the right is a polynomial in x of degree at most $n - 1$ that assumes the value 1 for each of $x = x_1, x_2, \ldots, x_n$ which is possible only if it is identically 1 . Setting $x = 0$ in (5.2), we see that

$$(5.3) \qquad 1 = \sum_{j=1}^{n} \prod_{\substack{i=1 \\ i \neq j}}^{n} \frac{1}{1 - \frac{x_i}{x_j}} .$$

Multiplying both sides of (5.3) by $F(a_1, \ldots, a_n)$, we find that

$$(5.4) \qquad F_n(a_1, \ldots, a_n) = \sum_{j=1}^{n} F_n(a_1, \ldots, a_j-1, \ldots, a_n) ,$$

and therefore

$$c_n(a_1, \ldots, a_n) = \sum_{j=1}^{n} c_n\,(a_1, \ldots, a_j - 1, \ldots, a_n)\ . \tag{5.5}$$

The two initial conditions on the $c_n(a_1, \ldots, a_n)$ and (5.5) uniquely define these numbers; since the multinomial coefficients $(a_1 + \ldots + a_n)!/(a_1! \ldots a_n!)$ also satisfy the same initial conditions and recurrence we see that $c_n(a_1, \ldots, a_n) = (a_1 + \ldots + a_n)!/(a_1! \ldots a_n!)$ as asserted in Dyson's conjecture. □

On the question of q-analogs for the Master Theorem and the Dyson conjecture, we make the following:

Conjecture. The constant term in the expansion of

$$\prod_{1 \leq i \neq j \leq n} (x_i \varepsilon_{ij}/x_j)_{a_i}$$

is $(q)_{a_1+a_2+\ldots+a_n}/((q)_{a_1}(q)_{a_2}\ldots(q)_{a_n})$, where $\varepsilon_{ij} = 1$ if $i < j$ and $\varepsilon_{ij} = q$ if $i > j$.

This conjecture is trivial if $n = 2$ and when $n = 3$ it reduces to

$$\sum_{k=-\infty}^{\infty} (-1)^k q^{\frac{1}{2}k(3k+1)} \begin{bmatrix} b+c \\ c+k \end{bmatrix} \begin{bmatrix} c+a \\ a+k \end{bmatrix} \begin{bmatrix} a+b \\ b+k \end{bmatrix} = \frac{(q)_{a+b+c}}{(q)_a (q)_b (q)_c} \tag{5.6}$$

where $\begin{bmatrix} A \\ B \end{bmatrix} = (q)_A/(q)_B(q)_{A-B}$ for $0 \leq B \leq A$ and $\begin{bmatrix} A \\ B \end{bmatrix} = 0$ otherwise. Identity (5.6) is the terminating form of the q-analog of Dixon's summation, a result due in part to F. H. Jackson (1941) and completed by L. Carlitz (1969).

For completeness we present a short proof of (5.6) that relies only on the q-analog of Saalschütz's theorem (Bailey (1935), p. 68); which we state in the following form:

$$q^{k^2} \begin{bmatrix} a+c \\ a+k \end{bmatrix} \begin{bmatrix} a+b \\ b+k \end{bmatrix} = \sum_{r=0}^{b} q^{(b-r)^2} \begin{bmatrix} b-k \\ r \end{bmatrix} \begin{bmatrix} c+k \\ c-b+r \end{bmatrix} \begin{bmatrix} a+c+r \\ c+b \end{bmatrix}\ . \tag{5.7}$$

Hence

$$\sum_{k=-\infty}^{\infty} (-1)^k q^{\frac{1}{2}k(3k+1)} \begin{bmatrix} b+c \\ c+k \end{bmatrix} \begin{bmatrix} c+a \\ a+k \end{bmatrix} \begin{bmatrix} a+b \\ b+k \end{bmatrix}$$

$$= \sum_{k=-\infty}^{\infty} (-1)^k q^{\frac{1}{2}k(k+1)} \sum_{r=0}^{b} \frac{q^{(b-r)^2} (q)_{a+c+r}}{(q)_r (q)_{b-k+r} (q)_{b+k-r} (q)_{a-b+r} (q)_{c-b+r}}$$

$$= \sum_{r=0}^{b} q^{(b-r)^2} \begin{bmatrix} a+c+r \\ b+c \end{bmatrix} \frac{(q)_{b+c}}{(q)_r (q)_{2b-2r} (q)_{c-b+r}} \sum_{k=-\infty}^{\infty} (-1)^k q^{\frac{1}{2}k(k+1)} \begin{bmatrix} 2b-2r \\ b+k-r \end{bmatrix}$$

$$= \sum_{r=0}^{b} (-1)^{b-r} q^{(b-r)^2 + \frac{1}{2}(r+1-b)(r-b)} \begin{bmatrix} a+c+r \\ b+c \end{bmatrix} \frac{(q)_{b+c} (q^{r+1-b})_{2b-2r}}{(q)_r (q)_{2b-2r} (q)_{c-b+r}} .$$

Now for $0 \leqq r < b$, $(q^{r+1-b})_{2b-2r}$ contains $1 - q^0 = 0$ as a factor and so vanishes. Hence the only nonvanishing term in the above sum occurs at $r = b$. Therefore

$$\sum_{k=-\infty}^{\infty} (-1)^k q^{\frac{1}{2}k(3k+1)} \begin{bmatrix} b+c \\ c+k \end{bmatrix} \begin{bmatrix} c+a \\ a+k \end{bmatrix} \begin{bmatrix} a+b \\ b+k \end{bmatrix}$$

$$= \begin{bmatrix} a+c+b \\ b+c \end{bmatrix} \frac{(q)_{b+c}}{(q)_b (q)_c} = \frac{(q)_{a+b+c}}{(q)_a (q)_b (q)_c} ,$$

which proves (5.6).

If our conjecture is true in general it would suggest strongly the importance of this type of theorem in the general theory of basic hypergeometric functions.

6. Saalschützian Series and Inversion Theorems.

We turn now to a topic that is related to our work in Section 2.

PROBLEM 6. What q-inversion theorems imply or are closely related to partition identities?

Göllnitz (1967) proved the following partition theorem:

Theorem 6. Let A(n) denote the number of partitions of n into parts $\equiv 2, 5$, or 11 (mod 12). Let B(n) denote the number of partitions of n into parts $\equiv 2, 4, 5 \pmod 6$. Let C(n) denote the number of partitions of n of the form $n = b_1 + b_2 + \ldots + b_s$, where $b_i - b_{i+1} \geqq 6$ with strict inequality if $b_i \equiv 0, 1, 3 \pmod 6$ and $b_s \neq 1, 3$. Then $A(n) = B(n) = C(n)$.

Göllnitz proved this theorem using a very intricate arithmetic argument patterned after Gleissberg's (1928) proof of Schur's (1926) theorem (Theorem 2 above).

Subsequently (Andrews (1969)) the following analytic theorem was proved which implies Theorem 6.

Theorem 7. If Λ_n is defined by

$$\prod_{n=0}^{\infty} (1+xq^{6n+2})(1+xq^{6n+4})(1+xq^{6n+5})(1-xq^{6n})^{-1} = \sum_{n=0}^{\infty} \frac{\Lambda_n x^n}{\prod_{j=1}^{n}(1-q^{6j})},$$

and if

$$d_m = 1 + \sum_{n=1}^{\infty} C_m(n) q^n$$

where $C_m(n)$ is the number of those partitions enumerated by C(n) with the added restriction that each part is $\leqq m$, then

$$d_{6n-1} = \sum_{0 \leqq 2j \leqq n} q^{6nj-3j^2+2j} \begin{bmatrix} n-j \\ j \end{bmatrix}_6 \Lambda_{n-2j}, \tag{6.1}$$

where

BASIC HYPERGEOMETRIC FUNCTIONS

$$\left[{m \atop n}\right]_r = \begin{cases} \dfrac{(1-q^{rm})(1-q^{r(m-1)})\ldots(1-q^{r(m-n+1)})}{(1-q^{rn})(1-q^{r(n-1)})\ldots(1-q^{r})}, & 0 \leqq n \leqq m , \\ 0 & \text{otherwise.} \end{cases}$$

H. Gould (1962), Gould and Hsu (1973) and L. Carlitz (1963b), (1973) have extensively studied inversion theorems clearly related to (6.1). In particular the methods of Carlitz (1963b), (1973) may be used to prove.

Theorem 8. For $\lambda \geqq 2$ each of the following identities implies the other

$$(6.2)\quad v_n = \sum_{0 \leqq \lambda s \leqq n} \frac{(-1)^s q^{(\lambda-1)s(s-1)/2+(3-\lambda)s}(q)_{n-\lambda s+s}\, u_{n-\lambda s}}{(q^{\lambda-1};q^{\lambda-1})_s (q)_{n-\lambda s}} \quad \text{for all } n \geqq 0 ;$$

$$(6.3)\quad u_n = \sum_{0 \leqq \lambda s \leqq n} \frac{q^{(3-\lambda)s}(q)_n (1-q^{n-\lambda s+1}) v_{n-\lambda s}}{(q^{\lambda-1};q^{\lambda-1})_s (q)_{n-s+1}} \quad \text{for all } n \geqq 0 .$$

When $\lambda = 2$, it is easy to show (Andrews (1969)) that (6.1) is a special case of (6.2). A natural follow up on Problem 6 is: Are there other cases of Theorem 8 related to partition theorems besides the case $\lambda = 2$? We should remark that one of Schur's partition theorems (Theorem 2 above, Schur (1926)) may be proved (Andrews (1967a), Theorem 5) using an identity that is precisely q-binomial inversion (see Rota and Goldman (1970), eq. (5), p. 244).

Concerning Theorem 7, it has recently been shown that

$$(6.4)\qquad \sum_{n\geq 0} d_{6n-1}\, x^n = \frac{(q^6;q^6)_\infty (-xq^2;q^6)_\infty (-xq^4;q^6)_\infty (-xq^5;q^6)_\infty}{(x;q^6)_\infty}$$

$${}_5\phi_4\left[\begin{matrix} x,\ ixq^{-1/2},\ -ixq^{-1/2},\ ixq^{5/2},\ -ixq^{5/2};q^6,\ q^6 \\ -x^2q^{-1},\ -xq^2,\ -xq^4,\ -xq^5 \end{matrix}\right]$$

The series here is a nonterminating Saalschützian series since the product of the denominator parameters divided by the product of the numerator parameters is q^6, the base. From the above identity Göllnitz's theorem follows immediately since

$$\sum_{n\geq 0} C(n)q^n = \lim_{n\to\infty} d_{6n-1} = \lim_{x\to 1} (1-x)\,\Sigma\, d_{6n-1}\, x^n$$

$$= (-q^2;q^6)_\infty (-q^4:q^6)_\infty (-q^5;q^6)_\infty \qquad \text{(by (6.4))}$$

$$= \sum_{n\geq 0} B(n)q^n .$$

The fact that $A(n) = B(n)$ is a consequence of a simple manipulation of the above infinite product.

PROBLEM 7. Are there other partition identities related to certain nonterminating Saalschützian series like that in (6.4)?

7. Conclusion. I hope this paper has presented topics in basic hypergeometric functions of some interest and has raised some problems worthy of further research. It has been my desire to show that there is a fruitful interaction between ordinary and basic hypergeometric functions while at the same time demonstrating that basic series possess a special character and have particular applications which clearly make them not

just a subtopic or curious generalization in the study of hypergeometric functions.

In ABHF applications have been considered at length. Also a short conference "Eulerian Series and Applications" (Andrews et al. (1974)) further explored basic hypergeometric functions and considered a number of problems that are not covered in this paper. These works should be consulted for further information and references on basic hypergeometric functions.

REFERENCES

W. A. Al-Salam (1965) Saalschützian theorems for basic double series, J. London Math. Soc., 40, 455-458.

G. E. Andrews (1966) An analytic proof of the Rogers-Ramanujan-Gordon identities, Amer. J. Math., 88, 844-846.

G. E. Andrews (1967a) On Schur's second partition theorem, Glasgow Math. J., 8, 127-132.

G. E. Andrews (1967b) A generalization of the Göllnitz-Gordon partition theorems, Proc. Amer. Math. Soc., 18, 945-952.

G. E. Andrews (1968a) On partition functions related to Schur's second partition theorem, Proc. Amer. Math. Soc., 19, 441-444.

G. E. Andrews (1968b) On q-difference equations for certain well-poised basic hypergeometric series, Quart. J. Math., 19, 433-447.

G. E. Andrews (1969) On a partition theorem of Göllnitz and related formulae, J. für die reine und angew. Math., 236, 37-42.

G. E. Andrews (1971a) A new property of partitions with applications to the Rogers-Ramanujan identities, J. Comb. Th., 10, 266-270.

G. E. Andrews (1971b) Number Theory, W. B. Saunders, Philadelphia.

G. E. Andrews (1972a) Partition identities, Advances in Math., 9, 10-51.

G. E. Andrews (1972b) Summations and transformations for basic Appell series, J. London Math. Soc. (2), 4, 618-622.

G. E. Andrews (1974a) (ABHF) Applications of basic hypergeometric functions, S. I. A. M. Review, 16, 441-484.

G. E. Andrews (1974b) An analytic generalization of the Rogers-Ramanujan identities for odd moduli, Proc. Nat. Acad. Sci., 71, 4082-4085.

G. E. Andrews (1974c) A general theory of identities of the Rogers-Ramanujan type, Bull. Amer. Math. Soc., 80, 1033-1952.

G. E. Andrews, R. Askey, L. Carlitz, N. J. Fine, H. W. Gould, K. W. Kadell, D. P. Roselle, and A. Verma (1974) Eulerian Series and Applications, Abstracts from the Conference at the Pennsylvania State University, May, 1974.

P. Appell and J. Kampé de Fériet (1926) Fonctions hypergeometriques et hypersphériques, Gauthier Villars, Paris.

W. N. Bailey (1935) Generalized Hypergeometric Series, Cambridge University Press (Reprinted: Hafner, New York, 1964).

L. Carlitz (1963a) A Saalschützian theorem for double series, J. London Math. Soc., 38, 415-418.

L. Carlitz (1963b) Some inversion formulas, Rend. Circ. Mat. Palermo (2), 12, 1-17.

L. Carlitz (1967a) Summation of a double hypergeometric series, Le Matematiche, 22, 138-142.

L. Carlitz (1967b) A summation theorem for double hypergeometric series, Rend. Sem. Mat. Padova, 37, 230-233.

L. Carlitz (1969) Some formulas of F. H. Hackson, Monat. für Math., 73, 193-198.

L. Carlitz (1970) Some applications of Saalschutz's theorem, Rend. Sem. Mat. Padova, 44, 91-95.

L. Carlitz (1973) Some inverse relations, Duke Math. J., 40, 893-901.

L. Carlitz and M. V. Subbarao (1972a) On a combinatorial identity of Winquist and its generalization, Duke Math. J., 39, 165-172.

L. Carlitz and M. V. Subbarao (1972b) On the quituple product identity, Proc. Amer. Math. Soc., 32, 42-44.

B. C. Carlson (1974) Inequalities for Jacobi polynomials and Dirichlet averages, S. I. A. M. J. Math. Anal., 5, 586-596.

BASIC HYPERGEOMETRIC FUNCTIONS

A. C. Dixon (1891) On the sum of the cubes of the coefficients in a certain expansion by the binomial theorem, Messenger of Math., 20, 79-80.

A. C. Dixon (1903) Summation of a certain series, Proc. London Math. Soc. (1), 35, 285-289.

F. J. Dyson (1962) Statistical theory of the energy levels of complex systems I, J. Math. Phys., 3, 140-156.

W. Gleissberg (1928) Über einen Satz von Herrn I. Schur, Math. Zeit., 28, 372-382.

H. Göllnitz (1967) Partitionen mit Differenzenbedingungen, J. für die reine und angew. Math., 225, 154-190.

I. J. Good (1962) Proofs of some 'binomial' identities by means of MacMahon's master theorem, Proc. Camb. Phil. Soc., 58, 161-162.

I. J. Good (1970) Short proof of a conjecture of Dyson, J. Math. Phys., 11, 1884.

B. Gordon (1961) A combinatorial generalization of the Rogers-Ramanujan identities, Amer. J. Math., 83, 393-399.

H. W. Gould (1962) A new convolution formula and some new orthogonal relations for inversion of series, Duke Math. J., 29, 393-404.

H. W. Gould and L. C. Hsu (1973) Some new inverse series relations, Duke Math. J., 40, 885-891.

J. Gunson (1962) Proof of a conjecture by Dyson in the statistical theory of energy levels, J. Math. Phys., 3, 752-753.

N. Hall (1936) An algebraic identity, J. London Math. Soc., 11, 276.

W. J. Holman, L. C. Biedenharn and J. D. Loucke (1975) On hypergeometric series well-poised in SU(n), SIAM J. Math. Anal. (to appear).

F. H. Jackson (1941) Certain q-identities, Quart. J. Math., 12, 167-172.

P. A. MacMahon (1894) A certain class of generating functions in the theory of numbers, Philos. Trans., 185, 111-160.

P. A. MacMahon (1902), The sums of powers of the binomial coefficients, Quart. J. Math., 33, 274-288.

P. A. MacMahon (1915) Combinatory Analysis, Vol. 1, Cambridge University Press, Cambridge (Reprinted: Chelsea, New York, 1960).

J. K. Percus (1971) Combinatorial Methods, Springer, New York.

G. C. Rota and J. Goldman (1970) On the foundations of combinatorial theory IV: Finite vector spaces and Eulerian generating functions, Studies in Appl. Math., 49, 239-258.

I. J. Schur (1926) Zur additiven Zahlentheorie, Sitzungsber. Akad. Wissensch. Berlin, Phys.-Math. Klasse, 488-495.

A. Selberg (1936) Über einige arithmetische Identitäten, Avh. Norsk. Vidensk. Akad. Oslo Mat. Naturvidensk. Kl. 8.

V. N. Singh (1957) Certain generalized hypergeometric identities of the Rogers-Ramanujan type, Pacific J. Math., 7, 1011-1014.

L. J. Slater (1952) Further identities of the Rogers-Ramanujan type, Proc. London Math. Soc. (2), 54, 147-167.

L. J. Slater (1966) Generalized Hypergeometric Functions, Cambridge University Press, Cambridge.

G. N. Watson (1929) A new proof of the Rogers-Ramanujan identities, J. London Math. Soc., 4, 4-9.

F. J. W. Whipple (1926) On well-poised series, generalized hypergeometric series having parameters in pairs, each pair with the same sum, Proc. London Math. Soc. (2), 24, 247-263.

K. Wilson (1962) Proof of a conjecture by Dyson, J. Math. Phys., 3, 1040-1043.

L. Winquist (1969) An elementary proof of $p(11m+6) \equiv 0 \pmod{11}$, J. Comb. Th., 6, 56-59.

Partially supported by National Science Foundation Grant MSP 74-97282.

7.6 The Well-Poised Thread: An Organized Chronicle of Some Amazing Summations and their Implications

THE RAMANUJAN JOURNAL 1, 7–23 (1997)

The Well-Poised Thread: An Organized Chronicle of Some Amazing Summations and their Implications

GEORGE E. ANDREWS* andrews@math.psu.edu
Department of Mathematics, Eberly College of Sciences, The Pennsylvania State University, University Park, Pennsylvania 16802

Received March 3, 1996; Accepted April 4, 1996

Abstract. This paper provides an organized history of well-poised hypergeometric series. The object is to reveal the process by which a rather narrow mathematical study blossomed into a topic of widespread importance. In addition, short biographies of the early contributors are included.

Key words: hypergeometric series, well-poised series

1991 Mathematics Subject Classification: Primary—33C20; Secondary—33D15, 11P83

Part I. Background

1. Introduction

This article arose from several motivating sources. The most immediate cause was an invitation by the American Mathematical Society to present the History Lecture at the 1994 Summer meetings in Minneapolis. This opportunity reminded me of an historical talk presented by Bailey to the London Mathematical Society in the late 1930's. I had obtained a xeroxed copy of Bailey's handwritten notes from Lucy Slater, and I always felt that Bailey's charming address merited a wider audience. The talk is untitled. The first paragraph reads as follows:

> "When I was asked to speak to this Society, I hesitated whether to choose as my subject "Well-Poised Hypergeometric Series" or some other subject which has interested me more recently. I did not, however, hesitate long. There is probably no other subject to which I have given so much time and thought, and in my opinion, it forms a most interesting corner of mathematics. The subject started just over a hundred years ago, but after Kummer's original formula, no further progress was made for about 65 years. Since 1900, numerous papers have been written on this and cognate subjects, and it is my purpose in this lecture to give a historical account of the development of the subject..."

I love this talk, and in this paper I shall rely on many of Bailey's insights. However the talk itself is primarily a chronological account of the main achievements in a century's

*Partially supported by National Science Foundation Grant DMS 951101 and by the University of Florida.

work on well-poised, hypergeometric series. This means it is aimed at a limited group of specialists.

In this paper I hope to extend Bailey's talk in two significant ways. First I want to place this topic in the elementary setting of reciprocal polynomials. In this way the seeming cumbersome nature of well-poised hypergeometric series will at least seem explicable in a more readily comprehended context. Here I shall follow the lead sketched in [6].

Second and more important, I wish to place in stark relief the moment in this history when well-poised series passed irreversibly from the introverted hobby of a few amateurs and specialists to a subject of broad interest which in the last three decades has found applications in orthogonal polynomials, statistical physics, additive number theory and combinatorics.

In the remainder of Part I (i.e., Sections 2 and 3), we shall discuss some elementary facts about reciprocal polynomials. Part II takes up the Well-Poised Thread and chronicles the history with some overview and guidance provided by reciprocal polynomials. Part III looks at q-analogs of Part II; this leads inexorably to Part IV, the moment when Watson showed that well-poised series are in fact powerfully applicable to questions in other branches of mathematics. Part V concludes with an all too brief look at current applications of well-poised series. To flesh out the subject I shall conclude many sections with brief biographies of mathematicians who have contributed to developments.

W.N. Bailey (1893–1961). Wilfrid Norman Bailey was born in Consett, county of Durham, England. He was educated at Consett Secondary School and Trinity College Cambridge. He was a Wrangler in Part II of the Tripos in 1915. *He served in the Royal Navy from* 1916 *to* 1919 *where he met the Rev. Dr. F.H. Jackson. He joined the University of Manchester in* 1920 *and in* 1944 *was appointed to the Chair of Mathematics at Bedford College of London University. He wrote* 73 *research papers and the classic text* [10] *Generalized Hypergeometric Series. A more extensive account of his life and work can be found in* [32].

2. Reciprocal polynomials

Reciprocal polynomials (over the complex numbers) are familiar to everyone. Succinctly put, they are polynomials whose coefficients read the same back to front as front to back. So

$$f(x) = \sum_{i=0}^{n} a_i x^i \tag{2.1}$$

is **reciprocal** provided $a_i = a_{n-i}$.

Note immediately from (2.1) that

$$f(x) = x^n f(1/x). \tag{2.2}$$

The number n is called the **reciprocal degree** of $f(x)$. This is the same as the **degree** of $f(x)$ if and only if $a_0 \neq 0$. In general, the reciprocal degree of a reciprocal polynomial is the lowest power of x actually appearing plus the highest power.

For example, $x^3 + 4x^4 + 4x^5 + x^6$ is a reciprocal polynomial of reciprocal degree 9.

Clearly the reciprocal polynomials of reciprocal degree n form an additive group (provided we adjoin 0 as the unit) because

$$f(x) \pm g(x) = x^n(f(1/x) \pm g(1/x)).$$

3. The Binomial Theorem and its enumerative corollaries

Perhaps the most famous reciprocal polynomial occurs in the Binomial Theorem:

$$(1+x)^n = \sum_{j=0}^{n} \binom{n}{j} x^j. \tag{3.1}$$

It is, of course, obvious that

$$(1+x)^n = x^n\left(1+\frac{1}{x}\right)^n. \tag{3.2}$$

Using this famous result for our prototypical example, we note several well-known enumerative facts.

1. **The coefficients are interesting.** Namely when multiplying out

$$\underbrace{(1+x)(1+x)\cdots(1+x)}_{n \text{ terms}}$$

we see that we will get x^j precisely as many times as we can find j terms (out of n) to choose x as a factor. In other words, $\binom{n}{j}$ is the number of ways j objects can be selected from n objects.

2. **$x = 1$ provides enumerative information.** Namely, with $x = 1$ in (3.2) we see that

$$2^n = \binom{n}{0} + \binom{n}{1} + \binom{n}{2} + \cdots + \binom{n}{n}.$$

In other words 2^n is the number of subsets of a set of n objects.

3. **$x = -1$ provides enumerative information.** Namely with $x = -1$ in (3.2) and $n > 0$, we see that

$$0 = \binom{n}{0} - \binom{n}{1} + \binom{n}{2} - \cdots + (-1)^n\binom{n}{n}.$$

In other words, there are exactly as many subsets with an even number of elements as there are with an odd number of elements.

All these observations are familiar from elementary, discrete mathematics. They are made here to illustrate the fact that the evaluation of reciprocal polynomials at ± 1 might well provide interesting information.

Part II. Hypergeometric series

4. Hypergeometric polynomials

The infinite series

$$\sum_{n=0}^{\infty} a_n$$

is called a **hypergeometric series** provided for each $n \geq 0$

$$\frac{a_{n+1}}{a_n} = R(n) \tag{4.1}$$

where $R(x)$ is a rational function of x. Equivalently we could require

$$a_{n+1} = R(n)R(n-1)\cdots R(0)a_0. \tag{4.2}$$

Now (4.2) is actually more useful as a definition than (4.1) in that if $R(x)$ has a positive integer zero, say N, then

$$a_j = 0 \quad \text{for } j > N. \tag{4.3}$$

Finally if we exhibit R in terms of its zeros and poles, and if we assume further that R has the same number of zeros as poles and also that 1 is a pole, we see that

$$R(n) = t\,\frac{(\alpha_0+n)(\alpha_1+n)\cdots(\alpha_r+n)}{(n+1)(\beta_1+n)\cdots(\beta_r+n)}. \tag{4.4}$$

Or equivalently

$$a_n = \frac{(\alpha_0)_n(\alpha_1)_n\cdots(\alpha_r)_n}{n!\,(\beta_1)_n\cdots(\beta_r)_n}\,t^n, \tag{4.5}$$

where

$$(A)_n = \begin{cases} A(A+1)\cdots(A+n-1) & \text{if } n>0, \\ 1 & \text{if } n=0. \end{cases} \tag{4.6}$$

Thus we arrive at the classical formulation of the hypergeometric series:

$${}_{r+1}F_r\left(\begin{matrix}\alpha_0, \alpha_1, \ldots, \alpha_r; t \\ \beta_1, \ldots, \beta_r\end{matrix}\right) = \sum_{n=0}^{\infty} \frac{(\alpha_0)_n(\alpha_1)_n\cdots(\alpha_r)_n t^n}{n!\,(\beta_1)_n\cdots(\beta_r)_n}. \tag{4.7}$$

By the comments leading up to (4.3), we see that

$${}_{r+1}F_r\left(\begin{matrix}-N, \alpha_1, \ldots, \alpha_r; t \\ \beta_1, \ldots, \beta_r\end{matrix}\right)$$

is a polynomial (hence **hypergeometric polynomial**) in t of degree N.

5. Reciprocal hypergeometric polynomials

Obviously we have been leading up to the

Question: What are the reciprocal hypergeometric polynomials?

This question is a little too general to be really useful. For example,

$$
{}_2F_1\left(\begin{matrix} -2, \dfrac{A(2B+A)}{4B-A^2}; x \\ \dfrac{2(2B+A)}{A^2-4B} \end{matrix}\right) = 1 + Ax + Bx^2. \tag{5.1}
$$

So in general if r is large enough with respect to N we wind up with a system of algebraic equations to solve to produce any polynomial whatsoever with unit constant term.

This leads us to the Revised Question. Given r, what conditions on $a_1, \ldots, a_r, b_1, \ldots, b_r$ are necessary in order that

$$
{}_{r+1}F_r\left(\begin{matrix} -N, a_1, \ldots, a_r; -(-1)^{rN}x \\ b_1, \ldots, b_r \end{matrix}\right) \tag{5.2}
$$

be a reciprocal polynomial for each integer $N \geqq 0$?

Answer: We first note that

$$
{}_{r+1}F_r\left(\begin{matrix} -N, a_1, \ldots, a_r; -(-1)^{rN}x \\ b_1, \ldots, b_r \end{matrix}\right) = \sum_{j=0}^{N} \binom{N}{j} \frac{(a_1)_j \cdots (a_r)_j}{(b_1)_j \cdots (b_r)_j} x^j (-1)^{rNj}. \tag{5.3}
$$

Hence for $0 \leqq j \leqq N$, we must have

$$
(-1)^{rNj} \frac{(a_1)_j \cdots (a_r)_j}{(b_1)_j \cdots (b_r)_j} = \frac{(a_1)_{N-j} \cdots (a_r)_{N-j}}{(b_1)_{N-j} \cdots (b_r)_{N-j}} (-1)^{rN(N-j)} \tag{5.4}
$$

In particular, with $j = 0$ we see that

$$
(b_1)_N \cdots (b_r)_N = (a_1)_N \cdots (a_r)_N (-1)^{rN} \tag{5.5}
$$

Furthermore since

$$
(A)_{N-j} = \frac{(A)_N}{(A+N-j)\cdots(A+N-1)} = \frac{(-1)^j (A)_N}{(1-N-A)_j}, \tag{5.6}
$$

we see that (in light of (5.5)) we may reduce (5.4) to

$$
\frac{(a_1)_j \cdots (a_r)_j}{(b_1)_j \cdots (b_r)_j} = (-1)^{rN} \frac{(1-N-b_1)_j \cdots (1-N-b_r)_j}{(1-N-a_1)_j \cdots (1-N-a_r)_j} \tag{5.7}
$$

for $0 \leqq j \leqq N$.

Now dividing the instance of (5.7) at $j+1$ by the instance at j, we find

$$\frac{(a_1+j)\cdots(a_r+j)}{(b_1+j)\cdots(b_r+j)} = \frac{(1-N-b_1+j)\cdots(1-N-b_r+j)}{(1-N-a_1+j)\cdots(1-N-a_r+j)} \tag{5.8}$$

for $0 \leqq j < N$.

If now $N > 2r+1$ (remember that we must have reciprocity for each $N > 0$) then the rational function identity (with variable j) in (5.8) implies that the set $\{a_1, \ldots, a_r\}$ and $\{1-N-b_1, \ldots, 1-N-b_r\}$ are identical.

In other words, we may (by reordering the b_i if necessary) deduce that the conditions required for our Revised Question are

$$b_i = 1 - N - a_i \quad \text{for } 1 \leqq i \leqq r$$

Note that at this moment we see the necessity of the $(-1)^{rN}$ that appears in (5.5).

Hence the hypergeometric polynomial

$${}_{r+1}F_r\left(\begin{matrix}-N, a_1, \ldots, a_r; \; -(-1)^{rN}x \\ 1-N-a_1, \ldots, 1-N-a_r\end{matrix}\right) \tag{5.9}$$

is a reciprocal polynomial.

6. The well-poised thread

The phenomenon discovered in Section 5 is called **well-poised** in the literature of hypergeometric series. Namely

$${}_{r+1}F_r\left(\begin{matrix}a_0, a_1, \ldots, a_r; t \\ b_1, \ldots, b_r\end{matrix}\right)$$

is called well-poised provided

$$1 + a_0 = a_1 + b_1 = a_2 + b_2 = \cdots = a_r + b_r.$$

With this prologue we may state:

The Well-Poised Thread
Evaluate reciprocal hypergeometric polynomials at $x = \pm 1$

We can make a start on this problem in the full generality of arbitrary reciprocal polynomials following the approach of [6].

Namely, if

$$f(x) = a_0 + a_1x + a_2x^2 + \cdots + a_2x^{n-2} + a_1x^{n-1} + a_0x^n, \tag{6.1}$$

then by the additive group structure mentioned at the end of Section 3

$$f(x) - a_0(1+x)^n = a_1'x + a_2'x^2 + \cdots + a_{n-1}'x^{n-1} \tag{6.2}$$

is again a reciprocal polynomial, and in like manner

$$f(x) - a_0(1+x)^n - a_1'x(1+x)^{n-2} = a_2''x^2 + \cdots + a_{n-2}''x^{n-2} \tag{6.3}$$

is as well, etc.

This process directly yields a second expansion for reciprocal polynomials (called a quadratic transformation in the literature), namely

$$f(x) = \sum_{0 \leqq 2\nu \leqq n} \overline{a_\nu}\, x^\nu (1+x)^{n-2\nu}. \tag{6.4}$$

Consequently

$$f(-1) = \begin{cases} 0 & \text{if } n \text{ odd} \\ (-1)^{n/2}\overline{a_{n/2}} & \text{if } n \text{ even,} \end{cases} \tag{6.5}$$

while

$$f(1) = 2^n \sum_{0 \leqq 2\nu \leqq n} \frac{\overline{a_\nu}}{4^\nu}. \tag{6.6}$$

Comparing (6.4) and (6.1), we see immediately that

$$a_j = \sum_{0 \leqq 2i \leqq n} \binom{n-2i}{j-i} \overline{a_i}, \tag{6.7}$$

and one of the classical "binomial inversions" of Gould [28; Table 2.2, Class 1, $q = -1$, p. 52] reveals

$$\overline{a_j} = \sum_{0 \leqq i \leqq j} (-1)^{j+i} \binom{n-j-i}{j-i} \frac{n-2i}{n-j-i}\, a_i. \tag{6.8}$$

7. The binomial theorem (again) and Kummer's theorem

Given the formulation of our project in Section 6, we are now in a position to proceed with the natural development wherein the flow of history correlates perfectly with increasing r in

$${}_{r+1}F_r\left(\begin{matrix} -N, a_1, \ldots, a_r; & -(-1)^{Nr}x \\ 1-N-a_1, \ldots, 1-N-a_r \end{matrix}\right). \tag{7.1}$$

The natural starting point is $r = 0$, in which case

$$\begin{aligned} {}_1F_0(-N; -x) &= \sum_{j=0}^{N} \frac{(-N)_j(-x)^j}{j!} \\ &= \sum_{j=0}^{N} \binom{N}{j} x^j \\ &= (1+x)^N, \end{aligned} \tag{7.2}$$

and we are back to Section 3 and the Binomial Theorem.

The next step is the case $r = 1$. This was treated by Kummer [22] in 1836. He proved (among many other things) that

$$ {}_2F_1\left(\begin{matrix} a, b; -x \\ 1+a-b \end{matrix}\right) = (1+x)^{-a}\, {}_2F_1\left(\begin{matrix} \frac{a}{2}, \frac{a+1}{2} - b; \frac{4x}{(1+x)^2} \\ 1+a-b \end{matrix}\right) \tag{7.3} $$

If a is a negative even integer this is (6.4) explicitly.

Furthermore, the case $x = -1$ is (with $a = b = -n$)

$$ \sum_{j=0}^{n} \binom{n}{j}^2 = \binom{2n}{n}, \tag{7.4} $$

while $x = 1$ yields (with $a = b = -n$) eventually

$$ \sum_{j=0}^{n} (-1)^j \binom{n}{j}^2 = \begin{cases} 0, & \text{if } n \text{ odd}, \\ (-1)^{n/2}\binom{n}{n/2}, & \text{if } n \text{ even}. \end{cases} \tag{7.5} $$

Equations (7.4) and (7.5) are two instances of the well-poised thread disguised as binomial coefficient identities.

E.E. Kummer (1810–1893). Ernst Eduard Kummer was born in Sorau, Germany (now Zary, Poland). He was privately tutored in early childhood; he entered the Gymnasium in Sorau in 1819 *and the University of Halle in* 1828. *In* 1831 *he was granted a doctorate for his prize essay on a problem posed by his mentor, Heinrich Ferdinand Scherk. He was appointed full professor at the University of Breslau in* 1842. *In* 1855, *he moved to a professorship in Berlin. He made major contributions to analysis, number theory and geometry. Among his accomplishments his invention of "ideals" in algebraic number theory (devised in his efforts to establish Fermat's Last Theorem) has perhaps had the most profound impact.*

8. Dixon's theorem

The next appearance of the Well-Poised Thread occurred in 1891 when Morley posed the following challenge to the mathematical community. Prove that

$$ \sum_{j=0}^{2p} (-1)^j \binom{2p}{j}^3 = \frac{(-1)^p (3p)!}{p!^3}, \tag{8.1} $$

or equivalently

$$ \sum_{j=-p}^{p} (-1)^j \binom{2p}{p+j}^3 = \frac{(3p)!}{p!^3}. \tag{8.2} $$

Note that (8.1) may be rewritten as

$$ {}_3F_2\left(\begin{matrix} -2p, -2p, -2p; 1 \\ 1, 1 \end{matrix}\right) = \frac{(-1)^p (3p)!}{p!^3}, \tag{8.3} $$

an $r = 2$ instance of the Well-Poised Thread.

Dixon [16] in 1891 managed to be the first to answer the challenge. Proofs of (8.1) were also published within the following decade by Morley [26], Richmond [27] and MacMahon [24].

A.C. Dixon (1865–1936). Alfred Carden Dixon was born in Northallerton. He was educated at the Quaker school in Kendal, Kingswood School near Bath, and Trinity College, Cambridge. He was Senior Wrangler in 1886. *In* 1893, *he was appointed to the Professorship at Queen's College, Galway. In* 1901 *he became professor at Queen's College, Belfast where he remained until his retirement in* 1930. *He published extensively in geometry, analysis, dynamics and algebra. A more complete account of his life and work appears in* [36].

9. The MacMahon-Whipple theorem

Obviously Dixon's theorem naturally raises the question of whether the full expansion (6.4) is possible for (5.9) when $r = 2$. This was answered affirmatively by MacMahon [24] in 1902. Unfortunately MacMahon couched his result in such obscure notation that it was not until 1927 that it resurfaced being rediscovered by Whipple [35].

The result in question is

$$
\begin{aligned}
&{}_3F_2\left(\begin{matrix}-n, b, c; -x\\ 1-b-n, 1-c-n\end{matrix}\right)\\
&\quad = (1+x)^n_3 F_2\left(\begin{matrix}-n/2, -n/2+1/2, 1-b-c-n; 4x/(1+x)^2\\ 1-b-n, 1-c-n\end{matrix}\right). \qquad (9.1)
\end{aligned}
$$

To be fair we should note that Whipple generalized this result to n being any complex number, and MacMahon had only the case of interest to us, namely n a nonnegative integer.

P.A. MacMahon (1854–1929). Percy Alexander MacMahon was born in Malta. He was educated at Cheltenham and as a gentleman cadet at Woolwich. His primary career was in the British Army first as Lieutenant in 1872, *then Captain in* 1881 *and finally Major in* 1889. *He was a mathematics professor in a variety of military schools, and received numerous honors including F.R.S. in* 1890, *the Royal Medal in* 1900, *the Sylvester Medal* 1919, *and the DeMorgan Medal in* 1923. *His full biography is contained in* [24].

10. Dougall's summations

Once one has proved (9.1), it is instructive to compare coefficients of x^j on each side, an easy exercise wherein one invokes the binomial series. The result is the following assertion:

$$
{}_5F_4\left(\begin{matrix}n-2j, n/2-j+1, n+b-j, n+c-j, -j; 1\\ n/2-j, 1-j-b, 1-j-c, n-j+1\end{matrix}\right) = \frac{(n-j)!(1-b-c-n)_j}{(n-2j)!(b)_j(c)_j}, \qquad (10.1)
$$

and we now have as a corollary of (9.1) an instance of the Well-Poised Thread for hypergeometric polynomials with $r = 4$.

MacMahon in 1902 did not observe this result; however just 5 years later [18] Dougall's study of the Vandermonde summation led him to many more general results among them:

$$
{}_5F_4\left(\begin{matrix} A, 1+A/2, C, D, E; 1 \\ A/2, 1+A-C, 1+A-D, 1+A-E \end{matrix}\right)
= \frac{\Gamma(1+A-C)\Gamma(1+A-D)\Gamma(1+A-E)\Gamma(1+A-C-D-E)}{\Gamma(1+A)\Gamma(1+A-D-E)\Gamma(1+A-C-E)\Gamma(1+A-C-D)}, \quad (10.2)
$$

where $\Gamma(z)$ is, of course, the standard gamma function of Euler.

John Dougall (1867–1960). John Dougall was born in Kippen, Stirlingshire. He attended the University of Glasgow beginning in 1881, *and received an M.A. in* 1886. *His work in elasticity won him the Makdougall-Brisbane Prize of* 1902–1904. *He also received a D.Sc. from Glasgow. He was President of the Edinburgh Mathematical Society in* 1925–1926 *and was a Fellow of the Royal Society of Edinburgh. In later life he was a reader for the publishers Blackie and Sons. A more extensive biography was published by I. Sneddon in the Year Book of the Royal Society of Edinburgh,* 1962.

11. Whipple's theorem

Once one sees all the results of Dougall [13] wherein the "well-poised" property (a term coined by Whipple [34]) is abundantly clear, it is natural to ask whether the Well-Poised Thread can be pushed further. The next natural result along this line is Whipple's Theorem [34]:

$$
{}_7F_6\left(\begin{matrix} A, 1+A/2, B, C, D, E, -m; 1 \\ A/2, 1+A-B, 1+A-C, 1+A-D, 1+A-E, 1+A+m \end{matrix}\right)
= \frac{(1+A)_m(1+A-D-E)_m}{(1+A-D)_m(1+A-E)_m}\,{}_4F_3\left(\begin{matrix} 1+A-B-C, D, E, -m; 1 \\ 1+A-B, 1+A-C, D+E-A-M \end{matrix}\right). \quad (11.1)
$$

It would be well at this point to step back a bit and examine the progress to date. Questions that first arose in the work of Kummer in the 19th century were taken up and generalized, at first, by professional mathematicians such as Dixon, Dougall and MacMahon. The final stages of generalization were accomplished by an ingenious amateur. Whipple, an astrophysicist, was, as we have mentioned before, the one who coined the term "well-poised."

Indeed, one might think the Well-Poised Thread had frayed into self-involved obscurity. Even charming binomial coefficient identities like (8.2) are sadly missing from Whipple's work.

F.J.W. Whipple (1876–1943). Francis John Welsh Whipple was the son of George Mathews Whipple who was Superintendent of Kew Observatory from 1878 *to* 1893. *Whipple was educated at Merchant Taylors' School and Trinity College, Cambridge. He was Second Wrangler in* 1897. *From* 1899 *to* 1912 *he was a mathematical master at Merchant Taylors' School. He then joined the Meteorological Office in London, and from* 1925 *to* 1939 *he served a Superintendent of Kew Observatory. Whipple wrote more than* 60 *papers in various topics in physics and its applications. He wrote 23 papers on mathematics, almost all on hypergeometric functions. A fuller account of his life and work is given in* [11].

To be fair it should be mentioned that Bailey, very much a professional mathematician, extended and generalized much of Whipple's work. Clearly there is enough order, symmetry and surprise to the Well-Poised Thread that it snared Bailey's interest for many years. Even so, in the talk alluded to in Section 1, Bailey felt compelled to provide the humble description of "a most interesting corner of mathematics" [emphasis added].

Fortunately another strand of the Well-Poised Thread was developing (in an even more disjointed fashion) in parallel with the work we have described so far.

Part III. q-analogs

12. Reciprocal polynomials in two variables and the q-binomial theorem

The one variable functional Eq. (2.2) can obviously be extended to polynomials in more than one variable. For example, we may define $f(x, q)$ to be a reciprocal polynomial of two variables provided.

$$f(x, q) = x^N q^M f(1/x, 1/q). \tag{12.1}$$

One of the first and most interesting instances of (12.1) is the q-Binomial Theorem [4, p. 36]

$$\begin{aligned} f(x, q) &:= (1+x)(1+xq)\cdots(1+xq^{n-1}) \\ &= \sum_{j=0}^{n} \begin{bmatrix} n \\ j \end{bmatrix}_q x^j q^{j(j-1)/2}, \end{aligned} \tag{12.2}$$

where

$$\begin{bmatrix} n \\ j \end{bmatrix}_q = \begin{cases} \dfrac{(1-q^n)(1-q^{n-1})\cdots(1-q^{n-j+1})}{(1-q^j)(1-q^{j-1})\cdots(1-q)} & \text{if } 0 < j < n \\ 1 & \text{if } j = 0 \text{ or } n \\ 0 & \text{if } j < 0 \text{ or } j > n. \end{cases} \tag{12.3}$$

Note that clearly

$$f(x, q) = x^n q^{n(n-1)/2} f(1/x, 1/q). \tag{12.4}$$

13. q-analogs of the well-poised thread

Many of the results found in the first eleven sections of this story have q-analogs, i.e., counterparts wherein rising factorials

$$A(A+1)\cdots(A+n-1) \tag{13.1}$$

are replaced by q-rising factorials

$$(1-q^A)(1-q^{A+1})\cdots(1-q^{A+n-1}), \tag{13.2}$$

and the binomial coefficients

$$\binom{n}{j} \tag{13.3}$$

are replaced by q-binomial coefficients

$$\begin{bmatrix} n \\ j \end{bmatrix}_q . \tag{13.4}$$

When I say the story of q-analogs is disjointed, I mean that these corresponding results do not follow the fairly strict chronology of the Well-Poised Thread, and the related questions about quadratic transformations are murky at best.

For example, there is a q-analog of (7.5), namely

$$\sum_{j=0}^{2n} (-1)^j q^{(n-j)^2} \begin{bmatrix} 2n \\ j \end{bmatrix}_q^2 = (-1)^n \begin{bmatrix} 2n \\ n \end{bmatrix}_{q^2} . \tag{13.5}$$

However this result was effectively proved first in 1942 by Daum [15].

There is a q-analog of Dixon's Theorem due to Jackson [21].

$$\sum_{j=-p}^{p} (-1)^j q^{j(3j-1)/2} \begin{bmatrix} 2p \\ p+j \end{bmatrix}^3 = \frac{(1-q)(1-q^2)\cdots(1-q^{3p})}{(1-q)^3(1-q^2)^3\cdots(1-q^p)^3} . \tag{13.6}$$

Jackson proved this result in 1941. As with some of the later workers in the Well-Poised Thread, we find Jackson to be another immensely talented amateur. He was (as noted below) a British Navy Chaplain, and some of his q-analog papers list his place of work as "H.M.S. Irresistible." How perfect for the first person to be genuinely beguiled by what Gian-Carlo Rota has irreverently called "the q-disease"!

Perhaps most surprising of all is the fact that Rogers [31; p. 29, second equation] found the q-analog of Dougall's ${}_5F_4$ summation (10.2) in 1894 thirteen years before Dougall's paper appeared. Rogers' biography appears at the end of Section 15.

F.H. Jackson (1870–1960). Frank Hilton Jackson was born in Hull and was educated at the Hull Grammar School and Peterhouse College in Cambridge. He took the Mathematical Tripos in 1889 *and achieved the rank of Wrangler. He was an ordained curate in the Church of England and served as a chaplain in the Royal Navy from* 1898 *to* 1908 *and* 1914 *to* 1918. *He wrote more than* 40 *papers in mathematics mostly on q-series. A more complete biography was by T.W. Chaundy published in London Math Soc.* 37 (1962), 126–128.

Part IV. The Moment

14. G.N. Watson's inspired discovery

While the Well-Poised Thread was becoming more and more introverted in the first few decades of the 20th century, the saga of the Indian genius, Ramanujan, unfolded. One of the most amazing discoveries of Ramanujan (a discovery brilliantly foreshadowed in the work of Rogers [30]) was the following pair of identities [4; p. 104]

$$1+\sum_{n=1}^{\infty}\frac{q^{n^2}}{(1-q)(1-q^2)\cdots(1-q^n)}=\prod_{n=1}^{\infty}\frac{1}{(1-q^{5n-4})(1-q^{5n-1})} \tag{14.1}$$

$$1+\sum_{n=1}^{\infty}\frac{q^{n^2+n}}{(1-q)(1-q^2)\cdots(1-q^n)}=\prod_{n=1}^{\infty}\frac{1}{(1-q^{5n-3})(1-q^{5n-2})}\,. \tag{14.2}$$

There is a grand story accompanying these formulae. It boils down to the fact that these two identities made the rounds in 1914–1915 as substantial unsolved problems conjectured by Ramanujan. After some time, Ramanujan ran across Rogers' 1894 paper [21] which contains full proofs of (14.1) and (14.2).

MacMahon was among those who found (14.1) and (14.2) intriguing. He noticed that they are equivalent to two assertions in additive number theory (i.e., the theory of partitions).

The first Rogers-Ramanujan partition theorem. The partitions of n into distinct nonconsecutive summands are equinumerous with the partitions of n into parts whose units digit is 1, 4, 6 or 9.

The second Rogers-Ramanujan partition theorem. The partitions of n into distinct nonconsecutive summands each >1 are equinumerous with the partitions of n into parts whose units digit is 2, 3, 7 or 8.

As an example of the first assertion, we note that for $n = 12$ the two relevant sets of partitions are $12, 11+1, 10+2, 9+3, 8+4, 8+3+1, 7+5, 7+4+1, 6+4+2$, and $11+1, 9+1+1+1, 6+6, 6+4+1+1, 6+1+1+1+1+1+1, 4+4+4, 4+4+1+1+1+1, 4+1+1+1+1+1+1+1+1$, and $1+1+\cdots+1$. Each set has 9 elements.

It was Watson, who first asked and answered the:

Question. Can one find a proof of the Rogers-Ramanujan identities that arises naturally among the q-analogs of hypergeometric series?

Watson [33] proved that there is a q-analog of Whipple's theorem (11.1). Namely,

$$\begin{aligned}&{}_8\phi_7\left(\begin{matrix}a,\ q\sqrt{a},\ -q\sqrt{a},\ b,\ \ c,\ \ d,\ \ e,\ \ q^{-m};q,\ \dfrac{a^2q^{m+2}}{bcde}\\ \sqrt{a},\ \ -\sqrt{a},\ \ \dfrac{aq}{b},\ \dfrac{aq}{c},\ \dfrac{aq}{d},\ \dfrac{aq}{e},\ aq^{m+1}\end{matrix}\right)\\ &=\frac{(aq;q)_m\left(\frac{aq}{de};q\right)_m}{\left(\frac{aq}{d};q\right)_m\left(\frac{aq}{e};q\right)_m}\,{}_4\phi_3\left(\begin{matrix}\dfrac{aq}{bc},\ d,\ \ e,\ \ q^{-m};\ \ q,\ q\\ \dfrac{aq}{b},\ \dfrac{aq}{c},\ \dfrac{deq^{-m}}{a}\end{matrix}\right),\end{aligned} \tag{14.3}$$

where

$$(A;q)_m=(1-A)(1-Aq)\cdots(1-Aq^{m-1}), \tag{14.4}$$

and

$$ {}_{r+1}\phi_r\left(\begin{matrix} a_0, a_1, \ldots, a_r; q, t \\ b_1, \ldots, b_r \end{matrix}\right) = \sum_{n=0}^{\infty} \frac{(a_0; q)_n (a_1; q)_n \cdots (a_r; q)_n t^n}{(q; q)_n (b_1; q)_n \cdots (b_r; q)_n} . \tag{14.5} $$

Note that the well-poised property for q-hypergeometric series is that the product (not the sum) of each column is a constant (in the case of (14.3) aq).

Watson easily deduced (14.1) from (14.3) by first setting $a = 1$, letting b, c, d, e and m tend to $+\infty$, and concluding with an invocation of Jacobi's Triple Product Identity. He then proved (14.2) in the same way only this time $a = q$.

This was indeed **The Moment**. The Well-Poised Thread and its q-analogs had previously seemed to its enthusiasts and observers to be increasingly introverted. However, Watson had tied the subject to the romance of the Rogers-Ramanujan story; but much more importantly he had shown that a subtle, esoteric looking identity, namely (14.3), had as corollaries surprising results in the theory of partitions.

I would be misleading if I suggested that applications gushed forth directly. However, Watson planted the seed that the well-poised hierarchy might well possess external applications and interest.

G.N. Watson (1886–1965). George Neville Watson was born at Westward Ho!. He was educated at St. Paul's and at Trinity College, Cambridge. Watson was Senior Wrangler in 1907*. He was awarded a Smith's Prize in* 1909 *and a Fellowship at Trinity in* 1910*. He moved to University College, London in* 1914*, and in* 1918 *he became Mason Professor of Mathematics at the University of Birmingham. He was elected a Fellow of the Royal Society in* 1919*. A more extensive account of his life and work can be found in* [37].

15. The result Ramanujan missed

It might well be objected that I am giving too much importance to Watson's achievement described in Section 14. Indeed, many discoveries by Rogers may, in hindsight be viewed at least as limiting cases of theorems for q-analogs of well-poised series; as we mentioned in Section 14, Rogers [31; p. 29, second equation] has the q-analog of Dougall's ${}_5F_4$ summation (10.2) thirteen years before Dougall. Bailey has made this clear in [12] and [13].

Even more to the point is Ramanujan's role in all this. In Ramanujan's Notebooks [14; p. 16, Entry 7] we find

$$ \frac{(\alpha; q)_\infty (\delta; q)_\infty}{(\beta; q)_\infty (\gamma; q)_\infty} \sum_{n=0}^{\infty} \frac{(\beta/\alpha; q)_n (\gamma/\alpha; q)_n \alpha^n}{(q; q)_n (\delta/\alpha; q)_n} $$

$$ = 1 + \sum_{n=1}^{\infty} \frac{q^{n^2-n}(1-\delta q^{2n-1})(\alpha; q)_n (\delta/q; q)_n (\delta/\beta; q)_n (\delta/\gamma; q)_n}{(1-\delta q^{-1})(q; q)_n (\delta/\alpha; q)_n (\beta; q)_n (\gamma; q)_n} \cdot \left(\frac{\beta\gamma}{\alpha}\right)^n, \tag{15.1} $$

If in (14.3) we take b and $m \to \infty$, $a = \delta/q$, $c = \alpha$, $d = \delta/\beta$ and $e = \delta/\gamma$, we find

$$1 + \sum_{n=1}^{\infty} \frac{q^{n^2-n}(1-\delta q^{2n-1})(\alpha;q)_n(\delta/q;q)_n(\delta/\beta;q)_n(\delta/\gamma;q)_n}{(1-\delta q^{-1})(\delta/a;q)_n(q;q)_n(\beta;q)_n(\gamma;q)_n} \cdot \left(\frac{\beta\gamma}{\alpha}\right)^n$$

$$= \frac{(\delta;q)_\infty(\beta\gamma/\delta;q)_\infty}{(\beta;q)_\infty(\gamma;q)_\infty} \sum_{n=0}^{\infty} \frac{(\delta/\beta;q)_n(\delta/\gamma;q)_n\left(\frac{\beta\gamma}{\delta}\right)^n}{(q;q)_n(\delta/\alpha;q)_n}. \tag{15.2}$$

This is not quite Ramanujan's result. However it is close, and a transformation due to Rogers [29]

$$\sum_{n=0}^{\infty} \frac{(\delta/\beta;q)_n(\delta/\gamma;q)_n\left(\frac{\beta\gamma}{\delta}\right)^n}{(q;q)_n(\delta/\alpha;q)_n} = \frac{(\alpha;q)_\infty}{(\beta\gamma/\delta;q)_\infty} \sum_{n=0}^{\infty} \frac{(\beta/\alpha;q)_n(\gamma/\alpha;q)_n\alpha^n}{(q;q)_n(\delta/\alpha;q)n} \tag{15.3}$$

would have provided Ramanujan with as much of (14.3) as he needed to prove his famous conjectures. Ramanujan did have Heine's transformation from which (15.3) is easily deduced; but Ramanujan apparently did not do so.

How close he came to proving the Rogers-Ramanujan identites before he came to England. What a surprise that he missed the opportunity.

S. Ramanujan (1887–1920). Srinivasa Ramanujan was born in Erode, India, in 1887; *he died in Madras in* 1920. *He attended elementary school and high school in Kumbakonam. In* 1904, *he entered Government College at Kumbakonam; however his neglect of certain nonmathematical subjects caused him to lose his scholarship, and he had to leave college. He contacted the star English analyst, G.H. Hardy in* 1913. *In* 1914 *he went to Cambridge where for four years, he and Hardy collaborated on a number of projects that were to profoundly influence much of the work on analytic number theory in the twentieth century. He became ill in* 1918, *and after convalescing for a year, returned to India where he died a year later. In* 1918 *he was elected Fellow of the Royal Society and was also elected to a fellowship at Trinity College.*

L.J. Rogers (1862–1933). Leonard James Rogers was born in Oxford. He was not sent to school due to a serious childhood illness but was tutored by J. Griffith, a mathematician at Jesus College, Oxford. At Balliol College, Oxford, Rogers did classical studies, mathematics and music. In 1888 *he went to Yorkshire College (now the University of Leeds) as Professor of Mathematics. He was elected a Fellow of the Royal Society in* 1924 *primarily due to the celebrity attending the Rogers-Ramanujan identities. Probably during his lifetime no one appreciated fully his many deep contributions. For a more detailed biography see* [17].

Part V. Conclusion

16. Applications

Given the lead up to Part 4 and Watson's proof of the Rogers-Ramanujan identities, we must start this section by mentioning the fact that well-poised q-hypergeometric series have played an extensive role in the theory of partitions; for particulars see [1], [4; Ch. 7], [7].

Indeed the main result in [1] is a three parameter generalization of the Rogers-Ramanujan identities, and the proof heavily relies on well-poised series. Subsequently questions raised in [1] have found application in the modular representations of the symmetric group [7].

In addition, other problems in number theory such as representations as sums of squares or congruences for the partition function can be viewed as corollaries of theorems on well-poised series [2].

In statistical mechanics, the extension of the Hard Hexagon model considered by Baxter et al. [5] relies in a substantial way on a theorem for well-poised series [3].

Even more substantial than anything listed above is the extensive application of well-poised series to problems in classical analysis. Askey [9] was the first to note the relation of Whipple's theorem (11.1) to the problem of connection coefficient formulas for Jacobi polynomials. Subsequently Askey, Gasper, Rahman [19] and many others have used well-poised series in the study of positivity questions, Poisson kernels, and a variety of related questions. In addition, an entire survey paper could be written just on integrals connected with well-poised series. For an introduction to related q-integrals see Askey's paper [8]. The book [19] Basic Hypergeometric Series by Gasper and Rahman is an excellent place to continue the study of The Well-Poised Thread.

Reaching beyond the series as described in this survey is an extensive and deep project undertaken by Steve Milne, Bob Gustafson and others wherein generalized well-poised series are attached to various classical groups. A good introduction to the Milne-Gustafson achievements is given in [25].

17. Summary

It should be emphasized that the Well-Poised Thread as it is described here is not part of the history of well-poised series but only a device to simplify the discussion of the history. The study of these series seems almost to have forced itself on the consciousness of several mathematicians, and the questions we have raised in our survey have only arisen in hindsight. Probably in the actual discovery of these results, the researchers were driven more by a love of formulas than anything else. With this in mind, we close with Richard Askey's comments [9; p. 89]

> "To wax philosophical for a minute, I would like to suggest that Hardy [1, p. 14] was wrong when he suggested that the age of great formulas may be over. Whipple's formula was found shortly after Ramanujan died, and it is clearly a very important formula. So is the addition formula of Šapiro and Koornwinder. Great formulas do not come along very often, but they never have. The rate at which important formulas are being found is probably decreasing, especially as a proportion of the mathematics being done, but there are still important formulas to be found. And when an explicit formula can be found there is nothing to beat it. In the rush to abstraction and generalization we often forget this."

References

1. G.E. Andrews, "On the general Rogers-Ramanujan theorem," *Memoirs of the Amer. Math. Soc.* (152) (1974).
2. G.E. Andrews, "Applications of basic hypergeometric functions," *S.I.A.M. Review* **16** (1974), 441–484.

3. G.E. Andrews, "Problems and Prospects for Basic Hypergeometric Functions," from *Theory and Applications of Special Functions* (R. Askey, ed.), Academic Press, New York, 1976, pp. 191–224.
4. G.E. Andrews, "The Theory of Partitions," *Encyclopedia of Math. and Its Applications*, Addison-Wesley, Reading, Mass. vol. 2 (Reissued: Cambridge University Press, Cambridge, 1985).
5. G.E. Andrews, R.J. Baxter, and P.J. Forrester, "Eight-vertex SOS model and generalized Rogers-Ramanujan-type identities," *J. Stat. Phys.* **35** (1984), 193–266.
6. G.E. Andrews, "Reciprocal polynomials and quadratic transformations," *Utilitas Math.* **28** (1985), 255–264.
7. G.E. Andrews, C. Bessenrodt, and J. Olsson, "Partition identities and labels for some modular characters," *Trans. Amer. Math. Soc.* **344** (1994), 597–615.
8. R. Askey, "Ramanujan's extension of the gamma and beta function," *Amer. Math.* (Monthly) **87** (1980), 346–359.
9. R. Askey, "Orthogonal Polynomials and Special Functions," *Regional Conf. Series in Applied Math.* 21, S.I.A.M., Philadelphia.
10. W.N. Bailey, *Generalized Hypergeometric Series*, Cambridge University Press, 1935 (Reissued: Hafner, New York, 1964).
11. W.N. Bailey, "Francis John Welsh Whipple," *J. London Math. Soc.* **18** (1943), 249–256.
12. W.N. Bailey, "Some identities in combinatory analysis," *Proc. London Math. Soc.* **49**(2) (1947), 421–435.
13. W.N. Bailey, "Identities of the Rogers-Ramanujan type," *Proc. London Math. Soc.* **50**(2) (1949), 1–10.
14. B. Berndt, *Ramanujan's Notebooks, Part III*, Springer-Verlag, New York, 1991.
15. J.A. Daum, "The basic analog of Kummer's theorem," *Bull. Amer. Math. Soc.* **48** (1942), 711–713.
16. A.C. Dixon, "On the sum of the cubes of the coefficients in a certain expansion by the binomial theorem," *Mess. of Math.* **20** (1891), 79–80.
17. A.C. Dixon, "Leonard James Rogers," *Obit. Notices of Fellows of the Royal Society* **1** (1932–35), 299–301.
18. J. Dougall, "On Vandermonde's theorem and some more general expansions," *Proc. Edinburgh Math. Soc.* **25** (1907), 114–132.
19. G. Gasper and M. Rahman, "Basic hypergeometric series," *Encyclopedia of Math. and Its Applications*, Cambridge University Press, Cambridge, 1990, vol. 35.
20. G.H. Hardy, *Ramanujan*, Cambridge University Press, Cambridge, 1940 (Reissued: Chelsea, New York, 1978).
21. F.H. Jackson, "Certain q-identities," *Quart. J. Math. Oxford Ser.* **12** (1941), 167–172.
22. E.E. Kummer, "Über die hypergeometrische Reihe...," *J. für die reine und angew. Math.* **15** (1836), 39–83, 127–172.
23. E.E. Kummer, *Collected Papers* (A. Weil, ed.), Springer-Verlag, Berlin, 1975, vols. I and II.
24. P.A. MacMahon, *Collected Papers* (G. Andrews, ed.), MIT Press, Cambridge, 1978, vol. 1.
25. S. Milne, "A q-analog of hypergeometric series well-poised in $SU(n)$ and invariant G-functions," *Adv. in Math.* **58** (1985), 1–60.
26. F. Morley, "On the series...," *Proc. London Math. Soc.* **34**(1) (1902), 397–402.
27. H.W. Richmond, "The sum of the cubes of the coefficients in $(1-x)^{2n}$," *Mess. of Math.* **21** (1892), 77–78.
28. J. Riordan, *Combinatorial Identities*, John Wiley, New York, 1968.
29. L.J. Rogers, "On a three-fold symmetry in the elements of Heine's series," *Proc. London Math. Soc.* **24** (1893), 171–179.
30. L.J. Rogers, "Second memoir on the expansion of certain infinite products," *Proc. London Math. Soc.* **25**(1) (1894), 318–343.
31. L.J. Rogers, "Third memoir on the expansion of certain infinite products," *Proc. London Math. Soc.* **26**(1), (1894), 15–32.
32. L.J. Slater, "W.N. Bailey," *J. London Math. Soc.* **37** (1962), 504–512.
33. G.N. Watson, "A new proof of the Rogers-Ramanujan identities," *J. London Math. Soc.* **4** (1929), 4–9.
34. F.J.W. Whipple, "On well-poised series, generalized hypergeometric series having parameters in pairs, each pair with the same sum," *Proc. London Math. Soc.* **23**(2) (1925), 104–114.
35. F.J.W. Whipple, "Some transformations of generalized hypergeometrics series," *Proc. London Math. Soc.* **26**(2), (1927), 257–272.
36. E.T. Whittaker, "Alfred Carden Dixon," *J. London Math. Soc.* **12** (1937), 145–155.
37. J.M. Whittaker, "George Neville Watson," *Biographical Memoirs of the Royal Soc.* **12** (1966), 521–530.

Chapter 8

Education, History, etc.

This chapter might well be titled, "Miscellaneous", or "Potpourri." It refers to a number of unrelated publications that do not quite fit into the first seven chapters.

One might suggest that the unearthing of Ramanujan's "Lost" Notebook should be included in the history chapter. As has been mentioned before, the exclusion of work on the "Lost" Notebook was made to save space, and in light of the fact that all my work on that subject has been collected in other books jointly written with Bruce Berndt.

I have chosen one purely historical paper [97] that is devoted to a letter that Rogers wrote to Jackson in 1917. I discovered this letter during my visit to Cambridge in 1976, when I also discovered Ramanujan's "Lost" Notebook. My wife, Joy, and I, and our small daughters, Amy and Katy, were house guests of Lucy Slater. One evening, Lucy announced: "When a mathematican visits Cambridge, his host takes him off to Hall (a formal hall dinner with the college Fellows), and his wife is left to babysit. Tonight I will be taking your wife to Hall, and you will be babysitting." I would note as explanation that the gender segregation of the Cambridge colleges in 1976 meant the Lucy could not invite me to Hall at Lady Lucy Cavendish College. At the time, I am sure Lucy thought that my wife was getting the best deal. My reaction however, was much like that of Brer Rabbit upon being tossed in the briar patch. After Amy and Katy were in bed, I got to go through Lucy's extensive collection of letters and reprints. By far the most appealing discovery of the evening was this illuminating letter of Rogers.

I have just a few comments about the section of my bibliography labelled "Book Reviews, Prefaces and Obituaries." The two obituaries allowed me to honor my friend and mentor, Fine, and the famous number theorist, Wright. Most of the book reviews express my admiration for the book in question; the only exceptions are the books by Exton [Ext83], Schoenfeld [Scho94], and McKnight *et al.* [McK-MMM00]. Indeed, the McKnight review might well have been chosen for inclusion. It contains heavy criticism of the idea that mathematics education is a science. The Essays in my bibliography mostly concern mathematics education and my skeptical view of many curriculum reforms, especially those that rely heavily on technology, and most especially on calculators. The only essay outside this topic is the first one on "the death of proof." This essay was written in response to a paper [Zei93] of my friend, and collaborator, Doron Zeilberger: a brilliant mathematician whose views on the future of technology in mathematics differ from mine.

Out of the papers on mathematics education, I have chosen [Essay4] "A Case for Balance." Originally, I thought "The Irrelevance of Calculus Reform" would be best. This latter paper caused more people to write to me than any other paper I have ever written. My correspondents ranged from famous algebraic geometers to assistant professors at rural community colleges. The vast majority of responses were quite positive: one letter began "Thank you! Thank you! Thank you!" I finally decided to omit this paper because it was fully reprinted in Krantz's book on teaching mathematics [Kran98].

8.1 L.J. Rogers and the Rogers–Ramanujan Identities

L.J. ROGERS AND THE ROGERS-RAMANUJAN IDENTITIES

George E. Andrews*

(received 29 January, 1981)

1. Introduction

The celebrated Rogers-Ramanujan identities are:

$$1 + \frac{q}{1-q} + \frac{q^4}{(1-q)(1-q^2)} + \frac{q^9}{(1-q)(1-q^2)(1-q^3)} + \cdots$$

$$= \frac{1}{(1-q)(1-q^4)(1-q^6)(1-q^9)(1-q^{11})(1-q^{14})(1-q^{16})(1-q^{19})\cdots} \tag{1.1}$$

and

$$1 + \frac{q^2}{1-q} + \frac{q^6}{(1-q)(1-q^2)} + \frac{q^{12}}{(1-q)(1-q^2)(1-q^3)} + \cdots$$

$$= \frac{1}{(1-q^2)(1-q^3)(1-q^7)(1-q^8)(1-q^{12})(1-q^{13})(1-q^{17})(1-q^{18})\cdots} \tag{1.2}$$

G.H. Hardy [10] provides us with the following fascinating account of their discovery.

> "They were found first in 1894 by Rogers, a mathematician of great talent but comparatively little reputation, now remembered mainly from Ramanujan's rediscovery of his work. Rogers was a fine analyst, whose gifts were, on a smaller scale, not unlike Ramanujan's; but no one paid much attention to anything he did, and the particular paper in which he proved the formulae was quite neglected.

---

* Partially supported by National Science Foundation Grant MCS-75-19162. I wish to thank Dr Lucy Slater of the Cambridge University Applied Economics Department for supplying me with the letters reprinted in Section 2.

Math. Chronicle 11(1982) Part 2 1-15.

"Ramanujan rediscovered the formulae sometime before 1913. He had then no proof (and knew that he had none), and none of the mathematicians to whom I communicated the formulae could find one. They are therefore stated without proof in the second volume of MacMahon's *Combinatory Analysis*.

"The mystery was solved, trebly, in 1917. In that year Ramanujan, looking through old volumes of the *Proceedings of the London Mathematical Society*, came accidentally across Rogers' paper. I can remember very well his surprise, and the admiration which he expressed for Rogers' work. A correspondence followed in the course of which Rogers was led to a considerable simplification of his original proof. About the same time I. Schur, who was then cut off from England by the war, rediscovered the identities again. Schur published two proofs, one of which is "combinatorial" and quite unlike any other proof known. There are now seven published proofs, the four referred to already, the two much simpler proofs found later by Rogers and Ramanujan and published in the *Papers*, and a much later proof by Watson based on quite different ideas. None of these proofs can be called both 'simple' and 'straightforward', since the simplest are essentially verifications; and no doubt it would be unreasonable to expect a really easy proof."

One of the questions that has gone unanswered over the years is: How did Rogers feel about all this? Lucy Slater has two letters written by L.J. Rogers to F.H. Jackson which recount not only the human aspect of Rogers' response but also contain some interesting mathematics of amazing simplicity which foreshadowed many subsequent discoveries. Section 2 is devoted completely to the reproduction of these letters. Section 3 considers the relationship of Rogers' observations to recent work and raises the possibility for further research. Section 4 is devoted to a half page of mathematics in Rogers' handwriting that was found mixed in with Ramanujan's "Lost" Notebook (cf. Andrews [8]).

2. The Letters from Rogers to Jackson

There are two changes in these letters from the original version. First I have inserted an equation number so that I can refer to this result subsequently, and I have inserted some references in square brackets. Second there are some minor misprints in the formulas of the second letter that have been corrected.

Feb. 13th The University Leeds

Dear Sir:

It is with a certain amount of amusement that a theorem which I proved nearly 24 years ago [Rogers [13]] should have remained in obscurity so long, and recently brought (sic) into prominence as a conjecture.

MacMahon wrote to me on the publication of his work regretting that he had overlooked my work before it was too late. Since then I have found out other ways of proving both the identities in a more direct way, which have been recently published in the London Math. Soc. [Rogers [14]]. My extra copies are at my home and I will send you one.

I have another proof besides which I sent to MacMahon in case he brings out a new edition which is as simple as one can make it I think. I enclose two old papers of mine. They have lain by many years and Leeds dirt penetrates everywhere, so I must apologize for their condition.

They contain too a disgraceful number of errata, which I have listed in my recent paper.

I have tried to extend the idea but without success viz. seeing that

$$1 - \frac{q}{1-q} + \frac{q^3}{(1-q)(1-q^2)} - \frac{q^6}{(1-q)(1-q^2)(1-q^3)} + \cdots$$

$$= (1-q)(1-q^2)(1-q^3) \cdots$$

$$= 1 - q - q^2 + q^5 + q^7 - \cdots$$

a θ-function of 3/2th order.

Also

$$\left.\begin{array}{l} 1+\dfrac{q^2}{1-q}+\dfrac{q^6}{(1-q)(1-q^2)}+\dots \\[2ex] 1+\dfrac{q}{1-q}+\dfrac{q^4}{(1-q)(1-q^2)}+\dots \end{array}\right\}\ \text{depend upon } \theta\text{-functions of order } 5/2 .$$

Will

$$\left.\begin{array}{l} 1+\dfrac{q^3}{1-q}+\dfrac{q^9}{(1-q)(1-q^2)}+\dots \\[2ex] 1+\dfrac{q^2}{1-q}+\dfrac{q^7}{(1-q)(1-q^2)}+\dots \\[2ex] 1+\dfrac{q}{1-q}+\dfrac{q^5}{(1-q)(1-q^2)}+\dots \end{array}\right\}\ \text{depend upon } \theta\text{-functions of order } 7/2\ ?$$

I do not get at anything, but as you will see from my last paper [Rogers [14]] certain simple basic series are connected with θ-functions of order 3/2,3,5/2,5,7/2,15/2,21/2, but I got no further.

I cannot just now lay my hands on a copy of my old "first memoir" which dealt with the operator

$$\eta f(x) = \frac{f(x)-f(xq)}{x}$$

and its affects on $(1-\lambda x)(1-\lambda xq)(1-\lambda xq^2)\ \dots$ to ∞, but you will find it in the London M.S. of about 1893 [Rogers [12]].

With many thanks for your paper and any others if you have them.

Yours truly,

L.J Rogers

Feb. 14th
6 Hollin Lane, Leeds

Dear Sir:

The proof of the identities which I referred to in my last letter to you has not been published, but is as follows: Let C_r denote $(1-x)(1-xq)\ \ldots\ (1-xq^{r-1})/(1-q)(1-q^2)\ \ldots\ (1-q^r)$. Take V_m to denote

$$(1-a_0x^m) - a_0a_1x^nq^{n+1-m}C_1(1-a_2x^mq^{2m}) + a_0a_1a_2a_3x^{2n}q^{4n+3-2m}C_2(1-a_4x^mq^{4m}) - \ldots$$

The general term being $+\ (-1)^r a_0a_1 \ldots a_{2r-1}x^{rn}q^{r^2n+r(r+1)/2-rm}C_r(1-a_{2r}x^mq^{2rm})$
$\vdots$

Then

$$V_m - V_{m-1} = a_0x^{m-1}(1-x) - a_0a_1x^nq^{n+1-m}C_1\{1-q+a_2x^{m-1}q^{2m-1}(1-xq)\} + \ldots$$

so that

$$\frac{V_m-V_{m-1}}{1-x} = a_0x^{m-1}\{(1-a_1x^{n-m+1}q^{n+1-m}) - a_1a_2x^nq^{n+m}(\eta C_1)(1-a_3x^{n+1-m}q^{3(n+1-m)}) + \ldots\}$$

where

$$\eta f(x) = f(xq)\ .$$

Then

$$\frac{V_m-V_{m-1}}{1-x} = a_0x^{m-1}\eta V_{n-m+1} \qquad (2.1)$$

with the suffixes of the a's increased by 1. This is unnecessarily general as we only want the case where every $a = 1$.

If

$$V_m = \nu_m(1-x)(1-xq) \ldots \text{ to } \infty$$

we have

$$\nu_m - \nu_{m-1} = x^{m-1}\eta\nu_{n-m+1} .$$

Let $n = 2$, then since evidently $\nu_0 = 0$

$$\nu_1 = \eta\nu_2 \qquad \text{when} \quad m = 1$$

$$\nu_2 - \nu_1 = x\eta\nu_1 \qquad \text{when} \quad m = 2$$

$$\therefore \quad \nu_1 = \eta\nu_1 + xq\eta^2\nu_1 .$$

Let

$$\nu_1 = 1 + c_1x + c_2x^2 + \ldots$$

then

$$1 + c_1x + c_2x^2 + \ldots = 1 + c_1qx + c_2q^2x^2 + \ldots + xq(1 + c_1q^2x + c_2q^4x^2 + \ldots)$$

$$\nu_1 = 1 + \frac{xq}{1-q} + \frac{x^2q^4}{(1-q)(1-q^2)} + \ldots = \frac{1}{\prod_{n=1}^{\infty}(1-xq^n)} V_1 .$$

Let $x = q$, then all the C's $= 1$ and

$$V_1 = 1 - q - q^4 + \ldots$$

the θ-series. [This leads to (1.2).] The other identity [i.e. (1.1)] comes from ν_2, when $x = q$. You will see that the idea of V_m is very widely general and when all the a's are unity leads to extensions for higher values of n .

It may be shown that V_m actually vanishes for all the factors $1-x$, $1-xq$, $1-xq^2$, ... equated to zero, but this property is not necessary for the above proof.

When all a's = 1 , or all = -1 , we also get

$$\frac{V_m - xV_{m-1}}{1-x} = \eta V_{n+m} \qquad \text{and} \qquad V_{-m} = -a_0 x^{-m} V_m$$

which accounts for all suffixes whether > or < m or negative.

When all the a's = -1 and $x = q$, V_m becomes a false θ-series like those in §9 p. 332 of the enclosed paper [Rogers [14]], which seem to me very remarkable. I have not considered them from the point of view of the V-method.

Yours truly,

L.J. Rogers

N.B. I find we have friends in common in the Gotts our present Lord Mayor and Lady Mayoress. I do not see them as much as I could wish as they nearly always go to Newby Wiske for weekends and other weekdays are most devoted to committees. Otherwise my common interests with Mr and Mrs Gott is in alpines.

3. The Implications of Rogers' Letters

It is interesting to see how calmly Rogers accepts the neglect of his work. Also observe the described attempt to extend Euler's result and the Rogers-Ramanujan identities into an infinite collection of such results. D.H. Lehmer [11] and H.L. Alder [1] came to much the same conclusions that Rogers did concerning the series

$$1 + \frac{q}{1-q} + \frac{q^5}{(1-q)(1-q^2)} + \frac{q^{12}}{(1-q)(1-q^2)(1-q^3)} + \dots$$

One notes with fascination that while the first letter describes a failure at generalization, the second letter represents a result which in fact contains the germ of some of the most involved and elaborate extensions of recent times.

The work in the second letter appeared in Rogers and Ramanujan [15] in the special case where all the $a_i = 1$. Indeed both A. Garsia and I have noted that Rogers [15] had the full functional equation developed independently by A. Selberg [16] and subsequently used in Andrews [2] to prove analytically Gordon's theorem. However Rogers never published this general elegant functional equation (2.1) with an arbitrary set of a_i.

It turns out that (2.1) contains also as special cases two of the most general functional equations known for generalized well-poised basic hyper-geometrical series. Indeed these general results correspond to the following specializations:

Equation (2.1) of Andrews [3] (restated as equation (2.5) of Andrews [6]) is the special case

$$\left.\begin{aligned} a_{2n} &= 1\,; \\ a_{2n-1} &= \frac{\left(\frac{1}{\alpha_1 q^{n-1}} - 1\right)\left(\frac{1}{\alpha_1 q^{n-1}} - 1\right)\cdots\left(\frac{1}{\alpha_\lambda q^{n-1}} - 1\right)}{\left(1 - \frac{xq^n}{\alpha_1}\right)\left(1 - \frac{xq^n}{\alpha_2}\right)\cdots\left(1 - \frac{xq^n}{\alpha_\lambda}\right)} \end{aligned}\right\} \qquad (3.1)$$

Several years after the above application was utilized in Andrews [3], [4], [5], a second related functional equation was required

(Theorem 4.1, p. 337 of Andrews [6]). This second equation is the case

$$\left.\begin{aligned} a_{2n} &= \frac{\left(\frac{1}{\alpha_1 q^n} - 1\right)\left(\frac{1}{\alpha_2 q^n} - 1\right)\cdots\left(\frac{1}{\alpha_\lambda q^n} - 1\right)}{\left(1 - \frac{xq^n}{\alpha_1}\right)\left(1 - \frac{xq^n}{\alpha_2}\right)\cdots\left(1 - \frac{xq^n}{\alpha_\lambda}\right)} \\ a_{2n-1} &= 1 \end{aligned}\right\} \tag{3.2}$$

To be more specific, the combination of (3.1) with (2.1) is one of the crucial functional equations used in the proof of

The General Rogers-Ramanujan Theorem (Andrews [4], [5]). *Let* $0 \leq \lambda \leq a \leq k$ *be integers. Define* $B_{\lambda,k,a}(n)$ *to be the number of partitions of* n *of the form* $n = b_1+b_2+\ldots+b_s$, *with* $b_i \geq b_{i+1}$, $b_i-b_{i+k-1} \geq \lambda+1$ *with strict inequality if* $(\lambda+1)|b_i$, *no parts* $\not\equiv 0 \pmod{\lambda+1}$ *are repeated, and at most* $a-1$ *parts not exceeding* $\lambda+1$ *appear. If* λ *is even, define* $A_{\lambda,k,a}(n)$ *to be the number of partitions of* n *into parts such that no part* $\not\equiv 0 \pmod{\lambda+1}$ *may be repeated and no part is* $\equiv 0, \pm(a - \frac{\lambda}{2})(\lambda+1) \pmod{(2k-\lambda+1)(\lambda+1)}$. *If* λ *is odd define* $A_{\lambda,k,a}(n)$ *to be the number of partitions of* n *into parts such that no part* $\not\equiv 0 \pmod{(\lambda+1)/2}$ *may be repeated, no part is* $\equiv \lambda+1 \pmod{2\lambda+2}$, *and no part is* $\equiv 0, \pm(2a-\lambda)(\lambda+1)/2 \pmod{(2k-\lambda+1)(\lambda+1)}$. *Then for all* $n \geq 0$,

$$A_{\lambda,k,a}(n) = B_{\lambda,k,a}(n) .$$

From the q-series point of view, we find that (3.2) was combined with (2.1) as an essential ingredient in proving [Andrews [6]]:

$$\prod_{n=1}^{\infty}(1+q^n+q^{2n}+q^{3n}) \prod_{\substack{n=1 \\ n\not\equiv 0,\pm 5 \pmod{11}}}^{\infty}(1-q^n)^{-1}$$

$$= \sum_{j,n\geq 0} \frac{q^{4n^2+12nj+8j^2+j}(-1)^j(q)_{4n+2j}}{(q^4;q^4)_n(q^2;q^2)_j(q^4;q^4)_{2n+2j}},$$

where $(A;q)_n = (A)_n = (1-A)(1-Aq)\ldots(1-Aq^{n-1})$.

Now we see that Rogers was able to state elegantly in 1918 this general extension of his method and that subsequent work has shown that special cases of his result have extensive applications in the theory of partitions and q-series.

It should also be pointed out that in a very real sense Rogers has seen through to the natural boundaries of his method. Indeed the complete generality of the a_i sequence makes clear that each term appearing in the V_m series is completely general.

4. A Further Note from Rogers

Here we reproduce a half page of mathematics found with Ramanujan's "Lost" Notebook [cf. Andrews [8]] and clearly written in Rogers' hand.

In the following the operator η is defined by

$$\eta f(x) \equiv f(xq) .$$

Also

$$(A)_\infty = \prod_{n=0}^{\infty}(1-Aq^n) ,$$

$$q_i = 1-q^i ,$$

$$q_i! = (1-q)(1-q^2)\ldots(1-q^i) .$$

"N.B. This transformation might interest you: In §10 [of Rogers [14]] (4) writing $x\eta$ for x we have since $(x\eta)^n = q^{n(n-1)/2}x^n\eta^n$

$$(1-x\eta) + x\eta(1-x\eta)(1-x\eta q) + qx^2\eta^2(1-x\eta)(1-x\eta q)(1-x\eta q^2) + \ldots$$
$$= 1 - q^2x^2\eta^2 - q^5x^3\eta^3 + q^{15}x^5\eta^5 + q^{22}x^6\eta^6 - \ldots \quad (4.1)$$

Now $\quad (1-x\eta)(-x)_\infty = (-x)_\infty - \dfrac{x(-x)_\infty}{1+x} = (-xq)_\infty = \eta(-x)_\infty ,$

$$(1-xq\eta)(-xq)_\infty = (-xq^2)_\infty = \eta^2(-x)_\infty .$$

Hence operating on $(-x)_\infty$ in (4.1) it is equivalent to

$$(\eta + x\eta^3 + qx^2\eta^5 + q^3x^3\eta^7 + \ldots)(-x)_\infty$$
$$= (1 - q^2x^2\eta^2 - \ldots)(-x)_\infty$$

Therefore

$$\frac{1}{1+x} + \frac{x}{(1+x)(1+xq)(1+xq^2)} + \frac{qx^2}{\cdots} + \ldots = 1 - \frac{q^2x^2}{(1+x)(1+xq)} - \ldots \quad (4.2)$$

or writing p_r for $(1+x)(1+xq)$ to r factors

$$\frac{1}{p_1} + \frac{x}{p_3} + \frac{qx^2}{p_5} + \frac{q^3x^3}{p_7} + \ldots = 1 - \frac{q^2x^2}{p_2} - \frac{q^5x^3}{p_3} + \frac{q^{15}x^5}{p_5} + \ldots . \quad (4.3)$$

Let $x = -q$:

$$\frac{1}{q_1} - \frac{q}{q_3!} + \frac{q^3}{q_5!} - \frac{q^6}{q_7!} + \ldots = 1 - \frac{q^4}{q_2!} + \frac{q^8}{q_3!} - \frac{q^{20}}{q_5!} + \frac{q^{28}}{q_6!} - \ldots \quad (4.4)$$

$x = -q^{1/2}$:

$$\frac{1}{q_1} - \frac{q}{q_1q_3q_5} + \frac{q^4}{q_1\cdots q_7} - \cdots = 1 - \frac{q^5}{q_1q_3} + \frac{q^{13}}{q_1q_3q_5} - \cdots \text{"} \tag{4.5}$$

Here again we see clearly illustrated the power of Rogers' technique in using the operator η to yield new q-identities. It is surely worth noting that identities (4.3), (4.4) and (4.5) have not appeared elsewhere in the literature to my knowledge; however they are formulas of great simplicity and elegance.

Also it is noteworthy that we now know how to extend (4.1), and consequently Rogers' technique is applicable to produce a large collection of new results.

For example, we know [Andrews [7], Th.1, p.290, with $y = xq^{-1}$] that if $D_{-1} = D_0 = 1$, and $D_n = D_{n-1} + q^n D_{n-2}$ for $n > 0$, then

$$\begin{aligned}1 &- \sum_{n=0}^{\infty}(1-xq)(1-xq^2)\ldots(1-xq^n)x^{2n+3}q^{2n+2}D_{n-1}\\ &- \sum_{n=1}^{\infty}(1-xq)(1-xq^2)\ldots(1-xq^n)x^{2n+2}q^{2n+1}D_{n-2}\\ &= 1 + \sum_{m=1}^{\infty}(-1)^m\{q^{m(5m-1)/2}x^{5m-2} + q^{m(5m+1)/2}x^{5m-1}\}.\end{aligned}$$

The same technique that Rogers applied above now yields

$$\left\{q - \sum_{n=0}^{\infty} D_{n-1} q^{(n+1)(2n+3)} x^{2n+3} \eta^{3n+3} - \sum_{n=1}^{\infty} D_{n-2} q^{(n+1)(2n+1)} x^{2n+2} \eta^{3n+2}\right\}(-x)_{\infty}$$

$$= \left\{q + \sum_{m=1}^{\infty} (-1)^m \left(q^{(m-1)(5m-6)/2+(5m-2)(5m-3)/2} x^{5m-2} \eta^{5m-2} \right.\right.$$

$$\left.\left. + \, q^{(m-1)(5m-4)/2+(5m-1)(5m-2)/2} x^{5m-1} \eta^{5m-1}\right)\right\}(-x)_{\infty} . \quad (4.6)$$

Hence

$$q - \sum_{n=0}^{\infty} \frac{q^{(n+1)(2n+3)} x^{2n+3} D_{n-1}}{(1+x)(1+xq)\dots(1+xq^{3n+2})} - \sum_{n=1}^{\infty} \frac{q^{(n+1)(2n+1)} x^{2n+2} D_{n-2}}{(1+x)(1+xq)\dots(1+xq^{3n+1})}$$

$$= q + \sum_{m=1}^{\infty} \frac{(-1)^m q^{15m^2-18m+6} x^{5m-2}}{(1+x)(1+xq)\dots(1+xq^{5m-3})} + \sum_{m=1}^{\infty} \frac{(-1)^m q^{15m^2-12m+3} x^{5m-1}}{(1+x)(1+xq)\dots(1+xq^{5m-2})} . \quad (4.7)$$

It would be interesting to have combinatorial explanations of Rogers' result (4.2) or extensions like (4.7).

5. Conclusion

Chapter III A of Andrews [9] is entitled "The Continuing Rediscovery of L.J. Rogers". Surely these few previously unpublished letters and notes substantiate the comments I made there. The elegant identity (2.1) undoubtedly contains further treasures besides those numerous results obtained from (3.1) and (3.2). Also the operator technique used in Section 4 is clearly applicable in many other identities. Thus again we find that Rogers foresaw further results that were to play a significant role in q-series and the theory of partitions long after his lifetime.

REFERENCES

1. H.L. Alder, *The nonexistence of certain identities in the theory of partitions and compositions*, Bull. Amer. Math. Soc. 54 (1948), 712-722.

2. G.E. Andrews, *An analytic proof of the Rogers-Ramanujan-Gordon identities*, Amer. J. Math. 88 (1966), 844-846.

3. G.E. Andrews, *On q-difference equations for certain well-poised basic hypergeometric series*, Quart. J. Math., Oxford (2) 19 (1968), 433-447.

4. G.E. Andrews, *A generalization of the classical partition theorems*, Trans. Amer. Math. Soc. 145 (1969), 205-221.

5. G.E. Andrews, *On the general Rogers-Ramanujan theorem*, Memoirs of the Amer. Math. Soc., No. 152 (1974).

6. G.E. Andrews, *On Rogers-Ramanujan type identities related to the modulus 11*, Proc. London Math. Soc. (3) 30 (1975), 330-346.

7. G.E. Andrews, *On identities implying the Rogers-Ramanujan identities*, Houston Journal of Math. 2 (1976), 289-298.

8. G.E. Andrews, *An introduction to Ramanujan's "lost" notebook*, Amer. Math. Monthly 86 (1979), 89-108.

9. G.E. Andrews, *Partitions: Yesterday and Today*, New Zealand Math. Soc., Wellington.

10. G.H. Hardy, *Ramanujan*, Cambridge University Press, London, 1940 (Reprinted: Chelsea, New York).

11. D.H. Lehmer, *Two nonexistence theorems on partitions*, Bull. Amer. Math. Soc. 52 (1946), 538-544.

12. L.J. Rogers, *On the expansion of some infinite products*, Proc. London Math. Soc., Ser. 1, 24 (1893), 337-352.

13. L.J. Rogers, *Second memoir on the expansion of certain infinite products*, Proc. London Math. Soc., Ser. 1, 25 (1894), 318-343.

14. L.J. Rogers, *On two theorems of combinatory analysis and some allied identities*, Proc. London Math. Soc., Ser. 2, 16 (1917), 315-336.

15. L.J. Rogers and S. Ramanujan, *Proof of certain identities in combinatory analysis*, Proc. Cambridge Phil. Soc. 19 (1919), 211-214.

16. A. Selberg, *Über einige arithmetische identitäten*, Avhandlinger Norske Akad., No. 8 (1936).

Pennsylvania State University

8.2 Mathematical Education: A Case for Balance

Mathematical Education: A Case for Balance

George E. Andrews

George Andrews received his BS and MA degrees from Oregon State University in 1960, and his Ph.D. in 1964 from the University of Pennsylvania. He joined the Pennsylvania State University and in 1981 became an Evan Pugh Professor. He has been a visiting faculty member at colleges and universities on four continents. He has published extensively on partitions and related areas of mathematics and is often a featured lecturer at professional meetings. In 1993 he received the MAA Distinguished Teaching Award for the Allegheny Mountain Section.

At the 1996 winter meetings of the AMS and MAA in Orlando, I participated in the Special Session on Calculus Reform. The most amazing feature of the many presentations by the well-intentioned reformers was the nearly complete abandonment of any of the standard means or devices of teaching. One particularly ardent reformer described in detail an expensive, made-to-order classroom deliberately designed to make lecturing *impossible*. As the only doubter to speak in the session, I started to wonder whether this really can be called "reform." Almost everything presented bespoke "revolution."

Taking this observation as a starting point, I wish to explore some of my concerns about the tendency toward imbalance in the current reforms of mathematics instruction at all educational levels.

Reformist Motivation

We have had intense periods of mathematics education reform before. The most recent was the "New Math," which in many ways seems quite opposite from the current reforms. Listen, for example, to James Russell, the Secretary of the Educational Policies Commission in Washington, speaking in 1965:

> And in the teaching of mathematics we need to give appropriate stress to its abstract character. The practice of teaching "two blocks plus two blocks equals four blocks" ought to be viewed as a possible impediment for some children. If we let the child learn that numbers are somehow "natural," or can be derived from experience, we are guilty of a real misrepresentation. Numbers are abstract and arbitrary definitions. They are not natural, or God-given. They are shapes which the human mind presses on events. **[13]**

In contrast, the recent *NCTM Assessment Standards* showcases the following vignette. Ms. Morris's first-grade students are studying arithmetic. A constructivist, Ms. Morris is encouraging the students to come up with "alternative solutions."

> *Ms. Morris*: The African elephant ate 37 peanuts. The Indian elephant ate 43 peanuts. How many fewer peanuts did the African elephant eat than the Indian elephant?
>
> The children worked on the problem for two or three minutes. Some of the children used stacking cubes that had been joined together in stacks of ten cubes. ...
>
> *Ms. Morris*: ... How did you figure it out, Mike?

Reprinted from the College Mathematics Journal
Vol. 27, No. 5, November 1996

Mike: Well, I had 43 here (pushing out four stacks of ten cubes and three additional cubes joined together), and I had 37 here (pushing out three stacks of ten cubes and a stack of seven). I put 30 on top of these 30. I took 3, and I put them here; there were 4 left, so I took 4 off, and there were 6 left. **[11]**

What can these two movements have in common? The first assures us that abstraction is so important that real-world instances are an impediment. The second is so firmly mired in politically correct reality (inclusive language has extended even to elephants!) that any abstraction seems impossible. However, upon reflection we find that each is energized by a vigorous *rejection of the past*—perhaps not the real past with all its complexities, but a simplified version, a caricature easy to ridicule.

This scorn for tradition among New Math reformers appears in the first few paragraphs of the 1964 book, *A Parent's Guide to the New Mathematics*. Its author announces a revolution in mathematics teaching, arising from an awareness of how outmoded the standard curriculum is:

> If you have a child in school you are beginning to feel the impact of change in mathematics—a change so far-reaching that it warrants the word revolution. . . .
>
> . . . If you dug up an old seventeenth-century don he could [until recently] walk right into any classroom and start teaching math. Nobody would notice the difference, since the content of the courses hadn't changed in the last 300 years.
>
> Furthermore, mathematics was the only subject of which this was true. The don couldn't teach history because he wouldn't know what had happened since the reign of William and Mary. In science he would obviously be lost. **[15]**

Today among the reformers, we hear similar voices railing against the dead hand of the past. Listen, for example, to Cal Moore:

> California has taken the lead in upgrading mathematics education for the 21st century. But while reforms are gradually taking hold, the majority of classrooms still rely on a traditional mathematics curriculum, that, as one cynical observer remarked, is largely composed of eight years of 15th century arithmetic, two years of 17th century algebra and one year of 3rd century B.C. geometry. **[10]**

Tom Tucker even reaches for wrestling metaphors to describe current struggles to liberate mathematics instruction from the past: "Just as calculus reform is unlocking the strangle-hold of the lecture method on pedagogy, it is also unlocking the strangle-hold of algebra on content" **[17]**.

I shall postpone for the moment a fuller description of the current reforms. Let us first turn our attention to some of the reasons that education is hard and that reform enthusiasm makes it harder.

Two Views of the Education Problem

Is respect for traditional teaching really worth preserving? E. F. Schumacher, a founder of the modern environmental movement, considered this question. He distinguished two types of problems, which he called *convergent* and *divergent* **[14]**. Convergent problems have a solution that most will agree is optimal:

> Take a design problem—say, how to make a two-wheeled, manpowered means of transportation. Various solutions are offered, which gradually and increasingly converge until, finally, a design emerges which is simply "the answer"—a bicycle, an answer that turns out to be amazingly stable in time. Why is this answer so stable? Simply because it complies with the laws of the Universe—laws at the level of inanimate nature.

Divergent problems are those that admit of multiple solutions, no one of which is "best," and Schumacher used education to illustrate such problems. Two groups present almost contradictory visions of education. Says the first group:

> Education is the process by which existing culture is passed on to the next generation. Those who have (or are supposed to have) knowledge and experience ***teach***, and those who as yet lack knowledge and experience ***learn***. This is quite clear, and implies that there must be a situation of authority and discipline.

Education, then, is the transmission of knowledge pure and simple; a structured curriculum and discipline are appropriate. However, another equally committed group takes an almost opposite view of education:

> Education is nothing more or less than the provision of a *facility*. The educator is like a good gardener, who is concerned to make available good, healthy, fertile soil in which a young plant can grow strong roots and then extract the nutrients it requires. The young plant will develop in accordance with its own laws of being, which are far more subtle than any human being can fathom, and will develop best when it has the greatest possible freedom to choose exactly the nutrient it needs.

Now the teacher should be a "facilitator," and students should be free to choose which topics to explore. Schumacher goes on to observe that a balance must be struck between the two:

> If our first group of advisers is right, discipline and obedience are "a good thing", and it can be argued with perfect logic that if something is "a good thing", more of it will be an even better thing; and this line of logic leads to the conclusion that perfect discipline and obedience would be a perfect thing... and the school would become a prison.
>
> Our second group of advisors, on the other hand, argues that in education freedom is "a good thing". If so, more freedom will be an even better thing, and perfect freedom would produce perfect education. The school would become a wilderness, even a kind of lunatic asylum.

Obviously reformers, devoted to throwing off the shackles of the past, will almost always fall into Schumacher's second group. They will often view all who oppose them as extremists of the traditionalist approach—"Back to the Basics" troglodytes. Such zeal makes it difficult for less radical colleagues to find common ground for discussing modest reforms that might improve our teaching.

Some Themes of Current Reform

Mike Reed **[12]** has given us succinct descriptions of reformed teaching. While he is addressing only calculus reform, his comments fairly accurately reflect themes in primary and secondary education as well:

> I believe that the most important aspect of Reform Calculus is the way the actual teaching is done. Specifically, I see an emphasis on classroom activities of four types: machine computation, cooperative learning in which students work together, investigative learning, and writing.

Professor Reed believes that these reforms require dedication, much more so than traditional approaches:

> ... teaching Reform Calculus requires more talent and training. The Old Calculus could be delivered by Professor Hohum, Mr. Part-time, and graduate student Littlenglish because lecturing on technique is easy. Getting them to deliver Reform Calculus will be much more difficult. **[12]**

Some of us may recall inspiring traditional courses and inspiring lecturers, but no one would ever guess from the reform literature that such ever existed. It would appear that formerly no one ever taught well and no one ever learned anything.

By now the reader may suspect that I am one of Schumacher's rigid traditionalists anxious to overpower the current revolutionaries so that I can go to the opposite extreme. Not so! I can still remember a rather free-form sophomore seminar in which I managed to solve the combinatorial derangement problem more or less from scratch. It was exhilarating and enlightening—all in all, a good thing! Of course the version of inclusion-exclusion I invented was the clunkiest one imaginable; it certainly is not how I think about that topic today.

What the "True Believers" Say

How can we gauge the tendency toward imbalance in the reform movement? One of the best ways is to listen to the radicals who express their philosophical positions most openly. Take Steven Leinwand, a mathematics supervisor for the Connecticut Department of Education, who is just fed up with arithmetic:

> It's time to confront those nagging doubts about continuing to teach our students computational algorithms for addition, subtraction, multiplication, and division.... And it's time to proclaim that, for many students, real mathematical power, on the one hand, and facility with multi-digit pencil-and-paper computational algorithms, on the other hand, may be mutually exclusive....
>
> Shouldn't we be as eager to end our obsessive love affair with pencil-and-paper computation as our move on from outhouses and sundials? In short, we know and should agree that the long division "gazinta" algorithm and its computational cousins are obsolete in light of everyday societal realities....
>
> In short, we now understand that teaching formal rules for adding and subtracting decimals, like their "gone-and-not-missed" cousin, the square root algorithms, remains a vestige of a sort-em-out approach that continues to fail both kids and society that so desperately needs a far more mathematically powerful citizenry. **[8]**

Sadly, there is worse. Like vultures waiting for poor old rational thought to give its last death twitch, the post-modernist buzzards circle ominously. They assure us that mathematics may be constructed in a variety of fairly arbitrary ways, and all we must watch is whether the constructing is done by the oppressors (the bad guys) or the oppressed (the good guys). For example, how shall we decide if there are infinitely many integers? We learn from Sandra Harding that the answer depends on our society and culture, not some abstract proof. *Social image replacement* holds the key:

> Common sense tells us that a part cannot be equal to the whole. Thus it is only relatively recently that mathematicians have been able to countenance the idea that the integers could be infinite in number. Earlier mathematicians' problem was as follows: one can match each sequential integer with an even integer (1-2, 2-4, 3-6, 4-8, ...), resulting in an infinite series in which there are as many even integers as there are integers—at first glance an absurdity. How was this paradox resolved? Mathematicians were willing to let go of the common sense truth that a part cannot be equal to the whole for this special circumstance in order to develop infinitesimal theory. They did so by replacing the social image of numbers as counting units with the social image of numbers as divisions of a line. These are *social* images because they reflect what people in historical cultures intentionally do. Not all cultures have been as preoccupied with measuring—dividing a line—as has ours for the last few centuries. A whole field of mathematical inquiry was made possible by the substitutions of a different kind of social image for thinking about what numbers are. As one commentator points out, such a process of socially negotiating cultural images in mathematics is similar to what we do when we exclude patriotic killing in wartime from the moral and legal category of murder. **[5]**

Unfortunately this paragraph is not a joke. It appears in a book that won the 1987 Jessie Bernard Award of the American Sociological Association.

It requires little imagination to conjure up the disaster awaiting mathematics education if it gets further infected with a relativistic philosophy like Harding's. Yet the colleges of education to whom politicians and administrators often turn for educational advice are filled with relativists (cf. **[7]**).

It is important to realize that these extreme views are intimately related to seemingly moderate and respectable reform ideas. Leinwand's views are the logical conclusion of the premature and indiscriminate use of calculators. Constructivism is the poor cousin of Harding's relativism. Relativism provides the philosophical clothing for constructivism—if reality is negotiable, we'd all better start negotiating early. On the other hand, if there truly is a real mathematical world out there, then discovery and transmission are both appropriate means of learning. This latter view was perhaps most memorably stated by Isaac Newton: "If I have seen further it is by standing on the shoulders of Giants."

Five Questions about Current Reforms

Numerous accounts laud the success of mathematics education reform in the schools and of calculus reform in colleges and universities. For example, with one exception [**2**], the February 1995 issue of *UME Trends* was devoted to tales of calculus reform triumphs. The one skeptical paper (mine) generated an avalanche of response, from which I have distilled five questions posed by traditionalists at all levels of mathematics teaching:

1. Are standards being maintained?
2. Has the use of calculators impaired basic skills?
3. Is constructivism the only valid basis for mathematics teaching?
4. Is homework being done?
5. Is group learning essential?

The reformers' answers to these questions are yes, no, yes, yes, and yes. Perhaps, so—maybe education isn't one of Schumacher's divergent problems after all! Nevertheless, these answers are at least debatable.

Let us turn for perspective on the first three questions to a news report from Great Britain, where the reform movement is farther advanced than in the United States:

> Calculators are to be banned in one of two maths papers to be sat by 11-year-olds next year because government curriculum advisers have identified a worrying lack of mathematical ability in tests taken this summer.
>
> The School Curriculum and Assessment Authority has also urged Gillian Shephard, the Education Secretary, to introduce a similar ban in tests for 14-year-olds by 1997. The results of this summer's tests will be published later this month, but officials analysing trends were worried by widespread confusion over decimal places and a lack of competence with fractions. Division is also a "particular weakness" among 11 and 14-year-olds.
>
> Teaching methods which encouraged children to discover mathematical rules for themselves were criticised for allowing children to "waste time" trying to work things out inefficiently.
>
> In one question this summer, 11-year-olds were asked to calculate 15 percent of 80. Just a quarter of the 600,000 pupils answered the question correctly. A very common wrong answer was 15, which "showed a lack of basic understanding of percentages." [**3**]

For calculus reform, we may consult Kerry Johnson:

> The overall trend seems to be that Harvard calculus grades are better than traditional students' grades, yet traditional students tend to be more successful in subsequent math courses. This is most likely due to Harvard students moving from an application based course to a more algebraically based course such as traditional Calculus 2, Differential Equations, or Linear Algebra. It appears that Harvard calculus does address the problem of students passing calculus, but may not fully prepare students for algebraically based subsequent courses. [**6**]

Concerning homework, reformer David Mathews **[9]** reports that students in the reform project C^4L study nine hours a week, whereas five hours is the number for traditional calculus. It seems "Professor Hohum, Mr. Part-time, and graduate student Littlenglish" are representing the traditionalists again. If in each contest the reform first-string varsity goes up against the traditionalist third-string junior-varsity injured list, it doesn't take Joe Paterno to predict the victor. Any calculus course that can be passed by an average student with five hours of work per week is a disgrace.

Is homework being done? Mathews' observations do not necessarily coincide with reports from other projects. For example, at Rensselaer Polytechnic Institute the Studio Calculus Project finds students studying 20 minutes more per week than in their traditional courses. Others report that their students spend many hours on special projects, but quiz and test performance indicates that they spend very little time on skill-building homework.

Proponents of group learning (question 5) have often turned to Uri Treisman for support. Treisman examined groups of black students and Chinese students at Berkeley. The former did not do well in calculus: "Unlike the Chinese students, black students rarely studied with classmates. Only two of the twenty students I interviewed reported that they regularly studied with other students ...." The Chinese students not only created informal study groups to augment individual study but put in longer hours:

> These students learned quickly, for example, that the often-quoted rule of thumb for estimating the number of hours that one should devote to study—two hours for each class hour—was seriously misleading. The blacks whom I had interviewed devoted approximately eight hours per week to homework and study for their four-unit math course; the Chinese students devoted roughly fourteen hours per week to these same tasks.
>
> The Chinese students spent roughly two hours studying alone for each of their hours together, and thus its was entirely possible that in their solitary study they had relearned any particular fact or method on their own. Correct mathematics has many possible sources; errors, on the other hand, are far easier to trace. **[16]**

Note that the group learning by the Chinese students was *informal* and was one-third of a 14-hour-per-week regimen!

Other studies of group learning are mixed. One review **[4]** reports: "Less than half of the studies comparing small group and traditional methods of mathematics instruction have shown a significant difference in student achievement; but when significant differences have been found, they almost always favored the small-group procedure."

To summarize, reformers have had some success, and they deserve a place on the stage. They have not, however, made an overwhelming case in support of the sweeping changes they advocate.

Turning the Tables

In my talk in Orlando, I tried to illustrate one subtle way that the reformers advance their agenda. Table 1 quotes the Guidelines for Pedagogy from the AMATYC Standards **[1]**.

Table 2 illustrates how a traditionalist could turn the tables by loading the language differently. Now the reformers look inane, and tradition seems the only way. The real challenge at all levels of education is to find a proper *balance* between the two "solutions" of the divergent problem of education, solutions reflected by the two columns in these tables.

Table 1. Guidelines for Pedagogy

Increased use	Decreased use
Active involvement of students	Passive listening
Technology to aid in concept development	Paper-and-pencil drill
Problem solving and multistep problems	One-step single-answer problems
Mathematical reasoning	Memorization of facts and procedures
Conceptual understanding	Rote manipulation
Realistic problems encountered by adults	Contrived exercises
An integrated curriculum with ideas developed in context	Isolated topic approach
Multiple approaches to problem solving	Requiring a particular method for solving a problem
Diverse and frequent assessment both in class and outside of class	Tests and a final exam as the sole assessment
Open-ended problems	Problems with only one possible answer
Oral and written communication	Requiring only short, numerical answers, or multiple choice responses
Variety of teaching strategies	Lecturing

Table 2. Guidelines for Pedagogy

Increased use	Decreased use
Focused homework	Chaotic self-instruction
Essential drill	Thoughtless, pervasive use of calculators
Problems designed to reinforce a clearly designated concept	Artificially contrived, multistep fiascoes
Appropriate memorization	Pretense that mathematical reasoning can precede mathematical facility
Drills fostering familiarity with necessary skills	Pretense that conceptual understanding can precede mathematical facility
Focused exercises	Contrived "realistic problems" supposedly encountered by adults
Focused topics	Mishmash of topics precluding conceptual organization
Clear elucidation of each mathematical method	Mishmash of problem solving techniques
Carefully prepared, objective tests	Variety of so-called "assessment instruments," each likely to hide the students' truly poor level of achievement
Focused problems	Problems so poorly posed that each possible answer is debatable
Clearly written answers to each problem	Allowing the student to explain away ignorance
Focused lectures	A phantasmagoria of so-called "teaching" so-called "strategies"

Schumacher pointed out that some teachers seem better able to find such a balance, and he described other gifts that a successful teacher must employ:

> Love, empathy, *participation mystique*, understanding, compassion—these are faculties of a higher order than those required for the implementation of any policy of discipline or of freedom. To mobilise these higher faculties or forces, to have them available not simply as occasional impulses but permanently; that requires a high level of self-awareness, and that makes a great educator. [**14**]

A call to balance the divergent requirements in education within a positive student-teacher interaction is not likely to produce strong enthusiasm or to fire revolutions, but at least it provides a realistic statement of our problem. Let us conclude with the words of Rita Kramer:

> The single most important factor in an individual's education is his teachers. All of us remember particular individuals who influenced and inspired us and gave direction to our lives. Not buildings, programs, curricula, philosophies of education, but men and women who by virtue of their personalities and their love of some discipline, some book, some kind of learning, opened the world to us, and showed us things we had not seen before, gave life a meaning it had not had before. What we need if we are to touch the minds of children, rescue the public school system and the democracy it should nourish, are inspiring teachers. They are precisely what the present system [i.e., colleges of education as currently constituted] militates against. Although they exist here and there, they do what they do in spite of the present system of preparation and in spite of all the odds against them which that system presents.
>
> Isn't it time we changed it? [7]

References

1. American Mathematical Association of Two-Year Colleges, *Crossroads in Mathematics: Standards for Introductory College Mathematics Before Calculus*, American Mathematical Association of Two-Year Colleges, Memphis, TN, 1995; p. 20.
2. George Andrews, The irrelevance of calculus reform: Ruminations of a sage-on-the-stage, *UME Trends* 6:6 (1995) 17, 23.
3. David Charter, Exams ban on calculators after maths' standards fall, *London Times*, December 7, 1995.
4. Neil Davidson and Diana Lambdin Kroll, An overview of research on cooperative learning related to mathematics, *Journal for Research in Mathematics Education* 22:5 (1991) 362–365; pp. 362–363.
5. Sandra Harding, *The Science Question in Feminism*, Cornell University Press, Ithaca, NY, 1986; p. 50.
6. Kerry Johnson, Harvard calculus at Oklahoma State University, *American Mathematical Monthly* 102:9 (November 1995) 794–797; p. 796.
7. Rita Kramer, *Ed School Follies: The Miseducation of America's Teachers*, Free Press/Macmillan, New York, 1991; p. 222.
8. Steven Leinwand, It's finally time to abandon computational algorithms, *Wisconsin Teacher of Mathematics* (Winter 1995) 8–9; p. 8.
9. David M. Mathews, Time to study: The C^4L experience, *UME Trends* 7:4 (September 1995) 14.
10. Calvin Moore, California's math reforms aim to inspire economy of thought, *Sacramento Bee*, October 18, 1995.
11. National Council of Teachers of Mathematics, *Assessment Standards for School Mathematics*, Reston, VA, 1995; pp. 32–33.
12. Michael C. Reed, Mainstreaming calculus reform, *SIAM News*, 27:7 (August/September 1994) 7.
13. James Russell, *Change and Challenge in American Education*, Houghton Mifflin, Boston, 1965; p. 56.
14. E. F. Schumacher, *A Guide for the Perplexed*, Sphere Books, London, 1978; pp. 140–143.
15. Evelyn Sharp, *A Parent's Guide to the New Mathematics*, Dutton, New York, 1964; pp. 11–12.
16. Philip Uri Treisman, A study of the mathematics performance of black students at the University of California, Berkeley, in Fisher, Keynes, and Wagreich, eds., *Mathematics and Education Reform, Proceedings of July 6–8, 1988, Workshop* (CBMS Issues in Mathematics Education Series, vol. 1), American Mathematical Society, Providence, RI, 1990, 33–46; pp. 38–39.
17. Thomas W. Tucker, Nonalgebraic approaches to calculus, in Roberts, ed., *Calculus: The Dynamics of Change* (MAA Notes Number 39), MAA, Washington, DC, 1996, 16–19; p. 16.

Permissions from Publishers and Copyright Holders

1. The Geometry of Numbers

1.1 Trans. Amer. Math. Soc. (2) 106 (1963) 270–279, A lower bound for the volume of strict convex bodies with many boundary lattice points, G. E. Andrews, copyright © 1963 American Mathematical Society, reprinted with permission from the American Mathematical Society.

2. q-Series

2.1.1 Andrews, George E., On the theorems of Watson and Dragonette for Ramanujan's mock theta functions. American Journal of Mathematics (2) 88 (1966) 454–490. © 1966 The Johns Hopkins University Press. Reprinted with permission of The Johns Hopkins University Press.

2.2.1 Proc. Amer. Math Soc. (2) 16 (1965) 333–334, A simple proof of Jacobi's triple product identity, G. E. Andrews, copyright © 1965 American Mathematical Society, reprinted with permission from the American Mathematical Society.

2.2.2 Andrews, George E., On the q-analog of Kummer's theorem and applications, Duke Math. J. (3) 40 (1973) 525–528, reprinted with permission of the publisher, Duke University Press.

2.3.1 Orthogonal Polynomials and Applications (Bar-le-Duc, 1984), Lecture Notes in Math N. 1171, 1985, 36–62, Classical orthogonal polynomials and applications, G. E. Andrews and R. Askey, copyright © 1985 Springer, reprinted with kind permission of Springer Science and Business Media.

2.4.1 George E. Andrews, On q-difference equations for certain well-poised basic hypergeometric series, Quart. J. Math. (2) 19 (1968) 433–447, by permission of the Oxford University Press.

2.5.1 Andrews, George E., An analytic generalization of the Rogers-Ramanujan identities for odd moduli, Proc. Nat. Acad. Sci. (10) 71 (1974) 4082–4085, © 1974 National Academy of Sciences, USA.

2.5.2 George E. Andrews. Multiple series Rogers-Ramanujan type identities, Pac. J. Math. (2) 114 (1984) 267–283, reprinted with permission.

2.5.3 Trans. Amer. Math. Soc. (1) 293 (1986) 113–134, The fifth and seventh order mock theta functions, G. E. Andrews, copyright © 1986 American Mathematical Society, reprinted with permission from the American Mathematical Society.

2.5.4 Invent. Math. 91 (1988) 391–407, Partitions and indefinite quadratic forms, G. E. Andrews and D. R. Hickerson, copyright © 1988 Springer, reprinted with kind permission of Springer Science and Business Media.

2.5.5 This article was published in *Journal of Combinatorial Theory Series A*, vol. 91, G. E. Andrews, Umbral calculus, Bailey chains and pentagonal number theorems, pages 464–475, Copyright © Elsevier 2000.

2.5.6 Special Functions 2000, J. Bustoz et al. eds., 1–22, Bailey's transform, lemma and tree, G. E. Andrews, copyright © 2001 Kluwer, reprinted with kind permission of Springer Science and Business Media.

2.6.1 J. Amer. Math. Society (3) 3 (1990) 653–669, Euler's "Exemplum Memorabile Inductionis Fallacis," G. E. Andrews, copyright © 1990 American Mathematical Society, reprinted with permission from the American Mathematical Society.

2.6.2 Analytic Number Theory, B. Berndt et al. eds., 1990 1–11, q-trinomial coefficients and Rogers-Ramanujan type identities, G. E. Andrews, copyright © 1990 Birkhäuser, reprinted with kind permission of Springer Science and Business Media.

2.6.3 This article was published in *Journal of Symbolic Computation*, vol. 20, G. E. Andrews, On a conjecture of Peter Borwein, pages 487–501, Copyright © Elsevier 1995.

2.7.1 George E. Andrews. An introduction to Ramanujan's "lost" notebook, Amer. Math Monthly (2) 86 (1979) 89–108.

2.8.1 J. of Stat. Phys., (3/4) 35 (1984) 193–266, Eight-vertex SOS model and generalized Rogers-Ramanujan-type identities, G. E. Andrews, R. J. Baxter, and P. J. Forrester, copyright © 1984 Springer, reprinted with kind permission of Springer Science and Business Media.

2.8.2 J. of Stat. Phys., (3/4) 47 (1987) 297–330, Lattice gas generalization of the hard hexagon model. III. q-trinomial coefficient, G. E. Andrews and R. J. Baxter, copyright © 1987 Springer, reprinted with kind permission of Springer Science and Business Media.

2.9.1 This article was published in *Journal of Computational and Applied Mathematics* vol. 68, G. E. Andrews, Pfaff's method (II) diverse applications, pages 15–23, Copyright © Elsevier 1996.

2.10.1 Annals of Combinatorics 4 (2000) 327–338, MacMahon's partition analysis: II Fundamental theorems, G. E. Andrews, © 2000 Springer, reprinted with kind permission of Springer Science and Business Media.

3. Partition Identities

3.1 Andrews, George E., An analytic proof of the Rogers-Ramanujan-Gordon identities. American Journal of Mathematics 88:4 (1966), 844–846. © 1966 The Johns Hopkins University Press. Reprinted with permission of The Johns Hopkins University Press.

3.2 J. Comb. Theory (4) 2 (1967) 431–436, Some new partition theorems, G. E. Andrews, © 1967 Springer, reprinted with kind permission of Springer Science and Business Media.

3.3 Proc. Amer. Math. Soc. (5) 18 (1967) 945–952, A generalization of the classical partition theorems, G. E. Andrews, copyright © 1967 American Mathematical Society, reprinted with permission from the American Mathematical Society.

3.4 Lecture Notes in Mathematics No. 251, (1972) 1–20. Sieves for theorems of Euler, Rogers and Ramanujan, G. E. Andrews, © 1972 Springer-Verlag, reprinted with kind permission of Springer Science and Business Media.

3.5 Combinatorics and Ramanujan's Lost Notebook, by George Andrews from Surveys in Combinatorics 1985: Invited Papers for the Tenth British Combinatorial Conference, edited by Ian Anderson. Copyright © 1985 Cambridge University Press. Reprinted with permission.

3.6 George E. Andrews. On a partition problem of H. L. Alder, Pacific J. Math. (2) 36 (1971) 279–284, reprinted with permission.

3.7 Andrews, George E., Partitions and Durfee dissection. American Journal of Mathematics (3) 101 (1979), 735–742. © 1979 The Johns Hopkins University Press. Reprinted with permission of The Johns Hopkins University Press.

3.8 This article was published in *Enumeration and Design*, G. E. Andrews, Use and Extension of Frobenius' representation of partitions, pages 51–65, Copyright © Elsevier 1984.

3.9 Bull. Amer. Math. Soc. (2) 18 (1988) 167–171, Dyson's crank of a partition, G. E. Andrews and F. G. Garvan, copyright © 1988 American Mathematical Society, reprinted with permission from the American Mathematical Society.

3.10 Andrews, George E. & Jorn B. Olsson: "Partition identities with an application to group representation theory." In: Journal für die reine und angewandte Mathematik, 413, Berlin, New York: de Gruyter 1991, pp. 198–212. Reprinted with permission from de Gruyter.

3.11 Andrews, George E. & Basil Gordon: "Generalizations and refinements of a partition theorem of Göllnitz." In: Journal für die reine und angewandte Mathematik, 469, Berlin, New York: de Gruyter 1995, pp. 165–188. Reprinted with permission from de Gruyter.

3.12 This article was published in *Discrete Mathematics* vol. 232, G. E. Andrews and R. P. Lewis, An algebriac identity of F. H. Jackson and its implications for partitions, pages 77–83, Copyright © Elsevier 2001.

3.13 Andrews, George E., Partitions with short sequences and mock theta functions, Proc. Nat. Acad. Sci. 102 (2005) 4666–4671, © 2005 National Academy of Sciences, USA.

3.14 Inventiones Math. 169 (2007) 37–73, Partitions, Durfee symbols, and the Atkin-Garvan moments of ranks, George E. Andrews, © 2007 Springer, reprinted with kind permission of Springer Science and Business Media.

3.15 Andrews, George E.: "The number of smallest parts in the partitions of n." In: Journal für die reine und angewandte Mathematik, 469, Berlin, New York: de Gruyter 624 2008, pp. 133–142. Reprinted with permission from de Gruyter.

3.16 Andrews, George E., On a partition theorem of N. J. Fine, J. Nat. Acad. Math., India, 1 (1983), 105–107. Reprinted with permission.

3.17 DIMACS Series in Discrete Mathematics and Theoretical Computer Science No. 64 (2004) M. B. Nathanson, Ed., 9–15, The Friedman-Joichi-Stanton monotonicity conjecture at primes, G. E. Andrews, copyright © 2004 American Mathematical Society, reprinted with permission from the American Mathematical Society.

3.18 Andrews, George E. The theory of compositions IV: Multicompositions, The Mathematics Student (Special Centenary Volume) (2007) pp. 25–31, reprinted with permission of the General Secretary of the Indian Mathematical Society.

4. Plane Partitions

4.1 Andrews, George E., MacMahon's conjecture on symmetric plane partitions, Proc. Nat. Acad. Sci. (2) 74 (1977) 426–429, © 1977 National Academy of Sciences, USA.

4.2 Inventiones Math. 53 (1979) 193–225, Plane partitions (III): The weak Macdonald conjecture, George E. Andrews, © 1979 Springer, reprinted with kind permission of Springer Science and Business Media.

4.3 Aequationes Mathematicae 33 (1987) 230–250, Plane partitions IV: A conjecture of Mills-Robbins-Rumsey, George E. Andrews, © 1987 Springer, reprinted with kind permission of Springer Science and Business Media.

4.4 George E. Andrews and W. H. Burge. Determinant identities, Pac. J. Math. (1) 158 (1993) 1–14, reprinted with permission.

5. Combinatorics, Fibonacci Numbers, and Computers

5.1 This article was published in *Journal of Combinatorial Theory Series A*, vol. 44, G. E. Andrews, q-Catalan numbers and hypergeometric series, pages 267–273 Copyright © Elsevier 1987.

5.2 George E. Andrews, Fibonacci numbers and the Rogers-Ramanujan identities, Fibonacci Quarterly (1) 42 (2004) 3–19, reprinted with permission.

5.3 George E. Andrews and R. J. Baxter, SCRATCHPAD explorations for elliptic theta functions, in Computers in Mathematics, Lecture Notes in Pure and Appl.

Math. No. 125, D. V. Chudnovsky and R. D. Jenks, eds. (M. Dekker, New York, 1990), 17–33, reprinted with permission from Taylor and Francis Group.

6. Number Theory

6.1 George E. Andrews, A note on the Bombieri-Selberg Formula for algebraic curves, Portugaliae Math. (2) 27 (1968) 75–81, reprinted with permission.

6.2 George E. Andrews, MacMahon's prime numbers of measurement, Amer. Math. Monthly (9) 82 (1975) 922–923.

6.3 This article was published in *Journal of Number Theory* vol. 23, G. E. Andrews, ΕΥΡΗΚΑ! num $= \Delta + \Delta + \Delta$, pages 285–293 Copyright © Elsevier 1986.

6.4 George E. Andrews and R. Roy, Ramanujan's method in q-series congruences, Electronic J. Combin. (Wilf Festschrift) (2) 4 (1997).

6.5 This article was published in *Discrete Mathematics* 204, George E. Andrews, q-analogs of the binomial coefficient congruences of Babbage, Wolstenholme and Glaisher, 15–25, copyright © 1999, with permission from Elsevier.

6.6 This article was published in *Journal of Number Theory* vol. 118, G. E. Andrews, A. Knopfmacher and B. Zimmerman, On the number of distinct values of multinomial coefficients, pages 15–30 Copyright © Elsevier 2006.

7. Surveys

7.1 This article was published in *Advances in Mathematics* vol. 9, G. E. Andrews, Partition Identities, pages 10–51, Copyright © Elsevier 1972.

7.2 Bull. Amer. Math. Soc. (6) 80 (1974) 1033–1052, A general theory of identities of the Rogers-Ramanujan type, G. E. Andrews, copyright © 1974 American Mathematical Society, reprinted with permission from the American Mathematical Society.

7.3 Ramanujan Journal 7 (2003) 385–400, Partitions: At the interface of q-series and modular forms, George E. Andrews, © 2003 Springer, reprinted with kind permission of Springer Science and Business Media.

7.4 George E. Andrews, Applications of basic hypergeometric functions, SIAM Rev. (4) 16 (1974) 441–484, copyright © 1974 Society for Industrial and Applied Mathematics. Reprinted with permission. All rights reserved.

7.5 This article was published in *Advances in Mathematics* vol. 9, G. E. Andrews, Partition Identities, pages 10–51, Copyright © Elsevier 1972.

7.6 Ramanujan Journal 1 (1997) 7–23, The well-poised thread: An organized chronicle of some amazing summations and their implications, George E. Andrews, © 1997 Springer, reprinted with kind permission of Springer Science and Business Media.

8. Education, History, etc.

8.1 George E. Andrews, L. J. Rogers and the Rogers-Ramanujan identities, Math. Chronicle (Part 2) 11 (1982) 1–15, reprinted with permission.

8.2 George E. Andrews, Mathematical education: A case for balance, The College Mathematics Journal (5) 27 (1996) 341–348.

Bibliography

[Alf-JO10] C. Alfes, M. Jameson and R. Lemke Oliver, *Proof of the Alder–Andrews conjecture*, Proc. Amer. Math. Soc., **139** (2011), 63–78.

[Ald56] H. L. Alder, *Research problem no. 4*, Bull. Amer. Math. Soc., **62** (1956), 76.

[Ald69] H. L. Alder, *Partition identities from Euler to the present*, Amer. Math. Monthly, **76** (1969), 733–746.

[Ask80] R. A. Askey, *Some basic hypergeometric extensions of the integrals of Selberg and Andrews*, S.I.A.M J. Math. Anal., **11** (1980), 938–951.

[Ask01] R. A. Askey, *The work of George Andrews: A Madison Perspective*, in The Andrews Festschrift: Seventeen Papers on Classical Number Theory and Combinatorics, eds. D. Foata and G.-N. Han, Springer, Berlin (2001).

[Bac02] P. G. H. Bachmann, *Niedere Zahlentheorie*, B. G. Teubner, Leipzig, 1902.

[Bár08] I. Bárány, *Random points and lattice points in convex bodies*, Bull. Amer. Math. Soc., **45** (2008), 339–365.

[Bel61] R. Bellman, *A Brief Introduction to Theta Functions*, Holt, Rinehart and Winston, New York (1961).

[Bri08] K. Bringmann, *On the explicit construction of higher deformations of partition statistics*, Duke Math. J., **144** (2008), 195–233.

[Bri-LO10] K. Bringmann, J. Lovejoy and R. Osburn, *Automorphic properties of generating functions for generalized rank moments and Durfee symbols*, Int. Math. Res. Not., (to appear).

[Bri-O06] K. Bringmann and K. Ono, *The $f(q)$ mock theta function conjecture and partition ranks*, Invent. Math., **165** (2006), 243–266.

[Ben88] D. Benson, *Spin modules for symmetric groups*, J. London Math. Soc. (2), **38** (1988), 250–262.

[Bai47] W. N. Bailey, *Some identities in combinatory analysis*, Proc. London Math. Soc. (2), **49** (1947), 421–435.

[Bai49] W. N. Bailey, *Identities of the Rogers–Ramanujan type*, Proc. London Math. Soc. (2), **50** (1949), 1–10.

[Berk09] A. Berkovich, *The tri-pentagonal number theorem and related identities*, Internat. J. Number Theory, **5** (2009), 1385–1399.

[Berk-MS96] A. Berkovich, B. M. McCoy and A. Schilling, *$N = 2$ supersymmetry and Bailey pairs*, Phys. A., **228** (1996), 33–62.

[Bes91] C. Bessenrodt, *A combinatorial proof of a refinement of the Andrews–Olsson partition identity*, Europ. J. Comb., **12** (1991), 271–276.

[Bes95] C. Bessenrodt, *Generalizations of the Andrews–Olsson partition identity and applications*, Discrete Math., **141** (1995), 11–22.

[Ben-K72] E. Bender and D. Knuth, *Enumeration of plane partitions*, J. Comb. Th., **13** (1972), 40–54.

[Bern85-98] B. C. Berndt, *Ramanujan's Notebooks*, Vols. I–V, Springer, New York, (1985), (1989), (1991), (1994), (1998).

[Bom63] E. Bombieri, *Sull'analogo della formula di Selberg nei corpi di funzioni*, Atti della Accademia Nazimale dei Lincei, Classe di Scienze, Fische, Mathematik Naturali (8), **35** (1963), 252–257.

[Bou-K11] C. Boulet and K. Kurşungöz, *Symmety of K-marked Durfee symbols*, Internat. J. Number Theory, **7** (2011), 215–230.

[Bou-E97] M. Bousquet-Mélou and K. Eriksson, *Lecture hall partitions I and II*, The Ramanujan J., **1** (1997), 101–111 and 165–185.

[Bre81] D. Bressoud, *Solution of Problem 74-12*, S.I.A.M. Rev., **23** (1981), 101–104.

[Bre99] D. Bressoud, *Proofs and Confirmations: The Story of the Alternating Sign Matrix Conjecture*, Cambridge University Press, New York (1999).

[Bri-OR08] K. Bringmann, K. Ono and R. Rhoades, *Eulerian series as modular forms*, J. Amer. Math. Soc., **21** (2008), 1085–1104.

[Bri09] K. Bringmann, *Asymptotics for rank partition functions*, Trans. Amer. Math. Soc., **361** (2009), 3483–3500.

[Cor-LS07] S. Corteel, S. Lee and C. D. Savage, *Five guidelines for partition analysis with applications to lecture hall-type theorems*, in Combinatorial Number Theory, de Gruyter, Berlin (2007), 131–155.

[Cor-L09] S. Corteel and J. Lovejoy, *Overpartitions and the q-Bailey identity*, Proc. Edinburgh Math. Soc., **52** (2009), 297–306.

[DeL-HTY04] J. A. DeLoera, R. Hemmecke, J. Tanzer and R. Yoshide, *Effective lattice point counting in rational convex polytopes*, J. Symb. comb., **38** (2004), 1273–1302.

[Dic19] L. E. Dickson, *History of the Theory of Numbers, Volume II: Diophantine Analysis*, Carnegie Institute of Washington, Washington (1919).

[Div76] B. Diviš, *On the lattice points on strictly convex surfaces*, J. Number Th., **8** (1976), 298–307.

[Dra52] L. A. Dragonette, *Some Asymptotic formulae for the mock theta series of Ramanujan*, Trans. Amer. Math. Soc., **72** (1952), 474–500.

[Dys44] F. J. Dyson, *Some guesses in the theory of partitions*, Eureka, **8** (1944), 10–15.

[Ext83] H. Exton, *Q-hypergeometric Functions and Applications*, E. Horwood, Chichester, U. K., (1983).

[Fok-FW95] R. Fokkink, W. Fokkink and B. Wang, *A relation between partitions and the number of divisors*, Amer. Math. Monthly, **102** (1995), 345–347.

[Fol08] A. Folsom, *A short proof of the mock theta conjecture using Maass Forms*, Proc. Amer. Math. Soc., **136** (2008), 4143–4149.

[Fol-O08] A. Folsom and K. Ono, *The spt-function of Andrews*, Proc. Nat. Acad. Sci. (USA), **105** (2008), 20152–20156.

[Gar-LZ06] S. Garoufalidis, T. Le and D. Zeilberger, *The quantum MacMahon master theorem*, Proc. Nat. Acad. Sci. USA, **103** (2006), 13928–13931.

[Gar08] S. Garthwaite, *Coefficients of the $\omega(q)$ mock theta function*, Internat. J. Number Theory, **6** (2008), 1027–1042.

[Göl60] H. Göllnitz, *Einfache Partitionen*, Diplomarbeit W. W. (1960), Göttingen, 65 pp.

[Göl67] H. Göllnitz, *Partitionen mit Differenzenbedingungen*, Journal für die reine und angewandte Mathematik **225** (1967), 154–190.

[Gor61] B. Gordon, *A combinatorial generalization of the Rogers–Ramanujan identities*, Amer. J. Math., **83** (1961), 393–399.

[Gor62] B. Gordon, *Some Ramanujan-like continued fractions*, Abstracts of Short Communications, Int. Congr. of Math., (1962), Stockholm, 29–30.

[Gor65] B. Gordon, *Some continued fractions of the Rogers–Ramanujan type*, Duke Math. J., **31** (1965), 741–748.

[Gor83] B. Gordon, *A proof of the Bender, Knuth conjecture*, Pac. J. Math., **108** (1983), 999–1013.

[Gold-R69] J. Goldman and G.-C. Rota, *The number of subspaces of a vector space*, from Recent Progress in Combinatorics, (W. T. Tutte, ed.), Academic Press, New York, (1969), 75–83.

[Hab88] L. Habsieger, *Une q-intégrale de Selberg–Askey*, S.I.A.M. J. Math. Anal., **19** (1988), 1475–1489.

[Hah49] W. Hahn, *Über orthogonalpolynome, die q-Differenzengleichungengenügen*, Mathematische Nachrichten, **2** (1949), 4–34.

[Hei1847] E. Heine, *Untersuchungen über die Reihe...*, Journal für die reine und angewandte Mathematik, **34** (1847), 285–328.

[Hic88a] D. Hickerson, *A proof of the mock theta conjectures*, Invent. Math., **94** (1988), 639–660.

[Hic88b] D. Hickerson, *On the seventh order mock theta functions*, Invent. Math., **94** (1988), 661–677.

[Hol-LR04] A. E. Holroyd, T. M. Liggett and D. Romils, *Integrals, parititions and cellular automata*, Trans. Amer. Math. Soc., **356** (2004), 3349–3368.

[Jor-WW79] J. H. Jordan, R. Walch and R. J. Wisner, *Triangles with integer sides*, Amer. Math. Monthly, **86** (1979), 686–689.

[Kim-Y10] S. Kim and A. J. Yee, *Göllnitz–Gordon identities and parity questions*, European J. Combin., **32** (2011) 288–293.

[Kad88] K. Kadell, *A proof of Askey's conjectured q-analog of Selberg's integral and a conjecture of Morris*, S.I.A.M. J. Math. Anal., **19** (1988), 969–986.

[Kan04] D. M. Kane, *Resolution of a conjecture involving cranks of partitions of Andrews and Lewis*, Proc. Amer. Math. Soc., **132** (2004), 2247–2256.

[Kei07] W. J. Keith, *Ranks of Partitions and Durfee Symbols*, Ph.D. thesis, Pennsylvania State University, (2007), viii+132 pp.

[Kit88] Y. Kitaoka, *Local densities of quadratic forms*, Adv. Stud. Pure Math., **13** (1988), 433–460.

[Kran98] S. Krantz, *How to Teach Mathematics*, 2nd ed., Amer. Math. Soc., Providence, (1998).

[Krat-CPR08] C. Krattenthaler, B. M. Chandrashekara and R. Raghavendra, *Analytic proof of the partition identity, $A_{5,3,3}(n) = B^0_{5,3,3}(n)$*, Ramanujan J., **15** (2008), 77–86.

[Krat-R07] C. Krattenthaler and T. Rivoal, *An identity of Andrews, multiple integrals, and very-well poised hypergeometric series*, Ramanujan J., **13** (2007), 203–219.

[Kur09] K. Kursungoz, *Parity Considerations in Andrews–Gordon Identities and k-Marked Durfee Symbols*, Ph.D. thesis, Pennsylvania State University, (2009).

[Lus77] G. Lusztig, *Irreducible representations of finite classical groups*, Invent. Math., **43** (1977), 125–175.

[Macd82] I. G. Macdonald, *Some conjectures for root systems*, S.I.A.M. J. Math. Anal., **13** (1982), 988–1007.

[MacM16] P. A. MacMahon, *Combinatory Analysis*, 2 Vols., Cambridge University Press, Cambridge, (1916).

[MacM78] P. A. MacMahon, *Collected Papers*, MIT Press, Cambridge, Vol. 1 (1978), Vol. 2 (1986).

[McK-MMM00] C. McKnight, M. McKnight, A. Magid and T. Murphy, *Mathematics Education Research; A Guide for the Research Mathematician*, Amer. Math. Soc., Providence, (2000).

[McL-S10] J. McLaughlin and A. V. Sills, *On a pair of identities from Ramanujan's lost notebook*, preprint, (2010).

[Men65] P. K. Menon, *On Ramanujan's continued fraction and related identities*, J. London Math. Soc., **40** (1965), 49–54.

[Mill-RR82] W. H. Mills, D. H. Robbins and H. Rumsey, *Proof of the Macdonald conjecture*, Invent. Math., **66** (1982), 73–87.

[Miln80] S. Milne, *Hypergeometric functions well-poised in SU(n) and generalizations of Biedenham's G-functions*, Adv. in Math., **36** (1980), 169–211.

[Miln85] S. Milne, *A q-analog of hypergeometric series well-poised in SU(n) and invariant G functions*, Adv. in Math., **58** (1985), 1–60.

[Miln02] S. Milne, *Infinite families of exact sums of squares formulas, Jacobi elliptic functions, continued fractions, and Schur functions*, Ramanujan J., **6** (2002), 7–149 (Preface by G. E. Andrews pages 5 and 6).

[Mon77] S. Montgomery *et al.* (eds.), *Selected Papers on Algebra*, Raymond W. Brink Selected Mathematical Papers, Vol. 3, Math. Assoc. of Amer., Washington, D. C., (1977).

[Mul-R70] R. Mullin and G.-C. Rota, *On the foundations of combinatorial theory, III: Theory of binomial enumeration, from Graph Theory and Its applications*, (B. Harris, ed.), Academic Press, New York, (1970), 167–213.

[Pak06] I. Pak, *Partition bijections, a survey*, Ramanujan J., **12** (2006), 5–75.

[Paule85] P. Paule, *On identities of the Rogers–Ramanjuan type*, J. Math. Anal. Appl., **107** (1985), 255–284.

[Paule87] P. Paule, *The concept of Bailey chains*, Séminaire Lotharingen de Combinatoire, B 18f (1987), 24.

[Por77] S. Porubský, *On MacMahon's segmented numbers and related sequences*, Nieuw Archief veer Wiskunde (3), **25** (1977), 403–408.

[Pre-S03] T. Prellberg and D. Stanton, *Proof of a monotonicity conjecture*, J. Comb. Th. (A), **103** (2003), 377–381.

[Rad73] H. A. Rademacher, *Topics in Analytic Number Theory*, Springer, Berlin (1973).

[Rob91] D. Robbins, *The story of* $1, 2, 7, 42, 429, 7436, \ldots$, Math. Intelligencer, **13** (1991), 12–19.

[Rog1893] L. J. Rogers, *On the expansion of some infinite products*, Proc. London Math. Soc., **25** (1893), 318–343.

[Rog17] L. J. Rogers, *On two theorems of combinatory analyses and some allied identities*, Proc. London Math. Soc. (2), **16** (1917), 315–336.

[San91] J. P. O. Santos, *Compurter Algebra and Identities of the Rogers–Ramanujan Type*, Ph.D. Thesis, Pennsylvania State University, (1991).

[Sav-S10] C. D. Savage and A. V. Sills, *On an identity of Gessel and Stanton and the new little Göllnitz identities*, Adv. Appl. Math., **46** (2011), 563–575.

[Schi-W98] A. Schilling and S. O. Warnaar, *Supernomial coefficients, polynomial identities and q-series*, Ramajuan J., **2** (1998), 759–494.

[Scho94] A. Schoenfeld, *Mathematical Thinking and Problem Solving*, Lawrence Erlbaum Associates, Hillsdale, New Jersey, Hove, UK, (1994).

[Schu17] I. Schur, *Ein Beitrag zur additiven Zahlentheorie und zur Theorie der Kettenbrüche*, Sitz, Preuss. Akad. Wiss. Phys.-Math Kl. (1917), 302–321.

[Schu26] I. Schur, *Zur additiven Zahlentheorie*, S.-B. Akad. Wiss, Berlin (1926), 488–495.

[Sel36] A. Selberg, *Über einige arithmetische Identitäten*, Avhandlinger Norske Akad., (1936), No. 8.

[Sil03] A. V. Sills, *Finite Rogers–Ramanujan type identities*, Electronic J. Combin., **10** (2003), #R13, 122 pages.

[Sil10] A. V. Sills, *Rademacher-type formulas for restricted partition and overpartition functions*, Ramanujan J., **23** (2010), 253–264.

[Sil-Z12] A. V. Sills and D. Zeilberger, *Rademacher's infinite partial fraction conjecture is (almost certainly) false*, J. Difference Eq. Appl., (to appear).

[Sla52] L. J. Slater, *Further Identities of the Rogers–Ramanujan type*, Proc. London Math. Soc. (2), **54** (1952), 147–167.

[Sla66] L. J. Slater, *Generalized Hypergeometric Functions*, Cambridge University Press, Cambridge, (1966).

[Slo73] N. J. A. Sloane, *Handbook of Integer Sequences*, Academic Press, New York, (1973).

[Spi02] V. P. Spiridonov, *An elliptic incarnation of the Bailey chain*, Int. Math. Res. Notices, (2002), No. 37, 1945–1977.

[Sta71] R. P. Stanley, *Theory and application of plane partitions I and II*, Studies in Appl. Math., **50** (1971), 167–188, 259–279.

[Ste90] J. Stembridge, *Nonintersecting paths Pfaffians, and plane partitions*, Adv. in. Math., **83** (1990), 96–131.

[Ste95] J. Stembridge, *The enumeration of totally symmetric plane partitions*, Adv. in Math., **111** (1995), 227–243.

[Swi74] H. P. F. Swinnerton-Dyer, *The number of lattice points on a convex curve*, J. Number Theory, **6** (1974), 128–135.

[Syl82] J. J. Sylvester, *A constructive theory of partitions arranged in three acts, an interact and an exodion*, Amer. J. Math., **5** (1882), 251–330, **6** (1884), 334–336 (Reprinted: Coll. Math. Papers of James Joseph Sylvester, Vol. 4, pp. 1–83, Cambridge University Press, Cambridge, (1912)).

[Wat29] G. N. Watson, *A new proof of the Rogers–Ramanujan identities*, J. London Math. Soc., **4** (1929), 4–9.

[Wat36] G. N. Watson, *The final problem: An account of the mock theta functions*, J. London Math Soc., **11** (1936), 55–80.

[Wat37] G. N. Watson, *The mock theta functions (2)*, Proc. London Math. Soc. (2), **42** (1937), 274–304.

[Xin04] G. Xin, *A fast algorithm for MacMahon's partitions analysis*, Electronic J. Combin, **11** (2004), R58.

[Yee04] A. J. Yee, *Partitions with difference conditions and Alder's conjecture*, Proc. nat. Acad. Sci. USA, **101** (2004), 91–109.

[Zei93] D. Zeilberger, *Theorems for a price: Tomorrow's semi-rigorous mathematical culture*, Notices Amer. Math. Soc., **40** (1993), No. 8, 978–981.

[Zei94] D. Zeilberger, *A constant term identity featuring the ubiquitous (and mysterious) Andrews–Mills–Robbins–Rumsey numbers* $1, 2, 7, 42, 429, \ldots$, J. Comb. Th. (A), **66** (1994), 17–27.

[Zei-B85] D. Zeilberger and D. Bressoud, *A proof of Andrews' q-Dyson conjecture*, Discr. Math., **54** (1985), 201–224.